Conversion Table (*Continued*)

	Dimensions	To Convert from	To	Multiply by
Specific heat	$L^2T^{-2}\theta^{-1}$	BTU/lbm °F	J/kg °C	4186.8
Specific volume	L^3M^{-1}	ft³/lbm	m³/kg	0.062428
Surface tension	MT^{-2}	lbf/ft	N/m	14.594
Temperature	θ	°F	°C	$t_C = (t_F - 32)/1.8$
		°R	K	1/1.8
Thermal conductivity	$MLT^{-3}\theta^{-1}$	BTU/h ft °F	W/m °C	1.7307
Thermal diffusivity	L^2T^{-1}	ft²/s	m²/s	0.092903
Torque	ML^2T^{-2}	lbf ft	N m	1.3558
Velocity	LT^{-1}	ft/s	m/s	0.3048
Viscosity (dynamic)	$ML^{-1}T^{-1}$	lbf s/ft²	N s/m²	47.88
		poise	N s/m²	0.1
Viscosity (kinematic)	L^2t^{-1}	ft²/s	m²/s	0.092903
Volume	L^3	ft³	m³	0.0283168

BTU in the units refers to International Steam Table (IT)
BTU (thermochemical) is also widely used.
1 BTU (IT) = 1055.056 J   1 BTU (thermochemical) = 1054.35 J
Dimensions: Length (*L*), Mass (*M*), Time (*T*), and Temperature (θ)

Physical Constants

Avogadro's number	6.024×10^{26} per kilomole
Boltzmann's constant (k)	1.3806×10^{-23} J/K molecule
Gravitational acceleration	9.807 m/s²
Planck's constant (h)	6.625×10^{-34} J s/molecule
Speed of light in vacuum (c_o)	2.998×10^8 m/s
Stefan-Boltzmann's constant	5.67×10^{-8} W/m² K⁴ = 0.1714×10^{-8} BTU/h-ft² °R⁴
Universal gas constant ($\bar{R}$)	8314.4 J/kmole K = 1545 ft-lbf/lbmole °R

ENGINEERING HEAT TRANSFER

■ ■ ■

N.V. Suryanarayana

Michigan Technological University

WEST PUBLISHING COMPANY

MINNEAPOLIS/ST. PAUL ■ NEW YORK ■ LOS ANGELES ■ SAN FRANCISCO

WEST'S COMMITMENT TO THE ENVIRONMENT

In 1906, West Publishing Company began recycling materials left over from the production of books. This began a tradition of efficient and responsible use of resources. Today, up to 95 percent of our legal books and 70 percent of our college and school texts are printed on recycled, acid-free stock. West also recycles nearly 22 million pounds of scrap paper annually–the equivalent of 181,717 trees. Since the 1960s, West has devised ways to capture and recycle waste inks, solvents, oils, and vapors created in the printing process. We also recycle plastics of all kinds, wood, glass, corrugated cardboard, and batteries, and have eliminated the use of Styrofoam book packaging. We at West are proud of the longevity and the scope of our commitment to the environment.

PRODUCTION CREDITS

Production, Prepress, Printing and Binding by West Publishing Company.

Text Design   Dianne Beasely
Artwork   Visual Graphic Systems, Ltd.
Copyediting   Betty O'Bryant
Composition   University Graphics, Inc.

 TEXT IS PRINTED ON 10% POST CONSUMER RECYCLED PAPER

LIBRARY OF CONGRESS CATALOGING-IN-PUBLICATION DATA

Suryanarayana, N. V. (Narasipur Venkataram), 1931-
    Engineering heat transfer / N.V. Suryanarayana.
       p.   cm.
    Includes index.
    ISBN 0-314-01093-9
    1. Heat—Transmission.   I. Title.
TJ260.S88 1994
621.402'2—dc20   94-9599
                                       CIP

*This book is dedicated to the memory of
my mother, Kamalamma and my father Venkataram,
whose immense faith and confidence in me shaped and encouraged my
educational endeavors.*

■   ■   ■

CONTENTS

3 CONDUCTION II   205

9 HEAT EXCHANGERS 739

10 MIXED-MODE HEAT TRANSFER 817

11 CONDUCTION—DIFFERENTIAL FORMULATION 869

PREFACE

We experience the results of all modes of heat transfer in our daily life. Yet, in my experience of over 20 years of teaching heat transfer and talking to my colleagues I find that students have difficulty in understanding the subject. Heat transfer is an abstract subject. We do not ''see'' any temperature profile although we experience the results of elevated temperatures. My main objective in writing this book is to help students understand heat transfer. To achieve this objective I have deviated from the traditional organization of the material and incorporated several features to assist the students to master the material in the book. The organization of the material is described in the following section on the ''Note to the Instructor.'' I have tried to emphasize physics retaining all the mathematical details that are appropriate for an introductory course on heat transfer. Recognizing that many students may not be quite comfortable in calculus, I have incorporated all the mathematical details in the development of the solutions.

I have tried to write this book from the perspective of the students and to help them understand heat transfer. I hope I have succeeded in that endeavor.

NOTE TO THE INSTRUCTOR

There are many books on heat transfer and there must be good reasons for adding one more book to the list. The main reason for writing this book is that during more than two decades of teaching heat transfer I found that it was difficult to follow the organization in the existing books. Many instructors may find the departure from the traditional sequence of presentation unusual. An explanation of the rationale behind the reordering of the material follows.

Organization:

In an introductory book it is necessary to present simple concepts in some detail before introducing more complex materials. The philosophy behind the organization is that one has to learn to walk before attempting to run. There are more than one chapter in each mode of heat transfer. In the first chapter in each mode, I have introduced simple concepts that the students can relate to from their experiences. The explanations are complete and all the details of the application of the laws of thermodynamics and the principles of fluid mechanics are presented. I have tried to synthesize all the disciplines as appropriate. The emphasis has been on the physics of the situations with appropriate mathematical rigor. As the required background in calculus is only ordinary differential equations, topics that need the solution of partial differential equations are presented in the last two chapters. A brief description of the arrangement of the materials in the various chapters is given below.

Chapter 1—Introduction gives an introduction to the three modes of heat transfer followed by several examples of the application of heat transfer and a brief review of the required background in thermodynamics and fluid mechanics.

Chapter 2—Conduction I introduces the student to one-dimensional, steady state problems, the application of thermal circuits, conduction shape factors, and a comprehensive discussion of extended surfaces. Complete statements of the physical laws, their translation to mathematical forms, and their solutions are presented. One can either derive the differential equation for each case or specialize a general form of the equation to specific cases. Different methods work for different students. Those who prefer the elegant route of starting with the general equation may start with the section on generalized one-dimensional, transient conduction equation with area change and internal energy generation (Section 2.7) and specialize it to the various cases. Overall heat transfer coefficient is defined and used in this chapter. Only the results of one-dimensional transient temperature distribution in plain slabs, cylinders and spheres and their application to multi-dimensional transient temperature problems are given. Details of the derivation of the partial differential equations and their solutions are deferred to Chapter 11.

Inclusion of internal energy generation in conduction problems makes the resulting equations amenable to a variety of mathematical solutions as the generation term can be made space or temperature dependent. But considering the low probability that a practicing engineer will be required to solve such problems this material is included in Chapter 3.

Chapter 3—Conduction II deals with internal thermal energy generation, variable thermal conductivity, and finite difference solutions for multi-dimensional problems. In the spirit of requiring only a background in ordinary differential equations, the approximate analytical technique of integral method for the solution of one-dimensional transient temperature distribution is included in this chapter.

Chapter 4—Convection I is the introductory chapter on convection. Instead of presenting each mode of convection in separate chapters, I have provided a unified presentation of forced external and internal convection and natural convection in this chapter. Boundary layer concepts are introduced by appealing to the intuition of the students without recourse to partial differential equations. The emphasis in this chapter is on determining the convective heat transfer coefficient and the use of the coefficient to determine the heat transfer rate or the surface temperature. Iterative solutions for convective problems are illustrated through several examples. A large number of correlations are included so that the students are aware that different correlations for the same configuration are available. I have also tried to emphasize the uncertainty in the computed value of the heat transfer coefficients.

Chapter 5—Convection II begins with tube banks combining internal and external flow correlations. A section on high speed flows is included. Integral technique for the approximate solutions of convection problems is presented.

Chapter 6—Heat Transfer with Change of Phase deals with heat transfer with phase change limited to boiling and condensation.

Chapter 7—Radiation I starts with an introduction to radiation with a discussion of the various terms and is essentially limited to gray, diffuse surfaces. View factors and view factor algebra are introduced. Finally solutions to radiative heat transfer in an enclosure by the electrical analogy, limited to gray, diffuse and opaque surfaces are given.

Chapter 8—Radiation II goes deeper into wavelength dependent properties, mathematical details of view factors, and includes the radiosity method for the solution of radiative heat transfer from surfaces. A proof of the equivalence of the electrical analogy and the radiosity method is given. The chapter ends with an introduction to gaseous radiation.

Chapter 9—Heat Exchangers is a classical treatment of heat exchangers. Both the LMTD and NTU methods are presented. The chapter includes a discussion of plate heat exchangers and regenerators.

Chapter 10—Mixed Mode Heat Transfer is a collection of problems to illustrate the solution of problems with more than one mode of heat transfer. No general solution methodology is possible for such problems and different methods of solutions are illustrated through examples. Solar collector is an application where all three modes of heat transfer are involved and it is dealt with in some detail.

Chapter 11—Conduction: Differential Formulation includes a complete derivation of the differential equation for the two-dimensional temperature distribution. Solutions to the partial differential equations by the separation of variables technique are given with all the details. The use of Laplace transforms for the solution of transient problems is illustrated.

Chapter 12—Convection: Differential Formulation starts with a complete derivation of the equations of balance of mass, momentum and energy for two-dimensional problems. Examples of exact solutions to the Navier Stokes equation are provided. The complete derivation of the boundary layer equations for momentum and energy and their solutions to the flat plate problem are included.

Appendixes inlcude extensive tables of properties, brief discussions of dimensional analysis, orthogonal functions, Laplace transforms with an extensive table of inverse transforms.

Pedagogy

I have tried to follow the accepted nomenclature. The one exception is the symbol i for specific enthalpy to avoid confusion with the symbol h for the convective heat transfer coefficient.

Physics and Mathematics:

I have tried to emphasize the physics retaining all the mathematical details. Sufficient details of the mathematics are given for the students to follow the manipulations. Having understood the basic principles of each mode the students will be in a better position to absorb the more complex heat transfer and mathematical aspects involved in the later chapters. The only mathematical requirement for following the book is familiarity with linear ordinary differential equations except in the last two chapters. The advanced analytical methods involving transforms and partial differential equations are presented in the last two chapters. By studying the material up to the last two chapters the students will have had a good grasp of heat transfer fundamentals and will find it easier to follow both the heat transfer and the mathematical aspects. In the last two chapters I have given a complete derivation of the partial differential equations. The derivations are limited to two-dimensional situations so that the students are not overwhelmed by the mathematical manipulations. After completing the derivations, complete solutions to some of the simpler problems are given. I have avoided the temptation to say ''after some manipulation it can be shown. . .''

In both conduction and convection I have emphasized that the total heat transfer rate is given by $q = \int q'' \, dA$ and how q'' is determined and used.

Conceptual Solutions:

The solutions to most problems, save those of the most elementary nature, begin with conceptual solutions. The conceptual solution is a road map where all the steps involved in the solution are verbally explained. Development of the conceptual solution requires that the students think of the various steps involved and builds their confidence in the subject. I consider the inclusion of the conceptual solutions as one of the more significant tools in understanding the material.

Projects:

The projects at the end of chapters 2 through 8 are meant to show how to model physical situations, to supply needed information which may be missing in the problem statement, and to synthesize the materials in the earlier chapters in the book and from other related subjects, mainly thermodynamics and fluid mechanics. Most of the projects are open-ended. The students may find other alternatives. In many cases, physical situations do not conform to the models for which solutions are available. In such cases I have explained how the bounds of the heat transfer rates can be established from the models for which solutions are available. The use of either the upper or lower value depends on the physical situation.

Examples and End-of-Chapter Problems:

I have included examples covering such ordinary engineering applications as domestic refrigerators and water heaters, the more classic engineering applications in power plants, computers, furnaces and so on and also such exotic applications that involve drop towers and space shuttles. I have also included the somewhat non-engineering applications such as finding the temperature at the end of the handle of a kitchen pot containing boiling water. Most of the examples and end-of-chapter problems are application based. Several design problems have been added in most chapters in line with the emphasis placed by ABET on design.

Charts:

In many cases charts have been provided to show the trends and, in some, cases, the maximum attainable values, for example, Heisler charts, view factors, and heat exchanger effectiveness. In the solution to problems the appropriate equations are used directly or through the software where the equations have been programmed. The students can also check the values obtained from the equations against the values in the charts. By using the software provided with the book accurate values from the equations are obtained quickly.

Chapter Introductions, Margin Notes, and Chapter Summaries:

The chapter introductions give an overview of the chapter. The margin notes, which indicate what the material is about, help to locate a particular topic in the chapter. The end-of-chapter summaries not only tell the students what they should have learned but also briefly give all the essential elements in the chapter. They help to refresh the memory of the students. The review questions are intended to encourage the students to think and understand the material.

Historical Vignettes:

To give the students some historical perspective I have included historical sketches of some of the people who have made seminal contributions to heat transfer. I hope the inclusion of the sketches makes the subject livelier.

Software:

Most students have access to personal computers and it is appropriate to make use of this powerful tool in the study of heat transfer. There are several ways of using the computer. I am a strong believer that the students should have a good understanding of the principles and must be able to develop the formulas that are used in heat transfer. The software is there to *assist in the rapid computations,* particularly repetitive computations involving complicated correlations *and to eliminate errors* in computations. By programming the many equations in the software, the tedium of having to compute values from these correlations and the possibility of errors are eliminated. All the decision making rests with the students—the software acts as a rapid calculator. Details of the software are given in the section on ''Note to the Student.''

    With the available software, the solution of problems requiring iterative solution, particularly those requiring repeated use of convection correlations (involving determining properties by linear interpolation from tabulated values), is no longer time consuming. Many engineering problems require iterations but many instructors avoided assigning problems requiring iteration. With the available software I hope the instructors are more comfortable in assigning problems that involve iteration.

Other Features:

Some of the features that are not normally found in other undergraduate books on heat transfer are

■ the introduction of the overall heat transfer coefficient in the first chapter on conduction

- concept of critical thickness of insulation introduced through problems 2.37, 2.45 and 2.56
- several features related to fins—suitable length of fins, when a fin can be considered to be infinite and so on
- rigorous derivation of the transient conduction equation
- justification for the one-term approximation for transient conduction
- discussion of variable thermal conductivity
- introduction of the tri-diagonal matrix algorithm
- explanation of stability criterion for the numerical solution of transient conduction problems
- inclusion of the integral method for the approximate solution of conduction and convection problems
- discussion of variation of the convective heat transfer coefficient in internal flows
- inclusion of the law of corresponding corners and radiative heat transfer coefficient
- inclusion of plate heat exchangers and regenerators
- many interesting application-oriented problems
- extensive heat transfer properties in the appendixes including an introduction to orthogonal functions; an extensive table of inverse Laplace transforms; and a brief introduction to dimensional analysis.

I have assumed that the students have had at least one course each on thermodynamics and fluid mechanics and an understanding of calculus including ordinary differential equations. Calculus beyond ordinary differential equations is not required except in Chapters 11 and 12. All the mathematical details are given and the students should have no difficulty in following the mathematical details.

There is enough material for a sequence of two three-semester credit courses. For the first course I would suggest Chapters 1, 2, 4, 6, 7, 9, and 10. Depending on the time available and the interests of the instructor additional topics from the remaining chapters may be included.

ACKNOWLEDGMENTS

Without the constant encouragement of my wife, Pramila, this book would not have been possible. She never lost patience with me even when night after night, for many years, I locked myself in the study. Thank you for your support.

My interest in heat transfer was sparked by the late H.S. Rao who was my supervisor in the shipyard in India where I worked. He not only spurred me on to study heat transfer but also encouraged me when I decided to come to this country for graduate education. I shall always be grateful to him.

Many people have helped me by reading the manuscript and offering valuable suggestions. I have incorporated most of the suggestions. I am particularly indebited to my colleague, Professor Oner Arici for his continuous encouragement from the beginning of the project and to Professor E.M. Sparrow who, in spite of his busy schedule, read the entire manuscript line by line and offered innumerable suggestions to improve the book. I thank the following reviewers for their useful comments.

B.K. Hodge
Mississippi State University

Costas P. Grigoropoulos
Univesity of California, Berkley

John W. Sheffield
University of Missouri, Rolla

B.P. Leonard
The University of Akron

Bakhtier Farouk
Drexel University

John G. Georgiadis
Duke University

C. Subba Reddy
Union College

Frederick Carlson
Clarkson University

M.M. Ohadi
University of Mayland at College Park

Roanald S. Mullisen
California Polytechnic State University, SLO

Allan Kirkpatrick
Colorado State University

John Biddle
California State Polytechnic University, Pomona

J.C. Han
Texas A & M Univesity

Mohamad Metghalchi
Northeastern University

Brian Vick
Virginia Poytechnic Institute and State University

M. Erol Ulucakli
Lafayette College

D.C. Look, Jr.
University of Missouri, Rolla

Maurice J. Marongiu
Illinois Institute of Technology

Howard L. Julien
New Mexico State University

Frank W. Schmidt
The Pennsylvania State University

Willard W. Pulkrabek
The University of Wisconsin, Platteville

Morton S. Isaacson
Boston University

Lary C. Witte
University of Houston

Patrick J. Burns
Colorado State University

Brian J. Savilonis
Worcester Polytechnic Institute

Sushil H. Bhavnani
Auburn University

Jack B. Chaddock
Duke University

Van P. Carey
University of California, Berkley

Kau-Fui V. Wong
University of Miami

James Klausner
Lehigh University and University of Florida, Gainesville

Emmanuel K. Glakpe
Howard University

O.A. Plumb
Washington State University

J. Edward Sunderland
University of Massachusetts at Amherst

David G. Briggs
The State University of New Jersey, Rutgers

James Hartley
Georgia Institute of Technology

Jacob N. Chung
Washington State University

Mason P. Wilson, Jr.
University of Rhode Island

L.N. Tao
Illinois Institute of Technology

J.D.A. Walker
Lehigh University

Ramendra P. Roy
Arizona State University

Deborah A. Kaminski
Rennselaer Polytechnic Institute

Shi-Chune Yao
Carnegie Mellon University

Seppo A. Korpella
The Ohio State University

David J. Kukulka
SUCB

K.V.C. Rao
Michigan Technological University

James L.S. Chen
University of Pittsburg

Melvyn C. Branch
University of Colorado at Boulder

I have had the good fortune to work with professional, friendly and helpful personnel at West Publishing Corporation. My special thanks to Mr. Peter C. Gordon, Editor; Ms. Lucy Paine Kezar, Editorial Assistant; Ms. Debra Aspengren Meyer, Assistant Production Editor; Ms. Andrea Peters, Ancillary Publications Editor, all of West Publishing. My thanks are extended to Ms. Betty O'Bryant, Copy Editor, and to Visual Graphics Systems for their assistance in copyediting the manuscript and preparing the art work.

NOTE TO THE STUDENT

The main objectives of this book are to get you interested in heat transfer. With that object, I have emphasized physics and included all the mathematical details to assist you to follow the subject. Applications varying from the mundane to the exotic have been included to make the book interesting to you and relevant in the context of current applications.

I would like to explain some of the features of the book.

I have tried to emphasize the physics of the situations. In many cases, simple explanations based on your day-to-day experiences have been given and the mathematical details have been included, and I have avoided the temptation of saying ''with a little manipulation it can be shown . . .''

When solving problems it is useful to write the solution in words before embarking on the detailed solution, *the conceptual solution*. Except for the drill type problems, I have included such conceptual solutions to the problems. I would urge you to write a similar conceptual solution before you start on the actual solution. I would also suggest that you carry out the solution in symbolic form as far as possible and substitute numerical values at the end.

To make computations easier and to eliminate the possibility of errors, program **HT.EXE** is provided in the disk accompanying the book. The program **HT.EXE** contains 10 modules. In the text the modules are referenced by the two letter abbreviations (in parentheses).

- **Transient Conduction 1 (TC)** provides the dimensionless transient temperature at a specified location and Fourier number (greater than 0.05) for a Biot number between 0.1 and 10 000 when a semi-infinite rectangular block, semi-infinite cylinder, or sphere is exposed to convection heat transfer.
- **Transient Conduction 2** is similar to Transient Conduction 1. It computes the dimensionless temperature (a) for specified values of the dimensionless distance and Biot nuber for Fourier numbers from 0 to 2 in steps of 0.1 or (b) for specified values of the Biot number and Fourier number for dimensionless distances of 0 to 1 in steps of 0.05.
- **Convection Correlations (CC)** computes convective heat transfer coefficients in forced and natural convection employing the correlations given in the text. Most of the correlations are programmed. You supply the type of convection, geometry, and fluid (properties of air, water and oil are programmed, properties of other fluids must be supplied by the user). After examining the Reynolds number or Raleigh number and the physical situation, you select the equation to be used.

The program computes the Nusselt number and the convective heat transfer coefficient.

■ **Radiation Function (RF)** calculates black body radiation functions for the defined temperature and the lower and upper limits of the wavelength.

■ **View Factor (VF)** computes view factors for four geometries—two identical rectangles directly opposite each other, two rectangles perpendicular to each other and sharing a common side, two coaxial disks, and two coaxial cylinders.

■ **Heat Exchangers (HE)** computes the LMTD correction factors, the effectiveness with known values of NTU and NTU for a defined value of the effectiveness.

■ **Simultaneous equations** solve simultaneous equations with more than two unknowns (essentially for the solution of radiative heat transfer in a N-surface enclosure).

■ **Bessel** computes the Bessel functions of the first kind of order 0 and 1.

■ **Error Functions** evaluates error functions.

BLVEL.EXE computes the velocity in the boundary layer for the flow of a fluid parallel to a flat plate. The functions obtained by the integration of the boundary layer equation for a fluid flowing parallel to a flat plate can be printed for any specified interval of the dimensionless distance.

Many heat transfer problems require iterative solutions. In programs on convection, iterative solutions call for repeated computations of convective heat transfer coefficients from complex correlations; such calculations can be tedious and there is the possibility of errors. The use of the software eliminates errors and the tedium. However, you have to make all the decisions, and unless you are familiar with the material the software may not be of much use; it is not intended to solve problems.

The margin notes are to assist you in locating the topics easily. The summary at the end of the chapters helps you to refresh your memory and to give you an idea of what you should have learned from the chapter. The review questions are intended to make you think.

If you find the the book useful and the subject interesting I will have achieved my objectives.

NOMENCLATURE

A, A_s	Surface area (m^2)
A_x	Area perpendicular to the x-direction
A_r	Area perpendicular to the r-direction
b, W	Width (m)
C.V.	Control Volume
c	Specific heat—solids and liquids (J/kg K); Sonic velocity (m/s)
C_F	Fanning friction factor
c_p	Specific heat at constant pressure (J/kg K)
c_v	Specific heat at constant volume (J/kg K)
D, d	Diameter (m)
D_h, d_h	Hydraulic mean diameter (m)
E	Emissive power (W/m^2)
E_b	Black body emissive power (W/m^2)
E_λ	Monochromatic emissive power (W/m^3 or W/m^2 μm)
$E_{b\lambda}$	Black body monochromatic emissive power (W/m^3 or W/m^2 μm)
$F_{1\text{-}2}$	View factor between surface 1 and surface 2
f	Darcy friction factor; LMTD correction factor
$f_{0\text{-}\lambda}$	Black body radiation function ($E_{b,0\text{-}\lambda}/E_b$)
G	Irradiation (W/m^2)
g	Gravitational acceleration (9.807 m/s^2)
H	Height
h, h_c	Convective heat transfer coefficient (W/m^2 K or W/m^2 °C)
h_r, h_R	Radiative heat transfer coefficient (W/m^2 K or W/m^2 °C)
I	Radiation intensity (W/m^2 sr), Enthalpy (J)
i	Specific enthalpy (J/kg)
I_b	Black body radiation intensity (W/m^2 sr)
i_{fg}	Specific enthalpy of vaporization (J/kg)
I_λ	Monochromatic radiation intensity (W/m^3 sr or W/m^2 sr μm)
J	Radiosity (W/m^2)
k	Thermal conductivity (W/m K or W/m° C)
L	Length (m)
m, M	Mass (kg)
$\dot{m}$	Rate of mass flow (kg/s)

$\dot{m}_c$	Mass rate of condensation (kg/s)
N	Number of surfaces; Number of tubes in a tube bank
NTU	Number of transfer units
P	Perimeter (m); Power (W)
p	Pressure (Pa, kPa)
Q	Rate of energy (W)
q	Heat transfer rate (W)
q_w	Heat transfer rate from wall surface (W)
q'	Heat transfer rate per unit length (W/m)
q''	Heat flux (W/m^2)
q'''	Internal energy generation rate per unit volume (W/m^3)
R, r	Radius (m)
r, θ, z	Cylindrical coordinates
r_i, r_o	Inner and outer radii (m)
r, θ, ϕ	Spherical coordinates
S	Conduction shape factor; Distance (m)
S_L	Distance between two adjacent rows of tubes (m)
S_T	Distance between two adjacent columns of tubes (m)
s	Distance (m)
T	Temperature (°C, K)
T_b	Bulk temperature (°C, K)
T_{bi}, T_{be}	Inlet and exit bulk temperatures (°C, K)
T_e	Exit temperature (°C, K)
T_i	Inner surface temperature; Also inlet temperature (°C, K)
T_m	Mean temperature (°C, K)
T_o	Outer surface temperature (°C, K)
T_s	Surface temperature (°C, K)
T_w	Wall temperature (°C, K)
T_∞	Free stream temperature (°C, K)
U	Overall heat transfer coefficient (W/m^2 K); Internal energy (J)
U_i, U_o	Overall heat transfer coefficients based on the inside and outside surface area, respectively, (W/m^2 K)
U_∞	Free stream velocity (m/s)
u	Specific internal energy (J/kg), x-component of velocity
u, v, w	Fluid velocity components—Cartesian coordinates
V	Velocity (m/s); Volume (m^3)
v	Specific volume (m^3/kg), y-component of velocity
V_∞	Free stream velocity (m/s)
v_x, v_y, v_z	Velocity components—Cartesian coordinates
W	Width (m)
$\dot{W}$	Work transfer rate (W)
x, y, z	Cartesian coordinates
x_c	Critical distance for transition to turbulence

Greek Letters

α	Thermal diffusivity (m^2/s); Absorptivity
β	Volumetric thermal expansion coefficient (K^{-1})
Γ	Mass rate of flow/unit width (Kg/m)
δ	Velocity boundary layer thickness
δ_T	Temperature boundary layer thickness
ϵ	Emissivity; Fin effectiveness; Heat exchanger effectiveness
η	Dimensionless distance; Fin efficiency; Similarity variable
ϕ	Viscous dissipation function (s^{-2})
Λ	Dimensionless length
λ	Wavelength (m, μm)
μ	Dynamic viscosity ($N\ s/m^2$)
ν	Kinematic viscosity $= \mu/\rho$ (m^2/s)
π	Dimensionless time
ρ	Density (kg/m^3); Reflectivity; Dimensionless distance (r/R)
σ	Stefan-Boltzman constant ($5.67 \times 10^{-8}\ W/m^2\ K^4$) Surface tension (N/m)
θ	Temperature difference (°C, K); Dimensionless temperature
τ	Time (s); Transmissivity; Shear stress (N/m^2)
ω	Solid angle (sr)

Dimensionless Parameters

Bi	Biot number, hL/k, hR/k
Ec	Eckert number, $v^2/c_p\ (T_s - T_\infty)$
Fo	Fourier number, $\alpha\ \tau/L^2$, $\alpha\ \tau/R^2$
Gr	Grashof number, $(g\ \beta\ \rho^2\ \Delta T\ L^3)/\mu^2$
Ja	Jakob number, $\dfrac{c_p\ (T_w - T_\infty)}{i_{fg}}$
Ma	Mach number, v/c
Nu	Nusselt number, hL/k, hx/k, hd/k
Pr	Prandtl number, $c_p\ \mu/k$, ν/α
Ra	Rayleigh number, Gr Pr
Re	Reynolds number, $\rho UL/\mu$, $\rho Ux/\mu$, $\rho Vd/\mu$
Pe	Peclet number, Re Pr
St	Stanton number, Nu/(Re Pr)

1

INTRODUCTION

The practice of heat transfer started when man learned to cool himself with water and dry himself in the sun. The science of heat transfer started when ancient man learned to conquer his fear of fire and strike a fire and maintain the flame at will. It is this desire to control and harvest the heat energy and transfer it most efficiently across interfaces that motivated research and yielded the astounding technological developments of air conditioning, refrigeration, transportation by land and air . . . .

From the inaugural address by Samuel Sideman at the Ninth International Heat Transfer Conference, Jerusalem, August 20, 1990.

In this chapter, we

- Define heat transfer
- Show the relationship between thermodynamics and heat transfer
- Define the three modes of heat transfer—conduction, convection, and radiation
- Give a brief introduction to each mode of heat transfer
- Briefly describe a few cases where heat transfer plays an important role
- Suggest a methodology for the solution of heat transfer problems
- Define some of the significant terms used in this book
- Give a brief review of thermodynamics and fluid mechanics with particular reference to heat transfer

Heat transfer is the energy transfer resulting only from temperature differences. Examples of heat transfer are many and within our day-to-day experiences—our bodies transfer heat to the surroundings; coolant in an automobile engine is the medium by which heat transfers from the engine cylinders to the surrounding air through the radiator; our houses get warmer in the summer and cooler in the winter partially as a result of heat transfer across the walls, doors, and windows; and the temperature of water in a pan kept on a stove initially increases until boiling begins, and thereafter, the water continues to boil without any increase in the temperature. A few cases where heat transfer plays a dominant role are briefly described in Section 1.7. In those cases where heat transfer plays a significant role, we need to know the details of the processes so that we may control them for our benefit.

1.1 RELATIONSHIP BETWEEN THERMODYNAMICS AND HEAT TRANSFER

Relationship between thermodynamics and heat transfer

Classical thermodynamics deals with equilibrium states. The First Law of Thermodynamics is a statement of the principle of conservation of energy relating the change in the energy of a fixed mass of a substance (or of a substance contained in a region in space) when it changes from one equilibrium state to another as a result of heat and work transfer across its boundaries. But thermodynamics does not attempt to answer such questions as how long it takes to accomplish the required heat transfer, or what conditions are to be satisfied to achieve the heat transfer in a given time. These are some of the questions that the study of heat transfer attempts to answer. The study of heat transfer also provides the heat flux (heat transfer rate per unit area) and the temperature distribution in a medium. These answers cannot be obtained from the balance laws alone (balance of mass, linear and angular momentum, energy, and entropy) but need supplementary relations. For example, consider a 2-kg spherical copper ball, initially at a uniform temperature of 100 °C, dropped into a light, insulated vessel containing 1-kg of water at 25 °C as shown in Figure 1.1.1. Applying the First Law of Thermodynamics, it is found that the final uniform temperature of the ball and the water is 36.7 °C. When the ball is being cooled by the water, the temperature at the center of the sphere is higher than the temperature at the surface. The temperature variation in the radial direction at different times or the time taken for the ball to reach the uniform final temperature cannot be determined by applying just the principles of thermodynamics. In the study of heat transfer, we attempt to answer the additional questions of temperature distribution and the heat transfer rate. Another difference is that in thermodynamics we generally deal with material at a uniform temperature, whereas in heat transfer we deal with materials at nonuniform temperatures; without temperature differences there is no heat transfer. In heat transfer, we explicitly satisfy the First Law of Thermodynamics and implicitly satisfy the

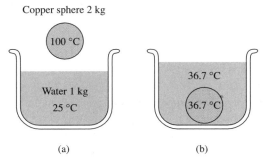

Copper sphere 2 kg

100 °C

Water 1 kg
25 °C

36.7 °C

36.7 °C

(a)   (b)

FIGURE 1.1.1 The Cooling of a Hot Spherical Ball Dropped in Cold Water
(a) A spherical copper ball initially at a uniform temperature is dropped in cold water. (b) After a sufficiently long time, the ball and the water reach the equilibrium temperature.

Second Law of Thermodynamics, i.e., that heat is transferred from a region of high temperature to a region of low temperature.

1.2 MODES OF HEAT TRANSFER

Two main modes of heat transfer—diffusion and radiation

Heat transfer has been defined as the energy transfer due to temperature differences. Heat transfer is a surface phenomenon; i.e., heat transfer occurs from (or to) a surface. We can recognize two modes of heat transfer—*diffusion*[1] and *radiation*. In diffusion from a surface, a material medium adjacent to the surface is required, and the effect of a temperature disturbance is propagated slowly (compared with radiation), first to the points in the immediate neighborhood of the disturbance, then to their neighboring points and so on. For example, consider the temperature history of a 1-cm diameter, 20-cm long plain carbon-steel rod, initially in equilibrium with the surrounding air at 20 °C, with one end of the rod immersed in water at 100 °C and held there. Figure 1.2.1 shows the temperature at four locations, 5 cm (A), 10 cm (B), 15 cm (C), and 20 cm (D), from the end that is in the water, at different times. At the instant the end is dipped, the temperature of the end increases but the rest of the bar remains at the initial temperature of 20 °C. The effect of the increase of the temperature at the end propagates to its neighbors due to diffusion of energy, and the

Diffusion

temperature of the neighboring material increases. This process of diffusion of energy towards regions of lower temperature continues and raises the temperature of the bar. Counting time from the instant the end is dipped in water, 20 seconds later the temperatures at A goes up to approximately 26 °C, but the temperatures at B, C, and D remain at 20 °C shown in Figure 1.2.1b. At the end of 3.5 minutes the temperatures at A, B, C, and D increase to 62 °C, 40 °C, 27 °C, and 24 °C, respectively (Figure 1.2.1c). After about 30 minutes the temperatures at the four points reach 78 °C, 64 °C, 57 °C, and 54 °C (Figure 1.2.1d), and there are no significant changes in the temperature at any part of the rod; i.e., steady state is reached.

Radiation

In contrast, radiation, with energy emitted by a surface or a gas, does not require a material medium for the transport of energy, and the effect of a change is felt over large distances almost instantaneously when compared with diffusion. The lighting of a lamp is a good example of radiation. As soon as the lamp is turned on, everyone in the room is aware of it. Another difference between diffusion and radiation is that in a diffusion process, the quantities of interest to us (for example, the heat transfer rate at a given location) can be determined from a knowledge of the state of the material in the immediate vicinity of the point. In the example of the steel rod immersed in hot water, it is only necessary to know the temperatures in the immediate vicinity of a point to determine the heat transfer rate per unit area in a given direction at that point. In radiation, we have to sum the effects of all points in the entire field that participates in radiation.

Although the heat transfer process has been divided into two categories, diffusion

---

[1]The term diffusion is generally used to denote mass transfer, but here it is used in a more general sense. Not only mass but also energy and momentum may be transferred by diffusion.

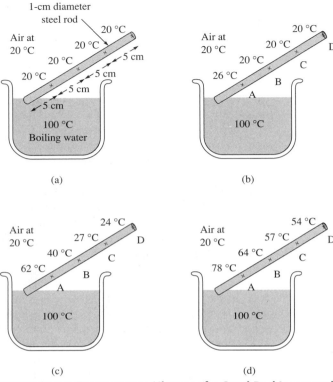

FIGURE 1.2.1 Temperature History of a Steel Rod Immersed in Water at 100 °C
The temperature of the rod at four locations at different times are shown. (a) Initial conditions, time: 0. (b) Temperature of the rod after 20 s. (c) Temperature after 210 s. (d) Temperature after 30 min.

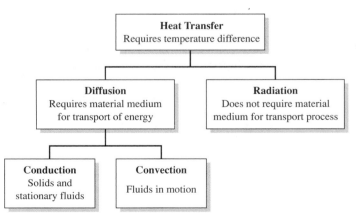

FIGURE 1.2.2 Diagram delineating conduction, convection, and radiation

Diffusion
subdivided—
conduction and
convection

and radiation processes, it is usual to subdivide diffusion further into *conduction* and *convection*. Heat transfer in solids and stationary fluids (fluids without bulk motion) is generally referred to as conduction, and heat transfer in fluids in bulk motion is referred to as convection. In conduction, energy transfer is by molecular diffusion, but in convection, energy is transferred by both molecular diffusion and bulk motion of the fluid, i.e., by the flow of the fluid at one temperature into a region at a different temperature. The various modes of heat transfer are shown in Figure 1.2.2.

1.3 CONDUCTION

• Conduction—heat
transfer in a
stationary medium

Conductive heat transfer is defined as heat transfer (diffusion of thermal energy) in solids and fluids without bulk motion. Conductive heat transfer generally takes place in solids, though it may occur in fluids without bulk motion or with rigid body motion. The basis for solving conduction problems is the First Law of Thermodynamics in conjunction with *Fourier's Law of Conduction*.

Fourier's Law of
Conduction

Fourier's Law of Conduction states that the heat flux in a given direction is proportional to the magnitude of the rate of change of temperature in that direction and is positive in the direction of decreasing temperature. If the temperature in a medium varies along a coordinate, say *x*, the local heat flux in the *x*-direction is given by

$$q''_x = -k \frac{\partial T}{\partial x} \tag{1.3.1}$$

Thermal
Conductivity

The coefficient, k, that relates the heat flux to the temperature gradient in Equation 1.3.1 is always positive, and is the *thermal conductivity* of the material. It is a property of the material. It may vary with spatial coordinates (nonhomogeneous materials) and it may depend on the temperature of the medium.

From Equation 1.3.1, it is clear that heat flux is a vector. In vector form it is given by

$$\mathbf{q}'' = -k \, \nabla \, T$$

In cartesian coordinates, $\nabla T = \hat{\imath} \, (\partial T/\partial x) + \hat{\jmath} \, (\partial T/\partial y) + \hat{k} \, (\partial T/\partial z)$, where $\hat{\imath}, \hat{\jmath},$ and $\hat{k}$ are the unit vectors along the axes.

Consider a rectangular slab, whose two opposite surfaces are maintained at uniform temperatures T_1 and T_2 as shown in Figure 1.3.1. When steady state is reached, i.e., when the temperature does not change with time, it is reasonable to assume that the temperature varies only in the *x*-direction. Such an assumption is valid if either the remaining surfaces are perfectly insulated, or the thickness of the slab, *L*, is much less than the length and the width of the slab so that the end effects (temperature variation due to heat transfer from the edge surfaces) are confined to a small region near the edges. From Fourier's law, the heat flux across any area perpendicular to the *x*-direction, from Equation 1.3.1, is

$$q''_x = -k \frac{dT}{dx} \tag{1.3.1a}$$

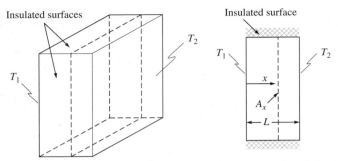

FIGURE 1.3.1 Two parallel surfaces of a rectangular slab are maintained at uniform temperatures.

In steady-state conditions, the heat transfer rate across every cross section perpendicular to the x-direction is constant. As the cross-sectional area is constant, the heat flux is also constant. Integrating Equation 1.3.1a,

$$\int_{T_1}^{T_2} k \, dT = -\int_0^L q''_x \, dx$$

Assuming constant thermal conductivity, we obtain

$$q''_x = k \frac{T_1 - T_2}{L} \tag{1.3.2}$$

The total heat transfer rate across a cross-sectional area of the slab is

$$q_x = \int_{A_x} q''_x \, dA_x = q''_x A_x$$

or

$$q_x = k A_x \frac{T_1 - T_2}{L} \tag{1.3.3}$$

where

q''_x = heat flux (W/m²) in the x-direction
q_x = heat transfer rate (W) in the x-direction
T_1 = temperature at $x = 0$ (°C or K)
T_2 = temperature at $x = L$ (°C or K)
L = thickness of the slab (m)
k = thermal conductivity of the material (W/m K or W/m °C)
A_x = cross-sectional area perpendicular to the x-direction (m²)

Note that the negative sign in Fourier's law implicitly satisfies the Second Law of Thermodynamics. For heat transfer in the x-direction to be positive, i.e., q''_x to be

TABLE 1.3.1 Thermal Conductivity of Some Materials at 300 K

Material	k (W/m K)
Carbon (type IIa diamond)	2620
Silver (pure)	429
Copper (pure)	401
Stainless steel (AISI 316)	13.4
Carbon (amorphous)	1.6
Glass (pyrex)	1.0
Concrete	0.663
Water	0.611
Sand	0.93
Fiberglass insulation (blanket)	0.046
Air	0.0262

positive, dT/dx must be negative (k is always positive) which means T must decrease in the x-direction. If T increases in the x-direction, i.e., dT/dx is positive, q''_x is negative indicating that the heat transfer is in the negative x-direction.

The thermal conductivities of materials vary greatly. The thermal conductivities of some materials are given in Table 1.3.1 to indicate the range of values of k. The values in the table are given only to show the relative magnitudes of the thermal conductivities of some common materials. For values of the thermal conductivities of common materials, refer to the tables in Appendix 1. The thermal conductivity varies from 0.0262 W/m K for air to 2620 W/m K for carbon (type IIa diamond), i.e., by a factor of 100 000.

EXAMPLE 1.3.1

The concrete walls and roof of a garage are 15-cm thick. The inside dimensions of the garage are 8-m long, 6-m wide, and 2.5-m high. The door is made of 12-mm thick hardwood and is 6-m wide and 2.5-m high. The garage is modeled in Figure 1.3.2. On a cold day, the garage is electrically heated. The inside and outside surface temperature are 5 °C and −10 °C, respectively. Determine the heat fluxes through the walls, roof, and the door, and the total heat transfer rate through the walls, roof, and the door.

Given

Dimensions of the garage: 8 m × 6 m × 2.5 m
Dimensions and material of the door: 12 mm × 6 m × 2.5 m, hardwood
Inside surface temperature (T_1) = 5 °C
Outside surface temperature (T_2) = −10 °C
Thickness of the wall = 15 cm

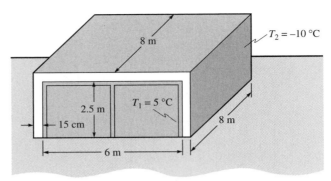

FIGURE 1.3.2 A garage with concrete walls and roof and a wooden door. The inner surfaces are at 5 °C and the outer surfaces at −10 °C.

Find

(a) The heat fluxes through the wall, roof, and the door
(b) The total heat transfer rate through the walls, roof, and the door

ASSUMPTIONS

1. Steady-state conditions
2. Constant properties
3. One-dimensional temperature distribution in the walls and the door

SOLUTION

(a) From Table A3B, the thermal conductivities of concrete and wood are 0.663 W/m K and 0.17 W/m K, respectively. As the heat transfer is by conduction, the heat flux under steady-state conditions is given by Equation 1.3.2. The heat flux across the concrete walls and roof is

$$q''_w = k_w \frac{T_1 - T_2}{L_w}$$

where

q''_w = heat flux across the concrete walls and roof (W/m²)
k_w = thermal conductivity of concrete (0.663 W/m K)
L_w = thickness of the concrete wall (0.15 m)
T_1, T_2 = temperatures of the inner and outer surfaces of the wall (°C)

Substituting the values,

$$q''_w = 0.663 \times \frac{5 - (-10)}{0.15} = \underline{66.3 \text{ W/m}^2}$$

Heat flux across the door is given by

$$q_d'' = k_d \frac{T_1 - T_2}{L_d}$$

where

q_d'' = heat flux across the door (W/m²)
k_d = thermal conductivity of hardwood (0.17 W/m K)
L_d = thickness of the door (0.012 m)
T_1, T_2 = temperatures of the inner and outer surfaces of the door (°C)

Substituting the values

$$q_d'' = 0.17 \times \frac{5 - (-10)}{0.012} = \underline{212.5 \text{ W/m}^2}$$

(b) The total heat transfer rate across the walls and the doors is obtained from

$q = q_w'' A_w + q_d'' A_d$

A_w = inner surface area of the walls and roof,
     $= 2 \times 8 \times 2.5 + 6 \times 2.5 + 8 \times 6 = 103 \text{ m}^2$

A_d = area of the inner surface of the door = $6 \times 2.5 = 15 \text{ m}^2$

$q = 66.3 \times 103 + 212.5 \times 15 = \underline{10,016 \text{ W}}$

COMMENTS

1. The total heat transfer across the walls and roof was computed on the basis of the inner surface area of the walls. But the surface area of the walls varies from 103 m² at the inner surface to 111.2 m² at the outer surface ($2 \times 8.15 \times 2.65 + 6.3 \times 2.65 + 8.15 \times 6.3$). A better estimate of the total heat transfer rate can be obtained by employing the arithmetic mean of the inner and outer surface areas, and is given by

$$q = 66.3 \times \frac{103 + 111.2}{2} + 212.5 \times 15 = \underline{10,288 \text{ W}}$$

2. The assumption of one-dimensional temperature distribution is not valid at the corners. But, as the thickness of the walls is much less than either the width or the height of the walls, one-dimensional approximation is valid for most of the walls except near the corners. The total heat transfer rate is slightly greater than the computed value due to this departure from the one-dimensional temperature distribution at the corners (why?).

3. The power input to the heater should be greater than the computed value of the total heat transfer rate as there is some heat transfer through the floor, which has been neglected, and due to the ingress of cold air through the cracks between the doors and the doors and the wall.

4. The heating load (10 kW) is large. It should be reduced by insulating the door and walls.

1.4 CONVECTION

Convection—heat transfer in a fluid in bulk motion

When there is a temperature gradient in a fluid (liquid or gas) in bulk motion, heat transfer occurs by two mechanisms, energy transfer by conduction, and by the bulk motion of the fluid. In addition to the random motion of the molecules, which results in conduction, there is the bulk motion of fluid packets. In the presence of temperature gradients, the bulk motion of the fluid packets results in heat transfer. The total heat transfer is then given by the superposition of the energy transfer due to molecular conduction and energy transfer, due to the bulk motion of the fluid packets. Generally, we are interested in the heat transfer rate between a fluid and its bounding solid surfaces. The heat flux from or to the solid surface is proportional to the difference between the surface temperature and a characteristic temperature of the fluid. The coefficient of proportionality is known as the *convective heat transfer coefficient* (Figure 1.4.1). The convective heat transfer coefficient, h, is defined by the equation

Convective heat transfer coefficient

$$q_c'' = h\,(T_s - T_f) \tag{1.4.1}$$

where

q_c'' = heat flux from the solid surface to the adjacent fluid (W/m^2)
T_s = temperature of the solid surface (°C or K)
T_f = temperature of the fluid with reference to which h is defined (°C or K)
h = convective heat transfer coefficient (W/m^2 °C or W/m^2 K)

The convective heat transfer coefficient is always positive. If $T_s < T_f$, the heat transfer is from the fluid to the solid surface. Thus, if $q_c'' < 0$ it indicates that the heat transfer is from the fluid to the solid. Equation 1.4.1 is also known as *Newton's Law of Cooling*. Note that when defining the convective heat flux from a solid surface to the adjacent fluid, the heat flux is always along the normal to the surface. *In convection problems, as we work only with this normal component of the heat flux, the vector notation is not generally used.* The convective heat flux may vary from

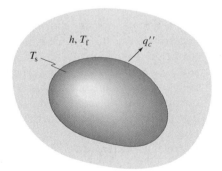

FIGURE 1.4.1 Definition of Convective Heat Transfer Coefficient
A solid surface at a temperature of T_s is adjacent to a fluid at T_f.

one location to another. In Equation 1.4.1, if q''_c is defined as the average heat flux (i.e., the total heat transfer rate divided by the surface area), then the average value of h is obtained. If q''_c is the local heat flux, then the local value of h is obtained. In this book the local heat transfer coefficient is denoted with a coordinate subscript, for example, h_x. The symbol h without any coordinate subscript denotes the average heat transfer coefficient. Sometimes, a subscript such as L or d is used to clarify the dimension over which the average heat transfer coefficient is determined. For example, h_L, when used with a plate of length L, denotes the average heat transfer coefficient over the length of the plate. Similarly, h_d is used to denote the average heat transfer coefficient associated with a cylinder or a sphere of diameter d.

Equation 1.4.1 defines the convective heat transfer coefficient. In the definition, the choice of the reference temperature is arbitrary; when a value of the convective heat transfer coefficient is given, the reference temperature for its definition must also be indicated. For example, with a fluid flowing in a tube whose surface temperature is different from that of the fluid, the temperature of the fluid varies across a cross section. In defining the convective heat transfer coefficient, one may choose the temperature at the axis of the tube, or the average temperature of the fluid at that section, as the reference temperature. As the heat flux is the same irrespective of the choice of the reference temperature, the value of the convective heat transfer coefficient is different for the two reference temperatures. In defining the convective heat transfer coefficient with heat transfer from (or to) fluids flowing inside ducts, the accepted reference temperature is the average temperature, also known as the bulk temperature. In external flows, i.e., the flow of a fluid over a surface, the reference temperature is the temperature of the fluid far away from the surface, also known as the free stream temperature.

We emphasize that *the convective heat transfer coefficient is not a property of the fluid but depends on a variety of factors, principally the velocity of the fluid, the flow pattern, and the geometry in addition to the properties of the fluid.* It is within our experience to realize that the convective heat transfer coefficient is much greater when we face wind velocities of 30 mph than in calm winds even when the air temperature is the same in both cases, say 0 °C. We feel the cold more at higher wind velocities because of greater heat transfer rates from our body to the air. From Equation 1.4.1 as the temperature difference in the two cases is the same, this increased heat transfer rate can take place only if the convective heat transfer coefficient is higher at higher wind velocities. Then again, with winds of 30 mph at 0 °C, we feel the cold more when we face the wind than when we turn away from it, a result of different flow patterns around our face in the two cases.

Convection is subdivided into *forced convection* and *natural convection*. In forced convection heat transfer, the motion of the fluid is caused by an external agency, such as a pump or a fan; the motion is not dependent on the difference in temperatures of the solid surface and the fluid. In natural convection the motion of the fluid is caused by the difference in the temperatures of the bounding surface and the fluid, such as occurs in water in a pan that is kept on a heated stove. In most cases, velocities of fluids in forced convection are greater than those in natural convection. As velocity is one of the parameters that determines the value of the convective heat transfer coefficient, the value of the convective heat transfer coefficient

TABLE 1.4.1 Magnitude of Convective Heat Transfer Coefficients

	h (W/m^2 K)
Natural convection (air)	5–15
Natural convection (water)	500–1000
Forced convection (air)	10–200
Forced convection (oil)	20–2000
Forced convection (water)	300–20 000
Water (boiling)	3000–100 000
Steam (condensing)	5000–10 000

in forced convection is usually much greater than in natural convection for the same fluid.

The range of values of convective heat transfer coefficients for some common cases are given in Table 1.4.1 These values are given to illustrate the range and magnitudes of the convective heat transfer coefficients, *and should not be treated as properties or used in the solution of problems.* Convective heat transfer coefficients vary from approximately 5 W/m^2 K in natural convection in gases to as high as 100 000 W/m^2 K in boiling of liquids, i.e., they vary by a factor of 20 000.

EXAMPLE 1.4.1 Lead is melted in a 9-mm thick cast-iron (4% C), rectangular tank (Figure 1.4.2). The inside surface of the tank is at the melting point of lead, 328 °C, and the outer surface is at 327 °C. The outer surface of the tank is exposed to air at 20 °C. Find the heat flux across the side walls of the tank and estimate the convective heat transfer coefficient associated with the outer surface of the tank.

Given

Thickness of the tank (L) = 9 mm
Inner surface temperature (T_1) = 328 °C

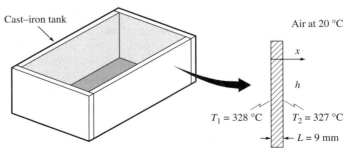

FIGURE 1.4.2 The sketch on the left shows the cast-iron tank. One of the side walls is represented on the right.

Outer surface temperature $(T_2) = 327\,°C$
Temperature of surrounding air $(T_\infty) = 20\,°C$

Find

(a) Heat flux through the tank walls
(b) The convective heat transfer coefficient h associated with the outer surface

ASSUMPTIONS

1. Steady state
2. The length, width, and height of the tank are much greater than the thickness of the tank, and end effects at the joints of the side walls are negligible for each wall. As the end effects are negligible, the temperature of the side wall varies only in the direction perpendicular to inner surfaces. Hence the temperature distribution is one-dimensional.

SOLUTION

(a) With the assumption of steady state, one-dimensional temperature distribution, and with the temperatures of the two parallel surfaces of the side walls known, the situation is identical to the case analyzed in Section 1.3. Denoting the temperatures of the inner and outer surfaces of the wall by T_1 and T_2, from Equation 1.3.2 we have

$$q''_k = \frac{k\,(T_1 - T_2)}{L} \tag{1}$$

where

q''_k = conductive heat flux
L = thickness of the wide wall

As for the properties of case iron, the side walls are at 327 °C, however, properties of cast iron (4% C) are available only at 27 °C in Table A1. We use the available value but recognize that our solution is approximate: $\rho = 7272\ kg/m^3$, $c_p = 420$ J/kg K, and $k = 52$ W/m K.

Substituting the known values into Equation 1,

$$q''_k = k\,\frac{T_1 - T_2}{L} = 52 \times \frac{328 - 327}{0.009} = \underline{5778\ W/m^2}$$

(b) A surface does not have any mass and cannot store energy. Therefore, from an energy balance on the outer surface, conductive heat flux to the surface equals the convective heat flux to the surrounding air.

$$q''_k = q''_c = h\,(T_s - T_\infty)$$

where

q''_c = convective heat flux from the surface to the surrounding fluid (W/m²)
h = convective heat transfer coefficient (W/m² K)

T_s = temperature of the outer surface (327 °C)

T_∞ = temperature of air (20 °C)

Employing the conductive heat flux from part a, the convective heat transfer coefficient is calculated.

$$5778 = h\,(327 - 20) \qquad h = 18.8 \text{ W/m}^2 \text{ K}$$

1.5 RADIATION

Radiative energy is energy emitted by matter as electromagnetic waves or photons resulting from the matter being at a finite temperature. The maximum radiative energy flux that can be emitted from a solid surface at a defined temperature is given by the *Stefan-Boltzmann Law:*

Stefan-Boltzmanns Law for maximum emitted radiative energy flux

$$E_b = \sigma T^4 \qquad\qquad (1.5.1)$$

where

E_b = maximum emitted radiative energy flux at temperature T, known as the black body emissive power of the surface (W/m²)

T = absolute temperature (K)

σ = Stefan-Boltzmann constant (5.67×10^{-8} W/m² K⁴)

Black Surface

A surface that emits the maximum amount of energy at any given temperature is an *ideal radiator,* also known as a *black surface.* The energy flux, E, emitted by real surfaces is less than this maximum, and the ratio, E/E_b, is the emissivity, ϵ, of the surface. Radiative heat exchange between surfaces, particularly if the medium separating the surfaces also participates in radiation (many gases do), is very complex. Some simple situations are considered in the chapters on radiative heat transfer.

Consider the radiative heat transfer from a convex surface (area A_s, emissivity ϵ,

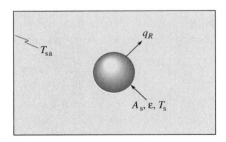

FIGURE 1.5.1 Radiative Heat Transfer
A solid with no concave surface, such as a sphere, is suspended in a large room.

and temperature T_s) to another much larger surface (temperature T_{sa}) that completely surrounds the first surface. The two surfaces are separated by a medium that does not participate in radiation (Figure 1.5.1). No part of the inner surface can see any other part of itself, i.e., no part of it is concave.

There is no restriction on the shape of the outer surface. In this particular case, the net radiative energy transfer rate from the inner surface to the outer surface is given by

$$q = \sigma A_s \epsilon \, (T_s^4 - T_{sa}^4) \tag{1.5.2}$$

$$q'' = \frac{q}{A_s} = \sigma \epsilon \, (T_s^4 - T_{sa}^4) \tag{1.5.3}$$

where

$\quad q$ = radiative heat transfer rate (W/m^2)
$\quad \sigma$ = Stefan-Boltzmann constant (5.67×10^{-8} W/m^2 K^4)
$\quad \epsilon$ = emissivity of the surface
$\quad T_s$ = temperature of the inner surface (K)
$\quad T_{sa}$ = temperature of the surrounding surface (K)
$\quad A_s$ = surface area of the inner surface (m^2)
$\quad q''$ = radiative heat flux (W/m^2)

EXAMPLE 1.5.1   The temperature of the outer surface of a small furnace is 40 °C when the surrounding air temperature is 25 °C, and the temperature of the walls of the room is 20 °C (Figure 1.5.2). If the emissivity of the furnace surface is 0.85 and the convective heat transfer coefficient is 10 W/m^2 K, determine the total heat transfer rate from the outer surface to the surroundings. The furnace is kept on a well-insulated surface and the outer surface area exposed to air is 0.8 m^2.

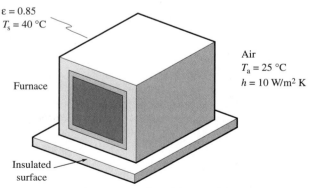

$\varepsilon = 0.85$
$T_s = 40\ °C$

Furnace

Air
$T_a = 25\ °C$
$h = 10$ W/m^2 K

Insulated surface

FIGURE 1.5.2 The outer surface of a small furnace in a large room is at 40 °C

Given

Outer surface temperature of the furnace walls $(T_s) = 40\,°C$
Surrounding air temperature $(T_a) = 25\,°C$
Temperature of the walls of the room $(T_{sa}) = 20\,°C$
Emissivity of the outer surface of the furnace $(\epsilon) = 0.85$
Convective heat transfer coefficient $(h) = 10\ W/m^2\,°C$
Area of the outer surface exposed to air $(A) = 0.8\ m^2$

Find

The total heat transfer rate from the outer surface of the furnace to the surroundings.

ASSUMPTIONS

1. Steady state
2. The surface area of the walls of the surroundings is much greater than that of the furnace
3. Air does not participate in radiation
4. As the furnace is placed on a well-insulated surface, the conductive heat transfer from the bottom surface is negligible in comparison with the convective and radiative heat transfer from the surface exposed to air.

SOLUTION

The total heat transfer rate from the outer surface is the sum of the convective and radiative heat transfer rates:

$$q_{tot} = q_c + q_R$$

The convective heat transfer rate is found from Equation 1.4.1, and as the surface of the surrounding walls of the room is much greater than the outer surface of the furnace, the radiative heat transfer rate is found from Equation 1.5.2.

$$q_c = hA\,(T_s - T_a) \qquad\qquad (1)$$

$$q_R = \sigma A \epsilon\,(T_s^4 - T_{s_a}^4) \qquad\qquad (2)$$

where

q_c = convective heat transfer rate from the outer surface of the furnace to the surrounding air (W)
h = convective heat transfer coefficient $(10\ W/m^2\ K)$
A = outer surface area of the furnace $(0.8\ m^2)$
T_s = temperature of the outer surface of the furnace (313.2 K)
T_a = temperature of the surrounding air (298.2 K)
q_R = radiative heat transfer rate from the outer surface (W)
σ = Stefan-Boltzmann constant $(5.67 \times 10^{-8}\ W/m^2\ K^4)$
ϵ = emissivity of the surface (0.85)
T_{sa} = temperature of the wall (293.2 K)

Substituting the values into Equations 1 and 2,

$$q_c = 10 \times 0.8 \times (313.2 - 298.2) = 120 \text{ W}$$
$$q_R = 5.67 \times 10^{4-8} \times 0.8 \times 0.85 \times (313.2^4 - 293.2^4) = 86.1 \text{ W}$$
$$q_{total} = q_c + q_R = 120 + 86.1 = \underline{206.1 \text{ W}}$$

COMMENTS

1. As the air temperature and the wall temperature are different, two different temperatures are used in the computation of the convective and radiative heat transfer rates.
2. The radiative heat transfer is 42% of the total heat transfer and is a significant part of the total heat transfer.

EXAMPLE 1.5.2

A thermocouple is a sensor to measure temperature. It consists of a junction of two wires of dissimilar materials. The junction of a thermocouple, which is in the form of a 2-mm diameter bead, is placed inside a long duct in which air flows. (Figure 1.5.3). The temperatures of the air and duct surface are 100 °C and 60 °C, respectively. The emissivity of the thermocouple junction surface is 0.7, and the convective heat transfer coefficient is 20 W/m² °C. Determine the equilibrium (steady-state) temperature of the junction.

Given

Temperature of the air in the duct (T_a) = 100 °C
Duct surface temperature (T_{sa}) = 60 °C
Diameter of the thermocouple bead (d) = 2 mm
Emissivity of the surface of the bead (ϵ) = 0.7
Convective heat transfer coefficient associated with the bead (h) = 20 W/m² °C

Find

The temperature of the bead (T_s)

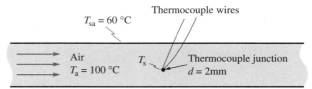

FIGURE 1.5.3 A thermocouple junction is placed in an air stream to measure the temperature of the air.

ASSUMPTIONS

1. The surface area of the duct walls is much greater than the surface area of the thermocouple bead
2. Air does not participate in radiative heat transfer
3. Radiative heat transfer is only between the junction and the duct walls
4. There is no concave surface on the bead
5. Conductive heat transfer through the thermocouple wires is negligible.

SOLUTION

Under equilibrium conditions, the bead does not store any energy and the net energy transfer to it is zero. As the only energy transfer is by heat transfer, the net heat transfer rate to the bead is zero. As conduction through the wires is neglected, heat transfer from the bead to the air is by convection and to the duct walls is by radiation.

$$q_c + q_R = 0 \tag{1}$$

where

$\quad q_c$ = convective heat transfer rate from the bead to the air (W)
$\quad q_R$ = radiative heat transfer rate from the bead to the duct walls (W)

From the definition of convective heat transfer coefficient,

$$q_c = hA \ (T_s - T_a) \tag{2}$$

where

$\quad h$ = convective heat transfer coefficient (20 W/m^2 K)
$\quad A$ = surface area of the bead (m^2)
$\quad T_s$ = temperature of the surface of the bead (K)
$\quad T_a$ = temperature of the air (373.2 K)

From assumptions 1–4, the radiative heat transfer rate from the bead to the duct walls is determined from Equation 1.5.2.

$$q_R = \sigma A \epsilon \ (T_s^4 - T_{sa}^4) \tag{3}$$

where

$\quad \sigma$ = Stefan-Boltzmann constant (5.67 $\times$ 10^{-8} W/m^2 K^4)
$\quad \epsilon$ = emissivity of the bead surface (0.7)
$\quad T_{sa}$ = temperature of the duct walls (333.2 K)

Introducing the expressions for convective and radiative heat transfer rate into Equation 1, and rearranging the equation, we obtain

$$T_s^4 + \frac{h}{\sigma \epsilon} T_s - (T_{sa}^4 + \frac{h}{\sigma \epsilon} T_a) = 0 \tag{4}$$

The solution of Equation 4 yields the value T_s.

Substituting the values of h, σ, ϵ T_{sa}, and T_a into Equation 4,

$$T_s^4 + 5.0391 \times 10^8 \times T_s - 2.0038 \times 10^{11} = 0 \tag{5}$$

Solving Equation 5 for T_s,

$$T_s = 363.1 \text{ K}$$

COMMENTS

1. Equation 5 has four roots, 363.1 K, $-$ 899.1 K, and two complex roots. But from the physics of the problem (and thermodynamic considerations, which forbid negative absolute temperatures) the appropriate value of T_s is 363.1 K.
2. If the thermocouple is used to measure the temperature of the air, it reads a temperature of 363.1 K (90 °C), significantly lower than the actual temperature of the air. Although it is unlikely that the duct wall temperature would be at 60 °C with the air at 100 °C, the example shows that if the duct wall temperature is different from the temperature of the air, a correction should be made to the temperature indicated by the thermocouple, or steps must be taken to reduce the effect of radiation heat transfer. Can you suggest any such step?

1.6 RELATIVE MAGNITUDES OF CONVECTIVE AND RADIATIVE HEAT TRANSFER

When heat transfer from a surface is by both radiation and convection, one of the modes is sometimes dominant; in such cases where there is a dominant mode, the heat transfer from the other mode can be neglected. To get an appreciation for the relative magnitudes in such cases, recall that the heat fluxes by the two modes are given by

$$q_c'' = h(T_s - T_a) \tag{1.4.1}$$

$$q_R'' = \sigma \, \epsilon \, (T_s^4 - T_{sa}^4) \tag{1.5.3}$$

where

q_c'' = convective heat flux (W/m²)
q_R'' = radiative heat flux (W/m²)
T_s = temperature of the surface (K)
T_a = temperature of the surrounding fluid (K)
T_{sa} = temperature of the surrounding surface (K)

Relative magnitudes of convective and radiative heat transfer rates

To estimate the relative magnitudes of convective and radiative heat fluxes in some common situations, consider surfaces exposed to a fluid that does not participate in radiation, such as air. The fluid temperature is T_a. The surrounding surfaces are also at the same temperature ($T_a = T_{sa}$). Most surfaces, excluding highly polished metallic surfaces, have an emissivity in the range of 0.6–0.8; we will use 0.8 for computing the radiative heat flux. The convective and radiative heat fluxes computed from Equations 1.4.1 and 1.5.3, respectively, are given in Table 1.6.1. The values of the convective heat transfer coefficients used in computing the convective heat fluxes are typical for the cases considered.

TABLE 1.6.1 Relative Magnitudes of Convective and Radiative Heat Fluxes

	q_C'' (W/m^2)	q_R'' (W/m^2)	q_R''/q_C
1. $T_S = 127\,°C$, $T_a = 27\,°C$ Air, natural convection, $h = 6\ W/m^2\,°C$	600	795	1.33
2. $T_S = 127\,°C$, $T_a = 27\,°C$ Air, forced convection, $h = 50\ W/m^2\,°C$	5000	795	0.159
3. $T_S = 127\,°C$, $T_a = 27\,°C$ Air, forced convection, $h = 200\ W/m^2\,°C$	2×10^4	795	0.0398
4. $T = 600\,°C$, $T_a = 500\,°C$ Air, forced convection, $h = 200\ W/m^2\,°C$	2×10^4	1.02×10^4	0.508

Table 1.6.1 shows that in natural convection in air, radiative heat transfer is significant even at low surface temperatures, but in forced convection it is not that significant, unless the temperature levels are high. Note that for the same temperature difference of 100 K and convective heat transfer coefficient of 200 W/m^2 K, radiative heat transfer becomes significant when the temperature level is raised from a mean of 350 K (case 3 in Table 1.6.1) to 823 K (case 4).

1.7   RELEVANCE OF HEAT TRANSFER

The relevance and importance of the study of heat transfer are illustrated by considering the role it plays in such varied cases as power plants, refrigeration equipment, solar heated homes, and transportation vehicles; all of which play a dominant role in the control of our modern-day life style, and in the human body itself.

1.7.1 Steam Power Plant

Heat transfer in a steam power plant

A steam power plant is schematically shown in Figure 1.7.1. The plant operates on the reheat cycle with two closed feedwater heaters. Steam from the boiler drum, superheated in the superheater, is admitted to the high pressure turbine. A part of the exhaust steam from the high pressure turbine is supplied to feedwater heater A. The remaining steam is reheated in the boiler and supplied to the low pressure turbine. After partial expansion, a part of the steam is extracted from the turbine and supplied to the second feedwater heater, B. The exhaust steam from the low pressure turbine is condensed in the condenser. The feedwater pump raises the pressure of the condensate from the condenser pressure to the boiler pressure. The feedwater is heated in the two feedwater heaters by the steam extracted from the turbines. The temperature of the feedwater is further raised in the economizer by the stack gases (combustion products) in the boiler before entering the boiler. The water is heated in the convection coils to produce steam, and the cycle is repeated. In addition to the economizer, steam drum, convection coils (tubes in which the water is heated in the boiler), superheater, and reheater, all of which are in the feedwater-steam

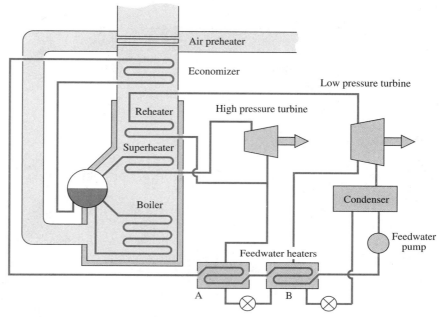

FIGURE 1.7.1 Schematic of a Steam Power Plant

Boiler

loop, the boiler has an air preheater where the combustion air is heated by the stack gases. Both the economizer and the air preheater reduce the fuel requirements.

Detailed views of a boiler, a condenser, and a feedwater heater are shown in Figures 1.7.2 through 1.7.5. Figure 1.7.2 is a dimensioned sketch of a boiler installation, and Figure 1.7.3. is a schematic view of a boiler. In the boiler, heat is transferred from the combustion products to the steam drum, convection coils, superheater, and reheater by forced convection and radiation. The combustion products contain significant amounts of particulates, moisture, and carbon dioxide, all of which participate in radiative heat transfer. As the temperature of the gases is quite high, radiative heat transfer from the gases to the boiler is quite significant, and in many cases, may be more than the convective heat transfer. In the economizer and air preheater, the temperature of the gases is much lower and forced convection heat transfer becomes more significant. In each case, heat is transferred to the water by conduction across the tube walls. In the economizer, superheater, reheater, and air preheater, heat transfer is by forced convection to a single phase fluid. In the convection coils, the heat transfer to the water is by forced convection; in parts of the convection coil, the heat transfer to the water results in a change of phase, from liquid phase to the vapor phase.

The boiler shown in Figure 1.7.2 is 73-m (238-ft) high and 70-m (230-ft) wide. The height of the boiler is approximately that of a building with 30 floors. A power plant has more than one boiler; you may imagine the size of the building to house these boilers.

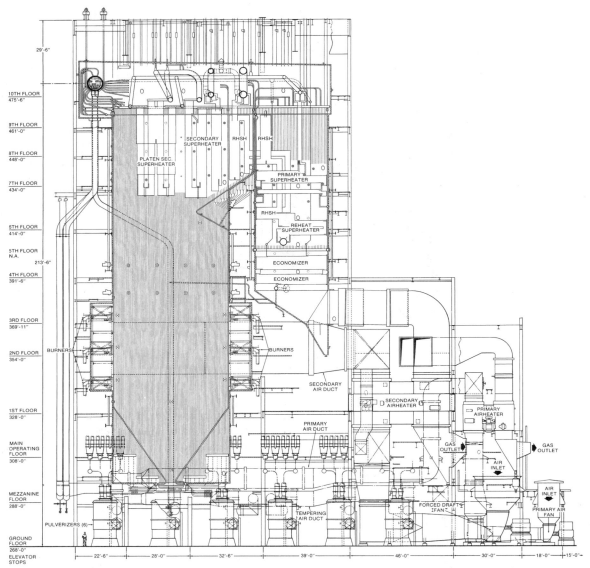

FIGURE 1.7.2 Dimensioned Sketch of a Boiler (Courtesy: Babcock & Wilcox, Barberton, OH)

In a modern power plant the temperature of the steam at inlet to the high pressure turbine is approximately 560 °C. The inner surface of the turbine is also at similar high temperatures. The turbine is well insulated to reduce the heat transfer to the surroundings and to reduce the temperature of the casing. The reduction in heat transfer from the turbine to the surroundings improves the efficiency of the turbine;

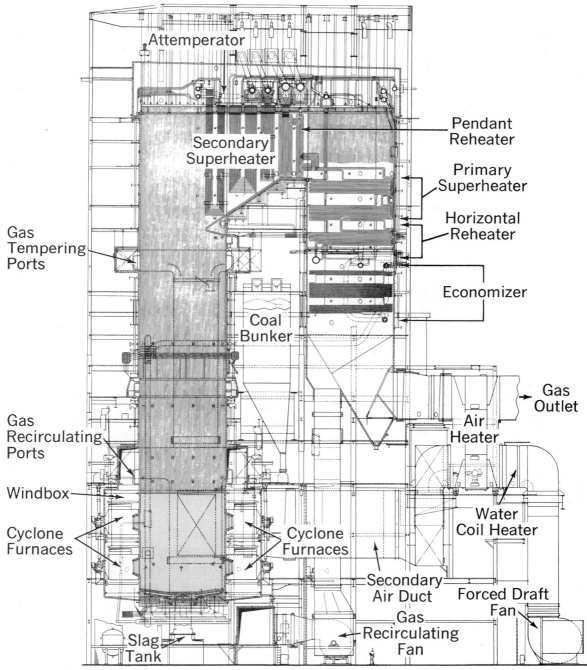

FIGURE 1.7.3 Schematic View of a Boiler (Courtesy: Babcock & Wilcox, Barberton, OH)

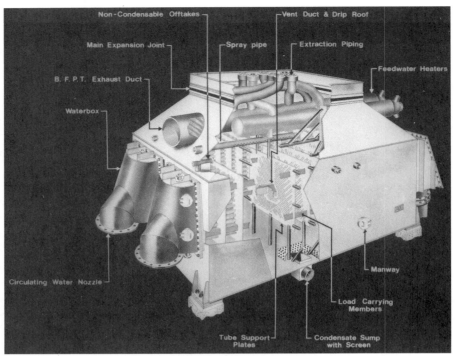

FIGURE 1.7.4 A Water-Cooled Steam Condenser (Courtesy: Senior Engineering Company, Los Angeles, CA)

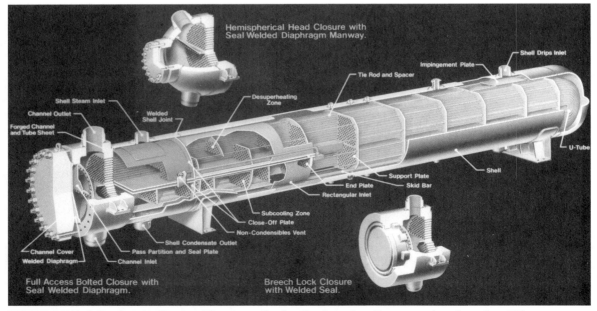

FIGURE 1.7.5 A Feedwater Heater (Courtesy: Senior Engineering Company, Los Angeles, CA)

the reduction in the temperature of the casing is necessary to make it safe for the operating personnel.

The exhaust steam from the turbines is condensed in a condenser. A typical condenser is shown in Figure 1.7.4. Steam is condensed by water flowing in the tubes in the condenser, with steam condensing on the outer surface of the tubes. Heat transfer to the steam is accompanied by a change in phase, from the vapor phase to the liquid phase. The heat transfer to the cooling water is by forced convection. A typical condenser in a large power plant is approximately 20-m long, 9-m wide, and 12-m high (70-ft long, 30-ft wide, and 40-ft high).

A feedwater heater is shown in Figure 1.7.5. In the feedwater heaters, heat transfer to the feedwater is by forced convection to a single phase fluid; heat transfer from the bleed steam (steam extracted from the turbines) results in a change of phase.

In all cases, conductive heat transfer takes place from one fluid to another across the solid surfaces that separate the two fluids.

1.7.2 Cooling System in a Transport Vehicle

A transport vehicle, such as an automobile or a truck, has a number of heat transfer components. A typical cooling system of a bus or a truck is shown in Figure 1.7.6. The engine is cooled by a coolant, typically a mixture of water and an antifreeze compound, circulating around the cylinders in the engine block. Heat is transferred to the coolant from the combustion products across the cylinder walls by conduction. Heat transfer to the coolant from the cylinder walls is by forced convection. The coolant passes through the radiator tubes. Atmospheric air, drawn over the radiator tubes, cools the coolant. In the radiator, there is forced convective heat transfer from the coolant to the tube walls, conductive heat transfer from the inner surface of the tubes to the outer surface and the thin metal sheets attached to the tubes, and forced convective heat transfer to the atmospheric air. If the engine is equipped with a supercharger, there is a charge-air cooler in which the pressurized air is cooled by atmospheric air. There is also a lubricating oil cooler and a heat exchanger for heating the air supplied to the truck cabin during cold days. The refrigerant in the air conditioning unit is cooled by an air-cooled condenser where the refrigerant is condensed by heat transfer to the atmospheric air. The refrigerant evaporates in the evaporator and cools the air supplied to the cabin on hot days.

1.7.3 Solar-Heated House

A solar-heated house makes use of the solar energy to supply a part or all of the space heating and hot water needs of a house. When solar energy is used for space heating, its availability is greatest when the need is least (on bright, sunny days), and its availability is least when the need is greatest (on cold nights). Due to the periodic availability of solar energy and the mismatch between the availability and the need, a large energy storage system is needed to store the solar energy. In most cases it is uneconomical to design a solar-heating system to meet the entire heating needs of a house. Hence, a solar-heated house usually has a conventional heating system, which uses oil, gas, or electricity. The system shown in Figure 1.7.7 consists

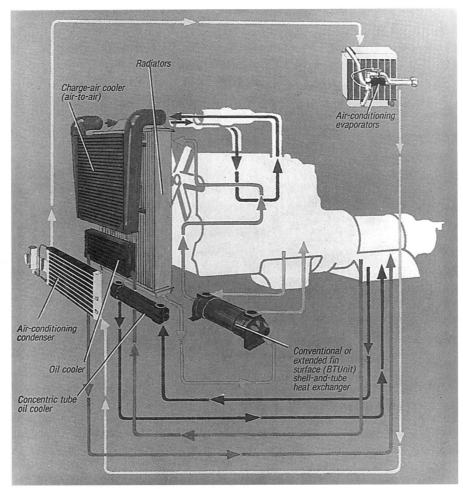

FIGURE 1.7.6 Diagram Showing the heat transfer equipment in a Truck (Courtesy: Modine Manufacturing Company, Racine, WI)

of a number of flat plate collectors, a storage tank, a hot water boiler for heating the house, and a domestic hot water heater for the supply of hot water.

flat plate collector

The flat plate collector, shown in Figure 1.7.8, consists of tubes brazed to a flat plate. One or two glass panes (known as cover plates) cover the plate. The collectors are usually mounted on the roof. Radiant energy from the sun heats the flat plate which, in turn, heats the liquid circulating through the tubes and the storage tank. Continuous circulation of the liquid in the storage tank through the collector, when solar energy is available, raises the temperature of the liquid in the storage tank. There are two heat transfer coils in the storage tank. The hot water return from the space heaters flows through one coil where it is heated by the liquid in the storage tank. If the temperature of the hot water at exit from the storage tank is not suffi-

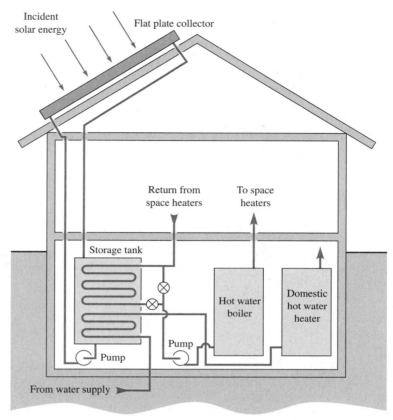

FIGURE 1.7.7 Solar-heated house

ciently high, the hot water is further heated in the hot water boiler. The water from the boiler is circulated through the space heaters located in different parts of the house. Water from the water mains flows through a second coil where it is heated by the liquid in the storage tank. If needed, the water is further heated in the domestic hot water heater to a temperature suitable for showers and kitchen use.

In the collector, the collector plate is heated by radiant energy from the sun. A major part of the energy is transferred to the working fluid. Heat is also transferred from the collector plate to the glass cover plates and from the glass plates to the surroundings by radiation and convection. To increase the efficiency of the collectors, i.e., the fraction of the incident solar energy that is transferred to the working fluid, such heat transfer from the collector to the surroundings must be reduced.

In the storage tank, heat is transferred from the working fluid to the water by forced convection. In the hot water heaters (space heating and domestic hot water supply) water is heated by the combustion products (oil or gas). The hot water for space heating is circulated through the radiators in different parts of the house. The air in different rooms is heated by the radiators by natural convection. From the

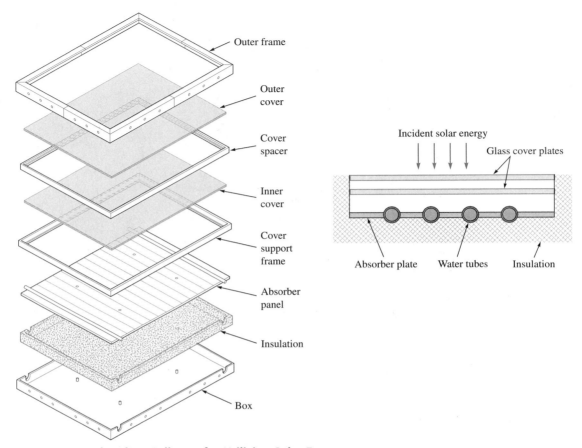

FIGURE 1.7.8 **Flat Plate Collector for Utilizing Solar Energy**

heated rooms, heat is transferred by natural convection to the inner surfaces of the outer walls, from the inner surfaces to the outer surfaces by conduction, and from the outer surfaces to the surroundings by convection and radiation. Convective heat transfer from the outer surfaces to the ambient air may be by either natural convection or forced convection depending on the wind velocity.

1.7.4 Refrigeration Plant

Refrigeration plants are used for air-conditioning, to cool warehouses, to produce ice, and so forth. A refrigeration plant is shown schematically in Figure 1.7.9. Refrigerant vapor from the evaporator is compressed in a reciprocating or rotary compressor. The high pressure vapor is condensed in the condenser either by water or air. The liquid refrigerant is throttled to a low pressure by an expansion valve in large plants and by a capillary tube in small systems, such as a domestic refrigerator. The throttling of the liquid to a low pressure results in a liquid-vapor mixture at a low

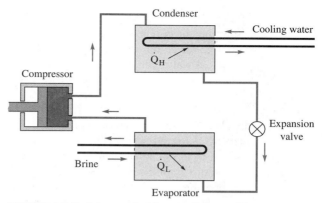

FIGURE 1.7.9 Schematic Diagram of a Refrigeration Plant

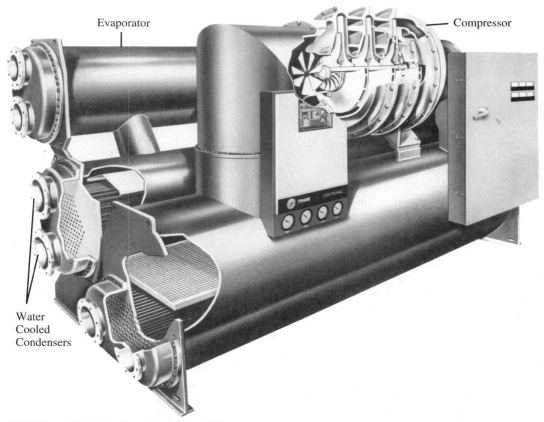

FIGURE 1.7.10 A Packaged Water Chiller (Courtesy: The Trane Company, La Crosse, WI)

temperature. The mixture enters the evaporator where heat is transferred to the refrigerant from the surrounding air, water, or brine. The vapor from the evaporator is compressed in the compressor and the cycle is repeated.

Heat transfer in a refrigeration unit

A packaged water chiller is shown in Figure 1.7.10. The unit has a three-state rotary compressor, a water chiller, and condensers, all mounted on a single frame. In the condensers, the refrigerant vapor is cooled by water flowing through the tubes. Forced convection heat transfer to the water in the tubes results in condensation of the vapor on the outer surface of the tubes. In the evaporator, the water in the tubes is cooled by the refrigerant, which evaporates on the outer surface of the tubes. Water in the tubes is cooled by forced convection. The unit shown in Figure 1.7.10 has one evaporator (adjacent to the control panel) and two water-cooled condensers. The unit operates with R-11 as the refrigerant. The larger units have refrigeration capacities of up to 1400 tons (12 000 BTU per hour of cooling capacity is one ton of refrigeration).

1.7.5 The Human Body

In the human body, there is continuous metabolic activity, equivalent to a source of energy, that tends to increase the temperature of the body. The rate of metabolism depends on many factors; for example, increased physical activity results in a higher metabolic rate. In cool climates and warm climates, in humid regions and arid regions, in deep sleep or when engaged in vigorous physical activity, the temperature of the human body is maintained at approximately 36.8 °C by a complex regulatory system involving both heat and mass transfer.

Maintaining human body temperature

When the ambient temperature is less than the body temperature, the skin temperature is essentially controlled by the blood circulation in the veins closest to the skin. The heat transfer from the body depends on the difference between the skin temperature and the temperature of the surroundings. To maintain a constant heat transfer rate from the body to the surroundings, as the ambient temperature decreases, the skin temperature must decrease. Such a reduction in the skin temperature is achieved by a reduction in the blood supply to the veins. If the surrounding air is much colder than the body, extended exposure may cause the temperature of the skin to decrease to the point of causing frost bite. The risk of frost bite may be reduced by increasing physical activity, which causes increased metabolic activity, necessitating increased heat transfer from the body. The need to increase heat transfer increases blood supply to the veins, which increases the skin temperature. As the surrounding temperature increases, or if there is increased physical activity, the blood supply to the veins increases and the heat transfer rate from the skin increases.

Heat transfer from the human body

When the ambient temperature is higher than the body temperature, an additional mechanism is required to control the body temperature. Evaporative cooling provides such a mechanism. Both high ambient temperature and increased physical activity activate the sweat glands. Water, along with other compounds, is discharged from the sweat glands on to the skin. As the surrounding air is generally not saturated, the water on the skin evaporates. The rate of evaporation of the water on the skin depends on the relative humidity, temperature, and movement of the surrounding air. The evaporation of water from the body transfers energy to the surroundings

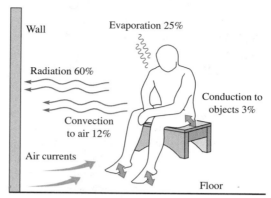

FIGURE 1.7.11 The Human Body and Heat Transfer Mechanisms

and cools the body. Additional evaporative cooling is provided by respiration with the humidity in the exhaled air being greater than the humidity in the inhaled air. The heat transfer from the human body is illustrated in Figure 1.7.11.

1.8 TERMINOLOGY

We often use terms such as pure, homogeneous, uniform, isotropic, transient and steady, one-dimensional, and uncertainty. It is necessary that we understand the meaning of these terms.

Pure substance

The term *pure substance* is generally used to indicate that the substance is made of only one type of molecule. For example pure water (in any state, ice, water, or steam) contains only H_2O molecules.

Homogeneous and heterogeneous substances

A *homogeneous substance* contains more than one pure substance or more than one phase of a pure substance so mixed that every volumetric element contains the same proportion of molecules of the different pure substances or of the phases. For example, air is a homogeneous mixture of nitrogen and oxygen (and a few other elements and compounds in small quantities), as every elemental volume of air contains these components in the same proportion. In a uniform state, temperature and density being the same everywhere, other properties of the substance are also the same everywhere. On the other hand, the composition of a material may be different at different locations; such materials are termed *inhomogeneous* or *heterogeneous substances*. Properties of inhomogeneous materials vary from location to location. An example of such a material is case-hardened steel, which has a higher percentage of carbon at the surface than in the interior.

Isotropic and anisotropic substances

If the properties of a substance of interest do not vary with direction, the substance is *isotropic*. For example, if the magnitudes of the temperature gradient at a point in two arbitrary directions in a block of material are equal, and the heat fluxes in these two directions have the same magnitude, the substance is isotropic, as the thermal conductivity, the property of interest to us, has the same value in two arbitrary directions. Most engineering substances may be treated as isotropic, but a few important exceptions exist. Examples of *anisotropic* substances in our daily experiences are wood (whose properties along and perpendicular to the grain are different), composites, and laminated plastics. Unless otherwise stated, we will assume that the material we deal with is isotropic.

One-dimensional

The term *one-dimensional* means that the variable of interest to us, say, temperature, can be expressed in terms of only one spatial coordinate with an appropriate choice of the coordinate system. Consider a hollow cylinder whose ends are perfectly insulated and whose inner and outer surfaces are maintained at uniform temperatures T_i and T_o, respectively (Figure 1.8.1). It is apparent that the temperature varies only in the radial direction; i.e., the temperature is a function of the radius only. As the temperature is a function of only one spatial coordinate, the radius, this is a one-dimensional temperature distribution. If for some reason, Cartesian coordinates are chosen, or if the z-axis is a line other than the axis of the cylinder, it is impossible to express the temperature distribution in terms of only one spatial coordinate. It is, therefore, important to choose an appropriate coordinate system so that the variable of interest can be expressed in terms of a minimum number of spatial coordinates.

Steady and transient states

The terms *steady-state* and *transient* (or *dynamic*) *state* are used to define if a variable is a function of time. By steady state we mean that the variable of interest to us *at given point* in space is independent of time, i.e., does not change with time in the region of interest. If the variable at a given point changes with time, it is in an unsteady or transient state.

To illustrate unsteady state, consider the heating of steel rods in a furnace. The heating may be accomplished by either a batch process or a continuous process. In the batch process, shown in Figure 1.8.2a, the rods are loaded into the furnace and left in it until the temperature of the rods reaches the desired value. In the furnace, the temperature of the rods, at a given location in space, changes with time, and hence, the rods are in an unsteady state.

Now consider the continuous process where the steel rods are fed to the furnace at a constant rate. The steel rods, of the same dimensions, are assumed to enter the furnace at the same temperature as shown in Figure 1.8.2b. If the conditions in the furnace do not change with time, the temperature of the rods at exit does not vary with time. What is more, if the temperature of the rods is measured at any fixed

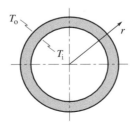

FIGURE 1.8.1 Example of one-dimensial temperature distribution
A hollow cylinder with the inner and outer surfaces is maintained at two different but uniform temperatures.

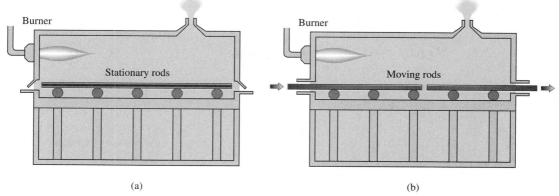

FIGURE 1.8.2 Examples of steady and transient states
(a) Steel rods are heated in a furnace in a batch process. (b) Steel rods are fed into a furnace at a steady state. The temperature of the rods at a fixed point in the furnace does not change.

point in the furnace, it will be the same at every instant. This, according to our definition, is a steady-state situation as the temperature, the variable of interest to us, does not change with time at a *fixed point in space*. Notice that if one follows a point fixed to the rod, the temperature changes with time.

Uncertainty

We will frequently use the term *uncertainty* to indicate that the value of a variable cannot be treated as exact. For example, to compute the convective heat flux from a surface, we may employ one of the correlations given in Chapter 4 to determine the convective heat transfer coefficient. However, the value of the convective heat transfer coefficient determined from a correlation is subjected to some uncertainty. The uncertainty is caused by many factors. Some of them are briefly described below.

Some causes of uncertainty

- Measurements made with instruments always have some uncertainty. For example, when we measure the temperature by a thermometer, the accuracy of the measurement depends on the calibration of the thermometer, the spacing between the division, the care exercised by the person reading the thermometer, and so on. The actual temperature may be different from the measured value.
- When values measured under controlled conditions are used in another situation, conditions in the two cases are rarely identical. For example, convective heat transfer coefficients in natural convection are determined in laboratories with great care to ensure that there is no bulk motion of the fluid. In most cases, it is difficult to avoid some bulk motion of the fluid. To determine the heat transfer rate from a water-heated radiator in a room (for space heating), we would assume that there is no bulk motion of the air and that the correlations developed from laboratory experiments are applicable. This is generally not true, as there is the possibility of some motion of the fluid due to the movement of people in the room, opening and closing of doors, and so forth. As another example, in determining the conductive heat flux in a slab of plain carbon steel, the thermal conductivity of the steel may be taken from Table A1. However, the value given in the table is for a particular sample (or the average of many samples) of carbon steel. The

slab for which the heat flux is being computed may not have exactly the same composition as the sample for which the value is given in the table; and hence, the thermal conductivity is different from the tabulated value.

■ Even when care is taken to determine the values of such quantities as convective heat transfer coefficient to ensure repeatability, several experimental values are determined under what one would consider as indentical conditions. However, the conditions are never identical, and there is generally a spread in the values determined even by the same person at different times. If similar experiments are conducted by different persons, one would find that the values reported have a mean value with some spread around the mean value. When experimental results are reported, in most cases, the spread of the measurements is indicated as uncertainty in the measurements.

The uncertainty in the value of a parameter, x, is generally expressed in one of two ways. One way is to express the value as $x \pm y$ to indicate that the measured or computed value is x but that the actual value may be between $(x + y)$ and $(x - y)$. The other way is to express it as $x \pm z$ % to indicate that the value may be between $(x + 100 \, (y/x))$ and $(x - 100 \, (y/x))$.

The uncertainty in some cases, such as the density of a substance, is quite small if the composition of the materials is nearly identical. But in some cases, such as convective heat transfer, the uncertainty may be quite significant. The convective heat transfer coefficient with single phase fluids generally has an uncertainty of about 10%. In the case of boiling heat transfer, the uncertainty may be quite high, 30% to 50%, or more.

Although most measurements are subject to uncertainties, one should minimize round-off errors. If only two significant digits are used at every stage of calculations, round-off errors may accumulate, and the final answer may have an uncertainty that is greater than the sum of the uncertainties of the various values used in the computations. It is suggested that computations be carried out with at least four significant figures at every stage. The final answer may then be rounded off in a manner that is appropriate to the variable.

1.9 METHODOLOGY IN THE SOLUTION OF HEAT TRANSFER PROBLEMS

Solution methodology

The starting point for the solution of problems in heat transfer is the balance laws—balance of mass, momentum (linear and angular), energy, and entropy (The Second Law of Thermodynamics). *For most problems in heat transfer considered in this book, the application of balances of mass and energy, with the Second Law implicitly satisfied, is sufficient. In many cases it may be appropriate to start with the energy equation and determine what other balance laws and auxiliary laws are needed.* The balance laws for a system (fixed mass) and control volume (fixed region in space)[2] are stated here.

---

[2]In some books *system* is referred to as closed system or control mass and *control volume* is referred to as an open system.

Balance of Mass

System: mass remains constant.

Mass balance   *Control Volume (C.V.):*

> Rate of increase of mass within the C.V.
> + Net rate of mass flow out of the C.V.
> = 0

Balance of Linear Momentum (Newton's Second Law)

System:

Balance of linear
momentum
> Rate of change of linear momentum in a given direction
> = Sum of all the forces on the system in that direction

Control Volume: in any given direction,

> Rate of change of linear momentum of the mass in the C.V.
> + Net rate of momentum flow out of the C.V.
> = Sum of all the forces on the material in the C.V.

Balance of Angular Momentum

System: about any given axis,

Balance of angular
momentum
> Rate of change of angular momentum
> = Sum of the moments of all the forces about the axis

Control Volume: about any given axis,

> Rate of change of angular momentum of the mass in the C.V.
> + Net rate of flow of angular momentum out of the C.V.
> = Sum of the moments of all the forces acting on the material in the C.V.

Balance of Energy (First Law of Thermodynamics)

Usually, in the type of applications that are considered here, changes in kinetic and potential energies are negligible. There are cases where it is not true, for example, flow of compressible fluids. *The law of balance of energy is stated here with the assumption that changes in kinetic and potential energies are negligible.*

System:

Balance of thermal
energy
> Rate of change of internal energy in the system
> = Net rate of heat transfer to the system
> − Net rate of work transfer from the system
> + Rate of internal energy generation

Control Volume:

> Rate of increase of internal energy of the mass in the C.V.
> + Net rate of enthalpy flow out of the C.V.
> = Net heat transfer rate to the C.V.
> − Net rate of work transfer from the C.V.
> + Rate of internal energy generation in the C.V.

In engineering texts on thermodynamics, internal energy generation is treated in different ways. Here, a variety of different cases are treated as internal energy generation. For example, resistance heating and exothermic or endothermic reactions without significant changes in properties of importance to heat transfer are all treated as internal energy generation. The phrase *internal heat generation* is also used to denote internal energy generation.

Internal energy generation

If the control volume is deformable, the work transfer from the control volume includes both shaft work and the work done by the control surface on the surroundings.

Depending on the problem, we choose either the system or the C.V. approach and apply the balance laws. But these balance laws alone may not be sufficient to give the required solutions. The balance laws are then supplemented by auxiliary (particular) laws. The difference between the balance laws and the auxiliary laws is that the balance laws are universally applicable whereas the auxiliary laws are applicable only to a specific class of problems. Examples of auxiliary laws are, $p = \rho RT$, which is applicable only to an ideal gas, and $q''_x = -k \, (\partial T/\partial x)$, which is applicable only for conductive heat transfer. Other auxiliary laws are pointed out later where appropriate.

Auxiliary laws

In the solution of heat transfer problems, we seek either the temperature distribution or the heat flux. From the temperature distribution, the temperature at any desired location can be found; for example, the maximum temperature in the medium. If the heat transfer is by conduction or convection, the temperature distribution also enables us to determine the heat flux by the application of Fourier's Law or a modified form of the law (in convection with turbulent flows). However, in convection problems, more often we seek the value of the convective heat transfer coefficient rather than the complete temperature distribution. The convective heat transfer coefficient relates the heat flux from the bounding surface to an appropriate temperature difference.

It should be emphasized that heat transfer is a surface phenomenon and that the total heat transfer across a surface is given by

Heat transfer is a surface phenomenon

$$q = \int \mathbf{q}'' \cdot \hat{\mathbf{n}} \, dA$$

where $\mathbf{q}''$ is the local heat flux vector, and $\hat{\mathbf{n}}$ is the unit vector along the outward normal to the infinitisimal area dA (Figure 1.9.1). In the case of conductive heat transfer $q''_n = -k \, (\partial T/\partial n)$, where n is the outward normal to the surface. If the temperature distribution is one-dimensional, say in the x-direction, $T = T(x,\tau)$.

FIGURE 1.9.1 Definitions of Unit Vector and Outward Normal

$\partial T/\partial x$ is uniform on A_x, and taking $n = x$, (Figure 1.9.1) we obtain

$$q_x = -kA_x \frac{\partial T}{\partial x}$$

If the surface is exposed to a moving fluid adjacent to it, the heat transfer to the fluid is by convection and the total heat transfer rate to the fluid is given by

$$q = \int q'' \, dA_s = \int h_x \, (T_s - T_{ref}) \, dA_s$$

where

  h_x = local convective heat transfer coefficient (W/m² K)
  T_s = surface temperature (°C or K)
  T_{ref} = reference temperature for the definition of the convective heat transfer
      coefficient (°C or K)
  q'' = local heat flux

Local and average heat transfer coefficient

Recall that in determining the convective heat transfer from a surface, we always deal with the normal component of the heat flux and the vector notation is not explicitly used. In general, both the local convective heat transfer coefficient, h_x, and the temperature difference, $T_s - T_{ref}$, may vary with A_s. In such cases, to find the total heat transfer rate, the relationship between h_x, $(T_s - T_{ref})$, and A_s must be known. If $(T_s - T_{ref})$ is constant, an average heat transfer coefficient, h, may be defined from the relation

$$q = \int h_x \, (T_s - T_{ref}) \, dA_s = hA_s \, (T_s - T_{ref})$$

or

$$h = \frac{1}{A_s} \int A_s \, h_x \, dA_s$$

Usually, the total heat transfer rate is computed, without employing the vector notation, as

$$q = \int q'' \, dA$$

In the above equation q'' is to be interpreted as the component of the heat flux vector along the normal to dA.

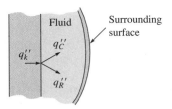

FIGURE 1.9.2 Surface Energy Balance

Surface energy
balance

In heat transfer, energy balance is frequently applied to a solid surface. As a surface has no mass, it cannot store energy. Referring to Figure 1.9.2, a solid surface receives energy from the interior of the solid by conduction. The surface transfers energy to the surrounding fluid by convection and to the surroundings by radiation. At any given instant the net energy transfer to the surface is zero and we have, for the surface,

$$q''_k - q''_c - q''_R = 0$$

where

q''_k = conductive heat flux from the interior of the solid to the surface
q''_c = convective heat flux from the surface to the surrounding fluid
q''_R = net radiative heat flux from the surface to the surroundings.

Note that the surface energy balance is true both for steady state and transient state (at any given instant of time) and also is true whether or not there is internal energy generation.

1.10 BRIEF REVIEW OF FLUID MECHANICS AND THERMODYNAMICS

In heat transfer, many results from fluid mechanics and thermodynamics are used frequently. Some of those results and concepts that are invoked in this book are briefly reviewed here.

The velocity of a fluid may be defined through the equation (Figure 1.10.1)

Fluid velocity

$$d\dot{m} = \rho\, \mathbf{V} \cdot \hat{\mathbf{n}}\, dA \qquad (1.10.1)$$

where

$d\dot{m}$ = rate of mass crossing an infinitesimal surface area dA (kg/s)
ρ = local density of the fluid (kg/m^3)
$\mathbf{V}$ = local velocity of the fluid (m/s)
$\hat{\mathbf{n}}$ = outward unit normal to dA

The total mass rate of flow of the fluid across a surface is then given by

$$\dot{m} = \int_A \rho\, \mathbf{V} \cdot \hat{\mathbf{n}}\, dA \qquad (1.10.2)$$

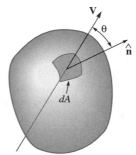

FIGURE 1.10.1 Definition of Fluid Velocity

Average velocity

If the density of the fluid is uniform on the surface, A, the average normal component of the velocity of the fluid across the surface is defined by $V_{av} = \dot{m}/\rho A$. Employing Equation 1.10.2, the average velocity is obtained from

$$V_{av} = \frac{1}{A} \int_A \mathbf{V} \cdot \hat{\mathbf{n}} \, dA$$

Balance of Mass For a stationary control volume with discrete inlets and outlets

Mass balance

$$\left\| \frac{dm_{cv}}{d\tau} = \sum_{in} \dot{m} - \sum_{out} \dot{m} \right\| \qquad \textbf{(1.10.4)}$$

In steady state, $dm_{cv}/d\tau = 0$.

Balance of energy

Balance of Energy (First Law of Thermodynamics) for a Control Volume

$$\left\| \sum_{in} \dot{m} \left(u + \frac{p}{\rho} + \frac{V^2}{2} + zg \right) + q \right.$$

$$\left. = \sum_{out} \dot{m} \left(u + \frac{p}{\rho} + \frac{V^2}{2} + zg \right) + \dot{W}_{cv} + \frac{dE_{cv}}{d\tau} \right\| \qquad \textbf{(1.10.5)}$$

where

$\dot{m}$ = mass rate of flow (kg/s)
u = specific internal energy (J/kg)
p = pressure (Pa)
ρ = density (kg/m³)
V = velocity (m/s)
z = elevation (m)
q = heat transfer rate to the C.V. (J/s or W)
$\dot{W}_{cv}$ = rate of work transfer from the C.V. to the surroundings (J/s or W)
E_{cv} = energy of mass in C.V. (J)
g = gravitational acceleration (m/s²)

The term $u + p/\rho$ is the specific enthalpy and is denoted by i. In most books on thermodynamics, the specific enthalpy is denoted by h, but in heat transfer, the convective heat transfer coefficient is denoted by h. Therefore, to avoid confusion, the specific enthalpy is denoted by i. In Equation 1.10.5, $\dot{W}_{cv}$ represents the work transfer by the control volume to the surroundings. The work transfer may be in the form of shaft work and due to the change in the volume of the control volume. If the control volume is deformable, the work transfer due to the motion of the control surface is given by $p \; (dV/d\tau)$, where p is the pressure at the control surface. The term $dV/d\tau$ represents the rate of change of volume of the control volume (and not the rate of change of velocity of the fluid).

Constant volume and constant pressure specific heats

Specific heats Constant volume and constant pressure specific heats are defined as

$$c_v = \left.\frac{\partial u}{\partial T}\right|_v \tag{1.10.6}$$

$$c_p = \left.\frac{\partial i}{\partial T}\right|_p \tag{1.10.7}$$

The two specific heats are related by

$$c_p - c_v = \frac{\beta^2 \, v \, T}{\beta_T} \tag{1.10.8}$$

where

β = volumetric thermal expansion coefficient = $(1/v) \; (\partial v/\partial T)|_p \; (K^{-1})$
v = specific volume (m^3/kg)
T = temperature (K)
β_T = isothermal compressibility = $-(1/v \; (\partial v/\partial p)|_T \; (Pa^{-1})$

If the density of a substance is strictly constant, $c_p = c_v$. For most solids and liquids, the density is substantially constant and $c_p \approx c_v$. Hence, for solids and liquids, only one specific heat, c_p, is given in the table of properties. Note that for ideal gases, Equation 1.10.8 reduces to, $c_p - c_v = R$ (gas constant).

For most substances the specific heats are weakly dependent on pressure and such dependence is generally neglected. For solid and liquids for which the densities are approximately constant,

$$di = c_p \; dT + v \; dp \tag{1.10.9}$$

$$i_2 - i_1 = \int_{T_1}^{T_2} c_p \; dT + \int_{p_1}^{p_2} v \; dp$$

For solids and liquids the changes in specific volume with pressure can be neglected and

$$i_2 - i_1 = \int_{T_1}^{T_2} c_p \; dT + v \; (p_2 - p_1) \tag{1.10.10}$$

If the specific heat is constant or varies linearly with temperature, Equation 1.10.10 can be recast as

$$i_2 - i_1 = c_{pm} \; (T_2 - T_1) + v \; (p_2 - p_1) \tag{1.10.11}$$

where c_{pm} is the mean specific heat = $[c_p \; (T_2) + c_p \; (T_1)]/2$ or $c_p \; [(T_1 + T_2)/2]$

In many engineering applications, the term $v \; (p_2 - p_1) \ll c_{pm} \; (T_2 - T_1)$ and it can be neglected. For example, consider water flowing in a 15-cm diameter, 60-m long tube with a velocity of 2 m/s, entering the tube at 55 °C at section 1, and existing at 65 °C at section 2, Figure 1.10.2. For this case, the friction factor is about 0.02 yielding a pressure drop, $p_1 - p_2 \approx 16$ kPa. From Table A5 the specific heat of water at 60 °C is 4181 J/kg °C and

$$c_{pm} (T_2 - T_1) = 4181 \times 10 = 41\,810 \text{ J/kg}$$
$$v (p_2 - p_1) = -0.001 \times 16\,000 = -16 \text{ J/kg}$$

If the water flows in a 20-m long, 1-cm diameter pipe with a velocity of 3 m/s and with the same change in temperature, the pressure drop is about 200 kPA and v $(p_2 - p_1) = -200$ J/kg. In both cases the change in the enthalpy due to the change in temperature is much greater than that due to the change in the pressure. Thus, in most heat transfer applications, the changes in temperature are significant and changes in specific enthalpy due to changes in pressure can be neglected. In such cases,

$$i_2 - i_1 \approx c_{pm} (T_2 - T_1) \tag{1.10.12}$$

For an ideal gas $pv = RT$, and, if c_p is constant or varies linearly with temperature, Equation 1.10.12 can be used. Thus, for both ideal gases and solids and liquids Equation 1.10.12 is used extensively. It should be recognized that Equation 1.10.12 is correct for an ideal gas with constant specific heats or specific heats varying linearly with temperature. For solids and liquids it is an acceptable engineering approximation.

An application that is of engineering significance is that of a fluid flowing in a duct and heated by the wall of the duct as shown in Figure 1.10.2. In steady state conditions combining Equations 1.10.4 and 1.10.5, we obtain

$$\dot{m} \left(u + \frac{p}{\rho} + \frac{V^2}{2} + zg\right)\Bigg|_1 + q = \dot{m} \left(\text{u} + \frac{p}{\rho} + \frac{V^2}{2} + zg\right)\Bigg|_2 + \dot{W}_{CV} \tag{1.10.13}$$

For the flow of a constant density fluid in a pipe (constant diameter) the velocity is constant. Further, in the absence of shaft work, Equation 1.10.13 simplifies to

$$\left(\frac{p_1}{\rho_1} + z_1 g\right) - \left(\frac{p_2}{\rho_2} + z_2 g\right) = u_2 - u_1 - \frac{q}{\dot{m}} \tag{1.10.14}$$

From thermodynamics we have

$$T \, ds = du + p \, dv \tag{1.10.15}$$

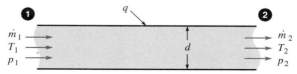

FIGURE 1.10.2 A fluid enters a duct with a uniform temperature T_1 and exits at at temperature T_2. The pressures at inlet and exit are p_1 and p_2.

For constant density,

$$T \, ds = du \qquad \text{(1.10.16)}$$

If the process is irreversible, $T \, ds > \delta(q/\dot{m})$, where $\delta(q/\dot{m})$ is the heat transfer to a unit mass of the fluid when there is an infinitesimal change in the state of the fluid. Therefore,

$$du > \frac{\delta q}{\dot{m}}$$

or

$$u_2 - u_1 > \frac{q}{\dot{m}}$$

The expression $u_2 - u_1 - q/\dot{m}$ represents the irreversible work transfer (w_{IR}) to the fluid and

$$u_2 - u_1 - \frac{q}{\dot{m}} = w_{\text{IR}}$$

The irreversible work transfer to the substance is commonly termed *losses*. The losses are generally not amenable to simple, analytical determination. Quite often they are calculated from empirical correlations. For example, in pipe flows of incompressible fluids,

$$\left(\frac{p_1}{\rho} + z_1 g \right) - \left(\frac{p_2}{\rho} + z_2 g \right) = h_f \, g \qquad \text{(1.10.18)}$$

where

$\begin{aligned}
h_f &= \text{fricton head} = f \, (L/d)(V^2/2g) \ (\text{m}) \\
f &= \text{friction factor} = f(\text{Re}_d, \epsilon/d) \\
\text{Re}_d &= \text{Reynolds number} = \rho V d / \mu \\
\epsilon/d &= \text{roughness factor} \\
V &= \text{mean velocity (m/s)}
\end{aligned}$

For flows in pipes without work transfer, if changes in potential and kinetic energies are small, Equation 1.10.13 reduces to

$$q = \dot{m} \, (i_2 - i_1)$$

For ideal gases or incompressible fluids with small changes in pressure, and without change in phase, $i_2 - i_1 \approx c_{\text{pm}} (T_2 - T_1)$ and

$$\| \quad q = \dot{m} c_{\text{pm}} (T_2 - T_1) \quad \| \qquad \text{(1.10.19)}$$

Equation 1.10.19 is extensively used to compute convective heat transfer rate in internal flows.

Note on Thermal Conductivity and Viscosity Both thermal conductivity and viscosity are temperature and pressure dependent. But their dependence on pressure is very weak compared with their dependence on temperature, and in engineering applications the pressure dependence is neglected so that

$$k = k(T)$$

$$\mu = \mu(T)$$

Thermal conductivity and viscosity—effect of temperature and pressure

The kinematic viscosity $\nu = \mu/\rho$ is also independent of pressure for an incompressible fluid for which the density is very weakly pressure dependent. For gases, however, the density is dependent on pressure, and the kinematic viscosity of gases is a function of both temperature and pressure. Thus, for gases

$$\nu(p,T) = \frac{\mu(T)}{\rho(p,T)} \tag{1.10.20}$$

Another property that is of significance in heat transfer is the thermal diffusivity, α, which is the ratio, $k/\rho c_p$. Here again, the constant pressure specific heat, c_p, may depend on temperature, but it is relatively insensitive to changes in pressure so that it can be considered either as a constant or, at most, dependent on temperature only. From the definition of thermal diffusivity it follows that the thermal diffusivity is dependent on temperature only for liquids and solids but is dependent on both temperature and pressure for gases.[3]

1.11 UNITS

In this book the SI units (Le System International d'Unites) are used. The basic units of time, mass, length, and temperature are second(s), meter(m), kilogram (kg), and degree celsius (°C) (or kelvin,K), respectively. A table of conversions from English units to S.I. units is given on front cover endsheets.

Time

The unit of time is the second, defined as the time required for 9,192,631,770 cycles of a cesium resonator using a beam of cesium-133 atoms.

Length

Until a few decades ago, the accepted standard of the meter was the distance between two marks on a platinum-iridium bar maintained at the International Bureau of Weights and Measures, Sevres, France. In 1983, the General Conference of Weights and Measures (GCPM) defined the meter as the length of the path traversed by light in vacuum during a time interval of 1/299,792,458 second.

Mass

The unit of mass, the kilogram, is the mass of a certain platinum-iridium cylinder maintained at the International Bureau of Weights and Measures.

In 1967, the GCPM defined the unit of temperature, the kelvin, as 1/273.16 of the temperature of the triple point of water. The relation between the kelvin and celsius scale is

$$K = {}^\circ C + 273.15$$

---

[3]For a more detailed explanation of the thermodynamic relations used in this section, refer to Van Wylen, G.J., and Sonntag, R.E., Fundamentals of Classical Thermodynamics, third ed. (New York: Wiley, 1994).

Temperature

In 1990, the International Committee of Weights and Measures adopted a temperature scale fixing the values of temperature of some substances in defined states. Some of the primary fixed-point temperatures in degree Celsius are

Triple point of equilibrium hydrogen	−259.347 °C
Triple point of oxygen	−218.792 °C
Triple point of water	0.01 °C
Normal freezing point of zinc	419.527 °C
Normal freezing point of gold	1064.18 °C

When quantities based on temperature differences are used, the magnitude of the quantity remains the same regardless of whether the kelvin or the celsius scale is used, as the difference in temperatures in both scales is the same. For example, the thermal conductivity is the heat flux per unit temperature gradient in a given direction. As the temperature gradient is the change in temperature per unit distance, the magnitude of the gradient is same in both the kelvin and the Celsius scales. *Hence, the thermal conductivity of a substance may be expressed either as W/m K or W/m °C, and the values in both cases are the same. Convective heat transfer coefficients and specific heats also have the same values in both temperature scales.* In this book both the celsius and the kelvin scales are used. However, when the actual temperature is used, the appropriate scale should be used. For example, both the ideal gas law ($p = \rho RT$) and the Stefan-Boltzmann's law ($E_b = \sigma T^4$) are valid only when the absolute scale of temperature, the kelvin scale (or the rankine scale in the British System of Units), is used.

The units for other quantities used in this book are given in the Conversion Table on the cover endsheets.

Although the unit of length is the meter, the derived units of centimeter (cm), milimeter (mm), and the kilometer (km) are widely used in engineering practice. At present, the British System of Units, the FPS system (foot, pound, and second), is the preferred system in both the building construction and heating, ventilating, and air-conditioning industries. Hence, in a few cases dealing with heat transfer in buildings, the FPS system of units is used.

1.12 SOME SUGGESTIONS FOR SOLVING PROBLEMS

Problem solving—
some suggestions

The majority of problems in this book can be worked in less than an hour each. In solving the problems you are urged to follow the steps given below. These steps will be found to be useful even during your career when you may be called upon to solve engineering problems. While working the first one or two problems on a specific topic, you may find it helpful to review the relevant sections in the book. Beyond about the second problem, you should be able to solve the problems without any reference to the book (except to obtain property values or to verify a particular equation) and certainly without reference to a similar worked problem.

The suggested steps in solving a problem are

1. Read the problem statement carefully and translate the statement into a sketch. Transfer all the information from the statement to the sketch. Identify what is known and what should be found.

2. List all assumptions. Some of the assumptions are clear before commencing the solution. A few of them will become necessary during the course of the solution.
3. Find and list all required properties. In most cases the properties that are required for the solution of the problems will be obvious.
4. Conceptualize the solution. Write the steps involved in getting the solution. For example, in Example 1.4.1, the numerical solution is preceded by such a conceptual solution.
5. Apply your analysis in symbolic form as far as possible, and substitute numerical values at the end. Many students find substitutions of numbers at every stage simpler, but much physical insight that can be obtained with the solution in symbolic form may be lost. Another advantage is that symbolic forms will assist you in determining if the solution is dimensionally correct.
6. Present your solution in a neat and orderly manner so that all the steps are clear not only to another person but to yourself at a later date.
7. It will be useful to write the solution to some problems completely with every step defended on the basis of well-established laws. This will not only help you to understand the subject clearly but will also enable you to synthesize the various subjects. This process will initially appear to be time consuming, but it will increase your confidence in your understanding of the subject.
8. Remember that the purpose of solving problems is to ensure a good understanding of analysis and not merely to pick a formula and substitute numbers.

SUMMARY

Heat transfer is energy transfer by virtue of temperature differences. Heat transfer may be either by *diffusion* or by *radiation*. For the transport of energy from a surface, diffusion heat transfer requires a material medium adjacent to the surface and is a much slower process than radiation. In diffusion the heat flux can be determined from a knowledge of the temperature in the immediate vicinity of the point. Radiative heat transfer does not require a material medium for the transport of energy and is a much faster process. To determine the radiative heat flux at a point, the effect of the entire region that participates in radiation must be considered.

Heat transfer is a *surface phenomenon*. The total heat transfer rate from a surface is obtained from

$$q = \int q'' \, dA$$

where q'' is the local heat flux along the normal to dA. Depending on the mode of heat transfer, the appropriate expression for q'' is used to obtain the total heat transfer rate.

Diffusion is subdivided into *conduction* and *convection*. Heat transfer in solids and stationary fluids (without bulk motion) is termed *conduction*. The conductive heat flux in a given direction, say the x-direction, is given by Fourier's Law,

$$q''_x = -k\frac{\partial T}{\partial x}$$

The coefficient of proportionality, k, is the *thermal conductivity* of the material and is a property of the material.

For steady, one-dimensional temperature distribution with constant properties, the total heat transfer rate is

$$q_x = - k A_x \frac{dT}{dx}$$

Heat transfer to a fluid in motion is termed *convective heat transfer. The convective heat transfer coefficient* is defined by the equation

$$q_c'' = h (T_s - T_{ref}),$$

where q_c'' is the convective heat flux to the fluid from the bounding surface, h is the convective heat transfer coefficient, and T_s and T_{ref} are the surface and reference temperature of the fluid, respectively. If the motion of the fluid is caused by an external force, such as due to a pump or a fan, and is not dependent on the difference in temperatures of the fluid and the solid surface, the heat transfer process is termed *forced convection.* If the fluid flow results from the difference in the temperatures of the fluid and the bounding surface, we have *natural convection.* With uniform h, T_s, and T_{ref}, the total heat transfer rate is

$$q_c = h A_s (T_s - T_{ref})$$

At a given temperature, the *maximum emitted radiative energy flux,* the *emissive power from an ideal surface,* also known as a black surface, is given by

$$E_b = \sigma T^4$$

where σ is the Stefan-Boltzmann constant and T is the absolute temperature of the surface. The ratio of the emitted energy flux of a surface to the maximum emitted energy flux at the same temperature is the *emissivity of the surface.* The radiative heat transfer rate from a convex or a plane surface at T_1 completely enclosed by a second much larger surface at T_2 is

$$q_R = \epsilon \sigma A_s (T_1^4 - T_2^4)$$

In natural convection with gases, radiative heat flux from a surface may be quite significant. In forced convection, the radiative heat flux from a surface is *usually* much less than the convective heat flux.

In this chapter we have illustrated the relevance of the study of heat transfer through several examples, developed the methodology for the solution of heat transfer problems, and defined various terms used in the book—pure, homogeneous, isotropic, one-dimensional temperature distribution, steady and unsteady states, and uncertainty. A brief review of fluid mechanics and thermodynamics, as required for an understanding of the material in this book has also been presented. We concluded the chapter with a brief discussion of units and suggestions for solving problems.

REFERENCES

There are many books on heat transfer and what follows is a short list of books that may be useful for further reading or as references.

INTRODUCTORY BOOKS

Bejan, A. (1993). *Heat Transfer*. New York: Wiley.

Chapman, A.J. (1987). *Fundamentals of Heat Transfer*. New York: Mcmillan.

Gebhart, B. (1971). *Heat Transfer*, 2d ed. New York: McGraw-Hill.

Holman, J.P. (1990). *Heat Transfer*, 7th ed. New York: McGraw-Hill.

Incropera, F.P., and De Witt, D.P. (1990). *Fundamentals of Heat and Mass Transfer*, 3d ed. New York: Wiley.

Janna, W.S. (1986). *Engineering Heat Transfer*. Boston: PWS Engineering.

Kreith, F., and Bohn, M.S. (1993). *Principles of Heat Transfer*, 5th ed. Minneapolis, MN: West.

Lienhard, J.H. (1987). *A Heat Transfer Text Book*, 2d ed., Englewood Cliffs, NJ: Prentice Hall.

Mills, A.F. (1992). *Heat Transfer*. Homewood, IL: Irwin.

Ozisik, M.N. (1985). *Heat Transfer—A Basic Approach*. New York: McGraw-Hill.

White, F.M. (1988). *Heat and Mass Transfer*. Reading, MA: Addison Wesley.

Whitaker, S. (1981). *Fundamental Principles of Heat Transfer*. Malabar, FL: Robert E. Krieger.

ADVANCED BOOKS

Arpaci, V. (1966). *Conduction Heat Transfer*. Reading, MA: Addison Wesley.

Arpaci, V., and Larsen, P.S. (1984). *Convection Heat Transfer*. Englewood Cliffs, NJ: Prentice Hall.

Bejan, A. (1984). *Convection Heat Transfer*. New York: Wiley-Interscience.

Bird, R.B., Stewart, W.E., and Lightfoot, E.N. (1960). *Transport Phenomena*. New York: Wiley.

Burmeister, L.C. (1994). *Convective Heat Transfer*. 2d ed. New York: Wiley.

Carslaw, H.S., and Jaeger, J.C. (1959). *Conduction of Heat in Solids*, 2d ed. London: Oxford.

Eckert, E.R.G., and Drake, R.M., Jr. (1972). *Analysis of Heat and Mass Transfer*. New York: McGraw-Hill.

Kays, W.M., and Crawford, M.E. (1993). *Convective Heat and Mass Transfer*, 3d ed. New York: McGraw-Hill.

Love, T.J. (1968). *Radiative Heat Transfer*. Columbus, OH: Merrill.

Siegel, R., and Howell, J.R. (1981). *Thermal Radiation Heat Transfer*, 3d ed. Washington, D.C.: Hemisphere.

Sparrow, E.M., and Cess, R.D. (1978). *Radiation Heat Transfer*. New York: McGraw-Hill.

Welty, J.R., (1974). *Engineering Heat Transfer*. New York: Wiley

REFERENCE BOOKS

ASHRAE Handbook of Fundamentals (1993). American Society of Heating, Refrigerating, and Air Conditioning Engineers. Atlanta.

C.R.C. (1973). *Handbook of Tables for Applied Engineering Science*. Cleveland, OH: CRC Press.

C.R.C. (1991). *Handbook of Chemistry and Physics*. Cleveland, OH: CRC Press.

Hewitt, G.F., ed. (1990). *Handbook of Heat Exchanger Design*. New York: Hemisphere.

Kakac, S., Shah, R.K., and Win Aung, eds (1987). *Handbook of Single Phase Convective Heat Transfer*. New York: Wiley-Interscience.

Raznjevic, K. (1976). *Handbook of Thermodynamic Tables and Charts.* New York: Hemisphere.

Rohsenow, W.M., Hartnett, J.P., and Ganic, E.N. (1985). *Handbook of Heat Transfer Fundamentals,* 2d ed. New York: McGraw-Hill.

Rohsenow, W.M., Hartnett, J.P., and Ganic, E.N. (1985). *Handbook of Heat Transfer Applications,* 2d ed. New York: McGraw-Hill.

Touloukian, Y.S., and Ho, C.Y., eds. (1972) *Thermophysical Properties of Matter,* Vols 1–9, New York: Plenum.

REVIEW QUESTIONS

(a) What are the differences between diffusion and radiation heat transfer?

(b) State Fourier's Law of conduction.

(c) Both thermodynamics and heat transfer deal with heat transfer. What are the differences in the study of heat transfer in the two cases?

(d) What is the thermodynamic significance of the negative sign in Fourier's Law?

(e) Define thermal conductivity.

(f) Define convective heat transfer coefficient.

(g) What is the principal difference between thermal conductivity and convective heat transfer coefficient?

(h) How is natural convection different from forced convection?

(i) Define a black surface.

(j) What is the emissivity of a surface?

(k) What is the range of values for the emissivity of a surface?

(l) Define the terms pure, homogeneous, isotropic, steady state, and one-dimensional temperature distribution. Give one example to illustrate each definition.

PROBLEMS

1.1 A 10-mm thick Teflon sheet is perfectly bonded to a large 9-mm thick, type 316 stainless steel sheet. The interface is at 119 °C and the outer surface of the steel sheet is at 120 °C. In steady state, determine

(a) The heat flux across the steel sheet.

(b) The heat flux across the Teflon sheet.

(c) The outer surface temperature of the Teflon sheet.

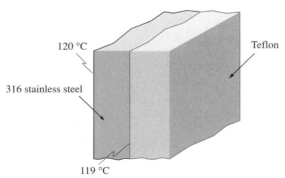

120 °C Teflon

316 stainless steel

119 °C

FIGURE P1.1

1.2 Consider two identical rectangular slabs, one made of copper and the other of type 316 stainless steel. If the temperature difference across the copper slab is 10 °C, what should be the temperature difference across the stainless steel slab for the heat transfer rates in the two cases to be equal?

1.3 Consider two slabs, one of copper and the other of AISI 316 stainless steel, as shown. The cross-sectional area of the copper slab is 0.2 m² and that of the stainless steel slab is 1.6 m². Both the slabs have the same thickness. One surface is maintained at 30 °C and the parallel surface at 20 °C in both cases. What is the ratio of the heat transfer rates $q_{Cu}/q_{st.st.}$?

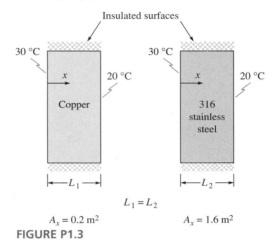

FIGURE P1.3

1.4 Two parallel surfaces of a 10-mm thick slab of stainless steel (AISI 302) are maintained at 30 °C and 20 °C. All other surfaces are perfectly insulated. The surface at 30 °C is exposed to a fluid at 50 °C. Determine the convective heat transfer coefficient on the fluid side.

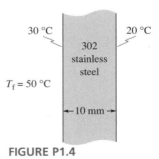

FIGURE P1.4

1.5 An ice chest, with inner dimensions of 50-cm length, 40-cm width, and 25-cm height is made of 25-mm thick Styrofoam. Estimate the heat transfer rate across the walls of the ice chest if its inner and outer surface temperatures (including the top lid) are 0 °C and 20 °C, respectively.

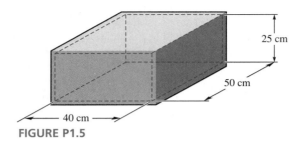

FIGURE P1.5

1.6 The heat flux across slab A in the sketch is 60 W/m². What are the heat fluxes across slabs B, C, and D.

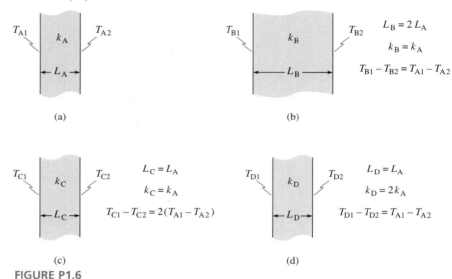

FIGURE P1.6

1.7 The heat flux across a plane wall is given by

$$q''_x = k \frac{T_1 - T_2}{L} = \frac{T_1 - T_2}{(L/k)}$$

The denominator, L/k, represents the resistance to heat transfer rate. Hence, the effectiveness of insulation materials for buildings is expressed by its R-value, which is L/k, where L is in ft. and k is the thermal conductivity in BTU/hr ft °F. Determine and verify the R-value of

(a) 3.5-in thick fiberglass blanket.
(b) 6-in thick fiberglass blanket.
(c) 2-in thick Styrofoam.
(d) 4-in thick Styrofoam.

1.8 For identical surface temperatures, compute the ratio of the heat fluxe across a 80-mm thick fiberglass blanket to the heat flux across a 50-mm thick Styrofoam board.

1.9 List some of the conditions that are required for a fluid adjacent to a heated vertical wall to be in motion when the temperature of the surface is increased above that of the fluid.

1.10 A solid cylinder is to be cooled rapidly. It can be cooled by a stream of air or by immersing it in a large body of water. Which is more likely to yield faster cooling? Both air and water are available at the same temperature. Justify your answer.

1.11 Water enters a 2.5-cm I.D. tube at a rate of 0.8 kg/s at 10 °C and leaves at 50 °C. Determine
(a) The average velocity of the water in the tube.
(b) The heat transfer rate to the water.

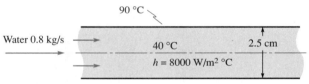

Water 0.8 kg/s

10 °C   2.5 cm   50 °C

FIGURE P1.11

1.12 In Problem 1.11, at the location where the water temperature is 40 °C, the temperature of the tube surface is 90 °C, and the convective heat transfer coefficient is 8000 W/m² °C. Determine the heat transfer rate per unit length of the tube at that location.

90 °C

Water 0.8 kg/s   40 °C   2.5 cm

$h = 8000$ W/m² °C

FIGURE P1.12

1.13 An immersion heater, fabricated by inserting an electric heating element inside a 1.2-cm O.D., 60-cm long tube, is immersed in water at 20 °C. When the power input is 2 kW, the temperature of the surface of the tube is 120 °C. Determine the convective heat transfer coefficient.

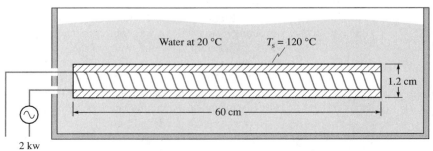

Water at 20 °C $T_s = 120$ °C

1.2 cm

60 cm

2 kw
FIGURE P1.13

1.14 An electric heating element is sandwiched between two vertical metal plates. When the power supply to the heating element is 1500 W/m², it is found that the mean surface temperature is 50 °C. The surrounding air temperature is 20 °C. Neglecting radiative heat transfer, determine the the convective heat transfer coefficient.

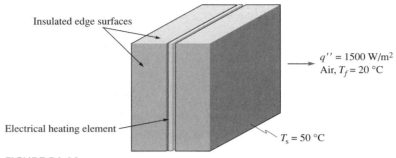

Insulated edge surfaces

$q'' = 1500$ W/m²
Air, $T_f = 20$ °C

Electrical heating element

$T_s = 50$ °C

FIGURE P1.14

1.15 The tungsten filament in a lamp has a diameter of 0.003 in and is 6-in long. When it is maintained at 2900 K, what is the amount of energy emitted by the filament if the emissivity of the filament is 0.3? If the surroundings are at 20 °C, what is the electrical power supply to the filament? Assume all heat transfer is by radiation only.

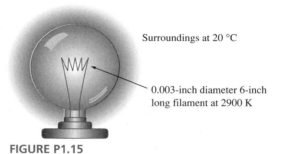

Surroundings at 20 °C

0.003-inch diameter 6-inch
long filament at 2900 K

FIGURE P1.15

1.16 To determine the emissivity of the surface of a copper surface, a 4-cm diameter copper sphere with an embedded heating element was placed inside a large evacuated vacuum chamber. To maintain a surface temperature of 200 °C, with the surroundings at 20 °C, it was found that the power supply to the heating element was 1.2 W. Estimate the emissivity of the surface of the sphere. (See figure, top of page 53).

1.17 An electrical cartridge heater is inserted into a hollow copper cylinder with the ends perfectly insulated. The temperature of the surface of the cylinder is 80 °C, and it is freely suspended in a large room. The temperature of the surrounding air and the walls is 20 °C. The outer diameter and the length of the copper cylinder are respectively 1 cm and 15 cm. Compute

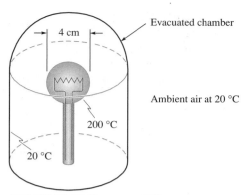

FIGURE P1.16 (See page 52)

(a) The heat transfer rate by convection if the convective heat transfer coefficient is 6 W/m² K.
(b) The heat transfer rate by radiation if the emissivity of the surface is 0.8.

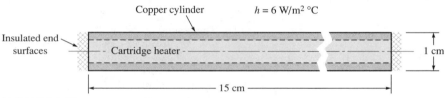

FIGURE P1.17

1.18 A 10-cm diameter sphere is freely suspended in a large room and is heated by an embedded electrical heater. When the power input to the heater is 30 W, the surface temperature of the sphere is 100 °C. If the convective heat transfer coefficient is 7 W/m² °C and the room is at 20 °C, determine the emissivity of the surface of the sphere.

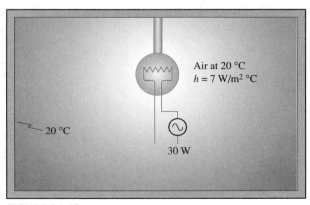

FIGURE P1.18

1.19 In Problem 1.13, instead of operating the heater in water it is operated in air which is at 20 °C. When the heater is operated in water, radiative heat transfer can be neglected. However, when it is operated in air, radiative heat transfer becomes significant and has to be considered. The total heat transfer rate from the heater is the sum of the radiative and convective heat transfer rates. In natural convection in air, the convective heat transfer coefficient may be estimated by

$$h = 1.4 \left(\frac{T_s - T_\infty}{d} \right)^{1/4} \ (\text{W/m}^2 \ °\text{C})$$

and the radiative heat transfer rate by

$$q_R = \sigma \epsilon A_s \ (T_s^4 - T_\infty^4) \ (\text{W})$$

where

T_s = temperature of the surface (K)
T_∞ = surrounding temperature (K)
σ = Stefan-Boltzman constant = $5.67 \times 10^{-8} \ \text{W/m}^2 \ \text{K}^4$
A_s = surface area of tube (m²)
ϵ = emissivity of the surface of the tube = 0.85
d = diameter of the tube (m)

Determine the temperature of the surface of the tube and the relative magnitude of the convective and radiative heat transfer rates. What do you expect would happen if the heating element is operated in air?

1.20 Liquid nitrogen (boiling point = 80 K, enthalpy of vaporization = 195.1 kJ/kg) is stored in a 30-cm diameter spherical container. The emissivity of the surface of the sphere is 0.1 and the convective heat transfer coefficient is 8 W/m² °C. The pressure inside the tank is maintained at a constant value (corresponding to the saturation temperature of 80 K) by venting it to the atmosphere. When the tank is full of liquid nitrogen, determine the rate of evaporation of the nitrogen assuming the temperature of the surface of the tank is 120 K and that of the surrounding is 20 °C.

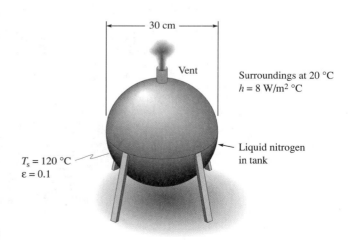

FIGURE P1.20

1.21 In Problem 1.20, to reduce the heat transfer rate, the tank is insulated with a 10-cm thick insulation material. The temperature of the outer surface of the insulation material is 285 K. The surrounding air is at 20 °C and the convective heat transfer coefficient is 8 W/m² °C. Determine the heat transfer rate and the rate of evaporation of nitrogen. Emissivity of the metal cladding on the insulation is 0.1.

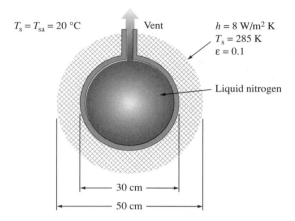

$T_s = T_{sa} = 20$ °C   Vent $h = 8$ W/m² K
$T_s = 285$ K
$\varepsilon = 0.1$

Liquid nitrogen

30 cm

50 cm

FIGURE P1.21

1.22 A simple method to estimate the solar insolation (solar energy incident on a surface per unit area of the surface perpendicular to the solar beam) is to measure the temperature of a sphere of known emissivity. In one experiment, the temperature of a 25-mm diameter sphere with an emissivity of 1, at different times are

Time	8 A.M.	10 A.M.	Noon
Air temperature	12 °C	24 °C	27 °C
Temp. of sphere	26.2 °C	39.4 °C	42.5 °C

(a) The convection heat transfer coefficient is estimated to be 8 W/m² °C. Taking into account both convective and radiative heat transfer from the sphere, estimate the solar insolation for each case.

(b) There is some uncertainty in the value of the convective heat transfer coefficient. It is estimated that the value of the convective heat transfer coefficient is 8 ± 0.8 W/m² °C. Determine the uncertainty in the value of the solar insolation. (Assume that all the incident solar energy is absorbed by the sphere.)

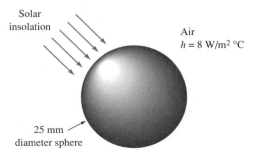

Solar insolation

Air
$h = 8$ W/m² °C

25 mm diameter sphere

FIGURE P1.22

1.23 Sometimes, it is useful to define a radiative heat transfer coefficient as the ratio of the radiative heat flux to the difference in temperatures of the surface and the surroundings. If a surface with an emissivity of 0.7 is enclosed in a much larger surface at 300 K (T_{sa}), determine

(a) The radiative heat transfer coefficient if the temperature of the surface (T_s) is 350 K, and 800 K.

(b) What conclusions do you draw from the values obtained?

1.24 When hot coffee that is kept in a vacuum flask cools with time, identify the heat transfer processes.

1.25 List the thermal conductivity, dynamic viscosity, kinematic viscosity, constant pressure specific heat, and thermal diffusivity of water and air at 200 kPa, 100 °C.

1.26 In the majority of engineering applications, it is assumed that the thermal conductivity is constant. But in some cases, the variation of the thermal conductivity of the material with temperature has to be considered, particularly if the temperature variation in the material is large. In most such cases it is adequate to express the thermal conductivity as

$$k = k_o (1 + \alpha T)$$

Plot the thermal conductivity of iron (99.75 % pure) as a function of temperature in the range of 400 K–1000 K. Employing linear regression analysis, determine the values of k_o and α.

1.27 The variation of viscosity of liquids with temperature is quite significant but, in most cases, cannot be expressed as a linear function with reasonable accuracy if the temperature range is large. Plot the viscosity of a few liquids as a function of temperature.

CHAPTER

2

CONDUCTION I

In Chapter 1, three modes of heat transfer were introduced. In this chapter on conductive heat transfer, we

- Show how to solve steady-state conduction problems involving one-dimensional temperature distribution
- Introduce the concept of thermal resistance for problems with one-dimensional temperature distribution and demonstrate the use of an electrical analogy to solve problems with layers of different materials
- Extend the application of thermal resistance to cases with two- and three-dimensional temperature distribution through conduction shape factor
- Introduce the concept of overall heat transfer coefficient
- Analyze problems involving extended surfaces
- Solve transient problems with uniform temperature
- Illustrate the use of one-dimensional transient temperature distribution solutions for the solution of one-dimensional and multidimensional transient temperature problems.

Recall that conductive heat transfer is heat transfer in solids and fluids without bulk motion; it is a diffusion process. In this chapter, solutions to one-dimensional, steady-state problems are derived, and the concepts of resistance to heat transfer and overall heat transfer coefficient are introduced. Conduction shape factor, an extension of the concept of thermal resistance to cases with two- and three-dimensional temperature distribution, are introduced. Heat transfer augmentation with extended surfaces is studied. The use of the available solutions for one-dimensional transient conduction problems, and their extension to multidimensional temperature distribution problems, is illustrated.

In diffusion problems related to heat transfer, we generally seek solutions to temperature as a function of time and spatial coordinates. The solution to the temperature distribution—temperature as a function of spatial and time coordinates—enables us to determine the conductive heat transfer rate across any surface in the medium through Fourier's Law. In many engineering applications, it is the heat transfer rate that needs to be determined; the solution to the temperature distribution (if it is necessary) is a means to that end. If thermal stresses are to be determined, the temperature distribution is required.

2.1 FOURIER'S LAW OF CONDUCTION

Fourier's Law of Conduction—Relation between heat flux and temperature gradient

The basis for the solutions to conduction heat transfer problems is the First Law of Thermodynamics in conjunction with *Fourier's Law of Conduction*. Fourier's Law states that at any location, *the heat transfer rate per unit area (the heat flux) in a given direction is proportional to the magnitude of the temperature gradient in that direction (space rate of change of temperature in that direction) and that the heat transfer takes place in the direction of decreasing temperature.* The heat flux can be treated as a vector by the above definition. Fourier's Law can be expressed as

$$q_x'' \propto -\frac{\partial T}{\partial x} \quad \text{or} \quad q_x'' = -k\frac{\partial T}{\partial x}$$

where

q_x'' = heat flux across an infinitesimal area perpendicular to the x-direction
x = any arbitrary direction
T = temperature

Thermal Conductivity

The partial derivative of T has been used as T can be a function of more than one spatial coordinate and also of time. The coefficient, k, is the *thermal conductivity* of the material. It is a property of the material and is always positive. Fourier's Law is

Joseph Fourier (1768–1830) proposed the law that bears his name in response to a prize question posed by the French Academy of Science: "To give the mathematical theory of the propagation of heat, and to compare this theory with exact observations." The complete exposition of the theory is contained in Fourier's book, *Theorie Analytique de la Chaleur,* published in 1822. Fourier was an applied mathemetician of great repute. He adapted a series expansion suggested by Daniel Bernoulli for the solution of a class of conduction problems, now commonly known as Fourier series.

valid for transient states also by evaluating the temperature gradient at any instant of time. The negative sign in Fourier's law implicitly satisfies the Second Law of Thermodynamics.

The thermal conductivities of substances vary with temperature. As a general rule (but not always true), for solids, the thermal conductivities of good electrical conductors (metals) decrease with an increase in temperature and those of poor conductors (nonmetals) increase with temperature. The thermal conductivities of gases increase with temperature but those of liquids do not show any regular pattern.

The thermal conductivity of a material may be a function of spatial coordinates and temperature. In most cases we deal with homogeneous materials, and the thermal conductivities of such materials depend only on the temperature. In this chapter, however, we consider those cases where the thermal conductivity can be treated as constant. The approximation is valid if the variation of the thermal conductivity in the medium, in which conductive heat transfer occurs, is not large. Cases where the thermal conductivity is a function of temperature are studied in Chapter 3. In Section 2.2, the differential equations for the temperature distribution for several cases are derived, and the solutions for specific boundary conditions are found. The derivation of a general differential equation for one-dimensional temperature distribution and a discussion of boundary conditions are presented in Section 2.7. We can also start with the general one-dimensional equation and specialize it to each case considered.

2.2 STEADY STATE, ONE-DIMENSIONAL CONDUCTION

The analytical determination of the temperature distribution involves the application of the balance and auxiliary laws, translating them to a mathematical form, and solving the resulting equations. If these laws are applied to an infinitesimal element in the medium, we get a differential equation that represents the physics of conduction. In this section we consider the solutions to steady state, one-dimensional conduction problems without internal energy generation.

When is temperature one-dimensional?

The conditions to be satisfied for the assumption of one-dimensional temperature distribution are that the geometry should be regular; i.e., the boundary conditions can be specified with one spatial coordinate, the material should be homogeneous, the temperatures on two surfaces separated by a finite distance along one coordinate should be uniform, and the end effects should be negligible. By end effects we mean the conditions at the edges that may lead to two- or three-dimensional temperature distribution. For example, even when the temperatures of the inner and outer surfaces of the walls of a house are uniform, the temperature distribution at the edges where two walls meet is two-dimensional. However, if such two-dimensional effects are confined to a small region near the edges, one-dimensional temperature distribution approximation is satisfied over a large part of the walls; neglecting such end effects, the resulting solution for the one-dimensional temperature distribution is a good approximation for the entire wall. We now proceed to find the one-dimensional temperature distribution in a rectangular slab and a hollow cylinder.

2.2.1 Rectangular Geometry—Rectangular Slab

Two parallel surfaces of a rectangular slab are maintained at temperatures T_1 and T_2 as shown in Figure 2.2.1. All other surfaces are perfectly insulated. Determine

(a) The temperature distribution.
(b) The heat transfer rate across a surface perpendicular to the x-direction.

ASSUMPTIONS

1. Steady state
2. Material is homogeneous with constant properties.
3. As the temperatures of the two parallel surfaces are uniform, and the material is homogeneous with constant properties, the temperature varies in the x-direction only; i.e., the temperature distribution is one-dimensional.

SOLUTION

One-dimensional temperature distribution— rectangular slab

As there is no mass flow involved in the problem, we choose the system approach and apply the appropriate laws to a fixed mass. The solution steps to the problem are to

(a) Choose an appropriate element.
(b) Apply the appropriate physical laws.
(c) Translate steps a and b into a mathematical form.
(d) Identify the needed auxiliary laws.
(e) Obtain the applicable equations.
(f) Solve the resulting equations.

(a) **Choose an Appropriate Element.** As the temperature varies only in the x-direction, choose an element of thickness Δx, perpendicular to the x-direction, as shown in Figure 2.2.1.

(b) **Apply the Appropriate Physical Laws.** As the focus is on a fixed mass, balance of mass is satisfied. For this problem the balance of momentum law is not needed. The balance of energy law is required. As there is no change in the

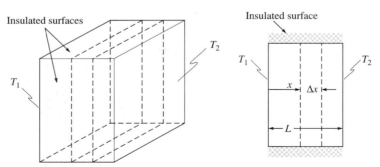

FIGURE 2.2.1 Determining One-Dimensional Temperature Distribution for a Rectangular Slab
The sketch shows the element chosen for analysis of the slab.

kinetic and potential energy, we apply the First Law of Thermodynamics to the element and obtain

> Rate of change of internal energy of the element
> = Net heat transfer rate to the element
> − Net rate of work done by the element
> + Net rate of internal energy generation

(c) Translate Steps a and b into a Mathematical Form. In steady state, there is no change in the internal energy of the element with time, and the left side of the equation is zero. To find the heat transfer rate to the element, as the temperature changes in the x-direction only, we recognize that heat transfer rates[1] q_x and $q_{x+\Delta x}$ take place across surfaces located at x and $x+\Delta x$, respectively. q_x is the heat transfer rate to the element. $q_{x+\Delta x}$ is the heat transfer rate from the element, and hence, it is negative. There is no heat transfer across the other insulated surfaces. The net rate of heat transfer to the element is given by

$$q_x - q_{x+\Delta x}$$

Both q_x and $q_{x+\Delta x}$ are assumed to be positive in the x-increasing direction. For this case there is neither work transfer (as there is no change in volume) nor any internal energy generation. We now have the energy equation,

$$q_x - q_{x+\Delta x} = 0$$

Dividing the equation by Δx and taking the limit as $\Delta x \rightarrow 0$ we have

$$-\frac{dq_x}{dx} = 0 \qquad \qquad \textbf{(2.2.1)}$$

(d) Identify the Needed Auxiliary Laws. As we are interested in the temperature distribution, we seek a law that relates the heat transfer rate to the temperature changes. Fourier's Law provides such a relation. This is the auxiliary law that is required in this case.

(e) Obtain the Applicable Equations. Now, as T is a function of x only, $q''_x = -k\,(dT/dx)$ and

$$q_x = \int_{A_x} q''_x\, dA_x = \int_{A_x} -k\frac{dT}{dx}\, dA_x$$

A_x represents the area of a surface perpendicular to the x-direction. As $T = T(x)$ only, dT/dx is the constant everywhere on A_x, (dT/dx has the same value at a fixed value of x regardless of the value of y and z). As the medium is assumed

---

[1]Strictly speaking, we should be using $q_x \mid_x$ and $q_x \mid_{x+\Delta x}$, the first subscript indicating the direction of heat transfer and the second defining the location of the surface. In one-dimensional problems, dropping the first subscript should not cause any confusion; it will make it easier to read the text.

to be homogeneous, the thermal conductivity is not dependent on the location. Hence,

$$q_x = -k \frac{dT}{dx} \int_{A_x} dA_x \quad \text{or} \quad q_x = -kA_x \frac{dT}{dx} \tag{2.2.2}$$

Substituting Equation 2.2.2 into Equation 2.2.1, we get

$$\frac{d}{dx}\left(kA_x \frac{dT}{dx}\right) = 0 \tag{2.2.3}$$

(f) Solve the Resulting Equations. To find T as a function of x, Equation 2.2.3 should be integrated twice. The solution contains two constants, which are evaluated from a knowledge of the conditions that must be satisfied for this particular problem—boundary conditions. There should be two independent conditions that T must satisfy (one for each order of the derivative). For this case T must satisfy

$$T(0) = T_1 \qquad T(L) = T_2 \tag{2.2.4}$$

Integrating Equation 2.2.3, we obtain

$$\frac{dT}{dx} = \frac{c_1}{kA_x} \tag{2.2.5}$$

With constant k and A_x integrating Equation 2.2.5, we obtain

$$T = \frac{c_1}{kA_x} x + c_2 \tag{2.2.6}$$

From the boundary conditions given by Equation 2.2.4, we evaluate the two constants, c_1 and c_2.

$$c_2 = T_1 \qquad c_1 = -\frac{T_1 - T_2}{L} kA_x$$

and

$$T = -\frac{T_1 - T_2}{L} x + T_1 \tag{2.2.7}$$

From Equation 2.2.7 it is seen that the temperature varies linearly with x, which is illustrated in Figure 2.2.2. To find the heat transfer rate across a surface located at $x = x_0$ $(0 \le x_0 \le L)$, by Fourier's Law

$$q_x = -kA_x \frac{dT}{dx}\bigg|_{x=x_0} = -kA_x \frac{-(T_1 - T_2)}{L}$$

or

$$q_x = kA_x \frac{T_1 - T_2}{L} \tag{2.2.8}$$

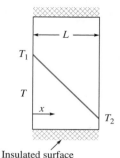

FIGURE 2.2.2 Linear Temperature Profile in a Rectangular Slab

Insulated surface

From Equation 2.2.8, it is evident that the heat transfer rate across every surface (perpendicular to the x-direction) is constant; this result is consistent with Equation 2.2.1.

COMMENTS

Note that q_x is positive for $T_1 > T_2$. Although heat transfer rates are computed across surfaces, for situations of this type where the heat transfer rate across every cross-sectional area is constant, we may say that the heat transfer rate across the slab is given by Equation 2.2.8.

The temperature distribution is independent of the thermal conductivity of the material, but the thermal conductivity determines the magnitude of the heat transfer rate.

It was assumed that the thermal conductivity of the material is constant. In most cases, the thermal conductivity is a function of the temperature. In the temperature range T_1 to T_2, if the change in the conductivity is small compared with its actual value, the thermal conductivity may be approximated as a linear function of the temperature. The arithmetic mean of the values of the thermal conductivity at T_1 and T_2, or its value at the arithmetic mean of the temperatures T_1 and T_2, may then be used to compute the heat transfer rate by Equation 2.2.8.

EXAMPLE 2.2.1   The outer wall of a house, constructed with common brick, is 4-m long, 2-m high, and 30-cm thick. The inner surface of the wall is at 20 °C and the outer surface at 0 °C (Figure 2.2.3). Find the heat transfer rate through the wall.

Given

Wall dimensions: 4-m long, 2-m high   Material of the wall: common brick
Thickness of wall (L) = 30 cm   Outer surface temperature (T_2) = 0 °C
Inner surface temperature (T_1) = 20 °C

Find

Heat transfer rate through the wall

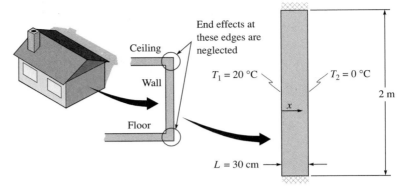

FIGURE 2.2.3 The sketch shows a wall with the inner surface at 20 °C, close to the room temperature, and the outer surface at 0 °C, a temperature that may be expected on a cold day.

ASSUMPTIONS

1. Steady state
2. Homogeneous material
3. One-dimensional temperature distribution. This is a reasonable assumption as the temperature on the inner and outer surfaces are uniform, and the material of the wall may be considered to be homogeneous. In the regions where the wall is attached to other walls, ceiling, and so forth, the temperature distribution may not be one-dimensional but we assume that such end effects are negligible.
4. Constant properties. In the limited range of temperature from 0 °C to 20 °C, we assume the properties to be constant.

SOLUTION

With the above assumptions the development of the solution for a rectangular slab leading to Equation 2.2.8 is valid.

$$q_x = kA_x \frac{T_1 - T_2}{L}, \qquad (A_x = 4 \times 2 = 8 \text{ m}^2)$$

From Table A3A the thermal conductivity of common building bricks is 0.72 W/m K. Substituting the values.

$$q_x = 0.72 \times 8 \times \frac{20 - 0}{0.3} = \underline{384 \text{ W}}$$

One-dimensional temperature distribution—hollow cylinder

2.2.2 Cylindrical Geometry—Hollow Cylinder

Now, consider a hollow, circular cylinder, whose end surfaces perpendicular to the axis are insulated, as shown in Figure 2.2.4a. The inner surface is maintained at T_i

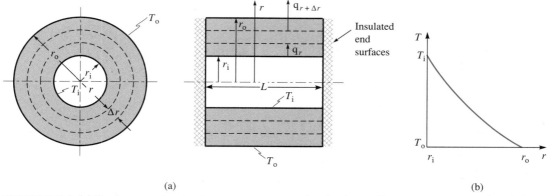

FIGURE 2.2.4 (a) The inner and outer temperatures are maintained at uniform temperatures of T_i and T_o. (b) Logarithmic temperature distribution given by Equation 2.2.11.

and the outer surface at T_o. The inner and outer radii are r_i and r_o, respectively, and the axial length is L. We wish to obtain the temperature distribution and the heat transfer rate.

ASSUMPTIONS

1. Steady state
2. Homogeneous material
3. Constant properties

SOLUTION

As the ends are insulated and the inner and outer surfaces are at uniform temperatures, the temperature varies with the radius only, i.e., $T = T(r)$. In cylindrical geometry, cylindrical coordinates are the appropriate choice for the solution of this problem. By taking a cylindrical element of radius r and thickness Δr and following the development for the rectangular slab, with r replacing x, we obtain

$$q_r - q_{r+\Delta r} = 0$$

Dividing by Δr and taking the limit as $\Delta r \to 0$, $-(dq_r/dr) = 0$. From Fourier's Law, as dT/dr is constant on A_r, $q_r = -kA_r (dT/dr)$ and

$$\frac{d}{dr}\left(kA_r \frac{dT}{dr}\right) = 0 \tag{2.2.9}$$

The two boundary conditions are

$$T(r_i) = T_i \qquad T(r_o) = T_o \tag{2.2.10}$$

A_r is the surface area perpendicular to the r-direction and has a magnitude of $2\pi rL$. Integrating Equation 2.2.9 once,

$$kA_r \frac{dT}{dr} = c_1$$

With $A_r = 2\pi r L$,

$$\frac{dT}{dr} = \frac{c_1}{2\pi k L r}$$

Integrating a second time (from the assumption of constant properties, k is constant),

$$T = \frac{c_1}{2\pi k L} \ln r + c_2$$

Employing the boundary conditions given by Equation 2.2.10, we get

$$T_i = \frac{c_1}{2\pi k L} \ln r_i + c_2 \qquad T_o = \frac{c_1}{2\pi k L} \ln r_o + c_2$$

Solving for c_1 and c_2 we obtain

$$c_1 = -2\pi k L \frac{T_i - T_o}{\ln(r_o/r_i)} \qquad c_2 = T_i + \frac{T_i - T_o}{\ln(r_o/r_i)} \ln (r_i)$$

Thus,

$$T = T_i + \frac{T_i - T_o}{\ln (r_o/r_i)} \ln r_i - \frac{T_i - T_o}{\ln (r_o/r_i)} \ln r$$

or

$$T = T_i - \frac{T_i - T_o}{\ln(r_o/r_i)} \ln r/r_i \qquad\qquad (2.2.11)$$

Figure 2.2.4b shows the temperature distribution in the cylinder.

To find the heat transfer rate across a cylindrical surface with radius r_1 ($r_i \leq r_1 \leq r_o$) by Fourier's Law,

$$q_{r_1} = -k A_r \left.\frac{dT}{dr}\right|_{r=r_1} \qquad\qquad (2.2.12)$$

In Equation 2.2.12, every quantity that depends on r should be evaluated at r_1.

$$A_r = 2\pi r_1 L, \quad \text{and} \quad \frac{dT}{dr} = -\frac{T_i - T_o}{r_1 \ln(r_o/r_i)}$$

$$q_{r_1} = 2\pi k L \frac{T_i - T_o}{\ln(r_o/r_i)} \qquad\qquad (2.2.13)$$

The heat transfer rate across every cylindrical shell in the cylinder is the same and is given by Equation 2.2.13.

EXAMPLE 2.2.2 An electrical heater is placed inside a thick, hollow (type AISI 316), stainless steel cylinder with inner and outer radii of 10 mm and 15 mm, respectively, and a length of 2 m (Figure 2.2.5). The power input to the heater is 30 kW. If the outer surface temperature of the cylinder is 70 °C, determine the inner surface temperature.

Given

Inner radius (r_i) = 10 mm
Length (L) = 2 m
Outer surface temperature (T_o) = 70 °C

Outer radius (r_o) = 15 mm
Power input to heater (P) = 30 000 W

Find

Inner surface temperature (T_i)

ASSUMPTIONS

1. Steady state
2. Homogeneous material
3. Constant properties
4. The power density (W/m²) on the inner surface is uniform. As the cylinder is 2-m long with a radial thickness of 5 mm, the heat transfer rate from the end surface is much smaller than the heat transfer rate from the outer cylindrical surface and is neglected. Putting this in another way, the end effects are assumed to be negligible.

SOLUTION

In steady state, applying the First Law of Thermodynamics with the cylinder and the heater as the system,

$$\begin{array}{c} \text{Heat transfer rate to the system} \\ - \text{ Work transfer rate from the system} \\ = 0 \end{array}$$

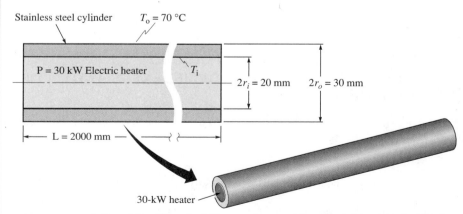

FIGURE 2.2.5 **A Cartridge Electrical Heater Encased in a Stainless Steel Cylinder**

The electrical power input to the heater is equivalent to work transfer to the system.[2] Therefore,

Heat transfer rate from the cylinder
= Power input to the heater

With the assumption of negligible heat transfer from the end surfaces,

Heat transfer rate from the outer surface
= Power input to the heater

In steady state, the heat transfer rate from the outer surface of a hollow cylinder, with end surfaces insulated, is related to the inner and outer surface temperature by Equation 2.2.13.

$$q_r = P = \frac{2\pi k L \ (T_i - T_o)}{\ln(r_o/r_i)} \tag{1}$$

where P is the power input to the heater.

From Table A1, k (316 stainless steel) = 13.4 W/m K. Substitution of the known values in Equation 1 yields

$$T_i = T_o + q_r \frac{\ln(r_o/r_i)}{2\pi k L} = 70 + \frac{30\ 000\ \ln(15/10)}{2\pi \times 13.4 \times 2} = \underline{142.2°C}$$

COMMENTS

The inner and outer surface temperatures of the cylinder are at 142.2 °C (415.4 K) and 70 °C (343.2 K). The thermal conductivity at 300 K was used in determining the inner surface temperature. From Table A1 we observed that the thermal conductivity varies from 13.4 W/m K at 300 K to 15.2 W/m K at 400 K. We may obtain a better estimate of the inner surface temperature by using the thermal conductivity at 106.1 °C, the arithmetic mean of 70 °C at 142.2 °C. From a linear interpolation the thermal conductivity at 106.1 °C is 14.8 W/m K. Using this value, the inner surface temperature is 135.4 °C.

2.3 CONDUCTION WITH BOUNDARIES EXPOSED TO CONVECTIVE HEAT TRANSFER

Convective boundary condition

So far, we considered cases where the *surface* temperatures (or temperatures at specified locations) were known. Such situations are easy to visualize, but difficult to attain in practice. More common situations involve boundaries with surfaces exposed to *fluids* whose temperatures are known. If the convective heat transfer coefficient between the surfaces and the fluids are fixed, it is evident that the tem-

---

[2]See Van Wylen, G.J., and Sonntag, R.E., *Fundamentals of Classical Thermodynamics,* 3rd ed. (New York: Wiley, 1985).

perature distribution in the wall and the heat transfer rate are also fixed; we should, therefore, be able to determine them.

As an illustration, the inner surface of the wall in Example 2.2.1 is likely to be exposed to room air maintained at T_{fi} and the outer surface to atmospheric air at T_{fo}, as shown in Figure 2.3.1a. If the convective heat transfer coefficients h_i and h_o on the inner and outer surfaces, respectively, are known, how do we determine the temperature distribution and heat transfer rate (Figure 2.3.1b)?

With the assumption of steady state, one-dimensional temperature distribution, and no internal energy generation, the differential equation is given by Equation 2.2.3,

$$\frac{d}{dx}\left(kA_x\frac{dT}{dx}\right) = 0$$

The boundary conditions given by Equation 2.2.4 are not applicable, as only the fluid temperatures T_{fi} and T_{fo}, and not the surface temperatures, are known. To obtain the boundary conditions, from an energy balance on the surface located at $x = 0$,

Heat transfer rate by conduction (in the x-direction) from the inner surface

= Heat transfer rate *from the inside fluid to the inner surface by convection*

$$-kA_x\frac{dT}{dx}\bigg|_{x=0} = h_iA_x(T_{fi} - T)\bigg|_{x=0} \qquad\qquad \textbf{(2.3.1)}$$

Similarly, at $x = L$,

Heat transfer rate to the outer surface by conduction (in the x-increasing direction)

= Convective heat transfer rate *from the surface to the outside fluid*

$$-kA_x\frac{dT}{dx}\bigg|_{x=L} = h_oA_x(T - T_{fo})\bigg|_{x=L} \qquad\qquad \textbf{(2.3.2)}$$

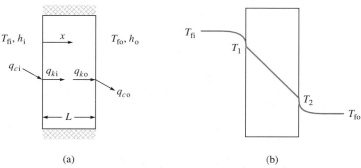

(a)   (b)

FIGURE 2.3.1 Determining Conductive Heat Transfer Rate with Boundaries Exposed to Convective Heat Transfer
(a) Model of a wall. (b) Temperature distribution in the wall and the fluids.

(Note the different signs in Equations 2.3.1 and 2.3.2. Both of them can be represented by

$$-k \frac{dT}{dn} = h(T - T_{fo})$$

where n is the direction of the outward normal to the surface. At $x = 0$, $dT/dx = -dT/dn$, and at $x = L$, $dT/dx = dT/dn$.)

The solution to Equation 2.2.3 is given by Equation 2.2.6.

$$T = \frac{c_1}{kA_x} x + c_2 \qquad (2.2.6)$$

From the boundary conditions given by Equations 2.3.1 and 2.3.2, we obtain

$$-KA_x \left[\frac{c_1}{kA_x} \right] = h_i A_x (T_{fi} - c_2) \qquad (2.3.3)$$

$$-kA_x \left[\frac{c_1}{kA_x} \right] = h_o A_x \left[\frac{c_1 L}{kA_x} + c_2 - T_{fo} \right] \qquad (2.3.4)$$

Solving Equations 2.3.3 and 2.3.4 for c_1 and c_2,

$$c_1 = -\frac{T_{fi} - T_{fo}}{[1/(h_i A_x)] + [L/(kA_x)] + [1/(h_o A_x)]} \qquad (2.3.5a)$$

$$c_2 = T_{fi} - \frac{T_{fi} - T_{fo}}{(h_i/h_o) + [(h_i L)/k] + 1} \qquad (2.3.5b)$$

To find the heat transfer rate across any cross section perpendicular to the x-direction, we have

$$q_x = -kA_x \frac{dT}{dx} = -c_1$$

or

$$q_x = \frac{T_{fi} - T_{fo}}{[1/(h_i A_x)] + [L/(kA_x)] + [1/(h_o A_x)]} \qquad (2.3.6)$$

COMMENT

With $q_x = h_i A_x (T_{fi} - T_1) = kA_x [(T_1 - T_2)/L]$, where T_1 and T_2 are the temperatures at $x = 0$ and $x = L$, respectively, we have

$$T_{fi} - T_1 = \frac{T_1 - T_2}{(h_i L)/k} \qquad (2.3.7)$$

If h_i is large, such that $(h_i L)/k \gg 1$, $(T_{fi} - T_1) \ll (T_1 - T_2)$; i.e., the difference between the temperatures of fluid 1 and the adjacent surface, is much smaller than the temperature change in the slab. In such cases, one may use the approximation that $T_1 \approx T_{f1}$ for computing the heat transfer rate and the temperature distribution.

With boiling and condensation of some fluids (for example, water) the convective heat transfer coefficient is very high, and the temperature of the adjacent surface will be nearly equal to the saturation temperature of the fluid. This is one way of maintaining the temperature of a surface at a defined value.

The heat transfer given by Equation 2.3.6 can be obtained by a simpler method—the electrical analogy.

2.4 ELECTRICAL ANALOGY

The expression for the heat transfer rate given by Equation 2.3.6 can be obtained by an alternate method. From an energy balance on the left surface of the slab,

Heat transfer rate by convection from the fluid at T_{fi} to the surface (q_{ci})
= Heat transfer rate by conduction in the x-direction by conduction (q_{ki})

Similarly, from an energy balance on the right surface,

Heat transfer rate to the surface in the x-direction by conduction (q_{ko})
= Heat transfer rate by convection from the surface to the fluid at T_{fo} (q_{co})

In steady state the net heat transfer rate to the slab is zero, and hence, the heat transfer rate across every cross section of the surface has the same value. Therefore, from the equalities given above

$$q_{ci} = q_{ki} = q_{ko} = q_{co} = q$$

and

$$q_{ci} = h_i A_x (T_{fi} - T_1)$$

$$q_{ki} = q_{ko} = \frac{kA_x (T_1 - T_2)}{L}$$

$$q_{co} = h_o A_x (T_2 - T_{fo})$$

T_1 and T_2 are the temperatures of the left and right surfaces of the solid. Rearranging the above three equations (with all heat transfer rates represented by q),

$$T_{fi} - T_1 = q \left(\frac{1}{h_i A_x} \right)$$

$$T_1 - T_2 = q \left(\frac{L}{kA_x} \right)$$

$$T_2 - T_{fo} = q \left(\frac{1}{h_o A_x} \right)$$

Adding the three temperature differences and rearranging the resulting equation, we obtain Equation 2.3.6.

The above three equations where the temperature differences are expressed as the

Electrical analogy—
thermal circuit

product of the heat transfer rate and a term containing the heat transfer coefficient or thermal conductivity, and the area of cross section are similar to the equations relating the voltage drop to the product of the current and the resistance in a resistance circuit. Further, the addition of the temperature differences resulting in the total temperature difference is similar to the addition of the voltage drop across each resistance to give the total voltage drop in a series resistance circuit. Therefore, the temperature differences are analogous to the voltage drop, the heat transfer rate is analogous to the current, and the coefficients of the heat transfer rate are analogous to the resistances. The similarity between the thermal circuit and the series electrical circuit is obvious.

In Figure 2.4.1 the heat transfer rates in steady state with one-dimensional temperature distribution without internal energy generation in a rectangular slab, hollow cylinder, and hollow sphere are given. The expressions for the heat transfer rates for a rectangular slab and a hollow cylinder were derived earlier in this chapter. The

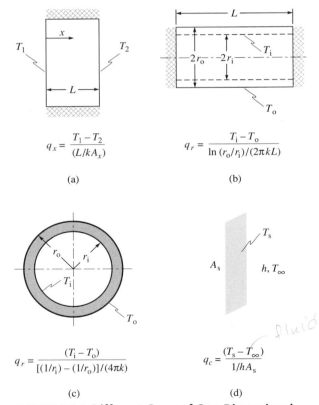

$$q_x = \frac{T_1 - T_2}{(L/kA_x)}$$

(a)

$$q_r = \frac{T_i - T_o}{\ln (r_o/r_i)/(2\pi kL)}$$

(b)

$$q_r = \frac{(T_i - T_o)}{[(1/r_i) - (1/r_o)]/(4\pi k)}$$

(c)

$$q_c = \frac{(T_s - T_\infty)}{1/hA_s}$$

(d)

FIGURE 2.4.1 Different Cases of One-Dimensional Temperature Distribution and Their Corresponding Thermal Resistances
(a) Plane slab. (b) Hollow cylinder. (c) Hollow sphere. (d) Convection.

derivation of the expression for the heat transfer rate in a hollow sphere is left as an exercise. The expression for the convective heat transfer rate follows from the definition of the convective heat transfer coefficient. In each case the heat transfer rate has been expressed as the ratio of an appropriate temperature difference to a quantity containing the geometrical parameters and the constant thermal conductivity of the material or the uniform convective heat transfer coefficient. In analogy with the electrical circuit, the denominator in each expression represents the thermal resistance in the circuit.

The analogy can be usefully employed to solve many heat transfer problems. For example, consider the wall shown in Figure 2.4.2. It is exposed to fluids on either side of the wall with known fluid temperatures and associated convective heat transfer coefficients (similar to Figure 2.3.1). With the assumptions of steady state, one-dimensional temperature distribution and constant properties, the thermal circuit identifies the convective resistances [$1/(h_iA_x)$ and $1/h_oA_x$] on either side of the wall and the conductive resistance [$L/(kA_x)$].

From the thermal circuit one can solve for any one of the parameters in the equation,

$$q_x = \frac{T_{fi} - T_{fo}}{[1/(h_iA_x)] + [L/(kA_x)] + [1/h_oA_x]}$$

If T_{fi} and T_{fo} are known, having found q_x, the inner and outer surface temperatures, T_1 and T_2, are easily found.

In applying the thermal circuit, continuity of the temperature across the interface of dissimilar materials is usually assumed; i.e., the interfacial thermal resistance has been assumed to be negligible. Such an assumption may not be justified in all cases. For a discussion of the interfacial resistance refer to Contact Resistance later in this section.

Overall Heat Transfer Coefficient

In the preceding example, the heat flux from one fluid to the other can be written as

$$q''_x = \frac{q_x}{A_x} = \frac{T_{fi} - T_{fo}}{(1/h_i) + (L/k) + (1/h_o)} = U\,(T_{fi} - T_{fo}) \qquad \textbf{(2.3.8)}$$

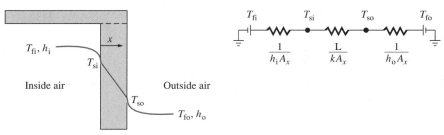

FIGURE 2.4.2 A Wall Exposed to Fluids on Either Side at Different Temperatures and the Corresponding Thermal Circuit

In Equation 2.3.8 the denominator represents the thermal resistance for a unit area of the surface. In many cases of one-dimensional, steady-state temperature distribution without internal energy generation, the heat flux can be expressed as the product of the reciprocal of the resistance for a unit area, and an appropriate temperature difference, as in Equation 2.3.8. In such cases, the reciprocal of the resistance for a unit area, U, is termed the *overall heat transfer coefficient*. In the example just considered,

Overall heat transfer coefficient

$$U = \frac{1}{(1/h_i) + (L/k) + (1/h_o)}$$

The concept of overall heat transfer coefficient is extensively used in cases of heat transfer from one fluid to another separated by a solid medium, particularly in heat exchanger applications and in estimating heating and cooling loads of buildings.

The use of the thermal circuit for the solution of problems involving one-dimensional temperature distribution is illustrated in Examples 2.4.1 through 2.4.3.

EXAMPLE 2.4.1   The exterior wall of a building is constructed of four materials: 12-mm thick gypsum board, 75-mm thick fiberglass insulation, 20-mm thick plywood, and 20-mm thick hardboard siding. For the purpose of determining the heat flux, the wall is modeled in Figure 2.4.3, with the four materials numbered 1 through 4. The inside and outside air temperatures are 20 °C and −10 °C, respectively. The convective heat transfer coefficient on the inner and outer surfaces of the wall are 6 W/m² °C and 10 W/m² °C, respectively. Determine the heat flux and the overall heat transfer coefficient.

Given

$L_1 = 12$ mm $L_3 = 20$ mm
$L_2 = 75$ mm $L_4 = 20$ mm
Inside air temperature $(T_i) = 20\,°C$ Outside air temperature $(T_o) = -10\,°C$
Inside heat transfer   Outside heat transfer
    coefficient $(h_i) = 6$ W/m² °C   coefficient $(h_o) = 10$ W/m² °C

Find

(a) The average heat flux through the wall
(b) The overall heat transfer coefficient

ASSUMPTIONS

1. Steady state
2. Each material is homogeneous with constant properties.
3. The temperature is continuous, i.e., there is no temperature change at the interfaces of the different materials.

SOLUTION

(a) The temperature is one-dimensional in each material and we construct the thermal circuit for the temperature distribution (Figure 2.4.3b). From Table A3A, the properties of the material are

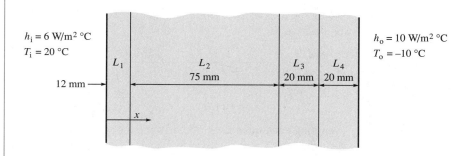

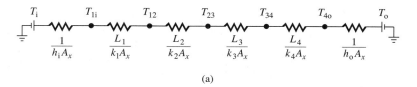

(a)

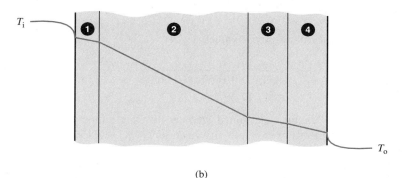

(b)

FIGURE 2.4.3 The Thermal Circuit of a One-Dimensional Temperature Distribution Problem
(a) A composite wall exposed to fluids on either side at different temperatures and the corresponding thermal circuit. (b) Temperature profile in the composite wall.

1. k_1 (gypsum, plasterboard) $= 0.176$ W/m K
2. k_3 (plywood) $= 0.115$ W/m K
3. k_2 (fiberglass) $= 0.036$ W/m K
4. k_4 (hardboard siding) $= 0.215$ W/m K

R_i and R_o represent the convective resistances at the inner and outer surfaces, respectively, and R_1, R_2, R_3, and R_4 the conductive resistances of the different materials. The double subscripts for the temperatures in the resistance diagram denote the temperature at the interface of two materials indicated by the two numerals of the subscript. The heat transfer rate across the wall is

$$q = \frac{T_i - T_o}{R_i + R_1 + R_2 + R_3 + R_4 + R_o} \tag{1}$$

where the heat transfer from the inner surface to the outer surface has been taken to be positive.

$$\frac{q}{A_x} = \frac{T_i - T_o}{A_x \, (R_i + R_1 + R_2 + R_3 + R_4 + R_o)}$$

$$= \frac{T_i - T_o}{(1/h_i) + (L_1/k_1) + (L_2/k_2) + (L_3/k_3) + (L_4/k_4) + (1/h_o)} \quad (2)$$

Denoting each quantity in the denominator of Equation 2 by R'', for example, $A_x R_i$ ($= 1/h_i$) by R_i'', with an appropriate subscript to indicate the thermal resistance for a unit area of each material and using the properties of the different materials from Table A3A, we have

$R_i'' = 1/h_i = 1/6 = 0.1667 \text{ m}^2 \text{ K/W}$
k_1 (gypsum) $= 0.176 \text{ W/m K};$ $R_1'' = 0.012/0.176 = 0.06818 \text{ m}^2 \text{ K/W}$
k_2 (fiberglass) $= 0.036 \text{ W/m K};$ $R_2'' = 0.075/0.036 = 2.0833 \text{ m}^2 \text{ K/W}$
k_3 (plywood) $= 0.115 \text{ W/m K};$ $R_3'' = 0.02/0.115 = 0.1739 \text{ m}^2 \text{ K/W}$
k_4 (hardwood siding) $= 0.215 \text{ W/m K};$ $R_4'' = 0.02/0.215 = 0.09302 \text{ m}^2 \text{ K/W}$
$R_o'' = 1/h_o = 1/10 = 0.1 \text{ m}^2 \text{ K/W}$

Substituting these values we get

$$q'' = \frac{q}{A_x} = \frac{20 - (-10)}{0.1667 + 0.06818 + 2.0833 + 0.1739 + 0.09302 + 0.1}$$
$$= \underline{11.17 \text{ W/m}^2}$$

Having determined the heat flux, the temperature at any location can be determined. For example, the value of T_{12} is obtained from

$$q'' = \frac{T_i - T_{12}}{R_i'' + R_1''}$$

or

$$11.17 = \frac{20 - T_{12}}{0.1677 + 0.06818} \qquad \underline{T_{12} = 17.4 \,°\text{C}}$$

(b) The overall heat transfer coefficient can be found either by the relation $U = q''/(T_i - T_o)$ or from the definition of the overall heat transfer coefficient as the reciprocal of the resistance for a unit area. As the latter definition is more useful (the overall heat transfer coefficient can be calculated without computing the heat flux), we determine U from that definition.

$$U = \frac{1}{R_i'' + R_1'' + R_2'' + R_3'' + R_4'' + R_o''}$$
$$= \frac{1}{0.1667 + 0.06818 + 2.0833 + 0.1739 + 0.09302 + 0.1}$$
$$= 0.372 \text{ W/m}^2 \,°\text{C}$$

The *R*-values that are listed on insulation materials for houses are the values of *L/k* for the insulations when British (FPS) units are used. For example, the thermal conductivity of fiberglass blanket insulation is 0.046 W/m K or 0.0266 BTU/hr.ft. °F. For a 3.5-in thick insulation, the *R*-value is $(3.5/12)/0.0266 = 10.96$ and is rounded off to 11. Similarly the *R*-value of a 6-in thick fiberglass blanket insulation is 19.

EXAMPLE 2.4.2   Pressurized hot water flows in a heavy walled carbon-manganese-silicon steel tube ($r_i = 25$ mm, $r_o = 40$ mm) as shown in Figure 2.4.4a. The temperature of the water at a particular location is 300 °C. The pipe is surrounded by air at 20 °C. The convective heat transfer coefficients are $h_i = 1500$ W/m² °C and $h_o = 6$ W/m² °C.

(a) Determine the heat transfer rate per meter length of the tube and the outer surface temperature of the bare pipe.
(b) Compute the overall heat transfer coefficient based on the inner surface area and on the outer surface area.

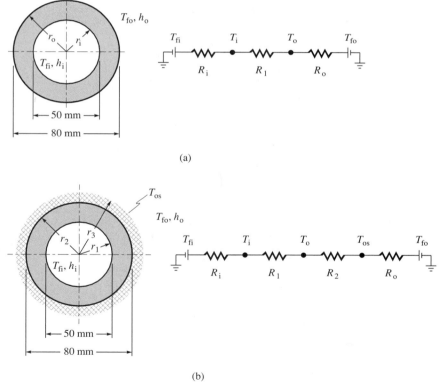

(a)

(b)

FIGURE 2.4.4 (a) A heavy, walled tube in which hot water flows exposed to a fluid on the outside, and the corresponding thermal circuit. $T_{fi} = 300$ °C; $h_i = 1500$ W/m² °C; $r_i = 25$ mm; $T_{fo} = 20$ °C; $h_o = 6$ W/m² °C; $r_o = 40$ mm. (b) The tube covered with insulation and its thermal circuit.

(c) It is proposed to insulate the tube with fiberglass insulation so that the outer surface temperature of the insulation is limited to 25 °C for reasons of safety and to reduce the heat transfer rate to the surroundings. Find the thickness of the insulation to be applied to the tube.

Given

Inside fluid temperature
(T_{fi}) = 300 °C

Surrounding air temperature
(T_{fo}) = 20 °C

Inside convective heat transfer
coefficient (h_i) = 1500 W/m^2 °C

Outside convective heat transfer
coefficient (h_o) = 6 W/m^2 °C

Inner radius of the tube (r_i) = 25 mm

Outer radius of the tube (r_o) = 40 mm

Find

(a) The heat transfer rate per meter length of the tube and the outer surface temperature of the bare tube (T_o)

(b) The overall heat transfer coefficients based on the inner and outer surface areas of the tube

(c) The thickness of fiberglass insulation to limit the outer surface temperature of the insulation (T_{os}) to 25 °C

ASSUMPTIONS

1. Steady state
2. Constant properties
3. When hot water flows inside the tube with heat transfer from the water to the surroundings, the temperature of the water decreases in the direction of flow, and the inner surface temperature of the tube also varies in the direction of flow. With the uniform outside conditions (air temperature and the convective heat transfer coefficient on the outside not varying in the direction of flow), this leads to some variation in the temperature in the tube in the direction of flow. Unless the flow rate is very small (leading to significant changes in the water temperature), the variation of temperature in the direction of flow is very much smaller than that in the radial direction and is neglected. The problem can then be treated as a one-dimensional problem with the temperature being a function of the radius only.

SOLUTION

(a) The applicable resistances are

$$R_i = \frac{1}{h_i A_i} = \frac{1}{h_i 2\pi r_i L} \qquad R_1 = \frac{\ln(r_o/r_i)}{2\pi k_{st} L} \qquad R_o = \frac{1}{h_o 2\pi r_o L}$$

All the conditions for the application of the thermal circuit are satisfied. Anticipating the average temperature of the tube to be close to 300 °C, we evaluate the thermal conductivity of the steel at 570 K. From Table A1 for C-Mn-Si steel at 570 K, k_{st} = 40 W/m K.

From the thermal circuit shown in Figure 2.4.4a,

$$q_r = \frac{T_{fi} - T_{fo}}{[1/(h_i 2\pi r_i L)] + [\ln(r_o/r_i)/(2\pi k_{st} L)] + 1/[h_o/(2\pi r_o L)]}$$

or

$$\frac{q_r}{L} = \frac{T_{fi} - T_{fo}}{[1/(h_i 2\pi r_i)] + [\ln(r_o/r_i)/(2\pi k_{st})] + [1/(h_o 2\pi r_o)]}$$

Denoting the thermal resistances for a unit length in the denominator by R', we have

$$R'_i = \frac{1}{h_i\, 2\pi r_i} = \frac{1}{1500 \times 2\pi \times 0.025} = 0.004244 \text{ m °C/W}$$

$$R'_1 = \frac{\ln(r_o/r_i)}{2\pi k_{st}} = \frac{\ln(40/25)}{2\pi \times 40} = 0.00187 \text{ m °C/W}$$

$$R'_o = \frac{1}{h_o 2\pi r_o} = \frac{1}{6 \times 2\pi \times 0.04} = 0.6631 \text{ m °C/W}$$

Substituting the values of the resistances,

$$\frac{q_r}{L} = \frac{300 - 20}{0.004244 + 0.00187 + 0.6631} = \underline{418.4 \text{ W/m}}$$

From the thermal circuit (or by the definition of the convective heat transfer coefficient) we have

$$\frac{q_r}{L} = \frac{T_{os} - T_{fo}}{[1/(h_o 2\pi r_o)]}$$

or

$$418.4 = \frac{T_{os} - 20}{0.6631} \qquad \underline{T_{os} = 297.4 \text{ °C}}$$

Note that the average temperature is ≈ 571 K. The thermal conductivity evaluated at 570 K is appropriate.

(b) From the expression for the heat transfer rate, we have

Heat transfer rate per unit area based on the inner surface area $= q''_i = \dfrac{q_r}{2\pi r_i L}$

Heat transfer rate per unit area based on the outer surface area $= q''_o = \dfrac{q_r}{2\pi r_o L}$

$$q''_i = \frac{T_{fi} - T_{fo}}{(1/h_i) + [(r_i/k_{st})\ln(r_o/r_i)] + [r_i/(h_o r_o)]}$$

By definition, the overall heat transfer coefficient based on the inner surface area is given by

$$U_i = \frac{q_i''}{T_{fi} - T_{fo}} = \frac{1}{(1/h_i) + [(r_i/k_{st})\ln(r_o/r_i)] + [r_i/(h_o r_o)]}$$

$$= \frac{1}{1/1500 + [(0.025/40)\ln(0.04/0.025)] + 0.025/(6 \times 0.04)} = \underline{9.51 \text{ W/m}^2\ {}^{\circ}\text{C}}$$

Similarly the overall heat transfer coefficient based on the outer surface area is given by

$$U_o = \frac{q_o''}{T_{fi} - T_{fo}} = \frac{1}{[r_o/(h_i r_i)] + [(r_o/k_{st})\ln(r_o/r_i)] + (1/h_o)}$$

$$= \frac{1}{0.04/(1500 \times 0.025) + [(0.04/40)\ln(0.04/0.025)] + 1/6} = \underline{5.95 \text{ W/m}^2\ {}^{\circ}\text{C}}$$

(c) Denoting the inner and outer radii of the tube by r_1 and r_2 and the outer radius of the insulation by r_3, the resistances for a unit length are

$$R_i' = \frac{1}{h_i 2\pi r_1}, \qquad R_1' = \frac{1}{2\pi k_{st}} \ln\left(\frac{r_2}{r_1}\right),$$

$$R_2' = \frac{1}{2\pi k_{ins}} \ln\left(\frac{r_3}{r_2}\right), \qquad R_o' = \frac{1}{h_o 2\pi r_3}$$

From Figure 2.4.4b,

$$\frac{T_i - T_{os}}{[1/(h_i 2\pi r_1)] + [1/(2\pi k_{st})\ln(r_2/r_1)] + [1/(2\pi k_{ins})\ln(r_3/r_2)]} = \frac{T_{os} - T_o}{1/h_o 2\pi r_3}$$

where

k_{ins} = thermal conductivity of fiberglass (Table A3A) = 0.036 W/m K
T_{os} = temperature of the outer surface of the insulation = 25 °C

Substituting the values,

$$\frac{275}{0.00602 + 4.421 \ln (25\ r_3)} = 188.5\ r_3$$

Solving this equation by one of the numerical methods (for example, by the method of bisection with the limits of 0.04 m as a minimum and 1 m as a maximum for r_3), we obtain $r_3 = 0.203$ m, giving an insulation thickness of

$$0.203 - 0.040 = 0.163 \text{ m} = \underline{16.3 \text{ cm}}$$

The heat transfer rate per meter length is given by

$$q_r = 2\pi r_3 h_o (T_{os} - T_o) = 2\pi \times 0.203 \times 6 \times (25 - 20) = \underline{38.3 \text{ W/m}}$$

COMMENT

Note the dramatic decrease in the heat transfer rate per unit length caused by the addition of the insulation. The insulation has not only made it safer for people who may be in the vicinity of the pipe, but has also resulted in a significant reduction in the heat transfer rate. It is thus possible to combine safety with other engineering and economic advantages.

Why is the overall heat transfer coefficient based on the outer surface area very nearly equal to h_o?

Thermal parallel circuit is an acceptable approximation in many cases

The thermal circuit works well with series circuits, but with parallel circuits it is not as accurate because thermal parallel circuits cannot be treated as one-dimensional problems. However, one may, in many cases, accept the approximation involved in applying the thermal circuit to parallel circuits. For example, consider the heat transfer rate across the wall of a residential building. One type of construction of the wall consists of dry wall followed by insulation between studs and an outer siding. The heat transfer across the wall has two parallel paths: one across the dry wall, the stud, and the siding; and another across the dry wall, the insulation, and the siding. In most cases, the heating and cooling loads of buildings are computed by such parallel circuits.

EXAMPLE 2.4.3

The wall of a house is constructed of 0.5-in thick wood siding, nominal 2 in × 4 in wood studs at 16-in centers and 0.5-in thick gypsum wallboard. The space between the studs is filled with 3.5-in thick fiberglass insulation. The convective heat transfer coefficients on the outer and inner surfaces are 6 BTU/hr ft^2 °F and 1.3 BTU/hr ft^2 °F, respectively. The inside air temperature is 68 °F and the outside air temperature

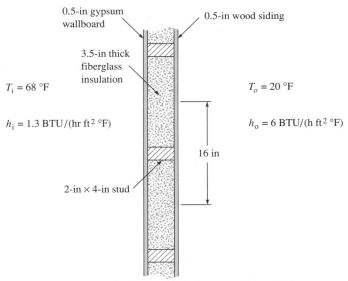

FIGURE 2.4.5 Details of the Wall Construction—A Cross Section of the Wall, Looking Down

is 20 °F. Compute the heat flux from the wall surface and the overall heat transfer coefficient. The details of the wall are shown in Figure 2.4.5.

Given

Construction details of the wall in
   Figure 2.4.5
Inside air temperature (T_i) = 68 °F
Outside air temperature (T_o) = 20 °F

Inside convective heat transfer
   coefficient (h_i) = 1.3 BTU/h ft^2 °F
Outside convective heat transfer
   coefficient (h_o) = 6 BTU/h ft^2 °F

Find

(a) The average heat flux through the wall
(b) The overall heat transfer coefficient

ASSUMPTIONS

1. Steady state
2. Constant properties
3. 2 in × 4 in studs actually measure approximately 1.5 in × 3.5 in.
4. Parallel thermal circuit is applicable

Solution

(a) To find the heat flux across the wall, choose a part of the wall 16-in wide with a stud in the center as shown in Figure 2.4.5. We may approximate the heat transfer rate to take place along parallel paths—one through that part of the wall where the stud is located and the other where there is insulation in the wall. For computing areas, we take one foot height of the wall (into the plane of the paper). If we consider the two parallel paths to be independent of each other (in line with the one-dimensional approximation), the equivalent resistance circuit is shown in Figure 2.4.6.

Values of thermal conductivities of the various building materials have been taken from the ASHRAE *Handbook of Fundamentals* (1989). The values in the Handbook are slightly different from those given in the tables in this book. The Handbook is an excellent source for properties and other information related to heat transfer in

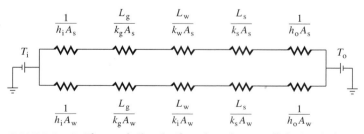

FIGURE 2.4.6 Thermal Circuit Showing the Parallel Circuit for the Wall Modeled in Figure 2.4.5

buildings. As FPS (British) units are widely used in the building construction industry, they are used in this problem also. With properties from the handbook, we have

L_g = thickness of gympsum board = 0.5 in
L_w = thickness of the insulation and stud = 3.5 in
L_s = thickness of siding = 0.5 in
k_w = thermal conductivity of pine studs = 0.07 BTU/h ft °F
k_s = thermal conductivity of siding = 0.07 BTU/h ft °F
k_i = thermal conductivity of insulation = 0.023 BTU/h ft °F
k_g = thermal conductivity of gypsum board = 0.248 BTU/h ft °F
h_i = inside convection heat transfer coefficient = 1.3 BTU/h ft^2 °F
h_o = outside convection heat transfer coefficient = 6 BTU/h ft^2 °F
A_s = area of cross section of stud/foot length = 1.5/12 = 0.125 ft^2
A_w = area of wall excluding studs/foot length = (16 − 1.5)/12 = 1.21 ft^2
T_i = 68 °F
T_o = 20 °F

The heat transfer rate per foot of length across the studs is given by

$$q_s = \frac{T_i - T_o}{[1/(h_iA_s)] + [L_g/(k_gA_s)] + [L_w/(k_wA_s)] + [L_s/(k_sA_s)] + [1/(h_oA_s)]}$$

$$= \frac{68 - 20}{[1/(1.3 \times 0.125)] + [(0.5/12)/(0.248 \times 0.125)] + [(3.5/12)/(0.07 \times 0.125)] \\ + [(0.5/12)/(0.07 \times 0.125)] + [1/(6 \times 0.125)]}$$

$$= \underline{1.02 \text{ BTU/h ft}}$$

The heat transfer rate per foot length across the insulation is given by

$$q_w = \frac{T_i - T_o}{[1/(h_iA_w)] + [L_g/(k_gA_w)] + [L_w/(k_iA_w)] + [L_s/(k_sA_w)] + [1/(h_oA_w)]}$$

$$= \frac{68 - 20}{[1/(1.3 \times 1.21)] + [(0.5/12)/(0.248 \times 1.21)] + [(3.5/12)/(0.023 \times 1.21)] \\ + [(0.5/12)/(0.07 \times 1.21)] + [1/(6 \times 1.21)]}$$

$$= \underline{4.04 \text{ BTU/h ft}}$$

(b) The total heat transfer rate across a wall area of 16/12 ft^2 is given by

$$q_s + q_w = 1.02 + 4.04 = 5.06 \text{ BTU/h.}$$

Average heat transfer rate/ft^2 of wall = 5.06/(16/12) = 3.8 BTU/h ft^2.
From the definition of the overall heat transfer coefficient,

$$U = \frac{q''}{T_i - T_o} = \frac{3.8}{68 - 20} = \underline{0.079 \text{ BTU/h ft}^2 \text{ °F} = 0.449 \text{ W/m}^2 \text{ °C}}$$

COMMENT

A comparison of the heat fluxes across the studs (1.02/0.125 = 8.16 W/m^2) and the insulation (4.04/1.21 = 3.34 W/m^2) shows the effectiveness of the insulation.

From the various cases discussed, and the introduction to the thermal circuit, we can conclude that use of the thermal circuit is limited to steady state, one-dimensional problems without internal energy generation and with constant thermal conductivity. In most practical cases, all these conditions are not totally satisfied, but one should ensure that they are reasonably satisfied if the circuit is to give results of acceptable accuracy.

Contact Resistance

Contact resistance
at the interface

In the application of the thermal circuit, we tacitly assumed that the temperature was continuous at the interface of two dissimilar materials (Figure 2.4.7a). In reality, because of the roughness of the interfaces, the materials are in contact at relatively few places, with the gaps filled with a fluid, usually the fluid surrounding the materials. The fluid in the gaps and the contact surfaces provide parallel paths for heat transfer. The resistance at the interface is confined to a very thin region on either side of the interface. As the resistance is confined to a very thin region, there is a sharp change in the temperature across the region. As the thickness of the region where such a change in temperature occurs is very small and unknown, it is usual to assume that there is a sudden change in the temperature on either side of the interface as shown in Figure 2.4.7b. The effective thermal conductivity across the interface is defined as $k_{eff} = q'' L_{eff}/\Delta T$, where L_{eff} is the effective thickness of the interface with a temperature drop of ΔT. Although L_{eff} is very small, the ratio L_{eff}/k_{eff} is finite and represents the contact resistance. The contact resistance is defined as

$$R''_c = \frac{\Delta T}{q''}$$

where

R''_c = contact resistance for a unit area (m^2 K/W)
q'' = heat flux across the interface (W/m^2)
ΔT = temperature drop across the interface (K)

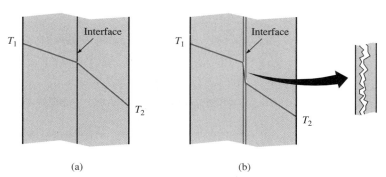

(a)   (b)

FIGURE 2.4.7 Contact resistance
(a) Two surfaces in perfect (ideal) contact with no interfacial resistance to heat transfer. Temperature is continuous at the interface. (b) Two surfaces in imperfect contact giving rise to contact resistance. Rough interface results in a finite thermal resistance in a very thin region and a consequent temperature drop.

The heat transfer at the interface is by conduction at the contact surfaces. Some of the factors that affect the thermal contact resistance are

■ **Surface finish:.** A smooth surface offers a greater area of contact, and hence, a lower resistance to heat flow.
■ **Pressure:.** Higher pressures result in greater surface contact, and hence, in lower resistance.
■ **Surface condition:.** If a surface is oxidized or rusted, the resistance to heat flow is increased and the contact resistance is higher.
■ **Fluid surrounding the materials..** The fluid surrounding the material will be trapped between the valleys at the interface. The path across the fluid is a parallel path for heat flow. A fluid with higher thermal conductivity offers lower thermal resistance.

The measured values of contact resistance under different conditions give an idea of the effect of some of the factors on contact resistance and its magnitude. The effects of surface conditions, contact pressure, and fluid between surfaces on contact resistance are illustrated in the values given below.

1. Effect of Surface. Contact resistance for different surfaces at moderate pressures

	R_c (m^2 K/W)
Aluminum surfaces	5×10^{-5}
Stainless steel surface	3×10^{-4}
Copper surfaces	10^{-5}

2. Effect of Pressure. Contact resistance at contact pressures of 100 kPa and 10 000 kPa (with vacuum interface)

	$R_c \times 10^4$ (m^2 K/W)	
Contact pressure	100 kPa	10,000 kPa
Stainless steel	6–25	0.7–4.0
Copper	1–10	0.1–0.5
Aluminum	1.5–5.0	0.2–0.4

3. Effect of Fluid Between Surfaces. Contact resistance between aluminum surfaces with different fluids at the interface surfaces

Interfacial fluid	$R_c \times 10^4$ (m^2 K/W)
Air	2.75
Hydrogen	0.720
Silicone oil	0.525
Glycerin	0.265

Contact resistances for a few cases are given in Table 2.4.1. For a more detailed discussion on contact resistance, refer to Schneider (1973) and Yovanovich (1986).

Other Types of Boundary Conditions

There are situations where neither the boundary temperatures nor the surrounding fluid temperatures and the associated heat transfer coefficients are known. For exam-

TABLE 2.4.1 Contact Resistance

Surface	Roughness (μm)	Temperature (°C)	Pressure (atm)	R'' (m² °C/W × 10⁴)
416 Stainless steel, ground, air	2.54	90–200	3–25	2.64
304 Stainless steel, ground air	1.14	20	40–70	5.28
416 Stainless steel, ground, with 0.001-in brass shims, air	2.54	30–200	7	3.52
Aluminum, ground, air	2.54	150	12–25	0.88
	0.25	150	12–25	0.18
Aluminum, ground, with 0.001-in brass shims, air	2.54	150	12–200	1.23
Copper, ground, air	1.27	20	12–200	0.07
Copper, milled, air	3.81	20	10–50	0.18
Copper, milled, vacuum	0.25	30	7–70	0.88

Extracted from Holman, J.P., *Heat Transfer,* 7th ed. (New York: McGraw-Hill, 1990).

Types of boundary conditions

ple, the heat transfer rate across a known surface may be specified, such as in the case of a resistance heater attached to the surface of a solid. If the surface of the heater not in contact with the solid is insulated and the heat transfer across the insulation is neglected, the power input to the heater, P, results in heat transfer across the solid. Denoting the direction of the heat flow by x, and the location of the surface to which the heater is attached by x_o, from Fourier's Law we have

$$P = q_x = -kA_x \left. \frac{dT}{dx} \right|_{x_o} \quad \text{or} \quad \left. \frac{dT}{dx} \right|_{x_o} = -\frac{P}{kA_x}$$

In this case, the derivative of the temperature at a given location is known. For an insulated surface, the heat transfer rate is zero, and the temperature gradient at the surface is also zero. This is a special case of known heat transfer rate.

If a surface is exposed to radiative heat transfer, the boundary condition contains the fourth power of the temperatures of the surface and the surroundings. With the differential equation in the first power of the temperature, such a boundary condition is nonlinear, and the solution to the resulting equation is more complex. See Section 2.7 for a more detailed discussion of boundary conditions.

2.5 CONDUCTION SHAPE FACTOR

There are many common cases with two- and three-dimensional temperature distributions where we would like to get a quick estimate of the heat transfer rate. For example, with a hot water pipe buried underground, the temperature in the soil is obviously two-dimensional. (Earth is some times used as a sink or source of heat.

Heat transfer with two- and three-dimensional temperature distributions

For example, it may be used as a source in heat pump applications for heating buildings.) The concept of resistance can be applied in such cases by relating the heat transfer rate to the geometry and thermal conductivity of the earth.

Consider the heat transfer from the inner surface to the outer surface of a long hollow block (Figure 2.5.1), satisfying the following conditions:

- Inner and outer surfaces are at uniform temperatures of T_1 and T_2, respectively
- Steady state
- No internal energy generation
- Constant properties
- Only one homogeneous material
- Constant thermal conductivity

The temperature distribution in the block is two-dimensional. The heat flow to the outer surface of the block is given by

$$q = \int q'' \, dA = -k \int \frac{\partial T}{\partial n} \, dA$$

where n represents the outward normal to the outer surface and the integration is over the outer surface. Defining a dimensionless temperature

$$\theta = \frac{T_1 - T}{T_1 - T_2}$$

the integral can be expressed as

$$-\int \frac{\partial T}{\partial n} \, dA = (T_1 - T_2) \int \frac{\partial \theta}{\partial n} \, dA = (T_1 - T_2) \, S$$

where θ is 0 on the inner surface and 1 at the outer surface. The value of $\partial\theta/\partial n$ depends only on the dimensions of the block and can be evaluated for different

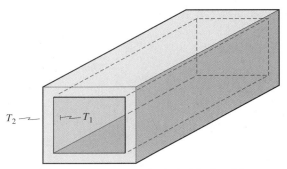

FIGURE 2.5.1 Heat Transfer in a Block with Two-Dimensional Temperature Distribution
A long, hollow block with the inner surface at T_1 and the outer surface at T_2.

dimensions. The magnitude of the integral, $\int (\partial\theta/\partial n) \, dA$, is known as the *conduction shape factor* and is denoted by S (with dimension of length).

The conduction shape factor for many cases with two- and three-dimensional temperature distribution have been determined, and some of them are given in Table 2.5.1. If the conduction shape factor for a particular geometry is known, the heat transfer rate is given by

$$q = Sk(T_1 - T_2)$$

The use of conduction shape factor is limited to those cases that satisfy the conditions stated at the beginning of this section.

Conduction shape factors can be determined by analytical or numerical solutions. An approximate value of the conduction shape factor for two-dimensional cases can be obtained by constructing flux plots consisting of isothermal lines and adiabatic lines, which are mutually perpendicular. For a discussion of flux plots, refer to Incropera and De Witt (1990) or Kreith and Bohn (1993).

TABLE 2.5.1 Conduction Shape Factors

Configuration	Shape Factor S(m)	Restrictions	
1. Edge of two adjoining walls	$0.54W$	$W > L/5$	
2. Conduction through corner of three adjoining walls (inner surface at T_1, outer surface at T_2)	$0.15L$	$L \ll$ length and width of wall	
3. Isothermal rectangular block embedded in a semi-infinite body with one face of the block parallel to the surface of the body	$2.756 \, L \left[\ln\left(1 + \dfrac{d}{W}\right) \right]^{-0.59}$ $\left(\dfrac{H}{d}\right)^{0.078}$	$L > W$	

TABLE 2.5.1 (continued)

Configuration	Shape Factor S(m)	Restrictions	
4. Thin rectangular plate buried in a semi-infinite medium having an isothermal surface	$\dfrac{\pi W}{\ln{(4W/L)}}$ $\dfrac{2\pi W}{\ln{(4W/L)}}$ $\dfrac{2\pi W}{\ln{(2W/L)}}$	$d = 0\ W > L$ $d \gg W\ W > L$ $d > 2W\ W \gg L$	
5. Cylinder centered inside a square of length L	$\dfrac{2\pi L}{\ln{(0.54W/R)}}$	$L \gg W$	
6. Isothermal cylinder buried in a semi-infinite medium	$\dfrac{2\pi L}{\cosh^{-1}{(d/R)}}$ $\dfrac{2\pi L}{\ln{(2d/R)}}$ $\dfrac{2\pi L}{\ln{(L/R)}\,\{1\,-\,\ln{(L/2d)}/\ln{(L/R)}\}}$	$L \gg R$ $L \gg R\ d > 3R$ $d \gg R\ L \gg d$	
7. Horizontal cylinder of length L midway between infinite parallel isothermal surfaces, both surfaces being at the same temperature	$\dfrac{2\pi L}{\ln{(4d/R)}}$		
8. Isothermal sphere in a semi-infinite medium	$\dfrac{4\pi R}{1\,-\,(R/2d)}$		

TABLE 2.5.1 (continued)

Configuration	Shape Factor S(m)	Restrictions	
9. Isothermal sphere in an infinite medium	$4\pi R$		
10. Isothermal sphere in a semi-infinite medium with an insulated surface	$\dfrac{4\pi R}{1 + (R/2d)}$		
11. Hemisphere on the surface of a semi-infinite medium	$2\pi R$		
12. Conduction between two isothermal spheres in an infinite medium	$\dfrac{4\pi d}{R_2/R_1 \, \{1 - (R_1/d)^4/[1 - (R_2/d)^2]\}}$	$d > 5R_{max}$	
13. Conduction between two isothermal cylinders of length, L, in an infinite medium	$\dfrac{2\pi L}{\cosh^{-1}\,[(d^2 - R_1^2 - R_2^2)/(2R_1 R_2)]}$	$L \gg R_1, R_2$ $L \gg d$	

EXAMPLE 2.5.1 A 10-mm, outer diameter (O.D.) hot water tube is embedded in a 150-mm thick concrete floor whose upper and lower surfaces are at 20 °C as shown in Figure 2.5.2. Determine the heat transfer rate if the tube is 2-m long and the surface of the tube is at 80 °C.

Given

Outer diameter of tube $(2R)$ = 0.01 m   Temperature of tube surface
Thickness of concrete $(2d)$ = 0.15 m (T_w) = 80 °C
Length of tube (L) = 2 m   Temperature of concrete surface
                                                     (T_s) = 20 ° C

Find

The heat transfer rate from the tube surface

ASSUMPTIONS

1. Steady state
2. Width of the concrete slab is much greater than the depth and the slab may be considered to be semi-infinite.
3. Constant properties

SOLUTION

From Table A3B, for air-dried concrete (2000 kg/m³), k = 0.896 W/m K. From the definition of conduction shape factor, $q = Sk (T_w - T_s)$. From Table 2.5.1, entry 7,

$$S = \frac{2\pi L}{\ln(4d/R)} = \frac{2 \times \pi \times 2}{\ln(4 \times 0.075/0.005)} = 3.069 \text{ m}$$
$$q = 3.069 \times 0.896 \times (80 - 20) = \underline{165 \text{ W}}$$

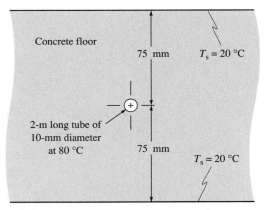

FIGURE 2.5.2 A Tube Used to Transport Hot Water Embedded in a Concrete Floor

EXAMPLE 2.5.2 Hot water flows in a 10-mm diameter hole in a rectangular block of stainless steel (type AISI 304) and maintains the temperature of the inner surface at 80 °C. The block is 50-mm wide, 50-mm high, and 5-m long (Figure 2.5.3). Air at 20 °C flows over the block, and the convective heat transfer coefficient associated with the outer surface is 40 W/m² °C. Determine the heat transfer rate from the water to the surroundings.

Given

Diameter of hole in block
$(2R) = 0.01$ m
Dimensions of block: 50-mm wide,
  50 mm high, and 5-m long
Inner surface temperature
$(T_i) = 80\,^{\circ}$ C

Outside air temperature
$(T_\infty) = 20\,^{\circ}$C
Outside heat transfer coefficient
$(h_o) = 40$ W/m² °C
Material of the block:
  304 stainless steel

Find

The heat transfer rate from the water in the block to the surrounding air

ASSUMPTION

Outer surface of the block is at a uniform temperature.

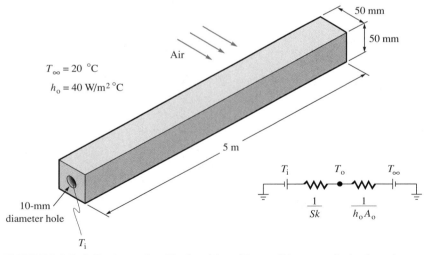

FIGURE 2.5.3 A Rectangular Block with a 10-mm Diameter hole Carrying Hot Water
The inner and outer surfaces of the block are at uniform temperatures. The thermal circuit is shown on the right.

SOLUTION

The heat transfer rate can be computed from the thermal circuit. The conduction resistance is computed through the appropriate conduction shape factor and added to the convective resistance associated with the outer surface. From the resistance circuit,

$$q = \frac{T_i - T_\infty}{[1/(Sk)] + [1/(h_o A_o)]}$$

where

k = thermal conductivity of stainless steel
h_o = convective heat transfer coefficient associated with the outer surface
A_o = outer surface area of the block

From Table A1 for type 304 stainless steel, $k = 14.9$ W/m K. Since $W = 0.05$ m and $L = 5$ m, $L \gg W$, and the conduction shape factor given in Table 2.5.1, entry 5, is used.

$$S = \frac{2\pi L}{\ln(0.54W/R)}$$

$$\frac{1}{Sk} = \frac{\ln(0.54W/R)}{2\pi Lk} = \frac{\ln(0.54 \times 0.05/0.005)}{2\pi \times 5 \times 14.9} = 3.603 \times 10^{-3}\ \text{K/W}$$

$$\frac{1}{h_o A_o} = \frac{1}{40 \times 4 \times (0.05 \times 5)} = 0.025\ \text{K/W}$$

$$q = \frac{80 - 20}{0.003603 + 0.025} = 2098\ \text{W}$$

Comment

The assumption that the outer surface is at a uniform temperature may not be valid, but the heat transfer rate is a good estimate.

2.6 EXTENDED SURFACES

Heat transfer augmentation by extended surfaces

One of the most common engineering application of heat transfer involves two fluids at different temperatures, separated by a solid wall. Examples of such applications are hot water space heaters; automobile radiators; evaporators and condensers of refrigeration systems; condensers, boilers and feedwater heaters in power plants; lawn mower, snow blower and motor bike engines.

Consider an application with heat transfer from one fluid to another separated by a rectangular plate as shown in Figure 2.6.1a. From the thermal circuit, the heat transfer rate is given by

$$q = \frac{T_i - T_o}{[1/(h_i A_i)] + [L/(kA)] + [1/(h_o A_o)]} \tag{2.6.1}$$

If we are interested in increasing the heat transfer rate, we should reduce the resistances shown in the denominator. By choosing a solid wall of a material of high thermal conductivity and small thickness, we can generally make the conductive resistance L/kA very much smaller than either of the other two resistances in the denominator; any further reduction in its value will not yield a significant increase in the heat flow. This is clarified by multiplying both the numerator and the denominator of Equation 2.6.1 by $h_i A_i$. The denominator is given by $1 + h_i A_i (L/kA) + [(h_i A_i)/(h_o A_o)]$. Consider the case of the plate being made of 2-mm thick copper with a thermal conductivity of 401 W/m K. If the inner fluid is a liquid with a convective heat transfer coefficient, h_i, of the order of 3000 W/m² °C and the outer fluid is air with a convective heat transfer coefficient, h_o, of 10 W/m² °C, the value of $(h_i A_i)/(h_o A_o)$ is 300 and that of $h_i A_i (L/kA)$ is 0.015. Even if we eliminate the conductive resistance entirely, the total value of the denominator is reduced by 0.015 from a total of 301.015; this reduction in the resistance is insignificant compared with the total resistance. From Equation 2.6.1 it is clear that, if we desire to increase the heat transfer rate, we must increase $h_i A_i$ and $h_o A_o$. In many cases the convective heat transfer coefficient on one side is much higher than on the other side, as illustrated by the values for h_i and h_o in this example. Hence, to increase the heat transfer rate, we should decrease $1/(h_o A_o)$ [as $1/(h_i A_i) \ll 1/(h_o A_o)$]. There are several methods of decreasing $1/(h_o A_o)$, and one such method is to increase the surface area associated with h_o, as in Figure 2.6.1b.

When is an extended surface useful?

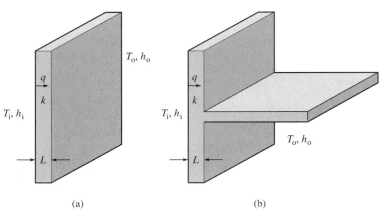

(a)   (b)

FIGURE 2.6.1 Heat Transfer Involving Two Fluids at Different Temperatures Separated by a Solid Surface
(a) Rectangular slab between two fluids. (b) One method of increasing the heat transfer by increasing the surface area.

(a)

(b)

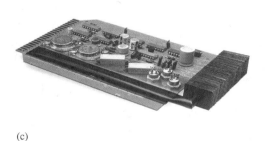
(c)

(e)

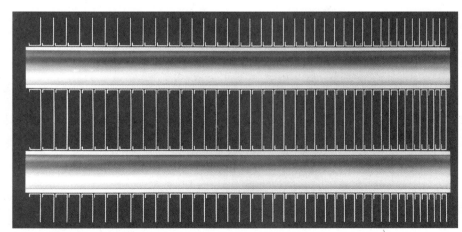
(d)

FIGURE 2.6.2 Some Applications of Extended Surfaces
(a) Furnace: Pin-stud and refractory construction (Courtesy: Babcock & Wilcox, Barberton, Ohio). (b) Pin fin heat sinks (Courtesy: Thermalloy Inc. Dallas, Texas). (c) Heat pipe with fins for a personal computer (Courtesy: Noren Products, Inc., Menlo Park, California). (d) Infinitely variable fin spacing in coils for refrigeration equipment (Courtesy: The Trane Company, La Crosse, Wisconsin). (e) An automotive heater core (Courtesy: Harrison Radiator Division, General Motors Corp. Lockport, New York).

Such an increase in the surface area to decrease the thermal resistance leads us to the concept of *extended surfaces*. An increase in the heat transfer rate can be achieved by adding ''surface'' to A_o. You may notice how such surfaces are added to baseboard heaters, automobile radiators, and refrigerator condensing coils. Examples of the use of extended surfaces in different applications are illustrated in Figure 2.6.2.

To analyze heat transfer from an extended surface, consider the surface shown in Figure 2.6.1b. It is clear that the temperature of the extended surface is not uniform, as its temperature varies with the distance from the plate to which it is attached. Hence, Equation 2.6.1 cannot be used to compute the heat transfer rate in such cases.

2.6.1 Analysis of Extended Surfaces

Analysis of fins with constant cross section and perimeter

In the extended surfaces shown in Figures 2.6.3a and b, the temperature of the fins vary with x, the distance from the plate to which the fins are attached. In the fins shown in Figure 2.6.3a, the areas of the cross sections (and their perimeters) in the direction in which the temperature is changing are constant. But the fins shown in Figure 2.6.3b do not satisfy this condition. In the case of the annular fin attached to

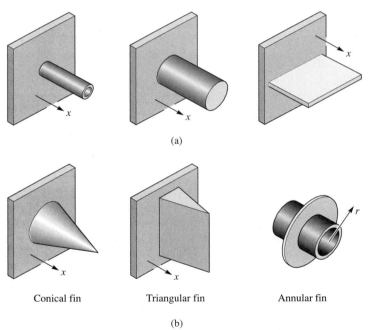

(a)

Conical fin   Triangular fin   Annular fin

(b)

FIGURE 2.6.3 Fins with Constant or Variable Area of Cross Section
(a) In these fins, the temperature is a function of x only. Cross-sectional areas perpendicular to the x-direction and their perimeters are constant. (b) In these fins the areas of cross section and perimeters are not constant.

a tube, the temperature changes in the radial direction. The area of a cross section of the fin in the radial direction is $2\pi rt$, where r is the radius at a given location and t is the thickness of the fin. Its perimeter is $4\pi r$. Both the area of the cross section and the perimeter are functions of the radius and are not constant. Such fins having variable cross sections or perimeters are not considered in this section. Here, analytical results are developed only for fins with constant areas of cross section and perimeters.

Consider a fin with constant perimeter and cross-sectional area in the direction in which the temperature changes, such as in a heat sink used in electronic equipment or an electric motor, as shown in Figure 2.6.4a and schematically shown in Figure 2.6.4b. The fin is attached to a surface that is maintained at T_b and is exposed to a fluid at T_∞, with a convective heat transfer coefficient h. The end surface is exposed to the fluid at T_∞, but the convective heat transfer coefficient h_e may be different from h. The temperature distribution in the fin and the heat transfer rate from such an extended surface to the surrounding fluid are to be found.

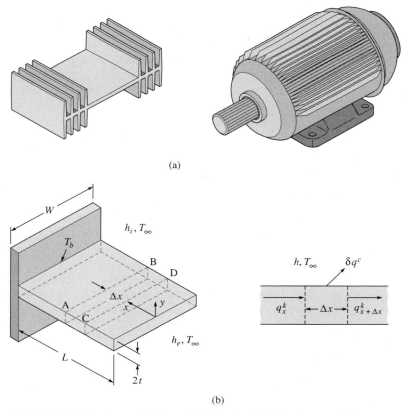

(a)

(b)

FIGURE 2.6.4 Rectangular Fin with Constant Perimeter and Cross-Sectional Area Perpendicular to the *x*-Direction
(a) A heat sink in an electronic equipment and an electric motor. (b) Schematic of a rectangular fin.

ASSUMPTIONS

1. Steady state
2. All material properties of the extended surface, the convective heat transfer coefficients, h and h_e, and the temperature of the surrounding fluid are constant.
3. Temperature variation in the direction perpendicular to the x-direction is negligible so that $T = T(x)$. This condition is satisfied if $ht/k \leq 0.1$. The dimensionless parameter ht/k is known as the *Biot number*. Its significance can be understood by recasting it as $(t/k)/(1/h)$. In this form, the numerator represents the conductive resistance in the y-direction, and the denominator represents the convective resistance. If the Biot number is very small, it means that the conductive resistance is much smaller than the convective resistance, and, when these are in series, the conductive resistance can be neglected in comparison with the convective resistance. If the conductive resistance in the y-direction is small, the temperature difference in the y-direction is also small compared with the difference in temperature of the surface at the same location (x) and the surrounding fluid. The fin may then be approximated as isothermal in the y-direction. (A more detailed justification for one-dimensional analysis, if $ht/k \ll 1$, is given in Section 2.6.6.)
4. No internal energy generation
5. Negligible radiative heat transfer (or radiative heat transfer is accounted for by an appropriate increase in the value of the convective heat transfer coefficient)

> If Biot number is less than 0.1, one-dimensional temperature distribution is a good approximation

ANALYSIS

Choose an element of length Δx as the system to be analyzed. Energy balance for the element yields

$$q_x^k - q_{x+\Delta x}^k - \delta q^c = 0 \qquad (2.6.2)$$

where

$q_x^k =$ conductive heat transfer rate to the element across surface AB
$q_{x+\Delta x}^k =$ conductive heat transfer rate from the element across surface CD
$\delta q^c =$ convective heat transfer rate to the surrounding fluid from the surfaces of the element between cross sections at AB and CD. In this case $\delta q^c = hP\Delta x\,(T - T_\infty)$, where P is the perimeter of the cross-sectional area, A_x, *exposed* to the surrounding fluid. $P\Delta x$ is the surface area of the element from which convective heat transfer to the surrounding fluid occurs.

Substituting the expression for δq^c in Equation 2.6.2, dividing by Δx, and taking the limit as $\Delta x \to 0$, we obtain

$$-\frac{dq_x^k}{dx} - hP\,(T - T_\infty) = 0$$

From Fourier's Law, with $T = T(x)$, $q_x^k = -kA_x\,(dT/dx)$;

$$\frac{d}{dx}\left(kA_x\frac{dT}{dx}\right) - hP\,(T - T_\infty) = 0$$

As k and A_x are constant, dividing the equation by kA_x, we obtain

$$\frac{d^2T}{dx^2} - m^2\,(T - T_\infty) = 0$$

where

$$m^2 = \frac{hP}{kA_x}$$

The differential equation is linear, of second order, and nonhomogeneous (What makes the equation nonhomogeneous?). Because it is easier to solve a homogeneous differential equation, the equation is made homogeneous by a change in the variable. Define $\theta = T - T_\infty$ so that θ represents the difference between the temperature at any location, x, in the extended surface and the temperature of the surrounding fluid. Substituting θ in the above equation,

$$\frac{d^2\theta}{dx^2} - m^2\,\theta = 0 \qquad (2.6.3)$$

Solution of Equation 2.6.3 is

$$\theta = c_1\,e^{-mx} + c_2\,e^{mx} \qquad (2.6.4)$$

or

$$\theta = c_3\,\cosh mx + c_4\,\sinh mx \qquad (2.6.5)$$

Two independent boundary conditions for θ are required to evaluate the constants c_1 and c_2, or c_3 and c_4 in the solution.

The boundary conditions are obtained from the known conditions that the temperature or its gradient must satisfy. At the base where the fin is attached, it must satisfy

$$\theta(x=L) = \theta_b = T_b - T_\infty \qquad (2.6.6)$$

From an energy balance on the surface at the free end, the temperature must satisfy the convective boundary condition,

Heat transfer rate by conduction in the x-direction

= Heat transfer rate by convection from the fluid to the surface

$$-kA_x\,\frac{dT}{dx}\,(0) = h_e A_x\,(T_\infty - T)_{x=0}$$

or

$$-k\,\frac{d\theta}{dx}\,(0) = -h_e\,\theta(0) \qquad (2.6.7)$$

Application of boundary conditions, Equations 2.6.6 and 2.6.7 to Equation 2.6.4 yields

$$\theta_b = c_1\,e^{-mL} + c_2\,e^{mL} \qquad (2.6.8)$$

$$-k\,(-c_1 m + c_2 m) = -h_e\,(c_1 + c_2) \qquad (2.6.9)$$

Solving Equations 2.6.8 and 2.6.9 for c_1 and c_2,

$$c_1 = \frac{\theta_b \, [1 - (h_e/km)]}{e^{-mL} \, [1 - (h_e/km)] + e^{mL} \, [1 + (h_e/km)]} \qquad (2.6.10)$$

$$c_2 = \frac{\theta_b \, [1 + (h_e/km)]}{e^{-mL} \, [1 - (h_e/km)] + e^{mL} \, [1 + (h_e/km)]} \qquad (2.6.11)$$

We thus obtain the temperature distribution as

$$\theta = \frac{\theta_b}{e^{-mL} \, [1 - (h_e/km)] + e^{mL} \, [1 + (h_e/km)]} \left[\left(1 - \frac{h_e}{km}\right) e^{-mx} + \left(1 + \frac{h_e}{km}\right) e^{mx} \right] \qquad (2.6.12)$$

The expression for the temperature distribution looks complicated, but if the known values in the expressions for the constants of integration are substituted, it is a simple expression like Equation 2.6.8, where the values of c_1 and c_2 are known. In fact, to reduce writing effort, hereafter, c_1 and c_2 are used as constants known from Equations 2.6.10 and 2.6.11 (although Equation 2.6.13 looks identical to Equation 2.6.4, it is given a different number to indicate that the constants in the equation are known).

$$\theta = c_1 \, e^{-mx} + c_2 \, e^{mx} \qquad (2.6.13)$$

As the main purpose of adding the extended surface is to increase the total heat transfer rate, the heat transfer rate from the fin to the surrounding fluid should be found. We may find the heat transfer rate by two methods: (1) by determining the convective heat transfer rate (the preferred method); and (2) from conductive heat transfer rate at the base ($x = L$). Though the value of h is assumed to be constant and known, the temperature varies along x. Because of the variation in $T - T_\infty$, the convective heat flux from the extended surface, given by $h(T - T_\infty)$, also varies with x. Hence, the total heat transfer rate from the fin to the surrounding fluid needs to be determined by integration. Starting with the heat transfer rate from an infinitesimal length dx, we obtain

$$q^c = \text{heat transfer rate from the fin to the surrounding fluid}$$

$$= \int q'' dA_s = \int_0^L hP \, (T - T_\infty) \, dx + h_e A_x \, (T - T_\infty) \,_{x=0}$$

where the integral gives the convective heat transfer rate from the surface associated with h, and the second term (associated with h_e) gives the heat transfer rate from the end surface. With $T - T_\infty$ given by Equation 2.6.13,

$$\int_0^L hP\theta \, dx = hP \int_0^L \left(c_1 e^{-mx} + c_2 e^{mx} \right) dx = \frac{hP}{m} \left(-c_1 e^{-mL} + c_2 e^{mL} + c_1 - c_2 \right)$$

and

$$h_e A_x \, (T - T_\infty)_{x=0} = h_e \, A_x \, (c_1 + c_2)$$

Hence,

$$q^c = \frac{hP}{m} \left[-c_1 e^{-mL} + c_2 e^{mL} + c_1 - c_2 \right] + h_e \, A_x \, (c_1 + c_2)$$

Substituting c_1 and c_2 and simplifying, we obtain

$$q_c = \sqrt{hPkA_x} \, \theta_b \, \frac{(e^{mL} - e^{-mL}) + [(h_e/km)(e^{mL} + e^{-mL})]}{(e^{mL} + e^{-mL}) + [(h_e/km)(e^{mL} - e^{-mL})]} \qquad (2.6.14)$$

Applying the energy balance to the entire fin, the heat transfer rate from the fin to the surroundings by convection must be equal to the heat transfer rate to the fin by conduction at the base of the fin ($x = L$). The heat transfer rate to the fin by conduction at the base is given by

$$q_b = kA_x \frac{dT}{dx}\bigg|_{x=L}$$

This gives the same result as before. (It is left as an exercise to show that the two expressions are equal to each other). If the temperature distribution is derived by analysis, the two values will be equal. But if the heat transfer rate is obtained by measured values of the temperature gradient, significant errors may be introduced. In general, if there is a choice between determining a particular quantity by integration or differentiation, integration is preferred. This should be clear if one tries to compute the temperature gradient from the measured values of the temperature along the fin. A small error, or uncertainty, in the measurement of the temperatures or the spacing between the temperature sensors may lead to a significant error in the gradient (see Problem 2.3). But integration will reduce such errors substantially.

The temperature distribution given by Equation 2.6.12 can be recast in terms of hyperbolic functions (see Appendix 2 for the definition of hyperbolic functions) and the result is

$$\theta = \frac{\theta_b}{\cosh(mL) + (h_e/km)\sinh(mL)} \left[\cosh(mx) + \frac{h_e}{km}\sinh(mx) \right] \qquad (2.6.15a)$$

Equation 2.6.15a can also be obtained by expressing the solution to the differential equation in terms of hyperbolic functions instead of exponential functions. This is left as an exercise. The heat transfer rate from the fin to the surrounding fluid is given by

$$q^c = \int_0^L hP\theta \, dx + h_e A_x \theta_{x=0}$$

Substituting for θ and $\theta_{x=0}$ from Equation 2.6.15a and performing the indicated operations, we obtain

$$q^c = \sqrt{hPkA_x} \, \theta_b \, \frac{\sinh mL + (h_e/km)\cosh mL}{\cosh mL + (h_e/km)\sinh mL} \qquad (2.6.15b)$$

Now consider the case of a fin with the free end perfectly insulated. Rarely is the free end of a fin insulated. Two examples when the end can be considered insulated are (1) the fin is enclosed inside a case made of an insulating material with no clearance between the fin and the case, and (2) a symmetric fin with no heat transfer across the plane midway between the ends of the fin. As the heat transfer rate from the surface of the free end is zero, we can obtain the solution by replacing the boundary conditions given in Equation 2.6.7 by $-kA \left. (d\theta/dx \right|_{x=0} = 0$ or by letting $h_e = 0$. Using Eq. 2.6.5 and the boundary condition $\left. d\theta/dx \right|_{x=0} = 0$,

$$\left. (c_3 m \sinh mx + c_4 m \cosh mx) \right|_{x=0} = 0$$

As $\sinh(0) = 0$ and $\cosh(0) = 1$, $c_4 = 0$.

The second boundary condition, $\theta_{x=L} = \theta_b$ gives, $c_3 = \dfrac{\theta_b}{\cosh(mL)}$

$$\theta = \frac{\theta_b}{\cosh(mL)} \cosh(mx) \qquad (2.6.16)$$

and

$$q^c = \int_0^L hP \, \theta_b \, \frac{\cosh mx}{\cosh mL} \, dx$$

or

$$q^c = \sqrt{hPkA_x} \, \theta_b \tanh(mL) \qquad (2.6.17)$$

Equation 2.6.16 can be recast as

$$\frac{\theta}{\theta_b} = \frac{\cosh[(mL)(x/L)]}{\cosh mL}$$

The variation of the dimensionless temperature θ/θ_b as a function of x/L for various values of the parameter mL is shown in Figure 2.6.5.

COMMENT

The choice of the origin for the solution is quite arbitrary but the algebra associated with different origins has different complexities. In general, it is more convenient to

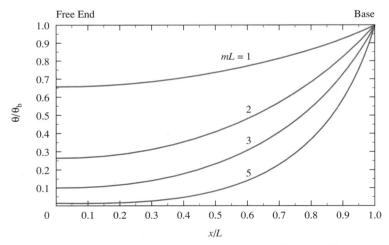

FIGURE 2.6.5 Dimensionless Temperature Variation in a Fin

choose the origin at the location where the boundary condition is homogeneous. For this case, the boundary condition at the free end, Equation 2.6.7 is homogeneous.

For some of the more common boundary conditions the constants of integration, the temperature distribution, and the heat transfer rates are obtained by solving Problem 2.75. Students are urged to obtain these solutions by following the method illustrated above.

EXAMPLE 2.6.1

Water flows in a tube which is exposed to air at a temperature of 20 °C. The temperature of the outer tube surface is 110 °C. To increase the heat transfer rate from the water to the air, 3-mm diameter and 60-mm long circular rods of brass are attached to the tube, with the axis of the rods perpendicular to the axis of the tube as shown in Figure 2.6.6. The convective heat transfer coefficient between the surface and air is 50 W/m² K. Determine the temperature at the tip of the rods, the heat transfer rate from one rod to the surrounding air by convection, and the tip temperature and convective heat transfer rate assuming the tip to be perfectly insulated.

Given

Tube surface temperature
  $(T_b) = 110\,°C$
Surrounding air temperature
  $(T_\infty) = 20\,°C$
Diameter of rods $(d) = 0.003$ m

Length of rods $(L) = 0.06$ m
Material of the rods: brass
Convective heat transfer coefficient
  $(h) = 50$ W/m² °C

Find

(a) The temperature of the tip of the rods
(b) The heat transfer rate from one rod to the surrounding air
(c) The tip temperature assuming the tip to be perfectly insulated.

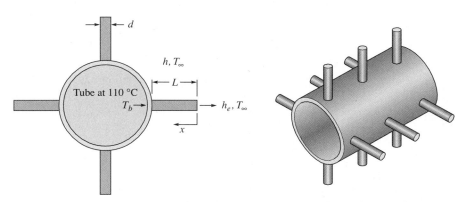

FIGURE 2.6.6 Cylindrical rods of diameter *d* attached to the surface of a tube.
The axes of the rods are perpendicular to the axis of the tube.

ASSUMPTIONS

1. Steady state
2. $h_e = h$
3. Constant properties
4. From Table A1 the thermal conductivity of brass is 110 W/m K. The Biot number, $hR/k = 50 \times 0.0015/110 = 0.0007$, is much less than 0.1 and one-dimensional temperature distribution is assumed.

SOLUTION

(a) With $\theta = T - T_\infty$, this is a fin with its tip exposed to the fluid and with the above assumptions, Equations 2.6.15a and b are applicable.

$$\frac{\theta}{\theta_b} = \frac{\cosh mx + (h_e/km)\sinh mx}{\cosh mL + (h_e/km)\sinh mL} \qquad \frac{\theta_{x=0}}{\theta_b} = \frac{1}{\cosh mL + (h_e/km)\sinh mL}$$

$$q_c = \sqrt{hkA_x P}\ \theta_b\ \frac{\sinh mL + (h_e/km)\cosh mL}{\cosh mL + (h_e/km)\sinh mL}$$

The values for the variables in the equation are

$$\theta_b = 110 - 20 = 90\,°C \qquad k = 110\ \text{W/m K}$$
$$L = 0.06\ \text{m} \qquad h = h_e = 50\ \text{W /m}^2\ \text{K}$$
$$d = 0.003\ \text{m} \qquad A_x = \pi d^2/4 = 7.069 \times 10^{-6}\ \text{m}^2$$
$$P = \pi d = 9.425 \times 10^{-3}\ \text{m} \qquad m^2 = hP/kA_x = 4h/kd = 606.1\ /\text{meter}^2$$
$$m = 24.62/\text{meter}; \qquad mL = 1.4772$$

Denoting the tip temperature $\theta_{x=0} = \theta_t$,

$$\frac{\theta_t}{\theta_b} = \frac{1}{\cosh(1.4772) + 50/(110 \times 24.62)\ \sinh(1.4772)} = 0.4268$$

$$\theta_t = 0.4268 \times 90 = 38.4\,°C; \qquad T_t = 38.4 + 20 = \underline{58.4\,°C}$$

(b) Substituting the values for q^c, with $h_e/km = 0.01846$, we get

$$q_c = (50 \times 110 \times 7.069 \times 10^{-6} \times 9.424 \times 10^{-3})^{0.5} \times 90 \times$$
$$\frac{\sinh(1.4772) + 0.01846 \times \cosh(1.4772)}{\cosh(1.4772) + 0.01846 \times \sinh(1.4772)}$$
$$= \underline{1.56\ \text{W}}$$

(c) For the insulated tip, from Equation 2.6.16 and 2.6.17,

$$\frac{\theta_t}{\theta_b} = \frac{1}{\cosh mL} \quad \text{and} \quad q^c = \sqrt{hPkA_x}\ \theta_b \tanh mL$$

$$\frac{\theta_t}{\theta_b} = 0.4339; \quad \theta_t = 39.1\,°C; \quad T_t = \underline{59.1\,°C}, \quad \text{and} \quad q^c = \underline{1.55\ \text{W}}$$

Note how close the values for the insulated tip fin are to those for the uninsulated tip. Can you explain the reasons on both physical and mathematical grounds?

EXAMPLE 2.6.2

Estimate the temperature at the free end of the handle of a pot used to boil water. Pots made either of stainless steel (type 316) or aluminum are available. The dimensions of the handle are 30-cm long, 25-mm wide, and 3-mm thick (Figure 2.6.7). The handle is exposed to air at 20 °C, and the convective heat transfer coefficient is 6 W/m^2 °C.

Given

Pot used for boiling water
Materials for the handle: Aluminum alloy (2024-T) and AISI 316 stainless steel
Dimensions of handle: 30-cm long, 25-mm wide, and 3-mm thick
Convective heat transfer coefficient (h) = 6 W/m^2 °C
Air temperature T_∞ = 20 °C

Find

The temperature of the free end of the handle

ASSUMPTIONS

1. Steady state
2. Constant properties
3. From Table A1, at 350 K, the thermal conductivity of aluminum alloy (2024-T) and AISI 316 stainless steel are, respectively, 181.5 W/m K and 14.3 W/m K. The higher value of the Biot number formed with the thermal conductivity of stainless steel is (6 × 0.003/2)/14.3 = 0.0006—very much less than 0.1. Hence, the temperature distribution in the handle is assumed to be one-dimensional.

SOLUTION

The temperature distribution in the handle can be determined by treating it as an extended surface attached to the pot. However, the temperature at the base, where the handle is attached to the pot, is not given. We know that the maximum temperature at the base of the handle is 100 °C (temperature of boiling water unless the pot is heated without water!). For simplicity (see Example 2.6.1), we neglect the heat transfer from the free end of the handle.

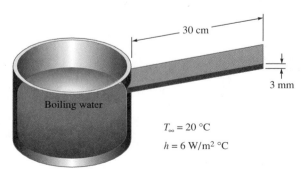

FIGURE 2.6.7 A Handle 30-cm Long Is Attached to a Pot in Which Water Boils

For the above model, the temperature distribution in the handle is given by Equation 2.6.16. With $\theta_{tip} = \theta_{x=0}$,

$$\theta_{tip} = \frac{\theta_b}{\cosh(mL)}, \quad \text{where } m = \sqrt{\frac{hP}{kA}} = \sqrt{\frac{6 \times 2 \times (0.003 + 0.025)}{k\,(0.003 \times 0.025)}} = \frac{66.9}{\sqrt{k}}$$

Results for the two pot materials are given in the following table.

	Aluminum	Stainless Steel
k (W/m K)	181.5	14.3
m (per meter)	4.966	17.69
mL	1.49	5.307
$\theta_{tip} = \dfrac{\theta_b}{\cosh mL} = \dfrac{80}{\cosh mL}$ (°C)	34.3	0.79
Tip temperature (°C) $= \theta_{tip} + T_\infty$	54.3	20.8

Having determined the temperatures, a more appropriate value of the thermal conductivity, evaluated at some mean temperature between 100 °C and the minimum temperature at the tip, can be used and the solution improved. This example illustrates how the principles of heat transfer may be applied in diverse problems not associated with engineering in the traditional sense.

2.6.2 Fin Effectiveness and Efficiency

Fin effectiveness and efficiency as figure of merit

To determine how good a fin is in increasing the heat transfer rate, a *figure of merit* is required. There are two definitions for the figure of merit. One is to compare the heat transfer rate from the fin with the heat transfer rate from the original area that the fin occupied. As fins are provided to increase the heat transfer rates, the figure of merit so defined should be greater than unity, and hence, we designate it as *effectiveness*, ϵ, of the fin (the term *efficiency* is usually associated with values of less than 1).

Fin effectiveness

$$\epsilon = \frac{q_c}{h_b\, A_c\, \theta_b}$$

where

ϵ = effectiveness of the fin
q_c = convective heat transfer rate from the fin
A_c = cross-sectional area of the fin
h_b = convective heat transfer coefficient in the absence of the fin

For example, for a fin with its free end insulated,

$$\epsilon = \frac{\sqrt{hPkA_c}\ \theta_b \tanh mL}{h_b\, A_c\, \theta_b}$$

and, if $h_b = h$,

$$\epsilon = \sqrt{\frac{Pk}{hA_c}} \tanh mL$$

Although the above definition of the figure of merit appears logical, the more widely accepted definition is to compare the fin with a fin made of an ideal conductor, i.e., a material of an infinite thermal conductivity; the figure of merit so defined is usually termed efficiency and is less than 1. The efficiency η, of a fin is given by

Fin efficiency

$$\eta = \frac{\text{Actual heat transfer rate from the fin}}{\text{Maximum possible heat transfer rate from a fin of the same dimensions}}$$

In the definition of the efficiency, the maximum heat transfer rate is interpreted as the heat transfer rate from a fin of the same geometry as the original fin but made of an ideal conductor. The conductive resistance of such an ideal conductor is zero and the temperature of the fin at every location is equal to the base temperature. For

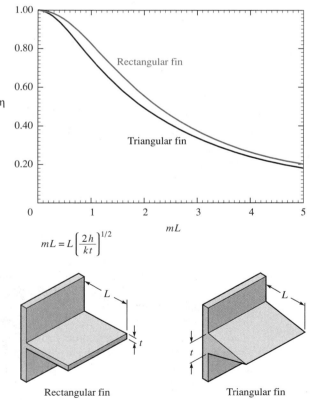

$$mL = L\left[\frac{2h}{kt}\right]^{1/2}$$

Rectangular fin

Triangular fin

FIGURE 2.6.8 Fin Efficiency—Rectangular and Triangular, Longitudinal Fins

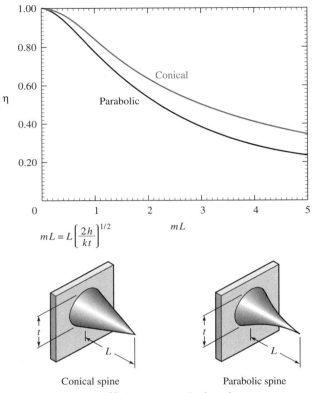

FIGURE 2.6.9 **Fin Efficiency—Conical and Concave Spine Fins**

a fin with the free end insulated, the heat transfer rate from an ideal conductor is given by $hA_s\theta_b$, where A_s is the surface area of the fin exposed to the surrounding fluid. Hence, for a fin with its free end insulated,

$$\eta = \frac{\sqrt{hPkA_c}\ \theta_b \tanh mL}{hPL\theta_b} = \frac{\tanh mL}{mL}$$

The fin efficiency has been determined for different geometries with cross-sectional areas and perimeters being functions of the coordinate x. Some of these are given in Figures 2.6.8 through 2.6.10, which can be used to determine the heat transfer rates for geometries that do not have simple solutions of the type we have seen so far. Knowing the geometry, we can determine the surface area of the fin exposed to the surrounding fluid. The maximum possible heat transfer rate, $q_{max} = hA_s\theta_b$ is then computed. The actual heat transfer rate is found by multiplying the maximum heat transfer rate by the fin efficiency, η, found from the graphs of η. The curves are plotted from numerical values given in Kern and Kraus (1972). These graphs are based on the concept of corrected length (see Section 2.6.4).

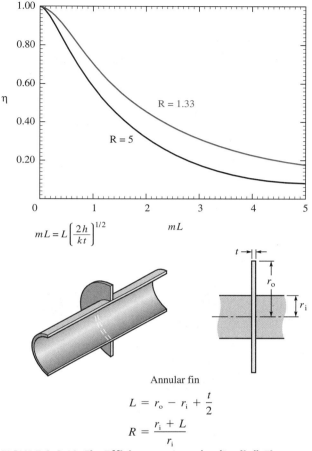

$$mL = L\left[\frac{2h}{kt}\right]^{1/2}$$

Annular fin

$$L = r_\mathrm{o} - r_\mathrm{i} + \frac{t}{2}$$

$$R = \frac{r_\mathrm{i} + L}{r_\mathrm{i}}$$

FIGURE 2.6.10 Fin Efficiency—Annular (Radial) Fins

EXAMPLE 2.6.3 A vapor flowing in a 25-mm O.D. tube is to be condensed. Air at 20°C flows perpendicular to the axis of the tube. As the heat transfer coefficient in condensation is much greater than the heat transfer coefficient in convection in air, it is proposed to attach type 2024-T6 aluminum fins in the form of circular disks 25-mm inner diameter (I.D.), 75-mm O.D., and 0.5-mm thick (Figure 2.6.11). The tube surface temperature is 47° C and the convective heat transfer coefficient on the outer surface is 50 W/m² K. Determine the heat transfer rate from one fin to the surrounding air.

Given

I. D. of annular fin $(2r_\mathrm{i})$ = 0.025 m   Tube surface temperature (T_b) = 47°C
O. D. of the fin $(2r_\mathrm{o})$ = 0.075 m   Surrounding air temperature
Thickness of fin (t) = 0.5 mm (T_∞) = 20°C
                                                           Convective heat transfer coefficient
                                                              (h) = 50 W/m² °C

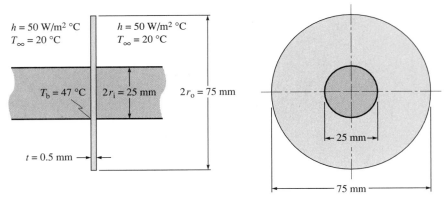

FIGURE 2.6.11 Annular Fin Attached to a 25-mm O.D. Tube

Find

The heat transfer rate from one fin to the surrounding air

ASSUMPTIONS

1. Steady state
2. Constant properties
3. From Table A1, for aluminum (2024-T6 alloy) $k = 177$ W/m K and Biot number $= ht/2k = 50 \times 0.00025/177 = 7.06 \times 10^{-5}$. As the Biot number is much less than 0.1, the temperature distribution is assumed to be one-dimensional, varying only in the radial direction.
4. Uniform convective heat transfer coefficient

SOLUTION

As the cross-sectional area in the radial direction and the perimeter are both functions of the radius, we cannot use the analytical relations developed so far. We, therefore, use the concept of fin efficiency. The heat transfer rate from the fin is calculated from the relation

$$q = \eta q_{max}$$

where

$\eta =$ efficiency of the fin to be determined from Figure 2.6.10
$q_{max} =$ maximum heat transfer rate from a fin of infinite thermal conductivity
$\quad = hA_s(T_b - T_\infty)$
$A_s =$ total surface area of the fin exposed to the fluid
$\quad = 2\pi(r_o^2 - r_i^2) + 2\pi r_o t$

The values are

$$A_s = 2 \times \frac{\pi(0.075^2 - 0.025^2)}{4} + \pi \times 0.075 \times 0.0005 = 7.972 \times 10^{-3} \text{ m}^2$$

$$q_{max} = hA_s\theta_b = 50 \times 7.972 \times 10^{-3}\,(47 - 20) = 10.76 \text{ W}$$

Referring to Figure 2.6.10,

$$L = r_o - r_i + t/2 = (0.075 - 0.025)/2 + 0.0005/2 = 0.02525 \text{ m}$$

$$R = \frac{r_i + L}{r_i} = \frac{0.025/2 + 0.02525}{0.025/2} = 3.02$$

$$mL = L\left(\frac{2h}{kt}\right)^{1/2} = 0.02525\left(\frac{2 \times 50}{177 \times 0.0005}\right)^{1/2} = 0.8488$$

For $R = 3$, $mL = 0.85$, from Figure 2.6.10, $\eta \approx 0.72$

$$q = \eta q_{max} = 0.72 \times 10.76 = \underline{7.74 \text{ W}}$$

Overall Heat Transfer Coefficient with Extended Surfaces

The concept of fin efficiency can be utilized in defining the overall heat transfer coefficient with extended surfaces.

Consider the heat transfer from one fluid to another separated by a solid boundary. One side of the solid is provided with a number of fins as shown in Figure 2.6.12. In determining the heat flow from the fin assembly depicted in Figure 2.6.12, we permit the heat transfer coefficient, h_o, associated with the unfinned outer surface to be different from h_f associated with the fins; the temperature of the fluid is assumed uniform over the entire outer surface. As the heat transfer rate in the region where the fins are located is greater than in the unfinned region (it is to increase the heat transfer rate that fins are provided), it is clear that the temperature distribution is not

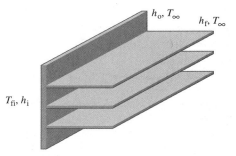

FIGURE 2.6.12 Overall Heat Transfer Coefficient with a Number of Fins Attached to a Surface

strictly one-dimensional as shown in Suryanarayana (1977). In many such cases, an engineering approximation of a one-dimensional temperature distribution can still be made if the temperature drop across the solid wall is not very great. In such cases, the concept of overall heat transfer coefficient can be applied to configurations with extended surfaces.

The nomenclature for the definition of the overall heat transfer coefficient is

h_o = convective heat transfer coefficient, outer surface
h_f = convective heat transfer coefficient, fin surface
h_i = convective heat transfer coefficient, inner surface
T_∞ = temperature of the fluid on the outer surface
T_{fi} = temperature of the fluid on the inner surface
A_f = total surface area of the fins exposed to the outer fluid at T_∞
A_s = area of unfinned surface exposed to fluid at T_∞
A_x = cross-sectional area of the solid wall
η = fin efficiency

With the assumption of one-dimensional temperature distribution in the solid wall and the concept of fin efficiency, the heat transfer rate is given by

$$q = (A_s h_o + \eta A_f h_f)(T_s - T_\infty)$$

where T_s is the temperature of the outer unfinned surface and the temperature at the base of the fins. We then have

$$T_{fi} - T_s = q\left(\frac{1}{h_i A_x} + \frac{L}{kA_x}\right) \qquad T_s - T_\infty = \frac{q}{A_s h_o + \eta A_f h_f}$$

Adding and rearranging, we have

$$q = \frac{T_{fi} - T_\infty}{1/(h_i A_x) + L/(kA_x) + 1/(A_s h_o + \eta A_f h_f)}$$

Overall heat transfer coefficient with fins

If the overall heat transfer coefficient based on the cross-sectional area A_x is defined as $U = (q/A_x)/(T_{fi} - T_\infty)$, we obtain

$$U = \frac{1}{1/h_i + L/k + A_x/(A_s h_o + \eta A_f h_f)}$$

As it is difficult to correctly determine the two different convective heat transfer coefficients, h_o and h_f, it is quite common to assume that they are equal.

2.6.3 Some Features of Analysis of Extended Surfaces

Our discussion of extended surfaces has been limited to those cases where the assumption of one-dimensional temperature distribution is a good engineering approximation. If a fin is infinitely long, we know intuitively that the free end is at the same temperature as the surrounding fluid. The question is how long should a fin be for it to be considered infinitely long. In discussing fin efficiency and the use of available charts for efficiency, the corrected length was utilized. A brief discussion follows of the corrected length and the suitable length of a fin; the conditions under which a fin can be considered to be infinite; and the validity of one-dimensional analysis.

2.6.4 Corrected Length

We have seen the simplicity of solution in terms of Equation 2.6.16 and 2.6.17 but as our intent in using the extended surface is to increase the heat transfer rate, it would be inappropriate to insulate the free end surface. (One case when there is no heat flow from the end surface is that of a fin attached to two plates, each maintained at the same temperature. Because of the symmetry, only one half of the fin need be analyzed. The cross-sectional surface midway between the two plates may be treated as the free end, and there is no heat flow across that surface.) However, these equations may be employed to get a very good approximation to the heat transfer rate by increasing the actual length by a small amount, and treating the free end of the extended fin as insulated. The length L_c to be added is determined by a simple approximation. Consider the two fins (a) and (b) shown in Figure 2.6.13. The actual fin is represented by (a) and the fin with the corrected length is represented by (b). The value of L_c has to be determined such that the heat transfer rates in both the configurations are approximately equal. It is anticipated that the extension is small. In most cases the value of mL is of the order of 1 or greater (see the succeeding section 2.6.5 on suitable length of fins); the magnitude of the temperature gradient near the free end is small. In such cases, close to the free end of (a), we expect the variation in the temperature to be small. The temperature in the extended part L_c may be considered to be reasonably constant at the value at the tip of (a), θ_t. We require that the heat transfer from the extended part in (b) should be equal to the heat transfer from the end surface of (a). Further, assuming that $h_e \cong h$,

$$h_e A_c \theta_t = hPL_c \theta_t \quad \text{or} \quad L_c = \frac{A_c}{P}$$

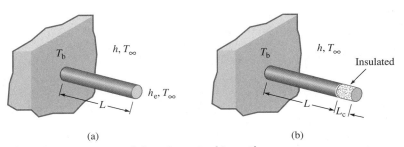

(a)   (b)

FIGURE 2.6.13 Determining Corrected Length
(a) Fin with free end exposed to the surrounding fluid. (b) Fin incorporating a corrected length.

Corrected length of
fins

where A_c is the cross-section area and P is the perimeter. For example, for a rectangular fin with heat transfer from all sides, denoting the thickness by t and the width by W,

$$L_c = \frac{Wt}{2(W + t)} = \frac{t}{2(1 + t/W)} \quad \text{If } t/W \ll 1, L_c = \frac{t}{2}$$

Similarly if the fin is cylindrical $L_c = R/2$, where R is the radius.

The correction to the length is based on the assumption that the temperature gradient near the free end is small. If the value of mL is of the order of 0.1 or less the temperature gradient at the free end is significant and the concept of corrected length should not be used.

2.6.5 Suitable Length of a Fin

In designing a fin, we may start by choosing an appropriate material for it. We may then specify the dimensions of the cross section. Having determined the material and the cross section of the fin, we next determine a suitable length for it. If it is too long, much of it is at the same temperature as the surrounding fluid, and little heat transfer takes place from a significant part of the fin, resulting in wastage of material and increased weight. If it is too short, we may not increase the heat transfer rate sufficiently. To determine an appropriate length, consider a fin whose cross-sectional dimensions and other parameters, except its length, are fixed. The heat transfer rate is a maximum if it is infinitely long. The answer to the question ''For what length does the fin behave as essentially infinite?'' can be given from two different viewpoints—one in terms of the temperature at the free end, and the other in terms of the total heat transfer rate. In both cases we assume the free end to be insulated (i.e., we use the concept of corrected length).

From Equation 2.6.16 the temperature at the free end of the fin is given by

$$\theta_t = \frac{\theta_b}{\cosh(mL)} \quad \text{or} \quad \frac{\theta_t}{\theta_b} = \frac{1}{\cosh(mL)}$$

When is a fin
infinitely long?

For a fin of infinite length, the free end temperature is equal to the surrounding fluid temperature, i.e., $\theta_t/\theta_b \approx 0$. The ratio of these two temperatures is dependent on the value of mL. The value of m is fixed by the cross-sectional dimensions, the convective heat transfer coefficient, and the thermal conductivity of the material. The values of the temperature ratio for different values of mL are given in the following table and shown in Figure 2.6.14 (see Figure 2.6.5 also).

mL	0.5	1.0	2.0	3.0	5.0
θ_t/θ_b	0.8868	0.6481	0.2658	0.0993	0.0135

The table shows the rapid decline in the free end temperature as the length is increased. At a length corresponding to a value of $mL = 3$, the difference in the temperature at the free end and the surroundings is about a tenth of the difference at the base. When the length is increased to $mL = 5$, this difference is only about 1.4% of that at the base. For all practical purposes we may consider the free end

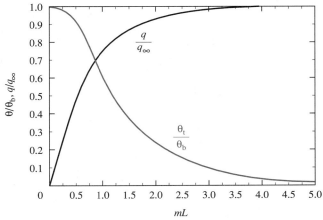

FIGURE 2.6.14 **Dimensionless Tip Temperature and Heat Transfer Rate as a Function of mL**

temperature to be that of the surroundings for $mL = 5$. Thus, from the point of view of the temperature of the free end, we may consider a fin to be of infinite length, when its length is greater than that corresponding to $mL = 5$.

From Equation 2.6.17 the heat transfer from the fin to the surroundings is given by

$$q_f = \sqrt{hPkA_x} \; \theta_b \tanh mL$$
$$q_\infty = \sqrt{hPkA_x} \; \theta_b$$

and

$$\frac{q_f}{q_\infty} = \tanh mL$$

where q_f represents the heat transfer rate from the fin of finite length L and q_∞ the heat transfer rate from an infinitely long fin. The ratio q_f/q_∞ represents the ratio of the actual heat transfer rate to the maximum possible heat transfer rate, with mL as the only variable. When this ratio is close to 1, the fin can be considered to be infinitely long. The values of q_f/q_∞ for some values of mL are given in the table below and shown in Figure 2.6.14.

mL	0.1	0.2	0.5	1.0	2.0	3.0
q_f/q_∞	0.0997	0.1974	0.4621	0.7616	0.9640	0.9951

The table shows how rapidly the heat transfer rate approaches that of an infinite fin when the length is increased. When the length is doubled from $mL = 0.1$ to $mL = 0.2$ (the value of m is constant), the heat transfer rate is essentially doubled. When the length is increased by a factor of 2.5 (from $mL = 0.2$ to $mL = 0.5$), the heat transfer rate is increased by a factor of 2.34, and at this increased length the heat transfer rate is 46.2% of the maximum possible heat transfer rate. From the values in the table, it is evident that for values of mL greater than about 2, there is no significant increase in the heat transfer.

One may, therefore, conclude that the fin behaves as essentially infinite if mL is greater than about 2.5. In actual practice, considering that the intent is to increase the heat transfer rate without ''wasting material,'' it would be appropriate to choose a length that corresponds to a value of around 1 for mL. In some cases, values of $mL < 1$ may be used and increased heat transfer rate obtained by increasing the number of fins (see Example 2.6.3). For example, consider two cases, A and B. Case A has a fin with $mL = 2$ and B has two fins with $mL = 1$ for each fin. The total lengths and masses are equal in both cases. In A the heat transfer rate from the fin is approximately $0.96\,q_\infty$, where q_∞ is the maximum heat transfer rate from a single fin of infinite length. In B the heat transfer rate from each fin is $0.76\,q_\infty$ and the total heat transfer rate from both fins is $1.52\,q_\infty$. Thus, with the same mass of the fins the heat transfer rate in B is 58% more than in A.

Determining how long a fin should be

2.6.6 Justification for One-Dimensional Analysis

In an extended surface there is conductive heat transfer not only in the x-direction but also in the y-direction (Figure 2.6.15). Therefore, there must be temperature variation in both directions. Yet, in the analysis of fins, such a variation of temperature in the y-direction was neglected. The condition for which the one-dimensional temperature distribution is a reasonable approximation needs to be known. The main intent is to determine the heat transfer rate from the surface to the surrounding fluid by convection and not so much to determine accurately the temperature itself. We wish to determine the condition for which the convective heat flux is approximately the same whether the temperature at the surface or the temperature in the interior of the fin at the same value of x is used. Denote the midplane temperature by T_c and the surface temperature by T_s at a given value of x. If $T_c - T_s$ is small compared with $T_s - T_\infty$, the variation of the temperature in the y-direction may be neglected. In other words, in the evaluation of the convective heat flux, $h(T_s - T_\infty)$, either the surface temperature or the midplane temperature can be used if

$$\frac{T_c - T_s}{T_s - T_\infty} \ll 1$$

FIGURE 2.6.15 Heat transfer in a rectangular fin of thickness $2t$; for $ht/k \ll 1$, $(T_c - T_s)/(T_s - T_\infty) \ll 1$

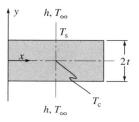

At the surface $h(T_s - T_\infty) = -k(\partial T/\partial y)|_{\text{surface}}$. Approximating $\partial T/\partial y|_{\text{surface}}$ by $(T_s - T_c)/t$, we obtain

Justification for one-dimensional temperature distribution—Biot number $\ll 1$

$$h(T_s - T_\infty) \approx k\frac{T_c - T_s}{t} \quad \text{or} \quad \frac{T_c - T_s}{T_s - T_\infty} \approx \frac{ht}{k}$$

If ht/k is much less than 1, we expect $(T_c - T_s)/(T_s - T_\infty)$ to be much less than 1. This is, indeed, the case for a rectangular fin. To obtain an acceptable level of engineering accuracy for the heat transfer rate from the fin from a one-dimensional analysis, it is sufficient if ht/k is less than 0.1. For a cylindrical fin Irey (1968) has shown that if the Biot number hR/k is less than 0.2 the difference between the heat transfer rates calculated with two- and one-dimensional temperature distributions, is less than 0.2%. The Biot number plays an important role in conduction problems with convective boundary conditions.

Now consider Equation 2.6.15a. If $h_e \approx h$, $(h_e/km) \approx (hA_x/kP)^{1/2}$. Therefore, (h_e/km) is equal to $(ht/2k)^{1/2}$ for a rectangular fin for which the width is much greater than the thickness, t, and equal to $(hR/2k)^{1/2}$ for a circular pin fin of radius R. In both cases, if the Biot number is much less than 1 for one-dimensional analysis to be appropriate, then (h_e/km) is also much less than 1. As $\sinh(mx) < \cosh(mx)$, Equation 2.6.15a can be approximated by Equation 2.6.16 and Equation 2.6.15b by Equation 2.6.17. But if $mL \ll 1$, $\sinh(mL) \ll \cosh(mL)$, and $\sinh(mL) \approx (h_ekm)/\cosh(mL)$ and Equation 2.6.15b cannot be approximated by Equation 2.6.17.

2.7 GENERALIZED ONE-DIMENSIONAL, TRANSIENT CONDUCTION EQUATION WITH AREA CHANGE AND INTERNAL ENERGY GENERATION

So far, the conduction equation for each one-dimensional case was derived. We can also derive a generalized, one-dimensional conduction equation with area change and internal energy generation and then specialize it to particular cases. We now derive such a generalized equation and illustrate the application of the generalized equation to some cases already considered. As an introduction to the transient cases to be considered in Section 2.8, the time dependency of the temperature is included.

Generalized one-dimensional transient conduction with internal energy generation

Consider a solid body in which the temperature is a function of time and only one spatial coordinate, $T(x, \tau)$. A heat flux of q_i'' reaches the surface parallel to the x-direction; the heat flux may be due to convective heat transfer from a surrounding fluid, radiant heat flux from surrounding surfaces, and so forth. For the geometry shown in Figure 2.7.1, in reality, the temperature is two-dimensional. But, if the rate of area change with respect to x is small or if the Biot number is small (as in the case of fins illustrated in Section 2.6), the changes in the temperatures in the direction normal to the x-direction can be neglected. The resulting one-dimensional temperature distribution represents the average temperature as a function of x. For the physical configuration shown in Figure 2.7.1 we define

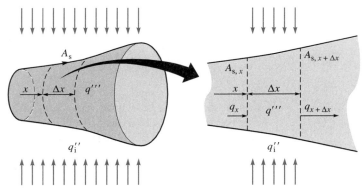

FIGURE 2.7.1 A solid body with internal energy generation, in which the temperature distribution is approximated as one-dimensional, $T = T(x, \tau)$. q_i'' is the heat flux from the surroundings

A_x = area perpendicular to the x-direction
A_s = surface area exposed to the heat flux q_i'', measured from $x = 0$
q''' = rate of internal energy generation per unit volume (for an explanation of internal energy generation see Section 3.1)

An energy balance for the element between x and $x + \Delta x$ (fixed mass) gives

> Rate of change of internal energy of the element
>
> = Net rate of heat transfer to the element
>
> + Internal energy generation
>
> − Work transfer from the element

Mass of element = $\rho A_x \Delta x$

Rate of change of internal energy of the element = $\dfrac{\partial}{\partial \tau}(\rho A_x u \, \Delta x)$

where u is the specific internal energy of the solid in the element. Heat transfer is by conduction into the element across the surface located at x, and out of the element across the surface located at $x + \Delta x$.

Net rate of heat transfer to the element by conduction = $q_x - q_{x+\Delta x}$

The surface area exposed to the heat flux q_i'' is measured from the origin ($x = 0$), and is given by $A_{s,x+\Delta x} - A_{s,x}$

Heat transfer rate from the surroundings = $q_i''(A_{s,x+\Delta x} - A_{s,x})$

Internal energy generation rate in the element = $q'''A_x \Delta x$

Rate of work transfer from the element = $p \dfrac{\partial V}{\partial \tau}$

where

V = volume of the element = $A_x \Delta x$
p = pressure

For solids the density changes due to temperature changes are small; if density changes are neglected then the work transfer is zero. However, here we account for the small changes in the density that lead to work transfer. In many conduction problems the pressure is constant or nearly constant and for such cases,

$$\text{Rate of work transfer from the element} = \frac{\partial(pA_x\Delta x)}{\partial \tau}$$

Substituting the terms in the energy equation, one obtains

$$\frac{\partial}{\partial \tau}(u\rho A_x\Delta x) = q_x - q_{x+\Delta x} + q_i''(A_{s,x+\Delta x} - A_{s,x}) + q'''A_x\Delta x - \frac{\partial(pA_x\Delta x)}{\partial \tau}$$

Rearranging the equation,

$$\frac{\partial}{\partial \tau}\left[\rho A_x\Delta x\left(u + \frac{p}{\rho}\right)\right] = q_x - q_{x+\Delta x} + q_i''(A_{s,x+\Delta x} - A_{s,x}) + q'''A_x\Delta x$$

But from mass conservation $\frac{\partial}{\partial \tau}(\rho A_x\Delta x) = 0$ and $u + p/\rho$ = specific enthalpy, i. Using these relations, we obtain

$$\rho A_x\Delta x \frac{\partial i}{\partial \tau} = q_x - q_{x+\Delta x} + q_i''(A_{s,x+\Delta x} - A_{s,x}) + q'''A_x\Delta x$$

Dividing the equation by $A_x\Delta x$ and taking the limit as $\Delta x \to 0$, we get

$$\rho \frac{\partial i}{\partial \tau} = -\frac{1}{A_x}\frac{\partial}{\partial x}(q_x) + \frac{1}{A_x}q_i''\frac{dA_s}{dx} + q''' = 0$$

From Fourier's Law $q_x = -kA_x(\partial T/\partial x)$. From the definition of specific enthalpy, at constant pressure, $\partial i/\partial \tau = c_p(\partial T/\partial \tau)$.

$$\rho c_p \frac{\partial T}{\partial \tau} = \frac{1}{A_x}\frac{\partial}{\partial x}\left(kA_x\frac{\partial T}{\partial x}\right) + \frac{1}{A_x}q_i''\frac{dA_s}{dx} + q''' \qquad (2.7.1)$$

For solids (and liquids with almost constant densities, the difference between the specific heats at constant volume and constant pressure is very small, and is generally neglected. In tables of properties, only one value of the specific heat, usually c_p, is given, and the same value may be used for c_v also.

Initial and Boundary Conditions

Boundary conditions

To solve Equation 2.7.1, we need one initial condition and two boundary conditions. The initial condition requires that we know the temperature distribution at some instant of time, i.e., $T(x, \tau_o) = T_o(x)$.

The boundary conditions for the temperatures may be one of several different types. Some of these are given below.

Specified temperature

1. **Specified Temperature.** $T(x_o, \tau) = T_o$.

Specified heat flux—insulated boundary is a special case of specified heat flux

2. **Specified Heat Flux.** At a known surface, the heat transfer rate is known, that is

$$\pm kA_x \frac{\partial T}{\partial x}(x_0, \tau) = q_0'' \quad \text{or} \quad \frac{\partial T}{\partial x}(x_0, \tau) = \pm \frac{q_0''}{kA_x}$$

An insulated boundary is a particular case of specified heat flux with $q'' = 0$.

Convective boundary condition

3. **Convective Boundary Condition.** At a boundary the surface is exposed to a fluid, and heat transfer occurs between the surface and the adjacent fluid by convection. From an energy balance on the surface the condutive heat flux to the surface from the interior of the solid must be equal to the convective heat flux from the surface to the surrounding fluid. That is,

$$\pm kA_x \frac{\partial T}{\partial x}(x_o, \tau) = hA_x[T(x_o, \tau) - T_\infty].$$

This equation can be recast as

$$\frac{\partial T}{\partial x}(x_o, \tau) \pm \frac{h}{k}[T(x_o, \tau) - T_\infty] = 0$$

Positive sign if the outward normal to the surface is in the positive x-direction and negative sign if the outward normal is in the negative x-direction.

4. **Matching Heat Flux and Temperature.** Is some cases, such as when the solid is made of two different materials or when one of the parameters in the differential equation is different in different parts of the solid body, the same differential equation cannot be used over the entire region. In such a case, solutions to each domain must be found separately. The two solutions should then satisfy certain conditions regarding the temperature and heat flux at the interface of the two domains.

Matching heat flux

(a) Matching Heat Flux. As a surface cannot store energy, the heat flux at the interface must be the same whether the interface is considered as belonging to one domain or the other. Referring to Figure 2.7.2, if A and B are different materials, or if the differential equations for the two domains are different, the solutions to each domain must satisfy the continuity of heat flux at the interface. Denoting the location of the interface by x_o, the matching heat flux condition requires

$$k_A \frac{dT_A}{dx}\bigg|_{x=x_o} = k_B \frac{dT_B}{dx}\bigg|_{x=x_o}$$

where T_A and T_B are the temperature distributions in the domains A and B, respectively.

Matching temperatures—Contact resistance

(b) Matching Temperatures. The heat flux is continuous at the interface but the temperature, in general, is not. If two solids are brought together, at the interface there is contact resistance leading to discontinuity in temperature. The contact resistance may be neglected if the joints are welded or brazed. Soldered joints

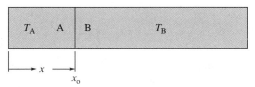

FIGURE 2.7.2 Conduction in two different domains, A and B.

may have non-negligible contact resistance. The contact resistance is not easily determined, and hence, in many cases continuity of temperature is assumed as an approximation. With such an approximation, referring to Figure 2.7.2,

$$T_A(x = x_o) = T_B(x = x_o)$$

If contact resistance is not negligible, and its value is known, then,

$$-k_A \left. \frac{dT_A}{dx} \right|_{x=xo} = (T_A - T_B)/R_c''$$

where R_c'' is the contact resistance for a unit area. If the material is homogeneous without any discontinuity the contact resistance is zero and the temperature is continuous.

From the general conduction equation to specific cases

We now apply Equation 2.7.1 to some specific cases that we considered earlier in this chapter. In all cases, the internal energy generation was zero, and hence, the last term in Equation 2.7.1 is zero. Only steady-state situations are considered; for such cases the left side of the equation is zero.

2.7.1. Plane Slab, Steady State, and Constant Thermal Conductivity

Refer to Figure 2.2.1. As the temperature is a function of only one variable, we replace the partial derivatives by total derivatives. Further, with no heat transfer to the surfaces parallel to the x-direction, the term containing q_i'' also drops out. With no internal energy generation ($q''' = 0$) the resulting equation is

$$\frac{d}{dx}\left(kA_x \frac{dT}{dx}\right) = 0$$

This is the same as Equation 2.2.3.

2.7.2. Cylinder, Steady State

Refer to Figure 2.2.4a. Replace the general coordinate x, in the equation by the radius in cylindrical coordinates, $T = T(r)$. $A_x = A_r$, area perpendicular to the radial direction. With no heat transfer to the end surfaces of the cylinder and internal heat generation being zero, we obtain

$$\frac{d}{dr}\left(kA_r \frac{dT}{dr}\right) = 0$$

This is the same as Equation 2.2.9. Note that $A_r = 2\pi rL$.

2.7.3. Extended Surface

Refer to Figure 2.6.4. In this case, the heat transfer from the surroundings is by convection and $q_i'' = -h(T - T_\infty)$. With a constant perimeter, the surface area is given by $A_s = Px$, and its derivative with respect to x is P. With constant A_x, and assuming constant thermal conductivity, after replacing $T - T_\infty$ by θ and hP/kA_x by m^2, we obtain

$$\frac{d^2\theta}{dx^2} - m^2 = 0$$

Thus, Equation 2.6.3 is obtained.

The general equation has been applied to cases where the internal heat generation is zero, and the simplified equations obtained. Applications with nonzero internal heat generation are discussed in Chapter 3.

2.8   TRANSIENT CONDUCTION—LUMPED ANALYSIS

So far, we have dealt with steady-state problems. We now study cases where the temperature is time dependent. In general, the temperature of a body may be expected to depend on both time and spatial coordinates. In this section, we consider a particularly simple case where the temperature can be considered to be time dependent but spatially uniform. As an example where such a simplification is a good approximation, consider a solid surrounded by a fluid at a temperature different from that of the solid. If the Biot number (hL_c/k), formed with the proper characteristic length, L_c, is small (<0.1), the approximation of spatially uniform temperature at any given instant gives acceptable results. An analysis where the temperature is considered spatially uniform, i.e., spatially lumped, is termed *lumped analysis*. Cases where lumped analysis is not applicable are considered in Section 2.9.

What is lumped analysis? When is lumped analysis valid?

As an illustration of lumped analysis, consider a solid body which is initially at a uniform temperature T_o. Starting at time $\tau = 0$ a heat transfer rate of q_i is imposed at the surface of the solid (Figure 2.8.1a). The internal energy generation rate is q''' per unit volume. Both q_i and q''' may be functions of the temperature and/or time.

From an energy balance on the solid

Rate of change of internal energy of the solid

= Net rate of heat transfer to the solid

− Rate of work done by the solid

+ Rate of internal energy generation

Rate of change of internal energy = $dU/d\tau$

where

$U = mu$ (m = mass of the solid, and u = specific internal energy)

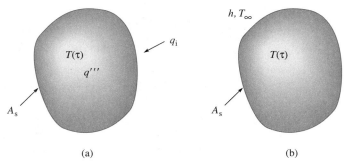

(a)   (b)

FIGURE 2.8.1 Transient conduction
(a) General case. (b) Heat transfer to surroundings by convection
and no internal energy generation.

Denoting the internal generation per unit volume by q''' and the total volume by V

$$\text{Internal energy generation} = q'''V$$

$$\text{Rate of work done by the solid} = p\,\frac{dV}{d\tau}$$

In the majority of transient conduction problems, the pressure is constant (or nearly so). For such cases $p(dV/d\tau) = m(dpv/d\tau)$, where v is the specific volume. In the absence of internal energy generation, we get

$$m\,\frac{du}{d\tau} = q_{\mathrm{i}} - m\,\frac{dp\,v}{d\tau} + q'''V$$

Replacing $u + pv$ by the specific enthalpy i, and with $di/d\tau = c_p(dT/d\tau)$, we obtain

$$mc_p\,\frac{dT}{d\tau} = q_{\mathrm{i}} + q'''V \tag{2.8.1}$$

Equation 2.8.1 can be integrated if the relations between q''' and τ or T, and between q_{i} and τ or T are known.

Consider the case where the solid is exposed to a fluid maintained at a constant temperature, T_∞, with a constant convective heat transfer coefficient, h, and no interal energy generation as shown in Figure 2.8.1b. Radiative heat transfer is negligible or it has been suitably compensated for by evaluating the value of the convective heat transfer coefficient[3] to include the effect of radiation.

Net rate of heat transfer to the solid (q_{i})
= Convective heat transfer rate from the surrounding fluid to the solid
= $-hA_{\mathrm{s}}(T - T_\infty)$

---

[3]In some cases when the temperature variation is not large, an approximate solution including the effect of radiation can be obtained by adding the radiative heat transfer coefficient to the convective heat transfer coefficient.

where

$$A_s = \text{surface area of the solid exposed to the fluid}$$
$$T = \text{temperature of the solid at any instant } \tau$$

$$mc_p \frac{dT}{d\tau} = -hA_s(T - T_\infty) \qquad \text{(2.8.1a)}$$

Initial condition: $T(\tau = 0) = T_o$.

Separating the variables, replacing m by $\rho V (\rho = $ density of the solid and $V = $ volume of the solid), and assuming constant specific heat,

$$\int_{T_o}^{T} \frac{dT}{T - T_\infty} = -\int_{0}^{\tau} \frac{hA_s}{\rho V c_p} d\tau \qquad \text{(2.8.2)}$$

$$\frac{T - T_\infty}{T_o - T_\infty} = \exp\left(-\frac{hA_s}{\rho V c_p} \tau\right)$$

Lumped analysis— time constant

In Equation 2.8.2, the term $(\rho V c_p / hA_s)$ can be thought of as the *time constant* for the transient case. It represents the time taken for the solid to change the temperature difference, $T - T_\infty$, to 37% $(1/e)$ of its initial value. Another interpretation may be given to the time constant. The rate of change of temperature at time $\tau = 0$, is $-[(hA_s)/(\rho V c_p)](T_o - T_\infty)$. The maximum change in temperature is $T_\infty - T_o$ (the final value of the temperature less the initial value). Hence, the magnitude of the rate of change of temperature of the body is a maximum at time $\tau = 0$. If the rate of change of temperature remained constant at the value at $\tau = 0$, the time taken for the temperature to change from T_o to T_∞ is given by the time constant.

Biot number should be less than ≈ 0.1 for lumped analysis

As already indicated, the analysis is valid only when the Biot number defined by hL_c/k ($k = $ thermal conductivity of the material of the solid) is less than about 0.1. The characteristic length, L_c, is the dimension along which the temperature variation can be neglected compared with some characteristic temperature difference. For example, the characteristic length for a slab with heat transfer from only two parallel surfaces (the remaining four surfaces being insulated) is half the thickness of the slab. For a cylinder, with heat transfer from only the cylindrical surface, and for a sphere, the radius is the characteristic length. However, for solids with irregular shapes an equivalent length is defined as the ratio of the volume to the surface area of the solid exposed to the fluid, i.e., $L_c = V/A_s$ ($V = $ volume and $A_s = $ surface area exposed to fluid).

EXAMPLE 2.8.1 Alloy tool steels can be hardened by cooling them in air or in oil. A rectangular block 10-cm long, 2-cm wide, and 2-cm thick is hardened by heating it to 1640 °C and cooling it (Figure 2.8.2). Estimate the time required to cool it from 1640 °C to 40 °C, and the rate of cooling when its temperature is 1000 °C if it is cooled (1) in air at 30 °C, with a convective heat transfer coefficient of 110 W/m² °C, and (2) by immersing it in a large bath of oil at 30 °C with a convective heat transfer coefficient of 200 W/m² °C. The properties of alloy steel are $\rho = 7850$ kg/m³, $c_p = 580$ J/kg K, and $k = 28$ W/m K.

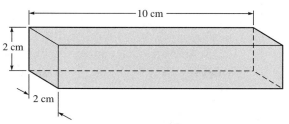

FIGURE 2.8.2 A block of high alloy tool steel is hardened by heating it to 1640 °C and cooling it in air or oil.

Given

Dimensions of steel block:
2-cm wide, 2-cm thick, and 10-cm
  long
Initial temperature of block
  $(T_o) = 1640\,°C$

Final temperature of the block =
  40 °C
Surrounding fluid temperature
  $(T_\infty) = 30\,°C$

Find

(a) The time to cool the block to 40 °C
(b) The rate of cooling when the block temperature is 1000 °C
if
  (1) the surrounding fluid is air with $h_a = 110\ W/m^2\ °C$
  (2) the surrounding fluid is oil with $h_o = 200\ W/m^2\ °C$

ASSUMPTIONS

1. Constant T_∞ and h
2. Constant properties
3. Negligible radiative heat transfer with oil. When cooled in air, radiative heat transfer has been compensated for by an appropriate value of h.

SOLUTION

(a) As the block is cooled by convection from all sides, a characteristic length for computing the Biot number is given by

$$L^* = \frac{\text{volume}}{\text{surface area}} = \frac{0.1 \times 0.02 \times 0.02}{2(0.1 \times 0.02 + 0.02 \times 0.02 + 0.1 \times 0.02)}$$

$$= 0.004545\ m$$

$$\text{Maximum value of Biot number} = h_{max}L^*/k = \frac{200 \times 0.004545}{28}$$

$$= 0.032$$

As the maximum value of the Biot number is much less than 1, lumped analysis is employed.

For convective heat transfer with constant T_∞ and h, the time required for cooling the block is given by Equation 2.8.2,

$$\theta = \frac{T - T_\infty}{T_o - T_\infty} = e^{-hA_s\tau/(\rho Vc_{sp})} \tag{1}$$

and the rate of change of temperature by Equation 2.8.1a.

$$\frac{d\theta}{d\tau} = -\frac{hA_s}{\rho Vc_p}\theta \tag{2}$$

To cool it from the initial temperature of 1640 °C to 40 °C, from Equation 1,

$$\tau = -\frac{\rho Vc_p}{hA_s}\ln\left(\frac{T - T_\infty}{T_o - T_\infty}\right) = -\frac{7850 \times 0.004545 \times 580}{h}\ln\frac{40 - 30}{1640 - 30}$$

$$\tau = \frac{105151}{h}\text{ s} \tag{3}$$

(b) With $\theta = (T - T_\infty)/(T_o - T_\infty)$, the rate of change of temperature of the block is given by

$$\frac{dT}{d\tau} = -\frac{hA_s}{\rho Vc_p}(T - T_\infty) = -\frac{h}{7850 \times 0.004545 \times 580}(1000 - 30)$$

$$\frac{dT}{d\tau} = -0.04687\ h\ \text{°C/s} \tag{4}$$

[Note that Equation 4 can also be obtained by directly applying the First Law of Thermodynamics to the block, leading to $mc_p(dT/d\tau) = -hA_s(T - T_\infty)$]

(1) For cooling in air, substituting $h_a = 110$ W/m² °C into Equations 3 and 4,

(a)
$$\tau = \frac{105151}{h_a} = \frac{105151}{110} = \underline{956\text{ s} = 16\text{ min}}$$

(b)
$$\frac{dT}{d\tau} = -0.04687 \times 110 = \underline{-5.2\text{°C /s}}$$

(2) For cooling in oil with $h_o = 200$ W/m² °C,

(a)
$$\tau = \frac{105151}{h_o} = \frac{105151}{200} = \underline{526\text{ s} = 8\text{ min }45\text{ s}}$$

(b)
$$\frac{dT}{d\tau} = -0.04687 \times 200 = \underline{-9.4\text{ °C /s}}$$

EXAMPLE 2.8.2 An electric cartridge heater fits snugly inside a hollow copper cylinder 1-cm I.D., 3-cm O.D., and 20-cm long (Figure 2.8.3). Initially, the cylinder is at 20 °C. At time $\tau = 0$, hot air at 100 °C is forced over the cylinder and at the same time the electric heater is turned on. The power supply to the heater is 1000 W and the convective heat transfer coefficient associated with the outside surface and the hot air is 100 W/m² °C. Determine the steady state temperature of the copper cylinder and the temperature of the cylinder 5 s, and 15 min after the electric power is turned on.

Given

Inside diameter $(2r_i) = 0.01$ m
Length $(L) = 0.2$ m
Initial temperature of cylinder
$(T_8) = 20$ °C

Outside diameter $(2r_o) = 0.03$ m
Convective heat transfer coefficient (h)
$= 100$ W/m² °C
Temperature of surrounding air
$(T_\infty) = 100$ °C
Power supply to heater $(P) = 1000$ W

Find

(a) The steady temperature of cylinder (T_{st})
(b) The temperature of the cylinder 5 s and 5 min after the power is turned on

ASSUMPTIONS

1. Mass of the heater is much less than that of the cylinder and $(mc_p)_{heater} \ll (mc_p)_{Cu}$.
2. Properties of copper from Table A1 are $k = 401$ W/m K, $c_p = 385$ J/kg K, $\rho = 8933$ kg/m³.
   Bi $= [h(r_o - r_i)]/k = (100 \times 0.01)/401 \ll 1$. Hence, lumped analysis is valid.
3. Radiation heat transfer is negligible.

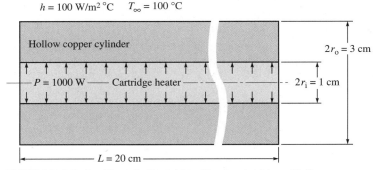

FIGURE 2.8.3 **An Electric Cartridge Heater Inside a Hollow Copper Cylinder**

SOLUTION

(a) In steady state conditions net energy transfer $= 0$. With P representing the electric power supply to the heater, $P + hA_s(T_\infty - T_{st}) = 0$, where T_∞ is the temperature of the air surrounding the cylinder and T_{st} is the steady-state temperature of the cylinder.

Substituting the values,

$$T_{st} = T_\infty + \frac{P}{hA_s}$$

$$= 100 + \frac{1000}{100[\pi \times 0.03 \times 0.2 + 2 \times \pi \times (0.015^2 - 0.005^2)]} = 597.4°C$$

(b) In transient conditions, from an energy balance with the copper cylinder and the heater as the system, from Equation 2.8.1,

$$\frac{dI}{d\tau} = P - hA_s(T - T_\infty)$$

where $I = U + pV$ is the total enthalpy of the system. With the assumption of uniform temperature, $I = (mi)_{Cu} + (mi)_{heater}$

$$\frac{dI}{d\tau} = [(mc_p)_{Cu} + (mc_p)_{heater}]\frac{dT}{d\tau} = (mc_p)_{Cu}\left(1 + \frac{(mc_p)_{heater}}{(mc_p)_{Cu}}\right)\frac{dT}{d\tau}$$

From assumption 1, $(mc_p)_{heater}/(mc_p)_{Cu} \ll 1$, and hence,

$$\frac{dI}{d\tau} \approx (mc_p)_{Cu}\frac{dT}{d\tau}$$

The energy equation is now written as

$$(mc_p)_{Cu}\frac{dT}{d\tau} = -hA_s\left(T - T_\infty - \frac{P}{hA_s}\right) = -hA_s(T - T_{st})$$

where T_{st} is the steady state temperature given by $T_\infty + \frac{P}{hA_s}$. The initial condition is $T(\tau = 0) = T_i$.

Separating the variables and integrating from $\tau = 0$ to $\tau = \tau$,

$$\ln\frac{T - T_{st}}{T_i - T_{st}} = -\frac{hA_s\tau}{(mc_p)_{Cu}}$$

or

$$T = T_{st} - (T_{st} - T_i)\exp\left(-\frac{hA_s\tau}{(mc_p)_{Cu}}\right)$$

$$\frac{hA_s}{(mc_p)_{Cu}} = \frac{100[\pi \times 0.03 \times 0.2 + 2 \times \pi(0.015^2 - 0.005^2)]}{8933 \times \pi \times [(0.03^2 - 0.01^2)/4] \times 0.2 \times 385} = 0.004652s^{-1}$$

With $T_i = 20°$ C, $T_{st} = 597.4$ °C,

$$T = 597.4 - 577.4e^{-0.004652\tau}$$

For $\tau = 5$ s, T = 332.8 °C.
For $\tau = 15$ min = 900 s, T = 588.6 °C.

COMMENT

Note that from the transient solution for the temperature, as $\tau \rightarrow \infty$, the temperature of the solid reaches its steady state temperature.

2.9   MULTIDIMENSIONAL TRANSIENT PROBLEMS—HEISLER CHARTS

In the previous section, solutions to lumped problems were presented. When the Biot number is not much less than 1, lumped analysis does not give acceptable results. Analytical solutions to multidimensional transient conduction problems for rectangular (plane) slabs, cylinders, and spheres are available, and some of them are derived in Chapter 11. An introduction to numerical solutions is given in Chapter 3. Here we confine ourselves to the use of available solutions to a class of one-dimensional transient problems, and their extension to the solutions to multidimensional transient problems. The class of problems considered is that of a solid mass of a rectangular block, cylinder, or sphere, initially at a uniform temperature T_i. At time $\tau = 0$, the solid is exposed to a fluid maintained at a constant temperature T_∞. The convective heat transfer coefficient h, between the surface of the solid and the fluid is constant. Cooling or quenching of steel in air or water for heat treatment is an example of such a case.

First, we study the solutions to one-dimensional transient problems, and then, we extend these solutions to multidimensional problems.

2.9.1 One-Dimensional Transient Problems

One-dimensional transient temperature distribution

Consider a rectangular slab with two parallel surfaces, $2L$ apart, exposed to a fluid (Figure 2.9.1), and with the remaining surfaces perfectly insulated. Initially, at time $\tau = 0$, the slab is at a uniform temperature T_o. It is then exposed to a fluid whose temperature is maintained at a constant value T_∞. The heat transfer coefficient h is constant. We wish to find $T(x, \tau)$.

As the temperature is a function of both the spatial coordinate, x, and time, τ, the resulting differential equation for T is a partial differential equation. From Equation 2.7.1, with no heat transfer to the surfaces parallel to the x-direction, and $q''' = 0$, we obtain

$$\frac{\partial^2 T}{\partial x^2} = \frac{1}{\alpha}\frac{\partial T}{\partial \tau} \tag{2.9.1}$$

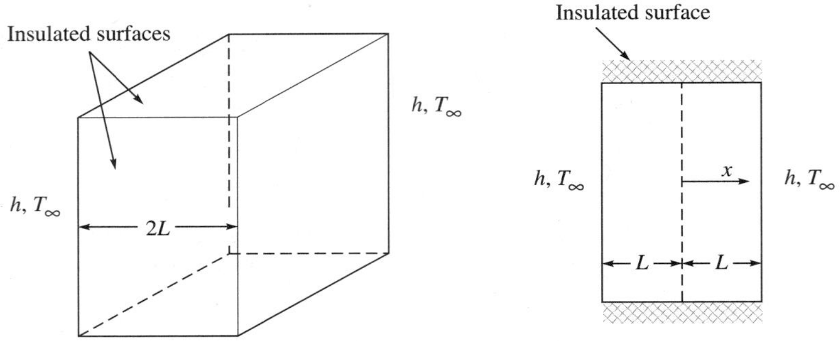

FIGURE 2.9.1 A Rectangular Block Is Initially at a Uniform Temperature T_o
Two parallel surfaces are exposed to a fluid at T_∞ and convective heat transfer coefficient h. All other surfaces are perfectly insulated.

The initial and boundary conditions are

$$T(x, 0) = T_o$$

$$k \frac{\partial T}{\partial x} (L, \tau) + h[T(L, \tau) - T_\infty] = 0$$

$$k \frac{\partial T}{\partial x} (-L, \tau) - h[T(-L, \tau) - T_\infty] = 0$$

α is the thermal diffusivity ($k/\rho c_p$), and x and τ are the dimensional x-coordinate and time, respectively.

We define dimensionless temperature, coordinate, and time as

$$\theta = \frac{T - T_\infty}{T_o - T_\infty} \qquad \eta = \frac{x}{L} \qquad F_o = \frac{\alpha \tau}{L^2}$$

The dimensionless form of Equation 2.9.1 is

$$\frac{\partial^2 \theta}{\partial \eta^2} = \frac{\partial \theta}{\partial Fo} \tag{2.9.2}$$

$$\theta(\eta, 0) = 1$$

$$\frac{\partial \theta}{\partial \eta} (1, Fo) + Bi \, \theta(1, Fo) = 0$$

$$\frac{\partial \theta}{\partial \eta} (-1, Fo) - Bi \, \theta(-1, Fo) = 0$$

where Bi is the Biot number (hL/k).

By employing dimensionless variables, the equations for all one-dimensional transient problems of a rectangular slab with different values at T_o, h, k, T_∞, and L have been reduced to one equation with only one parameter, Bi. This equation can be solved for different values of Bi. The differential equation and the boundary conditions for all cases with the same Biot number are the same, and hence, θ has the same solution, i.e., $\theta = \theta(x, Fo, Bi)$. The function $\theta(x, Fo)$ has been analytically determined for different values of Bi. The dimensionless time Fo $[=(\alpha\tau)/L^2]$ is known as the Fourier number. The analytical solution to Equation 2.9.2 is derived in Chapter 11. The solution is

$$\theta = 2 \sum_{n=1}^{n=\infty} \frac{\sin \mu_n}{\mu_n + \sin \mu_n \cos \mu_n} e^{-\mu_n^2 \text{Fo}} \cos \mu_n \eta$$

where μ_n are the roots of $\mu_n \tan \mu_n = \text{Bi}$. The first six values of μ_n are given in Appendix 2. Similar solutions have been obtained for cylinders (ends perfectly insulated) and spheres. Charts of temperature θ for a plane slab, cylinder (ends perfectly insulated), and sphere are presented in Figures 2.9.2 through 2.9.4. Heisler (1946) presented charts for the one-dimensional transient temperature distribution. Charts similar to Figures 2.9.2 through 2.9.4 are often referred to as Heisler charts.

In Figures 2.9.2 through 2.9.4, for Fo = 1.0, the value of θ evaluated from a single term of the series solutions is within 10^{-5} of the value obtained with 20 terms of the series (clearly indicated by the straight lines of θ against τ on the semilogarithmic plots). Thus, the temperature at any location is easily computed with a single term for values of Fo greater than 1. For a discussion on single term approximations see the next subsection.

The values of θ/θ_c represent the ratio of the temperature θ at different distances to the temperature θ_c at the midplane for the rectangular slab, at the axis for the cylinder, and at the center for the sphere. From the plot it is observed that the difference between the temperature at the surface and θ_c is less than 5% for Bi < 0.1 and, therefore, lumped analysis is valid for Bi < 0.1.

Approximation to Analytical Solution

One-term approximation to Series Solution for one-dimensional transient temperature distribution

The analytical solutions for each case—plane slab, cylinder with ends insulated, and sphere—is in the form of an infinite series. However, for sufficiently large values of the dimensionless time, Fo, the series converges rapidly, and a single term of the series (the first term) gives very good approximate results. To determine how large Fo should be to make such an approximation, consider the case of a plane slab. The solution for the dimensionless temperature for a plane slab is

$$\theta = 2 \sum_{n=1}^{\infty} \frac{\sin(\mu_n)}{\mu_n + \sin(\mu_n) \cos(\mu_n)} e^{-\mu_n^2 \text{Fo}} \cos\left(\mu_n \frac{x}{L}\right) \qquad (2.9.3)$$

where μ_n is the solution to $\mu_n \tan \mu_n = \text{Bi}$; and $\text{Bi} = (hL)/k$. This can be written in the form

$$\theta = 2 \frac{\sin(\mu_1)}{\mu_1 + \sin(\mu_1) \cos(\mu_1)} e^{-\mu_1^2 \text{Fo}} \cos\left((\mu_1 \frac{x}{L}\right) \times \\ \left\{ 1 + \sum_{n=2}^{\infty} a_n \frac{\cos[\mu_n(x/L)]}{\cos(\mu_1[(x/L)]} \right\} \qquad (2.9.4)$$

where

$$a_n = \frac{\mu_1 + \sin(\mu_1) \cos(\mu_1)}{\mu_n + \sin(\mu_n) \cos(\mu_n)} \left[\frac{\sin(\mu_n)}{\sin(\mu_1)} \right] e^{-(\mu_n^2 - \mu_1^2)\text{Fo}} \qquad n = 2, 3, \ldots, \infty$$

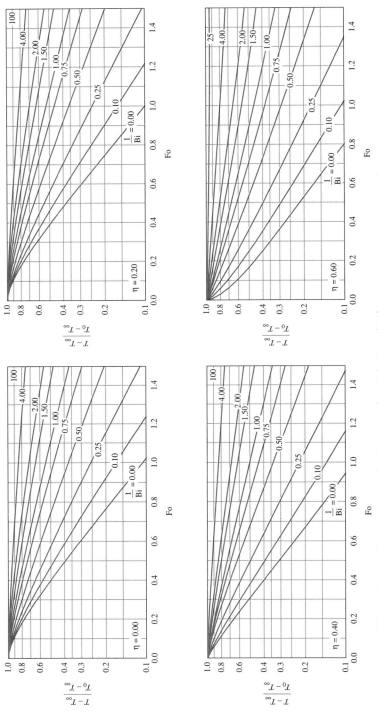

FIGURE 2.9.2 Dimensionless Temperature for Rectangular Slab: Bi = hL/k, $\eta = x/L$, $2L$ = thickness of plate, and Fo = $\alpha\tau/L^2$; and ratio of Dimensionless Temperatures θ/θ_c for Fo > 0.2.

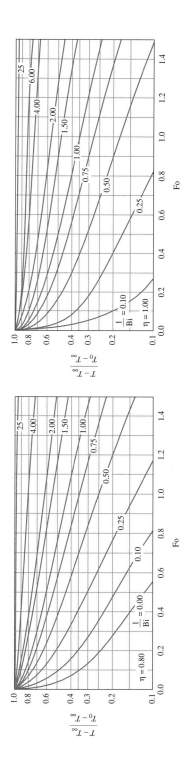

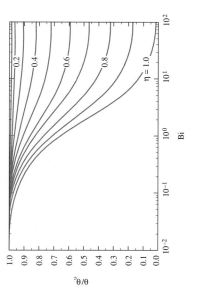

FIGURE 2.9.2 cont'd

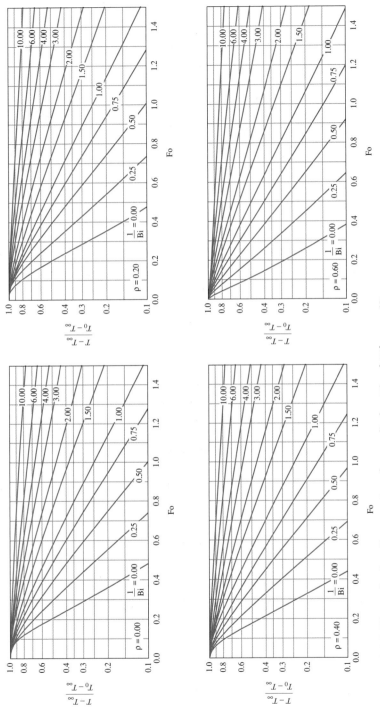

FIGURE 2.9.3 Dimensionless Temperature for a cylinder: $Bi = hR/k$, $\rho = r/R$, R = radius, and $Fo = \alpha\tau/R^2$; and ratio of Dimensionless Temperatures θ/θ_c for $Fo > 0.2$.

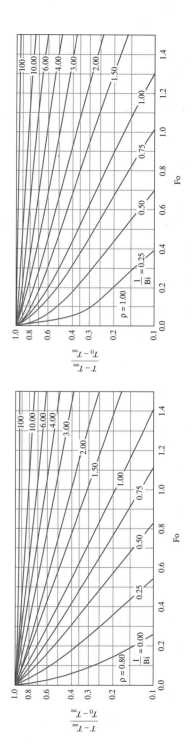

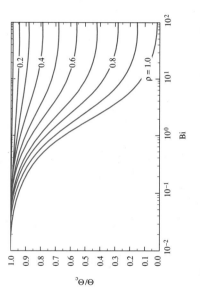

FIGURE 2.9.3 cont'd

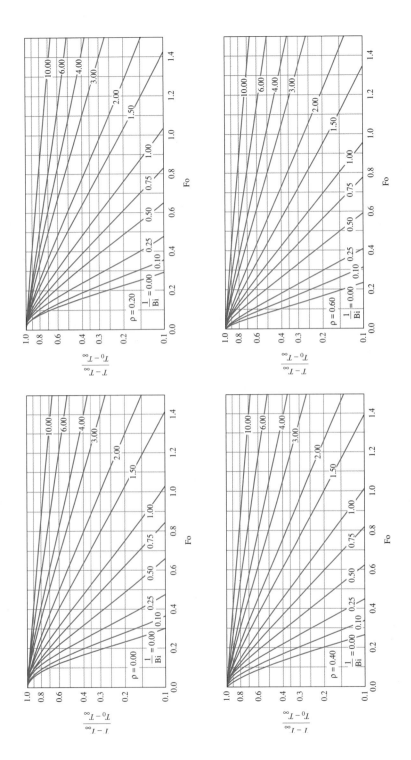

FIGURE 2.9.4 Dimensionless Temperature for a Sphere: $Bi = hR/k$, $\rho = r/R$, R = radius, and $Fo = \alpha\tau/R^2$; and ratio of Dimensionless Temperatures θ/θ_c for $Fo > 0.2$.

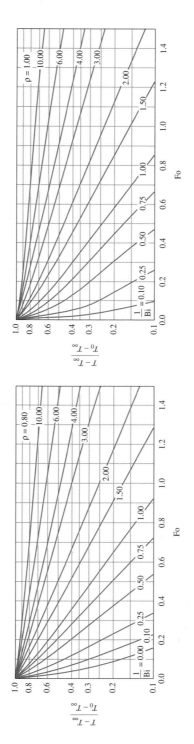

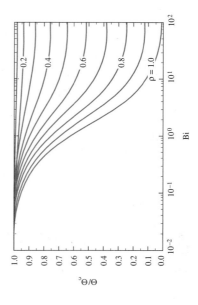

FIGURE 2.9.4 cont'd

The values of a_n $\{\cos[\mu_n(x/L)]/\cos[\mu_1(x/L)]\}$, denoted by b_n for Fo = 0.1 and 0.2, and $x/L = 0$ and 1 corresponding to the midplane and surface, respectively, are given in Table 2.9.1.

It is evident from Table 2.9.1 that for Bi > 0.1 and Fo ≥ 0.2, the second and subsequent terms in Equation 2.9.4 are significantly less than 1, and they contribute very little to the value of the series. Neglecting these terms introduces an error of approximately 2% and is generally acceptable in engineering calculations. However, it is more than likely that interpolating the values from the charts introduces errors greater than the error introduced by computing with one term of the series. Further, considering that the value of the convective heat transfer coefficient can be determined only within an uncertainty of ±10% in most cases, a 2% difference from the exact solution is acceptable in engineering practice. The same type of approximations can be made for cylindrical and spherical geometry. The one term approximation for Fo > 0.2 was suggested by Heisler. Considering that 80% to 90% of the heating/cooling occurs during periods of Fo > 0.2, the one term approximation is useful in a large number of cases.

The solutions to one-dimensional transient heat transfer from rectangular slabs, cylinders, and spheres have been programmed in the software (TC). The solutions are accurate for Fo ≥ 0.01 and 0.1 < Bi < 1000. (Computers without a math coprocessor take a bit of time to compute the solutions for cylinders.)

Specifying large Biot number is equivalent to specifying surface temperature

Note: Bi → ∞ is equivalent to h → ∞, implying that there is no difference between the fluid temperature and the surface temperature. The solution with Bi >> 1 (Bi ≈ 500) can be used for cases where the temperature of the surface changes at time $\tau = 0$.

TABLE 2.9.1 Values of $b_n = a_n \dfrac{\cos[\mu_n \, (x/L)]}{\cos[\mu_1 \, (x/L)]}$, Equation 2.9.4

Top rows are values for $x/L = 0$ and bottom rows are for $x/L = 1$.

	Bi	b_2	b_3	b_4	b_5	b_6
Fo = 0.1	0.1	−7.14E-3	9.45E-5	−3.03E-7	1.71E-10	−1.52E-11
		7.49E-3	9.92E-5	3.19E-7	1.79E-10	1.59E-11
	10.0	−5.99E-2	1.10E-3	−3.85E-6	2.24E-9	−2.01E-13
		1.67E-1	4.56E-3	1.95E-5	1.26E-8	1.21E-12
	100	−4.81E-2	6.00E-4	−1.29E-6	4.33E-10	−2.21E-14
		1.44E-1	2.99E-3	8.96E-6	3.86E-6	2.40E-13
Fo = 0.2	0.1	−2.63E-4	1.81E-6	−4.17E-11	2.34E-17	−2.89E-25
		2.77E-3	1.90E-6	4.38E-11	2.46E-17	3.04E-25
	10.0	−1.15E-2	7.27E-6	−1.43E-10	7.15E-17	−8.15E-25
		3.22E-2	3.01E-5	7.23E-10	4.03E-16	4.91E-24
	100.0	−6.95E-3	1.81E-6	−1.17E-11	1.71E-18	−5.45E-27
		2.08E-2	9.02E-6	8.12E-11	1.52E-17	5.91E-26

The main difference between the solutions for a plane slab (and a sphere), and a cylinder is that the solutions for a plane slab and sphere involve trigonometric functions and the solution for a cylinder involves the less common Bessel functions. The required values of the Bessel functions are tabulated in Appendix 2 and are also available in the software. For each geometry considered, the results of the one-term approximation are

$$\theta = \frac{T - T_\infty}{T_o - T_\infty} \qquad \theta_c = \frac{T_c - T_\infty}{T_o - T_\infty} = \text{center temperature}$$

■ **Plane Slab of Thickness 2L.**

$$\theta = c_1 e^{-\mu_1^2 Fo} \cos\left(\mu_1 \frac{x}{L}\right) \qquad \frac{\theta}{\theta_c} = \cos\left(\mu_1 \frac{x}{L}\right) \qquad Fo = \frac{\alpha\tau}{L^2} \qquad \textbf{(2.9.5a)}$$

■ **Cylinder (Ends Insulated) of Radius R.**

$$\theta = c_1 e^{-\mu_1^2 Fo} J_0\left(\mu_1 \frac{r}{R}\right) \qquad \frac{\theta}{\theta_c} = J_0\left(\mu_1 \frac{r}{R}\right) \qquad Fo = \frac{\alpha\tau}{R^2} \qquad \textbf{(2.9.5b)}$$

■ **Sphere of Radius R.**

$$\theta = c_1 e^{-\mu_1^2 Fo} \frac{\sin[\mu_1(r/R)]}{\mu_1(r/R)} \qquad \frac{\theta}{\theta_c} = \frac{\sin[\mu_1(r/R)]}{\mu_1(r/R)} \qquad Fo = \frac{\alpha\tau}{R^2} \qquad \textbf{(2.9.5c)}$$

For a sphere, note that as $r \to 0$, $\sin[\mu_1(r/R)]/[\mu_1(r/R)] \to 1$.

From the ratio of temperature at any location to that at the center given in each case, we may determine the value of the ratio of the temperature at the surface to the temperature at the center. For $Fo > 0.2$, θ_s/θ_c for a plane slab is given by $\cos(\mu_1)$.

From Table 2.9.2, for $Bi = 0.1$, $\mu_1 = 0.3111$ and θ_s/θ_c is 0.95, and for Bi of 0.06, μ_1 is 0.2425 and θ_s/θ_c is 0.97. Thus, as Bi gets smaller than 0.1, the difference between the surface and center line temperatures becomes less than 5%. Similar results are obtained in the case of cylinders and spheres. In such cases, lumped analysis is valid. Thus, the statement made in Section 2.8 that lumped analysis is valid if $Bi < 0.1$ is justified.

Computation of heat transfer

To find the heat transfer from the solid to the fluid, we proceed as follows: First, define Q_τ as the heat transfer from the solid to the surrounding fluid from $\tau = 0$ to any time τ. Then, for constant density, $c_p = c_v = c$, and with a reference temperature of T_r for computing the internal energy,

$$Q_\tau = \rho c \int (T_o - T_r)\, dV - \rho c \int (T - T_r)\, dV$$
$$= \rho c \int (T_o - T)\, dV$$

where V is the volume of the solid.

When the solid reaches the uniform temperature of the surroundings, the total heat transferred is the maximum and is given by $Q_o = \rho c (T_o - T_\infty)V$. We then have

$$\frac{Q_\tau}{Q_o} = \int_v \frac{T_o - T}{T_o - T_\infty} \frac{dV}{V} = \int_v (1 - \theta) \frac{dV}{V}$$

Q_τ/Q_o is evaluated by introducing the expressions for θ and V for each of the three cases. Knowing $Q_o = \rho c (T_o - T_\infty) V$, Q_τ can then be evaluated. The solutions with the one-term approximation (valid for Fo $>$ 0.2) follow.

■ **Plane Slab.**

$$\frac{dV}{V} = \frac{A_x dx}{A_x 2L} = \frac{dx}{2L}$$

Introducing Equation 2.9.5a for θ,

$$\frac{Q_\tau}{Q_o} = \int_{-1}^{1} (1 - c_1 e^{-\mu_1^2 Fo} \cos(\mu_1 \eta)) \frac{d\eta}{2} \qquad \text{where } \eta = \frac{x}{L}$$

$$\frac{Q_\tau}{Q_o} = 1 - \frac{c_1 e^{-\mu_1^2 Fo} \sin \mu_1}{\mu_1} \tag{2.9.6a}$$

■ **Cylinder.**

$$\frac{dV}{V} = \frac{2\pi rL\, dr}{\pi R^2 L} = 2 \frac{r\, dr}{R^2}$$

With θ given by Equation 2.9.5b, substituting $\eta = r/R$,

$$\frac{Q_\tau}{Q_o} = 2 \int_0^1 [1 - c_1 e^{-\mu_1^2 Fo} J_0(\mu_1 \eta)]\, \eta\, d\eta$$

With $z = \mu_1 \eta$,

$$\int_0^{\mu_1} z J_0(z)\, dz = z J_1(z) \Big|_0^{\mu_1} = \mu_1 J_1(\mu_1)$$

$$\frac{Q_\tau}{Q_o} = 1 - 2 \frac{c_1 e^{-\mu_1^2 Fo} J_1(\mu_1)}{\mu_1} \tag{2.9.6b}$$

■ **Sphere.**

$$\frac{dV}{V} = \frac{4\pi r^2\, dr}{(4/3)\pi\, R^3} = \frac{3r^2\, dr}{R^3}$$

With θ given by Equation 2.9.5c,

$$\frac{Q_\tau}{Q_o} = \int_0^R \left[1 - \frac{c_1 e^{-\mu_1^2 Fo} \sin[\mu_1 (r/R)]}{\mu_1 (r/R)} \right] 3 \frac{r^2}{R^3}\, dr$$

Changing the variable of integration to $\eta = r/R$ and performing the integration, we obtain

$$\frac{Q_\tau}{Q_o} = 1 - \frac{3c_1 e^{-\mu_1^2 Fo}}{\mu_1^3} (\sin \mu_1 - \mu_1 \cos \mu_1) \qquad \textbf{(2.9.6c)}$$

Values of μ_1 and c_1 to be used in Equations 2.9.5 and 2.9.6 are given in Table 2.9.2 for different values of Bi. Functions $J_0(x)$ and $J_1(x)$ are given in Appendix 2 and are available from the software. For values of Bi < 0.1, we can use lumped analysis.

Equations 2.9.5a, b, and c, and Equations 2.9.6a, b, and c give the approximate temperature distribution and the heat transferred for Fo > 0.2. These are obtained by using the first term of the exact series solution for the temperature distribution. Chen and Kuo (1979) have given an approximate solution by an integral method, which is presented below (see Chapter 3 for an introduction to integral methods). For brevity, only the temperature profile for Fo > 0.2 and the heat transferred for Fo > 0.1 are given here. The nomenclature for the approximate solution is

T_o = uniform initial temperature of solid
T_c = center line temperature
T_∞ = fluid temperature
k = thermal conductivity of solid
$Q_o = mc_p(T_o - T_\infty)$

Approximate integral solution to one-dimensional transient temperature distribution

■ **Plane Slab.** $\eta = x/L$, Bi = hL/k, $B = 1/$Bi, and Fo = $\alpha\tau/L^2$.

Fo > 0.2
$$\frac{T_c - T_\infty}{T_o - T_\infty} = \exp\left(\frac{a - Fo}{0.35 + B + \exp(-4B)}\right) \qquad \textbf{(2.9.7a)}$$

$$\frac{T(\eta) - T_\infty}{T_o - T_\infty} = 1 - \frac{(1 + c)\eta^2 - c\eta^3}{1 + (2 - c)B} \qquad \textbf{(2.9.7b)}$$

$$a = 0.167 - \frac{0.067}{1 + 6B} \qquad c = \frac{0.315}{1 + 2.5B}$$

Fo > 0.1: $$\frac{Q}{Q_o} = 1 - \frac{0.63 + 2B}{1 + 2B} \exp\left(\frac{a - Fo}{0.35 + B + \exp(-4B)}\right) \qquad \textbf{(2.9.7c)}$$

■ **Cylinder.** $\rho = r/R$, Bi = hR/k, $B = 1/$Bi, and Fo = $\alpha\tau/R^2$.

Fo > 0.2:
$$\frac{T_c - T_\infty}{T_o - T_\infty} = \exp\left(\frac{a - Fo}{0.13 + 0.5B + 0.04 \exp(-2B)}\right) \qquad \textbf{(2.9.8a)}$$

$$\frac{T(\rho) - T_\infty}{T_o - T_\infty} = 1 - \frac{(1 + c)\rho^2 - c\rho^3}{1 + (2 - c)B} \qquad \textbf{(2.9.8b)}$$

$$a = 0.14 - \frac{0.056}{1 + B} \qquad c = \frac{0.595}{1 + 3B}$$

Fo > 0.1: $$\frac{Q}{Q_o} = 1 - \frac{0.42 + 2B}{1 + 2B} \exp\left(\frac{a - Fo}{0.13 + 0.5B + 0.04 \exp(-2B)}\right) \qquad \textbf{(2.9.8c)}$$

TABLE 2.9.2 Values of c_1 and μ_1 in Equations 2.9.5 and 2.9.6.

For a plane slab, $Bi = (hL)/k$, where L = half thickness of slab. For a cylinder and a sphere, $Bi = (hR)/k$, where R = radius. In all cases, μ_1 is in radians.

Bi	Plane Slab		Cylinder		Sphere	
	μ_1	c_1	μ_1	c_1	μ_1	c_1
0.01	0.0998	1.0017	0.1412	1.0025	0.1730	1.0030
0.02	0.1410	1.0033	0.1995	1.0050	0.2445	1.0060
0.04	0.1987	1.0066	0.2814	1.0099	0.3450	1.0120
0.06	0.2425	1.0098	0.3438	1.0148	0.4217	1.0179
0.08	0.2791	1.0130	0.3960	1.0197	0.4860	1.0239
0.1	0.3111	1.0161	0.4417	1.0246	0.5423	1.0298
0.2	0.4328	1.0311	0.6170	1.0483	0.7593	1.0592
0.3	0.5218	1.0450	0.7465	1.0712	0.9208	1.0880
0.4	0.5932	1.0580	0.8516	1.0931	1.0528	1.1164
0.5	0.6533	1.0701	0.9408	1.1143	1.1656	1.1441
0.6	0.7051	1.0814	1.0184	1.1345	1.2644	1.1713
0.7	0.7506	1.0918	1.0873	1.1539	1.3525	1.1978
0.8	0.7910	1.1016	1.1490	1.1724	1.4320	1.2236
0.9	0.8274	1.1107	1.2048	1.1902	1.5044	1.2488
1.0	0.8603	1.1191	1.2558	1.2071	1.5708	1.2732
2.0	1.0769	1.1785	1.5995	1.3384	2.0288	1.4793
3.0	1.1925	1.2102	1.7887	1.4191	2.2889	1.6227
4.0	1.2646	1.2287	1.9081	1.4698	2.4556	1.7202
5.0	1.3138	1.2403	1.9898	1.5029	2.5704	1.7870
6.0	1.3496	1.2479	2.0490	1.5253	2.6537	1.8338
7.0	1.3766	1.2532	2.0937	1.5411	2.7165	1.8673
8.0	1.3978	1.2570	2.1286	1.5526	2.7654	1.8920
9.0	1.4149	1.2598	2.1566	1.5611	2.8044	1.9106
10.0	1.4289	1.2620	2.1795	1.5677	2.8363	1.9249
20.0	1.4961	1.2699	2.2880	1.5919	2.9857	1.9781
30.0	1.5202	1.2717	2.3261	1.5973	3.0372	1.9898
40.0	1.5325	1.2723	2.3455	1.5993	3.0632	1.9942
50.0	1.5400	1.2727	2.3572	1.6002	3.0788	1.9962
100.0	1.5552	1.2731	2.3809	1.6015	3.1102	1.9990
∞	1.5708	1.2732	2.4048	1.6021	3.1416	2.0000

■ **Sphere.** $\rho = r/R$, $Bi = hR/k$, $B = 1/Bi$, and $Fo = \alpha\tau/R^2$.

$$Fo > 0.2: \qquad \frac{T_c - T_\infty}{T_o - T_\infty} = \exp\left(\frac{3(a - Fo)}{0.2 + B + 0.11\exp(-1.5B)}\right) \qquad (2.9.9a)$$

$$\frac{T(\rho) - T_\infty}{T_o - T_\infty} = 1 - \frac{(1 + c)\rho^2 - c\rho^3}{1 + (2 - c)B} \qquad (2.9.9b)$$

$$a = 0.14 - \frac{0.08}{1 + B} \qquad c = \frac{0.838}{1 + 3B}$$

$$\text{Fo} > 0.1: \quad \frac{Q}{Q_o} = 1 - \frac{0.36 + 2B}{1 + 2B} \exp\left(\frac{3(a - \text{Fo})}{0.2 + B + 0.11 \exp(-1.5B)}\right) \quad \textbf{(2.9.9c)}$$

The approximate solutions have also been programmed in the software (TC).

EXAMPLE 2.9.1

Consider a potato wrapped in an aluminum foil and baked in an oven maintained at 204.5 °C (Figure 2.9.5). It is required to determine the maximum temperature difference within the potato 30 min and 60 min after it is put inside the oven. The potato is initially at a uniform temperature of 21 °C. The potato is approximately in the form of a cylinder 8-cm diameter and 16-cm long. The convective heat transfer coefficient is 12 W/m² K.

Given

Dimensions $2R = 8$ cm and $2L = 16$ cm

Initial temperature of potato $(T_o) = 21$ °C

Temperature of surrounding air $(T_\infty) = 204.5$ °C

Convective heat transfer coefficient $(h) = 12$ W/m² °C

Find

The maximum temperature difference in the potato 30 min and 60 min after it is put inside the oven.

ASSUMPTIONS

1. As the potato is wrapped in an aluminum foil, probably bright, radiative heat transfer is assumed to be small and that it has been approximately compensated by modifying the value of the convective heat transfer coefficient.
2. To get an appreciation for the temperature variation in the potato, the potato is modeled as a cylinder with its end surfaces insulated. The solution for the one-dimensional, transient temperature distribution is employed. The effect of heat transfer from the end surfaces will be examined in Example 2.9.2
3. Constant temperature of the air in the oven and constant convective heat transfer coefficient
4. Constant properties

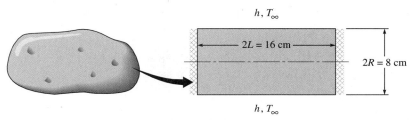

FIGURE 2.9.5 A Potato Modeled as a Cylinder with Insulated End Surfaces

SOLUTION

The maximum difference in the temperature is the difference in the temperatures at a point on the axis of the cylinder and a point on the surface. Properties of potatoes (derived from values available in the ASHRAE *Handbook of Fundamentals,* 1989) are thermal conductivity = 0.48 W/m K and thermal diffusivity = 1.33×10^{-7} m^2/s.

SOLUTION

Check if lumped analysis is suitable.

$$\mathrm{Bi} = \frac{hR}{k} = \frac{12 \times 0.04}{0.48} = 1$$

As the Biot number is not much less than 1, lumped analysis is inappropriate.

The temperatures can be obtained by calculating the various quantities that are required for using the charts in Fig. 2.9.3. In the table below the symbol ρ represents the dimensionless radial distance r/R. The values from the software (TC) and from the one-term approximation for Fo > 0.2 are given in the table. From Table 2.9.2, for Bi = 1,

$$\mu_1 = 1.2558$$
$$c_1 = 1.2071$$

Values of $J_0(\mu_1\rho)$ are available in Table A12 and in the software.

	From Software (or Figure 2.9.3)		From Equation 2.9.5b
	1800 s	3600 s	3600 s
$\mathrm{Fo} = \dfrac{\alpha\tau}{R^2}$	0.1496	0.2993	0.2993
$\theta = \dfrac{T - T_\infty}{T_\mathrm{o} - T_\infty}$ $(r = R, \rho = 1)$	0.6224	0.4849	0.4841
$\theta(r = 0, \rho = 0)$	0.9294	0.7509	0.7529
$T_\mathrm{s}(T$ at $r = R)\,°\mathrm{C}$	90.3	115.5	115.7
$T_\mathrm{c}(T$ at $r = 0)\,°\mathrm{C}$	34.0	66.7	66.3

COMMENTS

The solution illustrates the considerable difference between the temperatures at the surface and at points in the interior of the potato. This difference is due to the high value of the Bi. In this case, lumped analysis would not be valid. The ease with which the one-term approximation yields results for Fo > 0.2, compared with the use of the charts should also be obvious.

2.9.2 Application to Multidimensional Problems

The charts developed for one-dimensional temperature distribution can be extended to multidimensional problems for rectangular blocks and cylinders with heat transfer from all the surfaces to a surrounding fluid. The dimensionless temperature is expressed as the product of two or three dimensionless temperatures of one-dimensional problems. Consider the two-dimensional temperature distribution in a long, rectangular bar of cross-sectional dimensions $2L \times 2W$. The procedure for finding the two-dimensional temperature distribution follows.

Use of one-dimensional transient temperature distribution solution to multidimensional, transient temperature distribution problems Defining the dimensionless temperature $\theta = (T - T_\infty / T_o - T_\infty)$, the problem is split up into two problems as indicated in Figure 2.9.6. θ is now expressed as the product of dimensionless temperatures of two one-dimensional temperatures.

$$\theta = \theta_1 \left(\frac{x}{L}, \frac{\alpha\tau}{L^2}, \frac{h_1 L}{k} \right) \times \ \theta_2 \left(\frac{y}{W}, \frac{\alpha\tau}{W^2}, \frac{h_2 W}{k} \right)$$

θ_1 and θ_2 are found from the charts, from the software (TC) or from the one-term approximation (depending on the value of Fo). From this value of θ we determine the temperature T. The values of the convective heat transfer coefficients with each pair of parallel surfaces must be equal but may be different for different pairs. Note that h_1 and h_2 can have different values but that T_∞ must have the same value everywhere.

The use of one-dimensional temperature distribution solutions or charts can be extended to three-dimensional problems in Cartesian geometry and two-dimensional problems in a cylinder, where both the cylindrical and end surfaces participate in heat transfer. For example, if it is desired to determine the temperature distribution in a cylinder with all the surfaces exposed to a fluid maintained at a constant temperature (Figure 2.9.7—see following page), we proceed as follows.

Define θ_1 as the dimensionless temperature of a semi-infinite slab of thickness $2L$ and θ_2 as the dimensionless temperature of a semi-infinite cylinder of radius R. Both

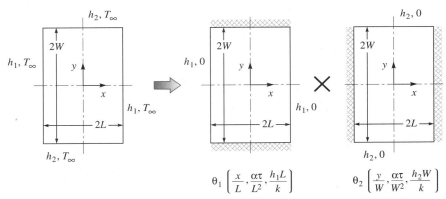

FIGURE 2.9.6 A Slab with Four Surfaces Exposed to a Fluid Considered as a Combination of Two Slabs
Only two parallel surfaces of each slab are exposed to the fluid.

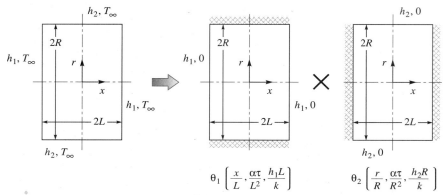

FIGURE 2.9.7 A Cylinder Exposed to a Fluid on all Sides Split into Two Cylinders
A cylinder with its cylindrical surface insulated (treated as a slab) and a cylinder with its end surfaces insulated.

the slab and cylinder are at a dimensionless initial temperature of 1 and exposed to convective heat transfer with Bi $= (h_1L)/k$ for the slab and $(h_2R)/k$ for the cylinder. If θ_1 is the solution to the one-dimensional transient dimensionless temperature distribution in a rectangular slab,[4] and θ_2 is the solution to the one-dimensional temperature distribution in a cylinder with insulated end surfaces, the final dimensionless temperature is the product of these two values, $\theta = \theta_1 \times \theta_2$.

In finding θ_1 and θ_2, the dimensionless distance, Fo and Bi must be evaluated separately. The procedure is illustrated through an example.

EXAMPLE 2.9.2

Reconsider Example 2.9.1 and include the effect of heat transfer from the end surfaces. Heating starts and after 3600 s the temperatures at two points on the surface, point A located on the cylindrical surface of one of the ends and point B located on the surface at the midplane, will be determined. See Figure 2.9.8.

Given

Dimensions: $2R = 8$ cm and $2L = 16$ cm

Initial temperature of potato $(T_o) = 21\ °C$

Surrounding air temperature $(T_\infty) = 204.5\ °C$

Convective heat transfer coefficient $(h) = 12\ \text{W/m}^2\ °C$

Find

The temperatures at two points, one on the periphery of an end surface and the other on the cylindrical surface at the midplane.

ASSUMPTION

The convective heat transfer coefficients associated with the cylindrical surface is likely to be different from the convective heat transfer coefficient associated with

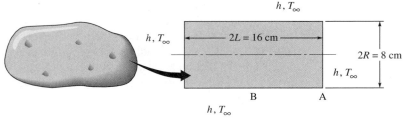

FIGURE 2.9.8 **The Potato in Example 2.9.1, Modeled as a Cylinder with All Its Surfaces Exposed to Hot Air in the Oven**

the end surfaces. As no specific information on the different values is available, it is assumed that the convective heat transfer coefficient has the same value on all surfaces.

SOLUTION

The problem is similar to that of Figure 2.9.7, where $R = 4$ cm and $L = 8$ cm. The final dimensionless temperature is $\theta = \theta_1 \times \theta_2$, where θ_1 is the dimensionless temperature of an infinitely long cylinder of radius 4 cm and θ_2 is the dimensionless temperature of a 16-cm thick semi-infinite slab.

From Example 2.9.1, for points A and B:

$$\rho_A = \rho_B = 1, \qquad \frac{\alpha\tau}{R^2} = 0.2993, \qquad \text{Bi} = \frac{hR}{k} = 1, \qquad \theta_{1,A} = \theta_{1,B} = 0.4849$$

The dimensionless temperature θ_2 at A and B are found in a manner similar to that in Example 2.9.1. To find θ_2,

$$\text{Bi} = \frac{hL}{k} = \frac{12 \times 0.08}{0.48} = 2$$

$$\frac{\alpha\tau}{L^2} = \frac{1.33 \times 10^{-7} \times 3600}{0.08^2} = 0.07481$$

$$\eta_A(=x/L) = 1, \text{ from the software (TC), } \theta_{2,A} = 0.5924$$

$$\eta_B = 0, \text{ from the software, } \theta_{2,B} = 0.9961$$

$$\theta_A = \theta_{1,A} \times \theta_{2,A} = 0.4849 \times 0.5924 = 0.2873$$

$$T_A = T_\infty - \theta_A \times (T_\infty - T_i) = 151.8\,°C$$

$$\theta_B = \theta_{1,B} \times \theta_{2,B} = 0.4849 \times 0.9961 = 0.483$$

$$T_B = T_\infty - \theta_B \times (T_\infty - T_i) = 115.9\,°C$$

COMMENT

Comparing these values with those obtained in Example 2.9.1, we can easily see the effect of heat transfer from the end surfaces on the temperature distribution on the cylindrical surface.

In using the charts for the solution of transient problems, you should ensure that all the assumptions made in obtaining the charts are reasonably satisfied. One of the assumptions is that the values of h and T_∞ are constant. It is unlikely that either of these will be strictly constant. Compared with an appropriate characteristic value, the changes in their values should not be significant. For example, if the change Δh in the value of the convective heat transfer coefficient compared with the initial (or final) value of h is small, we may treat it as being constant at some average value. Similarly, one may compare the change in the value of T_∞ with the difference in the initial temperatures of the solid and the fluid. Another factor to be satisfied is that in the actual case, the surfaces exposed to the fluid are the same as those for which the charts are developed. If only parts of a surface are exposed to the fluid, the charts or the solutions are not applicable. As already indicated, it is not necessary for the convective heat transfer coefficient to have the same value on all the surfaces. It is sufficient if the value of h is uniform for each simple problem. However, the temperature of the fluid adjacent to all surfaces must have the same value. Another important assumption is the there is no internal energy generation.

A word of caution. In employing the charts for the solution of multidimensional problems, the solutions found for one-dimensional problems can be multiplied only in dimensionless form to obtain the dimensionless temperature for the actual case. This dimensionless temperature can then be used to find the actual temperature.

2.10 PROJECTS

PROJECT 2.10.1

A 2500 W electrical heater is to be fabricated by sandwiching the heating element between two steel plates. The plates are 36-cm wide, 36-mm high and 2-mm thick. Air is forced parallel to the plate with a velocity of 6 m/s and the convective heat transfer coefficient is 12 W/m^2 °C. The plate temperature is to be no more than 100 °C when the temperature of the air at inlet is 20 °C.

The heat transfer rate from the plates is 248 W ($0.36 \times 0.36 \times 12 \times 80 \times 2$). To realize the heat transfer rate of 2500 W, it is proposed to add cylindrical fins to the plate (Figure 2.10.1). Aluminum and copper rods are available in diameters of 3 mm, 6 mm, and 10 mm. With the addition of fins, the average heat transfer coefficient is 60 W/m^2 °C. Design a system of fins with consideration for the volume and weight of the heater.

ASSUMPTIONS

1. The heater is completely enclosed with side plates, which are insulated. The clearance between the side plates and the fins is very small and the air flow through the clearance is also small. Therefore, the heat transfer from the end surfaces of the fins is negligible.

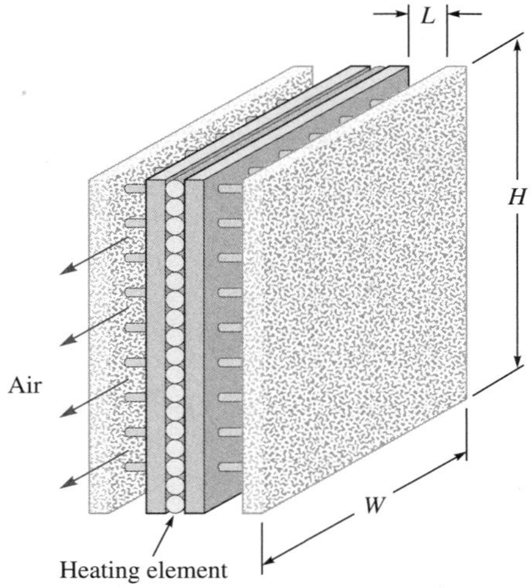

FIGURE 2.10.1 Conceptual Design of a Heater

2. The center-to-center spacing of the fins is $3d$ (d is the diameter) so that there is a minimum clear space of $2d$ between two adjacent fins.
3. The convective heat transfer coefficient has the same value for the fins and the unfinned part of the plates.

SOLUTION METHODOLOGY

Starting with one of the diameters of the rods, the heat transfer rate from one fin is given by Equation 2.6.17.

$$q_f = (hPkA_c)^{1/2} \, \theta_b \tanh mL$$

where

h = convective heat transfer coefficient
P = perimeter (πd)
k = thermal conductivity of the material of the fin
A_c = cross-sectional area ($\pi d^2/4$)
θ_b = the difference between the temperatures at the base and the air
m = $(hP)/(kA_c)^{1/2} = 4h/kd^{1/2}$

From the discussion in Section 2.6.5, the length of the fin should be such that the value of mL is less than 2.

Denoting the number of fins on each plate by n, the total heat transfer rate from the fins, q_{ft} is

$$q_{ft} = nq_f = n(hPkA_c)^{1/2} \, \theta_b \tanh mL$$

The heat transfer rate from the unfinned part of each plate q_{uf} is

$$q_{uf} = (A - n\pi d^2/4)h\,\theta_b$$

where A is area of each plate (WH).

The total heat transfer rate from both plates, q, which should be equal to the capacity of the heater, (2500 W) is

$$q = 2[(A - n\pi d^2/4)h\,\theta_b + n(hPkA_c)^{1/2}\,\theta_b \tanh mL]$$

Solving for n, we obtain

$$n = \frac{(q/2\theta_b) - Ah}{(hPkA_c)^{1/2}\tanh mL - h\pi d^2/4} \tag{1}$$

The mass of the fins is

$$M = n\,\frac{\pi d^2}{4}\,L\rho \tag{2}$$

The volume of the heater is directly proportional to the length of the fins. The number and mass of fins are calculated from Equations 1 and 2. The computations are repeated for the different diameters and materials.

As the air flows through the heater, the temperature of the air, and therefore, the temperature of the plates vary. For our computations we will assume that the heat flux (based on the plate area) is uniform and, therefore, θ_b is also uniform. We may use an appropriate value of θ_b recognizing that the air temperature increases towards the exit of the heater. The temperature of the air at exit is not yet available as the mass rate of flow of air is not known. We will assume a value of 40 °C for θ_b, perform the computations and check if the assumed value is appropriate.

The values computed from Equations 1 and 2 are given in Figure 2.10.2 and Table 2.10.1.

From Figure 2.10.1 and Table 2.10.1 it is observed that a large number of fin configurations satisfy the requirements. However, one would like to keep the volume

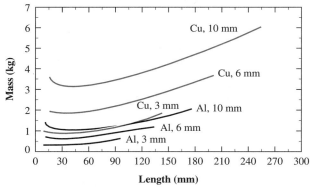

FIGURE 2.10.2 **Mass of Fins for Various Diameters of Aluminum and Copper Fins**

TABLE 2.10.1 Length, number, and mass of fins on each plate

Aluminum: Thermal conductivity = 177 W/m K   Density = 2770 kg/m³

$\frac{m}{mL}$	d = 3 mm 21.26m⁻¹			d = 6 mm 15.03m⁻¹			d = 10 mm 11.64m⁻¹		
	L(mm)	n	M(kg)	L(mm)	n	M(kg)	L(mm)	n	M(kg)
0.10	4.70	3525	0.32	6.65	1353	0.70	8.59	687	1.28
0.20	9.41	1626	0.30	13.30	597	0.62	17.18	288	1.08
0.30	14.11	1072	0.30	19.96	388	0.61	25.76	185	1.04
0.40	18.81	811	0.30	26.61	292	0.61	34.35	138	1.03
0.50	23.52	661	0.30	33.26	237	0.62	42.94	112	1.05
0.60	28.22	566	0.31	39.91	203	0.63	51.53	95	1.07
0.70	32.93	501	0.32	46.56	179	0.65	60.11	84	1.10
0.80	37.63	455	0.34	53.22	163	0.68	68.70	76	1.14
0.90	42.33	421	0.35	59.87	150	0.71	77.29	71	1.19
1.00	47.04	396	0.36	66.52	141	0.74	85.88	66	1.24
1.20	56.44	361	0.40	79.82	129	0.80	103.05	60	1.35
1.40	65.85	339	0.44	93.13	121	0.88	120.23	57	1.48
1.60	75.26	326	0.48	106.43	116	0.97	137.40	54	1.62
1.80	84.67	317	0.53	119.74	113	1.06	154.58	53	1.78
2.00	94.07	311	0.57	133.04	111	1.15	171.76	52	1.94

Copper: Thermal conductivity = 401 W/m K   Density = 8933 kg/m³

$\frac{m}{mL}$	d = 3 mm 14.12m⁻¹			d = 6 mm 9.99m⁻¹			d = 10 mm 7.74m⁻¹		
	L(mm)	n	M(kg)	L(mm)	n	M(kg)	L(mm)	n	M(kg)
0.10	7.08	2201	0.98	10.01	818	2.07	12.93	401	3.64
0.20	14.16	1050	0.94	20.02	380	1.92	25.85	181	3.28
0.30	21.24	698	0.94	30.04	251	1.90	38.78	118	3.22
0.40	28.32	531	0.95	40.05	190	1.92	51.70	89	3.24
0.50	35.40	434	0.97	50.06	155	1.96	64.63	73	3.30
0.60	42.48	372	1.00	60.07	133	2.01	77.56	62	3.39
0.70	49.56	330	1.03	70.09	118	2.08	90.48	55	3.50
0.80	56.64	300	1.07	80.10	107	2.16	103.41	50	3.63
0.90	63.72	278	1.12	90.11	99	2.25	116.33	46	3.77
1.00	70.80	261	1.17	100.12	93	2.35	129.26	43	3.94
1.20	84.96	238	1.28	120.15	85	2.57	155.11	40	4.31
1.40	99.12	224	1.40	140.17	80	2.82	180.97	37	4.72
1.60	113.28	215	1.54	160.20	76	3.09	206.82	36	5.18
1.80	127.44	209	1.69	180.22	74	3.39	232.67	35	5.67
2.00	141.60	206	1.84	200.25	73	3.69	258.52	34	6.19

and mass of the heater as low as possible. With a minimum center-to-center distance of $3d$, a total of 1600 fins of 3-mm diameter, 400 fins of 6-mm diameter, and 144 fins of 10-mm diameter (all of aluminum) can be accommodated on each plate. From Figure 2.10.1 it is also observed that 3-mm diameter aluminum fins have both low volume and mass. Thus, an appropriate choice of fins is

Material	Aluminum
Diameter	3 mm
Length	9.4 mm
Number of fins	1600
Mass of fins	0.3 kg

The actual number of fins required is 1625 but only 1600 have been suggested to obtain a square array of 40 × 40 fins and a center to center distance of 9 mm. A value of θ_b of 40 °C has been used. The air temperature increases in the direction of flow, and this increase in the air temperature results in an increase in the plate temperature. The maximum plate temperature occurs at exit. To find the maximum air temperature, the exit air temperature is needed. We now proceed to determine the exit temperature of the air. The properties of air at 20° C are

Density	1.204 kg/m^3
Specific heat (c_p)	1007 J/kg °C

The heat transfer rate to the air is

$$q = \dot{m}c_p(T_e - T_i)$$

where

$$\dot{m} = \text{mass rate of flow of air (kg/s)} = \rho V A$$
$$c_p = \text{specific heat at constant pressure (J/kg °C)}$$
$$T_e, T_i = \text{temperatures of air at exit of and inlet to the heater}$$

$$\dot{m} = 1.204 \times 6 \times 0.36 \times 0.0094 \times 2 = 0.04889 \text{ kg/s}$$

$$T_e = \frac{2500}{0.04889 \times 1007} + 20 = 70.8°C$$

With an exit temperature of air of 70.8 °C, the maximum plate temperature is 110.8 °C, which is higher than the specified temperature of 100 °C. To reduce the temperature of the plate to 100 °C, the exit temperature of air should be reduced by increasing the mass flow rate of the air without increasing the velocity of air. An increase in the mass flow rate requires that the length of the fins be increased such that the exit temperature of the air is 60 °C. To reduce the temperature rise of the air from 70.8 °C to 60 °C, the mass rate of flow and, therefore, the cross-sectional area for flow, should be increased by a factor of 70.8/60. This leads to an increase in the length of the fins to 11.1 mm. With an increase in the length of the fins, the heat transfer rate from each fin increases, and the total heat transfer rate also increases if the plate temperature is 40 °C higher than the air temperature. However, with the heat transfer rate fixed at 2500 W, the effect of increasing the length of the fins is to reduce the plate temperature. The plate temperature is computed from Equation 1,

$$\theta_b = \frac{q/2}{\{[Ah + n[(hPkA_c)^{1/2} \tanh mL - \pi d^2 h/4]\}}$$

Corresponding to the new value of 0.234 for mL, and a total of 1600 fins, the value of θ_b is 74 °C, which is slightly less than the prescribed maximum value of 80 °C. The given maximum value is a target and some leeway is permitted in the design of equipment. The final design is

Material	Aluminum
Diameter of fins	3 mm
Length of fins	11.1 mm
Number of fins	1600 on each plate
Velocity of air	6 m/s
Mass rate of flow of air	0.04889 kg/s
Inlet air temperature	20 °C
Exit air temperature	60 °C
Maximum plate temperature	94 °C
Mass of fins	0.7 kg

COMMENTS

Another factor to be considered is the fabrication costs. Fabrication costs can be decreased by reducing the number of fins. Therefore, in the final design one has to consider the mass, weight, and number of fins. Possible choices of reduced number of fins are

	3 mm	6 mm
Diameter	3 mm	6 mm
Number of fins	529 (23 × 23 matrix)	256 (16 × 16 matrix)
Approximate length of fins	30 mm	33 mm
Mass of fins	0.62 kg	1.32 kg

Note that if weight is the prime consideration, aluminum is preferable to copper in all cases, although the thermal conductivity of copper is higher than that of aluminum. Because of the significantly higher density of copper, the weight of copper fins is higher.

PROJECT 2.10.2 Propose a simple experiment to determine an approximate value of the thermal conductivity of a metallic alloy. There are several ways to determine the thermal conductivity of a solid. A few of the ways are described below. The suggested projects are simple and can be fabricated and used in the laboratory.

In some of the suggested experiments the thermal conductivity is found by determining the thermal diffusivity. To find the thermal conductivity from the value of the thermal diffusivity, the density and specific heat of the material are needed. The density of the material is found easily by finding the mass and volume of a sample of the substance. The volume of the sample is found by immersing it in water in a graduated jar and noting the increase in the volume of the water.

To find the specific heat of the alloy, a sample of the metal is heated uniformly to a known temperature. For example, the sample may be immersed in boiling water for sufficient time to ensure that the sample is at the same temperature as the water. It is then immersed in a known mass of water in a light, insulated metal container (Figure 2.10.3). The initial temperature of the water is measured. With the top of the container covered with a light insulating material, the water is continuously stirred and the rise in the temperature of the water monitored. The maximum, steady temperature of the water is noted. From an energy balance we have

$$mc(T_i - T_f) = (m_c c_c + m_w c_w)(T_f - T_\infty)$$

where

m = mass of the metal sample
c = specific heat of the metal sample
T_i = initial temperature of the sample
T_f = final equilibrium temperature of the metal and water
m_c = mass of the container
c_c = specific heat of the material of the container
m_w = mass of water
c_w = specific heat of water
T_∞ = initial uniform temperature of the water and the container

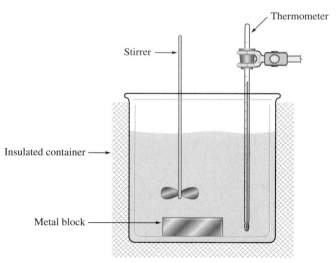

FIGURE 2.10.3 **Apparatus to Determine the Specific Heat of the Metal Alloy**

By starting with a sufficiently large mass of water, the increase in the temperature of the water can be limited so that the required mass of insulation material on the container (to reduce the heat transfer rate to the surroundings is small. However, the mass of water should be such that the increase in the temperature of the water can be measured accurately.

Having determined the specific heat of the material, we now briefly describe four methods for determining the thermal conductivity of the material.

1. Make use of the one-dimensional temperature distribution in a solid with a known heat transfer rate. Cartridge electrical heaters (cylindrical) are available in various sizes and capacities. If the available cartridge heater is 9 mm in diameter and 15 cm long, obtain a 15 cm long, hollow cylinder of the metal alloy with an inner diameter of 9 mm (so that the heater fits snugly into the hollow cylinder) and an outer diameter of approximately 50 mm. Install thermocouples (TC) at radial distances of 12 mm, 17 mm, and 22 mm by drilling axial holes, as shown in Figure 2.10.4, so that the thermocouple junctions are at least 4 cm from the end surface.

Insert the cartridge heater and insulate the end surfaces of the heater and the cylinder. Suspend the assembly in air with an enclosure to reduce movement of air around the heated cylinder. Regulate the power supply to the heater so that the temperature in the cylinder is around 60 °C–80 °C. When steady-state temperature has been reached, measure the power input to the heater and the temperatures at the three locations in the cylinder. Determine the thermal conductivity of the metal from the relation

$$q_r = \frac{2\pi k L (T_i - T_{i+1})}{\ln(r_{i+1}/r_i)}$$

where

$$q_r = \text{power supply to the heater}$$
$$T_{i+1}, T_i = \text{temperatures at radii } r_{i+1} \text{ and } r_i$$

With measured temperatures at three radii, we can obtain three values of the thermal conductivity from the three pairs of temperatures. The mean of the three values is the approximate value of the thermal conductivity of the material.

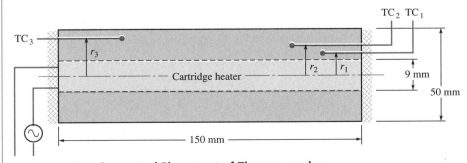

FIGURE 2.10.4 Suggested Placement of Thermocouples

The accuracy of the measured value depends on the accuracy of the measured distances between the thermocouple junctions, the measured temperatures, and the power supply to the heater.

2. Make use of the temperature distribution in an extended surface. The temperature distribution in an extended surface (Figure 2.10.5) is given by

$$\theta = c_1 e^{-mx} + c_2 e^{mx}$$

$$\theta = T - T_\infty \qquad m = \left(\frac{hP}{kA_x}\right)^{1/2}$$

If mL is greater than about 5, the fin behaves as an infinitely long one, and with θ at the base ($x = 0$) equal to θ_b

$$\theta = \theta_b e^{-mx}$$

If the convective heat transfer coefficient is known, then by determining the temperature distribution in a fin of the material at several locations, the value of m and k can be computed.

A simple geometry for the fin is a cylindrical rod of diameter approximately 5 mm and 50-cm long. Install about 10 thermocouples at intervals of 2 cm from one end of the rod. The rod is to be used as a fin in still air.

To determine the convective heat transfer coefficient associated with the horizontal rod surface, we use lumped transient analysis. Heat the rod to a uniform temperature of about 100 °C and cool it by freely suspending it in still air at room temperature of, say, 20 °C. To reduce the uncertainty in the value of the convective heat transfer coefficient due to air motion caused by moving people, opening of doors, and so on, enclose the rod in a large box as shown in Figure 2.10.6. Record the temperature of the rod and the surrounding air at several locations as a function of time.

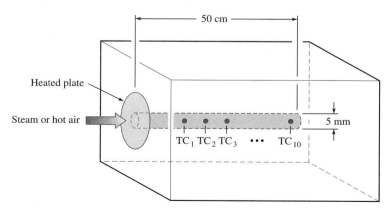

FIGURE 2.10.5 Using an Extended Surface to Determine the Thermal Conductivity

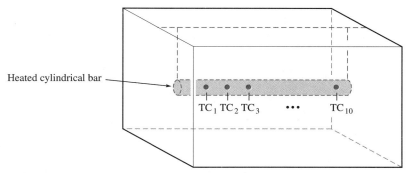

FIGURE 2.10.6 A Horizontal Rod Enclosed in a Large Box To Determine the Convective Heat Transfer Coefficient

From lumped analysis we have

$$mc\frac{dT}{d\tau} = -hA_s(T_s - T_\infty)$$

As the end-surface area is only 1.6% of the cylindrical surface area the error in determining the heat transfer coefficient by lumping the end surfaces and the cylindrical surface is quite small. Ensure that the air temperature is substantially constant. From the measured values of the temperature of the rod, the time rate of change of temperature of the rod at several values of T_s are calculated. Note that the value of the heat transfer coefficient determined in this manner includes the effect of radiative heat transfer. It is assumed that the specific heat of the material has already been determined.

Attach the rod to a metal disk approximately 3 cm in diameter and suspend the rod in the box previously used to determine the heat transfer coefficient. Heat the metal plate to approximately 100 °C. The heating may be accomplished by using steam or a hot air blower. Even a commercially available hair drier is satisfactory. You should ensure that the air inside the box is essentially free from disturbances. Record the temperatures until steady state is reached. A plot ln θ against x should produce a straight line as shown in Figure 2.10.7. Find the slope of the line that yields the value of m $[=(4h/kd)^{1/2}]$. From the measured values of h as a function of the temperature of the rod, find a reasonable average value of the convective heat

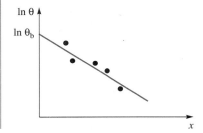

FIGURE 2.10.7 Temperature versus Distance from the Base

transfer coefficient over the temperature range in the rod. Employing the average value of h, determine the thermal conductivity.

3. Use two extended surfaces. An alternative method is to have two cylindrical rods of identical dimensions, say 5-mm diameter and 50-cm long, one rod made of the metal whose thermal conductivity is to be determined and the other rod made of a material whose thermal conductivity is known. Install two thermocouples at approximately the same locations in the two rods, say about 4 cm and 7 cm from one end of the rods. Attach a 3-cm diameter metal plate to the end closest to the thermocouples. Suspend the two rods in a horizontal position inside a large box of wood or some other material with a very low thermal conductivity, with the metal disk outside the box (Figure 2.10.8). Heat the metal disks to the same temperature with steam or hot air. Record the steady state temperatures in the two rods and the temperature of the air. From Figure 2.10.8, denoting the two rods by A and B, define $\theta = T - T_\infty$,

$$\ln (\theta_{A2}/\theta_{A1}) = -m_A(x_{A2} - x_{A1})$$

$$\ln (\theta_{B2}/\theta_{B1}) = -m_B(x_{B2} - x_{B1})$$

$$\frac{\ln (\theta_{A2}/\theta_{A1})}{\ln (\theta_{B2}/\theta_{B1})} = \frac{m_A(x_{A2} - x_{A1})}{m_B(x_{B2} - x_{B1})}$$

As the convection conditions for the two bars are the same, it is reasonable to assume that the convective heat transfer coefficients in the two cases are equal to each other. Thus,

$$\frac{\ln (\theta_{A2}/\theta_{A1})}{\ln (\theta_{B2}/\theta_{B1})} = \left(\frac{k_B}{k_A}\right)^{1/2} \frac{x_{A2} - x_{A1}}{x_{B2} - x_{B1}}$$

The accuracy of the measured value of the thermal conductivity is increased if the thermal conductivities of the two materials are approximately equal. It may be

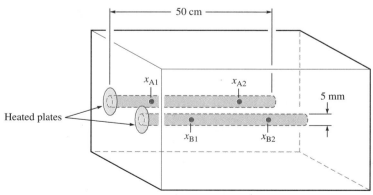

FIGURE 2.10.8 Two Fins Used To Determine the Thermal Conductivity of the Metal Alloy

advantageous to have several bars of different materials to cover a wide range of thermal conductivities. For example, stainless steel, brass, aluminum, and copper span a fairly large range of thermal conductivity.

4. Transient temperature distribution. Transient temperature distribution can be used in different ways to determine the thermal diffusivity and, hence, the thermal conductivity. In the method described in this section, the one-term approximation, Equation 2.9.5a, is used.

Take a 20-cm long, 1-cm diameter rod of the material and install thermocouples at the midplane and at known distances, say, at 2 cm and 4 cm, from the end surfaces, preferably on either side of the midplane. Attach 2-cm long end caps made of thin stainless steel sheet. Insulate the cylindrical surface of the rod (Figure 2.10.9). Heat the end surfaces with steam at atmospheric pressure or with hot air maintained at a constant temperature. Record the temperatures (indicated by the thermocouples and of steam) as a function of time. The temperatures at the midplane, θ_o, and at the other two locations are given by

$$\theta_o = c_1 \exp(-\mu_1^2 \, Fo_1) \cos \mu_1 \tag{1}$$

$$\theta_1 = c_1 \exp(-\mu_1^2 \, Fo_1) \cos(\mu_1 \eta_1) \tag{2}$$

$$\theta_2 = c_1 \exp(-\mu_1^2 \, Fo_1) \cos(\mu_1 \eta_2) \tag{3}$$

or

$$\frac{\theta_o}{\theta_1} = \frac{T_o - T_\infty}{T_1 - T_\infty} = \cos(\mu_1)/\cos(\mu_1 \eta_1) \tag{4}$$

$$\frac{\theta_o}{\theta_2} = \frac{T_o - T_\infty}{T_2 - T_\infty} = \cos(\mu_1)/\cos(\mu_1 \eta_2) \tag{5}$$

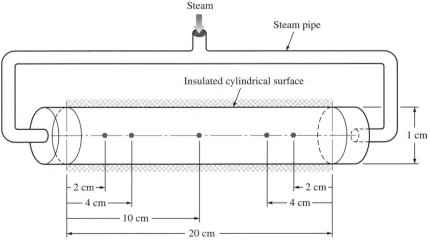

FIGURE 2.10.9 Transient Temperature Distribution To Determine the Thermal Conductivity of the Metal Alloy

where

$$0 < \mu_1 < \pi$$

$$\text{Fo} \geq 0.2$$

$$\eta = x/L$$

The value of μ_1 is found from Equations 4 and 5 by numerical methods. Ideally, the values from Equations 4 and 5 should be equal. But, because of the uncertainties in the measured values of the temperatures and the distances from the mid-plane, they will not be equal. An average value of μ_1 found from the two equations is used in subsequent calculations.

At a given location, find the temperatures at two different times. Denoting the two temperatures by θ_a and θ_b and the corresponding Fourier numbers by Fo_a and Fo_b,

$$\frac{\theta_a}{\theta_b} = \exp[-\mu_1^2(\text{Fo}_a - \text{Fo}_b)]$$

$$= \exp[\mu_1^2 \text{Fo}_a(\text{Fo}_b/\text{Fo}_a - 1)]$$

But

$$\text{Fo}_b/\text{Fo}_a = \tau_b/\tau_a$$

where τ_a and τ_b are real times. Knowing θ_a, θ_b, and μ_1, Fo_a is computed. From the relations

$$\text{Fo}_a = \alpha\tau_a/L^2 \quad \text{and} \quad \alpha = k/(\rho c_p)$$

the value of the thermal conductivity is determined.

Note: In all cases where the value of the convective heat transfer coefficient is required, radiative heat transfer is to be taken into account. If the heat transfer coefficient is determined experimentally without any correction for radiative heat transfer, the effect of radiative heat transfer is included in the experimentally determined value. To achieve the same radiative heat transfer coefficient, it is recommended to coat all the samples with a paint. If the convective heat transfer coefficient is calculated from correlations, the radiative heat transfer coefficient should be added to the convective heat transfer coefficient.

SUMMARY

The basis for the solution of conduction problems is the First Law of Thermodynamics in conjunction with Fourier's Law [$q_x'' = -k(\partial T/\partial x)$]

General One-Dimensional Conduction Equation

$$\rho c_p \frac{\partial T}{\partial \tau} = \frac{1}{A_x} \frac{\partial}{\partial x}\left(k A_x \frac{\partial T}{\partial x}\right) + \frac{1}{A_x} q_i'' \frac{dA_s}{dx} + q'''.$$

The steady, one-dimensional temperature conduction equation is obtained by setting $\partial T/\partial \tau = 0$. The boundary conditions for the temperature are

1. Specified temperature
2. Specified heat flux—insulated boundary condition is a special case of the heat flux being zero
3. Convective boundary condition
4. Matching temperature and heat flux

From the solution to $T(x, \tau)$, the heat transfer rate at any instant of time is obtained from

$$q_x = \int q''_x dA_x = \int - k \frac{\partial T}{\partial x} dA_x = -kA_x \frac{\partial T}{\partial x}$$

Steady, One-Dimensional Conduction with q''$_i$ = q''' = 0, Constant Properties

For specified temperatures at two locations in the x-direction (rectangular slabs) and r-direction (cylinders and spheres), the heat transfer rates are given by

$$\text{Rectangular slab } q_x = \frac{kA_x(T_1 - T_2)}{L}$$

$$\text{Hollow sylinder } q_r = \frac{2\pi kL(T_i - T_o)}{\ln (r_o/r_i)}$$

$$\text{Hollow sphere } q_r = \frac{4\pi k(T_i - T_o)}{1/r_i - 1/r_o}$$

The reciprocals of the coefficients of the temperature differences are the conductive thermal resistances. The thermal circuit with conductive and convective resistances in series can be employed for solving one-dimensional, steady-state conduction problems in solids with constant properties and without internal energy generations. The thermal circuit with parallel circuits is not as accurate as with series circuit because, with parallel thermal circuits, the temperature distribution is two-dimensional. However, the approximate solutions obtained by the application of the thermal circuit for parallel thermal circuits is widely used to determine the heating and cooling loads in buildings.

When solids are brought into thermal contact, there is a temperature discontinuity at the interface due to *contact resistance*. A brief discussion of such contact resistance was provided in Section 2.4.

Extended Surfaces

Extended surfaces are widely used to increase the heat transfer rate. For extended surfaces satisfying

■ Constant A_x (cross-sectional area) and P (perimeter)
■ Uniform h and T_∞
■ Constant properties

the resulting equation is a second order, linear, nonhomogeneous differential equation in $T(x)$. From the solution to $T(x)$ the heat transfer rate to the surrounding fluid is found from

$$q_{\text{fluid}} = \int h(T - T_\infty)dA_s$$

The heat transfer rate in the x-direction is given by $q_x = -kA_x(dT/dx)$. The one-dimensional temperature distribution in fins is valid if the Biot number (hL^*/k; L^* = characteristic length) is less than 0.1

Lumped Analysis

Spatial uniformity of the transient temperature in a solid body is valid if the Biot number (hL^*/k) is less than 0.1. The temperature distribution is found from

$$mc_p \frac{dT}{d\tau} = q_i + q'''V \qquad q_i = \text{heat transfer rate to the solid}$$

Transient Temperature Distribution

One-dimensional transient temperature distribution solutions for rectangular slabs, cylinders, and spheres exposed to a constant temperature fluid with uniform and constant heat transfer coefficient are available as a series solution and in charts as a function of dimensionless time and dimensionless distance.

The extension of the solution to one-dimensional transient temperature distribution to the solution of multi-dimensional transient temperature distribution is discussed and illustrated.

Software

The program HT on the disk contains the program (TC) to compute the transient, dimensionless temperatures for rectangular, cylindrical and spherical geometries. The program computes dimensionless temperature for Fo $\geq$ 0.01. Solutions by the approximate method, Equations 2.9.7 through 2.9.9 are also included.

REFERENCES

Chen, R.Y., and Kuo, L.T. (1970). "Closed form solutions for constant temperature heating of solids." *Mechanical Engineering News,* 16:20.

Heisler, M.P. (1946). "Temperature charts for induction and constant-temperature heating." ASME Semi-Annual Meeting, Detroit. Also in *TRANS. ASME.* 69 (1947):227.

Incropera, F.P., and De Witt, D.P. (1990). *"Fundamentals of Heat and Mass Transfer."* 3d ed. New York: Wiley.

Irey, R.K. (1968). "Errors in the one-dimensional fin solution." *J. Heat Transfer,* 99C:175.

Holman, J.P. (1990). *"Heat Transfer",* 7th ed. New York: McGraw-Hill.

Kern, D.Q., and Kraus, A.D. (1972). *"Extended Surface Heat Transfer."* New York: McGraw-Hill.

Kreith, F., and Bohn, M.S. (1993). *"Basic Heat Transfer,"* 5th ed. Minneapolis, MN: West.

Schneider, P.J. (1973). "Conduction interface resistance." In *Handbook of Heat Transfer,* eds. Rohsenow, W.M., and Hartnett, J.P. New York: McGraw Hill.

Suryanarayana, N.V. (1977). ''Two-dimensional effects on heat transfer rates from an array of straight fins.'' *J. Heat Transfer,* 99:129.

Yovanovich, M.M. (1986). ''Recent developments in thermal contact, gap and joint conductance.'' In *Heat Transfer,* vol. 1, eds. Tien, C.L., Carey, V.P., and Ferrel, J.K. New York: Hemisphere.

REVIEW QUESTIONS

(a) Define conductive heat transfer.

(b) State Fourier's Law of Conduction.

(c) What is the thermodynamic significance of the negative sign in Fourier's Law?

(d) What are the conditions to be satisfied for the application of the thermal circuit?

(e) What are the expressions for the thermal resistances for a rectangular slab, a hollow cylinder, and a hollow sphere?

(f) Will the thermal resistance of a rectangular slab increase or decrease if
  i. The thermal conductivity of the material is increased?
  ii. The cross sectional area is increased?
  iii. The thickness of the slab is increased?

(g) Define the ''R'' value of insulation material for residential use.

(h) What is conduction shape factor?

(i) What are the dimensions and units (S.I. system) for conduction shape factors?

(j) State the conditions that will lead to the use of extended surfaces.

(k) Define overall heat transfer coefficient.

(l) What is the corrected length of a fin?

(m) How will you fix the length of a fin? Will you make it as long as space permits?

(n) How long should a fin be before it can be considered as an infinite fin?

(o) State the condition to be satisfied to treat the temperature distribution in a fin as one-dimensional.

(p) Define and give a physical interpretation of the Biot number.

(q) What is a lumped system?

(r) When can the unsteady temperature in a solid body be considered to be spatially uniform?

(s) What is Fourier number?

PROBLEMS

2.1 Determine the heat flux across a slab of cartridge brass if two parallel surfaces, 25-mm apart, are at 40 °C and 39 °C.

2.2 The heat flux across a 10-mm stainless steel (AISI 316) plate is found to be 20 kW/m^2. If the temperature of the warmer surface is 28 °C, what is the temperature of the opposite surface?

2.3 In an experimental setup, the heat flux across a stainless steel (type 316) plate is determined by placing two thermocouples (temperature measuring devices) 6-mm apart. The measured temperature difference is 2.8 °C.

  (a) What is the heat flux?

  (b) It is known that the distance between the two thermocouples is 6 ± 0.5 mm. What are the maximum and minimum values of the heat fluxes if the measured temperature difference is 2.8 °C?

  (c) Similarly, the temperature difference is 2.8 ± 0.1 °C. What are the maximum and minimum values of the heat flux if the distance between the two thermocouples is 6 mm?

(d) If the measured distance is 6 ± 0.5 mm and the measured temperature difference is 2.8 ± 0.1 °C, what are the maximum and minimum possible values of
the heat flux?
What conclusions do you draw from your answers to parts a–d?

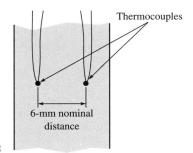

Thermocouples

6-mm nominal
distance

FIGURE P2.3

2.4 The temperature of air inside and outside a house are 20 °C and −5 °C, respectively. What are the inner and outer surface temperatures of a single pane glass
window 6-mm thick if the convective heat transfer coefficients on the inner and
outer surfaces are 6 W/m² °C and 10 W/m² °C, respectively?

2.5 The temperature at two locations of a stainless steel rod (type 316) are 40 °C and
30 °C, as shown. Determine the temperatures, T_{s1} and T_{s2}, of the two end surfaces
and heat transfer coefficient h_1. The cylindrical surfaces of the rod are perfectly
insulated.

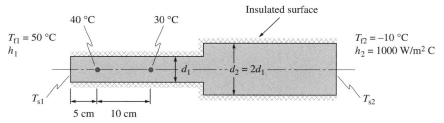

FIGURE P2.5

2.6 The inside dimensions of a laboratory furnace are 60-cm long, 40-cm high, and
80-cm wide. The walls are made of 2.5-cm thick firebrick, followed by 5-cm thick
glass wool, and covered by 3-mm thick steel plate. With a power supply to the
furnace of 1700 W, in steady state conditions, it is found that the outer surface
temperature of the steel plate is 60 °C with the surrounding fluid at 20 °C.

(a) Is the temperature distribution in the wall one-dimensional?. If it is not one-
dimensional, what are the reasons for the departure from one-dimensional temperature distribution?

(b) Assuming all outer surfaces are exposed to the surrounding air, estimate the
convective heat transfer coefficient between the steel plate and outside air.
Assume that the convective heat transfer coefficient is uniform on all surfaces.

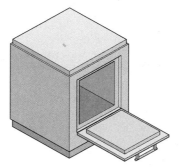

FIGURE P2.6

2.7 An experimental apparatus to measure the conductivity of metals consists of two identical plates of the metals, 1-cm thick, 20-cm wide, and 20-cm high. An electrical heater is sandwiched between the two plates. All the edge surfaces are heavily insulated. The temperature of the inner and outer surfaces are 29 °C and 27 °C, respectively, and the power input to the heater is 215 W. Estimate the thermal conductivity of the metal.

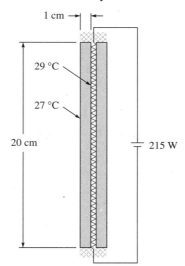

FIGURE P2.7

2.8 The temperature profile in a rectangular slab of a homogeneous material under steady state conditions is shown in the figure. For the material, does the thermal conductivity increase or decrease with an increase in the temperature?

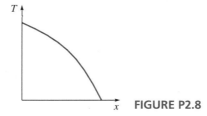

FIGURE P2.8

2.9 An ice box with inside dimensions of 60-cm length, 40-cm width, and 20-cm height, is made of 2.5-cm thick Styrofoam. It is filled with a mixture of crushed ice and water at 0 °C. Eighty percent of the mixture by mass is ice. The outer surfaces, including the top (also of 2.5-cm thick Styrofoam), are exposed to air at 20 °C, and the convective heat transfer coefficient between the outer surfaces and the surrounding air is 6 W/m² K. The inner surfaces may be assumed to be at 0 °C. Assuming one-dimensional temperature distribution in the walls of the ice box and all surfaces are exposed to air, estimate

    (a) The heat transfer rate to the ice box from the surroundings employing the arithmetic mean cross-sectional area.

    (b) The time taken for all the ice to melt assuming that the ice/water mixture is at a uniform temperature of 0 °C at all times. Assume that the density of ice and water ≈ 1000 kg/m³ and enthalpy of fusion of ice = 333.3 kJ/kg.

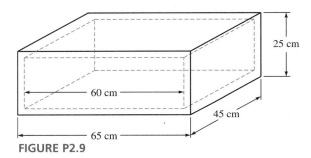

FIGURE P2.9

2.10 The wall of a house, 8-m long and 2.5-m high, is modeled as shown. Find

    (a) The total heat transfer rate from the inside air across the wall.

    (b) The inside surface temperature of the plasterboard.

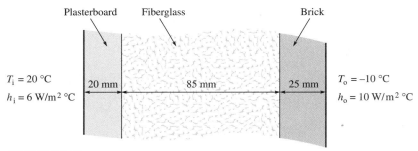

FIGURE P2.10

2.11 A wall of a cold storage facility is constructed with 15-cm thick common brick followed by cork insulation. If the outer surface of the wall is at 22 °C and the inner surface of the cork is at −10 °C, determine the thickness of the cork needed to limit the heat transfer rate to 8 W/m².

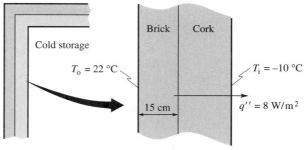

FIGURE P2.11

2.12 A long, 1-cm diameter aluminum rod carries a current of 1000 amps. The wire is covered with a 3-mm thick insulation ($k = 0.15$ W/m K). The temperature of the outer surface of the insulation is limited to 30 °C. The electrical resistance of the wire is 3.7×10^{-4} Ω/m. Determine the maximum temperature in the insulation material.

2.13 A window is made of 6-mm thick glass. The inner surface is exposed to air at 20 °C with a heat transfer coefficient of 6 W/m^2 K. The outer surface is exposed to air at -20 °C with an associated heat transfer coefficient of 30 W/m^2 K. What is the heat flux across the window?

2.14 The rear window of an automobile is defogged by an electrical heating element attached to the inner surface. When power is supplied to the heating element, it may be assumed that a condition of uniform heat flux exists. The inner surface is exposed to air at 20 °C with a convective heat transfer coefficient of 10 W/m^2 K. The outer surface is exposed to air at 0 °C with an associated convective heat transfer coefficient of 50 W/m^2 K. If the inner surface of the window is to be maintained at 12 °C, what should be the electrical power supply to the heater per unit area of the window? The window is made of 4-mm thick glass ($k = 1$ W/m K).

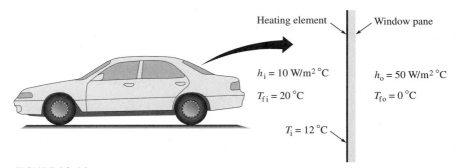

FIGURE P2.14

2.15 For the window in Problem 2.13, to reduce the heat loss, another glass pane of the same thickness is added to it with a 3-mm air gap between the two. Assuming that the heat transfer across the air gap is by pure conduction, estimate the heat flux across the window.

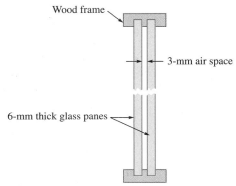

FIGURE P2.15

2.16 The thermal conductivity of a material is given by $k = k_o (1 + \alpha T)$, where α is a positive constant. Sketch the steady state temperature profile in a rectangular slab of the material whose two parallel surfaces are maintained at T_1 and T_2 $(T_1 > T_2)$.

2.17 A 2-cm diameter brass rod is attached to a 4-cm long, 2 cm-wide, and 30-cm high rectangular stainless steel plate (type 302). The free end of the brass rod is heated with a 2 W heating element. All the power input to the heater is transferred to the brass rod. If the temperature of the stainless steel plate 2 cm from the free end is 40 °C, what is the temperature of the outer surface of the stainless steel plate? All surfaces of the steel plate, except those with dimensions of 4 cm $\times$ 2 cm, are perfectly insulated.

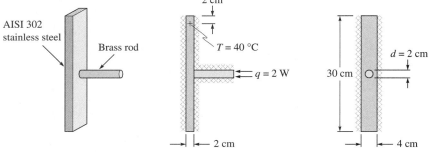

FIGURE P2.17

2.18 One method of measuring the thermal conductivity of a material is to heat a material of known conductivity in contact with a specimen of the material whose conductivity is to be determined, as shown in the figure. The specimens in a particular case are cylindrical pieces, each of 2-cm diameter. Thermocouples are located as shown. Material A is stainless steel with a thermal conductivity of 15.2 W/m K. In the experiment,

$$T_1 = 105 \text{ °C} \qquad T_2 = 98 \text{ °C} \qquad T_3 = 91 \text{ °C} \qquad T_4 = 62.1 \text{ °C} \qquad T_5 = 58.6 \text{ °C}$$

Determine the thermal conductivity of material B.

2.19 In Problem 2.18, for specimen A, which is used as a standard piece to determine the thermal conductivity of other materials, great care is taken to locate the ther-

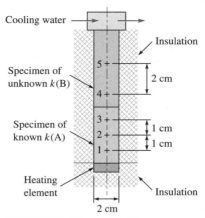

FIGURE P2.18 and P2.19

mocouples, which may be taken to be exactly 1-cm apart. In preparing the specimens whose thermal conductivities are to be determined, it is suspected that the spacing may be off by 0.5 mm. What are the limits of the thermal conductivity of material B? Given that a thermocouple can be located within 0.5 mm of the intended location, what is your recommendation to reduce the uncertainty in the measurement of the conductivity?

2.20 The outer surface of a furnace wall reaches a temperature of 400 °C. It is to be insulated so that the temperature of the outer surface of the insulation should not exceed 50 °C. It is proposed to insulate it with a high temperature insulation material followed by a more easily available and less expensive low temperature insulation material with a maximum working temperature of 200 °C. The outermost surface of the insulation material is exposed to air at 30 °C with a convective heat transfer coefficient of 10 W/m² °C. Determine the thicknesses of the insulation materials and the heat flux. The thermal conductivities of the high temperature insulation material and the low temperature insulation material are 0.18 W/m K and 0.036 W/m K, respectively.

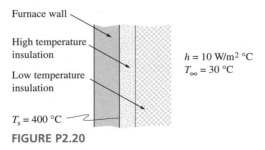

FIGURE P2.20

2.21 Refrigerant 113 at 47 °C flows over one surface of a 6-mm thick copper plate. The other side of the plate is cooled by water at 20 °C. The convective heat transfer coefficients on the refrigerant and water sides are 200 W/m² K and 500 W/m² K, respectively. Determine
(a) The heat flux across the copper plate.
(b) The temperature of each surface of the copper plate.

2.22 In an experiment, an electrical heater is sandwiched between two stainless steel (type 347) plates, each 20-cm wide, 20-cm high, and 1-cm thick. Air at 20 °C flows over the outer surfaces of the plates. The temperature of the inner surfaces of the plate (adjacent to the heater) is 80 °C. If the power input to the heater is 96 W, determine the convective heat transfer coefficient on the air side assuming that it is the same on both sides.

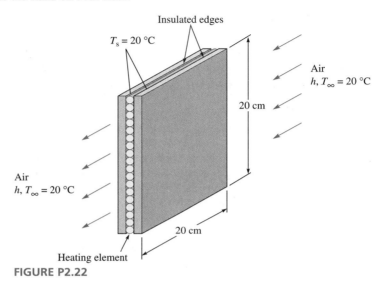

FIGURE P2.22

2.23 An electric heating element is bonded to one side of a 30-cm wide, 40-cm long, and 1.2-cm thick, stainless steel (type 316) plate. The opposite surface is exposed to air at 20 °C with a convective heat transfer coefficient of 20 W/m² °C. The

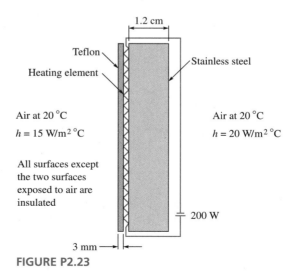

FIGURE P2.23

heating element is insulated with a 3-mm thick Teflon sheet, which is also exposed to air at 20 °C with a convection heat transfer coefficient of 15 W/m² °C. If the power input to the element is 200 W determine the maximum temperature in the assembly.

2.24 A boiler furnace is constructed with 300-mm thick fireclay brick. The inner surface temperature of the furnace is estimated to be 1100 °C. The outer surface is exposed to air at 30 °C with a convective heat transfer coefficient of 10 W/m² °C. Estimate the heat flux across the wall and the temperature of the outer surface.

2.25 The inside diameter of a fireplace chimney is 200 mm and made of very thin sheet steel (negligible conduction resistance) covered with 8-mm thick insulation material ($k = 0.05$ W/m K). The outer surface of the chimney is exposed to air at -10 °C with a convective heat transfer coefficient of 7 W/m² °C. The convective heat transfer coefficient on the inside surface is 20 W/m² °C. To avoid condensation, the inner surface temperature should be more than 50 °C. What should be the minimum temperature of the gases in the chimney?

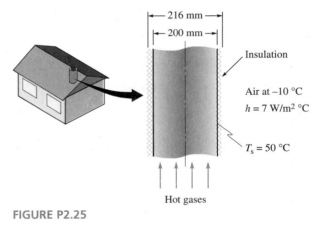

FIGURE P2.25

2.26 A wall is constructed with 18-mm thick Sheetrock (plasterboard) attached to 2 in $\times$ 4 in studs (actually measuring 1.5 in $\times$ 3.5 in) at 16-in centers. On the other side, 1.25-cm thick Celotex boards are attached to the studs, followed by 10-mm thick plywood sheet and 10-cm thick common brick. (for Celotex, $k = 0.048$ W/m K). The convective heat transfer coefficients associated with the inner and outer surfaces are 7 W/m² K and 20 W/m² K, respectively. The inside and outside air temperatures are 20 °C and -10 °C, respectively. The space between the studs is either a 3.5-in air gap or filled with fiberglass insulation. Determine, for each case (assume the thermal resistance per unit area of the air gap between the studs to be 6 m² K/W),
(a) The average heat flux.
(b) The inner surface temperature.

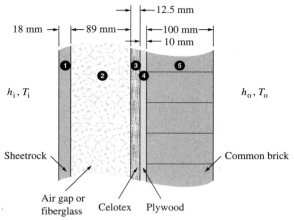

FIGURE P2.26 (See page 171)

2.27 The ceiling of a residential building consists of 16-mm (5/8-in) thick plasterboard (dry wall) supported by ceiling joists. The space between the joists is filled with cellulose insulation. Determine the average heat flux across the ceiling for the conditions given and the overall heat transfer coefficient.

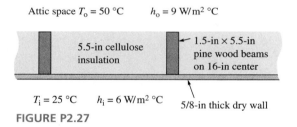

FIGURE P2.27

2.28 A cold storage room in a ship has walls as shown. Plain carbon steel flats 150 mm × 6 mm are welded to the 9-mm thick steel hull. The steel flats are 600-mm apart. Cork insulation 250-mm thick is applied to the hull. Determine the heat flux assuming that the thermal circuit for parallel paths is valid. (See Figure P2.28, top of next page).

2.29 Saturated steam at 200 °C flows inside a 5-cm O.D. carbon steel tube (wall thickness 3 mm). The tube is insulated with 5-cm thick fiberglass. The convective heat transfer coefficients on the steam and air side are 8000 W/m² °C and 7 W/m² °C, respectively. If the temperature of the outside air is 25 °C, determine
(a) The heat transfer rate per meter length of the tube.
(b) The temperature of the outer surface of the insulation.
(c) The rate of condensation of steam per meter length of the tube.

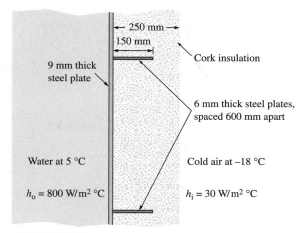

FIGURE P2.28

2.30 A 5-cm I.D., 3-mm thick Styrofoam cup is 7-cm high. The inner surface temperature of the cup is 70 °C, and the outer surface is exposed to air at 20 °C. The convective heat transfer coefficient between the outside surface and the air is 6 W/m^2 °C. Determine
(a) The heat transfer rate through the cylindrical surface of the cup.
(b) The outer surface temperature of the cup.

2.31 A cylindrical cartridge heater fits snugly inside a hollow 12-mm I.D and 18-mm O.D. stainless steel (AISI 302) tube. The power input to the heater is 0.25 kW/m length of the tube. Air at 20 °C flows over the tube. Neglecting the heat transfer from the end surfaces, determine
(a) The heat transfer coefficient associated with the outer surface if the outer surface temperature is 60 °C.
(b) The maximum temperature, and its location, in the tube.

2.32 Steam at 400 °C flows inside a type 316 stainless steel tube of inside diameter 2 cm and outside diameter 3 cm. The tube is insulated with molded cellular glass insulation. The radial thickness of the insulation is 1 cm, and its outer surface is exposed to air at 20 °C. The convective heat transfer coefficient between the outer surface and the air is 6 W/m^2 °C. The convective heat transfer coefficient on the steam side is very high. Determine
(a) The heat transfer rate per meter length of the tube.
(b) The temperature of the outer surface of the insulation.

2.33 The water in an electric water heater is maintained at 60 °C. The tank is made of 6-mm thick stainless (type 316) steel cylinder with 12-mm thick stainless steel end plates. The inside dimensions of the tank are 60-cm diameter and 100-cm high. The tank is covered with 5-cm thick fiberglass insulation. The surrounding air is at 15 °C with a convective heat transfer coefficient of 8 W/m^2 K. Assuming the inner surface temperature of the tank is 60 °C, estimate the electrical energy required during a 12-hr period when no water is drawn from the boiler.

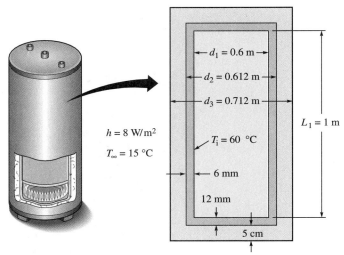

$d_1 = 0.6$ m

$d_2 = 0.612$ m

$d_3 = 0.712$ m

$L_1 = 1$ m

$h = 8$ W/m^2

$T_\infty = 15$ °C

$T_i = 60$ °C

6 mm

12 mm

5 cm

FIGURE P2.33

2.34 Consider two long, cylinders of an insulating material. Cylinder 1 has an inner diameter of 25 mm and an outer diameter of 50 mm. Cylinder 2 has an inner diameter of 100 mm and an outer diameter of 175 mm. The lengths of the two cylinders are equal. For the same values of the inner and outer surface temperatures in both cases, find the ratio of the heat transfer rates q_1/q_2.

2.35 The temperature distribution in a hollow cylinder of inner radius 5 cm, outer radius 5.5 cm, and length 1 m is given by

$$T = -625 \, r^2 + 91 \ln r + 284 \qquad (T \text{ in °C}, r \text{ in meters})$$

Water flows inside the cylinder and the cylinder is surrounded by air on the outside. The thermal conductivity of the material of the cylinder is 16 W/m K. Determine
(a) The heat transfer rate to water.
(b) The heat transfer rate to air.
(c) The net heat transfer rate from the cylinder to the air and water.

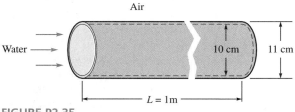

Air

Water

10 cm   11 cm

$L = 1$m

FIGURE P2.35

2.36 A long, 1.5-mm diameter electrical wire is covered with 1.5-mm thick Teflon. The convective heat transfer coefficient between the wire and the surrounding air (20 °C) is 8 W/m² K. Determine the maximum current in the wire if the resistance of the wire is 0.003 Ω/m and the maximum temperature in the insulation is not to exceed 50 °C.

2.37 The outer surface of a 1-cm O.D. tube is maintained at 100 °C and is surrounded by air at 20 °C. The convective heat transfer coefficient is 6 W/m² K.
  (a) Compute the heat transfer rate per meter length of the tube.
  (b) To reduce the heat transfer rate, it is proposed to cover the tube with 3-mm thick Teflon. What is the heat transfer rate per meter length of the pipe with this insulation, if the covective heat transfer coefficient has the same value of 6 W/m² K. (If your answer to part b is greater than part a, do not think your solution is incorrect—see Problem 2.45.)

2.38 The outer surface temperature of a 15-cm O.D. tube is 400 °C. It is to be insulated so that the outer surface temperature of the insulation is not to exceed 50 °C. It is proposed to insulate it with a high temperature insulation material followed by a more easily available and less expensive low temperature insulation material with a maximum working temperature of 200 °C. The outer surface of the insulation material is exposed to air at 30 °C with a convective heat transfer coefficient of 10 W/m² °C. Determine the thicknesses of the insulation materials. The thermal conductivity of the high temperature insulation material is 0.18 W/m K and that of the low temperature insulation material is 0.036 W/m K.

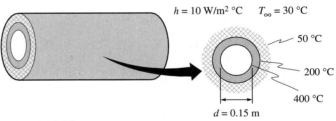

FIGURE P2.38

2.39 Water flows inside a plain carbon steel tube (12-mm I.D. and 1-mm wall thickness) with a velocity of 3 m/s. At a particular location, the temperature of the water is 60 °C and the convective heat transfer coefficient is 14,600 W/m² °C. If the temperature of the outer surface of the tube is 80 °C and the tube is surrounded by a hot fluid with a convective heat transfer coefficient of 8000 W/m² °C, determine
  (a) The temperature of the the fluid on the outside.
  (b) The rate of change of temperature of water per meter length of tube at that location.

2.40 A 12-mm I.D. copper tube is used to transport a liquid-vapor mixture of Refrigerant 12 at −20 °C. The tube has a wall thickness of 1 mm, and is covered with 50-mm thick cork insulation. It is exposed to air at 20 °C with a convective heat transfer coefficient of 10 W/m² °C. The covective heat transfer coefficient on the refrigerant side is 400 W/m² °C. Estimate the rate of vaporization of R-12 in a 15-m length of the tube. (The enthalpy of vaporization of R-12 at −20 °C is 161.7 kJ/kg).

2.41 Steam, flowing in a 15-mm I.D. × 21-mm O.D. plain carbon steel tube, maintains the inner surface of the tube at 300 °C. The tube is to be covered with a sufficient thickness of insulation such that the temperature of the outer surface is not to exceed 60 °C when exposed to air at 25 °C. If the convective heat transfer coefficient on the air side is 8 W/m² °C, propose a suitable insulation and determine the heat transfer rate per unit length of the tube.   **(Design)**

2.42 A liquid at 80 °C flows inside a type 316 stainless steel tube (18-mm I.D. × 24-mm O.D.). The temperature of the air surrounding the tube is 30 °C. The difference between the temperatures of the inner and outer surfaces of the tube is 1.8 °C. The convective heat transfer coefficient on the liquid side is 2700 W/m² °C. Determine the convective heat transfer coefficient on the air side.

2.43 In an experiment, a stainless steel tube (type 316, 25-mm I.D. with a wall thickness of 3 mm) is covered with 1-mm thick zirconia cloth. An electric heating element, made of a very thin metal, is placed around the cloth and is covered with 75-mm thick fiberglass insulation. Nitrogen at 40 °C flows inside the tube with a convective heat transfer coefficient of 50 W/m² °C. The surrounding air is at 20 °C with a convective heat transfer coefficient of 10 W/m² °C. If the power input to the heating element is 515 W/m length of the tube, determine
(a) The temperature of the heating element.
(b) The heat transfer rates to the nitrogen and air per meter length of the tube.

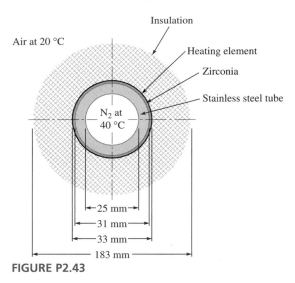

FIGURE P2.43

2.44 The circular edge of a thin metallic diaphragm of thickness, t, is attached to a heavy block of metal. The diaphragm is subjected to a uniform heat flux of q'' on one surface; the opposite surface is perfectly insulated.

(a) Derive the differential equation for the one-dimensional temperature distribution in the diaphragm.

(b) Find the functional relationship between the temperature T_c, at the center of the diaphragm, and T_R, at the periphery.

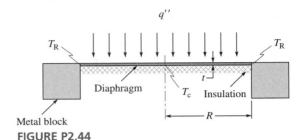

FIGURE P2.44

2.45 Consider a cylindrical surface of radius R_o whose surface temperature is maintained at T_s. It is exposed to a fluid at T_∞ with a convective heat transfer coefficient h. It is proposed to add insulation to the outer surface with a material of thermal conductivity k. If $R_o < k/h$, show that the heat transfer rate increases with insulation thickness and reaches a maximum at

$$R = \frac{k}{h}$$

where R is the outer radius of the insulation surface. Assume h remains constant. The radius of the insulation at which the heat transfer rate is the maximum is known as the critical radius of insulation.[5]

2.46 Steam, flowing in a 6-mm O.D. tube maintains the outer surface of the tube at 200 °C. The tube is to be insulated with a material having a thermal conductivity of 0.1 W/m K. The outer surface of the insulation is exposed to air at 20 °C with a convective heat transfer coefficient of 8 W/m² °C.

(a) Find the thickness of the required insulation so that the outer surface of the insulation is at 70 °C.

(b) Find the thickness of the insulation that will give the same heat transfer rate per unit length of the tube as with a bare pipe (without insulation) and the corresponding temperature of the outer surface of the insulation. Assume the value of the convective heat transfer coefficient remains at 8 W/m² °C.

2.47 An electric heating element is fabricated with 2.5-mm diameter nickel wire having a resistance of 0.22 Ω/m. It is covered with a ceramic tube ($k = 0.7$ W/m K) and a stainless steel (type 316) tube as shown. Determine the temperature of the surface of the nickel wire if the current in the wire is 20 A. The surrounding air temperature is 20 °C and the convective heat transfer coefficient is 10 W/m² °C. See Figure P2.47, next page.

---

[5]The current carrying capacity of wires can be increased by the addition of insulation if $R < k/h$. In some small-diameter steam pipes, the heat transfer rate may increase when they are insulated with a small thickness of insulation material.

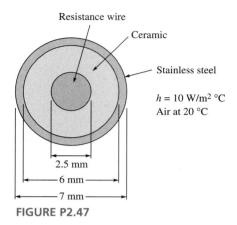

FIGURE P2.47

2.48 The steady state heat transfer rate across a hollow cylinder is given by

$$q = \frac{2\pi kL(T_i - T_o)}{\ln(r_o/r_i)}$$

A suggestion is made that the heat transfer rate be expressed as

$$q = \frac{kA(T_i - T_o)}{r_o - r_i}$$

where A is an appropriate mean area so that the latter equation gives the correct value of the heat transfer rate.

(a) Find an expression for the mean area A

(b) If instead of using the mean area found in part a, the arithmetic average $= \pi(r_i + r_o)L$ is used, what is the error in the heat transfer rate for values of $r_o/r_i = 2, 5,$ and 10?

2.49 Figure P2.49 shows a section of a hollow cylinder of length L. The cylindrical surfaces ($r = r_i$ and $r = r_o$) and the end surfaces perpendicular to the axis are perfectly insulated. The heat transfer is in the θ-direction only. $T(\theta = 0) = T_1$ and $T(\theta = \theta_o) = T_2$. Assuming that $T = T(\theta)$,

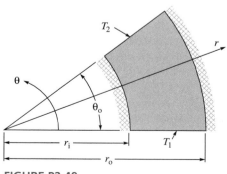

FIGURE P2.49

(a) Find an expression for $T(\theta)$.

(b) Show that the heat transfer rate is given by

$$kL \ln(r_o/r_i) \frac{T_1 - T_2}{\theta_o}$$

Note that $q_\theta'' = -k \dfrac{dT}{rd\theta}$

2.50 Saturated steam at 8 atm flows inside a 10-cm O.D. pipe. The original asbestos insulation is falling apart and is to be replaced. The primary concern is safety, and the outer surface temperature of the insulation is not to exceed 40 °C. As the convective heat transfer coefficient for condensing steam is very high, the convective resistance on the inside surface may be neglected. The ambient air temperature is 30 °C, and the associated convective heat transfer coefficient is 9 W/m^2 °C. Propose a suitable insulation.   **(Design)**

2.51 In an experiment, an electric heating element was introduced inside a 30-cm long heavy walled, stainless steel (type 316) tube, 10-mm I.D. and 15-mm O.D. Air at 20 °C flowed over the cylinder perpendicular to the axis of the tube. The end surfaces of the cylinder were perfectly insulated. The power input to the heating element was 100 W.

(a) The temperature of the outer surface of the tube was reported to be 98 °C. Determine the convective heat transfer coefficient.

(b) On closer examination, it was found that the thermocouple used to measure the surface temperature was actually located on the inner surface of the hollow tube, and the temperature of 98 °C reported was the inner surface temperature. Recompute the convective heat transfer coefficient.

2.52 To estimate the temperature of a bearing, it is modeled as the shaft being coaxial with the bearing, with the space between the shaft and the bearing filled with a lubricating oil. Consider a bearing for a 10-cm diameter shaft. The bearing is of cast iron and is 100-mm long with a radial thickness of 1 cm. The radial clearance between the shaft and the bearing is 0.1 mm. The viscosity of the lubricating oil is 0.01 N s/m^2. If the shaft rotates at 600 rpm, determine the maximum temperature in the bearing. Assume that the heat transfer rate to the shaft is negligible and that the surrounding air temperature is 20 °C with a convective heat transfer coefficient of 8 W/m^2 °C. Neglect the heat transfer from the end surfaces.

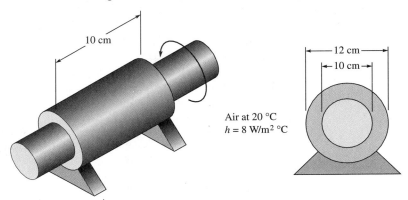

10 cm

12 cm

10 cm

Air at 20 °C
$h = 8$ W/m^2 °C

FIGURE P2.52

2.53 A 100-mm O.D. pipe is used to transport steam at 400 °C. To ensure safety of working personnel and to reduce heat transfer rate to the surroundings, it is required to insulate the surface of the pipe so that the temperature of the outer surface of the insulation is not to exceed 50 °C when exposed to air at 30 °C with a convective heat transfer coefficient of 10 W/m² °C. Assuming the temperature of the outer surface of the tube to be 400 °C, suggest a suitable insulation. As space is at a premium, it is desirable to keep the radial thickness of the insulation low.   **(Design)**

2.54 The inner and outer surfaces of a hollow sphere are maintained at temperatures T_i and T_o, respectively. If the inner and outer radii are r_i and r_o, show that the heat transfer rate through the sphere is given by

$$q_r = \frac{4\pi k(T_i - T_o)}{1/r_i - 1/r_o}$$

2.55 A spherical container has an inside diameter of 50 cm and is made of a 6-mm thick, type 316 stainless steel plate. Liquid nitrogen at 77.4 K is stored in the container. The convective heat transfer coefficient between the exterior surface and the surrounding air (20 °C) is 6 W/m² K. Assume that the inner surface is at 77.4 K.
(a) What is the heat transfer rate to the nitrogen?
(b) What is the heat transfer rate if the container is covered with 10-cm thick polyurethane insulation?
(Liquefied natural gas is stored in spherical containers for transportation in ships. Can you think of reasons for using spherical containers from the point of view of material, mechanical strength, and heat transfer?)

2.56 In a manner analogous to the cylinder as in Problem 2.45, show that the critical radius of insulation for a sphere is given by $R = 2k/h$.

2.57 The steady state heat transfer rate across a hollow sphere is given by

$$q = \frac{4\pi k(T_i - T_o)}{1/r_i - 1/r_o}$$

A suggestion is made to express the heat transfer rate as

$$q = \frac{kA\,(T_i - T_o)}{r_o - r_i}$$

where A is an appropriate mean area so that the latter expression gives the correct heat transfer rate.
(a) Find an expression for the mean area A
(b) If, instead of using the mean area found in part a, the arithmetic average value of the inner and outer surface areas is used, what is the error in the heat transfer rate for values of r_o/r_i = 1.1, 2, 5, and 10?

2.58 A frustum of a right circular aluminum cone has end surfaces of 5-cm and 8-cm diameter and is 10-cm long. One end surface is maintained at 80 °C and the other at 30 °C. The peripheral surface of the cone (excepting the two end surfaces) is insulated. Determine the temperature as a function of x, x being the axis perpendicular to the end surfaces, and the heat transfer rate through the cone. It will be convenient to choose the apex of the cone as the origin.

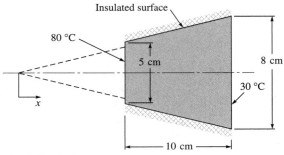

FIGURE P2.58

2.59 The conical section of a metal is shown in the sketch where $r = px^n$; $p > 0$, and $n > 0$.
   (a) Write the differential equation for the temperature in the solid.
   (b) Specify the boundary conditions in terms of the known quantities given in the sketch.
   (c) Derive an expression for the temperature distribution in the solid and for the heat transfer rate through it.
   (d) Find an expression for the resistance to heat transfer across the solid in a manner similar to that for a cylinder or a sphere.

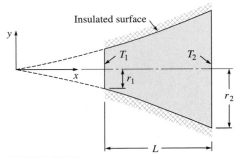

FIGURE P2.59

2.60 Re-work Problem 2.59 as if the right-hand side surface is exposed to a fluid maintained at T_∞ with a convective heat transfer coefficient h.

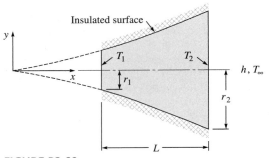

FIGURE P2.60

2.61 For both hollow cylinders and hollow spheres, show that the conductive resistance can be approximated as the conductive resistance of a plane slab,

$$(r_o - r_i)/(k\,A_r), \text{ if } \frac{r_o - r_i}{r_i} \ll 1$$

where A_r is the inner surface area.

2.62 It is proposed to construct a cold storage room making use of earth as insulation. The outer dimensions of the room are 20-m long, 10-m wide, and 3-m high. It is estimated that the outer surface of the room is at $-10\,°C$ and that the surface of the earth is at $10\,°C$. The top surface of the room is 3 m below the surface of the earth. Estimate the heat transfer rate to the room. For the soil, $k = 0.28$ W/m K.

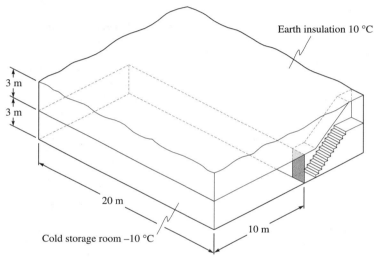

Earth insulation 10 °C

3 m

3 m

20 m

10 m

Cold storage room −10 °C

FIGURE P2.62

2.63 A sphere of radius R_1 and at T_1 is in close thermal contact with a concentric sphere of a material whose outer surface (radius R_2) is maintained at T_2. Find the conduction shape factor for the sphere. By letting R_2 become very large, derive the conduction shape factor for a sphere in an infinite medium and compare the result with the expression given in Table 2.5.1.

2.64 (See Figure P2.64, next page.) The inner dimensions of a small furnace, constructed with 10-cm thick fireclay bricks, are 0.5-m wide, 0.5-m high, and 0.5-m deep. If the inside surface temperature is 300 °C and the outer surface is exposed to air at 20 °C with a convective heat transfer coefficient of 7 W/m² °C, determine the heat transfer rate to the air. Employ conduction shape factor to find the total resistance to heat transfer.

2.65 Estimate the heat transfer rate to the surroundings in Problem 2.9 employing conduction shape factor.

2.66 Liquid nitrogen is stored in a 5-m diameter sphere buried below the surface of the earth, with its center 7 m below the surface. The surface of the sphere is maintained

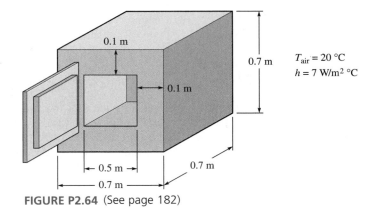

FIGURE P2.64 (See page 182)

at $-180\,°C$. Nitrogen vaporizes as a result of heat transfer to the tank. Constant pressure (and temperature) is maintained in the tank by venting the vapor generated to the atmosphere. If the surface of the earth is at $10\,°C$ and its thermal conductivity is $0.17\ W/m\ K$, estimate the heat transfer rate to the sphere and the percentage of nitrogen vaporized per hour when the tank is full of the liquid. The specific enthalpy and density of liquid nitrogen are $198.6\ kJ/kg$ and $808.6\ kg/m^3$, respectively.

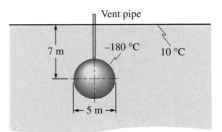

FIGURE P2.66

2.67 Instead of the spherical container in Problem 2.66, the same amount of liquid nitrogen is stored in a horizontal, 2-m diameter, cylindrical container with its axis 5 m below the surface of the earth. Determine the heat transfer rate and the fraction of the liquid evaporated per hour when the tank is full of liquid nitrogen. All other conditions are the same as in Problem 2.66.

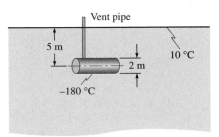

FIGURE P2.67

2.68 Hot water is transported in a 50-mm O.D. tube, buried 30 cm below the surface of the earth. The velocity of the water is 3 m/s. The surface of the tube is at 80 °C and the surface of the earth is at 0 °C. Determine the temperature drop of water per 100-m length of the tube. Thermal conductivity of the soil is 0.17 W/m K.

2.69 A hot water pipe is laid inside a 20-cm thick concrete floor with its axis at the midplane of the slab. The floor surfaces are at 20 °C. The pipe has an outside diameter of 5 cm and its surface is at 80 °C. Estimate the heat transfer rate per meter length of the pipe if the thermal conductivity of concrete is 1.1 W/m K.

2.70 Crude oil is very viscous at room temperature. The pumping power is reduced by heating it, and thus, reducing its viscosity. Crude oil is pumped in a 22-cm diameter pipe. The pipe surface is kept at 60 °C. The pipe is buried 1 m below the surface of the earth, which is at 0 °C. The thermal conductivity of the soil is 0.21 W/m K.
(a) Determine the heat transfer rate per meter length of the pipe.
(b) What is the reduction in the heat transfer rate if the pipe is covered with a 100-mm thick insulation material with a thermal conductivity of 0.05 W/m K?

Carbon

2.71 Steel rods of 2-mm diameter and 50-mm length, are attached to a tube (axis of the rods perpendicular to the surface of the tube) whose surface temperature is 100 °C. The convective heat transfer coefficient between the rods and ambient air (20 °C) is 20 W//m^2 K. Determine
(a) The heat transfer rate from a rod to the air by convection.
(b) The temperature at the free end of the rods.
(c) The heat transfer rate and the tip temperature assuming the tips of the rods to be perfectly insulated.

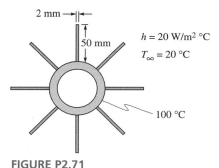

2 mm

50 mm

$h = 20$ W/m^2 °C

$T_\infty = 20$ °C

100 °C

FIGURE P2.71

2.72 A hollow aluminum alloy tube, 25-mm O.D., 15-mm I.D., and 300-mm long, is attached to a plate maintained at 300 °C. The tube is exposed to air (20 °C) and the convective heat transfer coefficient between the surface and air is 25 W/m^2 K. Assuming the heat transfer rate from the inside surface of the tube to be negligibly small, calculate the convective heat transfer rate from the tube to the surrounding air and the temperature of the tip of the tube.

2.73 Starting with Equations 2.6.12 and 2.6.14, derive Equations 2.6.15a and 2.6.15b, respectively.

2.74 Starting with Equations 2.6.5 and boundary conditions given by Equations 2.6.6 and 2.6.7, derive Equations 2.6.15a and 2.6.15b.

2.75 Consider an extended surface with different boundary conditions as indicated below and in Figure P2.75a-d. In each case, derive expressions for the temperature distribution and heat transfer rate from the fin to the surrounding fluid.

(a) One end at $\theta_b (x = L)$ and the other end $(x = 0)$ at θ_t:

$$\theta = \theta_t \cosh(mx) + \frac{\theta_b - \theta_t \cosh(mL)}{\sinh(mL)} \sinh(mx)$$

$$q = kA_x m \left[\theta_t \sinh(mL) + \frac{\theta_b - \theta_t \cosh(mL)}{\sinh(mL)} (\cosh mL - 1) \right]$$

(b) Infinitely long fin with $\theta \ (x = 0) = \theta_b$:

$$\theta = \theta_b e^{-mx} \qquad q = kA_x m \ \theta_b = \sqrt{hPkA_x} \ \theta_b$$

(c) One end surface $(x = 0)$ subjected to a heat flux of q_1'' and the other end surface to q_2'':

$$\theta = \frac{q_2'' + q_1'' \cosh(mL)}{km \ \sinh(mL)} \cosh(mx) - \frac{q_1''}{km} \sinh(mx)$$

(d) One end surface $(x = 0)$ subjected to a heat flux of q_1'' and the other end surface $(x = L)$ maintained at θ_2:

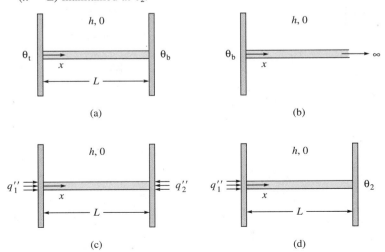

(a)   (b)

(c)   (d)

FIGURE P2.75

$$\theta = \left[\theta_2 + \frac{q_1''}{km}\sinh(mL)\right]\frac{\cosh(mx)}{\cosh(mL)} - \frac{q_1''}{km}\sinh(mx)$$

$$q = q_1'' A_x (1 - \cosh mL) + A_x \tanh(mL)(km\theta_2 + q_1'' \sinh mL)$$

2.76 A 6-cm long, 2-mm diameter solid rod of commercial bronze is attached to two plates maintained at 100 °C. The rod is surrounded by air at 20 °C with a convective heat transfer coefficient of 100 W/m² °C. Calculate the temperature of the rod midway between the plates and the heat transfer rate from the rod to the air.

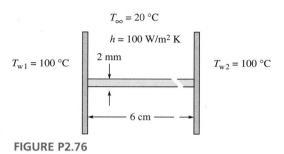

$T_\infty = 20$ °C

$h = 100$ W/m² K

$T_{w1} = 100$ °C   2 mm $T_{w2} = 100$ °C

6 cm

FIGURE P2.76

2.77 In Problem 2.76, if one plate is maintained at 100 °C and the other plate at 0 °C, find
(a) The temperature distribution in the rod.
(b) The heat transfer rate from the plate at 100 °C to the rod.
(c) The heat transfer rate from the plate at 0 °C to the rod.
(d) The total heat transfer rate from the rod to the surrounding fluid.
2.78 A steam pipe is supported by a plain carbon-steel flat plate strut. The strut is 75-mm wide, 150-mm long, and 3-mm thick. One end of the strut is welded to the pipe, which is at 200 °C, and the other end attached to a beam on the ceiling, which is at 20 °C. The strut is surrounded by air at 20 °C with a convection heat transfer coefficient of 8 W/m² °C.
(a) Estimate the heat transfer rate from the pipe to the strut.
(b) To reduce the heat transfer rate to the strut, it is proposed to insulate the strut so that the heat transfer rate to the air is negligible. Calculate the heat transfer rate to the insulated strut.

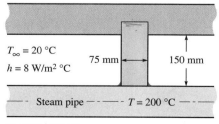

$T_\infty = 20$ °C
$h = 8$ W/m² °C

75 mm   150 mm

Steam pipe — $T = 200$ °C

FIGURE P2.78

2.79 To measure the temperature of a fluid flowing in a pipe a thermometer well is used. The analysis of extended surfaces can be applied to determine the error that may be caused in measuring the temperature of a fluid by using such a thermometer well.

Consider one such well made of brass tube 2-cm long, 4-mm I.D., and 2-mm wall thickness. A gas at 80 °C flows in the pipe with a convective heat transfer coefficient of 60 W/m^2 °C. The base of the tube (end attached to the pipe) is exposed to air at 20 °C (convective heat transfer coefficient on the air side is 7 W/m^2 °C). A thermocouple is soldered to the free end of the tube. Estimate the temperature of the thermocouple junction.

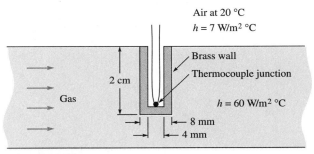

FIGURE P2.79

2.80 A solar collector is fabricated with a flat plate and a glass cover as shown. The space between the plate and the glass cover is completely evacuated. Heat is transferred to the water flowing in the tubes that are attached to the bottom surface of the plate. The bottom surface is perfectly insulated. The temperature of the plate where the tubes are attached is 30 °C. If the net radiant heat flux to the plate is 800 W/m^2, determine the maximum temperature in the plate.

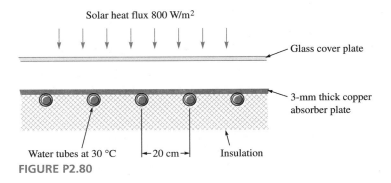

FIGURE P2.80

2.81 In Problem 2.80 the glass cover is removed. The convective heat transfer coefficient between the plate and the surrounding air is 6 W/m^2 °C, and the surrounding air temperature is 0 °C. Determine the heat transfer rate to the water per unit length of the tube.

2.82 A 2024-T6 aluminum alloy rod is attached to a plate as shown. Determine the heat transfer rate from the rod to the surrounding fluid by convection. Assume the free end to be insulated.

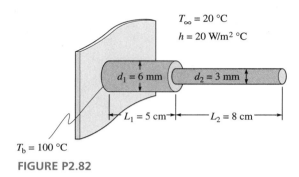

$T_\infty = 20$ °C
$h = 20$ W/m^2 °C
$d_1 = 6$ mm
$d_2 = 3$ mm
$L_1 = 5$ cm
$L_2 = 8$ cm
$T_b = 100$ °C

FIGURE P2.82

2.83 A rod is partially immersed in a liquid as shown. Find an expression for the temperature distribution in the rod. Assume the end surfaces of the rod to be perfectly insulated. The liquid temperature is T_f and the temperature of air is T_a. The heat transfer coefficients are h_f on the liquid side and h_a on the air side.

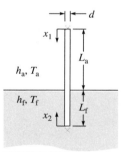

d
x_1
L_a
h_a, T_a
h_f, T_f
L_f
x_2

FIGURE P2.83

2.84 A copper tube of inner radius 4 mm, outer radius 6 mm, and length 200 mm is attached to two plates 200-mm apart. Plate A is maintained at 20 °C and plate B at 100 °C. The tube is surrounded by a fluid at 20 °C and the convective heat transfer coefficient between the outer surface of the tube and the fluid is 100 W/m^2 K. Find
(a) The heat transfer rate from plate A to the tube.
(b) The heat transfer rate from plate B to the tube.
(c) The heat transfer rate from the tube to the surrounding fluid.

2.85 Two power sources at 60 °C are connected by a plain carbon-steel rod. The convective heat transfer coefficient between the rod and the air that flows over it is 20 W/m² °C. The air temperature is 20 °C. Estimate the heat transfer rate from the rod to the surrounding air if the diameter of the rod is 6 mm and the length is 25 cm.

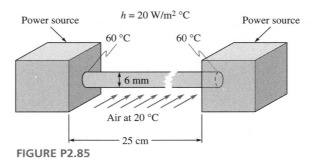

FIGURE P2.85

2.86 One end of a plain carbon steel rod is ground on a grinding wheel. The rod, 12-mm in diameter and 300-mm long, is surrounded by air at 20 °C with a convective heat transfer coefficient of 12 W/m² °C. To estimate the power dissipated in grinding, a thermocouple is installed at a distance of 15 mm from the end being ground. Under steady state conditions, the thermocouple indicates a temperature of 120 °C. The free end of the rod is pressed with a material of very low thermal conductivity. As the thermal conductivity of the material of the grinding wheel is much lower than that of the rod, the heat transfer rate to the grinding wheel may be neglected. Estimate the temperature of the surface of the rod in contact with the grinding wheel and the power dissipated due to grinding that shows as heat transfer rate.

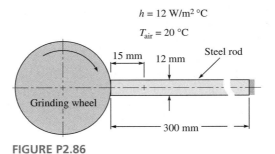

FIGURE P2.86

2.87 It is proposed to limit the temperature of an instrument package to 50 °C by providing 10 rectangular fins of aluminum (2024-T6), one of which is shown. Determine the total convective heat transfer rate from the fins to the air.

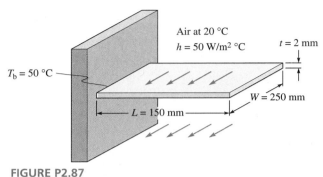

Air at 20 °C
$h = 50$ W/m² °C $t = 2$ mm

$T_b = 50$ °C

$W = 250$ mm

$L = 150$ mm

FIGURE P2.87

2.88 A fluid at 20 °C flows over a hollow, plain carbon-steel tube (3-mm I.D., 6 mm O.D., and 100-mm long) connected to two walls, one of which is maintained at 100 °C and the other at 50 °C. If the convective heat transfer coefficient is 70 W/m² °C, determine
(a) The temperature of the tube midway between the plates.
(b) The heat transfer rate from the plate at 100 °C.
(c) The heat transfer rate from the plate at 50 °C.
(d) The heat transfer rate from the tube to the surrounding fluid.

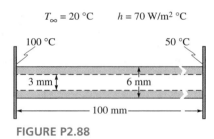

$T_\infty = 20$ °C $h = 70$ W/m² °C

100 °C   50 °C

3 mm   6 mm

100 mm

FIGURE P2.88

2.89 To increase the heat transfer rate from a surface maintained at 100 °C, a hollow brass tube 6-mm O.D., 3-mm I.D., and 10-cm long is attached to the surface. The convective heat transfer coefficient between the tube surface and the surrounding fluid (20 °C) is 100 W/m² K. Determine the convective heat transfer rate from the tube to the surrounding fluid. Assume that the heat transfer rate from the inner surface of the tube is negligible.

2.90 To estimate the thermal conductivity of a material, one end of a cylindrical rod of the material is placed in a furnace with a large part of it projecting into the room air. The diameter of the rod is 2.5 cm. In steady state the temperature at two points 10-cm apart are found to be 104 °C and 86 °C. The convective heat transfer coefficient between the surface of the rod and the air (20 °C) is 6 W/m² K. Estimate the thermal conductivity of the material of the rod.

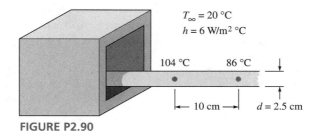

FIGURE P2.90

2.91 In an experiment, the convective heat transfer coefficient for air flowing over a cylinder is to be determined. To reduce end effects, the heat transfer rate from the end surfaces is to be reduced. Two methods are suggested.

One method is to insulate the end surface with 1-cm thick polyurethane as depicted on end A. The second method is to attach a hollow cylinder made of 0.05-mm thick stainless steel shim stock (end cap 1 cm long) as shown on end B. In both cases assume that one-dimensional temperature in the z-direction is valid. Power input to the 100-cm long cylinder is 500 W. Determine

(a) The convective heat transfer coefficient assuming the heat transfer rate from the end surfaces to be negligible.

(b) The heat transfer rates from the end surfaces. Assume that the convective heat transfer coefficient found in part a is valid every where and that contact resistance is negligible. The convective heat transfer rate from the inside surface and the free end of the end cap may be neglected. (Note that in the case of the insulated end, the temperature is two-dimensional. The assumption of the temperature being a function of the z-axis only neglects the heat transfer rate from the cylindrical surface of the insulation, thus, underestimating the total heat transfer rate from the insulation to the surroundings.)

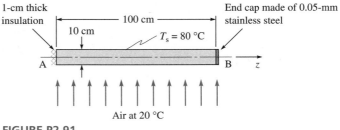

FIGURE P2.91

2.92 A power source has to dissipate its resistance losses as heat. The source is enclosed in a metal housing. To dissipate this heat, it is proposed to weld 3-mm thick plain carbon-steel plates to the housing as shown in Figure P2.92 and to blow air at 20 °C over the plates. The convective heat transfer coefficient between the plate and the flowing air is 20 W/m² K.

(a) Recommend a suitable length for the plate.

(b) For the recommended length determine the heat transfer rate from the extended surface.   **(Design)**

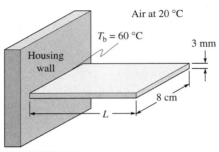

FIGURE P2.92

2.93 To estimate the thermal conductivity of a material, it is proposed to run an experiment as shown. Two very long rods of the same diameter, one of known thermal conductivity and the other whose conductivity is to be determined, are attached to a plate. The steady state temperatures at two points in the rods are measured as shown. For the measurements indicated, determine the unknown thermal conductivity. The value of the convective heat transfer coefficient for both rods may be assumed to be the same.

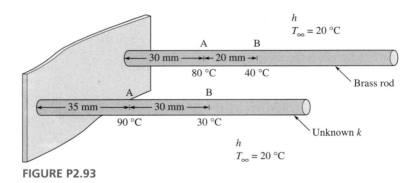

FIGURE P2.93

2.94 A handle made of 2 cm × 2 cm square steel bar is attached to the door of a furnace as shown. The temperature of the door is 100 °C and the convective heat transfer coefficient between the handle and the surrounding air (20 °C) is 8 W/m² K. Estimate the temperature at the midsection of the handle.

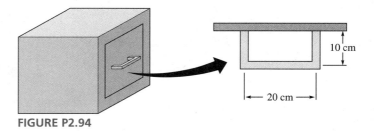

FIGURE P2.94

2.95 Two, long 1-cm diameter, plain carbon-steel rods are brought together and a current passed through the joint to raise the temperature of the joint to 200 °C. If the convective heat transfer coefficient between the rod and the surrounding air (20 °C) is 8 W/m² K, determine the minimum power input to the joint.

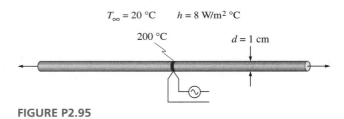

FIGURE P2.95

2.96 It is proposed to increase the heat transfer rate from two power sources whose maximum temperature is to be limited to 60 °C by connecting them with cylindrical rods. Rods of brass, copper, and steel are available. The convective heat transfer coefficient between the rods and the surrounding air is 20 W/m² °C. Suggest a suitable diameter for the rods and estimate the heat transfer rate from one rod for the suggested diameter.   **(Design)**

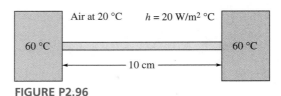

FIGURE P2.96

2.97 A steel rod 10-mm diameter is bent in the form of a semicircle of radius 40 cm and the ends are attached to a steel plate which is at 80 °C. If the surrounding air is at a temperature of 20 °C and the convective heat transfer coefficient is 6 W/m² K, estimate the temperature at the midpoint of the rod.

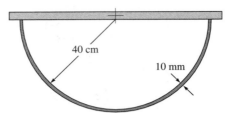

FIGURE P2.97 (see page 193)

2.98 Two very long rods, one of copper and the other of aluminum alloy, are attached to a surface maintained at T_w. The rods are exposed to air at T_∞ with a convective heat transfer coefficient of h.
(a) Find the ratio of q_{Cu}/q_{Al} if the diameters of the rods are equal.
(b) What is the ratio of heat transfer rates if the two rods have the same weight?
(c) Find the ratio of the diameters, d_{Al}/d_{Cu}, if the heat transfer rates from the two rods are equal.

2.99 For a straight fin of rectangular cross section, the heat transfer rate is given by Equation 2.6.15b. If the convective heat transfer coefficient is uniform and has the same value with and without fins, show that for $ht/2k > 1$, the use of a fin is not desirable as it will actually decrease the heat transfer rate. Assume that the width of the fin is large compared with the thickness so that $P/A_x = 2/t$, where t is the thickness of the fin.

2.100 Consider a rectangular fin with the width much greater than the thickness of the fin. For a fixed width, W, the mass of the fin is constant if the product $t \times L$ is constant (L = length of the fin). For a fin of fixed width and mass, show that the heat transfer rate is maximum if $\tanh(mL) = 3 \, mL \, \text{sech}^2(mL)$, where $m^2 = hP/kA_x = 2h/kt$. Neglect the heat transfer from the free end of the fin.

2.101 Derive (but do not solve) the differential equation for an annular fin of constant thickness.

2.102 Straight brass fins of triangular cross section are attached to a plate maintained at 80 °C. The fins are 6-mm high at the base and 10-cm long. They are exposed to air at 20 °C with a convective heat transfer coefficient of 50 W/m² °C. Assuming the width of the fins to be much greater than the height, determine
(a) Fin efficiency.
(b) The heat transfer rate from one fin per meter width of the fin.

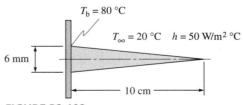

FIGURE P2.102

2.103 Annular fins of aluminum alloy are attached to the tubes of a space heater. The fins are 1-mm thick, 25-mm I.D., and 100-mm O.D. The tube surface is at 80 °C and is exposed to air at 20 °C. The convective heat transfer coefficient is 10 W/m² °C. The fins are spaced 5-mm apart and the tube is 2-m long. Assuming uniform convective heat transfer coefficient, determine the total heat transfer rate to the air.

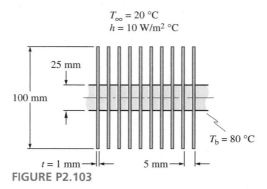

$T_\infty = 20\ °C$
$h = 10\ W/m^2\ °C$

25 mm

100 mm

$T_b = 80\ °C$

$t = 1\ mm$ 5 mm

FIGURE P2.103

2.104 In an experiment, air flows over a cylindrical copper tube (0.5-cm I.D., 1-cm O.D., and 15-cm long). The tube is attached to a plate, which is maintained at 100 °C, and is surrounded by air at 20 °C flowing over it. The heat transfer rate to the tube from the plate is found to be 11.1 W. Assuming the heat transfer rate from the free end to be negligible, estimate the convective heat transfer coefficient.

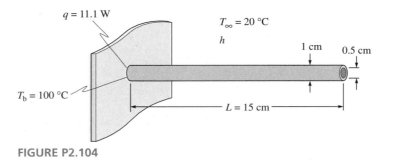

$q = 11.1\ W$

$T_\infty = 20\ °C$

h

1 cm   0.5 cm

$T_b = 100\ °C$

$L = 15\ cm$

FIGURE P2.104

2.105 A small refrigerant condenser consists of a cylindrical shell inside which water flows in a 2-cm diameter coaxial pipe. To increase the heat transfer rate, circular fins are attached to the pipe. The fins are in the form of hollow circular disks (2-cm I.D. and 8-cm O.D.) made of 1-mm thick brass. If the temperature of the base of the fin is 30 °C and the convective heat transfer coefficient between the surface and the refrigerant (40 °C) is 200 W/m² K, determine the heat transfer rate from one of the fins. (See figure.)

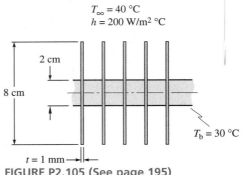

FIGURE P2.105 (See page 195)

2.106 Water, flowing in a 4-cm O.D. tube, is to be heated by combustion gases flowing over the the tube. Heat transfer to the water is increased with annular fins of plain carbon steel attached to the outer surface of the tube. The fins are 4-cm I.D., 12-cm O.D., and 1-mm thick. The convective heat transfer coefficient on the gas side is 20 W/m² °C. If the gases are at 300 °C and the tube surface at 60 °C, determine the heat transfer rate to a fin.

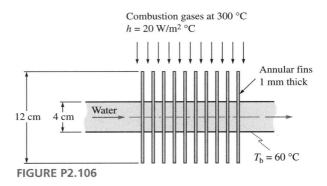

FIGURE P2.106

2.107 In Problem 2.106 the heat transfer rate is to be increased by 25%. Two possibilities are being considered:
(a) Increase the outer radius, keeping the thickness at 1 mm.
(b) Increase the thickness, keeping the outer radius at 6 cm.
Which would you recommend?

2.108 Rectangular fins of plain carbon-steel, 50-mm radial length and 2-mm thick, are attached to the outside surface of a 50-mm O.D. tube maintained at 80 °C. The fins are parallel to the axis of the tube and there are 18 fins around the tube. (Only 8 fins are shown.) Air at 20 °C flows parallel to the fins. The heat transfer rate from the tube at one location is found to be 1450 W/m length of the tube. Assuming uniform convective heat transfer coefficient on all surfaces, estimate the convective heat transfer coefficient. The heat transfer rate from the free ends of the fins may be neglected. (Hint: Solution is by iteration. Initially, assume a fin efficiency of 1 and determine h. With the computed value of h, determine the fin efficiency and repeat computations.)

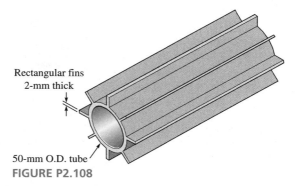

Rectangular fins
2-mm thick

50-mm O.D. tube
FIGURE P2.108

2.109 Rectangular fins are attached to a heavy plate as shown. Shown how you would compute the heat transfer rate, if
(a) The temperature T_i, of the inner surface is specified.
(b) The inner surface is exposed to a fluid at T_{fi} with a convective heat transfer coefficient of h_i.

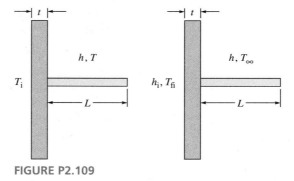

FIGURE P2.109

2.110 An electric heater is made by sandwiching a heating element between two thin plates 40-cm long and 40-cm high. When operating with air at 20 °C the heater is to transfer 2500 W with a maximum plate temperature of 120 °C. Propose a suitable system of cylindrical fins. Minimum diameter of the fins should be 3 mm with a minimum pitch of $4d$. Weight and volume are to be kept low. A fan forces air through the heater with a velocity of 5 m/s so that the convective heat transfer coefficient to the air is 80 W/m² °C.   **(Design)**

2.111 Water at 80 °C is separated from air at 20 °C by a vertical 5-mm thick plain carbon steel plate. The convective heat transfer coefficients on the water and air side are 1500 W/m² °C and 9 W/m² °C respectively. It is required to attain a heat transfer rate of 1000 W/m² (based on the area exposed to water) by providing extended surfaces on the air side. Assuming that the convective heat transfer coefficient remains at 9 W/m² °C, propose a suitable system of extended surfaces. Assume a one-dimensional temperature distribution in the solid wall, and use the concept of fin efficiency to determine the resistance to heat transfer. Acceptable materials are copper, steel, and aluminum. As heat transfer on the air side is by natural convection, you may use either vertical plates or pin fins (hollow or solid). Weight of the extended surfaces is to be kept low. **(Design)**

2.112 A copper cylinder, 1-cm diameter and 3-cm long, is initially at 20 °C. It is placed in a 100 °C air stream, which gives a convective heat transfer coefficient of 20 W/m² K over the entire cylinder including the end surfaces. How long will it take for the cylinder temperature to reach 95 °C?

2.113 A 1-cm diameter copper sphere is at a uniform temperature of 80 °C. At time $\tau = 0$, it is exposed to air at 20 °C. It is found that the convective heat transfer coefficient in air is approximated (for this case) by

$$h = 4.44 \ (T - T_\infty)^{1/4} \ \text{W/m}^2 \ °C$$

where T is the instantaneous surface temperature of the sphere and T_∞ is the temperature of the air (all temperatures in °C). Determine the time taken for the temperature of the sphere to fall from 80 °C to 60 °C and from 60 °C to 40 °C.

2.114 Coffee is served in a silver cup, which is modeled as a cylinder of 6-cm diameter and 5-cm height. The initial temperature of coffee is 80 °C. Assuming that the properties of coffee are the same as those of water and that the heat capacity of the cup (the product of the mass and specific heat of silver) is negligible compared with that of the coffee, estimate the temperature of the coffee 10 min later. The convective heat transfer coefficient between all the surfaces (including the top surface of the coffee) and the surrounding air is 6 W/m² K. Surrounding air temperature is 20 °C.

(In posing this problem, a number of simplifying assumptions have been made to get an estimate of the temperature without elaborate analysis. Discuss the differences between this simplified model and an actual case and how these differences may affect the solution obtained.)

2.115 In Problem 2.114, if the silver cup is replaced by a 3-mm thick Styrofoam cup with the same inner dimensions. Determine the temperature of the coffee as a function of time.

2.116 To sense the temperature of the air in a room, a transducer is made as a sphere of copper with a thermocouple located at the center of the sphere. The transducer is used to turn on the air conditioning system when the temperature of the air raises to 21 °C. When the temperature of the surrounding air (convective heat transfer coefficient 20 W/m² °C) increases from 20 °C to 21 °C, the temperature at the center of the sphere should increase from 20 °C to 20.1 °C in 1 min. What should be the diameter of the sphere?

2.117 A can of soft drink, 12-cm high and 6-cm diameter, is initially at 20 °C. It is placed in a refrigerator where the air is at 0 °C. The convective heat transfer coefficient associated with the outer surface of the can is 10 W/m² °C. Assuming the properties of the drink to be those of water and that the heat capacity of the container is negligible, estimate the time taken for the temperature of the drink to reach 5 °C.

2.118 A 3-cm thick plain carbon steel plate is initially at a uniform temperature of 50 °C. At time $\tau = 0$ one surface of the plate is subjected to a uniform heat flux of 5000 W/m². The other parallel surface is exposed to air at 20 °C with a convective heat transfer coefficient of 60 W/m² K. Assuming lumped analysis to be valid, determine

(a) The steady state temperature of the plate.

(b) The temperature of the plate 15 min. after the start of heating.

(c) The time taken to attain steady state. Assume that steady state is attained when the difference between the temperature of the plate and its steady state temperature is within 1% of the difference between the steady temperature of the plate and air.

2.119 A 10-cm diameter sphere of plain carbon-steel, coated with a special paint is subjected to solar radiation. The solar irradiation is 900 W/m^2 of surface perpendicular to the solar beam. The convective heat transfer coefficient between the surface and the surrounding air (20 °C) is 10 W/m^2 °C. All the solar energy is absorbed and the emitted energy is negligible. Assuming uniform temperature in the sphere, find

(a) The steady state temperature of the sphere.

(b) The time taken for the temperature to rise from the initial uniform tempeature of 20 °C to 35 °C.

2.120 The exhaust pipe of an automobile engine is initially at a uniform temperature of T_o. At time $\tau = 0$ exhaust gases at T_g flow through the tube. Assuming lumped analysis to be valid, find an expression for the temperature of the exhaust pipe as a function of time if the inside heat transfer coefficient is h_i and the outside heat transfer coefficient is h_o. The outside surface of the tube is exposed to surrounding air at T_a. Assume that the temperature of the exhaust gases is constant and uniform.

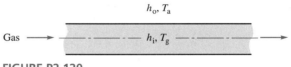

h_o, T_a

Gas $\longrightarrow$ h_i, T_g $\longrightarrow$

FIGURE P2.120

2.121 Stainless steel balls of 4-mm diameter, initially at 30 °C are heated for 1 min by blowing air at 95 °C (convective heat transfer coefficient of 40 W/m^2 °C) and then cooled in 25 °C air for 1 min (convective heat transfer coefficient of 20 W/m^2 °C). What is the temperature of the balls when they are removed from the cool air?

2.122 A metal cylinder, initially at a uniform temperature T_o is suspended with its axis vertical in air at T_∞. The heat transfer coefficient associated with the top, bottom, and cylindrical surfaces are h_T, h_B, and h_C, respectively. Find an expression for $T(\tau)$.

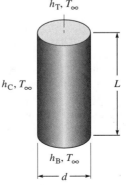

h_T, T_∞

h_C, T_∞

L

h_B, T_∞

d

FIGURE P2.122

2.123 A metal sphere, initially at a uniform temperature of T_o, is freely suspended in a large room. The temperature of the air and walls is T_∞. The emissivity of the surface of the sphere is ϵ and the convective heat transfer coefficient is h. Assuming a lumped analysis to be valid, derive the differential equation for the temperature of the sphere as a function of time. For radiative heat transfer, refer to Section 1.5, in Chapter 1.

2.124 A domestic iron is modeled as a hollow box as shown. The properties of the material of the iron may be assumed to those of plain carbon-steel. The mass of the iron is 0.5 kg. When the electrical connection is unplugged, the temperature of the iron is 90 °C. The iron is placed on an asbestos board, which may be treated as a nonconducting board. The surrounding air temperature is 20 °C and the convective heat transfer coefficient is 10 W/m² °C. Estimate the time required for the temperature of the iron to reach 40 °C (when it is be safe to handle and store it).

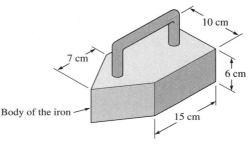

10 cm

7 cm

6 cm

Body of the iron

15 cm

FIGURE P2.124

2.125 An electrical current is passed through a stainless steel rod of diameter D and length L. The power supplied is P. If the convective heat transfer coefficient between the surface and the surrounding fluid (at T_∞) is h, assuming lumped capacity, derive an expression for the transient temperature of the rod.

2.126 A heated 2-cm diameter copper sphere is immersed in a liquid maintained at 20 °C. When the temperature of the sphere is 55 °C, it is found that the rate of decrease of the temperature of the sphere is 2.8 °C/s. Assuming constant heat transfer coefficient how long will it take for the temperature to drop from 80 °C to 30 °C?

2.127 To determine the convective heat transfer coefficient in a liquid, a 2-cm diameter copper sphere is heated to 60 °C and immersed in the liquid maintained at 20 °C. The temperature of the sphere 20 s after it was immersed in the liquid is 31 °C. Determine the average convective heat transfer coefficient.

2.128 A copper sphere of mass m_s, initially at a uniform temperature of T_{si} is immersed in a fluid of mass m_f, initially at a uniform temperature of T_{fi} ($T_{si} > T_{fi}$). As the copper sphere cools, the temperature of the fluid increases. The temperatures of the sphere T_s and the fluid T_f are both assumed to be spatially uniform and are functions of time only. Assuming constant convective heat transfer coefficient, find expressions for $T_s(t)$ and $T_f(t)$.

Hint

$$m_s\, c_{p_s} \frac{dT_s}{d\tau} = -\, m_f c_{pf} \frac{dT_f}{d\tau}; \qquad m_f c_{pf} \frac{dT_f}{d\tau} = hA_s(T_s - T_f)$$

Differentiate the second equation with respect to time and eliminate $dT_s/d\tau$ using the first equation. Solve for T_f with $T_f(0) = T_{fi}$ and at $\tau=0$, $mc_{pf}(dT_f/d\tau) = hA_s (T_{si} - T_{fi})$.

2.129 A large 25-mm thick ebonite sheet is to be heated to a temperature of 100 °C by exposing one surface of it to a stream of hot air at 150 °C. The convective heat transfer coefficient is 200 W/m² °C. Estimate the time for the minimum temperature in the sheet to rise from the initially uniform temperature of 20 °C to 100 °C.

2.130 A furnace is lined with 25-cm thick fireclay bricks. The heat transfer from the insulated outer surface of the brick is negligibly small. Initially, the bricks are at a uniform temperature of 20 °C. When the furnace is fired, the temperature of the combustion gases is maintained at 1000 °C. The convective heat transfer coefficient (including the effect of radiation) is 30 W/m² °C. Determine the temperature of the outer surface of the bricks 1 h, 2 h, and 4 h after combustion commences.

2.131 A 20-cm thick, large metallic slab, initially at a uniform temperature of 220 °C is cooled by a fluid at 20 °C. Five hundred seconds after initiating the cooling, it is found that the temperature at the surface and midplane are 43.7 °C and 187.8 °C, respectively. The density of the metal is 8200 kg/m³ and it specific heat is 489 J/kg K. Estimate the thermal conductivity of the metal and the convective heat transfer coefficient.

2.132 A long, 16-cm diameter, type 304 stainless steel cylinder, initially at a uniform temperature of 370 °C is cooled in a fluid at 20 °C. The exact value of the convective heat transfer coefficient is not known but it is known to be in the range of 1600–2000 W/m² K. Estimate the temperature at the axis of the cylinder 486 s after initiating the cooling, if the surface temperature at that instant is 35 °C.

2.133 A laccolith is a huge mass of igneous rock, approximately a sphere of radius $R = 1000$ m. The thermal diffusivity of the rock is 1.18×10^{-6} m²/s. The initial uniform temperature of the rock may be taken to be 1500 °C (between 1000 °C and 2000 °C). At time $\tau = 0$, the surface is assumed to be exposed to surroundings at a low temperature. The temperature of the surface is brought down to 20 °C and maintained at that value. This process is equivalent to exposing the rock to a fluid at 20 °C with a very high Biot number. Estimate the time taken for the center temperature to reach 100 °C. [Adapted from a problem in *Heat Conduction* by Ingersoll, L.R., Zobel, O.J., and Ingersoll, A.C., (Madiso, WI: The University of Wisconsin Press, 1954).]

2.134 Some people prefer to eat fruits that are approximately at room temperature. Consider grapefruits that are refrigerated to 4 °C. To warm them up before being served, they are taken out and exposed to room air (20 °C) for a sufficient time.

(a) Find the minimum temperature in a grapefruit 30, 60, and 180 min after it is taken out of the refrigerator and estimate the time required for the minimum temperature to reach 15 °C.

Convective heat transfer coefficient = 7 W/m² °C	Specific heat = 3900 J/kg K
Diameter of fruit = 10 cm	Thermal conductivity = 0.46 W/m K
Density = 900 kg/m³	

(b) If the fruit is immersed in water at 30 °C, with a convective heat transfer coefficient of 500 W/m² °C, estimate the time required for the minimum temperature to reach 15 °C. Note that not all of the fruit will be immersed in water, but to get an estimate of the temperature and time, assume that the convective heat transfer coefficient is uniform over the entire outer surface.

2.135 A long 10-cm diameter cylindrical bar of plain carbon steel is taken out of a furnace and quenched in a liquid maintained at 20 °C. The bar is initially at a uniform temperature of 300 °C. The convective heat transfer coefficient is 1200 W/m² °C. Determine
   (a) The temperature of the bar at the axis, 3 min after the bar is immersed in the liquid.
   (b) The time taken for the surface temperature to fall to 40 °C.

2.136 In Problem 2.135, the bar is 10-cm diameter and 20-cm long. Compute
   (a) The temperature at the midpoint on the axis of the bar 3 min after the start of quenching.
   (b) The time taken for the minimum temperature in the bar to reach 30 °C.

2.137 Before grinding glasses for a special application, they are stress relieved. Consider a disk of crown glass 50-cm diameter and 10-cm thick, initially at a uniform temperature of 20 °C. It is to be heat treated to relieve the stresses. It is placed in an oven with one of the flat surfaces on an insulating material. The convective heat transfer coefficient on the parallel surface is 9 W/m² °C. The surrounding air temperature is 300 °C. Assuming the heat transfer from the cylindrical surface to be negligible, find
   (a) The minimum temperature in the glass 90 min and 150 min after it is placed in the oven.
   (b) The time for the minimum temperature to reach 280 °C (c_p of crown glass = 840 J/kg K).

2.138 Re-solve problem 2.137, assuming the cylindrical surface to be exposed to air with a convection heat transfer coefficient of 12 W/m² °C.

2.139 A block of butter, 2.5-cm wide, 2.5-cm high, and 11-cm long, is taken out of a refrigerator and is at 2 °C. It is exposed to air at 20 °C with a convective heat transfer coefficient of 7 W/m² °C. The bottom surface may be considered to be perfectly insulated. Determine the time taken for the minimum temperature in the bar to reach 16 °C. Properties of butter: ρ = 980 kg/m³, c_p = 1040 J/kg K, and k = 0.2 W/m K.

2.140 Common bricks, at 1300 °C, are taken out of a kiln. One brick is removed from the kiln and allowed to cool in air at 30 °C. If the brick is 6-cm wide, 9-cm high, and 20-cm long, what is the minimum and maximum temperatures in the brick 30 min after it is taken out of the kiln. The convective heat transfer coefficient over all the surfaces is estimated to be 100 W/m² °C.

2.141 To estimate the thermal conductivity of a 4-cm thick steel slab, it is heated to a uniform temperature of 80 °C and cooled in a fluid at 20 °C. The temperature of the surface of the slab is found to be 43.7 °C after 34.5 s and 41.1 °C after 46 s. The density of steel is 8200 kg/m³ and the specific heat is 468 J/kg K. Estimate the thermal conductivity of the steel and the convective heat transfer coefficient employing the one-term approximate solution for the one-dimensional transient temperature distribution in a slab.

2.142 It is desired to determine the time taken for wooden beams to reach their ignition temperature when they are exposed to a fire. Consider a long, 5-cm wide and 10-cm deep maple beam, initially at a uniform temperature of 20 °C, exposed to a fire with the gases at 550 °C. The convective heat transfer coefficient is 20

W/m^2 °C. Determine, the maximum temperature in the beam 30 min after being exposed to combustion gases. The specific heat of wood is 1650 J/kg K. Assume that only three surfaces are exposed to the hot gases and that the fourth surface (5-cm wide) is insulated.

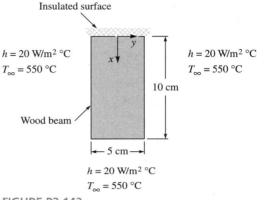

Insulated surface

$h = 20$ W/m^2 °C
$T_\infty = 550$ °C

$h = 20$ W/m^2 °C
$T_\infty = 550$ °C

10 cm

Wood beam

⊢— 5 cm —⊣

$h = 20$ W/m^2 °C
$T_\infty = 550$ °C

FIGURE P2.142

2.143 A block of ice, 25-cm thick, is placed in a box that is insulated on all sides except the top surface. Initially, it is a uniform temperature of -20 °C. It is then exposed to saturated air at -2°C. The convective heat transfer coefficient associated with the top surface of the ice block is 8 W/m^2 °C. At times 20 min and 2 h after the block is exposed to air, determine the temperature at
(a) The top surface.
(b) The bottom surface.

2.144 A block of common bricks, 1-m wide, 1-m high, and 3-m long, is heated to a uniform temperature of 320 °C. It is removed from the furnace and cooled in air at 20 °C with an associated convective heat transfer coefficient of 12 W/m^2 °C. Heat transfer occurs by convection from all surfaces except the bottom surface, which may be assumed to be insulated.
(a) Determine the maximum temperature in the block 24 h after it is removed from the furnace. Assume that the block behaves as a homogeneous material with the properties of common bricks.
(b) Find the time required for the maximum temperature in the block to be less than 60 °C.

CHAPTER

3

CONDUCTION II

In Chapter 2 we studied steady and transient thermal conduction with one-dimensional temperature distribution, constant properties, and without internal energy generation. We also presented an introduction to extended surfaces. In this chapter we

■ *Examine the effect of internal energy generation on temperature distribution and heat transfer*
■ *Study the effect of temperature dependence of thermal conductivity on the temperature profile and heat transfer in steady, one-dimensional problems*
■ *Introduce numerical methods for the solution of one- and two-dimensional steady-state problems and one-dimensional transient problems*
■ *Introduce an approximate analytical technique, the integral technique, and present some solutions to one-dimensional transient problems employing that technique*

3.1 INTERNAL ENERGY GENERATION

There are situations such as resistance heating, slow chemical reactions (exothermic or endothermic), and nuclear reactions where the temperature in a solid medium tends to increase (without heat transfer to the solid); but the material properties of interest in heat transfer—thermal conductivity, density, and specific heat—do not change significantly. Such cases are studied under *internal energy generation,* also referred to as *internal heat generation.* Internal energy generation is treated as a volumetric term. For example, if the total power dissipated in resistance heating in a volume V of a heating element is P, the internal energy generation rate per unit volume is P/V and is denoted by q'''. Without going into the thermodynamic details, we consider internal energy generation as a source of thermal energy. The effect of internal energy generation on the temperature distribution and heat transfer rate in a solid body is illustrated through several examples.

3.1.1 Rectangular Geometry

Consider the setting of cement (or concrete). During the setting, there is a slow exothermic reaction that tends to increase the temperature of the slab. The exothermic

One-dimensional temperature distribution—rectangular slab with internal energy generation

reaction is the equivalent of internal energy generation of q''' per unit volume. The steady-state temperature distribution in such a slab is to be determined (Figure 3.1.1).

ASSUMPTIONS

1. Length and width of the slab are much larger than the thickness of the slab, and end effects are negligible.
2. Temperatures on the two large parallel surfaces are uniform.
3. Rate of internal energy generation, q''', is uniform and constant.
4. Constant properties
5. Steady state

SOLUTION

Assumptions 1 through 4 imply one-dimensional temperature distribution. From an energy balance on the element of thickness Δx,

> Net heat transfer rate to the element
> + Rate of internal energy generation
> = 0

Therefore,

$$q_x - q_{x+\Delta x} + q'''A_x\Delta x = 0$$

Dividing by Δx and taking the limit as $\Delta x \to 0$,

$$-\frac{dq_x}{dx} + q''' A_x = 0$$

From Fourier's Law for a solid with one-dimensional temperature distribution, $q_x = -kA_x \, (dT/dx)$. Substituting for q_x and dividing by kA_x,

$$\frac{d^2T}{dx^2} + \frac{q'''}{k} = 0 \qquad\qquad (3.1.1)$$

(Equation 3.1.1 is a special case of Equation 2.7.1.) Integrating Equation 3.1.1 twice,

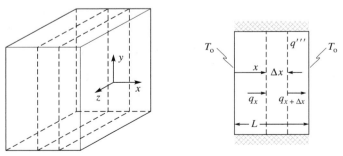

FIGURE 3.1.1 A Large Rectangular Slab with Internal Energy Generation

$$T = -\frac{q'''x^2}{2k} + c_1 x + c_2 \tag{3.1.2}$$

Constants c_1 and c_2 are to be determined from the known boundary conditions. If the temperature of the two surfaces are known and equal (T_o),

$$T(0) = T(L) = T_o$$

From the boundary conditions,

$$c_2 = T_o; \qquad c_1 = \frac{q'''L}{2k}$$

and

$$T = T_o + \frac{q'''L^2}{2k}\left(\frac{x}{L} - \frac{x^2}{L^2}\right) \tag{3.1.3}$$

Equation 3.1.3 can be recast in dimensionless form as

$$\theta = \frac{T - T_o}{(q'''L^2/2k)} = \frac{x}{L} - \frac{x^2}{L^2} \tag{3.1.4}$$

where θ is the dimensionless temperature. The temperature distribution represented by Equation 3.1.4 is shown in Figure 3.1.2.

The location where the maximum temperature occurs is found by differentiating Equation 3.1.3 with respect to x and equating the derivative to zero;[1] this gives the location of the maximum temperature as $x = L/2$ for this case.

The heat transfer rate from the surface at $x = 0$ is

$$q_{x=0} = -kA_x \frac{dT}{dx}\bigg|_{x=0} = -kA_x \frac{q'''L^2}{2k}\frac{1}{L} = -\frac{q'''LA_x}{2}$$

$q_{x=0}$ is the conductive heat transfer rate in the positive x-direction, from the surroundings to the surface. The negative sign indicates that the heat transfer is from the surface to the surroundings.

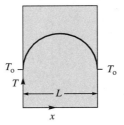

FIGURE 3.1.2 Temperature Distribution with Uniform Internal Energy Generation in a Rectangular Slab, Equation 3.1.4
Both end surfaces of the slab are at T_o.

T_o — T_o

T

L

x

---

[1]Depending on the boundary conditions, $dT/dx = 0$ may occur in the region $0 \le x \le L$ or outside that region. If it occurs outside that region, dT/dx is either positive or negative everywhere in the medium and the minimum and maximum temperatures occur at the boundaries; if $dT/dx > 0$, the temperature is a maximum at $x = L$, and if $dT/dx < 0$, the temperature is a maximum at $x = 0$.

You will observe that $(q'''LA_x)/2$ is half the total internal energy generation. In a similar manner the heat transfer rate from the surface located at $x = L$ is found to be $(q'''LA_x)/2$, and the total *net* heat transfer rate is $q'''LA_x$, the total internal energy generation. The heat transfer rate from each surface to the surroundings is half the total internal energy generation. This is only because the internal energy generation, and the boundary conditions are symmetric about the midplane of the slab. If either of them is not symmetric, the heat transfer rate from each surface would be different, but the total net heat transfer rate from the two surfaces to their surroundings would still be equal to the total internal energy generation rate.

3.1.2 Cylindrical Geometry

One-dimensional temperature distribution—hollow cylinder with internal energy generation

A current, I, flows through a long wire of diameter $2r_o$. The wire is surrounded by a fluid at T_∞ with a heat transfer coefficient of h (Figure 3.1.3). We wish to determine the radial temperature distribution in the wire.

ASSUMPTIONS

1. The wire is sufficiently long so that end effects are negligible.
2. Constant properties
3. Steady state
4. If the resistance of the wire is R, $q''' = I^2 R/(\pi r_o^2 L)$. q''' is uniform and constant.

SOLUTION

From assumptions 1 and 2 the temperature depends on the radius only. From an energy balance on the cylindrical element of radial thickness Δr (shown as the shaded area in Figure 3.1.3)

Net heat transfer rate to the element
+ rate of internal energy generation
= 0

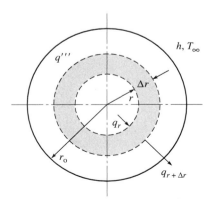

FIGURE 3.1.3 A Long Wire with a Uniform Internal Energy Generation Resulting from Resistance Heating

Therefore,

$$q_r - q_{r+\Delta r} + q''' \, 2\pi r \, \Delta r \, L = 0$$

Dividing by Δr and taking the limit as $\Delta r \to 0$,

$$-\frac{dq_r}{dr} + q''' \, 2\pi r \, L = 0$$

By applying Fourier's Law for the one-dimensional temperature distribution, $q_r = -kA_r \, (dT/dr)$. With $A_r = 2\pi rL$, after substituting the expression for q_r in the above equation and dividing by $2\pi kL$,

$$\frac{d}{dr}\left(r\,\frac{dT}{dr}\right) + \frac{q'''r}{k} = 0 \qquad \textbf{(3.1.5)}$$

(Note that Equation 3.1.5 is a special case of Equation 2.7.1.) Integrating Equation 3.1.5 twice,

$$T = -\frac{q'''r^2}{4k} + c_1 \ln r + c_2$$

The constants c_1 and c_2 are determined from the applicable boundary conditions. From the statement of the problem, there is only one boundary condition, a convective boundary condition at the surface, and a second independent boundary condition is not explicitly available. But from physical considerations, the statement of the problem is complete, and it should be possible to evaluate both the constants. The second boundary condition is provided by requiring that the temperature at the axis, $T(0)$ be finite. Therefore, the logarithmic term is not applicable, or $c_1 = 0$. From an energy balance on the outer surface of the wire,

Conductive heat transfer rate to the outer surface (from the interior of the wire)
= Convective heat transfer rate from the outer surface

$$-kA_r \left.\frac{dT}{dr}\right|_{r=r_o} = hA_r \left.(T - T_\infty)\right|_{r=r_o}$$

or

$$-k\left(-\frac{q'''r_o}{2k}\right) = h\left(-\frac{q'''r_o^2}{4k} + c_2 - T_\infty\right)$$

$$c_2 = T_\infty + \frac{q'''r_o^2}{4k}\left(1 + \frac{2k}{hr_o}\right)$$

and

$$T = \frac{q'''r_o^2}{4k}\left(1 + \frac{2k}{hr_o} - \frac{r^2}{r_o^2}\right) + T_\infty \qquad \textbf{(3.1.6)}$$

Equation 3.1.6 can be recast in dimensionless form as

$$\theta = \frac{T - T_\infty}{(q'''r_o^2)/4k} = 1 + \frac{2k}{hr_o} - \frac{r^2}{r_o^2} \qquad \textbf{(3.1.7)}$$

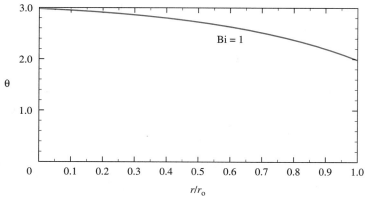

FIGURE 3.1.4 Dimensionless Temperature Distribution in a Wire with Uniform Internal Energy Generation, $hr_o/k = 1$

The dimensionless temperature distribution for Biot number $hr_o/k = 1$ is shown in Figure 3.1.4. The heat transfer from the outer surface, q_s, is found from

$$q_s = -kA_r \left. \frac{dT}{dr} \right|_{r=r_o} = -k2\pi r_o L \left(-\frac{q'''r_o}{2k} \right) = q'''\pi r_o^2 L$$

$q'''r_o^2 L$ is also the total internal energy generation.

COMMENT

Note that the dimensionless temperature is a function of the two dimensionless parameters, the dimensionless distance (r/r_o) and the Biot number hr_o/k and that ($q'''r_o^2/4k$) represents the characteristic temperature difference. As the Biot number increases, indicating increasing h, the difference between the temperatures of the surface and the surrounding fluid (for constant internal energy generation) decreases and for a Biot number of 100, the difference is very small; i.e., the temperature of the surface is very close to the temperature of the surrounding fluid. For low values of the Biot number (low values of h) the difference in temperatures is large. Also, the difference between the dimensionless temperatures at the axis and the surface is 1 or the difference in dimensinoal temperatures is ($q'''r_o^2/4k$) and is independent of the Biot number. The physical significance of the characteristic temperature difference, ($q'''r_o^2$)/4k, is that it represents the difference between the temperatures at the axis and the surface for any value of h or T_∞.

EXAMPLE 3.1.1 An electric current flows through the walls of a long, hollow stainless steel (type 304) tube of outer diameter 6 mm and radial thickness 1 mm (Figure 3.1.5). The convective heat transfer coefficient between the air and the outer surface of the tube is 15 W/m² °C. If the surface temperature is not to exceed 40°C above the temperature of the surrounding air, find the maximum current that the tube can carry and the maximum temperature in the tube.

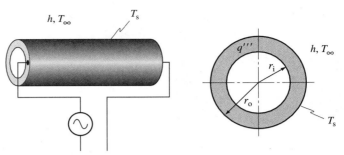

FIGURE 3.1.5 Internal Energy Generation Due To Resistance Heating

Given

$r_i = 0.002$ m $\qquad h = 15$ $W/m^2\,°C$

$r_o = 0.003$ m $\qquad T_s - T_\infty = 40\,°C$

Find

The maximum current in the wire (I) and the maximum temperature in the tube.

ASSUMPTIONS

1. Steady state
2. Constant properties
3. Uniform internal energy generation
4. From an energy balance on the air inside the tube, in steady state, the heat transfer rate to the air is zero. Therefore, the heat transfer rate to the air from the inner surface of the tube—the bounding surface for the air—is zero.
5. Negligible end effects. As the length of tube is much greater than the radial thickness of the tube, the end effects are restricted to a small region toward the ends of the tube and are neglected. If the tube is short, not very much greater than the thickness of the tube, the end effects are significant and lead to a two-dimensional temperature distribution.

SOLUTION

The electric power supply to the tube (P) is treated as work transfer to the tube. From an energy balance on the tube,

> Heat transfer rate to the tube
> \+ Work transfer rate to the tube (P)
> $= 0$

As the end effects are assumed to be negligible, and there is no heat transfer form the inner surface,

> Heat transfer rate from the outer surface $= P$

Therefore,

$$h2\pi r_o L\,(T_s - T_\infty) = P = q'''\pi(r_o^2 - r_i^2)L = I^2 R$$

where

T_s = surface temperature
R = electrical resistance of the wire

$$q''' = \frac{2hr_o\,(T_s - T_\infty)}{r_o^2 - r_i^2}$$

With $T_s - T_\infty = 40\,°C$, substituting the other known value we get

$$q''' = \frac{2 \times 15 \times 0.003 \times 40}{(0.006^2/4) - (0.004^2/4)} = \underline{720\,000\ \text{W/m}^3}$$

Also

$$I^2 R = q'''\,\pi(r_o^2 - r_i^2)L$$

$$I^2 = \frac{q'''\,\pi(r_o^2 - r^2 i)}{R/L}$$

R/L is the resistance per unit length. To find the value of R/L, for stainless steel (type 304), the electrical resistivity ρ is 72 $\mu\Omega$ cm or 0.72 $\times$ $\times$ $10^{-6}\Omega$ m (from CRC *Handbook of Physics and Chemistry,* 1991).

$$R/L = \frac{\rho}{A_c} = \frac{0.72 \times 10^{-6}}{(\pi/4)\,(0.006^2 - 0.004^2)} = 0.04584\ \Omega/\text{m}$$

$$I = \sqrt{\frac{q'''\pi\,(r_o^2 - r_i^2)}{R/L}} = \left(\frac{720\,000 \times \pi(0.003^2 - 0.002^2)}{0.04584}\right)^{0.5} = \underline{15.71\ \text{A}}$$

We need the temperature distribution to find the location of the maximum temperature and to evaluate it. The differential equation for the region $r_i \le r \le r_o$, where internal energy generation occurs is (Equation 3.1.5)

$$\frac{d}{dr}\left(r\,\frac{dT}{dr}\right) + \frac{q'''r}{k} = 0 \qquad (1)$$

The surface temperature T_s is known and forms one of the two required boundary conditions. From assumption 4, the inner surface is adiabatic. The two boundary conditions are, then given by

$$T(r_o) = T_s = T_\infty + 40 \qquad (2)$$

and

$$\frac{dT}{dr}\,(r_i) = 0 \qquad (3)$$

Integrating Equation 1 twice,

$$T = -\frac{q'''r^2}{4k} + c_1\,\ln r + c_2$$

From Equation 3,

$$c_1 = \frac{q'''r_i^2}{2k}$$

From Equation 2,

$$T_s = -\frac{q'''r_o^2}{4k} + \frac{q'''r_i^2}{2k} \ln r_o + c_2$$

or

$$c_2 = T_s + \frac{q'''r_o^2}{4k} - \frac{q'''r_i^2}{2k} \ln r_o$$

As $dT/dr = 0$ at the inner surface, the temperature is a maximum at that surface. With heat flow in the r-increasing direction, the temperature decreases in the r-increasing direction and the temperature at the inner surface can only be a maximum. This can also be verified by checking if the second derivative at r_i is positive or negative.

$$T_{max} = T(r_i) = -\frac{q'''r_i^2}{4k} + \frac{q'''r_i^2}{2k} \ln r_i + T_s + \frac{q'''r_o^2}{4k} - \frac{q'''r_i^2}{2k} \ln r_o \tag{4}$$

$$T_{max} - T_s = -\frac{q'''r_i^2}{4k} + \frac{q'''r_i^2}{2k} \ln r_i + \frac{q'''r_o^2}{4k} - \frac{q'''r_i^2}{2k} \ln r_o$$

From Table A1, for type 304 stainless steel, $k = 14.9$ W/m °C. Substituting all the values into Equation 4, we obtain $T_{max} - T_s = 0.021$ °C. Note the small difference between the temperatures at the inner and outer surfaces.

3.2 EFFECT OF VARIABLE THERMAL CONDUCTIVITY

Temperature dependent thermal conductivity

The thermal conductivity of a material is, in general, temperature dependent. In most cases, where the range of temperature occurring in a problem is not large, we may use a constant value for the conductivity, evaluated at a suitable average temperature. We now consider the effect of the temperature dependence of thermal conductivity on the heat transfer rate and the temperature distribution.

The temperature dependency of the thermal conductivity can be expressed as a polynomial in temperature, $k = k_o (1 + aT + bT^2 + \ldots)$, where k_o, a, b, and so on are constants determined by a regression analysis. The number of terms in the polynomial depends on the number of data points available, the range of temperature in the solid, and the accuracy desired. In many cases a linear relationship of the type $k = k_o(1 + aT)$ gives results that are acceptable in engineering applications. Such a linear relation is used to determine the temperature distribution and the heat transfer rates. The procedure for determining the desired quantities is illustrated through an example.

EXAMPLE 3.2.1 One surface of a 20-cm thick plain carbon-steel slab is maintained at 500 K and the parallel surface at 700 K (Figure 3.2.1). Determine the temperature at a point 5 cm from the surface at 500 K and the heat flux across the slab.

Given

$T_1 = 500\ K$ $L = 20$ cm
$T_2 = 700\ K$ $k = k(T)$

Find

(a) $T(x = 5$ cm$)$
(b) q''_x

ASSUMPTIONS

1. Steady state
2. Thermal conductivity is a linear function of the temperature.

SOLUTION

From the assumptions, the temperature is one-dimensional. From Table A1, the properties of plain carbon-steel are

Temperature (K)	400	600	800	1000
k (W/m K)	56.7	48.0	39.2	30.0

As the temperature in the slab is in the range of 500 K–700 K, we use the values of the conductivity at temperatures of 400 K, 600 K and 800 K to find the functional form of $k = k_o(1 + aT)$. From a linear regression analysis

$$k = 74.2 - 0.04375\ T = k_o(1 + aT)$$

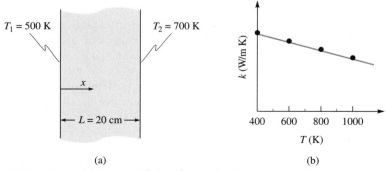

(a)   (b)

FIGURE 3.2.1 (a) Two parallel surfaces of a thick, plain carbon-steel slab are maintained at 500 K and 700 K. (b) Variation of thermal conductivity with temperature.

where

$$k_o = 74.2 \text{ W/m K}$$
$$a = -5.896 \times 10^{-4} \text{ K}^{-1}$$

From Figure 3.2.1b, showing the actual thermal conductivity (by circles and the linear approximation (solid line), it is clear that the thermal conductivity is very well approximated by the linear relationship. For a rectangular slab with one-dimensional temperature distribution, no internal energy, and with variable thermal conductivity,

$$\frac{d}{dx}\left(k\frac{dT}{dx}\right) = 0 \tag{1}$$

The boundary conditions are

$$T(0) = T_1 \qquad T(L) = T_2 \tag{2}$$

Integrating Equation 1 once,

$$k\frac{dT}{dx} = c_1 \tag{3}$$

As $-k(dT/dx) = $ heat flux $= q''$, the constant c_1 may be replaced by $-q''$. The magnitude of the heat flux, which has replaced the constant of integration, is still not known; it is to be evaluated from the known boundary conditions. With $k = k_o(1 + aT)$, integrating Equation 3,

$$k_o\left(T + \frac{aT^2}{2}\right) = -q'' x + c_2 \tag{4}$$

From the first boundary condition in Equation 2,

$$T(0) = T_1 \qquad k_o\left(T_1 + \frac{aT_1^2}{2}\right) = c_2 \tag{5}$$

Using the second boundary condition in Equation 2,

$$k_o\left(T_2 + \frac{aT_2^2}{2}\right) = -q'' L + c_2 = -q''L + k_o\left(T_1 + \frac{aT_1^2}{2}\right)$$

or

$$q'' = \frac{k_o\left[(T_2 + aT_2^2/2) - (T_1 + aT_1^2/2)\right]}{L} \tag{6}$$

Substituting the values

$$k_o = 74.2 \text{ W/m K}$$
$$T_1 = 500 \text{ K}$$
$$a = -5.896 \times 10^{-4}/\text{K}$$
$$T_2 = 700 \text{ K}$$
$$L = 0.2 \text{ m}$$

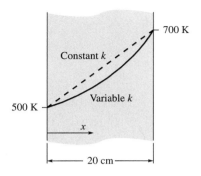

FIGURE 3.2.2 Temperature Distribution with Variable Thermal Conductivity

we obtain

$$c_2 = 31\,631 \text{ W/m} \quad \text{and} \quad q'' = -47\,951 \text{ W/m}^2$$

The temperature distribution is given by Equation 4,

$$k_o \left(T + \frac{aT^2}{2} \right) = -q''x + c_2$$

where the values of the constants q'' and c_2 are now known. Hence, for any given value of x, the resulting quadratic equation can be solved for the temperature.[2] At $x = 0.05$ m

$$\underline{T = 546.7 \text{ K}}$$

The temperature distribution is shown in Figure 3.2.2. The solid line represents the temperature with variable thermal conductivity and the dashed line is the temperature distribution with constant thermal conductivity.

Let us now compare the temperature and the heat transfer rate that we would have obtained by using a constant value for the thermal conductivity, evaluated at the average temperature of 600 K (from the equation obtained from linear regression). The thermal conductivity at 600 K is 47.95 W/m K. The value obtained from the linear regression is slightly different from the value at 600 K as the variation of the conductivity is not exactly linear with temperature. With a constant thermal conductivity of 47.95 W/m K, the temperature at $x = 0.05$ m is 550 K. For the heat flux we have

$$q'' = 47.95 \, \frac{500 - 700}{0.2} = -47,950 \text{ W/m}^2$$

---

[2]There are two roots to the quadratic equation. One value of the temperature will be out of the range of temperature in the solid.

In Example 3.2.1 there is some difference in the value of the temperature at 5 cm from the origin by the two methods, but the values of the heat fluxes are surprisingly close to each other. This is no accident. If there is no internal energy generation, it is easily shown that when the thermal conductivity varies linearly with temperature, the heat transfer rate can be calculated accurately by using a constant value of the thermal conductivity, evaluated at the mean value of the temperature of the two bounding surfaces. The expression for q'' in Equation 6 can be rearranged as

$$q'' = \frac{k_o}{L} \left(T_1 - T_2 + a \, \frac{T_1^2 - T_2^2}{2} \right)$$

$$= \frac{k_o}{L} (T_1 - T_2) \left(1 + a \, \frac{T_1 + T_2}{2} \right) = k_m \, \frac{T_1 - T_2}{L}$$

where

$$k_m = k_o \left(1 + a \, \frac{T_1 + T_2}{2} \right)$$

k_m is the value of the conductivity evaluated at the mean temperature $T_m = (T_1 + T_2/2)$. Although the correct value of the heat transfer rate is obtained by using the mean thermal conductivity, the temperature distribution is different, however.

In a more general sense, it can be shown that, for one-dimensional, steady-state conduction without internal energy generation, if the thermal conductivity varies linearly with temperature, the heat transfer rate can be determined by using a constant thermal conductivity evaluated at the mean temperature. Consider two cases:

$$k = k_o \, (1 + aT) \quad \text{and} \quad k = k_{const} = \text{constant}$$

For both cases, the differential equation is

$$\frac{d}{dx} \left(kA_x \frac{dT}{dx} \right) = 0$$

Integrating once

$$kA_x \frac{dT}{dx} = - q$$

where q is the heat transfer rate in the x-direction. Separating the variables and integrating again,

$$\frac{1}{T_1 - T_2} \int_{T_1}^{T_2} k \, dT = - \frac{1}{T_1 - T_2} \int_{x_1}^{x_2} \frac{q}{A_x} \, dx$$

For a given set of values of q, T_1, and T_2, the right side of the equation is a function of the geometry only and is the same for both cases whether k is a function of the temperature or it is constant.

For $k(T) = k_o(1 + aT)$.

$$k_o \left(1 + a \, \frac{T_1 + T_2}{2} \right) = - \frac{1}{T_1 - T_2} \int_{x_1}^{x_2} \frac{q}{A_x} \, dx$$

For $k = $ constant,

$$k_{\text{const}} = -\frac{1}{T_1 - T_2} \int_{x_1}^{x_2} \frac{q}{A_x} \, dx$$

Hence,

$$k_{\text{const}} = k_o \left(1 + a \frac{T_1 + T_2}{2} \right) = k_m$$

Mean thermal
conductivity

where k_m is the thermal conductivity evaluated at the arithmetic mean temperature. This relation applies to cases where A_x varies, as in cylindrical geometry, if the mean thermal conductivity is defined as

$$k_m = \frac{1}{T_1 - T_2} \int_{T_1}^{T_2} k \, dT$$

3.3 MULTIDIMENSIONAL TEMPERATURE DISTRIBUTION

Analytical solutions for multidimensional temperature distribution, with or without internal energy generation, are available for a number of cases. However, when the geometry is irregular, when the material is nonhomogeneous, or when the differential equation is nonlinear, analytical solutions are either very complex or do not exist. In such cases, numerical solutions are particularly useful. Numerical solutions to two-dimensional steady-state problems are presented in Sections 3.3.1 and 3.3.2 and to one-dimensional transient cases in Section 3.4.2.

3.3.1 Two-Dimensional, Steady-State Conduction Problems—Numerical Solutions

In the numerical solutions to conduction problems, the temperature gradients are approximated by values of the temperatures at finite intervals. For example, if the temperatures at two locations are $T_1(x_1)$ and $T_2(x_2)$, the temperature gradient may be approximated by the linear relationship $(T_2 - T_1)/(x_2 - x_1)$. The gradients may also be approximated by more realistic relationships, such as taking three points and passing a parabola through the three points. The equations representing the physical laws, such as the balance of energy, through such approximations to the derivatives are known as *difference equations*.

Difference
equation

The difference equations may be obtained either by starting with the differential equations or by applying the balance laws to a finite volume in the material. In conduction problems, the latter method is slightly less accurate than other more sophisticated differencing schemes, but it offers the advantages of a physical insight for the differencing scheme, ease of understanding, and simplicity of implementation—features that are particularly desirable at the introductory stage.

Numerical methods
in steady-state
conduction

To illustrate some of the possible methods of obtaining the difference equations, consider the two-dimensional, steady-temperature distribution in a rectangular slab. The slab is divided into a grid with lines parallel to the axes. The points of intersection of the lines are known as nodes. The problem is to find the temperatures at these nodes. If the temperatures at the nodes are determined, we have an approximate temperature distribution. The approximate temperature distribution also yields the heat flow in the slab. To obtain the temperatures at the nodes, difference equations are derived at every node where the temperature is to be determined. The solution to the system of difference equations yields the temperature distribution.

Differential equation for the two-dimensional steady-state temperature distribution

The differential equation for the two-dimensional temperature distribution is obtained by performing an energy balance on an element of dimensions Δx and Δy along the x- and y-directions, and W along the z-direction as shown in Figure 3.3.1

In steady state with no internal energy generation, the net heat transfer rate to the element is zero. Therefore,

$$\delta q_x|_{x,y} - \delta q_x|_{x+\Delta x,y} + \delta q_y|_{x,y} - \delta q_x|_{x,y+\Delta y} = 0$$

Substituting $\delta q_x = q_x'' \, W \, \Delta y$ and $\delta q_y = q_y'' \, W \, \Delta x$,

$$q_x'' \, W \, \Delta y|_{x,y} - q_x'' \, W \, \Delta y|_{x+\Delta x,y} + q_y'' \, W \, \Delta x|_{x,y} - q_x'' \, W \, \Delta x|_{x,y+\Delta y} = 0$$

Dividing by $W \, \Delta x \, \Delta y$ and taking the limit as $\Delta x \to 0$ and $\Delta y \to 0$.

$$-\frac{\partial q_x''}{\partial x} - \frac{\partial q_y''}{\partial y} = 0$$

From Fourier's Law, $q_x'' = -k \, (\partial T/\partial x)$ and $q_y'' = -k \, (\partial T/\partial y)$; with the assumption of uniform k,

$$\frac{\partial^2 T}{\partial x^2} + \frac{\partial^2 T}{\partial y^2} = 0 \qquad\qquad \textbf{(3.3.1)}$$

(A complete derivation of the general two-dimensional conduction equation is given in Chapter 11.)

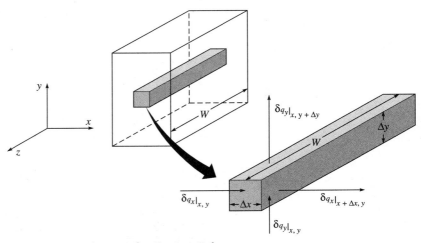

FIGURE 3.3.1 Element for Energy Balance

Linear
approximation for
derivatives

1. Linear Approximation of Derivatives. Referring to Figure 3.3.2, one method of arriving at the difference equation is to approximate the derivatives in Equation 3.3.1 as

$$\frac{\partial^2 T}{\partial x^2}(x,y) \approx \frac{\frac{\partial T}{\partial x}\left(x + \frac{\Delta x}{2}, y\right) - \frac{\partial T}{\partial x}\left(x - \frac{\Delta x}{2}, y\right)}{\Delta x}$$

With

$$\frac{\partial T}{\partial x}\left(x + \frac{\Delta x}{2}, y\right) \approx \frac{T(x + \Delta x, y) - T(x,y)}{\Delta x}$$

$$\frac{\partial T}{\partial x}\left(x - \frac{\Delta x}{2}, y\right) \approx \frac{T(x,y) - T(x - \Delta x, y)}{\Delta x}$$

we get

$$\frac{\partial^2 T}{\partial x^2}(x,y) = \frac{T(x - \Delta x,y) - 2T(x,y) + T(x + \Delta x,y)}{\Delta x^2}$$

Similarly,

$$\frac{\partial^2 T}{\partial y^2}(x,y) = \frac{T(x,y - \Delta y) - 2T(x,y) + T(x,y + \Delta y)}{\Delta y^2}$$

Substituting the expressions for the derivatives in Equation 3.3.1, we obtain the difference equation as

$$\frac{T(x - \Delta x, y) - 2T(x, y) + T(x + \Delta x, y)}{\Delta x^2}$$

$$+ \frac{T(x, y - \Delta y) - 2T(x, y) + T(x, y + \Delta y)}{\Delta y^2} = 0$$

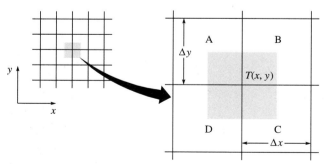

FIGURE 3.3.2 Sketch Showing the Grid and the Material Associated with a Node for the Development of Difference Equations for Rectangular Geometries

2. **Taylor Series Expansion.** Another way to obtain the difference equation is to expand $T(x + \Delta x, y)$ and $T(x - \Delta x, y)$ in a Taylor series about (x, y).

$$T(x + \Delta x, y) = T(x, y) + \Delta x \frac{\partial T}{\partial x} + \frac{\Delta x^2}{2!} \frac{\partial^2 T}{\partial x^2} + \frac{\Delta x^3}{3!} \frac{\partial^3 T}{\partial x^3} + \frac{\Delta x^4}{4!} \frac{\partial^4 T}{\partial x^4} + \cdots$$

$$T(x - \Delta x, y) = T(x, y) - \Delta x \frac{\partial T}{\partial x} + \frac{\Delta x^2}{2!} \frac{\partial^2 T}{\partial x^2} - \frac{\Delta x^3}{3!} \frac{\partial^3 T}{\partial x^3} + \frac{\Delta x^4}{4!} \frac{\partial^4 T}{\partial x^4} + \cdots$$

Adding the two equations, dividing by Δx^2, and rearranging, we obtain

Approximation of derivatives by Taylor series expansion

$$\frac{\partial^2 T}{\partial x^2} = \frac{T(x - \Delta x, y) - 2T(x, y) + T(x + \Delta x, y)}{\Delta x^2} - \frac{2\Delta x^2}{4!} \frac{\partial^4 T}{\partial x^4} - \cdots$$

Similarly, for $\dfrac{\partial^2 T}{\partial y^2}$ we get

$$\frac{\partial^2 T}{\partial y^2} = \frac{T(x, y - \Delta y) - 2T(x, y) + T(x, y + \Delta y)}{\Delta y^2} - \frac{2\Delta y^2}{4!} \frac{\partial^4 T}{\partial y^4} - \cdots$$

If we approximate $\dfrac{\partial^2 T}{\partial x^2}$ by $[T(x - \Delta x, y) - 2T(x, y) + T(x + \Delta x, y)]/\Delta x^2$, it is clear that we are neglecting $- (2\Delta x^2/4!)\,(\partial^4 T/\partial x^4) - (2\Delta x^4/6!)\,(\partial^6 T/\partial x^6) - (2\Delta x^8/8!)\,(\partial^8 T/\partial x^8) - \ldots$ If we assume that all derivatives are of the same order of magnitude as $\partial^2 T/\partial x^2$, the error in the approximation is of order of magnitude of Δx^2.

Adding the two second derivatives and equating the result to zero, we obtain the difference equation as

$$\frac{T(x - \Delta x, y) - 2T(x, y) + T(x + \Delta x, y)}{\Delta x^2}$$
$$+ \frac{T(x, y - \Delta y) - 2T(x, y) + T(x, y + \Delta y)}{\Delta y^2} = 0$$

Energy balance method

3. **Energy Balance Method.** The last method we consider, and the one we use to illustrate the numerical procedure, is based on the application of the balance of energy and a suitable approximation of the terms in the equation. For example, for the slab under consideration, the node (x, y) is associated with the material shaded in Figure 3.3.2. An energy balance applied to this material requires that the net heat transfer to the element be zero or

$$q^+_{x-\Delta x/2} + q^+_{x+\Delta x/2} + q^+_{y-\Delta y/2} + q^+_{y+\Delta y/2} = 0$$

where $q^+_{x-\Delta x/2}$ represents the heat transfer rate to the element across AD, $q^+_{x+\Delta x/2}$ represents the heat transfer rate to the element across BC, and so on (the superscript $+$ with q denotes that the heat transfer is to the element across each surface). The heat transfer rates across the boundaries are approximated by assuming a linear temperature profile between the node at $\{x, y\}$ and its neighboring nodes situated at $\{x - \Delta x, y\}$, $\{x + \Delta x, y\}$ $\{x, y - \Delta y\}$ and $\{x, y + \Delta y\}$. We then get, per unit width perpendicular to the plane of the paper,

$$q_{x-\Delta x/2}^{+} = k\,\Delta y\,\frac{T(x-\Delta x, y) - T(x, y)}{\Delta x} \qquad q_{x+\Delta x/2}^{+} = k\,\Delta y\,\frac{T(x+\Delta x, y) - T(x, y)}{\Delta x}$$

$$q_{y-\Delta y/2}^{+} = k\,\Delta x\,\frac{T(x, y-\Delta y) - T(x, y)}{\Delta y} \qquad q_{y+\Delta y/2}^{+} = k\,\Delta x\,\frac{T(x, y+\Delta y) - T(x, y)}{\Delta y}$$

With the assumption of constant thermal conductivity, after summing all the heat transfer rates from the neighboring nodes, and dividing by $\Delta x\,\Delta y$, we obtain the difference equation

$$\frac{T(x-\Delta x, y) - 2T(x, y) + T(x+\Delta x, y)}{\Delta x^2}$$

$$+ \frac{T(x, y-\Delta y) - 2T(x, y) + T(x, y+\Delta y)}{\Delta y^2} = 0$$

All three methods yield the same difference equation. The first two methods are purely a mathematical manipulation of the differential equation derived from the application of the balance of energy. The last method is based on the application of the balance of energy, which gives us a physical appreciation for the resulting difference equation. This last method is also easier to apply at the boundaries. However, it has the disadvantage that it offers no information on the accuracy of the difference equation, such as can be obtained by the application of the Taylor's series expansion.

3.3.2 Difference Equations for Two-Dimensional, Steady-State Problems

Consider a block shown in Figure 3.3.3 where a two-dimensional temperature distribution is expected. For example, it may represent one-quarter of the cross section of a long rectangular furnace or a chimney. The temperature distribution in such cases is obviously two- or three-dimensional. In this section, to illustrate the methodology, the temperature distribution is considered to be two-dimensional. Figure 3.3.3 shows the grid and the nodes. The nodes are identified by two indices, $\{i, j\}$, i running parallel to the x-axis and j running parallel to the y-axis. Each node also

Numerical procedure—steady state, two-dimensional temperature distribution

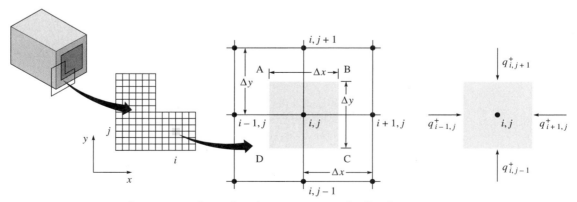

FIGURE 3.3.3 Determing a Two-Dimensional Temperature Distribution
The sketch represents a quadrant of the cross section of a long furnace.

identifies the material whose boundaries lie midway between the nodes. For example, the node $\{i, j\}$ identifies the material shaded in the figure. The distances between the neighboring nodes are Δx and Δy; these need not be equal to each other or be constant over the entire region. To keep the discussion simple, and to illustrate the methodology, the grid is so chosen the Δx and Δy are constant.

The starting point for the derivation of the difference equation is the balance of energy. In steady state with no internal energy generation the net rate of heat transfer to the material associated with a node is zero. Now, consider the element associated with the node i, j. The sum of the heat transfer rates from the neighboring elements must be zero.

$$q^+_{i-1,j} + q^+_{i+1,j} + q^+_{i,j-1} + q^+_{i,j+1} = 0$$

The subscripts indicate the neighboring elements of $\{i,j\}$. Our intent is to determine the temperatures at these nodes. The closer the nodes, the more accurate is the resulting solution. In the following discussion, the terms east, west, north, and south are used to facilitate identifying the location of the nodes relative to the node $\{i,j\}$. The heat transfer rate to the element across each surface is denoted by q^+ with the subscript denoting the node from which the heat transfer occurs. As the temperature distribution is two-dimensional, it is sufficient to consider a unit depth of the cross section in the z-direction. Consider the heat transfer rate to the element across its west boundary from the element denoted by $i-1,j$. The thermal conductivity of the element at the west boundary is k, and the area of the surface AD is Δy. The temperature gradient in the x-direction at this surface is approximated by $(T_{i,j} - T_{i-1,j})/\Delta x$. From Fourier's law applied to this surface, the heat transfer rate to the element i,j across AD is approximated by

$$q^+_{i-1,j} \approx k\Delta y \frac{T_{i-1,j} - T_{i,j}}{\Delta x}$$

In a similar manner, the heat transfer rates across the other boundaries are approximated by

Interior node— difference equation

$$q^+_{i+1,j} \approx k\Delta y \frac{T_{i+1,j} - T_{i,j}}{\Delta x}$$

$$q^+_{i,j-1} \approx k\Delta x \frac{T_{i,j-1} - T_{i,j}}{\Delta y}$$

$$q^+_{i,j+1} \approx k\Delta x \frac{T_{i,j+1} - T_{i,j}}{\Delta y}$$

If the material is such that its thermal conductivity is a strong function of space coordinates or temperature, you only need to determine its value appropriate to the particular boundary. For the present, assuming that the thermal conductivity is constant and summing the heat transfer rates to the node $\{i,j\}$, we get

$$\frac{\Delta y}{\Delta x}(T_{i-1,j} + T_{i+1,j}) + \frac{\Delta x}{\Delta y}(T_{i,j-1} + T_{i,j}) - 2\left(\frac{\Delta y}{\Delta x} + \frac{\Delta x}{\Delta y}\right)T_{i,j} = 0$$

Similar equations are written for every node where the temperature is not known. We then produce as many equations as the number of unknown temperatures at the different nodes. Solution to the resulting system of simultaneous, linear equations yields the temperature at every node. From the temperature distribution, the heat transfer rate across any surface can be determined.

Boundary Conditions The difference equation derived is applicable to an interior node. The difference equations at the boundaries take different forms depending on the boundary conditions. There are three common types of boundary conditions:

Specified temperature

1. **Specified Temperature.** If the temperature at a boundary is specified, there is no need to write the equation for the nodes placed on that surface.

Convective boundary condition— difference equation

2. **Convective Boundary Condition (Figure 3.3.4).** If a node is situated on a surface exposed to a fluid with heat transfer by convection, we have

$$q_{i-1,j}^+ + q_{i,j-1}^+ + q_{i,j+1}^+ + q_c^+ = 0$$

$$q_{i-1,j}^+ = k\,\Delta y\,\frac{T_{i-1,j} - T_{i,j}}{\Delta x} \qquad q_{i,j-1}^+ = k\,\frac{\Delta x}{2}\,\frac{T_{i,j-1} - T_{i,j}}{\Delta y}$$

$$q_{i,j+1}^+ = k\,\frac{\Delta x}{2}\,\frac{T_{i,j+1} - T_{i,j}}{\Delta y} \qquad q_c^+ = h\,\Delta y\,(T_\infty - T_{i,j})$$

Combining these equations we get (for constant k)

$$\frac{\Delta y}{\Delta x}\,T_{i-1,j} + \frac{\Delta x}{2\Delta y}\,(T_{i,j-1} + T_{i,j+1}) - \left(\frac{\Delta y}{\Delta x} + \frac{\Delta x}{\Delta y} + h\,\frac{\Delta y}{k}\right)T_{i,j} + h\,\frac{\Delta y}{k}\,T_\infty = 0$$

Imposed heat flux boundary condition— difference equation

3. **Specified Heat Flux (Figure 3.3.5).** If a surface is subject to a specified heat flux of magnitude q'', we replace the q_c^+ term in (2) by $q''\Delta y$ and obtain

$$\frac{\Delta y}{\Delta x}\,T_{i-1,j} + \frac{\Delta x}{2\Delta y}\,(T_{i,j-1} + T_{i,j+1}) - \left(\frac{\Delta y}{\Delta x} + \frac{\Delta x}{\Delta y}\right)T_{i,j} + \frac{q''\Delta y}{k} = 0$$

Note that for an insulated boundary q'' is zero.

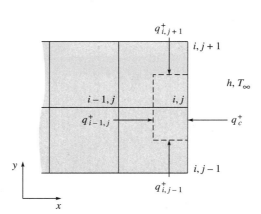

FIGURE 3.3.4 Convective Boundary Condition

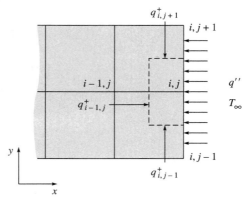

FIGURE 3.3.5 Imposed Heat Flux at the Boundaries

The system of linear, simultaneous equations is solved by any of the many available methods. Here we consider two such methods: The Gauss-Siedel method and the tri-diagonal matrix algorithm (TDMA). For the solution of two- and three-dimensional temperature distributions, both methods require iteration. For one-dimensional, transient temperature distribution, the TDMA gives the solution directly without iteration. Both methods are simple and easy to implement on a computer. The methods are illustrated in Examples 3.3.1, 3.3.2, and 3.3.3.

Irregular Boundaries The numerical procedure described for a regular geometry can be extended to irregular boundaries also, with some approximations. To illustrate the methodology involved in deriving the difference equation for irregular boundaries, consider the two-dimensional temperature distribution in an irregular solid as shown in Figure 3.3.6. In the Figure a and b represent the distance between the node and the boundaries as a fraction of the regular nodal spacing, Δx and Δy. Thus,

$$x_{i+a,j} - x_{i,j} = a\Delta x \qquad y_{i,j+b} - y_{i,j} = b\Delta y \qquad 0 < a < 1, \quad 0 < b < 1$$

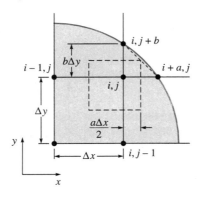

FIGURE 3.3.6 Difference Equations for Irregular Boundaries
A node is adjacent to an irregular boundary.

Irregular
boundary—
difference equation

The surface between the points represented by $(i + a, j)$ and $(i, j + b)$ is approximated by a straight plane as shown by the dotted line joining the two points. Applying the balance of energy and Fourier's Law of Conduction, we obtain

$$k \frac{\Delta y}{2} (1 + b) \frac{T_{i-1,j} - T_{i,j}}{\Delta x} + k \frac{\Delta y}{2} (1 + b) \frac{T_{i+a,j} - T_{i,j}}{a \Delta x} +$$

$$k \frac{\Delta x}{2} (1 + a) \frac{T_{i,j-1} - T_{i,j}}{\Delta y} + k \frac{\Delta x}{2} (1 + a) \frac{T_{i,j+b} - T_{i,j}}{b \Delta y} = 0$$

Dividing by $k (1 + a) (1 + b)$ and rearranging, we get

$$\frac{1}{1 + a} \frac{\Delta y}{\Delta x} T_{i-1,j} + \frac{1}{a(1 + a)} \frac{\Delta y}{\Delta x} T_{i+a,j} + \frac{1}{1 + b} \frac{\Delta x}{\Delta y} T_{i,j-1}$$

$$+ \frac{1}{b(1 + b)} \frac{\Delta x}{\Delta y} T_{i,j+b} - \left(\frac{1}{a} \frac{\Delta y}{\Delta x} + \frac{1}{b} \frac{\Delta x}{\Delta y} \right) T_{i,j} = 0$$

This form of the difference equation is appropriate where the boundary temperature are known. When the boundaries are exposed to convective heat transfer, the heat transfer rate to the node may be determined by the thermal circuit. For example, if the boundaries are exposed to a surrounding fluid at a temperature of T_∞ and a convective heat transfer coefficient, h, the convective heat transfer rate to $\{i,j\}$ is approximated by

$$\frac{T_\infty - T_{i,j}}{\{(2a \Delta x)/[k(1 + b)\Delta y]\} + \{2/[h(1 + b)\Delta y]\}}$$

$$+ \frac{T_\infty - T_{i,j}}{\{(2b \Delta y)/[k(1 + a)\Delta x]\} + \{2/[h(1 + a)\Delta x]\}}$$

Similarly, the difference equation can be obtained if the boundaries are subjected to known heat transfer rates.

We first consider a simple problem to show the salient points of the numerical procedure followed by a more realistic problem.

EXAMPLE 3.3.1 Four surfaces of a long rectangular bar, 6-cm thick and 8-cm wide, are maintained at different temperatures as shown in Figure 3.3.7. Determine the two-dimensional temperature distribution in the bar.

The grid with a total of 20 nodes is shown in the figure. The nodes are 2-cm apart in the x- and y-directions.

Consider a typical node, $\{i,j\}$. In steady state, the sum of the heat transfer rates from the neighboring nodes $\{i - 1, j\}$, $\{i + 1, j\}$, $\{i, j - 1\}$, and $\{i, j + 1\}$ is zero. As the temperature varies only in the x- and y-directions, we consider a unit depth in the z-direction.

$$q_{i-1,j}^+ = k \Delta y \frac{T_{i-1,j} - T_{i,j}}{\Delta x} \qquad q_{i+1,j}^+ = k \Delta y \frac{T_{i+1,j} - T_{i,j}}{\Delta x}$$

$$q_{i,j-1}^+ = k \Delta x \frac{T_{i,j-1} - T_{i,j}}{\Delta y} \qquad q_{i,j+1}^+ = k \Delta x \frac{T_{i,j+1} - T_{i,j}}{\Delta y}$$

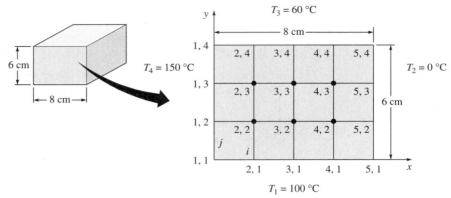

FIGURE 3.3.7 Numerical Procedure for Rectangular Geometry
A grid with 20 nodes; outer surfaces are at defined temperature.

Summing up all the terms, assuming constant thermal conductivity, and rearranging the equation, we get

$$T_{i,j} = \frac{\dfrac{\Delta y}{\Delta x}(T_{i-1,j} + T_{i+1,j}) + \dfrac{\Delta x}{\Delta y}(T_{i,j-1} + T_{i,j+1})}{2\,(\Delta y/\Delta x) + (\Delta x/\Delta y)} \tag{1}$$

Gauss-Siedel iteration

In Equation 1, Δx and Δy need not be equal.

In this example, we illustrate the application of the Gauss-Siedel iterative method for the solution of the equations. (The TDMA is illustrated in Example 3.3.3.) The procedure for the Gauss-Siedel method is as follows:

1. Assume any reasonable values for the unknown temperatures. The closer the assumed distribution to the actual solution, the faster the convergence.
2. Employing Equation 1, update the values for $T_{i,j}$. In Equation 1 use the latest available values of the temperatures for the terms on the right side.
3. Check for convergence of the solution. If the convergence criterion is satisfied, quit. Otherwise, go back to step 2.

CONVERGENCE CRITERIA

Different criteria are used to test for convergence. Three of them are given below.

Convergence criteria

1. **Absolute Convergence Criterion.** If the difference in the values at each node between two consecutive iterations is less than a predefined value, the solution is assumed to have converged.
2. **Relative Convergence Criterion.** At each node the difference in temperatures between two consecutive iterations is compared with some characteristic temperature difference and if the fraction is less than a predefined value at every node, the solution is assumed to have converged.

3. Absolute Sum of the Fractions Criterion. In yet another method, the absolute sum of the fractions obtained in the relative convergence criterion at all nodes is to be less than a predefined value for the solution to have converged.

The grid in Figure 3.3.7 has 5 nodes in the x-direction and 4 in the y-direction; set $\Delta x = \Delta y$ in Equation 1. The initial assumed temperature distribution is given in Table 3.3.1. In the table, the nodes are indicated by small-sized numerals and the temperatures by normal-sized numerals.

With $\Delta x = \Delta y$, Equation 1 is

$$T_{i,j} = \frac{T_{i-1,j} + T_{i+1,j} + T_{i,j-1} + T_{i,j+1}}{4} \tag{2}$$

Starting with node {2,2} proceed to other nodes in row 2 and then to row 3.

$$T_{2,2} = \frac{150 + 80 + 100 + 80}{4} = 102.5$$

$$T_{3,2} = \frac{102.5 + 60 + 100 + 70}{4} = 83.13$$

Note that in computing $T_{3,2}$, the latest available value for $T_{2,2}$, namely 102.5, has been used (and not the original value of 90).

$$T_{4,2} = \frac{83.13 + 0 + 100 + 50}{4} = 58.29$$

After updating the values in row 2, proceed to row 3 and repeat the procedure. The procedure is repeated until the criterion for convergence is satisfied. The values of the temperatures are given in Table 3.3.2. After the ninth iteration, there is no difference in the values (with two significant figures after the decimal point), and the solutions may be considered as having converged after 9 iterations.

TABLE 3.3.1 Initial Temperature Distribution

	60	60	60	
(1, 4)	(2, 4)	(3, 4)	(4, 4)	(5, 4)
150	80	70	50	0
(1, 3)	(2, 3)	(3, 3)	(4, 3)	(5, 3)
150	90	80	60	0
(1, 2)	(2, 2)	(3, 2)	(4, 2)	(5, 2)
	100	100	100	
(1, 1)	(2, 1)	(3, 1)	(4, 1)	(5, 1)

TABLE 3.3.2 Temperature Distribution After Iteration

Node	Number of Iterations						
	0	1	2	3	4	9	10
2, 2	90	102.5	107.2	107.94	108.04	108.1	108.1
3, 2	80	83.13	84.41	84.57	84.63	84.66	84.66
4, 2	60	58.29	58.01	58.04	58.06	58.07	58.07
2, 3	80	95.63	97.34	97.57	97.62	97.64	97.64
3, 3	70	72.19	72.34	72.43	72.47	72.48	72.48
4, 3	50	47.62	47.59	47.62	47.63	47.64	47.63

EXAMPLE 3.3.2 A very wide stainless steel ($k = 16$ W/m K) bar of rectangular cross section, 1-cm thick and 20-cm long, is attached to a surface maintained at 100 °C (T_b) (Figure 3.3.8). The bar is exposed to air at 20 °C (T_∞) and the convective heat transfer coefficient (h) is 6 W/m² °C. Determine the temperature distribution and the heat transfer rate from the bar to the surroundings per unit width. Assume that the convective heat transfer associated with the end surface (h_e) is also 6 W/m² °C.

Although analysis on the basis of one-dimensional temperature distribution is satisfactory in this case (why?), we consider it as a two-dimensional problem to illustrate the application of the numerical procedure. The results will also show if the assumption of one-dimensional temperature distribution is satisfactory.

As the bar is very wide in the z-direction, the temperature varies only in the x-

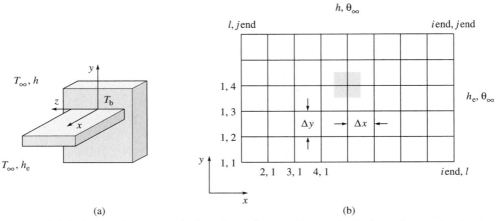

(a)   (b)

FIGURE 3.3.8 (a) A wide rectangular bar of stainless steel is attached to a surface maintained at 100 °C. (b) Grid for the slab and notation for the nodes.

and y-directions, except at the ends in the z-direction; such end effects are neglected. Because of the symmetry, only one-half of the thickness and one-half of the width of the bar need to be analyzed. To obtain the equations that yield the temperature distribution, for each node where the temperature is to be determined, the energy equation is written in the form

$$q_w^+ + q_e^+ + q_s^+ + q_n^+ = 0$$

where the subscripts denote regions to the west, east, south, and north of the boundaries of the element associated with the node $\{i,j\}$ and q^+ the heat transfer rate to the element. As the temperature does not vary in the z-direction, we consider unit depth of the material in the z-direction. As in the case of extended surfaces, *the difference between the temperature at any node and that of the surroundings is denoted by* θ. For $i = 1$ and $j = 1$ to $jend$, we have $\theta_{i,j} = \theta_o = 100 - 20 = 80\,°C$. As the temperatures at these nodes are known, we need not write the energy equation for these nodes. Also across the surface located at $j = 1$, there is no heat transfer (by symmetry). We now proceed to write the energy equation for the various nodes.

$i = 2$ to $iend - 1$, and $j = 1$, with $q_s^+ = 0$

$$q_w^+ = k\,\frac{\Delta y}{2}\,\frac{(\theta_{i-1,j} - \theta_{i,j})}{\Delta x}$$

In this equation, $\Delta y/2$ represents the area of the west border of the element and Δx the distance between the node and the west neighbor. Similarly, the heat transfer rates from the nodes to the east and north of $\{i,j\}$ are represented by

$$q_e^+ = k\,\frac{\Delta y}{2\Delta x}\,(\theta_{i+1,j} - \theta_{i,j}) \quad \text{and} \quad q_n^+ = k\,\frac{\Delta x}{\Delta y}\,(\theta_{i,j+1} - \theta_{i,j})$$

Summing up the heat transfer rates from the neighboring elements and equating the sum to zero, with some rearrangement we have

$$\frac{1}{2}\frac{\Delta y}{\Delta x}\,\theta_{i-1,j} + \frac{\Delta x}{\Delta y}\,\theta_{i,j+1} + \frac{1}{2}\frac{\Delta y}{\Delta x}\,\theta_{i+1,j} - \left(\frac{\Delta y}{\Delta x} + \frac{\Delta x}{\Delta y}\right)\theta_{i,j} = 0 \qquad \textbf{(1)}$$

$i = 2$ to $iend - 1$, and $j = 2$ to $jend - 1$

$$q_w^+ = k\,\frac{\Delta y}{\Delta x}\,(\theta_{i-1,j} - \theta_{i,j}) \qquad q_n^+ = k\,\frac{\Delta x}{\Delta y}\,(\theta_{i,j+1} - \theta_{i,j})$$

$$q_e^+ = k\,\frac{\Delta y}{\Delta x}\,(\theta_{i+1,j} - \theta_{i,j}) \qquad q_s^+ = k\,\frac{\Delta x}{\Delta y}\,(\theta_{i,j-1} - \theta_{i,j})$$

Summing up all the terms, we get

$$\frac{\Delta y}{\Delta x}\,(\theta_{i-1,j} + \theta_{i+1,j}) + \frac{\Delta x}{\Delta y}\,(\theta_{i,j-1} + \theta_{i,j+1}) - \left(2\frac{\Delta y}{\Delta x} + 2\frac{\Delta x}{\Delta y}\right)\theta_{i,j} = 0 \qquad \textbf{(2)}$$

The equations for the other nodes are obtained in a similar manner and they are

$i = 2$ to $iend - 1$, and $j = jend$

$$\frac{1}{2}\frac{\Delta y}{\Delta x}\left(\theta_{i-1,j} + \theta_{i+1,j}\right) + \frac{\Delta x}{\Delta y}\theta_{i,j-1} + \frac{h\,\Delta x}{k}\theta_\infty - \left(\frac{\Delta y}{\Delta x} + \frac{\Delta x}{\Delta y} + \frac{h\,\Delta x}{k}\right)\theta_{i,j} = 0$$

Note that in Equation 3, θ_∞, as defined, is zero.

$i = iend$, and $j = 1$

$$\frac{1}{2}\frac{\Delta y}{\Delta x}\theta_{i-1,j} + \frac{h_e\Delta y}{2k}\theta_\infty + \frac{1}{2}\frac{\Delta x}{\Delta y}\theta_{i,j+1} - \left(\frac{1}{2}\frac{\Delta y}{\Delta x} + \frac{h_e\Delta y}{2k} + \frac{1}{2}\frac{\Delta x}{\Delta y}\right)\theta_{i,j} = 0 \quad \textbf{(4)}$$

$i = iend$ and $j = 2$ to $jend - 1$

$$\frac{\Delta y}{\Delta x}\theta_{i-1,j} + \frac{h_e\Delta y}{k}\theta_\infty + \frac{1}{2}\frac{\Delta x}{\Delta y}\left(\theta_{i,j-1} + \theta_{i,j+1}\right) - \left(\frac{\Delta y}{\Delta x} + \frac{\Delta x}{\Delta y} + \frac{h_e\Delta y}{k}\right)\theta_{i,j} = 0$$

$i = iend$ and $j = jend$

$$\frac{1\Delta y}{2\Delta x}\theta_{i-1,j} + \frac{1}{2}\frac{h_e\Delta y}{k}\theta_\infty + \frac{1}{2}\frac{\Delta x}{\Delta y}\theta_{i,j-1} + \frac{h}{2}\frac{\Delta x}{k}\theta_\infty$$

$$-\left(\frac{1\Delta y}{2\Delta x} + \frac{1}{2}\frac{h_e\Delta y}{k} + \frac{1}{2}\frac{\Delta x}{\Delta y} + \frac{1}{2}\frac{h\,\Delta x}{k}\right)\theta_{i,j} = 0 \quad \textbf{(6)}$$

We have developed as many equations as there are unknown temperatures. The set of linear, simultaneous equations are now solved by the Gauss-Siedel iteration method. In solving the present problem, we use the relative convergence criterion. That is, when the magnitude of the temperature difference at every point between two consecutive iterations is less than a fraction of the maximum temperature difference, $\theta_o - \theta_\infty$, i.e., if $\left|(\theta_{i,j}^n - \theta_{i,j}^{n-1})/(\theta_o - \theta_\infty)\right| < \epsilon$, we assume that the correct temperature distribution is obtained. The actual value of ϵ to be employed is obtained by trial and error. For a further discussion of this procedure refer to a book on numerical methods. We now proceed to solve for the temperature at every node. The temperature $\theta_{i,j}$ at different nodes is given by the following set of equations, (after setting θ_∞ to zero).

$i = 2$ to $iend - 1$, $j = 1$

$$\theta_{i,j} = \frac{[(1/2)\,(\Delta y/\Delta x)\,(\theta_{i-1,j} + \theta_{i+1,j})] + [(\Delta x/\Delta y)\,\theta_{i,j+1}]}{(\Delta y/\Delta x) + (\Delta x/\Delta y)} \quad \textbf{(7)}$$

$i = 2$ to $iend - 1$, $j = 2$ to $jend - 1$

$$\theta_{i,j} = \frac{[(\Delta y/\Delta x\,(\theta_{i-1,j} + \theta_{i+1,j})] + [(\Delta x/\Delta y)\,(\theta_{i,j-1} + \theta_{i,j+1})]}{2\,(\Delta y/\Delta x) + 2\,(\Delta x/\Delta y)} \quad \textbf{(8)}$$

$i = 2$ to $iend - 1$, $j = jend$

$$\theta_{i,j} = \frac{[(1/2)\,(\Delta y/\Delta x)\,(\theta_{i-1,j} + \theta_{i+1,j})] + [(\Delta x/\Delta y)\,\theta_{i,j-1}]}{(\Delta y/\Delta x) + (\Delta x/\Delta y) + (h\,\Delta x/k)} \quad \textbf{(9)}$$

$i = i\text{end}, j = 1$

$$\theta_{i,j} \frac{(\Delta y/\Delta x)\,\theta_{i-1,j} + (\Delta x/\Delta y)\,\theta_{i,j+1}}{(\Delta y/\Delta x) + (h_e\Delta y/k) + (\Delta x/\Delta y)} \tag{10}$$

$i = i\text{end}, j = 2 \text{ to } j\text{end} - 1$

$$\theta_{i,j} = \frac{(\Delta y/\Delta x)\,\theta_{i-1,j} + [(1/2)(\Delta x/\Delta y)\,(\theta_{i,j-1} + \theta_{i,j+1})]}{(\Delta y/\Delta x) + (h_e\Delta y/k) + (\Delta x/\Delta y)} \tag{11}$$

$i = i\text{end}$ and $j = j\text{end}$

$$\theta_{i,j} = \frac{(\Delta y/\Delta x)\,\theta_{i-1,j} + (\Delta x/\Delta y)\,\theta_{i,j-1}}{(\Delta y/\Delta x) + (h_e\Delta y/k) + (\Delta x/\Delta y) + (h\,\Delta x/k)} \tag{12}$$

The heat transfer from one surface per unit width in the z-direction is then computed by

$$q' = \frac{h\,\Delta x}{2}\,\theta_o + \sum_{i=2}^{i\text{end}-1} h\,\Delta x\,\theta_{i,j\text{end}} + \frac{h_e\Delta y}{2}\,\theta_{i,j\text{end}}$$
$$+ \sum_{j=2}^{j\text{end}-1} h_e\Delta y\,\theta_{i\text{end},j} + \left(h\,\frac{\Delta x}{2} + h_e\,\frac{\Delta y}{2}\right)\theta_{i\text{end},j\text{end}} \tag{13}$$

The significance of the five terms in Equation 13 is shown in Figure 3.3.9; i.e., the first term corresponds to region 1 in the figure, and so on.

Analytical results with the assumption of one-dimensional temperature distribution are

Heat transfer rate: 104.7 W/m

Distance from base (m)	0.0	0.02	0.04	0.06	0.08	0.10
Temperature (°C)	80.0	68.05	58.16	50.01	43.36	38.01
Distance from base (m)	0.12	0.14	0.16	0.18	0.20	
Temperature (°C)	33.82	30.64	28.38	26.97	26.38	

For a very coarse network with $i\text{end} = 11$ and $j\text{end} = 5$, the initial guessed temperature field and the final solution with different values for ϵ for the convergence criterion are given in Table 3.3.3. The corresponding heat transfer rates are also given. The calculations proceed in the i direction and then in the j direction. The temperatures in the top row correspond to those at the surface, i.e., $j = j\text{end}$, and those in the first column correspond to the base temperature. *Although the tabulated values give temperatures to five significant figures, one should not expect this degree of accuracy in actual cases.* Rarely are temperatures measured to an accuracy of greater than 0.1 °C. Three significant figures to the the right of the decimal point are used to show how the iterations differ from one value to ϵ to another. For purposes of comparison, the analytical results based on one-dimensional temperature distribution are also given.

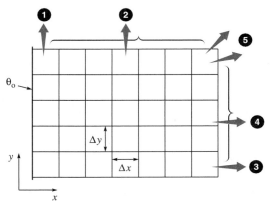

FIGURE 3.3.9 Significance of the Terms in Equation 13

The Biot number ht/k ($t =$ half thickness of the plate), is 0.001875. For this low value of the Biot number, we expect a one-dimensional analysis to be valid. The values in Table 3.3.3 show that compared with the difference in the temperatures of the surface and the surrounding fluid, the variation of the temperature in the y-direction is very small, confirming the validity of the one-dimensional approximation.

As mentioned earlier, the convergence criterion to ensure a defined measure of accuracy cannot be determined *a priori*. It is found by a trial and error procedure. For different values of the convergence criterion, ϵ, the temperature and heat transfer rates should be determined. In Table 3.3.3, values of the temperatures and heat transfer rates are given for ϵ varying from 10^{-4} to 10^{-7}. The differences in the temperature distribution and the heat transfer rates between $\epsilon = 10^{-6}$ and $\epsilon = 10^{-7}$ are very small. We expect the differences in the values for $\epsilon = 10^{-7}$ and lower values of ϵ to be negligibly small for engineering estimates, and the values given for $\epsilon = 10^{-7}$ are taken to represent an acceptable approximation to the correct values. Appendix 7 gives a FORTRAN program for the solution of this example by the Gauss-Siedel iteration.

Relaxation factor

The convergence of the iteration scheme can be speeded by employing a relaxation scheme. The scheme consists of using a value of $\theta_{i,j}$ that is slightly different from the value computed from the equations. If $\Delta\theta$ represents the difference between the computed value of the temperature from the equations after $n + 1$ iterations and $\theta_{i,j}^n$ (after n iterations) we write

$$\theta_{i,j}^{n+1} = \theta_{i,j}^n + R\,\Delta\theta$$

where

$R =$ relaxation factor,
$\Delta\theta =$ difference between the computed value of $\theta_{i,j}$ after $n + 1$ iterations and the previous value of $\theta_{i,j}$ after n iterations
$\theta_{i,j}^{n+1} =$ actual value of $\theta_{i,j}$ to be employed after $n + 1$ iterations for further computations

| TABLE 3.3.3 Temperature Distribution and Heat Transfer Rate-Example 3.2.2

Initial Guessed Temperature Distribution
80.000	72.000	64.000	56.000	48.000	40.000	32.000	24.000	16.000	8.000	0.000
80.000	72.000	64.000	56.000	48.000	40.000	32.000	24.000	16.000	8.000	0.000
80.000	72.000	64.000	56.000	48.000	40.000	32.000	24.000	16.000	8.000	0.000
80.000	72.000	64.000	56.000	48.000	40.000	32.000	24.000	16.000	8.000	0.000
80.000	72.000	64.000	56.000	48.000	40.000	32.000	24.000	16.000	8.000	0.000

$\epsilon = 10^{-4}$ Number of iterations = 735
$q' = 96$ W/m
80.000	69.263	60.154	51.983	44.333	37.014	30.053	23.701	18.431	14.860	13.534
80.000	69.792	60.180	52.006	44.353	37.031	30.066	23.710	18.436	14.862	13.535
80.000	69.313	60.200	52.024	44.368	37.043	30.075	23,716	18.440	14.863	13.535
80.000	69.326	60.212	52.035	44.378	37.051	30.081	23.720	18.441	14.862	13.535
80.000	69.329	60.214	52.036	44.379	37.052	30.083	23.722	18.444	14,866	13.538

$\epsilon = 10^{-5}$ Number of iterations = 4542
$q' = 100$ W/m
80.000	67.604	57.295	48.732	41.675	35.940	31.382	27.896	25.407	23.866	23.241
80.000	67.631	57.319	48.752	41.692	35.954	31.394	27.907	25.417	23.875	23.250
80.000	67.651	57.336	48.766	41.704	35.965	31.403	27.915	25.424	23.882	23.256
80.000	67.663	57.346	48.774	41.712	35.971	31.408	27.919	25.428	23.886	23.260
80.000	67.667	57.349	48.777	41.714	35.973	31.411	27.921	25.430	23.887	23.262

$\epsilon = 10^{-6}$ Number of iterations = 14261
$q' = 104$ W/m
80.000	67.975	58.040	49.851	43.167	37.790	33.560	30.354	28.076	26.660	26.064
80.000	68.003	58.063	49.872	43.185	37.805	33.574	30.366	28.088	26.671	26.074
80.000	68.022	58.080	49.886	43.198	37.816	33.584	30.375	28.096	26.679	26.082
80.000	68.034	58.091	49.895	43.205	37.823	33.590	30.381	28.101	26.683	26.086
80.000	68.038	58.094	49.898	43.208	37.825	33.592	30.383	28.103	26.685	26.088

$\epsilon = 10^{-7}$ Number of iterations = 24812
$q' = 104.9$ W/m
80.000	68.028	58.145	50.005	43.366	38.029	33.833	30.653	28.394	26.988	26.393
80.000	68.056	58.168	50.026	43.384	38.044	33.846	30.665	28.405	26.999	26.404
80.000	68.076	58.185	50.040	43.397	38.055	33.856	30.674	28.414	27.007	26.412
80.000	68.087	58.196	50.049	43.404	38.062	33.862	30.680	28.419	27.011	26.416
80.000	68.091	58.199	50.052	43.407	38.064	33.864	30.681	28.420	27.013	26.418

Relaxation factor = 1.5 $\epsilon = 10^{-7}$
Number of iterations = 11.693
$q^1 = 105$ W/m
80.000	68.031	58.151	50.015	43.379	38.044	33.850	30.672	28.414	27.009	26.415
80.000	68.059	58.175	50.036	43.397	38.060	33.864	30.685	28.426	27.020	26.426
80.000	68.079	58.192	50.050	43.410	38.071	33.874	30.694	28.434	27.028	26.433
80.000	68.091	58.202	50.059	43.417	38.078	33.880	30.699	28.439	27.033	26.438
80.000	68.095	58.206	50.062	43.420	38.080	33.882	30.701	28.441	27.035	26.440

Over-relaxation

If $R = 1$, we get the regular iteration scheme. If $R > 1$, the scheme is one of over-relaxation, and may speed up convergence. For $R = 1.5$, it is observed that for $\epsilon = 10^{-7}$, it takes 11 693 iterations, whereas with $R = 1$ ($\epsilon = 10^{-7}$) it takes 24 812 iterations, reducing the time taken for the computations by a factor of 2. In some iterative schemes, using a value of 1 for R does not result in a convergent solution, but, with R less than 1, convergent solutions are obtained.

The difference in the heat transfer rates obtained from the numerical procedure with $\epsilon = 10^{-7}$, $R = 1$, and from the one-dimensional analysis is less than 0.2%. The analytical solution is a one-dimensional approximation to a two-dimensional problem, and the numerical solution is an approximation to the differential equation. As both solutions are approximate, no conclusion as to which solution represents the correct value can be drawn. Note that there is considerable uncertainty in the value of the convective heat transfer coefficient ($\pm$ 10%) and you should not expect the predicted heat transfer rate to be closer than $\pm$ 10% to the realizable values.

The heat transfer rate to the air is determined by employing Equation 13. It can also be determined as the conductive heat transfer rate at $x = 0$, i.e., $-kA \times (d\theta/_x dx)|_{x=0}$. A crude approximation to the temperature derivative is $(\theta_{2j} - \theta_{1j})/\Delta x$. In this case, the derivative is approximately, $(68.09 - 80)/0.02 = 595.5$ K/m and $q_k = -16 \times 0.01 \times (-595.5) = 95.28$ W. This value of the heat transfer rate obtained from computing the slope from the numerical solution is approximately 9.2% lower than that obtained by integration (Equation 13). This shows that the values obtained by integration are more reliable than those obtained from derivatives in many cases—experimental and numerical. This reinforces the statement made in Section 2.6 on page 101 that in determining the heat transfer rate integration is the preferred mode.

3.3.3 Tri-Diagonal Matrix Algorithm

In many numerical solutions to heat transfer problems the equations to be solved can be cast in the form

$$a_i T_{i-1} + b_i T_i + c_i T_{i+1} = d_i$$

$$
\begin{bmatrix}
b_1 & c_1 & 0 & 0 & 0 & 0 & & \cdots & \\
a_2 & b_2 & c_2 & 0 & 0 & 0 & & \cdots & \\
0 & a_3 & b_3 & c_3 & 0 & 0 & & \cdots & \\
0 & 0 & a_4 & & & & & \cdots & 0 \\
& & & \ddots & & & & & \\
0 & 0 & & & & \ddots & a_{n-1} & b_{n-1} & c_{n-1} \\
0 & 0 & & & & & 0 & a_n & b_n
\end{bmatrix}
\begin{bmatrix}
T_1 \\ T_2 \\ \cdot \\ \cdot \\ \cdot \\ \cdot \\ T_n
\end{bmatrix}
=
\begin{bmatrix}
d_1 \\ d_2 \\ \cdot \\ \cdot \\ \cdot \\ \cdot \\ d_n
\end{bmatrix}
$$

Tri-diagonal matrix algorithm

This system of equations can be efficiently solved by the tri-diagonal matrix algorithm. The algorithm gives the values of the unknowns without iteration, however large the number of equations. A brief introduction to the algorithm follows.

Assume

$$T_{i-1} = p_{i-1}T_i + q_{i-1} \tag{3.3.2}$$

This ith line of the system of equations is

$$a_iT_{i-1} + b_iT_i + c_iT_{i+1} = d_i \tag{3.3.3}$$

Substituting Equation 3.3.2 in Equation 3.3.3, we get

$$(a_ip_{i-1} + b_i)\, T_i + c_iT_{i+1} = d_i - a_iq_{i-1} \tag{3.3.4}$$

Recast Equation 3.3.4 as

$$T_i = -\frac{c_i}{a_ip_{i-1} + b_i}\, T_{i+1} + \frac{d_i - a_i\, q_{i-1}}{a_i\, p_{i-1} + b_i} \tag{3.3.4a}$$

But from the assumed form of $T_{i-1} = p_{i-1}T_i + q_{i-1}$, T_i is given by

$$T_i = p_i\, T_{i+1} + q_i \tag{3.3.5}$$

Comparing Equations 3.3.4a and 3.3.5 we obtain

$$p_i = -\frac{c_i}{a_ip_{i-1} + b_i} \qquad q_i = \frac{d_i - a_iq_{i-1}}{a_ip_{i-1} + b_i} \tag{3.3.6}$$

Equation 3.3.6 contains the recursion formulae for p_i and q_i. From the first line, $p_1 = -c_1/b_1$ and $q_1 = d_1/b_1$. Having determined p_1 and q_1, the rest of $p_2, q_2, \ldots p_n, q_n$ are generated from Equation 3.3.6. From the nth row of the system of equations, we have

$$T_n = q_n$$

Starting with the computed value of T_n $(= q_n)$ and employing Equation 3.3.2, T_{n-1}, $T_{n-2}, T_{n-3}, \ldots, T_1$ are generated by back substitution.

The tri-diagonal matrix algorithm can be adapted to the solution of two-dimensional temperature distribution, but it will be a combination of direct solution and iteration. The equations are written in such a way that the left side of the equation contains temperatures in the same row, with the temperatures on the neighboring rows being transferred to the right-hand side and treated as constant. The temperatures in a particular row are then evaluated by employing the TDMA. Then, move to the next row and repeat the process. Thus, a combination of direct solution and iteration to the simultaneous equations is adopted to find the final temperature distribution.

To illustrate the application of the TDMA, consider the one-dimensional temperature distribution in a slab of thickness L as shown in Figure 3.3.10. Two parallel surfaces are maintained at 100 °C and 20 °C, respectively. It is required to find the temperature distribution within the slab employing the TDMA.

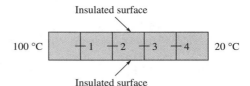

FIGURE 3.3.10 A bar whose two parallel surfaces are maintained at known temperatures.

The slab is divided into 5 segments of equal thickness with 4 interior nodes as shown. The difference equations for nodes 1 through 4 are

$$
\begin{aligned}
-2T_1 + T_2 &= -100 \\
T_1 - 2T_2 + T_3 &= 0 \\
T_2 - 2T_3 + T_4 &= 0 \\
T_3 - 2T_4 &= -20
\end{aligned}
\quad\text{or}\quad
\begin{bmatrix}
-2 & 1 & 0 & 0 \\
1 & -2 & 1 & 0 \\
0 & 1 & -2 & 1 \\
0 & 0 & 1 & -2
\end{bmatrix}
\begin{bmatrix}
T_1 \\ T_2 \\ T_3 \\ T_4
\end{bmatrix}
=
\begin{bmatrix}
-100 \\ 0 \\ 0 \\ -20
\end{bmatrix}
$$

The solution is shown in the following table.

i	a_i	b_i	c_i	d_i	p_i	q_i	T_i
1	0	−2	1	−100	1/2	50	84
2	1	−2	1	0	2/3	100/3	68
3	1	−2	1	0	3/4	25	52
4	1	−2	0	−20	0	36	36

The application of the TDMA for the solution of two-dimensional, steady state temperature distribution will be illustrated through an example.

EXAMPLE 3.3.3   Find the temperature distribution in the plate in Example 3.3.2 applying the TDMA. The difference equations obtained in the solution of Example 3.3.2 will be used here. The steps in obtaining the temperature distribution are

1. Obtain the difference equation for each row in the form

$$
a_{i,j}\,\theta_{i-1,j} + b_{i,j}\theta_{i,j} + c_{i,j}\theta_{i+1,j} = d_{i,j}
$$

In this form for Equation 2 of Example 3.3.2, we have, $i = 2$ to $iend - 1$, and $j = 2$ to $jend - 1$,

$$
a_{i,j} = \frac{\Delta y}{\Delta x} \quad b_{i,j} = -2\left(\frac{\Delta y}{\Delta x} + \frac{\Delta x}{\Delta y}\right) \quad c_{i,j} = \frac{\Delta y}{\Delta x} \quad d_{i,j} = -\frac{\Delta x}{\Delta y}(\theta_{i,j-1} + \theta_{i,j+1})
$$

Note that the temperature at nodes in the same row are on the left-hand side of the equation and those in the neighboring rows are on the right-hand side of the equation, where they are included in $d_{i,j}$.

2. Guess a reasonable temperature distribution.
3. Starting with row 1, solve for the nodal temperatures in each row employing TDMA. When solving for the temperatures in a particular row, use the latest available temperatures for the neighboring rows for computing $d_{i,j}$ on the right-hand side of the difference equation. For example, to compute $d_{i,j}$ for row 3, employ the temperatures obtained from the solution for row 2.
4. After solving for the temperatures in all the rows, test for convergence. If convergence criterion is not satisifed, go back to step 3 and iterate till a convergent solution is obtained.

A FORTRAN program to solve the equation by TDMA is given in Appendix 7. The coefficients used in the program are given below. Note that as the coefficients, a, b, and c do not depend on the temperatures, they need be computed only once at the beginning of the program. However, the coefficients $d_{i,j}$ depend on the temperatures at the nodes and have to be computed during each iteration. Hence, $d_{i,j}$s are shown separately.

$j = 1$, $i = 2$ to $iend - 1$

$$a_{i,j} = 0.5 \frac{\Delta y}{\Delta x} \quad (a_{i,j} = 0 \text{ for } i = 2)$$

$$b_{i,j} = -\frac{\Delta y}{\Delta x} - \frac{\Delta x}{\Delta y}$$

$$c_{i,j} = 0.5 \frac{\Delta y}{\Delta x}$$

$j = 2$ to $jend - 1$, $i = 2$ to $iend - 1$

$$a_{i,j} = \frac{\Delta y}{\Delta x} \quad (a_{2,j} = 0)$$

$$b_{i,j} = -2 \left(\frac{\Delta y}{\Delta x} + \frac{\Delta x}{\Delta y} \right)$$

$$c_{i,j} = \frac{\Delta y}{\Delta x}$$

$j = jend$, $i = 2$ to $iend - 1$

$$a_{i,j} = 0.5 \frac{\Delta y}{\Delta x} \quad (a_{2,j} = 0)$$

$$b_{i,j} = -\left(\frac{\Delta y}{\Delta x} + \frac{\Delta x}{\Delta y} + \frac{h \Delta x}{k} \right)$$

$$c_{i,j} = 0.5 \frac{\Delta y}{\Delta x}$$

$j = 1, i = i$end

$$a_{i,j} = \frac{\Delta y}{\Delta x}$$

$$b_{i,j} = -\left(\frac{\Delta y}{\Delta x} + \frac{h_e \Delta y}{k} + \frac{\Delta x}{\Delta y}\right)$$

$$c_{i,j} = 0$$

$j = 2$ to $j$end $- 1, i = i$end

$$a_{ij} = \frac{\Delta y}{\Delta x}$$

$$b_{i,j} = -\left(\frac{\Delta y}{\Delta x} + \frac{h_e \Delta y}{k} + \frac{\Delta x}{\Delta y}\right)$$

$$c_{i,j} = 0$$

$j = j$end, $i = i$end

$$a_{i,j} = \frac{\Delta y}{\Delta x}$$

$$b_{i,j} = -\left(\frac{\Delta y}{\Delta x} + \frac{h_e \Delta y}{k} + \frac{\Delta x}{\Delta y} + \frac{h \Delta x}{k}\right)$$

$$c_{i,j} = 0$$

The coefficient, $d_{i,j}$ are computed at each iteration.

$j = 1, i = 2$

$$d_{i,j} = -\frac{\Delta x}{\Delta y}\,\theta_{i,j+1} - 0.5\,\frac{\Delta y}{\Delta x}\,\theta_o$$

$j = 1, i = 3$ to $i$end

$$d_{i,j} = -\frac{\Delta x}{\Delta y}\,\theta_{i,j+1}$$

$j = 2$ to $j$end $- 1$

$i = 2$ $$d_{i,j} = -\frac{\Delta x}{\Delta y}\,(\theta_{i,j-1} + \theta_{i,j+1}) - \frac{\Delta y}{\Delta x}\,\theta_o$$

$i = 3$ to $i$end $- 1$ $$d_{i,j} = -\frac{\Delta x}{\Delta y}\,(\theta_{i,j-1} + \theta_{i,j+1})$$

$i = i$end $$d_{i,j} = -0.5 - \frac{\Delta x}{\Delta y}\,(\theta_{i,j-1} + \theta_{i,j+1})$$

$j = jend$

$i = 2$

$$d_{i,j} = -\frac{\Delta x}{\Delta y}\theta_{i,j-1} - 0.5\frac{\Delta y}{\Delta x}\theta_o$$

$i = 3$ to $iend$

$$d_{i,j} = -\frac{\Delta x}{\Delta y}\theta_{i,j-1}$$

A comparison between the solutions by the Gauss-Siedel iteration and by the TDMA show that the two methods are comparable in the number of iterations and accuracy.

3.3.4 Variable Thermal Conductivity

If the thermal conductivity is not constant, either because of the nonhomogeneity of the material or because of its dependence on temperature, it is easily handled in the numerical method. However, we should be careful about how we evaluate a suitable value of the thermal conductivity at the boundary between two elements. It is tempting to use the arithmetic mean of the two values at the neighboring nodes, but a little reflection will show that this mean is not the correct value. Consider a 3-mm thick copper sheet with a 3-mm thick Teflon backing, Figure 3.3.11. With thermal conductivities of 0.25 W/m K for Teflon and 401 W/m K for copper, the thermal resistance employing the arithmetic mean thermal conductivity is computed as follows:

Arithmetic mean thermal conductivity $= k_m = (401 + 0.25)/2 = 200.6$ W/m K
Thermal resistance $= (L_1 + L_2)/k_m = 0.006/200.6 = 2.991 \times 10^{-5}$ K m^2/W
Correct thermal resistance $= L_1/k_1 + L_2/k_2 = 0.003/401 + 0.003/0.25 = 1.2 \times 10^{-2}$ K m^2/W

Variable thermal conductivity

The correct resistance is very much higher than the resistance computed by employing the arithmetic mean value of the thermal conductivity. Now, consider the evaluation of the thermal conductivity at the boundary between two elements, Δx apart, with the nodes represented by subscripts w and e. If the thermal conductivity evaluated at w and e are k_w and k_e, respectively, what is the appropriate value at the boundary? From the thermal circuit, the heat transfer from w to e is given by $(T_w - T_e)/[(\Delta x/2k_w) + (\Delta x/2k_e)]$. If a suitable average value is k_m, $(T_w - T_e)/(\Delta x/k_m)$ should give the correct heat transfer rate. Equating the two expres-

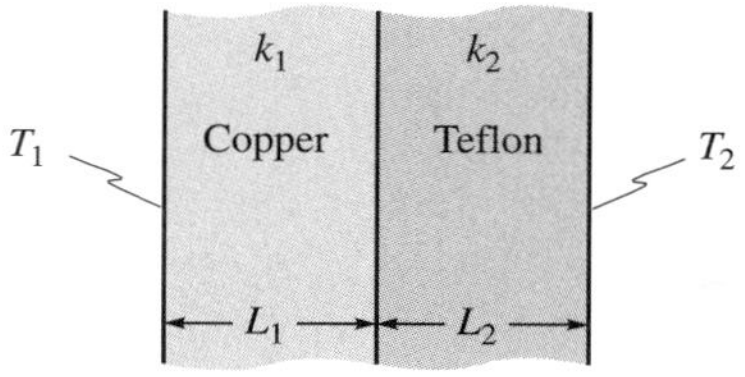

FIGURE 3.3.11 Variable Thermal Conductivity
Sheets of copper and Teflon of equal thickness.

sions for the heat transfer rate, the correct mean value of the thermal conductivity is given by

$$\frac{1}{k_m} = \frac{1}{2}\left(\frac{1}{k_w} + \frac{1}{k_e}\right)$$

If the spacing of the element is not uniform, it is easily shown that the correct mean value to be used is given by

$$\frac{1}{k_m} = \frac{1}{\Delta x_w + \Delta x_e}\left(\frac{\Delta x_w}{k_w} + \frac{\Delta x_e}{k_e}\right)$$

(A similar situation exists in fluid flows in evaluating shear stresses if the viscosity varies with location.)

3.4 TRANSIENT CONDUCTION

Analytical solutions for the transient temperature distribution are available for rectangular slabs, cylinders, and spheres. Approximate analytical solutions are discussed in Section 3.5 and exact analytical solutions in Chapter 11. In this section we prove the validity of the product solution employed in Section 2.9.2 and provide an introduction to the numerical solution of one-dimensional transient temperature problems.

3.4.1 Heisler Charts Revisited

In Chapter 2, the one-term approximation of analytical solutions for one-dimensional transient problems was given. The development of Heisler charts for one-dimensional, transient temperature distribution was explained. The use of the chart for the solution of two- and three-dimensional transient problems was illustrated.

Product solution—multi-dimensional transient temperature distribution

We now present the justification for using the products of one-dimensional transient solutions for two- and three-dimensional problems, without completely solving the applicable differential equations. As an illustration, we show that the solution to a two-dimensional, transient temperature distribution can be expressed as the product of the solutions to two one-dimensional transient problems. The extension to three-dimensional problems is straightforward.

Consider the two-dimensional, transient temperature distribution in a rectangular slab, shown in Figure 3.4.1. The slab is initially at a uniform temperature of T_o. At time $\tau = 0$, it is exposed to a fluid at a constant temperature T_∞ with constant convective heat transfer coefficients: h_1 on two parallel surfaces and h_2 on the other two parallel surfaces.

The differential equation (for derivation of the equations see Chapter 11) and the associated initial and boundary conditions are

$$\frac{\partial^2 T}{\partial x^2} + \frac{\partial^2 T}{\partial y^2} = \frac{1}{\alpha}\frac{\partial T}{\partial \tau} \tag{3.4.1}$$

Initial condition: $T(x, y, 0) = T_o$

FIGURE 3.4.1 Transient Conduction for Rectangular Geometries
A rectangular slab with two-dimensional, transient temperature Distribution.

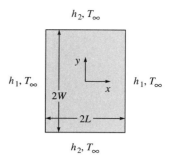

Boundary conditions: From an energy balance on each of the four surfaces, equating the conductive heat flux to the surface and the convective heat flux from the surface,

$$k \frac{\partial T}{\partial x} (L, y, \tau) + h_1 [T(L, y, \tau) - T_\infty] = 0$$

$$k \frac{\partial T}{\partial x} (-L, y, \tau) - h_1 [T(-L, y, \tau) - T_\infty] = 0$$

$$k \frac{\partial T}{\partial y} (x, W, \tau) + h_2 [T(x, W, \tau) - T_\infty] = 0$$

$$k \frac{\partial T}{\partial y} (x, -W, \tau) - h_2 [T(x, -W, \tau) - T_\infty] = 0$$

Defining a dimensionless temperature $\theta = (T - T_\infty)/(T_o - T_\infty)$, Equation 3.4.1 is recast as

$$\frac{\partial^2 \theta}{\partial x^2} + \frac{\partial^2 \theta}{\partial y^2} = \frac{1}{\alpha} \frac{\partial \theta}{\partial \tau} \qquad (3.4.2)$$

where the initial and boundary conditions are

$$\theta(x, y, 0) = 1$$

$$k \frac{\partial \theta}{\partial x} (L, y, \tau) + h_1 \theta(L, y, \tau) = 0$$

$$k \frac{\partial \theta}{\partial x} (-L, y, \tau) - h_1 \theta(-L, y, \tau) = 0$$

$$k \frac{\partial \theta}{\partial y} (x, W, \tau) + h_2 \theta(x, W, \tau) = 0$$

$$k \frac{\partial \theta}{\partial y} (x, -W, \tau) - h_2 \theta(x, -W, \tau) = 0$$

We will now show that a function formed by the product of two functions $\phi(x, \tau)$ and $\psi(y, \tau)$ satisfies the differential equation and the boundary conditions. It will also be shown that each function is the solution to a one-dimensional, transient conduction problem of the type discussed in Section 2.9.1. Substituting $\phi\psi$ for θ in the differential equation and dividing by $\phi\psi$, we obtain

$$\frac{1}{\phi}\frac{\partial^2\phi}{\partial x^2} + \frac{1}{\psi}\frac{\partial^2\psi}{\partial y^2} = \frac{1}{\alpha\phi}\frac{\partial\phi}{\partial\tau} + \frac{1}{\alpha\psi}\frac{\partial\psi}{\partial\tau}$$

We now require that

$$\frac{1}{\phi}\frac{\partial^2\phi}{\partial x^2} = \frac{1}{\alpha\phi}\frac{\partial\phi}{\partial\tau} \quad \text{and} \quad \frac{1}{\psi}\frac{\partial^2\psi}{\partial y^2} = \frac{1}{\alpha\psi}\frac{\partial\psi}{\partial\tau}$$

The above two equations are recast as

$$\frac{\partial^2\phi}{\partial x^2} = \frac{1}{\alpha}\frac{\partial\phi}{\partial\tau} \qquad (3.4.3)$$

$$\frac{\partial^2\psi}{\partial y^2} = \frac{1}{\alpha}\frac{\partial\psi}{\partial\tau} \qquad (3.4.4)$$

The initial and boundary conditions for ϕ and ψ are

$$\phi(x, 0) = 1$$

$$k\frac{\partial\phi}{\partial x}(L, \tau) + h_1\phi(L, \tau) = 0$$

$$k\frac{\partial\phi}{\partial x}(-L, \tau) - h_1\phi(-L, \tau) = 0$$

$$\psi(y, 0) = 1$$

$$k\frac{\partial\psi}{\partial y}(W, \tau) + h_2\psi(W, \tau) = 0$$

$$k\frac{\partial\psi}{\partial y}(-W, t) - h_2\psi(-W\ \tau) = 0$$

Equations 3.4.3 and 3.4.4 can now be made completely dimensionless:

$$\eta = \frac{x}{L}; \quad \text{Fo}_x = \frac{\alpha\tau}{L^2} \qquad \mu = \frac{y}{W}; \quad \text{Fo}_y = \frac{\alpha\tau}{W^2}$$

We then obtain

$$\frac{\partial^2\phi}{\partial\eta^2} = \frac{\partial\phi}{\partial\text{Fo}_x} \qquad\qquad\qquad \frac{\partial^2\psi}{\partial\mu^2} = \frac{\partial\psi}{\partial\text{Fo}_y}$$

$$\phi(\eta, 0) = 1 \qquad\qquad\qquad\qquad \psi(\mu, 0) = 1$$

$$\frac{\partial\phi}{\partial\eta}(1, \text{Fo}_x) + \text{Bi}_1\ \phi(1, \text{Fo}_x) = 0 \qquad \frac{\partial\psi}{\partial\mu}(1, \text{Fo}_y) + \text{Bi}_2\ \psi(1, \text{Fo}_y) = 0$$

$$\frac{\partial\phi}{\partial\eta}(-1, \text{Fo}_x) - \text{Bi}_1\ \phi(-1, \text{Fo}_x) = 0 \qquad \frac{\partial\psi}{\partial\mu}(-1, \text{Fo}_y) - \text{Bi}_2\ \psi(-1, \text{Fo}_y) = 0$$

$$\text{Bi}_1 = \frac{h_1 L}{k} \qquad\qquad\qquad\qquad \text{Bi}_2 = \frac{h_2 W}{k}$$

It is easily seen that the solutions to ϕ and ψ can be found from Figure 2.9.2. When Figure 2.9.2 is used, the appropriate values of $(x/L, \alpha\tau/L^2, h_1L/k)$ for ϕ and $(y/W, \alpha\tau/W^2, h_2W/k)$ for ψ should be used. Note that this procedure can be applied only when the fluid temperature is the same on all sides, and when the convective heat transfer coefficients associated with each pair of parallel surfaces are the same. However, the convective heat transfer coefficients of two perpendicular surfaces may be different.

3.4.2 Numerical Solutions

Transient
temperature
distribution—
numerical solution

In Chapter 2 we studied some transient problems (by lumped analysis) and the application of one-dimensional analytical solutions for rectangular, cylindrical, and spherical geometries to multidimensional problems. However, numerical solutions can be obtained for both regular and irregular geometries, for boundary conditions that are arbitrary functions of time, for bodies with internal energy generation, and for nonuniform and unsteady convective heat transfer coefficients. As an introduction to the numerical procedure, Examples 3.4.1 and 3.4.2 give numerical solutions to two one-dimensional transient conduction problems involving regular geometries.

EXAMPLE 3.4.1   A cylindrical rod attached to a surface (Figure 3.4.2) is initially at a uniform temperature T_o. At time $\tau = 0$, the temperature of the surface at $x = 0$ is changed to T_s and maintained at that value. The free end is exposed to a fluid at T_∞ with a convective heat transfer coefficient h. The cylindrical surface is insulated. We wish to determine the one-dimensional, transient temperature distribution in the rod.

As in the case of steady, two-dimensional problems, we divide the rod into a number of segments in the x-direction with the midpoint between the boundaries of each segment being a node. From an energy balance, for the material associated with

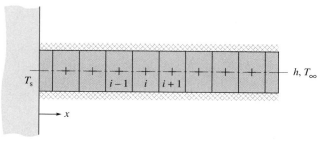

FIGURE 3.4.2 A cylindrical rod with one end maintained at a constant temperature T_s and the other end subjected to convective heat transfer.

the node i, in the time interval $(n - 1)\Delta\tau$ to $n\Delta\tau$, (where $\Delta\tau$ represents a time interval and n represents the number of time intervals),

> Increase in the internal energy of the element
> = Net heat transfer to the element from the nodes $i - 1$ and $i + 1$
> + Net work transfer to the element
> + Internal energy generation in the element

It has already been shown that, with constant pressure, the work term can be combined with the internal energy term (see Section 2.7). With internal energy generation being zero, the energy equation reduces to

> Increase in the enthalpy of the element
> = Net heat transfer from the the nodes $i - 1$ and $i + 1$ in time $\Delta\tau$

> Increase in the enthalpy of the element $= \rho\, c_p\, A\, \Delta x\, (T_i^n - T_i^{n-1})$

where T_i^{n-1} is the temperature of the element i at time $(n - 1)\Delta\tau$ and T_i^n is the temperature at $n\Delta\tau$. The temperature at a node is represented by a subscript to identify the node and a superscript to indicate the time in terms of the number of time intervals.

The heat transfer from the neighboring nodes $i - 1$ and $i + 1$ in time $\Delta\tau$ is approximated by

$$\left(kA_x \frac{T_{i-1}^{n-1} - T_i^{n-1}}{\Delta x} + kA_x \frac{T_{i+1}^{n-1} - T_i^{n-1}}{\Delta x} \right)\Delta\tau$$

Substituting the above expressions in the energy equation and rearranging we obtain

$$T_i^n - T_i^{n-1} = \frac{\alpha\Delta\tau}{\Delta x^2} (T_{i-1}^{n-1} + T_{i+1}^{n-1} - 2T_i^{n-1}) \tag{1}$$

where α is the thermal diffusivity, $k/\rho c_p$. If $\alpha\Delta\tau/\Delta x^2$ is chosen to be 1/2, Equation 1 simplifies to

$$T_i^n = \frac{T_{i-1}^{n-1} + T_{i+1}^{n-1}}{2} \tag{2}$$

Although it appears that we chose $\alpha\Delta\tau/\Delta x^2 = 1/2$ for convenience, it turns out that the maximum value of $\alpha\Delta\tau/\Delta x^2$ that we can use for meaningful results in this case is 1/2. Any value higher than 1/2 will give results that would violate the physics of the problem (see Section 3.4.3 for more details). Recall that $\alpha\Delta\tau/\Delta x^2$ is dimensionless; it is the nodal Fourier number.

If the temperature distribution is known at any instant, the temperature at every node after one time interval $\Delta\tau$ can be found from Equations 1 or 2, and subsequent times can be found by a marching process in time. The complete methodology of the solution is illustrated in Example 3.4.2.

EXAMPLE 3.4.2 Cylindrical rods are attached to a pipe to increase the heat tarnsfer rate. The rods, 3-mm diameter and 8-cm long are made of plain carbon-steel. At time $\tau = 0$, steam is admitted to the pipe so that the temperature of the pipe surface quickly increases to 100 °C and remains at that value. Air at 20 °C flows over the rods at 3 m/s. The convective heat transfer coefficient on the air side is 102 W/m² °C (Figure 3.4.3). Determine the transient temperature distribution in the rods by a numerical method.

We employ the marching process to determine the temperature distribution. To demonstrate the method, divide the fin into 16 equal parts, the midpoint of each element being a node (Figure 3.4.4). Subscript i denotes any node from 1 to 16. Apply the energy equation for the material associated with each node. For each element, in the time interval $\Delta\tau$, the balance of energy yields

$$\text{Increase in the enthalpy}$$
$$= \text{The net heat transfer to the element}$$

Denote the time steps by superscript n, and define $\theta = T - T_\infty$. For any node from 2 to 15, balance of energy in the time interval $n\Delta\tau$ to $(n + 1)\Delta\tau$ yields

$$\rho c_p\,\pi r^2\,\Delta x\,(\theta_i^{n+1} - \theta_i^n) = k\pi r^2\,\frac{(\theta_{i-1}^n - \theta_i^n)}{\Delta x}\,\Delta\tau$$
$$+ k\pi r^2\,\frac{(\theta_{i+1}^n - \theta_i^n)}{\Delta x}\,\Delta\tau - h2\pi r\,\Delta x\,\theta_i^n\,\Delta\tau$$

Dividing by $\rho c_p \pi r^2 \Delta x$ and rearranging the equation,

$$\theta_i^{n+1} - \theta_i^n = \frac{\alpha\Delta\tau}{\Delta x^2}\,(\theta_{i-1}^n + \theta_{i+1}^n - 2\theta_i^n) - \frac{2h}{\rho r c_p}\,\Delta\tau\theta_i^n$$

With Fo $= \alpha\Delta\tau/\Delta x^2$, Bi $= hr/k$, the equation is written as

$$\theta_i^{n+1} - \theta_i^n = \text{Fo}\left[\theta_{i-1}^n + \theta_{i+1}^n - \theta_i^n\left(2 + 2\text{Bi}\,\frac{\Delta x^2}{r^2}\right)\right] \qquad (1)$$

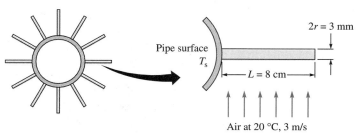

FIGURE 3.4.3 The temperature of the base of a cylindrical rod, initially at a uniform temperature, is changed to T_s.

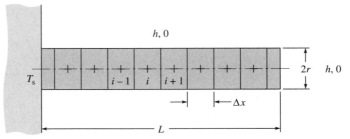

FIGURE 3.4.4 Nodes and Nodal Spacing for the Rod in Example 3.4.2

For $i = 1$,

$$\rho c_p \pi r^2 \frac{3}{4} \Delta x \, (\theta_i^{n+1} - \theta_i^n) = k\pi r^2 \frac{(\theta_{i-1}^n - \theta_i^n)}{\Delta x/2} \Delta \tau$$

$$+ k\pi r^2 \frac{(\theta_{i+1}^n - \theta_i^n)}{\Delta x} \Delta \tau - h2\pi r \frac{3}{4} \Delta x \, \theta_i^n \Delta \tau$$

Dividing by $\rho c_p \pi r^2 \frac{3}{4} \Delta x$ and rearranging with $\theta_{i-1}^n = \text{constant} = \theta_o$,

$$\theta_i^{n+1} - \theta_i^n = \text{Fo} \left[\frac{8}{3} \theta_o + \frac{4}{3} \theta_{i+1}^n - \theta_i^n \left(4 + 2\text{Bi} \frac{\Delta x^2}{r^2} \right) \right] \tag{2}$$

Lastly, for $i = 16$,

$$\rho c_p \pi r^2 \Delta x \, (\theta_i^{n+1} - \theta_i^n) = k\pi r^2 \frac{(\theta_{i-1}^n - \theta_i^n)}{\Delta x} \Delta \tau$$

$$- h2\pi r \Delta x \, \theta_i^n \Delta \tau - \frac{\theta_i^n}{(\Delta x/2\pi r^2 k) + (1/h\pi r^2)} \Delta \tau$$

where we have used the thermal circuit to model the heat transfer rate from the free end surface. It has also been assumed that the values of the convective heat transfer coefficients on the cylindrical surface and on the end surface are equal. Dividing by $\rho c_p \pi r^2 \Delta x$ and rearranging,

$$\theta_i^{n+1} - \theta_i^n$$

$$= \text{Fo} \left\{ \theta_{i-1}^n - \theta_i^n \left[\frac{2 + 3\text{Bi} \, (\Delta x/r) + 4\text{Bi} \, (\Delta x^2/r^2) + 2\text{Bi}^2 \, (\Delta x^3/r^3)}{2 + \text{Bi} \, (\Delta x/r)} \right] \right\} \tag{3}$$

The time step $\Delta \tau$ cannot be arbitrary. For a stable, meaningful solution, the value of $\Delta \tau$ should be below a certain value. If the time step is higher than this maximum value the solution will not converge. When the energy equation is written with $\theta_i^{n+1} - \theta_i^n$ on the left side (representing the increase in the temperature of the node in time $\Delta \tau$), it can be shown (see Section 3.4.3) that the maximum permissible value for Fo is the reciprocal of the coefficient of θ_i^n on the right side of the equation.

Hence, in this case, we use the minimum of the reciprocals of the coefficients of θ_i^n in Equations 1, 2, and 3:

$$r = 0.0015 \text{ m} \qquad L = 0.08 \text{ m} \qquad \Delta x = 80/16 = 5 \text{ mm} = 0.005 \text{ m}$$

$$\Delta x/r = 5/1.5 = 3.333 \qquad h = 102 \text{ W/m}^2 \text{ °C}$$

From Table A1, for plain carbon-steel, $k = 60.5$ W/m K

$$\text{Bi} = \frac{hr}{k} = \frac{102 \times 0.0015}{60.5} = 0.00253$$

$$\frac{1}{2 + 2\text{Bi} \, (\Delta x^2/r^2)} = 0.4863$$

$$\frac{1}{4 + 2\text{Bi} \, (\Delta x^2/r^2)} = 0.2465$$

$$\frac{2 + \text{Bi} \, (\Delta x/r)}{2 + 3\text{Bi} \, (\Delta x/r) + 4\text{Bi} \, (\Delta x^2/r^2) + 2\text{Bi}^2 \, (\Delta x^3/r^3)} = 0.9394$$

TABLE 3.4.1 Solution to Example 3.4.2

Number of time steps = 1   Fo × n = 0.2   Real time = 0.28 s
Temperature distribution

T(1) = 32.000000	T(2) = 6.400000	T(3) = 1.280000	T(4) = 0.256000
T(5) = 0.051200	T(6) = 0.010240	T(7) = 0.002048	T(8) = 0.000410
T(9) = 0.000082	T(10) = 0.000016	T(11) = 0.000003	T(12) = 0.000001
T(13) = 0.000000	T(14) = 0.000000	T(15) = 0.000000	T(16) = 0.000000

Heat transfer rate = 0.1923 W

Number of time steps = 6   Fo × n = 1.2   Real time = 1.69 s
Temperature distribution

T(1) = 60.220534	T(2) = 30.333754	T(3) = 13.401165	T(4) = 5.344601
T(5) = 1.966568	T(6) = 0.678428	T(7) = 0.222067	T(8) = 0.069593
T(9) = 0.021027	T(10) = 0.006158	T(11) = 0.001756	T(12) = 0.000489
T(13) = 0.000133	T(14) = 0.000036	T(15) = 0.000009	T(16) = 0.000001

Heat transfer rate = 0.5396 W

Number of time steps = 30   Fo × n = 6.0   Real time = 8.47 s
Temperature distribution

T(1) = 68.615247	T(2) = 50.131267	T(3) = 35.561028	T(4) = 24.436437
T(5) = 16.244406	T(6) = 10.439988	T(7) = 6.486498	T(8) = 3.897726
T(9) = 2.266827	T(10) = 1.277152	T(11) = 0.697856	T(12) = 0.370308
T(13) = 0.191249	T(14) = 0.096838	T(15) = 0.049796	T(16) = 0.025723

Heat transfer rate = 1.0613 W

Number of time steps = 505   Fo × n = 101.0   Real time = 142.66 s
Temperature distribution

T(1) = 70.594692	T(2) = 55.751398	T(3) = 44.041291	T(4) = 34.806308
T(5) = 27.527486	T(6) = 21.795793	T(7) = 17.289141	T(8) = 13.754287
T(9) = 10.992600	T(10) = 8.848897	T(11) = 7.202725	T(12) = 5.961587
T(13) = 5.055751	T(14) = 4.434322	T(15) = 4.062388	T(16) = 3.918978

Heat transfer rate = 1.6180 W

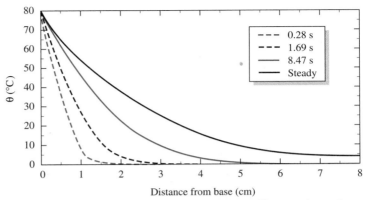

FIGURE 3.4.5 Temperature Distribution at Different Times for Example 3.4.2.

As the minimum value of the reciprocal of the coefficient of θ_i is 0.2465, choose Fo = 0.2 for the computations. With $\theta_i^\circ = 0$ for $i = 1$ to 16 and repeatedly applying Equations 1, 2, and 3, θ_i^n for $n = 1, 2, 3, \ldots$ are computed. The computations are easily programmed on a computer.

The heat transfer rate is computed from the equation,

$$q = \sum_{i=2}^{16} h2\pi r \Delta x\, \theta_i^n + h2\pi r\, \frac{\Delta x}{4}\, \theta_o + h2\pi r\, \frac{3\Delta x}{4}\, \theta_1^n + \frac{\pi r^2 \theta_{i=16}^n}{(\Delta x/2k) + (1/h)}$$

The second term represents the heat transfer rate from the element associated with the base and the third term the heat transfer rate from the element associated with the first node. The last term accounts for the heat transfer rate from the end surface of the fin, employing the thermal circuit.

Some of the results of the computations are given in Table 3.4.1 and shown in Figure 3.4.5. It is assumed that steady state is reached when the magnitude of the relative change in the temperature between two successive time steps defined as the absolute value of $[\theta_i^n - \theta_i^{n-1}]/\theta_i^{n-1}$ at every node is less than 10^{-6}. A comparison of the analytical solution for the steady-state temperature distribution and the numerical method after 142.7 s is given below.

Distance from base (cm)	Designation	Temperature (°C)	
		Analytical	Numerical
0.25	T(1)	71.0669	70.6
0.75	T(2)	56.0889	55.8
1.25	T(3)	44.2761	44.04
1.75	T(4)	34.9640	34.81
2.25	T(5)	27.6260	27.53
2.75	T(6)	21.8479	21.8
3.25	T(7)	17.3034	17.29
3.75	T(8)	13.7358	13.76

Distance from base (cm)	Designation	Temperature (°C)	
		Analytical	Numerical
4.25	T(9)	10.9438	10.99
4.75	T(10)	8.7697	8.85
5.25	T(11)	7.0908	7.20
5.75	T(12)	5.8122	5.96
6.25	T(13)	4.8618	5.06
6.75	T(14)	4.1859	4.43
7.25	T(15)	3.7463	4.06
7.75	T(16)	3.5183	3.92

The heat transfer rate from the analytical solution is 1.621 W and from the numerical solution the rate is 1.618 W. The values obtained from analysis and numerical integration are quite consistent even though a fairly crude net work of only 16 nodes was used.

The accuracy of the numerical solution can be increased by decreasing the size of the element. But from the definition of Fo and the limitation on its maximum value, it is evident that as the element size is decreased, the time step also has to be decreased. Hence, the computational time increases considerably. For example, if we reduce the element size by a factor of 2, the time step has to be decreased by a factor of 4 so that the total computations will increase by a factor of 8. The penalty for increased accuracy is the increase in computational time.

We resort to numerical computations when we do not have an exact solution. In such a case, the only way to ensure that the values obtained are within an acceptable level of approximation is to get a solution with smaller and smaller values of Δx and compare the results. It may be easier to look at the steady-state heat transfer rate. For example, we may assume that steady state has been reached when the ratio of the difference in heat transfer rate between two successive time steps to the heat transfer rate in the latest time step does not change by more than a defined small value, say 10^{-6}. If the difference between the steady-state heat transfer rate computed in this way for values of Δx and $\Delta x/2$ does not change by more than a predefined value of the order of 0.1%, we may assume that we have reached an acceptable level of approximation.

Stability criterion—transient temperature distribution, numerical method

3.4.3 One-Dimensional, Transient Temperature Distribution—Stability of Numerical Solutions

In example 3.4.2, it was pointed out that there was a maximum value for the nodal Fourier number. A physical interpretation for such a maximum value will now be given.

We write the finite difference form of the energy equation for any node, i, in the form

$$\theta_i^{n+1} - \theta_i^n = \text{Fo} \, (a \, \theta_{i-1}^n + b \, \theta_{i+1}^n - c \, \theta_i^n) \tag{3.4.5}$$

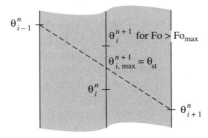

FIGURE 3.4.6 One-Dimensional, Transient Temperature Distribution
The maximum value of Fo must yield the steady-state temperature at node i.

It is obvious that the change in the magnitude of the temperature of the node at i on the left side of the equation can be made as large as desired by taking a sufficiently large value of Fo on the right side. We now ask, What is the maximum value of Fo that we can use to compute the change in the temperature of the node, given by the right side of the equation, without violating any physical laws? During any time step, we compute the change in the temperature of the node, i, assuming that the temperature of the neighboring nodes remain constant during this time step. With constant neighboring node temperatures, the maximum change in the value of the temperature at node i is when θ_i^{n+1} reaches its steady-state value; the corresponding maximum value of Fo is Fo_{max}. If a value of Fo greater Fo_{max} is used, the value of θ_i^{n+1} would shoot past the steady-state value, thus, violating the Second Law of Thermodynamics. Therefore, Fo must be less than Fo_{max}. This is illustrated in Figure 3.4.6.

From the energy equation for node i, the steady-state value of θ_i is obtained when the left side of Equation 3.4.5 is zero (with θ_{i-1}^n and θ_{i+1}^n fixed), that is,

$$\theta_{i,max}^{n+1} = \theta_{st} = \frac{a\,\theta_{i-1}^n + b\,\theta_{i+1}^n}{c}$$

$\theta_{i,max}^{n+1}$ is the value with Fo_{max}, which is computed from

$$\theta_i^{n+1}\big|_{max} - \theta_i^n = \theta_{i,st} - \theta_i^n = \text{Fo}_{max}\,(a\,\theta_{i-1}^n + b\,\theta_{i+1}^n - c\,\theta_i^n)$$

Substituting $(a\,\theta_i^n + b\,\theta_{i+1}^n)/c$ for $\theta_{i,max}^{n+1}$, we obtain

$$\frac{a\,\theta_{i-1}^n + b\,\theta_{i+1}^n}{c} - \theta_i^n = \text{Fo}_{max}\,(a\,\theta_{i-1}^n + b\,\theta_{i+1}^n - c\,\theta_i^n)$$

From the above equation, $\text{Fo}_{max} = 1/c$.

It should be clear by now, from the example problem, that the value of c may be different for different nodes. It is best to use the maximum value of c in determining the maximum value of Fo to be used in the computations. The reciprocal of the maximum value of, c, the coefficient of θ_i, gives the maximum value of Fo that can be used in the solution.

Implicit Method for Transient Problems So far we have considered the explicit method for the solution of transient problems; i.e., when the temperature distribution at a given instant is known, the temperature distribution after a defined time step is explicitly computed. In this method the time step is limited by the element size.

The limitation on the time step is eliminated in the *implicit method.* In the implicit method the temperatures at the nodes at a particular instant are expressed in terms of the temperatures at the end of the time step (backward differencing). The set of equations for all the time steps are then simultaneously solved. There are as many sets of simultaneous equations as there are time steps. Although the time step in this method is not limited by the element size, it should be small enough to give an acceptable solution—it can be larger than that required for the explicit method for the same element size. The number of simultaneous equations to be solved is now equal to the product of the number of nodes and the total number of time steps. The disadvantage of the method is that it requires a much larger number of equations to be solved simultaneously. For a more detailed discussion of the method, see Sucec (1985).

3.5   APPROXIMATE ANALYTICAL METHOD FOR TRANSIENT PROBLEMS—INTEGRAL METHOD

The exact solution to multidimensional transient problems requires the solution of partial differential equations. A few exact solutions are presented in Chapter 11. It is possible to find approximate analytical solutions to some one-dimensional transient conduction problems within the framework of ordinary differential equations. One such method is the *integral method.* The method can also be applied to the solution of some convective heat transfer problems; it is illustrated for transient conduction problems in this chapter and for convection problems in Chapter 5.

In differential methods the balance laws for mass, momentum, and energy are applied to an infinitesimal element, thus satisfying all the laws at every location in the region of interest. In integral methods the balance laws are satisfied over a finite region in one direction and an infinitesimal region in a perpendicular direction. As the physical laws are not satisfied at every point in the region of interest, the integral method yields approximate results. Such approximate solutions are satisfactory in many cases. As the solution is approximate, it is generally not possible to determine how good it is without comparing it with an exact solution (in which case there is no need for the approximate solution) or with experimental results. The integral method is illustrated through two examples.

EXAMPLE 3.5.1   A wide metal slab of thickness 2L, is cleaned by steam as it passes through a heating chamber, with a constant velocity V. In the chamber it is heated by the condensing steam as depicted in Figure 3.5.1. Determine the two-dimensional temperature distribution in the slab, $T(x, y)$.

ASSUMPTIONS

1. The slab enters the chamber at a uniform temperature T_i.
2. The dimension in the z-direction is much greater than the thickness 2L so that end effects may be neglected. The temperature distribution is two-dimensional, that is, $T = T(x, y)$.

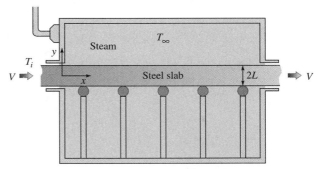

FIGURE 3.5.1 One-Dimensional Transient Conduction
A wide metal slab is heated by condensing steam.

3. Steady state
4. The heat transfer rate in the x-direction is much smaller than in the y-direction and can be neglected. This condition is satisfied if VL/α is much greater than 1. (See Section 12.5 for a detailed explanation.)
5. As the convective heat transfer coefficient in condensing steam is large, the Biot number on the steam side is large and the temperature of the surface is assumed to be equal to the temperature of the condensing steam, T_∞.
6. Constant properties (temperature independent)
7. The supporting rollers are also at T_∞ so that there is no conduction from the slab to the rollers.

SOLUTION

As the slab enters the chamber at a uniform temperature T_i, the condensing steam raises the surface temperature to T_∞. This tends to increase the temperature in the interior of the slab. The distance, δ, from the surface up to which this effect propagates depends on x, V, and the properties of the material of the slab. The distance δ is known as the penetration depth.

Now consider an elemental length of the slab shown in Figure 3.5.2. The control volume ABCD has a finite dimension in the y-direction and an elemental dimension in the x-direction. Because of symmetry, only one-half the thickness of the slab is considered for analysis. The temperature profile at a distance, x, from the entrance to the chamber is shown in Figure 3.5.2.

Consider the control volume ABCD. Applying the First Law of Thermodynamics (it has already been shown that with constant pressure, the work transfer term can be combined with the internal energy),

Rate of change of enthalpy in the C.V.
+ Net rate of enthalpy flow out of C.V.
= Net rate of heat transfer to the material in the C.V.
+ Net rate of internal energy generation

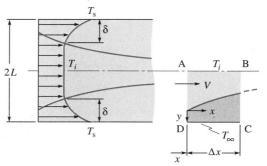

FIGURE 3.5.2 Temperature profile for Example 3.5.1
An elemental length of the solid showing the penetration of the temperature disturbance.

As the temperature does not vary in the z-direction, it is sufficient to consider a unit depth in that direction. For convenience, the origin is taken at a distance δ from the surface. As steady state is assumed, the rate of change of energy in the control volume $= 0$.

$$\text{Net rate of enthalpy flow out} = \int_{-(L-\delta)}^{\delta} V\rho i \, dy \bigg|_{x+\Delta x} - \int_{-(L-\delta)}^{\delta} V\rho i \, dy \bigg|_{x}$$

where $i = $ specific enthalpy of the slab.

As the axial conduction is negligible (assumption 4), the net rate of heat transfer is $q''_w \Delta x$, where q''_w is the heat flux to the surface exposed to the steam. As internal energy generation is zero we get

$$\int_{-(L-\delta)}^{\delta} V\rho i \, dy \bigg|_{x+\Delta x} - \int_{-(L-\delta)}^{\delta} V\rho i \, dy \bigg|_{x} = q''_w \Delta x$$

Dividing the equation by Δx and taking the limit as $\Delta x \to 0$, we obtain

$$\frac{d}{dx} \left(\int_{-(L-\delta)}^{\delta} V\rho i \, dy \right) = q''_w \tag{1}$$

Taking T_i as the reference temperature for computing changes in enthalpy,

$$i = i_i + c_p \, (T - T_i)$$

$$\frac{d}{dx} \left\{ \int_{-(L-\delta)}^{\delta} \rho V \, [i_i + c_p \, (T - T_i)] \, dy \right\} = \frac{d}{dx} \left(\int_{-(L-\delta)}^{\delta} \rho c_p V \, (T - T_i) \, dy \right) = q''_w$$

Rearranging the equation, we obtain

$$\frac{d}{dx} \left(\int_{-(L-\delta)}^{o} \rho c_p V \, (T - T_i) \, dy \right) + \frac{d}{dx} \left(\int_{0}^{\delta} \rho c_p V \, (T - T_i) \, dy \right) = q''_w$$

But, for $-(L - \delta) \leq y < 0$, $T = T_i$, and the first integral vanishes, and the energy equation simplifies to

$$\frac{d}{dx} \left(\int_0^\delta \rho c_p V \, (T - T_i) \, dy \right) = q''_w \tag{2}$$

The temperature T should satisfy the following conditions in the x- and y-directions:

$$T \, (x, y = 0) = T_i \tag{3a}$$

$$T \, (x, y = \delta) = T_\infty \tag{3b}$$

$$\frac{\partial T}{\partial y} \, (x, y = 0) = 0 \tag{3c}$$

The first condition is valid for values of x until $\delta = L/2$. Thereafter, the the midplane temperature begins to increase, and Equation 2 is no longer valid. The problem has to be solved in two domains: the first domain where $\delta \leq L/2$ and the second domain after δ reaches a value of $L/2$. Other boundary conditions, such as $(\partial^2 T/\partial y^2)(x, y = 0) = 0$, may be imposed but there is no certainty that such an increase in boundary conditions yields more accurate results.

The integral method consists of assuming a reasonable temperature profile, which involves the parameter δ. This assumed profile should satisfy the boundary conditions and the physics of the problem statement. The assumed profile is then introduced into Equation 2 and evaluated. This will then lead to an ordinary differential equation in δ. Details of the method now follow.

Assume a temperature profile of the form

$$T = a_0 + a_1 y + a_2 y^2 + a_3 y^3 + \dots$$

The logic behind assuming a polynomial profile is that, in most cases, any well-behaved function can be approximated by a polynomial with a sufficient number of terms. However, other functions, such as trigonometric or exponential functions, can be used for the profile if the physics of the problem suggests that they would be more appropriate. Reverting to the polynomial profile, the coefficients of y are functions of x so that the temperature profile is a function of x and y. The number of coefficients that can be retained in the function depends on the number of equations—balance equations and boundary conditions—that are available. For this problem, the four equations (Equation 2 and Equations 3a, 3b, and 3c) permit four unknowns to be evaluated. δ is one of the unknowns to be determined. Thus, only three coefficients of y in the profile can be retained;[3] such a profile has three terms:

$$T = a_0 + a_1 y + a_2 y^2$$

---

[3]It is not necessary that only the first three terms in the series be used. Any three terms of the series may be used. As the profile is not known, there is no reason to assume a more complex profile unless there is some indication that the temperature profile follows such a higher order polynomial. In this case there are no such reasons and the simplest profile that satisfies all the equations will be used.

From the boundary conditions, Equation 3,

$$T(x, 0) = T_i = a_0$$

$$\frac{\partial T}{\partial y}(x, 0) = 0 = a_1$$

$$T(y = \delta) = T_\infty = a_0 + a_2\delta^2 = T_i + a_2\delta^2$$

which gives

$$a_2 = \frac{T_\infty - T_i}{\delta^2}$$

Thus,

$$T - T_i = (T_\infty - T_i)\frac{y^2}{\delta^2} \qquad (4)$$

As δ is a function of x, Equation 4 gives the temperature distribution as a function of x and y. Employing Equation 4,

$$q''_w = k\left.\frac{\partial T}{\partial y}\right|_{y=\delta} = \frac{2k}{\delta}(T_\infty - T_i)$$

Therefore,

$$\frac{d}{dx}\left(\int_0^\delta V\frac{y^2}{\delta^2}\,dy\right) = \frac{2k}{\rho c_p \delta} = \frac{2\alpha}{\delta}$$

Performing the indicated integration we obtain,

$$\frac{1}{3}\frac{d\delta}{dx} = \frac{2\alpha}{\delta} \quad \text{or} \quad \frac{d\delta^2}{dx} = \frac{12\alpha}{V}$$

We have obtained a first order, ordinary differential equation in δ, requiring one boundary condition. The boundary condition is

$$\delta(x = 0) = 0$$

The solution to the differential equation with the boundary condition is

$$\delta = \sqrt{12\frac{\alpha x}{V}} \qquad (5)$$

Knowing the value of δ at any given location x, the temperature at any point y can be found from Equation 4. Equation 5 is valid for $0 < \delta < L$, i.e., for $x \leq x_o$, where

$$x_o = \frac{L^2 V}{12\alpha} \qquad (6)$$

For $x > x_o$, the temperature at the midplane is no longer constant at T_i but changes with x. The temperature profile for $x > x_o$ is shown in Figure 3.5.3 at three different

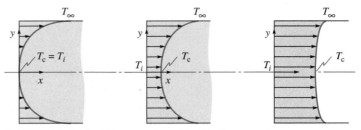

FIGURE 3.5.3 **The Temperature Profile for δ ≥ L**

locations. With increasing x, the midplane temperature increases. As x increases, T_c, the midplane temperature, should approach T_∞. The energy equation can be written as

$$\frac{d}{dx} \left(\int_0^L \rho c_p VT \, dy \right) = q_w''$$

From the above equation, subtract

$$\frac{d}{dx} \left(\int_0^L \rho c_p VT_i \, dy \right) = 0$$

and obtain

$$\frac{d}{dx} \left(\int_0^L \rho c_p V \, (T - T_i) \, dy \right) = q_w \tag{7}$$

Note that the origin is now at the midplane of the slab. Rewrite $T - T_i = (T - T_c)$ $+ (T_c - T_i)$ in Equation 7. $T - T_c$ represents the difference between the temperatures at any location and the midplane at the same location, and $T_c - T_i$ represents the increase in the midplane temperature over the initial temperature, T_i.

$$\frac{d}{dx} \int_0^L \rho c_p V \, [(T - T_c) + (T_c - T_i)] \, dy = q_w'' \tag{8}$$

Expressing the temperature as a polynominal,

$$T = a_0 + a_1 y + a_2 y^2 \tag{9}$$

The appropriate boundary conditions are now given by

$$T(0) = T_c = a_0$$

T_c represents the yet to be determined midplane temperature. By symmetry,

$$\frac{\partial T}{\partial y} (0) = 0 = a_1$$

$$T(L) = T_\infty = T_c + a_2 L^2$$

or

$$a_2 = \frac{T_\infty - T_c}{L^2}$$

Thus, the temperature profile is given by

$$T = T_c + \frac{T_\infty - T_c}{L^2} y^2 \tag{10}$$

In Equation 10 the midplane temperature T_c is now a function of x and this functional relationship between T_c and x is yet to be determined. The second term in Equation 10, $[(T_\infty - T_c)/L^2]y^2$ represents the difference between the temperature at any point (x, y) above the midplane temperature at the same value of x. If this midplane temperature is known, the temperature profile is specified by Equation 10. Thus, the problem reduces to finding T_c as a function of x. With $q_w'' = k(\partial T/\partial y)|_{y=L} = 2k[(T_\infty - T_c)/L$, substituting Equation 10 into Equation 8,

$$\frac{d}{dx} \int_0^L \rho c_p V \left[\frac{T_\infty - T_c}{L^2} y^2 + (T_c - T_i) \right] dy = 2k \frac{T_\infty - T_c}{L}$$

Performing the indicated integration and rearranging the equation,

$$\frac{d}{dx} \left[(T_\infty - T_c) \frac{L}{3} + (T_c - T_i) L \right] = 2 \frac{\alpha}{VL} (T_\infty - T_c)$$

As T_∞ and T_i are constant, we get

$$\frac{dT_c}{dx} = \frac{3\alpha}{VL^2} (T_\infty - T_c)$$

With $T_c(x_o) = T_i$, integrating the equation from x_o to x, $x > x_o$ (x_o is given by Equation 6),

$$\int_{T_i}^{T_c} \frac{dT_c}{T_\infty - T_c} = \frac{3\alpha}{VL^2} \int_{x_o}^{x} dx$$

or

$$T_\infty - T_c = (T_\infty - T_i) \exp - \left[\frac{3\alpha}{VL^2} (x - x_o) \right] \tag{11}$$

Equation 11 gives the midplane temperature at any value of $x > x_o$ and together with Equation 10, the temperature distribution at any location is defined. The surface heat flux can also be determined. Equation 11 predicts that as $x \to \infty$ $T_c \to T_\infty$, and the temperature profile becomes uniform. This is consistent with the physics of the problem.

The solution illustrates the methodology employed in integral methods. In this case, the problem is a two domain problem. In the first domain, defined by $x < x_o$, the penetration depth δ varies with x, with the temperature at the midplane being constant. In the second domain, defined by $x > x_o$, there is no penetration depth but the midplane temperature varies with x.

If one fixes the coordinates to the slab, the temperature distribution is obviously a function of time and the y-coordinate. The problem can also be solved in that frame with the solution being obtained in two time domains, the first one until the effect penetrates up to the midplane, and the second one thereafter. Such a transient temperature distribution may be obtained by starting from the basic principles. Alternatively, such a solution may also be obtained by recognizing that the time lapse is given by $\tau = x/V$ and substituting $V\tau$ for x in the solutions.

EXAMPLE 3.5.2

An electric resistance heating element is attached to one end of a very long cylinder, which is initially at a uniform temperature T_o (Figure 3.5.4). At time $\tau = 0$, the power to the electric heater is turned on, providing a constant heat flux, q_w''. The cylindrical surface is perfectly insulated. Find an expression for the transient temperature distribution in the cylinder.

ASSUMPTIONS

1. As the boundary condition is uniform and the material is assumed to be homogeneous, the temperature distribution is unsteady, one-dimensional
2. Constant properties

SOLUTION

The imposed heat flux at one end heats the rod, and the effect of the disturbance leads to an increase in the temperature. At any instant of time, δ represents the depth to which the effect of the imposed heat flux penetrates into the cylinder. Taking the rod as the system and applying the First Law of Thermodynamics, with $T(y, \tau)$ representing the temperature distribution (pressure being constant), we have

$$\frac{d}{d\tau}\left(\int_0^\infty \rho i A_y \, dy\right) = q_w'' A_y \tag{1}$$

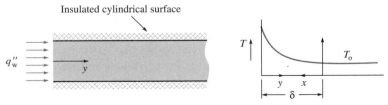

FIGURE 3.5.4 A Constant Heat Flux Imposed on One End of a Long Cylinder

where

$$i = \text{specific enthalpy} = \int_{T_{ref}}^{T} c_p \, dT$$

T_{ref} = the reference temperature for determining the specific enthalpy

A_y = area of cross section perpendicular to the y-direction

In heat transfer problems, it is the change in enthalpy that needs to be calculated and, hence, the choice of the reference temperature is quite arbitrary. For convenience, taking T_o as the reference temperature, and dividing Equation 1 by the constant area of cross section A_y, we obtain

$$\frac{d}{d\tau} \left[\int_0^\infty \rho c_p \, (T - T_o) \, dy \right] = q''_w$$

With δ representing the depth to which the effect of the heat flux penetrates, for $y > \delta$, $T = T_o$. As the integrand is 0 for $y > \delta$, the above equation can be written as

$$\frac{d}{d\tau} \left[\int_0^\delta \rho c_p \, (T - T_o) \, dy \right] = q''_w \tag{2}$$

For algebraic convenience, changing the origin as shown in Figure 3.5.4, with $x = (\delta - y)$, Equation 2 becomes

$$\frac{d}{d\tau} \left[\int_0^\delta \rho c_p \, (T - T_o) \, dx \right] = q''_w \tag{3}$$

In line with the logic in Example 3.5.1, a polynomial profile for the temperature T is assumed.

$$T = a_0 + a_1 x + a_2 x^2 + a_3 x^3 + \cdots$$

At the surface at $x = \delta$, the heat flux is related to the temperature gradient by Fourier's Law and

$$k \frac{\partial T}{\partial x} (\delta, \tau) = q''_w$$

The other boundary conditions are

$$T(0, \tau) = T_o \qquad \frac{\partial T}{\partial x} (0, \tau) = 0$$

The last boundary condition is obtained by requiring that the heat flux be continuous so that the temperature gradients evaluated at $x = 0^+$ and at $x = 0^-$ are equal. As the temperature for $x < 0$ is uniform, the gradient evaluated at $x = 0^+$ is zero.

With three boundary conditions and the energy equation, we can solve for four unknowns. As one of the unknowns to be determined is the penetration depth, δ, the maximum number of unknowns in the polynomial for T is 3. Choosing the first three terms in the polynomial,

$$T = a_0 + a_1 x + a_2 x^2 \tag{4}$$

Employing the boundary conditions,

$$T(0, \tau) = T_o = a_0 \qquad \frac{\partial T}{\partial x}(0, \tau) = 0 = a_1 \qquad k\frac{\partial T}{\partial x}(\delta, \tau) = q_w'' = 2ka_2\delta$$

or

$$a_2 = \frac{q_w''}{2\delta k}$$

The temperature profile is then given by

$$T = T_0 + \frac{q_w''}{2k}\frac{x^2}{\delta} \tag{5}$$

Substituting Equation 5 into Equation 3,

$$\frac{d}{d\tau}\int_0^\delta \rho c_p \frac{q_w''}{2k}\frac{x^2}{\delta}\,dx = q_w'' \tag{6}$$

$$\frac{d\delta^2}{d\tau} = 6\alpha \tag{7}$$

With $\delta(0) = 0$, integrating Equation 7, we obtain

$$\delta = \sqrt{6\alpha\tau} \tag{8}$$

Employing the value of δ given in Equation 8 in Equation 5, the one-dimensional transient temperature distribution is given by

$$\frac{T(x, \tau) - T_o}{(q_w''/k)} = \frac{x^2}{2\sqrt{6\alpha\tau}} \tag{9}$$

The temperature distribution given by Equation 9 is with the origin located at a distance δ from the end where the heat flux is imposed and is valid for

$$0 < x < \delta \qquad \delta = \sqrt{6\alpha\tau}$$

We now compare the temperature given by Equation 9 with that given by the exact solution (derived in Chapter 11). The exact solution is

$$\frac{T(x,t) - T_o}{q_w''/k} = 2\left[\sqrt{\frac{\alpha\tau}{\pi}}\exp\left(-\frac{y^2}{4\alpha\tau}\right) - \frac{y}{2}\,\mathrm{erfc}\left(\frac{y}{\sqrt{2\alpha\tau}}\right)\right]$$

where

$$\mathrm{erfc}\ z = \text{complimentary error function of } z = 1 - \mathrm{erf}\ z$$
$$\mathrm{erf}\ (z) = \text{error function of } z$$

Table A14 in Appendix 2 gives values of erf (z); the error function can also be found from HT(EF) software accompanying this book. Values obtained from Equation 9 are compared with the exact solution for the temperature at the surface where the heat flux is imposed, and at a distance $= \sqrt{\alpha\tau}$ from that surface.

Location	End Surface	$\sqrt{\alpha\tau}$ from End Surface
Equation 9	$1.225\ \sqrt{\alpha\tau}$	$0.4289\ \sqrt{\alpha\tau}$
Exact	$1.128\ \sqrt{\alpha\tau}$	$0.3993\ \sqrt{\alpha\tau}$
Error	$+\ 10\%$	$+\ 6\%$

The integral method predicts the same trend as the exact solution, that the temperature is proportional to $\sqrt{\alpha\tau}$. However, it overshoots the temperature excess, $T - T_o$, by about 10% at the end surface, and by about 6% at a distance $\sqrt{\alpha\tau}$ from the end surface. In many engineering applications, such a magnitude of error is acceptable. In this case, the integral method gave results that we may consider as acceptable. It may not always be so. The only way to determine if the results are reasonable is to compare them with some experimental results.

3.6 PROJECTS

PROJECT 3.6.1   A Liquid Level Measuring Device

Consider a thin resistance wire in a fluid through which an electric current flows. Because of the resistance heating, the temperature of the wire increases and, consequently, the resistance of the wire increases. The change in the resistance depends on the material of the wire, the current that causes the temperature to increase, and the fluid surrounding the wire.

Now consider a container partially filled with a liquid; the space above the free surface of the liquid is either a gas, such as air or the vapor of the liquid (or a mixture of gases) as shown in Figure 3.6.1. We assume the free surface is exposed to air. If a current flows through the wire, the temperature of the wire increases so that the electrical power supplied to the wire dissipates as convective heat transfer to the liquid and air. The approximate temperatures of the wire in the two sections are given by

$$q'_1 = h_1 \pi d\ (T_1 - T_\infty) \tag{1}$$

$$q'_2 = h_2 \pi d\ (T_2 - T_\infty) \tag{2}$$

q'_1 and q'_2 are approximately equal. Because of the difference in the values of h_1 and h_2 the temperatures in the two parts of the wire are different. The resistances of the wire in the two parts are related to the temperatures by

$$R_1 = R'_o\ (1 + \alpha T_1)\ L_1 \tag{3}$$

$$R_2 = R'_o\ (1 + \alpha T_2)\ L_2 \tag{4}$$

where R'_o is the resistance of the wire per unit length at 0 °C and α is the temperature coefficient of resistance of the material of the wire.

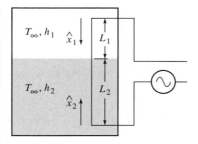

FIGURE 3.6.1 Liquid Depth Measuring Device
A resistance wire partially immersed in a liquid in a container.

Thus, the total resistance of the wire depends on the lengths L_1 and L_2 and temperatures T_1 and T_2. By relating the total resistance of the wire to the lengths and temperatures in the two sections, we can determine the depth of the liquid in the container. This principle is to be exploited to develop a liquid depth measuring instrument.

As the wire is partially immersed in the liquid, and the temperatures in the part exposed to air is higher than the temperature of the wire immersed in the liquid, the temperature varies continuously across the interface. The resistance in each part is to be determined by developing an expression for the temperatures as a function of the distance from the interface.

For a constant current through the wire, determine the total resistance of the wire as a function of the lengths of the wire in the air and liquid. The following parameters are known:

Air side heat transfer coefficient (W/m^2 °C)	120
Liquid side heat transfer coefficient (W/m^2 °C)	4500
Material of the wire	Copper clad steel
Wire gauge	30
Resistivity of wire material (Ω m)	35×10^{-8}
Temperature coefficient of resistance (K^{-1})	0.005
Thermal conductivity of wire material (W/m K)	52
Thermal diffusivity of wire material (m^2/s)	18×10^{-6}
Length of wire (m)	0.5

SOLUTION

The maximum Biot number based on the radius of the wire is $h_2 d/2k$. From the CRC *Handbook of Chemistry and Physics,* the diameter of 30-gauge wire is 2.548×10^{-4} m. The Biot number is 0.011 [$= (4500 \times 2.548 \times 10^{-4})/(2 \times 52)$]. As the Biot number is much less than 1, we assume one-dimensional temperature distribution in the wire.

The total resistance of the wire is given by

$$R = \int_0^{L_1} R_o' (1 + \alpha T_1) \, d\hat{x}_1 + \int_0^{L_2} R_o' (1 + \alpha T_2) \, d\hat{x}_2$$

T_1 and T_2 are the temperatures of the wire on the air and liquid side, respectively, and are functions of the distance from the ends of the wire. We proceed to determine $T_1(\hat{x}_1)$ and $T_2(\hat{x}_2)$.

Assume that the internal energy generation due to resistance heating is uniform (but different) in each section of the wire. Applying the steady state form of Equation 2.7.1 to each section of the wire,

$$\frac{d^2T_1}{d\hat{x}_1^2} - \frac{h_1P}{kA_x}(T_1 - T_\infty) + \frac{q_1'''}{k} = 0$$

$$\frac{d^2T_2}{d\hat{x}_2^2} - \frac{h^2P}{kA_x}(T_2 - T_\infty) + \frac{q_2'''}{k} = 0$$

The above two equations are recast as

$$\frac{d^2T_1}{d\hat{x}_1^2} - \frac{h_1p}{kA_x}\left(T_1 - T_\infty - \frac{q_1'}{h_1P}\right) = 0 \tag{5}$$

$$\frac{d^2T_2}{d\hat{x}_2^2} - \frac{h_2P}{kA_x}\left(T_2 - T_\infty - \frac{q_2'}{h_2P}\right) = 0 \tag{6}$$

$q_1'''A_x$ and $q_2'''A_x$ represent the internal energy generation per unit length and have been replaced by q_1' and q_2', respectively. The values of q_1' and q_2' are found from the values of the current and the resistances in each part. But the values of the resistances depend on the temperatures in the two parts and we do not, as yet, know the temperature distributions. But to start the computations, we calculate the resistances at T_∞ and update the values after determining the temperatures.

The boundary conditions are

$$\hat{x}_1 = 0: \qquad\qquad \frac{dT_1}{d\hat{x}_1} = 0 \tag{7}$$

$$\hat{x}_2 = 0: \qquad\qquad \frac{dT_2}{d\hat{x}_2} = 0 \tag{8}$$

$$\hat{x}_1 = L_1;\ \hat{x}_2 = L_2: \qquad\qquad T_1 = T_2 \tag{9}$$

$$\hat{x}_1 = L_1;\ \hat{x}_2 = L_2: \qquad\qquad \frac{dT_1}{d\hat{x}_1} = -\frac{dT_2}{d\hat{x}_2} \tag{10}$$

Define dimensionless variables,

$$\theta_1 = \frac{[T_1 - T_\infty - (q_1'/h_1P)]}{q_1'/(h_1P)} \qquad \theta_2 = \frac{[T_2 - T_\infty - (q_2'/h_2P)]}{q_2'/(h_2P)}$$

$$x_1 = \frac{\hat{x}_1}{L} \qquad x_2 = \frac{\hat{x}_2}{L} \qquad \frac{h_1P}{kA_x}L^2 = m_1^2 \qquad \frac{h_2P}{kA_x}L^2 = m_2^2$$

$$\frac{d^2\theta_1}{dx_1^2} - m_1^2\,\theta_1 = 0 \tag{11}$$

$$\frac{d^2\theta_2}{dx_2^2} - m_2^2\,\theta_2 = 0 \tag{12}$$

$$x_1 = 0: \qquad\qquad \frac{d\theta_1}{dx_1} = 0 \tag{13}$$

$$x_2 = 0: \qquad\qquad \frac{d\theta_2}{dx_2} = 0 \tag{14}$$

$$x_1 = f1 = L_1/L; \qquad x_2 = f2 = L_2/L$$

$$\theta_1 \frac{q_1'}{h_1 P} + T_\infty + \frac{q_1'}{h_1 P} = \theta_2 \frac{q_2'}{h_2 P} + T_\infty + \frac{q_2'}{h_2 P} \tag{15}$$

$$\frac{d\theta_1}{dx_1} = -\frac{h_1}{h_2} \frac{q_2'}{q_1'} \frac{d\theta_2}{dx_2} \tag{16}$$

Denoting $(h_1/h_2)(q_2'/q_1')$ by c, Equations 15 and 16 are written as

$$\theta_1 + 1 = c\,(\theta_2 + 1) \tag{17}$$

$$\frac{d\theta_1}{dx_1} = -c\frac{d\theta_2}{dx_2} \tag{18}$$

Solutions to Equations 11 and 12 are

$$\theta_1 = a_1 \cosh m_1 x_1 + a_2 \sinh m_1 x_1 \tag{19}$$

$$\theta_2 = b_1 \cosh m_2 x_2 + b_2 \sinh m_2 x_2 \tag{20}$$

From Equations 13 and 14, $a_2 = b_2 = 0$. From Equation 17,

$$a_1 \cosh m_1 f1 + 1 = c\,(b_1 \cosh m_2 f2 + 1) \tag{21}$$

From Equation 18,

$$a_1 m_1 \sinh m_1 f1 = -c b_1 m_2 \sinh m_2 f2 \tag{22}$$

Solving Equations 21 and 22,

$$a_1 = \frac{(c-1)\, m_2 \sinh m_2 f2}{Dr} \quad \text{and} \quad b_1 = -\frac{c-1}{c} \frac{m_1 \sinh m_1 f1}{Dr}$$

where $Dr = m_1 \sinh m_1 f1 \cosh m_2 f2 + m_2 \sinh m_2 f2 \cosh m_1 f1$.

Having found the expressions for the temperature distribution, we proceed to find the resistances in each part of the wire. Denoting the resistance of the wire exposed to the gas by R_1 and the resistance of the part exposed to the liquid by R_2,

$$R_1 = \int_0^{L_1} R_o'\,(1 + \alpha T_1)\,d\hat{x}_1 \qquad R_2 = \int_0^{L_2} R_o'\,(1 + \alpha T_2)\,d\hat{x}_2$$

where

$$T_1 = \theta_1 \frac{q_1'}{h_1 P} + T_\infty + \frac{q_1'}{h_1 P} \qquad T_2 = \theta_2 \frac{q_2'}{h_2 P} + T_\infty + \frac{q_2'}{h_2 P}$$

Introducing the expressions for θ_1 and θ_2 for T_1 and T_2 and the dimensionless distances x_1 and x_2 in the integrals for R_1 and R_2, we obtain, after performing the indicated integrations,

$$R_1 = R'_o \left[1 + \alpha \left(T_\infty + \frac{q'_1}{h_1 P} \right) \right] f1\ L + \frac{R'_o L \alpha q'_1 a_1}{h_1 P m_1} \sinh m_1 f1 \qquad (23)$$

$$R_2 = R'_o \left[1 + \alpha \left(T_\infty + \frac{q'_2}{h_2 P} \right) \right] f2\ L + \frac{R'_o L \alpha q'_2 b_1}{h_2 P m_2} \sinh m_2 f2 \qquad (24)$$

Equations 23 and 24 now permit the computations of the resistances R_1 and R_2. Recall that the values of q'_1 and q'_2 were computed from the values of the resistances evaluated at T_∞. After computing the resistances at the estimated values of the temperatures in each section, we now update the values of the internal energy generations and recompute the values of the resistances. The iteration is continued until two consecutive values of the total resistances are within a predefined limit. The computed values of the total resistance are shown in the table below as a function of the fraction of the wire exposed to the gas ($f1$) for the following conditions:

Length of wire   0.5 m
Current   1 A
T_∞ 25 °C

$f1$	0.0	0.1	0.2	0.3	0.4	0.5	0.6	0.7	0.8	0.9	1.0
$R(\Omega)$	3.88	4.08	4.29	4.5	4.71	4.92	5.13	5.35	5.56	5.77	6.0

If the current is reduced to 0.25 A, the resistance varies from 3.86 Ω at $f1 = 0$ to 3.95 Ω at $f1 = 1$.

Figure 3.6.2 shows the variation of the resistance of the wire as a function of the dimensionless depth of the liquid $f2$. The functional relationship between the wire resistance and the depth of the liquid can be computed for different lengths and

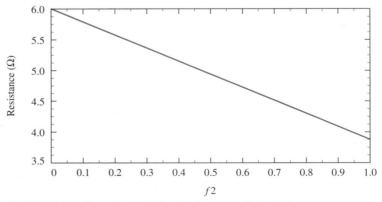

FIGURE 3.6.2 Variation of the Resistance of the Wire as a Function of the Depth of the Liquid in the Container ($f2$)

diameters of the wire from Equations 23 and 24. The linear relationship between the depth of the liquid and the resistance requires a simple electronic circuit, and the calibration is also simple.

COMMENTS

1. Depending on the calculator or computer program used, when the value of $m_1 f1$ or $m_2 f_2$ exceeds a certain value, you may run into numerical problems when Equations 23 and 24 are employed. In such cases the solutions can be obtained by using exponential functions instead of the hyperbolic functions with the origin at the liquid-gas interface. The boundary condition is modified by setting θ to zero as $m_1 f1$ (or $m_2 f2$) goes to infinity. The results when both $m_1 f1$ and $m_2 f2$ are large are

$$R_1 = R_o' \left[1 + \alpha \left(T_\infty + \frac{q_1'}{h_1 P} \right) \right] f1\, L + \frac{R_o' L \alpha q_1' c_1}{h_1 P m_1} [1 - \exp(-m_1\, f1)]$$

$$R_2 = R_o' \left[1 + \alpha \left(T_\infty + \frac{q_2'}{h_2 P} \right) \right] f2\, L + \frac{R_o' L \alpha q_2' c_2}{h_2 P m_2} [1 - \exp(-m_2\, f2)]$$

$$c_1 = \frac{(c - 1)\, m_2}{m_1 + m_2} \qquad c_2 = -\frac{(c - 1)}{c} \frac{m_1}{m_1 + m_2} \qquad c = \frac{q_2'}{q_1'} \frac{h_1}{h_2}$$

Note that for large values of mf, $\exp(-mf) \to 0$.

2. Although the values of h_1 and h_2 are given for a vertical wire, they are quite sensitive to departure of the wire from the vertical. The correct values should be obtained for the actual orientation at site. One method to obtain the values is to expose the entire wire either to the liquid or to the gas and pass a known current through the wire. By measuring the voltage drop across the wire, the heat dissipation (equal to the power input) is determined. From the measured value of current and voltage drop the resistance of the wire is found. From the resistance-temperature relationship the temperature of the wire is calculated. From the measured values of the heat dissipation and the temperature of the wire the heat transfer coefficient is determined.

3. A more direct method of calibrating the instrument is to measure the actual resistance as a function of the liquid depth at site. The measured resistance can be compared with the computed resistance based on the convective heat transfer coefficients determined as suggested in comment 2 above. The equations for the resistance assist us in choosing the appropriate size and material for the wire.

PROJECT 3.6.2   Project 3.6.1 Continued

There is some concern that the time required for the temperatures to reach their steady-state values may be excessive, requiring a long wait to obtain the liquid depth. It is required to estimate the time required to reach steady state.

The differential equations for the transient temperatures distribution in each domain, obtained from Equation 2.7.1, are

$$\frac{\partial \theta_1}{\partial \tau} = \frac{\partial^2 \theta_1}{\partial x_1^2} - m_1^2\, \theta_1 \qquad \frac{\partial \theta_2}{\partial \tau} = \frac{\partial^2 \theta_2}{\partial x_2^2} - m_2^2\, \theta_2$$

where τ is the dimensionless time $\alpha \hat{\tau}/L^2$.

The solutions to the equations can be obtained by employing Laplace transforms. But the inversion is involved and numerical procedures for the inversion may be required.

However, an estimate of the time required to achieve steady state can be obtained by neglecting the heat transfer at the interface so that the equations simplify to the form

$$\frac{\partial \theta}{\partial \tau} = -m^2\, \theta \qquad \theta\,(\tau = 0) = -1 \tag{1}$$

The solution to θ is

$$\theta = -\exp\,(-m^2 \tau) \tag{2}$$

The value of θ increases from -1 at $\tau = 0$, to 0 as $\tau \to \infty$. It is clear that the temperature reaches its steady-state value more rapidly for higher values of m. The higher value of the time for the temperature to reach its steady-state value is in domain 1, where the value of m is lower. We obtain an estimate of the time to reach steady state by assuming that steady state is reached when 99% of the maximum change in the value of θ has occurred, i.e., when θ reaches a value of -0.01. Employing the values in Project 3.6.1, we have

$$\alpha \hat{\tau}_{st}/L^2 = \frac{\ln 100}{m_1^2} \qquad \hat{\tau}_{st} = (\ln 100) \frac{kd}{4h\alpha}$$

For the material of the wire $\alpha = 18 \times 10^{-6}$ m^2/s. Substituting the various values,

$$\hat{\tau}_{st} = (\ln 100) \frac{52 \times 2.548 \times 10^{-4}}{4 \times 120 \times 18 \times 10^{-6}} = 7 \text{ s}$$

Assuming that the instrument is only turned on intermittently when the liquid level is to be measured, 7 s is not such a small time. If the operator reads the level immediately after turning on the instrument, erroneous values will be obtained.

PROJECT 3.6.3 Project 3.6.1 Continued

In Project 3.6.1, the internal energy generation was assumed to be uniform in each section of the wire. But as the internal energy generation is dependent on the resistance of the wire, which, in turn, depends on the temperature of the wire, for constant current, it is possible to directly express the internal energy generation as a function of the temperature. Thus,

$$q'_1 = = I^2R'_1 = I^2R'_o\,(1 + \alpha T_1) \qquad q'_2 = I^2R'_2 = I^2R'_o\,(1 + \alpha T_2)$$

Using the above expression in Equations 5 and 6 and modifying the boundary conditions appropriately, find the values of R_1 and R_2.

PROJECT 3.6.4   **Building Air Conditioning Load**

The refrigeration capacity required for the air conditioning of buildings depends on the cooling load. The cooling load depends, among other factors, on the heat transfer across the walls of the building (some of the other factors that contribute to the cooling load are the supply of fresh air; infiltration of warm air, people, appliances, and lighting in the building; solar infiltration through windows; and so forth). In this project we are interested in determining the heat transfer through one of the outer walls.

Consider the south facing wall of a building in which the inside air temperature is maintained at T_i (Figure 3.6.3).

An energy balance on the outer surface of the wall on a sunny day is given by

$$q'' = \alpha G_s + h_o\,(T_o - T_s) \tag{1}$$

where

q''	=	heat flux into the outer surface of the wall
α	=	absorptivity of the surface
G_s	=	solar insolation on a vertical south facing wall
h_o	=	convective heat transfer coefficient associated with the outer surface of the wall
T_o, T_s	=	the ambient air temperature and the outer surface temperature, respectively

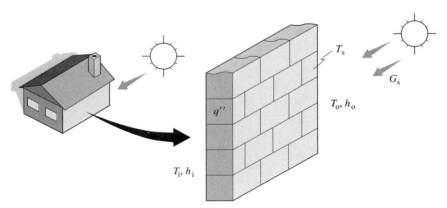

FIGURE 3.6.3 Determining Cooling Load
The south wall of a building.

The solar insolation is defined as the incident radiant, solar energy flux on a surface (W/m^2). α is the absorptivity of the surface and αG_s yields the radiant solar energy flux absorbed by the surface.

We now define an equivalent temperature, T_e, such that the convective heat flux from the ambient air temperature equals the total heat flux given by Equation 1. Recasting Equation 1,

$$h_o \left(T_e - T_s\right) = \alpha G_s + h_o \left(T_o - T_s\right)$$

or

$$T_e = T_o + \frac{\alpha G_s}{h_o} \tag{2}$$

The equivalent temperature T_e is known as the *sol-air temperature*. By using the sol-air temperature in conjunction with the convective heat transfer coefficient, the heat flux due both to convection and radiation is found without explicitly determining the two components separately. The sol-air temperatures have been computed for a large number of cases and are available in the ASHRAE *Handbook of Fundamentals* (1989).

Both the ambient air temperature and the solar insolation depend on the time of the year and time of day. Therefore, the sol-air temperature is a function of the time of the year and time of day.

The objective of the project is to determine the heat flux from the inner surface of the wall to the room, i.e., the heat transfer component of the cooling load, as a function of time. The heat flux from the inner surface of the wall depends on the heat flux to the outer surface of the wall. The heat flux to the outer surface of the wall can be computed if we know the convective heat transfer coefficient and the sol-air temperature. We will make use of the sol-air temperatures tabulated in the ASHRAE Handbook.

The south facing wall is modeled as consisting of 6-mm thick plywood siding, followed by 8-cm thick fiberglass insulation, and 1-cm thick dry wall (pearlite). The sol-air temperatures on a summer day (location 40 ° N latitude on July 21) are given below.

Time	2 a.m.	4 a.m.	6 a.m.	8 a.m.	10 a.m.	12 noon
T_e (°C)	24.4	23.3	23.8	27.2	31.6	41.6
Time	2 p.m.	4 p.m.	6 p.m.	8 p.m.	10 p.m.	12 midnight
T_e (°C)	50.5	50.0	39.4	29.4	27.2	25

Other known values are

$T_i = 25\,°C$
$h_i = 7\ W/m^2\,°C$
$h_o = 10\ W/m^2\,°C$

Assuming that the sol-air temperatures repeat themselves every day, determine the heat flux into the room and the total heat transfer per day per unit area. After the initial transients disappear, the temperatures become steady periodic. You may start with any reasonable temperature distribution in the wall and carry out the computations for a sufficient number of days until the temperatures become steady periodic.

SUMMARY

One-dimensional, steady temperature distribution with *internal energy generation* is obtained from the solution to

$$\frac{d}{dx}\left(kA_x \frac{dT}{dx}\right) + q''' A_x = 0$$

If the *thermal conductivity varies with temperature,* the heat transfer rate is correctly predicted by solving the conduction equation with a *constant mean thermal conductivity* defined by

$$k_m = \frac{1}{T_1 - T_2} \int_{T_1}^{T_2} k \, dT$$

However, for the correct temperature distribution, the temperature dependence of the thermal conductivity must be included in the differential equation. In most cases the use of an approximate linear relationship between the thermal conductivity and the temperature gives acceptable results for the temperature distribution and heat transfer.

When closed form solutions for the temperature distribution are not available, the temperature distribution can be obtained from *numerical procedures.* The difference equations are obtained from an energy balance. The resulting equations are solved by an iterative method. In the numerical solution to the transient temperature distribution, the nodal Fourier number, Fo, must satisfy the condition

$$\text{Fo}_{max} < \frac{1}{c}$$

where c is the coefficient of T_i^n in the nodal difference equation

$$T_i^{n+1} - T_i^n = \text{Fo} \, (aT_{i-1}^n + bT_{i+1}^n - cT_i^n)$$

Finite difference equations are derived for interior nodes, nodes at the boundaries with different boundary conditions, and at irregular boundaries. Two numerical methods—the Gauss-Siedel iteration and the tri-diagonal matrix algorithm (TDMA)—for the solution of the resulting linear, simultaneous equations are presented. Fortran programs for the two algorithms are given in Appendix 7.

The justification for the products of solutions of one-dimensional transient problems for the solutions of multidimensional, transient problems is provided in Section 3.4.

An approximate analytical solution, the integral method, for the solution of one-dimensional, transient conduction problems is introduced in Section 3.5. The technique satisfies the physical laws over finite regions. The technique is illustrated through problems that require the solution of ordinary differential equations.

REFERENCES

Adams, J.A., and Rogers, D.F. (1973). *Computer Aided Heat Transfer Analysis.* New York: McGraw-Hill.

Arpaci, V.S. (1966). *Conduction Heat Transfer.* Reading, MA: Addison-Wesley.

Dusinberre, G.M. (1961). *Heat Transfer Calculation by Finite Differences.* Scranton, PA: International TextBook.

Patankar, S.V. (1980). *Numerical Heat Transfer and Fluid Flow.* New York: McGraw-Hill.

Schneider, P.J. (1955). *Conductive Heat Transfer.* Reading, MA: Addison-Wesley.

Sucec, J. (1985). *Heat Transfer.* Dubuque, IA: Brown.

REVIEW QUESTIONS

(a) What is internal energy generation? Give some examples where internal energy generation occurs.

(b) When would you use temperature dependent thermal conductivity for the solution of heat transfer problems?

(c) When would you resort to numerical solutions for heat transfer problems?

(d) Give two convergence criteria for numerical solutions.

(e) To estimate the heat transfer rate across the boundary between two neighboring nodes, an appropriate value of the thermal conductivity at the boundary should be used. How is such an appropriate thermal conductivity related to the thermal conductivities evaluated at the temperature of the nodes?

(f) Discuss the constraints for using the product solution employing Hiesler charts.

(g) To use Hiesler charts for the solution of multidimensional transient problems, is it necessary that
   i. The convective heat transfer coefficients on all surfaces have the same value?
   ii. The fluid surrounding the various surfaces have the same temperature?

(h) In the numerical solution to two-dimensional, steady-state, problems, the element size $(\Delta x, \Delta y)$ can be fixed arbitrarily. In the solution to transient problems, can the time interval be selected arbitrarily?

(i) What do you understand by "stability criterion" for the solution of transient problems?

(j) Give the salient points of the integral method.

(k) What is penetration depth?

PROBLEMS

3.1 A plate is 1-cm thick, 6-cm wide and 20-cm long. A potential difference of 120 V is applied across two of the surfaces as shown. All end surfaces, except the two parallel surfaces, are perfectly insulated. The two uninsulated surfaces with dimensions of 6 cm × 20 cm are maintained at a uniform temperature of T_1. Determine $T_{max} - T_1$. Properties of the material of the plate are, $k = 15.2$ W/m K. Electrical resistivity $= 406 \times 10^{-5}$ Ω m.

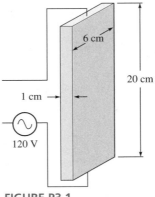

FIGURE P3.1

3.2 In Problem 3.1, one 6-cm surface (A) is maintained at 600 °C and the other 6-cm surface (B) at 400 °C. Find the heat transfer rate from surfaces A and B to the surroundings and the maximum temperature in the plate.

3.3 A potential difference of 120 V is applied to a 2-cm diameter, 1-m long cylindrical rod. Determine the difference between the temperatures at the axis and the surface of the rod. Properties of the material of the plate are, $k = 15.2$ W/m K. Electrical resistivity $= 406 \times 10^{-5} \Omega$ m.

3.4 A large 50-mm thick rectangular slab of type AISI 316 stainless steel, is heated by an induction coil. The surfaces are exposed to a fluid at 20 °C with a heat transfer coefficient of 20 W/m² °C. The temperature of the surfaces is to reach 300 °C. Assuming uniform internal heat generation due to the induction heating, determine
(a) The internal energy generation per unit volume.
(b) The maximum temperature in the slab.

3.5 A rectangular plate of thickness L, made of a semitransparent material, is subjected to laser irradiation. The absorption of energy (which is the equivalent of internal energy generation per unit volume) is given by $ae^{-\alpha x}$ per unit volume where a and α are known constants. Find the temperature distribution in the material of the plate if the surface temperatures are $T(x = 0) = T_1$ and $T(x = L) = T_2$. Also find the heat flux from each surface.

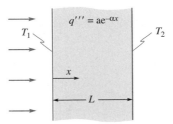

FIGURE P3.5

3.6 A 10-cm thick (AISI 316) steel slab is to be heated so that its surface exposed to air ($T_\infty = 20\,°C$, $h = 20\ \text{W/m}^2\,°C$) reaches a temperature of $400\,°C$ (Figure P3.6). Induction heating is employed. The internal energy generation due to the induction heating is represented by

$$q''' = q'_o\, x^2 \qquad x = \text{distance from the midplane}$$

Determine q'_o, the temperature at the midplane, and the temperature gradient at the surface.

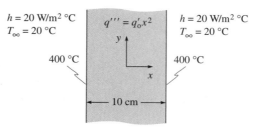

$h = 20\ \text{W/m}^2\,°C$
$T_\infty = 20\,°C$

$q''' = q'_o x^2$

$h = 20\ \text{W/m}^2\,°C$
$T_\infty = 20\,°C$

$400\,°C$ $400\,°C$

10 cm

FIGURE P3.6

3.7 When a 5-cm thick rectangular stainless steel (type AISI 316) slab is heated by induction heating, the internal energy generation is represented by $q''' = q_o\, e^{cx}$, for $x > 0$, and $q''' = q_o\, e^{-cx}$, for $x < 0$, where $q_o = 9.89 \times 10^{-3}$, W/m^3, $c = 800\ \text{m}^{-1}$, $x = $ distance from the midplane. If the slab is exposed to air at $20\,°C$ with a convective heat transfer coefficient of $20\ \text{W/m}^2\,°C$, determine
(a) The temperature at the surface.
(b) The temperature gradient at the surface.
(c) The temperature at the midplane.

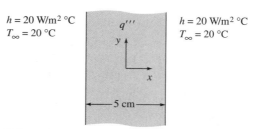

$h = 20\ \text{W/m}^2\,°C$
$T_\infty = 20\,°C$

q'''

$h = 20\ \text{W/m}^2\,°C$
$T_\infty = 20\,°C$

5 cm

FIGURE P3.7

3.8 In Problem 3.3, the outer surface of the rod is exposed to air at $20\,°C$ and the convective heat transfer coefficient is $40\ \text{W/m}^2\ \text{K}$. Find the temperatures at the surface and at the center line of the rod.

3.9 The uniform internal energy generation in a hollow cylinder of inner and outer radii r_i and r_o, respectively, is q''' per unit volume. The inner surface is cooled by a fluid at T_{fi} with a heat transfer coefficient of h_i, and the outer surface by a fluid at T_{fo} with a heat transfer coefficient of h_o. Find

(a) An expression for the temperature distribution in the cylinder.

(b) The heat transfer rate per unit length of the tube from the inner and outer surfaces.

3.10 A hollow tube, 8-mm I.D., 12-mm O.D., and 1-m long, is made of Inconel with an electrical resistivity of 122 $\mu\Omega$ cm. A potential difference of 10 V is applied across the end surfaces. The inside surface is perfectly insulated.

(a) Determine the difference between the inner and outer surface temperatures.

(b) If the outer surface is exposed to a fluid maintained at 100 °C and $h_o = 3000$ W/m^2 K, find the outer and inner surface temperatures.

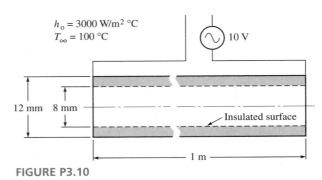

FIGURE P3.10

3.11 An 8-cm diameter nuclear fuel rod ($k = 55$ W/m k) is encased inside a type 316 stainless steel tube, which fits snugly over the fuel rod. The outer diameter of the steel tube is 10 cm. The steel tube is exposed to a fluid at 250 °C with a convective heat transfer coefficient of 3000 W/m^2 K. With uniform internal energy generation in the fuel rod, it is required that the temperature of the tube should not exceed 300 °C. Determine

(a) The outer surface temperature of the steel tube.

(b) The internal energy generation in the fuel rod (per unit length).

(c) The maximum temperature in the fuel rod.

(d) Redetermine part c if the thermal conductivity of the fuel rods is 4 W/m K.

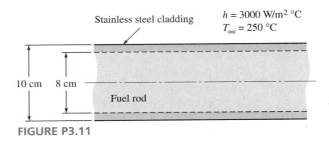

FIGURE P3.11

3.12 Nichrome V wire, made of an alloy containing 80% nickel and 20% chromium, is extensively used as an electrical heating element. It has a thermal conductivity of 17.1 W/m K and an electrical resistivity of 108 $\mu\Omega$ cm.

In an experimental setup, a hollow tube of nichrome, 25-mm I.D., 30-mm O.D., and 2-m long, is used as a heating element. A potential different of 5 V is impressed across the ends of the tube. Inside the tube a fluid flows at its saturation temperature of 50 °C with a convective heat transfer coefficient of 2000 W/m² K. Determine the inner and outer surface temperatures of the tube if the outer surface the tube is perfectly insulated.

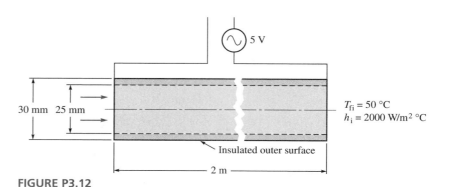

FIGURE P3.12

3.13 (a) From the values of the thermal conductivity of AISI 316 stainless steel, obtain a linear relationship between the thermal conductivity and temperature, $k = k_o (1 + \alpha T)$, where k_o and α are constants and T is in K ($k_o = 9.25$ W/m K; $\alpha = 1.622 \times 10^{-3}$ K^{-1}).

(b) One surface of a 25-mm thick stainless steel plate (type 316) is maintained at 800 K and the other parallel surface at 400 K. Employing the linear relationship between k and T found in part a, find

   i. An expression for the temperature distribution in the plate as a function of the distance from the surface maintained at 800 K.

   ii. The temperature at the midplane and compare it with that obtained by employing a constant thermal conductivity.

   iii. The heat flux.

3.14 (a) The inner and outer surface of a hollow stainless steel tube (type 316) of inner radius r_1 and outer radius r_2 are maintained at uniform temperatures of T_1 and T_2, respectively. The thermal conductivity is approximated as a linear function of the temperature (see Problem 3.13). Derive an expression for the temperature distribution and the heat transfer rate from the outer and inner surface per unit length of the cylinder.

(b) Employing the results of Problem 3.13a, find the heat transfer rate per unit length of the cylinder across the hollow cylinder, if $r_1 = 2$ cm, $r_2 = 6$ cm, $T_1 = 400$ K, and $T_2 = 800$ K.

(c) Find the temperature at $r = 3$ and 4 cm and compare them with the those obtained by assuming constant thermal conductivity evaluated at the arithmetic mean temperature.

3.15 Assuming $k = k_o (1 + \alpha T)$, determine the temperature distribution in a flat plate of thickness L, one surface of which is maintained at T_1 and the other parallel surface at T_2. A uniform internal energy generation of q''' per unit volume occurs in the plate. Also find the heat flux from each surface to the surroundings.

3.16. Re-solve Problem 3.14a if a uniform internal energy generation at q''' per unit volume takes place in the cylinder.

3.17 The thermal conductivity of magnesite (with 50% MgO) is approximated by

$$k = k_o (1 + \alpha T + \beta T^2) \qquad T \text{ in K}$$

$$k_o = 4.09 \text{ W/m K} \qquad \alpha = -1.742 \times 10^{-3} \text{ K}^{-1} \qquad \beta = 8.802 \times 10^{-7} \text{ K}^{-2}$$

It is found that the temperature in a slab of magnesite varies from 400 K to 1000 K. What is the mean value of the thermal conductivity to be used in determining the heat transfer rate?

3.18 A coarse network for a rectangular slab is shown in Figure P3.18 along with the assumed temperature distribution for the interior nodes. The temperatures of the surfaces are constant. Employing Gauss-Siedel iteration, write the temperatures at the nodes after the first and second iteration, if,

(a) $a = b = 4$ cm.

(b) $a = 4$ cm and $b = 8$ cm.

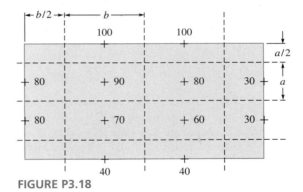

FIGURE P3.18

3.19 A plain carbon-steel plate has two cavities as shown. Surface A is maintained at 300 °C and the parallel surface B at 30 °C. All other surfaces, including those of the cavities, are perfectly insulated. Determine the heat transfer rate across the plate under steady-state conditions. The width of the plate perpendicular to the plane of the paper is 20 cm and the cavities extend over the entire width of the plate.

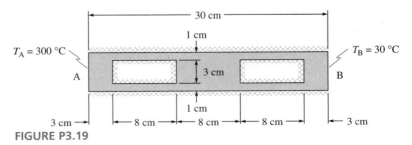

FIGURE P3.19

3.20 In Problem 3.19, if surface B is exposed to a fluid maintained at 20 °C with a heat transfer coefficient of 10 W/m² K, determine the heat transfer rate.

3.21 Re-solve Problem 3.20 taking into account the variation of thermal conductivity of the material with temperature.

3.22 Re-solve Problem 3.19 if, instead of the surfaces of the cavities being insulated, they are exposed to a fluid at 100 °C with a heat transfer coefficient of 20 W/m² K.

3.23 A 2-cm thick amorphous carbon plate is sandwiched between two AISI 316 stainless steel plates, each 0.5-cm thick. The assembly is 6-cm wide and 10-cm long. Determine the heat transfer rate assuming constant thermal conductivity for each material. Compare the heat transfer rate obtained by employing the electrical analogy for a parallel circuit. All surfaces, except the two end (parallel) surfaces, are perfectly insulated. Assume negligible contact resistance between the steel and carbon plates.

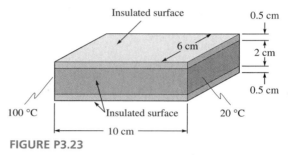

FIGURE P3.23

3.24 The two-dimensional temperature distribution in the block shown in Figure P3.24 is to be determined by Gauss-Siedel iteration. $\Delta x = 5$ mm, $\Delta y = 8$ mm. The temperatures after several iterations are as shown in the figure. Write the equation for the corner node and find its temperature if h = 30 w/m²°C and $T_\infty = 20$ °C.

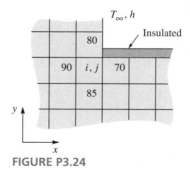

FIGURE P3.24

3.25 A handle is made of 20-mm diameter plain carbon-steel rod in the form of an *L* as shown. Part of the handle is insulated by Bakelite. The handle is attached to a door which is at 200 °C. The convective heat transfer coefficient is 10 W/m² °C. Determine the outer surface temperature of the Bakelite at different locations. Neglect the contact resistance between the Bakelite and steel.

3.26 The thickness at the base and tip of a very wide 50-mm long trapezoidal, rectangular fin are 8 mm and 3 mm, respectively. The nodal spacing is 10 mm. Write the nodal

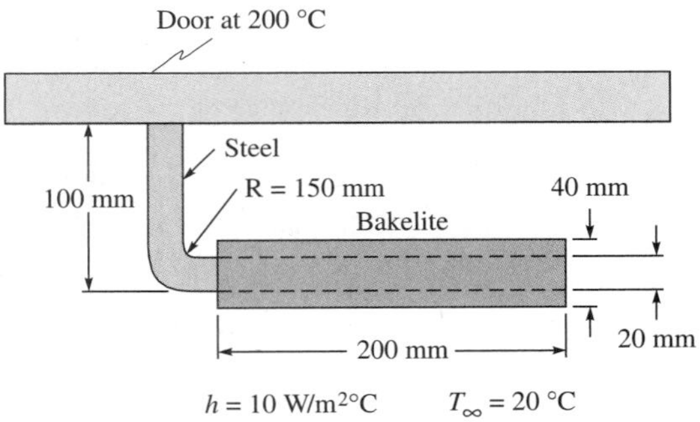

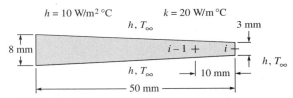

FIGURE P3.25 (See page 278)

equations for the interior node, $i-1$, and the end node i at the tip of the fin. Employ $\theta = T - T_\infty$.

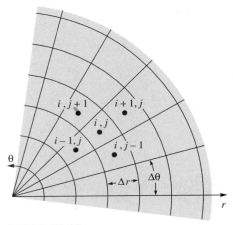

FIGURE P3.26

3.27 Show that the two-dimensional nodal equation in cylindrical coordinates for $T(r, \theta)$ is given by

$$T_{i,j} = \frac{[(r_{i,j} - \Delta r/2)\,(\Delta\theta/\Delta r)\,T_{i-1,j}] + [(r_{i,j} - \Delta r/2)\,(\Delta\theta/\Delta r)\,T_{i+1,j}]}{2\,[r_{i,j}\,(\Delta\theta/\Delta r) + \Delta r/(r_{i,j}\Delta\theta)]} \atop {- [(\Delta r/r_{i,j}\Delta\theta)\,(T_{i,j-1} + T_{i,j+1})]}$$

Note that the heat flux in the θ-direction is given by $q''_\theta = -k\,\dfrac{\partial T}{r\,\partial\theta}$.

FIGURE P3.27

3.28 A rectangular plain carbon-steel plate, A, 6-cm thick and 8-cm long is attached to a second plain carbon-steel plate, B, 10-cm thick and 6-cm long. The width perpendicular to the plane of the paper is large and two-dimensional temperature distribution may be assumed. The end surface of A is maintained at 80 °C and the parallel end surface of B at 20 °C. All other surfaces are perfectly insulated. Find the two-dimensional temperature distribution and the heat transfer rate per unit width of the plate.

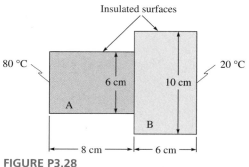

FIGURE P3.28

3.29 Re-solve Problem 3.28 if plate B is type 316 stainless steel.

3.30 The inside dimensions of a furnace are 15-cm wide, 15-cm high, and 2-m long. It is lined with 10-cm thick fireclay brick followed by 15-cm thick insulation ($k = 0.04$ W/m K). The temperatures in the walls may be assumed to be two-dimensional. The inside surfaces are maintained at 400 °C. The outside surfaces, except the bottom surface, are exposed to air at 30 °C with a convective heat transfer coefficient of 10 W/m² °C. The bottom surface is perfectly insulated. Determine the heat transfer rate to the air from the 2-m long walls.

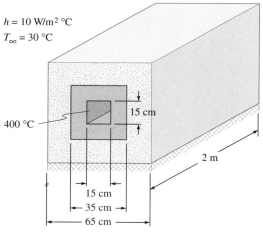

FIGURE P3.30

3.31 An annular fin is attached to a pipe maintained at 200 °C. The material of the fin is AISI 316 stainless steel. Find the temperature distribution in the fin employing numerical methods. The heat transfer coefficient is 10 W/m² K.

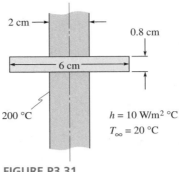

FIGURE P3.31

3.32 A type 316 stainless steel rod, 2-cm diameter and 14-cm long, is attached to a plate maintained at 400 °C. The convective heat transfer coefficient between the surfaces of the rod and the surrounding air (at 30 °C) is 20 W/m² K. Estimate the tip temperature and heat transfer rate from the rod to the air
   (a) Assuming constant thermal conductivity employing analytical techniques.
   (b) Assuming the thermal conductivity to vary linearly with temperature ($k = 9.25 + 0.015 \, T$ W/m K, T in K) employing numerical methods.

3.33 In Problem 2.91, to estimate the heat transfer rate from the insulated end, the one-dimensional approximation was made. Employing numerical methods determine the heat transfer rate from the insulated end assuming the surface in contact with the cylinder to be at 80 °C and the heat transfer coefficient associated with the surfaces to be 26 W/m² K. Note that Bi = $hR/k > 1$ and the one-dimensional approximation is not appropriate.

3.34 A rectangular duct, 10-cm wide and 15-cm high, is insulated with 10-cm thick cork insulation. The surface temperature of the duct is −20 °C and the surrounding air is at 30 °C. If the convective heat transfer coefficient on the outer surface is 8 W/m² K, compute the heat transfer rate per unit length of the duct. Assume two-dimensional temperature distribution.

3.35 A long I-beam of plain carbon-steel is shown in the Figure P3.35. The top surface is maintained at 35 °C and the bottom surface at 20 °C. All other surfaces are perfectly insulated. Assuming two-dimensional temperature distribution, determine the heat transfer rate across the beam per unit length of the beam.

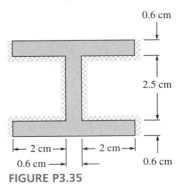

FIGURE P3.35

3.36 The steel hull of a cold storage room in a ship is insulated with 200-mm thick cork insulation. The hull is strengthened by welding steel bulb angles at intervals of 600 mm as shown in Figure P3.36. The outer surface of the hull is exposed to water at 15 °C with a convective heat transfer coefficient of 3000 W/m K. The inner surface is exposed to air at −18 °C with a convective heat transfer coefficient of 20 W/m² K. Estimate the heat flux and compare it with that obtained by ignoring the bulb angles. Assume properties of steel to be those of plain carbon-steel.

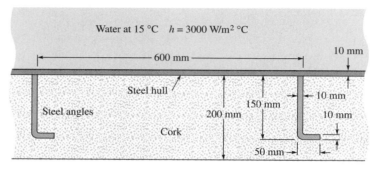

Water at 15 °C $h = 3000$ W/m² °C

600 mm

Steel hull

Steel angles

Cork

10 mm

10 mm

150 mm

10 mm

200 mm

50 mm

Air at −18 °C $h = 20$ W/m² °C

FIGURE P3.36

3.37 A rectangular slab is subjected to a radiant heat flux of q'' (radiant energy absorbed by the surface) on one surface. The parallel surface is perfectly insulated. Write the nodal equation for the nodes at the corners A and B.

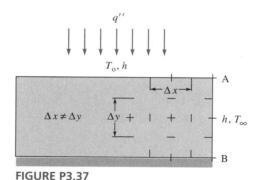

q''

T_o, h

Δx

A

$\Delta x \neq \Delta y$ Δy +   +

h, T_∞

B

FIGURE P3.37

3.38 A solid block of plain carbon-steel has two rectangular cavities (along the entire length of the block). The surfaces of the cavities are maintained at 100 °C with outside surfaces of the block exposed to air at 20 °C (with a convective heat transfer coefficient of 20 W/m² K). Determine the heat transfer rate from the cavities per unit length of the block.

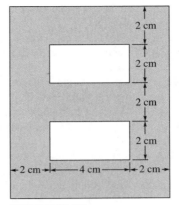

FIGURE P3.38

3.39 A very wide straight fin of cartridge brass has a triangular profile. It is 1-cm thick at the base, and is 3-cm long. The base of the fin is maintained at 80 °C, and the surrounding air is at 20 °C with a convective heat transfer coefficient of 60 W/m² °C. Determine the temperature distribution and the heat transfer rate from the fin under steady state conditions. Assume one-dimensional temperature distribution.

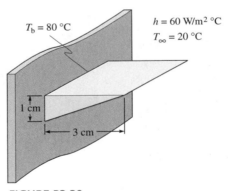

FIGURE P3.39

3.40 Consider the fin in Problem 3.39. Initially the fin is at a uniform temperature of 20 °C. At time $\tau = 0$, the temperature of the base of the fin is raised to 80 °C and maintained at that value. Determine the transient temperature distribution and the heat transfer rates. Compare the steady-state values with those obtained in Problem 3.39.

3.41 To detect unusually high temperatures (caused by hot gases from a fire) a thermocouple is attached to one side of a solid, 40-mm thick hardwood door. The temperature on the other side of the door is raised to 400 °C at time $\tau = 0$. The initial uniform temperature of the door is 20 °C. Employing numerical methods, determine the time taken for the temperature of the cool side of the door to increase by 5 °C assuming

(a) The cool side is perfectly insulated.

(b) The cool side is exposed to air at 20 °C with a convective heat transfer coefficient of 10 W/m² K.

3.42 To reduce the response time of the thermocouple in Problem 3.41, a 6-mm diameter brass plug is embedded in the door. The heat transfer rate from the cylindrical surface may be neglected as the surrounding wood acts as an insulator. Estimate the time for the cool side temperature of the brass plug to increase by 5 °C for both cases indicated in Problem 3.41.

3.43 A vertical plate, 30-cm wide, 40-cm high, and 3-mm thick, made of plain carbon-steel is attached to a pipe along one edge. The plate is exposed to air at 20 °C. At time $\tau = 0$, steam enters the pipe and maintains the edge of the plate connected to the pipe at 100 °C. The convective heat transfer coefficient on the air side is 7 W/m² °C. Determine

(a) The transient temperature distribution.

(b) The steady-state temperature distribution and heat transfer rate and compare them with those obtained by analytical methods.

(c) Repeat part a and include radiation heat transfer. The plate is in a large room at 20 °C and the emissivity of the surfaces is 0.8.

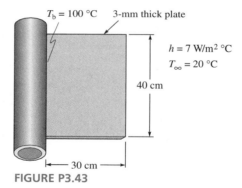

$T_b = 100$ °C   3-mm thick plate

$h = 7$ W/m² °C

$T_\infty = 20$ °C

40 cm

30 cm

FIGURE P3.43

3.44 Following the logic given in Section 3.4.3 find the maximum time step for the following transient cases to ensure stable numerical solutions:

(a) Prescribed boundary temperature, unequal grids (one-dimensional)

$$\Delta\tau_{max} = \frac{\Delta x^- \, \Delta x^+}{2\alpha}$$

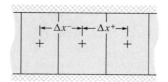

Δx^- Δx^+

FIGURE P3.44a

(b) Two-dimensional temperature distribution, specified boundary temperatures, equal grid spacing.

$$\Delta\tau_{max} = \frac{\Delta x^2}{4\alpha}$$

(c) Two-dimensional temperature distribution, corner exposed to convection heat transfer, equal grid spacing.

$$\Delta\tau_{max} = \frac{1}{\alpha} \frac{\Delta x \, \Delta y}{\{(\Delta y/\Delta x) + (\Delta x/\Delta y) + [(h/k)(\Delta x + \Delta y)]\}}$$

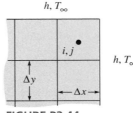

FIGURE P3.44c

3.45 The one-dimensional transient temperature distribution in a pin fin is to be determined numerically. Show that for stability, the maximum value of the time step is given by

$$\frac{\alpha \, \Delta\tau}{\Delta x^2} = \frac{1}{2 + m^2}$$

where

$\Delta\tau$ = maximum time step
$m^2 = 4h/kd$
h = convective heat transfer coefficient
α = thermal diffusivity
k = thermal conductivity
d = diameter

3.46 Consider the transient, one-dimensional temperature distribution in a slab of thickness L. The block is initially at a uniform temperature T_o. At time $\tau = 0$, the two parallel surfaces are exposed to a fluid maintained at T_∞ with a convective heat transfer coefficient of h. All the other surfaces are perfectly insulated.

(a) Write the finite difference form of the transient conduction equation in dimensionless form in terms of

$$\theta \text{ (dimensionless temperature)} = \frac{T_o - T}{T_o - T_\infty}$$

$$x \text{ (dimensionless distance)} = \frac{x}{L}$$

$$\text{Biot number} = \frac{hL}{k} \qquad \text{Fo (dimensionless time)} = \frac{\alpha\tau}{L^2}$$

(b) Find $\theta(x, \text{Fo})$, for $x = 0$, 0.5 and 1, as a function of Fo and plot it for Bi of 1.0 and 5.0. Compare your results with the plot in Figure 2.9.2

3.47 A 1-cm diameter, 30-cm long brass rod is attached to a metallic door. The convective heat transfer coefficient is 10 W/m² °C in an atmosphere of air at 20 °C. The radiant heat transfer rate per unit surface area of the rod is given by

$$q_R'' = \sigma \, \epsilon \, (T^4 - T_\infty^4) \qquad \sigma = 5.67 \times 10^{-8} \text{ W/m}^2 \text{ K}^4$$

$$\epsilon = \text{emissivity of the surface} = 0.8$$
$$T = \text{temperature of the surface in K}$$
$$T_\infty = \text{ambient temperature in K (293.2 K)}$$

The temperature of the rod at the end attached to the door is 300 °C. Employing a numerical method, find the steady-state temperature of the rod and the heat transfer rate from the rod to the surroundings.

3.48 For the rod in Problem 3.47, find the transient, one-dimensional temperature distribution and the heat transfer rate if the rod is initially at 20 °C, and the temperature of the end attached to the door is raised from 20 °C to 300 °C at time $\tau = 0$. Compare the steady-state solution with that of Problem 3.47.

3.49 The construction of a wall is shown in Figure P3.49. Initially, the temperature distribution in the wall is steady. The effect of solar insolation on the heat transfer rate across the wall is to be determined. As an approximation, assume that at time $\tau = 0$, a heat flux of 500 W/m² is imposed on the outer surface of the wall. Determine the heat flux q_x'' from the inner surface of the wall, 2 h and 8 h after the heat flux is imposed.

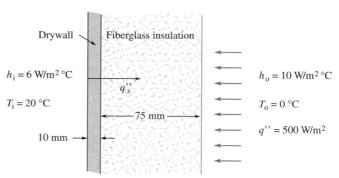

FIGURE P3.49

3.50 A more realistic solar insolation is $q'' = 800 \sin (\tau\pi/8)$ for $0 < \tau < 8$, τ in hours. Re-solve Problem 3.49.

3.51 A high temperature furnace is constructed with 20-cm thick firebricks. The inside dimensions of the furnace are 0.5-m wide, 1-m deep, and 0.8-m high. The thermal conductivity of the firebricks is

T (°C)	260	540	815
k (W/m K)	0.12	0.14	0.17

The temperature of the inner surface of the furnace walls is 800 °C. The ambient air is at 35 °C with an associated convective heat transfer coefficient of 10 W/m² °C. Find the three-dimensional temperature distribution in the furnace and the heat transfer rate to the surroundings. Comment on the desirability of insulating the outer surface.

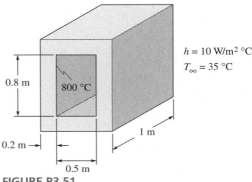

$h = 10$ W/m² °C
$T_\infty = 35$ °C

0.8 m

800 °C

0.2 m

1 m

0.5 m

FIGURE P3.51

3.52 A 1000 W electric heating element is inserted into a 30-cm long, horizontal, 1-cm I.D, 2-cm O.D., type 316 stainless steel tube. The tube is freely suspended in a large room, which is at 20 °C. Employing numerical methods, determine the transient temperature distribution in the cylinder. The convective heat transfer coefficient on the air side is 8 W/m² °C. The heat transfer rate from the end surfaces may be neglected but the radiative heat transfer from the cylindrical surface is to be included.

3.53 An annular fin of inner radius r_i, outer radius r_o, and constant thickness t, is attached to a steam pipe and exposed to ambient air. The convective heat transfer coefficient on the air side is h. Assuming a one-dimensional, transient temperature distribution, write the finite difference form of the energy equation for an internal node, the node closest to the pipe surface, and the node closest to the free end. Assume uniform Δr.

3.54 For the conditions given below, find the transient temperature distribution and heat transfer rate for the fin in Problem 3.53.

Material of fin: Aluminum alloy

$$t = 3 \text{ mm} \qquad r_i = 1 \text{ cm} \qquad r_o = 4 \text{ cm}$$
$$T_b = 80 \text{ °C} \qquad T_\infty = 20 \text{ °C} \qquad h = 50 \text{ W/m}^2 \text{ °C}$$

3.55 A cylindrical rod of plain carbon-steel 1-cm diameter and 30-cm long is attached to a furnace door. The furnace door is at a temperature of 120 °C. The rod is in air at 30 °C. It is desired to find the temperature distribution in the rod as a function of time. Radiative heat transfer is to be included. The rod is in a large room, which is at 30 °C. The convective heat transfer coefficient is 10 W/m² °C.

(a) Assuming one-dimensional temperature distribution write the finite difference form of the energy equation to determine the transient temperature distribution.

(b) Find the transient temperature distribution assuming the temperature of the door increases from 30 °C to 120 °C at $\tau = 0$.

(c) Repeat parts a and b if the door temperature increases linearly with time from 30 °C to 120 °C in 30 min and is constant at 120 °C thereafter.

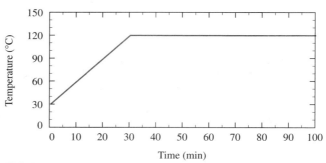

FIGURE P3.55

3.56 In an extended surface of rectangular cross section, the heat flux q_x'' decreases towards the free end. Hence, the material of the extended surface is not fully utilized. To make better use of the material, it is proposed to make it of variable thickness such that q_x'' is constant.
 (a) Determine $\theta(x)$ in terms of q_x'' and x.
 (b) Employing $\theta(x)$ determined in part a, find the relationship between the thickness t and x. Assume $P = 2$ W, where W is the width of the fin and P is the perimeter.

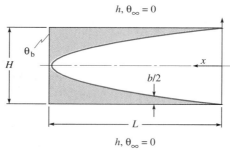

FIGURE P3.56

The exact solution for Problems 3.57 and 3.58 is

$$\frac{T_o - T}{T_o - T_i} = \text{erf}\left[\frac{y}{2\,(\alpha\tau)^{0.5}}\right]$$

The heat flux at $y = 0$ is given by

$$q_{eo}'' = -k\,\frac{\partial T}{\partial y}\,(0,\tau) = \left(\frac{k\rho c_p}{\pi\tau}\right)^{0.5}(T_o - T_i)$$

After obtaining the solutions determine q_o''/q_{eo}'' for each case. q_o'' is the approximate heat flux at $y = 0$ and q_{eo}'' is the exact heat flux.

3.57 For the rod in Example 3.5.2, find the transient temperature distribution, if the temperature of the surface at $y = 0$ is raised to T_o at $\tau = 0$, by integral methods employing a linear temperature profile; $T(0,\tau) = T_o$ and $T(\delta, \tau) = T_i$.

3.58 Re-solve Problem 3.57 employing

(a) A second-order temperature profile, adding the boundary condition, $\partial T/\partial y = 0$ at (δ,τ).

(b) A third-order temperature profile adding the boundary conditions $\partial T/\partial y = 0$ at (δ,τ) and $\partial^2 T/\partial y^2 = 0$ at (δ,τ).

(c) A fourth-order temperature profile with the boundary condition $\partial^2 T/\partial y^2 = 0$ at $(0,\tau)$ in addition to those in part b. The boundary condition is obtained by evaluating the differential equation $\partial T/\partial \tau = \alpha(\partial^2 T/\partial y^2)$ at $y = 0$, where the temperature is constant and $(\partial T/\partial \tau) = 0$.

3.59 An internal energy generation of q''' occurs in a thick slab. Two parallel surfaces are maintained at T_o. At time $\tau = 0$, the internal energy generation ceases and the parallel surfaces continue to be maintained at T_o. Find the transient, one-dimensional temperature distribution by the integral method.

3.60 (a) Employing the integral method, find the one-dimensional transient temperature distribution for the brass plug in Problem 3.42 if the temperature at one end is raised to 400 °C at $\tau = 0$, and the other end is insulated.

(b) A safe is to be made to withstand the effects of fire for 90 min. The walls of the safe are insulated with a sufficient thickness of fiberglass board such that the inside surface temperature will not rise above 150 °C in 90 min when the outside surface of the insulation is raised from 20 °C to 800 °C. Estimate the thickness of the insulation required, assuming the heat transfer rate from the inside surface to be negligible.

3.61 One end of an aluminum alloy rod, initially at a uniform temperature of 20 °C, is immersed in boiling water at 100 °C. The rod is 0.5-cm diameter and 20-cm long and is surrounded by air at 20 °C. The convective heat transfer coefficient on the air side is 10 W/m² K. Employing the integral method, find the transient temperature distribution assuming the free end to be perfectly insulated. Find the time taken for the free end to reach a temperature of 2 °C less than its steady-state temperature. The convective heat transfer coefficient for the end in boiling water is very large.

3.62 Assume that the surface of the earth in a particular location has a uniform temperature of 12 °C. It is required to estimate the depth at which water pipes are to be laid without freezing the water. For this purpose it is assumed that the surface temperature is reduced to −10 °C and maintained at that value. It is further assumed that if the temperature at depth, d, does not fall below 4 °C in 30 days after the surface temperature is reduced to −10 °C, it is safe to lay the water line at that depth. Determine the depth d. The properties of soil are $k = 2.3$ W/m K and $\alpha = 7.75 \times 10^{-7}$ m²/s.

CHAPTER

4

CONVECTION I

In solving problems on conduction in solid bodies with surfaces exposed to convective heat transfer, we assumed that the convective heat transfer coefficient was known. We now show how to determine the convective heat transfer coefficient and how to use it. To effectively use the convective heat transfer coefficient, an introduction to some of the associated fluid mechanics phenomena is useful. In this chapter, we

- *Distinguish laminar and turbulent flows*
- *Introduce the concept of boundary layers*
- *Define fully developed temperature and velocity profiles in internal flows*
- *Study how to determine the convective heat transfer coefficient in forced and natural convection for external and internal flows*
- *Show how to use the convective heat transfer coefficient*

4.0 INTRODUCTION

We have defined *convective heat transfer as heat transfer in a fluid in motion.* In most engineering applications involving convective heat transfer, the heat transfer rate from a solid surface to an adjacent fluid in motion is needed. When discussing conductive heat transfer in solid bodies with the boundaries exposed to a fluid, we assumed that the convective heat transfer coefficient was known. In this chapter, we show how to determine the convective heat transfer coefficient and how to use it.

Recall that the convective heat transfer coefficient, h is defined by Equation 1.4.1:

$$q'' = h(T_s - T_f) \tag{1.4.1}$$

where

q'' = heat flux *from* the solid surface *to* the surrounding fluid
T_s = surface temperature, also denoted by T_w (wall temperature)
T_f = reference temperature of the fluid

In Equation 1.4.1, if q'' is the average heat flux from a surface at a uniform temperature (the total heat transfer rate divided by the surface area), h is the average convective heat transfer coefficient. If the local heat flux is used in the equation, then h is the local heat transfer coefficient. To distinguish the two, the average heat transfer coefficient is denoted by h and the local heat transfer coefficient with a coordinate subscript, for example, h_x. Further, the average heat transfer coefficient is sometimes denoted with a subscript showing the dimension over which the average value is obtained. For example, for a flat plate, h_L indicates that it is the average value over a length L of the plate in the direction of the flow; for a cylinder with cross flow, h_d indicates that it is the average value over the perimeter of a cylinder of diameter d.

With convective heat transfer from a solid surface to an adjacent fluid, the temperature of the fluid varies with distance from the solid surface. When defining the convective heat transfer coefficient, any appropriate fluid temperature of the fluid can be used as the reference temperature; as the reference temperature is arbitrary in defining h, the reference temperature should be specifically stated. One condition for choosing the reference temperature is that, if the heat transfer is from the plate to the fluid, the reference temperature must be less than the surface temperature; i.e., h is always positive. With heat transfer from a solid plane surface to a fluid flowing adjacent to it, the reference temperature is usually the fluid temperature far away from the surface; with heat transfer from the walls of a tube in which a fluid flows, the reference temperature is the mean fluid temperature at the section.

It should be emphasized that the convective heat transfer coefficient is not a property of the fluid. Equation 1.4.1 only defines it. The convective heat transfer coefficient is dependent on a variety of factors, involving flow conditions, geometry

Equation 1.4.1 is popularly referred to as Newton's Law of Cooling. Sir Issac Newton is credited with the idea of the convective heat transfer coefficient. He also worked extensively in many other fields. Newton developed the binomial theorem, and simultaneously with Leibnitz, he laid the foundations of modern calculus. However, he is best known for his discovery of the gravitational force and the three Laws of Motion bearing his name. Fourier is also credited with putting forward the concept of heat flux and heat transfer coefficient.

Convective heat
transfer coefficient
is not a property of
the fluid

of the solid surface, and the physical properties of the fluid. For example, on a cool day with air at 5 °C (41 °F), a person experiences a greater heat transfer rate from the body to the air when the wind velocity increases. If you immerse your hand in still water at 5 °C (41 °F), the feeling of cold is much more than in air as a result of the higher heat transfer rate that occurs in water. Thus, it is evident that the flow velocity and the properties of the fluid determine the value of the convective heat transfer coefficient. The effect of the geometry is not so obvious, but with identical conditions of the fluid (its temperature and velocity being the same) the heat flux from a flat plate is different from the heat flux from a cylinder.

4.0.1 Classification

Convection is a complex phenomenon. The derivation of the governing differential equations to determine the convective heat transfer coefficient and the solutions to the equations are quite involved. However, it is possible to gain an understanding of the physics involved in the determination of the convective heat transfer coefficient through approximate methods. An introduction to one such approximate method—the integral method—is given in Chapter 5. An introduction to the differential equations and their solutions is given in Chapter 12.

In this section, *the emphasis is on how to determine the convective heat transfer coefficient and, having determined it, on how to use it to solve problems involving convective heat transfer.* Many dimensionless, empirical correlations for finding the values of the convective heat transfer coefficients are available.

In heat transfer and fluid mechanics, the functional relationships between the relevant variables are expressed in dimensionless form; i.e., the variables are combined into several groups such that each group is dimensionless. Each dimensionless group now becomes a single variable. Thus, when dimensionless variables are used, the number of independent variables is reduced. Another advantage is that the constants that may be involved in the correlations are independent of the system of units. A brief discussion of dimensional analysis is given in Appendix 6. The dimensionless parameters have physical significance that gives an insight into the nature of the problems. For example, for flows inside circular tubes, the Reynolds number ($\rho U d / \mu$) can be interpreted as the ratio of inertia forces (ρU^2) to viscous forces ($\mu U / d$); the Biot number (hL/k) can be interpreted as the ratio of the conductive resistance (L/k) to the convective resistance ($1/h$).

Convection
classification

Many correlations are available for determining the convective heat transfer coefficient in various physical configurations. It would be appropriate to classify the convective configurations in some suitable manner so that the appropriate correlation can be chosen. Convective heat transfer can be studied under four categories.

1. *Driving Force*

Forced Convection. Fluid flow is caused by external forces, such as by a fan or a pump, and some characteristic fluid velocity is available. The flow is not dependent on the difference between the temperatures of the fluid and the solid surface.

Natural Convection. Flow results solely from temperature differences in the fluid in the presence of a body force. No characteristic fluid velocity is readily available. The density of a fluid decreases with an increase in the temperature. In a gravitational field, such density differences caused by the temperature differences result in buoyancy forces, which lead to fluid motion.

2. *Geometry*

   External Flow. Flow of a semi-infinite fluid over a surface.

   Internal Flow. Flow confined between solid surfaces (inside tubes or ducts).

3. *Type of flow*

   Laminar Flow

   Turbulent Flow

4. *Number of Phases of Fluid*

   Single Phase. The fluid exists in only one phase—either as a liquid or as a gas.

   Two Phases. The fluid exists as a mixture of liquid and vapor phases, as in boiling and condensation, or as a mixture of solid and liquid, as in freezing and melting.

   In this chapter we confine ourselves to the study of heat transfer in single-phase fluids. Boiling and condensation heat transfer are studied in Chapter 6.

4.0.2 Laminar and Turbulent Flow

Reynold's
experiments

The terms laminar and turbulent flow used in the above classification of convection need some explanation. To understand laminar and turbulent flows, it is appropriate to refer to the classic experiments of Osborne Reynolds in 1883. Figure 4.0.1 shows the apparatus Reynolds used and sketches of his observations. He observed the flow of water in a long, horizontal glass tube, which was attached to a tank, by injecting a dye into the water with different rates of mass flow of water. Figure 4.0.2 shows photographs of flow patterns. When the flow rate was small, the dye line remained parallel to the surface of the tube, Figure 4.0.2a. As the flow rate was increased, the dye initially remained parallel to the tube wall but, at some distance downstream,

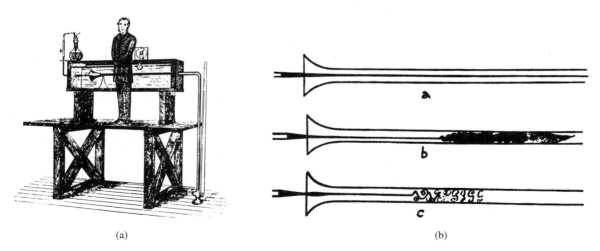

(a)　　　　　　　　　　　　　　　　　　(b)

FIGURE 4.0.1 The Classic Experiments of Osborne Reynolds
(a) A sketch of the apparatus Reynolds used to study the transition from laminar to turbulent flows in pipes. (b) Reynolds' sketches of his observations. From *Journal of Heat Transfer,* 103 (1981). Reproduced with permission from the American Society of Mechanical Engineers.

(a)

(b)

(c)

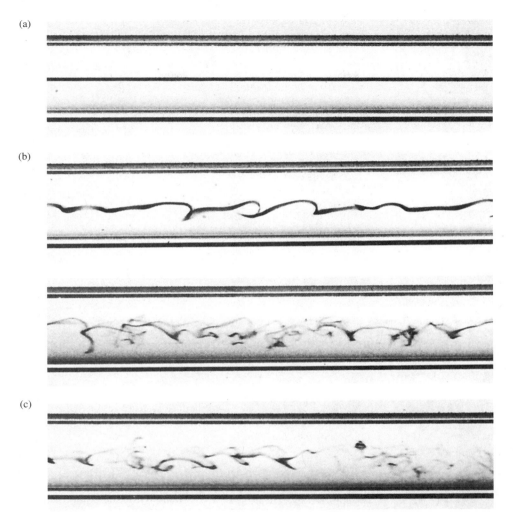

FIGURE 4.0.2 Photos Indicating Transition from Laminar to Turbulent Flow
Reynolds himself kept only sketches of flows. These photos were taken with the
apparatus used by Reynolds. The experimental apparatus is still in use at the University
of Manchester. Reproduced with permission from *An Album of Fluid Motion*,
assembled by van Dyke, Milton, Stanford, CA: Parabolic Press, 1982).

the dye line became wavy, as shown in Figure 4.0.2b. When the flow rate was
increased further, the waviness increased, and beyond a certain mass flow rate, the
flow pattern of the dye line became chaotic or random to such an extent that the dye
line lost its identity at a short distance from the point of injection (Figure 4.0.2c).
The type of flow shown in Figure 4.0.2a, where the streamlines[1] are aligned more
or less parallel to the bounding solid surfaces, is termed *laminar flow*. The flow

---

[1]A streamline is a line such that at every point the tangent to it has the same unit vector as the velocity
vector at that point.

Osborne Reynolds (1842–1912) contributed in many fields such as mechanics, thermodynamics, electricity, navigation and rolling friction. In 1868, he was appointed to the Chair of Engineering at Owens College (subsequently named Victoria University) in England. In the early years, he conducted his experiments at home as there was no laboratory in the college. He is well known for his experiments on transition to turbulent flows in pipes and development of the dimensionless number, $\rho V d/\mu$, named after him. He is also known for modifying the differential equations for laminar flows to account for turbulence.

shown in Figure 4.0.2b, where departure from laminar flow sets in, is termed *transitional flow*. Finally, when the fluid motion becomes chaotic, as in Figure 4.0.2c, it is termed *turbulent flow*.

4.0.3 Heat Transfer in Laminar and Turbulent Flows

In laminar flows, heat transfer is essentially by the random motion of the molecules. In turbulent flows, heat transfer occurs by random motion not only at the microscopic level (molecular) but also at the macroscopic level with finite-sized fluid packets of varying sizes (eddies) participating in energy transfer by their chaotic motion.

 Consider the flow of a fluid in a tube whose surface temperature is higher than that of the fluid. If the flow is laminar (Figure 4.0.3a), the fluid flows along streamlines that are parallel to the tube wall. In this case, the transfer of energy from one layer of the fluid to the next is by molecular diffusion (conduction) only. If the flow is turbulent (Figure 4.0.3b), the fluid motion is no longer only along lines parallel to the surface. There is motion perpendicular to the axis of the tube also. Because of this motion perpendicular to the axis, eddies at higher (or lower) temperature move toward fluid at a lower (or higher) temperature and then mix. Such mixing of the fluid results in a transfer of energy in addition to the heat transfer by conduction. Indeed, in turbulent flows, energy transfer by the mixing of fluid packets is much greater than that by molecular motion, and quite often, it is the predominant mode of energy transfer.

Enhancement of heat transfer by turbulent eddies

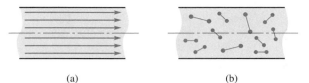

FIGURE 4.0.3 Laminar and Turbulent Flows
(a) Laminar flow: Streamlines are parallel to the
bounding surface. (b) Turbulent flow: Fluid packets
(eddies) are in random motion.

For internal flows in a circular tube, Reynolds found that the velocity above which the flow ceased to be laminar was a function of the density of the fluid (ρ), the diameter (d) of the tube, and the viscosity (μ) of the fluid. From dimensional reasoning, it can be shown that the dimensionless group $\rho V d / \mu$ determines when the flow is laminar or turbulent. This dimensionless group is known as the *Reynolds number,* in his honor. We should not conclude that the Reynolds number is always formed with the diameter as the characteristic length. Reynolds number is defined in terms of a characteristic velocity and a characteristic length, both of which depend on the configuration considered. When using the Reynolds number, the characteristic length and velocity should be clearly stated. In the case of a circular tube, if the mean velocity—defined as $\dot{m}/(\rho \pi d^2/4)$—is employed in computing the Reynolds number, the flow is laminar for Reynolds numbers less than about 2100. (You will find that different values, ranging from 2000–2300, are used as the transitional Reynolds number.) The Reynolds number below which the flow remains laminar is known as the *critical Reynolds number,* and the velocity corresponding to the critical Reynolds number is known as the *critical velocity.* Just above the critical Reynolds number, the flow is in a transitional mode as shown in Figure 4.0.2b. Above a Reynolds number of about 10 000, the flow becomes fully turbulent. Most flows are turbulent. For example, consider the flow of water in a tube of 12-mm inside diameter (the size of pipes used in most houses for water supply). With properties of water at 20 °C (Table A5) the critical velocity in this case is

$$\frac{\rho V_{\text{cr}} d}{\mu} = 2100 \qquad V_{\text{cr}} = \frac{2100 \times 985.3 \times 10^{-6}}{998.2 \times 0.012} = 0.173 \text{ m/s}$$

A velocity of 0.173 m/s corresponds to a volumetric flow rate of approximately 0.3 gpm (gallon per minute). You may like to get an appreciation of how small this flow rate is by adjusting the flow in a faucet to yield about one gallon every three minutes.

You may also observe laminar flow in natural convection, such as that of smoke from a chimney on a calm day. Initially, in calm air, the smoke rises in vertical lines, later develops waviness, and ultimately exhibits random motion.

Heat Transfer in Laminar Flows

Referring to the classification of convective heat transfer and taking one type from each of the first three categories—cause of flow, geometry, and type of flow—we can identify the various possible situations. For example, heat transfer could be by natural convection, external flow, and turbulent flow, or by forced convection, internal flow, and laminar flow. After identifying the physical configuration, an appropriate correlation is chosen from among the many available, and the heat transfer coefficient is determined. The correlations are based sometimes on analyses but, in many cases, on a combination of analyses and experiments. Quite often, the correlations are obtained from the available experimental results by regression analysis. For convenience, many of the correlations in this chapter are listed at the end of the chapter (pp. 385–391).

Determining convective heat transfer coefficient from analytical solutions

To understand how the convective heat transfer coefficient is determined from analytical solutions, consider a fluid flowing parallel to a flat plate as shown in Figure 4.0.4. The solution of the applicable differential equations for balance of mass, momentum, and energy (derivation of the steady, two-dimensional equations for an incompressible fluid is given in Chapter 12) yield the velocity and temperature as functions of the spatial coordinates x and y (Figure 4.0.4). Adjacent to the plate, the

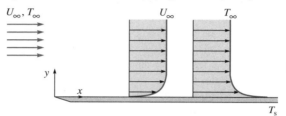

FIGURE 4.0.4 Fluid Flow Over a Flat Plate
The velocity and temperature distribution of the fluid are known.

fluid is at rest relative to the plate (no slip) and the heat transfer across this layer is by conduction. The local heat flux is given by

$$q_w'' = q_y''|_{y=0} = -k \left. \frac{\partial T}{\partial y} \right|_{y=0}$$

The local convective heat transfer coefficient for flow over a flat plate is defined by

$$q_w'' = h_x(T_s - T_\infty)$$

so that

$$h_x = \frac{-k(\partial T/\partial y)|_{y=0}}{T_s - T_\infty}$$

With the temperature distribution $T(x, y)$ known from the solution to the differential equations, the numerator in the expression for h_x and, hence, the local convective heat transfer coefficient are evaluated. The local convective heat transfer coefficient varies with x, but in Chapters 2 and 3, an average, uniform value of the convective heat transfer coefficient was used.

Heat Transfer in Turbulent Flows

If the fluid is in turbulent flow, the situation is more complicated as there are random fluctuations superimposed on the mean velocity components in the x-, y-, and z-directions. Although we speak of time independent mean velocity components, the differential equations defining the flow field involve the fluctuating components, which are time dependent. Similarly, we should also consider the fluctuating component of the temperature. Figure 4.0.5 shows the time dependence of the velocity components and temperature at some point in the flow. We denote the instantaneous

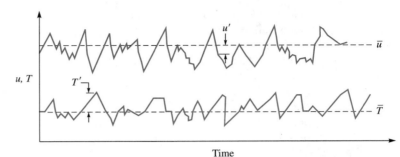

FIGURE 4.0.5 **Variation of Velocity Components and Temperature with Time at a Defined Point in the Flow**

velocities and temperature as the sum of their mean (time independent) and fluctuating (time dependent) components.

$$u = \bar{u} + u' \qquad v = \bar{v} + v' \qquad T = \bar{T} + T'$$

where the overbars denote the mean values and the primes denote the instantaneous fluctuating values. By substituting the instantaneous values (sum of the mean and fluctuating components) in the time dependent differential equations for mass, momentum, and energy balance, and taking the time-averaged form of the equations, it can be shown that

$$\frac{\partial \bar{u}}{\partial x} + \frac{\partial \bar{v}}{\partial y} = 0 \tag{4.0.1}$$

$$\bar{u}\frac{\partial \bar{u}}{\partial x} + \bar{v}\frac{\partial \bar{u}}{\partial y} = \frac{1}{\rho}\frac{\partial}{\partial y}\left(\tau_{yx} - \rho\overline{u'v'}\right) \tag{4.0.2}$$

$$\bar{u}\frac{\partial \bar{T}}{\partial x} + \bar{v}\frac{\partial \bar{T}}{\partial y} = -\frac{1}{\rho c_p}\frac{\partial}{\partial y}\left(q''_y - \rho c_p\overline{v'T'}\right) \tag{4.0.3}$$

Effect of turbulence on heat transfer

The term $\rho\overline{u'v'}$ represents the momentum exchange, and $\rho c_p\overline{v'T'}$ represents the energy transfer due to the motion of the eddies. It is also known from experimental results that both $\overline{u'v'}$ and $\overline{v'T'}$ are negative. τ_{yx} represents the shear stress $[\mu(\partial\bar{u}/\partial y)]$, and q''_y represents the heat flux in the y-direction $[-k(\partial\bar{T}/\partial y)]$ in laminar flow. In a formal way, we define the terms involving the fluctuating components as

$$-\rho\overline{u'v'} = \mu_t\frac{\partial\bar{u}}{\partial y} \qquad -\rho c_p\overline{v'T'} = k_t\frac{\partial\bar{T}}{\partial y}$$

where μ_t and k_t may be interpreted as the effective turbulent viscosity and turbulent thermal conductivity, respectively. The total heat flux, q''_y, is then expressed as

$$q''_y = -(k + k_t)\frac{\partial\bar{T}}{\partial y}$$

From the expression for the heat flux, it is clear that turbulence has an effect equivalent to increasing the thermal conductivity of the fluid. We should, therefore, expect higher heat transfer rates in turbulent flows than in laminar flows. Such is actually the case. This can also be visualized in another way. Because of the fluctuating velocity and temperature components, the eddies are moving both in the positive and negative y-direction and these carry energy. However, the net result (consistent with the Second Law of Thermodynamics) is that more energy is carried from a region of higher temperature to a region of lower temperature than in the opposite direction, and hence, there is an apparent increase in the thermal conductivity of the fluid. It should be emphasized, that *unlike* μ *and* k, *the turbulent viscosity* μ_t, *and the turbulent thermal conductivity* k_t, *are introduced as a matter of convenience, and that they are not properties of the fluid.* Considerable work is going on to relate these turbulent quantities to flow variables that can be measured.

One of the early models, due to Prandtl, is the *Prandtl mixing length model.* He hypothesized that fluid packets are in random motion and that they travel a certain distance before they mix with the rest of the fluid and lose their identity. He called this distance the ''mixing length.'' It is analogous to the mean free path in gases. Different hypotheses have been put forward to relate the mixing length to the flow variables, the turbulent viscosity, and thermal conductivity, but none of them is satisfactory in all situations.

Fluctuations increase the heat transfer rate. If the spatial distribution of $\overline{T}(x, y)$ is known, the flow being stationary at the boundaries, the heat flux from the solid boundary is given by Fourier's Law,

$$q''_w = -k \left. \frac{\partial \overline{T}}{\partial y} \right|_{y=0} \qquad h_x = \frac{q''_w}{T_s - T_\infty} = \frac{-k(\partial \overline{T}/\partial y)|_{y=0}}{T_s - T_\infty}$$

4.0.4 Empirical Correlations

The analytical determination of the convective heat transfer coefficient is limited to a few simple geometries, such as a flat plate with a fluid flowing parallel to it. Even in such simple cases, the solutions to the differential equation are quite involved. To get engineering estimates of the convective heat transfer coefficients in applications of practical significance, we resort to the presentation of experimental data in the form of empirical equations relating the convective heat transfer coefficient to the variables that affect it. However, instead of relating the convective heat transfer coefficient directly to the dimensional variables, the equations are presented in terms of dimensionless parameters.

Use of dimensionless parameters in correlations

In most correlations in convective heat transfer, there are three dimensionless parameters. One dimensionless parameter contains variables that are related to the motion of the fluid, another one has the convective heat transfer coefficient in it, and the third one couples the motion with the heat transfer rate. The parameter that contains the heat transfer coefficient is the Nusselt number. The Prandtl number is the coupling parameter.

To undersatnd how empirical correlations are developed, consider forced convection. In forced convection the Reynolds number characterizes the flow. For heat transfer from a cylinder of diameter d, to a fluid flowing perpendicular to the axis of the cylinder, the characteristic velocity is the velocity U far away from the cylinder, the free stream velocity. For this case the definition of the three dimensionless parameters is

Nusselt number: $Nu = hd/k$
Reynolds number: $Re = \rho Ud/\mu$
Prandtl number: $Pr = c_p\mu/k$

If the experimentally determined Nusselt numbers for different Reynolds numbers at different but discrete Prandtl numbers are plotted on a log-log paper, we may get curves such as shown in Figure 4.0.6a for three different values of the Prandtl num-

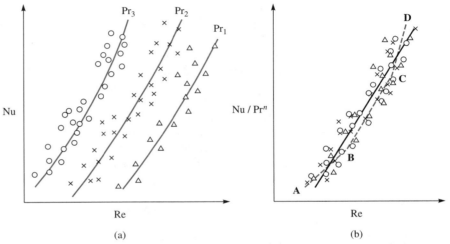

FIGURE 4.0.6 (a) Qualitative variation of Nu for discrete values of Pr. (b) Nu/Pr^n is plotted against Re.

ber. From the curves we may attempt to find the dependence of the Nusselt number on the Prandtl number in the form of $Nu/Pr^n = f(Re)$ from intuition and numerical experimentation. If we choose a suitable value of n, the three distinct plots in Figure 4.0.6a may collapse into one curve such as shown in Figure 4.0.6b. We may now attempt to find an equation that relates the Nusselt number to the Reynolds number and Prandtl number. Recognizing that the variation of Nu on the log-log paper is almost (but not quite) a straight line, we attempt to represent the data by an equation of the type

$$Nu = c\, Re^m\, Pr^n \qquad (4.0.4)$$

Why there are different correlations for the same configuration

We may try to represent all the experimental data by one equation over the entire range of Re shown by the straight line. However, it is obvious that as the plot is not quite linear on the log-log plot, the resulting equation may not predict the values with acceptable accuracy over the entire range of Re. We may then choose to represent the data with several equations similar to Equation 4.0.4 over limited ranges of Re, the values of c and m being different in the different ranges of Re as shown by the three dashed lines (AB, BC, CD). We may also represent the functional relationship over the entire range of Re and Pr by a more complex function.

Depending on the chosen functional relationship and the experimental data used for the correlation, different correlations are obtained. The values of the convective heat transfer coefficient determined from the different correlations will not be exactly equal to one another. Further, as there is a spread of experimental data, as shown in Figure 4.0.6, the convective heat transfer coefficient predicted from a given correlation has an uncertainty associated with it.

4.1 FORCED CONVECTION

Forced convection has been defined as the heat transfer to a fluid whose motion is caused by an external agency, such as a pump or a fan, and the motion of the fluid is not dependent on temperature differences in the fluid. In forced convection, we expect the heat transfer coefficient to be a function of the fluid velocity, density, viscosity (which influences the motion of the fluid close to the solid boundary), specific heat, thermal conductivity, and a characteristic length. The choice of the characteristic length depends on the geometry and the flow pattern. The characteristic length may be the diameter of the tube for flows inside tubes, the diameter of a cylinder when the fluid flow is perpendicular to the axis of the cylinder, or the distance from the leading edge of a plate for flows parallel to the flat plate. Representing the characteristic length by L, the functional relationship between the average heat transfer coefficient h and the variables that affect it is indicated by

$$h = h(L, k, \rho, V, \mu, c_p)$$

Dimensionless parameters for forced convection

From dimensional analysis, we obtain three dimensionless groups, (hL/k), $(\rho VL/\mu)$, and $(c_p\mu/k)$. Recall that the Reynolds number is used in connection with a friction factor in pipe flows and in our discussion on Reynolds' experiments on the onset of turbulence. The other two parameters are the Nusselt number (hL/k) and the Prandtl number, $(c_p\mu/k)$. An equivalent definition of Prandtl number is the ratio, ν/α. Thus, from dimensional analysis we obtain

$$\mathrm{Nu} = \mathrm{Nu}(\mathrm{Re}, \mathrm{Pr})$$

where the nomenclature is

Nu = Nusselt number = hL/k
Re = Reynolds number = $\rho VL/\mu$
Pr = Prandtl number = $c_p\mu/k = \nu/\alpha$
h = convective heat transfer coefficient (W/m^2 °C)
L = characteristic length (m)
k = thermal conductivity of the fluid (W/m K)
ρ = density of the fluid (kg/m^3)
V = velocity of the fluid (m/s)
μ = dynamic viscosity of the fluid (N s/m^2)
ν = kinematic viscosity of the fluid (m^2/s) = μ/ρ
α = thermal diffusivity of the fluid (m^2/s) = $k/(\rho c_p)$

The functional relationship between the Nusselt number, the Reynolds number, and the Prandtl number can be expressed in the form of either an equation or a graph. Also the relationships may be developed either for the local heat transfer coefficient with a spatial coordinate as the characteristic length or the average heat transfer coefficient with an appropriate dimension of the solid surface as the characteristic dimension.

The experimental data, from which the correlations are obtained, have a spread around the correlating equations. The realized value of the convective heat transfer coefficient in an actual case may be higher or lower than the value predicted from the correlations. Many correlations are accompanied by the value of the uncertainty giving the probable spread of experimental measurements on either side of the correlating equation. Generally, the uncertainty in the convective heat transfer coefficient computed from the correlations for single-phase fluids is about $\pm 10\%$.

4.1.1 Forced Convection Correlations—External Flow Parallel to a Flat Plate

Heat transfer rate from a plane surface to a fluid flowing parallel to it has been studied extensively, both analytically and experimentally. The simplicity of the geometry and correlations is conducive to an understanding of the nature of the correlations.

Consider the forced flow of a fluid parallel to a flat plate as shown in Figure 4.1.1. The fluid approaches the plate with a uniform velocity U_∞ (free stream veloc-

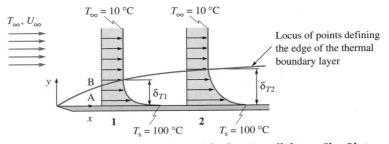

FIGURE 4.1.1 Forced Convection with Flow Parallel to a Flat Plate
A fluid approaches a flat plate with a uniform velocity and temperature. The temperature of the fluid changes from T_s to T_∞ in a distance δ_T. δ_T increases in the direction of flow.

ity) and a uniform temperature T_∞ (free stream temperature). The plate is at a uniform temperature T_s. Suppose the plate is at 100 °C (T_s) and the approaching fluid is at 10 °C (T_∞). At section 1 of the plate, at point A very near the plate, the temperature of the fluid will be close to 100 °C; at a point far away from the plate the fluid will be at 10 °C. The fluid temperature changes from 100 °C at the plate to 10 °C at some point B, which is at a distance δ_{T1} from the plate. Downstream of section 1, at section 2, a similar situation exists except that the thickness of the layer of the fluid δ_{T2}, across which the fluid temperature changes from 100 °C to 10 °C, is greater than δ_{T1}. The region where the temperature of the fluid changes from the surface temperature to the free stream temperature is known as the *temperature* or *thermal boundary layer*. The temperature gradient at the plate in the y-direction, $\partial T/\partial y|_{y=0}$, is a function of δ_T and $T_\infty - T_s$. The magnitude of the heat flux is proportional to the product of the temperature gradient at the plate and the thermal conductivity of

Temperature boundary layer

the fluid. In the boundary layer the temperature changes from T_s to T_∞ in a distance δ_T. We anticipate the temperature profiles at different locations to be similar and, therefore, the temperature gradient at the wall is proportional (not necessarily linearly) to $(T_s - T_\infty)/\delta_T$. As $\delta_{T1} < \delta_{T2}$, with constant $T_s - T_\infty$, the magnitude of the temperature gradient at the wall is greater at section 1 than at section 2. Hence, the heat flux at section 1 is greater than at section 2. The local convective heat transfer coefficient, being the ratio of the heat flux to $T_s - T_\infty$, is greater at section 1 than at 2; the convective heat transfer coefficient is, therefore, a function of the x-coordinate.

Velocity boundary layer

A similar situation exists regarding changes in velocity, as shown in Figure 4.1.2. At section 1, the velocity of the fluid changes from 0 at the plate to the free stream

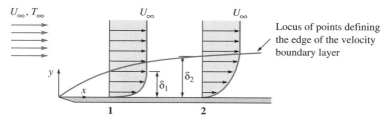

FIGURE 4.1.2 Fluid Velocity Profile for Flow Over a Flat Plate
A fluid approaches a flat plate with a uniform velocity U_∞. The velocity of the fluid changes from 0 to U_∞ in a short distance δ. δ increases with x.

velocity U_∞ at some distance δ_1. The effect of decelerating the fluid adjacent to the plate penetrates through a greater distance as we proceed downstream. At section 2, downstream of section 1, the velocity changes from 0 at the plate to U_∞ in a distance δ_2, which is greater than δ_1. The region where the velocity changes from 0 at the plate to the free stream velocity is known as the *velocity* or *momentum boundary layer*; its thickness is usually represented by δ. The velocity profiles in the velocity boundary layer at different locations are also similar, and the velocity gradient at the plate, $\partial u/\partial y|_{y=0}$, is proportional (not necessarily linearly) to U_∞/δ. The wall shear stress, τ_w, is given by $\mu(\partial u/\partial y)|_{y=0}$. As $\delta_1 < \delta_2$, it follows that $\tau_{w1} > \tau_{w2}$. It is evident that the Fanning friction factor, c_f, defined as $\tau_w(\rho U_\infty^2/2)$, is a function of the x-coordinate.

Transition from laminar to turbulent boundary layer

The velocity boundary layer may be either laminar or turbulent. If the boundary layer is disturbed by an external force, the velocity fluctuations caused by the disturbance may be either dampened or amplified. If the fluctuations are dampened, the boundary layer will be laminar. If the fluctuations are amplified, they lead to chaotic motion and the boundary layer may become turbulent. In any flow, disturbances are always present, for example, due to vibration of the plate; however, the magnitude of the disturbances may be so small that we may not be aware of them. For some distance from the leading edge, the disturbances are dampened and the boundary layer is laminar as shown in Figure 4.1.3. Beyond a certain distance, the disturbances begin to be amplified in the direction of flow—the *transition region*. Further downstream, the motion becomes chaotic resulting in a turbulent boundary layer. Even in the turbulent region, close to the plate, the flow is similar to that in the transition region; very close to the plate, the flow appears to be laminar. This region is usually referred to as the *viscous sublayer*.

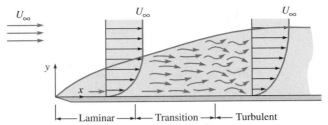

FIGURE 4.1.3 Transition from Laminar to Turbulent Boundary Layer
A fluid approaches a flat plate with a uniform velocity. Near the leading edge, the boundary layer is laminar. At some distance downstream from the leading edge the boundary layer becomes unstable, and further downstream it becomes turbulent.

It has been established analytically and experimentally that for laminar boundary layers, $\delta/x \sim 1/\mathrm{Re}_x^{1/2}$ ($\mathrm{Re}_x = \rho U_\infty x/\mu$), so that if $\mathrm{Re}_x \gg 1$, $\delta/x \ll 1$. When the boundary layer thickness is much smaller than the distance from the leading edge, the boundary layer is characterized as thin. If a variable changes by the same amount across the boundary layer (at a defined value of x) and from the leading edge to the specified location, the magnitude of the gradient of the variable in the y-direction is much greater than the gradient in the x-direction. For a laminar boundary layer, it has also been established that $\delta_T/x \sim 1/(\mathrm{Re}_x^{1/2}\mathrm{Pr}^{1/3})$ and, therefore, if $\mathrm{Re}_x^{1/2}\,\mathrm{Pr}^{1/3} \gg 1$, $\delta_T/x \ll 1$.

Relative magnitudes of velocity and temperature boundary layers— effect of Prandtl number in laminar boundary layers

The temperature and velocity boundary layers, in general, have different thicknesses as shown in Figure 4.1.4. The kinematic viscosity ν ($= \mu/\rho$) indicates how rapidly the velocity boundary layer grows in the laminar region. For two fluids A and B, if $\nu_A > \nu_B$, the effect of the wall is propagated more rapidly in fluid A than in fluid B; in fluid A the boundary layer grows more rapidly in the x-direction than in fluid B. Similarly, the thermal diffusivity α ($= k/\rho c_p$), is an indication of how rapidly the thermal boundary layer grows in a fluid. If $\alpha_A > \alpha_B$, the thermal boundary layer in fluid A grows more rapidly than in fluid B. If $\nu = \alpha$ ($\mathrm{Pr} = 1$; for example,

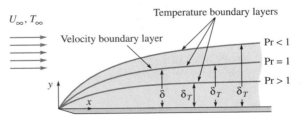

FIGURE 4.1.4 Temperature and Velocity Boundary Layers
A fluid with a uniform velocity and temperature approaches a flat plate whose temperature is different from that of the fluid. The relative magnitudes of the velocity and temperature boundary layers depend on the Prandtl number.

air has a Pr ≈ 0.7) for a fluid, both the velocity and temperature boundary layers grow at the same rate, and $\delta = \delta_T$. If $\nu/\alpha > 1$ (oils have Pr >> 1), the velocity boundary layer grows more rapidly than the thermal boundary layer, and $\delta > \delta_T$; if $\nu/\alpha < 1$ (liquid metals have Pr << 1) then, $\delta < \delta_T$. Thus, the ratio δ/δ_T is a function of the ratio ν/α. By analysis and experiments, it has been established that for laminar boundary layers,

$$\delta/\delta_T = Pr^n$$

where $n = 1/3$ for fluids with Prandtl number in the range 0.6 to 50. If Pr << 1, as for liquid metals, $n = 1/2$. In oils (Pr >> 1) $\delta >> \delta_T$; in liquid metals (Pr << 1) $\delta << \delta_T$. For laminar boundary layers over a flat plate,

$$\frac{\delta}{x} = \frac{5}{Re_x^{1/2}} \qquad Re_x = \frac{\rho U_\infty x}{\mu}$$

where x is measured along the plate from the leading edge in the direction of flow.

In the turbulent region, as a result of the chaotic motion of the eddies, the boundary layer grows more rapidly than in the laminar region. The transfer of momentum and energy by the eddies is more significant than the transfer of momentum or energy by molecular diffusion (except for liquid metals with Pr << 1, where energy transfer by molecular diffusion is much greater than by the eddies). As the transfer of both momentum and energy are by the same eddies, the thickness of the velocity boundary layer is approximately equal to the thickness of the temperature boundary layer, $\delta \approx \delta_T$. In turbulent boundary layers,

Velocity and temperature boundary layer thicknesses are approximately equal in turbulent boundary layers

$$\frac{\delta}{x} = \frac{0.37}{Re_x^{0.2}} \qquad Re_x = \frac{\rho U_\infty x}{\mu} \qquad \delta \approx \delta_T$$

Useful engineering solutions for the wall shear stress and the convective heat transfer coefficient have been obtained for both the laminar and turbulent boundary layers. It is very difficult to obtain such useful relations for the transition boundary layer. Because of this difficulty, it is generally assumed that the transition from laminar to turbulent boundary layer is abrupt somewhere in the transition region. Various values ranging from 10^5 to 10^6 are used for Re_{cr}, below which the boundary layer is laminar. A commonly used value for Re_{cr} is 5×10^5. In this book we use 5×10^5 for Re_{cr} unless specifically stated otherwise. We assume that the boundary layer becomes turbulent when the Reynolds number is greater than the critical value. If a device to induce disturbances is introduced in the laminar boundary layer, the boundary layer downstream from that location may become turbulent. A tripping wire at the leading edge is used to obtain turbulent boundary layer from the leading edge.

In the problems solved in this text, the properties of air, water, and engine oil have been evaluated from the CC module of the HT software provided. Properties in the software are found from a regression analysis of the tabulated values in Appendix 1. The values found from such a regression analysis may differ slightly from the tabulated values. The differences are generally within 0.5%. Also, many of the values given in the example problems are from the values generated by computer from the HT programs on the disk; the values in the text have been rounded off to four significant figures. In some cases, where the final results are transferred from the computer, they may differ slightly from the values computed by using the intermediate, rounded values (round-off errors).

EXAMPLE 4.1.1

A fluid at 20 °C flows parallel to a flat plate with a velocity of 5 m/s (Figure 4.1.5). Find the velocity boundary layer thickness at distances of 10 cm and 2 m from the leading edge, if the fluid is air at atmospheric pressure, water, and engine oil.

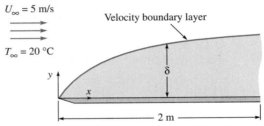

$U_\infty = 5$ m/s

Velocity boundary layer

$T_\infty = 20$ °C

y

δ

x

2 m

FIGURE 4.1.5 A fluid approaches a flat plate with a uniform velocity of 5 m/s at 20 °C. The velocity boundary layer thickness increases in the direction of flow.

Given

$U_\infty = 5$ m/s $T_\infty = 20$ °C

Find

δ at $x = 0.1$ m and 2 m for air, water, and engine oil as the fluid

SOLUTION

The boundary layer thickness is given by

$$\frac{\delta}{x} = \frac{c}{\mathrm{Re}_x^m}$$

where Re_x is $\rho U_\infty x/\mu$. The values for c and m are

$\mathrm{Re}_x < \mathrm{Re}_{cr}$	Laminar boundary layer	$c = 5.0$	$m = 0.5$
$\mathrm{Re}_x > \mathrm{Re}_{cr}$	Turbulent boundary layer	$c = 0.37$	$m = 0.2$

Properties of air, water, and engine oil at 20 °C from Tables A7, A5, and A4, respectively, are

	ρ (kg/m³)	μ (N s/m²)	ν (m²/s)
Air	1.209	181.6×10^{-7}	15.12×10^{-6}
Water	998.2	985.3×10^{-6}	98.71×10^{-8}
Engine oil	888	800.0×10^{-3}	901×10^{-6}

Values of Re_x, δ/x, and δ are given in the following table.

x (m)	Fluid	Re_x		δ/x	δ (mm)
0.1	Air	3.329×10^4	$5.0/Re_x^{0.5}$	2.74×10^{-2}	2.74
	Water	5.065×10^5	$0.37/Re_x^{0.2}$	2.675×10^{-2}	2.67
	Engine oil	5.55×10^2	$5.0/Re_x^{0.5}$	2.122×10^{-1}	21.2
2.0	Air	6.657×10^5	$0.37/Re_x^{0.2}$	2.53×10^{-2}	50.6
	Water	1.013×10^7	$0.37/Re_x^{0.2}$	1.47×10^{-2}	29.4
	Engine oil	1.11×10^4	$5.0/Re_x^{0.5}$	4.75×10^{-2}	94.9

COMMENT

As the kinematic viscosity of engine oil is much higher than that of air or water, the boundary layer in engine oil grows more rapidly and is much thicker than that of air or water. It also remains laminar for a much greater distance. The higher viscosity of the oil dampens the disturbances and maintains a laminar boundary layer for a much longer distance from the leading edge.

Heat Transfer

The heat transfer coefficient for a laminar boundary layer has been determined analytically both for uniform surface temperature and uniform surface heat flux. The results have also been experimentally verified. The correlations in forced convection involve three dimensionless groups—the Nusselt number, the Reynolds number, and the Prandtl number.

Uniform Surface Temperature, Laminar Boundary Layer ($Re < Re_{cr}$) A simple case of heat transfer from a flat plate is that of a fluid flowing parallel to the plate maintained at a uniform temperature. The fluid approaches the plate with a uniform velocity, U_∞, and a uniform temperature, T_∞. We have already concluded from boundary layer concepts that the heat flux at the wall is a function of the x-coordinate (Figure 4.1.6). As the fluid adjacent to the plate is stationary, the local heat flux q_x'' at the wall is given by

$$q_x'' = -k \left.\frac{\partial T}{\partial y}\right|_{y=0}$$

In the laminar boundary layer,

$$\left.\frac{\partial T}{\partial y}\right|_{y=0} \sim \frac{T_\infty - T_s}{\delta_T}$$

With $\delta/\delta T \sim Pr^{1/3}$ and $\delta/x \sim Re_x^{-1/2}$

$$\frac{1}{\delta_T} \sim \frac{Re_x^{1/2}\, Pr^{1/3}}{x}$$

Therefore,

$$q_x'' \sim -k\,\frac{T_\infty - T_s}{x}\, Re_x^{1/2}\, Pr^{1/3}$$

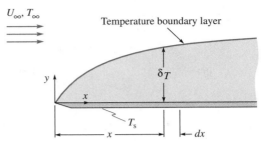

FIGURE 4.1.6 Heat Transfer from a Flat Plate
A fluid flows parallel to a flat plate, approaching it at a uniform velocity and temperature. The plate is maintained at a uniform temperature and heat is transferred from the plate to the fluid.

Correlations for local convective heat transfer coefficient for flat plate—laminar boundary layers

With the local convective heat transfer coefficient given by

$$h_x = \frac{q_x''}{T_s - T_\infty}$$

we obtain

$$\frac{h_x x}{k} \sim \text{Re}_x^{1/2} \, \text{Pr}^{1/3}$$

The dimensionless group $h_x x/k$ is the Nusselt number based on the local convective heat transfer coefficient. The value of the constant of proportionality is 0.332. Thus, the correlation valid for

$\text{Pr} \gtrsim 0.6$, $\text{Re}_x < \text{Re}_{cr}$, and properties evaluated at $T_f = (T_s + T_\infty)/2$

$$\text{Nu}_x = 0.332 \, \text{Re}_x^{1/2} \, \text{Pr}^{1/3} \qquad \textbf{(4.1.1)}$$

where

$$\text{Nu}_x = \frac{h_x x}{k} \qquad \text{Re}_x = \frac{\rho U_\infty x}{\mu} = \frac{U_\infty x}{\nu} \qquad \text{Pr} = \frac{\nu}{\alpha} = \frac{c_p \mu}{k}$$

Laminar boundary layer—uniform surface temperature

The properties of the fluid used in evaluating the different dimensionless parameters in Equation 4.1.1 are temperature dependent. For the best agreement of Equation 4.1.1 with experimental measurements, the properties are evaluated at the arithmetic mean of the free stream temperature, T_∞, and the wall temperature, T_s. The mean temperature is also known as the film temperature, denoted by T_f; $T_f = (T_s + T_\infty)/2$.

We can determine the local heat transfer coefficient from Equation 4.1.1. Then, employing the local convective heat transfer coefficient, we can determine the local heat flux. We are also interested in determining the total heat transfer rate. If the length of the plate in the direction of the flow of the fluid is L and the width is W, to find the total heat transfer rate from the plate to the fluid, we proceed as follows:

$$q_w = \int q_x'' \, dA = \int_0^L h_x (T_s - T_\infty) W \, dx$$

From Equation 4.1.1,

$$h_x = 0.332 \, \mathrm{Pr}^{1/3} \frac{k(U_\infty/v)^{1/2}}{x^{1/2}}$$

$$q_w = \int_0^L W(T_s - T_\infty) \, 0.332 \, \mathrm{Pr}^{1/3} \, k \left(\frac{U_\infty}{v}\right)^{1/2} \frac{dx}{x^{1/2}}$$

$$= W(T_s - T_\infty) \, 0.332 \, \mathrm{Pr}^{1/3} \, k \left(\frac{U_\infty}{v}\right)^{1/2} 2(L)^{1/2} \tag{4.1.2}$$

We define the average convective heat transfer coefficient, h_L, over the entire length of the plate, L, as

$$h_L = \frac{(q_w/A)}{(T_s - T_\infty)}$$

The numerator q_w/A is the average surface heat flux. The convective heat transfer coefficient used in Chapters 2 and 3 is the average heat transfer coefficient defined above.

Average convective heat transfer coefficient— laminar boundary layer, uniform surface temperature

We now define the Nusselt number based on the average heat transfer coefficient over the length L of the plate as $h_L L/k$. Using Equation 4.1.2 to compute h_L (with $A = WL$) and simplifying, we obtain the correlation valid for

$\mathrm{Pr} \gtrsim 0.6$, $\mathrm{Re}_L < \mathrm{Re}_{cr}$ and properties evaluated at $T_f = (T_s + T_\infty)/2$

$$\mathrm{Nu}_L = \frac{h_L L}{k} = 0.664 \, \mathrm{Re}_L^{1/2} \, \mathrm{Pr}^{1/3} \tag{4.1.3}$$

where

$$\mathrm{Nu}_L = \frac{h_L L}{k} \qquad \mathrm{Re}_L = \frac{\rho U_\infty L}{\mu}$$

From an examination of Equations 4.1.1 and 4.1.3, it is evident that $h_L = 2h_{x=L}$, i.e., the average convective heat transfer coefficient over a length, L from the leading edge is twice the local heat transfer coefficient at $x = L$ (Figure 4.1.7). Nu_L is the

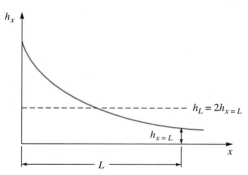

FIGURE 4.1.7 Variation of Local Heat Transfer Coefficient with the Distance from the Leading Edge
The average heat transfer coefficient over a length L is twice the local heat transfer coefficient at $x = L$.

Nusselt number based on the average heat transfer coefficient (and not the average Nusselt number).

Local heat transfer coefficient—Pr ≪ 1, laminar boundary layers, uniform surface temperatures

Equation 4.1.1, can be used for fluids with high Prandtl numbers, such as oils, with reasonable accuracy (for very high Prandtl numbers, the constants 0.332 in Equation 4.1.1, and 0.664 in Equation 4.1.3 are replaced by 0.339 and 0.678, respectively, for slightly better accuracy). However, for liquid metals that have Pr ≪ 1, Equations 4.1.1 and 4.1.3 are not valid. For such cases (see Section 12.5.2), the local Nusselt number is given by

$$\| \quad Nu_x = 0.564\ Re_x^{1/2}\ Pr^{1/2} \qquad Pr < 0.05,\ Re_x\ Pr > 100 \quad \| \qquad (4.1.4)$$

Churchill (1976) recommends the following correlations for all values of Pr:

$$\| \quad Nu_x = \frac{0.3387\ Pr^{1/3}\ Re_x^{1/2}}{[1 + (0.0468/Pr)^{2/3}]^{1/4}} \qquad Re_x < Re_{cr} \quad \| \qquad (4.1.5a)$$

$$\| \quad Nu_L = \frac{0.6774\ Pr^{1/3}\ Re_L^{1/2}}{[1 + (0.0468/Pr)^{2/3}]^{1/4}} \qquad Re_L < Re_{cr} \quad \| \qquad (4.1.5b)$$

Local heat transfer coefficient—laminar boundary layers, uniform heat flux

Uniform Heat Flux, Laminar Boundary Layer (Re < Re_cr) Equations 4.1.1 through 4.1.5 are for uniform wall temperature. For uniform heat flux the correlation for laminar boundary layers is

$$\| \quad Nu_x = 0.453\ Re_x^{1/2}\ Pr^{1/3} \qquad Pr > 0.1 \quad \| \qquad (4.1.6a)$$

But for fluids with very low Prandtl numbers, the correlation is (see Problem 12.15),

$$\| \quad Nu_x = 0.886\ Re_x^{1/2}\ Pr^{1/2} \qquad Pr < 0.05 \quad \| \qquad (4.1.6b)$$

Churchill (1973a) recommends the following single correlation for all Prandtl numbers.

$$\| \quad Nu_x = \frac{0.464\ Re_x^{1/2}\ Pr^{1/3}}{[1 + (0.0207/Pr)^{2/3}]^{1/4}} \qquad Re_x < Re_{cr} \quad \| \qquad (4.1.6c)$$

With uniform heat flux there is no need to find the average heat transfer coefficient as the heat transfer rate is already known. Equations 4.1.6a, b, and c are useful to find the wall temperature at any desired location through the relation

$$T_s - T_\infty = \frac{q_x''}{h_x}$$

where T_s is the local surface temperature, q_x'' is the local wall heat flux, and h_x is the local heat transfer coefficient.

Local heat transfer coefficient—turbulent boundary layers, uniform surface temperature

Uniform Surface Temperature, Turbulent Boundary Layer (Re > Re_cr) When the boundary layer becomes turbulent (Re_x > Re_cr), the relations in Equations 4.1.1 and 4.1.3 for laminar boundary layers are no longer valid. For a flat plate at a uniform temperature, from Reynolds analogy (Section 5.2) and the friction factor recommended by Schliching (1979), the correlations for the local convective heat transfer coefficient for a turbulent boundary layer are

$0.6 \le \mathrm{Pr} \le 60$ and properties at $T_\mathrm{f} = (T_\mathrm{w} + T_\infty)/2$

$$\mathrm{Nu}_x = \frac{h_x x}{k} = 0.0296\ \mathrm{Re}_x^{4/5}\ \mathrm{Pr}^{1/3} \qquad \mathrm{Re}_{\mathrm{cr}} < \mathrm{Re}_x < 10^7 \qquad \textbf{(4.1.7a)}$$

$$\mathrm{Nu}_x = 1.596\ \mathrm{Re}_x (\ln\ \mathrm{Re}_x)^{-2.584}\ \mathrm{Pr}^{1/3} \qquad 10^7 < \mathrm{Re}_x < 10^9 \qquad \textbf{(4.1.7b)}$$

To find the total heat transfer rate from a plate when $\mathrm{Re}_L > \mathrm{Re}_{\mathrm{cr}}$, Equation 4.1.1 is employed for $0 < x \le x_{\mathrm{cr}}$, where the boundary layer is laminar, and Equation 4.1.7a or 4.1.7b is employed for $x \ge x_{\mathrm{cr}}$, where the boundary layer is turbulent. Thus, for $\mathrm{Re}_{\mathrm{cr}} < \mathrm{Re}_L < 10^7$,

$$q_\mathrm{w} = \int_0^{x_{\mathrm{cr}}} h_x (T_\mathrm{s} - T_\infty) W\ dx + \int_{x_{\mathrm{cr}}}^L h_x (T_\mathrm{s} - T_\infty) W\ dx$$

Substituting Equation 4.1.1 in the first integral and Equation 4.1.7a in the second, we obtain

$$q_\mathrm{w} = 0.664 \left(\frac{U_\infty}{\nu}\right)^{1/2} \mathrm{Pr}^{1/3}\ kW(T_\mathrm{s} - T_\infty) x_{\mathrm{cr}}^{1/2}$$
$$+\ 0.037 \left(\frac{U_\infty}{\nu}\right)^{4/5} \mathrm{Pr}^{1/3}\ kW(T_\mathrm{s} - T_\infty)(L^{4/5} - x_{\mathrm{cr}}^{4/5})$$

With $h_L = q_\mathrm{w}/[WL(T_\mathrm{s} - T_\infty)]$ and $\mathrm{Nu}_L = h_L L/k$ we obtain

Average heat transfer coefficient— boundary layer laminar for a part of the length and turbulent for the remaining part, uniform surface temperature

$$\mathrm{Nu}_L = \frac{h_L L}{k} = [0.664\ \mathrm{Re}_{\mathrm{cr}}^{1/2} + 0.037(\mathrm{Re}_L^{4/5} - \mathrm{Re}_{\mathrm{cr}}^{4/5})]\mathrm{Pr}^{1/3} \qquad \textbf{(4.1.8)}$$

If $\mathrm{Re}_{\mathrm{cr}} = 5 \times 10^5$, Equation 4.1.8 simplifies to

$$\mathrm{Nu}_L = (0.037\ \mathrm{Re}_L^{4/5} - 871)\mathrm{Pr}^{1/3} \qquad 5 \times 10^5 < \mathrm{Re}_L < 10^7 \qquad \textbf{(4.1.9a)}$$

For $\mathrm{Re}_L > 10^7$, employing Equation 4.1.7b and Equation 4.1.1 based on the friction factor recommended by Schlichting (1979), the Nusselt number with the average heat transfer coefficient is

$10^7 < \mathrm{Re}_L < 10^9$, $0.6 \le \mathrm{Pr}$, and properties at $T_\mathrm{f} = (T_\mathrm{s} + T_\infty)/2$

$$\mathrm{Nu}_L = [1.967\ \mathrm{Re}_L (\ln\ \mathrm{Re}_L)^{-2.584} - 871]\mathrm{Pr}^{1/3} \qquad \textbf{(4.1.9b)}$$

Equation 4.1.8 can be recast as

$$\mathrm{Nu}_L = 0.037\ \mathrm{Re}_L^{4/5}\ \mathrm{Pr}^{1/3} \left[1 - \left(\frac{\mathrm{Re}_{\mathrm{cr}}}{\mathrm{Re}_L}\right)^{4/5} \left(1 - \frac{0.664}{0.037\ \mathrm{Re}_{\mathrm{cr}}^{3/10}}\right)\right]$$

For $\mathrm{Re}_{\mathrm{cr}} = 5 \times 10^5$ the above equation simplifies to

$$\mathrm{Nu}_L = 0.037\ \mathrm{Re}_L^{4/5}\ \mathrm{Pr}^{1/3} \left[1 - 0.65 \left(\frac{\mathrm{Re}_{\mathrm{cr}}}{\mathrm{Re}_L}\right)^{4/5}\right] \qquad \textbf{(4.1.8a)}$$

For $\mathrm{Re}_{\mathrm{cr}}/\mathrm{Re}_L \ll 1$ (i.e., $x_{\mathrm{cr}}/L \ll 1$), we may set the term within the square brackets to 1. In some cases, such as when the fluid approaches a blunt plate or when the free stream turbulence is very high, the boundary layer becomes turbulent at the leading edge. Thus, either when $\mathrm{Re}_{\mathrm{cr}}/\mathrm{Re}_L \ll 1$ or when $x_{\mathrm{cr}} \approx 0$,

$$\mathrm{Nu}_L = 0.037\ \mathrm{Re}_L^{4/5}\ \mathrm{Pr}^{1/3} \qquad \mathrm{Re}_{\mathrm{cr}} = 0,\ \mathrm{Re}_L < 10^7 \qquad \textbf{(4.1.10a)}$$

If $Re_L > 10^7$, Equation 4.1.9b leads to

$$Nu_L = 1.967\ Re_L(\ln Re_L)^{-2.584}\ Pr^{1/3} \quad Re_{cr} = 0,\ 10^7 < Re_L < 10^9 \quad \textbf{(4.1.10b)}$$

Equation 4.1.10a overpredicts h_L by 11% for $Re_L = 10\ Re_{cr}$ and by 6% for $Re_L = 20\ Re_{cr}$, compared with Equation 4.1.9a. Equations 4.1.10a and b are valid when the boundary layer can be treated as turbulent from the leading edge.

For the local heat transfer coefficient for fluids with a large variation in viscosity, based on experimental and analytical results, Whitaker (1985) recommends that the exponent for the Prandtl number in Equation 4.1.7a should be 0.43 and that the correlation should be modified by the factor $(\mu_\infty/\mu_s)^{1/4}$, where μ_∞ and μ_s are the dynamic viscosities of the fluid at the free stream and surface temperatures, respectively. The recommended correlations are

$$Nu_x = 0.332\ Re_x^{1/2}\ Pr^{1/3} \left(\frac{\mu_\infty}{\mu_s}\right)^{1/4} \quad Re_x < Re_{cr} \quad \textbf{(4.1.11a)}$$

$$Nu_x = 0.0296\ Re_x^{4/5}\ Pr^{0.43} \left(\frac{\mu_\infty}{\mu_s}\right)^{1/4} \quad Re_x > Re_{cr} \quad \textbf{(4.1.11b)}$$

Employing Equations 4.1.11a and b, and setting $Re_{cr} = 5 \times 10^5$, the correlation for the average heat transfer coefficient is

$$Nu_L = 0.664\ Re_L^{1/2}\ Pr^{1/3} \left(\frac{\mu_\infty}{\mu_s}\right)^{1/4} \quad Re_L < 5 \times 10^5 \quad \textbf{(4.1.11c)}$$

$$ \textbf{(4.1.11d)} $$

$$Nu_L = (0.037\ Re_L^{4/5} - 871)Pr^{0.43} \left(\frac{\mu_\infty}{\mu_s}\right)^{1/4} \quad Re_L > 5 \times 10^5$$

In Equations 4.1.11a–d, the dynamic viscosities, μ_∞ and μ_s, are evaluated at the free stream temperature, T_∞, and wall temperature, T_s, respectively. All other properties are evaluated at the film temperature, $T_f = (T_s + T_\infty)/2$.

Uniform Heat Flux, Turbulent Boundary Layer ($Re > Re_{cr}$) For turbulent boundary layers, the local heat transfer coefficient is slightly higher than the local heat transfer coefficient for uniform wall temperature. The correlation suggested by Kays and Crawford (1980) is

$$ Nu_x = 0.03\ Re_x^{0.8}\ Pr^{0.6} \quad \textbf{(4.1.12)}$$

Local convective
heat transfer
coefficient—
turbulent boundary
layers, uniform
heat flux

Correlations, by Thomas and Al-Sharifi (1985) for uniform heat flux for turbulent boundary layers are

$0.5 < Pr < 10$

$$Nu_x = \frac{\sqrt{c_{fx}/2}\ Re_x\ Pr}{2.21\ \ln(Re_x\sqrt{c_{fx}/2}) - 0.232\ \ln Pr + 14.9\ Pr^{0.623} - 15.6} \quad \textbf{(4.1.13a)}$$

$10 < Pr < 500$

$$Nu_x = \frac{\sqrt{c_{fx}/2}\ Re_x\ Pr}{2.21\ \ln(Re_x\sqrt{c_{fx}/2}) - 0.232\ \ln Pr + 10\ Pr^{0.741} - 6.21} \quad \textbf{(4.1.13b)}$$

$$c_{fx} \approx 0.0592\ Re_x^{-0.2} \quad \textbf{(4.1.13c)}$$

Equations 4.1.13a and b can be used for uniform wall temperatures also. c_{fx} is the local friction factor, which is approximated by Equation 4.1.13c.

Flow Over a Plate with Unheated Starting Length, Uniform Surface Temperature So far, we have considered a plate heated from the leading edge. In some cases, the heating may start some distance downstream from the leading edge as in Figure 4.1.8. The velocity boundary layer always starts from the leading edge. The heat transfer coefficients where there is an unheated starting length is different from that when the heat transfer starts from the leading edge. Solutions for the case of an unheated starting length have been obtained by integral methods by Kays and Crawford (1980).

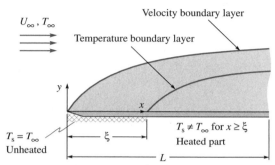

FIGURE 4.1.8 Heat Transfer from a Plate with an Unheated Starting Length
A fluid flows parallel to a flat plate with a uniform free stream velocity and temperature. For $x > \xi$, the plate is heated and maintained at a uniform temperature, $T_s > T_\infty$.

For laminar boundary layer, the local Nusselt number, for $x > \xi$, is given by

$$\mathrm{Nu}_x = \frac{0.332 \ \mathrm{Re}_x^{1/2} \ \mathrm{Pr}^{1/3}}{[1 - (\xi/x)^{3/4}]^{1/3}} \tag{4.1.14}$$

For $\xi = 0$, Equation 4.1.14 reduces to Equation 4.1.1.

For turbulent flow, Kays and Crawford (1980) show that the local Nusselt number, for $x > \xi$, is

$$\mathrm{Nu}_x = \frac{0.0296 \ \mathrm{Re}_x^{4/5} \ \mathrm{Pr}^{3/5}}{[1 - (\xi/x)^{9/10}]^{1/9}} \tag{4.1.15}$$

Local and average heat transfer coefficient— uniform surface temperature, unheated starting length

(In Kays and Crawford, the constant in Equation 4.1.15 is 0.0287; consistent with the expression for c_{fx} in Equation 4.1.13c the constant has been changed to 0.0296.) The Nusselt number, based on the average heat transfer coefficient over the entire heated length, can be found by integrating Equations 4.1.14 and 4.1.15, depending on the Reynolds number. Although it appears as if no closed form solution is possible and that the integration has to be performed by numerical methods, these expressions have been integrated and the results given in a closed form by Thomas (1977). The average value of the convective heat transfer coefficient from $x = \xi$ to $x = L$ for uniform surface temperature and for laminar boundary layers is given by

$$h_L = 2 \frac{[1 - (\xi/L)^{3/4}]}{1 - \xi/L} h_{x=L} \tag{4.1.16}$$

If the boundary layer is turbulent starting from the leading edge,

$$h_L = \frac{5[1 - (\xi/L)^{9/10}]}{4(1 - \xi/L)} h_{x=L} \tag{4.1.17}$$

If the boundary layer is partly laminar and partly turbulent, the expression for the average heat transfer coefficient,

$$h_L = \frac{1}{L - \xi} \int_{\xi}^{L} h(x)\, dx$$

is found by integration by substituting Equation 4.1.14 for the laminar boundary layer and Equation 4.1.15 for the turbulent boundary layer.

Methodology for Determining the Convective Heat Transfer Coefficient
1. Identify the type of convection—forced or natural.
2. Identify the geometry—external or internal—and the geometry of the bounding surfaces—flat plate, cylinder, or other.
3. Evaluate fluid properties at the appropriate temperatures. Note that although viscosity, thermal conductivity, and specific heats are functions of both pressure and temperature, their dependence on pressure is weak and is usually neglected. Thus, their values at a given temperature can be used at all pressures at the same temperature.
4. Calculate the Reynolds number based on the appropriate characteristic length dimension. The characteristic length dimension is usually the length of the plate for flow over a flat plate, the diameter of the cylinder or sphere, and so forth.
5. Identify other conditions—unheated length, uniform surface temperature, uniform surface heat flux, local or average convective heat transfer coefficient, and so on.
6. Select a correlation appropriate for the type of convection and Reynolds number and other conditions.

There may be more than one correlation given for a particular configuration. All the correlations are based on experimental data (not necessarily the same set) and sometimes in combination with analysis. The differences in the different sets of experimental data arise as a result of the many variables that affect the heat transfer coefficient, some of which are difficult to measure. For example, the free stream turbulence has an effect on the transition Reynolds number but is not easily measurable. Every correlation has a band of uncertainty, and different correlations give slightly different values of the convective heat transfer coefficient.

EXAMPLE 4.1.2 The viscosity of a liquid, flowing parallel to a flat plate (Figure 4.1.9), adjacent to the plate can be reduced by heating the liquid. Such a reduction in the viscosity leads to a reduction in the drag force on the plate. It is proposed to determine experimentally the reduction in the drag force that can be achieved by heating a liquid.

The experimental apparatus consists of two flat plates, 4-mm apart, 1-m long in the direction of the flow, and 0.6-m wide. Water at 20 °C flows parallel to the plate with a free stream velocity of 6 m/s. The plates are to be heated to a uniform temperature of 40 °C by supplying steam into the space between the two plates. Determine the heat flux at a distance of 1 cm and 50 cm from the leading edge of the plate and the total heat transfer rate.

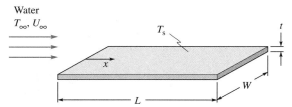

FIGURE 4.1.9 Water at 20 °C flows parallel to flat plates, 1-m long and 0.6-m wide and maintained at 40 °C. Free stream velocity of water is 6 m/s.

Given

$U_\infty = 6$ m/s $T_\infty = 20$ °C $T_s = 40$ °C
$L = 1$ m $W = 0.6$ m $t = 4$ mm

Find

(a) q_x'' (local convective heat flux from the plate) at $x = 1$ cm and 50 cm
(b) q_w (total heat transfer rate from the plates)

ASSUMPTIONS

1. Steady state
2. Effect of blunt edges at the leading edge is negligible and the flow is considered to be parallel to the flat plate.
3. $Re_{cr} = 5 \times 10^5$

SOLUTION

The local heat flux is computed from $q_x'' = h_x(T_s - T_\infty)$. h_x is found from one of the correlations depending on the value of Re_x. The total heat transfer rate is given by $q = h_L A(T_s - T_\infty)$; the average heat transfer coefficient is computed from one of the correlations depending on the value of Re_L. In all the correlations the properties are evaluated at the film temperature.
From Table A5, properties of water at the film temperature of $T_f = (40 + 20)/2 = 30$ °C are
$\rho = 995.6$ kg/m^3 $c_p = 4176$ J/kg K $k = 0.618$ W/m K
$\mu = 778.6 \times 10^{-6}$ N s/m^2 $\nu = 78.2 \times 10^{-8}$ m^2/s   Pr $= 5.26$
(a) The convective heat flux from the plate is computed as follows.

$x = 1$ cm

$$Re_x = \frac{U_\infty x}{\nu} = \frac{6 \times 0.01}{78.2 \times 10^{-8}} = 7.673 \times 10^4$$

For $Re_x < 5 \times 10^5$, from Equation 4.1.1,

$$Nu_x = 0.332\ Re_x^{1/2}\ Pr^{1/3} = 0.332(7.673 \times 10^4)^{1/2}(5.26)^{1/3} = 159.9$$

$$h_x = Nu_x\ \frac{k}{x} = 159.9 \times \frac{0.618}{0.01} = 9882\ W/m^2\ {}^\circ C$$

$$q''_x = h_x(T_s - T_\infty) = 9882(40 - 20) = \underline{1.976 \times 10^5\ W/m^2}$$

$x = 0.5$ m

$$Re_x = \frac{U_\infty x}{\nu} = \frac{6 \times 0.5}{78.2 \times 10^{-8}} = 3.836 \times 10^6$$

For $5 \times 10^5 < Re_x < 10^7$, employing Equation 4.1.7a,

$$Nu_x = 0.0296\ Re_x^{4/5}\ Pr^{1/3} = 0.0296(3.836 \times 10^6)^{4/5}(5.26)^{1/3} = 9522$$

$$h_x = Nu_x\ \frac{k}{x} = 9522 \times \frac{0.618}{0.5} = 11\ 770\ W/m^2\ {}^\circ C$$

$$q''_x = h_x(T_s - T_\infty) = 11\ 770(40 - 20) = \underline{2.354 \times 10^5\ W/m^2}$$

(b) The total heat transfer rate from *both* the plates is q. Total heat transfer rate from *one* plate is given by $q/2 = h_L WL(T_s - T_\infty)$, where W is width of the plate and L is length of the plate:

$$Re_L = \frac{U_\infty L}{\nu} = \frac{6 \times 1}{78.2 \times 10^{-8}} = 7.673 \times 10^6$$

For $5 \times 10^5 < Re_L < 10^7$, from Equation 4.1.9a,

$$Nu_L = (0.037\ Re_L^{4/5} - 871)Pr^{1/3} = [0.037(7.673 \times 10^6)^{4/5} - 871](5.26)^{1/3}$$

$$= 19\ 211$$

$$h_L = Nu_L\ \frac{k}{L} = 19\ 211 \times \frac{0.618}{1} = 11\ 870\ W/m^2\ {}^\circ C$$

$$q = 2WLh_L(T_s - T_\infty) = 2 \times 0.6 \times 1 \times 11\ 870(40 - 20)$$

$$= 2.849 \times 10^5\ W = \underline{284.9\ kW}$$

COMMENT

Note the large amount of power required to heat the plate. Also, both in the laminar and turbulent boundary layer regions, the convective heat transfer coefficient decreases in the direction of flow, but the values are significantly higher in the turbulent region.

EXAMPLE 4.1.3 Atmospheric air, at 0 °C, flows parallel to the vertical wall of a building with a velocity of 10 m/s as shown in Figure 4.1.10. The wall is 20-m long (in the direction of flow), 2.5-m high, and its surface is at 10 °C. Estimate

(a) The convective heat transfer rate from the wall to the air.

(b) The convective heat transfer rate from that part of the wall where the boundary layer is turbulent.

(c) The average convective heat transfer coefficient in the part where the boundary layer is turbulent.

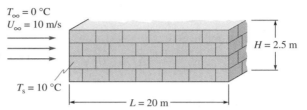

FIGURE 4.1.10 Air flows parallel to a vertical wall. The wall is at a uniform temperature.

Given

$U_\infty = 10$ m/s $\qquad T_\infty = 0\,°C \qquad L$ (length) $= 20$ m

H (height) $= 2.5$ m $\qquad T_s = 10\,°C$

Find

(a) q_w (heat transfer rate from wall)

(b) heat transfer rate from the wall in the turbulent boundary layer region

(c) h_{tur} (the average heat transfer coefficient in the turbulent boundary layer region)

ASSUMPTIONS

1. Steady state
2. The flow over the wall is considered as flow parallel to a thin plate. The effects of the adjoining walls are neglected.

SOLUTION

(a) The heat transfer rate from the wall to the air is calculated from $q_w = h_L\,A(T_s - T_\infty)$. The average heat transfer coefficient is calculated from one of the applicable correlations depending on the value of Re_L. The properties are evaluated at the film temperature.

Film temperature of air $= (0 + 10)/2 = 5\,°C$. From the software CC (or from Table A7), properties of air at $5\,°C$ are

$\rho = 1.269$ kg/m^3 $\qquad c_p = 1006$ J/kg $°C \qquad \mu = 1.745 \times 10^{-5}$ N s/m^2

$k = 0.0245$ W/m K $\qquad Pr = 0.717$

For a wall at a uniform temperature we have

$$q_w = h_L HL(T_s - T_\infty)$$

The value of the convective heat transfer coefficient has to be determined from an appropriate correlation. To identify the correlation to be used, evaluate Re_L.

$$Re_L = \frac{\rho U_\infty L}{\mu} = \frac{1.269 \times 10 \times 20}{1.745 \times 10^{-5}} = 1.455 \times 10^7$$

Assuming a critical Reynolds number of 5×10^5, as $Re_L > 5 \times 10^5$, the boundary layer is laminar near the leading edge, becoming turbulent at some point downstream. For $Re_L > 10^7$, we employ Equation 4.1.9b to determine the average convective heat transfer coefficient.

$$Nu_L = \frac{h_L L}{k} = [1.967\ Re_L(\ln Re_L)^{-2.584} - 871]Pr^{1/3}$$

$$= \{1.967 \times 1.455 \times 10^7 \times [\ln(1.455 \times 10^7)]^{-2.584} - 871\}(0.717)^{1/3}$$

$$= 17\ 543$$

$$h_L = Nu_L \frac{k}{L} = 17\ 543 \times \frac{0.0245}{20} = 21.5\ \text{W/m}^2\ \text{K}$$

$$q_w = h_L A(T_s - T_\infty) = 21.5 \times 20 \times 2.5(10 - 0) = \underline{10\ 750\ \text{W}}$$

As the wall and the terrain may be rough, assuming that the boundary layer becomes turbulent from the leading edge itself and using Equation 4.1.10b,

$$Nu_L = 1.967\ Re_L(\ln Re_L)^{-2.584}\ Pr^{1/3}$$

$$= 1.967 \times 1.455 \times 10^7 \times [\ln(1.455 \times 10^7)]^{-2.584}(0.717)^{1/3} = 18\ 323$$

$$h_L = Nu_L \frac{k}{L} = 18\ 323 \times \frac{0.0245}{20} = 22.45\ \text{W/m}^2\ ^\circ\text{C}$$

$$q_w = h_L A(T_s - T_\infty) = 22.45 \times 20 \times 2.5(10 - 0) = \underline{11\ 225\ \text{W}}$$

(b) If we assume that the boundary layer becomes turbulent at $Re_{cr} = 5 \times 10^5$, then the critical length $L_{cr} = 20 \times 5 \times 10^5/(1.455 \times 10^7) = 0.6873$ m. This length, over which the boundary layer is laminar, is indicated by L_{lam}. We can now determine the average heat transfer coefficient and the heat transfer rate from this length of the plate. Subtracting the heat transfer rate in the laminar region from the total heat transfer rate found in part a will give us the heat transfer rate from that part of the wall where the boundary layer is turbulent. That is,

$$q_{lam} = h_{lam}L_{lam}H(T_s - T_\infty)$$

$$q_{tur} = q_w - q_{lam}$$

Compute h_{lam}, q_{lam}, and q_{tur}

$$Nu_{lam} = 0.664\ Re_{lam}^{1/2}\ Pr^{1/3} = 0.664(5 \times 10^5)^{1/2} \times 0.717^{1/3} = 420.2$$

$$h_{lam} = Nu_{lam} \frac{k}{L_{lam}} = 420.2 \times \frac{0.0245}{0.6873} = 14.98\ \text{W/m}\ ^\circ\text{C}$$

$$q_{lam} = h_{lam}L_{lam}H(T_s - T_\infty) = 14.98 \times 0.6873 \times 2.5 \times 10 = 257\ \text{W}$$

$$q_{tur} = q_{total} - q_{lam} = 10\ 750 - 257 = \underline{10\ 493\ \text{W}}$$

(c) The average heat transfer coefficient in the part where the boundary layer is turbulent is

$$h_{tur} = \frac{q_{tur}}{A_{tur}(T_s - T_\infty)} = \frac{10\ 493}{(20 - 0.6873) \times 2.5 \times 10} = \underline{21.73\ \text{W/m}^2\ ^\circ\text{C}}$$

COMMENT

As $\text{Re}_{cr}/\text{Re}_L = 5 \times 10^5/1.455 \times 10^7 = 0.03436$, the condition approximating Equation 4.1.9b by Equation 4.1.10b is satisfied, and the heat transfer rate found by using Equation 4.1.10b is only 4.4% more than that obtained from the use of Equation 4.1.9b.

It is not always necessary to find the heat transfer rate to determine the average heat transfer coefficient, as in part c. The average heat transfer coefficient from $L = L_1$ to $L = L_2$ can be computed from

$$h_{1-2} = \frac{q_2 - q_1}{H(L_2 - L_1)(T_s - T_\infty)} = \frac{(h_2 L_2 - h_1 L_1)H(T_s - T_\infty)}{H(L_2 - L_1)(T_s - T_\infty)}$$

$$= \frac{k}{L_2 - L_1}(\text{Nu}_2 - \text{Nu}_1)$$

where the subscripts 1 and 2 denote that the variable is evaluated over L_1 or L_2. Employing the appropriate correlations for Nu_2 and Nu_1, the average heat transfer coefficient from L_1 to L_2, h_{1-2} is directly computed.

Variation of the Local and Average Heat Transfer Coefficient, Flat Plate To show the variation of the local heat transfer coefficient with the distance from the leading edge, Equations 4.1.1 and 4.1.7a are recast as

$$c_1 h_x = 0.332 \, \text{Re}_x^{-1/2} \qquad \text{Re}_x < 5 \times 10^5$$

$$c_1 h_x = 0.0296 \, \text{Re}_x^{-1/5} \qquad \text{Re}_x > 5 \times 10^5$$

$$c_1 = \left(k \frac{U}{\nu} \text{Pr}^{1/3}\right)^{-1}$$

where U is the free stream velocity.

For a given flow over a flat plate, $c_1 h_x$ represents the local heat transfer coefficient (multiplied by a constant c_1), and Re_x is proportional to the distance from the leading edge. Similarly, from Equations 4.1.3 and 4.1.9a, the average heat transfer coefficient is given by

Local and average heat transfer coefficient for a flat plate, uniform surface temperature

$$c_1 h_L = 0.664 \, \text{Re}_L^{-1/2} \qquad \text{Re}_L < 5 \times 10^5$$

$$c_1 h_L = \frac{0.037}{\text{Re}_L^{0.2}} - \frac{871}{\text{Re}_L} \qquad \text{Re}_L > 5 \times 10^5$$

Figure 4.1.11 shows the variation of the local and average heat transfer coefficient (multiplied by c_1) with Re_x, which represents the dimensionless distance from the leading edge. Both decrease with x when the boundary layer is laminar. When transition to turbulent boundary layer occurs, the predicted value of the local convective heat transfer coefficient jumps to a higher value and then decreases with x. The average heat transfer coefficient decreases in the laminar region. It begins to increase for $\text{Re}_L > 5 \times 10^5$, reaches a maximum at $\text{Re}_L = 2.18 \times 10^6$ (see Problem 4.38), and then decreases again.

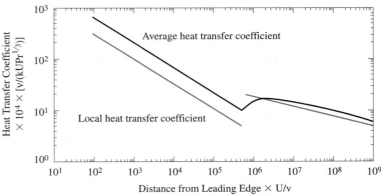

FIGURE 4.1.11 **Variation of Local and Average Convective Heat Transfer Coefficient with Distance from the Leading Edge for a Flat Plate**

4.1.2 Forced Convection Correlations—External Flow Over Cylinders and Spheres

Boundary layer flows perpendicular to the axis of a cylinder and over spheres are different from the boundary layer flow over a flat plate. With uniform flow over a flat plate, the pressure is uniform; there is no pressure gradient in the direction of flow. For cylinders and spheres, if the pressure increases in the direction of flow (adverse pressure gradient), the velocity profile in the boundary layer tends to flatten resulting in decreasing values of the velocity gradient $\partial u/\partial y|_{y=0}$ (Figure 4.1.12). At some point, if the magnitude of the adverse pressure gradient is sufficiently high,

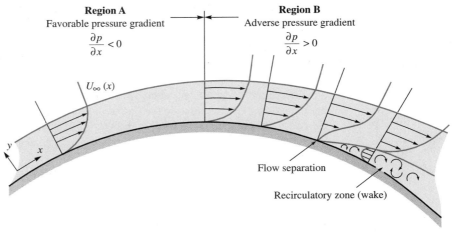

FIGURE 4.1.12 **Effect of Adverse Pressure Gradient on Velocity Profile and Boundary Layer Separation, on a Curved Surface**
(a) Flow over a cylinder: $Re_d = 9.6$. (b) Flow over a cylinder: $Re_d = 2000$.

$\partial u / \partial y |_{y=0}$ becomes zero and the boundary layer detaches from the surface. An adverse pressure gradient exists if the free stream velocity in the direction of flow decreases, for example, due to the curvature of the surface as shown in region B in Figure 4.1.12. In region A, the free stream velocity increases and the pressure decreases. But in region B, the velocity decreases and the pressure increases. In this region of adverse pressure gradient, the velocity profile flattens and the possibility of boundary layer separation exists.

Flows over spheres and cylinders have different flow regimes

In flows over cylinders and spheres, for the part of the solid facing the oncoming fluid, the pressure gradient is favorable (pressure decreasing in the direction of flow). But for the part that faces away from the oncoming fluid, the pressure gradient is adverse. At very low Reynolds numbers ($\mathrm{Re}_d = \rho V d / \mu$, where d is the diameter of the cylinder), the streamlines follow the surface of the cylinder, and the boundary layer is laminar, Figure 4.1.13a. As the fluid velocity is increased, due to the adverse pressure gradient toward the rear of the cylinder, the boundary layer separates. The boundary layer is laminar with a wake, which may be turbulent, Figure 4.1.13b. As the velocity is increased, the point of *boundary layer separation moves upstream.* For $\mathrm{Re}_d \approx 2 \times 10^5$ the boundary layer becomes turbulent at some point, and the *separation point moves downstream.* With a further increase in the velocity, a greater part of the boundary layer becomes turbulent, and the *separation point moves upstream* again.[2] Because of the complexities in the flows over cylinders and spheres, only numerical solutions have been successful in predicting many (but not all) observed phenomena.

It should now be evident that with flow across a cylinder the boundary layer cannot be classified as laminar or turbulent in the same sense as flows over a flat plate. The boundary layer can be laminar followed by a turbulent wake, or it can be partly laminar and partly turbulent followed by a turbulent wake. Delineating the boundary layer as laminar or turbulent does not completely describe the flow and the flow is characterized only in terms of a range of Reynolds numbers. For the same reason, some of the correlations for the convective heat transfer coefficient have constants that are functions of the Reynolds number. Because of the complexities in determining the local convective heat transfer coefficient, only correlations for the average heat transfer coefficient are given.

Forced convection correlations for cross flow over cylinders

Cylinders For cross flow over cylinders the characteristic dimension for forming the Nusselt and Reynolds numbers is the diameter of the cylinder. As in the case of flat plates, correlations relating the Nusselt number (based on the average heat transfer coefficient over the entire cylindrical surface), the Reynolds number, and the Prandtl number are available in different forms. One such form is

$$\left\| \quad \mathrm{Nu}_d = \frac{h\,d}{k} = c\,\mathrm{Re}_d^m\,\mathrm{Pr}^n \quad \right\| \qquad \textbf{(4.1.18)}$$

---

[2]The drag on the cylinder and sphere increases with velocity when the boundary layer is laminar, with the point of separation moving upstream with increasing velocity; but, if the point of separation moves downstream for the same velocity, the drag force is reduced. Turbulence has the effect of delaying separation. For a particular velocity, if the boundary layer is laminar with turbulent wake, the point of separation can be moved downstream by making the boundary layer turbulent. This will result in a reduction in the drag force. Golf balls are given dimples to promote such a transition to turbulent flow and a reduction in the drag force.

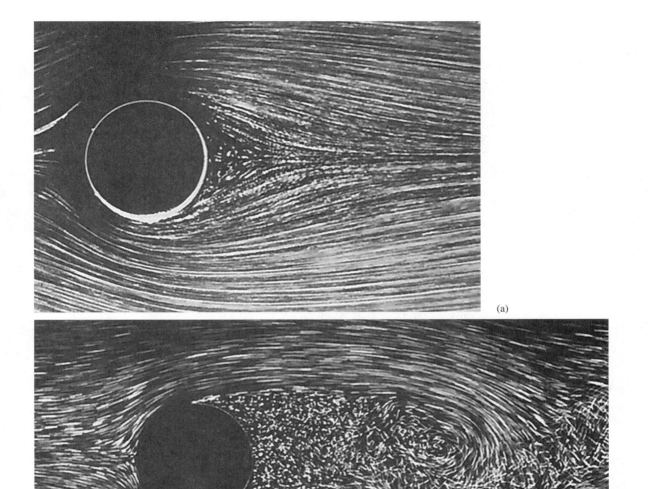

FIGURE 4.1.13 Flow over a Cylinder at Different Reynolds Numbers (a) $Re_d = 9.5$ (b) $Re_d = 2000$
Reproduced with permission from *An Album of Third Motion,* assembled by van Dyke, Milton, (Stanford, CA: Parabolic Press, 1982).

where

$$Nu_d = \text{Nusselt number}$$
$$h = \text{average heat transfer coefficient}$$
$$k = \text{thermal conductivity of the fluid}$$
$$d = \text{diameter of the cylinder}$$
$$Re_d = \text{Reynolds number} = \rho U_\infty \, d/\mu$$

c, m, n = constants that are functions of Re_d (Table 4.1.1)
Properties at $T_f = (T_s + T_\infty)/2$

Based on extensive experimental measurements with air in cross flow over cylinders of different diameters, Hilpert (1933) determined the value of $c\,Pr^n$ in Equation 4.1.18. Morgan (1975) recalculated the values of cPr^n for air. For most of the experimental measurements, the Prandtl number for air is close to 0.7. Using the value of 0.7 for Prandtl number of air, the constants in Equation 4.1.18 (based on the values of cPr^n recommended by Morgan) are given in Table 4.1.1.

TABLE 4.1.1 Constants in Equation 4.1.18 for Flow Over Cylinders, Air at $Pr \gtrsim 0.7$

Re_d	$c\,Pr^{1/3}$ ($Pr = 0.7$)	c	m
4–35	0.795	0.895	0.384
35–5000	0.583	0.657	0.471
5000–50 000	0.148	0.167	0.633
50 000–230 000	0.0208	0.0234	0.814

For $Re_d\,Pr > 0.2$, based on the paper by Churchill and Bernstein (1977), the following correlations are recommended:

$Re_d\,Pr > 0.2$, properties at $T_f = (T_s + T_\infty)/2$
$Re_d > 400\ 000$

$$Nu_d = 0.3 + \frac{0.62\,Re_d^{1/2}\,Pr^{1/3}}{[1 + (0.4/Pr)^{2/3}]^{1/4}} \left[1 + \left(\frac{Re_d}{282\ 000}\right)^{5/8}\right]^{4/5} \tag{4.1.19}$$

$20\ 000 < Re_d < 400\ 000$

$$Nu_d = 0.3 + \frac{0.62\,Re_d^{1/2}\,Pr^{1/3}}{[1 + (0.4/Pr)^{2/3}]^{1/4}} \left[1 + \left(\frac{Re_d}{282\ 000}\right)^{1/2}\right] \tag{4.1.20}$$

$Re_d < 20\ 000$

$$Nu_d = 0.3 + \frac{0.62\,Re_d^{1/2}\,Pr^{1/3}}{[1 + (0.4/Pr)^{2/3}]^{1/4}} \tag{4.1.21}$$

Differences in predictions of Nusselt numbers between Equation 4.1.18 (along with the constants from Table 4.1.1), and Equations 4.1.19, 20, and 21 are within about 10% for $Re_d < 5000$ and $Re_d > 75\ 000$ but as high as 13% to 30% for Re_d in the range 10 000 to 50 000. Equation 4.1.18 has been developed mainly from data with air, whereas Equation 4.1.19 through 4.1.21 are based on a much broader data base with a wider range of Prandtl numbers. It is, therefore, recommended that Equation 4.1.18 be restricted to Prandtl numbers in the range of 0.5 to 10. Equations 4.1.19 through 4.1.21 can be used for all Prandtl numbers.

Forced convection correlation for flows over spheres

Spheres For a sphere, Whitaker (1972) recommends

$3.5 < Re_d < 76\ 000, \quad 0.71 < Pr < 380, \quad 1 < \mu_\infty/\mu_s < 3.2,$
Properties at T_∞, except μ_s at T_s

$$Nu_d = 2.0 + (0.4\,Re_d^{1/2} + 0.06\,Re_d^{2/3})Pr^{2/5}\left(\frac{\mu_\infty}{\mu_s}\right)^{1/4} \tag{4.1.22}$$

TABLE 4.1.2 Values of c and m for Use in Equation 4.1.18 for Long Bars

Evaluate all properties at the arithmetic mean temperature $(T_s + T_\infty)/2$; $n = \frac{1}{3}$ in all cases.

Geometry	Re_D	c	m
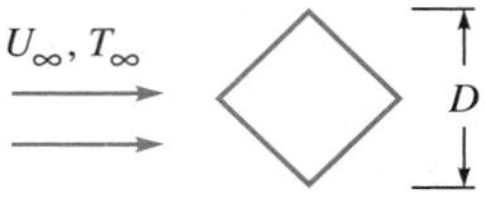 U_∞, T_∞ ◇ D	5×10^3–10^5	0.246	0.588
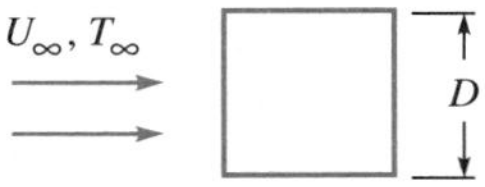 U_∞, T_∞ □ D	5×10^3–10^5	0.102	0.675
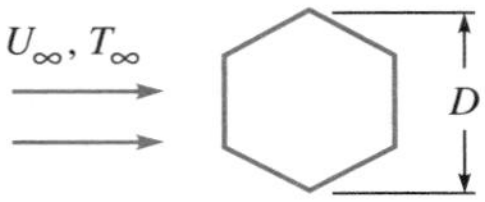 U_∞, T_∞ ⬡ D	5×10^3–1.95×10^4 1.95×10^4–10^5	0.10 0.085	0.638 0.782
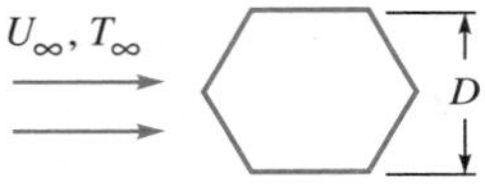 U_∞, T_∞ ⬡ D	5×10^3–10^5	0.153	0.638
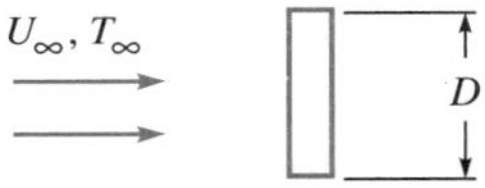 U_∞, T_∞ ▯ D	4×10^3–1.5×10^4	0.228	0.731

Forced convection correlations for flows over various geometric shapes

Geometries Other Than Cylinders and Spheres For geometries other than cylinders and spheres, the characteristic dimensions and the values of the constants to be used in Equation 4.1.18 recommended by Jacob (1949) are given in Table 4.1.2.

EXAMPLE 4.1.4 It is proposed to fabricate a space heater with 15-mm O.D. tubes with steam flowing inside the tubes (Figure 4.1.1). Air flows perpendicular to the tube axis at 3 m/s. If the temperatures of the surface of the tube and air are 100 °C and 10 °C, respectively, determine the heat transfer rate to the air per meter length of a tube.

Given

Air in cross flow over a tube

$U_\infty = 3$ m/s $T_\infty = 10\,°C$ $d = 15$ mm $T_s = 100\,°C$

Find

q/L (heat transfer rate per meter length of the tube)

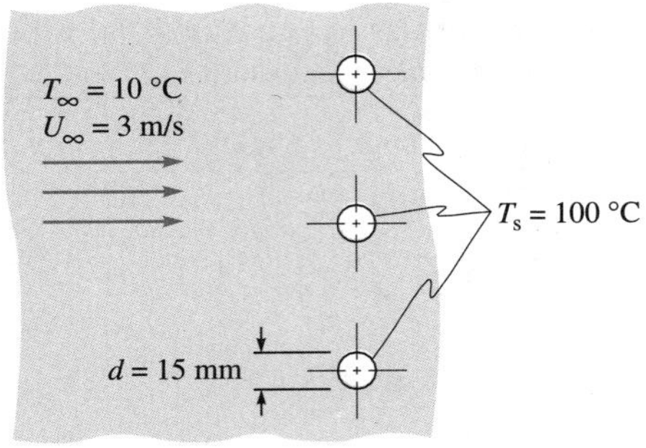

FIGURE 4.1.14 Heat Transfer with Air Flowing Over a Bank of Tubes Heated by Steam

ASSUMPTIONS

1. Steady state
2. The tubes are sufficiently far apart that the fluid flow over one tube is not affected by the presence of the neighboring tubes. If the spacing between the tubes is s, then $s \gg d$ so that there is no significant increase in the velocity of the air between the tubes.

SOLUTION

The heat transfer rate from one tube is calculated from $q = h\pi dL(T_s - T_\infty)$, where h is the average heat transfer coefficient to be found from one of the correlations for flow over cylinders. The heat transfer rate per meter length of one tube is computed from $q/L = h\pi d(T_s - T_\infty)$. The correlation for h depends on the value of Re_d. The properties for the correlations are evaluated at the film temperature.

From the software (CC) (or from Table A7), properties of air at the arithmetic average temperature of 55 °C are

$$\rho = 1.076 \text{ kg/m}^3 \qquad c_p = 1008 \text{ J/kg °C} \qquad \mu = 1.98 \times 10^{-5} \text{ N s/m}^2$$
$$k = 0.0282 \text{ W/m K} \qquad \mathrm{Pr} = 0.709$$

$$\mathrm{Re}_d = \frac{\rho d V_\infty}{\mu} = \frac{1.076 \times 0.015 \times 3}{1.98 \times 10^{-5}} = 2445$$

In Equation 4.1.18, from Table 4.1.1, $c = 0.657$ and $m = 0.471$; thus,

$$\mathrm{Nu}_d = \frac{hd}{k} = 0.657 \ \mathrm{Re}_d^{0.471} \ \mathrm{Pr}^{1/3} = 0.657 \times 2445^{0.471} \times 0.709^{1/3} = 23.1$$

$$h = \mathrm{Nu}_d \frac{k}{d} = 23.1 \times \frac{0.0282}{0.015} = 43.37 \text{ W/m}^2 \text{ K}$$

$$q/L = h\pi d(T_s - T_\infty) = 43.37 \times \pi \times 0.015(100 - 10) = \underline{184 \text{ W/m}}$$

If, instead of using Equation 4.1.18, we use Equation 4.1.21, then

$$\mathrm{Nu}_d = 0.3 + \frac{0.62\ \mathrm{Re}_d^{1/2}\ \mathrm{Pr}^{1/3}}{[1\ +\ (0.4/\mathrm{Pr})^{2/3}]^{1/4}} = 0.3 + \frac{0.62 \times 2445^{1/2} \times 0.709^{1/3}}{[1\ +\ (0.4/0.709)^{2/3}]^{1/4}} = 24.3$$

$$h = \mathrm{Nu}_d\,\frac{k}{d} = 24.3 \times \frac{0.0282}{0.015} = 45.68\ \mathrm{W/m^2\,{}^\circ C}$$

$$\frac{q}{L} = h\,\pi d(T_s - T_\infty) = 45.68 \times \pi \times 0.015(100 - 10) = \underline{194\ \mathrm{W}}$$

The heat transfer rate obtained by using Equation 4.1.21 is 5.4% higher than that obtained by using Equation 4.1.18. Such differences should be expected if different correlations are used.

4.2 FORCED CONVECTION CORRELATIONS—INTERNAL FLOWS

We now turn our attention to another common engineering application, heat transfer to fluids flowing inside tubes and ducts. In external flows the boundary layer growth is not limited by any confining surface. However, in internal flows, such as in tubes, there is a limit to the velocity and temperature boundary layer thicknesses—the radius of the tube. There are rapid changes in the value of the convective heat transfer coefficient in the region where the boundary layer thickness increases but much smaller changes (due only to the properties being temperature dependent) in the region where the boundary layer thickness has reached its maximum value. A brief discussion of the two regions follows.

Fully developed velocity and temperature profile defined

When a fluid enters a tube from a large reservoir, the velocity profile varies in the direction of flow, but, after some distance, the profile is invariant. The region where the velocity profile varies is known as the *hydrodynamically developing region* and the region where the velocity profile is invariant is known as the *hydrodynamically fully developed region*. Similarly, just downstream of the location where the heating of the fluid starts, the temperature profile varies in the direction of flow, but, after some distance downstream, the dimensionless temperature profile is also invariant. Thus, there are *thermally developing* and *thermally fully developed regions*. Our discussion of heat transfer will be generally limited to the hydrodynamically and thermally fully developed regions.

Fully developed velocity profile in internal flows

To appreciate the meaning of fully developed region, consider the flow of an incompressible fluid in a tube attached to a large reservoir as shown in Figure 4.2.1a. At the entrance to the tube, section A, the velocity is uniform at U_o. As the fluid flows in the tube, the shear stress at the wall decelerates the fluid close to the wall. As a consequence of the deceleration, the velocity profile at section B is different from that at A. At section B, the velocity varies from zero at the wall to some value U_c at a distance δ from the surface of the tube; the velocity from that point to the axis is uniform at U_c. Because the velocity close to the wall is less than that at section A, mass balance requires that $U_c > U_o$. Further downstream, the distance δ across which the velocity changes from 0 to a uniform velocity U_c (U_c increases along the

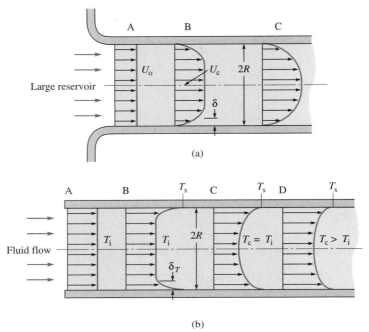

FIGURE 4.2.1 Development of Velocity and Temperature Boundary Layers in Internal Flows
(a) A fluid enters a circular tube with a uniform velocity. The fluid near the surface is decelerated and the fluid near the axis is accelerated. (b) A fluid enters the heated section of a tube with fully developed velocity profile and uniform temperature. The temperature of the fluid close to the heated surface increases.

Fully developed temperature profile in internal flows

axis) increases. Although δ increases in the direction of the flow, there is limit to its growth, the limit of its thickness being the tube radius R, as shown at section C. From section C onwards, the velocity profile remains unaltered. The unchanging velocity profile that exists beyond C is known as the *fully developed velocity profile;* the region of the fully developed velocity profile is known as the *hydrodynamically fully developed region.* The region where the velocity profile is developing, up to section C, is the *hydrodynamically developing* or *the entrance region.*

Now consider the region where heating begins, with the velocity profile being fully developed (Figure 4.2.1b). Denote the surface temperature by T_s, and the temperature of the fluid entering the heated part by T_i ($T_i < T_s$). A short distance downstream, at B, the fluid adjacent to the tube is heated by the tube surface and the temperature of the fluid varies from T_s, immediately adjacent to the tube, to T_i at a small distance δ_T from the tube surface. The distance δ_T across which the temperature changes from T_s to T_i increases in the direction of flow until, at some location, C, it equals the radius of the tube. Up until section C, the center-line temperature remains constant at T_i but beyond this section, the center-line temperature begins to change in the direction of the flow. If we plot the radial profile of the dimensionless temperature $(T - T_c)/(T_s - T_c)$, where T_c is the center-line temperature, the profile changes in shape in the region AC but does not change from section C onwards (although the value of T_c changes). This region downstream of C, where the profile of the dimensionless temperature does not change, is known as the *thermally fully developed region*. The region where the profile is changing (AC in Figure 4.2.1b) is the *thermally developing region* or the *thermal entrance region*. In region AC the velocity profile is fully developed and the temperature profile is developing. Downstream of section C both the velocity and temperature profiles are fully developed. In most cases, we will consider heat transfer only in the region of fully developed velocity and temperature profiles. (Depending on where the heating of the fluid starts, we may have developing temperature profile with fully developed velocity profile or developing velocity and developing temperature profiles.) When both temperature and velocity profiles are developing, they may not develop at the same rate and we may have different entrance lengths for velocity and temperature in such cases.

Entrance length for circular pipes

The entrance length depends on the Reynolds number. An estimate of the length required to yield a fully developed velocity profile and a fully developed temperature profile can be obtained from the value of the Reynolds number and Prandtl number. For laminar flows, from the results presented by Shah and Bhatti (1987), the hydrodynamic entrance length is given by

$$\frac{L_e}{d} \approx 0.0565 \, \mathrm{Re}_d$$

The thermal entrance length for simultaneously developing velocity and temperature profiles depends on whether the tube surface is at a uniform temperature or the surface is subjected to a uniform heat flux.

$$\frac{L_{e,th}}{d} \approx 0.037 \, \mathrm{Re}_d \, \mathrm{Pr} \qquad \text{Uniform surface temperature}$$

$$\frac{L_{e,th}}{d} \approx 0.053 \, \mathrm{Re}_d \, \mathrm{Pr} \qquad \text{Uniform heat flux}$$

where

$$
\begin{aligned}
L_e &= \text{hydrodynamic entrance length} \\
\mathrm{Re}_d &= \text{Reynolds number} = \rho V \, d/\mu \\
\mathrm{Pr} &= \text{Prandtl number} = \nu/\alpha \\
L_{e,th} &= \text{thermal entrance length} \\
V &= \text{mean velocity of the fluid} \\
d &= \text{diameter of the tube}
\end{aligned}
$$

For turbulent flows Bhatti and Shah (1987) recommend that the hydrodynamic entrance length be computed from the formula of Zhi-qing (1982):

$$L_e/d = 1.359 \, \mathrm{Re}_d^{1/4}$$

In most engineering applications Re_d is less than 250 000 and $L_e < 30d$.

Variation of heat transfer coefficient in the entrance region

For simultaneously developing velocity and temperature profiles, calculated values of the local Nusselt number ($\mathrm{Nu}_d = h_x \, d/k$) are shown in Figure 4.2.2. $\mathrm{Nu}_{x,T}$ represents the local Nusselt number for uniform surface temperature and $\mathrm{Nu}_{x,H}$ represents the local Nusselt number for uniform surface heat flux. The figure shows the significantly higher heat transfer coefficients near the entrance to the tubes. For

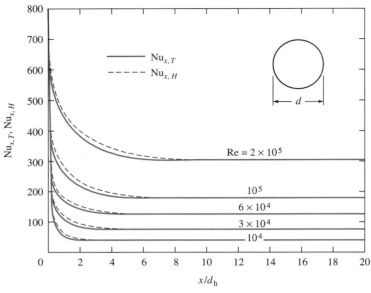

FIGURE 4.2.2 Variation of Local Nusselt Number in the Developing Velocity and Temperature Profile Region.
Nusselt numbers for both uniform surface temperature ($Nu_{x,T}$) and uniform surface heat flux ($Nu_{x,H}$) are shown for Pr = 0.73. [After Deissler (1953).] Reproduced with permission from *Handbook of Single Phase Convective Heat Transfer,* by Kakac, S., Shah, R. K., and Aung, W., eds. (New York: Wiley Interscience, 1987).

Re_d = 10 000, the Nusselt number reaches a constant value (fully developed Nu) for $x/d \approx 6$. For $Re_d = 2 \times 10^5$, x/d for fully developed Nusselt number is less than 10. Hence, we may assume fully developed velocity and temperature profiles for $x/d >$ 10. Although Figure 4.2.2 is for a fluid with Pr = 0.73, in turbulent flows the mechanisms for transfer of momentum and energy are nearly independent of the Prandtl number and we may expect the values in Figure 4.2.2 for the entrance length to be approximately valid for fluids with higher Prandtl numbers. Hereafter, for turbulent flows, we will assume fully developed conditions to exist if $x/d >$ 10. The criterion for estimating the entrance length for turbulent flows in circular tubes can also be used for noncircular ducts by replacing the diameter by the hydraulic mean diameter, d_h. The *hydraulic mean diameter* is defined as the ratio of 4 times the area of cross section to the wetted perimeter of the duct:

$$d_h = 4A_c/P$$

where A_c is the area of cross section of the flow perpendicular to the direction of flow and P is the wetted perimeter, i.e., the length of the line common to the solid surface and the fluid in the cross section.

From Figure 4.2.2, it is clear that, in turbulent flows, the Nusselt numbers for uniform surface temperature and uniform heat flux are equal in the fully developed region. Furthermore, in the developing region, for uniform surface heat flux, the Nusselt numbers and, hence, the heat transfer coefficients are only slightly higher than those for uniform surface temperature. Therefore, *all correlations for turbulent flows may be used for both uniform surface temperature and uniform heat flux conditions.* Recall that the flow in tubes may be either laminar or turbulent with a critical Reynolds number of 2100, i.e., if $Re_d < 2100$, the flow is laminar.

Similar to convective heat transfer coefficient for flows over an external surface, the local convective heat transfer coefficient in internal flows is defined as

$$q''_{zw} = h_z(T_s - T_{ref})$$

where

q''_{zw} = local heat flux from the surface to the fluid
T_s = local surface (wall) temperature
T_{ref} = local fluid reference temperature
h_z = local convective heat transfer coefficient
z = axial coordinate

Reference temperature for heat transfer coefficient in internal flows

For external flows, the reference temperature of the fluid is the free stream temperature (T_∞), which is constant. But in internal flows, there is no such constant temperature of the fluid as the temperature of the fluid varies in the direction of flow due to heat transfer.

The temperature of the fluid varies not only in the direction of flow but also across a cross section perpendicular to the direction of flow. The choice of a reference temperature in the definition of the heat transfer coefficient is arbitrary. Referring to Figure 4.2.1b, we could use the local center-line temperature or the arithmetic mean

Bulk temperature

or some combination of the center-line and surface temperatures. The accepted reference temperature for computing the convective heat transfer coefficient is the fluid bulk temperature, also known as the mixing cup or mean temperature. It is defined as

$$T_b = \frac{\int_{A_c} \rho u c_p T \, dA_c}{\int_{A_c} \rho u c_p \, dA_c}$$

where u is the axial velocity of the fluid and A_c is the cross-sectional area perpendicular to the flow direction. The integration is performed over the cross section where the reference temperature is to be determined. The numerator gives the rate of enthalpy flow across the cross section and T_b represents the uniform temperature that the fluid should have at that location to give the same enthalpy flow rate.[3] The specific heat is assumed to be uniform over the cross section. With this assumption,

$$T_b = \frac{\int_{A_c} \rho u T \, dA_c}{\int_{A_c} \rho u \, dA_c}$$

---

[3]Some people define the reference temperature on the basis of internal energy flow rate instead of the enthalpy flow rate. With the assumption of constant specific heat over the cross section the two definitions give the same value for the mean temperature.

The denominator represents the mass flow rate of the fluid across the cross section. If the fluid at a section of the pipe is collected, the temperature that it would attain after thorough mixing (without any heat transfer) is the bulk temperature; the reason for naming it the mixing cup temperature is obvious.

To illustrate the concept of bulk temperature, consider the fully developed (both hydrodynamically and thermally) laminar flow of a fluid in a circular tube as shown in Figure 4.2.1b. The velocity and temperature distributions are functions of the radius at any given cross section. In laminar flows with uniform surface heat flux, in the hydrodynamically and thermally fully developed region, the velocity and temperature profiles are given by (see Example 12.4.1)

$$u = u_c \left(1 - \frac{r^2}{R^2} \right)$$

$$T = T_c + (T_s - T_c) \frac{4}{3} \frac{r^2}{R^2} \left(1 - \frac{r^2}{4R^2} \right)$$

where

u = axial velocity of the fluid
u_c = center-line velocity
T = temperature of the fluid
T_c = center-line temperature
T_s = surface temperature
R = radius of tube
r = radial coordinate

For a circular tube $dA_c = 2\pi r\, dr$. With the further assumption that the density of the fluid is uniform over the cross section, the bulk temperature is given by

$$T_b = \frac{\int_0^R uT\, 2\pi r\, dr}{\int_0^R u\, 2\pi r\, dr} = \frac{\int_0^R uTr\, dr}{\int_0^R ur\, dr}$$

Substituting the profiles for u and T and recognizing that u_c, T_c, and T_s are not functions of the radius, we obtain

$$T_b = T_c + \frac{7}{18} (T_s - T_c)$$

When we speak of the *temperature of a fluid flowing in a tube*, we generally refer to its bulk temperature.

Correlation for laminar flow in circular pipes— entrance region

Circular Pipes As in external flows, the correlations for the convective heat transfer coefficients in internal flows are also given in terms of the Nusselt, Reynolds, and Prandtl numbers. In laminar flows, two cases are considered—uniform surface temperature and uniform heat flux.[4] In the entrance region for laminar flows, for uniform

[4]In most cases, the condition of uniform heat flux exists on the outer surface of the tube if it is electrically heated. The same condition may not exist on the inner surface of the tube due to the effect of conduction in the axial direction in the material of the tube. Such axial conduction is generally neglected and it is assumed that a uniform heat flux condition exists on the inner surface of the tube.

surface temperature, the average heat transfer coefficient in the tube can be obtained from the correlations suggested by Sieder and Tate (1936):

$$\begin{Vmatrix} 0.48 < \text{Pr} < 16\,700,\ 0.0044 < (\mu/\mu_s) < 9.75,\ \text{uniform } T_s, \\ L/d < 8\ (\mu_s/\mu)^{0.42}/(\text{Re}_d\,\text{Pr}),\ \text{and properties at } (T_i + T_e)/2, \\ \text{except } \mu_s \text{ at } T_s \\ \overline{\text{Nu}}_d = 1.86\left(\dfrac{\text{Re}_d\,\text{Pr}}{L/d}\right)^{1/3}\left(\dfrac{\mu}{\mu_s}\right)^{0.14} \end{Vmatrix} \qquad \textbf{(4.2.1)}$$

The overbar on the Nusselt number indicates that it is formed with the average heat transfer coefficient over the length of the tube. The fluid temperature for determining the total convective heat transfer rate is the arithmetic mean of the inlet and exit temperatures.

$$q = h\,\pi dL\left(T_s - \frac{T_i + T_e}{2}\right)$$

Heat transfer coefficient correlations for fully developed region in internal, laminar and turbulent flows

For $L/d > 8\ (\mu_s/\mu)^{0.42}/(\text{Re}_d\,\text{Pr})$, the fully developed region predominates, and the correlations for fully developed flow are applicable.

Fully developed laminar flows:

$$\begin{Vmatrix} \text{Evaluate properties at the bulk temperature,} \\[4pt] \text{Uniform surface temperature} \quad \text{Nu}_d = 3.66 \\[4pt] \text{Uniform surface heat flux} \quad \text{Nu}_d = 4.36 \end{Vmatrix}$$

(4.2.2a) for uniform surface temperature, (4.2.2b) for uniform surface heat flux.

Fully developed turbulent flows (smooth circular pipes): Uniform wall temperature and uniform surface heat flux,

$$\text{Nu}_d = 0.023\ \text{Re}_d^{4/5}\,\text{Pr}^{1/3} \qquad \textbf{(4.2.3a)}$$

The Dittus-Boelter (1930) relation gives slightly better results:

$$\begin{Vmatrix} \text{Fluid properties evaluated at the bulk temperature, } T_b \\ 0.7 \leq \text{Pr} \leq 160,\ \text{Re}_d \geq 10\,000,\ L/d \geq 10 \\ n = 0.4 \text{ for heating } (T_s > T_b) \\ n = 0.3 \text{ for cooling } (T_s < T_b) \\ \text{Nu}_d = 0.023\ \text{Re}_d^{4/5}\,\text{Pr}^{n} \end{Vmatrix} \qquad \textbf{(4.2.3b)}$$

For large variations in viscosity, Sieder and Tate (1936) suggest

$$\begin{Vmatrix} \text{Fluid properties at the bulk temperature, except } \mu_s \text{ at } T_s \\ 0.7 \leq \text{Pr} \leq 16\,700,\ \text{Re}_d \geq 10\,000,\ L/d \geq 10 \\ \text{Nu}_d = 0.027\ \text{Re}_d^{4/5}\,\text{Pr}^{1/3}\left(\dfrac{\mu}{\mu_s}\right)^{0.14} \end{Vmatrix} \qquad \textbf{(4.2.3c)}$$

Correlation for fully developed and developing turbulent internal flows

Equations 4.2.3a, b, and c have significant uncertainties, as high as 20%. For internal turbulent flows the correlations suggested by Gnielinsky (1976, 1990) have much better accuracy, with uncertainties of around 6%. The recommended correlations are

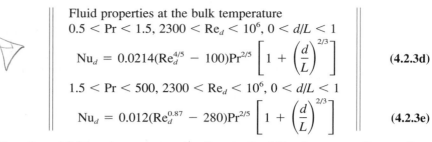

Fluid properties at the bulk temperature
$0.5 < \text{Pr} < 1.5, 2300 < \text{Re}_d < 10^6, 0 < d/L < 1$

$$\text{Nu}_d = 0.0214(\text{Re}_d^{4/5} - 100)\text{Pr}^{2/5}\left[1 + \left(\frac{d}{L}\right)^{2/3}\right] \qquad (4.2.3d)$$

$1.5 < \text{Pr} < 500, 2300 < \text{Re}_d < 10^6, 0 < d/L < 1$

$$\text{Nu}_d = 0.012(\text{Re}_d^{0.87} - 280)\text{Pr}^{2/5}\left[1 + \left(\frac{d}{L}\right)^{2/3}\right] \qquad (4.2.3e)$$

Equations 4.2.3d and e are approximations to the following composite equation, also by Gnielinsky (1976, 1990):

$0 < d/L < 1, 0.6 < \text{Pr} < 2000, 2300 < \text{Re}_d$

$$\text{Nu}_d = \frac{(f/8)(\text{Re}_d - 1000)\text{Pr}}{1 + 12.7(f/8)^{1/2}(\text{Pr}^{2/3} - 1)}[1 + (d/L)^{2/3}] \qquad (4.2.3f)$$

$$f = (0.79 \ln \text{Re}_d - 1.64)^{-2} \qquad (4.2.3g)$$

To take into account the property variation of fluids due to temperature, Gnielinsky (1990) suggests that the Nusselt number in Equation 4.2.3d, e, and f be multiplied by

$$\left(\frac{T_b}{T_s}\right)^{0.45} \quad \text{for gases and} \quad \left(\frac{\text{Pr}}{\text{Pr}_s}\right)^{0.11} \quad \text{for liquids}$$

The absolute temperature and Prandtl number with subscript s are evaluated at the surface temperature.

Equations 4.2.3d, e, and f can be used to account for the higher heat transfer coefficient in the developing region; in such cases the correlation yields the average heat transfer coefficient over the length of the tube. Properties are to be evaluated at the mean of the inlet and exit temperatures. To use the equations in the fully developed region set d/L to zero. If the equations are used for circular tubes, d is the diameter of the tube. For noncircular tubes, both the Nusselt number and Reynolds number are based on, d, the hydraulic mean diameter.

The foregoing correlations are for uniform surface temperature of the tube, but they can also be used for uniform heat flux in turbulent flows. For more accurate results, based on experimental data with uniform heat flux, Petukhov (1970) recommends

Properties at the bulk temperature, except μ_s
at the surface temperature
$n = 0.11$ for liquids, heating
$n = 0.25$ for liquids, cooling
$n = 0$ for gases
$0.5 < \text{Pr} < 200$ 6% uncertainty
$200 < \text{Pr} < 2000$ 10% uncertainty
$10^4 < \text{Re}_d < 5 \times 10^6, 0.08 < (\mu/\mu_s) < 40$

$$\text{Nu}_d = \frac{(f/8)\text{Re}_d \text{Pr}}{1.07 + 12.7(f/8)^{1/2}(\text{Pr}^{2/3} - 1)}\left(\frac{\mu}{\mu_s}\right)^n \qquad (4.2.4a)$$

where f is the friction factor from Equation 4.2.3g.

Petukhov's correlation, Equation 4.2.4a agrees with the most reliable experimental data on heat and mass transfer. Gnielinsky's correlations, Equations 4.2.3d and e predict Nusselt numbers that are close to Equation 4.2.4a over a wide range of variables and are easier to use.

The correlations presented so far do not apply to liquid metals, which have very low Prandtl numbers. For liquid metals (Pr < 0.1) Sleicher and Rouse (1975) recommend

$$\text{Uniform surface temperature: Pr} \ll 1$$
$$\text{Nu}_{d,b} = 4.8 + 0.0156\ \text{Re}_{d,f}^{0.85}\ \text{Pr}_s^{0.93} \tag{4.2.4b}$$
$$\text{Uniform heat flux: Pr} \ll 1$$
$$\text{Nu}_{d,b} = 6.3 + 0.0167\ \text{Re}_{d,f}^{0.85}\ \text{Pr}_s^{0.93} \tag{4.2.4c}$$

The subscripts b, f, and s denote that properties are to be evaluated at the bulk temperature, film temperature (arithmetic mean temperature), and surface temperature, respectively.

EXAMPLE 4.2.1

Steam is generated on the surface of tubes (surrounded by water) with pressurized water flowing inside the tubes of a heat exchanger. At a particular section, the velocity of the water in the tubes is 3 m/s. The inside diameter of the tubes is 25 mm and the tube surfaces are at 250 °C. Find the convective heat transfer coefficient by different correlations at a section where the bulk temperature of the pressurized water is 280 °C. This section is 2.5 m from the entrance of the water to the tube (Figure 4.2.3).

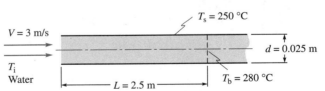

FIGURE 4.2.3 Pressurized water enters a tube with a uniform velocity and temperature. The tube surface temperature is 250 °C and that of the pressurized water at a section 2.5 m from the entrance to the tube is 280 °C.

Given

$V = 3$ m/s $T_b = 280$ °C $T_s = 250$ °C
$d = 25$ mm $L = 2.5$ m

Find

h (from different correlations)

ASSUMPTION

Steady state

SOLUTION

The correlations for the convective heat transfer coefficient depend on the Reynolds number and Prandtl number. From Table A5, properties of water at the bulk temperature of 280 °C are

$\rho = 750.7 \text{ kg/m}^3$ $c_p = 5234 \text{ J/kg K}$ $k = 0.574 \text{ W/m K}$
$\mu = 96.0 \times 10^{-6} \text{ Ns/m}^2$ $\nu = 12.79 \times 10^{-8} \text{ m}^2/\text{s}$
$\text{Pr} = 0.88$ $\text{Pr}_s = 0.85$

Thus,

$$\text{Re}_d = \frac{V d}{\nu} = \frac{3 \times 0.025}{12.79 \times 10^{-8}} = 5.864 \times 10^5$$

At a distance of 2.5 m from the entrance to the tube, $L/d = 2.5/0.025 = 100$. As the flow is turbulent and $L/d > 10$, fully developed conditions exist. From Equation 4.2.3a,

$$\text{Nu}_d = 0.023 \text{ Re}_d^{4/5} \text{ Pr}^{1/3} = 0.023 \times (5.864 \times 10^5)^{4/5} \times (0.88)^{1/3} = 907.4$$

$$h = \text{Nu}_d \frac{k}{d} = 907.4 \times \frac{0.574}{0.025} = \underline{20\,830 \text{ W/m}^2 \text{ °C}}$$

From Equation 4.2.3b, for cooling of the fluid ($T_s < T_b$), the exponent $n = 0.3$

$$\text{Nu}_d = 0.023 \text{ Re}_d^{0.8} \text{ Pr}^{0.3} = 0.023 \times (5.864 \times 10^5)^{4/5}(0.88)^{0.3} = 911.2$$

$$h = \text{Nu}_d \frac{k}{d} = 911.2 \times \frac{0.574}{0.025} = \underline{20\,920 \text{ W/m}^2 \text{ °C}}$$

From Equation 4.2.3c, where $\mu = 96 \times 10^{-6}$ N s/m^2 and $\mu_s(250 \text{ °C}) = 107.7 \times 10^{-6}$ N s/m^2,

$$\text{Nu}_d = 0.027 \text{ Re}_d^{4/5} \text{ Pr}^{1/3} (\mu/\mu_s)^{0.14}$$

$$= 0.027(5.864 \times 10^5)^{4/5}(0.88)^{1/3} \left(\frac{96}{107.7}\right)^{0.14} = 1048$$

$$h = \text{Nu}_d \frac{k}{d} = 1048 \times \frac{0.574}{0.025} = \underline{24\,070 \text{ W/m}^2 \text{ °C}}$$

From Equation 4.2.3d, and setting $d/L =$ for fully developed conditions,

$$\text{Nu}_d = 0.0214(\text{Re}_d^{0.8} - 100) \text{ Pr}^{0.4} \left(\frac{\text{Pr}}{\text{Pr}_s}\right)^{0.11}$$

$$= 0.0214[(5.864 \times 10^5)^{0.8} - 100)](0.88)^{0.4} \left(\frac{0.88}{0.85}\right)^{0.11} = 838.2$$

$$h = \text{Nu}_d \frac{k}{d} = 838.2 \times \frac{0.574}{0.025} = \underline{19\,246 \text{ W/m}^2 \text{ °C}}$$

COMMENT

The value of h varies from a low of 19 246 W/m² °C to a high of 24 070 W/m² °C. If we take the average of the highest and lowest values, 21 658 W/m² °C, the variation in the value of h is ±11%. Equations 4.2.3a, b, and d predict h that are within about 5% of their mean value. Such variation in the value of h, computed by different correlations, is not unusual.

In finding the solution to a specific problem, more than one correlation may be used to determine the range of the values of the convective heat transfer coefficient. An appropriate value of h is then chosen.

Internal flows— dealing with temperature dependent properties of the fluid

In the computations of the Nusselt number for internal flows, the properties of the fluid are evaluated at its bulk temperature. When a fluid flowing in a tube is heated or cooled, its bulk temperature varies in the direction of flow. Because of the temperature dependence of the properties used in the correlations, the convective heat transfer coefficient varies in the direction of flow. In many cases, we would like to determine an approximate average convective heat transfer coefficient for the entire length of the tube. Such an average convective heat transfer coefficient may be determined in one of two ways.

1. Determine the arithmetic average of the heat transfer coefficient based on the bulk temperature at inlet and exit, i.e., with h_i at T_{bi} and h_e at T_{be},

$$h_m = \frac{h_i + h_e}{2}$$

2. Determine the heat transfer coefficient based on the arithmetic mean of the fluid bulk temperatures at inlet and exit,

$$h_m = h \text{ at } (T_{bi} + T_{be})/2$$

In most cases the difference between the values determined by the two methods will not be great. It may be advantageous to determine the mean value by method 1. If the heat transfer coefficients at inlet and exit are reasonably close to each other, an arithmetic mean of the two values can be used over the entire tube. If the two values are not close to each other, it may be advisable to divide the tube into several smaller sections such that the values of the convective heat transfer coefficient at inlet and exit at each of the sections are close to each other. The mean of the values at inlet and exit (or at the mean temperature) can then be used for each section. In all cases, it is well to remember that the convective heat transfer coefficients, as determined from the correlations, have an uncertainty of ±10%. In some cases the exit bulk temperature may not be readily available to evaluate the mean value of the heat transfer coefficient. In such cases, assume any reasonable value for the exit bulk temperature as a starting point. After obtaining the solution to the problem, verify if the assumed value is appropriate. If it is not, use the newly computed value of the bulk temperature at exit to update the convective heat transfer coefficient, and repeat the computations if necessary.

Noncircular Ducts For turbulent flows in noncircular ducts, the correlations for turbulent flows for circular tubes can be used with the hydraulic mean diameter as the characteristic length for both the Reynolds and Nusselt numbers.

Total Heat Transfer Rate, Uniform Wall Temperature As the bulk temperature is not uniform, the temperature difference to be used in the computation of $q''[= h(T_s - T_b)]$ varies in the direction of flow. We now examine how to use the convective heat transfer coefficient to find the heat transfer rate. To determine the total heat transfer rate in steady flow, we proceed as follows.

Consider an elemental length Δz of a tube with a surface area ΔA_s, given by $A_{s,z+\Delta z} - A_{s,z}$; A_s is the total surface area from the location where the heating starts (Figure 4.2.4). In this elemental length for a single phase fluid,

Finding exit temperature of fluid and heat transfer rate— internal flow, uniform surface temperature

Heat transfer rate to the fluid $= h\,\Delta A_s(T_s - T_b)$

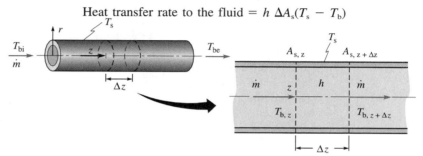

FIGURE 4.2.4 A fluid enters a tube at a temperature of T_{bi} and exits the tube at T_{be}. The surface of the tube is at a uniform temperature of T_s.

Assuming constant specific heats, the heat transfer rate is also related to the change in the bulk temperature of the fluid (see Equation 1.10.19),

$$\text{Heat transfer rate to the fluid} = \dot{m}c_p(T_{b,z+\Delta z} - T_{b,z})$$

Hence,

$$\dot{m}c_p(T_{b,z+\Delta z} - T_{b,z}) = h\,\Delta A_s(T_s - T_b)$$

Dividing by Δz and taking the limit as $\Delta z \to 0$, we obtain

$$\dot{m}c_p \frac{dT_b}{dz} = h\frac{dA_s}{dz}(T_s - T_b)$$

Assuming a suitable average convective heat transfer coefficient over the entire length of the tube, h can be treated as a constant. Separate the variables and integrate from $z = 0$ (inlet to the heated section, $A_s = 0$) to $z = L$ (exit, $A_s = A_{s,t}$, total heating surface area):

$$\int_{T_{bi}}^{T_{be}} \frac{dT_b}{T_s - T_b} = \int_0^{A_{s,t}} \frac{h}{\dot{m}c_p}\, dA_s$$

Performing the integration,

$$\ln \frac{T_s - T_{be}}{T_s - T_{bi}} = -\frac{hA_{s,t}}{\dot{m}c_p} \tag{4.2.5a}$$

or

$$T_s - T_{be} = (T_s - T_{bi}) \exp\left(-\frac{hA_{s,t}}{\dot{m}c_p}\right) \tag{4.2.5b}$$

Equation 4.2.5a (or b) gives the exit temperature. The heat transfer rate is determined from Equation 1.10.19. The total heat transfer rate is obtained from

$$q = \dot{m}c_p(T_{be} - T_{bi})$$

With

$$\dot{m}c_p = \frac{q}{T_{be} - T_{bi}}$$

Equation 4.2.5a can be recast as

$$q = hA_{s,t}\frac{(T_s - T_{be}) - (T_s - T_{bi})}{\ln[(T_s - T_{be})/(T_s - T_{bi})]} \tag{4.2.6}$$

Logarithmic mean temperature difference (LMTD) defined

In Equation 4.2.6, the coefficient of $hA_{s,t}$, $[(T_s - T_{be}) - (T_s - T_{bi})]/\ln[(T_s - T_{be})/(T_s - T_{bi})]$, can be interpreted as an appropriate mean temperature difference to determine the heat transfer rate when the bulk temperature varies. Because of the presence of the logarithmic term, it is termed the logarithmic mean temperature difference (LMTD).

COMMENT

It is tempting to determine the heat transfer rate by employing the arithmetic mean of the temperature differences at inlet and exit through the relation

$$q = hA_{s,t}\frac{(T_s - T_{be}) + (T_s - T_{bi})}{2} \tag{4.2.7}$$

However, except when $(hA_s/(\dot{m}c_p) \ll 1$, Equation 4.2.7 does not give correct results and in some cases it may lead to the violation of the Second Law of Thermodynamics. Equating the heat transfer rate given by $q = \dot{m}c_p(T_{be} - T_{bi})$ to the heat transfer rate given by Equation 4.2.7, and rearranging the result,

$$\frac{hA_{s,t}}{2\dot{m}c_p}[(T_s - T_{be}) + (T_s - T_{bi})] = (T_s - T_{bi}) - (T_s - T_{be})$$

Denoting $(hA_{s,t})/2\dot{m}c_p$ by c, and solving for $T_s - T_{be}$,

$$(T_s - T_{be})(1 + c) = (T_s - T_{bi})(1 - c)$$

It is evident that if $(hA_{s,t})/(2\dot{m}c_p) > 1, T_s - T_{bi}$ and $T_s - T_{be}$ have opposite signs; if T_{bi} is less than T_s, T_{be} is greater than T_s. This violates the Second Law.

As the Second Law of Thermodyanmics is violated by using Equation 4.2.7 if $(hA_s/\dot{m}c_p) > 2$, do not use Equation 4.2.7.

Equations 4.2.5a and b can be used to determine T_{be} and q if A_s (hereafter, we will indicate the total surface area by A_s) is known or to determine A_s when the inlet and exit temperatures are known.

Axial Temperature Profiles with Uniform Surface Temperature and Uniform Surface Heat Flux For the two cases, uniform surface temperature and uniform surface heat flux, the variation of the fluid temperature in the direction of flow is different. For fluids with constant properties these two cases are shown in Figures 4.2.5a and b for heat transfer to the fluid.

With a uniform surface temperature and constant h, the heat transfer rate per unit area (and, hence, per unit length for a constant diameter tube) is initially high as the magnitude of the difference in temperatures of the surface and the fluid is high near the inlet. This results in rapid changes in the bulk temperature of the fluid. As the bulk temperature increases, the magnitude of the difference between the surface temperature and the fluid temperature decreases resulting in a decrease in the heat transfer rate per unit length of the tube; this decrease leads to a decrease in the magnitude of the rate of change of the fluid temperature. The fluid temperature asymptotically reaches the surface temperature as the surface area of the tube increases, i.e., for large values of z.

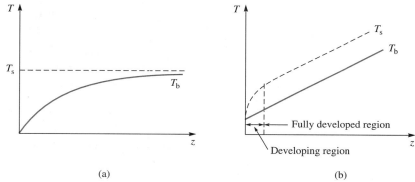

(a)   (b)

FIGURE 4.2.5 Variation of Fluid Temperature in a Tube with Uniform Temperature and Uniform Heat Flux
(a) Fluid temperature variation in a tube with uniform surface temperature. (b) Fluid temperature variation in a tube with uniform surface heat flux.

Variation of fluid
bulk temperature—
internal flows,
uniform heat flux

With uniform heat flux (Figure 4.2.5b) the difference between the bulk temperature and the surface temperature is given by

$$T_s - T_b = \frac{q''}{h}$$

where q'' is the heat flux from the tube surface to the fluid. In the fully developed region, the value of h is constant (except for the variation of the properties of the fluid with temperature). As both q'' and h are constant, $T_s - T_b$ is constant. But, as the value of the convective heat transfer coefficient is not constant in the entrance region, $T_s - T_b$ is not constant. Close to the inlet (see Figure 4.2.2) the value of h is very high and, hence, $T_s - T_b$ is very small. As the value of h approaches that of fully developed flow, $T_s - T_b$ approaches the constant value given by q''/h. The fluid bulk temperature varies linearly because the heat transfer rate per unit length is constant. In the fully developed region, where $T_s - T_b$ is constant, both the tube surface temperature and fluid temperature vary linearly.

In many cases of internal flow the subscript b may not be used to indicate that the temperature is the bulk temeprature; if there is no indication otherwise, the temperature should be interpreted as the bulk temperature.

EXAMPLE 4.2.2

A water heater is fabricated by a resistance wire wound uniformly over a 10-mm diameter and 4-m long tube (Figure 4.2.6). The resistance element maintains a uniform heat flux of 1000 W/m². The mass flow rate of water is 12 kg/h, and its inlet temperature is 10 °C. Estimate the surface temperature of the tube at exit.

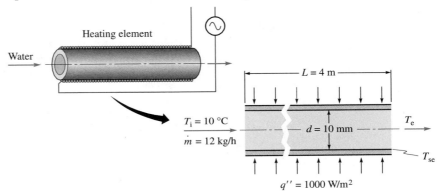

FIGURE 4.2.6 Water at 10 °C centers a 10-mm I.D. tube subjected to a uniform heat flux of 1000 W/m².

Given

Fluid: water $T_{bi} = 10\ °C$
$\dot{m} = 12$ kg/h $d = 10$ mm
$q'' = 1000$ W/m² $L = 4$ m

Find

T_{se} (the exit surface temperature)

ASSUMPTION

Steady state

SOLUTION

With a known heat flux the local surface temperature at the exit of the tube is given by

$$q'' = h_e(T_{se} - T_{be}) \tag{1}$$

where h_e is the local heat transfer coefficient at exit. The exit bulk temperature is found from

$$q = q''\pi\, dL = \dot{m}c_p(T_{bi} - T_{be}). \tag{2}$$

Depending on the conditions at exit (developed or developing profiles) and the Reynolds number at exit, an appropriate correlation for h_e is chosen. To determine the heat transfer coefficient at exit, properties of the fluid are evaluated corresponding to the conditions at exit; in particular, properties of the fluid are evaluated at T_{be} [and not at $(T_{bi} + T_{be})/2$]. To find T_{be}, c_p is evaluated at any reasonable temperature $> T_{bi}$. Having found the value of T_{be}, a more appropriate value of c_p is used to improve the value of T_{be}. Note that the value of c_p is not very sensitive to temperature.

For an initial estimate of T_{be}, evaluate c_p at any reasonable value of the temperature $> 10\,°C$, say, $20\,°C$ as a first approximation: $c_p = 4182$ J/kg K.

$$T_{be} - T_{bi} = \frac{1000 \times \pi \times 0.01 \times 4}{(12/3600) \times 4182} = 9\,°C \qquad T_{be} = 19\,°C$$

The average of the bulk temperatures at inlet and exit is $14.5\,°C$. The value of c_p at $14.5\,°C$ is 4187 J/kg K. If this value of c_p is used, $T_{be} = 19\,°C$. Now, determine the entrance length to check if the flow is fully developed. Evaluate the properties of the fluid at $12\,°C$, as we expect the temperature in the entrance region to be closer to $10\,°C$. From the software (CC), the properties of water at $12\,°C$ are

$\rho = 999.8$ kg/m^3 $k = 0.590$ W/m K
$\mu = 1219 \times 10^{-6}$ N s/m^2 Pr $= 8.67$

$$\text{Re}_d = \frac{\rho V d}{\mu} \qquad \dot{m} = \rho V \frac{\pi d^2}{4}$$

which gives

$$\rho V d = \frac{4\dot{m}}{\pi d} \qquad \text{Re}_d = \frac{4\dot{m}}{\pi\, d\, \mu}$$

$$\text{Re}_d = \frac{4 \times 12/3600}{\pi \times 0.01 \times 1219 \times 10^{-6}} = 348$$

The flow is laminar. For laminar flow in tubes with uniform heat flux, the entrance length is

$$L_{e,th} \cong 0.053\, \text{Re}_d\, \text{Pr}\, d = 0.053 \times 348 \times 8.67 \times 0.01 = 1.6\text{ m}$$

Hence, at exit of the 4-m long tube fully developed conditions exist. For laminar flow with uniform heat flux and for fully developed conditions, from Equation 4.2.2b,

$$\text{Nu}_d = 4.36$$

From the software (CC), thermal conductivity of water at $19\,°C$ is 0.602 W/m °C. At the exit, the local convective heat transfer coefficient is

$$h_e = \text{Nu}_d \frac{k}{d} = 4.36 \times \frac{0.602}{0.01} = 262.5\text{ W/m}^2\,°C$$

$$T_{se} - T_{be} = \frac{q''}{h_e} = \frac{1000}{262.5} = 3.8\,°C$$

$$T_{se} = 19 + 3.8 = \underline{22.8\,°C}$$

EXAMPLE 4.2.3 A water heater consists of a 25-mm diameter tube inside a second, coaxial tube (Figure 4.2.7). Water at 10 °C enters the inner tube at 0.8 kg/s. Condensing steam in the annulus maintains the temperature of the inner tube at 90 °C.

(a) Determine the exit temperature and the heat transfer rate if the tube is 10-m long.

(b) What should be the length of the tube for the exit temperature of the water to be 70 °C?

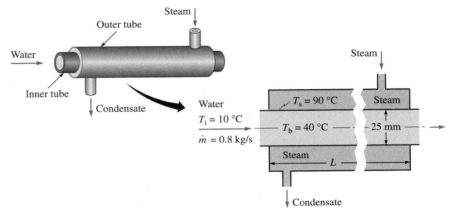

FIGURE 4.2.7 Water enters a tube at 10 °C. The tube surface is maintained at a uniform temperature of 90 °C.

Given

Fluid: Water $\qquad T_{bi} = 10\,°C$
$\dot{m} = 0.8$ kg/s $\qquad T_s = 90\,°C$
$d = 25$ mm

Find

(a) T_e (exit temperature) if $L = 10$ m
(b) L for $T_e = 70\,°C$

ASSUMPTIONS

1. Steady state
2. Heat transfer begins at inlet to the tube.
3. Entrance length is small compared with the total length of the tube, and the correlations for fully developed conditions are applicable. This assumption will be checked after computing the length of the tube.
4. The convective heat transfer coefficient, determined at the mean of the bulk temperatures at inlet and exit, is uniform.

SOLUTION

(a) There are two aspects to this problem: Determining the exit temperature and identifying a suitable temperature for evaluating the properties of the fluid. An exit temperature will be assumed for evaluating the properties to compute the convective heat transfer coefficient. With the computed heat transfer coefficient, the exit temperature, T_{be}, will be determined from Equation 4.2.5b. If the properties of the fluid at $(T_{bi} + T_{be})/2$ are different from those used in the computations, the computations will be repeated using the computed value of the exit temperature.

We begin the computations by assuming the exit temperature to be 70 °C. Mean bulk temperature is $(10 + 70)/2 = 40$ °C. From the software CC (or Table A5), properties of water at 40 °C are

$$\rho = 992.2 \text{ kg/m}^3 \qquad c_p = 4175 \text{ J/kg K} \qquad Pr = 4.19$$
$$\mu = 633.7 \times 10^{-6} \text{ N s/m}^2 \qquad k = 0.631 \text{ W/m K}$$

$$Re_d = \frac{\rho V d}{\mu} \qquad \dot{m} = \rho V \frac{\pi d^2}{4} \qquad \rho V d = \frac{4\dot{m}}{\pi d}$$

$$Re_d = \frac{4\dot{m}}{\pi d \mu} = \frac{4 \times 0.8}{\pi \times 0.025 \times 0.0006337} = 64\ 295$$

The flow is turbulent. For a $Pr = 4.19$, employing Equation 4.2.3b, with $n = 0.4$ as water is heated,

$$Nu_d = 0.023 \times 64\ 295^{0.8} \times 4.19^{0.4} = 286.6$$

$$h = Nu_d \frac{k}{d} = 286.6 \times \frac{0.631}{0.025} = 7234 \text{ W/m}^2 \text{ K}$$

For a uniform surface temperature, from Equation 4.2.5b,

$$T_s - T_{be} = (T_s - T_{bi}) \exp\left[-\frac{hA_s}{(\dot{m}c_p)}\right]$$

$$= (90 - 10) \exp\left(-\frac{7234 \times \pi \times 0.025 \times 10}{0.8 \times 4175}\right) = 14.6 \text{ °C}$$

$$T_{be} = 90 - 14.6 = 75.4 \text{ °C}$$

$$q = \dot{m}c_p(T_{be} - T_{bi}) = 0.8 \times 4175 \times (75.4 - 10) = 218\ 436 \text{ W}$$

The exit temperature of 75.4 °C is slightly higher than the assumed value of 70 °C for determining the convective heat transfer coefficient. An improved estimate of the average convective heat transfer coefficient can be obtained by using an exit temperature of 75.4 °C, i.e., at a mean bulk temperature of 42.7 °C. The new values are

$$h = 7406 \text{ W/m}^2 \text{ °C} \qquad c_p = 4174 \text{ J/kg °C}$$

Employing the new values of h and c_p, we obtain $T_{be} = 76$ °C. This changes the value of the heat transfer rate to 220 387 W. Further iteration is not required.

As the flow is turbulent, the entrance length is less than $10d$ or 0.25 m. For a length of 10 m the entrance length is quite small. The assumption that correlations for fully developed flows are applicable is reasonable. As the convective heat transfer coefficient in the developing region is higher than in the fully developed region, the total heat transfer rate will be slightly greater than the computed value.

$$T_{be} = \underline{76 \text{ °C}} \qquad q = \underline{220\ 387 \text{ W}}$$

If Equation 4.2.3e (setting $d/L = 0$) is used, the convective heat transfer coefficient is 8960 W/m² K, which is 21% higher than the value obtained from Equation 4.2.3b.

(b) The required surface area will be determined from Equation 4.2.5a in three different ways (Figure 4.2.8), on the basis of an average convective heat transfer coefficient computed by

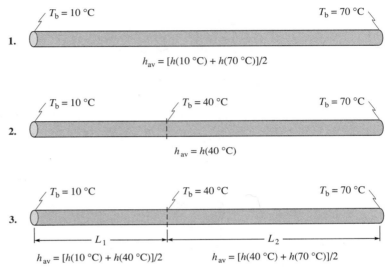

FIGURE 4.2.8 Three Methods for Calculating the Average Heat Transfer Coefficient and Length of Tube

1. Determining the convective heat transfer coefficients corresponding to bulk temperatures of 10 °C and 70 °C and taking the mean of the two values.
2. Determining the convective heat transfer coefficient at the average of the bulk temperatures at inlet and exit, i.e., at 40 °C.
3. Dividing the tube into two sections, the first one where the water is heated from 10 °C to 40 °C and the second one where it is heated from 40 °C to 70 °C. In each section the average of the heat transfer coefficients at inlet and exit will be employed.

The computations are shown in the following table.

Bulk temperature	10 °C	40 °C	70 °C
c_p (J/kg K)	4199	4175	4185
k (W/m K)	0.587	0.631	0.662
μ (N s/m²)	1291×10^{-6}	633.7×10^{-6}	393.3×10^{-6}
Pr	9.24	4.19	2.49
$\text{Re}_d \ [= (4\dot{m})(\pi d \mu)]$	31 562	64 295	103 598
$\text{Nu}_d \ (= 0.023 \ \text{Re}_d^{0.8} \ \text{Pr}^{0.4})$	222.5	286.6	340.5
$h \ [= \text{Nu}_d(k/d) \ (\text{W/m}^2 \ \text{K}]$	5222	7234	9021

1. Taking the average heat transfer coefficient at inlet and exit, $h = (5222 + 9021)/2 = 7122$ W/m² K. The average value of $c_p = (4199 + 4185)/2 = 4192$ J/kg K. From Equation 4.2.5a,

$$A_s = \frac{\dot{m}c_p}{h} \ln\left(\frac{(T_s - T_b)_i}{(T_s - T_b)_e}\right) = \frac{0.8 \times 4192}{7122} \ln\left(\frac{80}{20}\right) = 0.653 \text{ m}^2$$

$$L = \frac{A_s}{\pi d} = \frac{0.653}{\pi \times 0.025} = \underline{8.31 \text{ m}}$$

2. Employing $h = h(40\,°C) = 7234 \text{ W/m}^2 \text{ K}$ and $c_p = 4175 \text{ J/kg.K}$, and repeating the steps in 1 above,

$$A_s = 0.64 \text{ m}^2 \quad \text{and} \quad L = \underline{8.15 \text{ m}}$$

3. Split the tube into two sections with the inlet and exit temperatures in the first section being 10 °C and 40 °C, and those in the second section being 40 °C and 70 °C. In each case the mean of the heat transfer coefficients evaluated at the inlet and exit conditions of the tube will be used.

	Section 1	Section 2
h_i (W/m^2 K)	5222	7234
h_e (W/m^2 K)	7234	9021
$h = (h_i + h_e)/2$ (W/m^2 K)	6228	8128
$c_{p,i}$ (J/kg.K)	4199	4175
$c_{p,e}$ (J/kg K)	4175	4185
$c_{p,av}$ (J/kg K)	4187	4180
A_s (m^2)	0.253	0.377
L (m)	3.22	4.8

Total length $= 3.22 + 4.8 = \underline{8.02 \text{ m}}$

COMMENTS

Several features can be observed from the solutions to this problem.

1. The value of the heat transfer coefficient found from Equation 4.2.3b and Equation 4.2.3e differ by approximately 21%.

2. The average of the convective heat transfer coefficients at inlet (10 °C) and at exit (70 °C) differs by only 1.6% from the value at the mean of these two temperatures (40 °C).

3. The average heat transfer coefficient can also be computed as the arithmetic mean of the values evaluated at different bulk temperatures. For example, in the current example, the average h can be computed as $h_{av} = [h(10\,°C) + h(40\,°C) + h(70\,°C)]/3 = 7159 \text{ W/m}^2 \text{ °C}$. This value is within 1% of $h(40\,°C)$.

4. In method 3, the heat transfer rates $[\dot{m}c_p(T_e - T_i)]$ in section 1 (100.5 kW) and section 2 (100.4 kW) are nearly equal. The average heat transfer coefficient in section 1 (6228 W/m^2 K) is significantly less than that at section 2 (8128 W/m^2 K). Yet, the length of the tube in section 1 is less than that in section 2. This is because the heat flux at any section is related to the product of the heat transfer coefficient and the difference in temperatures of the surface and the fluid at that section. The difference in the temperatures in section 1 is significantly greater than that in a similar location in section 2. Hence, the heat transfer rate per unit length, even with a lower value of the heat transfer coefficient, is greater in section 1 than in section 2. The higher temperature difference in section 1 more than offsets the effect of the lower heat transfer coefficient compared with those in section 2.

EXAMPLE 4.2.4 In the miniaturization of modern computers, one of the limiting factors is the rise in the temperatures of the chips due to internal energy generation in the chips. The reliability of the chips decreases rapidly when their tempertures go above a certain value, typically between 85 °C and 100 °C. Some high capacity components may require high thermal dissipation, which may reach 50 W/cm² and, in some cases, even 200 W/cm².

One of the methods to cool a computer is to mount the circuit boards on a metal plate. An array of such metal plates is attached to two end-plates, which are cooled by a liquid.

Consider a plate 100-cm wide, 50-cm deep, and 6-mm thick on which a circuit board is mounted (Figure 4.2.9). Several such plates are attached to two heavy end plates through each of which four, 8-mm diameter passages are drilled. Cold water (with an additive to suppress the freezing point) flows through the passages and cools the end plates, which, in turn, cool the plates on which the circuit boards are mounted as shown in the figure. Water enters each passage at 0 °C at 0.15 kg/s. The end plates, which are 1.5-m high, are at 30 °C. Estimate the heat transfer rate from the end plates to the cooling water.

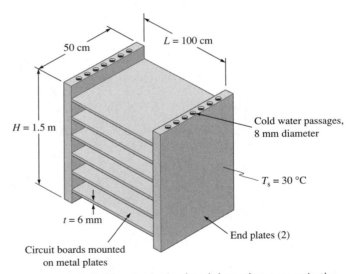

FIGURE 4.2.9 Cold water is circulated through passages in the end plates of a computer to cool the circuit boards mounted on metal plates attached to the end plate.

Given

Fluid: Water $\quad T_i = 0\,°C \quad T_s = 30\,°C$

$\dot{m} = 0.15$ kg/s $\quad H = 1.5$ m

$d = 8$ mm

Find

q (heat transfer rate to the cooling water)

ASSUMPTIONS

1. Steady state
2. The end plates are at a uniform temperature.
3. For the water in the passages, fully developed velocity and temperature profiles exist.
4. Properties of water with the additive are the same as those of water.

SOLUTION

The convective heat transfer rate is computed from Equation 1.10.19.

$$q = \dot{m}c_p(T_e - T_i) \tag{1}$$

The exit temperature, T_e, is computed from Equation 4.2.5b.

$$T_s - T_e = (T_s - T_i) \exp\left(-\frac{hA_s}{\dot{m}c_p}\right) \tag{2}$$

A reasonable average value of h can be evaluated at a mean bulk temperature of $(T_e + T_i)/2$. As T_e is not known, it is determined by iteration.

1. Assume a value of T_e.
2. Compute h at $T_b = (T_i + T_e)/2$.
3. Compute T_e from Equation 2.
4. If the computed value T_e is close to the assumed value of T_e in step 1, the correct value of T_e has been found; otherwise, starting with the computed value of T_e, repeat steps 2 through 4.

Assume $T_e = 10\,°C$; $T_m = (T_i + T_e)/2 = 5\,°C$. From the software (CC), the properties of water at 5 °C are

$\rho = 1000.3 \text{ kg/m}^3$ $c_p = 4207 \text{ J/kg °C}$ $\mu = 1.4996 \times 10^{-3} \text{ N s/m}^2$
$k = 0.578 \text{ W/m °C}$ $Pr = 10.91$ $Pr_s = 5.26$

$$Re_d = \frac{4\dot{m}}{\pi d\mu} = \frac{4 \times 0.15}{\pi \times 0.008 \times 1.4996 \times 10^{-3}} = 15\,920$$

Employing Equation 4.2.3e, with $d/L = 0$,

$$Nu_d = 0.012 \times (15\,920^{0.87} - 280) \times 10.91^{0.4} \left(\frac{10.91}{5.26}\right)^{0.11} = 143.6$$

$$h = 143.6 \times \frac{0.578}{0.008} = 10\,375 \text{ W/m}^2 \text{ °C}$$

From Equation 2,

$$30 - T_e = (30 - 0) \exp\left(-\frac{10\,375 \times \pi \times 0.008 \times 1.5}{0.15 \times 4207}\right)$$

$$T_e = 13.9\,°C$$

The assumed value of $T_e = 10\,°C$ is lower than the computed value of 13.9 °C. Repeating the calculations with $T_e = 13.9\,°C$, $T_m \approx 7\,°C$,

$$Re_d = 16\,924 \qquad Pr = 10.2$$

From Equation 4.2.3e,

$$\text{Nu}_d = 146.8, \qquad h = 10\,671 \text{ W/m}^2 \,^{\circ}\text{C}$$

With $h = 10\,671$ W/m² °C and c_p at 7 °C = 4203 J/kg °C, from Equation 2, $T_e = 14.2$ °C, which is close to the assumed value of 13.9 °C. The correct value of T_e is 14.2 °C.

q = Heat transfer rate from each passage to the cooling water

$$= \dot{m}c_p(T_e - T_i) = 0.15 \times 4203 \times (14.2 - 0) = 8952 \text{ W/passage}$$

Total heat transfer
rate to the water   = 8952 × 8 = 71 616 W
in 8 passages

$$\approx \underline{72 \text{ kW}}$$

COMMENT

As the flow is turbulent, the entrance length is less than 10d or 0.08 m. As the total length of the passages is 1.5 m, the effect of the higher heat transfer coefficient in the entrance region is to increase the heat transfer rate slightly. The calculated value of q is on the conservative side.

4.3 NATURAL CONVECTION

In Section 4.0.1, we explained the difference between forced and natural convection. In natural convection the motion of the fluid adjacent to a surface is caused by the difference in temperature between the surface and the fluid whereas in forced convection an external force (pump or blower) sets up the motion. In forced convection the motion of the fluid is not dependent on the difference between the temperatures of the bounding solid surface and the fluid, and a characteristic velocity is available; in natural convection a characteristic velocity is not readily available. Examples of natural convection heat transfer are the cooling of coffee in a covered cup, the heat transfer from baseboard heaters, the cooling of electronic components in computers without a cooling fan, and the heat transfer from the human body when a person is at rest. The term free convection is also used, but quite often its use is restricted to denoting natural convection in an unbounded fluid.

Fluid motion caused by temperature differences

One of the necessary conditions for the motion of a fluid due to temperature differences only is the existence of a body force; without a body force there is no natural convection. To understand why convective motion occurs as a result of temperature differences, consider a fluid adjacent to a vertical surface as shown in Figure 4.3.1a. The initial uniform temperature of the surface and the fluid is T_{∞}. Consider two fluid elements AB and CD of the same dimensions. AB is adjacent to the solid surface but CD is horizontally displaced and far away from AB. The relationship between the pressures at the various points are given by

$$p_A = p_C \qquad p_B = p_D$$

$$p_D - p_C = \rho_{\infty}gH \qquad p_B - p_A = \rho gh$$

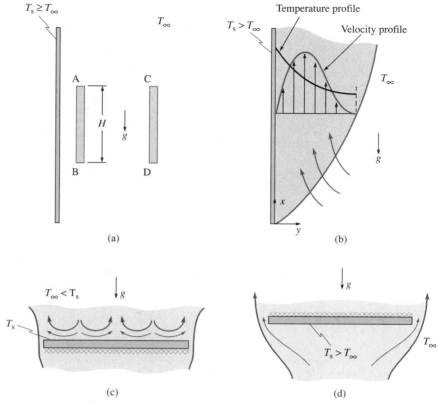

FIGURE 4.3.1 Convective Currents Adjacent to Vertical and Horizontal Plates

(a) A fluid and an adjacent vertical surface are at a uniform temperature. (b) The temperature of the vertical surface is raised and convective currents are set up. (c) Heated horizontal surface is facing up. (d) Heated horizontal surface is facing down.

where

ρ_{∞} = density of the fluid at T_{∞}
ρ = density of the fluid column AB
H = height of the elements AB and CD

Initially $\rho = \rho_{\infty}$. As the fluid column AB is in equilibrium, from a force balance on the fluid column, we have

$$\rho g A_c H = (p_B - p_A)A_c = \rho_{\infty} g H A_c$$

where A_c is the cross-sectional area of the fluid column AB.

Now let the temperature of the solid surface be raised to T_s. The temperature of the fluid column AB increases as a result of the heat transfer from the heated surface. The densities of fluids generally decrease with an increase in the temperature and, therefore, the weight of the fluid column AB is less than the original weight of the fluid column. But the upward force due to the pressure at B and A remains at

$\rho_\infty gHA_c$, which is greater than the weight of the fluid column. There is, therefore, a net upward force on the column AB and the fluid moves up. The cooler fluid flows into the region originally occupied by AB. Thus, convective currents are set up. Recall from fluid mechanics that when the upward force due to the pressure is greater than the weight of the substance in the fluid, the net resultant upward force is termed *buoyancy force*. If the surface temperature is less than the fluid temperature, the fluid adjacent to the surface becomes heavier and downward currents are set up.

The fluid immediately adjacent to the solid surface is at rest (no slip condition). The velocity of the fluid increases as we go farther away from the surface. The motion of the fluid is resisted by viscous forces. As we move away from the surface, the temperature of the fluid approaches T_∞ and the buoyancy force decreases; the velocity of the fluid decreases. Thus, the velocity of the fluid increases with y, reaches a maximum and gradually decreases. A typical velocity profile is shown in Figure 4.3.1b.

Differences in convection currents for vertical and horizontal surfaces

The natural convection currents associated with heated horizontal surfaces are different from those next to a vertical surface. Consider a horizontal heated surface at a uniform temperature T_s, which is greater than the ambient fluid temperature T_∞, shown in Figure 4.3.1c. The heated surface is facing up and the gravitational force is perpendicular to the surface (the gravitational force in the case of natural convection in vertical plates is parallel to the surface). The heated, lighter fluid adjacent to the surface, tends to rise but its motion is hindered by the heavier, cooler fluid above it. The situation is unstable (''top heavy''). A slight disturbance, such as imperceptible vibrations, will disturb the equilibrium and convection currents are set up. The heated, lighter fluid moves up and the cooler, heavier fluid moves down to occupy the space vacated by the heated fluid. The motion of the fluid is influenced by such factors as minor inclinations of the horizontal surface, end effects, and so forth. The flow pattern of the fluid if the heated face is facing down is shown in Figure 4.3.1d. In this case, the lighter, heated fluid moves towards the edges and then rises. The flow patterns with a cold horizontal surface facing up is similar to that of a heated plate facing down, and the flow with a horizontal, cold surface facing down is similar to that of a heated surface facing up.

In natural convection, the forces on the fluid are much smaller, the velocities are usually much lower, and the heat transfer rates much lower than in forced convection. But in many cases it accounts for a major part of the heat transfer from a solid surface. In most cases of forced convection, unless the temperature levels are high, the radiative heat transfer from a surface is much smaller than that by convection and can be neglected. Also, with natural convection heat transfer to liquids, radiative heat transfer can be neglected. However, for common surfaces (excluding highly polished surfaces) radiative heat transfer from a surface to the surroundings is generally of the same magnitude as heat transfer by natural convection to an adjacent gas. To find the total heat transfer rate from a surface adjacent to a stationary gas, both natural convection and radiation should be considered.

The photographs in Figure 4.3.2 shows the isotherms in natural convection adjacent to a heated vertical plate. The flow near the leading edge is steady, laminar. At some height above the bottom edge of the plate, the flow becomes unstable and waves appear in much the same way as the waves in internal flows in Reynolds experiments (Section 4.0).

FIGURE 4.3.2 Photograph Showing Isotherms in Natural Convection in Vertical Plates
Photograph courtesy of E.R.G. Eckert and E. Soehngen.

4.3.1 Natural Convection Correlations

As in forced convection, the correlations relating the convective heat transfer coefficient to other parameters are expressed in dimensionless form. Consider the heat transfer from a vertical plate of height L maintained at a temperature T_s and surrounded by a quiescent fluid at T_∞. From the discussion on the cause and nature of flow in natural convection, it is evident that the flow and, therefore, the average convective heat transfer coefficient in natural convection are related to the following variables:

Variables affecting motion and heat transfer rate in natural convection

- Gravitational acceleration g: Without a body force there is no natural convection.
- Rate of change of density of the fluid with respect to temperature, $\partial\rho/\partial T$, at constant pressure: For a given temperature difference $\Delta T = T_s - T_\infty$, a fluid that has a greater value of $|\partial\rho/\partial T|$ experiences a greater buoyancy force and has higher velocities.
- Temperature difference ΔT: A higher temperature difference results in greater density changes and greater buoyancy forces.
- Viscosity of the fluid, μ, which results in a decelerating force
- Specific enthalpy. Due to heat transfer, the enthalpy of the fluid changes. The enthalpy is a function of the density, ρ, of the fluid and its constant pressure specific heat, c_p.
- Thermal conductivity of the fluid, k
- Height L of the plate

Thus, denoting the average heat transfer coefficient over the length L by h_L',

$$h_L = h_L\left(g, \frac{\partial\rho}{\partial T}, \Delta T, L, \rho, \mu, k, c_p\right).$$

From dimensional analysis, the functional relationship can be expressed in terms of three dimensionless variables.

$$\frac{h_L L}{k} = f\left(\frac{g\,\beta\rho^2\,\Delta T L^3}{\mu^2}, \frac{c_p\mu}{k}\right)$$

where

$\Delta T = |T_s - T_\infty|$
β = volumetric thermal expansion coefficient $= 1/v(\partial v/\partial T)|_p = -(1/\rho)(\partial\rho/\partial T)|_p$
v = specific volume

or

$$\mathrm{Nu}_L = f(\mathrm{Gr}_L, \mathrm{Pr})$$

$$\mathrm{Nu}_L = \frac{h_L L}{k} \qquad\qquad \text{Nusselt number}$$

$$\mathrm{Gr}_L = \frac{g\,\beta\rho^2\,\Delta T L^3}{\mu^2} \qquad\qquad \text{Grashof number}$$

$$\mathrm{Pr} = \frac{c_p\mu}{k} = \frac{\mu/\rho}{k/(\rho c_p)} = \frac{\nu}{\alpha} \qquad \text{Prandtl number}$$

For an ideal gas, $pv = RT$, and it is easily shown that $\beta = 1/T$, T in absolute temperature.

The Grashof number represents the ratio of the buoyancy force to the viscous force in the fluid, just as the Reynolds number represents the ratio of the inertia force to the viscous force. The Grashof number plays a role in natural convection similar to the Reynolds number in forced convection.

If natural convection heat transfer occurs from a surface other than a vertical surface, the height L is replaced by an appropriate characteristic length to form the Nusselt and Grashof numbers. For example, in the case of a sphere or a horizontal cylinder, the characteristic length is the diameter.

In natural convection, many correlations contain the product of the Grashof number and the Prandtl number; the product of the two is the Rayleigh number. The correlations are then expressed in the form

$$\mathrm{Nu} = f(\mathrm{Ra}, \mathrm{Pr})$$

$$\mathrm{Ra} = \mathrm{Gr}\,\mathrm{Pr} \qquad \text{Rayleigh number}$$

As the Rayleigh number is a product of two dimensionless numbers, it is also dimensionless. Correlations to determine the average convective heat transfer coefficient for some common geometries are given in this section. A few more correlations for different geometries are given in Chapter 5. Except for correlations for vertical plates, most correlations of engineering significance are for the average heat transfer coefficient.

Vertical Plates, Uniform Wall Temperature Velocity and temperature profiles for natural convection on a heated vertical plate are shown in Figure 4.3.3. Analytical solutions, leading to an ordinary differential equation, and numerical solutions for

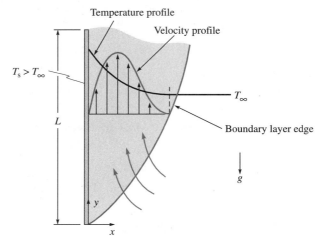

FIGURE 4.3.3 Velocity and Temperature Profiles for a Vertical Plate
A vertical plate is heated. Natural convection is set up with the velocity = 0 at the plate and at the edge of the boundary layer. The velocity reaches a maximum inside the boundary layer. The temperature of the fluid decreases from T_s at the wall to T_∞ at the edge of the boundary layer.

natural convection on vertical plates have been developed for laminar and turbulent flows. An approximate solution for laminar convection is presented in Chapter 5. But most of the engineering correlations, derived from experimental measurements, are more reliable than those obtained by analysis, as the simplifying assumptions made in analysis are not completely satisfied. Churchill and Chu (1975) recommend the following correlations:

Evaluate β at T_∞ and all other properties at $T_f = (T_s + T_\infty)/2$
L = characteristic length = height of the plate

$$\mathrm{Nu}_L = 0.68 + \frac{0.67\,\mathrm{Ra}_L^{1/4}}{[1 + (0.492/\mathrm{Pr})^{9/16}]^{4/9}} \qquad 10^{-1} < \mathrm{Ra}_L < 10^9 \qquad \textbf{(4.3.1a)}$$

$$\mathrm{Nu}_L = \left\{ 0.825 + \frac{0.387\,\mathrm{Ra}_L^{1/6}}{[1 + (0.492/\mathrm{Pr})^{9/16}]^{8/27}} \right\}^2 \qquad 10^{-1} < \mathrm{Ra}_L < 10^{12} \qquad \textbf{(4.3.1b)}$$

The lower limit of 10^{-1} for Ra_L is based on the experimental data used by Churchill and Chu. But they have concluded that the equations can be used for much lower values of Ra_L. Equation 4.3.1b may be used for all values of Ra_L, but for $\mathrm{Ra}_L < 10^9$, Equation 4.3.1a gives better results.

Natural convection correlation—vertical surface, uniform heat flux

Vertical Plates, Uniform Heat Flux The difference between the correlations for uniform wall temperature and uniform heat flux is small. Churchill and Chu (1975) recommend that Equations 4.3.1a and b be used by replacing the constant 0.492 in the denominator by 0.437 for a uniformly heated plate. The temperature at the midheight of the plate, $T_{L/2}$, is used to evaluate the film temperature for determining the properties of the fluid, and $T_{L/2} - T_\infty$ is used to form the Raleigh number. As $T_{L/2}$ is not known, it is determined by iteration so that the value of the convective heat transfer coefficient satisfies the relation

$$q'' = h_{y=L/2}(T_{L/2} - T_\infty)$$

Evaluate β at T_∞ and all other properties at $(T_{y=L/2} + T_\infty)/2$

$$Nu_L = 0.68 + \frac{0.67 \, Ra_L^{1/4}}{[1 + (0.437/Pr)^{9/16}]^{4/9}} \qquad\qquad Ra_L < 10^9 \qquad\qquad \textbf{(4.3.2a)}$$

$$Nu_L = \left\{0.825 + \frac{0.387 \, Ra_L^{1/6}}{[1 + (0.437/Pr)^{9/16}]^{8/27}}\right\}^2 \qquad 10^{-1} < Ra_L < 10^{12} \quad \textbf{(4.3.2b)}$$

When a known heat flux is imposed on the plate, the heat transfer rate is already known. In such cases, our interest is to determine the surface temperature. As the surface temperature varies along the plate, it is more useful to determine the local temperature. The local temperature can be determined if the heat flux and the convective heat transfer coefficient at that location are known. To determine the local heat transfer coefficient from the known value of the uniform heat flux, a modified Grashoff number is defined as

$$Gr_x^* = Gr_x \, Nu_x = \left(\frac{g \, \beta \rho^2 (T_s - T_\infty)x^3}{\mu^2}\right)\left(\frac{h_x x}{k}\right) = \frac{g \, \beta \rho^2 q_w'' x^4}{\mu^2 k}$$

where

T_s = local surface temperature
q_w'' (local heat flux) = $h_x(T_s - T_\infty)$

Based on the works of Sparrow and Gregg (1956) and Vliet and Liu (1969), the following correlations between the local Nusselt number and the modified Grashoff number are suggested.

Properties at the film temperature $T_f = (T_s + T_\infty)/2$
x = coordinate from the leading edge along the plate

$$Nu_x = 0.6(Gr_x^* \, Pr)^{0.2} \qquad\qquad 10^5 < Gr_x^* \, Pr < 10^{13} \qquad\qquad \textbf{(4.3.2c)}$$

$$Nu_x = 0.568(Gr_x^* \, Pr)^{0.22} \qquad\qquad 10^{13} < Gr_x^* \, Pr < 10^{16} \qquad\qquad \textbf{(4.3.2d)}$$

There is considerable uncertainty in the value of h computed from Equation 4.3.2d in the range $10^{13} < Ra_x^* < 10^{14}$ due to the transition of the boundary layer from laminar to turbulent flow.

The local wall temperature is not known *a priori* and it has to be determined by iteration. To find the local temperature at a given location, a wall temperature is assumed for determining the properties to be used in the correlations for the local value of h_x. From the value of h_x so determined, the wall temperature is found. If the wall temperature is different from the originally assumed value of T_s, the latest available value of T_s is used to update the value of h_x. The process is repeated until two consecutive values of T_s are sufficiently close to each other.

Vertical and Inclined Cylinders Equations 4.3.1 and 4.3.2 can be used for a vertical cylinder if the diameter of the cylinder is sufficiently large compared with its axial length. The criterion for employing the correlations for vertical plates to vertical cylinders is

$$\frac{d}{L} > \frac{35}{Gr_L^{1/4}}$$

where

d = diameter of the cylinder
L = axial length of the cylinder

For vertical and inclined cylinders with end surfaces insulated Churchill (1990) proposed

Evaluate β at T_∞ and all other properties at $(T_s + T_\infty)/2$
θ = angle of inclination of the cylinder axis from the vertical
L = length of the cylinder
d = diameter of the cylinder
$Ra_L = \{[g\,\beta\rho^2(T_s - T_\infty)L^3]/\mu^2\}Pr \qquad 10^5 < Ra_L < 10^9$

$$Nu_L = \frac{0.518(Ra_L \cos \theta)^{1/4}\{1 + [2.8(d/L)\tan \theta]^{3/2}\}^{1/6}}{[1 + (0.559/Pr)^{9/16}]^{4/9}} \qquad \textbf{(4.3.2e)}$$

EXAMPLE 4.3.1

A space heater is modeled as a series of 20 vertical plates, each 400-mm high, 150-mm wide, and 15-mm thick (Figure 4.3.4). If the temperatures of the surface of the plates and air are 60 °C and 10 °C, respectively, estimate the heat transfer rate to the air by natural convection from the vertical surfaces.

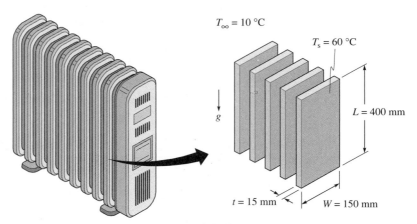

$T_\infty = 10$ °C
$T_s = 60$ °C
g
$L = 400$ mm
$t = 15$ mm
$W = 150$ mm

FIGURE 4.3.4 A Space Heater Modeled as a Series of Vertical Plates

Given

$T_s = 60$ °C $L = 400$ mm
$T_\infty = 10$ °C $W = 150$ mm $t = 15$ mm

Find

q (heat transfer rate from the vertical surfaces)

ASSUMPTIONS

1. Steady state
2. The flow adjacent to the vertical surface is not affected by the bottom surface.
3. The flow next to a vertical plate is not affected by the presence of the neighboring plates. The spacing of the plates should be greater than the maximum boundary layer thickness. This assumption will not be valid if the plates are closely spaced.
4. Air is an ideal gas.

SOLUTION

Evaluate β at T_∞ (283.15 K) and all other properties of air at $T_f = (60 + 10)/2 = 35\,°C = 308.15$ K. From the software CC (or Table A7) properties of air at 35 °C are

$$\rho = 1.146 \text{ kg/m}^3 \qquad \mu = 1.888 \times 10^{-5} \text{ N s/m}^2 \qquad k = 0.0267 \text{ W/m °C}$$
$$\beta = 1/283.15 = 3.532 \times 10^{-3} \text{ K}^{-1} \qquad Pr = 0.711$$

$$Ra_L = Gr_L \, Pr = \frac{g\,\beta\rho^2\,\Delta T L^3}{\mu^2} \, Pr$$

For a vertical surface the characteristic length is the height of the surface and $L = 0.4$ m

$$\Delta T = T_s - T_\infty = 60 - 10 = 50\,°C$$

$$Ra_L = \frac{9.807 \times 3.532 \times 10^{-3} \times 1.146^2 \times 50 \times 0.4^3 \times 0.711}{(1.888 \times 10^{-5})^2} = 2.903 \times 10^8$$

Employing Equation 4.3.1a,

$$Nu_L = 0.68 + \frac{0.67 \times (2.903 \times 10^8)^{1/4}}{[1 + (0.492/0.711)^{9/16}]^{4/9}} = 67.82$$

$$h = Nu_L \frac{k}{L} = 67.82 \times \frac{0.0267}{0.4} = 4.53 \text{ W/m}^2 \text{ K}$$

$$q = hA(T_s - T_\infty) = 4.53 \times 2 \times (0.150 + 0.015) \times 0.4 \times 50$$

$$= \underline{29.9 \text{ W/plate}}$$

If Equation 4.3.1b is employed, $Nu = 84.16$, $h = 5.63$ W/m² °C, and $q = 37.2$ W/plate.

Natural convection correlations— inclined plates, more complex flows, critical Grashof number

Inclined Plates For a downward-facing heated or an upward-facing cooled inclined isothermal surface (Figures 4.3.5a and b) Equations 4.3.1a and b may be used by using the component of the gravitational force parallel to the plate to evaluate Ra_L. If the angle of inclination of the surface to the vertical is θ, $g\cos\theta$ is used in place

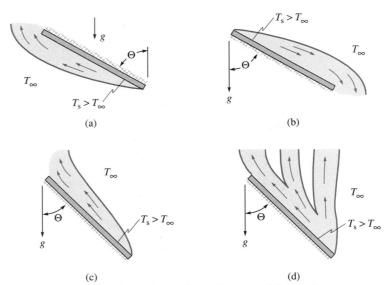

FIGURE 4.3.5 Heat Transfer with Different Configurations of Inclined Plates
(a) Downward-facing heated plate. (b) Upward-facing cooled plate.
(c) Upward-facing heated plate, $Gr_L < Gr_c$. (d) Upward-facing heated plate, $Gr_L > Gr_c$.

of g to compute the Ra_L in Equations 4.3.1a and b. The characteristic length is the length along the plate.

$$Gr_L = \frac{g\, \cos\, \theta \beta\, \Delta T\, L^3}{\nu^2} \qquad Ra_L = Gr_L\, Pr \qquad Nu_L = \frac{h_L L}{k}$$

If $\theta > 88°$, it is suggested that the correlations for horizontal plates be used.

The fluid flow in natural convection with an upward-facing heated plate is more complex than with a downward-facing heated plate. The following discussion on upward-facing heated plate applies to downward-facing cooled plates also. With an upward-facing heated plate, for small values of Gr_L (formed with $g\, \cos\, \theta$), the fluid motion is parallel to the plate similar to the flow with a downward-facing heated plate as shown in Figure 4.3.5c. But when the value of $Gr_L Pr$ exceeds a critical value, $Gr_c Pr$, the boundary layer detaches itself from the plate due to the component of the buoyant force perpendicular to the plate. The value of $Gr_c Pr$ depends on θ. The flow pattern for $Gr_L\, Pr > Gr_c\, Pr$ is shown in Figure 4.3.5d. It is evident that, as the flow pattern with $Gr_L\, Pr < Gr_c\, Pr$ is similar to the one with the downward-facing heated plate (Figure 4.3.5a), Equations 4.3.1a and b are applicable. But if $Gr_L > Gr_c$ those equations are no longer applicable.

The value of $Gr_c\, Pr$ depends on the angle θ. Fujii and Imura (1972) experimentally determined the values of Gr_c and developed a correlation for the Nusselt number for $Gr > Gr_c$. The values of Gr_c and the correlation are

θ (°)	15	30	60	70
$Gr_c\, Pr$	5×10^9	2×10^9	10^8	10^6

Gr_c Pr is approximated by

$$\ln(Gr_c \ Pr) = 14.223 + 8.3945 \cos \theta \qquad 15° < \theta < 60°$$

$$\ln(Gr_c \ Pr) = 3.848 + 29.146 \cos \theta \qquad 60° < \theta < 70°$$

> Properties at $T_r = T_s - 0.25(T_s - T_\infty)$; β at $T_\infty + 0.25(T_s - T_\infty)$
> θ = angle of inclination from the vertical
> L = length of the plate in the flow direction
> $Gr = [g\beta \cos \theta \rho^2(T_s - T_\infty)L^3]/\mu^2$
> $10^6 < Gr \ Pr < 10^{11} \qquad Gr > Gr_c$
>
> $$Nu_L = 0.14[(Gr \ Pr)^{1/3} - (Gr_c \ Pr)^{1/3}] + 0.56(Gr_c \ Pr \cos \theta)^{1/4}$$ **(4.3.3a)**

For $Gr < Gr_c$ employ Equations 4.3.2a or b with g replaced by $g \cos \theta$.

Equation 4.3.3a was obtained from experimental measurements with uniform heat flux. They may be used for uniform surface temperature also but the uncertainty in the predicted value of h will be higher. The validity of Equation 4.3.3a for large surfaces has not been established as the correlation was derived from experiments on plates that which were 5-cm and 30-cm wide.

For an inclined plate subjected to a uniform heat flux, with the heated surface facing down, Fussey and Warneford (1978) recommend

> Evaluate all properties at the film temperature $(T_s + T_\infty)/2$
> Ra_x^* = modified Raleigh number = $Gr_x^* \ Pr$;
> $Gr_x^* = g\beta q'' x^4/(kv^2)$
> θ = angle of inclination of the surface to the vertical
> (Figure 4.3.5)
> $Ra_{x,cr}^* = 6.31 \times 10^{12} e^{0.0705\theta}$ (θ in degrees)
>
> $$Nu_x = 0.592(Ra_x^* \cos \theta)^{0.2} \quad 0 < \theta < 86.5° \quad Ra_x^* < Ra_{x,cr}$$ **(4.3.3b)**
>
> $$Nu_x = 0.889(Ra_x^* \cos \theta)^{0.205} \qquad \theta < 31° \qquad Ra_x^* > Ra_{x,cr}^*$$ **(4.3.3c)**

Natural convection correlations— horizontal plates

Horizontal Plates, Uniform Surface Temperature Fujii and Imura (1972) measured average heat transfer coefficients with downward-facing heated, horizontal, rectangular plates of width L (smaller dimension) with an imposed uniform heat flux. Conditions corresponding to infinite strips were simulated. Though the experiments were based on uniform heat flux, the correlations obtained may be used for uniform surface temperatures also. The terms downward-facing heated surface and upward-facing cooled surface are defined in Figure 4.3.6. The suggested correlations are

> Properties (except β) at $T_r = T_s - 0.25(T_s - T_\infty)$
> Evaluate β at $T_\infty + 0.25(T_s - T_\infty)$
> Characteristic length = width (smaller dimension)
>
> Downward-facing heated plates and upward-facing cooled plates (Figures 4.3.6a and b)
>
> $$Nu_L = 0.58 \ Ra_L^{1/5} \qquad 10^6 < Ra_L < 10^{11}$$ **(4.3.4)**

Upward-facing heated plates and downward-facing cooled plates (Figures 4.3.6c and d)

$$\text{Nu}_L = 0.16 \, \text{Ra}_L^{1/3} \qquad 6 \times 10^6 < \text{Ra}_L < 2 \times 10^8 \qquad \textbf{(4.3.5a)}$$

$$\text{Nu}_L = 0.13 \, \text{Ra}_L^{1/3} \qquad 5 \times 10^8 < \text{Ra}_L < 10^{11} \qquad \textbf{(4.3.5b)}$$

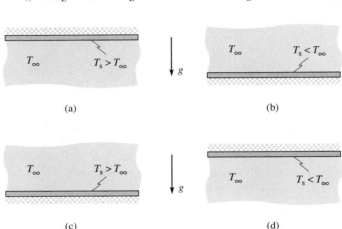

(a)   (b)

(c)   (d)

FIGURE 4.3.6 Definition of Downward-Facing Heated Surface, Upward-Facing Cooled Surface, Upward-Facing Heated Surface and Downward-Facing Cooled Surface
(a) Downward-facing heated surface. (b) Upward-facing cooled surface. (c) Upward-facing heated surface. (d) Downward-facing cooled surface.

Based on the experimental results of McAdams (1954), Lloyd and Moran (1974), and Goldstein et al. (1973), the following correlations for horizontal surfaces are recommended.

Properties of the fluid at the film temperature, $T_f = (T_s + T_\infty)/2$

Characteristic length $= A/P$, where A is the area of the surface and P is the perimeter of the surface

Downward-facing cooled plates and upward-facing heated plates

$$\text{Nu}_L = 0.54 \, \text{Ra}_L^{1/4} \qquad 2.2 \times 10^4 < \text{Ra}_L < 8 \times 10^6 \qquad \textbf{(4.3.5c)}$$

$$\text{Nu}_L = 0.15 \, \text{Ra}_L^{1/3} \qquad 8 \times 10^6 < \text{Ra}_L < 1.5 \times 10^9 \qquad \textbf{(4.3.5d)}$$

$$\text{Nu}_L = 0.96 \, \text{Ra}_L^{1/6} \qquad 1 < \text{Ra}_L < 200 \qquad \textbf{(4.3.5e)}$$

$$\text{Nu}_L = 0.59 \, \text{Ra}_L^{1/4} \qquad 200 < \text{Ra}_L < 10^4 \qquad \textbf{(4.3.5f)}$$

Downward-facing heated plate and upward-facing cooled plate

$$\text{Nu}_L = 0.27 \, \text{Ra}_L^{1/4} \qquad 10^5 < \text{Ra}_L < 10^{10} \qquad \textbf{(4.3.5g)}$$

EXAMPLE 4.3.2

The case of a personal computer is 40-cm wide, 50-cm deep, and 10-cm high (Figure 4.3.7). The top surface is to dissipate 25 W to the surrounding air, which is at 20 °C. Determine the temperature of the top surface.

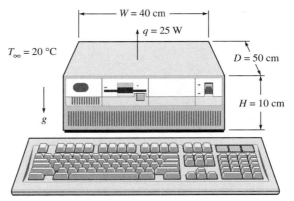

FIGURE 4.3.7 The top surface of a computer is to dissipate 25 W.

Given

$T_\infty = 20 \,°\text{C}$ $W = 40$ cm

$q = 25$ W $D = 50$ cm

$H = 10$ cm

Find

T_s (temperature of the top surface)

ASSUMPTIONS

1. Steady state

2. Uniform surface temperature

3. Conduction between the top surface and the side plates is negligible.

4. There is no monitor on the computer case.

5. Air is an ideal gas.

6. Negligible effect of the side plates on the flow.

SOLUTION

The surface temperature is evaluated from the relation

$$q = hA(T_s - T_\infty) \tag{1}$$

The surface temperature T_s is needed to determine $T_s - T_\infty$ and to evaluate the properties. But T_s is yet to be determined. It is found by the following iterative procedure.

1. Assume a value of T_s.

2. Determine h.

3. Calculate T_s from Equation 1. If the calculated value of T_s is significantly different from the value used to determine h in step 2, go to step 2 and evaluate h from the calculated value of T_s. Repeat the process till the values of T_s used to compute h and the value found from Equation 1 are sufficiently close to each other.

The heated horizontal surface is facing up; therefore, the characteristic length is

$$L = \frac{A}{P} = \frac{0.4 \times 0.5}{2(0.4 + 0.5)} = 0.1111 \text{ m}$$

Assume 40 °C for T_s. $T_f = 30$ °C. From the software CC (or Table A7), properties of air at 30 °C are

$\rho = 1.165$ kg/m^3 $c_p = 1007$ J/kg K
$\mu = 1.865 \times 10^{-5}$ N s/m^2 $k = 0.0264$ W/m K
$\beta = 1/303.2$ K^{-1} Pr $= 0.712$

$$\text{Ra}_L = \frac{g \, \beta \rho^2 \, \Delta T \, L^3}{\mu^2} \, \text{Pr}$$

$$= \frac{9.807 \times (1/303.2) \times 1.165^2 \times (313.2 - 293.2) \times 0.1111^3 \times 0.712}{(1.865 \times 10^{-5})^2}$$

$$= 2.464 \times 10^6$$

For a horizontal plate facing up, employing Equation 4.3.5c for $\text{Ra}_L = 2.464 \times 10^6$

$$\text{Nu}_L = 0.54 \times (2.464 \times 10^6)^{1/4} = 21.39$$

$$h = \text{Nu}_L \frac{k}{L} = 21.39 \times \frac{0.0264}{0.1111} = 5.08 \text{ W/m}^2 \text{ °C}$$

$$q = hA(T_s - T_\infty) \qquad 25 = 5.08 \times 0.5 \times 0.4(T_s - 20) \qquad T_s = 44.6 \text{ °C}$$

The computed value of $T_s = 44.6$ °C is greater than the originally assumed value of 40 °C. With $T_s = 44.6$ °C, find the new value of h. With properties of air from the software CC (or Table A7) at $T_f = 32.3$ °C,

$$\text{Ra}_L = 2.927 \times 10^6 \qquad \text{Nu}_L = 22.34 \qquad h = 5.34 \text{ W/m}^2 \text{ °C}$$

$$T_s = T_\infty + \frac{q}{Ah} = 20 + \frac{25}{0.4 \times 0.5 \times 5.34} = 43.4 \text{ °C}$$

The new value of 43.4 °C is quite close to the previous value of 44,6 °C and

$$T_s \cong \underline{43.4 \text{ °C}}$$

COMMENT

The value of the convective heat transfer coefficient has been computed to two decimal places. The realizable value is not likely to be within $\pm 10\%$ of the computed value. Hence, further iteration to find the "exact" value of the convective heat transfer coefficient is not warranted.

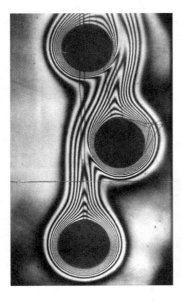

FIGURE 4.3.8 Heated Horizontal Cylinders
Flow pattern around the top cylinder is affected by
the flows around the cylinders below it. Correlations
for horizontal cylinders are applicable to the
bottommost cylinder if it is not too close to the
cylinder above it. Photograph courtesy of
E.R.G. Eckert and E. Soehngen.

Long Horizontal Cylinders (See Figure 4.3.8) For horizontal cylinders, the correlation between Nu_d and Ra_d given by Morgan (1975) is in the form

Properties at $T_f = (T_s + T_\infty)/2$
c and m from Table 4.3.1.

$$Nu_d = c Ra_d^m \tag{4.3.6a}$$

Natural convection
correlations—
horizontal cylinders

The diameter of the cylinder is the characteristic length for both the Nusselt and
Rayleigh numbers.

$$Nu_d = \frac{hd}{k} \qquad Ra_d = \frac{g \beta \rho^2 \, \Delta T \, d^3}{\mu^2} \, Pr$$

The values of c and m are functions of Ra_d and are given in Table 4.3.1. All properties
are evaluated at the film temperature.

Comprehensive correlations for a wide range of Ra_d are given by Churchill and
Chu (1975). For slightly better accuracy in the lower ranges of Ra they recommend
Equation 4.3.6c

TABLE 4.3.1 Values of c and m in Equation 4.3.6a

Ra	c	m
10^{-10}–10^{-2}	0.675	0.058
10^{-2} –10^{2}	1.02	0.148
10^{2} –10^{4}	0.85	0.188
10^{4} –10^{7}	0.48	0.25
10^{7} –10^{12}	0.125	0.333

Evaluate β at T_∞; all other properties at $T_f = (T_s + T_\infty)/2$
Characteristic length = diameter

$$\text{Nu}_d = \left[0.6 + \frac{0.387 \, \text{Ra}_d^{1/6}}{[1 + (0.559/\text{Pr})^{9/16}]^{8/27}} \right]^2 \qquad 10^{-3} < \text{Ra}_d < 10^{13} \qquad \textbf{(4.3.6b)}$$

$$\text{Nu}_d = 0.36 + \frac{0.518 \, \text{Ra}_d^{1/4}}{[1 + (0.559/\text{Pr})^{9/16}]^{4/9}} \qquad 10^{-6} < \text{Ra}_d < 10^9 \qquad \textbf{(4.3.6c)}$$

Spheres A simple correlation for heat transfer from spheres (Figure 4.3.9) recommended by Yuge (1960) is

Properties at $T_f = (T_s + T_\infty)/2$
Characteristic dimension = diameter

$$\text{Nu}_d = 2 + 0.426 \, \text{Ra}_d^{1/4} \qquad 1 < \text{Ra}_d < 10^5 \qquad \text{Pr} \approx 1 \qquad \textbf{(4.3.7a)}$$

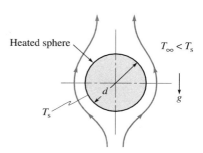

FIGURE 4.3.9 Heated Sphere

Natural convection correlation— spheres

For a wider range of Ra_d Churchill (1983) recommends

Evaluate β at T_∞; all other properties at $T_f = (T_s + T_\infty)/2$

$$\text{Nu}_d = 2 + \frac{0.589 \, \text{Ra}_d^{1/4}}{[1 + (0.469/\text{Pr})^{9/16}]^{4/9}} \qquad \text{Ra}_d < 10^{11} \quad \text{Pr} \gtrsim 0.7 \qquad \textbf{(4.3.7b)}$$

Note that when the fluid is stationary (in the absence of a body force), $\text{Ra}_d = 0$ and heat transfer is by conduction only. In such a case the limiting value of Nu_d is 2.

Natural convection heat transfer coefficient estimates for bodies for which correlations are not available—a "generalized" correlation

Estimate for Other Geometries For laminar convection without boundary layer separation for those cases where a correlation is not readily available, Lienhard (1987) suggests that a rough estimate for Nu can be obtained from the relation

$$\text{Nu}_\tau = 0.52 \, \text{Ra}_\tau^{1/4} \qquad \text{(Not valid for } \text{Pr} \ll 1\text{)} \qquad \textbf{(4.3.8a)}$$

The characteristic dimension for Nu_τ and Ra_τ is the length of travel, τ, of the boundary layer.

In the case of spheres and horizontal cylinders, for example, the length of travel of the boundary layer is $\pi \, d/2$, the distance from the bottom to the top along the surface. For those cases, Equation 4.3.8a yields

$$\frac{h(\pi d/2)}{k} = 0.52 \left[\frac{g \beta \, \Delta T \, (\pi d/2)^3}{\nu^2} \, \text{Pr} \right]^{1/4} \qquad \text{or} \quad \text{Nu}_d = 0.464 \, \text{Ra}_d^{1/4}$$

Sparrow and Stretton (1985) performed detailed natural convection experiments with cubes in many different orientations. From their experimental data and those of others with spheres and cylinders (radius equal to the axial length), they concluded that Lienhard's method underpredicts the heat transfer coefficient. They proposed the following prediction method.

Evaluate β at T_∞; all other properties at $(T_s + T_\infty)/2$

$$\text{Nu**} = 5.748 + 0.752\,[\text{Ra**}/F(\text{Pr})]^{0.252} \qquad \textbf{(4.3.8b)}$$

where

$\text{Nu**} = (hL**)/k$
$\text{Ra**} = [g\beta\rho^2(T_s - T_\infty)L**^3\,\text{Pr}]/\mu^2$
$F(\text{Pr}) = [1 + (0.492/\text{Pr})^{9/16}]^{16/9}$
$L** = A/d**$
$\pi d**^2/4 = A_H$
$\quad A_H$ = projected area of the body on a horizontal plane

Equation 4.3.8b was proposed for bodies with unity aspect ratios (cubes with equal sides, cylinders with radius equal to the axial length, and spheres). The validity of the equation for other bodies is not yet established.

Note on the Correlation for Natural Convection Most of the correlations presented in this section are based on laboratory experiments, where great care is taken to suppress forced motion of the surrounding fluid. In many engineering applications, some forced motion of the fluid is inevitable. Though such forced motion may result in small velocities, it may lead to higher or lower convective heat transfer rates than predicted by the correlations for natural convection. For a further discussion, see Section 4.4 on mixed convection.

A second observation is that considerable uncertainties exist in the values of h predicted by the correlations; the uncertainties are generally within $\pm 10\%$, though in some cases they may be as high as $\pm 20\%$.

A third observation is that, with natural convection in gases, radiative heat transfer may be quite significant and it should be included in computing the total heat transfer rates.

EXAMPLE 4.3.3 An ornamental space heater is in the form of a 60-cm diameter sphere, which is freely suspended in a large room (Figure 4.3.10). The surface of the sphere is maintained at 100 °C and the room is at 20 °C. Determine the convective heat transfer rate.

Given

Fluid: Air $T_\infty = 20\,°C$
$T_s = 100\,°C$ $d = 0.6$ m

Find

q (convective heat transfer rate)

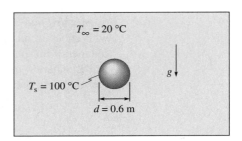

FIGURE 4.3.10 A sphere is freely suspended in a large room.

ASSUMPTIONS

1. Steady state
2. Air is an ideal gas.

SOLUTION

Evaluate properties of air at the film temperature. From the software CC (or Table A7), properties of air at $T_f = 60\,°C = 333.15\,K$

$\rho = 1.0597\ \text{kg/m}^3 \qquad \mu = 2.003 \times 10^{-5}\ \text{N s/m}^2 \qquad k = 0.0285\ \text{W/m}\,°C$
$\beta(293.15\ K) = 1/293.15\ K^{-1} \qquad Pr = 0.708$

$$Ra_d = Gr_d\ Pr = \frac{g\,\beta\rho^2\ \delta T\ d^3}{\mu^2}\ Pr$$

$$= \frac{9.807 \times (1/293.15) \times 1.0597^2 \times 80 \times 0.6^3 \times 0.708}{(2.003 \times 10^{-5})^2} = 1.147 \times 10^9$$

Employing Equation 4.3.7b,

$$Nu_d = 2 + \frac{0.589\ Ra_d^{1/4}}{[1 + (0.469/Pr)^{9/16}]^{4/9}} = 2 + \frac{0.589 \times (1.147 \times 10^9)^{1/4}}{[1 + (0.469/0.708)^{9/16}]^{4/9}} = 85.61$$

$$h = \frac{85.61 \times 0.0285}{0.6} = 4.07\ \text{W/m}^2\,°C$$

$$q = hA\,(T_s - T_\infty) = 4.07 \times 4\pi\,(0.6/2)^2 \times (100 - 20) = \underline{368.2\ W}$$

EXAMPLE 4.3.4 Hot water flows at the rate of 0.5 l/s in a horizontal, 2-cm I.D. copper pipe in the basement of a house (Figure 4.3.11). Determine the rate of change of temperature of the water per meter length of the pipe at a section where the water is at 50 °C. Ambient air is at 15 °C.

Given

Fluid: Water in the tube and air surrounding the tube
$\dot{V}$ (volumetric flow rate of water) = 0.5 l/s
$T_b = 50\,°C \qquad T_\infty$ (air temperature) = 15 °C
$d_i = 2\ \text{cm} \qquad d_o \approx d_i$

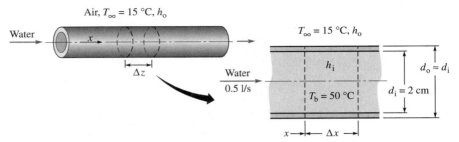

FIGURE 4.3.11 Hot water flows in a copper pipe. The ambient air is at 15 °C.

Find

dT/dx (rate of temperature change per meter length at 50 °C)

ASSUMPTIONS

1. Steady state
2. The effect of any wall or ceiling next to the pipe is negligible.
3. Air is an ideal gas.
4. The wall thickness of the tube is negligible and the conductive resistance is negligible.
5. Negligible radiative heat transfer (true only if the copper pipe is highly polished)
6. Fully developed velocity and temperature profiles for the water

SOLUTION

From the First Law of Thermodynamics, the rate of change of temperature of the water per unit length of the pipe is given by

$$\dot{m}c_p \frac{dT}{dx} = -q' \tag{1}$$

where

$\dot{m}$ = mass rate of flow of water
q' = heat transfer rate from the water to the surrounding air per unit length of the pipe

From the thermal circuit we have

$$q' = \frac{T_b - T_\infty}{(1/h_i \, \pi \, d_i) + (1/2 \, \pi \, k) \, \ln(d_o/d_i) + (1/h_o \, \pi \, d_o)} \tag{2}$$

where

T_b = bulk temperature of the water
T_∞ = ambient air temperature

In Equation 2, h_i and h_o are the convective heat transfer coefficient associated with the inner and outer surfaces of the tube, respectively. As the pipe is made of copper, which has a very high thermal conductivity, and the thickness of the copper tube is small, the conductive resistance in the above equation can be neglected. Further assume $d_o/d_i \approx 1$ and obtain

$$q' = \frac{(T_b - T_\infty)\pi\, d}{(1/h_i) + (1/h_o)} \tag{3}$$

Substituting Equation 3 into Equation 1 we get

$$\frac{dT^.}{dx} = -\left[\frac{\pi d(T_b - T_\infty)}{(1/h_i) + (1/h_o)}\right]\frac{1}{\dot{m}c_p} \tag{4}$$

All the quantities on the right hand side of Equation 4, except the convective heat transfer coefficients, are known. To determine h_i, we employ correlations for forced convection and to determine h_o we will employ correlations for natural convection.

To determine h_o, we need the surface temperature, T_s to compute Ra_d and to evaluate the properties of air. Many of the correlations for h_i require the surface temperature. T_s is determined by the following iterative procedure:

1. Assume T_s.
2. Compute h_i and h_o.
3. From an energy balance, $h_i(T_b - T_s) = h_o(T_s - T_\infty)$; therefore,

$$T_s = \frac{h_i T_b + h_o T_\infty}{h_i + h_o} \tag{5}$$

Compute T_s from Equation 5.
4. If the value of T_s obtained in step 3 is close to the value used to determine the heat transfer coefficients, the correct values for h_i and h_o have been obtained. If the computed and assumed values of the surface temperature are not close to each other, go to step 2 and compute the heat transfer coefficients with the value of T_s obtained in step 3. Repeat the computations as necessary.

Guess T_s

As the value of h_i with forced convection with water is expected to be much higher than the value of h_o with natural convection in air, the surface temperature is expected to be close to T_b. Assume 47 °C for T_s.

Evaluate h_i

Compute Re_d.

$$Re_d = \frac{\rho V d}{\mu} = \frac{4\dot{m}}{\pi d\mu}$$

Evaluate properties of water at the bulk temperature of 50 °C. From the software CC (or Table A5) properties of water at 323.2 K are

$\rho = 988$ kg/m^3 $c_p = 4175$ J/kg.K
$k = 0.643$ W/m K $\mu = 529 \times 10^{-6}$ N.s/m^2
$Pr = 3.44$ $Pr_s(47\ °C) = 3.637$
$\dot{m}$ = volumetric flow rate × density = (0.5/1000) × 988 = 0.494 kg/s

$$Re_d = \frac{4\dot{m}}{\pi d\mu} = \frac{4 \times 0.494}{\pi \times 0.02 \times 529 \times 10^{-6}} = 59\,430$$

For Pr > 1.5, employing Equation 4.2.3e for fully developed flows,

$$\text{Nu}_d = 0.012 \, (\text{Re}_d^{0.87} - 280) \, \text{Pr}^{0.4} \, (\text{Pr}/\text{Pr}_s)^{0.11}$$

$$= 0.012 \, (59\,430^{0.87} - 280) \, 3.437^{0.4} \, (3.437/3.637)^{0.11} = 272.7$$

$$h_i = 272.7 \times 0.643/0.02 = 8765 \text{ W/m}^2 \, ^\circ\text{C}$$

Evaluate h_o

With $T_s = 47 \, ^\circ\text{C}$, film temperature = 31 °C. From the software CC (or Table A7), properties of air at 31 °C are

$$\rho = 1.161 \text{ kg/m}^3 \qquad c_p = 1007 \text{ J/kg K} \qquad \mu = 186.9 \times 10^{-7} \text{ N s/m}^2$$
$$\nu = 16.1 \times 10^{-6} \text{ m}^2/\text{s} \qquad k = 0.0264 \text{ W/m K} \qquad \text{Pr} = 0.712$$
$$\beta(T_\infty = 288.15 \text{ K}) = 1/288.15 \text{ K}^{-1} = 3.470 \times 10^{-3} \text{ K}^{-1}$$

$$\text{Ra}_d = \text{Gr}_d \, \text{Pr} = \frac{g \, \beta \rho^2 \, \Delta T \, d^3}{\mu^2} \, \text{Pr}$$

$$= \frac{9.807 \times 3.470 \times 10^{-3} \times 1.161^2 \times (47 - 15) \times 0.02^3 \times 0.712}{(186.9 \times 10^{-7})^2}$$

$$= 23\,910$$

For $\text{Ra}_d = 23\,910$, from Equation 4.3.6c,

$$\text{Nu}_d = 0.36 + \frac{0.518 \times (23\,910)^{1/4}}{[1 + (0.559/0.712)^{9/16}]^{4/9}} = 5.23$$

$$h_o = \text{Nu}_d \frac{k}{d} = 5.23 \times \frac{0.0264}{0.02} = 6.92 \text{ W/m}^2 \text{ K}$$

Compute T_s from Equation 5

$$T_s = \frac{h_i T_b + h_o T_\infty}{h_i + h_o} = \frac{8765 \times 50 + 6.92 \times 15}{8765 + 6.92} = 50 \, ^\circ\text{C}$$

Recompute h_i and h_o with $T_s = 50 \, ^\circ\text{C}$. Following the procedure outlined above,

$$\text{Re}_d = 59\,430 \qquad\qquad \text{Pr} = 3.437 \qquad h_i = 8820 \text{ W/m}^2 \, ^\circ\text{C}$$
$$c_p(50 \, ^\circ\text{C}) = 4175 \text{ J/kg} \, ^\circ\text{C}$$
$$\text{Ra} = 25\,696 \qquad\qquad \text{Pr} = 0.712 \qquad h_o = 7.07 \text{ W/m}^2 \, ^\circ\text{C}$$

$$T_s = \frac{8820 \times 50 + 7.07 \times 15}{8820 + 7.07} = 50 \, ^\circ\text{C}$$

As the assumed and computed values of T_s are equal, the correct values of the convective heat transfer coefficients are

$$h_i = 8820 \text{ W/m}^2 \, ^\circ\text{C} \qquad h_o = 7.07 \text{ W/m}^2 \, ^\circ\text{C}$$

From Equation 4,

$$\frac{dT}{dx} = -\left[\frac{\pi d(T_b - T_\infty)}{(1/h_i) + (1/h_o)}\right] \frac{1}{\dot{m} c_p} = -\frac{\pi \times 0.02 \times (50 - 15)}{(1/8820) + (1/7.07)} \times \frac{1}{0.494 \times 4175}$$

$$= -0.008 \, ^\circ\text{C/m}$$

COMMENT

It is clear that with flowing water, the temperature drop of the water resulting from the heat transfer to the ambient air is very small; for a 20-m long pipe, the temperature drop is only 0.16 °C. Insulating the pipe will not result in any significant changes in the exit temperature of the water. But, with insulation and no water flow, it will take much longer for the temperature of the water to reach the temperature of the ambient air.

4.4 MIXED CONVECTION

Simultaneously existing forced and natural convection

In forced convection, the heat transfer is by a combination of forced convection and natural convection. If the velocity in forced convection is sufficiently high, the effect of natural convection is negligible in comparison with forced convection. But when the fluid velocity in forced convection is small, natural convection heat transfer becomes significant. The effect of natural convection depends on the direction of the forced flow and the direction of the body force that results in natural convection. For example, consider the fluid flow parallel to a plate. Four different configurations of the directions of the flow and the gravitational force are shown in Figure 4.4.1 for $T_s > T_\infty$. In every case, the flow pattern due to natural convection alone is shown.

Effect of forced flow on natural convection

Figure 4.4.1a depicts a heated vertical plate. If there is a slight upward motion of the fluid due to an external force, there is an increase in the effective velocity of the fluid, which leads to an increase in the heat transfer coefficient. If the motion of the fluid due to the external force is downwards (Figure 4.4.1b), the effective velocity is decreased. If the surface is horizontal (Figure 4.4.1c), and the flow due to the external force is parallel to the plate, the flow pattern is altered; the change in the flow pattern when the heated surface faces up is different from the change in the flow pattern when the heated surface faces down (Figure 4.4.1d).

Relative magnitude of Grashof and Reynolds numbers can be used to estimate the relative magnitudes of natural and forced convection heat transfer coefficients for vertical plates

In some cases, the relative magnitudes of natural convection and forced convection heat transfer rates can be assessed from the magnitudes of the Grashof and Reynolds numbers. For example, consider the situations shown in Figure 4.4.1a. The velocity of the fluid due to forced convection is in the upward direction and with a heated plate the motion of the fluid due to natural convection is also in the same direction. In this case forced convection heat transfer is enhanced due to natural convection. (If the plate is cooled, the motion due to natural convection is opposite to that due to forced convection, and the heat transfer is impeded.) The Nusselt number with the average heat transfer coefficient for forced convection in laminar flow of a fluid is given by Equation 4.1.3.

$$\text{Nu}_{Lf} = 0.664 \ \text{Re}_L^{1/2} \ \text{Pr}^{1/3}$$

The Nusselt number for natural convection for $\text{Ra}_L < 10^9$ is given by Equation 4.3.1a, which can be approximated by

$$\text{Nu}_{Ln} \approx \frac{0.67 \ \text{Ra}_L^{1/4}}{[1 + (0.492/\text{Pr})^{9/16}]^{4/9}}$$

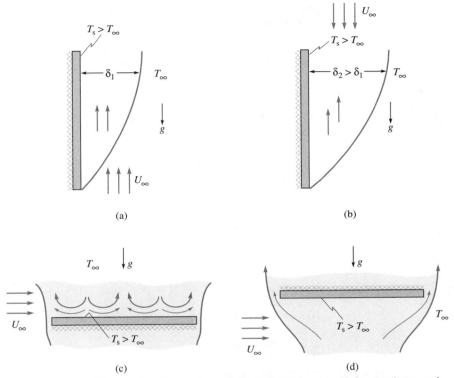

FIGURE 4.4.1 Different Configurations with Fluid Flow Parallel to a Plate and the Direction of Gravity

If the flow is perpendicular to the plate, four similar configurations are possible. Flow patterns due only to natural convection are shown. (a) Vertical plate, forced convection aids natural convection. (b) Vertical plate, forced convection impedes natural convection. (c) Upward-facing heated plate, forced convection flows parallel to plate. (d) Downward-facing heated plate, forced convection flows parallel to plate.

For $Pr \approx 1$, the Nusselt numbers in forced and natural convection can be approximated by

$$\mathrm{Nu}_{Lf} = 0.664 \ \mathrm{Re}_L^{1/2} \qquad \mathrm{Nu}_{Ln} \approx 0.533 \ \mathrm{Gr}_L^{1/4}$$

In both cases, the characteristic length dimension is the length of the plate, L. Hence, the ratio of the convective heat transfer coefficient is given by the ratio of the Nusselt numbers. As the temperature difference for computing the heat transfer rate in both cases is the same, the ratio of the heat transfer rate due to natural convection (q_n) to the heat transfer rate due to forced convection (q_f) is given by the ratio of the Nusselt numbers.

$$\frac{q_n}{q_f} = \frac{\mathrm{Nu}_{Ln}}{\mathrm{Nu}_{Lf}} \approx \frac{0.533 \ \mathrm{Gr}_L^{1/4}}{0.664 \ \mathrm{Re}_L^{1/2}} \approx \frac{\mathrm{Gr}_L^{1/4}}{\mathrm{Re}_L^{1/2}}$$

We may conclude that for this case of laminar convection (with both forced and natural convection), natural convection heat transfer is the dominant mode if

$$\frac{\text{Gr}_L^{1/4}}{\text{Re}_L^{1/2}} \gg 1 \quad \text{or} \quad \frac{\text{Gr}_L^{1/2}}{\text{Re}_L} \gg 1$$

Forced convection heat transfer will be the dominant mode if $\text{Gr}_L^{1/2}/\text{Re}_L \ll 1$ and the two modes of heat transfer rates will be of the same order of magnitude if $\text{Gr}_L^{1/2}/\text{Re}_L \approx 1$.

From the works of Lloyd and Sparrow (1961), Mori (1960), Yuge (1960), and Fand and Keswani (1973), the criteria for neglecting the effect of natural convection in three cases of laminar flows without separation are summarized in Table 4.4.1.

TABLE 4.4.1 Criteria for Neglecting Natural Convection Heat Transfer

Configuration	Criterion	h_n/h_f*	Pr
Heated vertical plate, upward flow	$\text{Gr}_L^{1/2}/\text{Re}_L < 0.4$	0.05	0.03–100
Horizontal plates, upward-facing heated surface, flow parallel to plate	$\text{Gr}_L^{1/2}/\text{Re}_L^{1.25} < 0.3$	0.1	Pr $\approx$ 1
Horizontal cylinders, cross flow	$\text{Gr}_d/\text{Re}_d^2 < 0.5$	0.05	

*h_n and h_f are the heat transfer coefficients in natural and forced convection, respectively.

Mixed convection in vertical and horizontal tubes

Laminar flows in horizontal and vertical tubes with low velocities are important in engineering applications. The effect of natural convection in such flows may be significant. Metais and Eckert (1964) developed a map to delineate the different regimes where one or the other mode of convection is dominant and where both are significant. Figures 4.4.2a and b show the relative significance of natural and forced convection heat transfer in horizontal and vertical tubes. These maps are applicable for $10^{-2} < \text{Pr}(d/L) < 1$, where d and L are the diameter and axial lengths of the tube, respectively.

The maps show the limits of forced and natural convection regimes. The limits are defined ''in such a way that the actual heat flux under the combined influence of the forces does not deviate by more than 10 percent from the heat flux that would be caused by the external forces alone or by the body forces alone,'' [Metais and Eckert (1964)]. The Grashof number is based on the diameter of the tube.

For flows in horizontal tubes, when both natural and forced convections are of approximately the same magnitude, correlations have been proposed by Depew and August (1971) for isothermal tube surface and by Morcos and Bergles (1975) for uniform heat flux.

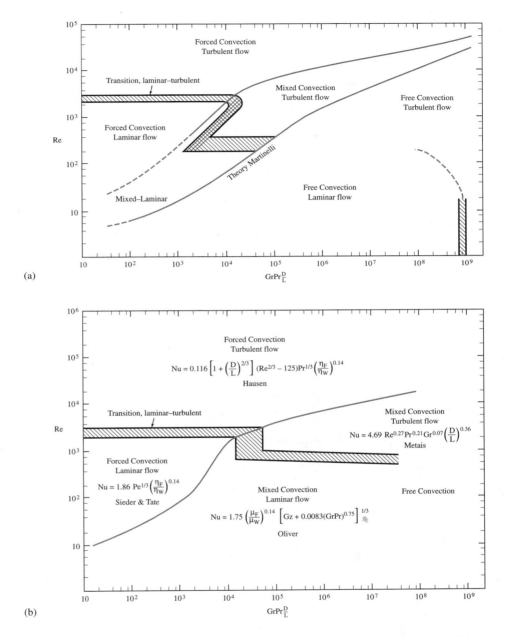

FIGURE 4.4.2 Maps Delineating Forced, Mixed, and Natural Convection for Laminar Flows in Vertical and Horizontal Tubes

(a) Vertical tubes. (b) Horizontal tubes. In both (a) and (b), $10^{-2} < \Pr(d/L) < 1$. Aiding flow means forced and natural convection flows in the same direction. Opposing flow means the flows in each mode are in opposite directions. UWT represents uniform surface temperature and UHT represents uniform heat flux.

Uniform Surface Temperature

$L/d < 28.4$

μ_s = dynamic viscosity, evaluated at the wall temperature
All other properties at the average bulk temperature of
the fluid

$$\text{Nu}_d = 1.75\,[\text{Gz} + 0.12\,(\text{Gz}\,\text{Gr}^{1/3}\,\text{Pr}^{0.36})^{0.88}]^{1/3}(\mu_b/\mu_s)^{0.14} \qquad \textbf{(4.4.1)}$$

where

Nu_d = Nusselt number = hd/k
Gz = Graetz number = $\dot{m}c_p/(kL)$
Gr = Grashof number = $g\beta\,\Delta T\,d^3/v^2$
Pr = Prandtl number = v/α
d = diameter of the tube
L = length of the tube

Equation 4.4.1 correlates the experimental data within about $\pm 40\%$.

Uniform Heat Flux

Properties at $(T_s + T_b)/2$

$3 \times 10^4 < \text{Ra} < 10^6, \quad 4 < \text{Pr} < 175, \quad 2 < h\,d^2/(k_w t) < 66$

$$\text{Nu}_d = \left\{ (4.36)^2 + \left[0.145 \left(\frac{\text{Gr}_d^*\,\text{Pr}^{1.35}}{P_w^{0.25}} \right)^{0.265} \right]^2 \right\}^{0.5} \qquad \textbf{(4.4.2)}$$

where

Nu_d = Nusselt number = $h\,d/k$
Pr = Prandtl number = v/α
Gr_d^* = modified Grashof number = $g\beta\,d^4 q_w''/(v^2 k)$
P_w = tube wall parameter = $k\,d/(k_w t)$
q_w'' = heat flux at the outer surface of the tube
k_w = tube wall thermal conductivity
Ra_d = Raleigh number = $g\beta\,\Delta T\,d^3\,\text{Pr}/v^2$
t = tube wall thickness

All properties are evaluated at the fluid film temperature = $(T_w + T_b)/2$, T_w and T_b being the inside wall temperature and bulk temperature, respectively. Equation 4.4.2 is for a tube having an inside diameter of d and wall thickness of t. For correlations in mixed convection for different flow geometries refer to Kakac et al. (1987).

4.5 PROJECTS

Project 4.5.1   Temperature Distribution in a Linoleum Sheet Moving in a Water Bath

Square floor tiles, 30 cm on each side, are made by cutting each tile from a continuous 3-mm thick sheet of linoleum (density 535 kg/m³). The optimal cutting temperature is 65 °C. The sheet emerges from a curing furnace at 100 °C and is passed through a water bath to reduce the temperature to 65 °C prior to the cutting operation. The schematic shown in Figure 4.5.1 is supplied by the systems engineer to the thermal engineer.

The systems engineer requests the thermal engineer for the specifications of both the water temperature T_w in the bath, and the length L of immersion of the sheet in the bath. The following additional information is available: The linoleum consists of 60% limestone and 40% vinyl resin, and the lowest temperature at which water is available is 5 °C.

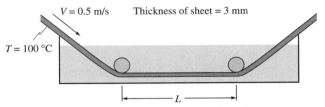

V = 0.5 m/s   Thickness of sheet = 3 mm

T = 100 °C

L

FIGURE 4.5.1 Schematic of the Setup for Cooling Linoleum Sheets Prior to Cutting Them into Tiles

SOLUTION

To determine the length L and the water temperature T_w, we need to assess if there is a significant temperature variation across the thickness of the sheet by computing the Biot number. If there is symmetric heating from the top and bottom surfaces, the Biot number is computed as

$$\text{Bi} = \frac{ht}{k2} \tag{1}$$

If, however, the heat transfer from the bottom surface is very much less than the heat transfer from the top surface because of the limited depth of water below the sheet (as depicted in Figure 4.5.1) the appropriate Biot number is

$$\text{Bi} = \frac{ht}{k} \tag{2}$$

If Bi < 0.1, the spatial variation across the thickness of the sheet can be neglected. As the convective heat transfer is to water, it is expected that the convective heat transfer coefficient will be high and consequently the value of the Biot number will be greater than 0.1.

To obtain a reasonable estimate for h, two models are examined. One of these is the forced convection to water flowing parallel to the sheet with a free stream velocity of U_∞ (= 0.5 m/s) and a free stream temperature of T_∞ (= 5 °C) as shown in Figure 4.5.2.

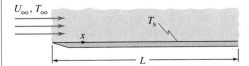

U_∞, T_∞

T_s

x

L

FIGURE 4.5.2 First Model for Estimating h, with Water Flowing Parallel to the Sheet

In the second model, shown in Figure 4.5.3, the sheet moves in an otherwise stagnant water bath. The second model is closer to the physical situation depicted in Figure 4.5.1. But, because the convective heat transfer results are more readily available for the first model shown in Figure 4.5.2, the first model will be adopted for estimating L and T_w ($= T_\infty$).

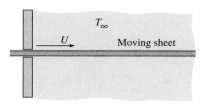

T_∞

U Moving sheet

FIGURE 4.5.3 Second Model for Estimating h, with the Sheet Moving Through an Otherwise Stagnant Water Bath

The strategy to be used in computing the Biot number is based on the assumption that the correctly computed Bi will significantly exceed 0.1. If Bi were computed with a value of h that is less than the correct value and if, even then, the computed value of Bi exceeded 0.1, the correct Bi would be even more in excess of 0.1.

The lowest value of h is that of Figure 4.5.2 for laminar boundary layer flows. For laminar boundary layer flows, the average value h for a plate length of L in the direction of flow of the water, is computed from Equation 4.1.3.

$$\text{Nu}_L = 0.664 \ \text{Re}_L^{1/2} \ \text{Pr}^{1/3} \tag{3}$$

In Equation 3, U_∞ is set equal to U_s, the speed of the sheet. The plate length will be guessed and computations of Bi and other parameters performed. Depending on the value of the temperature of the sheet at exit from the bath, the value of L will be revised and the computations repeated.

With the assumed value of 5 °C for the water (the convective heat transfer coefficient decreases with a decrease in the water temperature), we commence our computations with an initial guess of 1.5 m for L. Equation 3 is valid for uniform surface temperature. But the surface temperature is not only not known but also varies in the direction of the motion of the sheet due to the heat transfer from the sheet. If, as anticipated, Bi is much greater than 0.1, the surface temperature quickly drops close to T_∞. We, therefore, estimate h with both water and the sheet at 5 °C.

$$L = 1.5 \text{ m} \qquad U_\infty = 0.5 \text{ m/s} \qquad T_\infty = 5 \,°\text{C} \qquad T_s \text{ (approximate)} = 5 \,°\text{C}$$

From the software CC at $T_f = 5$ °C, the properties of water are

$\rho = 1000 \text{ kg/m}^3 \qquad \mu = 1.5 \times 10^{-3} \text{ N s/m}^2$
$k = 0.578 \text{ W/m °C} \qquad \text{Pr} = 10.9$
$\text{Re}_L = 5 \times 10^5$

From Equation 3

$$\text{Nu}_L = 1041.7 \qquad h = 401 \text{ W/m}^2 \,°\text{C}$$

To obtain a lower bound of Bi use Equation 1. From Table A3B, the thermal conductivity of linoleum (with a density of 535 kg/m³) is 0.081 W/m °C.

$$\text{Bi} = \frac{ht/2}{k} = \frac{401 \times 0.0015}{0.081} = 7.43$$

The actual value of Bi will be greater than 7.43 as a lower than the realizable value of h is used. It has now been established that Bi $\gg$ 0.1. Consequently, the temperature variation across the thickness of the sheet must be considered.

To take into account the timewise and spacewise variation of the temperature within the sheet, one more property of the material of the sheet, the thermal diffusivity, α, is needed. As the values of k and ρ for the material of the sheet are already known, we need either α or c_p. The specific heat or thermal diffusivity of linoleum is not available in Table A3B. From the ASHRAE *Handbook of Fundamentals* (1989) the specific heat of limestone is 909 J/kg °C. From Table A3B, the specific heat of polyvinyl chloride (PVC) is 840 J/kg °C. A reasonable approximate value of the specific heat of linoleum, which is a mixture of limestone and PVC, is 875 J/kg °C. For linoleum, with a thermal conductivity of 0.081 W/m °C and density of 535 kg/m³,

$$\alpha = \frac{k}{\rho c} = \frac{0.081}{535 \times 875} = 1.73 \times 10^{-7} \text{ m}^2/\text{s}$$

CONDUCTION ANALYSIS

The high value of Bi suggests that the convective resistance between the sheet surface and the water is negligible in comparison with the conductive resistance in the sheet. It is, therefore, reasonable to assume that when the sheet enters the water bath, the surface temperature of the sheet quickly reaches a value close to the water temperature. If the sheet is cooled symmetrically from the top and bottom, then the temperature distribution across the thickness of the sheet will evolve as the sheet moves through the water bath. The conceptual temperature variation across a section as a function of time in the water bath is shown in Figure 4.5.4. The figure shows that the attainment of 65 °C at the midplane of the sheet is the critical issue in the required length of the water bath. The time required for the midplane temperature to reach 65 °C will be found from the software TC with the assumption that the sheet is uniformly at 100 °C at time equal to zero, and it is immersed in a water bath with a convective heat transfer coefficient of 401 W/m² °C (this value of h is lower than the value for the moving sheet). At the midplane, the value of the dimensionless distance $y/(t/2)$ is zero and at that location,

$$\theta(0, \tau) = \frac{T - T_\infty}{T_i - T_\infty} = \frac{65 - 5}{100 - 5} = 0.632$$

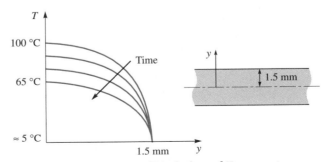

FIGURE 4.5.4 Conceptual Variation of Temperature across a Cross Section, with Time

From the software, with Bi = 7.43, after a few trials,

$$Fo = 0.356$$

$$\tau = \frac{FoL^2}{\alpha} = \frac{0.356 \times 0.0015^2}{1.73 \times 10^{-7}} = 4.6 \text{ s}$$

The sheet is moving at 0.5 m/s and in 4.6 s the sheet moves 2.3 m. The first estimate of L is, therefore, 2.3 m. But the computed value of L (2.3 m) is significantly higher than the originally assumed value of 1.5 m. Computations are repeated with a value of 2.3 m for L. With $L = 2.3$ m the Reynolds number is greater than 5×10^5 and we employ Equation 4.1.9a for computing h. From the software the convective heat transfer coefficient is 568 W/m^2 °C. Repeating the computations, we obtain Bi = 10.5, $\theta(0, \tau) = 0.632$, Fo = 0.335, $\tau = 4.4$ s and $L = 2.2$ m. From these computations, the value of L should be 2.2 m.

Now suppose the water temperature is 25 °C instead of 5 °C. Then,

$h = 973$ W/m^2 °C   Bi = 18
$\theta = 0.533$ Fo = 0.39
$\tau = 5$ seconds $L = 2.5$ m

As a water temperature of 25 °C is not unusual, it is desirable to provide a travel length of 2.5 m:

$$L = \underline{2.5 \text{ m}}$$

It may not be necessary that the temperature at all points across the thickness should be less than 65 °C. With a lower value of L, the temperature at some location between the surface and the midplane may reach 65 °C with the points between that location and the midplane being at a temperature higher than 65 °C. For example, with $L = 2$ m, $T_\infty = 25$ °C, the travel time is 4 s and

Temperature midplane = 73 °C.
Temperature at 0.6 mm from the midplane (0.9 mm from the surface) = 65 °C.

The thermal engineer supplies the following information to the systems design engineer:

	5 °C	25 °C
Water temperature		
L for midplane temperature to reach 65 °C	2.2 m	2.5 m

Water temperature $T_w = 25$ °C, $L = 2$m
Temperature at midplane = 73 °C
Temperature at 0.6 mm from the midplane (0.9 mm from the surface) = 65 °C

The system design engineer has the necessary information to choose an appropriate value of L and T_w.

COMMENT

In the analysis it is tacitly assumed that the axial conduction is negligible compared with the conduction in the y-direction. It is appropriate that we check if such an assumption is justified. The conductive heat flux in the x- and y-directions are proportional to the magnitudes of $\partial T/\partial x$ and $\partial T/\partial y$. At the trailing end (exit from the water bath) the temperature changes from 65 °C to approximately 5 °C (or 25 °C depending on the water temperature) in a distance of 0.0015 mm. The magnitude of $\partial T/\partial y$ is approximately, 60/0.0015 or 40 000 °C/m. In the x-direction, the temperature changes from 100 °C to 65 °C in a distance of 2.5 m. The magnitude of $\partial T/\partial x$ is approximately 35/2.5 or 14 °C/m. As the magnitude of $\partial T/\partial y$ is vastly greater than the magnitude of $\partial T/\partial x$, neglecting axial conduction is quite appropriate.

Project 4.5.2 Estimation of Performance of a Space Heater Under Off-Design Conditions

Quite often the numerical value of a key parameter or variable needed to obtain a desired result is not known. It may then be necessary to estimate the upper and lower bounds of the parameter and find the corresponding upper and lower bounds of the desired result.

To illustrate such a physical situation, consider the design of a space heating system to maintain the temperature in a room at a prescribed comfort level. A key part of the design is the determination of the steady-state convective heat transfer rate from the space heater to the surrounding air.

Suppose the design calls for an unfinned, horizontal tube carrying hot water. For ease of maintenance and control, a valve is installed at the inlet to the tube as depicted in Figure 4.5.5. The valve is fully accessible and may be adjusted at will by the

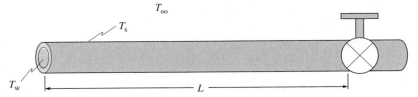

FIGURE 4.5.5 **Conceptual Design of a Space Heater**

occupants of the room. At any given air temperature (around 22 °C) some people find it too warm and some find it too cool. The tendency is, then, for some people to partially close the valve with the intent of reducing the heating capacity and for some to open the valve to increase the heating capacity. Thus, the extent to which the valve is open is quite random. As a consequence, the designer of the heating system does not know the water velocity in the tube. An estimate of the lower and upper bounds of the heating capacity is required.

The design parameters (sketched in Figure 4.5.6) are

Surrounding air temperature	20 °C
Barometric pressure	730 mm Hg
Inner diameter of tube	40 mm
Wall thickness of tube	3 mm
Material of the tube	Plain carbon-steel
Inlet temperature of water	75 °C
Length of tube	5 m

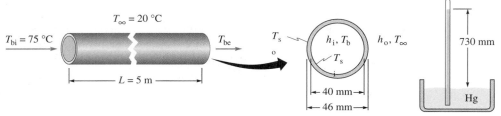

FIGURE 4.5.6 Preliminary Design of Heating System

Employing the concept of thermal resistance, the heat transfer rate to the air per meter length of the tube is given by

$$q' = \frac{T_b - T_\infty}{(1/h_i\pi \, d_i) + (1/2\pi k) \ln(d_o/d_i) + (1/h_o\pi d_o)} \tag{1}$$

From an examination of Equations 4.3.6b and c it is clear that the natural convective heat transfer coefficient on the air side, h_o, varies as $\frac{1}{3}$ or $\frac{1}{4}$ power of $(T_s - T_\infty)$ and, therefore, does not vary very much with the temperature of the surface. Calculations show that h_o varies from 6.2 W/m² °C to 5.3 W/m² °C at surface temperatures of 75 °C and 50 °C, respectively. Therefore, the resistance per unit length of the tube on the air side, $R_a' \, [= (1/h_o\pi d_o)]$, does not vary much. The conductive resistance per unit length, $R_T' \, [= (1/2\pi k) \ln(d_o/d_i)]$ is constant. The convective heat transfer coefficient on the water side varies significantly with a minimum at the lowest velocity and a maximum at the highest velocity, and the only thermal resistance that changes with the velocity of water is the convective resistance, $R_w' \, [= (1/h_i\pi \, d_i)]$. To find an upper bound for q' we set R_w' to zero.

From Table A1, the thermal conductivity of plain carbon-steel is 60.5 W/m °C. The conductive resistance R_T' is

$$R_T' = \frac{1}{2\pi k} \ln\left(\frac{d_o}{d_i}\right) = \frac{1}{2\pi \times 60.5} \ln (46/40) = 3.677 \times 10^{-4} \text{ m °C/W}$$

To find h_o, we need the outer surface temperature T_{so}. Since T_{so} depends on h_o both are found by iteration (see Example 4.3.4). Tentatively assume h_w is high and, therefore, $T_{so} = 75$ °C.

$$p_{atm} = \rho_{Hg}gh \qquad h = \text{barometric height} = 730 \text{ mm Hg}$$

From Table A4, density of mercury at 20 °C is 13 579 kg/m³.

$$p_{atm} = 13\ 579 \times 9.807 \times 0.73 = 9.7214 \times 10^4 \text{ Pa}$$

From the software CC, for $T_{so} = 75$ °C, $d_o = 0.046$ m, $T_\infty = 20$ °C,

$T_f = 47.5$ °C $p_{atm} = 97.214$ kPa
$\mu = 1.946 \times 10^{-5}$ N s/m² $\rho = 1.0564$ kg/m³
Pr = 0.71 $\beta(293.15 \text{ K}) = 3.411 \times 10^{-3} \text{ K}^{-1}$

$$Ra_d = \frac{g\,\beta\rho^2(T_{so} - T_\infty)\,d_o^3}{\mu^2} \text{ Pr} = 3.746 \times 10^5$$

From Equation 4.3.6c,

$$\text{Nu}_d = 10.05 \qquad h_o = 6.04 \text{ W/m}^2\,^\circ\text{C}$$

Now check if the assumed value of T_{so} for computing h_o is reasonable. Equating the heat transfer rate from the water to the surrounding air to the heat transfer rate from the outer surface to the surrounding air we obtain

$$q' = \frac{T_b - T_\infty}{(1/h_i\pi d_i) + (1/2\pi k)\,\ln(d_o/d_i) + (1/h_o\pi d_o)} = h_o\pi d_o(T_{so} - T_\infty)$$

From the assumption of negligible convective resistance on the water side,

$$\frac{75 - 20}{[1/(6.04 \times \pi \times 0.046)] + 3.677 \times 10^{-4}} = 6.04 \times \pi \times 0.046 \times (T_{so} - 20)$$

$$T_{so} \approx 75\,^\circ\text{C}$$

The computed and assumed values of T_{so} are equal and, therefore, T_{so} is 75 °C. The upper bound of the heat transfer rate per unit length, q'_u, is given by

$$q'_u = \frac{75 - 20}{[1/(6.04 \times \pi \times 0.046)] + 3.677 \times 10^{-4}} = \underline{48 \text{ W/m}}$$

The upper bound for the heat transfer rate for the 5-m long tube (assuming that the water temperature does not change significantly, an assumption to be verified later) is

$$q_u = q'_u L = 48 \times 5 = \underline{240 \text{ W}}$$

To find the lower bound, q'_L, the convective resistance on the water side should be a maximum; i.e., the velocity of the water should be zero. But water that is stationary in the tube contradicts the steady-state assumption as the temperature of the stagnant water in the tube would decrease with time towards the room temperature. Therefore, assume a reasonable minimum velocity of 0.25 m/s. Further assume a bulk temperature of 75 °C. As the convective heat transfer coefficient on the water side is much higher than on the air side, the inner surface temperature will be approximately equal to the bulk temperature of 75 °C. Therefore, assume $T_{si} = 75$ °C. The assumptions of T_b and T_{si} being equal to 75 °C are to be verified later. For $V = 0.25$ m/s, $T_b = 75$ °C, $T_{si} \approx 75$ °C, and $d_i = 0.04$ m we obtain from the software CC,

$T_f = 75\,^\circ\text{C}$ $\rho = 974.8 \text{ kg/m}^3$

$k = 0.666 \text{ W/m}\,^\circ\text{C}$ $\mu = 3.693 \times 10^{-4} \text{ N s/m}^2$

$\text{Re}_d = 2.64 \times 10^4$ $\text{Pr} = 2.322$

The flow is turbulent. As L/d_i (= 5/0.04 = 125) is significantly greater than 10 (as the Prandtl number increases the entrance length decreases for turbulent flows), use a correlation for fully developed velocity and temperature profiles. From Equation 4.2.3e,

$$\text{Nu}_d = 113.4 \qquad h_i = 1889 \text{ W/m}^2\,^\circ\text{C}$$

Assuming T_{so} is approximately 75 °C (to be verified later), then $h_o = 6.04 \text{ W/m}^2\,^\circ\text{C}$ (computed when calculating the upper bound of heat transfer rate).

$$q'_L = \frac{T_b - T_\infty}{(1/h_i \pi d_i) + (1/2\pi k) \ln (d_o/d_i) + (1/h_o \pi d_o)}$$

$$= \frac{75 - 20}{[1/(1889 \times \pi \times 0.04)] + 3.677 \times 10^{-4} + [1/(6.04 \times \pi \times 0.046)]}$$

$$= 47.8 \text{ W/m}$$

With the assumption of negligible change in the water temperature,

$$q'_L = 47.8 \times 5 = 239 \text{ W}$$

Check if the assumption of the outer surface temperature being approximately 75 °C is appropriate.

$$T_{so} = \frac{q'_L}{h_o \pi d_o} + T_\infty = \frac{47.8}{6.04 \times \pi \times 0.046} + 20 = 74.8 °C$$

The assumption that the outer surface temperature is approximately 75 °C for computing h_o is appropriate.

Check if the exit temperature of the water is close to 75 °C.

$$q = \dot{m} c_p (T_{be} - T_{bi})$$

$$\dot{m} = \frac{\rho V \pi \, d_i^2}{4} = \frac{974.8 \times 0.25 \times \pi \times 0.04^2}{4} = 0.3062 \text{ kg/s}$$

$$c_p = 4189 \text{ J/kg} °C$$

$$T_{be} = T_{bi} - \frac{q}{\dot{m} c_p} = 75 - \frac{239}{0.3062 \times 4189} = 74.8 °C$$

There is only a 0.2 °C temperature drop in the water and it is negligible compared with the difference in the temperatures of the water and the surrounding air.

Thus, the lower and upper bounds of the heat transfer rate from the 5-m long tube is

$$239 \text{ W} < q < 240 \text{ W}$$

From the upper and lower bounds of the heat transfer rate the conclusion is that the valve setting has no effect on the heat transfer rate as long as it is sufficiently open to give a minimum water velocity of 0.25 m/s. People who try to close or open the valve in an attempt to increase or decrease the temperature in the room will only have a psychological satisfaction.

SUMMARY

In this chapter the concepts of boundary layer, laminar and turbulent flows, and developing and fully developed regions in internal flows, are introduced. Correlations to determine the convective heat transfer coefficient in many situations of forced and natural convection are presented. In both external and internal flows, if a uniform heat flux is imposed on the surface, the surface temperature is to be determined. If the surface is maintained at a uniform temperature, the heat transfer rate is required. The computation of the total heat transfer rate is different in external and internal flows. In external flows the reference temperature for defining the convective heat transfer coefficient is known and is usually constant; in internal flows the reference temperature varies in the direction of flow. If property variations are small, the heat transfer coefficient is approximately constant in fully developed internal flows; in external flows the heat transfer coefficient varies with the distance from the leading edge because of the nature of the boundary layer (velocity and temperature) adjacent to the surface.

Values of the convective heat transfer coefficient predicted by the correlations have uncertainties of the order of $\pm 10\%$. Correlations for natural convection are based upon experiments in laboratories where precautions are taken to suppress the bulk motion of the fluid. In the majority of ''real-life'' situations, there is some bulk motion of the fluid; even when the velocity of the fluid due to such bulk motion is small, the convective heat transfer coefficient may be substantially higher than the value predicted by the correlations. With heat transfer to gases by natural convection, even at room temperatures, radiative heat transfer from the surface may be significant. The total heat transfer rate is then given by the sum of the convective and radiative heat transfer rates. To facilitate choosing an appropriate equation, the various correlations in this chapter have been compiled at the end of this section. To use the correlations effectively, the reader should be familiar with the material in the text.

Convection heat transfer coefficient, h, is defined by

$$q_c'' = h(T_s - T_f)$$

T_f is the reference temperature of the fluid. If q_c'' is the local heat flux, the local heat transfer coefficient is obtained. If q_c'' is the average heat flux, we get the average heat transfer coefficient.

The physical situations for determining the convective heat transfer coefficient can be grouped as follows:

Driving force	Forced or natural convection
Geometry	External or internal flows
Type	Laminar or turbulent
Number of phases	Single- or two-phase flows

In *forced convection* the motion of the fluid is not dependent on the existence of a difference in the temperatures of the surface and fluid. In *natural convection* the motion of the fluid is caused by the difference in temperatures of the surface and the fluid.

For *external flows (flow parallel to a flat plate)* the reference temperature is the free stream temperature of the fluid, T_∞. The thin layer of the fluid where the velocity changes from 0 to U_∞ is the *velocity boundary layer*. The layer of the fluid where the temperature changes from the temperature of the surface, T_s, to the free stream temperature, T_∞, is the *temperature boundary layer*. The critical Reynolds number, Re_{cr}, is the Reynolds number ($\rho U_\infty x / \mu$) below which the boundary layer is laminar.

A commonly accepted value for Re_{cr} is 5×10^5. Correlations for the local convective heat transfer coefficients are available for both laminar and turbulent boundary layers. For uniform T_s and T_∞ the total heat transfer rate is obtained from

$$q = \int h_x (T_s - T_\infty) \, dA$$

The correlations between Nu_x and (Re_x, Pr) are different for laminar and turbulent boundary layers. Depending on whether the boundary layer is laminar or turbulent, the appropriate correlation is used to find the total heat transfer rate. The average heat transfer coefficient is defined as

$$h_L = \frac{q}{A(T_s - T_\infty)}$$

The correlation between Nu_L (based on the average heat transfer coefficient) and (Re_L, Pr), when the boundary layer is partly laminar and partly turbulent, is derived as a function of Re_{cr}.

With uniform heat flux, the heat transfer rate is known. The local convective heat transfer coefficient *defines the local surface temperature.*

Correlations for the Nusselt number based on the average heat transfer coefficient for *cylinders* and *spheres* are presented. For *internal flows,* of constant density fluids and ideal gases, the heat transfer rate to the fluid is given by

$$q = \dot{m} c_p (T_{be} - T_{bi})$$

where T_{be} and T_{bi} are the bulk temperatures at exit and inlet respectively. The *bulk temperature of the fluid* is the temperature that the fluid would attain if it is collected at the location and mixed. For internal flows the bulk temperature is the reference temperature for the definition of the convective heat transfer coefficient. Internal flows in circular tubes are assumed to be *laminar for $Re_d < 2100$ and turbulent for $Re_d > 10\,000$.* For a certain distance from the location where the fluid enters the tube or heating of the fluid starts, the velocity and temperature profiles vary in the direction of flow. Beyond the *entrance length,* the velocity and temperature profiles do not vary; the region where the profiles do not vary is known as the *fully developed region.* The entrance length for turbulent flows is less than 10 diameters. Many correlations relating Nu_d and (Re_d, Pr) are presented for both uniform surface temperature and uniform heat flux. For *laminar flows the convective heat transfer coefficient varies with the thermal conductivity of the fluid (Nu_d is constant).* In *turbulent flows the convective heat transfer coefficient* is also a function of other temperature dependent properties of the fluid and, hence, varies in the direction of flow. The reference temperature (the bulk temperature) varies in the direction of flow. For uniform surface temperature the heat transfer can be calculated with acceptable accuracy by employing an average heat transfer

$$h_{av} = \frac{h(T_{bi}) + h(T_{be})}{2} \quad \text{or} \quad h_{av} = h\left(\frac{T_{bi} + T_{be}}{2}\right)$$

For *uniform surface temperature*

$$\ln \frac{T_s - T_{be}}{T_s - T_{bi}} = -\frac{hA_s}{\dot{m} c_p}$$

For *uniform heat flux* the heat transfer rate is known and the local convective heat transfer coefficient defines the surface temperature. The bulk temperature at any location is obtained from the relation

$$q = q''A_s = \dot{m}c_p(T_b - T_{bi})$$

In *natural convection* the reference temperature for defining the convective heat transfer coefficient is the temperature of the stagnant fluid. In natural convection the *Grashof number* plays a role similar to the role of Reynolds number in forced flows in defining the flow regimes. Correlations usually involve the *Rayleigh number, the product of the Grashof number and Prandtl number.*

When the fluid velocity is low, *both forced and natural convection occur.* Forced convection may enhance or decrease the fluid velocity caused by natural convection and the resulting convective heat transfer is higher or lower than with natural convection alone.

Software

The convection correlations segment of the program HT computes convective heat transfer coefficient in forced and natural convection employing correlations given in the text. Most of the correlating equations in the text are included in the program. The program computes the properties of water, air, and oil at the temperature appropriate for the correlating equation. For properties of water, air, and oil, the program uses equations found by regression analysis. For other fluids the user should supply the properties.

To use the convection correlations program, supply the type of convection, geometry, fluid, and any other information required to compute the heat transfer coefficient. The program then computes the Reynolds number or Grashof number and Prandtl number. Based on these values, select one of the correlation equations listed in the program. The program computes the Nusselt and number and the convective heat transfer coefficient. This program is particularly useful when the convective heat transfer coefficient has to be repeatedly computed, such as in cases where an iterative solution is required.

CONVECTION CORRELATIONS

I FORCED CONVECTION—EXTERNAL FLOWS

Evaluate properties of fluid at the film temperature, $T_f = (T_s + T_\infty)/2$, unless stated otherwise.

Flat Plate

Properties at $T_f = (T_w + T_\infty)/2$.

Laminar Boundary Layers (Re_x and Re_L and Re_{cr})

Uniform surface temperature

Local. $\text{Pr} \gtrsim 0.6$ $\qquad\qquad\qquad$ $\text{Nu}_x = \dfrac{h_x x}{k} = 0.332\ \text{Re}_x^{1/2}\ \text{Pr}^{1/3}$ $\qquad$ **(4.1.1)**

$\qquad$ $\text{Pr} < 0.05$, $\text{Re}_x\ \text{Pr} > 100$ $\qquad$ $\text{Nu}_x = 0.564\ \text{Re}_x^{1/2}\ \text{Pr}^{1/2}$ $\qquad$ **(4.1.4)**

For large variation in properties $\quad \mathrm{Nu}_x = 0.332 \, \mathrm{Re}_x^{1/2} \, \mathrm{Pr}^{1/3} (\mu_\infty/\mu_s)^{1/4}$ **(4.1.11a)**

All values of Pr $\quad \mathrm{Nu}_x = \dfrac{0.3387 \, \mathrm{Pr}^{1/3} \, \mathrm{Re}_x^{1/2}}{[1 + (0.0468/\mathrm{Pr})^{2/3}]^{1/4}}$ **(4.1.5a)**

Average. $\mathrm{Pr} \gtrsim 0.6 \quad \mathrm{Nu}_L = \dfrac{h_L L}{k} = 0.664 \, \mathrm{Re}_L^{1/2} \, \mathrm{Pr}^{1/3}$ **(4.1.3)**

$\mathrm{Nu}_L = 0.664 \, \mathrm{Re}_L^{1/2} \, \mathrm{Pr}^{1/3} (\mu_\infty/\mu_s)^{1/4}$ **(4.1.11c)**

All values of Pr $\quad \mathrm{Nu}_L = \dfrac{0.6774 \, \mathrm{Pr}^{1/3} \, \mathrm{Re}_L^{1/2}}{[1 + (0.0468/\mathrm{Pr})^{2/3}]^{1/4}}$ **(4.1.5b)**

Uniform heat flux

$\mathrm{Pr} \gtrsim 0.1 \quad \mathrm{Nu}_x = 0.453 \, \mathrm{Re}_x^{1/2} \, \mathrm{Pr}^{1/3}$ **(4.1.6a)**

$\mathrm{Pr} < 0.05 \quad \mathrm{Nu}_x = 0.886 \, \mathrm{Re}_x^{1/2} \, \mathrm{Pr}^{1/2}$ **(4.1.6b)**

All Pr $\quad \mathrm{Nu}_x = \dfrac{0.464 \, \mathrm{Re}_x^{1/2} \mathrm{Pr}^{1/3}}{[1 + (0.0207/\mathrm{Pr})^{2/3}]^{1/4}}$ **(4.1.6c)**

Turbulent Boundary Layers

Uniform surface temperature

Local. $0.6 \leq \mathrm{Pr} \leq 60$

$\mathrm{Nu}_x = \dfrac{h_x x}{k} = 0.0296 \, \mathrm{Re}_x^{4/5} \, \mathrm{Pr}^{1/3} \qquad \mathrm{Re}_{cr} < \mathrm{Re}_x < 10^7$ **(4.1.7a)**

$\mathrm{Nu}_x = 1.596 \, \mathrm{Re}_x (\ln \mathrm{Re}_x)^{-2.584} \, \mathrm{Pr}^{1/3} \qquad 10^7 < \mathrm{Re}_x < 10^9$ **(4.1.7b)**

$\mathrm{Nu}_x = 0.0296 \, \mathrm{Re}_x^{4/5} \, \mathrm{Pr}^{0.43} (\mu_\infty/\mu_s)^{1/4} \qquad \mathrm{Re}_x > \mathrm{Re}_{cr}$ **(4.1.11b)**

Average convective heat transfer coefficient. $\mathrm{Pr} \gtrsim 0.6$

$\mathrm{Nu}_L = \dfrac{h_L L}{k} = [0.664 \, \mathrm{Re}_{cr}^{1/2} + 0.037(\mathrm{Re}_L^{4/5} - \mathrm{Re}_{cr}^{4/5})] \, \mathrm{Pr}^{1/3}$ **(4.1.8)**

For $\mathrm{Re}_{cr} = 5 \times 10^5$ Equation 4.1.8 reduced to

$5 \times 10^5 < \mathrm{Re}_L < 10^7$

$\mathrm{Nu}_L = \dfrac{h_L L}{k} = (0.037 \, \mathrm{Re}_L^{4/5} - 871) \, \mathrm{Pr}^{1/3}$ **(4.1.9a)**

$\mathrm{Re}_{cr} = 5 \times 10^5 \qquad 10^7 < \mathrm{Re}_L < 10^9$

$\mathrm{Nu}_L = [1.967 \, \mathrm{Re}_L (\ln \mathrm{Re}_L)^{-2.584} - 871] \, \mathrm{Pr}^{1/3}$ **(4.19b)**

$\mathrm{Nu}_L = (0.037 \, \mathrm{Re}_L^{4/5} - 871) \, \mathrm{Pr}^{0.43} (\mu_\infty/\mu_s)^{1/4}$ **(4.1.11d)**

Uniform heat flux

$\mathrm{Nu}_x = 0.03 \, \mathrm{Re}_x^{0.8} \, \mathrm{Pr}^{0.6}$ **(4.1.12)**

$0.5 < \mathrm{Pr} < 10$ **(4.1.13a)**

$\mathrm{Nu}_x = \dfrac{\sqrt{c_{fx}/2} \, \mathrm{Re}_x \, \mathrm{Pr}}{2.21 \ln (\mathrm{Re}_x \sqrt{c_{fx}/2}) - 0.232 \ln \mathrm{Pr} + 14.9 \, \mathrm{Pr}^{0.623} - 15.6}$

$$10 < \text{Pr} < 500 \tag{4.1.13b}$$

$$\text{Nu}_x = \frac{\sqrt{c_{fx}/2}\ \text{Re}_x\ \text{Pr}}{2.21\ \ln(\text{Re}_x\sqrt{c_{fx}/2}) - 0.232\ \ln \text{Pr} + 10\ \text{Pr}^{0.741} - 6.21}$$

$$c_{fx} \approx 0.0592\ \text{Re}_x^{-0.2} \tag{4.1.13c}$$

Cylinders

$$\text{Nu}_d = \frac{h\,d}{k} = c\text{Re}_d^m\ \text{Pr}^n \tag{4.1.18}$$

For values of c, m, and n see Table 4.1.1

For $\text{Re}_d\ \text{Pr} > 0.2$

$\text{Re}_d > 400\ 000$

$$\text{Nu}_d = 0.3 + \frac{0.62\ \text{Re}_d^{1/2}\ \text{Pr}^{1/3}}{[1 + (0.4/\text{Pr})^{2/3}]^{1/4}}\left[1 + \left(\frac{\text{Re}_d}{282\ 000}\right)^{5/8}\right]^{4/5} \tag{4.1.19}$$

$20\ 000 < \text{Re}_d < 400\ 000$

$$\text{Nu}_d = 0.3 + \frac{0.62\ \text{Re}_d^{1/2}\ \text{Pr}^{1/3}}{[1 + (0.4/\text{Pr})^{2/3}]^{1/4}}\left[1 + \left(\frac{\text{Re}_d}{282\ 000}\right)^{1/2}\right] \tag{4.1.20}$$

$\text{Re}_d < 20\ 000$

$$\text{Nu}_d = 0.3 + \frac{0.62\ \text{Re}_d^{1/2}\ \text{Pr}^{1/3}}{[1 + (0.4/\text{Pr})^{2/3}]^{1/4}} \tag{4.1.21}$$

Spheres

Evaluate all properties at T_∞, except μ_s at T_s.

$$3.5 < \text{Re}_d < 76\ 000 \qquad 0.71 < \text{Pr} < 380 \qquad 1 < \mu_\infty/\mu_s < 3.2$$

$$\text{Nu}_d = 2.0 + (0.4\ \text{Re}_d^{1/2} + 0.06\ \text{Re}_d^{2/3})\ \text{Pr}^{2/5}\left(\frac{\mu_\infty}{\mu_s}\right)^{1/4} \tag{4.1.22}$$

II FORCED CONVECTION—INTERNAL FLOWS

Evaluate all properties at the bulk temperature, except μ_s at T_s.

Laminar Flows

Uniform wall temperature

$$0.48 < \text{Pr} < 16\ 700 \qquad 0.0044 < \mu/\mu_s < 9.75$$

$$\overline{\text{Nu}_d} = 1.86\left(\frac{\text{Re}_d\ \text{Pr}}{L/d}\right)^{1/3}\left(\frac{\mu}{\mu_s}\right)^{0.14} \qquad \frac{L}{d} < \frac{8}{\text{Re}_d\ \text{Pr}}\left(\frac{\mu_s}{\mu}\right)^{0.42} \tag{4.2.1}$$

Equation 4.2.1 yields the average heat transfer coefficient over a length L of the tube in the entrance region.

Fully developed flow

| Uniform surface temperature | $\text{Nu}_d = 3.66$ | (4.2.2a) |

Uniform surface temperature $\text{Nu}_d = 3.66$ (4.2.2a)

Uniform heat flux $\text{Nu}_d = 4.36$ (4.2.2b)

Turbulent Flows

Uniform surface temperature and uniform heat flux

$$\text{Nu}_d = 0.023 \, \text{Re}_d^{4/5} \, \text{Pr}^{1/3} \tag{4.2.3a}$$

$0.7 \leq \text{Pr} \leq 160 \qquad \text{Re}_d \geq 10\,000 \qquad L/d \geq 10$

$$\text{Nu}_d = 0.023 \, \text{Re}_d^{4/5} \, \text{Pr}^n \tag{4.2.3b}$$

$n = 0.4$ for heating ($T_s > T_b$) and $n = 0.3$ for cooling ($T_s < T_b$)
For large variations in viscosity:

$0.7 \leq \text{Pr} \leq 16\,700 \qquad \text{Re}_d \geq 10\,000 \quad L/d \geq 10$

$$\text{Nu}_d = 0.027 \, \text{Re}_d^{4/5} \, \text{Pr}^{1/3}(\mu/\mu_s)^{0.14} \tag{4.2.3c}$$

$0.5 < \text{Pr} < 1.5 \qquad 2300 < \text{Re}_d < 10^6 \qquad 0 < d/L < 1$

$$\text{Nu}_d = 0.0214(\text{Re}_d^{4/5} - 100) \, \text{Pr}^{2/5} \left[1 + \left(\frac{d}{L} \right)^{2/3} \right] \tag{4.2.3d}$$

$1.5 < \text{Pr} < 500 \qquad 2300 < \text{Re}_d < 10^6$

$$\text{Nu}_d = 0.012(\text{Re}_d^{0.87} - 280) \, \text{Pr}^{2/5} \left[1 + \left(\frac{d}{L} \right)^{2/3} \right] \tag{4.2.3e}$$

Modify Equations 4.2.3d and e by multiplying them by $(T_b/T_s)^{0.45}$ for gases and by $(\text{Pr}/\text{Pr}_s)^{0.11}$ for liquids

$$\text{Nu}_d = \frac{(f/8)(\text{Re}_d - 1000) \, \text{Pr}}{1 + 12.7(f/8)^{1/2}(\text{Pr}^{2/3} - 1)} \left[1 + \left(\frac{d}{L} \right)^{2/3} \right] \tag{4.2.3f}$$

$$f = (0.79 \ln \text{Re}_d - 1.64)^{-2} \tag{4.2.3g}$$

The friction factor f is given by Equation 4.2.3g. $n = 0.11$ for liquids heating, $n = 0.25$ for liquids cooling, and $n = 0$ for gases.

$10^4 < \text{Re}_d < 5 \times 10^6 \qquad 0.08 < \mu/\mu_s < 40$

$$\text{Nu}_d = \frac{f/8 \, \text{Re}_d \, \text{Pr}}{1.07 + 12.7(f/8)^{1/2}(\text{Pr}^{2/3} - 1)} \left(\frac{\mu}{\mu_s} \right)^n \tag{4.2.4a}$$

For $\text{Pr} \ll 1$,

Uniform surface temperature

$$\text{Nu}_{d,b} = 4.8 + 0.0156 \, \text{Re}_{d,f}^{0.85} \, \text{Pr}_s^{0.93} \tag{4.2.4b}$$

Uniform heat flux

$$\text{Nu}_{d,b} = 6.3 + 0.0167 \, \text{Re}_{d,f}^{0.85} \, \text{Pr}_s^{0.93} \tag{4.2.4c}$$

The subscripts b, f, and s denote that properties are to be evaluated at the bulk temperature, film temperature (arithmetic mean temperature) and wall temperature, respectively

Uniform wall temperature $\qquad \ln \dfrac{T_s - T_{be}}{T_s - T_{bi}} = -\dfrac{hA_s}{\dot{m}c_p}$ $\tag{4.2.5a}$

III NATURAL CONVECTION

Evaluate all properties at the film temperature $T_f = (T_s + T_\infty)/2$. For some correlations β is evaluated at T_∞.

$$Gr = g\beta \; \Delta T \; L^3/\nu^2 \qquad Nu = hL/k \qquad\qquad Ra = Gr \; Pr$$

$$Gr^* = g\beta q_w'' \; L^4/(\nu^2 k) \qquad L = \text{characteristic length}$$

Vertical Plates

Uniform wall temperature. Evaluate β at T_∞ and all other properties at $(T_s + T_{()})/2$.

$$Nu_L = 0.68 + \frac{0.67 \; Ra_L^{1/4}}{[1 + (0.492/Pr)^{9/16}]^{4/9}} \qquad 10^{-1} < Ra_L < 10^9 \tag{4.3.1a}$$

$$Nu_L = \left\{ 0.825 + \frac{0.387 \; Ra_L^{1/6}}{[1 + (0.492/Pr)^{9/16}]^{8/27}} \right\}^2 \quad 10^{-1} < Ra_L < 10^{12} \tag{4.3.1b}$$

Uniform heat flux. Evaluate β at T_∞ and all other properties at $(T_{y=L/2} + T_\infty)/2$.

$$Nu_L = 0.68 + \frac{0.67 \; Ra_L^{1/4}}{[1 + (0.437/Pr)^{9/16}]^{4/9}} \qquad Ra_L < 10^9 \tag{4.3.2a}$$

$$Nu_L = \left\{ 0.825 + \frac{0.387 \; Ra_L^{1/6}}{[1 + (0.437/Pr)^{9/16}]^{8/27}} \right\}^2 \quad 10^{-1} < Ra_L < 10^{12} \tag{4.3.2b}$$

Evaluate all properties at $T_f = (T_s + T_\infty)/2$.

$$Nu_x = 0.6(Gr_x^* \; Pr)^{0.2} \qquad 10^5 < Gr_x^* \; Pr < 10^{13} \tag{4.3.2c}$$

$$Nu_x = 0.568(Gr_x^* \; Pr)^{0.22} \qquad 10^{13} < Gr_x^* \; Pr < 10^{16} \tag{4.3.2d}$$

$$Gr_x^* = (g\beta\rho^2 q_w'' x^4)/(\mu^2 k)$$

Inclined Plates

Uniform surface temperature. For downward-facing heated plates or upward-facing cooled plates, use equations for vertical plates, replacing g by $g \cos \theta$ to compute Ra_L (θ is the angle of inclination of the plate to the vertical).

Uniform heat flux. For downward-facing heated plates and upward-facing cooled plates, evaluate properties (except β) at $T_r = T_s - 0.25(T_s - T_\infty)$; evaluate β at $T_\infty + 0.25(T_s - T_\infty)$.

$Gr_c \; Pr = e^{14.223 + 8.3945\cos\theta} \qquad 15° < \theta < 60°$
$Gr_c \; Pr = e^{3.848 + 29.146\cos\theta} \qquad 60° < \theta < 70°$
θ = angle of inclination of the plate from the vertical.
For $Gr < Gr_c$ use Equations 4.3.2a or 4.3.2b with g replaced by $g \cos \theta$.
For $Gr > Gr_c$,

$$Nu_L = 0.14[(Gr \; Pr)^{1/3} - (Gr_c \; Pr)^{1/3}] + 0.56(Gr_c \; Pr \cos \theta)^{1/4} \tag{4.3.3a}$$

Evaluate properties at $(T_s + T_\infty)/2$.
Ra_x^* is the modified Raleigh number $= Gr_x^* \; Pr$
θ is the angle of inclination of the surface to the vertical.
$Ra_{x,cr}^* = 6.31 \times 10^{12} e^{0.07050} \qquad$ (θ in degrees)

$$Nu_x = 0.592(Ra_x^* \cos \theta)^{0.2} \qquad Ra_x^* < Ra_{x,cr} \qquad 0 < \theta < 86.5° \tag{4.3.3b}$$

$$Nu_x = 0.889(Ra_x^* \cos \theta)^{0.205} \qquad Ra_x^* > Ra_{x,cr}^* \qquad \theta < 31° \tag{4.3.3c}$$

Horizontal Plates

Downward-facing heated plates and upward-facing cooled plates, rectangular plates, L is the width of the plate (smaller dimension). Evaluate properties at $T_s - 0.25(T_s - T_\infty)$, except β at $T_\infty + 0.25(T_s - T_\infty)$.

$$\mathrm{Nu}_L = 0.58 \ \mathrm{Ra}_L^{1/5} \qquad 10^6 < \mathrm{Ra}_L < 10^{11} \tag{4.3.4}$$

Evaluate properties at T_f, $L = A/P$.

$$\mathrm{Nu}_L = 0.27 \ \mathrm{Ra}_L^{1/4} \qquad 10^5 < \mathrm{Ra}_L < 10^{10} \tag{4.3.5g}$$

Upward-facing heated plates and downward-facing cooled plates, rectangular plates, L is the width of the plate (smaller dimension). Evaluate properties at $T_s - 0.25(T_s - T_\infty)$, except β at $T_\infty + 0.25(T_s - T_\infty)$.

$$\mathrm{Nu}_L = 0.16 \ \mathrm{Ra}_L^{1/3} \qquad 6 \times 10^6 < \mathrm{Ra}_L < 2 \times 10^8 \tag{4.3.5a}$$

$$\mathrm{Nu}_L = 0.13 \ \mathrm{Ra}_L^{1/3} \qquad 5 \times 10^8 < \mathrm{Ra}_L < 10^{11} \tag{4.3.5b}$$

Evaluate properties at $(T_s + T_\infty)/2$; $L = A/P$.

$$\mathrm{Nu}_L = 0.54 \ \mathrm{Ra}_L^{1/4} \qquad 2.2 \times 10^4 < \mathrm{Ra}_L < 8 \times 10^6 \tag{4.3.5c}$$

$$\mathrm{Nu}_L = 0.15 \ \mathrm{Ra}_L^{1/3} \qquad 8 \times 10^6 < \mathrm{Ra}_L < 1.5 \times 10^9 \tag{4.3.5d}$$

$$\mathrm{Nu}_L = 0.96 \ \mathrm{Ra}_L^{1/6} \qquad 1 < \mathrm{Ra}_L < 200 \tag{4.3.5e}$$

$$\mathrm{Nu}_L = 0.59 \ \mathrm{Ra}_L^{1/4} \qquad 200 < \mathrm{Ra}_L < 10^4 \tag{4.3.5f}$$

Long Horizontal Cylinders

Characteristic length is the diameter. Evaluate properties at $(T_s + T_\infty)/2$. The values of c and m are functions of Ra_d and are given in Table 4.3.1.

$$\mathrm{Nu}_d = c \ \mathrm{Ra}_d^m \tag{4.3.6a}$$

Evaluate β at T_∞ and all other properties at $(T_s + T_\infty)/2$.

$$\mathrm{Nu}_d = \left[0.6 + \frac{0.387 \ \mathrm{Ra}_d^{1/6}}{[1 + (0.559/\mathrm{Pr})^{9/16}]^{8/27}} \right]^2 \qquad 10^{-3} < \mathrm{Ra}_d < 10^{13} \tag{4.3.6b}$$

$$\mathrm{Nu}_d = 0.36 + \frac{0.518 \ \mathrm{Ra}_d^{1/4}}{[1 + (0.559/\mathrm{Pr})^{9/16}]^{4/9}} \qquad 10^{-6} < \mathrm{Ra}_D < 10^9 \tag{4.3.6c}$$

Spheres

Evaluate β at T_∞ and all other properties at $(T_s + T_\infty)/2$.

$$\mathrm{Nu}_d = 2 + 0.426 \ \mathrm{Ra}_d^{0.25} \qquad 1 < \mathrm{Ra}_d < 10^5 \qquad \mathrm{Pr} \approx 1 \tag{4.3.7a}$$

$$\mathrm{Nu}_d = 2 + \frac{0.589 \ \mathrm{Ra}_d^{1/4}}{[1 + (0.469/\mathrm{Pr})^{9/16}]^{4/9}} \qquad \mathrm{Ra}_d < 10^{11} \qquad \mathrm{Pr} \gtrsim 0.7 \tag{4.3.7b}$$

IV COMBINED FORCED AND NATURAL CONVECTION

Horizontal Tubes

Uniform surface temperature. $L/d \leq 28.4$; evaluate properties at the average bulk temperature, except μ_s at T_s.

$$\mathrm{Nu}_d = 1.75[\mathrm{Gz} + 0.12(\mathrm{Gz} \ \mathrm{Gr}^{1/3} \ \mathrm{Pr}^{0.36})^{0.88}]^{1/3}(\mu/\mu_s)^{0.14} \tag{4.4.1}$$

where

$\mathrm{Nu}_d = hd/k$ $\mathrm{Gz} = \text{Graetz number} = \dot{m}c_p/(kL)$
$\mathrm{Gr} = g\beta\,\Delta T\,d^3/\nu^2$ $\mathrm{Pr} = \nu/\alpha$
$d = \text{diameter}$ $L = \text{Length}$

Uniform heat flux. Evaluate properties at $(T_s + T_b)/2$.

$$3 \times 10^4 < \mathrm{Ra} < 10^6 \qquad 4 < \mathrm{Pr} < 175 \qquad 2 < hd^2/(k_w t) < 66$$

$$\mathrm{Nu}_d = \left\{ (4.36)^2 + \left[0.145 \left(\frac{\mathrm{Gr}_d^* \, \mathrm{Pr}^{1.35}}{P_w^{0.25}} \right)^{0.265} \right]^2 \right\}^{0.5} \tag{4.4.2}$$

where

$\mathrm{Nu}_d = hd/k$ $\mathrm{Pr} = \nu/\alpha$

$\mathrm{Gr}_d^* = g\beta\,d^4 q_w''/(\nu^2 k)$ $q_w'' = \text{wall heat flux}$

$k_w = \text{tube wall thermal conductivity}$ $P_w = kd/(k_w t)$

$\mathrm{Ra}_d = g\beta\,\Delta T\,d^3\,\mathrm{Pr}/\nu^2$ $t = \text{tube wall thickness}$

REFERENCES

Bhatti, M.S., and Shah, R.K. (1987). Turbulent and transition flow convective heat transfer in ducts. In *Handbook of Single-Phase Convective Heat Transfer,* eds. Kakac, S., Shah, R.K., and Aung, W. New York: Wiley Interscience.

Churchill, S.W. (1976). A comprehensive correlation equation for forced convection from flat plate. *AIChE J.* 22(2):264.

———(1990). Free convection around immersed bodies. In *Handbook of Heat Exchanger Design,* ed. Hewitt, G.F. New York: Hemisphere Publishing.

Churchill. S.W., and Bernstein, M. (1977). A correlating equation for forced convection from gases and liquids to a circular cylinder in cross flow. *J. Heat Transfer* 99:300.

Churchill, S.W., and Chu, H.H.S. (1975a). Correlating equations for laminar and turbulent free convections from a horizontal cylinder. *Int. J. Heat Mass Transfer* 18:1049.

———(1975b). Correlating equations for laminar and turbulent free convection from a vertical plate. *Int. J. Heat Mass Transfer* 18:1323.

Churchill, S.W., and Ozoe, H. (1973a). Correlation for laminar forced convection with uniform heating in flow over a plate and in developing and fully developed flow in a tube. *J. Heat Transfer* 95C:78.

———(1973b). Correlations for laminar and forced convection in flow over an isothermal flat plate and in developing and fully developed flow in an isothermal tube. *J. Heat Transfer* 95C:416.

Deissler, R.G. (1953). Analysis of turbulent heat transfer and flow in the entrance regions of smooth passages. NACA TN 3016. Quoted in *Handbook of Single-Phase Convective Heat Transfer,* eds. Kakac, S., Shah, R.K., and Aung, W. New York: Wiley Interscience, 1987.

Depew, C.A., and August, S.E. (1971). Heat transfer due to combined free and forced convection in a horizontal and isothermal tube. *TRANS. ASME* 93C:380.

Dittus, F.W., and Boelter, L.M.K. (1930). Heat transfer in automobile radiators of the tubular type. *Univ. Calif. Pub. Eng.* 13:443.

Fand, R.M., and Keswani, K.K. (1973). Combined natural and forced convection heat transfer from horizontal cylinders to water. *Int. J. Heat Mass Transfer* 16:1175.

Fussey, D.E., and Warneford, I.P. (1978). Free convection from a downward facing inclined flat plate. *Int. J. Heat Mass Transfer* 21:119.

Fujii, T., and Imura, H. (1972). Natural convection heat transfer from a plate with arbitrary inclination. *Int. J. Heat Mass Transfer* 15:755.

Gnielinsky, V. (1976). New equations for heat and mass transfer in turbulent pipe channel flow. *Int. Chem. Eng.* 16:359.

Gnielinsky, V. (1990). Forced convection in ducts. In *Handbook of Heat Exchanger Design,* ed. Hewitt, G.F. New York: Hemisphere Publishing.

Goldstein, R.J., Sparrow, E.M., and Jones, D.C. (1973). Natural convection mass transfer adjacent to horizontal plates. *Int. J. Heat Mass Transfer* 16:1025.

Haaland, S.E. (1983). Simple and explicit formulas for the friction factor in turbulent pipe flow. *J. Fluids Eng.* 105:89.

Hilpert, R. (1933). Warmeabgabe von geheizen Drahten und Rohren, *Forsch. Geb. Ingenieurwes* 4:220.

Jacob, M. (1949). *Heat Transfer,* vol. 1. New York: Wiley.

Kakac, S., Shah, R.K., and Aung, W. (1987). *Handbook of Single-phase Convective Heat Transfer.* New York: Wiley Interscience.

Kays, W.M., and Crawford, M.E. (1980). *Convective Heat and Mass Transfer.* New York: McGraw-Hill.

Lienhard, J.H. (1987). *A Heat Transfer Textbook,* 2d ed. Englewood Cliffs, NJ: Prentice Hall.

Lloyd, J.R., and Moran, W.R. (1974). Natural convection adjacent to horizontal surfaces of various plan forms, *J. Heat Transfer* 96:443.

Lloyd, J.R., and Sparrow, E.M. (1961). Combined forced and free convection flow on vertical surfaces. *Int. J. Heat Mass Transfer* 13:434.

Metais, B., and Eckert, E.R.G. (1964). Forced, mixed, and free convection regimes. *TRANS. ASME* 86C:295.

Morcos, S.M., and Bergles, A.E. (1975). Experimental investigation of combined forced and free laminar convection in a horizontal tube. *TRANS. ASME* 97C:212.

Morgan, V.T. (1975). The overall convective heat transfer from smooth circular cylinders.'' In *Advances in Heat Transfer,* vol. 11, eds. Irvine, T.F., and Hartnett, J.P. New York: Academic Press, 1975.

Mori, Y. (1960). ''Buoyancy effects in forced laminar convection flow over a horizontal plate.'' *TRANS. ASME.* 82C:214.

Petukhov, B.S. (1970). Heat transfer and friction in turbulent pipe flow with variable physical properties. In *Advances in Heat Transfer,* vol. 6. eds. Irvine, T.F., and Hartnett, J.P. New York: Academic Press, 1970.

Shah, R.K. and Bhatti, M.S. (1987). Laminar convective heat transfer in ducts. In *Handbook of Single-Phase Convective Heat Transfer,* eds. Kakac, S., Shah, R.K., and Aung, W. New York: Wiley Interscience.

Schlichting, H. (1979). *Boundary Layer Theory,* 7th ed. New York: McGraw-Hill.

Sieder, E.N., and Tate, C.E. (1936). Heat transfer and pressure drop of liquids in tubes. *Ind. Eng. Chem.* 28:1429.

Sleicher, C.A., and Rouse, M.W. (1975). A convenient correlation for heat transfer to constant and variable property fluids in turbulent pipe flow. *Int. J. Heat Mass Transfer* 18:677.

Sparrow, E.M., and Gregg, J.L. (1956). Laminar free convection from a vertical plate. *TRANS. ASME* 78C:435.

Sparrow, E.M., and Stretton, A.J. (1985). Natural convection from variously oriented cubes and other bodies of unity aspect ratios. *Int. J. Heat Mass Transfer* 28:747.

Thomas, L.C., and Al-Sharifi, M.M. (1981). An integral analysis for heat transfer to turbulent incompressible boundary layer flow. *J. Heat Transfer* 103:772.

Thomas, W.C. (1977). Note on the heat transfer equation for forced-convection flow over a flat plate with an unheated starting length. *Mechanical Engineering News (ASEE)* 9(1):19.

Vliet, G.C., and Liu, C.K. (1969). An experimental study of natural convection boundary layers. *TRANS. ASME* 91C:517.

Whitaker, S. (1972). Forced convection heat transfer correlations for flow in pipes, past plates, single cylinders, single spheres, and flow in packed beds and tube bundles. *AIChE J.* 18:361.

Whitaker, S. (1985). *Fundamental Principles of Heat Transfer.* Malabar, FL: Robert E. Krieger.

Yuge, T. (1960). Experiments on heat transfer from spheres including combined natural and forced convection. *TRANS. ASME* 82C:214.

Zhi-qing, W. (1982). Study on correction coefficients of laminar and turbulent entrance region effects in round pipes. *Appl. Math. Mech.* 3(3):433.

REVIEW QUESTIONS

(a) Define convective heat transfer coefficient. How does it differ from the thermal conductivity of the fluid?

(b) Both the Nusselt number and the Biot number have the same form. What are the differences between the two in terms of the variables employed in defining them and in their significance?

(c) What is the difference between forced and natural convection?

(d) Define the bulk temperature of a fluid.

(e) How would you use the convective heat transfer coefficient in internal, forced convection when the tube surface is at a uniform temperature? When a uniform heat flux is imposed on the tube surface?

(f) What is the effect of the Prandtl number of a fluid on the relative thicknesses of velocity and temperature boundary layers when the fluid flow is parallel to a flat plate?

(g) When a fluid flow is parallel to a flat plate, does the drag force increase or decrease by heating the plate if the fluid is air? If the fluid is oil?

(h) Two fluids, with different properties, flow with equal free stream velocities parallel to a flat plate. What property of the fluids determines if the velocity boundary layer thickness of one is greater than that of the other?

(i) What do you understand by the terms fully developed velocity and temperature profile regions in internal flows?

(k) Do you expect the convective heat transfer coefficient in the thermally developing region to be higher or lower (or equal to) than the convective heat transfer coefficient in the fully developed temperature profile region? Support your answer with qualitative logic.

(l) Fluids in space ships are subjected to a higher effective body force (during launch) and lower effective body force (orbiting) than on earth. How would you simulate such conditions on earth?

(m) How would you achieve very low Ra?

PROBLEMS
Assume only convective heat transfer and standard atmospheric pressure for air unless stated otherwise.

4.1 Properties of a liquid at 30 °C are
$$\rho = 1396 \text{ kg/m}^3 \qquad c_p = 935 \text{ J/kg °C}$$
$$\alpha = 5.59 \times 10^{-8} \text{ m}^2/\text{s} \qquad \nu = 21.4 \times 10^{-8} \text{ m}^2/\text{s}$$
Determine the dynamic viscosity, thermal conductivity, and Prandtl number of the fluid.

4.2 Water flows parallel to a flat plate at a free stream temperature of 40 °C. At a particular section, the plate temperature is 20 °C and the temperature gradient in water perpendicular to the plate surface is 120 000 °C/m at the surface. Calculate the heat flux at the wall and the convective heat transfer coefficient at that location.

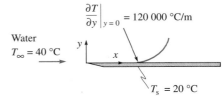

FIGURE P4.2

4.3 Air at 60 °C flows parallel to a flat plate which is at 20 °C. The heat flux at a particular location is 1440 W/m². Calculate the convective heat transfer coefficient and the temperature gradient in the air perpendicular to the plate at that location.

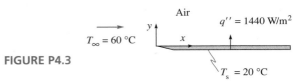

FIGURE P4.3

4.4 A fluid at 40 °C flows parallel to a flat plate with a uniform velocity. Estimate the distance from the leading edge over which the boundary layer is laminar for the following free stream velocities.
(a) Air   10 m/s
(b) Water   4 m/s
(c) Oil   1 m/s

4.5 A fluid at 40 °C flows parallel to a flat plate with a uniform free stream velocity. Calculate the velocity and temperature boundary layer thicknesses at distances of 1 cm, 50 cm, and 100 cm from the leading edge for the following conditions:
(a) Air   3 m/s
(b) Air   15 m/s
(c) Water   3 m/s
(d) Engine oil   1 m/s

4.6 A fluid at 40 °C flows parallel to a flat plate with a uniform velocity of 3 m/s. The plate is maintained at a uniform temperature of 60 °C. Plot the velocity and temperature boundary layer thicknesses as a function of the distance from the leading edge if the fluid is
(a) Air
(b) Water
(c) Engine oil

4.7 Atmospheric air at 300 K flows parallel to a 2-m long, thin flat plate which is maintained at 400 K. The free stream velocity of the air is 15 m/s. Assuming a critical Reynolds number of 500 000, determine (per meter width of one side of the plate)

(a) The total heat transfer rate from the plate to the air.

(b) The heat transfer rate from that part of the plate where the boundary layer is turbulent.

4.8 A fluid with constant properties, flows over two plates of the same width but of different lengths as shown. Determine the ratio of q_1/q_2.

$$U_{\infty 2} = U_{\infty 1} \qquad L_2 = 2L_1 \qquad Re_{L1} = 2 \times 10^5$$
$$T_{\infty 2} = T_{\infty 1} \qquad T_{s2} = T_{s1}$$

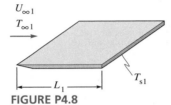
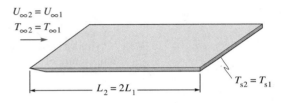

FIGURE P4.8

4.9 A fluid flows parallel to a flat plate of length L in the direction of flow and width W. It is known that $h_x = cx^{-1/n}$, where c and n are positive constants and x is the coordinate in the direction of the flow with its origin at the leading edge. Find an expression for the total heat transfer rate from one side of the plate if

(a) $T_s - T_\infty = $ constant

(b) $T_s - T_\infty = ax^{(1/p)}$ (a and p are positive constants.)

4.10 As shown in Figure P4.10, heat is transferred from a flat plate to a fluid at a Reynolds number, $Re_L = 2 \times 10^5$. To increase the heat transfer rate, two proposals are made: (1) To double the velocity of the fluid keeping all other conditions the same (Figure P4.10b), and (2) To double the length keeping all other conditions the same (Figure P4.10c). Determine q_A/q_B.

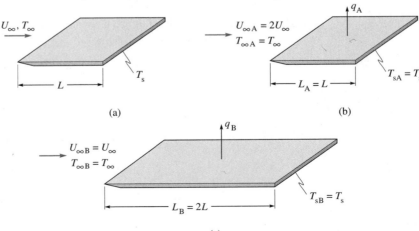

FIGURE P4.10

4.11 The surface temperature of a bar exposed to a fluid is approximated by

$$T = T_2 + (T_1 - T_2) \cos\left(\frac{\pi x}{2L}\right)$$

Estimate the convective heat transfer rate from the bar to the surrounding fluid, which is at T_∞, if the convective heat transfer coefficient is h.

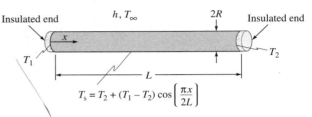

$$T_s = T_2 + (T_1 - T_2) \cos\left[\frac{\pi x}{2L}\right]$$

FIGURE P4.11

4.12 A cylindrical bar is attached to a plate and the bar exposed to a fluid. The surface temperature of the bar is:

$$T = T_\infty + (T_b - T_\infty)e^{-cx}$$

where c is a constant. Determine the convective heat transfer rate from the bar to the fluid if

$T_\infty = 20\ °C \qquad T_b = 100\ °C \qquad c = 15\ m^{-1}$
$L = 20\ cm \qquad h = 20\ W/m^2\ °C \qquad d = 1\ cm$

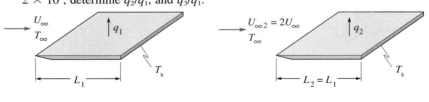

FIGURE P4.12

4.13 Three plates of equal width are maintained at the same temperature. If $Re_{L1} = 2 \times 10^5$, determine q_2/q_1, and q_3/q_1.

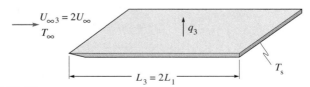

FIGURE P4.13

4.14 A thin flat plate of length, $L = 1.75$ m, separates two streams of a gas that flow parallel to the plate in opposite directions on either side of the plate. On one side of the plate, $U_{\infty 1} = 30$ m/s, $T_{\infty 1} = 200$ °C, while on the other side $U_{\infty 2} = 10$ m/s, $T_{\infty 2} = 25$ °C. Assuming isothermal correlations apply and conduction in the plate in the direction of flow is negligible, determine the heat flux and surface temperature at the location A if

(a) The fluid is a gas with constant properties $\nu = 18 \times 10^{-6}$ m²/s, $k = 0.028$ W/m °C, and Pr $= 0.7$.

(b) The fluid is air.

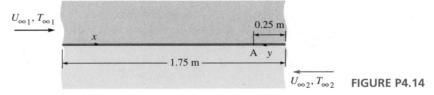

FIGURE P4.14

4.15 Atmospheric air at 300 K flows parallel to a 3-m long flat plate with a free stream velocity of 10 m/s. The plate is at a uniform temperature of 450 K. Determine (per meter width of the plate)

(a) The total heat transfer rate from the plate to the air.

(b) The heat transfer rate from that part of the plate where the boundary layer is turbulent.

4.16 A fluid flows parallel to a thin flat plate of length L, maintained at a uniform temperature. At $x_1 = L/2$, the local heat transfer coefficient is $h_{x1} = 20$ W/m² °C. Find the average heat transfer coefficient, h_L, over the entire length of the plate if

(a) $Re_{x1} = 100\,000$

(b) $Re_{x1} = 300\,000$

(c) $Re_{x1} = 600\,000$

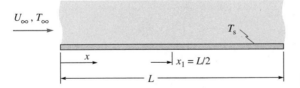

FIGURE P4.16

4.17 Oil at 100 °C is cooled by a column of 2-cm wide (in the direction of flow of the oil) strips that are maintained at 60 °C. The velocity of the oil is 1.5 m/s. What is the heat transfer rate per unit area of the strip?

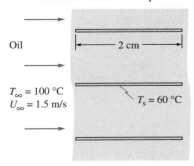

FIGURE P4.17

4.18 A flat plate heater heats engine oil flowing parallel to the plate. The plate is 60-cm long in the direction of the flow of the oil and 80-cm wide. The oil flows with a free stream velocity of 2 m/s and a free stream temperature of 60 °C. The heat transfer rate is to be 6 kW. The plate is heated by condensing steam that maintains both surfaces of the plate at its saturation temperature. Determine the temperature of the plate.

4.19 A copper plate is 1-m long, 50-cm wide, and 1-cm thick. It is at a uniform temperature of 90 °C. It is placed in a stream of water that is at 20 °C and flowing at 2 m/s. Estimate the time taken for the temperature of the plate to reach 30 °C if

(a) The 50-cm dimension is parallel to the flow direction.

(b) The 1-m wide dimension is parallel to the flow direction.

(Assume flat plate correlations are valid)

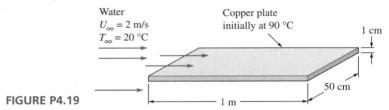

FIGURE P4.19

4.20 A thin, 0.8-m long copper plate separates a stream of air at 100 °C flowing at 15 m/s from a stream of water (flowing in the opposite direction) at 20 °C flowing at 3 m/s on the other side of the plate. Determine the heat transfer rate from the air to the water per meter width of the plate. Assume that the correlations for average heat transfer coefficients for isothermal surfaces are valid.

FIGURE P4.20

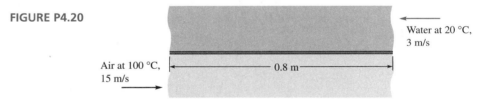

4.21 Thin strip heaters, 10-mm wide and 50-cm wide, are placed on a 50-cm wide flat plate. The heaters are placed such that there is no space between two adjacent heaters and the top surface of the elements form a smooth surface. Each strip is electrically and thermally insulated from its neighboring strips. Air at 20 °C flows over the plate with a free stream velocity of 10 m/s. By controlling the power input to each of the elements, the surfaces of strips are maintained at a uniform temperature of 80 °C. Determine the heat transfer rate from strips 1, 5, 10, 80, 151, and 200 from the leading edge and the total heat transfer rate.

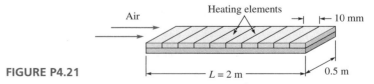

FIGURE P4.21

4.22 The bottom surface of a crank case of an automobile engine is 50-cm wide (parallel to the wheel axis) and 30-cm long. The hot oil from the engine keeps the bottom surface at 70 °C when the automobile is going at 65 mph.

(a) Determine the heat transfer rate from the surface to the air (air temperature is 35 °C).

(b) Because of the nature of the placement of the crank case and free stream turbulence, the boundary layer is turbulent from the leading edge. What is the heat transfer rate?

FIGURE P4.22

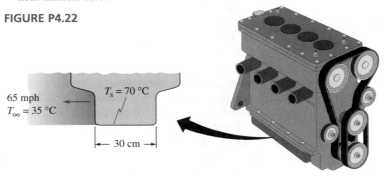

65 mph
$T_\infty = 35\ °C$ $T_s = 70\ °C$

30 cm

4.23 The crank case bottom surface in Problem 4.22 is aligned with another plate as shown in Figure P4.23. Assuming the boundary layer is turbulent from the leading edge, estimate the heat transfer rate from the bottom surface. Assume that the extension plate is a perfect nonconductor.

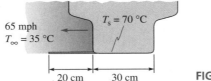

65 mph
$T_\infty = 35\ °C$ $T_s = 70\ °C$

20 cm   30 cm   **FIGURE P4.23**

4.24 The boundary layer on a flat plate is turbulent from the leading edge. A uniform heat flux is imposed on the plate. If the fluid free stream temperature is T_∞ and the free stream velocity is U_∞ find an expression for the temperature of the surface of the flat plate as a function of the distance from the leading edge.

4.25 Air at 20 °C and 10 m/s flows parallel to a flat plate, which is electrically heated. The uniform heat flux from the plate to the air is 1200 W/m². A tripping wire placed at the leading edge of the flat plate ensures that the boundary layer is turbulent from the leading edge. Estimate the temperature of the plate at distances of 50 cm and 100 cm from the leading edge.

4.26 The hood of an automobile is approximated as a flat plate, 1.22-m (4 ft) long and 1.52-m (5 ft) wide. It is anticipated that the temperature of the hood will reach 65.6 °C (150 °F). Estimate the convective heat transfer rate from the hood to the surrounding air at 35 °C (95 °F) when the automobile is travelling at 60 mph in calm winds.

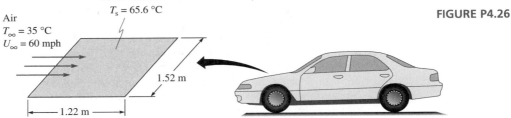

$T_s = 65.6\ °C$ **FIGURE P4.26**

Air
$T_\infty = 35\ °C$
$U_\infty = 60\ mph$

1.52 m

1.22 m

4.27 Plain carbon-steel plates, 1-m long, 2.5-m wide, and 10-mm thick are initially at 200 °C. They are placed on a material of low thermal conductivity and cooled by blowing air over the other parallel surface. The temperature of the air is 20 °C and its velocity 10 m/s parallel to the 1-m dimension. Determine
  (a) The average convective heat transfer coefficient when the plate is at 200 °C and at 60 °C.
  (b) The rate of initial cooling of the plates.
  (c) The time taken for the plate temperature to reach 60 °C, assuming the convective heat transfer coefficient to be constant at the arithmetic mean of the values at 200 °C and 60 °C.
  Repeat parts a, b, and c if the air flows parallel to the 2.5-m dimension.

4.28 An electric heater, 25-cm wide and 60-cm long, is fabricated by sandwiching a resistance heating element between two thin metal plates. When the heating element is switched on, the heat flux from the heater is uniform. Water at 20 °C flows parallel to the 60-cm dimension with a velocity of 3 m/s. If the temperature of the surface of the heating element is not to exceed 60 °C, what is the maximum power supply to the heating element?

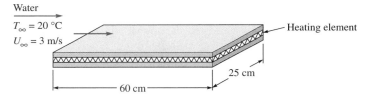

FIGURE P4.28

4.29 The average temperature of the roof of a house is 50 °C. The roof is 20-m long and 8-m wide. On a windy day, air at 30 °C flows parallel to the 20-m dimension with a velocity of 10 m/s. Determine the heat transfer rate to the air by convection if
  (a) $Re_{cr} = 5 \times 10^5$
  (b) $Re_{cr} = 0$ (As the roof is jagged, it is reasonable to assume that the boundary layer is turbulent from the leading edge.)

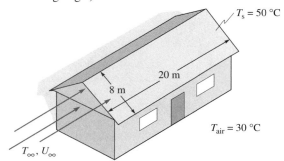

FIGURE P4.29

4.30 An array of computer chips is mounted on a board 0.3-m wide and 0.3-m deep. The total amount of energy to be dissipated is 25 W; the dissipation is uniform. To limit the temperature of the chips to 60 °C, it is proposed to install a fan to draw air over the chips. The maximum air temperature is expected to be 40 °C. The top surface of the chips may be considered to form a plane surface; the bottom surface is perfectly insulated.

(a) Estimate the minimum air velocity required if the critical Reynolds number is 5×10^5.

(b) Determine the maximum temperature for the velocity determined in part a if the boundary layer is turbulent from the leading edge.

4.31 Consider Problem 4.30. Some computers are now cooled by special electrically insulating liquids. The properties of one such liquid are

$$\rho = 1730 \text{ kg/m}^3 \qquad c_p = 1040 \text{ J/kg K}$$
$$\mu = 0.55 \times 10^{-3} \text{ N s/m}^2 \qquad k = 0.06 \text{ W/m K}$$

The liquid velocity is limited to 1 m/s, and the liquid temperature is 10 °C. Determine the heat flux if the maximum temperature of the board is limited to 60 °C, and the length of the board is 0.6 m.

4.32 Air, at 20 °C, flows parallel to a 1-m long plate with a velocity of 20 m/s. The plate is at 80 °C. To increase the heat transfer rate from the plate it is proposed to replace the plate with 50 1-cm wide strips placed 2-cm apart in a vertical column as shown. Estimate the heat transfer rate (per meter width) from the continuous plate and from the 50 strips also maintained at 80 °C.

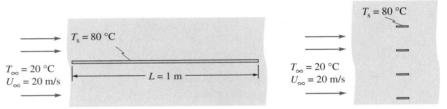

FIGURE P4.32

4.33 Re-solve Problem 4.32 if the boundary layer is turbulent from the leading edge.

4.34 In Problem 4.32, besides air, the use of argon and hydrogen are being considered. Which gas would you recommend for maximum heat transfer?

4.35 Atmospheric air at 0 °C flows with a velocity of 2 m/s over a triangular plate with its apex facing the air stream. The length of a plate in the direction of flow is 2 m and its width at the base is 1.3 m as shown in Figure P4.35. The plate is at 40 °C. Assuming that the correlations for forced convection over a heated semi-infinite plate are applicable to every elemental strip parallel to the direction of flow, determine the total heat transfer rate from one surface of the plate to the air.

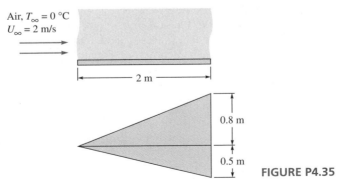

FIGURE P4.35

4.36 Atmospheric air at 20 °C is forced over a plain carbon-steel plate connecting two blocks which are at 100 °C. The free stream velocity of the air is 5 m/s. The plate is 30-cm long in the direction of flow and 20-cm wide.

(a) Compute the average convective heat transfer coefficient for a film temperature of 45 °C.

(b) Assuming that the convective heat transfer coefficient computed in part a is uniform over the entire plate, determine the thickness of the plate so that the temperature of the plate midway between the two blocks is 40 °C.

(c) What is the heat transfer rate from the plate to the air for the plate thickness determined in part b?

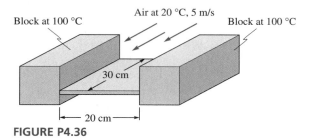

Block at 100 °C   Air at 20 °C, 5 m/s   Block at 100 °C

30 cm

20 cm

FIGURE P4.36

4.37 Plot the average heat transfer coefficient (on log-log paper) as a function of the length for the flow of water parallel to a flat plate for Re_L up to 10^9.

Temperature of water	10 °C
Temperature of plate	40 °C
Velocity of water	3 m/s

4.38 Equations 4.1.3 and 4.1.9a can be recast as

$$\frac{h_L}{k(U_\infty/\nu)\,\mathrm{Pr}^{1/3}} = \frac{0.664}{Re_L^{0.5}} \qquad \text{for } Re_L < 5 \times 10^5$$

$$\frac{h_L}{k(U_\infty/\nu)\,\mathrm{Pr}^{1/3}} = \frac{0.037}{Re_L^{0.2}} - \frac{871}{Re_L} \qquad \text{for } 5 \times 10^5 < Re_L$$

For a given fluid, T_∞, and U_∞, the left-hand side represents the average heat transfer coefficient divided by a constant. The equation relates h_L with L as Re_L is directly proportional to L. Plot $h_L/[k(U_\infty/\nu)\,\mathrm{pr}^{1/3}]$ against Re_L for $Re_L = 10^4$ to 2×10^8. You will notice that h_L has a local minimum, at $Re_L = 5 \times 10^5$ and a local maximum at $Re_L > 5 \times 10^5$ (see Figure 4.1.11).

(a) Show that the local maximum value of h_L occurs at $Re_L = 2.1801 \times 10^6$.

(b) Show that the same value of h_L as in part a occurs at $Re_L = 1.726 \times 10^5$.

(c) A local minimum value of h_L occurs at $Re_L = 5 \times 10^5$. Show that h_L has the same value at $Re_L = 9.0237 \times 10^7$.

(d) Determine two values of Re_L for which $h_L = h_{2L}$. A plate with dimensions $L \times 2L$ will have the same heat transfer rate when the flow is aligned parallel to either side for the value of the Reynolds number determined.

Solve the problem assuming that Equation 4.1.9a is valid for all $Re_L > 5 \times 10^5$ as it is a good approximation for $Re_L > 10^7$.

4.39 Supply of fresh water to arid zones by towing icebergs has been seriously considered. An iceberg, 1-km long and 300-m wide, is to be towed in water whose average temperature is 10 °C. Determine the average rate at which the flat bottom of the iceberg will melt (mm/h) when it is towed at
(a) 1 km/h
(b) 2 km/h
(Latent heat of fusion of ice is 333.4 kJ/kg and its density is 917 kg/m³.)

4.40 It is proposed to mount a solar panel, 50-cm wide and 200-cm long, on the wing of a plane flying at an altitude of 2000 m at a speed of 300 km/h. Solar energy at the rate of 900 W/m² (net) is absorbed by the panel. Determine the equilibrium temperature at the trailing edge of the panel if
(a) The longer side is parallel to the flow of the air.
(b) The shorter side is parallel to the flow of the air.
   (In both cases assume that only convection heat transfer between the panel and the surrounding air and that the leading edge of the panel is also the leading edge of the wing).
(c) In part a the leading edge of the panel is 50 cm from the leading edge of the wing.
(For the solution to this problem, you need to find the density and temperature of the air at an altitude of 2000 m from a book on fluid mechanics.)

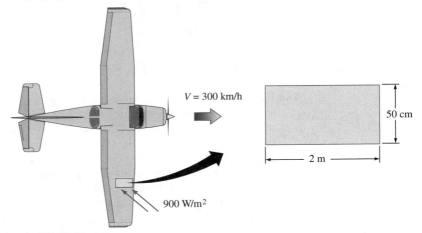

FIGURE P4.40

4.41 The net solar energy absorbed by the top, flat surface of a metal shed is 700 W/m². The top surface is 2-m wide and 4-m long. A breeze at 20 mph blows across the shed. The air temperature is 10 °C. Determine the temperature of the trailing edge of the top surface (assume that the heat transfer rate from the bottom surface of the roof is negligible) if the air flows parallel to the
(a) 2-m dimension
(b) 4-m dimension

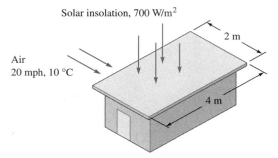

FIGURE P4.41

4.42 Glycerin is to be heated by passing it over a flat plate maintained at 60 °C. The plate is 30-cm long in the direction of flow and 50-cm wide. The glycerin approaches the plate at 20 °C with a velocity of 3 m/s. Compute the heat transfer rate from one side of the plate.

4.43 Silcon chips are mounted on an insulated surface which is 20-cm long and 20-cm wide. Air at 20 °C is forced over the uninsulated top surface with a velocity of 15 m/s. The top surface can be treated as a rough, plane surface. Each chip dissipates the same amount of energy to the air. If the temperature of the chips is to be limited to 60 °C, what is the maximum energy dissipation from the surface. Assume that the boundary layer is turbulent from the leading edge.

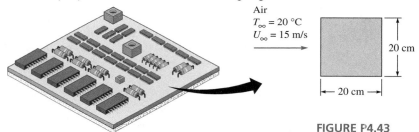

FIGURE P4.43

4.44 Electronic components are mounted on the underside of a heavy copper plate. The temperature of the copper plate is nearly uniform. Water at 20 °C flows over the plate (parallel to the surface of the plate) with a velocity of 4 m/s. The plate is 30-cm long in the direction of the flow.

(a) If the temperature of the plate is not to exceed 50 °C, determine the heat transfer rate to the water per meter width of the plate.

(b) Compare the heat transfer rate per meter width if, instead of water, air at 15 m/s, 20 °C is used to cool the copper plate.

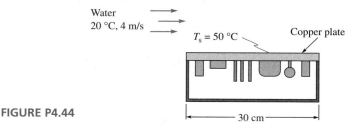

FIGURE P4.44

4.45 It is desired to dissipate heat from flat plates to water flowing at 10 °C with a velocity of 4 m/s. The plates, which are at a uniform temperature of 90 °C, are aligned parallel to the flow of water. Two possible alternatives are being considered.

(a) To make each plate 1-m long in the direction of flow.

(b) To double the number of plates with each plate 0.5 m in the direction of flow. Which arrangement would you suggest to maximize the heat transfer rate? Support your answer with quantitative results. The width of the plates is the same in both cases.

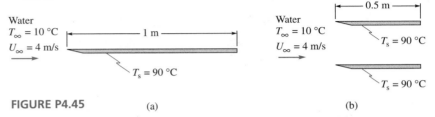

FIGURE P4.45   (a)   (b)

4.46 Air flows over a 30-cm long, 10-mm O.D. tube. The air is at 200 kPa, 20 °C and is in cross flow with a velocity of 10 m/s. The tube surface is at 90 °C. Determine the heat transfer rate to the air.

4.47 In Problem 4.46, the heat transfer rate is to be 125 W. What should be the velocity of the air?

4.48 Hollow, cylindrical brass tubes, 6-mm O.D., 3-mm I.D., and 10-cm long, are attached to the surface of a power source of an instrument package. Air at 20 °C is drawn across the tubes at 5 m/s. If the temperature of the end of the tubes attached to the power source is not to exceed 60 °C, find the heat transfer rate from one tube. Assume correlations for a single tube are applicable.

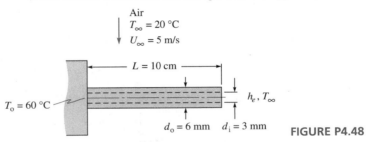

FIGURE P4.48

4.49 In the instrument package in Problem 4.48 the brass tubes are replaced by aluminum alloy (2023-T6) plates, 10-cm wide, 15-cm long, and 1-mm thick, with the 10-cm dimension perpendicular to the surface of the package. Air is drawn parallel to the 15-cm dimension. Estimate the heat transfer rate from one plate.

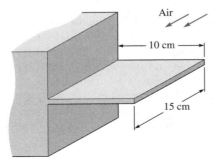

FIGURE P4.49

4.50 Air flows perpendicular to the axes of two cylinders (ends insulated) as shown. Estimate q_A/q_B if

(a) $Re_{d,A} = 20\,000$
(b) $Re_{d,A} = 40\,000$

$$d_B = 2d_A \qquad T_{sA} = T_{sB} \qquad L_B = L_A$$

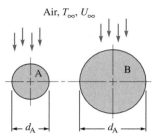

FIGURE P4.50

4.51 A 2.5-cm diameter, 30-cm long cylindrical rod is rotated at 600 rpm about a parallel axis. The distance between the axis of the rod and the axis of rotation is 30 cm. The temperature of the surface of the rod is 127 °C and that of the surrounding air is 27 °C. Neglecting the heat transfer from the end surfaces, estimate the convective heat transfer rate from the rod to the surroundings.

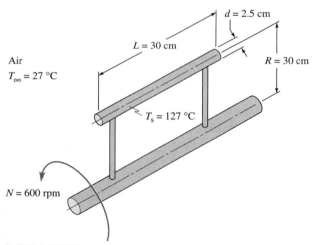

FIGURE P4.51

4.52 A hot wire/film anemometer is an instrument used to measure velocities of fluids. In one version of the instrument, a resistance wire is maintained at a constant temperature. The heat transfer rate is a function of the velocity of the fluid. By calibrating the voltage across the element (which is related to the power dissipation, i.e., heat transfer) as a function of the fluid velocity, the instrument can be used to measure the velocity of fluids. In one such instrument, the resistance wire is made of a platinum wire, 0.2-mm diameter and 0.5-cm long. The wire is maintained at 50 °C when it is used in water (at 20 °C) flowing perpendicular to the axis of the element. Estimate the voltage across the element when the velocity of the water is 0.5 m/s, 1 m/s, 3 m/s, and 5 m/s. Resistivity of platinum $= 0.17$ $\mu\Omega$ m. It is convenient to employ Equation 4.1.18 and Table 4.1.1.

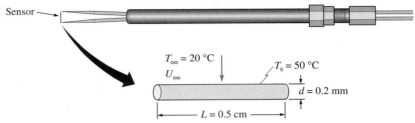

FIGURE P4.52

4.53 In an air dryer, air at 30 °C flows with a velocity of 4 m/s perpendicular to the axis of an 18-gauge nichrome wire that is at 400 °C. The length of the wire is 50 cm. Determine the current in the wire. The diameter of an 18-gauge nichrome wire is 1.024 mm (0.0403 in) and its resistance is 1.384 Ω/m.

4.54 In Problem 4.53, the current in the wire is 15 A. The velocity of the air is increased to 6 m/s. Determine the temperature of the surface of the wire.

4.55 A 4-cm diameter 2024-T6 aluminum alloy rod, initially at 200 °C, is placed in a stream of engine oil at 20 °C. The oil flows perpendicular to the axis of the rod with a velocity of 1 m/s. Assuming constant convective heat transfer coefficient evaluated at a surface temperature of 150 °C, determine the time required for the temperature of the rod to decrease from 200 °C to 100 °C if
(a) The rod is long (i.e., $L \gg d$) and the heat transfer from the end surfaces is negligible compared with the heat transfer from the cylindrical surface.
(b) The rod is 6-cm long. Evaluate separately, h_1 for the cylindrical surface and h_2 for the end surfaces. Employ any reasonable characteristic length to determine h_2.

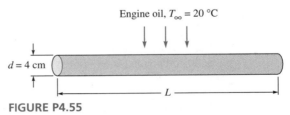

FIGURE P4.55

4.56 In cooling towers, water drops fall vertically down with air flowing up. The cooling of the water is partially by evaporation, and partially by convection due to the difference in the temperatures of the water and air. In a part of the cooling tower, the air may be saturated. In one cooling tower, there are a large number of spherical water drops with an average diameter of 1.5 mm. The velocity of the air relative to the drops is 1.5 m/s. If the temperature of the air is 25 °C and water drops are at 70 °C, determine the convective heat transfer rate from a drop to the surrounding air. Neglect evaporation.

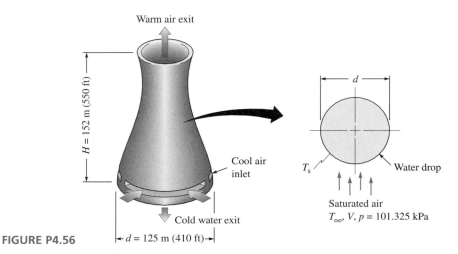

FIGURE P4.56

4.57 The temperature of the surfaces of a duct transporting cold air is 5 °C. The duct is 20-cm wide and 20-cm high. Air at 20 °C flows perpendicular to the axis of the duct with a velocity of 5 m/s. Estimate the heat transfer rate per meter length of the duct if the flow of air is
   (a) Parallel to the top and bottom surfaces of the duct.
   (b) Parallel to one of the diagonal surfaces of the duct.

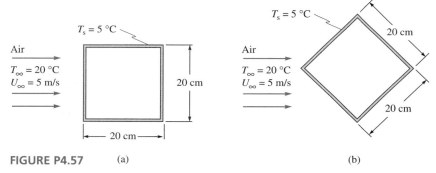

FIGURE P4.57   (a)   (b)

4.58 Long, square stainless steel (type 347) bars are to be cooled in a stream of water at 20 °C flowing with a velocity of 2 m/s. The bars are 2 cm on each side. Assuming that the convective heat transfer coefficient evaluated at a surface temperature of 60 °C represents an appropriate average value, estimate the time for the temperature at the axis of the bars to decrease from 80 °C to 40 °C if the water flow is parallel to one of the surfaces and perpendicular to the axis of the bar.

4.59 Long, hollow bars of plain carbon-steel, 3 cm on each side and 2-mm thick, are to be heated in an oven. In the oven, hot air at 100 °C flows perpendicular to the axis of the bars and parallel to two opposite surfaces with a velocity of 10 m/s. When the temperature of the bars is 70 °C determine
   (a) The heat transfer rate to the bars per meter length.
   (b) The rate of increase of the temperature of the bars.
   (c) The time taken for the temperature of the bars to increase from 60 °C to 80 °C assuming that the value of the convective heat transfer coefficient, found in part a, represents a suitable average value during the heating process.

4.60 A thin, square plate, 20 cm on each side, is maintained at 80 °C. Air at 10 °C is forced over the plate at 10 m/s. The plate can be placed with the air flow
(a) Parallel to one of the edges.
(b) Parallel to a diagonal.
Determine the heat transfer rate from one surface in each case.

4.61 A 10-cm diameter sphere of ice at 0 °C is placed in a stream of air at 20 °C flowing with a velocity of 10 m/s. Assuming that the convective heat transfer coefficient is constant at the arithmetic mean of the values evaluated at sphere diameters of 10 cm and 4 cm, calculate the time taken for the diameter of the sphere to reach 4 cm. Assume that all the water formed is removed as it forms and that evaporation is negligible. Enthalpy of fusion of ice is 333.4 kJ/kg. (The assumption of negligible evaporation will lead to an overestimate of the time.)

4.62 A 10-cm diameter copper sphere at 80 °C is placed in a stream of air at 20 °C. The free stream velocity of the air is 10 m/s. Estimate
(a) The initial heat transfer rate.
(b) The time taken for the temperature of sphere to drop to 40 °C.

4.63 A fluid at 30 °C flows in a 2-cm I.D. tube with a velocity of 1 m/s. Determine whether the flow is laminar or turbulent if the fluid is
(a) Atmospheric air
(b) Water
(c) Engine oil

4.64 A fluid at 30 °C enters a 2-cm I.D. tube with a uniform velocity of 1 m/s. Determine the hydrodynamic and thermal entrance lengths if the fluid is
(a) Atmospheric air
(b) Water
(c) Engine oil
The tube is at a uniform temperature.

4.65 A number of correlations for the convective heat transfer coefficient for internal flows have been given. Equation 4.2.4a is one of the most reliable correlations and Equations 4.2.3d, e, and f predict Nusselt numbers that are close to Equation 4.2.4a. To get an appreciation for the variation of the convective heat transfer coefficient predicted by different correlations, evaluate the convective heat transfer coefficient employing different correlations for fully developed velocity and temperature profiles in the following case.
Water flows in a long, 2.5-cm I.D. tube at a rate of 2000 kg/h. The tube surface is at 150 °C. Calculate the heat transfer coefficients at sections where the water temperature is 20 °C, 70 °C, and 120 °C.

4.66 In Problem 4.65, determine the length of the tube between the sections where the water temperatures are 20 °C and 120 °C.

4.67 In a refrigeration system, using Refrigerant-12, the liquid leaves the condenser at 20 °C. It is to be further subcooled to 10 °C by the incoming vapor, in a 2-cm I.D. tube. The tube surface is at −5 °C. If the velocity of the liquid refrigerant is 0.5 m/s, determine the length of tube required.
(Refrigerant-12 is a fluoro-carbon, one of a family of fluids, widely used in refrigeration and air-conditioning plants. Some of those fluids were also used as carriers in aerosols. There is strong evidence that the presence of such fluids in the atmosphere has caused a depletion in the ozone layer. Alternative fluids to replace the fluoro-carbons are being developed.)

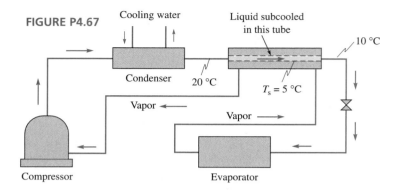

FIGURE P4.67

4.68 Engine oil is to be cooled from 85 °C to 65 °C in a 1-cm I.D. tube whose surface is at 40 °C with water flowing over the tube. Determine the length of the tube, if the velocity of the oil is 1 m/s.

4.69 When air is compressed in a compressor, its temperature increases. Between the compressor and the air reservoir, an after-cooler is used to cool the air. In a proposed design of the after-cooler, air from the compressor enters a 15-mm I.D. tube and leaves at 25 °C. The tube's surface is maintained at 18 °C. In the compressor, atmospheric air at 20 °C is compressed to 800 kPa. The compression process is approximated by $pv^{1.3}$ = constant. The velocity of the air entering the cooling tube is 15 m/s. Determine the length of the tube required.

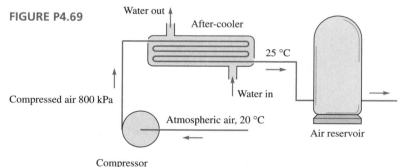

FIGURE P4.69

4.70 Water flows at a rate of 0.1 kg/s through a 1-cm diameter tube. The water is to be heated from 60 °C to 90 °C with the tube wall maintained at a uniform temperature of 100 °C by steam condensing on the outside of the tube. Estimate
(a) The average heat transfer coefficient between the inside surface and the water.
(b) The length of the tube required.

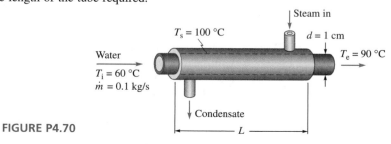

FIGURE P4.70

4.71 Water flows in a 1-cm diameter tube with an average velocity of 1 m/s. The temperature of the water at a particular location is 60 °C and a uniform heat flux of 200 000 W/m² is imposed on the tube surface. At this location, assuming fully developed condition, estimate
(a) The convective heat transfer coefficient.
(b) The tube surface temperature

4.72 200 kg/h of water are to be heated from 10 °C to 80 °C. The water flows through a 8-mm I.D. tube, which is heated electrically by a resistance wire wrapped around it. With this arrangement a uniform heat flux of 200 kW/m² is imposed on the tube surface. Determine
(a) The length of the tube required.
(b) The temperature of the tube surface at exit.

4.73 A fluid flows in a 12-mm I.D., 6-m long tube. A uniform heat flux of 2000 W/m² is imposed on the tube surface by an electrical heating element. The temperature of the fluid at exit is 80 °C. Estimate the surface temperature of the tube at exit.

Mass flow rate = 15 kg/h ρ = 1500 kg/m³   Pr = 5.2
μ = 4.22 × 10⁻⁴ N s/m² k = 0.0516 W/m K

4.74 Water enters a tube at 10 °C with a velocity of 4 m/s. The inside diameter of the tube is 3 cm. The tube is wrapped with a resistance heating wire so that when the wire is energized, a uniform heat flux is imposed on the tube. At the exit of the tube the temperature of the water is to be 80 °C. The pressure of the water in the tube corresponds to a saturation temperature of 140 °C. If there should be no boiling of water in any part in the tube, what is the minimum length of the tube?

4.75 In diesel-engine propelled ships, exhaust gases from the engines are utilized to produce steam in an exhaust gas boiler. The steam is used for space heating and to heat the fuel oil for the engines. In one such boiler, exhaust gases enter 2-cm I.D. tubes at 300 °C. To prevent condensation of the acid bearing moisture in the gases, the temperature of the gases should not be less than 200 °C at the exit of the boiler. The velocity of the gases in the tubes is 10 m/s. The temperature of the tube surface is 165 °C. Assuming that the properties of the exhaust gases are approximately those of air, determine the length of each tube.

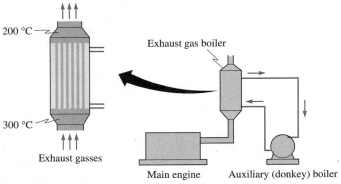

FIGURE P4.75

4.76 1.2 kg/s of water are to be heated from 10 °C to 80 °C. Tubes of 4-cm I.D. and 3-cm I.D. are available. The temperature of the surface of the tubes is 100 °C. either one 4-cm diameter tube or two 3-cm diameter tubes (in parallel) can be used. Determine the configuration that will yield the smaller length.

4.77 The Reynolds number of a liquid flowing in a tube of diameter, d, is 80 000. For the same mass flow rate and bulk temperature of the liquid, and tube surface temperature, determine the percentage change in the heat transfer rate per unit length of the tube if the diameter of the tube is doubled?

4.78 UCARTHERM is a special heat transfer fluid that is manufactured by Union Carbide. A 50% mixture of the fluid with water enters a 1-cm diameter, 10-m long tube at 10 °C with a velocity of 3 m/s. The tube surface is at 90 °C. Determine the heat transfer rate. Properties of the liquid are

Specific gravity   1.06
Specific heat   3600 J/kg °C
Thermal conductivity   0.406 W/m °C
Viscosity   0.0012 N s/m²

4.79 In Problem 4.78 what should be the tube surface temperature if the heat transfer rate is 80 000 W?

4.80 In an experimental program, the blood from a patient is withdrawn, heated from 37.5 °C to 45 °C, cooled to 37.5 °C, and returned to the patient. The blood is withdrawn at a rate of 3 l/min in a 1.5-cm I.D. stainless steel tube, which is subjected to a uniform heat flux. The temperature of the surface of the tube is not to exceed 55 °C at any section. The blood is cooled by passing it through another section of 1.5-cm I.D. tube, which is maintained at 25 °C. Determine the length of tubes required for heating and cooling. Properties of blood are

$\rho = 1060 \text{ kg/m}^3$ $\mu = 4 \times 10^{-3} \text{ N s/m}^2$
$k = 0.52 \text{ W/m K}$ $c_p = 3850 \text{ J/kg °C}$

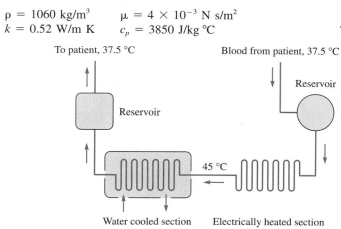

FIGURE P4.80

4.81 Air flows in a 2-cm dia. tube that is at 100 °C. At the entrance to the tube the temperature of the air is 0 °C. The tube is 10-m long. The pressure of the air in the tube is 1000 kPa and the velocity is 20 m/s.
    (a) Estimate the convective heat transfer coefficient between the air and the tube surface where the temperature of the air is 27 °C.

(b) Assuming the convective heat transfer coefficient to be constant at some suitable value, find the exit temperature of the air.

(c) Compute the heat transfer rate from the tube to the air. (Repeat computations if necessary.)

4.82 Milk is to be heated from 20 °C to 72 °C in a 1.5-cm I.D. tube that is at 100 °C. Determine the length of the tube required if the mass flow rate of the milk is 0.5 kg/s. Properties of milk

$\rho = 1030$ kg/m³ $c_p = 3850$ J/kg K
$k = 0.6$ W/m K $\mu = 2.12 \times 10^{-3}$ N s/m²

4.83 Water at 0.5 kg/s and 20 °C enters a 5-m long, 2-cm I.D. tube maintained at a uniform surface temperature. If the total heat transfer rate to the water is 200 kW, what should be the surface temperature of the tube?

4.84 Engine oil flows inside a 20-m long, 2-cm I.D. coiled tube with a velocity of 0.5 m/s. Determine the uniform surface temperature of the tube to heat the oil from 60 °C to 90 °C.

4.85 Water is to be heated from 10 °C to 60 °C. The water flows at a rate of 200 kg/h through a 8-mm diameter, 3-m long tube whose surface is subjected to a uniform heat flux. Determine

(a) The value of the uniform heat flux on the tube.

(b) The temperature of the tube at exit.

4.86 It is proposed to build a steam condenser as a double pipe heat exchanger. Water flows in the 15-mm I.D. and 5-m long inner tube of a thin, high conductivity material. The inner tube is surrounded by saturated steam at atmospheric pressure. Velocity of the water is 2.5 m/s.

(a) Determine the convective heat transfer coefficient at a section where the water temperature is 30 °C.

(b) Assuming that the average convective heat transfer coefficient over the entire length of the tube (5 m) is equal to that found in part a, compute the heat transfer rate to the water and the mass rate of condensation of steam, if the water enters the tube at 20 °C. (Assume that the convective heat transfer coefficient on the steam side is very large compared with that on the water side.)

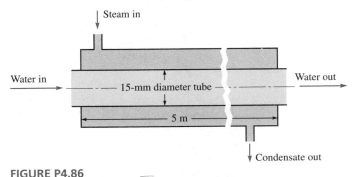

Steam in

Water in   15-mm diameter tube   Water out

5 m

Condensate out

FIGURE P4.86

4.87 To form ice in an ice rink, an array of parallel pipes is installed in a 20-mm thick water layer. A secondary refrigerant, is circulated through the pipes. During the formation of ice it may be assumed that the tubes are at a uniform temperature of 0 °C.

An ice rink is 61-m long, and 27-m wide. The array of pipes is parallel to the width of the rink. An aqueous solution of ethylene glycol (50% by mass) is used as the secondary refrigerant (coolant). The operating conditions are

Diameter of pipes (I.D.)	10 mm
Inlet temperature of coolant	$-15\,°C$
Exit temperature of coolant	$-10\,°C$

The properties of coolant are

Density	$1080\ kg/m^3$
Dynamic viscosity	$0.02\ N\ s/m^2$
Specific heat	$3010\ J/kg\,°C$
Thermal conductivity	$0.42\ W/m\ °C$
Latent heat of fusion of ice	$333.4\ kJ/kg$

(a) Determine the mass rate of flow of coolant.
(b) Determine the pitch of the tubes (spacing), S, if all the water is to be frozen (from and at $0\,°C$) in 2 hours. Enthalpy of fusion of ice (latent heat of fusion) is 333.4 kJ/kg. (After freezing commences, heat transfer from the water to the pipes is more complex. The solution, with the assumption of the tube remaining at $0\,°C$ during the entire solidification process, will give a rough estimate of the pitch of the tubes.)

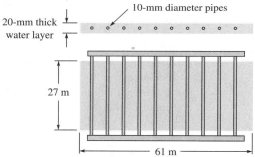

FIGURE P4.87

4.88 Air, at atmospheric pressure and a temperature of $5\,°C$, enters a rectangular duct with a velocity of 6 m/s. The duct is 20-cm wide, 8-cm high, and 20-m long. The temperature of the surface of the duct is $15\,°C$. Estimate the temperature of the air at exit.

4.89 Effective use of solar energy (for space heating) needs a means of storing energy during periods of sunshine. The stored energy is used during nights and cloudy periods. Some of the media used for such thermal storage are water, rock pebbles, and phase change materials. Phase change materials offer the advantage of high energy storage per unit volume at a set temperature because of the latent heat (heat of fusion). Waxes at different melting temperatures have been developed as phase change materials for thermal energy storage.

In one thermal energy storage system, hot water from the solar collectors is circulated through a bank of coaxial tubes. The annulus is filled with a phase change material. Air is forced over the bank of tubes when energy is required.

The arrangement of one such tube and properties of the phase change material are

Melting point (°C)	33
Density (kg/m³)	1460
Latent heat (kJ/kg)	238
Diameter of inner tube (mm)	25
Diameter of outer tube (mm)	150
Length of tube (m)	4
Velocity of water (m/s)	0.2
Inlet temperature of water	50 °C

Estimate
(a) The heat transfer rate from the water per tube.
(b) The time taken to melt all the phase change material in a tube.

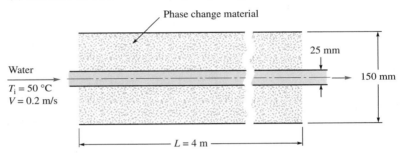

FIGURE P4.89

4.90 A Compact heat exchanger is constructed as a series of triangular ducts with hot and cold fluids flowing in adjacent channels as shown. The hot fluid is air at 200 °C, 150 kPa, and with a velocity of 12 m/s. The duct surfaces are at 110 °C. Each side of the triangular duct is 12 mm and the ducts are 3-m long. Estimate the exit temperature of the air.

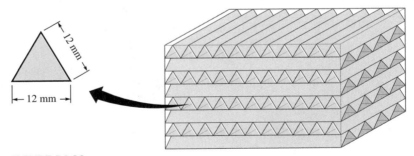

FIGURE P4.90

4.91 Atmospheric air at 300 °C flows at 20 m/s perpendicular to the axis of a thin, 2-cm diameter tube carrying water at 80 °C. The velocity of the water in the tube is 1 m/s. Determine
(a) The convective heat transfer coefficient on the water and air sides. (Calculating the convective heat transfer coefficient on the air side requires iteration on the outer surface temperature.)
(b) The heat transfer rate per unit length.
(c) The minimum pressure of the water to prevent boiling.

4.92 Consider an incompressible fluid flowing radially outwards between two coaxial disks of radius r_2; the gap between the disks is g. One of the plates is maintained at a uniform temperature T_s for $r_1 < r < r_2$ and insulated for $0 < r < r_1$. Such a flow is quite complex and no simple correlations are available. From the limited available experimental information the convective heat transfer coefficient is approximated as

$$h = c_1 e^{-c_2 r} + c_3$$

where c_1, c_2, and c_3 are constants. The convective heat transfer coefficient is based on the local bulk mean temperature of the fluid. The radial velocity of the fluid decreases in the direction of the flow and the convective heat transfer coefficient may be expected to decrease with increasing radius.

Derive an expression for the heat transfer rate to the fluid and exit temperature of the fluid if the mass flow rate of the fluid is $\dot{m}$ and its inlet temperature is T_i.

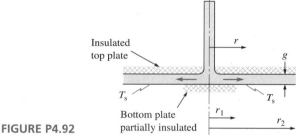

Insulated top plate

T_s T_s

Bottom plate partially insulated

FIGURE P4.92

4.93 In an experiment to determine the convective heat transfer coefficient, 0.062 kg/s of water enter a type 304, 10-mm I.D. stainless steel tube at 20 °C. The wall thickness of the tube is 1 mm. The water exits the tube at 60 °C and the maximum surface temperature is 90 °C. A uniform heat flux is imposed on the tube by passing a current in the tube.
(a) Determine the length of the tube.
(b) Estimate the current and voltage across the tube if the electrical resistivity of stainless steel is 72×10^{-8} Ω m.

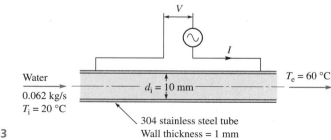

Water
0.062 kg/s
$T_i = 20\ °C$

$d_i = 10$ mm

$T_e = 60\ °C$

304 stainless steel tube
Wall thickness = 1 mm

FIGURE P4.93

4.94 A vertical plate, 30-cm wide and 50-cm high, is suspended in a fluid maintained at 20 °C. Determine the temperature of the plate for a heat transfer rate of 220 W from both surfaces of the plate by natural convection if the fluid is
(a) Air
(b) Water
(c) Oil

4.95 Show that from Equations 4.3.1a and b, for natural convection from vertical surfaces to atmospheric air at film temperature of 300 K, the average heat transfer coefficient can be approximated as

$$h_L = 1.33 \left(\frac{T_s - T_\infty}{L} \right)^{1/4} \text{W/m}^2 \text{ K} \qquad \text{for } 10^5 < Ra_L < 10^9$$

$$h_L = 1.26 \, (T_s - T_\infty)^{1/3} \text{ W/m}^2\text{K} \qquad \text{for } 10^9 < Ra_L < 10^{13}$$

4.96 An electric transformer is cooled by immersing it in transformer oil. The vertical cylinder containing the oil is 0.3-m diameter and 0.5-m high and is at 60 °C. Estimate
   (a) The convective heat transfer rate from the cylindrical surface when it is surrounded by air at 40 °C.
   (b) The radiative heat transfer rate (employing Equation 1.5.2) if the emissivity of the surface is 0.8.

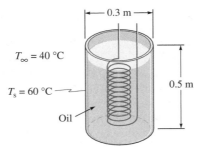

$T_\infty = 40\ °\text{C}$

$T_s = 60\ °\text{C}$

0.3 m

0.5 m

Oil

FIGURE P4.96

4.97 To increase the heat transfer rate from the transformer in Problem 4.96, 20 vertical strips of plain carbon-steel are welded to the transformer case. Each strip is 70-mm wide and 3-mm thick. Estimate the heat transfer rate from each strip if the cylindrical surface is at 60 °C and the ambient air is at 40 °C.

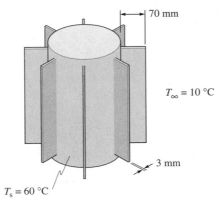

70 mm

$T_\infty = 10\ °\text{C}$

3 mm

$T_s = 60\ °\text{C}$

FIGURE P4.97

4.98 The outer dimensions of a refrigerator are 0.7 m (width), 0.6 m (depth), and 1.5 m (height). The outer surface is at 17 °C when the ambient air is at 22 °C. Estimate the heat gain to the vertical surfaces from the air by natural convection and radiation. The emissivity of the surface is 0.8.

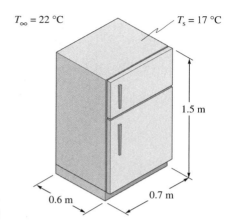

$T_\infty = 22\ ^\circ C$

$T_s = 17\ ^\circ C$

1.5 m

0.6 m

0.7 m

FIGURE P4.98

4.99 A vertical steam pipe is anchored to a wall with a plain carbon-steel plate. The plate is 30-cm long, 20-cm high, and 3-mm thick. The temperature of the steam pipe is 100 °C and the ambient air is at 20 °C. The temperature of the plate varies in the x-direction.

(a) Determine the convective heat transfer coefficient for a 20-cm vertical plate for surface temperatures of 100 °C, 60 °C, and 40 °C.

(b) Assuming the convective heat transfer coefficient to be uniform at the arithmetic mean of the three values in part a, determine the convective heat transfer rate from the plate to the surroundings.

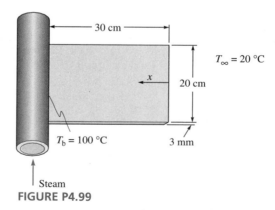

30 cm

$T_\infty = 20\ ^\circ C$

x

20 cm

$T_b = 100\ ^\circ C$

3 mm

Steam

FIGURE P4.99

4.100 An electronic (telephone) switch box is 71-cm wide, 86-cm deep, and 20-cm high. It has a card carrier, which carries 10 printed wire boards on which discrete elements are mounted. The card carrier is 58-cm wide and 15-cm high. The printed wire board is 15-cm high and 80-cm deep. The total heat dissipation from the boards is 100 W (10 W/board). If the case is completely enclosed, determine the case temperature assuming that heat transfer is by natural convection from all the surfaces except the back surface, which is assumed to be insulated as it is against a wall. The ambient air temperature is 45 °C. Suggest a method to limit the temperature of the cards to 5 °C above the ambient air temperature.

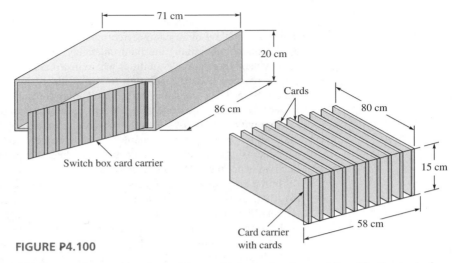

FIGURE P4.100

71 cm

20 cm

86 cm

Cards

80 cm

15 cm

58 cm

Switch box card carrier

Card carrier
with cards

4.101 Space ships are launched with very high acceleration and the effective body force inside the space ship is many times the gravitational force on the earth. An experimental rig to simulate high body forces consists of a cylindrical vessel, 0.3-m diameter and 0.3-m long (inside dimensions), which is rotated about an axis perpendicular to the axis of the vessel as shown. The mean distance between the axis of rotation and the vessel is 2 m.

The container is filled with air at 200 kPa, 20 °C. To determine the natural convection heat transfer rate experimentally, a 10-cm long and 5-cm wide thin plate is placed inside the vessel with the center line (parallel to the 10-cm dimension) of the plate along the axis of the vessel. The plate is heated by passing an electric current through it. Estimate the power required to maintain the plate at 40 °C for rotational speeds of 300 rpm and 600 rpm and compare them with power requirements if the container is stationary with its axis vertical.

At 300 rpm the effective body force per unit mass along the radius ($= \omega^2 r$) is 1974 N. The earth's gravitational force is only 4.968^{-3} times the force along the radius and can be neglected. If you do not wish to neglect it, the axis of the package should be inclined at an angle of 0.28° to the horizontal plane. At 600 rpm the effect of the earth's gravitational force is even smaller.

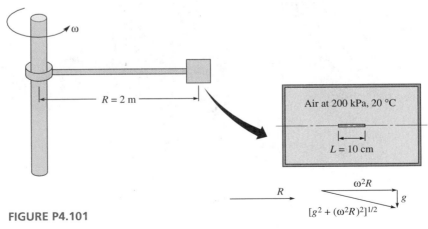

ω

$R = 2$ m

Air at 200 kPa, 20 °C

$L = 10$ cm

R

$\omega^2 R$

g

$[g^2 + (\omega^2 R)^2]^{1/2}$

FIGURE P4.101

4.102 In space ships in orbit, the body force is very small, termed microgravity conditions. One method of simulating microgravity conditions on a fluid on earth is to let a container containing the fluid fall freely. At NASA-Lewis Research Center in Cleveland there are two facilities where instrumented packages fall freely, simulating reduced gravity conditions. In one of them (right) the free fall is approximately 10 m. The second one (left), a well approximately 4.6-m diameter and 152-m deep (15-ft diameter and 500-ft deep), is used to conduct experiments under microgravity conditions. When the well is evacuated, the effective body force on the fluid in a freely falling container is approximately 10^{-5} g per unit mass. (The container is brought to rest at the bottom by a bed of expanded polystyrene beads.) Such an experimental facility is known as a drop tower.

(a) In the drop tower (152-m deep), what is the duration of free fall?

(b) Although the duration of the free fall is short, in many experiments steady-state conditions are reached during the fall. An experimental setup to determine the heat transfer coefficient in natural convection with vertical plates consists of a container 1-m diameter and 1.5-m high. The test piece is a thin plate, 15-cm high and 10-cm wide. The plate is heated by passing an electric current through it. The power dissipation is the heat transfer rate from both surfaces of the plate. Steps are taken to ensure that the radiative heat transfer is negligible compared with convective heat transfer. The fluid in the container is at 20 °C and the power dissipation is adjusted to maintain the plate at 30 °C prior to releasing the container for a free fall. Assuming steady conditions are reached

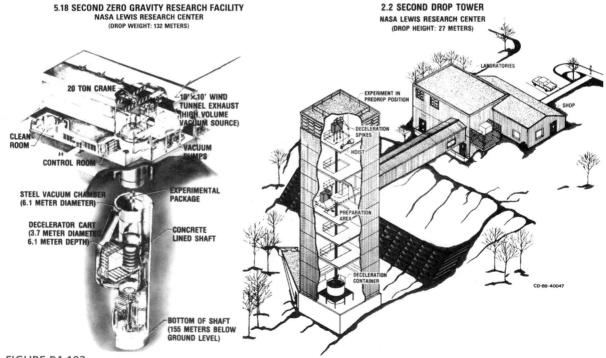

FIGURE P4.102
Courtesy of NASA-Lewis Research Center, Cleveland.

during the free fall, and that the effective body force on the fluid is 10^{-4} g per unit mass, estimate the temperature of the plate during free fall, if the fluid in the container is (1) Air at 100 kPa, and (2) Hydrogen at 10 kPa.

(c) There is some concern that the temperature of the gases will not remain constant during the experiments due to the heat transfer from the plate. Determine the temperature rise in the gases due to the heat transfer during free fall.

4.103 In a workshop, the vertical insulated chimney of a furnace is 50 cm in diameter and 5-m high. The temperature of the outer surface of the chimney is 50 °C. Estimate the heat transfer rate from the chimney if the emissivity of the metal cladding on the insulation is 0.1 and the ambient air is at 30 °C.

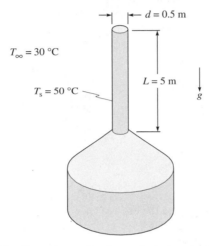

FIGURE P4.103

4.104 The inside dimensions of a domestic hot water heater are 60 cm (diameter) and 1.6 m (height). The heater is insulated with 6-cm thick fiberglass and covered with a thin metal cladding. The measured temperature of the cladding is 22 °C when the ambient air is at 21 °C. If the emissivity of the cladding surface is 0.8, estimate the heat transfer rate from the outer cylindrical, bottom, and top surfaces and the temperature of the water in the boiler. Assume that the metal adjacent to the water is at the same temperature as the water.

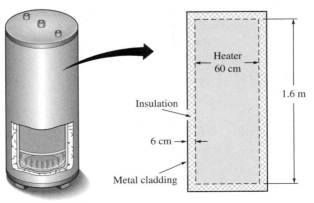

FIGURE P4.104

4.105 The surface temperature of a 3-m high vertical wall is 25 °C with ambient air at 20 °C. It is desired to determine experimentally the natural convection heat transfer coefficient from laboratory experiments on a shorter plate. Two different experiments are to be designed: (1) A 0.8-m high plate maintained at 40 °C in an atmosphere of pressurized air at 20 °C, and (2) a vertical plate maintained at 25 °C in still water at 20 °C. Assume that the Nusselt number is a function of the Raleigh number only. To achieve similarity between the laboratory experiment and the actual case, determine

(a) The pressure of air.
(b) The height of the plate in water.
(c) The ratio of the experimentally determined convective heat transfer coefficient to the actual convective heat transfer coefficient in each case.

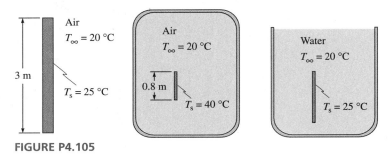

FIGURE P4.105

4.106 The net radiant energy flux absorbed (due to direct and diffuse incident solar energy, and the emitted energy) by a 2.5-m high, well-insulated vertical wall of a building is 60 W/m². Determine the surface temperature at midheight of the wall if the ambient air temperature is 30 °C.

4.107 An 80 W heater in the form of a freely suspended vertical plate operates at a surface temperature of 80 °C in oil at 40 °C. Determine the height of the plate if its width is 10 cm.

4.108 An oil-filled space heater is modeled as a rectangular block 60-cm high, 100-cm long, and 10-cm wide. If the surface temperature of the block is 60 °C in a room at 20 °C, estimate the convective heat transfer rate to the air. (Assume that the natural convection heat transfer rate from each surface is not affected by the presence of the other surfaces. The total heat transfer rate from the heater is the sum of the convective and radiative heat transfer rates.)

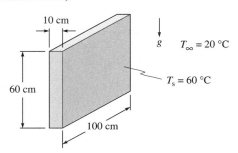

FIGURE P4.108

4.109 The average temperature of the roof of a house heated by the sun is found to be 50 °C. The roof is 20-m long and 8-m wide and is inclined at 45° to the vertical plane. Estimate the heat transfer rate from the roof to the atmosphere on a calm day if the surrounding air is at 30 °C.

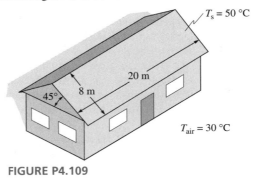

FIGURE P4.109

4.110 On a clear day, the net radiant energy absorbed by the roof in Problem 4.109 is 50 W/m². If the surrounding air temperature is 30 °C, estimate the temperature of the roof. Assume the heat transfer to the attic space is negligible.

4.111 A square, thin plate, 0.3 m × 0.3 m, is maintained at 100 °C. One surface of the plate is perfectly insulated and the other surface is exposed to air at 30 °C. Determine the orientation of the plate that results in the highest heat transfer rate by computing the heat transfer rate if
(a) The plate is vertical.
(b) The plate is horizontal with the heated surface facing up.
(c) The plate is horizontal with the heated surface facing down.

4.112 A thin, 16-cm diameter electrically heated circular, horizontal plate is at 70 °C in a large body of water at 10 °C. Determine the power input to the heater
(a) If the bottom surface of the heater is perfectly insulated.
(b) If the top surface of the heater is perfectly insulated.

4.113 Re-solve Problem 4.112 if the fluid surrounding the plate is air at atmospheric pressure.

4.114 An AM/FM receiver is 40-cm wide, 30-cm deep, and 10-cm high. The heat dissipation from the electronics is 10 W. Assume the heat transfer is from the top surface only. Estimate the temperature of the top surface when the ambient air temperature is 40 °C.

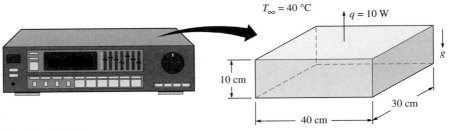

FIGURE P4.114

4.115 A cylinder, 15-cm long and 12-cm in diameter that is at a uniform temperature of 100 °C is to be slowly cooled by suspending it in a large room. The ambient air is at 20 °C. Determine the orientation of the cylinder (axis vertical or horizontal) that will yield the slower rate of initial cooling.

4.116 An electric current passes through a 15-cm long, horizontal, 24-gauge nichrome wire (1.024-mm diameter, 1.384 Ω/m) such that its temperature is 80 °C when kept in water at 20 °C. Estimate the current in the wire.

4.117 An immersion heater is made of 15-mm diameter, 60-cm long tube. It is used to heat water in a large tank and is placed horizontally in the water. The pressure of the water in the tank is 500 kPa. The surface temperature of the heater is 120 °C. Determine the heat transfer rate when the temperature of the water is 60 °C. What will be the consequence if the heater designed to operate in water is operated in air at 20 °C?

4.118 The oil in a 40-cm diameter spherical container is maintained at 60 °C by a 1.5-cm diameter, horizontal, cylindrical electric heater. The total length of the heating element (U-shape) is 60 cm. The ambient air temperature is 0 °C. Determine the capacity of the heater, and the temperature of the heater surface. Assume the surface of the sphere is at 60 °C. The emissivity of the container is 0.8.

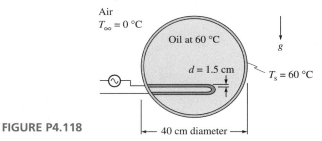

Air
$T_\infty = 0\ °C$

Oil at 60 °C

g

$d = 1.5$ cm

$T_s = 60\ °C$

FIGURE P4.118

← 40 cm diameter →

4.119 Heavy fuel oil, used in diesel-engine ships, must be heated prior to using it in the engine. Steam coils are used to heat the oil in the fuel tanks. The fuel oil in one such tank is heated by horizontal 2.5-cm O.D. tubes whose surfaces are at 70 °C. The oil in the tank is at 10 °C. Determine the heat transfer rate per meter length of the tube. Properties of the oil at 40 °C are

$\rho = 920\ \text{kg/m}^3 \qquad k = 0.12\ \text{W/m °C} \qquad \nu = 8.5 \times 10^{-6}\ \text{m}^2/\text{s}$
$c_p = 2100\ \text{J/kg °C} \qquad \beta = 7 \times 10^{-4}\ \text{K}^{-1}$

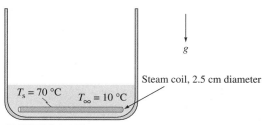

g

Steam coil, 2.5 cm diameter

$T_s = 70\ °C$ $T_\infty = 10\ °C$

FIGURE P4.119

4.120 A condenser for a refrigeration system consists of a horizontal row of 20 tubes, each 2-m long and 1-cm diameter. The tubes are at 60 °C and the ambient air at 30 °C. Estimate the heat transfer rate if the emissivity of the tube surfaces is 0.8.

4.121 Two horizontal tubes of diameters d_1 and d_2 ($= 2d_1$) are freely suspended in a fluid. *Estimate* the ratio of the heat transfer rates per unit length, q_2'/q_1' if both are at the same temperature. $\mathrm{Ra}_{d_1} = 6 \times 10^6$.

4.122 One technique for experimentally determining the convective heat transfer coefficient is to record the temperature of a solid when it is cooled (or heated) by exposing it to a fluid. If lumped analysis is valid, Equation 2.8.1a is used to determine h.

An experiment is to be designed to determine the natural convective heat transfer coefficient of horizontal cylinders in water by the transient technique. Horizontal cylinders of different diameters are heated to a uniform temperature and immersed in a large body of water at 20 °C. Their temperatures are recorded as a function of time. There is some concern about the validity of lumped analysis if the maximum diameter of the plain carbon-steel cylinders (solid) is 5 cm.
(a) Determine if lumped analysis is valid.
(b) If lumped analysis is not valid for the 5-cm diameter rod, propose a remedial measure.

4.123 The hood of an automobile is 1.8-m long and 1.1-m wide. On a sunny day the net radiant energy absorbed by the hood (radiant energy reaching the hood $-$ radiant energy leaving the hood) is estimated at 220 W/m^2 when the ambient air temperature is 30 °C. Modeling the hood as a horizontal plate, and neglecting the heat transfer from the lower surface of the hood (facing the engine), estimate the temperature of the hood.

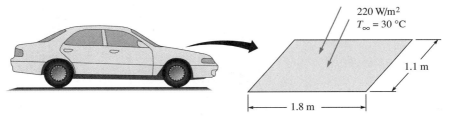

FIGURE P4.123

4.124 An incandescent lamp is approximated as a 75-mm diameter sphere. The temperature of the glass bulb of a 60 W lamp is 110 °C when the ambient air is at 20 °C. Estimate the heat transfer rate from the bulb by natural convection.

4.125 A thin circuit board is placed horizontally in a large cabinet. The temperature of the air in the cabinet is 20 °C. The board is 40-cm long and 30-cm wide. The hot surface is facing up and the heat transfer from the lower surface is negligible. The surface temperature of the board is not to exceed 70 °C. If the heat generation in the board is assumed to be uniform, what is the maximum heat dissipation rate from the board by natural convection?

4.126 A tube transporting a fluid is surrounded by another fluid. Show that the heat flux based on the inside surface area of the tube is given by

$$q'' = U(T_i - T_o)$$

$$U = \left[\frac{1}{h_i} + \frac{d_i}{2k} \ln \left(\frac{d_o}{d_i} \right) + \left(\frac{d_i}{d_o} \right) \left(\frac{1}{h_o} \right) \right]^{-1}$$

where

T_i = bulk temperature of the fluid in the tube
T_o = temperature of the fluid surrounding the tube
h_i, h_o = convective heat transfer coefficients on the inside and outside surfaces of the tube
d_i, d_o = inside and outside diameters of the tube
k = thermal conductivity of the material of the tube
U = overall heat transfer coefficient

(If the tube is insulated, it is only necessary to add the conductive resistance of the insulation in the expression for U.)

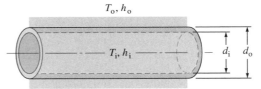

FIGURE P4.126

4.127 The temperature of the fluid surrounding a tube and the value of U (Problem 4.126) are uniform. Following the development in Section 4.2 for uniform temperature of the tube surface show that

$$\ln \left[(T_{fe} - T_o)/(T_{fi} - T_o) \right] = -U \pi \, d_i L/(\dot{m} c_p)$$

where

T_{fe} = temperature of the fluid at exit of the tube
T_{fi} = temperature of the fluid at inlet of the tube
L = length of the tube

4.128 Water flows in 1.5-cm diameter, thin copper tube with a velocity of 4 m/s. The tube is surrounded by air at 20 °C. At the section of the tube where the water temperature is 80 °C, determine
(a) The heat transfer rate per unit length.
(b) The overall heat transfer coefficient (see Problem 4.126) defined by

$$q' = U \pi d (T_b - T_o)$$

where

T_b = bulk temperature of the water in the tube
T_o = air temperature

(c) The water enters the tube at 85 °C. The tube located in the basement of a house is 15-m long. Assuming the overall heat transfer coefficient is constant at the value determined in part b above, calculate the exit water temperature.

4.129 Re-solve Problem 4.128 if the air is in cross flow with a velocity of 20 m/s.

4.130 Crude oil is pumped through buried pipes over large distances. To reduce the pumping power, the oil in the pipes is heated; without such heating it would be almost impossible to pump the oil in such places as Alaska. Consider the transport of oil between two pumping stations, 60-km apart. Oil is heated to 110 °C before it enters the pumps. The flow rate of oil through the 1-m I.D. pipe is 400 kg/s. The pipe is insulated with a 20-cm thick insulation material (k_{ins} = 0.06 W/m K) and buried with its axis 3-m below the surface of the earth. The temperature of the surface of the earth is estimated to be −35 °C, and the thermal conductivity of the soil is 0.6 W/m °C. The radial thickness and conductive resistance of the pipe material are negligible. Properties of the oil may be treated as constant at the values given below.

ρ = 890 kg/m^3 c_p = 1980 J/kg °C
k = 0.15 W/m °C ν = 9 × 10^{-6} m^2/s

Equivalent roughness of the pipe material = 0.046 mm.
Estimate
(a) The exit temperature of the oil.
(b) The pumping power required. The pressures at inlet at both pumping stations are equal.
(To solve this problem, use the appropriate conduction shape factor for the resistance of the soil to determine the overall heat transfer coefficient as in Problem 4.126 and the result of Problem 4.127 to determine the exit temperature of the oil. To find the pumping power, find the pressure drop using the friction factor for laminar or turbulent flow depending on the Reynolds number.)

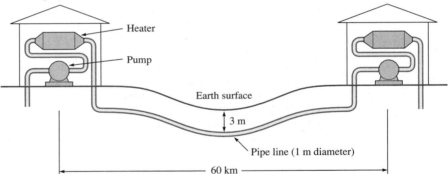

FIGURE P4.130

4.131 An ornamental space heater, in the form of a 2-ft diameter sphere, is freely suspended in a large room. The temperature of the air in the room is 20 °C. If the surface of the sphere is at 100 °C, determine the heat transfer rate by convection.

4.132 Saturated steam at 120 °C flows in a 10-cm I.D. steel pipe with a wall thickness of 0.5 cm. The pipe is exposed to air at 25 °C. Propose a suitable insulation to limit the temperature of the outer surface of the insulation to 40 °C. Because of the heat transfer rate to the surrounding atmosphere, steam condenses inside the tube. The condensate is to be removed in a steam trap. For a 10-m length of the pipe, determine the mass of condensate to be removed per hour. As the condensation heat transfer coefficient is large, the inside surface temperature of the tube may be assumed to be 120 °C. (The solution to this problem needs iteration on the outer surface temperature) **Design**

4.133 At a section of a thin, horizontal 1.5-cm diameter tube, water at 80 °C flows with a velocity of 4 m/s. It is surrounded by air at 20 °C. Determine the heat transfer rate per meter length of the tube, if

(a) The bare tube is exposed to air.

(b) The tube is insulated with 1-cm thick rockwool insulation.

In each case determine the rate of decrease of the water temperature in the direction of flow.

4.134 Design an experiment to determine the forced convection heat transfer coefficient with uniform heat flux, both in the developing and developed regions, with air or water as the fluid. The Reynolds numbers should cover the range 10 000 to 100 000. Air flow rate is limited to 0.05 kg/s.   **Design**

4.135 Design a forced convection space heater with a capacity of 1500 W to supplement central heating. The heater should use electric power available at 110 V. Identify the criteria for a successful design and propose a design.   **Design**

4.136 To supply domestic hot water, a 25-mm diameter, thin copper tube is used in the basement of a house. The 15-m long tube is horizontal. The temperature of the water is 60 °C and the flow rate of water is 0.2 l/s. The ambient air is at 12 °C.

(a) Compute the drop in temperature of the water per meter length of the tube.

(b) If the water flow ceases, compute the time taken for the water temperature to fall from 60 °C to 40 °C.

(c) Considering the drop in temperature of the water per meter length when hot water is being used, is any insulation of the tube necessary? Is insulation necessary to extend the time for the water temperature to drop from 60 °C to 40 °C when there is no flow?

(d) Propose a suitable insulation so that the outer surface temperature is less than 40 °C. Compute the time for the water temperature to fall from 60 °C to 40 °C when there is no flow.   **Design**

4.137 Design an electric hot water heater with a capacity of 40 gallons. The heater should have sufficient capacity to heat the water from 8 °C to 60 °C in 2 hours. Propose a design for the horizontal heating element and for the insulation of the boiler. **Design**

4.138 Propose a suitable design for the baseboard heater of a room. The capacity of the heater should be 3000 W when supplied with 0.08 kg/s of water at 65 °C. **Design**

CHAPTER

CONVECTION II

In Chapter 4 correlations for convective heat transfer coefficient for a few common geometries were presented. The use of the convective heat transfer coefficients for different situations was discussed. In this chapter we

- *Present correlations for a few more configurations of engineering significance*
- *Study forced convection in high speed flows where viscous dissipation is significant*
- *Show the similarity between momentum and heat transfer in laminar flows, and prove Reynolds analogy, which relates fluid friction factor to heat transfer parameters (Nusselt number, Reynolds number, Prandtl number)*
- *Introduce an approximate analytical technique, the integral technique, for the solutions of momentum and energy equations and obtain solutions for a few cases*

5.1 CONVECTION HEAT TRANSFER CORRELATIONS—SPECIAL CASES

In Chapter 4, convective heat transfer correlations for many simple geometries were presented. In this section we present correlations for a few other cases—forced convection over tube banks, natural convection in enclosures and with thin wires, natural convection from closely spaced vertical surfaces, and forced convection in high speed flows where viscous dissipation is significant. All the aforementioned situations have engineering significance.

5.1.1. Forced Convection—Flow Over Tube Banks

Forced convection correlations for the heat transfer coefficient for flows over single tubes were given in Chapter 4. In many applications, such as in heat exchangers, forced convection heat transfer occurs with a fluid flowing over a bank of tubes. The tubes may be aligned as in Figure 5.1.1a or staggered as in Figure 5.1.1b. In such configurations, the flow over each tube is different from the flow over a single cylinder surrounded by an infinite fluid medium. The maximum velocity of the fluid between two columns is given by $U_\infty S_T/(S_T - d)$. If $S_T \gg d$, the maximum velocity is close to the free stream velocity U_∞, and the flow is similar to that over a single

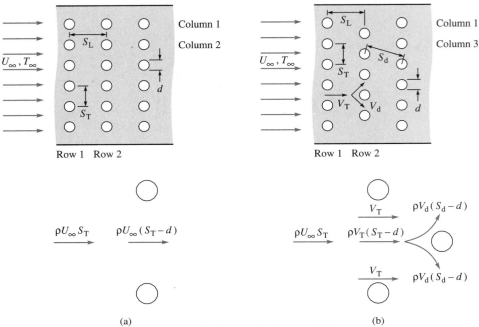

FIGURE 5.1.1 Two Arrangements of Tube Banks ($a = S_T/d$; $b = S_L/d$)
(a) In-line arrangement. (b) Staggered arrangement.

Heat transfer with flow over tube banks

cylinder. Otherwise, there is a significant increase in the velocity between the tubes in each row. Unless the rows are very far apart ($S_L \gg d$), the flow over the tubes in subsequent rows is affected by the presence of the tubes in the previous row(s) and by the neighboring columns. The extent of the effect of the neighboring columns and rows depends on the separation distance between the columns and rows. As a result of such an effect on the flow, the heat transfer coefficient in one row may not be the same as the heat transfer coefficient in another row. This effect is more pronounced in the first few rows. It has been found that, after about the first ten rows, the flow pattern repeats itself and the average value of the convective heat transfer coefficient after about 16 rows is not affected very much by increasing the number of rows. If the average heat transfer coefficient for all the tubes is known, the heat transfer rate from a bank of tubes maintained at a uniform temperature may be determined using the result obtained for internal flows with uniform tube surface temperature.

Convective heat transfer correlation, tube banks

For the average convective heat transfer coefficient, from experimental data, Zukauskas (1987) recommends correlations of the form

$$\mathrm{Nu}_d = c\,\mathrm{Re}_d\mathrm{Pr}_\infty^n \left(\frac{\mathrm{Pr}}{\mathrm{Pr_s}}\right)^{0.25} \tag{5.1.1}$$

$$\mathrm{Nu}_d = \frac{hd}{k}$$

h = average convective heat transfer coefficient with 16 or more rows of tubes, where the plane of a row of tubes is perpendicular to the free stream and the plane of a column of tubes is parallel to the free stream

$$\text{Re}_d = (\rho U_{\max} d)/\mu$$

$\text{Pr} = $ Prandtl number of the fluid

$\text{Pr}_s = $ Prandtl number evaluated at the tube surface temperature T_s

All properties of the fluid, except Pr_s, are evaluated at the mean bulk temperature, $(T_{bi} + T_{be})/2$; Pr_s is evaluated at T_s.

$T_{bi}, T_{be} = $ inlet and exit bulk temperatures of the fluid, respectively.

The complete form of Equation 5.5.1 for different configurations is given in Table 5.1.1. The uncertainty in the values of Nu_d is $\pm$ 15%.

TABLE 5.1.1 Correlations for Banks of Tubes

$0.7 < \text{Pr} < 500$ Number of rows ≥ 16 $a = S_T/d$, $b = S_L/d$ (Figure 5.1.1)

Evaluate all properties, except Pr_s, at the mean of the inlet and exit temperature of the fluid; evaluate Pr_s at T_s.

Correlation	Range of Re_d
In-line arrangement	
$\text{Nu}_d = 0.9\text{Re}_d^{0.4}\text{Pr}^{0.36}(\text{Pr}/\text{Pr}_s)^{0.25}$	$10^0 - 10^2$
$\text{Nu}_d = 0.52\text{Re}_d^{0.5}\text{Pr}^{0.36}(\text{Pr}/\text{Pr}_s)^{0.25}$	$10^2 - 10^3$
$\text{Nu}_d = 0.27\text{Re}_d^{0.63}\text{Pr}^{0.36}(\text{Pr}/\text{Pr}_s)^{0.25}$	$10^3 - 2 \times 10^5$
$\text{Nu}_d = 0.033\text{Re}_d^{0.8}\text{Pr}^{0.4}(\text{Pr}/\text{Pr}_s)^{0.25}$	$2 \times 10^5 - 2 \times 10^6$
Staggered Arrangement	
$\text{Nu}_d = 1.04\text{Re}_d^{0.4}\text{Pr}^{0.36}(\text{Pr}/\text{Pr}_s)^{0.25}$	$10^0 - 5 \times 10^2$
$\text{Nu}_d = 0.71\text{Re}_d^{0.5}\text{Pr}^{0.36}(\text{Pr}/\text{Pr}_s)^{0.25}$	$5 \times 10^2 - 10^3$
$\text{Nu}_d = 0.35(a/b)^{0.2}\text{Re}_d^{0.6}\text{Pr}^{0.36}(\text{Pr}/\text{Pr}_s)^{0.25}$	$10^3 - 2 \times 10^5$
$\text{Nu}_d = 0.031(a/b)^{0.2}\text{Re}_d^{0.8}\text{Pr}^{0.36}(\text{Pr}/\text{Pr}_s)^{0.25}$	$2 \times 10^5 - 2 \times 10^6$

In computing Re_d, the maximum average velocity between tubes is to be used. For the in-line arrangement from conservation of mass, assuming constant density, we have, $\rho U_\infty S_T = \rho U A_c$, where A_c is the cross-sectional area perpendicular to the flow direction. As the velocity is a maximum where the area of cross section is a minimum, the maximum velocity is given by

$$U_{\max} = \frac{U_\infty S_T}{S_T - d} \tag{5.1.2}$$

For a staggered arrangement, the minimum cross-sectional area may be either between two tubes in a row or between one tube and a neighboring tube in the succeeding row. From conservation of mass we have (Figure 5.1.1b),

$$\rho U_\infty S_T = \rho V_T(S_T - d) = 2\rho V_d(S_d - d)$$

where

$V_T = $ the average velocity between two tubes in a column

$V_d = $ the average velocity between one tube and a neighboring tube in the succeeding row

It is evident that if $S_T - d < 2(S_d - d)$, V_T is the maximum velocity. If $(S_T - d) > 2(S_d - d)$, V_d is the maximum velocity. Thus, for a staggered arrangement

For $S_T - d < 2(S_d - d)$ or $S_d > (S_T + d)/2$

$$U_{max} = \frac{U_\infty S_T}{S_T - d} \tag{5.1.3}$$

For $S_T - d > 2(S_d - d)$ or $S_d < (S_T + d)/2$

$$U_{max} = \frac{U_\infty S_T}{2(S_d - d)} \tag{5.1.4}$$
$$S_d = [S_L^2 + (S_T/2)^2]^{1/2}$$

Tube banks, the effect of number of rows

The correlations given in Table 5.1.1 are for tube banks with 16 or more rows. When there are less than 16 rows, the heat transfer coefficient is reduced. The ratio of the convective heat transfer coefficient with less than 16 rows to the heat transfer coefficient with 16 rows is given by a correction factor defined by

$$\frac{h_N}{h_{16}} = c_1 \tag{5.1.5}$$

where h_N is the convective heat transfer coefficient with N rows, $N < 16$, and h_{16} is the value obtained from the appropriate correlation in Table 5.1.1. The value of the correction factor in Equation 5.1.5 is shown in Figure 5.1.2. The correction factor is a function of the Reynolds number, but for $Re_d > 1000$, it is independent of Re_d. The values of c_1 for $Re_d > 1000$ are given in Table 5.1.2.

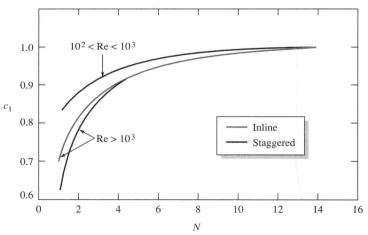

FIGURE 5.1.2 Correction Factor, c_1 in Equation 5.1.5
Reproduced by permission from Convection heat transfer in cross flow by Zukawskas, A., in *Handbook of Single Phase Convective Heat Transfer,* eds. Kakac, S., Shah, R. K., and Aung, W. (New York: Wiley Interscience, 1987).

TABLE 5.1.2 Value of Correction Factor, c_1, Equation 5.1.5 ($Re_d > 1000$)

N	1	2	3	4	5	7	10	13
In-line	0.70	0.80	0.86	0.90	0.93	0.96	0.98	0.99
Staggered	0.64	0.76	0.84	0.89	0.93	0.96	0.98	0.99

Determining
pressure drop in
tube banks

Pressure Drop   With flow over tube banks there is a significant pressure drop in the fluid. To compute the fan or pumping power required to maintain the flow, the pressure drop is needed. Zukauskas (1987) recommends that the pressure drop be computed by the relation

$$\Delta p = p_{in} - p_{out} = N\chi \frac{\rho U_{max}^2}{2} f \tag{5.1.6}$$

where p_{in} and p_{out} are the pressures at the inlet and outlet of the tube bank. The values of χ and f are presented in Figures 5.1.3. In Figure 5.1.3a, the friction factor, f, is presented graphically for $S_L = S_T$ and for different ratios of $b\ (= S_L/d)$. For ratios of $S_L/S_T \neq 1$, a correction factor χ is given in the inset of Figure 5.1.3a for different values of $(a - 1)/(b - 1)$, where $a = S_T/d$ and $b = S_L/d$. Figure 5.1.3b gives the values of f for a staggered arrangement with the neighboring tubes forming an equilateral triangle. For other arrangements ($S_T/S_d \neq 1$) the correction factor χ is given in the inset. The value of f is for the pressure drop with one row of tubes; it has to be multiplied by the number of rows to get the total pressure drop.

Tube banks—heat
transfer
computation with
fluid temperature
changing in the
direction of flow

When a fluid flows over a bank of tubes, the temperature of the fluid changes. This change in the temperature of the fluid should be considered when computing the heat transfer rate. In so far as heat transfer is concerned, the configuration is similar to that of heat transfer to a fluid flowing in a tube with a uniform surface temperature and convective heat transfer coefficient (see Section 4.2). Equations 4.2.5a and b are also applicable for flows over tube banks.

$$\frac{T_s - T_{be}}{T_s - T_{bi}} = \exp\left(-\frac{hA_s}{\dot{m}c_p}\right) \tag{4.2.5b}$$

where

T_s = temperature of the surface of the tubes
T_{bi} = temperature of the fluid at inlet to the tube bank
T_{be} = temperature of the fluid at exit from the tube bank
A_s = total surface area of the tubes in contact with the fluid
$\dot{m}$ = mass rate of flow of the fluid
h = average convective heat transfer coefficient at $(T_{bi} + T_{be})/2$
c_p = specific heat of the fluid

After finding the exit temperature of the fluid from Equation 4.2.5b, the heat transfer rate is calculated from Equation 1.10.19

$$q = \dot{m}c_p(T_{be} - T_{bi}) \tag{1.10.19}$$

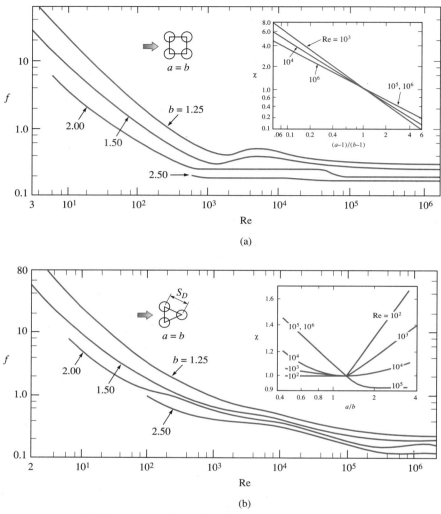

(a)

(b)

FIGURE 5.1.3 Friction Factors for Tube Banks
Reproduced by permission from Convection heat transfer in cross flow, by
Zukawskas, A., in *Handbook of Single Phase Convective Heat Transfer*, eds. Kakac,
S., Shah, R. K., and Aung, W. (New York: Wiley Interscience, 1987).

EXAMPLE 5.1.1 A heat exchanger, with staggered tubes, is used to heat atmospheric air entering at
5 °C. The temperature of the surface of the tubes is maintained at 100 °C with steam
condensing inside the tubes. Other details of the heat exchanger are

Diameter of the tubes = 25 mm	$U_\infty = 15$ m/s
Number of columns = 20	Number of rows = 20
$S_L = S_T = 50$ mm	Length of tubes = $L = 3$ m

Determine

(a) The exit temperature of air and the heat transfer rate.
(b) The mass rate of condensation of steam assuming the steam condenses from saturated dry vapor state at 125 kPa.
(c) The fan power required to maintain the air flow.

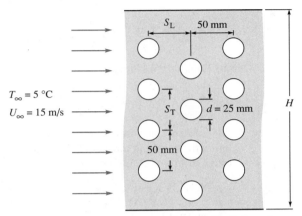

FIGURE 5.1.4 Arrangement of tubes in Example 5.1.1 (only 4 columns are shown)

Given

Fluid: air $d = 0.025$ m $L = 3$ m
$T_s = 100\ °C$ $T_{bi} = 5\ °C$
$U_\infty = 15$ m/s   Number of rows = 20   Number of columns = 20

Find

(a) T_{be} (the exit bulk temperature of air)
(b) $\dot{m}_{st}$ (mass rate of condensation of steam)
(c) Fan power

Assumptions

1. Steady state
2. Uniform surface temperature

SOLUTION

(a) The exit temperature is found from Equation 4.2.5b. But it requires the evaluation of convective heat transfer coefficient at the arithmetic mean of the inlet and exit temperatures of the air. As the exit air temperature is not known, it is determined by iteration. The iteration steps are

1. Guess a reasonable exit temperature.
2. Evaluate h from an appropriate correlation in Table 5.1.1.
3. Determine T_{be} from Equation 4.2.5b.

4. If the value of T_{be} is different from the value used in step 2, recompute h using the computed value of T_{be} and repeat computations until the value of T_{be} used to compute h and the value obtained from Equation 4.2.5b agree within an acceptable limit.

For the geometry shown in Figure 5.1.4,

$$S_d = \left[S_{\text{L}}^2 + \left(\frac{S_{\text{T}}}{2} \right)^2 \right]^{1/2} = (50^2 + 25^2)^{1/2} = 55.9 \text{ mm}$$

$$\frac{S_{\text{T}} + d}{2} = \frac{50 + 25}{2} = 37.5 \text{ mm} < S_d.$$

From Equation 5.1.3,

$$U_{\text{max}} = \frac{U_\infty S_{\text{T}}}{S_{\text{T}} - d} = \frac{15 \times 50}{50 - 25} = 30 \text{ m/s}$$

1. Assume an exit temperature of 60 °C. The arithmetic mean of the inlet and exit temperature of the air is 32.5 °C. From the software CC. (or Table A7) the properties of air at 32.5 °C and 101.3 kPa are

$\rho = 1.155 \text{ kg/m}^3$ $c_p = 1007 \text{ J/kg °C}$
$\mu = 1.877 \times 10^{-5} \text{ N s/m}^2$ $k = 0.0266 \text{ W/m °C}$
$\text{Pr} = 0.712$ $\text{Pr}_s(100 \text{ °C}) = 0.705$

$$\text{Re}_d = \frac{1.155 \times 30 \times 0.025}{1.877 \times 10^{-5}} = 46\,151$$

2. From Table 5.1.1,

$\text{Nu}_d = 0.35(a/b)^{0.2}\text{Re}_d^{0.6}\text{Pr}^{0.36}(\text{Pr}/\text{Pr}_s)^{0.25}$
$a = S_{\text{T}}/d = 50/25 = 2$ $b = S_{\text{L}}/d = 50/25 = 2$ $a/b = 1$
$\text{Nu}_d = 0.35 \times 46\,151^{0.6} \times 0.712^{0.36}(0.712/0.705)^{0.25} = 195.2$
$h = (195.2 \times 0.0266)/0.025 = 207.7 \text{ W/m}^2 \text{ °C}$
$\dot{m} = \rho U_\infty A$ $A = H \times L$
$H = 20 \times S_{\text{T}} + S_{\text{T}}/2 = 20 \times 0.05 + 0.05/2 = 1.025 \text{ m}$ $L = 3 \text{ m}$
$\rho(5 \text{ °C}, 101.3 \text{ kPa}) = 1.269 \text{ kg/m}^3$
$\dot{m} = 1.269 \times 15 \times 1.025 \times 3 = 58.53 \text{ kg/m}^3$
$A_s = $ surface area of tubes $= 400 \times \pi \times 0.025 \times 3 = 94.25 \text{ m}^2$

3. From Equation 4.2.5b,

$$T_s - T_{\text{be}} = (T_s - T_{\text{bi}}) \exp\left(-\frac{hA_s}{\dot{m}c_p} \right)$$

$$100 - T_{\text{be}} = (100 - 5) \exp\left(-\frac{207.7 \times 94.25}{58.53 \times 1007} \right) \Rightarrow T_{\text{be}} = 31.8 \text{ °C}$$

4. The initially guessed value of T_{be} $(= 60 \text{ °C})$ for computing h is too high. Recompute h with $T_{be} = 31.8 \text{ °C}$. Assuming $T_{\text{be}} = 31.8 \text{ °C}$, the arithmetic mean of the inlet and exit temperatures of the air is 18.4 °C. Properties of air at 18.4 °C and 101.3 kPa are

$$\rho = 1.211 \text{ kg/m}^3 \qquad c_p = 1007 \text{ J/kg °C}$$
$$\mu = 1.81 \times 10^{-5} \text{ N s/m}^2 \qquad k = 0.0255 \text{ W/m °C}$$
$$\text{Pr} = 0.714 \qquad \text{Pr}_s (100 \text{ °C}) = 0.705$$
$$\text{Re}_d = 50\ 180 \qquad \text{Nu}_d = 205.6$$
$$h = 209.7 \text{ W/m}^2 \text{ °C}$$

From Equation 4.2.5b,

$$100 - T_{be} = (100 - 5) \exp\left(-\frac{209.7 \times 94.25}{58.53 \times 1007}\right) \Rightarrow T_{be} = 32.1 \text{ °C}$$

The computed exit temperature of the air, 32.1 °C, is quite close to the assumed value of 31.8 °C for calculating h. Therefore,

$$T_{be} = \underline{32.1 \text{ °C}}$$

$$q = \dot{m}c_p(T_{be} - T_{bi}) = 58.53 \times 1007(32.1 - 5) = \underline{1.597 \times 10^6 \text{ W}}$$

(b) Mass rate of condensation of steam is given by

$$\dot{m}_c = \frac{q}{i_{fg}(125 \text{ kPa})}$$

From steam tables, i_{fg} for water at 125 kPa = 2.241×10^6 J/kg

$$\dot{m}_c = \frac{q}{i_{fg}(125 \text{ kPa})} = \frac{1.597 \times 10^6}{2.241 \times 10^6} = \underline{0.713 \text{ kg/s}}$$

(c) The fan power required is obtained from the energy equation with a control volume around the fan as shown by the dotted line in Figure 5.1.5. Denoting the

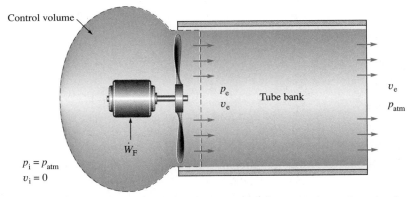

FIGURE 5.1.5 Control Volume for Computing the Fan Power Required to Overcome the Frictional Resistance in the Tube Banks

power input to the fan by $\dot{W}_F$, from the First Law of Thermodynamics, Equation 1.10.5,

$$\dot{W}_F = \dot{m}\left(u_e - u_i + \frac{p_e}{\rho_e} - \frac{p_i}{\rho_i} + \frac{v_e^2 - v_i^2}{2} + z_e - z_i - \frac{\dot{Q}}{\dot{m}}\right) \qquad (1)$$

where the subscripts i and e denote the states of air at inlet and exit to fan, respectively. $\dot{Q}$ in Equation 1 is the heat transfer rate to the control volume. The minimum fan power required is obtained when there are no irreversible effects in the flow through the fan. For such a reversible flow,

$$u_e - u_i - \frac{\dot{Q}}{\dot{m}} = 0$$

Further, v_i the velocity at inlet to the control volume may be made very small by taking the control surface sufficiently far away from the fan. It is only at such a distance from the fan that free stream conditions (in terms of pressure and temperature) exist. The change in the potential energy may also neglected. Thus, the energy equation reduces to

$$\dot{W}_F = \dot{m} \left(\frac{p_e}{\rho_e} - \frac{p_i}{\rho_i} + \frac{v_e^2}{2} \right)$$

p_i = atmospheric pressure = pressure at exit of the heat exchanger

p_e = pressure at exit of the fan

$p_e - p_i = p_e - p_{atm}$ = pressure drop across the heat exchanger. The pressure drop is obtained from Equation 5.1.6 in conjunction with Figure 5.1.3.

From the arrangement of the tubes,

$a = S_T/d = 50/25 = 2$ $N = 20$ $Re_d = 54\ 180$
$a/b = S_L/S_T = 1$ $\rho = 1.211\ \text{kg/m}^3$ $U_{max} = 30\ \text{m/s}$
$\chi = 1$ $f = 0.21$

$$p_e - p_{atm} = \text{pressure drop} = 20 \times 1.0 \times \frac{1.211 \times 30^2}{2} \times 0.21 = 2288\ \text{Pa}$$

$$\dot{W}_F = 58.53 \left(\frac{2288}{1.211} + \frac{15^2}{2} \right) = 1.172 \times 10^5\ \text{W} \approx \underline{117\ \text{kW}}$$

COMMENT

The required minimum fan power is quite large. The high mass flow rate of air and the considerable pressure drop across the tube banks require the high power. The power required to increase the kinetic energy is about 5% of the power required to overcome the pressure drop. The computed fan power is the minimum power required, assuming no irreversible effects. In an actual situation, irreversibilities do exist and they result in a positive value for $u_e - u_i - (\dot{Q}/\dot{m})$ in the energy equation. But such irreversibilities depend on the design of the fan, and even if the design is known, it is very difficult to compute such irreversibilities. The actual required power is estimated by knowing the efficiency of similar fans, defined by $\eta = \dot{W}_{F,min}/\dot{W}_{F,actual}$.

5.1.2 Natural Convection—Rectangular Cavities

In Chapter 4, natural convection correlations are given for surfaces when they are surrounded by an extensive fluid medium. In such engineering applications as solar collectors and double-glazed windows (with the space between two glass panes filled with a dry gas), there is heat transfer from one surface to another surface in the enclosure by natural convection. Such rectangular cavities may be either horizontal, vertical, or inclined to the horizontal plane at an arbitrary angle.

When a fluid is enclosed between two surfaces maintained at temperatures T_1 and T_2, the average heat flux is expressed as

$$q'' = h(T_1 - T_2) \tag{5.1.7}$$

In the following sections, correlations for determining the average heat transfer coefficient, h, in Equation 5.1.7 are given for some configurations similar to Figure 5.1.6.

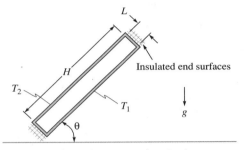

FIGURE 5.1.6 A Rectangular Cavity Inclined to the Horizontal Plane

In all the correlations the properties of the fluid are evaluated at the arithmetic mean temperature $(T_1 + T_2)/2$, and the characteristic length L in Ra_L is as indicated for each configuration.

Natural convection—vertical rectangular cavities

Rectangular, Vertical Cavities: $\theta = 90°$ Correlations for rectangular, vertical cavities (Figure 5.1.7) are expressed in the form

$$Nu_L = c \, Ra_L^m \, Pr^n \tag{5.1.8}$$

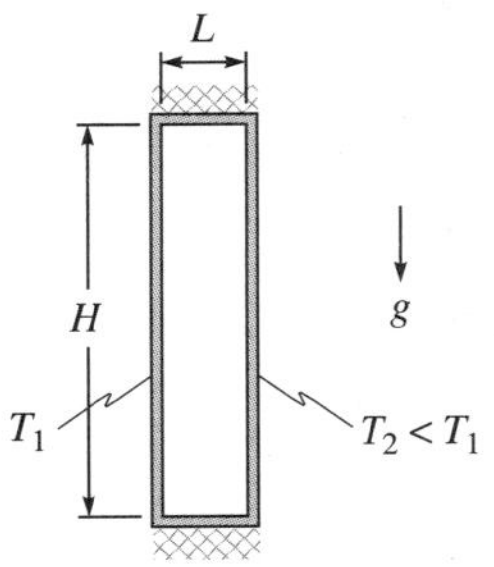

FIGURE 5.1.7 Vertical Cavity

For aspects ratios of $1 < H/L < 40$, correlations recommended by Catton (1986) and McGregor and Emery (1969) are

$$\left\| \begin{array}{ccc} 1 < H/L < 2 & 10^{-3} < \text{Pr} < 10^5 & 10^3 < (\text{Ra}_L\text{Pr})/(0.2 + \text{Pr}) \end{array} \right.$$

$$\text{Nu}_L = 0.18 \left(\frac{\text{Pr}}{0.2 + \text{Pr}} \text{Ra}_L \right)^{0.29} \tag{5.1.9}$$

$$\begin{array}{ccc} 2 < H/L < 10 & \text{Pr} < 10^5 & \text{Ra}_L < 10^{10} \end{array}$$

$$\text{Nu}_L = 0.22 \left(\frac{\text{Pr}}{0.2 + \text{Pr}} \text{Ra}_L \right)^{0.28} \left(\frac{L}{H} \right)^{0.25} \tag{5.1.10}$$

$$\begin{array}{ccc} 10 < H/L < 40 & 1 < \text{Pr} < 2 \times 10^4 & 10^4 < \text{Ra}_L < 10^7 \end{array}$$

$$\text{Nu}_L = 0.42 \, \text{Ra}_L^{0.25}\text{Pr}^{0.012} \left(\frac{L}{H} \right)^{0.3} \tag{5.1.11}$$

$$\begin{array}{ccc} 1 < H/L < 40 & 1 < \text{Pr} < 20 & 10^6 < \text{Ra}_L < 10^9 \end{array}$$

$$\left. \text{Nu}_L = 0.046 \, \text{Ra}_L^{1/3} \right\| \tag{5.1.12}$$

Horizontal Cavities: $\theta = 0$ Even when $T_1 > T_2$, for $\text{Ra}_L < 1708$, heat transfer is essentially by conduction if end effects are insignificant. For $\text{Ra}_L > 1708$, convection currents set in. For the lower surface hotter than the upper surface with air between the plates (Figure 5.1.8), Jakob (1949) recommends

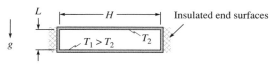

FIGURE 5.1.8 Horizontal Cavity

$$\text{Nu}_L = 0.195 \, \text{Gr}_L^{1/4} \qquad 10^4 < \text{Gr}_L < 4 \times 10^5 \tag{5.1.13a}$$

$$\text{Nu}_L = 0.068 \, \text{Gr}_L^{1/3} \qquad 4 \times 10^5 < \text{Gr}_L < 10^7 \tag{5.1.13b}$$

From measurements with liquids in a horizontal cavity, bottom surface heated, Globe and Dropkin (1959) suggest

$$\text{Nu}_L = 0.069 \, \text{Ra}_L^{1/3}\text{Pr}^{0.074} \qquad 3 \times 10^5 < \text{Ra}_L < 7 \times 10^9 \tag{5.1.13c}$$

From experimental measurements, Hollands et al. (1976) recommend

$$\text{Nu}_L = 1 + 1.44 \left(1 - \frac{1708}{\text{Ra}_L} \right)^* + \left[\left(\frac{\text{Ra}_L}{5830} \right)^{1/3} - 1 \right]^* \tag{5.1.13d}$$

In Equation 5.1.13d, the terms with the brackets and asterisk, [ ]*, should be set equal to zero if they are negative. All properties are evaluated at the arithmetic mean temperature.

An interesting feature of Equations 5.1.13b and c (for $\text{Ra}_L > 3 \times 10^5$) is that Nu_L is proportional to L, which makes the heat transfer coefficient independent of the spacing L.

Natural
convection—
Inclined rectangular
cavities

Inclined Rectangular Cavities As a result of the oil shortages in the 1970s and possibilities of long-term shortages in oil supplies, solar energy utilization was extensively studied. One of the components associated with solar energy utilization for heating is the flat plate solar collector, which has a cavity between a glass cover and an absorber plate. As the collectors are inclined to the horizontal plane, convection in inclined cavities received a great deal of attention (see Figure 5.1.9). Based on

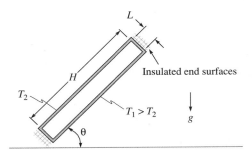

FIGURE 5.1.9 A Rectangular Cavity Inclined to the Horizontal Plane

the available analytical and experimental works by Ayyaswamy and Catton (1973), Arnold et al. (1974), and Hollands et al. (1976), Catton (1986) recommends the correlations given below. Evaluate properties in the table at $(T_1 + T_2)2$.

Angle of Inclination	Aspect Ratio H/L	Equation Number
$0 < \theta < \theta*$	>10	5.1.14
	<10	5.1.15
$\theta* < \theta < 90°$	All	5.1.16
$90° < \theta < 180°$	All	5.1.17

For values of $\theta*$ see Table 5.1.3.

$$\text{Nu}_L = 1 + 1.44 \left[1 - \frac{1708}{\text{Ra}_L \cos \theta}\right]^* \left[1 - \frac{1708(\sin 1.8\theta)^{1.6}}{\text{Ra}_L\cos \theta}\right] + \left[\left(\frac{\text{Ra}_L \cos \theta}{5830}\right)^{1/3} - 1\right]^* \quad \textbf{(5.1.14)}$$

The asterisk on the brackets, [ ]*, indicates that if the quantity in brackets is negative, it should be set equal to zero. Equation 5.1.14 is valid for angles of inclination less than $\theta*$, the critical angle, given in Table 5.1.3.

TABLE 5.1.3 Critical Angle $\theta*$ for Inclined Cavities

H/L	1	3	6	12	>12
$\theta*$ (deg)	25	53	60	67	70

$0 < \theta < \theta*$

$$\text{Nu}_L = \text{Nu}_L(\theta = 0) \left[\frac{\text{Nu}_L(\theta = 90°)}{\text{Nu}_L(\theta = 0)}\right]^{\theta/\theta*} (\sin \theta*)^{\theta/(4\theta*)} \quad \textbf{(5.1.15)}$$

$\theta^* < \theta < 90°$

$$Nu_L = Nu_L(\theta = 90°)(\sin\theta)^{1/4} \tag{5.1.16}$$

$90° < \theta < 180°$

$$Nu_L = 1 + [Nu_L(\theta = 90°) - 1]\sin\theta \tag{5.1.17}$$

In horizontal, rectangular cavities, as Ra_L is reduced, the motion of the fluid also decreases and below a certain critical value of Ra_L heat transfer is essentially by pure conduction. For example, for a horizontal cavity heated from below, the critical Rayleigh number is 1708, and below that value, heat transfer is by conduction across the fluid layer. Thus, as $Ra_L \to 0$ the limiting value of $Nu_L \to 1$. Also for a horizontal cavity heated from above, heat transfer is by conduction. There may be end effects, which may become significant as the aspect ratio H/L becomes small. Because of the end effects, the realized value of Nu_L is greater than the limiting value of 1.

5.1.3 Natural Convection—Coaxial Cylinders and Concentric Spheres

Natural convection—heat transfer rate coaxial cylinders and concentric spheres

The heat transfer rate by natural convection between two long, horizontal, coaxial, isothermal cylinders, and between two concentric, isothermal spheres was studied by Raithby and Hollands (1975). The heat transfer rate is expressed in a form that does not involve the convection heat transfer coefficient in the usual sense. They expressed the heat transfer rate per unit length of the inner cylinder as

$$q' = \frac{2\pi k_{eff}\,(T_i - T_o)}{\ln(d_o/d_i)} \tag{5.1.18}$$

where

$\quad T_i$ = temperature of the inner cylindrical surface
$\quad T_o$ = temperature of the outer cylindrical surface
d_i, d_o = diameters of the inner and outer cylinders, respectively
$\quad k_{eff}$ = thermal conductivity of a stationary medium with identical inner and outer surface temperatures for the same heat transfer rate

The value of the k_{eff} is determined from the relation

$$\frac{k_{eff}}{k} = 0.386 \left(\frac{Pr}{0.861 + Pr}\right)^{1/4} Ra_c^{1/4} \tag{5.1.19a}$$

where

$$Ra_c = \frac{[\ln(d_o/d_i)]^4}{L^3(d_i^{-0.6} + d_o^{-0.6})^5} Ra_L \tag{5.1.19b}$$

$$Ra_L = \frac{g\,\beta\rho^2\Delta T L^3}{\mu^2} Pr$$

Evaluate properties at $(T_i + T_o)/2$
$10^2 < Ra_c < 10^7$
$L = (d_o - d_i)/2$

k = thermal conductivity of the fluid
If $k_{eff}/k < 1$, set $k_{eff} = k$

For $Ra_c < 100$, heat transfer is effectively by conduction and $k_{eff} = k$.
   For heat transfer between two concentric spheres, the suggested correlation is

$$q = k_{eff}\pi \frac{D_i D_o}{L} (T_i - T_o) \tag{5.1.20a}$$

$$\frac{k_{eff}}{k} = 0.74 \left(\frac{Pr}{0.861 + Pr}\right)^{1/4} Ra_s^{1/4} \tag{5.1.20b}$$

$$Ra_s = \frac{L}{(D_o D_i)^4} \frac{Ra_L}{(D_i^{-1.4} + D_o^{-1.4})^5} \tag{5.1.20c}$$

$$Ra_L = \frac{g\beta\rho^2\Delta T L^3}{\mu^2} Pr \qquad \Delta T = |T_i - T_o|$$

Evaluate properties at $(T_i + T_o)/2$
$10^2 < Ra_s < 10^4$
$L = (D_o - D_i)/2$
If $\dfrac{k_{eff}}{k} < 1$, set $k_{eff} = k$

EXAMPLE 5.1.2   A 10-cm diameter sphere is maintained at 120 °C (Figure 5.1.10). It is enclosed in a 12-cm diameter concentric spherical surface maintained at 100 °C. The space between the two surfaces is filled with air at 200 kPa. Determine the convective heat transfer rate from the inner sphere.

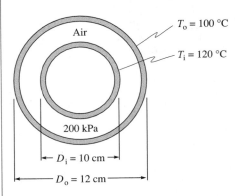

T_o = 100 °C
T_i = 120 °C
Air
200 kPa
D_i = 10 cm
D_o = 12 cm

FIGURE 5.1.10 Two Concentric Isothermal Spheres with Air Between Them

Given

Fluid: air p = 200 kPa
D_i = 0.1 m T_i = 120 °C
D_o = 0.12 m T_o = 100 °C

Find

q (convective heat transfer rate from the inner sphere)

SOLUTION

The heat transfer rate is computed from Equation 5.1.20a in conjunction with Equation 5.1.20c. From Table A7 the properties of air at $(120 + 100)/2 = 110\,°C$ and 200 kPa are

$$\rho = \frac{p}{RT} = \frac{200}{0.287 \times 383.15} = 1.819 \text{ kg/m}^3 \qquad k = 0.0319 \text{ W/m K}$$

$$\mu = 2.22 \times 10^{-5} \text{ N s/m}^2 \qquad\qquad \text{Pr} = 0.703$$

$$c_p = 1012 \text{ J/kg K} \qquad\qquad \beta = \frac{1}{383.15} \text{ K}^{-1}$$

$$D_o = 0.12 \text{ m}$$

$$\qquad\qquad\qquad\qquad L = (D_o - d_i)/2 = 0.01 \text{ m}$$

$$D_i = 0.1 \text{ m}$$

$$\text{Ra}_L = \frac{g\,\beta\rho^2(T_i - T_o)L^3}{\mu^2}\,\text{Pr}$$

$$= \frac{9.807 \times \dfrac{1}{383.15} \times 1.819^2 \times (120 - 100) \times 0.01^3}{(2.22 \times 10^{-5})^2} \times 0.703 = 2416$$

From Equation 5.1.20c,

$$\text{Ra}_s = \frac{L}{(D_o D_i)^4}\frac{\text{Ra}_L}{(D_i^{-1.4} + D_o^{-1.4})^5} = \frac{0.01 \times 2416}{(0.s12 \times 0.1)^4(0.12^{-1.4} + 0.1^{-1.4})^5} = 6.62$$

From Equation 5.1.20b,

$$k_{\text{eff}} = 0.74\left(\frac{\text{Pr}}{0.861 + \text{Pr}}\right)^{1/4}\text{Ra}_s^{1/4}\,k$$

$$= 0.74\left(\frac{0.703}{0.861 + 0.73}\right)^{1/4} \times 6.62^{1/4} \times 0.0319 = 0.031 \text{ W/m K}$$

As $k_{\text{eff}} < k$, set $k_{\text{eff}} = k = 0.0319$ W/m K. From Equation 5.1.20a,

$$q = 0.031 \times \pi\,\frac{0.12 \times 0.1}{0.01}\,(120 - 100) = \underline{2.4 \text{ W}}$$

5.1.4 Natural Convection from Thin Wires

Correlations for natural convection from thin wires

In Section 4.3, for natural convection from vertical cylinders, it is recommended that the correlations for vertical surfaces be used provided that $L/d < 35/\text{Gr}_L^{1/4}$. If the length of the cylinder is less than given by that relation, the boundary layer thickness is much less than the radius of the cylinder, and Equation 4.3.2e is recommended for $10^5 < \text{Ra}_L < 10^9$. But for very thin wires, Ra_L is very small and the condition that the boundary layer thickness be much less than the radius of the cylinder is not satisfied. Specific correlations have been proposed for such wires. From studies on convection from thin wires with arbitrary angles of inclination to the horizontal plane (Figure 5.1.11), Fujii et al. (1986), recommend

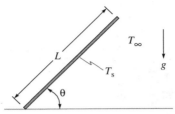

FIGURE 5.1.11 A Thin, Isothermal Wire Inclined to the Horizontal

Evaluate all properties at $(T_s + T_\infty)/2$.
The diameter, d, is the characteristic length for all orientations of the wire.
Horizontal wire: $10^{-8} < \mathrm{Gr}_d \mathrm{Pr} < 10^6$

$$\frac{2}{\mathrm{Nu}_d} = \ln\left[1 + \frac{3.3}{c\,(\mathrm{Gr}_d\mathrm{Pr})^n}\right] \tag{5.1.21a}$$

Vertical wire: $c\,[\mathrm{Gr}_d\mathrm{Pr}(d/L)]^{1/4} > 2 \times 10^{-3}$

$$\mathrm{Nu}_d = c\left(\mathrm{Gr}_d\mathrm{Pr}\,\frac{d}{L}\right)^{1/4} + 0.763\,c^{1/6}\left(\mathrm{Gr}_d\mathrm{Pr}\,\frac{d}{L}\right)^{1/24} \tag{5.1.21b}$$

Inclined wires: θ = angle of inclination to the horizontal plane.
$2 \times 10^{-5} < \mathrm{Gr}_d\mathrm{Pr}\cos\theta < 10^{-3}$ and $0 < \theta < 85°$.
For $85° < \theta < 90°$, use Equation 5.1.21b for vertical wires.
Replace g by $g\cos\theta$ in computing Ra_L.

$$\frac{2}{\mathrm{Nu}_d} = \ln\left[1 + \frac{3.3}{c\,(\mathrm{Gr}_d\mathrm{Pr}\cos\theta)^n}\right] \tag{5.1.21c}$$

c and n in Equations 5.1.21a, b, and c, are defined as

$$c = \frac{0.671}{[1 + (0.492/\mathrm{Pr})^{9/16}]^{4/9}} \qquad n = 0.25 + \frac{1}{10 + 5(\mathrm{Gr}_d\mathrm{Pr})^{0.175}}$$

5.1.5 Natural Convection from Closely Spaced Vertical Plates

Optimum spacing of vertical fins

In many cases, arrays of parallel plates are used as a bank of extended surfaces to increase the heat transfer rates. In some cases, such as electronic cooling, they are closely spaced as shown in Figure 5.1.12. When these plates are vertical, cool air flows between the plates and cools the equipment. With natural convection a boundary layer develops. If the spacing between two adjacent fins is more than $2\delta_{\max}$ ($\delta_{\max}$ is the maximum thickness of the boundary layer at the trailing edge of the fin), T_∞ can still be used as the reference temperature for the definition of the heat transfer coefficient. However, the number of fins is limited, as is the heat transfer rate. If the spacing is less than $2\delta_{\max}$, the number of fins is increased but parts of the fins are exposed to heated air only and this reduces the heat transfer rate. Bar-Cohen and Rohsenow (1984) modified the available analytical correlations to get better agree-

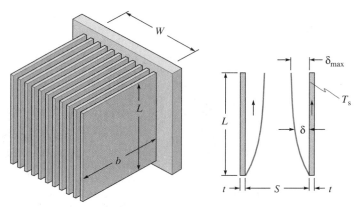

FIGURE 5.1.12 An Array of Vertical Fins

ment with experimental measurements. To obtain the maximum heat transfer rate for a given base width, W, they recommend that the optimum spacing, S, be computed from the following equation,

$$S + 1.5t - 0.0025P^{1.5}S^7 = 0 \tag{5.1.22}$$

$$P = \frac{g\beta\rho^2(T_s - T_\infty)}{\mu^2} \frac{\text{Pr}}{L} = \frac{\text{Ra}_L}{L^4}$$

For the optimum spacing obtained from Equation 5.1.22,

$$\text{Nu}_S = \frac{hS}{k} = \left[\frac{576}{P^2S^8} + \frac{2.873}{P^{1/2}S^2} \right]^{-1/2} \tag{5.1.23}$$

The heat transfer rate from the array of fins is then given by

$$q = h2NLb(T_s - T_\infty) \tag{5.1.24}$$

where N is the number of fins and h is the average heat transfer coefficient.

As $t \to 0$, Equations 5.1.22 and 5.1.23 simplify to

$$S = \frac{2.714}{P^{0.25}} \qquad \text{Nu}_S = 1.31$$

The equations given for closely spaced vertical fins are based on the assumption of nearly isothermal fins, i.e., for short fins with mb not much greater than 0.1 [$(m = (hP/kA)^{1/2}$].

5.1.6 Forced Convection with High Speed Flows over Flat Plates

High speed flows, Mach number greater than 0.2

In computing the forced convection heat transfer from a plate to the surrounding fluid, it is usual to neglect the effect of the work transfer due to viscous stresses. This is justified in many cases, but when the velocity of a fluid is high—when the Mach number (ratio of the velocity of the fluid to the sonic velocity) is greater than about 0.2, viscous work (the work done against the shear stresses) becomes significant. In this section the conditions under which such effects are important and the modifications to the existing correlations in such flows are given.

Just as rubbing of two solid surfaces in contact tends to increase the temperature of the surfaces, in fluid flows the shear stresses (similar to the friction forces between the solid plates) combined with the motion of the fluid layers tend to increase the temperature of the fluid. The magnitude of the increase in temperature depends on the shear stress and the velocity gradient in addition to the properties of the fluid. The effect of shear work is termed viscous dissipation. In a boundary layer with flow over a flat plate, the viscous dissipation is proportional to $(\partial u/\partial y)^2$, where u is the velocity parallel to the plate and y is the distance along the normal to the plate. As this gradient is a maximum at the surface, the viscous dissipation is also greatest at the boundary. Viscous dissipation results in an increase in the temperature of the fluid; the increase in the temperature of the fluid caused by viscous dissipation is greatest at the boundary. The adiabatic wall temperature is defined as the temperature of the fluid at the solid insulated surface. At the solid surface the velocity of the fluid is zero and, if the effect of viscous work and heat transfer from the fluid element is neglected, from the First Law of Thermodynamics, the increase in the specific enthalpy of the fluid element adjacent to the insulated surface (due the reduction in the kinetic energy) is $U_\infty^2/2$. However, in high speed flows, the increase in the specific enthalpy is further augmented by viscous work but decreased by heat transfer. For a fluid with Prandtl number less than 1 (low viscosity, high thermal conductivity) the effect of heat transfer is greater than the effect of viscous work and the actual enthalpy rise is less than $U_\infty^2/2$. For a fluid with Prandtl number greater than 1 (high viscosity, low thermal conductivity) the effect of viscous work is greater than that of heat transfer and the enthalpy increase is greater than $U_\infty^2/2$. The ratio of the actual difference in the specific enthalpy (due to the combined effect of reduced kinetic energy, viscous work, and heat transfer) to the increase in the specific enthalpy due to reduced kinetic energy alone is known as the *recovery factor,* denoted by r. From the foregoing discussion, it is evident that the recovery factor is a function of the Prandtl number. Assuming constant specific heats, the adiabatic wall temperature is obtained from the relation,

Viscous dissipation— Adiabatic wall temperature

$$T_{\text{aw}} = T_\infty + r\, \frac{U_\infty^2}{2c_p} \tag{5.1.25}$$

where

r = recovery factor
T_{aw} = adiabatic wall temperature
T_∞ = free stream temperature
U_∞ = free stream velocity

For flat plates, for $0 < \text{Pr} < 15$, the value of r is well approximated by

Laminar flow: $\text{Re} < \text{Re}_{\text{cr}}$ $\qquad\qquad r = \text{Pr}^{1/2}$ $\qquad\qquad$ **(5.1.26)**

Turbulent flow: $\text{Re} > \text{Re}_{\text{Cr}}$ $\qquad\qquad r = \text{Pr}^{1/3}$ $\qquad\qquad$ **(5.1.27)**

Eckert number and viscous dissipation

Consider the ratio $(T_{\text{aw}} - T_\infty)/(T_s - T_\infty)$, where T_s is temperature of the wall. From Equation 5.1.25, this ratio is given by the dimensionless number,

$$\frac{r}{2}\, \frac{U_\infty^2}{c_p(T_s - T_\infty)}$$

This dimensionless group, $U_\infty^2/c_p(T_s - T_\infty)$, is known as *Eckert number*, Ec. Thus,

$$\frac{T_{aw} - T_\infty}{T_s - T_\infty} = r \frac{U_\infty^2}{2c_p(T_s - T_\infty)} = \frac{r}{2} \text{Ec} \tag{5.1.28}$$

In Equation 5.1.28, if $(r/2)\text{Ec} \ll 1$ [$(\text{Pr}^{1/2}/2)\text{Ec}$ for laminar boundary layers and $(\text{Pr}^{1/3}/2)\text{Ec}$ for turbulent boundary layers], the difference between the adiabatic temperature and the free stream temperature as a fraction of the difference between the surface temperature and the free stream temperature is very small; in such cases viscous dissipation may be neglected. For gases with $\text{Pr} \approx 1$, the condition simplifies to $\text{Ec}/2 \ll 1$. If $(r/2)\text{Ec} \approx 1$, viscous dissipation should not be neglected. For example, consider laminar flow of air with

$U_\infty = 20$ m/s $T_\infty = 20\,°C$ $T_s = 80\,°C$
$c_p = 1007$ J/kg K $\text{Pr} = 0.7$

$$\frac{r}{2}\text{Ec} = \frac{\text{Pr}^{1/2}}{2}\text{Ec} = 0.003$$

In the above case viscous dissipation can be neglected. Now for supersonic flow of air at, say 500 m/s, the value of $(r/2)\text{Ec}$ is 1.73, and viscous dissipation cannot be neglected. For oils, which have very high Prandtl numbers, and high speed flows of gases, viscous dissipation becomes significant. From Equation 5.1.28 it is clear that if $(\text{Pr}^n/2)\text{Ec} \ll 1$ ($n = 1/2$ for laminar boundary layers and 1/3 for turbulent boundary layers) viscous dissipation can be neglected.

E.R.G. Eckert, originally from Germany, is now Professor Emeritus at the University of Minnesota. Immediately after the war in 1945, Eckert came to Wright-Patterson Air Force Base from Germany. Subsequently, he worked at the NACA Lewis Laboratory (now NASA-Lewis Research Center) and in 1952 took a teaching position at the University of Minnesota where he established a very productive heat transfer laboratory. His book on heat transfer in 1946 was one of two modern U.S. heat transfer texts (the other was by Max Jakob). He has worked extensively on heat transfer in high speed flows.

Another parameter that indicates when a flow is to be considered as high speed flows is the Mach (Ma) number, defined as the ratio of the velocity of the fluid to the sonic velocity of the fluid. If the ratio is much less than 1, viscous dissipation is not usually significant. For an ideal gas the sonic velocity is given by $(kRT)^{1/2}$, where k is the ratio of the specific heat c_p/c_v, R is the gas constant, and T is the absolute temperature. In the case considered above, the sonic velocity at 20 °C is 343.2 m/s. The Mach number at a velocity of 20 m/s is only 0.058 and viscous dissipation is negligible. With a velocity of 500 m/s, the Mach number is 1.46 and we should expect viscous dissipation to be significant.

The temperature profiles for subsonic flows, when viscous dissipation can be neglected, and in supersonic flows, when viscous heating is significant, are shown in Figure 5.1.13. From Equation 5.1.28, if $(r/2)\text{Ec} > 1$, $T_{aw} > T_s$. This means that

Temperature profiles for low and high Mach number flows

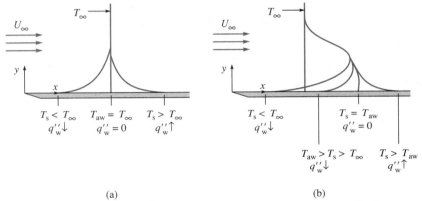

(a)   (b)

FIGURE 5.1.13 Temperature Profiles in Subsonic and Supersonic Flows
(a) Low speed, Ma $\ll$ 1. (b) High speed, Ma > 0.2.

the temperature of the fluid adjacent to the plate surface is greater than that of the plate itself as shown in Figure 5.1.13; the heat transfer is from the fluid to the plate even when $T_s > T_\infty$. This suggests that the appropriate reference temperature for defining the convective heat transfer coefficient in high speed flows is not the free stream temperature but the adiabatic wall temperature T_{aw}. Thus, the convective heat transfer coefficient is defined as

$$q'' = h(T_s - T_{aw}) \qquad (5.1.30)$$

where q'' is the heat flux from the plate to the fluid. For low speed flows, $T_{aw} \approx T_\infty$ and the definition of the convective heat transfer coefficient given by Equation 5.1.30 is valid for low speed flows also. With convective heat transfer coefficient defined by Equation 5.1.30, correlations developed in Chapter 4 can be used for high speed flows.

Incompressible Flows (Liquids, Large Pr)
Laminar flows: $r = \text{Pr}^{1/2}$
$0.6 < \text{Pr} < 50 \qquad \text{Re} < \text{Re}_{cr} \qquad \text{Re}_{cr} = 5 \times 10^5 \qquad$ Properties at $T_f = (T_s + T_\infty)/2$

$$\text{Nu}_x = 0.332\ \text{Re}^{1/2}\text{Pr}^{1/3} \qquad (5.1.31a)$$

$$\text{Nu}_L = 0.664\ \text{Re}_L^{1/2}\text{Pr}^{1/3} \qquad (5.1.31b)$$

Turbulent flows: $r = Pr^{1/3}$
$0.6 < Pr < 50$ Properties at $T_f = (T_s + T_\infty)/2$

$$Nu_x = 0.0296\ Re_x^{4/5} Pr^{1/3} \qquad 5 \times 10^5 < Re_x < 10^7 \qquad \textbf{(5.1.32a)}$$

$$Nu_x = 1.596\ Re_x\ (\ln Re_x)^{-2.584}\ Pr^{1/3} \qquad Re_x > 10^7 \qquad \textbf{(5.1.32b)}$$

As the recovery factor r is different in the laminar and turbulent regions, the reference temperature (T_{aw}) for the definition of the convective heat transfer coefficient is different in the two regimes. Hence, a meaningful average Nusselt number cannot be obtained when the boundary layer is partly laminar and partly turbulent. The laminar and turbulent regions will have to be treated separately.

Convective heat transfer correlations, for high speed flows

Compressible Flows (Gases) Equations similar to Equations 5.1.31 and 5.1.32 can be used but much better results are obtained if, instead of using the film temperature T_f, a reference temperature T^* is used for evaluating the properties. T^* is defined as

$$T^* = 0.5(T_s + T_\infty) + 0.22(T_{aw} - T_\infty) \qquad \textbf{(5.1.33)}$$

The correlations are then given by

Laminar flows: $r = Pr^{1/2}$
$0.6 < Pr < 50$ $Re < Re_{cr}$, $Re_{cr} = 5 \times 10^5$ Properties at T^* (Equation 5.1.33)

$$Nu_x = 0.332\ Re_x^{1/2} Pr^{1/3} \qquad\qquad \textbf{(5.1.34a)}$$

$$Nu_L = 0.664\ Re_L^{1/2} Pr^{1/3} \qquad\qquad \textbf{(5.1.34b)}$$

Turbulent flows: $r = Pr^{1/3}$
$0.6 < Pr < 50$ Properties at T^* (Equation 5.1.33)

$$Nu_x = 0.0296\ Re_x^{4/5} Pr^{1/3} \qquad 5 \times 10^5 < Re_x < 10^7 \qquad \textbf{(5.1.35a)}$$

$$Nu_x = 1.596\ Re_x (\ln Re_x)^{-2.584} Pr^{1/3} \qquad Re_x > 10^7 \qquad \textbf{(5.1.35b)}$$

Again, to reiterate, the adiabatic wall temperatures in the laminar and turbulent regions are different, and it is meaningless to find an average value for the convective heat transfer coefficient when the boundary layer is partly laminar and partly turbulent. However, an average value of the convective heat transfer coefficient can be found and used in a meaningful way when the boundary layer is wholly laminar. Otherwise, each region has to be treated separately.

EXAMPLE 5.1.3

A supersonic aircraft flies at 1400 mph at an altitude of 13 000 m, where the temperature of the air is $-56\ °C$ and the pressure is 18.5 kPa. Estimate the heat transfer rate per meter width, from a section of the wing 4-m long in the direction of air flow maintained at 70 °C

Given

Fluid: air	$T_\infty = -56\ °C$
$U_\infty = 1400$ mph	$T_s = 70\ °C$
$p = 18.5$ kPa	$L = 4$ m

Find

q/W (Heat transfer rate per meter width of the wing)

SOLUTION

Determine whether the boundary layer is laminar or turbulent at the trailing edge (Figure 5.1.14).

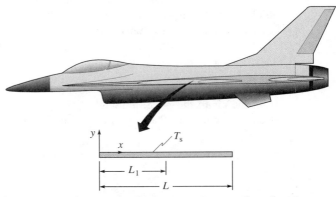

FIGURE 5.1.14 Heat Transfer Rate for an Aircraft Wing at High Speeds

$$U_\infty = \frac{1400 \text{ m/h} \times 5280 \text{ ft/m}}{3600 \text{ s/h} \times 3.2808 \text{ ft/m}} = 625.9 \text{ m/s}$$

At the free stream temperature of -56 °C the sonic velocity of air (considered as an ideal gas) is $(kRT_\infty)^{1/2}$, where k is the ratio of specific heats and R is the gas constant. The values of k and R for air are 1.4 and 287 J/kg K, respectively. Substituting the values, the sonic velocity is 295.4 m/s and the local Mach number (the ratio of the velocity of the air to the sonic velocity) is 2.12. At such high Mach numbers, we expect viscous dissipation to be significant.

To find the reference temperature to be used in Equation 5.1.33, assume a laminar boundary layer. Determine the adiabatic wall temperature. For the purpose of determining the adiabatic wall temperature, the Prandtl number can be assumed to be constant at 0.7 and c_p at 1009 J/kg K.

$$T_{\text{aw}} = \text{Pr}^{1/2} \frac{U_\infty^2}{2 \times c_p} + T_\infty$$

$$= 0.7^{1/2} \frac{625.9^2}{2 \times 1009} + (-56) = 106 \text{ °C}$$

$$T^* = 0.5 \ (T_s + T_\infty) + 0.22(T_{\text{aw}} - T_\infty)$$

$$= 0.5(70 - 56) + 0.22(106 + 56) = 42.6 \text{ °C}$$

From the software (CC) properties of air at 42.6 °C and 18.5 kPa are

$\rho = 0.204 \text{ kg/m}^3$ $\mu = 1.923 \times 10^{-5} \text{ N s/m}^2$
$k = 0.0273 \text{ W/m K}$ $\text{Pr} = 0.71$

The Prandtl number used in calculating the reference temperature is very slightly less than 0.71 at 42.6 °C and no refinement in determining the reference temperature is needed.

$$\text{Re}_L = \frac{0.204 \times 625.9 \times 4}{1.923 \times 10^{-5}} = 2.656 \times 10^7$$

As $\text{Re}_L > 5 \times 10^5$, the boundary layer is partly laminar and partly turbulent. The adiabatic wall temperature is different in the laminar and turbulent boundary layer regions. Hence, the heat transfer rate has to be determined in each region separately.

Laminar Region

In the laminar region, $T_{aw} = 106\ °C$ and $T_s = 70\ °C$. As $T_{aw} > T_s$, heat transfer is from the air to the plate, although the surface temperature is greater than the free stream temperature. The heat transfer rate in the laminar region is given by

$$\frac{q_1}{W} = h_1 L_1 (T_{aw} - T_s)$$

where

q_1/W = heat transfer rate from the fluid to the plate per unit width of the plate
h_1 = average convective heat transfer coefficient in the laminar region
L_1 = the length of the plate from the leading edge, where the boundary layer is laminar
T_{aw} = adiabatic wall temperature in the laminar region
T_s = temperature of the surface in the laminar region

$$h_1 L_1 = \text{Nu}_{L1} k = 0.664\ \text{Re}_{L1}^{1/2} \text{Pr}^{1/3}\ k$$
$$= 0.664(5 \times 10^5)^{1/2}\ 0.71^{1/3} \times 0.0273 = 11.4\ \text{W/m K}$$

$$\frac{q_1}{W} = 11.4(106 - 70) = \underline{410.4\ \text{W/m}}$$

Turbulent Region

To find the heat transfer rate in the turbulent region we proceed as follows.

$$\frac{q_2}{W} = \int_{L_1}^{L} h_x (T_{aw2} - T_s) dx \tag{1}$$

where

q_2/W = heat transfer rate per meter width in the turbulent region
h_x = local heat transfer coefficient in the turbulent region
L = total length of the plate
L_1 = length of the plate where the boundary layer is laminar
T_{aw2} = adiabatic wall temperature in the turbulent region
T_s = plate temperature

Rearranging Equation 1 with a change of variables,

$$x = \text{Re}_x\ v/U_\infty \qquad dx = v/U_\infty d\text{Re}_x \qquad h_x = \text{Nu}_x k/x = \text{Nu}_x k U_\infty/(\text{Re}_x v)$$

$$\frac{q_2}{W} = k(T_{aw2} - T_s) \int_{\text{Re}_{cr}}^{\text{Re}_L} \frac{\text{Nu}_x}{\text{Re}_x} d\text{Re}_x$$

In the turbulent boundary layer region $Nu_x = 0.0296\ Re_x^{4/5}Pr^{1/3}$. Therefore,

$$\frac{q_2}{W} = k(T_{aw2} - T_s) \int_{Re_{cr}}^{Re_L} 0.0296 Re_x^{-1/5}Pr^{1/3}dRe_x$$

$$= k(T_{aw2} - T_s)0.037(Re_L^{4/5} - Re_{cr}^{4/5}) \tag{2}$$

$$T_{aw2} = 0.71^{1/3}\frac{625.9^2}{2 \times 1007} - 56 = 118\ °C$$

$$T^* = 0.5(70 - 56) + 0.22(118 + 56) = 45.3\ °C$$

From the software (CC), properties of air at 45.3 °C, 18.5 kPa are

$\rho = 0.202\ kg/m^3$ $c_p = 1008\ J/kg$ $\mu = 1.936 \times 10^{-5}\ N\ s/m^2$
$k = 0.0275\ W/m\ °C$ $Pr = 0.71$

$$Re_L = \frac{0.202 \times 625.9 \times 4}{1.936 \times 10^{-5}} = 2.612 \times 10^7$$

Substituting $Re_L = 2.612 \times 10^7$ and $Re_{cr} = 5 \times 10^5$ in Equation 2

$$\frac{q_2}{W} = 0.0275(118 - 70) \times 0.71^{1/3} \times 0.037[(2.612 \times 10^7)^{4/5} - (5 \times 10^5)^{4/5}]$$

$$= 35\ 810\ W/m$$

Total heat transfer rate per meter width of the plate is

$$\frac{q_1}{W} + \frac{q_2}{W} = 410.4 + 35\ 810 = \underline{36\ 220\ W/m}$$

COMMENT

Equation 5.1.35a was used even though $Re_L > 10^7$. A slightly more accurate result will be obtained by using Equation 5.1.35a for $Re_{cr} \leq Re_x \leq 10^7$ and Equation 5.1.35b for $10^7 < Re_x \leq Re_L$. If Equation 5.1.35b is employed, the integral has to be evaluated numerically.

5.2 REYNOLDS ANALOGY

In Chapter 4, the local convective heat transfer coefficient correlations for a fluid flowing over a flat plate at a uniform surface temperature for laminar and turbulent boundary layers are given as (Figure 5.2.1)

Laminar boundary layer $Nu_x = 0.332\ Re_x^{1/2}Pr^{1/3}$ **(4.1.1)**

Reynolds analogy—Relationship between fluid flow and heat transfer parameters

Turbulent boundary layer $Nu_x = 0.0296\ Re_x^{4/5}Pr^{1/3}$ **(4.1.7a)**

The concept of velocity and temperature boundary layers was also introduced. In this section an important relation between the fluid flow parameters and the heat transfer parameters will be established.

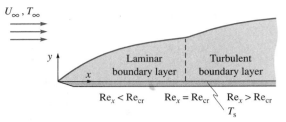

FIGURE 5.2.1 On a flat plate, the boundary layer is initially laminar and becomes turbulent at a certain distance downstream.

In laminar boundary layers, for $Pr < 1$, the thermal boundary layer thickness is greater than the velocity boundary layer thickness and, for $Pr = 1$, the two boundary layers have the same thickness. In a conceptual sense, it was shown that the convective heat transfer coefficient decreases in the direction of flow. It was also demonstrated (Section 4.1.1) that the wall shear stress decreases in the direction of flow. The shear stress in the x-direction, τ_{yx}, for an incompressible fluid is given by

$$\tau_{yx} = \mu \left(\frac{\partial u}{\partial y} + \frac{\partial v}{\partial x} \right)$$

where

μ = viscosity of the fluid
u = velocity component in the x-direction
v = velocity component in the y-direction

At the wall, $v = 0$, and the wall shear stress is given by

$$\tau_w = \mu \left. \frac{\partial u}{\partial y} \right|_{y=0} \tag{5.2.1}$$

The heat flux is related to the difference in temperatures of the wall and the fluid (free stream), and the thermal boundary layer thickness through the thermal conductivity. Similarly, the wall shear stress is also related to the difference in the x-components of the fluid velocity at the wall and the free stream velocity, the viscosity of the fluid, and the velocity boundary layer thickness. Denoting the x-component of the velocity by u,

$$\tau_w \sim \mu \frac{U_\infty - u_{y=0}}{\delta}$$

where δ is the velocity boundary layer thickness.

For a fluid with constant viscosity and free stream velocity, as $u_{y=0} = 0$, it is evident that the wall shear stress also decreases in the direction of flow as the velocity boundary layer thickness increases in that direction. If the velocity distribution in the fluid is known, the wall shear stress can be evaluated from Equation 5.2.1. Solutions (see Example 5.3.1 and Chapter 12 for solutions to laminar boundary layers) for the velocity distribution for both laminar and turbulent boundary layers are available. From these analytical solutions, correlations for wall shear stress and boundary layer thickness have been obtained.

The friction factor defined as $C_f = \tau_w/(\rho U_\infty^2/2)$ is given by

Laminar boundary layers
$$C_f = 0.664 \ \text{Re}_x^{-1/2} \qquad \qquad (5.2.2)$$

$$\frac{\delta}{x} = \frac{5.0}{\sqrt{\text{Re}_x}}$$

Turbulent boundary layers
$$C_f = 0.0592 \ \text{Re}_x^{-1/5} \qquad \qquad (5.2.3)$$

$$\frac{\delta}{x} = 0.37 \ \text{Re}_x^{-1/5}$$

From Equations 4.1.1 and 4.1.7a for uniform surface temperature, the Nusselt number can be recast as

Laminar boundary layer
$$\frac{\text{Nu}_x}{\text{Re}_x\text{Pr}} = 0.332 \ \text{Re}_x^{-1/2}\text{Pr}^{-2/3}$$

Turbulent boundary layer
$$\frac{\text{Nu}_x}{\text{Re}_x\text{Pr}} = 0.0296 \ \text{Re}_x^{-1/5}\text{Pr}^{-2/3}$$

For both laminar and turbulent boundary layers for Pr = 1, we obtain

Laminar boundary layer
$$\frac{\text{Nu}_x}{\text{Re}_x\text{Pr}} = 0.332 \ \text{Re}_x^{-1/2} = \frac{C_f}{2}$$

Turbulent boundary layer
$$\frac{\text{Nu}_x}{\text{Re}_x\text{Pr}} = 0.0296 \ \text{Re}_x^{-1/5} = \frac{C_f}{2}$$

Relation between Stanton number and friction factor

The dimensionless group Nu/(Re) is known as the *Stanton number* and denoted by St. Thus, for both laminar and turbulent boundary layers, for Pr = 1

$$\text{St}_x = \frac{C_{fx}}{2} \qquad \qquad (5.2.4)$$

The relation represented by Equation 5.2.4 is known as *Reynolds analogy*. For laminar boundary layers, the validity of Equation 5.2.4 can be proved without obtaining the detailed solutions for the velocity and temperature profiles.

To prove Equation 5.2.4, we start with the boundary layer differential equations for the balance of mass, momentum, and energy for flow over a flat plate maintained at a uniform temperature. The derivations of these equations are given in Chapter 12. Only the final form of the two-dimensional equations for laminar, steady flow of an incompressible fluid over a flat plate are given here.

Balance of mass

$$\frac{\partial u}{\partial x} + \frac{\partial v}{\partial y} = 0 \qquad \qquad (5.2.5)$$

Balance of Momentum

x-momentum
$$u\frac{\partial u}{\partial x} + v\frac{\partial u}{\partial y} = \nu\frac{\partial^2 u}{\partial y^2} \qquad \qquad (5.2.6)$$

y-momentum
$$0 = \frac{\partial p}{\partial y} \qquad \qquad (5.2.8)$$

Balance of Energy   Neglecting viscous dissipation, the balance of thermal energy is

$$u \frac{\partial T}{\partial x} + v \frac{\partial T}{\partial y} = \alpha \frac{\partial^2 T}{\partial y^2} \qquad (5.2.8)$$

$$v = \frac{\mu}{\rho} = \text{kinematic viscosity} \qquad \alpha = \frac{k}{\rho c_p} = \text{thermal diffusivity}$$

Boundary conditions are

$$u(x = 0, y) = U_\infty \qquad u(x, y \to \infty) = U_\infty \qquad u(x, y = 0) = 0$$
$$v(x = 0, y) = 0 \qquad v(x, y \to \infty) = 0 \qquad v(x, y = 0) = 0$$
$$T(x = 0, y) = T_\infty \qquad T(x, y \to \infty) = T_\infty \qquad T(x, y = 0) = T_s$$

To prove the validity of Reynolds analogy, we use the dimensionless form of the balance equations. Referring to Figure 5.2.2, as Pr = 1, the boundary layer thickness

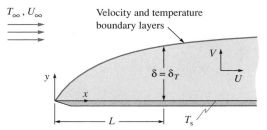

FIGURE 5.2.2 Velocity and Temperature Boundary Layer Thicknesses for Pr ≈ 1

for both velocity and temperature boundary layers are equal everywhere. If we seek the solutions to the balance equations at some point $x = L$, we can take the characteristic length as L, the characteristic velocity is U_∞, and characteristic temperature difference is $T_s - T_\infty$. The dimensionless variables are defined as

$$u^* = \frac{u}{U_\infty} \qquad v^* = \frac{v}{U_\infty} \qquad x^* = \frac{x}{L} \qquad y^* = \frac{y}{L} \qquad \theta = \frac{T_s - T}{T_s - T_\infty}$$

Introducing the dimensionless variables into Equations 5.2.5 through 5.2.8,

Balance of mass
$$\frac{\partial u^*}{\partial x^*} + \frac{\partial v^*}{\partial y^*} = 0 \qquad (5.2.9)$$

Balance of linear momentum:

Laminar Boundary layer equations— dimensionless form

x-momentum
$$u^* \frac{\partial u^*}{\partial x^*} + v^* \frac{\partial u^*}{\partial y^*} = \frac{1}{\text{Re}_L} \frac{\partial^2 u^*}{\partial y^{*2}} \qquad (5.2.10)$$

y-momentum
$$0 = -\frac{\partial p^*}{\partial y^*} \qquad (5.2.11)$$

Balance of energy
$$u^* \frac{\partial \theta}{\partial x^*} + v^* \frac{\partial \theta}{\partial y^*} = \frac{1}{\text{Re}_L \text{Pr}} \frac{\partial^2 \theta}{\partial y^{*2}} \qquad (5.2.12)$$

where Re_L is $(U_\infty L)/v$ and Pr is v/α.

Boundary conditions are

$$u^*(x^* = 0, y^*) = 1 \qquad \theta(x^* = 0, y^*) = 1$$
$$u^*(x^*, y^* = 0) = 0 \qquad \theta(x^*, y^* = 0) = 0$$
$$u^*(x^*, y^* \to \infty) = 1 \qquad \theta(x^*, y^* \to \infty) = 1$$
$$v^*(x^*, y^* = 0) = 0$$

After Equations 5.2.9 and 5.2.10 are solved for u^* and v^*, Equation 5.2.12 can be solved for θ. Notice the similarity between Equation 5.2.10 and 5.2.12 for Pr = 1. Further, for Pr = 1, not only are Equations 5.2.10 and 5.2.12 similar but also the boundary conditions are identical for u^* and θ. Thus, if u^* satisfies Equation 5.2.10 and the boundary conditions associated with it, it will also satisfy Equation 5.2.12 and its boundary conditions. Hence,

$$\theta = u^*$$

Thus,

$$\left.\frac{\partial \theta}{\partial y^*}\right|_{y^*=0} = \left.\frac{\partial u^*}{\partial y^*}\right|_{y^*=0} \tag{5.2.13}$$

Reverting to dimensional variables, Equation 5.2.13 becomes,

$$-\frac{1}{T_s - T_\infty}\left.\frac{\partial T}{\partial y}\right|_{y=0} = \frac{1}{U_\infty}\left.\frac{\partial u}{\partial y}\right|_{y=0} \tag{5.2.14}$$

The heat flux to the fluid from the wall is given by $q'' = -k(\partial T/\partial y)|_{y=0}$, and the wall shear stress by $\tau_w = \mu(\partial u/\partial y)|_{y=0}$. From these relations, we obtain

$$-\left.\frac{\partial T}{\partial y}\right|_{y=0} = \frac{q''}{k} \qquad \left.\frac{\partial u}{\partial y}\right|_{y=0} = \frac{\tau_w}{\mu}$$

Substituting these values of $-(\partial T/\partial y)|_{y=0}$ and $(\partial u/\partial y)|_{y=0}$ into Equation 5.2.14,

$$\frac{q''}{k(T_s - T_\infty)} = \frac{\tau_w}{\mu U_\infty}$$

Introducing $h = q''/(T_s - T_\infty)$ and $C_f = \tau_w/(\rho U^2_\infty/2)$

$$\frac{h}{k} = \frac{C_f}{2}\frac{\rho U_\infty}{\mu}$$

Multiplying both sides by L, and with $\mathrm{Nu}_L = hL/k$, $\mathrm{Re}_L = \rho U_\infty L/\mu$

$$\frac{\mathrm{Nu}_L}{\mathrm{Re}_L} = \frac{C_f}{2}$$

This relation was derived for Pr = 1. To account for small deviations of Pr from 1, it was proposed

$$\frac{\mathrm{Nu}_L}{\mathrm{Re}_L \mathrm{Pr}} = \mathrm{St} = \frac{C_f}{2} \tag{5.2.15}$$

The relation given by Equation 5.2.15 is the well-known Reynolds analogy. It must be noted that this is strictly valid for Pr = 1. This analogy was developed by Reynolds (1874). Colburn (1933) modified Reynolds analogy to accommodate a range of Prandtl numbers and showed that a more appropriate analogy is

$$\text{St Pr}^{2/3} = \frac{C_f}{2} \tag{5.2.16}$$

Reynolds analogy improved—Colburn's j factor

Equation 5.2.16 is known as Colburn's analogy; Colburn gave the dimensionless group, St Pr$^{2/3}$, a special symbol, now commonly referred to as Colburn's *j*-factor,

$$j = \text{St Pr}^{2/3}$$

Equation 5.2.15 is valid for laminar boundary layers with Pr = 1. In turbulent boundary layers, the turbulent Prandtl number is nearly equal to 1 for all fluids and the Reynolds analogy is applicable even when Pr is not 1.

The importance of Equation 5.2.16 is that values of the friction factor are available for a vast number of cases and those values can be used to determine the heat transfer coefficient. Equation 5.2.16 is valid for Prandtl numbers in the range of 0.5 to 1000, but it does not give very good values for liquid metals, which have very low Prandtl numbers. It must be emphasized that the analogy is not applicable in all cases.

5.3 APPROXIMATE ANALYTICAL SOLUTIONS—INTEGRAL METHOD

Integral methods for convection problems

Approximate solutions to transient, one-dimensional conduction problems by integral methods are presented in Chapter 3. Analytical solutions to the partial differential equations in conduction problems are generally easier to find than analytical solutions for problems involving fluid flows. As the differential equations of fluid flows are more complex than those in conduction in solids, solutions by approximate methods are even more important. With the widespread availability of computers and development of numerical methods for the solution of problems involving fluid flows, some people are of the opinion that finding analytical solutions is no longer important. However, there are many cases where approximate analytical solutions may be obtained faster than numerical solutions. Moreover, such analytical solutions show trends that cannot be easily found from numerical methods. It is in this context that one approximate technique, the integral technique, is presented here.

The basic elements of integral methods were developed in Chapter 3. The methodology involves defining a control volume (C.V.) that is finite in one direction and infinitesimal in another. The balance laws are then applied to the element as a whole. Depending on the number of boundary conditions available for the variable of interest, the variable is represented as a polynomial or some other suitable function of the coordinates. Introduction of the function into the balance equations results in an ordinary differential equation. The solution to the differential equation is the solution to the problem. The method will be illustrated through some examples. The examples involve the concept of boundary layers, which was introduced in Chapter 4; the concepts are elaborated on before presenting approximate solutions to a few boundary layer problems.

5.3.1 Differential Equations and Boundary Layers in Fluid Flows

Solutions to problems in convective heat transfer require that the fluid flow problems be first (or simultaneously) solved. The solutions to the differential equations for the fluid flow are complex and, in many cases, there are no known solutions unless some approximations are made. One such approximation that led to great advances in the field of fluid mechanics is that of the boundary layer concept. Because of the importance of this concept in fluid mechanics and, therefore, in convective heat transfer, it is appropriate to give some historical background to its development.

Consider the flow over a flat plate with heat transfer from the plate to the fluid, as shown in Figure 5.3.1. The boundary layer shown represents either the velocity

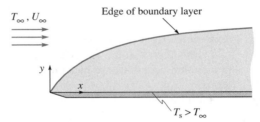

FIGURE 5.3.1 A fluid flows parallel to and over a flat plate with a uniform velocity. The plate is maintained at a uniform temperature.

boundary layer or the temperature boundary layer; their thicknesses may not be equal. The simplified, steady, two-dimensional equation for the temperature distribution in an incompressible fluid (neglecting viscous dissipation) is

$$u \frac{\partial T}{\partial x} + v \frac{\partial T}{\partial y} = \alpha \left(\frac{\partial^2 T}{\partial x^2} + \frac{\partial^2 T}{\partial y^2} \right) \tag{5.3.1}$$

where

u = local fluid velocity in the x-direction
v = local fluid velocity in the y-direction
T = temperature of the fluid
α = thermal diffusivity of the fluid

Equation 5.3.1 represents the First Law of Thermodynamics applied to an infinitesimal C.V., after subtracting the changes in kinetic and potential energies. In deriving Equation 5.3.1 it has been assumed that there is no work transfer due to external forces. But, actually, there is work transfer to a differential element, caused by the viscous forces on the surfaces of the C.V. and the fluid velocities at those surfaces. This ''frictional work'' transfer, known as viscous dissipation, is neglected in many engineering applications. A brief discussion of when viscous dissipation is not significant is given in Section 5.1.6. The effect of viscous work has been neglected in Equation 5.3.1. The left side of the equation represents the net rate of enthalpy flow out of the C.V., and the right side represents the net heat transfer rate into it by conduction.

Boundary layer
differential
equations

To solve the equation, the velocity components are needed. They are obtained as the solution to the following system of equations,

Balance of mass
$$\frac{\partial u}{\partial x} + \frac{\partial v}{\partial y} = 0 \tag{5.3.2}$$

Balance of momentum
$$u\frac{\partial u}{\partial x} + v\frac{\partial u}{\partial y} = -\frac{1}{\rho}\frac{\partial p}{\partial x} + v\left(\frac{\partial^2 u}{\partial x^2} + \frac{\partial^2 u}{\partial y^2}\right) \tag{5.3.3}$$

y-momentum
$$u\frac{\partial v}{\partial x} + v\frac{\partial v}{\partial y} = -\frac{1}{\rho}\frac{\partial p}{\partial y} + v\left(\frac{\partial^2 v}{\partial x^2} + \frac{\partial^2 v}{\partial y^2}\right) \tag{5.3.4}$$

In Equations 5.3.3 and 5.3.4 the terms on the left side represent the net rate of momentum flow out of the C.V. and those on the right side represent the external forces in the same direction along which the momentum flow is evaluated. The first term on the right side is due to the pressure variation in the fluid and the remaining two terms represent the viscous forces. The terms on the left side in the energy and momentum equations are generally referred to as inertia, convection, or advection terms.

Until the early part of the this century, the flow equations for a number of cases were solved assuming that the viscous forces were negligibly small. The solutions gave good results where pressure forces were dominant. An example of such a case is the prediction of the lift force caused by pressure variation on an airfoil. However, the solutions failed to predict the drag forces caused by viscous effects. The main reason for not including the viscous terms was that, with their inclusion, the equations could not be solved. Early in this century, by a combination of intuition and insight (and daring), Prandtl suggested that not all the terms in the equations were of equal importance and that the less important terms be neglected to get a solution with viscous forces. The resulting set of equations are

$$\frac{\partial u}{\partial x} + \frac{\partial v}{\partial y} = 0 \tag{5.3.5}$$

$$u\frac{\partial u}{\partial x} + v\frac{\partial u}{2y} = -\frac{1}{\rho}\frac{\partial p}{\partial x} + v\frac{\partial^2 u}{\partial y^2} \tag{5.3.6}$$

$$0 = -\frac{\partial p}{\partial y} \tag{5.3.7}$$

For flow parallel to a semi-infinite, flat plates, $\partial p/\partial x = 0$. By changing the variables, the system of partial differential equations was reduced to an ordinary, nonlinear equation. One of his graduate students, Blasius, obtained a numerical solution to the ordinary differential equation. Just as the momentum equations are approximated, the energy equation may also be simplified to

Determining
convective heat
transfer coefficient
from the analytical
solutions for the
temperature field

$$u\frac{\partial T}{\partial x} + v\frac{\partial T}{\partial y} = \alpha\frac{\partial^2 T}{\partial y^2} \tag{5.3.8}$$

In arriving at Equation 5.3.8, the heat transfer in the x-direction [represented by $\alpha(\partial^2 T/\partial x^2)$] is neglected in comparison with the heat transfer in the y-direction [represented by $\alpha(\partial^2 T/\partial y^2)$]. The solutions for u and v from the momentum equations

Ludwig Prandtl (1875–1953) is considered as one of the founders of present-day fluid mechanics. He studied mechanical engineering and did his doctoral thesis at the University of Munich. He presented his first paper outlining the concepts of the boundary layer in 1904. The concepts were so novel that there were no publications on the subject, except by his group, for twenty years. The quantitative results were presented by his student, Blasius, in 1907. Prandtl also developed the mixing length theory of turbulence. The dimensionless number named after him, Prandtl number, was formulated by another eminent researcher in the field of heat transfer, Wilhelm Nusselt. Prandtl contributed in solid mechanics also.

are used in Equation 5.3.8 to find the solution to the temperature field. From the temperature field $T(x, y)$, the local wall heat flux from the wall is found from the Fourier Law

$$q''_w = -k \left. \frac{\partial T}{\partial y} \right|_{y=0}$$

The local convective heat transfer coefficient is defined as

$$h_x = \frac{q''_w}{T_s - T_\infty}$$

From the analytical solution to $T(x, y)$,

$$h_x = \frac{-k (\partial T/\partial y)|_{y=0}}{T_s - T_\infty}$$

Some illustrative examples of the application of integral methods will now be presented.

EXAMPLE 5.3.1   An incompressible fluid flows over a flat plate with a uniform free stream velocity U_∞ (Figure 5.3.2). Find an expression for the velocity distribution $u(x, y)$ and the friction factor C_f.

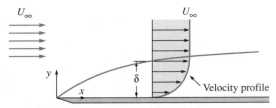

FIGURE 5.3.2 A fluid approaches a flat plate with a uniform velocity and flows parallel to it. A boundary layer develops adjacent to the plate

ASSUMPTIONS

1. Boundary layer concepts are valid. In particular, the pressure outside the boundary layer being uniform, $\partial p/\partial x = 0$. As $\partial p/\partial y = 0$ inside the boundary layer, the pressure inside the boundary layer is also uniform.
2. Steady state
3. As the plate extends to a large distance in the z-direction (perpendicular to the x-y plane), the analysis is performed for a unit depth in the z-direction.

SOLUTION

Choose a differential-integral C.V. ABCD—differential in the x-direction and finite in the y-direction—as shown in Figure 5.3.3. Apply the balance and applicable particular laws to this C.V.

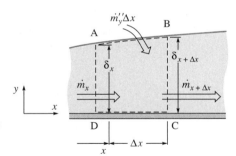

FIGURE 5.3.3 Control Volume, Finite in the y-Direction and Elemental in the x-Direction

The boundary layer, the region where the velocity in the x-direction changes from 0 at the plate to U_∞ at the edge of the boundary layer, has a thickness $\delta(x)$. For $y > \delta$, $u = U_\infty$.

Mass Balance

In steady state,

$$\text{Net rate of mass flow out of the C.V.} = 0$$

or

$$\begin{array}{l}\text{Rate of mass flow out of the C.V.}\\ -\ \text{Rate of mass flow into C.V.}\\ =\ 0\end{array}$$

Mass enters the C.V. across surfaces AD and AB and leaves across BC as shown in Figure 5.3.3. The mass flow rates across AD and BC are denoted by $\dot{m}_x$ and $\dot{m}_{x+\Delta x}$, respectively. $\dot{m}_y''$ represents the mass flow rate entering the C.V. per unit area perpendicular to the y-direction and $\dot{m}_y'' \Delta x$ is the mass flow rate entering the C.V. across AB. Balance of mass results in

$$\dot{m}_{x+\Delta x} - \dot{m}_x - \dot{m}_y'' \Delta x = 0$$

Dividing by Δx and taking the limit as $\Delta x \to 0$,

$$\dot{m}_y'' = \frac{d\dot{m}_x}{dx} \qquad \dot{m}_x = \int_0^\delta \rho u \, dy$$

$$\dot{m}_y'' = \frac{d}{dx} \left(\int_0^\delta \rho u \, dy \right) \tag{1}$$

Balance of Linear Momentum, x-Direction

The steady-state balance of linear momentum in the x-direction for the C.V. is

Net rate of momentum flow out in the x-direction
= Sum of the forces on the material in the C.V. in the x-direction

Denote the momentum flow rates across AD and BC by $\dot{M}_x$ and $\dot{M}_{x+\Delta x}$ (Figure 5.3.4). Mass $\dot{m}_y'' \Delta x$ enters the C.V. per unit time across AB with a velocity U_∞ in the

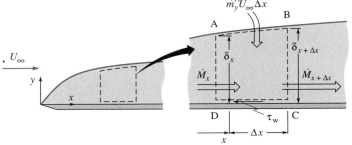

FIGURE 5.3.4 Control Volume for Analysis of Momentum in the Boundary Layer

x-direction. Therefore, $\dot{m}_y'' \Delta x U_\infty$ is the rate of momentum flow into the C.V. across AB. The pressure being uniform, the net force due to the pressure of the fluid is zero. The x-direction shear stress on the control surface is τ_w in the negative x-direction. As viscous forces outside the boundary layer are negligible, shear stress on AB is negligible. Thus,

$$\dot{M}_{x+\Delta x} - \dot{M}_x - \dot{m}_y'' U_\infty \Delta x = -\tau_w \Delta x$$

Dividing by Δx and taking the limit as $\Delta x \to 0$, we obtain

$$\frac{d\dot{M}_x}{dx} - \dot{m}_y'' U_\infty = -\tau_w$$

$$\dot{M}_x = \int_0^\delta \rho u^2 dy$$

Substituting $(d/dx) (\int_0^\delta \rho u \, dy)$ for $\dot{m}_y''$ from Equation 1 we obtain

$$\frac{d}{dx} \left(\int_0^\delta \rho u^2 \, dy \right) - U_\infty \frac{d}{dx} \left(\int_0^\delta \rho u \, dy \right) = -\tau_w$$

or

$$\frac{d}{dx} \left(\int_0^\delta \rho u^2 \, dy \right) - \frac{d}{dx} \left(\int_0^\delta \rho U_\infty u \, dy \right) = -\tau_w \qquad (2)$$

Dividing Equation 2 by ρU_∞^2 and rearranging,

$$\frac{d}{dx} \int_0^\delta \frac{u}{U_\infty} \left(\frac{u}{U_\infty} - 1 \right) dy = -\frac{\tau_w}{\rho U_\infty^2} \qquad (3)$$

Now, choose a reasonable velocity profile for u to satisfy the boundary conditions. With the chosen velocity profile, the wall shear stress τ_w can also be evaluated. Introducing the velocity profile into Equation 3, an ordinary differential equation in δ is obtained. The solution to δ establishes the velocity profile and the wall shear stress.

In selecting an expression for the velocity profile, one may choose to identify and satisfy different boundary conditions, and, depending on the number of boundary conditions that are satisfied and the functional form of the profile chosen, the results may be somewhat different. Any reasonable profile may be used.

First, consider the boundary conditions to be satisfied by u. Some of the known boundary conditions that one would like to satisfy are

$$u(x, 0) = 0 \qquad (4a)$$

$$u(x, \delta) = U_\infty \qquad (4b)$$

$$\frac{\partial u}{\partial y} (x, \delta) = 0 \qquad (4c)$$

$$\frac{\partial^2 u}{\partial y^2} (x, 0) = 0 \qquad (4d)$$

Boundary condition 4d comes from evaluating the differential equation for u at $y = 0$. From Equation 5.3.6, with a uniform pressure, $(\partial p/\partial x) = 0$,

$$u \frac{\partial u}{\partial x} + v \frac{\partial u}{\partial y} = v \frac{\partial^2 u}{\partial y^2} \qquad (5)$$

Evaluating Equation 5 at $y = 0$, with $u = v = 0$, $(\partial^2 u/\partial y^2) = 0$. We have chosen four boundary conditions to be satisfied by the velocity profile. More conditions may be used. For example, we may use the continuity of the second and higher order derivatives at $y = \delta$. If the derivatives are required to be continuous at $y = \delta$, they may be evaluated at either δ^+ or δ^- (i.e., on either side of $y = \delta$). At δ^+, as U_∞ is constant, its derivatives of all orders are zero and, hence, at $y = \delta$,

$$\frac{\partial^2 u}{\partial y^2} = \frac{\partial^3 u}{\partial y^3} = \cdots = \frac{\partial^n u}{\partial y^n} = 0$$

But there is no guarantee that using more boundary conditions improves the resulting solution. Hence, only four boundary conditions given in Equation (4) will be employed.

As discussed in Section 3.5, any functional form may be chosen for the profile. We choose a polynomial in y. Equation 1 has already been used in arriving at Equation 3, leaving us with a total of 5 independent equations: Equation 3 and the four boundary conditions given by Equation 4. As the boundary layer thickness, δ, is one of the unknowns to be determined, the chosen polynomial may have a maximum of four coefficients. These coefficients may be functions of x. Choosing the simplest polynomial in y with four coefficients,

$$u = a + by + cy^2 + dy^3 \tag{6}$$

Employing the boundary conditions given by Equation 4,

$$u(x, 0) = 0 = a$$

$$\frac{\partial^2 u}{\partial y^2}(x, 0) = 0 = 2c$$

$$u(x, \delta) = U_\infty = b\delta + d\delta^3$$

$$\frac{\partial u}{\partial y}(x, \delta) = 0 = b + 3d\delta^2$$

Solving for b and d, and introducing them into Equation 6,

$$\frac{u}{U_\infty} = \frac{3}{2}\frac{y}{\delta} - \frac{1}{2}\left(\frac{y^3}{\delta^3}\right) \tag{7}$$

From Equation 7,

$$\tau_w = \mu \left.\frac{\partial u}{\partial y}\right|_{y=0} = \frac{3\mu}{2\delta} U_\infty$$

Substituting Equation 7 for the velocity profile and $(3\mu/2\delta)U_\infty$ for τ_w into Equation 3, and performing the indicated integration, we get

$$\frac{39}{280}\frac{d\delta}{dx} = \frac{3\nu}{2\delta U_\infty} \qquad \delta\frac{d\delta}{dx} = \frac{140}{13}\frac{\nu}{U_\infty}$$

$$\frac{d\delta^2}{dx} = \frac{280}{13}\frac{\nu}{U_\infty} \tag{8}$$

Solving Equation 8 with $\delta(x = 0) = 0$, we obtain

$$\delta = \sqrt{\frac{280}{13}\frac{\nu x}{U_\infty}} \tag{9}$$

or

$$\frac{\delta}{x} = 4.64\left(\frac{\nu}{U_\infty x}\right)^{1/2} = 4.64\,\mathrm{Re}_x^{-1/2} \tag{10}$$

Substituting $\delta(x)$ from Equation 9 into Equation 7, the velocity distribution $u(x, y)$ is obtained. The wall shear stress and the friction factor can then be determined from the velocity profile. The result is

$$\tau_w = \mu \left.\frac{\partial u}{\partial y}\right|_{y=0} = \frac{3}{2}\,\mu\,\frac{U_\infty}{\delta}$$

and

$$C_f = \frac{\tau_w}{\rho U_\infty^2/2} = 2\left(\frac{3}{2}\frac{\nu}{\delta}\frac{1}{U_\infty}\right) = \frac{3\nu}{U_\infty}\sqrt{\frac{13}{280}\frac{U_\infty}{\nu x}}$$

With $\text{Re}_x = (U_\infty x)/\nu$

$$C_f = \frac{0.646}{\text{Re}_x^{1/2}} \tag{11}$$

From the solution to the boundary layer differential equations, the value of C_f is $0.664\,\text{Re}_x^{-1/2}$. The approximate solution is within 2.7% of the exact solution. We may not always be that fortunate.

When the boundary layer differential equations are solved, one of the boundary conditions is that the x-component of the velocity reaches U_∞ as the distance from the plate becomes infinite. With such a definition, the boundary layer thickness is not defined. For practical purposes, the boundary layer thickness in the exact solution is determined as the distance from the plate where the x-component of the velocity reaches $0.99\,U_\infty$. The boundary layer thickness from the exact solution defined in that way is $\approx 5/\text{Re}_x^{1/2}$.

It should be emphasized that the technique of integral methods has been demonstrated for a case for which an exact solution is available. It shows the comparative simplicity and ease with which approximate solutions by integral methods can be found. But there are cases for which even integral methods are not that easily applied. *In practice, the integral method is employed to solve problems for which exact solutions are not easily available or to get a quick solution.* In such a case there is no way to tell how good the approximate solution is. In most cases the approximate solutions should be validated by experiments.

Equation 11 can be used to determine the heat transfer coefficient through Reynolds or Colburn's analogy (Equation 5.2.16). Employing Colburn's analogy,

$$\text{Nu}_x = 0.323\text{Re}_x^{1/2}\text{Pr}^{1/3} \tag{12}$$

This is again 2.7% less than the value obtained from exact solutions.

Instead of employing Colburn's analogy, we may use the integral techniques to determine the thermal boundary layer thickness, δ_T. If $\text{Pr} > 1$, then $\delta > \delta_T$ and if $\text{Pr} < 1$, $\delta < \delta_T$. If $\text{Pr} \ll 1$, as in the case of liquid metals, $\delta \ll \delta_T$ and we may analyze the heat transfer to liquid metals by neglecting the momentum boundary layer thickness. In the next example, we demonstrate the application of such an approximation for liquid metals.

EXAMPLE 5.3.2 Consider a liquid metal, such as sodium, with Pr $\ll$ 1 flowing parallel to a thin flat plate with a free stream velocity U_∞ and temperature T_∞. The underside of the plate is electrically heated by a uniformly spaced resistance wire. The electrical heating imposes a uniform heat flux q_w'' on the plate. The temperature of the plate is to be determined.

ASSUMPTIONS

1. The plate width is large and two-dimensional analysis is applicable.
2. Heating starts from the leading edge.
3. Steady state
4. Constant properties
5. The conduction in the plate in the direction of flow is negligible.
6. Negligible viscous dissipation

SOLUTION

The plate width is large and the temperature distribution is two-dimensional, $T = T(x, y)$. The analysis is performed for a unit width of the plate in the z-direction.

As the Prandtl number is very small, the thermal boundary layer thickness is much greater than the velocity boundary layer thickness as shown in Figures 5.3.5 and 5.3.6. To obtain the plate temperature by integral methods, consider the C.V. ABCD, of unit width perpendicular to the x-y plane (Figure 5.3.6)

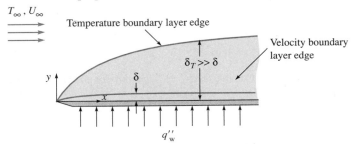

FIGURE 5.3.5 A liquid metal flows over a flat plate. The velocity boundary layer is much thinner than the temperature boundary layer.

Mass Balance

Denoting the mass rate of flow per unit area perpendicular to the y-direction by $\dot{m}_y''$

$$\dot{m}_{x+\Delta x} - \dot{m}_x - \dot{m}_y''\Delta x = 0$$

$\dot{m}_x$ and $\dot{m}_{x+\Delta x}$ represent the mass flow rates across AD and BC, respectively. $\dot{m}_y''\Delta x$ is the mass rate of flow entering the C.V. across AB.

Dividing by Δx and taking the limit as $\Delta x \rightarrow 0$,

$$\dot{m}_y'' = \frac{d\dot{m}_x}{dx} = \frac{d}{dx}\left(\int_0^{\delta_T} \rho u \, dy\right) \tag{1}$$

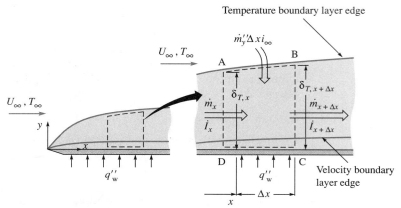

FIGURE 5.3.6 Control Volume Used in the Analysis for Example 5.3.2

Energy Balance

$$\dot{I}_{x+\Delta x} - \dot{I}_x - \dot{m}_y''\Delta x i_\infty = q_w''\Delta x$$

$\dot{I}_{x+\Delta x}$ and $\dot{I}_x$ are the rates of enthalpy flow across BC and AD, respectively. The specific enthalpy of the fluid in the free stream is i_∞ and $\dot{m}_y''\Delta x i_\infty$ is the rate of enthalpy flow into the C.V. across AB. The heat flux at the wall is q_w'' and $q_w''\Delta x$ is the heat transfer rate to the C.V. at the wall. Dividing by Δx and taking the limit as $\Delta x \to 0$,

$$\frac{d\dot{I}_x}{dx} - \dot{m}_y''i_\infty = q_w'' \tag{2}$$

$$\dot{I}_x = \int_0^{\delta_T} i\,\rho u\,dy$$

Substituting $(d/dx)\,(\int_0^{\delta_T} \rho u\,dy)$ for $\dot{m}_y''$ from Equation 1 into Equation 2, and rearranging (i_∞ is constant),

$$\frac{d}{dx}\int_0^{\delta_T} \rho u\,(i - i_\infty)dy\Bigg] = q_w''$$

Assuming constant specific heats,

$$i - i_\infty = c_p(T - T_\infty)$$

and

$$\frac{d}{dx}\left[\int_0^{\delta_T} \rho u c_p(T - T_\infty)dy\right] = q_w'' \tag{3}$$

To evaluate the integral, the functional relation between u, T, and y should be known. But as the Prandtl number is very small, the kinematic viscosity is much less than the thermal diffusivity. Therefore, the thermal boundary layer grows much more rapidly than the velocity boundary layer. Consequently, at any given location, if the heating starts from the leading edge, $\delta_T \gg \delta$. For $\delta < y < \delta_T$ the velocity is uniform at U_∞. Equation 3 is written as

$$\frac{d}{dx}\left[\int_0^{\delta_T} \rho u c_p (T - T_\infty)\, dy\right]$$

$$= \frac{d}{dx}\left[\int_0^{\delta} \rho u c_p (T - T_\infty) dy\right] + \frac{d}{dx}\left[\int_\delta^{\delta_T} \rho u c_p (T - T_\infty) dy\right]$$

$$\approx \frac{d}{dx}\left[\int_0^{\delta} \rho u c_p (T - T_\infty) dy\right] + \frac{d}{dx}\left[\int_\delta^{\delta_T} \rho U_\infty c_p (T - T_\infty) dy\right]$$

The region $0 < y < \delta$ is much smaller than the region $\delta < y < \delta_T$ and the mass flow rate within the velocity boundary layer is much smaller than the mass flow rate in the remaining part of the control surface. Replacing u with any value between 0 and U_∞ in evaluating the integral makes very little difference to the total integral and, therefore, replacing u with U_∞ is an acceptable approximation. With such an approximation, the energy equation takes the form,

$$\frac{d}{dx}\left(\int_0^{\delta_T} \rho U_\infty c_p (T - T_\infty) dy\right) = q_w'' \tag{4}$$

If the temperature profile is known with δ_T as a parameter, it can be substituted in the integral on the left side. Equation 4 then leads to an ordinary differential equation in δ_T. The solution of the differential equation yields the temperature profile, which permits the evaluation of the wall temperature.

With the same rationale as in the previous example, the temperature profile is assumed to be a polynomial in y. The number of terms to be used in the polynomial is to be determined from the number of boundary conditions to be used. For the boundary conditions for T we choose

$$T(x, \delta_T) = T_\infty \tag{5a}$$

$$\frac{\partial T}{\partial y}(x, \delta_T) = 0 \tag{5b}$$

$$-k \frac{\partial T}{\partial y}(x, 0) = q_w'' \tag{5c}$$

$$\frac{\partial^2 T}{\partial y^2}(x, 0) = 0 \tag{5d}$$

The last boundary condition is obtained by evaluating Equation 5.3.8 at $y = 0$. At $y = 0$, $u = v = 0$ and, therefore, $\partial^2 T/\partial y^2 = 0$. As there are five equations—four boundary conditions (represented by Equation 5) and Equation 4—a third-order temperature profile containing four coefficients can be used. With four boundary conditions and Equation 4, the four coefficients in the polynomial, and the thermal boundary layer thickness δ_T can be determined. In the region $0 < y < \delta_T$, we assume the temperature profile to be

$$T = a + by + cy^2 + dy^3$$

To satisfy the boundary conditions, we require

$$T(x, \delta_T) = T_\infty = a + b\delta_T + c\delta_T^2 + d\delta_T^3$$

$$\frac{\partial T}{\partial y}(x, \delta_T) = 0 = b + 2c\delta_T + 3d\delta_T^2$$

$$\frac{\partial^2 T}{\partial y^2}(x, 0) = 0 = 2c$$

$$-k\frac{\partial T}{\partial y}(x, 0) = -kb = q_w''$$

Solving for the four coefficients,

$$c = 0 \qquad b = -\frac{q_w''}{k} \qquad d = \frac{q_w''}{3\delta_T^2 k} \qquad a = T_\infty + \frac{2}{3}\frac{q_w''\delta_T}{k}$$

Introducing the coefficients into the temperature profile and rearranging it,

$$T - T_\infty = \frac{q_w''}{k}\left(\frac{2}{3}\delta_T - y + \frac{1}{3\delta_T^2}y^3\right) \tag{6}$$

Substituting Equation 6 into Equation 4,

$$\frac{d}{dx}\int_0^{\delta_T}\left[\rho U_\infty c_p \frac{q_w''}{k}\left(\frac{2}{3}\delta_T - y + \frac{1}{3\delta_T^2}y^3\right)\right]dy = q_w''$$

Evaluating the integral and simplifying,

$$\frac{d\delta_T^2}{dx} = 4\frac{k}{\rho c_p U_\infty} = 4\frac{\alpha}{U_\infty}$$

With $\delta_T(0) = 0$, the solution to the equation is

$$\delta_T = \sqrt{\frac{4\alpha x}{U_\infty}}$$

which can be recast as

$$\frac{\delta_T}{x} = \left(\frac{4\alpha}{U_\infty x}\right)^{0.5} = \left(4\frac{\alpha}{\nu}\frac{\nu}{U_\infty x}\right)^{0.5} = \frac{2}{(Re\ Pr)^{0.5}} \tag{7}$$

With δ_T from Equation 7, the temperature profile is known from Equation 6. The wall temperature may be found from Equations 6 and 7.

$$T_s - T_\infty = \frac{q_w''}{k}\frac{2}{3}\delta_T = \frac{2}{3}\frac{q_w''}{k}\sqrt{\frac{4\alpha x}{U_\infty}} \tag{8}$$

The heat flux is already known. The heat transfer coefficient may be determined from Equation 8. With the local heat transfer coefficient defined as

$$h_x = \frac{q_w''}{T_s - T_\infty} = \frac{3k}{2}\sqrt{\frac{U_\infty}{4\alpha x}}$$

$$Nu_x = \frac{h_x x}{k} = \frac{3}{4}\left(\frac{U_\infty x}{\alpha}\right)^{0.5} = 0.75\left(\frac{U_\infty x}{\nu}\frac{\nu}{\alpha}\right)^{0.5} = \underline{0.75\ (Re\ Pr)^{0.5}}$$

The exact solution for the boundary layer equations for small Prandtl numbers for uniform heat flux is

$$Nu_x = 0.886(Re_x)^{0.5} \tag{9}$$

While the integral method correctly predicts the Nusselt number dependence on $(Re_x\,Pr)^{0.5}$, it underestimates the heat transfer coefficient by as much as 15%. Hence, solutions obtained by integral methods should be validated by some other evidence of their correctness, usually by experiments.

The next example illustrates the use of the integral method for the solution of a problem involving phase change. The original solution was given by Nusselt (1916).

EXAMPLE 5.3.3 One surface of a vertical plate is exposed to a semi-infinite medium of a quiescent dry, saturated vapor. The plate is maintained at a temperature $T_s < T_{sat}$, and the vapor condenses on the vertical surface. The motion of the vapor far away from the plate is negligible; the vapor motion is caused by the condensation of the vapor on the plate. Determine the heat flux to the wall as a function of the distance from the top edge of the plate.

The physical configuration is shown in Figure 5.3.7a. As a result of heat transfer from the vapor to the wall, the vapor adjacent to the wall condenses on the plate and the condensed liquid flows down the plate. With heat transfer across the condensate film, more vapor condenses on the film. As the density of the liquid is greater than that of the vapor, the condensate film flows down the plate. As the condensate film flows down, it picks up more and more condensate, and its thickness increases in the direction of flow.

If the velocity profile in the condensate film, and its thickness $\delta(x)$ are known, the mass flow rate of the condensate can be determined. The product of the enthalpy

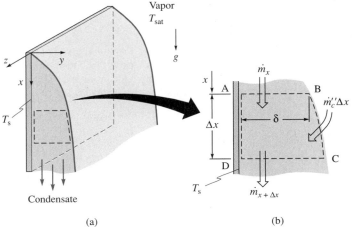

(a)   (b)

FIGURE 5.3.7 Condensation of a Vapor on a Vertical, Cooled Plate
(a) Condensation on a vertical plate. (b) Control volume.

of vaporization and the mass flow rate of the condensate (which is a function of the distance x) is the rate of heat transfer to the wall from the top edge to x. The heat transfer coefficient is determined as the ratio of the heat flux to the difference in the wall and vapor temperatures.

ASSUMPTIONS

1. The condensate film is laminar with constant properties.
2. The vapor is at its saturation temperature.
3. The energy transfer in the direction of flow, the x-direction, is negligible compared with the heat transfer rate in the y-direction. The temperature gradient $\partial T/\partial x$, and the y-component of the velocity, v, are very small and the terms $u(\partial T/\partial x)$, and $v(\partial T/\partial y)$ in the energy equation (Equation 5.3.3) are very small compared with the conduction heat transfer term, $\alpha(\partial^2 T/\partial y^2)$ and can be neglected. Similarly, the terms $(\partial u/\partial x)$ and $v(\partial u/\partial y)$ in Equation 5.3.6 can be neglected. Both the above assumptions are valid if the condensate film thickness is very small.
4. The shear stress in the x-direction at the condensate-vapor interface is negligible.
5. From assumption 3, the temperature profile in the condensate film is linear. This is reasonable, because with $u(\partial T/\partial x)$, $v(\partial T/\partial y)$, and $\alpha(\partial^2 T/\partial x^2)$ being negligible, from Equation 5.3.8, $(\partial^2 T/\partial y^2) \cong 0$. This means that the temperature profile is almost linear. The heat transfer across the condensate film being by conduction only, the heat flux to the wall at any location is given by $q''_w = k(T_{sat} - T_s)/\delta$, where δ is the local film thickness.

ANALYSIS

Consider the C.V. ABCD in Figure 5.3.7b. It is assumed that the plate is very wide in the z-direction, and that the variables do not change in that direction. It is then sufficient to analyze the flow per unit width in the z-direction.

Mass Balance

$$\dot{m}_{x+\Delta x} - \dot{m}_x - \dot{m}''_c \Delta x = 0 \tag{1}$$

where

$\dot{m}_{x+\Delta x}$ = mass rate of flow of the condensate out of C.V. across CD
$\dot{m}_x$ = mass rate of flow of condensate into the C.V. across AB
$\dot{m}''_c$ = condensation rate from the vapor on the condensate film per unit area (parallel to the plate) across BC

Dividing Equation 1 by Δx and taking the limit as $\Delta x \to 0$,

$$\frac{d\dot{m}_x}{dx} - \dot{m}''_c = 0 \tag{2}$$

Balance of Linear Momentum

To obtain the velocity profile, Equation 5.3.6 is employed with the addition of the term representing the gravitational force on the liquid condensate that causes the liquid flow. Neglecting the convection terms and including the gravitational force, (weight of the condensate) Equations 5.3.6 and 5.3.7 become

$$v \frac{\partial^2 u}{\partial y^2} - \frac{1}{\rho} \frac{\partial p}{\partial x} + g = 0 \qquad \text{(3a)}$$

$$\frac{\partial p}{\partial y} = 0 \qquad \text{(3b)}$$

In Eqs. 3a and 3b, ρ is the density of the condensate. (All properties of the liquid are denoted without any subscript. Properties of the vapor are denoted by the subscript v.) v is the kinematic viscosity of the condensate.

As $\partial p/\partial y = 0$, $p = p(x)$; that is, the pressure is a function of x only and, hence, dp/dx can be evaluated at any value of y. Evaluating dp/dx in the vapor, which is stationary,

$$\frac{dp}{dx} = \rho_v g$$

Substituting $dp/dx = \rho_v g$ into Equation 3a, and replacing $v\rho$ by μ,

$$\frac{\partial^2 u}{\partial y^2} = -\frac{(\rho - \rho_v)g}{\mu} \qquad \text{(4)}$$

The boundary conditions for u are

$$u(x, 0) = 0$$

From assumption 4

$$\frac{\partial u}{\partial y}(x, \delta) = 0$$

Integrating Equation 4 twice,

$$u = -\frac{(\rho - \rho_v)g}{\mu} \frac{y^2}{2} + ay + b$$

From the boundary conditions, $u(x, 0) = 0 = b$

$$\frac{\partial u}{\partial y}(x, \delta) = 0 = -\frac{(\rho - \rho_v)g}{\mu} \delta + a \quad \text{and} \quad a = \frac{(\rho - \rho_v)g}{\mu} \delta$$

giving the velocity profile

$$u = \frac{(\rho - \rho_v)g}{\mu} \delta y - \frac{(\rho - \rho_v)g}{\mu} \frac{y^2}{2} = \frac{(\rho - \rho_v)g}{\mu} \delta^2 \left(\frac{y}{\delta} - \frac{y^2}{2\delta^2} \right) \qquad \text{(5)}$$

Energy Balance

As the enthalpy flow into and out of the C.V. has been neglected (the advection terms $u(\partial T/\partial x)$ and $v(\partial T/\partial y)$ in the energy equation are neglected), the heat transfer rate to the wall results only from the condensation of the vapor. The temperature of the condensate varies from T_s at the wall to T_{sat} at the edge of the film. In reality, part of the heat transfer to the wall results in condensation of the vapor and part in cooling the condensate below the saturation temperature. The heat transfer required to cool the condensate below the saturation temperature is much less than the heat transfer to condense the vapor. At this stage, we make the approximation that all the heat transfer rate to the wall results in condensation of the vapor. The heat transfer rate from the vapor due to condensation is $\dot{m}_c'' i_{fg} \Delta x$, where $\dot{m}_c \Delta x$ is the rate of vapor crossing the surface BC and condensing on the liquid surface and i_{fg} is the enthalpy of vaporization.

The heat transfer rate to the plate is also given by $q_w'' \Delta x$. Thus, $\dot{m}_c'' i_{fg} = q_w''$. From Equation 2, $\dot{m}_c'' = d\dot{m}_x/dx = d/dx \left(\int_0^\delta \rho u \, dy \right)$ and from assumption 5, $q_w'' = k(T_{sat} - T_s)/\delta$. Substituting the expressions for $\dot{m}_c$ and q_w'' in $\dot{m}_c'' i_{fg} = q_w''$,

$$\frac{d}{dx} \left(i_{fg} \int_0^\delta \rho u \, dy \right) = k \frac{T_{sat} - T_s}{\delta} \tag{6}$$

Substituting Equation 5 into Equation 6, and evaluating the integral,

$$\frac{d}{dx} \left[i_{fg} \rho \frac{(\rho - \rho_v)g}{3\mu} \delta^3 \right] = k \frac{T_{sat} - T_s}{\delta}$$

$$\delta \frac{d\delta^3}{dx} = \frac{3\mu(T_{sat} - T_s)k}{\rho(\rho - \rho_v)g i_{fg}}$$

Rewriting,

$$\delta \frac{d\delta^3}{dx} = 3\delta^3 \frac{d\delta}{dx} = \frac{3}{4} \frac{d\delta^4}{dx}$$

and with $\delta(0) = 0$, we obtain

$$\delta = \left[\frac{4\mu(T_{sat} - T_s)kx}{\rho(\rho - \rho_v)g i_{fg}} \right]^{1/4} \tag{7}$$

With

$$q_w'' = \frac{k(T_{sat} - T_s)}{\delta} \qquad h_x = \frac{q_w''}{(T_{sat} - T_s)} = \frac{k}{\delta}$$

and

$$h_x = \left[\frac{\rho(\rho - \rho_v)g i_{fg} k^3}{4\mu(T_{sat} - T_s)x} \right]^{1/4} = 0.707 \left[\frac{\rho(\rho - \rho_v)g i_{fg} k^3}{\mu(T_{sat} - T_s)x} \right]^{1/4} \tag{8}$$

The Nusselt number is found from

$$Nu_x = \frac{h_x x}{k} = 0.707 \left[\frac{\rho(\rho - \rho_v)g i_{fg} x^3}{\mu(T_{sat} - T_s)k} \right]^{1/4}$$

Having found the local heat transfer coefficient, we find the average heat transfer coefficient over a height, L, from

$$q = \int q''dA = \int_0^L h_x(T_{sat} - T_s)W \, dx$$

where W is the width of the plate. With both W and $T_{sat} - T_s$ constant, defining the average heat transfer coefficient h_L by $q/[WL(T_{sat} - T_s)]$, we have

$$h_L = \frac{1}{L} \int_0^L h_x \, dx = \frac{4}{3} \times 0.707 \left[\frac{\rho(\rho - \rho_v)gi_{fg}k^3}{\mu(T_{sat} - T_s)L} \right]^{1/4}$$

The Nusselt number based on the average heat transfer coefficient is given by

$$\text{Nu}_L = \frac{h_L L}{k} = 0.943 \left[\frac{\rho(\rho - \rho_v)gi_{fg}L^3}{\mu(T_{sat} - T_s)k} \right]^{1/4} \tag{9}$$

Recall that in arriving at Equation 9, the heat transfer that results in subcooling the condensate was neglected. The actual heat transfer, being the sum of the heat transfer to achieve the condensation of the vapor and to cool it below the saturation temperature, is greater than that obtained by using the convective heat transfer coefficient computed from Equation 9. To account for the effect of cooling the condensate below the saturation temperature, Rohsenow (1956) showed that i_{fg} in Equation 9 should be replaced by i'_{fg}

$$i'_{fg} = i_{fg} \left[1 + 0.68 \frac{c_p \Delta T}{h_{fg}} \right]$$

(Rohsenow's modification is discussed in Section 6.2.) Equations 8 and 9 are derived assuming a smooth, interfacial surface exposed to the vapor. It has been observed that as the condensate film thickness increases, corresponding to $\text{Re}_\delta \approx 7$, ($\text{Re}_\delta = \rho u_m \delta/\mu$, where u_m is the mean velocity of the condensate film), interfacial waves set in. The interfacial waves tend to increase the heat transfer rate to the wall. The increase in the heat transfer is approximately 20%. Incorporating the 20% increase in the heat transfer coefficient, we obtain

$$\text{Nu}_x = 0.848 \left[\frac{\rho(\rho - \rho_v)gi'_{fg}k^3}{\mu(T_{sat} - T_s)x} \right]^{1/4} \tag{10}$$

$$\text{Nu}_L = 1.13 \left[\frac{\rho(\rho - \rho_v)gi'_{fg}k^3}{\mu(T_{sat} - T_s)L} \right]^{1/4} \tag{11}$$

(The increase in the heat transfer coefficient due to the interfacial waves depends on the Reynolds number, Re. For a discussion of such effects and the resulting correlations, see Section 6.2.2)

The application of integral methods has been illustrated for a flow problem, a forced convection heat transfer problem, and a heat transfer problem with phase change. We now conclude with an example on natural convection. The problem to be discussed has many interesting features and will be treated in some detail.

EXAMPLE 5.3.4 One surface of a vertical plate maintained at a uniform temperature of T_s is exposed to a semi-infinite, quiescent fluid at $T_\infty < T_s$. Determine the convective heat transfer coefficient as a function of the distance from the bottom edge.

The physical configuration considered is depicted in Figure 5.3.8a. For fluid motion due to temperature differences without any external force, there should be *gravitational force* and *temperature dependent density*.

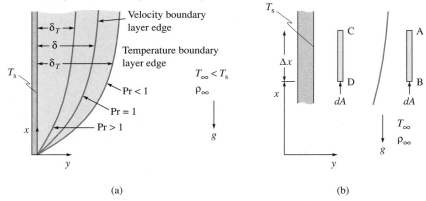

(a)   (b)

FIGURE 5.3.8 Natural Convection Resulting from Heating a Vertical Plate
(a) A vertical plate is at a uniform temperature surrounded by a fluid at a lower temperature. (b) Two vertical columns of the fluid, one far away from the plate and the other adjacent to the plate.

We anticipate a boundary layer developing adjacent to the plate as shown in Figure 5.3.8a. From the boundary layer equations, as $\partial p/\partial y = 0$, the pressure is a function of x only. The hydrostatic pressure variation is given by

$$\frac{dp}{dz} = -\rho_\infty g \qquad p_{x+\Delta x} - p_x = -\rho_\infty g\,\Delta x$$

where the density of the fluid at T_∞ is denoted by ρ_∞.

Now consider two elements of the fluid, each with the same cross-sectional area dA and height Δx, one far away from the surface and another close to the surface as shown in Figure 5.3.8b. Each element experiences an upward force given by $(p_x - p_{x+\Delta x})dA$. In the stationary fluid, far away from the surface, this upward force is balanced by the weight of the element of the fluid and

$$(p_x - p_{x+\Delta x})dA = \rho_\infty g\,\Delta x\,dA$$

For the element, CD, close to the surface, as the pressure is a function of x only, the upward force on this element due to the pressure is also given by $(p_x - p_{x+\Delta x})dA = \rho_\infty g\,\Delta x\,dA$.

The weight of the element, which represents the downward force on the element due to gravity, is $\rho g\,\Delta x\,dA$. But the fluid adjacent to the surface is at a higher temperature than the fluid far away from the surface, and $\rho < \rho_\infty$. The downward force on the element, $\rho g\,\Delta x\,dA$, is less than the upward force due to the pressure, $\rho_\infty g\,\Delta x\,dA$. Hence, there is a net upward force acting on the element close to the surface, $(\rho_\infty - \rho)g\,\Delta x\,dA$. This net upward force causes the element close to the surface to move up and sets up a convection current. (If the temperature of the surface is less than that of the fluid, a downward current is set up as the density close to the plate is higher than that far away from the plate.) Such convective motions can be observed in our daily life. On a sunny day the air adjacent to a window pane gets warmer than the room air and rises; when a hot solid is immersed in a cold fluid, convective currents can be observed around the solid; the chimney effect, which causes the hot gases to rise, is another example of the same phenomenon.

The upward force on the fluid adjacent to the vertical plate is opposed by viscous forces. The fluid motion is determined by the combination of the viscous and pressure forces; note that the pressure forces are related to the gravitational force. With an increase in the value of y, the upward velocity increases. But as the distance in the y-direction increases further, the temperature of the fluid approaches that of the fluid far away, and the net upward force on the fluid decreases. This causes the velocity to decrease. It essentially becomes zero at some distance $y = \delta$. The velocity profile in the boundary layer is sketched in Figure 5.3.9. At $y = \delta$ you should expect that $u = 0$ and $\partial u / \partial y = 0$ as the fluid for $y > \delta$ is stationary.

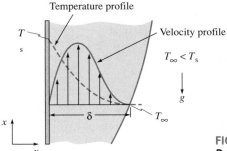

FIGURE 5.3.9 Velocity and Temperature Profiles in the Fluid

Some comments on the relative magnitudes of the thickness of the velocity boundary layer δ and temperature boundary layer δ_T are in order. For $Pr > 1$, it is expected that $\delta > \delta_T$. Physically, this is a plausible situation, as, when the forces due to pressure and gravity are balanced, the shear forces caused by the motion of the fluid in $0 < y < \delta_T$ tend to drag the fluid up. But for $Pr < 1$, δ_T should be greater than δ. Is this possible? If the temperature of the fluid is greater than that of the stationary fluid, the fluid should move upwards; i.e., it should be within the velocity boundary layer. This leads us to the conclusion that δ_T cannot be greater than δ. On the other hand, if the changes in temperature are small, the velocity may decrease more rapidly than the temperature, and the velocity may become insignificantly small for $\delta < y < \delta_T$. In this sense, one can conceive of $\delta < \delta_T$. In the analysis we permit the two boundary layer thicknesses to be different. With these introductory comments, we proceed to the analysis.

ASSUMPTIONS

1. The upward motion of the fluid is confined to the velocity boundary layer, $0 < y < \delta$ and the temperature variation is confined to the temperature boundary layer, $0 < y < \delta_T$. For $y > \delta$, $u = 0$ and for $y > \delta_T$, $T = T_\infty$.
2. Density differences are due to temperature differences only. With this assumption the variation of density with temperature can be determined by employing the coefficient of thermal expansion, β.

$$\beta = -\frac{1}{\rho}\frac{\partial\rho}{\partial T}\bigg|_p$$

As the pressure variation is expected to be small, changes in pressure can be neglected even for gases in the computation of the density changes. The density of the fluid at any location is then found from

$$d\rho = \frac{\partial\rho}{\partial T}\bigg|_p dT = -\rho\beta dT \qquad (1)$$

Changes in density are significant only when they cause the motion of the fluid; in all other cases the density can be considered to be constant. [This type of approximation, where a property (density in this case) is treated as constant except where differences in the property cause the phenomenon of interest (motion due to differences in density in this case) is known as the Boussinesq approximation.] Therefore, the density in Equation 1 can be replaced by ρ_∞ if the analysis is restricted to small values of $T_s - T_\infty$. The density at temperature T is evaluated as

$$\rho = \rho_\infty + \int_{T_\infty}^{T}\frac{\partial\rho}{\partial T}\bigg|_p dT = \rho_\infty - \int_{T_\infty}^{T}\rho_\infty B\, dT$$

With the further assumption that β is constant (or using a suitable average value of β),

$$\rho = \rho_\infty - \rho_\infty\beta(T - T_\infty) \qquad (2)$$

3. The plate is very wide in the z-direction (perpendicular to the x-y plane). End effects are negligible. The analysis is carried out for a unit depth in the z-direction.
4. Boundary layer equations are applicable with gravitational forces included in the equations.

SOLUTION

Mass Balance

Referring to Figure 5.3.10, for both velocity and temperature boundary layers the mass crossing EBC is found from (see Example 5.3.3),

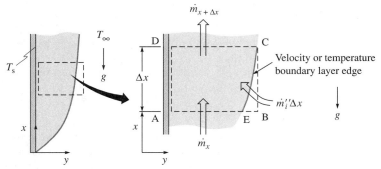

FIGURE 5.3.10 **Control Volume for the Application of Mass Balance**
Both velocity and temperature boundary layer are represented by EC. However, it does not mean that δ equals δ_T.

$$\frac{d}{dx}\int_0^* \rho u\ dy - \dot{m}_i'' = 0 \tag{3}$$

where, $\dot{m}_i''$ is the rate of mass crossing the boundary EBC per unit area perpendicular to the y-direction. The total mass flow rate crossing EBC is $\dot{m}_i''\Delta x$. The upper limit, denoted by *, represents δ for the velocity boundary layer and δ_T for the temperature boundary layer.

Balance of Momentum, x-Direction

Referring to Figure 5.3.11,

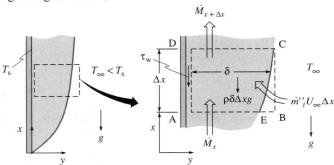

FIGURE 5.3.11 Control Volume for the Application of Momentum Balance

$$\dot{M}_{x+\Delta x} - \dot{M}_x - \dot{m}_i'' U_\infty \Delta x = -\tau_w \Delta x + (p\delta)_x$$
$$- (p\delta)_{x+\Delta x} + p(\delta_{x+\delta x} - \delta_x) - \int_{CV} \rho g\ dy\,\Delta x$$

where

$$\dot{M}_x = \text{rate of momentum flow in the } x\text{-direction crossing AE}$$
$$\dot{M}_{x+\Delta x} = \text{rate of momentum flow in the } x\text{-direction crossing CD}$$
$$(p\delta)_x = \text{force due to pressure on AE in the } x\text{-direction}$$
$$(p\delta)_{x+\Delta x} = \text{force due to pressure on CD in the negative } x\text{-direction}$$
$$p(\delta_{x+\delta x} - \delta_x) = \text{force due to pressure on EB}$$
$$\int_{CV} \rho g\ dy\ \Delta x = \text{weight of the fluid in the C.V.}$$

Dividing by Δx and taking the limit as $\Delta x \to 0$,

$$\frac{d\dot{M}}{dx} = -\tau_w - \delta\frac{dp}{dx} - \int_0^\delta \rho g\ dy \tag{4}$$

In arriving at Equation 4, $\dot{m}_i U_\infty$ has been set equal to 0, as $U_\infty = 0$ for $y > \delta$. Also as boundary layer equations are applicable, from assumption 4, $\partial p/\partial y = 0$ and the pressure is a function of x only. As discussed in the introductory comments on this problem,

$$\frac{dp}{dx} = -\rho_\infty g \quad \text{and} \quad -\delta\frac{dp}{dx} = \delta\rho_\infty g = \int_0^\delta \rho_\infty g\ dy$$

With $\dot{M}_x = \int_0^\delta \rho u^2 \, dy$, Equation 4 is recast as

$$\frac{d}{dx} \int_0^\delta \rho u^2 \, dy = -\tau_w + \int_0^\delta (\rho_\infty - \rho)g \, dy$$

Dividing by ρ (which can be treated as constant except in the last term of the equation, which causes the motion) and invoking Equation 2,

$$\frac{d}{dx} \int_0^\delta u^2 \, dy = -\frac{\tau_w}{\rho} + \int_0^\delta g\,\beta(T - T_\infty)dy \qquad (5)$$

In arriving at Equation 5, as the density changes are small, ρ_∞/ρ has been set equal to 1.

Any reasonable velocity profile can be chosen for use in Equation 5. The boundary conditions are

$$u(x, 0) = 0 \qquad u(x, \delta) = 0 \qquad \frac{\partial u}{\partial y}(x, \delta) = 0$$

A fourth boundary condition is obtained from boundary layer equations. The boundary layer equation, including the gravitational force, applied at $y = 0$ gives

$$0 = -\frac{dp}{dx} - \rho_w g + \mu \left.\frac{\partial^2 u}{\partial y^2}\right|_{y=0}$$

where ρ_w is the density of the fluid at T_s (at the wall). With $dp/dx = -\rho_\infty g$

$$-\frac{dp}{dx} - \rho_w g = g(\rho_\infty - \rho_w) = g \int_{T_s}^{T_\infty} \left.\frac{\partial \rho}{\partial T}\right|_p dT = -g \int_{T_s}^{T_\infty} \beta\rho \, dT$$

To evaluate the last integral, set $\rho = \rho_{av} = (\rho_\infty + \rho_w)/2$ and denote $(T_s - T_\infty)$ by ΔT.

$$-g \int_{T_s}^{T_\infty} \beta\rho \, dT = g\,\beta\rho_{av}\Delta T$$

Therefore,

$$\left.\frac{\partial^2 u}{\partial y^2}\right|_{y=0} = -\frac{g\,\beta\rho_{av}\Delta T}{\mu}$$

The first three boundary conditions are satisfied by

$$u = u_c \frac{y}{\delta}\left(1 - \frac{y}{\delta}\right)^2$$

where u_c is yet to be determined. Employing the last boundary condition,

$$\left.\frac{\partial^2 u}{\partial y^2}\right|_{y=0} = -\frac{g\,\beta\rho_{av}\Delta T}{\mu} = -\frac{4u_c}{\delta^2}$$

or

$$u_c = a\delta^2 \qquad a = \frac{g\,\beta\rho_{av}\Delta T}{4\mu}$$

and

$$u = a\delta y \left(1 - \frac{y}{\delta}\right)^2 \tag{6}$$

The temperature profile is chosen to satisfy the boundary conditions

$$T(x, 0) = T_x \qquad T(x, \delta) = T_\infty \qquad \frac{\partial T}{\partial y}(x, \delta_T) = 0$$

From the boundary layer equations, $(\partial^2 T/\partial y^2)(x, 0) = 0$. These conditions are satisfied by the profile

$$\theta = \frac{T - T_\infty}{T_s - T_\infty} = 1 - \frac{3}{2}\frac{y}{\delta_T} + \frac{1}{2}\frac{y^3}{\delta_T^3} \tag{7}$$

The temperature profile given by Equation 7 is shown by dashed lines in Figure 5.3.9.

Energy Balance

Referring to Figure 5.3.12,

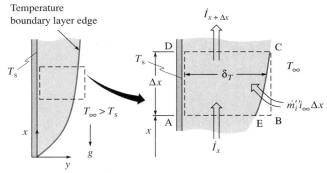

FIGURE 5.3.12 Control Volume for the Application of Energy Balance

Mass conservation $$\frac{d}{dx}\int_0^{\delta_T} \rho u \, dy - \dot{m}_i = 0 \tag{8}$$

Energy equation $$\dot{I}_{x+\Delta x} - \dot{I}_{x} - i_\infty \dot{m}_i \Delta x = q_w'' \Delta x$$

where

$\dot{I}_x$ = rate of enthalpy flow across AB = $\int_0^{\delta_T} \rho u i \, dy$

$\dot{I}_{x+\Delta x}$ = rate of enthalpy flow across DC

i and i_∞ are the specific enthalpies at T and T_∞, respectively

$q_w'' = -k \left.\frac{\partial T}{\partial y}\right|_{y=0}$

Substituting d/dx ($\int_0^{\delta_T} \rho u \, dy$) for $\dot{m}_i$ from Equation 8, dividing the energy equation by Δx, and taking the limit as $\Delta x \to 0$, we obtain

$$\frac{d}{dx} \left(\int_0^{\delta_T} \rho u i \, dy \right) - \frac{d}{dx} \left(\int_0^{\delta_T} \rho u i_\infty \, dy \right) = -k \frac{\partial T}{\partial y} \bigg|_{y=0}$$

or

$$\frac{d}{dx} \left(\int_0^{\delta_T} \rho u (i - i_\infty) dy \right) = -k \frac{\partial T}{\partial y} \bigg|_{y=0}$$

The pressure does not change in the y direction. By the definition of enthalpy, assuming constant specific heats, $i - i_\infty = c_p(T - T_\infty)$. The energy equation can be simplified by replacing

$$T - T_\infty = \theta \, \Delta T \qquad \Delta T = T_s - T_\infty \qquad k \frac{\partial T}{\partial y} \bigg|_{y=0} = k \Delta T \frac{\partial \theta}{\partial y} \bigg|_{y=0}$$

The final form of the energy equation is

$$\frac{d}{dx} \left(\int_0^{\delta_T} u \theta \, dy \right) = -\alpha \frac{\partial \theta}{\partial y} \bigg|_{y=0} \tag{9}$$

Employing Equations 6 and 7, the integrals in Equations 5 and 9 can be evaluated. Two cases arise, $\delta_T < \delta$ and $\delta_T > \delta$. The case of $\delta_T = \delta$ can be treated as the limiting case of either.

Case 1 $\delta_T < \delta$. With $u = u_c(y/\delta)[1 - (y/\delta)]^2$

$$\frac{d}{dx} \left(\int_0^{\delta} u^2 \, dy \right) = \frac{d}{dx} \left(\int_0^{\delta} \left[u_c \frac{y}{\delta} \left(1 - \frac{y}{\delta} \right)^2 \right]^2 dy \right)$$

$$= \frac{1}{105} \frac{d}{dx} (u_c^2 \delta) = \frac{1}{105} \frac{d}{dx} (a^2 \delta^5)$$

The integral $\int_0^{\delta} g\beta(T - T_\infty) \, dy$ in Equation 5 needs to be evaluated only between 0 and δ_T, as $T - T_\infty = 0$ for $y > \delta_T$.

$$\int_0^{\delta} g \beta(T - T_\infty) \, dy = \int_0^{\delta_T} g \beta(T - T_\infty) dy$$

$$= \int_0^{\delta_T} g \beta \Delta T \left(1 - \frac{3}{2} \frac{y}{\delta_T} + \frac{1}{2} \frac{y^3}{\delta_T^3} \right) dy$$

$$= \frac{3}{8} g \beta \Delta T \delta_T$$

Having evaluated the integrals in Equation 5, we get

$$\frac{1}{105} \frac{d}{dx} (a^2 \delta^5) = -\nu a \delta + \frac{3}{8} g \beta \Delta T \delta_T \tag{10}$$

where

$$\frac{\tau_w}{\rho} = \frac{\mu}{\rho} \frac{\partial u}{\partial y}\bigg|_{y=0} = \nu \frac{u_c}{\delta} = \nu a \delta$$

Assuming δ_T/δ to be constant and denoting the ratio by r, with $\delta(x = 0) = 0$, Equation 10 can be integrated to yield

$$\frac{\delta}{x} = \left[\frac{42\nu}{a}(3r - 2)\frac{1}{x^3}\right]^{1/4}$$

Substituting $a = g\beta\Delta T \, \rho/(4\mu)$ and $Gr_x = g\beta\Delta T x^3/\nu^2$, we obtain

$$\frac{\delta}{x} = \left(168 \frac{3r - 2}{Gr_x}\right)^{1/4} \quad \text{or} \quad \frac{\delta}{x} = 3.6 \left(\frac{3r - 2}{Gr_x}\right)^{1/4} \tag{11}$$

Now turning to Equation 9, evaluating the first integral,

$$\int_0^{\delta_T} u_c \frac{y}{\delta}\left(1 - \frac{y}{\delta}\right)^2 \left(1 - \frac{3}{2}\frac{y}{\delta_T} + \frac{1}{2}\frac{y^3}{\delta_T^3}\right) dy = u_c r \delta_T F_1 = a\delta^3 r^2 F_1$$

where

$$F_1 = \frac{1}{10} - \frac{1}{12}r + \frac{3}{140}r^2 \qquad r = \delta_T/\delta$$

Equation 9 now yields,

$$\frac{d}{dx}(a\delta^3 r^2 F_1) = \frac{3}{2}\frac{\alpha}{\delta_T} = \frac{3}{2}\frac{\alpha}{r\delta}$$

With the assumption that r is constant and with $\delta(x = 0) = 0$,

$$\delta \frac{d\delta^3}{dx} = \frac{3}{2}\frac{\alpha}{ar^3 F_1}$$

or

$$\delta^4 = \frac{8}{Pr}\frac{\nu^2}{g\beta\Delta T}\frac{x}{r^3 F_1}$$

$$\frac{\delta}{x} = \left(\frac{8}{Pr}\frac{\nu^2}{g\beta\Delta T x^3}\frac{1}{r^3 F_1}\right)^{1/4}$$

With $\delta = \delta_T/r$, we get

$$\frac{\delta_T}{x} = 1.682 \left(\frac{r}{Pr \, Gr_x F_1}\right)^{1/4} \tag{12}$$

For $r = 1$, equating Equations 11 and 12, we obtain $Pr = 1.25$. For this analysis, the velocity and thermal boundary layers have the same thickness for $Pr = 1.25$.

To evaluate the Nusselt number, the value of r for a given Pr is found by equating Equations 11 and 12 after substituting $\delta = \delta_T/r$, and solving the resulting equation, which is

$$21 \, Pr \, F_1 r^3 (3r - 2) - 1 = 0$$

The Nusselt number is, then, obtained from

$$q_w'' = -k \left.\frac{\partial T}{\partial y}\right|_{y=0} = -k\Delta T \left.\frac{\partial \theta}{\partial y}\right|_{y=0} = -k\Delta T \left(-\frac{3}{2\delta_T}\right)$$

$$\mathrm{Nu}_x = \frac{h_x x}{k} = \frac{q_2'' x}{\Delta T k} = \frac{3x}{2\delta_T} = \frac{3}{2(1.682)} \left(\frac{F_1}{r}\right)^{1/4} (\mathrm{Gr}_x\mathrm{Pr})^{1/4} \tag{13}$$

$$\mathrm{Nu}_x = 0.8918 \left(\frac{F_1}{r}\right)^{1/4} (\mathrm{Gr}_x\mathrm{Pr})^{1/4}$$

Equation 13 is valid for $\mathrm{Pr} \geq 1.25$. ($r < 1$).

Case 2 $\delta_T > \delta$ In this case, the left-hand side of Equation 9 yields

$$\frac{1}{105} \frac{d}{dx} (a^2\delta^5)$$

The integral on the right-hand side gives

$$\int_0^\delta g\beta(T - T_\infty)dy = g\beta\Delta T \int_0^\delta \left(1 - \frac{3}{2}\frac{y}{\delta_T} + \frac{1}{2}\frac{y^3}{\delta_T^3}\right) dy$$

$$= g\beta\Delta T\delta \left(1 - \frac{3}{4r} + \frac{1}{8r^3}\right) = g\beta\Delta T\delta f$$

where

$$f = \left(1 - \frac{3}{4r} + \frac{1}{8r^3}\right)$$

Equation 9 now yields

$$\frac{1}{105} \frac{d}{dx} (a^2\delta^5) = g\beta\Delta Tf\delta - va\delta$$

or

$$\frac{1}{\delta} \frac{d\delta^5}{dx} = 105 \frac{g\beta\Delta Tf - va}{a^2}$$

With $\delta(x = 0) = 0$, integration of the equation gives

$$\delta^4 = 84 \left(g\beta\Delta T \frac{f}{a^2} - \frac{v}{a}\right) x$$

Substituting $a = g\beta\rho_{av}\Delta T/(4\,\mu)$, we obtain

$$\frac{\delta}{x} = 4.281(4f - 1)^{1/4}\mathrm{Gr}_x^{-1/4} \tag{14}$$

Turning to the energy equation, as $\delta_T > \delta$ and $u = 0$ for $y > \delta$, Equation 9 needs to be integrated between 0 and δ only. The result of the integration is

$$\frac{d}{dx}(a\delta^3 F_2) = \frac{3}{2}\frac{\alpha}{\delta r}$$

where $F_2 = 1/12 - 1/20r + 1/210r^3$. With $\delta(0) = 0$, the solution to this equation is

$$\delta^4 = \frac{2\alpha x}{raF_2} = \frac{2\alpha x}{r[g\beta\rho_{av}\Delta T/(4\mu)]F_2}$$

or

$$\frac{\delta}{x} = 1.682(F_2 r Gr_x Pr)^{-1/4} \tag{15}$$

With $\delta = \delta_T/r$, Equation 15 yields

$$\frac{\delta_T}{x} = 1.682(F_2 Pr\ Gr_x/r^3)^{-1/4} \tag{16}$$

Comparing Equation 14 and 15,

$$Pr = \frac{1}{42(4f - 1)F_2 r}$$

For $r = 1$, $Pr = 1.25$, which is the same as for $r < 1$.

To compute the Nusselt numbers, the ratio r, for a given Prandtl number has to be obtained by solving

$$Pr < 1.25 \qquad\qquad 42\ Pr\ F_2 r(4f - 1) - 1 = 0$$

The Nusselt number is, then, given by

$$Nu_x = -k\frac{\partial T}{\partial y}\bigg|_{y=0} = k\Delta T\frac{3}{2\delta_T}$$

Substituting Equation 16 for δ_T, we obtain

$$Nu_x = 0.8918(F_2/r^3)^{1/4}(Gr_x Pr)^{1/4} \tag{17}$$

Equation 17 is valid for $Pr \le 1.25$.

From these expressions, the Nusselt number based on the average heat transfer coefficient is easily found as $h_L = (4/3)h_{x=L}$.

Results of numerical integration of the differential equation are available in Ede (1967). To compare the results from the approximate analysis with those from the numerical integration, Equations 13 and 17 are recast as

$$\frac{Nu_x}{(Gr_x Pr)^{1/4}} = c \tag{18}$$

The constant c in Equation 18 depends on the Prandtl number. c equals 0.8918 $(F_1/r)^{1/4}$ for $Pr \ge 1.25$ and 0.8918 $(F_2/r^3)^{1/4}$ for $Pr \le 1.25$. The values of r and c are given in Table 5.3.1. A graph of r versus Pr is shown in Figure 5.3.13.

TABLE 5.3.1 Comparison of Results Obtained by the Integral Technique and Solution by Numerical Integration

Pr	r	c Integral Technique	c Numerical Integration	% Difference
0.01	11.07	0.07788	0.1802	56.8
0.03	4.636	0.1465	0.2311	36.7
0.09	2.378	0.233	0.2828	17.6
0.5	1.24	0.3505	0.3717	5.7
0.72	1.13	0.3693	0.3874	4.7
0.733	1.125	0.3702	0.3881	4.6
1.0	1.048	0.3845	0.4010	4.1
1.5	0.9647	0.4013	0.4163	3.6
2.0	0.9151	0.412	0.4260	3.3
3.5	0.8374	0.43	0.4424	2.8
5.0	0.7989	0.4395	0.4511	2.6
7.0	0.7692	0.4471	0.4583	2.4
10.0	0.744	0.4538	0.4644	2.3
100	0.6763	0.4728	0.4900	3.5
1000	0.6677	0.4754	0.4992	4.8
10 000	0.6668	0.4757	0.5014	5.12

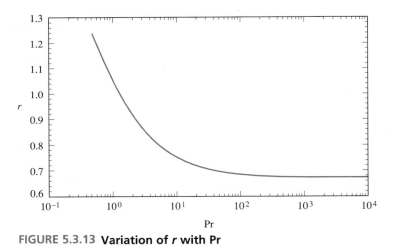

FIGURE 5.3.13 Variation of *r* with Pr

For a large range of Prandtl numbers, $0.5 \leq \text{Pr} \leq 10\,000$, the agreement between the results of the approximate solution and numerical integration of the differential equation is within 5%. However, the method fails for liquid metals that have $\text{Pr} < 0.03$.

The development given here for natural convection is more detailed than those given in most cases. Although this leads to considerable algebraic complexities, it brings out many of the salient points of approximate analysis by the integral technique.

5.4 PROJECTS

PROJECT 5.4.1 The inside of a building (20-m long, 10-m wide and 2.5-m high) is to be maintained at 20 °C in the winter months. The building, located in Minneapolis, has no partitions inside (it is one large room). The construction details of the walls and ceiling are shown in Figure 5.4.1. The following additional information is taken from the ASHRAE *Handbook of Fundamentals* (1989).

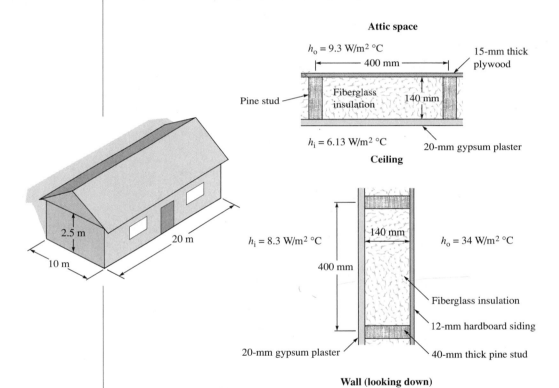

FIGURE 5.4.1 Details of Walls and Ceiling

Outside convective heat transfer coefficient	34 W/m² °C
Inside heat transfer coefficient	
Vertical surfaces	8.3 W/m² °C
Horizontal surface facing down	6.13 W/m² °C
Horizontal surface facing up	9.3 W/m² °C
Air infiltration through windows and doors	Approximately 0.6 times the volume of the building per hour

During the winter months the ambient temperature is equal to or greater than −27 °C for 99% of the time, and equal to or greater than −24 °C for 97.5% of the time.

Saturated steam at 100 °C is available for heating. A space heating system is to be designed to keep the inside of the building at 20 °C. The system should ensure reasonably uniform heating. Propose a suitable system.

SOLUTION

The main steps in the design are
1. Calculating the heating load—the required heat transfer rate from the heater to the room.
2. Considering the different possible configurations for the heater, selecting a desirable configuration, and sizing the heater, i.e., determining the required heat transfer area.

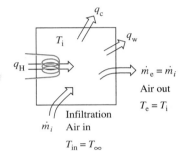

FIGURE 5.4.2 Components of Heating Load

Heating Load

The different components of the heating load are shown in Figure 5.4.2. With the air-space in the room as the control volume, in steady state, the balance equations are as follows:

Mass balance $\qquad \dot{m}_i = \dot{m}_e$

Energy balance Denoting $\dot{m}_i$ and $\dot{m}_e$ by $\dot{m}$ (mass rate of flow of air through the room)

$$\dot{m}i_\infty + q_H = \dot{m}i_i + q_w + q_c \tag{1}$$

where

$\quad i_\infty, i_i$ = specific enthalpies of the air at T_∞ and T_i, respectively
$\quad q_w, q_c$ = heat transfer rate from the room to the ambient air across the walls and ceiling respectively
$\quad q_H$ = required heat transfer rate from the heater

Treating air as an ideal gas,

$$i_i - i_\infty = c_p(T_i - T_\infty) \tag{2}$$

q_w and q_c are determined from the thermal circuit and are given by

$$q_w = \frac{T_i - T_\infty}{R_w} \quad \text{and} \quad q_c = \frac{T_i - T_{at}}{R_c}$$

where

$\quad R_w, R_c$ = thermal resistance of the walls and ceiling, respectively
$\quad T_{at}$ = attic air temperature

The attic temperature will be between T_∞ and T_i, closer to T_∞ than to T_i. How close T_{at} is to T_∞ depends on how well the attic space is ventilated. We assume that the attic is well ventilated and that T_{at} is approximately equal to T_∞. The actual value of T_{at} being higher than T_∞, the computed value of q_c will be greater than the actual value resulting in a conservative estimate for q_H.

In writing Equation 1, the windows and doors have been neglected. Usually the heat fluxes across windows and doors are much greater than the heat fluxes across the walls and ceiling. In the absence of any details of windows and doors, the computed value of q_H will be increased by 10% to allow for the higher heat fluxes across the windows and doors.

The value of T_∞ is not directly available. For determining the heating load the 97.5% value of $-24\,°C$ will be used. During those periods when the temperature falls below $-24\,°C$, one of two things will happen: (1) Usually, the computed heat transfer area is increased to allow for a safety margin. The increase in the required heat transfer rate when the outside temperature falls below $-24\,°C$ is accommodated within the safety margin. (2) If the actual and required values are equal, during those periods when the ambient air temperature falls below the value used for the design, the temperature of the room falls below $20\,°C$ by the same amount. For example, if the ambient air temperature falls to $-26\,°C$ for an extended period, the inside air temperature falls to $18\,°C$. However, if the ambient air temperature falls below the design value for brief periods (usually in the early hours of the morning), the effect on the inside air temperature is moderated by the energy storage in the walls, furniture, and other materials. Therefore, we choose $-24\,°C$ as the design ambient air temperature.

As no information is available regarding the details of the floor, we assume that it is very well insulated and the heat transfer rate across the floor is negligible.

Heater Configuration

There are many different configurations that will satisfy the requirements. A brief description of a few of them follow.

- A tube bank heat exchanger might be used, where the room air is forced over a bank of tubes inside which steam flows and maintains the surface temperature at close to $100\,°C$. The rate of flow of air is determined by the imposed increase in temperature and the heating load. The inlet temperature is the temperature of the room air and the exit temperature is high enough that it results in the least mass flow rate of air (lowest fan power) but low enough that it is not uncomfortable to the people.
- A steam pipe may be run along the walls slightly above the floor. Depending on the required heating surface area, the length of the tube is determined. If the length is greater than the perimeter of the room either a second tube row may be installed or the tube may be finned.
- A discrete number of steam heaters, much like the hot water heaters used for space heating, might be suitably distributed around the room to achieve uniform heating.

Determining the Heating Load

The heat transfer rate is determined from

$$q_H = \dot{m}c_p(T_i - T_\infty) + q_w + q_c$$

where

$\dot{m}$ = (0.6 × volume of the room × density of the air)/3600 kg/s
Volume of the room = 20 × 10 × 2.5 = 500 m³
$\rho = p/(RT)$ = 101 325/(287 × 293.15) = 1.204 kg/m³

Thus,

$$\dot{m} = \frac{0.6 \times 500 \times 1.204}{3600} = 0.1003 \text{ kg/s}$$

From Table A7, c_p = 1007 J/kg °C

$$\dot{m}c_p(T_i - T_\infty) = 0.1003 \times 1007 \times [20 - (-24)] = 4444 \text{ W}$$

Heat Transfer Across Walls   To compute the heat transfer rates across the walls and ceiling, employ the thermal circuit. For the heat transfer across the walls and the ceiling determine the heat transfer rate q'_w, from an area that is 1 m along the

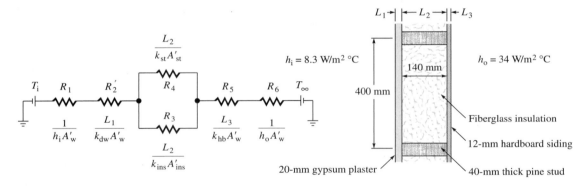

Wall (looking down)

FIGURE 5.4.3 **Details of Wall with the Thermal Circuit**

studs and 400-mm wide (Figure 5.4.3). The nomenclature and computations are as follows:

A'_w = area of the wall = 1 × 0.4 = 0.4 m²
k_{dw} (gypsum plaster dry wall, light weight aggregate, Table A3A) = 0.23 W/m °C
L_1 = 0.02 m L_2 = 0.14 m L_3 = 0.012 m
k_{ins} (fiberglass blankets, Table A3A) = 0.046 W/m °C
A'_{ins} = 1 × (0.4 − 0.04) = 0.36 m²
A'_{st} = pine stud = 0.04 m²
k_{st} (pine studs, Table A3A) = 0.12 W/m °C
k_{hb} (hard board, Table A3A) = 0.215 W/m °C

$$R_1 = \frac{1}{h_i A'_w} = \frac{1}{8.3 \times 0.4} = 0.3012 \text{ °C/W}$$

$$R_2 = \frac{L_1}{k_{dw}A'_w} = \frac{0.02}{0.23 \times 0.4} = 0.2174 \text{ °C/W}$$

$$R_3 = \frac{L_2}{k_{ins}A'_{ins}} = \frac{0.14}{0.046 \times 0.36} = 8.454 \text{ °C/W}$$

$$R_4 = \frac{L_2}{k_{st}A'_{st}} = \frac{0.14}{0.12 \times 0.04} = 29.17 \text{ °C/W}$$

$$R_5 = \frac{L_3}{k_{hb}A'_w} = \frac{0.012}{0.215 \times 0.4} = 0.1395 \text{ °C/W}$$

$$R_6 = \frac{1}{h_oA'_w} = \frac{1}{34 \times 0.4} = 0.07353 \text{ °C/W}$$

$$R_{eq} = R_1 + R_2 + \frac{R_3R_4}{R_3 + R_4} + R_5 + R_6 = 0.3012 + 0.2174 + \frac{8.454 \times 29.17}{8.454 + 29.17}$$
$$+ 0.1395 + 0.07353 = 7.286 \text{ °C/W}$$

$$q'_w = \frac{T_i - T_\infty}{R_{eq}} = \frac{20 - (-24)}{7.286} = 6.04 \text{ W}$$

$$q''_w = \text{heat flux across the wall} = \frac{q'_w}{A'_w} = \frac{6.04}{0.4} = 15.1 \text{ W/m}^2$$

$A_w = \text{total area of the walls} = 2 \times (20 + 10) \times 2.5 = 150 \text{ m}^2$

$q_w = \text{total heat transfer rate across the walls} = q''_wA_w = 15.1 \times 150 = 2265 \text{ W}$

Heat Transfer Across the Ceiling   The ceiling heat transfer computations are as follows:

k_{pw} (thermal conductivity of plywood, Table A3A) $= 0.115$ W/m °C

All other values are the same as for the walls.

$$R_1 = \frac{1}{h_iA'_w} = \frac{1}{6.13 \times 0.4} = 0.4078 \text{ °C/W}$$

$$R_2 = \frac{L_1}{k_{dw}A'_w} = \frac{0.02}{0.23 \times 0.4} = 0.2174 \text{ °C/W}$$

$$R_3 = \frac{L_2}{k_{ins}A'_{ins}} = \frac{0.14}{0.046 \times 0.36} = 8.454 \text{ °C/W}$$

$$R_4 = \frac{L_2}{k_{st}A'_{st}} = \frac{0.14}{0.12 \times 0.04} = 29.17 \text{ °C/W}$$

$$R_5 = \frac{L_3}{k_{pw}A'_w} = \frac{0.015}{0.115 \times 0.4} = 0.3261 \text{ °C/W}$$

$$R_6 = \frac{1}{h_oA'_w} = \frac{1}{9.3 \times 0.4} = 0.2688 \text{ °C/W}$$

$$R_{eq} = R_1 + R_2 + \frac{R_3R_4}{R_3 + R_4} + R_5 + R_6 = 0.4078 + 0.2174 + \frac{8.454 \times 29.17}{8.454 + 29.17}$$
$$+ 0.3261 + 0.2668 = 7.773 \text{ °C/W}$$

$$q'_c = \frac{T_i + T_\infty}{R_{eq}} = \frac{20 - (-24)}{7.773} = 5.66 \text{ W}$$

$$q''_c = \frac{q'_c}{A'_c} = \frac{5.66}{0.4} = 14.15 \text{ W/m}^2$$

$$q_c = q''_c A_c = 14.15 \times 20 \times 10 = 2830 \text{ W}$$

$$q_H = \dot{m}c_p(T_i - T_\infty) + q_w + q_c = 4444 + 2265 + 2830 = 9539 \text{ W}$$

To compensate for the increased heat flux across the windows and doors, the designed heat load is increased by 10%. Hence,

$$\text{Designed heat load} = 1.1q_H = 1.1 \times 9539 = 10\,492 \text{ W} \approx 10\,500 \text{ W}$$

Heater Design

Tube Bank A heater with a bank of tubes in which the steam flows is proposed (Figure 5.4.4). The specifications of the heater should include

- Diameter of the tubes
- Length of each tube
- Number of rows and columns
- Spacing between the tubes
- Arrangement of tubes (aligned or staggered)
- Mass rate of flow of air

A large number of combinations of the different variables will satisfy the design requirements. As one of the possibilities, fix the diameter of tubes, number of columns, and length of each tube. Further, choose aligned arrangement with $3d$ as the center-to-center distance between two consecutive tubes. Even after fixing several variables, some iteration may be necessary. Start with

Diameter of tubes	20 mm
Length of tube	1 m
Number of columns	12
Center-to-center distance of tubes	60 mm

The required mass rate of flow of air is determined from (see Figure 5.4.4)

$$q_H = \dot{m}c_p(T_e - T_i)$$

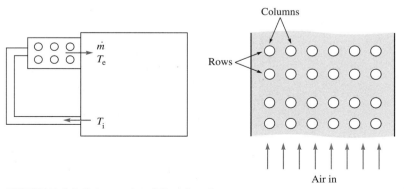

FIGURE 5.4.4 Schematic of Heating Arrangement

To keep the mass rate of flow of air to a minimum (to reduce the required fan power) the value for T_e should be high but should not be so high as make it uncomfortable for the people near the air registers (outlets). A reasonable value is 50 °C. With T_i = 20 °C and q_H = 10 500 W,

$$\dot{m} = \frac{10\ 500}{1007(50 - 20)} = 0.3476 \text{ kg/s}$$

The required heating surface area is found from Equation 4.2.5b.

$$\ln\left(\frac{T_s - T_e}{T_s - T_i}\right) = -\frac{hA_s}{\dot{m}c_p}$$

To find the total surface area, determine h at $(T_i + T_e)2$ = 35 °C. The heat transfer coefficient in condensation of steam is quite high and the inside convective resistance is negligible. To prevent corrosion, select copper tubes. The thermal conductivity of copper being very high and the thickness of the tubes being small, the conductive resistance is also very small compared with the convective resistance on the air side. Therefore, the tube surface temperature is approximately equal to the steam temperature, 100 °C.

To find the convective heat transfer coefficient, employ the appropriate correlation from Table 5.1.1. With the initial dimensions of

$$W = \text{width of the heater} = 1 \text{ m}$$
$$H = \text{height} = \text{number of columns} \times \text{center-to-center distance}$$
$$= 12 \times 0.06 = 0.72 \text{ m}$$

For air ρ = 101 325/(287 × 293.15) = 1.204 kg/m³. Thus,

$$U_\infty = \frac{\dot{m}}{\rho A} = \frac{0.3476}{1.204 \times 1.2} = 0.401 \text{ m/s}$$

$$S_T = S_L = 0.06 \text{ m} \qquad U_{max} = \frac{U_\infty S_T}{S_T - d} = \frac{0.401 \times 0.06}{0.06 - 0.02} = 0.6015 \text{ m/s}$$

From the software CC properties of air at 35 °C are

ρ = 1.146 kg/m³ $\qquad c_p$ = 1007 J/kg °C
μ = 1.888 × 10^{-5} N s/m² $\qquad k$ = 0.0267 W/m °C
Pr = 0.711 $\qquad Pr_s$(100 °C) = 0.705

$$Re_d = \frac{\rho U_{max} d}{\mu} = \frac{1.146 \times 0.6015 \times 0.02}{1.888 \times 10^{-5}} = 730.2$$

From Table 5.1.1, for Re_d = 730.2,

$$Nu_d = 0.52\ Re_d^{0.5}\ Pr^{0.36} \left(\frac{Pr}{Pr_s}\right)^{0.25} = 0.52 \times 730.2^{0.5} \times 0.711^{0.36} \times \left(\frac{0.711}{0.705}\right)^{0.25}$$

$$= 12.45$$

$$h = Nu_d \frac{k}{d} = 12.45 \frac{0.0267}{0.02} = 16.62 \text{ W/m}^2 \text{ °C}$$

From Equation 4.2.5b

$$A_s = -\frac{\dot{m}c_p}{h} \ln\left(\frac{T_s - T_e}{T_s - T_i}\right) = -\frac{0.3476 \times 1007}{16.62} \ln\left(\frac{100 - 50}{100 - 20}\right) = 9.9 \text{ m}^2$$

With 12 tubes, 1-m long, in each row, the number of rows of tubes is found from

$$\text{Number of rows} = \frac{\text{Total heat transfer area}}{\text{Surface area of tubes per row}}$$

$$= \frac{9.9}{\pi \times 0.02 \times 12} = 13.1, \text{ say } 13$$

In computing the heat transfer coefficient the correlation applicable for 16 or more rows of tubes was used, but the required number of rows is 13. With 13 rows the heat transfer coefficient is only slightly different from the heat transfer coefficient with 16 rows. We choose to neglect the difference.

The heater consists of 13 rows of tubes with 12 tubes in each row. The complete specification of the heater is

Diameter of tubes	20 mm
Length of tubes	1 m
Width of heater	0.72 m
Number of rows	13
Number of columns	12
Arrangement	Aligned

Steam Pipe   A simpler alternative is to run a pipe along the perimeter of the room slightly above the floor level. There should be sufficient clearance for air movement and a suitable cover should be provided to prevent people touching the hot pipe. The required heat transfer rate determined in the preceding section is 10 500 W.

An estimate of the required length of the pipe can be made without detailed calculations. Recall that in natural convection in air the convective heat transfer coefficient is approximately 6 W/m^2 °C to 10 W/m^2 °C. Employing a value of 8 W/m^2 °C, the approximate length of 25-mm diameter pipe is 208 m. But the perimeter of the room is 80 m. Provide the details of the calculations and arrangement of the tubes.

Steam Heaters   Using a discrete number of heaters, many combinations are possible. Develop the specifications for this heating system.

PROJECT 5.4.2   A flat plate solar collector, shown in Figure 5.4.5, consists of an absorber plate, which absorbs the major part of the incident radiant energy. The absorber plate is covered with one or more glass cover plates. In water heating solar collectors wter flows in tubes brazed to the absorber plate. The solar collector is usually tilted from the horizontal plane. The angle of tilt depends on the latitude of the location.

Consider a solar collector 1-m wide and -2 m long with one glass cover plate. The absorber plate is at 60 °C and the ambient air is at −5 °C. The air gap between

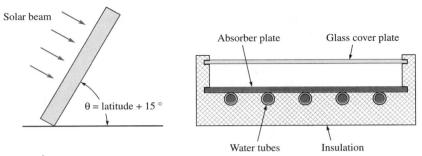

FIGURE 5.4.5 Schematic of a Flat Plate Collector

the absorber plate and the glass cover plate is 25 mm. The collector is tilted at an angle that is equal to the latitude plus 15°.

1. Determine the convective heat transfer rate from the absorber plate to the glass cover plate.
2. Suggest suitable methods to reduce the convective heat transfer from the absorber plate and compute the reduced heat transfer rate.

As an approximation, neglect the radiative heat transfer from the plates. Two possible methods are (1) add one more cover plate and (2) introduce a horizontal plate perpendicular to the absorber plate at the midheight of the collector. You may think of other methods but you should be able to estimate the reduction in the heat transfer rate.

SUMMARY

Forced convection correlations for tube banks, natural convection correlations for enclosures and thin wires, and forced convection correlations in high speed flows are presented.

In *tube banks* with uniform surface temperature for evaluating the heat transfer coefficient the arithmetic mean of the temperature of the fluid at exit and inlet to the bank is used. From the following relations the heat transfer rate and the surface area for the heat transfer rate are found.

$$q = \dot{m}c_p(T_e - T_i) \qquad \ln\frac{T_s - T_e}{T_s - T_i} = -\frac{hA_s}{\dot{m}c_p}$$

For *natural convection in enclosures* the heat transfer coefficient is defined on the basis of the difference in the temperatures of the parallel surfaces,

$$q_c'' = h(T_1 - T_2)$$

In *high speed flows* (Mach > 0.2) viscous dissipation is significant. For flows parallel to a flat plate, the fluid reference temperature for the definition of the convective heat transfer coefficient is the adiabatic wall temperature T_{aw}.

$$T_{aw} = T_\infty + r\,\frac{U_\infty^2}{2c_p} \qquad q_c'' = h(T_s - T_{aw})$$

where r is the recovery factor ($= \text{Pr}^{1/2}$ for laminar boundary layers and $= \text{Pr}^{1/3}$ for turbulent boundary layers). For uniform surface temperature the correlations for flows parallel to a flat plate given in Chapter 4 for laminar and turbulent boundary layers can also be used for high speed flows with T_{aw} replacing T_∞ in the definition of q_c''.

Reynolds analogy relates the Stanton number to the friction factor for $\text{Pr} \approx 1$ and is

$$\text{St}_x = \frac{C_{fx}}{2}$$

For $\text{Pr} \neq 1$, Colburn's analogy is

$$\text{St}_x\,\text{Pr}^{2/3} = \frac{C_{fx}}{2}$$

The integral method, introduced in Chapter 3, is extended to the solution of several convection problems. Solutions are obtained for a flow problem, a forced convection problem, a problem with a change in phase, and a problem on natural convection. To better understand the physics involved, the different cases are discussed in some detail. The cases show that the integral method gives results that are acceptable in engineering practice.

REFERENCES

Arnold, J.N., Catton, I., and Edwards, D.K. (1975). Experimental investigation of natural convection in a finite rectangular region inclined at various angles from 0° to 180°. *Proceedings 1974 Heat Transfer and Fluid Mechanics Institute.* Stanford, CA: Stanford University Press. p. 321.

Ayyaswamy, P.S., and Catton, I. (1973). The boundary layer regime for natural convection in a differently heated tilted rectangular cavity. *TRANS. ASME* 95C: 543.

Bar-Cohen, A., and Rohsenow, W.M. (1984). Thermally optimum spacing of vertical, natural convection cooled parallel plates. *J. Heat Transfer* 106: 116.

Catton, I. (1978). Natural convection in enclosures. *Proceedings Sixth International Heat Transfer Conference.* vol 6. Toronto, p. 13.

Colburn, A.P. (1933). A method of correlating forced convection heat transfer data and a comparison with fluid friction data. *TRANS. AIChE* 29: 174.

Ede, A.J. (1967). Advances in free convection. In *Advances in Heat Transfer,* vol. 4, eds. Irvine, T.F., and Hartnett, J.P. New York: Academic Press.

Fujii, T., Koyama, Sh., and Fujii, M. (1986). Experimental study of free convection heat transfer from an inclined fine wire to air. *Proceedings VIII International Heat Transfer Conference,* vol. 3. San Francisco, p. 1323.

Globe, S., and Dropkin, D. (1959). Natural convection heat transfer in liquids confined between two horizontal plates. *TRANS. ASME* 81C:24.

Hollands, K.G.T., Unny, T.E., Raithby, G.D., and Konicek, L. (1976). Free convective heat transfer across inclined air layers. *TRANS. ASME* 98C: 189.

Jakob, M. (1949). *Heat Transfer.* New York: Wiley.

Le Fevre, E.J. (1956). 9th International Proceedings Congress of Applied Mechanics, vol. 4. paper I-168. Brussels.

MacGregor, R.K., and Emery, A.P. (1969). Free convection through vertical plane layers: moderate and high Prandtl number fluids. *TRANS. ASME.* 91C:391.

McAdams, W.H. (1954). *Heat Transmission.* New York: McGraw-Hill.

Nusselt, W. (1916). Die Oberflachenkondensation des Wasserdampfes. *Z.V.D.I.* 60.

Raithby, G.D., and Hollands, K.G.T. (1975). A general method of obtaining solutions to laminar and turbulent tree convection problems. In *Advances in Heat Transfer,* vol. 11, eds. Irvine, T.F., and Hartnett, J.P. New York: Academic Press.

Reynolds, O. (1874). On the Extent and Action of the Heating Surface for Steam Boilers.'' *Proceedings Manchester Lit. Phil. Soc.* 14.

Rohsenow, W.M. (1956). ''Heat transfer and temperature distribution in laminar film condensation.'' *TRANS. ASME* 78:1645.

Zukauskas, A. (1972). Heat transfer from tubes in cross flow. In *Advances in Heat Transfer,* vol. 8, Irvine, T.F., and Hartnett, J.P. New York: Academic Press.

————(1987). Convection heat transfer in cross flow. In *Handbook of Single Phase Convective Heat Transfer,* New York: eds. Kakac, S., Shah, R.K., and Win Aung. Wiley Interscience.

REVIEW QUESTIONS

(a) Is the average heat transfer coefficient in a tube bank higher or lower than in a single tube?

(b) Explain why Equation 4.2.5b developed for internal flows can be used for flow in tube banks.

(c) What is pumping power?

(d) What is the criterion to indicate pure conduction for heat flow across a fluid between two concentric spheres?

(e) What is the difference in the definition of the convective heat transfer coefficient in natural convection in rectangular enclosures and in natural convection in vertical surfaces?

(f) What is the criterion to indicate that viscous dissipation is significant?

(g) Define recovery factor in high speed flows.

(h) What is the adiabatic wall temperature in high speed flows?

(i) If a fluid is brought to rest in a reversible and adiabatic process, its temperature rise is given by $U^2/2c_p$. In high speed flows, the difference between the fluid free stream temperature and the adiabatic wall temperature may be higher or lower than $U^2/2c_p$. Explain.

(j) Why is the convective heat transfer coefficient in high speed flows over a flat plate defined by $q_w'' = h(T_s - T_{aw})$ instead of $q_w'' = h(T_s - T_\infty)$?

(k) You are told that with flows over a flat plate, under certain conditions, even when the plate temperature is greater than the fluid free stream temperature, heat flows from the fluid to the plate. Do you agree? If you agree, under what condition is it possible? Explain.

(l) What do you understand by the terms velocity and temperature boundary layer?

(m) How would you determine if the velocity boundary layer is thicker or thinner than the temperature boundary layer in forced convection? Explain.

(n) For flow over a flat plate, what is the necessary condition for the boundary layer equations to be valid?

(o) What is Reynolds analogy?

(p) Explain why the temperature boundary layer grows much more rapidly than the velocity boundary layer in liquid metals.

(q) You are told that in a particular case of fluid flow over a flat plate the temperature boundary layer thickness is much smaller than the velocity boundary layer thickness. What can you conclude about the nature of the fluid?

(r) State the fundamental laws used in integral analysis for temperature distribution in forced convection.

(s) Solutions by integral methods are not exact and should be verified by experiments. Why?

(t) In boundary layer flows over flat plate, at $y = 0$, $\partial^2 u/\partial y^2 = 0$. Justify.

(u) In integral analysis of flows over flat plates, at $y = \delta$, $\tau_{yx} = 0$. Justify.

PROBLEMS

5.1 An air preheater is made of 12-mm O.D. tubes each of which is 3-m long, arranged in 20-rows with 16 tubes in each row (16 columns). The tubes are aligned. $S_T = S_L = 24$ mm. Atmospheric air enters the preheater at 20 °C with a velocity of 5 m/s. The surfaces of the tube are maintained at 100 °C. Determine

(a) The average convective heat transfer coefficient.

(b) The exit temperature of the air and the heat transfer rate from the tubes to the air.

(c) The pumping power.

5.2 Re-solve Problem 5.1 if the tubes are staggered.

5.3 Re-solve Problem 5.1 if the number of tubes is changed to 16 tubes per row with 6 tubes in each column.

5.4 A brief description of thermal storage for solar energy application is given in Problem 4.89. Now consider a bank of tubes of such double-pipe heat exchangers with the annulus filled with a phase change material. For properties of phase change material refer to Problem 4.89. Details of the bank of tubes to heat the air are

Outer diameter of tubes (m)	0.15
Length of tubes (m)	4
Inner tube diameter (m)	0.025
Velocity of air (m/s)	3
Inlet temperature of air (°C)	15
Spacing between columns, S_T (m)	0.25
Spacing between rows, S_L (m)	0.25
Arrangement of tubes	Staggered
Number of rows	20
Number of columns	10

Determine the exit temperature of air. Estimate the time taken to freeze all the phase change material and the fan power. Note that with the assumption of uniform convective heat transfer coefficient the rate of freezing of the phase change material is the highest in the first row and lowest in the last row. Hence, the assumption that the freezing is at the same rate in all the tubes will give an approximate solution for the estimate of the time to freeze all the material.

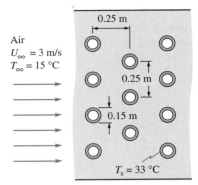

Air
$U_\infty = 3$ m/s
$T_\infty = 15\ °C$

0.25 m

0.25 m

0.15 m

$T_s = 33\ °C$

FIGURE P5.4

5.5 Hollow, cylindrical, brass tubes, 6-mm O.D., 3-mm I.D., and 20-cm long are attached to the surfaces of two power sources in a staggered arrangement with 10 columns and 5 rows. The temperature of the surfaces of the power source is 60 °C. The rows are 24-mm apart and the columns are 18-mm apart. Air at 20 °C is drawn over the bank with a free stream velocity of 5 m/s. Making suitable assumptions, estimate the heat transfer rate to the air.

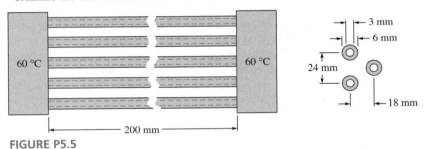

60 °C

60 °C

3 mm

6 mm

24 mm

18 mm

200 mm

FIGURE P5.5

5.6 A vertical wall in a house has an air gap of 7.5 cm as shown. The inner surfaces of the wall are at 14 °C and 10 °C, respectively. The height of the wall is 2.5 m. Determine the heat transfer rate across the air gap per unit width of the wall.

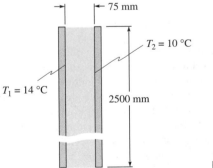

75 mm

$T_2 = 10\ °C$

$T_1 = 14\ °C$

2500 mm

FIGURE P5.6

5.7 In the wall in Problem 5.6, if a horizontal baffle is introduced in the air gap at a height of 1.25 m from the bottom of the wall determine the heat transfer rate across the air gap per unit width of the wall.

5.8 A solar collector has an air gap of 5 cm between the glass cover plate and the absorber plate. The collector is 1.3-m long and 1.2-m wide. The absorber plate is at 45 °C and the glass cover plate is at 30 °C. Determine the convective heat transfer rate across the gap if the angle of inclination θ, of the absorber plate with the horizontal is

(a) θ = 0°
(b) θ = 30°
(c) θ = 75°

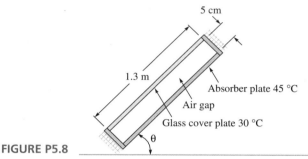

FIGURE P5.8

5.9 Re-solve Problem 5.8 if a baffle is introduced across the air gap at a distance of 0.65 m from one end of the collector.

5.10 A glass window, 1.2-m high and 1-m wide, is separated from a glass storm window by an air gap of 6 cm. If the temperatures of the window panes are 18 °C and −10 °C, respectively, determine the heat transfer rate across the air gap.

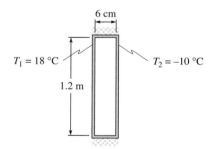

FIGURE P5.10

5.11 In Problem 5.10, the temperatures of the glass panes were specified. More realistically, only the air temperature on either side of the window are known. In such a case, finding the heat flux across the window involves significant iterations to find the temperatures of the glass panes, which are required to determine the heat transfer coefficients. To illustrate the iterations, consider the case when the inside glass temperature and the outside air temperature are specified at 18 °C and −10 °C, respectively. Find the heat flux across the window. Neglect the conduction resistance of the glass panes and radiative heat transfer.

5.12 A solar collector is made of a 5-cm diameter horizontal tube placed inside a coaxial 10-cm diameter glass tube. The space between the tubes is filled with nitrogen at 100 kPa. If the temperatures of the inner and outer tubes are 60 °C and 10 °C, respectively, compute the convective heat transfer rate per meter length of the tube.

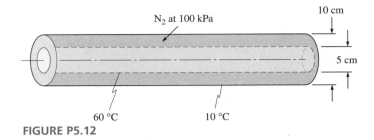

FIGURE P5.12

5.13 To reduce the convective heat transfer in the solar collector in Problem 5.12, a suggestion is made to decrease the pressure of nitrogen to 5 kPa. Calculate the percentage reduction in the heat transfer rate.

5.14 A solar collector is 2-m long and 1.3-m wide. The absorber plate is at 70 °C and the glass cover plate at 35 °C. The gap between the absorber plate and the glass cover is 3 cm and the space is filled with air at atmospheric pressure. For an optimum winter performance of the collector it is suggested that the angle between the collector plate and the horizontal plane should be (latitude + 15°). Determine the convective heat transfer rate from the absorber plate by natural convection if the collector is located in

(a) Edmonton, Alberta (Canada), latitude ≅ 54°.
(b) Miami, Florida (USA), latitude ≅ 26°.

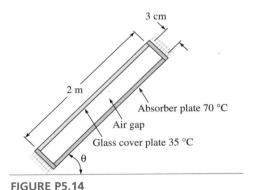

FIGURE P5.14

5.15 Two plates, each 0.6-m square, are separated by an air gap of 4 cm and form two surfaces of a rectangular enclosure. One plate is maintained at 90 °C and the other at 10 °C. All the side surfaces are perfectly insulated. Determine the convective heat transfer rate from the heated plate to the cooled plate if,

(a) The plates are horizontal with the heated plate at the top.
(b) The plates are vertical.
(c) The heated plate faces upwards, tilted at an angle of 30° to the vertical.
(d) The heated plate faces downwards, tilted at an angle of 60° to the vertical.

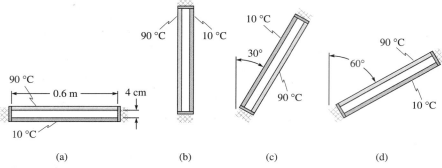

(a)   (b)   (c)   (d)

FIGURE P5.15

5.16 A 10-cm long 30-gauge nichrome wire is immersed in water at 20 °C. The wire is maintained at 80 °C by passing an electric current through it. Determine the current through the wire if the wire is
(a) Vertical
(b) Horizontal
(c) Inclined at an angle of 60° to the horizontal
Diameter of wire = 0.2548 mm. Resistance of wire = 0.2214 Ω/cm.

5.17 In Problem 5.16 a potential difference of 8 V is impressed across the wire. Determine the temperature of the wire if it is horizontal.

5.18 A heat sink is made of a 4-cm wide, 6-cm plate to which vertical fins are to be attached (see Figure 5.1.12). The fins are made of 1-mm thick 2024-T6 aluminum which are 4-cm high. Propose a suitable system of fins, and determine the total heat transfer rate if the base plate is at 60 °C, and the surrounding air temperature is 25 °C   **Design**

5.19 A room heater consists of a rectangular, hollow block 60-cm wide, 60-cm high, and 10-cm deep, filled with a fluid. The fluid is heated by an electrical heater. The fluid maintains the surfaces at a uniform temperature of 60 °C. It is proposed to augment the heat transfer rate from the heater by incorporating vertical fins of rectangular cross section. The fins are to be made of 1-mm thick copper. Compute
(a) The optimum spacing for the fins.
(b) The total heat transfer rate to the air.   **Design**

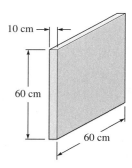

FIGURE P5.19

5.20 An airplane flies at an altitude of 12,000 m at a speed corresponding to a local Mach number of 2.2. If the wing of the plane is 3-m long in the direction of flow, determine the maximum temperature that the wing will attain. At an altitude of 12,000 m the temperature of air is −56.5 °C.

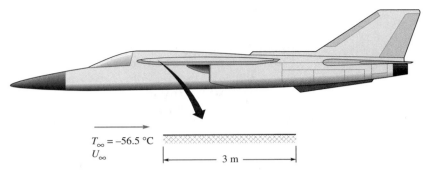

$T_\infty = -56.5\ °C$
U_∞

| 3 m |

FIGURE P5.20

5.21 A scale model of an airplane is to be tested in a supersonic wind tunnel. At the section where the plane is located in the wind tunnel, the air pressure is 10 kPa, and the temperature is $-40\ °C$. If the Mach number is 2.5, determine the heat transfer rate from the wing if it is modeled as a plate 70-cm long (in the direction of flow) and 1-m wide. The wing is maintained at $20\ °C$.

5.22 An 80-cm long and 1-m wide flat plate is placed in an air stream. The air, at 80 kPa and $20\ °C$, flows at Mach 2.

   (a) Find the heat transfer rate to one side of the plate if the plate is maintained at $40\ °C$ by cooling it with a liquid.

   (b) Estimate the temperature of the plate at distance of 4 cm from the leading edge if the coolant flow ceases.

5.23 During re-entry, the velocity of space shuttles is around 3660 m/s at an altitude of approximately 10 km. At such high speeds the temperature of the surface of the shuttle will reach very high values unless steps are taken to reduce the surface temperature. The space shuttle is approximately 37-m long. At an altitude of 10 km the temperature of air is $-56\ °C$. Modeling the space shuttle as a flat plate, estimate the adiabatic temperature of the shuttle surface. Identify at least one method of effectively reducing the temperature of the surface to $1000\ °C$. (You may recall that special tiles are glued to the space shuttle to reduce the skin temperature.)

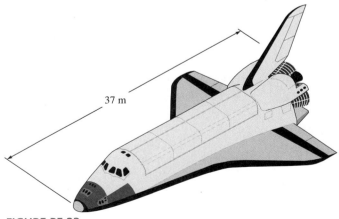

37 m

FIGURE P5.23

5.24 In Example 5.3.1, a third order velocity profile was used. Find the boundary layer thickness and the friction factor by employing
(a) A linear velocity profile with $u(x, 0) = 0$ and $u(x, \delta) = U_\infty$.
(b) A second order velocity profile by adding $(\partial u/\partial y)(x, \delta) = 0$ to the boundary conditions in part a.
(c) $u = U_\infty \sin [(\pi/2)(y/\delta)]$

5.25 In Example 5.3.2, a solution was obtained for uniform heat flux. Find an expression for the Nusselt number if a fluid with very low Prandtl number is heated by a plate maintained at a uniform temperature.

5.26 Consider the flow of a fluid over a flat plate at a uniform temperature. For an incompressible fluid with $\Pr > 1.0$, the thermal boundary layer thickness is less than the velocity boundary layer thickness. Denoting the ratio δ/δ_T by r show that for a flat plate at a uniform temperature,

$$f(r) \frac{d\delta^2}{dx} = \text{constant} \frac{\alpha}{U_\infty}$$

where $f(r)$ is a constant that depends on the temperature profile chosen. Employing Equation 5.3.15 for $d\delta^2/dx$ show that $f(r) = c/\Pr$. Choosing any reasonable temperature profile find $f(r)$ and c.

5.27 Consider the heat transfer to an incompressible fluid flowing parallel to a plate with a free stream velocity of U_∞ and a free stream temperature of T_∞. To find the heat transfer rate from the plate, the momentum and energy equations are to be solved by integral methods. Derive:

x-momentum:
$$\frac{d}{dx} \int_0^\delta u(U_\infty - u) \, dy = \nu \left. \frac{\partial u}{\partial y} \right|_{y=0}$$

Energy:
$$\frac{d}{dx} \int_0^{\delta_T} u(T - T_\infty) \, dy = -\alpha \left. \frac{\partial T}{\partial y} \right|_{y=0}$$

Assume:
$$\frac{u}{U_\infty} = \frac{3}{2}\frac{y}{\delta} - \frac{1}{2}\frac{y^3}{\delta^3}$$

$$\frac{T - T_\infty}{T_s - T_\infty} = 1 - \frac{3}{2}\frac{y}{\delta_T} + \frac{1}{2}\frac{y^3}{\delta_T^3}$$

Solve the momentum equation.

For $\Pr < 1$, $\delta < \delta_T$, and for $\Pr > 1$, $\delta > \delta_T$. Assuming $\delta_T/\delta = r = $ constant, simplify the energy equation for both cases. For $\Pr > 1$, show that

$$(14 - r^2)r^3 \Pr - 13 = 0$$

From the above expressions, the value of r can be found for different values of $\Pr > 1$. In a similar manner, find r for $\Pr > 1$.

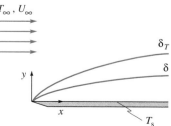

FIGURE P5.27

Problems 5.28 through 5.32 are two-domain problems. In Problem 5.28, with x as the axial coordinate (the origin being the location where the heating starts), for $0 < x < L_{Te}$, the temperature boundary layer thickness, δ_T, grows and the temperature profile is a function of x. At $x = L_{Te}$, $\delta_T = b$. For $x > L_{Te}$, the temperature profile is invariant and the centerline temperature increases.

5.28 An incompressible fluid is in laminar flow between two parallel plates with a fully developed velocity profile. Both the plates are at the same uniform temperature (different from the fluid inlet temperature). Employing a second order temperature profile find an expression for the temperature boundary layer thickness (in the thermal entrance region) as a function of the axial distance. Estimate the thermal entrance length. Establish the temperature profile as a function of the axial distance in both the thermally developing and fully developed regions. Note that this is a two domain problem.

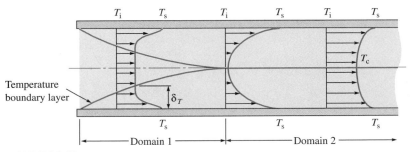

FIGURE P5.28

5.29 Re-solve Problem 5.28 when a uniform heat flux q_w'' is imposed on both the plates.

5.30 In Problem 5.29, the fluid is atmospheric air entering the heated region at 27 °C. The plates are 20-mm apart and the mass flow rate of air per unit width of the plates is 0.01 kg/s. A heat flux of 1000 W/m² is imposed on each plate. Determine
   (a) The axial distance to the location where the thermal boundary layer thickness is 5 mm and the wall temperature at that location.
   (b) The thermal entrance length L_{Te} and T_s at $x = L_{Te}$.

5.31 Re-solve Problem 5.28 if one of the plates is maintained at a uniform temperature and the other plate is perfectly insulated,

5.32 Re-solve Problem 5.28 if one plate is subjected to a uniform heat flux and the other plate is perfectly insulated.

5.33 Can you think of using the results of Problems 5.28 and 5.31 to solve a problem where the two plates are maintained at different but uniform temperatures or the two plates are subjected to different but uniform heat fluxes?

6

HEAT TRANSFER WITH CHANGE OF PHASE

In the previous two chapters on convection, heat transfer to single phase fluids was studied. In this chapter we

- *Introduce the different regimes in surface boiling and condensation—nucleate and film boiling and film-wise and drop-wise condensation*
- *Consider heat transfer that results in change of phase from liquid to vapor (boiling) and from vapor to liquid (condensation)*
- *Present correlations to estimate heat transfer rates in nucleate and film boiling and film-wise condensation and maximum heat flux under nucleate boiling conditions*
- *Provide a brief introduction to forced convection boiling and condensation inside tubes*

So far, heat transfer to (or from) single-phase fluids has been considered. Many devices, such as steam boilers and condensers in power plants and condensers and evaporators in refrigerating and air-conditioning plants, involve heat transfer to or from fluids that undergo phase changes. Another application of heat transfer with phase change is in thermal storage. In some cases with electrical load fluctuations, power plants with a lower peak capacity can meet such demands by working for longer periods (at peak capacity) with thermal storage. For example, power demand for air conditioning is much higher during afternoons than during early mornings. Without thermal storage the air-conditioning plant should have sufficient capacity to meet the afternoon cooling loads, but with thermal storage, a smaller air-conditioning plant, running for a longer period, can satisfy the cooling needs. When the air-conditioning demand is less than the capacity of the unit, as during nights, a suitable thermal storage material is cooled. In times of high air-conditioning demand the air is cooled by the cool material. In such cases thermal storage with phase-change material is advantageous. For a unit volume of the material, energy storage with phase-change materials is significantly greater than the energy storage with single-phase materials working within reasonable temperature ranges. Another application is in solar energy systems. As the availability of solar energy is periodic (both short term during each 24-hour period and long term as when cloudy days intervene between sunny days) thermal storage is necessary. In such cases phase-change materials offer an attractive alternative to storage in single-phase material. In this chapter heat transfer involving phase change is considered. Only changes between liquid and vapor phases are studied here because of their engineering importance.

6.1 BOILING

Difference between boiling and evaporation

First we consider the liquid-to-vapor phase change. Liquid-to-vapor phase change can occur either at the liquid-vapor interface or at a solid-liquid interface. When the vapor pressure is less than the saturation pressure, phase change from liquid to vapor occurs at the liquid-vapor interface. Such a phase change is termed *evaporation.* The liquid may be exposed to either its vapor or a gas. In either case, if the vapor pressure (partial pressure if the gas contains components other than the vapor of the liquid) is less than the saturation pressure at the temperature of the gas, evaporation results. Examples of evaporation are drying of wet clothes exposed to air and cooling of our body on hot days by the evaporation of perspiration. When a solid surface in contact with a liquid is at a temperature sufficiently higher than the saturation temperature of the liquid, vapor is produced at the interface. Such a process is known as *boiling.* In this chapter we restrict our discussion to boiling and condensation on solid surfaces.

Boiling is a complex phenomenon and no satisfactory analytical solution to predict heat transfer rates in boiling is available, although there are a very large number of published papers (in the hundreds) dealing with this topic. Many correlations to predict heat transfer rates in boiling have been proposed, but none of them gives results that are satisfactory in more than a few specific situations. The uncertainties in the correlations for boiling are even greater than those in single-phase correlations—on the order of $\pm 20\%$, and much more than that in many cases of boiling. To be able to use the available correlations intelligently, it is necessary to have a knowledge of the macroscopic physical process of boiling. Figure 6.1.1 shows three regimes of boiling.

Pool boiling and forced convection boiling

Boiling in an otherwise stationary liquid is known as *pool boiling.* If boiling occurs in a liquid already in motion it is termed *forced convection boiling.*

Pool Boiling A good way to understand boiling is to refer to the classic experimental results of Nukiyama (1966) in 1934. He placed a horizontal wire (silver- or nickel-plated copper wire, 20-cm long and 3-mm in diameter) in a pool of water at its saturation temperature and heated the wire electrically. The resistance of the wire was used to determine its temperature. He measured the heat flux as a function of ΔT ($= T_s - T_{sat}$), the difference in the temperature of the wire, and the saturation temperature of the water (also known as the excess temperature) and obtained a graph similar to the one shown in Figure 6.1.2. When this difference in temperature was slightly greater than zero, boiling did not begin. [Boiling begins at the saturation temperature only when the liquid, in contact with the saturated vapor, is heated. When a submerged surface is heated a little above the saturation temperature of the surrounding liquid, boiling does not begin. For details, see Janna (1986)]. Heat transfer occurred as natural convection in a single-phase fluid. This is represented by the part AB in Figure 6.1.2. When the temperature difference was increased

Pool boiling regimes

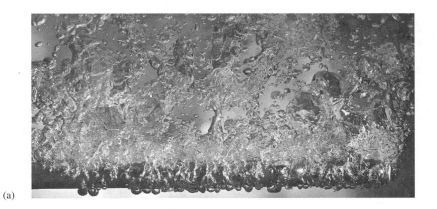

(a)

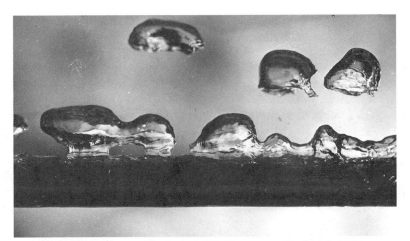

(b)

(c)

FIGURE 6.1.1 Different Regimes of Boiling on a Horizontal Wire
(a) Nucleate boiling. (b) Transition boiling. (c) Film boiling. Photographs
courtesy of J.W. Westwater.

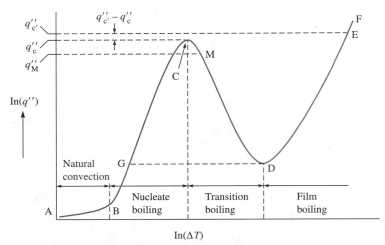

FIGURE 6.1.2 Boiling Curve Showing Different Regimes
B, Inception of boiling; D, Liedenfrost point (minimum film boiling heat
flux); C, Critical heat flux, CHF (maximum nucleate boiling heat flux).

Incipience of
boiling

beyond about 5 °C, a few small bubbles of vapor were observed on the surface of
the wire, to detach from the wire surface, and then to rise. The initiation of boiling,
also known as *incipience of boiling,* is represented by point B. Beyond B boiling
occurred. Just beyond B, an increasing number of vapor bubbles formed and
detached themselves from the surface. As seen from the graph, for a defined increase
in ΔT, the required increase in the heat flux q'' was much greater in the nucleate
boiling region (BGC) than in the natural convection region (AB). Beyond B, with
an increase in q'', not only did the number of bubbles increase but also their size.
More and bigger bubbles began to appear. During this stage, there was considerable
agitation in the liquid, and as the bubbles left the wire, ''cool'' liquid reached the
surface promoting heat transfer from the wire. But beyond a certain point, the rate
of increase of q'' with ΔT began to fall. The heat flux reached a maximum value as
shown by point C. When the heat flux was increased beyond this value, the wire
burned out! However, when similar experiments were tried with a platinum resistance wire, it did not melt when the heat flux was raised beyond this maximum value
but became white hot, jumped to point E, and followed another part of the boiling
curve EF. When the heat flux was reduced, instead of going from point E to C it
followed the path FED. In this region it was observed that the entire wire was
surrounded by a film of vapor. When the heat flux was reduced below a value
corresponding to point D, the temperature jumped to G and the boiling curve GB
was obtained again. In Nukiyama's experiments, the part of the curve between D
and C was missing. This part cannot be easily obtained by controlling the heat flux;
however, it can be obtained by controlling the temperature. Another way to obtain
the entire boiling curve ABCDEF is to monitor the temperature of a solid, which is
cooled by immersing it in a liquid. If the initial temperature of the solid corresponds
to a point in the region DF, the solid undergoes all the temperature regimes in the
boiling curve. By recording the temperature as a function of time, the heat flux is

determined as a function of ΔT for all values. (By ensuring that the Biot number is much less than 1, spatial uniformity of temperature can be assumed and computations of the heat transfer rate become simple.) Heat transfer coefficients in boiling cryogenic fluids, such as liquid nitrogen, are easily determined by such transient techniques.

A careful observation of boiling over surfaces provides a physical explanation of the boiling curve. After the incipience of boiling (point B), the number and size of bubbles grow. As these bubbles leave the surface, cool liquid replaces them at the surface and promotes heat transfer. In this region, heat transfer from the surface is directly to the liquid at the perimeter of the bubbles (changing the liquid to vapor), and indirectly through the vapor in the bubbles to the liquid. Initially, with an increase in ΔT, the number and frequency of the bubbles increase with a large increase in the heat flux. But beyond a certain value of ΔT, the number of nucleation sites (locations where vapor bubbles initiate) decreases, and the bubbles become larger. Some of the bubbles coalesce to form large bubbles, and more and more of the surface is occupied by the vapor bubbles. The initial formation of small bubbles, their frequent removal from the surface, and the consequent stirring of the liquid gives way to the formation of larger bubbles that remain on the surface for longer periods than the small bubbles. As the thermal conductivity of the vapor is less than that of the liquid, the heat flux from the portions of the surface in contact with the vapor is less than that from the surface in contact with the liquid. As the area of the surface exposed to the vapor increases (and the area exposed to the liquid decreases) the average heat flux from the surface decreases. Thus, beyond a certain value of ΔT, we should expect a drop in the rate of increase of the average heat flux with ΔT. As ΔT is increased further, the heat transfer rate reaches a maximum. This is what happens at point C. Beyond point C, to transfer the increased power, the temperature must drastically increase, jumping to point E. With water as the boiling fluid, the temperature corresponding to point E is higher than the melting points of many materials of resistance wires and tubes, and this will cause them to melt. This phenomenon is known as *burn-out* of the wire or pipes.

Why did an increase in the heat flux above the value corresponding to point C in Nukiyama's experiments lead to the burn-out of the wire? To answer this question, consider what happens when the heat flux is raised slightly above q_c'' to $q_{c'}''$ in Figure 6.1.2. Heat transfer to the fluid is limited to q_c'' and, hence, the difference, $q_{c'}'' - q_c''$, results in an increase in the internal energy of the wire. Consequently, the temperature of the wire increases above T_c. If the new temperature is T_M, the heat flux to the fluid q_M'' is even less than q_c'' and the difference $q_{c'}'' - q_M''$ further raises the temperature of the wire forcing the heat flux curve to go along CDE until the equilibrium point E is reached. Unless the material of the wire is such that its melting point is above T_E, the wire will melt before the equilibrium point is reached. The time taken for the surface temperature to go from C to E depends on the heat capacity of the wire—the product of the mass of the wire and the specific heat of the material of the wire. The smaller this value, the shorter is the time. If the resistance wire, such as the one used in the experiment, has a very small mass, its heat capacity is very small and it takes very little time for the temperature to reach the melting point of the wire (which is below T_E). However, if the heat capacity of the heating element is large, it will take a much longer time to burn out.

Boiling may occur with or without bulk motion of the fluid. When the boiling liquid is in bulk motion due to an external force, it is termed *forced convection boiling*. Examples of forced convection boiling are the boiling of water flowing in the tubes of a boiler in a steam power plant and the evaporation of the refrigerant in an evaporator of a refrigeration plant. Boiling of a stationary liquid, such as in a pot of water on a stove, is termed pool boiling.

The several distinct regimes in pool boiling are indicated in Figure 6.1.2. The maximum heat flux in the nucleate boiling region, point C, is known as the *critical heat flux* (CHF) or the peak heat flux. The region DEF, where the surface is completely blanketed by the vapor, is the *film boiling region*. Between point C and D, even with an increase in the temperature of the heated surface, the heat flux decreases. In the region CD there is a combination of nucleate boiling and film boiling—the region where transition from nucleate to film boiling occurs. With an increase in the temperature, the area of the surface with film boiling increases and the area with nucleate boiling decreases with a net decrease in the average heat flux.

The region CD is termed the *transition boiling*. When the transition from nucleate to film boiling is completed, the point D corresponding to the minimum heat flux is known as the *Leidenfrost point*.

Phenomena of Nucleate Boiling In engineering applications of boiling heat transfer, we would like to operate the equipment in the nucleate boiling region, where high heat fluxes are achieved at low values of ΔT. How does nucleation start? Machined surfaces have a large number of surface irregularities, some of which are in the form of cavities, Figure 6.1.3a. When a liquid comes in contact with such a surface, due to surface tension forces, the liquid is unable to displace the gas or vapor in such cavities and the gas or vapor is trapped in the cavities. When the temperature of the surface is increased, due to the heat transfer across the gas or vapor in the cavities, more liquid evaporates and the bubbles become larger, Figure

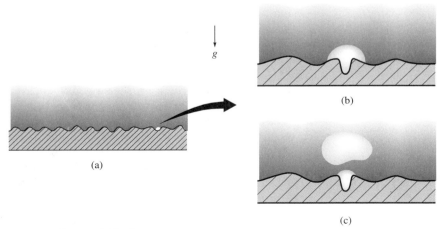

(a)

(b)

(c)

FIGURE 6.1.3 Bubble formation
Imperfections in a machined surface with vapor or gas in the cavities cause bubble formation. (a) Imperfections in a surface. (b) Bubble around a cavity. (c) Vapor bubble detaches from the surface and a new bubble forms.

6.1.3b. When the buoyancy force on a bubble is large enough, the bubble detaches from the surface, and a new bubble starts around the residual vapor in the cavity, Figure 6.1.3c. In the regime of nucleate boiling, initially isolated bubbles appear. With an increase in ΔT, more and more bubbles are generated and the heat flux increases (region BC, Figure 6.1.2). But when the value of ΔT is higher than ΔT corresponding to point C, many bubbles coalesce, the heated surface area available for the cool liquid decreases, and the heat flux begins to decrease—region CD. The effect of the lack of an adequate number of cavities as nucleation sites can be observed with boiling from a smooth surface, for example, boiling of water in a glass (pyrex) container. Lienhard (1987) has given an interesting discussion of nucleation of bubbles.

Regimes in forced convection boiling

Forced Convection Boiling In forced convection boiling also there are several regimes. Consider a liquid entering a tube with an imposed heat flux on the surface of the tube as shown in Figure 6.1.4. Initially, the heat transfer is by forced convec-

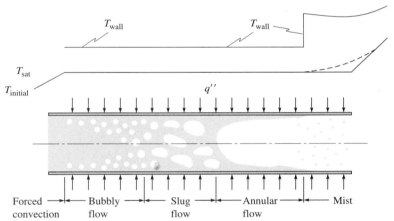

FIGURE 6.1.4 Different Regimes in Forced Convection Boiling and Profiles of Fluid and Tube Surface Temperatures

tion, until the temperature of the liquid reaches its saturation temperature. When the temperature of the liquid increases beyond its saturation temperature, bubbles of vapor begin to appear and we have the bubbly region. When the number of bubbles grow, two or more bubbles coalesce and large bubbles are formed in the interior of the tube. This is the slug flow region. If the mixture of vapor and liquid is further heated, there is a vapor core around the axis of the tube with a liquid film adjacent to the tube surface. This flow is termed *annular flow*. Further downstream, with more heating, entrained liquid drops appear inside the annular liquid film—the annular flow with drop entrainment. With continued heating, the annular liquid film evaporates and there are only small liquid drops—the drop-flow or misty regime. Note that the phase change at the liquid free surface is evaporation. Finally, when all the drops evaporate, we again have a single-phase, forced convection heat transfer to the vapor phase. Figure 6.1.4 also shows the temperature profile as a function of the axial distance of the tube.

6.1.1 Pool Boiling Correlations

Equipment involving boiling should operate in the region BC close to C in Figure 6.1.2. This will ensure high heat transfer rates at low values of ΔT. It is also necessary that the heat flux be less than the CHF—the maximum heat flux corresponding to point C—to prevent possible burn-out of the tubes. Because of our interest in operating engineering equipment in the nucleate boiling region below the CHF, a large amount of work has been devoted to developing correlations to predict the heat transfer rate in the nucleate boiling region and to predict the critical heat flux.

Factors affecting heat flux in pool boiling

Consider the factors that affect the heat flux in pool boiling. It is clear that the heat flux is affected by ΔT. The frequency of bubble removal is determined by a combination of the buoyancy force on the bubble, which is related to $(\rho - \rho_v)g$, the surface tension, σ, and the enthalpy of vaporization i_{fg}. The heat flux may be affected by a length dimension of the surface L, for example, the height of a vertical surface. It is also affected by the thermophysical properties of the liquid, k, c_p, and μ. Thus, the functional relation between the heat flux and the other variables may be expressed as (all properties are those of liquid except ρ_v, which represents the density of the vapor),

$$q'' = q''[\Delta T, L, (\rho - \rho_v)g, i_{fg}, \sigma, k, c_p, \mu]$$

From dimensional analysis we obtain

$$\frac{q''L}{\mu i_{fg}} = f\left(\sqrt{\frac{g(\rho - \rho_v)}{\sigma}}\, L, \frac{c_p\,\Delta T}{i_{fg}}, \frac{c_p\mu}{k}\right)$$

where

$$g(\rho - \rho_v)L^2/\sigma = \text{Bond number}$$
$$(c_p\,\Delta T)/i_{fg} = \text{Jakob number}$$
$$(c_p\mu)/k = \text{Prandtl number}$$

The Bond number represents the ratio of the buoyant force [proportional to $g(\rho - \rho_v)L^3$] to the surface tension force (capillary force proportional to σL). The Jakob number is the ratio of the sensible heat to the latent heat (sensible heat is the increase in the enthalpy of a substance resulting from an increase in the temperature—the increase in the enthalpy is ''sensed'' by the increase in the temperature). The significance of the Prandtl number as the parameter coupling the heat transfer and flow variables has already been discussed (Section 4.1). In addition to the three named dimensionless variables, there is also an unnamed dimensionless variable, $[(q''L)/\mu i_{fg})]$. Note that by combining the dimensionless variables, i.e., multiplying or dividing one by another, we can obtain other dimensionless variables. For example,

$$\frac{q''L}{\mu i_{fg}} \times \frac{c_p\mu}{k} \times \frac{i_{fg}}{c_p\,\Delta T} = \frac{q''}{\Delta T} \times \frac{L}{k} = \frac{hL}{k} = \text{Nusselt number}$$

The dimensionless variables have been obtained for the most general case. In any particular case, only some of the dimensionless variables may be significant.

In predicting the heat transfer rate in the natural convection region, AB in Figure 6.1.2, the correlations for natural convection are applicable. In the nucleate boiling region, BC, an alternate form of correlation, obtained by manipulating the different dimensionless variables, is used. The correlation is

$$\frac{c_p\,\Delta T}{i_{fg}} = f\left(\frac{q''}{\mu i_{fg}}\sqrt{\frac{\sigma}{g(\rho - \rho_v)}}, \frac{c_p\mu}{k}\right) \tag{6.1.1}$$

The prediction of the CHF, and ΔT at which CHF occurs, is of great importance because they establish the limits of operation of the devices in the nucleate boiling region. If the imposed heat flux is greater than the CHF, it may lead to film boiling and possible burn-out. Nuclear reactors must operate well below the CHF as an accidental heat flux greater than the CHF may lead to melt-down. No satisfactory correlation is available for the transition region CD. The transition region is of little practical significance as boilers are not designed to operate in this region; in operation, boilers operate either in the nucleate boiling region or in the film boiling region. While one would like to avoid operating the devices in the film boiling region, film boiling is inevitable in some cases. For example, in cryogenic applications involving such liquids as liquid hydrogen and liquid nitrogen (LH_2 is used as a fuel in space applications, and LN_2 is used in low temperature applications, such as cooling of some types of equipment, and in some medical procedures), film boiling may occur. We now give some of the available correlations for predicting the heat flux in the nucleate boiling and film boiling regions, the CHF, and the minimum heat flux in the film boiling region.

Nucleate Boiling In the nucleate boiling region, Rohsenow (1952) developed the correlation

Nucleate boiling correlation

$$\frac{c_p\, \Delta T}{i_{fg}\, Pr^n} = C_{sf} \left[\frac{q''}{\mu i_{fg}} \left(\frac{\sigma}{g\,(\rho - \rho_v)} \right)^{1/2} \right]^{1/3} \tag{6.1.2a}$$

$$q'' = \mu i_{fg} \left(\frac{g\,(\rho - \rho_v)}{\sigma} \right)^{1/2} \left(\frac{c_p\, \Delta T}{i_{fg}\, Pr^n\, C_{sf}} \right)^3 \tag{6.1.2b}$$

TABLE 6.1.1 Values of n and C_{sf} in Equation 6.1.2

Fluid	Surface	C_{sf}	n
Benzene	Chromium	0.101	1.7
Carbon tetrachloride	Polished copper	0.007	1.7
Ethyl alcohol	Chromium	0.0027	1.7
Isopropyl alcohol	Copper	0.0025	1.7
n-Butyl alcohol	Copper	0.003	1.7
n-Pentane	Polished copper	0.0154	1.7
	Polished nickel	0.0127	1.7
	Emery-rubbed copper	0.0074	1.7
	Chromium	0.015	1.7
Water	Brass	0.006	1.0
	Copper polished	0.0128	1.0
	Copper, lapped	0.0147	1.0
	Copper, scored	0.0068	1.0
	Stainless steel, ground and polished	0.008	1.0
	Stainless steel, Teflon pitted	0.0058	1.0
	Stainless steel, chemically etched	0.0133	1.0
	Stainless steel, mechanically polished	0.0132	1.0
	Nickel	0.006	1.0
	Platinum	0.013	1.0

where

$$\Delta T = T_s - T_{sat}$$

i_{fg} = enthalpy of vaporization of the liquid

Pr = Prandtl number of the liquid

n = constant (generally equal to 1.7; for water, equal to 1)

C_{sf} = constant dependent on the combination of the liquid and the material of the surface and its surface finish

c_p = specific heat of the liquid

μ = dynamic viscosity of the liquid

ρ = liquid density

σ = surface tension of the liquid

ρ_v = vapor density

All properties refer to the liquid unless it has a subscript v to denote a property of the vapor. Values of n and C_{sf}, suggested by Rohsenow (1952) and Vachon et al. (1968), for some of the liquid-surface combinations are given in Table 6.1.1.

It should be emphasized that the values of C_{sf} are sensitive to surface finish. Hence, depending on the surface finish, we can expect significant deviations from

---

TABLE 6.1.2 Definition of L* and Values of C for Some Configurations for Use in Equation 6.1.3

L = characteristic geometric length

$$\text{Characteristic length} = \left[\frac{\sigma}{g(\rho - \rho_v)} \right]^{0.5}$$

$$L^* = \text{dimensionless length} = L \left[\frac{g(\rho - \rho_v)}{\sigma} \right]^{0.5}$$

where σ = surface tension ρ = liquid density ρ_v = vapor density

Configuration	C	Characteristic Dimensional Length (L)	Range of L^*
1. Infinite long, horizontal flat heater	0.149	Width	$L^* \geq 27$
2. Small horizontal heater[1]	$0.159\lambda_d^2/A_{heater}$	Width	$9 < L^* < 20$
3. Horizontal cylinder	0.118	Radius	$L^* > 1.2$
4. Horizontal cylinder	$0.123/(L^*)^{0.25}$	Radius	$0.15 \leq L^* \leq 1.2$
5. Horizontal ribbon, oriented vertically	$0.155/(L^*)^{0.25}$	Height	$0.15 \leq L^* \leq 2.96$
6. Horizontal ribbon, oriented vertically, one side insulated	$0.183/(L^*)^{0.25}$	Height	$0.15 \leq L^* < 5.86$
7. Sphere	0.11	Radius	$L^* > 4.26$
8. Sphere	$0.227/(L^*)^{0.5}$	Radius	$0.15 < L^* < 4.26$
9. Small slender cylinder of any cross section[2]	$0.183/P^{0.25}$	Perimeter	$0.15 \leq L^* < 5.86$

[1] $\lambda_d = 10.9 \sqrt{\dfrac{\sigma}{g(\rho - \rho_v)}}$

[2] P = transverse perimeter

the predicted values. Because of the dependence of q'' on ΔT^3 in Equation 6.1.2b, the uncertainty in the predicted heat flux for a given value of ΔT may be as high as 100%; but the uncertainty in ΔT for a given value of q'' is much lower—approximately 25%. In boilers the operating value of the heat flux, in general, is significantly less than the CHF. The value of ΔT corresponding to the peak heat flux is not high, and the actual operating temperature is not a major concern if the heat flux does not exceed the CHF. The surface tensions of some of the fluids are given in Table 6.1.3.

Critical Heat Flux Based on dimensional arguments, Kutateladze (1963) derived the following relation for the critical heat flux for boiling from an infinite horizontal plate.

Critical heat flux correlation

$$\text{CHF} = q''_{\text{max}} = C\rho_v^{0.5}i_{\text{fg}}[\sigma g(\rho - \rho_v)]^{0.25} \qquad (6.1.3)$$

From the experimental measurements of several workers, Kutateladze demonstrated that in Equation 6.1.3 the average value of C is 0.13 for a horizontal surface derived from the experimental measurements is in agreement with the analytical result of $[\pi/(128 \times 3^{0.5})]^{0.5} < C < (\pi/128)^{0.5}$ by Zuber (1959a). From their own and others'

TABLE 6.1.3 Surface Tension of Some Fluids (Liquid-Vapor), $\sigma = a - bT$

Fluid	T	$a \times 10^4$ (N/m)	$b \times 10^4$ (N/m °C)
Ammonia	−75 °C−−40 °C	264	−2.228
Acetone	25 °C–50 °C	262.6	1.12
Benzene	10 °C–80 °C	314.5	1.291
Butane	−70 °C−−20 °C	148.7	1.206
Butyl alcohol	10 °C–100 °C	271.8	0.898
Carbon dioxide	−30 °C–20 °C	43.4	1.6
Carbon tetrachloride	15 °C–105 °C	294.9	1.224
Ethyl alcohol	10 °C–70 °C	240.5	0.832
Ethylene glycol	20 °C–140 °C	502.1	0.890
Isopropyl alcohol	10 °C–80 °C	229.0	0.789
Mercury	5 °C–200 °C	4906	2.049
Methyl alcohol	10 °C–60 °C	240.0	0.773
Nitrogen	77 K–90 K	264.2	2.265 (T in K)
Octane	10 °C–120 °C	235.2	0.951
Oxygen	−202 °C−−184 °C	−337.2	−2.56
Pentane	10 °C–30 °C	182.5	1.102
Propyl alcohol	10 °C–90 °C	252.6	0.777
Propane	−90 °C–10 °C	92.2	0.874
Toluene	10 °C–100 °C	309.0	1.189
Water	10 °C–100 °C	758.3	1.477
Water	100 °C–350 °C	822.4	2.246

Surface tension of methane: 0.01888 N/m at 90 K and 0.01237 N/m at 115 K

(a) For all fluids except water, values of a and b are found by linear regression from the values given in Jasper (1972).
(b) For water, values of a and b are found by linear regression from the values given in Kakac et al. (1987).

experimental measurements, the value of C has been obtained for different geometries by Sun and Lienhard (1970), Ded and Lienhard (1972), Lienhard and Dhir (1973), and Lienhard et al. (1973). The constant is a function of the characteristic geometric dimension, L, of the heating surface. A dimensionless length L^* is formed by dividing the characteristic geometric dimension by the characteristic length $\{\sigma/[g(\rho - \rho_v)]\}^{0.5}$. The results are summarized in Table 6.1.2. Surface tensions of some fluids, approximated as linear functions of the temperature by Jasper (1972), are given in Table 6.1.3. Equation 6.1.3 in combination with Table 6.1.2 is expected to yield peak heat flux values within an uncertainty of ±20%.

EXAMPLE 6.1.1

A horizontal copper electric heater is 30-cm long and 1-cm in diameter as shown in Figure 6.1.5. Determine

(a) The maximum power input to the heater for nucleate pool boiling in saturated water at 100 °C.

(b) The heat flux corresponding to a $\Delta T = 10$ °C.

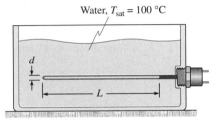

Water, $T_{sat} = 100$ °C

FIGURE 6.1.5 Horizontal Electric Heater Immersed in Water

Given

Fluid: Water $T_{sat} = 100$ °C
$L = 30$ cm $d = 1$ cm

Find

(a) CHF
(b) q'' for a $\Delta T = 10$ °C

SOLUTION

(a) The maximum power input corresponds to the maximum heat transfer rate from the heater to the water in nucleate boiling. The maximum heat transfer rate occurs when the heat flux corresponds to the CHF. Employing Equation 6.1.3

$$q''_{max} = C\rho_v^{0.5}i_{fg}[g(\rho - \rho_v)\sigma]^{0.25}$$

To use this equation, we need the value of C, which requires the value of L^*. From Table 6.1.2,

$$L^* = R\sqrt{g(\rho - \rho_v)/\sigma}$$

where R is 0.005 m.

From Tables A5 and A6, properties of saturated water and vapor at 100 °C are

$\rho = 958.3$ kg/m^3 $\rho_v = 0.5977$ kg/m^3 $i_{fg} = 2.257 \times 10^6$ J/kg

From Table 6.1.3, for water

$$\sigma = a - bT = 0.07583 - 1.477 \times 10^{-4} \times 100 = 0.06106 \text{ N/m}$$

$$\text{Characteristic length} = \left[\frac{\sigma}{g(\rho - \rho_v)} \right]^{1/2} = \left[\frac{0.06106}{9.807(958.3 - 0.5977)} \right]^{1/2}$$

$$= 0.00255 \text{ m}$$

$$L^* = \frac{\text{Radius}}{\text{Characteristic length}} = \frac{0.005}{0.00255} = 1.961$$

For a horizontal cylinder, for $L^* = 1.961$, from Table 6.1.2, $C = 0.118$. Employing Equation 6.1.3,

$$q''_{max} = 0.118 \times 0.5977^{0.5} \times 2.257 \times 10^6 [0.06106 \times 9.807(958.3 - 0.5977)]^{0.25}$$

$$= 1.01 \times 10^6 \text{ W/m}^2$$

$$\text{Maximum power input} = q''_{max} \pi \, dL = 1.01 \times 10^6 \times \pi \times 0.01 \times 0.3$$
$$= \underline{9520 \text{ W}}$$

Keeping in mind that the peak heat flux computed from Equation 6.1.3 is subject to an uncertainty of $\pm 20\%$, it may be prudent to limit the maximum power input to the heater to 80% of the calculated value, $0.8 \times 9520 \cong \underline{7600 \text{ W}}$.

(b) heat transfer rate corresponding to a $\Delta T = 10 \,°\text{C}$ can be determined through Equation 6.1.2b in conjunction with Table 6.1.1.

$$q'' = \mu i_{fg} \left[\frac{g(\rho - \rho_v)}{\sigma} \right]^{0.5} \left[\frac{c_p \, \Delta T}{i_{fg} \, \text{Pr}^a \, C_{sf}} \right]^3$$

From Table A5, for saturated water at $100 \,°\text{C}$,

$$c_p = 4211 \text{ J/kg K} \qquad \mu = 276 \times 10^{-6} \text{ N s/m}^2 \qquad \text{Pr} = 1.71$$

From Table 6.1.1, $n = 1$. A conservative value for q'' is obtained by employing the value of C_{sf} for polished copper; $C_{sf} = 0.0128$. Substituting these values

$$q'' = 276 \times 10^{-6} \times 2.257 \times 10^6 \left[\frac{9.807(958.3 - 0.5977)}{0.06106} \right]^{1/2}$$

$$\times \left[\frac{4211 \times 10}{2.257 \times 10^6 \times 1.71 \times 0.0128} \right]^3 = 1.513 \times 10^5 \text{ W/m}^2$$

$$\text{Power} = q'' \pi \, dL = 1.513 \times 10^5 \times \pi \times 0.01 \times 0.3 = \underline{1426 \text{ W}}$$

EXAMPLE 6.1.2

The heater for a steam boiler to produce saturated steam at $170 \,°\text{C}$ is made of an electrical heating element inside a 15-mm O.D. mechanically polished stainless steel tube (Figure 6.1.6). The power input to the heater is 5 kW. If the surface temperature of the heater is not to exceed $175 \,°\text{C}$, find the length of the heater.

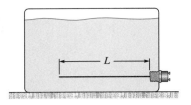

FIGURE 6.1.6 Electric Heating Element Inside a Boiler

Given

Fluid: Water   Mechanical polished stainless steel
$T_{\text{sat}} = 170\ \text{°C}$ $d = 15\ \text{mm}$
$P = 5\ \text{kW} = q$ $\Delta T = 5\ \text{°C}$

Find

L (length of heating element)

ASSUMPTION

Pool boiling

SOLUTION

The length of the heater is obtained from

$$q = q'' \pi d L$$

where

q = heat transfer rate from the heater = power input
q'' = heat flux
d = diameter of the heating element
L = length of the heating element

q'' is obtained from Equation 6.1.2b.
From Tables A5 and A6 properties of saturated water and steam at 170 °C are

$\rho = 897.3\ \text{kg/m}^3$ $c_p = 4340\ \text{J/kg °C}$ $k = 0.681\ \text{W/m °C}$
$\mu = 157.6 \times 10^{-6}\ \text{N s/m}^2$ $\text{Pr} = 1.0$
$i_{\text{fg}} = 2.0495 \times 10^6\ \text{J/kg}$ $\rho_v = 4.119\ \text{kg/m}^3$

From Table 6.1.3 the surface tension of water at 170 °C is

$$\sigma = 822.4 \times 10^{-4} - 2.246 \times 10^{-4} \times 170 = 0.04406\ \text{N/m}$$

From Table 6.1.1 for mechanically polished stainless steel $C_{\text{sf}} = 0.0132$ and $n = 1.0$. With $\Delta T = 5\ \text{°C}$, from Equation 6.1.2b

$$q'' = 157.6 \times 10^{-6} \times 2.0495 \times 10^6 \left[\frac{9.807 \times (897.3 - 4.119)}{0.04406} \right]^{1/2}$$

$$\times \left(\frac{4340 \times 5}{2.0495 \times 10^6 \times 1 \times 0.0132} \right)^3 = 74\,325\ \text{W/m}^2$$

$$L = \frac{q}{q'' \pi\, d} = \frac{5000}{74\,325 \times \pi \times 0.015} = \underline{1.43\ \text{m}}$$

Check if the heater operates in the nucleate boiling region. Find CHF from Equation 6.1.3. If the heat flux is significantly less than the CHF (say less than 80% of the CHF), it is reasonable to assume nucleate boiling conditions.

Referring to Table 6.1.2, the characteristic length is given by

$$L = \left[\frac{\sigma}{g(\rho - \rho_v)}\right]^{0.5} = \left[\frac{0.04406}{9.807(897.3 - 4.119)}\right]^{0.5} = 0.002243 \text{ m}$$

$$L^* = \frac{0.015/2}{0.002243} = 3.344$$

For a horizontal cylinder with $L^* = 3.344$, C in Equation 6.1.3 is 0.118. From Equation 6.1.3

$$\text{CHF} = 0.118 \times 4.119^{0.5} \times 2.0495 \times 10^6[0.04406 \times 9.807(897.3 - 4.119)]^{0.25}$$
$$= 2.175 \times 10^6 \text{ W/m}^2$$

As the heat flux is only 3.4% of the CHF, nucleate boiling is assured.

Film Boiling From an analysis similar to the analysis of film condensation on vertical surfaces (Example 5.3.3) the following expression for film boiling (region DEF in Figure 6.1.2) on horizontal cylinders and spheres is employed.

$$\frac{h\,d}{k_v} = c\left[\frac{\rho_v(\rho - \rho_v)gi'_{fg}\,d^3}{k_v\mu_v(T_s - T_{sat})}\right]^{1/4} \tag{6.1.4}$$

Film boiling correlations

The value of the constant c in Equation 6.1.4 is 0.62 for cylinders [Bromley (1950)] and 0.67 for spheres [Dhir and Lienhard (1971)]. The liquid properties are evaluated at T_{sat} and the vapor properties at $(T_s + T_{sat})/2$. i'_{fg} is the corrected enthalpy of vaporization to take into account the superheating of the vapor above the saturation temperature. The suggested correction for i_{fg} is

$$i'_{fg} = i_{fg} + 0.4c_{pv}\,\Delta T \tag{6.1.5}$$

Equation 6.1.4 is recast as

$$\frac{h\,d}{k_v} = c(\text{Ra}')^{1/4} \tag{6.1.6}$$

where Ra' represents the modified Rayleigh number,

$$\text{Ra}' = \frac{\rho_v(\rho - \rho_v)gi'_{fg}\,d^3}{\mu_v k_v\,\Delta T}$$

For film boiling on spheres with cryogenic fluids, Frederking and Clark (1963) propose

$$\frac{h\,d}{k_v} = 0.14\,\text{Ra}'^{1/3} \tag{6.1.7}$$

Experimental results of Lewis et al. (1965) for spheres and Suryanarayana and Merte (1970) for vertical surfaces (height replacing the diameter in the definition of Ra') with liquid nitrogen and liquid hydrogen are in good agreement with Equation 6.1.7. From the definition of Ra′ and Equation 6.1.7, it is evident that the film boiling heat transfer coefficients of both spheres and vertical surfaces are independent of the linear dimension. This conclusion is supported by Kalinin et al. (1975), who conclude that the film boiling heat transfer coefficient is independent of the linear dimensions if the characteristic dimension in the vertical plane is greater than about 15 mm.

For film boiling from large (infinite) horizontal surfaces, Berenson (1961) arrived at

$$\frac{hL^*}{k_v} = 0.425 \left[\frac{i'_{fg}\rho_v g (\rho - \rho_v)L^{*3}}{k_v \mu_v \, \Delta T} \right]^{1/4} \tag{6.1.8}$$

where

$$L^* = \left[\frac{\sigma}{g(\rho - \rho_v)} \right]^{1/2}$$

With many liquids, such as water, the surface temperatures in the film boiling region are quite high and radiative heat transfer is significant. In natural convection in gases, radiative heat transfer rate (when one surface is completely enclosed by a much larger second surface) computed from Equation 1.5.2 can be added to the natural convective heat transfer rate to get the total heat transfer rate. However, it is more complicated in film boiling; radiative heat transfer from the surface to the liquid generates vapor, making the vapor layer thicker than what it would have been with natural convection alone. The thicker vapor layer results in reduced convective heat transfer. To account for this interaction between radiative and convective heat transfer, Bromley (1950) suggested

$$q'' = q''_{con} + 0.75 q''_{rad} \tag{6.1.9}$$

where q''_{con} is $h(T_s - T_{sat})$ and q''_{rad} is $\sigma\varepsilon(T_s^4 - T_{sat}^4)$, and h is found from Equations 6.1.4, 6.1.7, or 6.1.8. The radiative correction term is generally not significant in film boiling in cryogenic fluids.

Another complication in film boiling is the presence of interfacial waves. Except for thin cylinders or vertical surfaces with very small heights, film boiling is accompanied by large amplitude interfacial waves. The interfacial waves increase the heat transfer rate significantly. Hence, equations based on a smooth vapor-liquid interface, such as Equation 6.1.4, are likely to underestimate the heat transfer rates. Unfortunately, there is yet no simple, satisfactory way to account for the effect of the interfacial waves.

Minimum Heat Flux The expression for the minimum heat flux in the film boiling region, point D in Figure 6.1.2, derived by Zuber (1959b) is

$$q''_{min} = \text{const. } \rho_v i'_{fg} \left[\frac{\sigma g (\rho - \rho_v)}{(\rho + \rho_v)^2} \right]^{1/4}$$

where

$$i'_{fg} = i_{fg} + 0.4 c_{pv} \, \Delta T$$

From his analysis for infinite horizontal surfaces Zuber determined the constant to be $\pi/24$. Berenson (1961) suggested a value of 0.09 for the constant and the correlation for the minimum heat flux for infinite horizontal surfaces in the film boiling region is

$$q''_{min} = 0.09\rho_v i'_{fg} \left[\frac{g\sigma(\rho - \rho_v)}{(\rho + \rho_v)^2} \right]^{1/4} \quad \text{(6.1.10)}$$

Combining Equation 6.1.8 and Equation 6.1.10, we can solve for ΔT corresponding to q''_{min}. The result is

$$\Delta T_{min} = 0.127 \frac{\rho_v i'_{fg}}{k_v} \left[\frac{g(\rho - \rho_v)\mu_v L^{*3}}{(\rho + \rho_v)^2} \right]^{1/3} \quad \text{(6.1.11)}$$

where

$$L^* = \left[\frac{\sigma}{g(\rho - \rho_v)} \right]^{1/2}$$

For a horizontal wire of radius R, Lienhard and Wong (1964) suggest

$$q''_{min} = 0.0464\rho_v i'_{fg} \left[\frac{g\sigma(\rho - \rho_v)}{(\rho + \rho_v)^2} \right]^{1/4} \left[\frac{18}{R'^2(2R'^2 + 1)} \right]^{1/4} \quad \text{(6.1.12)}$$

where

$$R' = R \left[\frac{g(\rho - \rho_v)}{\sigma} \right]^{1/2}$$

EXAMPLE 6.1.3 Long, 3-cm diameter, plain carbon-steel cylindrical rods at 300 °C are rapidly cooled by immersing them (one at a time) horizontally in a water bath as shown in Figure 6.1.7. Determine

(a) The heat flux when the surface temperature of the cylinders is 300 °C.
(b) The minimum heat flux in the film boiling region and the temperature at which it occurs.
(c) The maximum heat flux.

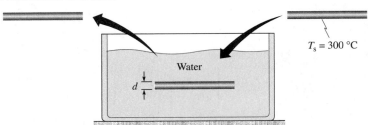

FIGURE 6.1.7 Cooling Horizontal Cylindrical Rods

Given

Fluid: Water at atmospheric pressure $d = 3$ cm
Material: Plain carbon-steel

Find

(a) q'' when $T_s = 300\,°C$
(b) q''_{min} and T_s
(c) CHF

ASSUMPTIONS

1. After quenching the first few pieces, the water temperature rises to the saturation temperature at atmospheric pressure (approximately 101.35 kPa), 100 °C
2. Emissivity of the rolled steel, from Table A9 is 0.66.

SOLUTION

(a) To find the heat flux at 300 °C, we first determine whether nucleate or film boiling occurs at 300 °C by finding the temperature corresponding to the minimum heat flux in the film boiling region. As the diameter of the rods, 30 mm, is greater than 15 mm above which the heat flux is insensitive to the linear dimension, we use Equations 6.1.10 and 6.1.11 to estimate q''_{min} and ΔT_{min}. Thus, we start by solving part b of the problem

(b) To use Equation 6.1.10, we need the properties of the vapor at $(T_s + T_{sat})/2$. But T_s is not yet available. We assume a value of T_s, determine ΔT_{min}, and verify if the assumed value of T_s is correct.

For the purpose of evaluating properties of the vapor, we assume $\Delta T = 80\,°C$, i.e., $T_s = 180\,°C$.

$$(T_s + T_{sat})/2 = (180 + 100)/2 = 140\,°C$$

From Table A6, properties of vapor at 140 °C (density from steam tables), are

$\rho_v = 0.5296\ \text{kg/m}^3$ $i_{fg} = 2.257 \times 10^6\ \text{J/kg}$
$c_{pv} = 2243\ \text{J/kg °C}$ $k_v = 0.0258\ \text{W/m °C}$
$\mu_v = 13.54 \times 10^{-6}\ \text{N s/m}^2$ $Pr_v = 1.177$

$$i'_{fg} = i_{fg} + 0.4 c_{pv}\,\Delta T = 2.257 \times 10^6 + 0.4 \times 2243 \times 80$$
$$= 2.329 \times 10^6\ \text{J/kg °C}$$

From Table 6.1.3, surface tension of water at 100 °C is

$\sigma = 0.07583 - 1.477 \times 10^{-4} \times 100 = 0.06106\ \text{N/m}$
$\rho = 958.3\ \text{kg/m}^3$

From Equation 6.1.11,

$$L^* = \left[\frac{\sigma}{g(\rho - \rho_v)}\right]^{0.5} = \left[\frac{0.06106}{9.807 \times (958.3 - 0.5296)}\right]^{0.5} = 0.00255\ \text{m}$$

$$\Delta T_{min} = \frac{0.127 \times 0.5296 \times 2.329 \times 10^6}{0.0258}$$

$$\times \left[\frac{9.807(958.3 - 0.5296)13.54 \times 10^{-6} \times 0.00255^3}{(958.3 + 0.5296)^2}\right]^{1/3}$$

$$\Delta T_{min} = 80\,°C$$

$$T_{s,min} = \underline{180\,°C}$$

The computed value of ΔT (= 80 °C) is also the assumed value for determining the properties and no iteration is required.

From Equation 6.1.10,

$$q''_{min} = 0.09 \times 0.5296 \times 2.329 \times 10^6 \left[\frac{9.807 \times 0.06106(958.3 - 0.5296)}{(958.3 + 0.5296)^2} \right]^{1/4}$$

$$= \underline{17\ 544\ \text{W/m}^2}$$

(a) Now, we know that at 300 °C film boiling conditions exist. We use Equation 6.1.4 to compute the heat flux at 300 °C. From Table A6, properties of vapor at (100 + 300)/2 = 200 °C (density from steam tables at 200 °C, 101.35 kPa) are

$$\rho_v = 0.4156\ \text{kg/m}^3 \qquad c_{pv} = 2837\ \text{J/kg °C}$$
$$k_v = 0.0317\ \text{W/m °C} \qquad \mu_v = 15.65 \times 10^{-6}\ \text{N s/m}^2$$

Values of ρ and σ at 100 °C have been found in part b.

$$i'_{fg} = 2.257 \times 10^6 + 0.4 \times 2837 \times 200 = 2.484 \times 10^6\ \text{J/kg}$$

$$\text{Ra}' = \frac{\rho_v(\rho - \rho_v)gi'_{fg}\ d^3}{\mu_v k_v\ \Delta T}$$

$$= \frac{0.4156(958.3 - 0.4156)9.807 \times 2.484 \times 10^6 \times 0.03^3}{15.65 \times 10^{-6} \times 0.0317 \times 200} = 2.639 \times 10^9$$

$$\text{Nu}_d = 0.62(2.639 \times 10^9)^{1/4} = 140.5$$

$$h = \frac{140.5 \times 0.0317}{0.03} = 148.5\ \text{W/m}^2\ \text{°C}$$

From Equation 6.1.9,

$$q'' = q''_{con} + 0.75q''_{rad}$$

$$= 148.5(300 - 100) + 0.75 \times 5.67 \times 10^{-8} \times 0.66(573.15^4 - 373.15^4)$$

$$= \underline{32\ 185\ \text{W/m}^2}$$

(c) q''_{max} corresponds to the CHF. CHF is found from Equation 6.1.3 in conjunction with Table 6.1.2. From Table 6.1.2, ρ_v (100 °C, 100 kPa) = 0.5977 kg/m³. From Table 6.1.2, for a horizontal cylinder

$$L^* = R \left[\frac{g(\rho - \rho_v)}{\sigma} \right]^{0.5} = 0.015 \left(\frac{9.807(958.3 - 0.5977)}{0.06106} \right)^{0.5} = 5.88$$

From Table 6.1.2, for a horizontal cylinder with $L^* = 5.88$, $C = 0.118$.

$$q''_{max} = 0.118 \times 0.5977^{0.5} \times 2.257 \times 10^6[0.06106 \times 9.807(958.3 - 0.5977)]^{1/4}$$
$$= \underline{1.008 \times 10^6\ \text{W/m}^2}$$

COMMENT

Cooling of the rods starts with film boiling with decreasing heat flux until the minimum heat flux is reached. Thereafter, there is an increase in the heat flux, in the transition regime, until the CHF is reached. Nucleate boiling is established and the heat flux then reduces, ending with natural convection in a single-phase liquid.

Incipience of
boiling correlation

Incipience of Boiling The temperature at which boiling is initiated depends on the heat flux imposed on the surface. A simple, general correlation relating the excess temperture ΔT ($= T_s - T_{sat}$) to the incipience of boiling at a defined heat flux is not available. But, for water, Rohsenow (1973) suggests the following dimensional correlation:

$$q_i'' = 5.28 p^{1.156}(1.8\ \Delta T)^{2.406/p^{0.00234}} \tag{6.1.13}$$

where

q_i'' = imposed heat flux (W/m^2)
p = pressure (kPa)
$\Delta T = T_s - T_{sat}$ (°C) at which boiling is initiated

In pool boiling, for a given ΔT, if q_{NC}'' for natural convection is greater than q_i'' computed from Equation 6.1.13, boiling does not occur. However, if $q_i'' > q_{NC}''$ boiling occurs.

Forced convection
boiling
computation

Forced Convection Boiling As shown in Figure 6.1.4, different regimes of boiling occur in forced convection boiling. The heat transfer coefficient depends on the boiling regime and, hence, has different values in different parts of the tube. Although different regimes are shown in the figure, the exact definition of each regime is somewhat subjective and the criteria for determining the different regimes are not well established. Because of these complications, an exact determination of the heat flux in forced convection boiling is quite involved. But for low quality boiling of water (quality less than 10%) Rohsenow (1985) suggests that an estimate of the heat flux can be obtained by one of the following relations (both results follow experimental results reasonably well).

$$q'' = (q_{FC}''^2 + q_B''^2 - q_{Bi}''^2)^{1/2} \tag{6.1.14a}$$

$$q'' = q_{FC}'' + q_B'' - q_{Bi}'' \tag{6.1.14b}$$

where

q'' = heat flux
q_{FC}'' = forced convection heat flux computed from Equation 4.3.2b replacing 0.023 by 0.019
q_B'' = heat flux determined from pool boiling experiments or from Equation 6.1.2b
q_{Bi}'' = boiling heat flux found from a plot of q'' versus ΔT from the following procedure

On a log-log paper plot q_{FC}'' for the given velocity of water, curve A in Figure 6.1.8. Plot the pool boiling curve, B, either from experimental measurements or computed from Equation 6.1.2b. Find the point of intersection I, of curve A and Equation 6.1.13. Directly below point I, locate q_{Bi}'' on the boiling curve. Then, for the given velocity of the water in the tube, Equation 6.1.14a or b gives the relation between q'' and ΔT. To the left of I, there is only forced convection and no boiling. Boiling occurs only if ΔT is to the right of I.

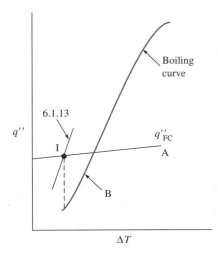

FIGURE 6.1.8 **Procedure for Finding q_{Bi}''**

Quite often, $q_{Bi}'' \ll q_B''$, and q_{Bi}'' can be neglected in Equation 6.1.14 giving much simpler correlations:

$$q'' = (q_{FC}''^2 + q_B''^2)^{1/2} \qquad \text{(6.1.14c)}$$

$$q'' = q_{FC}'' + q_B'' \qquad \text{(6.1.14d)}$$

When the quality of the steam exceeds about 10%, annular flow occurs. For correlations in the annular flow regime, and for a more general correlation for forced convection, refer to Rohsenow (1985) and the literature cited therein.

6.2 CONDENSATION

Condensation is also a common engineering occurrence. It occurs in such applications as condensers and feedwater heaters in steam power plants, condensers in refrigeration units, and in many chemical plants. Condensation can occur in different ways. If the vapor pressure is higher than the saturation pressure, condensate liquid drops, suspended in the vapor, are formed. For example, if saturated vapor is expanded in an insulated nozzle, the temperature falls below the saturation temperature at exit of the nozzle and liquid drops form throughout the vapor. Such condensation is known as bulk or homogeneous condensation. Another type of condensation, surface condensation, occurs when the vapor comes in contact with a surface at a temperature below the saturation temperature of the vapor.

Filmwise and dropwise condensation

Just as there are two modes of boiling—nucleate and film boiling—there are two modes of condensation—*filmwise* and *dropwise,* as shown in Figure 6.2.1. But, unlike in boiling, the mode of condensation is dependent on the surface characteristics and not on the temperature difference. If the liquid wets a surface—when a drop of the liquid on the surface tends to spread to the surface—film condensation takes place. In film condensation, the cooled surface is completely covered with a

FIGURE 6.2.1 Photograph Showing Film Condensation (right) and Dropwise Condensation Obtained by a Special Surface Coating (left)
Photograph courtesy of J.W. Westwater.

film of the condensate. On the other hand, if the liquid does not wet the surface (a drop on the surface remains as a drop and does not spread) drops of condensate form on the surface and roll off; such condensation is known as dropwise condensation. Dropwise condensation leads to much higher values of the heat transfer coefficient than filmwise condensation. But, to maintain dropwise condensation, special surface preparation, which is generally expensive, is required. In many cases, even with such treatment, dropwise condensation is not sustained for extended periods. It ultimately reverts to some form of filmwise condensation. Although it is easy to obtain filmwise condensation with wetting fluids, such as some refrigerants in the Freon family, it may not be that common with water. Quite often, condensation of steam on a surface is neither in the form of drops nor a film but in the form of rivulets closer to filmwise condensation. This form of condensation leads to heat transfer rates higher than filmwise condensation but significantly lower than dropwise condensation. With condensation of pure steam for extended periods, filmwise condensation results.

6.2.1 Film Condensation Correlations

In a classic analysis in 1916, Nusselt (in whose honor we have the Nusselt number in convection) analyzed film condensation on a vertical surface at a uniform temperature, surrounded by a stagnant, pure, single-phase, saturated vapor. The salient points of his model and analysis are explained in Example 5.3.3. Here, we give the results of his analysis. Assuming a parabolic velocity profile with zero interfacial shear stress and linear temperature profile for the condensate (Figure 6.2.2), his analysis yields

$$u = \frac{\rho - \rho_v}{\mu} g \, \delta^2 \left(\frac{y}{\delta} - \frac{y^2}{2\delta^2} \right) \tag{6.2.1}$$

where u is local velocity.

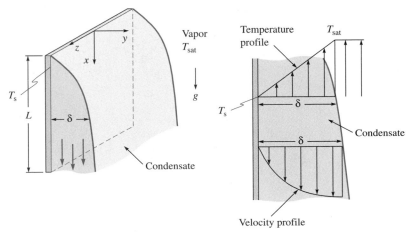

FIGURE 6.2.2 A Vertical Surface at a Temperature Less Than the Surrounding Vapor Temperature.
The vapor condenses on the surface and the condensate flows down the plate with increasing thickness

Denoting the mass rate of condensate per unit width (along the z-direction) of the condensing surface by Γ

$$\Gamma = \rho v \delta = \int_0^\delta \rho u \, dy = \frac{g \rho (\rho - \rho_v) \, \delta^3}{3\mu} \tag{6.2.2}$$

where v is average condensate velocity,

$$\delta = \left[\frac{4k \, \Delta T \mu x}{\rho (\rho - \rho_v) g i_{fg}} \right]^{1/4} \tag{6.2.3}$$

where

$$\Delta T = T_{sat} - T_s$$
$$\rho_v = \text{density of the vapor}$$

All other properties are those of the condensate.

In most cases δ/L is much less than 1. With the assumption of linear temperature profile, the condensation heat transfer rate to the wall per unit area, the heat transfer coefficient, and the local Nusselt number are given by

$$q'' = \frac{k \, \Delta T}{\delta} \qquad h_x = \frac{q''}{\Delta T} = \frac{k}{\delta} \qquad Nu_x = \frac{h_x x}{k} = \frac{x}{\delta}$$

Filmwise condensation correlation— vertical plates

Wilhelm Nusselt (1882–1957) obtained his doctoral degree from the Technisce Hochshule Munchen with a dissertation on the thermal conductivity of insulating materials. Nusselt was an engineer who used his considerable mathematical prowess to formulate and obtain experimental results. His seminal paper on "The fundamental law of heat transfer" may be said to form the foundation for the science of heat transfer. His paper on condensation was published in 1916. His use of dimensionless parameters prior to the publication of the concepts of Buckingham and Rayleigh on dimensional analysis indicates that Nusselt independently formulated his ideas on dimensional analysis. He also contributed in combustion and mass transfer.

With δ given by Equation 6.2.3, we have

$$\text{Nu}_x = 0.707 \left[\frac{\rho(\rho - \rho_v)g i_{\text{fg}} x^3}{\Delta T \mu k} \right]^{1/4} \tag{6.2.4}$$

In arriving at Equation 6.2.4, it was assumed that all the heat transfer to the wall results in condensation of the saturated vapor to saturated liquid; this is equivalent to assuming that all the condensate is at the saturation temperature. Obviously, this is not the case, as with a linear temperature profile, the temperature of the condensate adjacent to the wall is equal to the wall temperature T_s, which is less than the saturation temperature T_{sat}. A part of the heat transfer to the wall results in condensation of the vapor, and a much smaller part in cooling the condensate to some temperature between T_s and T_{sat}. This leads to a condensation rate that is slightly less than that obtained by assuming that all the heat transfer results in condensation only. The decrease in the condensation rate can be accounted for by modifying the value of i_{fg}. The heat transfer rate resulting in subcooling of the condensate should be a function of ΔT and the specific heat of the condensate. Rohsenow (1956) showed that the heat transfer due to the cooling of the liquid below the saturation temperature is $0.68 c_p \, \Delta T$. Thus, the total heat transfer per unit mass of the condensate is $i_{\text{fg}} + 0.68 c_p \, \Delta T$. Therefore, to reflect the subcooling of the condensate in the analysis, i_{fg} should be modified to

$$i'_{\text{fg}} = i_{\text{fg}} + 0.68 c_p \, \Delta T \tag{6.2.5}$$

With this modified enthalpy of vaporization, Equation 6.2.2 becomes

$$\text{Nu}_x = 0.707 \left[\frac{\rho(\rho - \rho_v)g i'_{fg} x^3}{\Delta T \mu k} \right]^{1/4} \qquad (6.2.6)$$

Equation 6.2.4 gives the local heat transfer coefficient. The average heat transfer coefficient over a length L from the leading edge can be obtained from the relation, $h = q_w/(\Delta TWL)$, where W is the width of the condensing surface.

$$q_w = \int_0^L q''_w W \, dx = \int_0^L h \, \Delta TW \, dx = \int_0^L \Delta TW \left[\frac{\rho(\rho - \rho_v)g i'_{fg}}{4\Delta T \mu k} \right]^{1/4} \frac{k}{x^{1/4}} \, dx$$

$$q_w = W \, \Delta T \frac{4}{3} \left[\frac{\rho(\rho - \rho_v)g i'_{fg}}{4\Delta T \mu k} \right]^{1/4} k L^{3/4}$$

$$h = q_w/(W \, \Delta TL) = \frac{4}{3} \left[\frac{\rho(\rho - \rho_v)g i'_{fg}}{4\Delta T \mu k L} \right]^{1/4} k = \frac{4}{3} \frac{k}{\delta_L}$$

δ_L is the condensate film thickness at $x = L$. The Nusselt number based on the average heat transfer coefficient is

$$\text{Nu}_L = \frac{hL}{k} = 0.943 \left[\frac{\rho(\rho - \rho_v)g i'_{fg} L^3}{\Delta T \mu k} \right]^{1/4} \qquad (6.2.7a)$$

An alternate form of Equation 6.2.7a in terms of the Reynolds number is obtained from Equation 6.2.2.

$$\text{Re} = \frac{\rho v \delta}{\mu} = \frac{\Gamma}{\mu} = \frac{g \rho(\rho - \rho_v) \delta_L^3}{3\mu^2}$$

$$h = \frac{4}{3} \frac{k}{\delta_L} = \frac{4}{3} k \left[\frac{g \rho(\rho - \rho_v)}{3\mu^2 \, \text{Re}} \right]^{1/3} \qquad (6.2.7b)$$

$$\frac{h}{k} \left[\frac{\mu^2}{g \rho(\rho - \rho_v)} \right]^{1/3} = 0.924 \, \text{Re}^{-1/3}$$

The dimensionless group $h/k \{ \mu^2/[g \rho(\rho - \rho_v)] \}^{1/3}$ is known as the condensate number and is denoted by Co.

The mass flow rate of the condensate per unit width of the condensing surface, Γ, can also be expressed in terms of the heat transfer rate per unit width and the corrected enthalpy of vaporization.

$$\Gamma = \frac{h \, \Delta TL}{i'_{fg}} \qquad (6.2.8a)$$

$$\text{Re} = \frac{\Gamma}{\mu} = \frac{h \, \Delta TL}{\mu i'_{fg}} \qquad (6.2.8b)$$

In evaluating Equations 6.2.2 through 6.2.8, properties of the condensate are evaluated at the arithmetic mean temperature, $0.5(T_{sat} + T_s)$.

The relations developed for vertical surfaces can be used for inclined surfaces (with the condensation on the upper surface) by replacing g by $g \cos \theta$ (θ = angle of inclination of the condensing surface with the vertical) and vertical cylinders if the radius of the cylinder is much greater than the maximum value of the condensate film thickness, i.e., $R \gg \delta_{max}$.

Effect of interfacial waves

It has been observed that when the Reynolds number ($= \Gamma/\mu$) exceeds a critical value, the interface becomes wavy, but the condensate film is laminar. Kutateladze (1963) computed the critical Reynolds number of several fluids and found that ripples appear if the Reynolds number exceeds about 7.5. When such ripples appear, the heat transfer coefficient increases. Based on experimental results, Kutateladze concluded that when these ripples appear, the heat transfer coefficient can be approximated by

$$h = 0.8 \, \mathrm{Re}^{0.11} \, h_0 \qquad (6.2.9)$$

where

h = average heat transfer coefficient with ripples
h_0 = average heat transfer coefficient computed from Equation 6.2.7a

White (1988) recommends the following correlation attributed to Kutateladze (1963) for $7.5 < \mathrm{Re} < 450$:

$$\frac{h}{k} \left[\frac{\mu^2}{g \, \rho (\rho - \rho_v)} \right]^{1/3} = \frac{\mathrm{Re}}{1.47 \, \mathrm{Re}^{1.22} - 1.3} \qquad (6.2.10)$$

Equation 6.2.9 predicts higher heat transfer coefficients than Equation 6.2.10. In this book we use Equation 6.2.10. Equations 6.2.1 through 6.2.10 are valid if the condensate film is laminar or laminar wavy; they are not valid if the condensate film is turbulent. When $\mathrm{Re} > 450$, effects of turbulence in the condensate film become significant.

EXAMPLE 6.2.1

Dry, saturated steam at 100 °C condenses on a 3-cm O.D., 0.8-m long vertical tube, Figure 6.2.3. The tube surface is at 60 °C. Determine

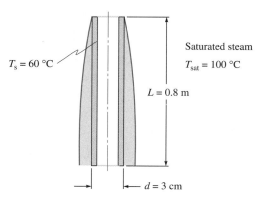

$T_s = 60$ °C

Saturated steam
$T_{sat} = 100$ °C

$L = 0.8$ m

$d = 3$ cm

FIGURE 6.2.3 Condensation on a Vertical Tube

(a) The heat transfer rate.
(b) The mass rate of condensation of steam.
(c) The condensate film thickness at the bottom of the tube.

Given

Fluid: Steam $\qquad L = 0.8$ m $\qquad T_s = 60\,°C$
$d = 3$ cm $\qquad T_{sat} = 100\,°C$

Find

(a) q (heat transfer rate)
(b) $\dot{m}_c$ (mass rate of condensation of steam)
(c) δ (condensate film thickness) at the bottom of the rube

SOLUTION

(a) The total heat transfer rate is found from $q = hA\,\Delta T$. The average convective heat transfer coefficient, h, is found either from Equation 6.2.7a or 6.2.10 depending on the value of Re. Re can be evaluated only after evaluating h; hence h is found by iteration, if necessary.
From Table A6, the properties of steam at $T_{sat} = 100\,°C$ are

$$\rho_v = 0.5977\ \text{kg/m}^3 \qquad i_{fg} = 2.257 \times 10^6\ \text{J/kg}$$

From Table A5, properties of water at $(100 + 60)/2 = 80\,°C$, are

$$\rho = 971.8\ \text{kg/m}^3 \qquad c_p = 4194\ \text{J/kg}\,°C$$
$$k = 0.67\ \text{W/m}\,°C \qquad \mu = 348.1 \times 10^{-6}\ \text{N s/m}^2$$
$$\text{Pr} = 2.18$$

$$\begin{aligned} i'_{fg} &= i_{fg} + 0.68 c_p\,\Delta T \\ &= 2.257 \times 10^6 + 0.68 \times 4194(100 - 60) = 2.371 \times 10^6\ \text{J/kg} \end{aligned}$$

From Equation 6.2.7a,

$$\begin{aligned} h &= 0.943 \left[\frac{\rho(\rho - \rho_v)g i'_{fg} k^3}{\Delta T \mu L} \right]^{1/4} \\ &= 0.943 \left[\frac{971.8(971.8 - 0.5977)9.807 \times 2.371 \times 10^6 \times 0.67^3}{(100 - 60)348.1 \times 10^{-6} \times 0.8} \right]^{1/4} \\ &= 4653\ \text{W/m}^2\,°C \end{aligned}$$

Check if interfacial ripples occur. From Equation 6.2.8b and the definition of Re,

$$\begin{aligned} \text{Re} &= \frac{\Gamma}{\mu} = \frac{h\,\Delta T L}{i'_{fg}\mu} = \frac{h \times 40 \times 0.8}{2.371 \times 10^6 \times 348.1 \times 10^{-6}} \\ &= 0.03877 h = 0.03877 \times 4653 = 180.4 \end{aligned}$$

As Re > 7.5 interfacial waves appear. We can solve for Re by eliminating h between Equation 6.2.8b and 6.2.10. From Equation 6.2.8b,

$$h = \frac{\text{Re}\,\mu i'_{fg}}{\Delta T L}$$

From Equation 6.2.10

$$h = k \left[\frac{g\rho(\rho - \rho_v)}{\mu^2} \right]^{1/3} \frac{\text{Re}}{1.47\,\text{Re}^{1.22} - 1.3}$$

Equating the two expression for h and solving for Re,

$$\text{Re} = \left[\frac{(a\,\Delta TL)/(i'_{fg}\mu) + 1.3}{1.47} \right]^{1/1.22}$$

where

$$a = k \left[\frac{g\rho(\rho - \rho_v)}{\mu^2} \right]^{1/3} = 0.67 \left[\frac{9.807 \times 971.8(971.8 - 0.5977)}{(348.1 \times 10^{-6})^2} \right]^{1/3}$$

$$= 2.843 \times 10^4 \text{ W/m}^2\,°\text{C}$$

Substituting the values,

$$\text{Re} = \left(\frac{(2.843 \times 10^4 \times 40 \times 0.8)/(2.371 \times 10^6 \times 348.1 \times 10^{-6}) + 1.3}{1.47} \right)^{1/1.22}$$

$$= 227.5$$

$$h = \frac{\text{Re}\,\mu i'_{fg}}{\Delta TL} = \frac{227.5 \times 348.1 \times 10^{-6} \times 2.371 \times 10^6}{40 \times 0.8} = 5868 \text{ W/m}^2\,°\text{C}$$

$$q = h\pi\,dL\,\Delta T = 5868 \times \pi \times 0.03 \times 0.8 \times 40 = \underline{1.77 \times 10^4 \text{ W}}$$

(b) Rate of condensation =

$$\dot{m}_c = q/i'_{fg} = \frac{1.77 \times 10^4}{2.371 \times 10^6} = \underline{0.00747 \text{ kg/s } (\approx 27 \text{ kg/h})}$$

(c) Condensate film thickness at the bottom of the tube =

$$h_x = \frac{k}{\delta_x} \qquad \delta_L = \frac{k}{h_{x=L}} \qquad h_{x=L} \approx \frac{3}{4} h = \frac{3}{4} \times 5868 = 4401 \text{ W/m}^2\,°\text{C}$$

$$\delta_L = \frac{0.67}{4401} = 1.52 \times 10^{-4} \text{ m} = 0.152 \text{ mm}$$

$$R = \text{radius of the tube} = 15 \text{ mm} \qquad \frac{\delta_L}{R} = 0.0101$$

As the maximum value of $\delta/R \ll 1$, the use of correlations for vertical surfaces for the vertical cylinder in this case is valid.

COMMENT

In the wavy region δ is not defined and the value found is an acceptable average value.

Filmwise condensation—effect of turbulence

Recall that with flow over a flat plate in forced convection, when the Reynolds number (based on the distance from the leading edge) exceeds about 5×10^5, the boundary layer becomes turbulent. We, therefore, anticipate that when the condensate Reynolds number based on the film thickness exceeds a certain value the condensate film becomes turbulent. Critical Reynolds numbers in the range of 100 to 500 have been quoted by different workers. Here we assume $Re_{cr} = 450$. For $Re > 450$, Labuntsov (1957) recommends

$$\left\| \quad \frac{h}{k}\left[\frac{\mu^2}{g\rho(\rho - \rho_v)}\right]^{1/3} = \frac{Re}{2188 + 41(Re^{0.75} - 89.5)\,Pr^{-0.5}} \quad \right\| \quad \textbf{(6.2.11)}$$

To summarize, for condensation on vertical surfaces,

$$Re < 10 \qquad Nu_L = 0.943\left[\frac{\rho(\rho - \rho_v)gi'_{fg}L^3}{\Delta T\mu k}\right]^{1/4} \qquad \textbf{(6.2.7a)}$$

$$\frac{h}{k}\left[\frac{\mu^2}{g\rho(\rho - \rho_v)}\right]^{1/3} = 0.924\,Re^{-1/3} \qquad \textbf{(6.2.7b)}$$

$$10 < Re < 450 \qquad \frac{h}{k}\left[\frac{\mu^2}{g\rho(\rho - \rho_v)}\right]^{1/3} = \frac{Re}{1.47\,Re^{1.22} - 1.3} \qquad \textbf{(6.2.10)}$$

$$Re > 450 \qquad \frac{h}{k}\left[\frac{\mu^2}{g\rho(\rho - \rho_v)}\right]^{1/3} = \frac{Re}{2188 + 41(Re^{0.75} - 89.5)\,Pr^{-0.5}} \qquad \textbf{(6.2.11)}$$

The values of $Co = (h/k)\{\mu^2/[g\rho(\rho - \rho_v)]\}^{1/3}$ are shown as a function of Re in the laminar, laminar wavy, and turbulent condensate film for discrete values of Pr in Figure 6.2.4.

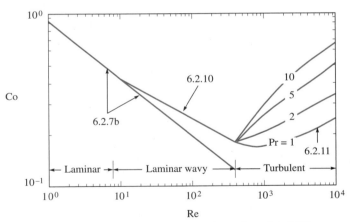

FIGURE 6.2.4 Values of Co as a Function of Re for Different Values of Pr

---

EXAMPLE 6.2.2 If the tube in Example 6.2.1 is 5-m long, determine the heat transfer rate and the mass rate of condensation.

Given

Fluid: Saturated steam $d = 3$ cm $L = 5$ m
$T_{sat} = 100\,°C$ $T_s = 60\,°C$

Find

q and $\dot{m}_c$ (mass rate of condensation)

SOLUTION

From Table A6, at $T_{sat} = 100\ °C$,

$\rho_v = 0.5977\ kg/m^3 \qquad i_{fg} = 2.257 \times 10^6\ J/kg\ °C$

From Table A5 properties of water at $(100 + 60)/2 = 80\ °C$, are

$\rho = 971.8\ kg/m^3 \qquad c_p = 4194\ J/kg\ °C \qquad k = 0.67\ W/m\ °C$
$\mu = 348.1\ N\ s/m^2 \qquad Pr = 2.18$

In Example 6.2.1, at a distance of 0.8 m from the top Re equals 227.5. Therefore, at a distance of 5 m from the top, we expect the Reynolds number to be greater than 450 and the condensate film towards the bottom of the tube to be turbulent. From Equation 6.2.8b

$$h = \frac{Re\ \mu i'_{fg}}{\Delta TL} \tag{1}$$

From Equation 6.2.11,

$$h = \frac{a\ Re}{2188 + 41(Re^{0.75} - 89.5)\ Pr^{-0.5}} \tag{2}$$

where

$$a = k\left[\frac{g\rho(\rho - \rho_v)}{\mu^2}\right]^{1/3} = 0.67\left[\frac{0.807 \times 971.8(971.8 - 0.5977)}{(348.1 \times 10^{-6})^2}\right]^{1/3}$$

$$= 2.843 \times 10^4\ W/m^2\ °C$$

Eliminating h between Equations 1 and 2 and solving for Re,

$$Re = \left[\frac{(a\ \Delta T\ L)/(i'_{fg}\mu) - 2188}{41\ Pr^{-0.5}} + 89.5\right]^{4/3} \tag{3}$$

Substituting all the values into Equation 3,

$$Re = \left[\frac{(2.843 \times 10^4 \times 40 \times 5)/(2.371 \times 10^6 \times 348.1 \times 10^{-6}) - 2188}{41 \times 2.18^{-0.5}}\right.$$

$$\left.+ 89.5\right]^{4/3} = 1649$$

$$h = \frac{Re\ i'_{fg}\mu}{\Delta TL} = \frac{1649 \times 2.371 \times 10^6 \times 348.1 \times 10^{-6}}{40 \times 5} = 6805\ W/m^2\ °C$$

$$q = 6805 \times \pi \times 0.03 \times 5 \times 40 = \underline{1.283 \times 10^5\ W}$$

Mass rate of condensation $= q/i'_{fg} = 1.283 \times 10^5/(2.371 \times 10^6) = \underline{0.0541\ kg/s}$

Film Condensation on Horizontal Cylinders and Spheres From an analysis similar to that for vertical surfaces, the average condensation heat transfer coefficient on a horizontal cylinder at a uniform surface temperature is correlated by (Figure 6.2.5).

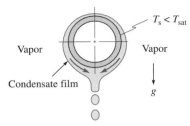

FIGURE 6.2.5 **Condensation on a Horizontal Cylinder**
The condensate flows along the surface of the cylinder.

Filmwise condensation on spheres and horizontal cylinders

$$\mathrm{Nu}_d = \frac{h\,d}{k} = 0.729 \left[\frac{\rho(\rho - \rho_v)gi'_{fg}\,d^3}{\Delta T\mu k} \right]^{1/4} \tag{6.2.12}$$

Dhir and Lienhard (1971) have shown that for condensation on spheres at a uniform temperature, the constant 0.729 in Equation 6.2.12 should be replaced by 0.815.

To account for the effect of ripples, it is suggested that the heat transfer coefficient computed from Equation 6.2.12 be increased by 20% if the Reynolds number [$\Gamma/\mu = (h\pi\,d\,\Delta T)/(2i'_{fg}\mu)$] is greater than 10. It is unlikely that the diameter of tubes used in engineering applications will be large enough for the condensate Reynolds number to exceed 450.

Condensation on horizontal tube banks

Film Condensation on Tube Banks In many types of heat transfer equipment condensation occurs on the outside of horizontal tube banks. As shown in Figure 6.2.6, each tube except for the top tube, is exposed to a condensate film whose

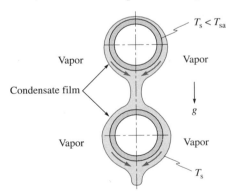

FIGURE 6.2.6 **Film Condensation on a Column of Tubes**
The condensate film thickness on one row of tubes is increased by the condensate falling from the row above it. The condensate film is also disturbed by such dripping.

average film thickness is greater than that of the tube directly above it. This increase in the condensate film thickness is caused by the dripping of the condensate from the tubes above. The increase in the thickness of the film leads to an increase in the resistance to the heat transfer to the surface, which in turn leads to a decrease in the value of the heat transfer coefficient. On the other hand, the condensate film on one row is disturbed by the condensate falling on it from the row above it, and such a disturbance results in an increase in the heat transfer rate. There is, however, a net decrease in the heat transfer rate. From experimental results, Kutateladze (1963) determined the ratio of the average heat transfer coefficient with N tubes in a column (h_N) to the average heat transfer coefficient (h) determined from Equation 6.2.12. The ratio of h_N/h is given in Table 6.2.1.

TABLE 6.2.1 Ratio of h_N/h

Configuration	N	2	3	5	7	10	14
In-line	h_N/h	0.91	0.85	0.74	0.66	0.57	0.5
Staggered	h_N/h	0.99	0.95	0.9	0.82	0.74	0.66
With in-line arrangement for $N > 14$, $h_N/h = 1/N^{1/4}$							

h_N = average heat transfer coefficient with N tubes in a column
h = heat transfer coefficient from Equation 6.2.12

EXAMPLE 6.2.3   A heat exchanger has forty, 12-mm diameter tubes in each column as shown in Figure 6.2.7. The tube surfaces are at 20 °C when surrounded by saturated steam at 40 °C. Compute the average heat transfer coefficient.

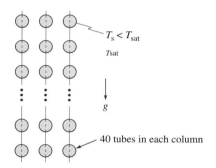

FIGURE 6.2.7 A Tube Bank with 40 Tubes in a Column

Given

Fluid: Steam $T_{sat} = 40\,°C$
$d = 12$ mm $T_s = 20\,°C$
Number of tubes in a column = 40

Find

h (average heat transfer coefficient)

Solution

Assume that heat transfer from tubes in each column is independent of the heat transfer from the neighboring columns. To find the average heat transfer coefficient with a bank of tubes, the heat transfer coefficient with a single horizontal tube, found from Equation 6.2.12, is modified with the correction factor from Table 6.2.1. As there are more than 14 tubes,

$$h_{40} = \frac{h}{40^{1/4}}$$

where h is the heat transfer coefficient from Equation 6.2.12.

From Table A5, properties of water at $T_f = (20 + 40)/2 = 30\,°C$ are

$\rho = 995.6 \text{ kg/m}^3 \qquad c_p = 4176 \text{ J/kg }°C$
$k = 0.618 \text{ W/m }°C \qquad \mu = 778.6 \times 10^{-6} \text{ N s/m}^2$

From Table A6, properties of saturated vapor at $40\,°C$ are

$\rho_v = 1/19.52 = 5.123 \times 10^{-2} \text{ kg/m}^3 \qquad i_{fg} = 2.407 \times 10^6 \text{ J/kg}$

Thus,

$$\rho - \rho_v = 995.6 - 5.123 \times 10^{-2} \approx 995.6 \text{ kg/m}^3$$

$$i'_{fg} = 2.407 \times 10^6 + 0.68 \times 4176(40 - 20) = 2.464 \times 10^6 \text{ J/kg}$$

From Equation 6.2.12,

$$h = 0.729 \left[\frac{g\rho(\rho - \rho_v)i'_{fg}k^3}{\mu\, \Delta T d} \right]^{1/4}$$

$$= 0.729 \left[\frac{9.807 \times 995.6^2 \times 2.464 \times 10^6 \times 0.618^3}{778.6 \times 10^{-6} \times 20 \times 0.012} \right]^{1/4} = 9614 \text{ W/m}^2\,°C$$

Check for the effect of ripples if any.

$$\text{Re} = \frac{h\pi\, d/2\, \Delta T}{i'_{fg}\mu} = \frac{9614 \times \pi \times 0.006 \times 20}{2.464 \times 10^6 \times 778.6 \times 10^{-6}} = 1.89$$

As Re < 10, ripples do not appear.
  With 40 columns per tube,

$$h = h/N^{1/4} = 9614/40^{1/4} = \underline{\underline{3823 \text{ W/m}^2\,°C}}$$

COMMENT

Although ripples do not appear on the first row, it is possible that they will appear in the lower rows as Re increases towards the lower rows.

Condensation Inside Horizontal Tubes When a vapor condenses inside a horizontal tube, there are different flow regimes (as in boiling). The condensate flow is strongly influenced by the vapor velocity and gravity.

If the vapor velocity is very small, the condensate that is formed on the tube wall flows towards the bottom due to gravity (along the tube wall in the same vertical plane). The condensate collected in the bottom of the tube flows out of the tube by the hydraulic gradient, the condensate thickness decreasing in the direction of flow. The correlation for the condensation on the inside wall of the tube can be determined by an analysis similar to the one for condensation on the outside of horizontal tubes. This type of condensation, where gravity is the only driving force for the condensate, shown in Figure 6.2.8a, was studied by Chaddock (1957) and Chato (1962).

As the vapor velocity is increased, the effect of the shear stress at the interface and of the momentum of the condensing vapor (inertial effects) as the driving force for the condensate film become significant. The condensate flow is by the combined

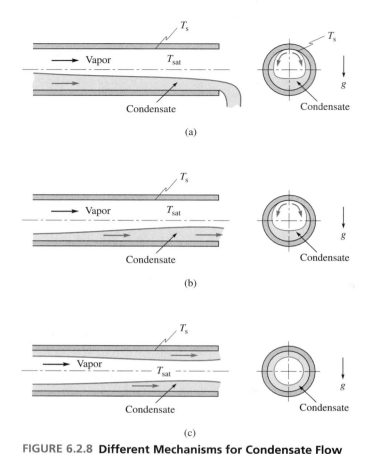

FIGURE 6.2.8 **Different Mechanisms for Condensate Flow**
(a) Condensation inside a horizontal tube at very low vapor velocities. Condensate flow is by hydraulic gradient with film thickness decreasing in the direction of flow. Gravity is the driving force for draining the condensate. (b) Condensation inside a horizontal tube with the condensate draining to the bottom by gravity and the collected condensate drained by interfacial shear stress and the inertia of the condensing vapor. (c) Annular film condensation with a uniform condensate film thickness around the tube, the thickness of the film increasing in the direction of flow. Interfacial shear stress is the driving force for draining the condensate.

action of the interfacial shear stress and the inertial effects of the condensing vapor, with the thickness of the film increasing in the direction of flow as shown in Figure 6.2.8b.

For low vapor velocities Chato (1962) recommends

Condensation
inside horizontal
tubes

$$h_{\mathrm{m}} = 0.556 \left[\frac{g \rho (\rho - \rho_{\mathrm{v}}) i_{\mathrm{fg}}' k^3}{\mu \, d \, \Delta T} \right]^{1/4} \tag{6.2.13}$$

where

h_m = average heat transfer coefficient based on the total inner surface area of the tube

$i'_{fg} = i_{fg} + 0.68 c_p\,\Delta T$

$q = h_m \pi\, dL\, \Delta T$

L = length of the tube

Equation 6.2.13 is valid for $\mathrm{Re}_v = (4\dot{m})/(\pi\, d\mu_v) < 35\,000$. $\dot{m}$ is the mass rate of flow of vapor and condensate inside the tube.

With increasing vapor velocities, the effect of the interfacial shear stress becomes predominant and the condensate does not flow down the wall of the tube as in the previous two cases. The absence of flow of the condensate down the wall of the tube leads to an approximately uniform condensate film thickness around the periphery of the tube. The thickness of the film increases in the direction of flow. This type of condensation where the condensate film thickness is uniform around the periphery, as shown in Figure 6.2.8c, is known as *annular film condensation* with a smooth interface. With increasing thickness of the film, interfacial waves appear resulting in wavy annular regime. Further downstream other regimes exist.

For annular condensation, Akers and Rosson (1960) formulated empirical correlations for the average heat transfer coefficients. The correlations depend on the vapor and liquid Reynolds numbers. In the condensing section, the mass rates of flow of the vapor and condensate vary. Taking the arithmetic mean of the mass flow rates of the vapor and condensate at inlet and exit, vapor and condensate Reynolds numbers, Prandtl number, and Jakob number are defined as follows.

$$\mathrm{Re}_v = \frac{4\dot{m}_v}{\pi d\mu}\,(\rho/\rho_v)^{1/2} \qquad \mathrm{Re}_L = \frac{4\dot{m}_L}{\pi d\mu} \qquad \mathrm{Nu} = \frac{h\,d}{k}$$

$$\mathrm{Pr} = \frac{c_p \mu}{k} \qquad\qquad \mathrm{Ja} = \frac{c_p\,\Delta T}{i_{fg}}$$

where

$\dot{m}_v$ = arithmetic mean of the mass flow rates of the vapor at inlet and exit of the tube

$\dot{m}_L$ = arithmetic mean of the mass flow rates of the condensate at inlet and exit of the tube

h = average heat transfer coefficient = $[q/(\pi\, dL)]/\Delta T$; $\Delta T = T_{sat} - T_s$

i_{fg} = enthalpy of vaporization

$1000 < \mathrm{Re}_v < 20\,000$

$$\mathrm{Nu} = 13.8\,\mathrm{Pr}^{1/3}\,\mathrm{Re}_v^{1/5}\,\mathrm{Ja}^{-1/6} \tag{6.2.14a}$$

$20\,000 < \mathrm{Re}_v < 100\,000 \quad \text{and} \quad \mathrm{Re}_L < 5000$

$$\mathrm{Nu} = 0.1\,\mathrm{Pr}^{1/3}\,\mathrm{Re}_v^{2/3}\,\mathrm{Ja}^{-1/6} \tag{6.2.14b}$$

$\mathrm{Re}_v > 20\,000 \quad \text{and} \quad \mathrm{Re}_L > 5000$

$$\mathrm{Nu} = 0.026\,\mathrm{Pr}^{1/3}(\mathrm{Re}_v + \mathrm{Re}_L)^{0.8} \tag{6.2.14c}$$

Note the definition of the vapor Reynolds number Re_v. It is based on the vapor velocity and *liquid viscosity*. Properties are those of the liquid; properties of vapor are indicated by the subscript v.

The correlations are valid in the annular regime. Most of the condensation heat transfer occurs in the annular flow region and Equation 6.2.14 can be used even when only condensate leaves the tube. For more elaborate correlations, refer to Soliman et al. (1968) and Traviss et al. (1973).

EXAMPLE 6.2.4

Dry, saturated steam at 80 °C enters a 25-mm diameter tube with a velocity of 20 m/s as shown in Figure 6.2.9. The tube surface is at 70 °C. Estimate the length of the tube to condense the steam.

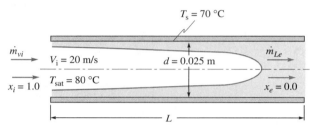

FIGURE 6.2.9 Annular Condensation of Steam Inside a Horizontal Tube

Given

Fluid: Steam
$d = 25$ mm
V_i (vapor velocity at inlet) = 20 m/s
x_i (quality at inlet) = 1.0

$T_{sat} = 80$ °C
$T_s = 70$ °C
x_e (quality at exit) = 0.0

Find

L (length of tube)

ASSUMPTIONS

1. Condensate leaves the tube at a temperature less than T_{sat} and the heat transfer rate is determined from $q = \dot{m}_v i'_{fg}$. The modified enthalpy of vaporization is determined from $i'_{fg} = i_{fg} + 0.68c_p\,\Delta T$.
2. The average heat transfer coefficient is determined from Equation 6.2.14
3. As most of the condensation occurs in the annual condensation region, the average heat transfer coefficient in the annular regime is assumed to be applicable to the entire condensing length.

SOLUTION

From the relations

$$q = \dot{m}_v i'_{fg} \qquad i'_{fg} = i_{fg} + 0.68c_p\,\Delta T$$

determine the heat transfer rate. From Equation 6.2.14, determine the average heat transfer coefficient. Determine the length of the tube for complete condensation from the relation

$$q = h\pi \, dL \, \Delta T$$

From Table A5, properties of water at $(80 + 70)/2 = 75\,°C$ are

$\rho = 974.9 \text{ kg/m}^3 \qquad c_p = 4190 \text{ J/kg }°C$
$k = 0.666 \text{ W/m }°C \qquad \mu = 369.3 \times 10^{-6} \text{ N s/m}^2$
$\text{Pr} = 2.32$

From Table A6, properties of saturated steam at $80\,°C$ are

$$\rho_v = 1/3.407 \text{ kg/m}^3 \qquad i_{fg} = 2.309 \times 10^6 \text{ J/kg}$$

As the quality is 1 at inlet and 0 at exit,

$$\dot{m}_{Le} = \dot{m}_{vi} = (1/3.407)20 \times \pi(0.025^2/4) = 2.882 \times 10^{-3} \text{ kg/s}$$

As $\dot{m}_{ve} = \dot{m}_{Li} = 0$, the average mass flow rates of the condensate and the vapor are equal and are given by

$$\dot{m}_v = \dot{m}_L = 0.5 \times 2.882 \times 10^{-3} = 1.441 \times 10^{-3} \text{ kg/s}$$

$$\text{Re}_L = \frac{4\dot{m}_L}{\pi \, d\mu} = \frac{4 \times 1.441 \times 10^{-3}}{\pi \times 0.025 \times 369.3 \times 10^{-6}} = 198.7$$

$$\text{Re}_v = \frac{4\dot{m}_v}{\pi \, d\mu} (\rho/\rho_v)^{1/2} = 198.7 \times (974.9 \times 3.407)^{1/2} = 11\,452$$

$$i'_{fg} = 2.309 \times 10^6 + 0.68 \times 4190 \times 10 = 2.337 \times 10^6 \text{ J/kg}$$

$$\text{Ja} = \frac{4190 \times 10}{2.337 \times 10^6} = 0.01793$$

For $\text{Re}_L = 198.7$ and $\text{Re}_v = 11\,452$, from Equation 6.1.14a,

$$\text{Nu} = 13.8 \times 2.32^{1/3} \times 11\,452^{1/5} \times 0.01793^{-1/6} = 231.5$$

$$h = \frac{231.5 \times 0.666}{0.025} = 6167 \text{ W/m}^2\,°C$$

$$q = \dot{m}_{Le}i'_{fg} = 2.882 \times 10^{-3} \times 2.337 \times 10^6 = 6735 \text{ W}$$

$$L = \frac{q}{h\pi \, d \, \Delta T} = \frac{6735}{6167 \times \pi \times 0.025 \times 10} = \underline{1.39 \text{ m}}$$

6.3 PROJECTS

PROJECT 6.3.1 Design of an Electric Steam Boiler

Design an electric steam boiler to produce 60 kg/h of saturated steam at $150\,°C$.

SOLUTION

The design of any component or system requires some preliminary choices. From the design of the boiler we have to decide whether the steam should be produced by pool boiling or forced convection boiling, the material of the heating element, the shape of the boiler, and so forth. The starting point for the design is making the preliminary choices. In the course of executing the design, it may sometimes be necessary to modify the original choice.

Preliminary Selections

Pool Boiling or Forced Convection Boiling The steam may be produced either by pool boiling with an immersion electric heater or by forced convection boiling with water flowing in an electrically heated tube. For forced convection boiling the units have to be custom built. But immersion heaters are commercially available and a suitable available unit can be chosen for pool boiling. Therefore, pool boiling is preferred to forced convection boiling. Choose pool boiling.

Material of Heater Two commonly used materials for the fabrication of heaters are copper and stainless steel. Copper is less expensive but stainless steel, while more expensive, is much more resistant to chemical action. Choose stainless steel.

Operating Conditions The physical dimensions of the heater can be minimzied by operating the heater at the highest, safe, heat flux, i.e., slightly below the critical heat flux. However, there is significant uncertainties in the values predicted by the correlations. If, for any reason, the critical heat flux in the actual conditions is less than the predicted value, the heater will burn out. To avoid such an undesirable consequence, choose a heat flux that is only 50% of the predicted CHF. From the computed heat flux and the required heat transfer rate, the heater surface area is determined.

Physical Dimensions There is considerable latitude in defining the dimensions of the heater, which depend on the shape of the heating element and the dimensions of the boiler. The most common shape for an immersion heater is a cylinder. The dimensions of the boiler are partially determined from the dimensions of the heating element.

We now begin the design of the boiler with the dimensions of the boiler. Boilers are usually cylindrical. To determine the volume of the boiler, assume that the boiler volume should be sufficient to operate the boiler for four hours without any supply of feedwater. Therefore, it should hold 240 kg of water. With the density of water being approximately 1000 kg/m³, the minimum boiler volume should be 0.24 m³. In addition to the water there should be some steam space above the liquid. Assume a minimum steam space of 40% of the liquid volume. The total volume of the boiler should be approximately 0.34 m³. For a cylindrical boiler assume a length/diameter ratio of 1.5. Then,

$$\frac{\pi D^2}{4} \times 1.5D = 0.34$$

Therefore,

$$D = \left(\frac{4 \times 0.34}{1.5\pi}\right)^{1/3} = 0.66 \text{ m}$$

Choose a diameter of 70 cm. Length of the boiler is 1.05 m (= 1.5 × 0.7). Choose a length of 1m. The volume of the cylinder with a diameter of 0.7 m and a length of 1 m is 0.385 m^2, which is 13% higher than the originally estimated volume of 0.34 m^3. The boiler may be operated with the axis vertical (Figure 6.3.1a) or horizontal (Figure 6.3.1b).

Determine the Heater Capacity The power supply should be sufficient to raise the temperature of 60 kg/h of water from the temperature at which it is supplied to 150 °C and change its phase from water to steam. The required power is given by

$$P = \dot{m}_{st}[c_p(T_{sat} - T_{in}) + i_{fg}]$$

For the inlet temperature of the water we assume a room temperature of 20 °C. From Table A6 at a saturation temperature of 150 °C, i_{fg} = 2114.3 kJ/kg. From Table A5, c_p(85 °C) = 4198 J/kg °C. Thus,

$$P = \frac{60}{3600} [4198(150 - 20) + 2114.3 \times 1000] = 44\,334 \text{ W}$$

The power supplied is also the heat transfer rate from the heater.

Determine the Heat Transfer Area of the Heating Element Denoting the operating heat flux by q'', the heat transfer area of the element is P/q''. To determine q'', the critical heat flux is needed. From Equation 6.1.3,

$$\text{CHF} = q''_{max} = C\rho_v^{0.5}i_{fg}[\sigma g(\rho - \rho_v)]^{0.25}$$

To determine C from Table 6.1.2, the diameter of the heating element is required. The diameters of immersion heater range from 5 mm to 40 mm. Choose a diameter of 10 mm. From Table 6.1.2 for a horizontal cylinder,

$$L^* = L\left(\frac{g(\rho - \rho_v)}{\sigma}\right)^{0.5}$$

At the saturation temperature of 150 °C from Tables A5 and A6

$$\rho = 916.9 \text{ kg/m}^3 \qquad \rho_v = \frac{1}{0.3928} = 2.546 \text{ kg/m}^3$$

At 150 °C the surface tension of water in contact with its own vapor is (822.4 − 2.246 × 150)10^{-4} = 0.04855 N/m. L is the radius of the heating element = 0.005 m. Thus,

$$L^* = 0.005\left(\frac{9.807(916.9 - 2.546)}{0.04855}\right)^{0.5} = 2.15$$

From Table 6.1.2 for L^* = 2.15, C = 0.118.

$$\text{CHF} = 0.118 \times 2.546^{0.5} \times 2114.3 \times 10^3[0.04855 \times 9.807(916.9 - 2.546)]^{0.25}$$
$$= 1.818 \times 10^6 \text{ W/m}^2$$

$$\text{Operating heat flux (equal to 50\% of the CHF)} = q'' = 0.5 \times 1.818 \times 10^6 = 9.09 \times 10^5 \text{ W}$$

As an additional piece of information, determine the operating temperature of the heating element. From Equation 6.1.2a,

$$\frac{c_p \, \Delta T}{i_{fg} \, \mathrm{Pr}^n} = C_{sf} \left\{ \frac{q''}{\mu i_{fg}} \left[\frac{\sigma}{g(\rho - \rho_v)} \right]^{1/2} \right\}^{1/3}$$

From Table A5 at a saturation temperature of 150 °C

$$\mu = 179.9 \times 10^{-6} \text{ N s/m}^2 \qquad \mathrm{Pr} = 1.12$$

From Table 6.1.1 for water-stainless steel (mechanically polished)

$$C_{sf} = 0.0132 \qquad n = 1 \qquad c_p(150 \text{ °C}) = 4270 \text{ J/kg °C}$$

Substituting the values, we obtain $\Delta T = 13$ °C and the operating temperature is 150 + 13 = 163 °C.

Determine the Length of the Heating Element With $q'' = 9.09 \times 10^5$ W/m^2 and q = 44 334 W

$$A = \frac{\text{surface area}}{\text{of the heating element}} = 44 \ 334/9.09 \times 10^5 = 0.0488 \text{ m}^2$$

$$L = \text{length of the element} = \frac{A}{\pi \, d} = \frac{0.0488}{\pi \times 0.01} = 1.55 \text{ m} \approx 1.5 \text{ m}$$

The diameter of the boiler is only 0.7 m. A design with the heating element mounted on the cylindrical surface of the boiler is shown in Figure 6.3.1a; a design with the heating element mounted on one end surface is shown in Figure 6.3.1b.

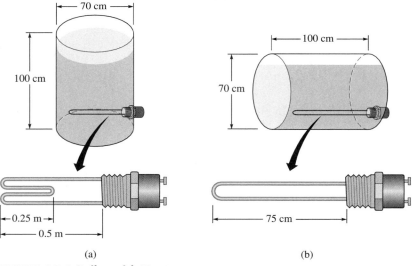

(a)   (b)

FIGURE 6.3.1 Boiler with Heater
(a) Vertical boiler. (b) Horizontal boiler.

COMMENT

Correlation for pool boiling of saturated water has been employed in the computations. Two effects need to be considered. First, with the introduction of feedwater there is some forced convection effect but, since the velocity of the feedwater is low, the effect of forced convection is neglected. Second, with feedwater entering the boiler there is some subcooling of the water in the boiler but with the large amount of water in the boiler the degree of subcooling is small. Further, in the nucleate boiling region, subcooling has the effect of slightly increasing the heat flux. Therefore, we may expect the heater to operate at a slightly lower temperature than the design temperature.

SUMMARY

When a phase change from liquid to vapor occurs at the liquid-vapor interface it is termed *evaporation*. If the phase change occurs at a solid surface, which is at a temperature higher than the liquid saturation temperature, the phase change is termed *boiling*. Boiling in an otherwise stagnant liquid is *pool boiling*. Boiling of a liquid in motion is *forced convection boiling*.

If the surface superheat (the difference between the surface and liquid temperatures) is small, the heat transfer is by natural convection. As the surface superheat is increased *nucleate boiling*, with vapor bubbles leaving the surface, occurs with a significant increase in the heat flux. There is a maximum limit to the heat flux in the nucleate boiling regime. The maximum heat flux is termed the *critical heat flux* (CHF). An increase in the heat flux beyond the CHF leads to an unstable regime, *transition boiling*—a mix of nucleate and film boiling—and then to *film boiling*. In the transition boiling region the heat flux decreases with an increase in the surface superheat and reaches a minimum value. For heat fluxes greater than the minimum value film boiling is established.

Correlations for nucleate boiling, CHF, and film boiling are provided. A brief introduction to forced convection boiling is also given.

There are two modes of condensation—*filmwise condensation* and *dropwise condensation*. Heat transfer coefficients with dropwise condensation are much greater than with film condensation but surface preparation for dropwise condensation is expensive and in many cases dropwise condensation cannot be sustained for long periods.

Correlations for filmwise condensation on vertical surfaces and horizontal tubes are given. With increasing condensate Reynolds number ($7.5 <$ Re < 450), *interafacial waves* appear; the waves result in an increase in the heat transfer coefficient. At high condensate Reynolds numbers (Re > 450) turbulence sets in in the condensate film. Correlations for smooth condensate-vapor interface, with interfacial waves, and for turbulent condensate film are provided.

A brief introduction to forced condensation inside horizontal tubes is also included.

REFERENCES

Akers, W.W., and Rosson, H.F. (1960). Condensation inside horizontal tubes. *Chem. Engg. Prog. Symp.* Series 3, 56:145.

Berenson, P.J. (1961). Film boiling heat transfer from a horizontal surface. TRANS. ASME 83C:351.

Bromley, L.A. (1950). Heat transfer in stable film boiling. *Chem. Eng. Pog.* 46:221.

Chaddock, J.B. (1957). Film condensation of a vapor in a horizontal tube. *Refrigerating Engineering* 65:36.

Chato, J.C. (1962). Laminar condensation inside horizontal and inclined tubes. *ASHRAE J.* 4:52.

Ded, J.S., and Lienhard, J.H. (1972). The peak pool boiling heat flux from a sphere. *AIChE J.* 18:337.

Dhir, V.K., and Lienhard, J.H. (1971). Laminar film condensation on plane and axisymmetric bodies in non-uniform gravity. *TRANS. ASME* 93C:97.

Frederking, T.H.K., and Clark, J.A. (1963). Natural convection film boiling on a sphere. *Advances in Cryogenic Engineering,* 8:501.

Janna, W.S. (1986). *Engineering Heat Transfer.* Boston: PWS Engineering.

Jasper, J.J. (1972). The surface tensions of pure liquid compounds. *J. Physical and Chemical Reference Data* 1(4):841.

Kakac, S., Shah, R.K., and Aung, W. (1987). *Handbook of Single-Phase Convective Heat Transfer,* New York: Wiley Interscience.

Kalinin, E.K., Berlin, I.I., and Kostyuk, V.V. (1975). Film-boiling heat transfer. In *Advances in Heat Transfer,* vol. 11, eds Irvine, T.F., and Hartnett, J.P. New York: Academic Press.

Kutateladze, S.S. (1963). *Fundamentals of Heat Transfer.* New York: Academic Press.

Lewis, E.W., Merte, H., Jr. and Clark, J.A. (1965). "Heat transfer at zero gravity." Symposium paper, The 55th National Meeting, American Institute of Chemical Engineers, Houston.

Lienhard, J.H. (1987). *A Heat Transfer Text Book.* 2nd ed. Englewood Cliffs, NJ: Prentice-Hall.

Lienhard, J.H., and Dhir, V.K. (1973). Hydrodynamic predictions of peak pool-boiling heat fluxes from finite bodies. *TRANS. ASME* 95C:152.

Lienhard, J.H., Dhir, V.K., and Riherd, D.M. (1973). Peak pool boiling heat-flux measurements on finite horizontal flat plates. TRANS. ASME 95C:477.

Labuntsov, D.A. 91957). Heat transfer in film condensation of pure steam on vertical surfaces and horizontal tubes. *Teploenergetika* 4:72. Quoted in White (1988).

McAdams, W.H. (1954). *Heat Transmission.* New York: McGraw-Hill.

Nukiyama, S. (1966). The maximum and minimum values of the heat transmitted from metal to boiling water at atmospheric pressure. *Int. J. Heat Mass Transfer* 9:1419. Originally published in *J. Soc. Mech. Engrs.,* Japan, 1934.

Rohsenow, W.M. (1952). A method of correlating heat transfer data for surface boiling of liquids. *TRANS. ASME* 74:969.

———(1956). Heat transfer and temperature distribution in laminar film condensation. *TRANS. ASME* 78:1645.

———(1973). Boiling. In *Handbook of Heat Transfer,* eds. Rohsenow, W.M. and Hartnett, J.P. New York: McGraw-Hill.

———(1985). Boiling. In *Handbook of Heat Transfer Fundamentals.* eds. Rohsenow, W.M., Harnett, J.P., and Ganic, E.N. New York: McGraw-Hill.

Rufer, C.E., and Kezios, S.P. (1966). Analysis of two-phase, one-component stratified flow with condensation. *TRANS. ASME* 88C:265.

Soliman, M., Scuster, J.R., and Berenson, P.J. (1968). A general heat transfer correlation for annular flow condensation. *TRANS. ASME* 90C:267.

Suryanarayana, N.V., and Merte, H., Jr. (1970). Film boiling on vertical surfaces. ORA Report 07461, Department of Mechanical Engineering, The University of Michigan, Ann Arbor.

———(1972). Film boiling on vertical surfaces. *TRANS. ASME* 94C:377.

Sun, K., and Linehard, J.H. (1970). The peak pool boiling heat flux on horizontal cylinders. *Int. J. Heat Mass Transfer* 13:1425.

Traviss, D.P., Rohsenow, W.M., and Baron, A.B. (1973). Forced-convection condensation inside tubes: A heat transfer equation for condenser design. *TRANS. ASHRAE* 79(1):157.

Vachon, R.I., Nix, G.H., and Tanger, G.E. (1968). Evaluation of constants for the Rohsenow pool boiling correlation. *TRANS. ASME* 90C:239.

White, F.M. (1988). *Heat and Mass Transfer.* Reading, MA: Addison-Wesley.

Zuber, N. (1959a). Hydrodyanmic aspects of boiling heat transfer. AEC Report AECU-4439, Physics and Mathematics. Quoted by Kutateladze (1963).

———(1959b). On the stability of boiling heat transfer. *TRANS. ASME* 80:711.

REVIEW QUESTIONS

(a) What are the different regimes in pool boiling?

(b) What is the difference between pool boiling and forced convection boiling?

(c) What is the difference between evaporation and boiling?

(d) What are critical heat flux and burnout?

(e) What is the consequence of imposing a heat flux greater than the CHF on a surface immersed in a saturated liquid?

(f) If the temperature of a highly polished surface immersed in a liquid is greater than the saturation temperature of the liquid, do you anticipate a large number of small bubbles or a few large bubbles due to nucleate boiling?

(g) Describe one method by which all regimes of boiling can be observed.

(h) What is the difference between film condensation and dropwise condensation?

(i) Generally, radiative heat transfer is significant in film boiling. Why?

(j) Is radiative heat transfer always significant in film boiling?

(k) Define Reynolds number used in the correlations for film condensation.

(l) In condensation on the outer surface of a horizontal cylinder, the Reynolds number was defined as $(h\pi d\,\Delta T)/2\mu$. Why is only half the mass rate of flow considered in the definition of the Reynolds number?

(m) What is the significance of the Reynolds number in film condensation?

PROBLEMS

6.1 Water is heated in a flat bottom, 20-cm diameter, stainless steel kettle. It is placed on a 2-kW flat electric heating element. Estimate the temperature of the bottom surface of the kettle during boiling of water.

FIGURE P6.1

6.2 Saturated water at 100 °C is to be evaporated by boiling in a 30-cm diameter pot at the rate of 3.5 kg/h. Estimate the surface temperature of the copper-plated bottom surface of the pan.

6.3 In Problem 6.2, what is the maximum heat transfer rate to sustain nucleate boiling?

6.4 Re-solve Problem 6.3 if the bottom surface is stainless steel.

6.5 Calculate the CHF on a horizontal surface with saturated water at 100 °C, saturated mercury at 195 °C, and saturated nitrogen at atmospheric pressure. Assume a large horizontal surface in each case. Properties of mercury and nitrogen are

Saturation Temperature	Mercury 195 °C	Nitrogen 77.35 K
ρ_v (kg/m^3)	0.1011	4.613
ρ (kg/m^3)	13 123	808.61
i_{fg} (J/kg)	3×10^5	1.986×10^5

6.6 A horizontal, 10-cm long, 14-gauge platinum wire is immersed in saturated water at 100 °C. It is heated by passing an electric current through it. Estimate
(a) The power dissipated by the wire if its temperature is 120 °C.
(b) The maximum current in the nucleate boiling region.
(c) The heat transfer rate if its temperature is 300 °C.
(Diameter of 14-gauge wire = 1.628 mm; resistance of wire = 4.81×10^{-4} Ω/cm)

6.7 Steam is produced in a boiler at the rate of 0.01 kg/s with a 10-mm diameter, horizontal heater immersed in saturated water at 100 °C. The heating element is made of copper. What should be the length of the heating element if its surface temperature is not to exceed 120 °C?

6.8 A boiler is to produce 50 kg/h of steam at atmospheric pressure. The heating element consists of 1.5-cm diameter horizontal brass tubes; the length of the tubes should not exceed 1 m. The tubes are to be operated at 60% of the critical heat flux. Determine
(a) The surface temperature of the tubes.
(b) The number of tubes needed (if more than one is needed).

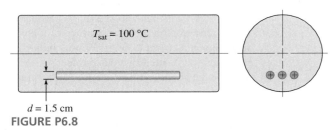

$d = 1.5$ cm
FIGURE P6.8

6.9 In an experiment to investigate the effect of material on boiling heat transfer from thin wires in different fluids, the maximum power supply to the wires is to be estimated. In one experiment, it is proposed to use a 10-cm long, 20-gauge platinum wire (0.8118-mm diameter 0.00193 Ω/cm) in saturated water at 100 °C. Estimate the maximum power required.

6.10 A 10-mm diameter, horizontal heating element immersed in saturated water at 100 °C is to produce 0.01 kg/s of steam. The heating element is made of copper and is maintained at 110 °C. Determine the length of the element required.

6.11 Saturated water at 100 °C is boiled in a 20-cm diameter and 10 cm high pot. Determine
(a) The heat transfer rate if the bottom surface is at 105 °C.
(b) The maximum heat transfer rate for nucleate boiling.
(c) The time required for all the water to evaporate if the bottom surface temperature is 105 °C.

6.12 To find the efffect of reduced gravity, Lewis et al. (1965) conducted boiling experiments in a drop tower (see Problem 4.102 for a sketch of a drop tower). The experimentally determined values of the minimum heat flux in film boiling with liquid nitrogen at atmospheric pressure on a 25-mm diameter sphere are

g_{eff}/g	1.0	0.33	0.02
q''_{min} (W/m²)	5360–6625	4100–4420	2750–3470

Compare the experimental values with those obtained from the correlation for q''_{min}. For properties of liquid nitrogen at atmospheric pressure see Problem 6.5.

6.13 Compare the heat flux from a brass surface at $\Delta T = 3$ °C when immersed in saturated water at 100 °C and 180 °C.

6.14 With the trend towards very large scale integration of electronic devices resulting in high heat dissipation rates, there is a need for improving the cooling of the chips. An effective method for dissipating large amounts of heat is to boil a liquid. To ensure reliability, the maximum heat flux is limited to 40% of the CHF as the chips will fail if (even locally) the heat flux exceeds the CHF. Determine the CHF if the fluid is saturated R-113 at atmospheric pressure. Estimate the temperture of the surface if the heat flux is 10% of the CHF. For boiling of R-113, assume $C_{sf} = 0.02$ and $n = 1.7$. Properties of saturated R-113 at atmospheric pressure are

$T_{sat} = 47.6$ °C $\rho = 1507.3$ kg/m³ $c_p = 984$ J/kg °C
$\mu = 516 \times 10^{-6}$ N s/m² $k = 0.0705$ W/m °C $\rho_v = 7.464$ kg/m³
$i_{fg} = 1.438 \times 10^5$ J/kg $\sigma(40 °C) = 0.015$ N/m

6.15 Design an electrically heated boiler to produce 100 kg/h of saturated steam at 800 kPa. **Design**

6.16 Plain carbon-steel spheres, 25-mm diameter ($\varepsilon = 0.25$), are heated to 500 °C and quenched in water (one at a time) at 101.35 kPa. Estimate
(a) The initial rate of cooling.
(b) The maximum rate of cooling.
(c) The minimum rate of cooling in film boiling.
Assume water is at its saturation temperature.

6.17 To investigate the effect of gravity on boiling, Lewis et al. (1965) conducted boiling experiments in a 32-ft deep drop tower (see Problem 4.102 for a brief description of a drop tower). During the fall of the drop package, they recorded the time-temperature history of a 25-mm diameter copper sphere immersed in liquid nitrogen at atmospheric pressure. From the recorded temperature-time curve in film boiling, they computed the heat flux ($q'' = (\dot{m}c_p/A_s)(dT/d\tau)$) at different temperatures. Some of the results are given below.

ΔT (°C)	93	91	99
g_{eff}/g	1.0	0.33	0.2
q'' (W/m²)	12 600	9780	9460

Compare the experimental values with those found from Equation 6.1.7. For properties of liquid nitrogen refer to Problem 6.5.

6.18 In Nukiyama's experiments it was found that when the heat flux exceeded the CHF, the wire burned out. To understand why it happened, consider a 1-mm diameter horizontal wire in saturated water at 100 °C. Employing Equation 6.1.3,
(a) Estimate the CHF.
(b) Estimate the temperature of the wire in film boiling that will yield the same q'' as the CHF, if the emissivity of the surface of the wire is 0.7.

6.19 Forced convection boilers with heated tubes can be made more compact than boilers with immersion heaters. In one such forced convection boiler saturated water at 180 °C flows in an electrically heated 2-cm diameter tube. The imposed heat flux is 2×10^5 W/m². Determine the surface temperature at which boiling is initiated. Assume a very low velocity for water.

6.20 In Problem 6.19, what should be the velocity of water to suppress boiling if the excess temperature (ΔT) is 0.6 °C.

6.21 Saturated water at 180 °C flows in a 2-cm diameter brass tube with a velocity of 3 m/s. Compute the heat flux if the tube surface is at 185 °C. Neglect q''_{Bi}.

6.22 Saturated water at 100 °C flows in a 15-mm diameter brass tube with a velocity of 4 m/s. Estimate the tube surface temperature if a heat flux of 2×10^5 W/m² is imposed. Neglect q''_{B1}.

6.23 Saturated water at 100 °C enters a 25-mm diameter copper tube with a velocity of 3 m/s. The tube surface is at 110 °C. Determine the heat flux and estimate the length of the tube for complete evaporation of water. Neglect q''_{B1}.

6.24 Saturated steam at 140 °C condenses on a 0.5-m high vertical plate whose surface is maintained at 80 °C. Determine the heat transfer rate to the plate per meter width of the plate.

6.25 Re-solve Problem 6.24 if the plate is inclined at 30° to the vertical with condensation on the upper surface.

6.26 Saturated steam at 60 °C condenses on the outside surface of a 2-m long, 2-cm O.D. vertical tube. The temperature of the outer surface of the tube is 50 °C.
(a) Compute the average heat transfer coefficient.
(b) Estimate the condensate film thickness employing correlations for vertical, plane surfaces, and verify that those correlations are valid for this case.
(c) Determine the mass rate of condensation of steam.
(d) Estimate the average condensate velocity at the bottom of the tube.

6.27 Show that for laminar film condensation on a vertical surface (without ripples),

$$\text{Re} = \frac{\Gamma}{\mu} = 0.943 \left[\frac{g\rho(\rho - \rho_v)L^3 \, \Delta T^3 k^3}{\mu^5 i'^3_{fg}} \right]^{1/4}$$

Show that $[g\rho(\rho - \rho_v)L^3 \, \Delta T^3 k^3]/(\mu^5 i'^3_{fg})$ is dimensionless.

6.28 Show that the condensate number

$$\text{Co} = \frac{h}{k} \left[\frac{\mu^2}{g\rho(\rho - \rho_v)} \right]^{1/3}$$

is dimensionless.

6.29 Saturated Refrigerant 12 at 40 °C condenses on the outer surface of 10-mm diameter, horizontal tube. The surface of the tube is at 30 °C. Determine the length of the tube required to condense 0.1 kg/s. Properties are of R-12

i_{fg} (40 °C) = 129.4 kJ/kg ρ_v (40 °C) = 54.57 kg/m³

6.30 A double pipe heat exchanger is made of 15-mm diameter, 1.5-m long horizontal tube inside a 20-cm diameter coaxial tube. The heat exchanger is to condense 60 kg/h of Refrigerant 12 (in the annulus) at a saturation temperature of 40 °C. What should be the surface temperature of the tube? Properties of R-12 are

$$i_{fg} \ (40 \ °C) = 129.4 \ \text{kJ/kg} \qquad \rho_v \ (40 \ °C) = 54.47 \ \text{kg/m}^3$$

6.31 A Refrigerant 12 condenser consists of an inline bank of 300, 15-mm diameter, 2-m long tubes. The surfaces of the tubes are at 30 °C when condensing saturated Refrigerant 12 at 50 °C. The tubes can be arranged as a rectangular, horizontal bank with 20 tubes or 15 tubes in each column. Which arrangement will yield the higher rate of condensation? If they are arranged with 20 tubes in each column, compute the condensation rate in kg/h. At $T_{sat} = 50 \ °C$, properties of R-12 are

$$i_{fg} = 122.3 \ \text{kJ/kg} \qquad \rho_v = 69.7 \ \text{kg/m}^3$$

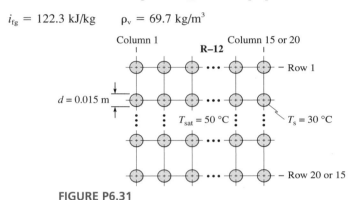

FIGURE P6.31

6.32 For condensation on a horizontal tube, the Reynolds number of the condensate at the bottom of the tube is defined as

$$\text{Re} = \frac{\Gamma}{\mu} \qquad \Gamma = \frac{h \pi d \ \Delta T}{2 i'_{fg}}$$

Employing Equation 6.2.12 for h, show that

$$\frac{h}{k} \left[\frac{\mu^2}{g \rho (\rho - \rho_v)} \right] = 0.763 \ \text{Re}^{-1/3}$$

6.33 Equation 6.2.9 is an empirical relation for the ratio of the average heat transfer coefficient with interfacial ripples to the average heat transfer coefficient computed from Equation 6.2.7a. Assume that a similar relation for the local heat transfer coefficient exists.

$$h_x/h_{xo} = C \ \text{Re}^n \qquad \text{Re} > \text{Re}_{cr}$$

where

h_x = local heat transfer coefficient with ripples
h_{xo} = local heat transfer coefficient computed from Equation 6.2.4
Re = local Reynolds number = Γ_x/μ
Re_{cr} = Reynolds number at which ripples appear

Employing Equation 6.2.2 for Γ and Equation 6.2.3 for δ, show that the average heat transfer coefficient with ripples is

$$\frac{h}{k}\left[\frac{\mu^2}{g\rho(\rho - \rho_v)}\right]^{1/3} = \frac{4}{3^{4/3}\,\mathrm{Re}^{4/3}}\left[\mathrm{Re_{cr}} + \frac{C}{n+1}\,(\mathrm{Re}^{n+1} - \mathrm{Re_{cr}^{n+1}})\right]$$

$\mathrm{Re} = \mathrm{Re}_{x=L}$

6.34 Saturated steam condenses on the inner vertical surface of a cylindrical container of diameter d and height H. The height H is such that the condensate is in laminar flow with a smooth interface. Find an expression for $dz/d\rho$, $z = $ height of the condensate in the container at time τ. Show that the total time for the tank to be filled by the condensate is

$$\tau_o = 1.06d\left[\frac{\mu i_{fg}^{\prime 3}\rho^3 H}{g(\rho - \rho_v)k^3\,\Delta T^3}\right]^{1/4}$$

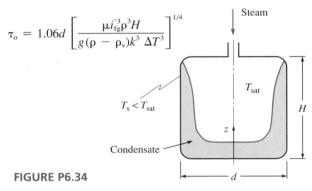

FIGURE P6.34

6.35 Saturated Refrigerant 12 vapor at 40 °C condenses on the outer surface of a 3-cm O.D. horizontal tube. The tube is maintained at 30 °C. Find the heat transfer rate per meter length of the tube. Properties of R-12 at $T_{sat} = 40$ °C are

$\rho_v = 54.57$ kg/m^3 $i_{fg} = 129.4$ kJ/kg

6.36 Saturated Refrigerant 12 vapor at 40 °C (quality $= 1$) enters a 20-mm diameter, horizontal tube. The surface temperature of the tube is 30 °C. The mass rate of the refrigerant flow is 10 kg/h. What should be the length of the tube for complete condensation of the refrigerant? Properties of R-12 are

$i_{fg}\,(40\,°C) = 129.4$ kJ/kg $\rho_v\,(40\,°C) = 54.47$ kg/m^3

6.37 Discussing the possibility of a turbulent condensate with condensation on a horizontal tube, it was stated that it is unlikely that for horizontal tubes used in engineering applications of condensation, the Reynolds number will exceed 450. For condensation of steam at 100 °C on a horizontal tube maintained at 60 °C, find the diameter for which $\mathrm{Re_{max}}\,[= (h\pi\,d\,\Delta T)/(2i_{fg}^{\prime}\mu)]$ is 450.

6.38 In a small steam condenser, there are 36 horizontal, 2-cm diameter tubes arranged in a square array. The condenser is to condense saturated steam at 50 °C. The temperature of the surface of the tubes is 40 °C. Determine
(a) The average heat transfer coefficient.
(b) The heat transfer rate to the tubes if they are 2-m long.
(c) The mass rate of condensation of steam.

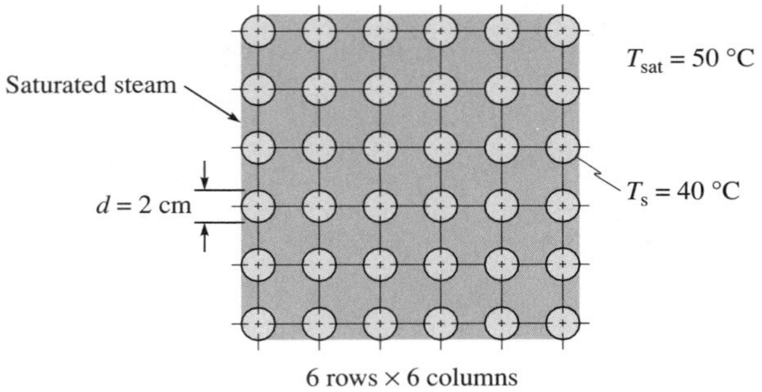

6 rows × 6 columns

FIGURE P6.38

6.39 Re-solve Problem 6.38 if the condenser tubes are vertical.

6.40 A 2.5-cm diameter horizontal tube is used to supply cold water. The temperature of the tubes is 10 °C. The surrounding air is at 40 °C with a relative humidity of 57%. Assuming that the vapor in the air condenses as saturated steam at its partial pressure,

   (a) Estimate the convective heat transfer coefficient.

   (b) Due to the presence of air, the actual heat transfer coefficient is expected to be reduced to 50% of the value calculated in part a. Determine the heat transfer rate per meter length of the tube and mass rate of condensation.

6.41 A 1-kW steam turbine demonstration plant is to be designed to operate on saturated steam at 160 °C. Find the mass rate of flow of the required steam, assuming an internal turbine efficiency of 60% and exhausting at atmospheric pressure. Design the boiler and condenser for the unit.   **(Design)**

CHAPTER

7

RADIATION I

In the preceding chapters we studied heat transfer by conduction and convection, both classified as diffusion of thermal energy processes. We now study heat transfer by radiation, which is different from heat transfer by diffusion. In this chapter we

■ *Discuss features that distinguish radiation from diffusion*

■ *Define emissive power, blackbody, and monochromatic emissive power—terms that are specific to radiative heat transfer*

■ *Introduce Planck's distribution for blackbody radiation and radiation functions*

■ *Define emissivity, absorptivity, transmissivity, reflectivity, and radiosity*

■ *Introduce view factor and view factor algebra*

■ *Solve problems involving radiative heat transfer among gray, opaque surfaces separated by a medium that does not participate in radiative heat transfer*

■ *Give an introduction to radiation intensity*

7.0 INTRODUCTION

The main differences between heat transfer by radiation and heat transfer by conduction and convection were touched upon in Section 1.2. We review those differences now as, having studied conductive and convective heat transfer, the differences become clearer.

There are two main differences between radiative heat transfer and conductive and convective heat transfer. The first difference is that in conduction and convection, the heat flux at a point is uniquely determined by the state of the medium immediately adjacent to the point. To determine the radiative heat flux at a point, we should know the state of the entire medium surrounding the point. Second, the time scale in radiation is much smaller than the time scale in conduction or convection. We will now elaborate on these two differences.

In conduction a knowledge of the temperatures of the medium in the immediate vicinity of a point is sufficient to determine the heat flux at that point. In convection the thermal energy transfer across any infinitesimal area can be determined if the local velocities and the temperatures in the immediate vicinity of the point are known. Thus, in both conduction and convection a knowledge of the parameters (temperatures and velocities) in the immediate vicinity of the point is sufficient to determine the energy flux at that location.

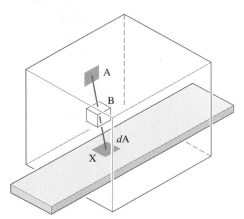

FIGURE 7.0.1 A Steel Slab in a Furnace
Radiant energy reaches the steel slab from
both the walls of the furnace and
combustion gases.

Differences
between radiative
and conductive and
convective heat
transfer

Now, consider the radiative heat transfer to a steel slab in a furnace, Figure 7.0.1. If the furnace is fired by oil or natural gas, the products of combustion contain carbon dioxide and water vapor, both of which participate in radiative heat transfer; i.e., the effect of emission and absorption of radiant energy by the gases is significant. In the figure, A represents an infinitesimal area on the furnace wall, B an infinitesimal volume of the combustion products, and X an infinitesimal area on the steel slab. The infinitesimal volume B is between X and A along a line connecting X and A. Radiant energy reaches X both from the furnace walls and combustion gases. Examine the energy reaching the element X from the walls. Radiant energy leaves the infinitesimal surface, A, on one of the walls of the furnace and reaches X. The energy that reaches X is less than the energy leaving A, since a part of the energy leaving A is absorbed by some of the gases, such as carbon dioxide and water vapor. The amount of energy that is absorbed depends on the length of the path between A and X and the density and temperature of the gases. To determine the energy that leaves the walls of the furnace and reaches the element X, we sum the energy reaching X from all the elements, such as A, that make up the walls and can "see" X. Similarly, a part of the energy emitted by each infinitesimal volume of the gases surrounding X, such as volume B, reaches X after attenuation by the gases lying between the element and X. The emission from the element depends on the composition and temperature of the gases in the element. The attenuation depends on the distance between the element and X and the composition and temperature of the gases between the element and X. Thus, the calculation of the total amount of radiant energy received by the element X, involves the summation of the radiant energy from every element of the walls and every packet of gases that can see X. It is evident that a mere knowledge of the state of the medium in the immediate vicinity of X on the gaseous side is not sufficient to find the radiant heat flux to X.

The second difference is the time scale involved in the two processes. Energy transfer by conduction and convection is much slower than by radiation. Figure 7.0.2 shows a slab of material of thickness L. An electrical resistance heating element is

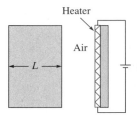

FIGURE 7.0.2 A heater, separated from the surface of a slab by air, transfers energy to the surface facing it by radiation. Heat transfer from the surface facing the heater to the opposite surface of the slab is by conduction.

placed at a small distance from the slab with air between the slab and the element. If the heating element is energized, the slab surface facing the element is immediately affected by the energy reaching the surface by radiation, and the temperature of the surface increases. But the effect of the change in the temperature of the surface is not immediately transmitted to the opposite surface of the slab. The time required for the thermal disturbance at one surface of the slab to affect the parallel surface can be calculated by the method presented in Example 3.5.1. The time for a temperature disturbance to propagate across a copper slab by conduction is about 3 s, if the thickness of the slab is 2 cm, and 12 s, if the thickness is 4 cm. Even if the distance between the heating element and the slab surface is much greater than 4 cm, the time for the effect of a change in the temperature of the heating element to reach the slab by radiation is negligibly small since radiation travels at the speed of light. The time scales in convection are similar to those in conduction.

In this chapter, we restrict ourselves to radiative heat transfer in a few simple cases with a large number of engineering applications. We consider radiative heat transfer between simple surfaces (defined here as diffuse and gray) separated by a medium that does not participate in radiation (emission and absorption in the medium being negligible). Radiative properties (reflectivity and emissivity, which are defined later) are independent of direction for a diffuse surface, and independent of wavelength for a gray surface. Emission and absorption by air are negligible unless the air is at a very high temperature. A medium containing gases such as carbon dioxide, water vapor, and other complex molecules even in small amounts may lead to absorption and emission of radiant energy. Some cases that involve a participating medium or nongray surfaces are considered in Chapter 8.

7.1 RADIATION PHENOMENON

A single theory to explain all the observed results of radiation is not yet available. Some results are explained by electromagnetic wave theory and others by quantum concepts. For our purposes, the electromagnetic wave theory is more appropriate. Different mechanisms in the electromagnetic phenomena lead to different types of radiation such as γ-rays, x-rays, radio waves, and thermal radiation. If the radiation is solely due to the temperature of the medium, it is termed *thermal radiation*.

Radiation spectra and thermal wave band

Various spectra of radiation are given different names as shown in Table 7.1.1 and Figure 7.1.1. The classification of the different wave bands is somewhat arbitrary. Different sources give different ranges for the various radiation spectra. The highest temperature in engineering applications is about 6000 K (temperature of the

TABLE 7.1.1 Electromagnetic Frequency Spectra

Application	Typical Frequency (Hz)	Typical Wavelength	Approximate Wave Band
Electric AC power	60	5×10^6 m	1.2×10^7–5×10^6 m
Induction furnace	2000	1.5×10^5 m	6×10^5–10^5 m
RF heating—metals	10^4	3×10^4 m	3×10^5–300 m
AM radio	10^6	300 m	550–180 m
Shortwave radio	2×10^7	15 m	120–13 m
Microwave	2.7×10^7	11 m	
FM radio	10^8	3 m	3.4–2.8 m
TV channels 2–13	1.8×10^8	1.67 m	5.55–1.4 m
Radar	10^{10}	0.03 m	0.25–0.015 m
Far infrared	3×10^{12}	100 μm	25–1000 μm
Infrared	1.25×10^{14}	2.4 μm	0.8–25 μm
Infrared heaters	1.5×10^{14}	2 μm	
Visible light	5×10^{14}	0.6 μm	0.4–0.7 μm
Solar—maximum intensity	7.1×10^{14}	0.42 μm	
Thermal radiation	10^{14}	3 μm	0.1–1000 μm
X rays	10^{16}	3×10^{-4}	10^{-5}–2×10^{-2} μm
Gamma rays	10^{21}	3×10^{-7} μm	4×10^{-7}–1.4×10^{-4} μm
Cosmic rays	3×10^{23}	10^{-9} μm	

Adapted from *C.R.C. Handbook of Tables for Applied Engineering Science,* and also from Bolz, R. E., Tuve, G. L., and Chapman, A. J., eds., *Heat Transfer* (New York: MacMillan, 1984).

c, the velocity of propagation of electromagnetic waves in vacuum is given by $c = \eta \lambda = 2.997925 \times 10^8$ m/s.

RF: Radiofrequency   FM: Frequency modulation   AM: Amplitude modulation

Notes: μm (micrometer or micron) $= 10^{-6}$ m. Another unit for wavelength is the Angstrom, which is 10^{-4} μm or 10^{-10} m.

surface of the sun). The lowest temperature, below which radiative energy transfer is negligible in most cases, is 50 K. At any temperature between 50 K and 6000 K the radiant energy in the wave band 0.1 μm to 1000 μm has a direct effect on the temperature of the medium participating in radiation and on the radiative heat transfer. The thermal radiant energy in the wave band 0 to 0.1 μm and above 1000 μm is negligibly small compared with the radiant energy in the band 0.1 μm to 1000 μm. Hence, the wave band 0.1 μm to 1000 μm is usually referred to as the thermal radiation wave band. The visible spectrum is in the wave band 0.4 μm ~ 0.7 μm. Radiation in the wavelengths less than 0.4 μm is termed *ultraviolet radiation* and radiation in the wavelengths greater than in the visible spectrum is termed *infrared radiation*.

In dealing with radiation, we introduce several terms that are defined in the succeeding paragraphs. To understand radiative heat transfer, a good grasp of these definitions is essential. *All these definitions are given in relation to solid surfaces.*

In this chapter *only radiant energy will be considered,* unless stated otherwise. For example, the term heat transfer rate should be interpreted as *radiative heat transfer rate.* This applies to other terms involving energy.

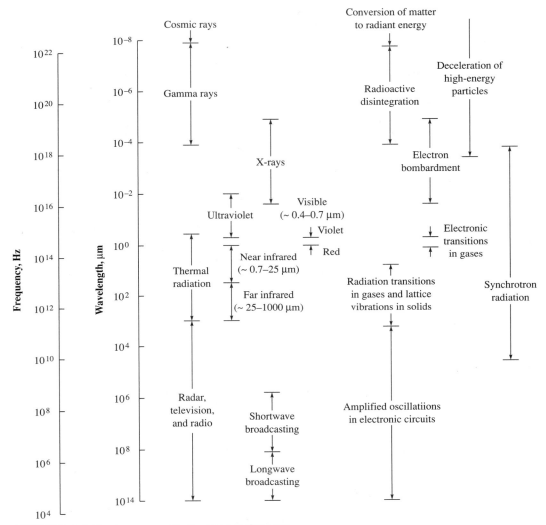

FIGURE 7.1.1 Electromagnetic Frequency Spectra

Emissive power

Emissive Power E Every surface emits energy. *The total amount of energy emitted by a surface in all directions, i.e., in a hemisphere surrounding it, over all wavelengths per unit surface area per unit time is its emissive power at the temperature of the surface.* The emissive power of a surface is dependent on the characteristics and temperature of the surface.

Blackbody—Stefan-Boltzmann Law for blackbody emissive power

Blackbody At any given temperature, there is an upper limit to the emissive power of a surface whatever be the nature of the surface. The surface that has the maximum emissive power at a given temperature is known as a *black surface*. As a black surface emits the maximum amount of radiant energy, it is also known as an ideal radiator or emitter. Not all ''black'' surfaces appear black at all temperatures; but a

In 1879 Josef Stefan pointed out that at high temperatures the radiative heat transfer rate from a surface was proportional to the fourth power of the absolute temperature of the surface. In 1884 Ludwig Boltzmann, pictured here, presented a derivation of the T^4 radiation law based on thermodynamic considerations. Boltzmann's contributions in thermodynamics are significant. The epitaph on his tombstone in Vienna is "$S = k \ln \omega$" to commemorate his contribution to the theory of entropy.

A blackbody has the highest emissive power at a defined temperature

true black surface appears black at room temperatures ($\approx$ 300 K). The emissive power of a black surface is given by the Stefan-Boltzmann law

$$\| \quad E_b = \sigma T^4 \quad \| \tag{7.1.1}$$

where

σ = Stefan-Boltzmann constant = 5.67×10^{-8} W/m^2 K^4
T = temperature of the surface (K)

Monochromatic emissive power

Monochromatic Emissive Power All surfaces emit radiation at many wavelengths. The radiation from many surfaces is continuous in wavelengths from 0 to ∞ μm. The monochromatic emissive power of a surface is defined by the equation

$$dE = E_\lambda \, d\lambda$$

where

dE = emissive power in the infinitesimal wave band between λ and $\lambda + d\lambda$ (W/m^2)
E_λ = monochromatic emissive power (W/m^3)

The monochromatic emissive power of a black surface is the highest at every wavelength (at any given temperature) and is given by *Planck's Distribution*

$$\| \quad E_{b\lambda} = \frac{c_1}{\lambda^5[\exp(c_2/\lambda T) - 1]} \quad \| \tag{7.1.2}$$

where

$$c_1 = 3.74047 \times 10^8 \text{ W } \mu\text{m}^4/\text{m}^2$$
$$c_2 = 1.43868 \times 10^4 \text{ } \mu\text{m K}$$

(λ is in μm). Curves of the monochromatic emissive power of a blackbody are shown as functions of the wavelength λ, at a few temperatures, in Figure 7.1.2.

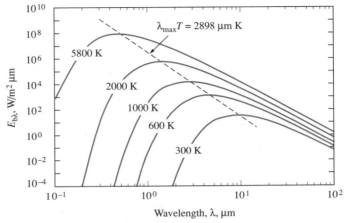

FIGURE 7.1.2 Monochromatic Emissive Power of a Blackbody
The wavelength, at which $E_{b\lambda}$ is a maximum, increases with a decrease in temperature.

From Figure 7.1.2 we observe that

1. At any given wavelength, the blackbody monochromatic emissive power increases with temperature.
2. The wavelength λ_{max} at which $E_{b\lambda}$ is a maximum decreases as the temperature increases.

The logarithmic scale used in Figure 7.1.2 masks the relative magnitude of the monochromatic emissive powers at different temperatures. If the maximum value of $E_{b\lambda}$ at a room temperature of 300 K is represented by a height of 1 cm (slightly less than 0.5 in) on a regular graph paper, the maximum value of $E_{b\lambda}$ at 5800 K (the approximate temperature of the surface of the sun) is 27 km ($\approx$ 17 miles) high!

The wavelength, at which the blackbody monochromatic emissive power is maximum is determined by setting $dE_{b\lambda}/d\lambda = 0$ (holding the temperature constant). At a fixed temperature, differentiating $E_{b\lambda}$ with respect to λ (from Equation 7.1.2), and setting it equal to zero, we obtain

$$\frac{dE_{b\lambda}}{d\lambda} = -\frac{5c_1}{\lambda^6[\exp(c_2/\lambda T) - 1]} + \frac{c_1 \exp(c_2/\lambda T)}{\lambda^5[\exp(c_2/\lambda T) - 1]^2} \frac{c_2}{\lambda^2 T} = 0$$

Simplifying and rearranging, we obtain, $x = 5(1 - e^{-x})$, where $c_2/\lambda T = x$. The solution to the equation is

$$x = \frac{c_2}{\lambda T} = 4.9651$$

Until Max Planck (1858–1947) proposed his formula for monochromatic emissive power, two formulae existed. One was good for short wavelengths but failed at long wavelengths—Wien's distribution, $E_{b\lambda} = c_1/[\lambda^5 e^{(c_2/\lambda T)}]$—and the other was good for long wavelengths but failed at short wavelengths—Raleigh-Jean's distribution, $E_{b\lambda} = c_1\lambda T/(c_2\lambda^5)$. In October 1900, based on his prior work on entropy, a knowledge of mathematical functions, and, in his own words, "lucky intuition", he modified Wien's distribution by adding (-1) to the denominator. This modification accurately predicted the monochromatic emissive power at all wavelengths. Planck was not satisfied with his empirical approach to the formula. He worked furiously and, in the brief span of a few weeks, proposed the famous Quantum Theory in December 1900. The theory was so revolutionary that it took the scientific community seventeen years before he was awarded the Nobel Prize.

With $c_2 = 1.43868 \times 10^4$ µm K, we get

Wein's Displacement Law

$$\| \quad \lambda_{max}T = 2897.6 \text{ µm K} \quad \| \tag{7.1.3}$$

where λ_{max} denotes the wavelength at which the monochromatic emissive power, $E_{b\lambda}$, is a maximum (and not the maximum value of λ, which is ∞) at the defined temperature T. The relation given by Equation 7.1.3 is known as *Wien's Displacement Law*.

We can obtain the Stefan-Boltzmann relation (Equation 7.1.1) from Equation 7.1.2 as follows:

$$E_b = \int_0^\infty E_{b\lambda}\, d\lambda = \int_0^\infty \frac{c_1}{\lambda^5[\exp(c_2/\lambda T) - 1]}\, d\lambda$$

$$= T^4 \int_0^\infty \frac{c_1\, d\lambda\, T}{(\lambda T)^5[\exp(c_2/\lambda T) - 1]} = c_1 T^4 \int_0^\infty \frac{dx}{x^5[\exp(c_2/x) - 1]}$$

where $\lambda T = x$. With a change of variable $c_2/x = y$, the integral becomes

$$E_b = \frac{c_1}{c_2^4} T^4 \int_0^\infty \frac{y^3}{e^y - 1} \, dy$$

From a table of integrals

$$\int_0^\infty \frac{y^3}{e^y - 1} \, dy = 6 \sum_{n=1}^\infty \frac{1}{n^4} \quad \text{and} \quad \sum_{n=1}^\infty \frac{1}{n^4} = \frac{\pi^4}{90}$$

which gives

$$E_b = \frac{c_1 \pi^4}{c_2^4 15} T^4 = \sigma T^4 \tag{7.1.1}$$

where

$$\sigma = \frac{c_1 \pi^4}{c_2^4 15} = \frac{3.74047 \times 10^8}{(1.43868 \times 10^4)^4} \frac{\pi^4}{15}$$
$$= 5.6699 \times 10^{-8} \text{ W/m}^2 \text{ K}^4 \approx 5.67 \times 10^{-8} \text{ W/m}^2 \text{ K}^4$$

Equation 7.1.1 gives the maximum amount of energy that can be emitted by a surface at a temperature T over all wavelengths. In some cases, the energy emitted in a finite wave band λ_1 to λ_2 may be required. For example, we may wish to find the energy emitted in the visible spectrum when a blackbody is maintained at a defined temperature. In such cases, we need

$$E_{b,\lambda_1-\lambda_2} = \int_{\lambda_1}^{\lambda_2} E_{b\lambda} \, d\lambda \tag{7.1.4}$$

While the integral with the lower and upper limits of 0 and ∞ could be evaluated, Equation 7.1.4 cannot be evaluated in a closed form for finite values of the limits of integration. It needs to be evaluated numerically. Instead of evaluating the integral for different values of wavelengths and different temperatures, a table of the computed integrals may be made and used for all possible combinations of the limits of integration and temperature by changing the independent variable from λ to λT. Now,

$$E_{b,\lambda_1-\lambda_2} = \int_{\lambda_1}^{\lambda_2} E_{b\lambda} \, d\lambda = \int_0^{\lambda_2} E_{b\lambda} \, d\lambda - \int_0^{\lambda_1} E_{b\lambda} \, d\lambda$$

Changing the variable from λ to λT, and noting that the integral is evaluated at a fixed value of T, after a little manipulation, we obtain (with the upper limit of the wavelength denoted by λ),

$$f_{0-\lambda} = \frac{E_{b,0-\lambda}}{\sigma T^4} = \frac{c_1}{\sigma} \int_0^{\lambda T} \frac{dx}{x^5 (e^{c_2/x} - 1)} \tag{7.1.5}$$

TABLE 7.1.2 Blackbody Radiation Functions

λT (μm K)	$f_{0-\lambda}$	λT (μm K)	$f_{0-\lambda}$	λT (μm K)	$f_{0-\lambda}$
100	0.000000	4800	0.607589	9500	0.903124
200	0.000000	4900	0.620937	9600	0.905490
300	0.000000	5000	0.633777	9700	0.907782
400	0.000000	5100	0.646127	9800	0.910002
500	0.000000	5200	0.658001	9900	0.912153
600	0.000000	5300	0.669417	10000	0.914238
700	0.000002	5400	0.680392	11000	0.931929
800	0.000016	5500	0.690940	12000	0.945138
900	0.000087	5600	0.701079	13000	0.955179
1000	0.000321	5700	0.710824	14000	0.962938
1100	0.000911	5800	0.720192	15000	0.969021
1200	0.002134	5900	0.729196	16000	0.973855
1300	0.004317	6000	0.737852	17000	0.977741
1400	0.007791	6100	0.746173	18000	0.980901
1500	0.012850	6200	0.754174	19000	0.983494
1600	0.019720	6300	0.761869	20000	0.985643
1700	0.028535	6400	0.769268	21000	0.987437
1800	0.039344	6500	0.776386	22000	0.988947
1900	0.052111	6600	0.783234	23000	0.990227
2000	0.066733	6700	0.789823	24000	0.991319
2100	0.083058	6800	0.796164	25000	0.992256
2200	0.100895	6900	0.802268	26000	0.993064
2300	0.120037	7000	0.808144	27000	0.993765
2400	0.140266	7100	0.813803	28000	0.994376
2500	0.161366	7200	0.819253	29000	0.994911
2600	0.183132	7300	0.824504	30000	0.995381
2700	0.205370	7400	0.829563	35000	0.997044
2800	0.227904	7500	0.834439	40000	0.998008
2900	0.250577	7600	0.839139	45000	0.998605
3000	0.273248	7700	0.843671	50000	0.998994
3100	0.295797	7800	0.848042	55000	0.999259
3200	0.318120	7900	0.852258	60000	0.999444
3300	0.340130	8000	0.856325	65000	0.999579
3400	0.361755	8100	0.860251	70000	0.999678
3500	0.382937	8200	0.864040	75000	0.999754
3600	0.403628	8300	0.867699	80000	0.999812
3700	0.423794	8400	0.871232	85000	0.999857
3800	0.443406	8500	0.874645	90000	0.999893
3900	0.462446	8600	0.877943	95000	0.999922
4000	0.480902	8700	0.881129	100000	0.999946
4100	0.498767	8800	0.884210		
4200	0.516040	8900	0.887188		
4300	0.532723	9000	0.890068		
4400	0.548823	9100	0.892853		
4500	0.564348	9200	0.895548		
4600	0.579309	9300	0.898156		
4700	0.593718	9400	0.900680		

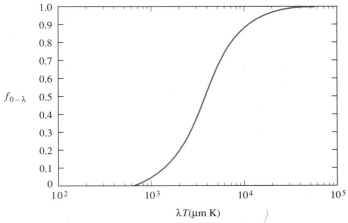

FIGURE 7.1.3 Graph of $f_{0\text{-}\lambda}$ as a Function of λT

where λT has been replaced by x. The values of the integral on the right-hand side of Equation 7.1.5 are given in Table 7.1.2 for various values of the upper limit and are shown in Figure 7.1.3. From Table 7.1.2 it is clear that the emitted energy as a fraction of the total emitted energy is negligible for $\lambda T < 1000$ μm K.

Wiebelt (1960) has given the following approximations to the values in the table

$$x \geq 2: \qquad f_{0\text{-}\lambda} = \frac{15}{\pi^4} \sum_{m=1,2,\ldots} \frac{e^{-mx}}{m^4} \{[(mx + 3)mx + 6]mx + 6\}$$

$$x < 2: \qquad f_{0\text{-}\lambda} = 1 - \frac{15}{\pi^4} x^3 \left(\frac{1}{3} - \frac{x}{8} + \frac{x^2}{60} - \frac{x^4}{5040} + \frac{x^6}{272\,160} - \frac{x^8}{13\,305\,600} \right)$$

where $x = c_2/\lambda T$. The values from the curve fits are within 0.01% of the values in Table 7.1.2. In the HT software, subprogram RF, $f_{0\text{-}\lambda}$ is computed from Wiebelt's polynomial approximations.

EXAMPLE 7.1.1

A blackbody is maintained at 2800 K (approximate temperature of the filament of an incandescent lamp) and another blackbody at 5760 K (approximate effective blackbody temperature of the surface of the sun). Determine the energy in the visible spectrum as a fraction of the total energy emitted in each case.

From Table 7.1.1 the visible spectrum is in the range $\lambda_1 = 0.4$ μm to $\lambda_2 = 0.7$ μm. With values of $f_{0\text{-}\lambda}$ from Table 7.1.2 (or from the software RF) we construct the following table.

	2800 K	5760 K
$\lambda_2 T$ (μm K)	1960	4032
$\lambda_1 T$ (μm K)	1120	2304
$\dfrac{E_{b,0\text{-}\lambda_2 T}}{\sigma T^4} = f_2$	0.06088	0.4867
$\dfrac{E_{b,0\text{-}\lambda_1 T}}{\sigma T^4} = f_1$	0.00116	0.1208
$f_2 - f_1$	0.0597	0.3659

f_1 and f_2 give the energy emitted in the wave band 0 to λ_1 and 0 to λ_2, respectively, at the specified temperatures as a fraction of the total energy emitted. The last row gives the difference between these two values and represents the energy in the visible spectrum as a fraction of the total energy radiated at these two temperatures. At 2800 K about 6% of the emitted energy is in the visible spectrum, whereas at 5760 K 36.6% of the emitted energy is in the visible spectrum. The sun is a lot more efficient as a light source than an incandescent lamp.

Table 7.1.2 shows that even at such a low temperature as 100 K, the energy emitted by a blackbody in the wave band 0.1 μm to 1000 μm is more than 99.99% of the blackbody emissive power. At higher temperatures, the energy emitted in the thermal radiation band (0.1 μm to 1000 μm) is even greater. Hence, the wave band 0.1 μm to 1000 μm is designated as the thermal radiation band.

We now define some of the radiative properties of surfaces. In the most general case the radiative properties of surfaces are dependent on the direction (the angle between the normal to the surface and the line indicating the direction), wavelength, surface temperature, material, and condition of the surface. The integrated mean value of a property at a specified wavelength over all directions is the spectral hemispherical property. The integrated mean value of the property over all directions and over the entire wave band is the total hemispherical property, often referred to simply as the total property, for example, total emissivity.

Emissivity

Emissivity ε A blackbody is an ideal emitter, just as an inviscid fluid is an ideal fluid, or a frictionless surface is an ideal surface. The energy emitted by any real surface is less than the energy emitted by a blackbody at the same temperature. The ratio of the monochromatic emissive power E_λ to the monochromatic blackbody emissive power $E_{b\lambda}$ at the same temperature is the spectral hemispherical emissivity of the surface

$$\varepsilon_\lambda = \frac{E_\lambda}{E_{b\lambda}}$$

From the above relation we obtain the total emissive power of a surface (over all wavelengths) as

$$E = \int_0^\infty E_\lambda \, d\lambda = \int_0^\infty \varepsilon_\lambda E_{b\lambda} \, d\lambda$$

Total hemispherical emissivity

To determine E, we should know the functional relation between ε_λ and λ. We define the *total hemispherical emissivity, ε*, as

$$\left\| \quad \varepsilon = \frac{E}{E_b} = \frac{1}{E_b} \int_0^\infty \varepsilon_\lambda E_{b\lambda} \, d\lambda \quad \right\| \tag{7.1.6}$$

Gray surface—a surface with wavelength-independent radiation properties

Here, ε can be interpreted either as the emissivity of a body, which is wavelength independent, i.e., ε_λ is constant, or as the average emissivity of a surface at that particular temperature. *A surface whose radiation properties are independent of the wavelength is known as a gray surface and for such a surface, the emissive power is given by*

$$E = \varepsilon E_b$$

A black surface is a gray surface with $\varepsilon = 1$. The spectral hemispherical emissivity, ε_λ, is a function of both the temperature of the surface and wavelength. Even if ε_λ is weakly dependent on the temperature but varies significantly with the wavelength, the total hemispherical emissivity ε is still a function of the temperature because $E_{b\lambda}$ is temperature dependent. In many cases ε_λ is a weak function of both the temperature and wavelength, and in such cases, ε defined by Equation 7.1.6 is reasonably constant over a wide range of temperatures.

For a black surface, $\varepsilon = 1$. Even for a gray surface, while the emissivity is independent of the wavelength, it may vary with temperature. In most cases considered in this chapter, properties of surfaces are assumed to be wavelength independent. We also restrict ourselves to only those surfaces whose properties are assumed to be uniform over the surface and assumed to be temperature independent.

Absorbtivity, reflectivity, and transmissivity

Absorptivity α, Reflectivity ρ, and Transmissivity τ Consider a semitransparent surface that receives incident radiant flux of G units (Figure 7.1.4). The incident

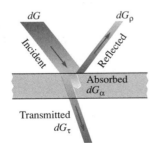

FIGURE 7.1.4 A Radiant Beam Reaching a Semitransparent Surface
Part of the beam is transmitted, a part absorbed, and the rest reflected.

radiant flux is also known as *irradiation*. Let dG represent the incident energy in the wave band λ to $\lambda + d\lambda$. Part of it may be transmitted, part of it absorbed, and the rest reflected by the surface. We define monochromatic properties, absorptivity α_λ, reflectivity ρ_λ, and transmissivity τ_λ as

$$\alpha_\lambda = \frac{dG_\alpha}{dG} \qquad \rho_\lambda = \frac{dG_\rho}{dG} \qquad \tau_\lambda = \frac{dG_\tau}{dG}$$

where dG_α, dG_ρ, and dG_τ are, respectively, the absorbed, reflected, and transmitted part of the incident energy dG. With $dG = dG_\alpha + dG_\rho + dG_\tau$, we have

$$\alpha_\lambda + \rho_\lambda + \tau_\lambda = 1 \tag{7.1.7a}$$

The total hemispherical properties, absorptivity α, reflectivity ρ, and transmissivity τ, are defined by

$$\alpha = \frac{G_\alpha}{G} \qquad \rho = \frac{G_\rho}{G} \qquad \tau = \frac{G_\tau}{G}$$

where G_α, G_ρ, and G_τ are, respectively, the absorbed, reflected, and transmitted part of the incident energy flux (in all wavelengths from all directions). As $G = G_\alpha + G_\rho + G_\tau$, it follows that

$$\alpha + \rho + \tau = 1 \tag{7.1.7b}$$

Opaque surface

A surface with zero transmissivity ($\tau = 0$) is termed an opaque surface. For opaque surfaces $\alpha + \rho = 1$.

Blackbody radiation characteristics

Blackbody Radiation We have already indicated that a blackbody has the highest emissive power at a defined temperature. A blackbody also absorbs all the incident energy (in every wave band). Thus, for a black surface, $\alpha_\lambda = 1$ and $\alpha = 1$. From Equations 7.1.7a and b it follows that for a black, opaque surface, $\tau = 0$ and $\rho = 0$. Further, the emission from a black surface is direction independent, i.e., it is a diffuse emitter. We now summarize the radiative characteristics of a blackbody.

1. It is a perfect emitter. At any prescribed temperature no surface emits more energy in any wave band than a blackbody at the same temperature and in the same wave band.
2. A blackbody absorbs *all* the incident radiant energy.
3. A blackbody is a diffuse emitter.

Why is a blackbody termed black?

Our perception of the color of a surface depends on the spectral reflectivity of the surface. If the reflectivity of the surface is such that it reflects mainly in a narrow band, say around 0.65 μm, which represents the color red, the surface appears red if the incident energy has a red component. If the incident energy does not have any red component and the surface reflectivity is mainly in the red band, then the surface appears black. For a black surface $\varepsilon = \alpha = 1$, and $\rho = 0$. Because a black surface does not reflect any incident energy, and at room temperatures the emitted energy in the visible spectrum is insignificant, the surface appears black, and hence, the term black. But at higher temperatures, the emitted energy in the visible spectrum becomes significant; even if there is no reflected energy, depending on the temperature, a black surface will appear red, orange, white, and so forth, because of the emitted energy. Although a truly blackbody appears black at room temperature, many surfaces have an absorptivity of very nearly 1 but reflect light. Such bodies may appear white, red, and so on because of reflection of incident energy but behave as black bodies. Therefore, it should not be concluded that the color of a surface indicates if a surface has a high emissivity.

No surface is a perfect emitter (just as no surface is frictionless). By utilizing the property of a blackbody that it absorbs all incident radiant energy, a very close

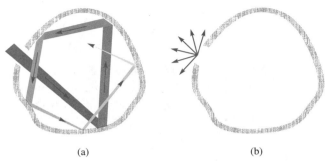

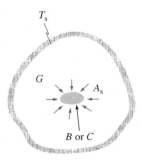

FIGURE 7.1.5 (a) A large cavity with a small opening. Incident energy undergoes repeated reflection in the cavity and very little of the incident energy leaves the cavity. (b) The emitted energy leaving the cavity is direction independent.

A small opening in a large cavity is a close approximation to a blackbody

approximation to a blackbody can be achieved by a large cavity with a small opening as shown in Figure 7.1.5. (Such a cavity with a small opening is often referred to as a *hohlrum*.) The surface of the cavity is isothermal. Any radiant energy entering the cavity through the small opening undergoes repeated reflection. Every time the radiant energy strikes the cavity surface, a part of the energy is absorbed and the rest is reflected. Since every ray of the incident energy experiences many reflections before it re-emerges, most of the incident energy is absorbed by the cavity. The larger the cavity and the smaller the opening, the smaller the amount of incident energy that is reradiated through the opening. Almost all the energy emerging from the opening is due to emission from the cavity surface and such emission is direction independent as shown in Figure 7.1.5b. Since almost all the incident energy is absorbed by the cavity surface and the energy emerging from the opening is diffuse, the cavity has the characteristics of a black surface even if the absorptivity of the surface is low; the cavity is a better approximation to a blackbody if its absorptivity is high, however.

Kirchhoff's Law Consider a small blackbody, B, with a surface area A_s, inside a large isothermal cavity whose surface is at a uniform temperature T_s (Figure 7.1.6). After a sufficient time, the temperature of B must equal T_s. In such an equilibrium state the net heat transfer rate from B is zero.

The total energy emitted by the blackbody is $A_s \sigma T_s^4$. Denoting the incident energy per unit area of B from the surface of the cavity to B by G, the total incident energy

FIGURE 7.1.6 A Small Body in Thermal Equilibrium Inside a Large Cavity Whose Surface Is at Uniform Temperature

on B is $A_s G$. Since B is black, it absorbs all the incident energy. Thus, with no heat transfer from B, in equilibrium conditions,

$$A_s \sigma T_s^4 = A_s G$$

Therefore, the incident energy flux on a surface enclosed by the cavity is equal to the blackbody emissive power at T_s, the temperature of the cavity surface.

Equality of emissivity and absorptivity in an isothermal enclosure in thermal equilibrium

Now replace B by an opaque body C with a surface area A_s whose emissivity is ε and whose absorptivity is α. When equilibrium conditions are established, the temperature of C equals T_s. The irradiation on C is σT_s^4 as shown in the preceding discussion. The total incident energy is $A_s \sigma T_s^4$ and the absorbed energy is $\alpha A_s \sigma T_s^4$. The emitted energy is $\varepsilon A_s \sigma T_s^4$. In equilibrium conditions the net heat transfer from C is zero and the total incident energy equals the total energy leaving surface C. From an energy balance on the surface C,

Incident energy
= Emitted energy
+ Reflected energy

But for an opaque surface the difference between the incident energy and the reflected energy is the absorbed energy and

Absorbed energy
= Emitted energy

or

$$\alpha A_s \sigma T_s^4 = \varepsilon A_s \sigma T_s^4$$

$$\| \quad \varepsilon = \alpha \quad \| \qquad \textbf{(7.1.8a)}$$

Kirchhoff's Law— equality of spectral emissivity and absorptivity

The preceding concepts can be applied to spectral conditions and it can be shown that

$$\| \quad \varepsilon_\lambda = \alpha_\lambda \quad \| \qquad \textbf{(7.1.8b)}$$

The relation expressed by Equations 7.1.8b is known as Kirchhoff's Law. Equation 7.1.8b is derived with the assumption of thermodynamic equilibrium and isothermal surfaces; i.e., both the surfaces are at the same temperature and there is no net heat transfer from the surfaces. In real situations, in most cases there is heat transfer from the surfaces and the question arises whether Equation 7.1.8b can be used when such a nonequilibrium condition exists. It also raises the question whether the monochromatic emissivity and absorptivity are properties of surfaces and, if so, how they are influenced by the incoming radiation. In most cases even when there is heat transfer from a surface, the surface can be assumed to be in local thermodynamic equilibrium (we can then speak of the temperature and other properties of the surface) and Equation 7.1.8b is valid and Kirchhoff's Law can be expected to hold. There is also experimental evidence that Equation 7.1.8a is valid even when there is heat transfer from a surface. We will, therefore, assume the validity of Equation 7.1.8b in all cases. When Kirchhoff's Law is invoked, it should be interpreted in that sense. Furthermore, for a gray surface, since ε_λ and α_λ are wavelength independent, it follows that $\varepsilon = \alpha$. For a more detailed discussion on Kirchhoff's Law refer to Chandrasekhar (1960), Sparrow and Cess (1978), and Siegel and Howell (1981).

Validity of Kirchhoff's Law when the surfaces are not in thermodynamic equilibrium

For a gray surface $\alpha = G$

Total hemispherical absorptivity   We define the *total hemispherical absorptivity* (similar to Equation 7.1.6 for emissivity) as

$$\alpha = \frac{1}{G} \int_0^\infty \alpha_\lambda G_\lambda \, d\lambda \qquad (7.1.9)$$

where $G_\lambda \, d\lambda$ represents dG, the incident energy in the wave band λ to $\lambda + d\lambda$, and the integral is the energy absorbed by the surface over all the wavelengths. By Kirchhoff's Law, $\alpha_\lambda = \varepsilon_\lambda$, but, in general, the average emissivity and absorptivity as defined by Equations 7.1.6 and 7.1.9 will not be equal *if they are functions of the wavelength*. This is because E_λ is dependent on the temperature of the surface, but G_λ is dependent on the properties of the source of the incident energy, the temperature of the source, and a few other parameters. E_λ and G_λ are not usually equal. Only in the special case of the surface being surrounded by an identical surface at the same temperature will the two quantities be equal.[1]

For a nongray surface there is no unique value of total hemispherical absorptivity

Gray surface   **Gray Surface** For a gray surface the absorptivity, transmissivity, and reflectivity, are independent of the wavelength. For such a surface, we have

$$\alpha + \tau + \rho = 1$$

$$\varepsilon = \alpha$$

Diffuse surface   **Diffuse Surface** In general, the radiative properties of a surface, absorptivity, reflectivity, and emissivity are direction dependent. Surfaces for which these properties are direction independent are known as *diffuse surfaces*.

EXAMPLE 7.1.2

A plane, gray, diffuse, opaque surface (absorptivity = 0.7) with a surface area of 0.5 m^2, is maintained at 500 °C and receives radiant energy at a rate of 10 000 W/m^2 (Figure 7.1.7). Determine per unit time

(a) The energy absorbed.
(b) The radiant energy emitted.
(c) The total energy leaving the surface per unit area.
(d) The radiant energy emitted by the surface in the wave band 0.2 μm to 4 μm.
(e) The net radiative heat transfer from the surface.

Given

Plane, gray, diffuse, opaque surface $\alpha = 0.7$
A_s (surface area) = 0.5 m^2 $G = 10\ 000 \text{ W/m}^2$
$T_s = 500 \text{ °C}$

Find

(a) Rate of energy absorbed
(b) Emitted radiant energy
(c) Total radiant energy leaving the surface per unit area

---

[1]In applications involving solar energy it is usual to refer to solar absorptivity in the sense that the average absorptivity is defined by Equation 7.1.9 for incident energy from the sun. The emissivity for most surfaces in solar applications is emissivity around room temperature (≈ 300 K).

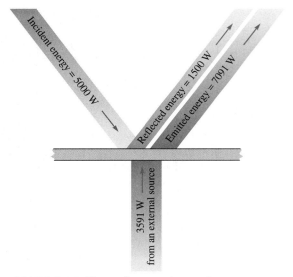

FIGURE 7.1.7 The surface receives radiant energy at the rate of 5000 W. Radiant energy leaves the surface at the rate of 8591 W. The net radiant energy transfer from the surface is 3591 W. To maintain the surface at 500 °C, the surface has to be supplied with 3591 W from an external source.

(d) Emitted radiant energy in the wave band 0.2 μm–4 μm

(e) q (net radiative heat transfer rate from the surface)

SOLUTION

Denoting the incident radiant flux by G,

(a) Rate of energy absorbed $= \alpha AG = 0.7 \times 0.5 \times 10\,000 = \underline{3500\text{ W}}$

(b) Rate of radiant energy emitted $= A\varepsilon\sigma T^4$. For a gray surface, $\varepsilon = \alpha = 0.7$.

$$A\varepsilon\sigma T^4 = 0.5 \times 0.7 \times 5.67 \times 10^{-8} \times 773.15^4 = \underline{7091\text{ W}}$$

(c) Total energy flux leaving the surface is the sum of the emitted energy flux and the reflected energy flux $= \varepsilon E_b + \rho G$. With $\rho = 1 - \alpha = 0.3$, the total energy flux leaving the surface is

$$0.7 \times 5.67 \times 10^{-8} \times 773.15^4 + 0.3 \times 10\,000 = \underline{17\,182\text{ W/m}^2}$$

(d) Rate of radiant energy emitted in the wave band 0.2 μm to 4 μm is given by

$$AE_{\lambda 1\text{-}\lambda 2} = AE\,\frac{E_{0\text{-}\lambda_2} - E_{0\text{-}\lambda_1}}{E} = A\varepsilon E_b\,\frac{\varepsilon E_{b,0\text{-}\lambda_2} - \varepsilon E_{b,0\text{-}\lambda_1}}{\varepsilon E_b}$$

$$= A\varepsilon E_b\left(\frac{E_{b,0\text{-}\lambda_2}}{E_b} - \frac{E_{b,0\text{-}\lambda_1}}{E_b}\right) = A\varepsilon E_b(f_{0\text{-}\lambda_2} - f_{0\text{-}\lambda_1})$$

From Table 7.1.2 (or from the software RF)

$$\lambda_2 T = 4 \times 773.15 = 3092.6 \text{ μm K} \quad f_{0\text{-}\lambda_2} = 0.294$$

$$\lambda_1 T = 0.2 \times 773.15 = 154.6 \text{ μm K} \quad f_{0\text{-}\lambda_1} \approx 0$$

Hence,

$$AE_{\lambda_2-\lambda_1} = A\varepsilon E_b(f_{0\text{-}\lambda_2} - f_{0\text{-}\lambda_1})$$

$$= 0.5 \times 0.7 \times 5.67 \times 10^{-8} \times 773.15^4(0.294 - 0) = \underline{2085 \text{ W}}$$

(e) From an energy balance on the surface

| Net radiative heat transfer rate from the surface | = | Total energy leaving the surface | − | Energy reaching the surface |

$$= 0.5 \times 17\ 182 - 0.5 \times 10\ 000 = \underline{3591 \text{ W}}$$

To maintain the temperature of the surface at 500 °C, the surface should be supplied with 3591 W of energy from an external source as shown in Figure 7.1.7.

EXAMPLE 7.1.3

A specially coated diffuse, opaque surface whose absorptivity is 1 for $0 < \lambda < 2$ μm and 0.1 for 2 μm $< \lambda < \infty$ is exposed to solar radiation in the outer reaches of the atmosphere (Figure 7.1.8). The incident solar energy is 1353 W/m². Determine

(a) The heat flux by radiation from the surface to the surroundings if the surface is maintained at 100 °C by a coolant.
(b) The equilibrium temperature of the surface, if the coolant flow stops and the surface is insulated on the side that does not receive solar radiation.
(c) Compare the values in parts a and b if the surface is black.

Assume the sun behaves as a blackbody at 5760 K. The sky is at 0 K.

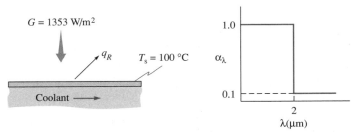

FIGURE 7.1.8 A Specially Coated Surface with Wavelength Dependent Absorptivity

Given

Surface: specially coated
$0 < \lambda < 2$ μm $\alpha_\lambda = 1$
2 μm $< \lambda < \infty$ $\alpha_\lambda = 0.1$
G (incident solar energy flux) $= 1353$ W/m²

Find

(a) q'' if $T_s = 100 \, °C$ with a coolant cooling the surface
(b) T_s if the coolant flow ceases.
(c) Compare the values in a and b if $\alpha_\lambda = 1$ for all wavelengths (black surface)

SOLUTION

From an energy balance on the surface exposed to solar radiation

> Heat transfer rate from the surface
> = Rate of radiant energy leaving the surface
> − Rate of radiant energy reaching the surface
> = Rate of (emitted energy + reflected energy − incident energy)
> = Rate of (emitted energy − absorbed energy)

On a unit area basis,

$$q'' = \text{Heat flux from the surface}$$
$$= \text{Emissive power} - \text{Absorbed energy flux}$$

Emissive power $= E = \int_0^\infty \varepsilon_\lambda E_{b\lambda} \, d\lambda$
Absorbed energy flux $= \int_0^\infty \alpha_\lambda G_\lambda \, d\lambda$

$$q'' = \int_0^\infty \varepsilon_\lambda E_{b\lambda} \, d\lambda - \int_0^\infty \alpha_\lambda G_\lambda \, d\lambda \tag{1}$$

By Kirchhoff's Law, $\varepsilon_\lambda = \alpha_\lambda$. The heat flux is found by evaluating the right-hand side of Equation 1. If the coolant flow stops and $q'' = 0$,

$$\int_0^\infty \varepsilon_\lambda E_{b\lambda} \, d\lambda = \int_0^\infty \alpha_\lambda G_\lambda \, d\lambda \tag{2}$$

As the characteristics of the incident radiation are known, the right-hand side of Equation 2 can be evaluated. The temperature for which the left-hand side of Equation 2 is equal to the right-hand side is to be found.

(a) Emissive power $= E = \int_0^\infty \varepsilon E_{b\lambda} \, d\lambda = \int_0^{\lambda_1} \varepsilon_1 E_{b\lambda} \, d\lambda + \int_{\lambda_1}^\infty \varepsilon_2 E_{b\lambda} \, d\lambda$ where $\lambda_1 = 2 \, \mu m$, $\varepsilon_1 = \alpha_1 = 1$ and $\varepsilon_2 = \alpha_2 = 0.1$, and $E_{b\lambda} = E_{b\lambda}(T_s)$.

$$E = \varepsilon_1 E_{b,0-\lambda_1} + \varepsilon_2 E_{b,\lambda_1-\infty}.$$

But $E_{b,\lambda_1-\infty} = E_b - E_{b,0-\lambda_1}$; therefore,

$$\text{Emissive power} = \varepsilon_1 E_{b,0-\lambda_1} + \varepsilon_2 (E_b - E_{b,0-\lambda_1})$$
$$= \varepsilon_1 \frac{E_{b,0-\lambda_1}}{E_b} E_b + \varepsilon_2 \frac{E_b - E_{b,0-\lambda_1}}{E_b} E_b$$
$$= [\varepsilon_1 f_{0-\lambda_1} + \varepsilon_2 (1 - f_{0-\lambda_1})] E_b$$

Similarly, the absorbed energy flux per unit area is given by

$$\int_0^\infty \alpha_\lambda G_\lambda \, d\lambda = \int_0^{\lambda_1} \alpha_1 G_\lambda \, d\lambda + \int_{\lambda_1}^\infty \alpha_2 G_\lambda \, d\lambda$$

where $\lambda_1 = 2 \, \mu m$, $\alpha_1 = 1$, and $\alpha_2 = 0.1$.

Rate of absorbed energy per unit area $= \alpha_1 G_{0\text{-}\lambda_1} + \alpha_2 G_{\lambda_1\text{-}\infty}$

$$= \alpha_1 \frac{G_{0\text{-}\lambda_1}}{G} G + \alpha_2 \frac{G - G_{0\text{-}\lambda_1}}{G} G$$

But as the source of incident energy is the sun, assumed to be black (subscript s indicating the sun),

$$\frac{G_{0\text{-}\lambda_1}}{G} = \left. \frac{E_{b,0\text{-}\lambda_1}}{E_b} \right|_s = f_{0\text{-}\lambda_1,s}$$

Rate of absorbed energy $= \alpha_1 f_{0\text{-}\lambda_1,s} + \alpha_2 (1 - f_{0\text{-}\lambda_1,s}) G$

Thus,

$$q'' = [\varepsilon_1 f_{0\text{-}\lambda_1} + \varepsilon_2 (1 - f_{0\text{-}\lambda_1})] E_b - [\alpha_1 f_{0\text{-}\lambda_1,s} + \alpha_2 (1 - f_{0\text{-}\lambda_1,s}) G] \qquad (3)$$

Note that $f_{0\text{-}\lambda_1,s}$ is evaluated at the temperature of the sun and $f_{0\text{-}\lambda_1}$ at the temperature of the surface. The value of $f_{0\text{-}\lambda}$ is evaluated from the software RF (or from Table 7.1.2).

$T(K)$	373.15	5760
$\lambda_1 \ (\mu m)$	2	2
$\lambda_1 T \ (\mu m \ K)$	746.3	11 520
$f_{0\text{-}\lambda_1}$	0.0	0.9392

$E_b = 5.67 \times 10^{-8} \times 373.15^4 = 1099 \ W/m^2$

$q'' =$ Radiant heat flux from the surface to the surrounding
$= 0.1 \times 1099 - [1 \times 0.9392 + 0.1(1 - 0.9392)] \times 1353$
$= 109.9 - 1279 = \underline{-1169 \ W/m^2}$

The emitted energy is 109.9 W/m^2 and the absorbed energy is 1279 W/m^2. The negative sign indicates that the net radiative heat transfer is to the surface at the rate of 1169 W/m^2. To keep the surface at 100 °C, the surface should be cooled by heat transfer to an external medium, in this case by convection to the coolant (a fluid such as air or water).

(b) Heat transfer rate to the surface is zero and, hence,

Emissive power = Rate of energy absorbed = 1279 W/m^2

or

$$[\varepsilon_1 f_{0\text{-}\lambda_1} + \varepsilon_2 (1 - f_{0\text{-}\lambda_1})] E_b = 1279 \qquad (4)$$

Both $f_{0\text{-}\lambda_1}$ and E_b depend on the temperature of the surface, which is to be determined. Though $f_{0\text{-}\lambda_1}$ and E_b depend on the temperature, no explicit functional relationship between them in an equation form is available. The temperature of the surface has to be determined by iteration. Assume an arbitrary temperature for the surface and determine the emissive power. If the emissive power of the surface is less than the absorbed energy of 1279 W/m^2, then the assumed temperature is too low; if the emissive power is higher than the absorbed energy flux, the assumed temperature is too high. A new temperature is assumed and the calculations are repeated.

The computed values for this problem are given in the following table. The emissive power of the surface is denoted by E. In arriving at the values in the table, the principles of bisection for finding the roots of Equation 4 have been employed.

T(K)	$\lambda_1 T$	$f_{0-\lambda_1}$	E_b	E	Value of T
500	1000	0.000321	3544	355.4	Too low
800	1600	0.019726	23 224	2735	Too high
650	1300	0.004318	10 121	1052	Too low
725	1450	0.010110	15 665	1709	Too high
688	1376	0.006825	12 704	1348	A little too high
670	1340	0.005536	11 426	1200	A little too low
680	1360	0.006228	12 123	1280.3	

At a temperature of 680 K the emitted energy (1280.3 W/m²) is very nearly equal to the absorbed energy (1279 W/m²). Hence, under equilibrium conditions

$$T = 680 \text{ K}$$

(c) For a black surface we have

$q'' = $ Emitted energy $-$ Absorbed energy $= 5.67 \times 10^{-8} \times 373.15^4 - 1353$
$= -253.7 \text{ W/m}^2$

Under equilibrium conditions, $E_b = G$ and

$$T = \left(\frac{1353}{5.67 \times 10^{-8}} \right)^{1/4} = \underline{393 \text{ K}}$$

COMMENTS

It is clear that the special surface is much more effective as a solar collector than a black surface. The special surface transfers 1169 W/m² to the coolant whereas the black surface transfers only 253.7 W/m². The reason for this difference is that the major part of the radiation from the sun is in the wave band 0–2 μm (94%), and all of it is absorbed by the special surface. Thus, the energy absorbed by the special surface is slightly less than the energy absorbed by the black surface. However, the major part of the blackbody radiation at the surface temperature (373 K) is in the wave band 2 μm–∞. In this wave band, the blackbody emits all the energy, whereas the special surface emits only 10% of the energy emitted by the blackbody. Therefore, there is not much difference in the energy absorbed by the two surfaces, but the emitted energy from the special surface is significantly less than the emitted energy from the black surface, which leads to a greater heat transfer rate from the special surface to the coolant.

7.2 RADIOSITY AND VIEW FACTOR

To understand how the radiative heat transfer rate from a surface is determined, consider surface i, which is completely surrounded by other surfaces, such as surface k, as illustrated in Figure 7.2.1. The heat transfer rate from surface i is obtained from the relation

q_i = Radiative heat transfer rate from surface i
= Total radiant energy leaving i
− Total radiant energy reaching i

FIGURE 7.2.1 **An Enclosure with *N* surfaces**

The energy leaving surface i consists of both the emitted energy and the reflected part of the incident energy. The total rate of radiant energy leaving a surface per unit area of the surface is its *radiosity, J*.

To determine the energy reaching surface i from other surfaces, such as k, we first need to know the total energy leaving surface k. But, as only a fraction of the energy leaving k reaches i, we also need the fraction of the energy leaving k that arrives at i. The ratio of the energy leaving k and reaching i to the total energy leaving surface k is the *view factor, F_{k-i}*, of surface i with respect to surface k.

It is now evident that to determine the heat transfer rate from a surface we need the radiosities and the view factors between surfaces.

7.2.1 Radiosity

Radiosity defined

Radiosity, J, is the total radiant energy leaving a surface per unit area per unit time. The total radiant energy leaving a surface consists of the emitted energy and the reflected part of the incident energy shown in Figure 7.2.2. Thus, radiosity is defined by the relation

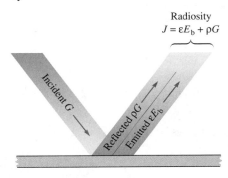

Radiosity
$J = \varepsilon E_b + \rho G$

FIGURE 7.2.2 **Definition of Radiosity of a Surface (Specular Reflection Not Implied)**

$$\| \quad J = \varepsilon E_b + \rho G \quad \| \tag{7.2.1}$$

where

J = radiosity of the surface (W/m^2)
ε = emissivity of the surface
E_b = blackbody emissive power at the temperature of the surface (W/m^2)
ρ = reflectivity of the surface
G = incident radiant flux (W/m^2)

For a black surface
radiosity equals the
emissive power

Although the concept of radiosity can be extended to include the transmitted energy in the case of a semitransparent material, we restrict our discussion to opaque surfaces only. In Example 7.1.2, the solution to part c is the radiosity of the surface. For a *black surface* $\varepsilon = 1$, $\rho = 0$, and $J = E_b$.

For a gray, diffuse surface with uniform radiosity, its radiosity, emissive power, and the heat transfer rate from the surface are mutually related. To obtain the relation consider a surface with surface area A, whose radiosity (assumed to be uniform) is J:

Total rate of energy leaving the surface = AJ
Total rate of energy arriving at the surface = AG
Net rate of radiant energy leaving the surface = $AJ - AG$

Since the net rate of radiant energy leaving the surface is also equal to the net radiant heat transfer rate from the surface, q, it follows that

$$q = AJ - AG = A(J - G) \tag{7.2.2}$$

Solving for G from Equation 7.2.1,

$$G = \frac{J - \varepsilon E_b}{\rho} = \frac{J - \varepsilon E_b}{1 - \varepsilon}$$

where, for an opaque, gray surface, the relations $\rho = 1 - \alpha$, and $\alpha = \varepsilon$, have been used. Substituting this expression in Equation 7.2.2,

$$\left\| \quad q = \frac{A\varepsilon}{1 - \varepsilon}(E_b - J) \quad \right\| \tag{7.2.3a}$$

Equation 7.2.3a is used to determine the radiant heat transfer rate from a surface whose emissivity is less than 1. *Equation 7.2.3a is not applicable to a black surface for which* $J = E_b$. For a black surface, $\alpha = \varepsilon = 1$, and $\rho = 0$. Thus,

$$q = A(E_b - G) \tag{7.2.3b}$$

EXAMPLE 7.2.1   A gray, diffuse, opaque surface (absorptivity = 0.8) is at 100 °C and receives an irradiation of 1000 W/m² (Figure 7.2.3). If the area of the surface is 0.1 m², compute

(a) The radiosity of the surface.
(b) The net radiative heat transfer rate from the surface.

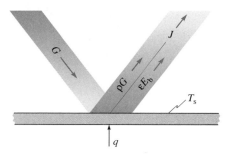

**FIGURE 7.2.3 Surface Energy Balance
with Radiant Energy Transfer**

Given

$\alpha = 0.8$ $\qquad$ $G = 1000 \text{ W/m}^2$
$T_s = 100\,°\text{C}$ $\qquad$ $A_s = 0.1 \text{ m}^2$

Find

(a) J
(b) q

SOLUTION

(a) From the definition of radiosity, Equation 7.2.1,

$$J = \varepsilon E_b + \rho G$$

For a gray, diffuse surface, $\varepsilon = \alpha = 0.8$. For an opaque surface, $\tau = 0$.

$$\alpha + \rho = 1 \qquad \rho = 1 - \alpha = 0.2 \qquad E_b = \sigma T^4$$

$$J = 0.8 \times 5.67 \times 10^{-8} \times 373.15^4 + 0.2 \times 1000 = \underline{1079.4 \text{ W/m}^2}$$

(b) From an energy balance on the surface,

Net radiative heat transfer rate
= Total radiant energy leaving the surface
$\quad$ −Total rate of radiant energy reaching the surface
= $A(J - G)$

$$q = A(J - G) = 0.1(1079.4 - 1000) = \underline{7.94 \text{ W}}$$

Alternatively, from Equation 7.2.3a,

$$q = \frac{A\varepsilon}{1 - \varepsilon}(E_b - J) = \frac{0.1 \times 0.8}{1 - 0.8}(5.67 \times 10^{-8} \times 373.15^4 - 1079.4)$$
$$= \underline{7.96 \text{ W}}$$

EXAMPLE 7.2.2 Re-solve Example 7.2.1 if the surface is black.

SOLUTION

(a) For a black surface, $\rho = 0$, $\varepsilon = 1$ and

$$J = E_b = 5.67 \times 10^{-8} \times 373.15^4 = \underline{1099.3 \text{ W/m}^2}$$

(b) From Equation 7.2.3b

$$q = A(E_b - G) = 0.1(1099.3 - 1000) = \underline{9.93 \text{ W}}$$

7.2.2 View Factor

Consider two diffuse surfaces i and k, Figure 7.2.4a, whose radiosities (assumed to be uniform on each surface) are J_i and J_k, respectively, separated by a nonpartici-pating medium (the medium neither emits nor absorbs radiant energy). The rate of

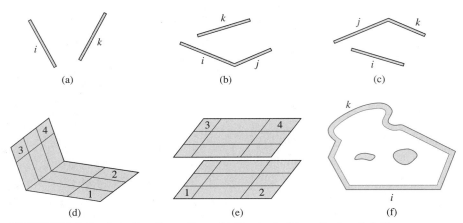

FIGURE 7.2.4 Surface Configurations to Illustrate View Factor Relations

radiant energy leaving surface i is A_iJ_i. Only a fraction of the total energy leaving surface i reaches surface k. *The view factor* F_{i-k} *of surface k with respect to surface i is defined as the ratio of the radiant energy leaving surface i and reaching surface k, to the total radiant energy leaving surface i.*

$$F_{i-k} = \frac{\text{Radiant energy leaving surface i and reaching surface k}}{\text{Total radiant energy leaving surface i}}$$

Thus, the radiant energy reaching surface k from i is $A_iJ_iF_{i-k}$

Although the view factor is defined here in terms of radiant energies, it can be shown (see Chapter 8) that it is purely a function of the geometry of the two surfaces, the orientation of one with respect to the other, and the spacing between the two. View factors can be evaluated from purely mathematical and/or numerical methods without any reference to radiative heat transfer. However, in the expression for the energy reaching surface k from surface i ($= A_iJ_iF_{i-k}$), it is evident that the view factor can be used only with a surface of uniform radiosity.

View factor is also known by other terms such as geometric factor, angle factor, configuration factor, and shape factor (not the same as conduction shape factor).

Some of the significant and useful relations concerning view factors are given by Equation 7.2.4. In these relations the subscripts {i, j, k} refer to three surfaces, and a combination such as ij means that the two surfaces i and j are taken together. For example, ij represents the two surfaces i and j taken together and F_{ij-k} means the view factor of surface k with respect to the combined surface ij. The term F_{ij-k} represents the energy leaving surface ij and reaching surface k as a fraction of the total energy leaving surface ij. The proof of the relations given in Equation 7.2.4 can be found in Chapter 8. Some of the relations can also be derived in a simpler way with the assumption of uniform radiosity.

Reciprocity Relation For two surfaces i and k (Figure 7.2.4a),

$$A_iF_{i-k} = A_kF_{k-i} \qquad\qquad \textbf{(7.2.4a)}$$

Decomposition Relations For three surfaces i, j and k (Figures 7.2.4b and c, respectively),

$$A_{ij}F_{ij\text{-}k} = A_iF_{i\text{-}k} + A_jF_{j\text{-}k} \tag{7.2.4b}$$

$$A_iF_{i\text{-}jk} = A_iF_{i\text{-}j} + A_iF_{i\text{-}k} \tag{7.2.4c}$$

Law of Corresponding Corners Two rectangular surfaces (Figures 7.2.4d and e), which have a common side or are directly opposite to each other, are divided into segments as shown. For the segments of the type shown in the figures,

$$A_1F_{1\text{-}4} = A_2F_{2\text{-}3} \tag{7.2.4d}$$

Summation Relation For an enclosure with N surfaces (Figure 7.2.4f),

$$\sum_{k=1}^{N} F_{i\text{-}k} = 1 \tag{7.2.4e}$$

Relations given by Equations 7.2.4a, b, c, and e will be shown to be valid on the basis of uniform radiosity of each surface involved in these relations. More rigorous proofs of these equations and Equation 7.2.4d are given in Chapter 8.

Equation 7.2.4a: $A_iF_{i\text{-}k} = A_kF_{k\text{-}i}$ Consider two surfaces i and k (Figure 7.2.4a) at uniform surface temperatures T_i and T_k, respectively. The net energy transfer from surface i to k, q_{ik}, is

$$q_{ik} = A_iJ_iF_{i\text{-}k} - A_kJ_kF_{k\text{-}i}$$

The above relation is valid for all cases. In particular, it is valid when both i and k are black and at the same uniform temperature T. As the surfaces are black $J = E_b$ for each surface. As the temperatures of the surfaces are equal, $E_{bi} = E_{bk}$. Further, if the temperatures of the two surfaces are equal, the net energy exchange should be zero. Therefore, we have

$$q_{ik} = 0 = E_{bi}(A_iF_{i\text{-}k} - A_kF_{k\text{-}i})$$

Since $E_{bi} \neq 0$,

$$\underline{A_iF_{i\text{-}k} = A_kF_{k\text{-}i}}$$

Equation 7.2.4b: $A_{ij}F_{ij\text{-}k} = A_iF_{i\text{-}k} + A_jF_{j\text{-}k}$ Referring to Figure 7.2.4b, with the assumption of uniform radiosity for each surface i and j, and with the further assumption that $J_i = J_j = J$, we have

Energy leaving surface ij and reaching surface k $= A_{ij}JF_{ij\text{-}k}$

Total energy leaving surface ij $= A_{ij}J$

Energy leaving surface ij and reaching surface k
$=$ Energy leaving surface i and reaching surface k
$+$ Energy leaving surface j and reaching surface k
$= A_iJF_{i\text{-}k} + A_jJF_{j\text{-}k}$

Thus,

$$A_{ij}JF_{ij\text{-}k} = A_iJF_{i\text{-}k} + A_jJF_{j\text{-}k}$$

or

$$A_{ij}F_{ij\text{-}k} = A_iF_{i\text{-}k} + A_jF_{j\text{-}k}$$

Equation 7.2.4c: $A_iF_{i\text{-}jk} = A_iF_{i\text{-}j} + A_iF_{i\text{-}k}$ Referring to Figure 7.2.4c, from the definition of the view factor, denoting the radiosity of surface i by J_i,

Energy leaving surface i and reaching surface jk = $A_iJ_iF_{i\text{-}jk}$

Energy leaving i and reaching jk
= Energy leaving i and reaching j
+ Energy leaving i and reaching k
= $A_iJ_iF_{i\text{-}j} + A_iJ_iF_{i\text{-}k}$

Equating the two expressions, we obtain Equation 7.2.4c.

Equation 7.2.4e: $\sum_{k=1}^{N} F_{i\text{-}k} = 1$, for an enclosure with N surfaces. Referring to Figure 7.2.4f, the energy leaving surface i and reaching any surface k = $A_iJ_iF_{i\text{-}k}$. The total energy leaving surface i and reaching all the surfaces in the enclosure is then given by the sum

$$\sum_{k=1}^{N} A_iJ_iF_{i\text{-}k}$$

But since surface i is a part of this enclosure, the sum of the energy leaving surface i and reaching each of the surfaces k (including the surface i itself) in the enclosure should also be equal to the total energy leaving surface i = A_iJ_i. Thus,

$$A_iJ_i = \sum_{k=1}^{N} A_iJ_iF_{i\text{-}k}$$

or

$$\sum_{k=1}^{N} F_{i\text{-}k} = 1$$

Though view factor has been defined in terms of two distinct surfaces, the view factor $F_{i\text{-}i}$ for a surface i can be either zero or a positive value depending on the geometry of the surface. If no part of the surface "sees" itself (a plane or a convex surface), no part of the energy leaving the surface falls on itself and $F_{i\text{-}i} = 0$. If a part of the surface can see another part of the surface (concave surfaces), a part of the energy leaving the surface falls on itself and $F_{i\text{-}i} \neq 0$. For example, in Figure 7.2.4b, if surfaces i and j are taken together $F_{ij\text{-}ij} \neq 0$ as i (a part of ij) sees surface j (also a part of ij).

To clarify the concept of view factor, the view factors for a few simple configurations will be determined.

EXAMPLE 7.2.3 A flat surface, 1, is completely enclosed by a second surface, 2 (Figure 7.2.5). Determine the view factors $F_{1\text{-}2}$, $F_{2\text{-}1}$, and $F_{2\text{-}2}$.

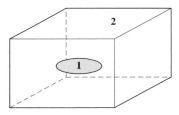

FIGURE 7.2.5 A Flat Surface, 1, Completely Enclosed by a Second Surface, 2

SOLUTION

As no part of surface 1 ''sees'' itself, all the energy leaving surface 1 reaches surface 2, and from the definition of view factor,

$$F_{1\text{-}2} = 1$$

To determine $F_{2\text{-}1}$, from Equation 7.2.4a,

$$A_2 F_{2\text{-}1} = A_1 F_{1\text{-}2}$$

$$F_{2\text{-}1} = \frac{A_1 F_{1\text{-}2}}{A_2} = \frac{A_1}{A_2}$$

From Equation 7.2.4e,

$$F_{2\text{-}1} + F_{2\text{-}2} = 1$$

and

$$F_{2\text{-}2} = 1 - F_{2\text{-}1} = 1 - \frac{A_1}{A_2}$$

EXAMPLE 7.2.4 The cross section of a very long duct (Figure 7.2.6) is an isosceles triangle. Surface 1 is the base; areas of surfaces 2 and 3 are equal. Determine $F_{1\text{-}2}$ and $F_{2\text{-}3}$.

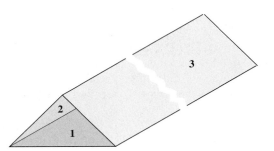

FIGURE 7.2.6 A Long Duct Whose Cross Section Is an Isosceles Triangle
Surface 1 is the base, and the area of surface 2 is equal to the area of surface 3.

SOLUTION

As it is a very long duct, it can be assumed, as a good approximation, that very little energy leaving any one of the surfaces passes through the end openings. This is equivalent to saying that the three surfaces form a complete enclosure. For such an enclosure, from Equation 7.2.4e,

$$F_{1-1} + F_{1-2} + F_{1-3} = 1$$

As surface 1 is a plane surface, no part of it can "see" any other part of itself and $F_{1-1} = 0$. Further, by symmetry and definition of view factor,

$$F_{1-2} = F_{1-3}$$

or

$$F_{1-2} = 0.5 = F_{1-3}$$

To find F_{2-3} for surface 2, for the three-surface enclosure,

$$F_{2-1} + F_{2-2} + F_{2-3} = 1$$

For the plane surface 2, $F_{2-2} = 0$ and $F_{2-1} + F_{2-3} = 1$. From the reciprocity relation, Equation 7.2.4a,

$$A_1 F_{1-2} = A_2 F_{2-1}$$

or

$$F_{2-1} = \frac{A_1 F_{1-2}}{A_2} = 0.5 \frac{A_1}{A_2}$$

giving

$$F_{2-3} = 1 - F_{2-1} = 1 - 0.5 \frac{A_1}{A_2}$$

View factors for some common configurations are presented in Figures 7.2.7 through 7.2.10 in graphical form and in Table 7.2.1 by algebraic expressions. The use of these view factors can be extended by employing the view factor relations given by Equations 7.2.4a through e.

Examples 7.2.5 and 7.2.6 illustrate how the values of view factors available in Figures 7.2.7 through 7.2.10 can be used to determine view factors of configurations for which the view factors are not directly available. In all problems dealing with view factors, the geometries of the surfaces should be known.

In the software HT, subprogram VF, the view factors are computed by using the algebraic expressions given in Table 7.2.1 for the different geometries.

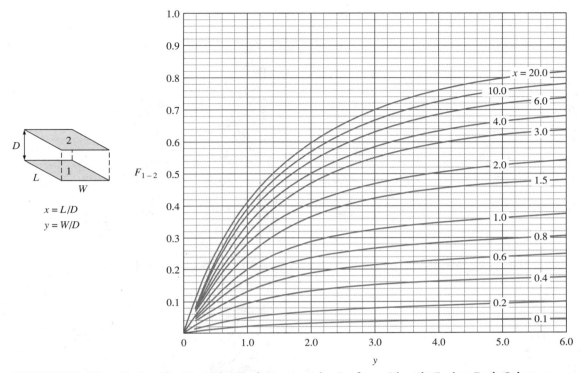

FIGURE 7.2.7 View Factors for Two Identical, Rectangular Surfaces Directly Facing Each Other

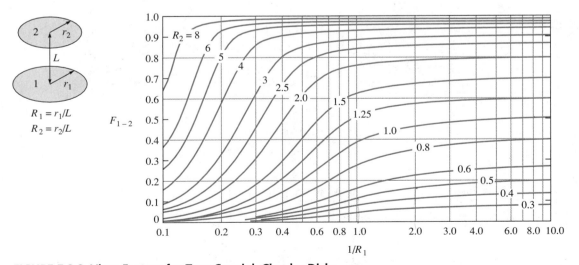

FIGURE 7.2.8 View Factors for Two Coaxial, Circular Disks

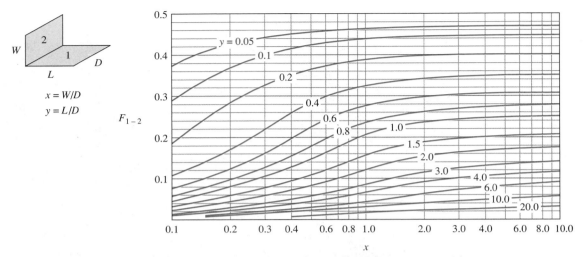

FIGURE 7.2.9 **View Factors for Two Rectangular Surfaces Sharing a Common Side**

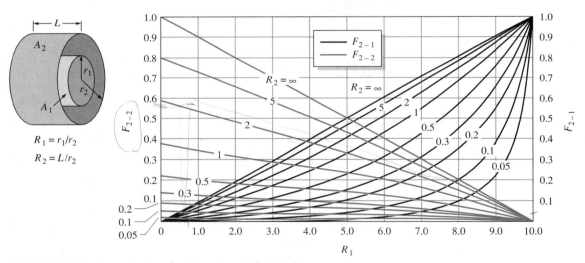

FIGURE 7.2.10 **View Factors for Two Coaxial Cylinders**

TABLE 7.2.1 View Factor Relations for Some Configurations

1. **Identical, aligned rectangles directly opposite to each other** (see Figure 7.2.7)

$$F_{1\text{-}2} = \frac{2}{\pi xy} \left\{ \ln \left[\frac{(1 + x^2)(1 + y^2)}{1 + x^2 + y^2} \right]^{1/2} + x(1 + y^2)^{1/2} \tan^{-1} \left[\frac{x}{(1 + y^2)^{1/2}} \right] \right.$$

$$\left. + y(1 + x^2)^{1/2} \tan^{-1} \left[\frac{y}{(1 + x^2)^{1/2}} \right] - x \tan^{-1}x - y \tan^{-1}y \right\}$$

| **TABLE 7.2.1** (continued)

2. Two coaxial circular disks
    See Figure 7.2.8: $R_1 = r_1/L$ $R_2 = r_2/L$ $x = 1 + [(1 + R_2^2)/R_1^2]$

$$F_{1\text{-}2} = \frac{1}{2}\left[x - \left(x^2 - 4\,\frac{R_2^2}{R_1^2}\right)^{1/2}\right]$$

3. Two, perpendicular rectangles sharing a common side
    See Figure 7.2.9: $x = W/D$ $y = L/D$ $a = x^2$ $b = y^2$

$$F_{1\text{-}2} = \frac{1}{\pi y}\left[y\,\tan^{-1}\left(\frac{1}{y}\right) + x\,\tan^{-}\left(\frac{1}{x}\right) - (a + b)^{1/2}\tan^{-1}\left[\frac{1}{(a + b)^{1/2}}\right]\right.$$

$$\left. + \frac{1}{4}\ln\left\{\frac{(1 + a)(1 + b)}{1 + a + b}\left[\frac{b(1 + a + b)}{(1 + b)(a + b)}\right]^b\left[\frac{a(1 + a + b)}{(1 + a)(a + b)}\right]^a\right\}\right]$$

4. Two Coaxial Cylinders
    See Figure 7.2.10:

$$R = r_2/r_1 \quad M = L/r_1 \quad A = M^2 + R^2 - 1 \quad B = M^2 - R^2 + 1 \quad x = \frac{4(R^2 - 1) + (M^2/R^2)(R^2 - 2)}{M^2 + 4(R^2 - 1)}$$

$$F_{2\text{-}1} = \frac{1}{R} - \frac{1}{\pi R}\left[\cos^{-1}\left(\frac{B}{A}\right) - \frac{1}{2M}\{[(A + 2)^2 - 4R^2]^{1/2}\cos^{-1}\left(\frac{B}{RA}\right) + B\,\sin^{-1}\left(\frac{1}{R}\right)\right.$$

$$\left. - \frac{\pi A}{2}\right]$$

$$F_{2\text{-}2} = 1 - \frac{1}{R} + \frac{2}{\pi R}\tan^{-1}\left[\frac{2(R^2 - 1)^{1/2}}{M}\right] - \frac{M}{2\pi R}\left\{\frac{(4R^2 + M^2)^{1/2}}{M}\sin^{-1}(x)\right.$$

$$\left. - \sin^{-1}\left(\frac{R^2 - 2}{R^2}\right) + \frac{\pi}{2}\left[\frac{(4R^2 + M^2)^{1/2}}{M} - 1\right]\right\}$$

$$F_{2\text{-}3} = \frac{1}{2}(1 - F_{2\text{-}1} + F_{2\text{-}2})$$

For any argument y $-\pi/2 \le \sin^{-1} y \le \pi/2$ $0 \le \cos^{-1} y \le \pi$

EXAMPLE 7.2.5   Figure 7.2.11 depicts a window in the wall of a room. Find the view factor $F_{\text{B-A}}$, where A is the window and B is the floor.

SOLUTION

For convenience, divide the floor into two parts, 1 and 2 as shown. The window is represented by 4 and the remaining part of the wall by 3. We need $F_{12\text{-}4}$. From Equation 7.2.4b we have

$$A_{12}F_{12\text{-}4} = A_1 F_{1\text{-}4} + A_2 F_{2\text{-}4} \tag{1}$$

$F_{1\text{-}4}$ can be found from entry 3, Table 7.2.1, or Figure 7.2.9. We now need $F_{2\text{-}4}$. To find $F_{2\text{-}4}$ we proceed as follows. From Equations 7.2.4b and c

$$A_{12}F_{12\text{-}34} = A_{12}F_{12\text{-}3} + A_{12}F_{12\text{-}4} = A_1 F_{1\text{-}3} + A_2 F_{2\text{-}3} + A_1 F_{1\text{-}4} + A_2 F_{2\text{-}4}$$

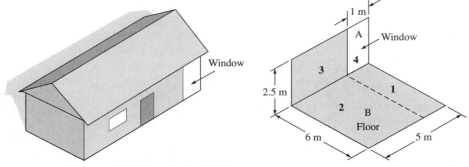

FIGURE 7.2.11 Model for Example 7.2.5

From the Law of Corresponding Corners, Equation 7.4.2d, $A_2F_{2\text{-}4} = A_1F_{1\text{-}3}$. Therefore,

$$A_2F_{2\text{-}4} = \frac{A_{12}F_{12\text{-}34} - A_2F_{2\text{-}3} - A_1F_{1\text{-}4}}{2}$$

Substituting the above expression in Equation 1 we obtain

$$A_{12}F_{12\text{-}4} = \frac{A_{12}F_{12\text{-}34} - A_2F_{2\text{-}3} + A_1F_{1\text{-}4}}{2}$$

The view factors, $F_{12\text{-}34}$, $F_{2\text{-}3}$, and $F_{1\text{-}4}$ can now be found from entry 3, Table 7.2.1, or Figure 7.2.9. To find $F_{12\text{-}34}$, referring to Figure 7.2.9 and the dimensions given in Figure 7.2.11,

$L = 6$ m, $D = 5$ m, and $W = 2.5$ m
$y = L/D = 6/5 = 1.2$
$x = W/D = 2.5/5 = 0.5$

For $x = 0.5$ and $y = 1.2$, from Figure 7.2.9, $F_{12\text{-}34} = 0.13$.
 In a similar manner, we find, $F_{2\text{-}3} = 0.12$ and $F_{1\text{-}4} = 0.06$. Thus,

$$F_{12\text{-}4} = \frac{1}{2A_{12}} (A_{12}F_{12\text{-}34} - A_2F_{2\text{-}3} + A_1F_{1\text{-}4})$$

$$= \frac{1}{2 \times 30} (30 \times 0.13 - 24 \times 0.12 + 6 \times 0.06) = \underline{0.023}$$

COMMENT

More accurate values of the view factors can be found quickly from the software VF.

EXAMPLE 7.2.6   Figure 7.2.12 shows the floor and roof of a house. The roof has a skylight in one corner, shown as 1. The house has radiant heating with heated floors. To determine the radiant energy that reaches the skylight from the floor, the view factor of the floor to the skylight is needed. Determine the view factor.

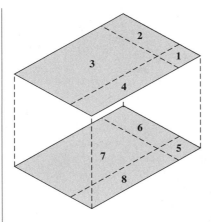

FIGURE 7.2.12 A Room with the Floor Represented by 5,6,7,8 and the Skylight by 1

SOLUTION

The floor and the ceiling are divided into four parts each to facilitate the solution and to take advantage of the view factor algebra. From Equation 7.2.4b

$$A_{5678}F_{5678\text{-}1} = A_5F_{5\text{-}1} + A_6F_{6\text{-}1} + A_7F_{7\text{-}1} + A_8F_{8\text{-}1} \tag{1}$$

The view factors $F_{i\text{-}1}$ ($i = 5, 6, 7,$ and 8) have to be determined. $F_{5\text{-}1}$ can be found directly from Figure 7.2.7.

To find $F_{6\text{-}1}$, we use Equations 7.2.4b and c.

$$A_{56}F_{56\text{-}12} = A_5F_{5\text{-}1} + A_5F_{5\text{-}2} + A_6F_{6\text{-}1} + A_6F_{6\text{-}2} \tag{2}$$

From Equation 7.2.4d, $A_5F_{5\text{-}2} = A_6F_{6\text{-}1}$. Substituting this relation,

$$A_6F_{6\text{-}1} = \frac{1}{2}\left(A_{56}F_{56\text{-}12} - A_4F_{5\text{-}1} - A_6F_{6\text{-}2}\right) \tag{3}$$

All the view factors on the right-hand side can be found from Figure 7.2.7. In the same manner,

$$A_8F_{8\text{-}1} = \frac{1}{2}\left(A_{58}F_{58\text{-}14} - A_5F_{5\text{-}1} - A_8F_{8\text{-}4}\right) \tag{4}$$

$F_{7\text{-}1}$ can be found in a similar manner but the algebra is a little bit more involved. With some patience and perseverance, it can be determined. Now,

$$A_{67}F_{67\text{-}14} = A_6F_{6\text{-}1} + A_6F_{6\text{-}4} + A_7F_{7\text{-}1} + A_7F_{7\text{-}4} \tag{5}$$

From Equation 7.2.4d, $A_6F_{6\text{-}4} = A_7F_{7\text{-}1}$. The term $A_6F_{6\text{-}1}$ has already been determined (Equation 3) and in a similar manner,

$$A_7F_{7\text{-}4} = \frac{1}{2}\left(A_{78}F_{78\text{-}34} - A_7F_{7\text{-}3} - A_8F_{8\text{-}4}\right) \tag{6}$$

Only $A_{67}F_{67\text{-}14}$ remains to be determined to find $A_7F_{7\text{-}1}$ in Equation 5. From Equation 7.2.4b

$$A_{5678}F_{5678\text{-}1234} = A_{58}F_{58\text{-}14} + A_{58}F_{58\text{-}23} + A_{67}F_{67\text{-}14} + A_{67}F_{67\text{-}23} \tag{7}$$

From Equation 7.2.4d, $A_{67}F_{67\text{-}14} = A_{58}F_{58\text{-}23}$ and

$$A_{67}F_{67\text{-}14} = \frac{1}{2}\left(A_{5678}F_{5678\text{-}1234} - A_{58}F_{58\text{-}14} - A_{67}F_{67\text{-}23}\right) \tag{8}$$

Now we have all the view factors required to determine $F_{5678\text{-}1}$.

Examples 7.2.5 and 7.2.6 show how the values of view factors that are available for simple geometries can be utilized to determine view factors in other geometries. The method of finding the view factors by using the available view factors and view factor relations (Equation 7.2.4) is known as *view factor algebra.*

Figures 7.2.7–7.2.10 cover some of the common configurations. View factors for a very large number of orientations of surfaces are available in the literature. For example, Siegel and Howell (1992) catalog 42 view factors and give references for the view factors for 142 configurations.

7.3 RADIATIVE HEAT TRANSFER AMONG GRAY, DIFFUSE, AND OPAQUE SURFACES

We are now equipped to determine the radiative heat transfer rates from surfaces. We restrict ourselves to

- Diffuse, gray, and opaque surfaces
- Complete enclosures with N surfaces
- A medium in the enclosure that does not participate in radiation

Note that steady state is not required to compute radiative heat transfer as the heat transfer is practically instantaneous. If the surfaces are not in steady state, the computed radiative heat transfer rates are applicable at the instant at which the surface conditions (temperature or heat flux) are utilized to compute the heat transfer rates.

N-Surface enclosure

To understand the restriction of complete enclosure, consider surfaces i, j, and k (Figure 7.3.1) in a room. To determine the heat transfer rate from i to k, or the net

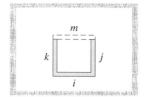

FIGURE 7.3.1 An Incomplete Enclosure, with Three Surfaces

heat transfer rate from surface i, we will have to complete the enclosure with an imaginary surface, m, whose properties represent those of the surfaces of the room that enclose surfaces i, j, and k. The last restriction is generally satisfied if the surfaces are separated by air. But even small quantities of gases such as carbon monoxide, carbon dioxide, and water vapor will affect the radiative heat transfer between surfaces.

What is a radiation
surface?

One more point needs emphasis. When we speak of an *N*-surface enclosure, we do not necessarily mean that there are *N* different geometric surfaces. More appropriately, we are assuming the radiosity of each of the *N* surfaces to be uniform. Even if a surface has a uniform temperature, it may not form one surface in the radiation sense, as it may not have uniform radiosity; recall that radiosity consists of both emitted and reflected energy. A surface with a uniform temperature (and uniform properties) has uniform emissive power, but may not have uniform incident (and hence, reflected) energy. To illustrate this point, consider a room whose floor and ceiling are maintained at two different temperatures, and the four walls are insulated (Figure 7.3.2a). We can think of this as a three-surface problem, the ceiling, the

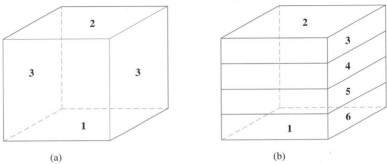

FIGURE 7.3.2 A Room with a Heated Ceiling, a Cooled Floor, and Insulated Walls
(a) Three-surface enclosure. (b) Six-surface enclosure.

floor, and the walls forming the three surfaces. If we wish to consider the walls to be more than one radiative surface, we may be tempted to consider each of the four walls as a separate surface but a little thought will convince us that this is not appropriate. Incident radiant flux from the ceiling (and the floor) on the walls decreases with distance and, hence, the radiosity of the walls varies with the distance from the ceiling (and the floor). Therefore, instead of considering the four walls to be four different surfaces, it is more appropriate to divide the walls into four different surfaces by three horizontal planes and consider the surfaces of all the walls between two consecutive horizontal planes as one radiation surface as shown in Figure 7.3.2b.

Some of the causes of nonuniform radiosity of a surface are

Causes of
nonuniform
radiosity

- Nonuniform temperature of a surface—emitted energy is different at different locations.
- Nonuniform emissivity and, therefore, nonuniform reflectivity. In this case neither the emitted energy nor the reflected energy is uniform.
- Nonuniform incident energy and, hence, nonuniform reflected energy.

In practice, very few surfaces have uniform radiosity. The one exception is a perfectly black surface with a uniform surface temperature. In all other cases, we have to start with a certain number of surfaces, each assumed to have a uniform radiosity. The solution found for this number of surfaces is compared with the solution when the bounding walls of the enclosure are further subdivided into a larger number of surfaces. If the two solutions are nearly equal to each other in terms of heat transfer rate and temperature of each surface, the assumed number of surfaces reasonably satisfies the assumption of uniform radiosity of each surface. Otherwise, the number of surfaces is increased again and the process repeated until the solutions from two consecutive configurations are within acceptable limits of the desired accuracy.

When the surfaces of an enclosure exchange energy by radiation, for each surface we know either the temperature at which the surface is maintained or the imposed heat transfer rate. If we know the temperature of the surface, we wish to determine the heat transfer rate from it; if we know the heat transfer rate, we wish to determine its temperature. For example, we may wish to construct an electrical heater so that the heater surface transfers a certain amount of energy by radiation. We would like to determine the heater surface temperature to ensure that it does not exceed the safe operating temperature of the material of the surface.

Now, consider an N-surface enclosure (Figure 7.3.3). Note that the enclosure may include one or more surfaces inside it. Two surfaces are represented by i and k. Radiative properties of every surface and the view factors between every pair of surfaces are known.

FIGURE 7.3.3 An Enclosure with _N_ Surfaces, Separated by a Medium That Does Not Participate in Radiation

From Equation 7.2.3a, the net radiative heat transfer rate from surface i is given by

$$\left\| \quad q_i = \frac{A_i \varepsilon_i}{1 - \varepsilon_i}(E_{bi} - J_i) \quad \right\| \tag{7.3.1}$$

However, Equation 7.3.1 cannot be used for a black surface for which $\varepsilon = 1$. When $\varepsilon = 1$, the denominator is zero but the numerator is also zero as $E_{bi} = J_i$. To find a more general expression for the radiative heat transfer from a surface, we proceed as follows. With $A_i F_{i-k} J_i$ representing the energy reaching surface k from surface i, and, $A_k F_{k-i} J_k$ representing the energy reaching surface i from surface k, the net energy transfer from surface i to surface k is represented by

$$q_{ik} = A_i F_{i-k} J_i - A_k F_{k-i} J_k$$

With $A_i F_{i-k} = A_k F_{k-i}$

$$\| \quad q_{ik} = A_i F_{i-k}(J_i - J_k) \quad \| \tag{7.3.2}$$

As q_{ik} represents the net energy transfer from surface i to surface k, if we sum up such net energy exchanges from surface i to each of the surfaces that form the enclosure, we get the total net radiant energy transfer rate from surface i to its surroundings, i.e., $q_i = \sum_{k=1}^{N} q_{ik}$.

$$\left\| \quad q_i = \sum_{k=1}^{N} A_i F_{i-k}(J_i - J_k) \quad \right\| \tag{7.3.3}$$

On the right-hand side of Equation 7.3.3, for surface i, even if F_{i-i} is nonzero, the net energy transfer from i to i is zero ($J_{k=i} = J_i$). When using Equations 7.3.1, 7.3.2, and 7.3.3, remember that

■ Both Equation 7.3.1 and Equation 7.3.3 yield the net radiant heat transfer from surface i.
■ Equation 7.3.1 cannot be used for a black surface as $\varepsilon = 1$ for such a surface.
■ Equation 7.3.3 is more general and can be used for all surfaces, including black surfaces.

From Equations 7.3.1, 7.3.2, and 7.3.3, it is evident that, once the radiosities of the surfaces are determined, the temperature of a surface with a known heat transfer rate or the heat transfer rate from a surface with a known temperature can be determined. For a black surface, if the temperature is known, its radiosity, being equal to the emissive power, is also known. Thus, the problem reduces to one of determining the radiosities of the surfaces of the enclosure.

Equations 7.3.1 and 7.3.2 can be recast as

$$q_i = \frac{E_{bi} - J_i}{(1 - \varepsilon_i)/A_i\varepsilon_i} \tag{7.3.4}$$

$$q_{ik} = \frac{J_i - J_k}{1/A_iF_{i-k}} \tag{7.3.5}$$

Electrical analogy   Equations 7.3.4 and 7.3.5 are of the form $I = E/R$ in electrical circuits. Equation 7.3.4 is similar to that of a battery where the denominator represents the internal resistance, with q_i representing the current from the battery. Equation 7.3.5 is the equivalent of a resistance $1/A_iF_{i-k}$, which is subjected to a potential difference, $J_i - J_k$, and q_{ik} is the current in that branch. The resistances and the radiosities of two surfaces of an enclosure are shown in Figure 7.3.4.

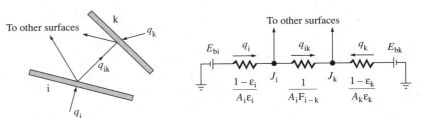

FIGURE 7.3.4 Radiative Heat Exchange Between Two Surfaces with Its Electrical Analog

Each surface is at a potential E_{bi} relative to ground. The term $A_i\varepsilon_i/(1 - \varepsilon_i)$ represents the internal resistance of the battery and, with a current of q_i from the surface, the potential at the surface is reduced to J_i. The term $1/A_iF_{i-k}$ is the resistance between two batteries with a potential difference of $J_i - J_k$. If surface k is insulated, there is no heat transfer to the surface; i.e., the ground connection to the battery at E_{bk} is cut. We may use the electrical analogy to solve the heat transfer problem. The circuit is solved by any one of the established methods. Kirchhoff's Laws for the solution of electrical networks may also be used here. The net energy transfer to any surface must be equal to zero, which is the equivalent of the statement that the algebraic sum of the currents to any junction must be zero. By applying this rule to each junction in the circuit, we produce as many equations as the number of unknown radiosities. The solution of the resulting system of simultaneous, linear equations yields the radiosities of the various surfaces. Radiative properties of each surface, such as emissivity, should be known and the various view factors computed prior to solving for the radiosities. The electrical network was proposed by Oppenheim (1956).

EXAMPLE 7.3.1

A surface 1 maintained at T_1 is completely enclosed by a surface 2, which is maintained at T_2. Determine the net radiative heat transfer rate from surface 1 to surface to 2.

ASSUMPTIONS

1. Each surface has a uniform radiosity; the configuration can be considered as an enclosure with two surfaces.
2. The surfaces are gray, diffuse, and opaque.
3. The medium between the surfaces does not participate in radiation.

SOLUTION

The corresponding electrical circuit is shown in Figure 7.3.5, where the resistances are given by

$$R_1 = \frac{1 - \varepsilon_1}{A_1 \varepsilon_1} \qquad R_2 = \frac{1 - \varepsilon_2}{A_2 \varepsilon_2} \qquad R_{12} = \frac{1}{A_1 F_{12}} = \frac{1}{A_2 F_{21}}$$

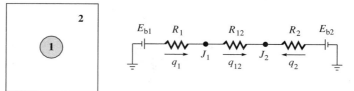

FIGURE 7.3.5 A convex surface, 1, is enclosed by a surface 2. The corresponding electrical circuit for the radiation heat transfer is also shown.

From the circuit we have

$$q_1 = q_{12} = -q_{21} = -q_2 = \frac{E_{b1} - E_{b2}}{(1 - \varepsilon_1)/(A_1 \varepsilon_1) + 1/(A_1 F_{1-2}) + (1 - \varepsilon_2)/(A_2 \varepsilon_2)} \quad \textbf{(1)}$$

Thus, if we know the properties and view factors, given any two values of E_{b1}, E_{b2}, and q_1 or q_2 (equivalent to giving any two values of T_1, T_2, and q_1 or q_2), we can solve for the third.

If surface 1 is convex towards surface 2, $F_{1-2} = 1$. Then Equation 1 simplifies to

$$\frac{q_1}{A_1} = \frac{E_{b1} - E_{b2}}{1/\varepsilon_1 + A_1/A_2(1 - \varepsilon_2/\varepsilon_2)} \quad \textbf{(2)}$$

Equation 2 can be applied to any convex or plane surface 1 in a two-surface enclosure. Surface 2 may be of any shape, as long as it completely encloses surface 1. *If surface 1 is not convex or plane, i.e.,* $F_{1\text{-}2} \neq 1$, *Equation 1 with the correct value of* $F_{1\text{-}2}$ *should be used.*

If we are dealing with two parallel surfaces of equal areas so close to each other that $F_{1\text{-}2} \approx 1$, the two surfaces can be considered to form an enclosure. Then, with $A_1 = A_2$, Equation 2 simplifies to

$$\frac{q_1}{A_1} = \frac{E_{b1} - E_{b2}}{1/\varepsilon_1 + 1/\varepsilon_2 - 1}$$

Lastly, if surface 1 is plane or convex, $A_1 \gg A_2$ or $\varepsilon_2 = 1$ (black surface), then $A_1/A_2(1 - \varepsilon_2/\varepsilon_2) \ll 1$, and Equation 2 reduces to

$$q_1 \approx A_1\varepsilon_1(E_{b1} - E_{b2}) \tag{3}$$

Equation 3 is Equation 1.5.2 given under radiation in Section 1.5. The restrictions imposed on that formula are now clear. Equation 3 can be used for a plane or a convex surface enclosed by a much larger surface. *Equations 1, 2, and 3 are useful relations for two-surface enclosures.*

EXAMPLE 7.3.2 A 3-cm diameter painted copper sphere is suspended in a large room where the walls and air are at 20 °C (Figure 7.3.6). If the emissivity of the painted surface is 0.9, determine the net radiative heat transfer rate from the surface and the rate of change of temperature of the sphere when its temperature is 600 °C,

(a) Neglecting convection.
(b) Including convection.

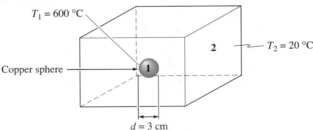

FIGURE 7.3.6 A copper sphere, 1, is freely suspended in a large room. The walls are represented as surface 2.

Given

Material: Copper $T_2 = 20\,°C$
$d_1 = 3$ cm $\varepsilon_1 = 0.9$
$T_1 = 600\,°C$

Find

q and $dT/d\tau$ for the sphere
(a) Neglecting convective heat transfer
(b) Including convective heat transfer

ASSUMPTIONS

1. The sphere is completely enclosed by the walls of the room.
2. Assuming the radiosities of the spherical surface and of the walls to be uniform, the configuration is treated as a two-surface enclosure.
3. The surfaces are gray, diffuse, and opaque.
4. At any instant of time the heat transfer rate from the surface can be evaluated by the relations valid for steady state at the instantaneous temperatures of the surfaces.
5. Lumped analysis is applicable to determine the rate of change of temperature.

SOLUTION

This problem is similar to Example 7.3.1. All the results from that problem are utilized here. From Equation 1, Example 7.3.1, the heat transfer rate q_1 from the sphere is given by

$$q_1 = \frac{E_{b1} - E_{b2}}{(1 - \varepsilon_1)/(A_1\varepsilon_1) + 1/(A_1F_{1\text{-}2}) + (1 - \varepsilon_2)/(A_2\varepsilon_2)}$$

As no part of the sphere can see another part of itself, $F_{1\text{-}1} = 0$. Since the sphere is completely enclosed by the room surface, which is considered as surface 2, $F_{1\text{-}1} + F_{1\text{-}2} = 1$ and, hence, $F_{1\text{-}2} = 1$. As the room is large, we assume that $A_2 \gg A_1$. For the wall surfaces, the emissivity is around 0.8. As long as the emissivity is not very small (close to zero, corresponding to a highly reflective surface such as we would obtain if all the walls were good mirrors!), we can approximate

$$\frac{A_1\varepsilon_1(1 - \varepsilon_2)}{A_2\varepsilon_2(1 - \varepsilon_1)} \ll 1$$

The heat transfer rate, obtained from Equation 3, Example 7.3.1, is

$$q_1 \approx A_1\varepsilon_1(E_{b1} - E_{b2})$$
$$= \pi \times 0.03^2 \times 0.9 \times 5.67 \times 10^{-8}(873.15^4 - 293.15^4) = \underline{82.8 \text{ W}}$$

(a) Rate of change of temperature neglecting convection.

To check the validity of the assumed lumped analysis, recall that with convective heat transfer lumped analysis is applicable if $\text{Bi} \ll 1$ (Section 2.8). The Biot number with convective heat transfer can be recast as

$$\text{Bi} = \frac{hL}{k} = \frac{q_c''}{T_s - T_2} \frac{L}{k}$$

where

$q_c'' = $ convective heat flux
$T_s = $ surface temperature
$T_2 = $ surrounding fluid temperature
$L = $ characteristic length dimension
$k = $ thermal conductivity of the solid

A similar criterion can be applied with radiative heat transfer by replacing the convective heat flux q_c'', with the radiative heat flux q_R'' to form an effective Biot number.

With the radius of the sphere as the characteristic dimension, we have

$q_R'' = 82.8/(\pi \times 0.03^2) = 29\ 285\ \text{W/m}^2$

$L = \text{radius of the sphere} = 0.015\ \text{m}$

$T_s - T_2 = 580\ ^\circ\text{C}$

$k_{\text{Cu}}\ (873\ \text{K}) = 361\ \text{W/m}\ ^\circ\text{C}$

$$\text{Bi} = \frac{29\ 285 \times 0.015}{580 \times 361} = 0.002$$

As Bi $\ll$ 1, lumped analysis is applicable. (Note that $q_R''/(T_s - T_2) = 50.5\ \text{W/m}^2$ can be interpreted as a radiative heat transfer coefficient. For a discussion of radiative heat transfer coefficient see Section 7.4.)

Now, to find the rate of change of temperature of the sphere, from Equation 2.8.1

$$mc_p \frac{dT}{d\tau} = \rho V c_p \frac{dT}{d\tau} = -q_i$$

where ρ is the density of the material, V is the volume of the sphere, and q_i is the heat transfer rate from the sphere. From Table A1, the properties of copper are $\rho = 8933\ \text{kg/m}^3$, $c_p(873\ \text{K}) \approx 361\ \text{J/kg K}$.

$$\frac{dT}{d\tau} = \frac{-q_i}{(4/3)\pi R^3 \rho c_p} = \frac{-82.8}{(4/3) \times \pi \times (0.03/2)^3 \times 8933 \times 361} = \underline{-1.82\ ^\circ\text{C/s}}$$

The negative sign indicates that the temperature decreases with time. The computed rate of change of temperature of the sphere is valid only when the sphere temperature T_s is 600 $^\circ$C. As the temperature of the sphere changes, the heat transfer rate and, therefore, the rate of change of temperature also change.

(b) Rate of change of temperature including convection.

To include the convective heat transfer, determine the convective heat transfer coefficient through an appropriate correlation. The convective heat transfer rate is then added to the radiative heat transfer rate determined in part a to get the total heat transfer rate. Assuming that the air in the room is not moving with any significant velocity, the convective heat transfer is by natural convection. Determine the convective heat transfer coefficient for natural convection from a sphere by Equation 4.3.7b.

Film temperature is 310 $^\circ$C. From the software CC (or Table A7) properties of air at 310 $^\circ$C are

$\rho = 0.6054\ \text{kg/m}^3$ $c_p = 1048\ \text{J/kg K}$

$\mu = 2.973 \times 10^{-5}\ \text{N s/m}^2$ $k = 0.0446\ \text{W/m K}$

$\text{Pr} = 0.699$ $\beta\ (293.15\ \text{K}) = 3.411 \times 10^{-3}\ \text{K}^{-1}$

$$\begin{aligned}
\text{Ra}_d &= \frac{g\beta\rho^2(T_s - T_\infty)d^3}{\mu^2}\ \text{Pr} \\
&= \frac{9.807 \times 3.411 \times 10^{-3} \times 0.6054^2 \times (600 - 20) \times 0.03^3}{(2.973 \times 10^{-5})^2} \times 0.699 \\
&= 1.517 \times 10^5
\end{aligned}$$

From Equation 4.3.7b,

$$\text{Nu}_d = 2 + \frac{0.589(1.517 \times 10^5)^{1/4}}{[1 + (0.469/0.699)^{9/16}]^{4/9}} = 10.95$$

$$h = \text{Nu}_d \frac{k}{d} = \frac{10.95 \times 0.0446}{0.03} = 16.3 \text{ W/m}^2 \text{ K}$$

$$q_c = 16.3 \times 4 \times \pi \times 0.015^2(600 - 20) = 26.7 \text{ W}$$

Total heat transfer rate = 82.8 + 26.7 = 109.5 W.

The addition of convective heat transfer coefficient increases the total heat transfer rate and also the effective Biot number. The effective Biot number is now given by using the total heat flux in forming the Biot number.

$$q''_\text{tot} = \frac{109.5}{\pi \times 0.03^2} = 38\,728 \text{ W/m}^2$$

$$\text{Bi} = \frac{38\,728 \times 0.015}{580 \times 361} = 0.0028$$

As Bi ≪ 1, lumped analysis is appropriate.

The magnitude of the rate of temperature change of the sphere is directly proportional to the total heat transfer rate and, therefore, the rate of change of temperature of the sphere is given by

$$\frac{dT}{d\tau} = -1.82 \times \frac{109.5}{82.8} = \underline{-2.41 \text{ °C/s}}$$

COMMENT

The radiative heat transfer rate is 82.8 W whereas the convective heat transfer rate is 26.7 W. This example shows that when heat transfer is by natural convection in gases, radiative heat transfer is as significant as convective heat transfer and, in many cases, greater than the convective heat transfer.

EXAMPLE 7.3.3 Consider a long enclosure with three radiation surfaces (Figure 7.3.7). Surfaces 1 and 2 are maintained at T_1 and T_2, respectively, and surface 3 is perfectly insulated. Determine the net radiative heat transfer rate from surface 1 and the temperature of surface 3.

ASSUMPTIONS

1. Each surface has uniform radiosity and the enclosure can be treated as a three-surface enclosure.
2. Diffuse, gray, and opaque surfaces
3. Medium separating the surfaces does not participate in radiation.
4. Negligible convection inside the enclosure

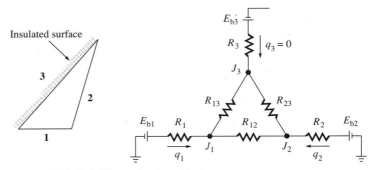

FIGURE 7.3.7 A Three-Surface Enclosure
Two surfaces are maintained at uniform temperatures, and the third
is perfectly insulated. The electrical circuit for the problem is shown at
right.

SOLUTION

The electrical circuit for the problem is shown in Figure 7.3.7, where the resistances
are

$$R_1 = \frac{1 - \varepsilon_1}{A_1 \varepsilon_1} \qquad R_2 = \frac{1 - \varepsilon_2}{A_2 \varepsilon_2} \qquad R_3 = \frac{1 - \varepsilon_3}{A_3 \varepsilon_3}$$

$$R_{12} = \frac{1}{A_1 F_{12}} \qquad R_{13} = \frac{1}{A_1 F_{13}} \qquad R_{23} = \frac{1}{A_2 F_{23}}$$

The battery connection of E_{b3} to the ground has been cut so that $q_3 = 0$, representing
the insulated surface. This is a series-parallel circuit and the equivalent resistance
between the E_{b1} and E_{b2} is

$$R_{eq} = R_1 + R_2 + \frac{R_{12}(R_{13} + R_{23})}{R_{12} + R_{13} + R_{23}}$$

and

$$q_1 = -q_2 = \frac{E_{b1} - E_{b2}}{R_{eq}}$$

Having determined q_1 we determine J_1 and J_2, and then J_3 by the relation

$$\frac{J_1 - J_3}{R_{13}} = \frac{J_3 - J_2}{R_{23}}$$

For surface 3,

$$q_3 = 0 = \frac{A_3 \varepsilon_3}{1 - \varepsilon_3} (E_{b3} - J_3)$$

and for all values of ε_3, $E_{b3} = J_3$. With $E_{b3} = \sigma T_3^4$, T_3 is evaluated.

EXAMPLE 7.3.4 The inside dimensions of a hollow cylinder are 6-cm diameter and 12-cm long. The cylindrical surface 2, is perfectly insulated. End surface 1 is maintained at 300 °C, and end surface 3 at 100 °C. Determine the total radiative heat transfer rate from surface 1 to surface 3, if each surface has an emissivity of 0.8.

Given

$L = 12$ cm	$r_1 = r_3 = 3$ cm	$T_1 = 300\,°C$
$q_2 = 0$	$\varepsilon_1 = \varepsilon_2 = \varepsilon_3 = 0.8$	$T_3 = 100\,°C$

Find

q_1 (total radiative heat transfer rate from surface 1)

ASSUMPTIONS

1. All surfaces are diffuse, gray, and opaque.
2. The medium inside the cylinder does not participate in radiation.
3. Uniform radiosity for each of the three surfaces—the cylindrical surface, and the two end surfaces—so that the cylinder can be treated as a three-surface enclosure
4. Negligible convection

SOLUTION

Figure 7.3.8 shows the physical configuration and the electrical analog. To find q_1, we need the view factors $F_{1\text{-}2}$ and $F_{1\text{-}3}$. Note that by symmetry, $F_{1\text{-}2} = F_{3\text{-}2}$. The term

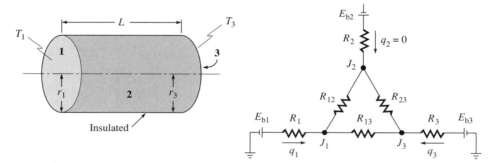

FIGURE 7.3.8 A Hollow Cylinder
The two end surfaces are maintained at uniform temperatures and the cylindrical surface is perfectly insulated. The electrical analog is shown at right.

$F_{1\text{-}3}$ is found from view factor graphs for two coaxial disks, Figure 7.2.8, or from the formula for coaxial disks in Table 7.2.1. Employing the formula

$$F_{1\text{-}3} = \frac{1}{2}\left[x - \left(x^2 - 4\frac{r_3^2}{r_1^2}\right)^{1/2}\right] \qquad x = 1 + \frac{1 + (r_3/L)^2}{(r_1/L)^2}$$

where $r_1/L = 0.25 = r_3/L$ (dimensionless radii) and $x = 18$. Substituting the values,

$$F_{1\text{-}3} = 0.05573 = F_{3\text{-}1}$$

From view factor algebra, $F_{1-1} + F_{1-2} + F_{1-3} = 1$ and as $F_{1-1} = 0$,

$$F_{1-2} = 1 - 0.05573 = 0.94427 = F_{3-2}$$

$$A_1 = \pi \times 0.03^2 = 0.0028274 \text{ m}^2 = A_3$$

$$A_2 = \pi \times 0.06 \times 0.12 = 0.02262 \text{ m}^2$$

$$R_1 = \frac{1 - \varepsilon_1}{A_1 \varepsilon_1} = \frac{1 - 0.8}{0.0028274 \times 0.8} = 88.42 \text{ m}^{-2} = R_3$$

$$R_{13} = \frac{1}{A_1 F_{1-3}} = \frac{1}{0.0028274 \times 0.05573} = 6346 \text{ m}^{-2}$$

$$R_{12} = \frac{1}{A_1 F_{1-2}} = \frac{1}{0.0028274 \times 0.94427} = 374.6 \text{ m}^{-2} = R_{32}$$

From Figure 7.3.8, for the series-parallel circuit, we have

$$R_{eq} = R_1 + R_3 + \frac{R_{13}(R_{12} + R_{32})}{R_{13} + R_{12} + R_{32}} = 88.42 + 88.42 + \frac{6346(374.6 + 374.6)}{6346 + 374.6 + 374.6}$$

$$= 846.9 \text{ m}^{-2}$$

$$q_1 = \frac{E_{b1} - E_{b3}}{R_{eq}} = \frac{5.67 \times 10^{-8}(573.15^4 - 373.15^4)}{846.9} = \underline{5.93 \text{ W}}$$

We can find the temperature of surface 2 by finding its radiosity. Noting that $J_2 = E_{b2}$ and $R_{12} = R_{23}$,

$$J_2 = E_{b2} = \frac{J_1 + J_3}{2}$$

From $q_1 = (E_{b1} - J_1)/R_1$,

$$5.93 = \frac{5.67 \times 10^{-8} \times 573.15^4 - J_1}{88.42} \qquad J_1 = 5594.3 \text{ W/m}^2$$

From $-q_1 = q_3 = (E_{b3} - J_3)/R_3$,

$$-5.93 = \frac{5.67 \times 10^{-8} \times 373.15^4 - J_3}{88.42} \qquad J_3 = 1623.6 \text{ W/m}^2$$

$$J_2 = (5594.3 + 1623.6)/2 = 3609 \text{ W/m}^2 = E_{b2} = \sigma T_2^4$$
$$= 5.67 \times 10^{-8} T_2^4$$
$$T_2 = 502.3 \text{ K} = 229.1 \text{ °C}$$

COMMENTS

Although surface 2 is perfectly insulated, it participates in heat transfer from surface 1 to surface 3 by providing an alternate path for the heat transfer. First there is a net energy transfer from surface 1 to surface 2, and an equal amount from surface 2 to surface 3. Can you think of an analogy in fluid flow?

Note on view factor $F_{2\text{-}2}$: Because surface 2 is concave, a part of the energy leaving that surface reaches itself and $F_{2\text{-}2} \neq 0$. We may compute $F_{2\text{-}2}$ with the known values of the view factors. From the reciprocity relation, Equation 7.2.4a,

$$A_2 F_{2\text{-}1} = A_1 F_{1\text{-}2} \qquad F_{2\text{-}1} = \frac{A_1 F_{1\text{-}2}}{A_2} = \frac{\pi \times 6^2 \times 0.94427}{4 \times \pi \times 6 \times 12} = 0.11803 = F_{2\text{-}3}$$

For the three-surface enclosure, $F_{2\text{-}1} + F_{2\text{-}2} + F_{2\text{-}3} = 1$.

$$F_{2\text{-}2} = 1 - 0.11803 - 0.11803 = \underline{0.7639}$$

Heat Transfer with Black Surfaces If a surface i is black, its radiosity is equal to its emissive power. When summing the energies to a black surface, the term $(E_{bi} - J_i)/R_i$ is replaced by the heat transfer rate q_i. In the remaining equations the radiosity of the black surface is replaced by its emissive power. Thus, if the temperature of a black surface is known, q_1 is the unknown. If q_1 is known, E_{bi} is the unknown. We then have as many equations as the number of unknown quantities. When an enclosure contains a black surface, the manner in which the solution is obtained is illustrated in Examples 7.3.5 and 7.3.6.

EXAMPLE 7.3.5

Reconsider the configuration in Example 7.3.4, where $\varepsilon_1 = \varepsilon_3 = 0.8$, $T_1 = 300\,°C$, and $T_3 = 100\,°C$.

(a) Determine q_1, q_2, and q_3 if $T_2 = 500\,°C$ and $\varepsilon_2 = 0.8$.
(b) Determine q_1, q_2, and q_3 if $T_2 = 500\,°C$ and $\varepsilon_2 = 1$.
(c) Determine q_1, q_3, and T_2 if $q_2 = 30\,W$ and $\varepsilon_2 = 1$.

Given

$d = 6$ cm	$L = 12$ cm	$T_1 = 300\,°C$
$\varepsilon_1 = \varepsilon_3 = 0.8$	$T_3 = 100\,°C$	

Find

(a) q_1, q_2, and q_3 if $T_2 = 500\,°C$, $\varepsilon_2 = 0.8$
(b) q_1, q_2, and q_3 if $T_2 = 500\,°C$, $\varepsilon_2 = 1$
(c) q_1, q_3, and T_2 if $q_2 = 30\,W$, $\varepsilon_2 = 1$

ASSUMPTIONS

1. All surfaces are diffuse, gray, and opaque.
2. The medium inside the cylinder does not participate in radiation.
3. Uniform radiosity for each of the three surfaces—the cylindrical surface and the two end surfaces—so that the enclosure can be treated as having three surfaces

SOLUTION

(a) The equivalent electrical circuit is shown in Figure 7.3.9. In addition to the resistance computed in Example 7.3.4, the value of one more resistance, R_2, is needed.

$$R_2 = \frac{1 - \varepsilon_2}{A_2 \varepsilon_2} = \frac{1 - 0.8}{0.02262 \times 0.8} = 11.1 \text{ m}^{-2}$$

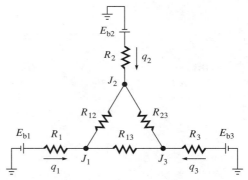

FIGURE 7.3.9 Electrical Analog for Example 7.3.5

From an energy balance the net heat transfer rate to each surface is zero. By repeated application of the energy balance for each surface, we obtain three equations in the three unknown radiosities.

$$\frac{E_{b1} - J_1}{R_1} + \frac{J_2 - J_1}{R_{12}} + \frac{J_3 - J_1}{R_{13}} = 0 \tag{1}$$

$$\frac{E_{b2} - J_2}{R_2} + \frac{J_1 - J_2}{R_{12}} + \frac{J_3 - J_2}{R_{23}} = 0 \tag{2}$$

$$\frac{E_{b3} - J_3}{R_3} + \frac{J_1 - J_3}{R_{13}} + \frac{J_2 - J_3}{R_{23}} = 0 \tag{3}$$

$$E_{b1} = 6118.7 \text{ W/m}^2 \qquad E_{b2} = 20\,260 \text{ W/m}^2 \qquad E_{b3} = 1099.3 \text{ W/m}^2$$

From the results of Example 7.3.4,

$$R_1 = 88.42 \text{ m}^{-2} = R_3 \qquad R_{13} = 6346 \text{ m}^{-2} \qquad R_{12} = 374.6 \text{ m}^{-2} = R_{32} = R_{23}$$

Substituting the various values and simplifying the equations, we get

$$-0.014137J_1 + 0.0026695J_2 + 0.00015758J_3 = -69.2 \tag{4}$$

$$0.0026695J_1 - 0.095429J_2 + 0.0026695J_3 = -1825.2 \tag{5}$$

$$0.00015758J_1 + 0.0026695J_2 - 0.014137J_3 = -12.43 \tag{6}$$

Solving for J_1, J_2, and J_3 we obtain

$$J_1 = 8628.7 \text{ W/m}^2 \qquad J_2 = 19\,497.9 \text{ W/m}^2 \qquad J_3 = 4657.2/\text{m}^2$$

The heat transfer rate from each surface can now be calculated from Equation 7.3.4.

$$q_1 = \frac{E_{b1} - J_1}{R_1} = \frac{6118.7 - 8628.7}{88.42} = \underline{-28.39 \text{ W}}$$

$$q_2 = \frac{E_{b2} - J_2}{R_2} = \frac{20\,260 - 19\,497.9}{11.1} = \underline{68.66 \text{ W}}$$

$$q_3 = \frac{E_{b3} - J_3}{R_3} = \frac{1099.3 - 4657.2}{88.42} = \underline{-40.24 \text{ W}}$$

COMMENTS

The qs represent the net heat transfer rates from each surface. Thus, surface 2 is to be supplied with 68.66 W from an external source, and surfaces 1 and 3 should be cooled by removing 28.39 W and 40.24 W, respectively. Also the net radiative heat transfer rate from surface 2 (68.66 W) should be equal to the sum of the radiative heat transfer rates to surfaces 1 and 3 ($28.39 + 40.24 = 68.63$ W). The difference of 0.03 W between these two values is caused by round-off errors in the numerical solution for the radiosities. (If a computer is used to compute the coefficients with 8 significant figures and solve for the radiosities, the results with 6 significant figures are, $J_1 = 8628.88$ W/m^2, $J_2 = 19\,498.1$ W/m^2, $J_3 = 4657.57$ W/m^2, $q_1 = -28.39$ W, $q_2 = 68.63$ W, and $q_3 = -40.24$ W. Note that $q_1 + q_2 + q_3 = 0$.)

(b) In this case, $J_2 = E_{b2} = 20\,260$ W/m^2. Equations 1 and 3 in part a can be used as they stand. But in Equation 2, the first term (where both the numerator and the denominator are zero) is replaced by q_2, the radiant heat transfer rate from surface 2. We then have three equations in the three unknowns, J_1, q_2, and J_3.

$$\frac{E_{b1} - J_1}{R_1} + \frac{J_2 - J_1}{R_{12}} + \frac{J_3 - J_1}{R_{13}} = 0$$

$$q_2 + \frac{J_1 - J_2}{R_{12}} + \frac{J_3 - J_2}{R_{23}} = 0$$

$$\frac{E_{b3} - J_3}{R_3} + \frac{J_1 - J_3}{R_{13}} + \frac{J_2 - J_3}{R_{23}} = 0$$

Substituting the values of E_{b1}, E_{b2}, and E_{b3} (noting that $J_2 = E_{b2}$) and the various resistances we obtain (after simplification),

$$-0.014137J_1 + 0.00015758J_3 = -123.28 \tag{7}$$

$$0.0026695J_1 + q_2 + 0.0026695J_3 = 108.17 \tag{8}$$

$$0.00015758J_1 - 0.014137J_3 = -66.52 \tag{9}$$

Solving for J_1, q_2, and J_3,

$$J_1 = 8773.9 \text{ W/m}^2 \qquad q_2 = \underline{71.93 \text{ W}} \qquad J_3 = 4803.2 \text{ W/m}^2$$

Employing Equation 7.3.4 for surfaces 1 and 3,

$$q_1 = \frac{6118.7 - 8773.6}{88.42} = \underline{-30.03 \text{ W}}$$

$$q_3 = \frac{1099.3 - 4803.2}{88.42} = \underline{-41.89 \text{ W}}$$

COMMENT

The sum of q_1 and q_3 is equal to q_2 in magnitude ($q_1 + q_3 = -71.92$ W) but is of opposite sign. The slight difference is due to round-off errors in the computations. The heat transfer from surface 2 increased from 68.66 W to 71.93 W as a result of increasing its emissivity from 0.8 to 1.0.

(c) Heat transfer rate q_2 is known but E_{b2} is not known. Equations 1 and 3 of part a can be used as they are after replacing J_2 by E_{b2}. The first term in Equation 2, which represents q_2, is replaced by its known value of 30 W. After simplifying, we obtain

$$-0.014137J_1 + 0.0026695E_{b2} + 0.00015758J_3 = -69.2 \qquad (10)$$

$$0.0026695J_1 + 0.005339E_{b2} + 0.0026695J_3 = -30 \qquad (11)$$

$$0.00015758J_1 + 0.0026695E_{b2} - 0.014137J_3 = -12.43 \qquad (12)$$

Solving for J_1, E_{b2}, and J_3,

$$J_1 = 6920.8 \text{ W/m}^2 \qquad E_{b2} = 10\,554.1 \text{ W/m}^2 \qquad J_3 = 2949.3 \text{ W/m}^2$$

Employing Equation 7.3.4 for surfaces 1 and 3, we obtain

$$q_1 = \frac{6118.7 - 6920.8}{88.42} = \underline{-9.1 \text{ W}}$$

$$E_{b2} = 10\,554.1 \text{ W/m}^2 = \sigma T_2^4 \qquad T_2 = \underline{657 \text{ K}}$$

$$q_3 = \frac{1099.3 - 2949.3}{88.42} = \underline{-20.9 \text{ W}}$$

EXAMPLE 7.3.6 Reconsider Example 7.3.4. Determine q_1 and T_2 if all surfaces are black.

Given

$d = 6$ cm $\qquad L = 12$ cm $\qquad r_1 = r_3 = 3$ cm $\qquad T_1 = 300\,°C$

$T_3 = 100\,°C \qquad q_2 = 0 \qquad\qquad \varepsilon_1 = \varepsilon_2 = \varepsilon_3 = 1$

Find

q_1 (heat transfer rate from surface 1)

ASSUMPTION

The cylinder is a three-surface enclosure.

SOLUTION

First, we obtain the solution employing the electrical analogy. We then show an alternate method of getting the solution for a black surface enclosure.

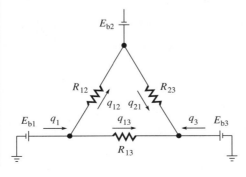

FIGURE 7.3.10 **Electrical Analog for an All Black Three-Surface Enclosure**

The electrical circuit for the three-surface enclosure with one insulated surface is obtained by setting $R_1 = R_2 = R_3 = 0$ in Figure 7.3.8 and is shown in Figure 7.3.10. From the results in Example 7.3.4,

$$R_{12} = R_{32} = 374.6 \text{ m}^{-2} \qquad R_{13} = 6346 \text{ m}^{-2}$$

$$E_{b1} = 6118.7 \text{ W/m}^2 \qquad E_{b3} = 1099.3 \text{ W/m}^2$$

$$q_1 = \frac{E_{b1} - E_{b3}}{R_{eq}}$$

$$R_{eq} = \frac{R_{13}(R_{12} + R_{32})}{R_{13} + R_{12} + R_{32}} = \frac{6346 \times 2 \times 374.6}{6346 + 2 \times 374.6} = 670.1 \text{ m}^{-2}$$

$$q_1 = \frac{6118.7 - 1099.3}{670.1} = \underline{7.5 \text{ W}}$$

ALTERNATE SOLUTION

As the surfaces are black, the only energy leaving each surface is due to emission from the surface and all the incident energy is absorbed.

$$\begin{aligned} &\text{Net heat transfer rate} \\ &= \text{Rate of emitted energy} \\ &- \text{Rate of incident energy} \end{aligned}$$

Now consider surface 1.

Emitted energy $= A_1 E_{b1}$
Incident energy $=$ absorbed energy $= A_1 E_{b1} F_{1-1} + A_2 E_{b2} F_{2-1} + A_3 E_{b3} F_{3-1}$. As $F_{1-1} = 0$,

$$q_1 = A_1 E_{b1} - A_2 E_{b2} F_{2-1} - A_3 E_{b3} F_{3-1} \tag{1}$$

Similarly, for surface 2 (note that $F_{2-2} \neq 0$)

$$q_2 = 0 = A_2 E_{b2} - A_1 E_{b1} F_{1-2} - A_2 E_{b2} F_{2-2} - A_3 E_{b3} F_{3-2}$$

$$E_{b2}(1 - F_{2-2})A_2 = A_1 E_{b1} F_{1-2} + A_3 E_{b3} F_{3-2} \tag{2}$$

With $F_{3-3} = 0$, for surface 3

$$q_3 = A_3 E_{b3} - A_1 E_{b1} F_{1-3} - A_2 E_{b2} F_{2-3} \tag{3}$$

As $q_3 = -q_1$, we need to determine only q_1 and E_{b2}. From Equation 2

$$E_{b2} = \frac{A_1 E_{b1} F_{1-2} + A_3 E_{b3} F_{3-2}}{(1 - F_{2-2})A_2}$$

From the results of Example 7.3.4,

$$A_1 = 0.002827 \text{ m}^2 = A_3 \qquad A_2 = 0.02262 \text{ m}^2$$

$$F_{1-3} = 0.05573 = F_{3-1} \qquad F_{1-2} = 0.94427 = F_{3-2}$$

$$F_{2-1} = 0.11803 \qquad F_{2-2} = 0.7639$$

Substituting the values, and noting $A_1 F_{1\text{-}2} = A_3 F_{3\text{-}2}$

$$E_{b2} = \frac{A_1 F_{1\text{-}2}}{(1 - F_{2\text{-}2})A_2}(E_{b1} + E_{b3}) = \frac{0.002827 \times 0.94427}{(1 - 0.7639) \times 0.02262}(6118.7 + 1099.3)$$

$$= 3608 \text{ W/m}^2 = \sigma T_2^4$$

$$T_2 = (3608/5.67 \times 10^{-8})^{1/4} = \underline{502.3 \text{ K}}$$

Substituting the values in Equation 1,

$$q_1 = 2.8274 \times 10^{-3} \times 6118.7 - 2.2619 \times 10^{-2} \times 3608 \times 0.11803$$
$$- 2.8274 \times 10^{-3} \times 1099.3 \times 0.05573 = \underline{7.5 \text{ W}}$$

7.3.1 Radiation Shields

Radiation shields to reduce radiative heat transfer from a surface

In some cases it is necessary to reduce the radiative heat transfer from a surface to its surroundings. For example, we wish to reduce the heat transfer rate from the top surface of a furnace that may be at a much higher temperature than its surroundings. From every day experience, we know that placing a surface that absorbs radiant energy (such as an opaque surface, a glass, or wire mesh screen) between a high temperature source, such as a fireplace, and ourselves reduces the radiant energy reaching us. The same principle is applied to reduce the radiant energy transfer from one surface to another. A surface that is introduced between two surfaces to reduce

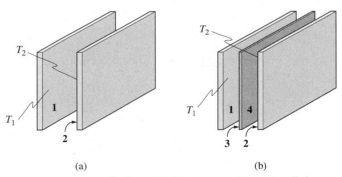

(a)   (b)

FIGURE 7.3.11 Radiation Shield Between Two Parallel Surfaces to Reduce Radiative Heat Transfer
(a) A plane surface 1, close to a parallel surface 2. (b) Radiation shield between the two surfaces.

the radiative heat transfer rate is known as a *radiation shield*.

Consider a plane surface 1, close to a second surface 2, as shown in Figure 7.3.11a. The two surfaces are directly opposite to each other and $A_1 = A_2$. The two surfaces are so close to each other that $F_{12} \approx 1$. Furthermore, the *net* radiative heat transfer rate, q_0 (the subscript indicates the number of radiation shields), from surface 1 to surface 2 is given by (see Example 7.3.1),

$$q_0 = \frac{A_1(E_{b1} - E_{b2})}{1/\varepsilon_1 + 1/\varepsilon_2 - 1} \tag{7.3.8}$$

Now, let a radiation shield, i.e., a thin flat plate, be placed between surfaces 1 and 2. To illustrate the effect of introducing the shield, we neglect the convective heat transfer. The shield has two surfaces, surface 3 facing surface 1 and surface 4 facing surface 2. The following assumptions are made in determining the heat transfer rate.

1. The shield is thin, the conductive resistance across the shield is negligible, and the temperature drop across it is also negligible so that $T_3 = T_4$ and $E_{b3} = E_{b4}$.
2. The temperature of surfaces 1 and 2 are not affected by the introduction of the shield.
3. As $F_{12} \approx 1$, $F_{13} = F_{42} \approx 1$.
4. Convection is negligible.

With these assumptions the electrical analog with the radiation shield is shown in Figure 7.3.12. As $T_3 = T_4$, $E_{b3} = E_{b4}$ but ε_3 and ε_4 may not be equal to each other

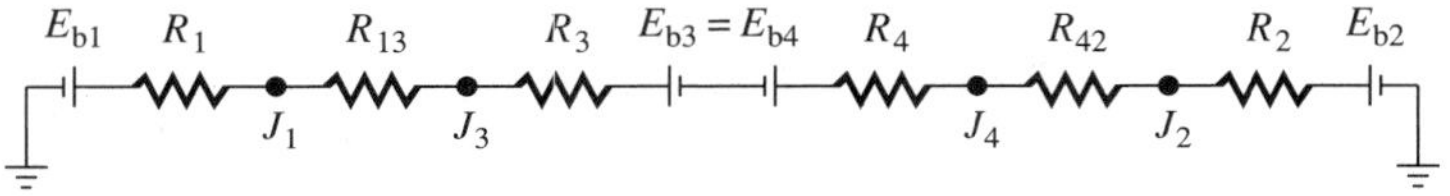

FIGURE 7.3.12 Electrical Analog with Radiation Shield

as the two surfaces may have different emissivities. The various resistances are given by

$$R_1 = \frac{1 - \varepsilon_1}{A_1 \varepsilon_1} \qquad R_{13} = \frac{1}{A_1 F_{1\text{-}3}} = \frac{1}{A_1} \qquad R_3 = \frac{1 - \varepsilon_3}{A_3 \varepsilon_3}$$

$$R_4 = \frac{1 - \varepsilon_4}{A_4 \varepsilon_4} \qquad R_{42} = \frac{1}{A_4 F_{4\text{-}2}} = \frac{1}{A_4} = \frac{1}{A_1} \qquad R_2 = \frac{1 - \varepsilon_2}{A_2 \varepsilon_2}$$

Since the area of each surface is equal to A_1, the heat transfer rate q_1 with one shield is given by

$$
\begin{aligned}
q_1 &= \frac{A_1(E_{b1} - E_{b2})}{(1 - \varepsilon_1)/\varepsilon_1 + 1 + (1 - \varepsilon_3)/\varepsilon_3 + (1 - \varepsilon_4)/\varepsilon_4 + 1 + (1 - \varepsilon_2)/\varepsilon_2} \quad \textbf{(7.3.9)} \\
&= \frac{A_1(E_{b1} - E_{b2})}{1/\varepsilon_1 + 1/\varepsilon_3 + 1/\varepsilon_4 + 1/\varepsilon_2 - 2}
\end{aligned}
$$

If all the surfaces have the same emissivity, say ε ($\varepsilon_1 = \varepsilon_3 = \varepsilon_4 = \varepsilon_2 = \varepsilon$), we have

$$q_0 = \text{Heat transfer without radiation shield} = \frac{A_1(E_{b1} - E_{b2})}{2/\varepsilon - 1}$$

$$q_1 = \text{Heat transfer with one radiation shield} = \frac{A_1(E_{b1} - E_{b2})}{4/\varepsilon - 2} = \frac{q_0}{2}$$

With all surfaces having the same emissivity it is easily shown that the heat transfer rate with N radiation shields is given by

$$q_N = \frac{q_0}{n + 1}$$

The above relation shows that, considering the increased use of material and fabrication complexity, the reduction in the radiative heat transfer rate beyond two radiation shields may not be significant in most applications.

From an examination of the resistances it is evident that the total resistance can be increased either by decreasing the emissivity of the shield or increasing the number of radiation shields. Example 7.3.7 illustrates the reduction in radiation heat transfer that can be achieved by introducing a radiation shield.

EXAMPLE 7.3.7 The ceiling of many furnaces is in the shape of an arch. The arch is constructed with suspended firebricks (Figure 7.3.13). The bricks deteriorate due to cracking and disintegration, which leads to significantly higher temperature of the surface of the arch exposed to the surroundings. A reduction in the heat transfer rate from the surface of the arch to the surroundings reduces the fuel requirements of the furnace and also results in a more comfortable working environment. In one such furnace it is found that the average surface temperature of the top surface is 200 °C. The emissivity of the surface is 0.9. To reduce the radiative heat transfer rate from the surface, it is proposed to add a radiation shield of steel ($\varepsilon = 0.85$) close to the top surface. The temperature of the surroundings is 35 °C. Determine, per unit surface area,

(a) The radiative heat flux without and with the radiation shield.
(b) The temperature of the shield.

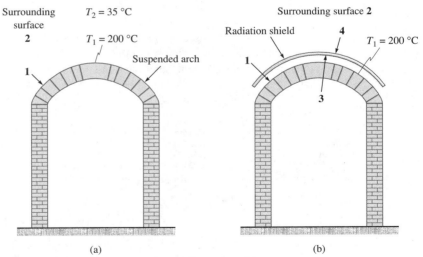

FIGURE 7.3.13 Furnace with Suspended Arch
(a) Furnace without shield. (b) Furnace with a radiation shield to reduce the radiant heat transfer rate from the top surface.

Given

$T_1 = 200 \,°\text{C}$ $\varepsilon_1 = 0.9$
$T_2 = 35 \,°\text{C}$ $\varepsilon_3 = \varepsilon_4 = 0.85$

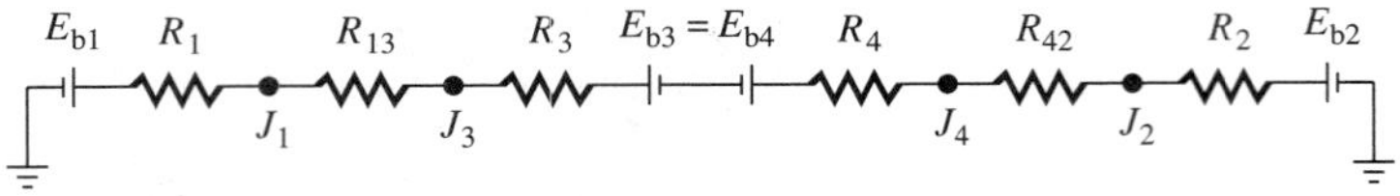

FIGURE 7.3.14 Electrical Analog with the Radiation Shield

Find

(a) q_0'' (without shield)
(b) q_1'' (with one radiation shield)

ASSUMPTIONS

1. All surfaces are diffuse, gray, and opaque.
2. The air surrounding the furnace does not participate in radiation.
3. Negligible convective heat transfer
4. The temperature of the arch is 200 °C after the insertion of the shield. This restriction is relaxed in Example 10.1.3.
5. The surrounding surface area is much greater than the arch surface area.

SOLUTION

The arch surface is denoted by 1 and the surrounding surface by 2. Only radiative heat transfer is considered.

(a) With the assumptions made, Equation 3 of Example 7.3.1 is applicable.

$$\frac{q_0}{A_1} = \varepsilon_1 \sigma(T_1^4 - T_2^4) = 0.9 \times 5.67 \times 10^{-8}(473.15^4 - 308.15^4) = \underline{2097 \text{ W/m}^2}$$

It is reasonable to assume that the shield has the same dimensions as the arch top and that areas of the arch top and the shield are equal. It is also reasonable to assume that the gap between the shield and the arch is very small and that all energy from the arch surface reaches the shield. This is equivalent to assuming that $F_{1\text{-}3} \approx 1$. With $F_{1\text{-}3} = F_{4\text{-}2} \approx 1$ and $E_{b3} = E_{b4}$,

$$\frac{q_1}{A_1} = \frac{E_{b1} - E_{b2}}{1/\varepsilon_1 + 1/\varepsilon_3 + 1/\varepsilon_4 - 1 + A_1/A_2[(1 - \varepsilon_2)/\varepsilon_2]}$$

With $A_1/A_2 \ll 1$ and $\varepsilon \sim 1$ for all the surfaces, the last term in the denominator can be neglected in comparison with the other terms.

$$\frac{q_1}{A_1} = \frac{E_{b1} - E_{b2}}{1/\varepsilon_1 + 1/\varepsilon_3 + 1/\varepsilon_4 - 1}$$

With $\varepsilon_1 = 0.9$, $\varepsilon_3 = \varepsilon_4 = 0.85$, the heat transfer rate from the arch surface is

$$\frac{q_1}{A_1} = \frac{5.67 \times 10^{-8}(473.15^4 - 308.15^4)}{1/0.9 + 1/0.85 + 1/0.85 - 1} = \underline{945.8 \text{ W/m}^2}$$

(b) The temperature of the shield is determined by the expression for the heat transfer rate from surface 1 to surface 3.

$$\frac{q_1}{A_1} = \frac{E_{b1} - E_{b3}}{[(1 - \varepsilon_1)/\varepsilon_1] + 1 + [(1 - \varepsilon_3)/\varepsilon_3]}$$

Substituting the various values,

$$945.8 = \frac{5.67 \times 10^{-8}(473.15^4 - T_3^4)}{[(1 - 0.9)/0.9] + 1 + [(1 - 0.85)/0.85]}$$

$$T_3 = 411.4 \text{ K} = 138.2 \,°\text{C}$$

COMMENT

By using one radiation shield, the radiative heat transfer rate is reduced from 2097 W/m^2 to 945.8 W/m^2, a reduction of 55%. The heat transfer rate can be further reduced by decreasing the emissivity of the shield. If, instead of employing a plane carbon sheet steel, a heat resistant, oxidized steel with $\varepsilon = 0.65$ is used, the heat transfer rate is reduced to 731 W/m^2.

Insertion of the radiation shield changes the surface temperature of the arch surface. Convection also changes the heat transfer rate from the surface. The effects of the change in the surface temperature and convective heat transfer are considered in Example 10.1.3.

7.4 RADIATIVE HEAT TRANSFER COEFFICIENT

In analogy with the convective heat transfer coefficient we formally define a radiative heat transfer coefficient associated with a surface completely surrounded by a second surface as

$$q_R'' = h_R(T_1 - T_2) \tag{7.4.1}$$

where

q_R'' = radiant heat flux from surface 1 to surface 2
h_R = radiative heat transfer coefficient
T_1 = temperature of surface 1
T_2 = temperature of surrounding surface 2

Radiative heat transfer coefficient

Radiative heat transfer coefficient is generally used only in a two-surface enclosure. For a convex or a plane surface completely enclosed by a second surface, from Equation 2, Example 7.3.1,

$$q_R'' = \frac{\sigma(T_1^4 - T_2^4)}{1/\varepsilon_1 + A_1/A_2[(1 - \varepsilon_2)/\varepsilon_2]} \tag{7.4.2a}$$

$$h_R = \frac{q''}{T_1 - T_2} = \frac{\sigma(T_1^2 + T_2^2)(T_1 + T_2)}{1/\varepsilon_1 + A_1/A_2[(1 - \varepsilon_2)/\varepsilon_2]}$$

If $(A_1/A_2)[(1 - \varepsilon_2)/\varepsilon_2] \ll 1$,

$$h_R = \sigma\varepsilon_1(T_1^2 + T_2^2)(T_1 + T_2) \tag{7.4.2b}$$

From the definition of h_R it is clear that the coefficient is dependent on the temperatures of the two surfaces. However, it is reasonably constant over a limited range of values of $(T_1 - T_2)$. In such cases, the radiative heat transfer coefficient can be used to get a good estimate of the heat transfer rate. When the heat transfer from a surface is both by radiation and convection, and both the surrounding fluid and surface are at the same temperature T_2, the total heat transfer rate from the surface is obtained by computing the two heat transfer rates and adding them. Thus, we have

$$q = (h_c + h_R)A(T_1 - T_2)$$

where A is the area of the surface, and T_1 is the temperature of the surface. An estimate of the variation of h_R with T_1, and the temperature difference $(T_1 - T_2)$ is obtained by taking the ratio of the radiative heat transfer coefficients at T_1 ($\neq T_2$) and at $T_1 = T_2$ for different values and T_1/T_2.

$$h_R(T_1, T_2) = \sigma\varepsilon(T_1^2 + T_2^2)(T_1 + T_2) \qquad h_R(T_1 = T_2, T_2) = \sigma\varepsilon 4T_2^3$$

As $T_1 \to T_2$, $h_R(T_1 = T_2, T_2)$ is the limiting value of h_R. Defining the ratio of the two heat transfer coefficients by R,

$$R = \frac{h_R(T_1, T_2)}{h_R(T_2, T_2)} = \frac{1}{4}\left[1 + \left(\frac{T_1}{T_2}\right) + \left(\frac{T_1}{T_2}\right)^2 + \left(\frac{T_1}{T_2}\right)^3\right]$$

Figure 7.4.1 shows the variation of R with T_1/T_2. For $T_1/T_2 = 1.05$, $R = 1.08$. In this small range of the temperatures, the value of the radiative heat transfer coefficient

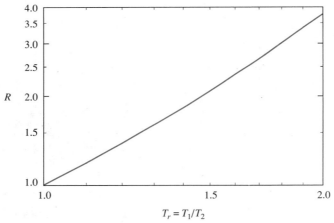

FIGURE 7.4.1 **Variation of the Ratio of Radiative Heat Transfer Coefficients**

has a maximum variation of 8% and the use of the average of the minimum and maximum values of h_R will yield results with acceptable accuracy in most engineering problems. The effect of the uncertainty in value of the surface emissivity is likely to be greater than the variation of h_R with the temperature if $T_1/T_2 < 1.05$.

The utility of the radiative heat transfer coefficient is illustrated through two examples.

EXAMPLE 7.4.1

In the computation of cooling loads of buildings, the heat transfer rate from each wall of a room is needed. Consider a room with only one exterior wall exposed to the outside atmosphere (Figure 7.4.2). In winter, the temperature of the inner surface

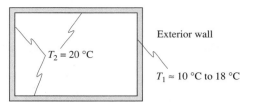

FIGURE 7.4.2 The temperature of the inner surface of the exterior wall varies with the conditions on the outside.

of the exterior wall is expected to be in the range 10 °C to 18 °C when the room air is maintained at 20 °C. Estimate the radiative heat transfer coefficient associated with the inner surface of the exterior wall.

Given

$T_1 = 10\,°C{-}18\,°C \qquad T_2 = 20\,°C$

Find

h_R (associated with the inner surface of the exterior wall)

ASSUMPTIONS

1. All surfaces are gray, diffuse, and opaque.
2. The emissivity of the plastered wall surfaces ≈ 0.9.

SOLUTION

The radiative heat flux from the exterior wall is found by considering it as one surface with area A_1 and temperature T_1 and the rest of the surfaces of the room as a second, much larger surface with area A_2 and temperature T_2. From Equation 3 of Example 7.3.1, we have

$$q_1'' = \varepsilon_1 \sigma(T_1^4 - T_2^4)$$

and from Equation 7.4.2b

$$h_R = \varepsilon_1 \sigma(T_1^2 + T_2^2)(T_1 + T_2)$$

The temperature of surface 2 is $T_2 = 20\,°C = 293.15\,K$ and T_1 varies from 10 °C to 18 °C (283.15 K to 291.15 K). Hence, the minimum and maximum values of h_R are

$$h_R(\text{min}) = 0.9 \times 5.67 \times 10^{-8}(283.15^2 + 293.15^2)(283.15 + 293.15)$$
$$= 4.89 \text{ W/m}^2 \text{ K}$$

$$h_R(\text{max}) = 0.9 \times 5.67 \times 10^{-8}(291.15^2 + 293.15^2)(291.15 + 293.15)$$
$$= 5.1 \text{ W/m}^2 \text{ K}$$

In the limited range of temperature of the surface, an average value $(4.89 + 5.1)/2 \approx 5 \text{ W/m}^2 \text{ K}$ may be used for h_R. The variation from the average value of 5 W/m^2 K is less than 2.2%.

COMMENT

There is some uncertainty in the value of the emissivity of the wall surfaces, which may lead to a greater uncertainty in the computed heat transfer rate than the error involved in using a constant value of h_R in this case.

   The natural convective heat transfer coefficient for a 2.5-m high wall is 3 W/m² °C for $T_1 = 283.2$ K and 1.8 W/m² °C for $T_1 = 292.2$ K. Even at room temperature, radiative heat transfer is more significant than the convective heat transfer in this case. The combined heat transfer coefficient (including both radiative and convective heat transfer) is 8 W/m² °C at $T_1 = 281.2$ K. ASHRAE (1989) recommends 8.3 W/m² °C for $\varepsilon = 0.9$. Natural convection correlations are for still air but in houses there is some movement of air.

EXAMPLE 7.4.2

Figure 7.4.3 shows a hot water heater, 60-cm outside diameter and 1.8-m high, which is insulated with 25-mm thick fiberglass insulation. The insulation is covered with a thin sheet of metal cladding. The water in the boiler is maintained at 80 °C. The temperature of the surrounding air and surfaces is 10 °C. The painted metal cladding has an emissivity of 0.95. To reduce the heat transfer to the surroundings, it is proposed to replace the cladding with a polished one having an emissivity of 0.1. If the convective heat transfer coefficient is 8 W/m² °C, determine the reduction in the heat transfer rate from the cylindrical surface by replacing the cladding.

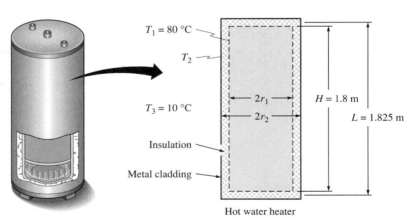

FIGURE 7.4.3 **A Hot Water Boiler with Fiberglass Insulation and a Metal Cladding**

Given

$r_1 = 30$ cm	$r_2 = 32.5$ cm	$H = 1.8$ m	$h_{conv} = 8$ W/m² °C
$T_1 = 80$ °C	$T_3 = 10$ °C	$\varepsilon_2 = 0.95$	

Find

The reduction in heat transfer rate if $\varepsilon_2 = 0.1$

ASSUMPTIONS

1. One-dimensional temperature distribution in the cylindrical part of the insulation; i.e., the two-dimensional effects at the corners are negligible
2. Negligible thermal resistance between the water and the inner surface of the insulation, and in the metal cladding
3. Steady state

SOLUTION

In steady state, from an energy balance,

Heat transfer rate across the insulation
= Heat transfer rate from the cladding to the surroundings

The heat transfer from the cladding to the surroundings is by convection to the surrounding air and by radiation to the surrounding surfaces.

$$2\pi kL \frac{T_1 - T_2}{\ln(r_2/r_1)} = (h_c + h_R)A_2(T_2 - T_3) \tag{1}$$

In Equation 1, the left side represents the conduction heat transfer rate across the insulation, and the right side represents the sum of the heat transfer rates by convection and radiation from the metal cladding to the surroundings.

With $A_2 = 2\pi r_2 L$, solving for T_2 we obtain

$$T_2 = \frac{kT_1/\ln(r_2/r_1) + (h_c + h_R)r_2 T_3}{k/\ln(r_2/r_1) + (h_c + h_R)r_2} \tag{2}$$

where

T_1 = temperature of the inner surface of the insulation = 80 °C (assumption 2)

T_2 = temperature of the outer surface of the insulation = temperature of the outer surface of the cladding (assumption 2)

T_3 = temperature of the surrounding air and surfaces = 10 °C

r_1, r_2 = inner and outer radii of the insulation

k = thermal conductivity of fiberglass insulation = 0.036 W/m K (Table A3A)

h_c, h_R = convective and radiative heat transfer coefficients associated with the outer surface of the cladding

After determining T_2 from Equation 2, iteratively if necessary (since h_R depends on T_2), the heat transfer rate from the cladding is determined from either the left side or the right side of Equation 1.

As the cladding is completely enclosed by a second, much larger surface,

q_2 = radiation heat transfer rate from the cladding
$= A_2 \varepsilon_2 \sigma(T_2^4 - T_3^4)$

and

$$h_R = \frac{q_2/A_2}{(T_2 - T_1)} = \varepsilon_2 \sigma(T_2^2 + T_3^2)(T_2 + T_3)$$

To estimate the average value of h_R, treated as constant in Equation 2, an upper limit of $T_2 (= T_{2max})$ is found by setting $h_R = 0$, i.e., neglecting radiative heat transfer. The lower limit of T_2 is T_3.

$$T_{2max} = \frac{kT_1/\ln(r_2/r_1) + h_c r_2 T_3}{k/\ln(r_2/r_1) + h_c r_2}$$

$$= \frac{0.036 \times 353.15/\ln(0.325/0.3) + 8 \times 0.325 \times 283.15}{0.036/\ln(0.325/0.3) + 8 \times 0.325} = 293.5 \text{ K}$$

$$\frac{h_{Rmax}}{\varepsilon_2} = \sigma(T_{2max}^2 + T_3^2)(T_{2max} + T_3)$$

$$= 5.67 \times 10^{-8}(293.5^2 + 283.15^2)(293.5 + 283.15) = 5.44 \text{ W/m}^2 \text{ K}$$

$$\frac{h_{Rmin}}{\varepsilon_2} = \sigma(T_{2min}^2 + T_3^2)(T_{2min} + T_3)$$

$$= 5.67 \times 10^{-8}(283.15^2 + 283.15^2)(283.15 + 283.15) = 5.15 \text{ W/m}^2 \text{ K}$$

$$\frac{h_{Rav}}{\varepsilon_2} = \frac{5.44 + 5.15}{2} = 5.295 \text{ W/m}^2 \text{ K}$$

As the maximum variation of h_R is within 3% of the mean value, the effect of the variation of h_R with T_2 can be neglected.

With $\varepsilon_2 = 0.95$,

$$h_R = h_{Rav} = 5.295 \times 0.95 = 5.03 \text{ W/m}^2 \text{ K}$$

From Equation 2,

$$T_2 = \frac{0.036 \times 353.15/\ln(0.325/0.3) + (8 + 5.03) \times 0.325 \times 283.15}{0.036/\ln(0.325/0.3) + (8 + 5.03) \times 0.325} = 289.9 \text{ K}$$

$$q_2 = (h_c + h_R)A_2(T_2 - T_3)$$

$$= (8 + 5.03) \times 2 \times \pi \times 0.325 \times 1.825 \times (289.9 - 283.15) = \underline{327.8 \text{ W}}$$

With $\varepsilon_2 = 0.1$,

$$h_R = 5.295 \times 0.1 = 0.5295 \text{ W/m}^2 \text{ K}$$

$$T_2 = \frac{0.036 \times 353.15/\ln(0.325/0.3) + (8 + 0.5295) \times 0.325 \times 283.15}{0.036/\ln(0.325/0.3) + (8 + 0.5295) \times 0.325}$$

$$= 292.9 \text{ K}$$

$$q_2 = (8 + 0.5295) \times 2 \times \pi \times 0.325 \times 1.825 \times (292.9 - 283.15) = \underline{309.9 \text{ W}}$$

The heat transfer rate is reduced from 327.8 W to 309.9 W, a reduction of 5.5%.

COMMENTS

The difference between the heat transfer rates is only 5.5%. But with $\varepsilon_2 = 0.95$, the magnitude of the radiative heat transfer rate is comparable to that due to convection. If radiation is neglected, $T_2 = 293.5$ K and $q_2 = 312.8$ W by convection only. This is about 5.9% less than that obtained by including radiation. With a reduction in the heat transfer rate, T_2 increases leading to a higher natural convective heat transfer rate.

The problem can be solved by employing the correct expression for the radiative heat transfer. The resulting equation is

$$k \frac{T_1 - T_2}{\ln(r_2/r_1)} = r_2 h_c (T_2 - T_3) + r_2 \varepsilon_2 \sigma (T_2^4 - T_3^4)$$

For $\varepsilon_2 = 0.95$, the value of T_2 found numerically is 289.8 K. By substituting 289.8 K in the right-hand side of the above equation, the heat transfer rate is 326.7 W, which is very close to the value obtained by using the radiative heat transfer coefficient. By employing the radiation heat transfer coefficient, the need for using numerical methods to determine the value of T_2 is eliminated. The simplicity of the procedure is obvious.

7.5 RADIATION INTENSITY

In Section 7.1 a brief introduction to wavelength dependent emissivity and absorptivity is provided. But, in addition to being wavelength dependent, radiation properties may also be direction dependent. An understanding of the concepts introduced in Sections 7.1 and 7.2 and of the directional dependency of the radiative properties is facilitated by a knowledge of the radiation intensity. For the purpose of introducing radiation intensity we limit our discussion to surface radiation. Radiation in gases is considered in Chapter 8.

Radiation intensity is an indication of the brightness of radiation to the "eye," which is sensitive to radiation at all wavelengths. As the human eye perceives radiation only in the limited wave band of 0.4 μ–0.7 μm, consider a plate at 800 °C. The brightness of the plate remains almost unaltered whether the viewer is directly in front of it (Figure 7.5.1a) or is at an angle (Figure 7.5.1b). But the amount of

Radiation intensity is an indication of the brightness of radiation

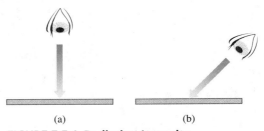

(a)   (b)

FIGURE 7.5.1 Radiation Intensity

radiant energy reaching the eye is greater when the viewer is directly in front than when the viewer is at an angle to the plate. As an understanding of radiation intensity requires a knowledge of a solid angle, a brief introduction to solid angle follows.

Solid Angle In two dimensions, an infinitesimal angle is defined as

$$d\theta = \frac{ds}{r}$$

where ds is an infinitesimal arc of a circle of radius r (Figure 7.5.2a). If the radius vector joining ds to the center of the circle is not perpendicular to ds but makes an angle α with ds (Figure 7.5.2b), then

$$d\theta = \frac{ds \cos \alpha}{r}$$

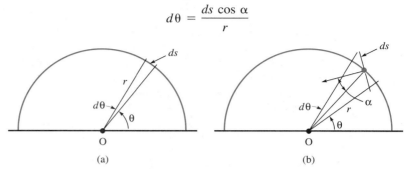

FIGURE 7.5.2 Definition of an Angle

Definition of solid angle

We may consider a solid angle as an extension of our concept of an angle in two dimensions to an angle in three dimensions. Consider an infinitesimal area dA_2 of a hemisphere of radius r centered around O as shown in Figure 7.5.3. The infinitesimal solid angle $d\omega$, subtended by dA_2 at O is given by

$$d\omega = \frac{dA_2}{r^2}$$

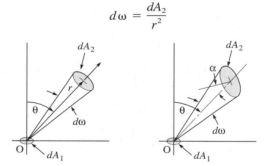

FIGURE 7.5.3 Definition of Solid Angle

If the normal to dA_2 does not pass through O but makes an angle α with the radius vector joining O to dA_2, then

$$d\omega = \frac{dA_2 \cos \alpha}{r^2}$$

In terms of spherical coordinates shown in Figure 7.5.4, $dA_2 = r \sin \theta \, d\phi \, r \, d\theta$ and

$$d\omega = \frac{dA_2}{r^2} = \sin \theta \, d\phi \, d\theta$$

The unit of a solid angle is steradian (sr). The total solid angle subtended by a sphere at its center is 4π sr.

Radiation intensity is defined through Equation 7.5.1. Consider an infinitesimal area dA at temperature T located at O (Figure 7.5.4). It emits radiation in all directions over all wavelengths. Consider another infinitesimal area dA_2 located at a distance

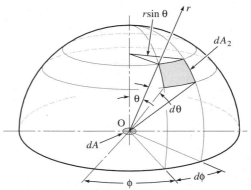

FIGURE 7.5.4 Radiative Energy from an Infinitesimal Area *dA* Reaching a Hemisphere with *dA* at the Center

of r and normal to the radius vector r. Denoting the emissive power of dA by E, we denote the energy emitted by dA in all directions over all wavelengths by dE. The energy emitted by dA in the wave band λ to $\lambda + d\lambda$ in the solid angle $d\omega$ is denoted by d^4E (to account for the infinitesimal wavelength $d\lambda$ and the product of two infinitesimal lengths in dA_2 used in defining $d\omega$). The energy emitted by dA and reaching dA_2 in the wave band λ to $\lambda + d\lambda$ is given by

$$d^4E = I_{\lambda,e}\, d\lambda\, d\omega\, dA \cos\theta \qquad\qquad \textbf{7.5.1}$$

where

$$
\begin{aligned}
E &= \text{emissive power of } dA \ (=dE/dA)\\
I_{\lambda,e} &= \text{spectral radiation intensity emitted by } dA \text{ at } T\\
d\omega &= \text{solid angle subtended by } dA_2 \text{ at O}\\
dA \cos\theta &= \text{projected area of } dA \text{ in the direction } \theta
\end{aligned}
$$

Spectral intensity (wavelength dependent) is direction dependent. By introducing this relation for the infinitesimal solid angle into Equation 7.5.1, the energy emitted by an infinitesimal surface in the wave band λ to $\lambda + d\lambda$ in the direction (ϕ, θ) in the solid angle $d\omega$ is expressed as

$$
\begin{aligned}
d^4E &= I_{\lambda,e}(\lambda, \phi, \theta)\, dA \cos\theta\, d\omega\, d\lambda\\
&= I_{\lambda,e}(\lambda, \phi, \theta)\, dA \cos\theta \sin\theta\, d\phi\, d\theta\, d\lambda
\end{aligned}
$$

where $I_{\lambda,e}(\lambda, \phi, \theta)$ is the spectral radiation intensity emitted from the surface at λ in the direction (ϕ, θ). The emitted energy flux in all directions over all wavelengths in the hemisphere surrounding dA is then given by

$$\frac{dE}{dA} = \int_{\lambda=0}^{\infty} \int_{\theta=0}^{\pi/2} \int_{\phi=0}^{2\pi} I_{\lambda,e}(\lambda, \phi, \theta) \cos\theta \sin\theta\, d\phi\, d\theta\, d\lambda$$

dE/dA is the emissive power of dA.

$$\left\| \quad E = \int_{\lambda=0}^{\infty} \int_{\theta=0}^{\pi/2} \int_{\phi=0}^{2\pi} I_{\lambda,e}(\lambda, \phi, \theta) \cos\theta \sin\theta\, d\phi\, d\theta\, d\lambda \quad \right\| \qquad \textbf{7.5.2}$$

(The reason for employing the fourth order infinitesimal d^4E is explained by the triple integral in Equation 7.5.2.)

From the definition of monochromatic emissive power the energy emitted in the wave band λ to $\lambda + d\lambda$ in all directions is given by

$$E_\lambda \, d\lambda = \int_{\theta=0}^{\pi/2} \int_{\phi=0}^{2\pi} I_{\lambda,e}(\lambda, \phi, \theta) \cos \theta \sin \theta \, d\phi \, d\theta \, d\lambda$$

or

$$\left\| \; E_\lambda = \int_{\theta=0}^{\pi/2} \int_{\phi=0}^{2\pi} I_{\lambda,e}(\lambda, \phi, \theta) \cos \theta \sin \theta \, d\phi \, d\theta \; \right\| \qquad \textbf{7.5.3}$$

Relation Between Radiation Intensity and Emissive Power for a Diffuse Surface In Equation 7.5.3 the emitted intensity, $I_{\lambda,e}$, is a function of direction and wavelength. For a diffuse emitter, the emitted intensity is independent of direction, and for such a surface,

Relation between
radiation intensity
and emissive power
for a gray surface

$$I_{\lambda,e}(\lambda, \phi, \theta) = I_{\lambda,e}(\lambda)$$

For a diffuse emitter Equation 7.5.2 simplifies to

$$E = \int_{\theta=0}^{\pi/2} \int_{\phi=0}^{2\pi} \cos \theta \sin \theta \, d\phi \, d\theta \int_{\lambda=0}^{\infty} I_{\lambda,e} \, d\lambda$$

Defining the total intensity of emitted radiation, $I_e = \int_0^\infty I_{\lambda,e} \, d\lambda$ and with

$$\int_{\theta=0}^{\pi/2} \int_{\phi=0}^{2\pi} \cos \theta \sin \theta \, d\phi \, d\theta = \pi$$

we obtain

$$\underline{E = \pi I_e} \qquad \textbf{7.5.4}$$

Radiation intensity is a fundamental quantity in radiation. Although only emitted intensity has so far been considered, the concept of intensity can be extended to any region in space where radiant energy is being transmitted—fluids, transparent and translucent solids, and vacuum. The total intensity from a surface, which consists of the intensities due to emission and reflection of incident energy, may then be related to the total energy leaving the surface. The total radiant energy flux leaving a surface in all directions into a hemisphere with the infinitesimal surface at the center of the base of the hemisphere is the radiosity of the surface, J, and

$$J = \int_{\theta=0}^{\pi/2} \int_{\phi=0}^{2\pi} I \cos \theta \sin \theta \, d\phi \, d\theta$$

where I is the total intensity equal to the sum of the emitted and reflected intensities. For a diffuse surface, both the emitted and reflected intensities are independent of direction and

$$\| \quad J = \pi I \quad \| \qquad \textbf{7.5.5}$$

Radiation Properties of a Surface The spectral properties of an opaque surface—absorptivity, reflectivity, and emissivity—are defined in terms of the incident or emitted intensity. These are discussed in Chapter 8.

7.6 PROJECTS

PROJECT 7.6.1 **Energy Resource Conservation—A Preheating Furnace for Iron Ore**

A gas-fired pellet preheating furnace for processing iron ore is 64-m long and 6-m wide. A 5.6-m wide traveling grate is used to transport the ore through the furnace. The grate moves at a rate of 12 cm/s. The material of the grate is predominantly stainless steel and has a mass of 2300 kg/lineal meter. The grate exits the furnace at 360 °C at one end of the furnace and reenters at the other end as shown in Figure 7.6.1. The 2.6-m high side plates are at an average temperature of 125 °C. The ambient air is at 35 °C.

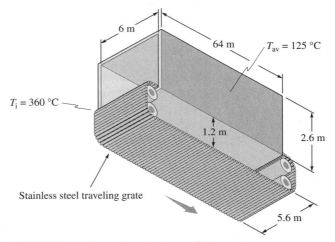

FIGURE 7.6.1 Traveling Grate and Furnace

Determine the heat transfer rate from the exposed part of the traveling grate and suggest a method to reduce the heat transfer to the surroundings. All the available information has been included in the statement of the problem. If any additional information is required, make appropriate assumptions.

Heat Transfer from the Grate

The heat transfer from the grate is by convection and radiation to the surroundings. To determine the convective heat transfer from the grate, we assume a 1 m/s air motion across the 5.6-m dimension of the grate. The radiative heat transfer from the lower surface of the grate, which is exposed to the surroundings, is computed by modeling it as a surface completely enclosed by a second much larger surface. The upper surface of the grate close to the floor of the furnace sees both the floor of the furnace and, to a lesser extent, the surrounding surfaces.

To compute the convective heat transfer coefficient and to determine the heat transfer rate, the temperature of the grate is needed. The only temperature that is available is the grate temperature at exit from the furnace. As the grate moves, its temperature decreases due to the heat transfer but the temperature at the other end of the grate is not known. The estimated total heat transfer rate by convection and radiation must satisfy the balance of energy applied to the moving grate as the control volume as shown in Figure 7.6.2. Therefore,

$$q = \dot{m}c_p(T_i - T_e) \tag{1}$$

where

$$q = \text{total heat transfer rate from the grate} = q_{1R} + q_{1c} + q_{2R} + q_{2c}$$

$q_{1R}, q_{2R} = $ radiative heat transfer rates from the upper and lower surfaces of the grate, respectively

$q_{1c}, q_{2c} = $ convective heat transfer rates from the upper and lower surfaces of the grate, respectively

$\dot{m} = $ rate of mass of the grate entering the C.V.

$c_p = $ specific heat of the material of the grate

$T_i, T_e = $ temperatures of the grate at inlet and exit of the C.V.

T_i is the temperature of the grate at exit from the furnace (360 °C) and T_e is the temperature at which the grate reenters the furnace. T_e is yet to be determined.

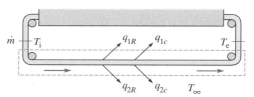

FIGURE 7.6.2 Control Volume for Balance of Energy

From the problem statement, the material of the grate is predominantly stainless steel. It is reasonable to assume that most of the material is some type of steel. From Table A1, it is observed that at 600 K the specific heats of the different types of alloy steels and stainless steels vary from a low of 550 J/kg K to a high of 582 J/kg K, with the specific heats of different types of stainless steel being in the range of 550 J/kg K–559 J/kg K. It is, therefore, reasonable to assume that the average specific heat of the materials is 557 J/kg K at 600 K—the specific heat of type 304 stainless steel at 600 K.

The exit temperature of the grate (from the C.V.), T_e, is found by iteration. An exit temperature is assumed and the heat transfer rate determined from Equation 1. It is compared with the total heat transfer rate by convection and radiation. When the two heat transfer rates are equal, the correct exit temperature has been determined.

To determine the convective and radiative heat transfer rates the following models are employed.

Convective Heat Transfer As already indicated, there is likely to be a significant movement of air around the grate. Because of the low velocity of the air, we anticipate both forced convection and natural convection to be significant. We assume an air speed of 1 m/s across the grate and compute the forced convective heat transfer coefficient. The model for forced convection is shown in Figure 7.6.3. No correla-

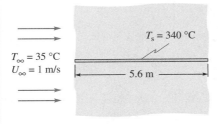

$T_\infty = 35\ °C$
$U_\infty = 1\ m/s$

$T_s = 340\ °C$

5.6 m

FIGURE 7.6.3 Model for Forced Convection

tions are available to determine the natural convective heat transfer coefficient from the upper surface of the grate (assumed to be stationary). The lower limit of the natural convective heat transfer coefficients are computed by treating the upper and lower surfaces as stationary horizontal surfaces surrounded by stationary air. An arithmetic mean of T_i and T_e will be used to determine the convective heat transfer coefficients. After determining the heat transfer coefficients by natural and forced convection, the higher value will be suitably modified to account for the combined effect of the two modes of convection.

Radiative Heat Transfer The radiative heat transfer from the top surface is computed by considering it as one surface of a three-surface enclosure—the upper grate surface, the floor of the furnace, and the surrounding surfaces. As the temperature of the furnace floor is not known, it is assumed that the temperature of the floor is equal to the temperature of the grate surface. To determine the radiative heat transfer rate (proportional to the differences between the fourth powers of the absolute temperatures), the arithmetic mean of the temperature defined as $[(T_i^4 + T_e^4)/2]^{1/4}$ will be used. The radiative heat transfer from the bottom surface will be computed by considering it as a surface enclosed by a much larger second surface (the surroundings).

Employing the above models we begin the computations.

Assume $T_e = 320\ °C$.
The specific heat of type 304 stainless steel at a mean temperature of 340 °C is 557 J/kg °C.
$\dot{m} = 2300\ kg/m \times 0.12\ m/s = 276\ kg/s$

From Equation 1

$$q = 276 \times 557 \times (360 - 320) = 6.149 \times 10^6\ W$$

Convective Heat Transfer Rate Computations
(a) *Forced convection.* From the software (CC), the properties of atmospheric air at $T_f = (340 + 35)/2 = 187.5\ °C$ are

$$\rho = 0.766 \text{ kg/m}^3 \qquad \mu = 2.531 \times 10^{-5} \text{ N s/m}^2$$
$$k = 0.037 \text{ W/m °C} \qquad Pr = 0.7$$

$$Re_L = \frac{\rho U_\infty L}{\mu} = \frac{0.766 \times 1 \times 5.6}{2.531 \times 10^{-5}} = 1.695 \times 10^5$$

Although $Re_L < 5 \times 10^5$, based on the construction of the grate with many moving links, we assume that the boundary layer is turbulent from the leading edge. With $Re_{cr} = 0$, from Equation 4.1.8,

$$Nu_L = 0.037(1.695 \times 10^5)^{0.8} \times 0.7^{1/3} = 501$$

$$h_L = \frac{501 \times 0.037}{5.6} = 3.31 \text{ W/m}^2 \text{ °C}$$

(b) *Natural convection.* As the grate is 64-m long and 5.6-m wide, we choose the characteristic length as the width of the grate (= 5.6 m) and employ Equation 4.3.4, 4.3.5a, or 4.3.5b. The reference temperature $= 340 - 0.25(340 - 35) = 264 \text{ °C} = 537 \text{ K}$. From Table A7, properties of air at 537 K are

$$\rho = 0.659 \text{ kg/m}^3 \qquad \mu = 281 \times 10^{-7} \text{ N s/m}^2$$
$$k = 0.0418 \text{ W/m °C} \qquad Pr = 0.698$$

$$\beta[\text{at } 35 + 0.25(340 - 35) = 111.3 \text{ °C}] = 2.602 \times 10^{-3} \text{ K}^{-1}$$

$$Ra_L = \frac{9.81 \times 2.602 \times 10^{-3} \times 0.659^2 \times (340 - 35) \times 5.6^3 \times 0.698}{(281 \times 10^{-7})^2}$$
$$= 5.249 \times 10^{11}$$

No correlation is available for $Ra_L > 10^{11}$. However, the upper limit of Equation 4.3.5b is 10^{11}, so we will use it to *estimate* the convective heat transfer coefficient for the upper grate surface (with the expectation that it is unlikely that the actual value will be very different from the estimated value).

$$Nu_L = 0.13 \times (5.249 \times 10^{11})^{1/3} = 1049$$

$$h_1 = \frac{1049 \times 0.0418}{5.6} = 7.83 \text{ W/m}^2 \text{ °C}$$

For the lower surface from Equation 4.3.4,

$$h_2 = 0.58 \times (5.249 \times 10^{11})^{0.2} \times \frac{0.0418}{5.6} = 0.96 \text{ W/m}^2 \text{ °C}$$

For the combined heat transfer coefficients due to forced and natural convection, increasing the higher heat transfer coefficient by 10%, we use

$$h_1 \approx 8.6 \text{ W/m}^2 \text{ °C} \qquad h_2 = 3.6 \text{ W/m}^2 \text{ °C}$$

Then heat transfer by convection is

$$q_c = (8.6 + 3.6) \times 64 \times 5.6 \times (340 - 35) = 1.334 \times 10^6 \text{ W}$$

Radiative Heat Transfer Computations From Table A9 the emissivity of highly oxidized stainless steel is 0.8 and for oxidized steel also it is ≈ 0.8. Therefore, for the upper surface, modeled as one surface of a three-surface enclosure (Figure 7.6.4)

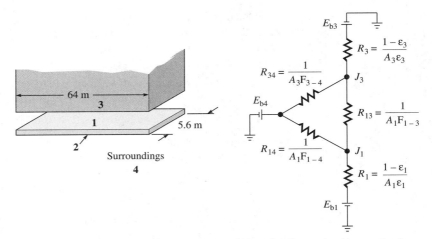

FIGURE 7.6.4 Model for Radiative Heat Transfer from the Upper Surface

$$\varepsilon_1 = \varepsilon_2 = 0.8 \qquad T_4 = 308 \text{ K}$$

$$T_1 = \left(\frac{633^4 + 593^4}{2}\right)^{1/4} = 614 \text{ K} = T_3$$

As the arithmetic mean temperature is 613 K and is only one kelvin lower than the computed mean, we may use the arithmetic mean temperature in the computations. From the software (VF)

$$F_{1\text{-}3} = 0.796 = F_{3\text{-}1} \qquad F_{1\text{-}4} = 1 - F_{1\text{-}3} = 0.204 = F_{3\text{-}4}$$

From the symmetry of the electrical circuit ($E_{b3} = E_{b1}$, $R_1 = R_3$, $R_{14} = R_{34}$), the heat transfer (current) in the branch $J_1 \to J_3$ is zero. The heat transfer rate is determined from

$$q_{1R} = \frac{E_{b1} - E_{b4}}{R_1 + R_{14}} = \frac{A_1\sigma(T_1^4 - T_4^4)}{(1 - \varepsilon_1)/\varepsilon_1 + 1/F_{1\text{-}4}}$$

where $A_1 = 64 \times 5.6 = 358.4 \text{ m}^2$. For the upper grate surface the heat transfer rate is

$$q_{1R} = \frac{358.4 \times 5.67 \times 10^{-8} \times (614^4 - 308^4)}{(1 - 0.8)/0.8 + 1/0.204} = 5.251 \times 10^5 \text{ W}$$

The radiative heat transfer rate from the lower surface of the grate is

$$q_{2R} = \sigma A_2 \varepsilon_2 (T_2^4 - T_4^4)$$
$$= 5.67 \times 10^{-8} \times 358.4 \times 0.8(614^4 - 308^4) = 2.164 \times 10^6 \text{ W}$$

The total convective and radiative heat transfer from the grate is

$$q = (1.334 + 0.525 + 2.164) \times 10^6 = 4.023 \times 10^6 \text{ W}$$

The value computed from Equation 1 ($= 6.149 \times 10^6$ W) is considerably higher than the combined convective-radiative heat transfer rates ($= 4.023 \times 10^6$ W). The assumed value of T_e is, therefore, too low and a higher value is needed. With a higher value of T_e the heat transfer rate from Equation 1 is reduced and the heat transfer rates by convection and radiation increased. A reasonable estimate is obtained by using a value slightly higher than 4.023×10^6 W, say 4.1×10^6 W for q in Equation 1,

$$T_e = T_i - \frac{q}{\dot{m}c_p} = 360 - \frac{4.1 \times 10^6}{276 \times 557} = 333 \,°C$$

The calculations are repeated with $T_e = 333\,°C$. From Equation 1, with $q = 4.1 \times 10^6$ W,

$$T_1 = T_2 = \frac{360 + 333}{2} = 346.5\,°C = 619.5\,K$$

Convective and radiative heat transfer rates are

$$q_c = (8.6 + 3.6) \times 358.4 \times (346.5 - 35) = 1.362 \times 10^6\,W$$

$$q_{1R} = \frac{358.4 \times 5.67 \times 10^{-8}(619.5^4 - 308^4)}{(1 - 0.8)/0.8 + 1/0.204} = 0.5455 \times 10^6\,W$$

$$q_{2R} = 358.4 \times 5.67 \times 10^{-8} \times 0.8 \times (619.5^4 - 308^4) = 2.248 \times 10^6\,W$$

Heat transfer by convection and radiation is

$$q = (1.362 + 0.5455 + 2.248) \times 10^6 = \underline{4.16 \times 10^6\,W\ (4160\ kW)}$$

The computed heat transfer by convection-radiation is only 1.5% higher than the heat transfer computed from Equation 1. No further iteration is necessary.

Note: The heat transfer to the surroundings is a heat loss in the sense that resources (oil, gas, or coal) have to be unnecessarily used. In the summer months such a heat transfer makes the surroundings uncomfortably warm, necessitating further use of energy to provide comfortable working conditions. Even in the winter months such heat transfer may lead to locally uncomfortable conditions.

$$\text{Heat loss} = 4.16 \times 10^6\,W \times \frac{BTU}{1055.1\,W/s} = 3943\ BTU/s$$

The heating value of natural gas is approximately 1000 BTU/cft. Assuming an efficiency of 85% for the fuel utilization, the amount of natural gas required is 4.64 cft/s. Assuming the furnace is operated 92% of the time (11 months in the year) the yearly gas consumption is 1.346×10^8 cft. The annual gas consumption for heating an average-sized 3-bedroom house in Chicago is approximately 60 000 cft. The computed gas used to provide the heat loss from the grate is enough to heat 2200 such houses!

Reducing the Heat Loss

An obvious method to reduce the heat loss is to use a radiation shield around the grate. The introduction of a metal radiation shield significantly reduces the heat loss. But the heat loss can be further minimized by insulating the radiation shield. Consider providing a steel radiation shield insulated with a 6-cm thick insulation material with a thermal conductivity of 0.1 W/m °C as shown in Figure 7.6.5.

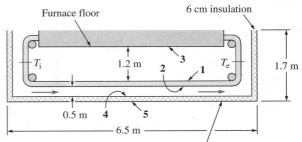

FIGURE 7.6.5 Grate with an Insulated Radiation Shield

An upper bound for the heat loss can be found by computing the radiative and convective heat transfer with T_4 set at 360 °C (the temperature of the grate at exit from the furnace).

The conductive heat flux is given by

$$q''_k = \frac{k(T_4 - T_5)}{L_5} \tag{2}$$

where L_5 is the thickness of the insulation, 6 cm. The heat flux from the outer surface by convection and radiation is given by

$$q''_5 = h(T_5 - T_\infty) + \sigma \varepsilon_5 (T_5^4 - T_\infty^4) \tag{3}$$

Equating Equations 2 and 3,

$$\frac{k(T_4 - T_5)}{L_5} = h(T_5 - T_\infty) + \sigma \varepsilon_5 (T_5^4 - T_\infty^4) \tag{4}$$

Using the following values

$k = 0.1$ W/m °C $\quad L_5 = 0.06$ m $\quad h = 3.6$ W/m² °C

$T_4 = 633$ K $\quad \varepsilon_5 = 0.8$ $\quad T_\infty = 308$ K

and solving Equation 4 for T_5, $T_5 = 353$ K. Then from Equation 2,

$$q''_k = \frac{0.1(633 - 353)}{0.06} = 467 \text{ W/m}^2$$

The area of the shield is $A = 65 \times 6.6 + 2 \times 65 \times 1.7 + 2 \times 6.6 \times 1.7 = 672$ m² and

$$q = 467 \times 672 = \underline{3.138 \times 10^5 \text{ W}}$$

The heat loss of 3.138×10^5 W represents the upper limit. Compared with the heat loss of 4.16×10^6 W without the shield, the introduction of the insulated radiation shield reduces the heat loss by a minimum of 92%.

To determine the actual heat transfer, the following assumptions are made:

1. The furnace floor and the grate are at the same temperature.
2. Because of the high levels of temperatures of the grate and the shield surface 4, the heat transfer from the grate is predominantly by radiation, and convective heat transfer is negligible.

3. The radiative heat transfer from the grate is found by considering the upper surface of the grate, the furnace floor, and the shield as one enclosure and the lower surface of the grate and the shield as another enclosure.
4. As the 66 m × 6.6 m horizontal surface of the shield is much greater than the other surfaces, the convective heat transfer coefficient associated with the lower surface of the shield (exposed to the ambient air) is applicable to all the surfaces.

For the radiative heat transfer the following surface properties and view factors are used (Figure 7.6.5).

$$\varepsilon_1 = \varepsilon_2 = \varepsilon_3 = \varepsilon_4 = \varepsilon_5 = 0.8$$
$$F_{1-3} = 0.796 = F_{3-1} \qquad F_{1-4} = 0.204 = F_{3-4} \qquad F_{2-4} = 1$$

The electrical analog is shown in Figure 7.6.6.

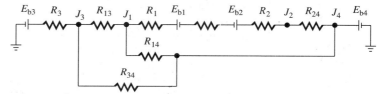

FIGURE 7.6.6 The Electrical Analog for Radiative Heat Transfer

As $R_1 = R_3$, $E_{b1} = E_{b3}$, and $R_{34} = R_{14}$, the net heat transfer between surfaces 1 and 3 is zero ($J_1 = J_3$). The following equations are solved simultaneously, where

q_1 = heat transfer rate from surface 1
q_2 = heat transfer rate from surface 2

$$q_{12} = q_1 + q_2 = \dot{m}c_p(T_i - T_e) \tag{5}$$

The radiative heat transfer rates q_1 and q_2 are determined from the following set of equations.

$$\frac{E_{b3} - J_3}{R_3} + \frac{J_4 - J_3}{R_{34}} = 0 \tag{6}$$

$$\frac{E_{b1} - J_1}{R_1} + \frac{J_4 - J_1}{R_{14}} = 0 \tag{7}$$

$$\frac{E_{b2} - J_2}{R_2} + \frac{J_4 - J_2}{R_{24}} = 0 \tag{8}$$

$$\frac{E_{b4} - J_4}{R_4} + \frac{J_2 - J_4}{R_{24}} + \frac{J_1 - J_4}{R_{14}} + \frac{J_3 - J_4}{R_{34}} = 0 \tag{9}$$

Solving for the radiosities, we obtain

$$q_{12} = \frac{E_{b1} - J_1}{R_1} + \frac{E_{b2} - J_2}{R_2} \tag{10}$$

$$q_4 = \text{radiative heat transfer rate to surface 4} = \frac{J_4 - E_{b4}}{R_4} \tag{11}$$

The conductive heat transfer rate across the radiation shield is given by

$$q_4 = \frac{kA_4(T_4 - T_5)}{L_5} \tag{12}$$

$$q_4 = q_5 = hA_5(T_5 - T_\infty) + \sigma A_5 \varepsilon_5 (T_5^4 - T_\infty^4) \tag{13}$$

where q_5 is the heat transfer rate from the outer surface of the shield to the surroundings by convection and radiation.

Conceptually, there are 9 unknowns (T_e, T_4, T_5, q_{12}, q_4, J_1, J_2, J_3, and J_4) and 9 equations—Equations 5 through 13. However, the solution has to be obtained by iteration.

Combining Equations 12 and 13,

$$\frac{k(T_4 - T_5)}{L_5} = h(T_5 - T_\infty) + \sigma\varepsilon_5(T_5^4 - T_\infty^4) \tag{14}$$

The solution procedure is

1. Assume T_e and determine T_1, T_2, and T_3.
2. (a) Assume T_4 and solve for J_1, J_2, J_3, and J_4.
   (b) Solve Equation 14 for T_5 and compute q_4 from Equation 12. Find the value of T_4 for which q_4 from Equations 11 and 12 are equal.
3. Find the value of T_e for which q_{12} from Equations 5 and 10 are equal.

The procedure requires double iteration.

1. An appropriate initial guess is obtained by equating the upper bound of q_{12} ($= 3.138 \times 10^5$ W) to q in Equation 1

$$T_e = T_i - \frac{q_{12}}{\dot{m}c_p} = 360 - \frac{3.138 \times 10^5}{276 \times 557} = 358\,°C$$

$$T_1 = T_2 = 359\,°C = 632\ K$$

2. (a) Assume $T_4 = 630$ K

$$R_1 = 6.9754 \times 10^{-4}\ \mathrm{m}^{-2} = R_2 = R_3 \qquad R_4 = 3.7202 \times 10^{-4}\ \mathrm{m}^{-2}$$
$$R_{14} = 1.36774 \times 10^{-2}\ \mathrm{m}^{-2} = R_{34} \qquad R_{24} = 2.7902 \times 10^{-3}\ \mathrm{m}^{-2}$$
$$E_{b1} = E_{b2} = E_{b3} = 9045.9\ \mathrm{W/m^2} \qquad E_{b4} = 8931.9\ \mathrm{W/m^2}$$

Solving for the radiosities,

$$J_1 = J_3 = 9045.88\ \mathrm{W/m^2} \qquad J_2 = 9045.84\ \mathrm{W/m^2} \qquad J_4 = 9045.62\ \mathrm{W/m^2}$$

$$q_4 = \frac{J_4 - E_{b4}}{R_4} = \frac{9045.62 - 8931.93}{3.7202 \times 10^{-4}} = 3.056 \times 10^5\ \mathrm{W}$$

(b) Solving Equation 14 for T_5,

$$\frac{0.1(630 - T_5)}{0.06} = 3.6(T_5 - 308) + 5.67 \times 10^{-8} \times 0.8 \times (T_5^4 - 308^4)$$

$$T_5 = 353.3\ K \qquad q_5 = \frac{0.1(630 - 353.2) \times 672}{0.06} = 3.099 \times 10^5\ \mathrm{W}$$

3. The value used for q_{12} (= 3.138×10^5 W) and the computed values for q_4 (= 3.056×10^5 W) and q_5 (= 3.099×10^5 W) are within 2% of each other and no further iteration is needed.

$$q_{12} \approx \underline{3.1 \times 10^5 \text{ W}}$$

Note that the radiative heat flux to surface 4 is 461 W/m². Even if the convective heat transfer coefficient among surfaces 1, 2, and 4 is as high as 10 W/m² °C, the convective heat flux is only 20 W/m² or just about 4% of the radiative heat flux. Neglecting the convective heat flux is, therefore, justified.

PROJECT 7.6.2 Project 7.6.1 Continued

Estimate the heat loss from the side plates. Suggest methods to reduce the heat loss and determine the reduction in the heat loss. Note that any reduction in the heat transfer rate from the side plates is likely to result in an increase in the temperature of the side plates. The conditions inside the furnace are required to determine the temperature of the side plates. In the absence of such information, only an estimate based on the side plate temperature being fixed can be obtained.

PROJECT 7.6.3 Consider an extended surface (Figure 7.6.7) where heat transfer from the surface is by radiation only (as in space applications).

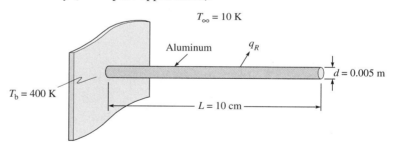

FIGURE 7.6.7 Extended Surface with Heat Transfer by Radiation Only

(a) Show that the differential equation for the one-dimensional temperature distribution is

$$\frac{d^2T}{dx^2} - \frac{\sigma P \varepsilon}{k A_x} (T^4 - T_\infty^4) = 0$$

To solve the differential equation, linearize the radiative heat transfer term by employing the radiative heat transfer coefficient

$$h_R = \sigma \varepsilon (T^2 + T_\infty^2)(T + T_\infty)$$

Evaluate h_R at a reasonable value of T and solve the differential equation. Find the heat transfer rate and plot the temperature distribution.

(b) A more accurate solution is obtained by solving the differential equation by numerical method. The nodal equation is (Figure 7.6.8)

$$\frac{kA_x(T_{i-1} - T_i)}{\Delta x} + \frac{kA_x(T_{i+1} - T_i)}{\Delta x} + P\Delta x \, \sigma\varepsilon(T_\infty^4 - T_i^4) = 0$$

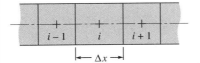

FIGURE 7.6.8 **Nodes for Numerical Solution**

To solve the resulting system of simultaneous equations linearize the radiative heat transfer term by employing the radiative heat transfer coefficient evaluated at the temperature of the node (T_i). Assume a reasonable temperature distribution and evaluate h_R at every node. Solve the equations by TDMA or any other suitable numerical method. Update the values of h_R and repeat the computations until two successive sets of T_i differ by less than a predetermined value. Compare the results of the numerical solution with the analytical solution found in part a.

SUMMARY

Radiative heat transfer differs from conductive and convective heat transfer in two respects. The transport of energy by radiation is much faster than by conduction or convection. The heat flux by conduction or convection at a point is defined by the state of the medium in the immediate vicinity of the point. In radiation, however, the effect of the entire field that participates in radiation has to be considered.

For our purposes the electromagnetic wave theory is appropriate to explain the radiation phenomenon. The surface that emits the maximum amount of energy at a given temperature is a *black surface* or *an ideal emitter*. The *energy flux* emitted by a body *in all wavelengths and directions* is its *emissive power, E,* at the temperature of the surface. The *blackbody emissive power, E_b,* is given by the Stefan-Boltzmann Law:

$$E_b = \sigma T^4 \qquad \sigma = 5.67 \times 10^{-8} \text{ W/m}^2 \text{ K}^4$$

The ratio of the emissive power of a surface to the blackbody emissive power at the same temperature is the *emissivity* of the surface.

$$\varepsilon = \frac{E(T)}{E_b(T)}$$

The *monochromatic emissive power of a surface* is defined by

$$dE = E_\lambda \, d\lambda$$

The surface *spectral emissivity,* ε_λ, is the ratio of the monochromatic emissive power of the surface to the monochromatic emissive power of a blackbody at the same temperature and wavelength.

$$\varepsilon_\lambda = \frac{E_\lambda(T)}{E_{b\lambda}(T)}$$

The ratio of the emissive power of a blackbody in the wave band 0 to λ to blackbody emissive power at the same temperature is the blackbody radiation function $f_{0\text{-}\lambda}$.

$$f_{0\text{-}\lambda} = \frac{E_{b,0\text{-}\lambda}(T)}{E_b(T)}$$

Tabulated values of $f_{0\text{-}\lambda}$ are given in Table 7.1.2 and are available in the software RF.

The *absorptivity,* α, of a surface is the ratio of the absorbed energy flux to the incident energy flux.

$$\alpha = \frac{G_{abs}}{G}$$

Similarly, the *reflectivity,* ρ, and *transmissivity,* τ, are defined as

$$\rho = \frac{G_{ref}}{G} \qquad \tau = \frac{G_{trans}}{G}$$

Denoting the incident energy flux in the wave band λ to $\lambda + d\lambda$ by dG, the *spectral absorptivity, reflectivity,* and *transmissivity* are given by

$$\alpha_\lambda = \frac{dG_{abs}}{dG} \qquad \rho_\lambda = \frac{dG_{ref}}{dG} \qquad \tau_\lambda = \frac{dG_{trans}}{dG}$$

Kirchhoff's Law states that $\varepsilon_\lambda = \alpha_\lambda$. From energy conservation,

$$\alpha + \rho + \tau = 1 \qquad \alpha_\lambda + \rho_\lambda + \tau_\lambda = 1$$

If the radiative properties of a surface are *direction independent* the surface is *diffuse.* If the radiative properties of a surface are *wavelength independent* the surface is *gray.* A surface that does not transmit radiant energy (i.e., $\tau = 0$) is an *opaque surface.* For a *gray, diffuse surface*

$$\varepsilon = \alpha$$

A black surface is an *ideal emitter* and is *gray, diffuse,* and *opaque* ($\alpha = \varepsilon = 1$, $\rho = \tau = 0$).

The following definitions are restricted to diffuse, gray, and opaque surfaces.

The *total radiant energy* flux from a surface is the *radiosity* of the surface.

$$J = \varepsilon E_b + \rho G$$

The ratio of the radiant energy leaving a surface, i, and reaching a surface, k, to the total radiant energy leaving surface i is the *view factor,* $F_{i\text{-}k}$. That is,

$$F_{i\text{-}k} = \frac{\text{Radiant energy leaving i and intercepted by k}}{\text{Total radiant energy leaving surface i}}$$

View factor relations are

Reciprocity relation $A_i F_{i-k} = A_k F_{k-i}$

Decomposition relations $A_{ij} F_{ij-k} = A_i F_{i-k} + A_j F_{j-k}$

$F_{i-jk} = F_{i-j} + F_{i-k}$

Law of corresponding corners   *See page 583.*

Summation rule For an N-surface enclosure $\displaystyle\sum_{k=1}^{N} F_{i-k} = 1$

View factors for four surface configurations are given in Figures 7.2.7 through 7.2.10 and are available in the software VF.

Radiant energy exchange from surfaces in an N-surface enclosure with all surfaces being gray, diffuse, and opaque is given by

$$q_{ik} = A_i F_{i-k}(J_i - J_k)$$

$$q_i = \sum_{k=1}^{n} A_i F_{i-k}(J_i - J_k)$$

For a *nonblack surface* q_i is also given by

$$q_i = \frac{A_i \varepsilon_i}{1 - \varepsilon}(E_{bi} - J_i)$$

For a *two-surface enclosure*

$$q_1 = \frac{E_{b1} - E_{b2}}{(1 - \varepsilon_1)/(A_1 \varepsilon_1) + 1/(A_1 F_{1-2}) + (1 - \varepsilon_2)/(A_2 \varepsilon_2)}$$

For a convex or plane surface 1, completely enclosed by surface 2, the *radiation heat transfer coefficient* is defined as

$$h_R = \frac{q_R''}{T_1 - T_2} = \frac{\sigma(T_1^2 + T_2^2)(T_1 + T_2)}{1/\varepsilon_1 + A_1/A_2(1 - \varepsilon_2)/\varepsilon_2}$$

If $\dfrac{A_1}{A_2}\dfrac{1 - \varepsilon_2}{\varepsilon_2}\varepsilon_1 \ll 1$

$$h_R = \sigma\varepsilon_1(T_1^2 + T_2^2)(T_1 + T_2)$$

Software

In the software, HT, there are three programs that are useful in the solution of problems with radiative heat transfer.

Radiation Function (RF) computes blackbody radiation function, $f_{0-\lambda}$. The user supplies the temperature of the surface and the upper and lower values of the wavelengths λ_2 and λ_1, respectively. The program computes $f_{0-\lambda_1}$, $f_{0-\lambda_2}$, and $f_{\lambda_1-\lambda_2} = f_{0-\lambda_2} - f_{0-\lambda_1}$ from Wiebelt's polynomial approximations.

View Factors (VF) computes the view factors for the following geometries:

■ Two rectangular surfaces perpendicular to each other and sharing a common side
■ Two equal rectangular surfaces directly opposite to each other
■ Two coaxial disks
■ Two coaxial cylinders

The user supplies the geometric information. The lengths can be in any consistent units; that is, they should all be in meters, millimeters, inches, and so forth.

Simultaneous Equations (SE) solves linear simultaneous equations with up to 15 unknowns. This program is useful in the solution of radiation heat exchange with more than three surfaces in an enclosure. It also simplifies computations when the assumption of uniform radiosity has to be justified with computations by increasing the number of surfaces.

REFERENCES

Chandrasekhar, S. (1960). *Radiative Heat Transfer*. New York: Dover.

Clark, J.A., and Korybalski, M.E. (1974). Algebraic methods for the calculation of radiation exchange in an enclosure. *Warme -und Stoffubertragung,* 7:31.

Edwards, D.K. (1981). *Radiation Heat Transfer Notes*. Washington, D.C.: Hemisphere Publishing.

Howell, J.R. (1982). *A Catalog of Radiation Configuration Factors*. New York: McGraw-Hill.

Hottel, H.C., and Sarofim, A.F. (1967). *Radiative Transfer*. New York: McGraw-Hill.

Love, T.J. (1968). *Radiative Heat Transfer*. Colombus, OH: Merrill.

Planck, M. (1959). *The Theory of Heat Radiation*. New York: Dover.

Siegel, R., and Howell, J.R. (1992). *Thermal Radiation Heat Transfer,* 3rd ed. Washington, D.C.: Hemisphere.

Sparrow, E.M. (1963). On the calculation of radiant interchange between surfaces. In *Modern Development in Heat Transfer,* ed. Ibele, W. New York: Academic Press.

Sparrow, E.M., and Cess, R.D. (1978). *Radiation Heat Transfer,* aug. ed. New York: McGraw-Hill.

Wiebelt, John A. (1960). *Engineering Radiation Heat Transfer*. New York: Holt, Rinehart and Winston.

REVIEW QUESTIONS

(a) What are the differences between heat transfer by radiation and by conduction?

(b) What is Planck's Distribution of monochromatic emissive power?

(c) What is Wien's Displacement Law?

(d) Define

Monochromatic emissive power

Emissive power

Emissivity

Absorptivity

Transmissivity

Radiosity

(e) State Kirchhoff's Law (radiation).

(f) What is a gray surface?

(g) What is a black surface?

(h) What is a diffuse surface?

(i) What is required for the total absorptivity of a surface to be different for irradiation from a high temperature black source and a low temperature black source?

(j) Define view factor.

(k) What is the ratio of the radiosity to the emissive power of a black surface?

(l) State the reciprocity relation for view factors.

(m) If a surface emits 200 W at a temperature of T_1 K, how much energy will it emit at $2T_1$ K?

(n) You might have observed early morning frost on a clear day even when the minimum air temperature during the night was above 0 °C. On a clear day, the effective sky temperature can be as low as -45 °C. Explain how such frost formation takes place.

PROBLEMS

Neglect convective heat transfer unless stated otherwise.

7.1 The radiant energy from the sun on the skylight of a house, perpendicular to the solar beam, is 900 W/m². The area of the skylight is 0.2 m². The reflectivity and transmissivity of the material of the skylight are, respectively, 0.1 and 0.2. Determine the amount of energy

(a) Absorbed

(b) Reflected

(c) Transmitted

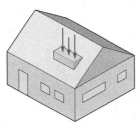

FIGURE P7.1

7.2 Re-solve Problem 7.1 if the normal to the surface is at 30° to the solar beam.

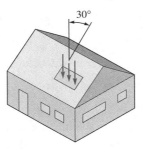

FIGURE P7.2

7.3 Heat rejection from space shuttles can be achieved by radiation from the surface to the sky. To cool a liquid in a space ship, it is proposed to circulate it through tubes embedded in the surface exposed to the sky. The liquid maintains the temperature of the exposed surface at 500 K. The incident energy is 1200 W/m². The emissivity of the surface is 0.9. Determine, per unit area,

(a) The absorbed energy.

(b) The reflected energy.

(c) The emitted energy.

(d) The radiant energy leaving the surface.

(e) The net radiant heat transfer from the surface.

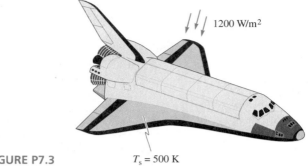

FIGURE P7.3 $T_s = 500$ K

7.4 Determine the energy flux emitted from a blackbody in the wave band $\lambda_{max} - 0.5$ μm to $\lambda_{max} + 0.5$ μm if the temperature of the surface is
(a) 500 K
(b) 1000 K
(c) 2000 K
(d) 4000 K
The wavelength at which the monochromatic emissive power is a maximum is λ_{max}.

7.5 The maximum monochromatic emissive power of a gray surface is found to be at a wavelength of 1.4 μm. What is the emissive power of the surface if the emissivity of the surface is 0.8?

7.6 The solar constant is the incident solar energy flux on a surface normal to the solar beam placed in the outer reaches of the atmosphere. Its mean value has been determined to be 1353 W/m^2. If the distance between the centers of the earth and the sun is 1.5×10^{11} m, determine the apparent blackbody temperature of the surface of the sun. Assume the sun to be a sphere with a diameter of 1.39×10^9 m.

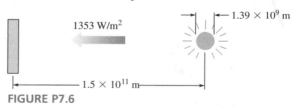

FIGURE P7.6

7.7 The temperature of a heating element is 800 °C. Determine the emitted energy flux from the surface in the spectrum 0.4 μm–0.8 μm and in the infrared wave band 0.8 μm–2 μm if the
(a) Surface is black.
(b) Emissivity is 0.8.

7.8 In Problem 7.7 find the wavelength at which the blackbody monochromatic emissive power is a maximum and determine the emitted energy flux in the wave band that is 1 μm on either side of that wavelength, as a fraction of the total emitted energy.

7.9 The transmissivity of two sheets of glass are approximated as follows:
Plain glass $\tau_\lambda = 0.9$ for 0.2 μm $< \lambda <$ 2 μm
Tinted glass $\tau_\lambda = 0.8$ for 0.4 μm $< \lambda <$ 1 μm
Outside the specified wave bands the transmissivity is zero for both cases. What is the ratio of solar energy transmitted through the sheets as a fraction of the total incident solar energy? What is the ratio of the solar energy transmitted in the spec-

trum 0.4 μm–0.8 μm to the total incident solar energy? Assume that the sun behaves as a blackbody at 5800 K.

7.10 In a flat plate solar collector it is required to choose the glass cover plate that transmits the maximum amount of solar energy. Three types of glass sheets, A, B, and C, with transmissivities as shown in the figure, are available. Which type of glass will you choose? Assume the sun behaves as a blackbody at 5800 K.

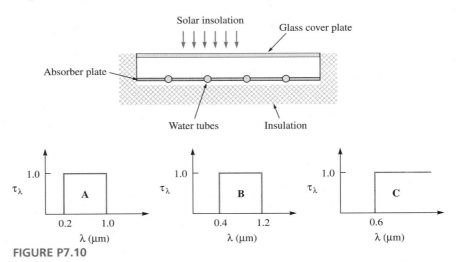

FIGURE P7.10

7.11 Re-solve Problem 7.10 for three types of glass having the following properties:

Glass A: $\tau = 0$ for $0 < \lambda < 0.2$ μm
$\tau = 0.8$ for 0.2 μm $< \lambda < 2$ μm
$\tau = 0$ for $\lambda > 2$ μm
Glass B: $\tau = 0$ for $0 < \lambda < 0.4$ μm
$\tau = 1$ for 0.4 μm $< \lambda < 1$ μm
$\tau = 0$ for $\lambda > 1$ μm
Glass C: $\tau = 0.7$ for all λ

7.12 A furnace is equipped with a 1-cm square quartz window. The furnace walls are at 1000 °C. The quartz has a transmissivity of 0.9 to radiation in the wave band 0.3 μm–1.0 μm and is opaque at other wavelengths. Assuming the furnace walls to be black, determine the radiant energy transmitted through the window.

7.13 A special surface on a space shuttle has an absorptivity of 0.9 in the wave band $0 < \lambda < 2$ μm and zero at other wavelengths. It receives solar radiation at a rate of 900 W/m^2 and is maintained at 800 K. Consider the sun as a blackbody at 5800 K. Determine, per unit area of the surface,
(a) The rate of radiant energy absorbed by the surface.
(b) The rate of radiant energy emitted by the surface.
(c) The rate of radiant energy transfer to the surface.
Assume that the surroundings are at 0 K.

7.14 Solar irradiation on a nonreflecting glass plate 1 m × 2 m is 1000 W/m^2. The transmissivity of the glass is 0.9 from 0 to 4 μm and 0.1 beyond 4 μm. Assuming the sun to behave as a blackbody at 5800 K, determine

(a) Solar energy transmitted through the plate.

(b) Energy absorbed by the plate.

7.15 A special type of glass has a transmissivity of 0.9 in the wave band 0–1.8 μm, and 0.1 in the wave band 1.8 μm–∞. It is used in a solar collector that has a black absorber plate maintained at 60 °C. Estimate the incident solar energy flux on the plate if the incident solar energy flux on the glass is 900 W/m². Assume the sun behaves as a blackbody at 5800 K.

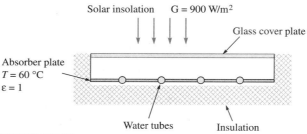

FIGURE P7.15

7.16 Color photographs are made with chemicals that are sensitive to light in different colors. The films reproduce colors if the incident energy has components in different colors in a proportion defined by a temperature. Consider a film that is designated as "daylight", which is designed for radiation from a blackbody at 5800 K. If this film is used in sunlight, it will reproduce true colors. The same film is used indoors lighted by incandescent lamps with a filament temperature of 3000 K. The filaments are gray. Explain the possible results. Support your answer with calculations. Wavelength of blue color is 0.47 μm and that of red is 0.65 μm.

7.17 Two surfaces, one with $\varepsilon_1 = 0.9$ and the other with $\varepsilon_2 = 0.2$ are available as sources for an incandescent lamp. Material 1 can be maintained at 2800 K, whereas material 2 can be maintained at 3000 K. For the same area identify the surface that emits a higher amount of energy in the spectrum 0.4 μm $< \lambda <$ 0.8 μm.

7.18 From Planck's Distribution, we have

$$E_{b\lambda} = \frac{c_1}{\lambda^5[\exp(c_2/\lambda T) - 1]}$$

For long wavelengths, show that $E_{b\lambda} = c_1\lambda T/(c_2\lambda^5)$. This is the Raleigh-Jean's formula. It is a good approximation for long wavelengths but fails at short wavelengths. For short wavelengths, show that Planck's distribution may be approximated by

$$E_{b\lambda} = \frac{c_1}{\lambda^5 \exp(c_2/\lambda T)}$$

This is Wien's formula. Planck combined the two approximate formulae into a single one, which, in turn, resulted in his Quantum Theory.

7.19 A very long duct is constructed in the form of an isosceles triangle with the two equal sides 0.3-m long and perpendicular to each other. Find the view factors $F_{1\text{-}2}$ and $F_{1\text{-}3}$.

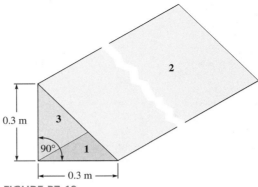

0.3 m   **3**

90°   **1**

2

0.3 m

FIGURE P7.19

7.20 A long circular cylinder 1 of diameter d_1 is enclosed by a second, long, circular cylinder 2 of diameter d_2. Find F_{2-1} and F_{2-2}.

7.21 A kiln is in the form of a hemisphere of radius R. Considering the base as surface 1 and the inside surface of the hemisphere as surface 2, find F_{2-2}.

7.22 A circular disk 1 of 20-cm diameter is placed 40 cm from a coaxial ring 2 of 20-cm I.D. $\times$ 40-cm O.D. Determine F_{2-1}.

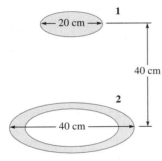

1

20 cm

40 cm

2

40 cm

FIGURE P7.22

7.23 A 10-cm diameter $\times$ 10 cm long cylinder 1 is enclosed in a hollow coaxial cylinder 2 of 20-cm I.D. and the same length. Designating the end surfaces as 3 and 4, determine all the view factors for the four surfaces.

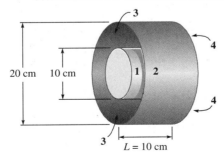

3

4

20 cm   10 cm   **1 2**

4

3 $L = 10$ cm

FIGURE P7.23

7.24 End surface 1, of a conical duct is 20-cm diameter and end surface 2 is 10-cm diameter. The axial length of the duct is 10 cm. Denoting the conical surface as 3, determine, F_{1-2}, F_{1-3}, and F_{3-3}.

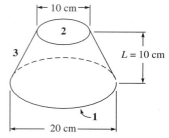

FIGURE P7.24

7.25 Determine the required view factors in each case represented in Figure P7.25. For case a, determine $F_{1\text{-}3}$, $F_{4\text{-}2}$. For case b, determine $F_{2\text{-}4}$, $F_{1\text{-}3}$, $F_{1\text{-}4}$.

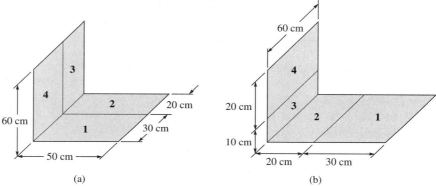

(a)   (b)

FIGURE P7.25

7.26 A 6-cm diameter circular disk 1 is 5 cm away from a coaxial ring 2, which is 6-cm I.D. × 10-cm O.D. Determine the view factors $F_{1\text{-}2}$ and $F_{2\text{-}1}$.

7.27 A picture window, 3-m long and 1.8-m high, is installed in a wall. The bottom edge of the window is on the floor, which is 6 m × 10 m as shown. Determine $F_{1\text{-}2}$.

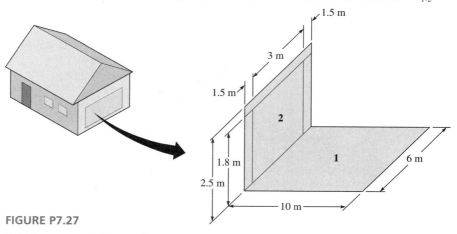

FIGURE P7.27

7.28 Two rectangular, parallel surfaces, 1 and 2, 10 cm × 20 cm are 5-cm apart. Determine $F_{1\text{-}2}$ if
(a) The two surfaces are directly opposite to each other.
(b) They are aligned but with the 10-cm edge offset by 10 cm.

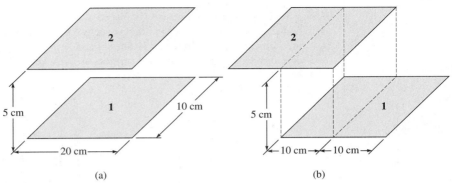

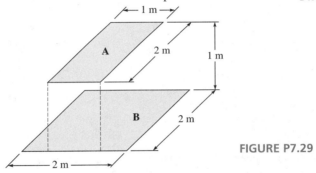

FIGURE P7.28

7.29 A 1 m × 2 m rectangular plate A, is directly opposite a 2 m × 2 m plate B. The distance between the two plates is 1 m. Determine F_{B-A}.

FIGURE P7.29

7.30 A block has a rectangular slot, 2-cm high and 1-cm wide. Denoting the slot surface as 1, determine F_{1-1} if

(a) The slot is very long and the view factor of the slot surface with respect to the ends (1 cm × 2 cm) is negligibly small.

(b) The length of the slot is 5 cm.

FIGURE P7.30

7.31 A long plate, of height L is placed in front of a semicylindrical surface of diameter d ($L < d$) as shown. Denoting the plate as 1, the semicylinder as 2, and the surroundings surfaces as 3, determine F_{2-1}, F_{2-3}, and F_{2-2}.

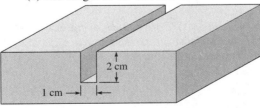

FIGURE P7.31

7.32 A long cylinder of diameter D is placed at the axis of a semicylindrical surface of diameter d. Denoting the cylinder as surface 1, the surrounding surfaces as 2, and the semicylindrical surface as 3, determine F_{3-2}, F_{3-1}, and F_{3-3}. The diameters of the cylinder and the semicylindrical surfaces are, respectively, 10 cm and 1 m.

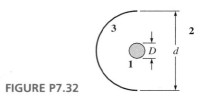

FIGURE P7.32

7.33 A circular disk 1, of diameter d, is placed in front of a hemisphere 2, of diameter D ($d < D$). Denoting the surrounding surfaces as 3, find F_{2-1}, F_{2-2}, and F_{2-3}.

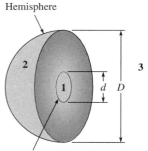

FIGURE P7.33   Circular disk

7.34 A 0.5-m diameter sphere is suspended in a large room. If the temperature of the walls of the room is 20 °C and that of the surface of the sphere is 80 °C, estimate the radiative heat transfer rate from the sphere ($\varepsilon = 0.9$).

7.35 In Problem 7.34 the sphere is placed inside a 1-m diameter spherical shell maintained at 20 °C. Determine the heat transfer rate from the sphere if the emissivity of the surface of the outer shell is
(a) 0.8
(b) 0.2

7.36 A block of metal at 120 °C has a 3-cm diameter hemispherical cavity. If the emissivity of the surface is 0.6 and the surroundings are at 20 °C, determine the radiative heat transfer rate from the cavity.

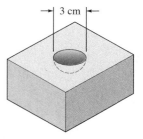

FIGURE P7.36

7.37 The base A, of a 20-cm diameter hemisphere is at 300 °C and the hemispherical surface B, is at 200 °C. The emissivities are $\varepsilon_A = 0.4$ and $\varepsilon_B = 1.0$. Determine

(a) The total amount of radiant energy leaving surface B.

(b) The values of $F_{B\text{-}A}$ and $F_{B\text{-}B}$.

(c) The fraction of the radiative energy leaving B and reaching A.

(d) For surface A, the total emitted energy and reflected part of the incident energy.

(e) The radiosity of surface A.

(f) The radiative heat transfer rate from surface A.

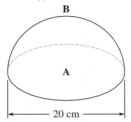

FIGURE P7.37

7.38 Consider a cylindrical cavity of depth L and diameter d with an emissivity of ε. Based on the energy passing through the circular opening of the cavity to an environment at 0 K, show that the effective emissivity of the aperture is given by

$$\frac{1}{A_1/A_2[(1 - \varepsilon_2)/\varepsilon_2] + 1}$$

where A_1 is the area of the aperture and A_2 is the surface area of the cavity.

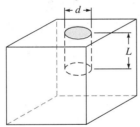

FIGURE P7.38

7.39 The temperature of a sheet of oxidized steel, 1 m $\times$ 2 m, is 750 K. The sheet is suspended inside a furnace whose walls are maintained at 1500 K. The emissivity of the sheet is 0.8 and the walls of the furnace behave as black surfaces. Determine, per unit time,

(a) The radiant energy emitted by the steel per unit area of the sheet.

(b) The energy incident on the steel per unit area of the sheet steel.

(c) The radiosity of the steel surface.

(d) The total radiative energy leaving the steel surface.

(e) The radiant energy incident on the steel in the wave band 2 μm to 4 μm, per unit area of the sheet.

7.40 A 100 W heating element is inserted into a hollow copper cylinder. The cylinder, 2-cm O.D. $\times$ 15-cm long, is freely suspended in a room that is at 20 °C. Neglecting convective heat transfer and the heat transfer from the end surfaces, determine the temperature of the copper cylinder if

(a) The emissivity of the polished outer surface is 0.1.

(b) The emissivity of the outer surface increases to 0.6 due to oxidation.

(c) Comment on the validity of neglecting convective heat transfer.

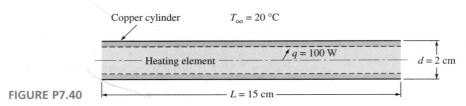

7.41 A carbon-steel block, 10-cm long, 10-cm wide, and 15-cm deep, is heated to 1000 °C and freely suspended in a large room that is at 30 °C. Assuming that lumped analysis is valid, determine the time taken for the slab to cool to 500 °C, neglecting convective heat transfer. Verify that lumped analysis is valid.

7.42 Liquid nitrogen at 77 K is to be stored in a small cylindrical container made of sheet steel. The container is 10-cm diameter and 30-cm high. It is enclosed in a coaxial cylinder with a radial clearance of 0.5 cm. The space between the two cylinders is completely evacuated. Assume the inner cylinder to be at 77 K. For an outer cylinder temperature of 10 °C, determine the radiative heat transfer rate to the inner cylinder if

(a) The surfaces are all black.

(b) The surfaces are coated with a highly reflecting material with a reflectivity of 0.95.

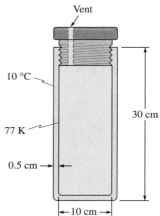

FIGURE P7.42

7.43 The long triangular duct shown is considered as a three-surface enclosure. Surfaces 1 (BC) and 2 (AB) are maintained at 400 K and 325 K, respectively, and surface 3 (AC) is insulated. Determine the radiative heat transfer rate from surface 1 per unit length of the duct and the temperature of surface 3 ($\varepsilon_1 = 0.8$, $\varepsilon_2 = 0.9$, $\varepsilon_3 = 0.1$).

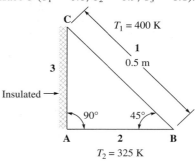

FIGURE P7.43

7.44 Re-solve Problem 7.43 if all surfaces are black.

7.45 A long, cylindrical rod 1 ($\varepsilon_1 = 0.7$, diameter = 10 mm) radiates energy to its surroundings 2 ($T_2 = 300$ K). An insulated coaxial, semicylindrical reflector 3 ($\rho_3 = 0.95$, diameter = 10 cm) is placed behind the heating element. The heater output is 1000 W/m. Considering radiative heat transfer only, and assuming all surfaces to be gray and diffuse, determine the temperatures of the heating rod and the semi-cylindrical surface. ($F_{3\text{-}1} = 0.1$, $F_{3\text{-}2} = 0.5398$, $F_{1\text{-}3} = F_{1\text{-}2} = 0.5$.)

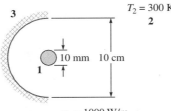

FIGURE P7.45

7.46 A thermocouple, placed in a stream of hot gases, indicates a temperature of 300 °C. The gases flow in a 10-cm diameter, 5-m long duct whose surface temperature is 200 °C. The gray thermocouple junction has a diameter of 2 mm and an emissivity of 0.65. If the convective heat transfer coefficient between the thermocouple junction and the gas stream is 130 W/m² K, find the correct temperature of the gas. The gas does not participate in radiation.

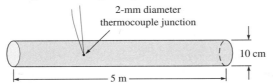

FIGURE P7.46

7.47 Consider a long evacuated enclosure whose cross section is a quarter circle with a radius of 10 cm. All interior surfaces are diffuse and gray. Surfaces 1 and 2 are at 1200 K, and have an emissivity of 0.5. Surface 3 is at 500 K, and has an emissivity of 0.9. Per unit area of surface 3,
   (a) Develop an expression for the energy that is incident on surface 3 in terms of the view factors, the radiosities of surfaces 1 and 2, and their areas.
   (b) Determine the view factors $F_{1\text{-}3}$ and $F_{1\text{-}2}$.
   (c) Determine the net rate of radiative heat transfer to surface 3.

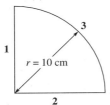

FIGURE P7.47

7.48 A cylinder 1, 15 cm in diameter and 30 cm in length, is at 200 °C. It is enclosed inside a coaxial cylinder 2 (25-cm I.D. and 30-cm long) at 20 °C. The end surfaces 3 and 4 are perfectly insulated. If all the surfaces are black, determine,
   (a) The radiative heat transfer rate from the inner cylinder.
   (b) The temperature of the end surfaces.

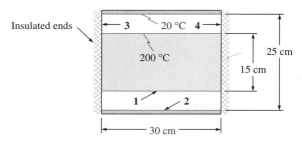

FIGURE P7.48

7.49 Re-solve Problem 7.48 if $\varepsilon_1 = 0.8$, $\varepsilon_2 = 0.7$, and $\varepsilon_3 = \varepsilon_4 = 0.9$.

7.50 In Problem 7.48, find the heat transfer rate from each surface if the end surfaces (instead of being insulated) are at 150 °C and all surfaces are black.

7.51 Re-solve Problem 7.50, if $\varepsilon_1 = 0.8$, $\varepsilon_2 = 0.7$, and $\varepsilon_3 = \varepsilon_4 = 0.9$.

7.52 Wood chips are dried by radiative heating in a furnace. The chips are fed to the furnace on a conveyor belt that forms the bottom surface of the a rectangular furnace, 2-m wide, 1-m high, and 20-m long. The side and end walls are perfectly insulated and the top surface ($\varepsilon_1 = 0.8$) is maintained at 200 °C. If the temperature of the wood chips ($\varepsilon_2 = 0.9$) is 50 °C, find the heat transfer rate from the top surface. Assume that moisture is rapidly removed and the medium surrounding the chips does not participate in radiation.

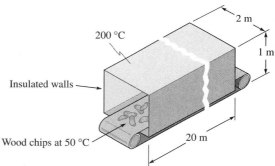

FIGURE P7.52

7.53 Instead of the furnace being rectangular as in Problem 7.52, the furnace has a heated top which is 1-m wide × 20-m long and the bottom conveyor belt, on which the wood chips are transported, is 2-m wide. The side and end walls are perfectly insulated. The emissivity of the heated surface is 0.8 and that of the wood chips is 0.9. If the heated top is at 200 °C and the wood chips are at 50 °C, what is the radiative heat transfer rate to the wood chips? ($F_{1\text{-}2} = 0.33$.)

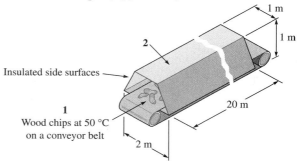

FIGURE P7.53

7.54 Reconsider the furnace in Problem 7.52. If the ends of the furnace are open to the surroundings at 20 °C, determine
(a) The heat transfer rate from the top surface.
(b) The heat transfer rate to the wood chips.

7.55 A vacuum flask is modeled as a cylinder 5-cm I.D. $\times$ 25-cm high. To reduce heat transfer, it is enclosed in another, thin 6-cm diameter coaxial cylinder. The reflectivity of the highly polished surfaces of the cylinders facing each other is 0.95. The annular space is completely evacuated. The temperatures of the inner and outer surfaces are 80 °C and 20 °C, respectively. Assume that the two cylinders form a complete enclosure.
(a) Determine the rate of radiative energy transfer from the inner cylinder.
(b) Instead of the vacuum space, it is suggested that the inner cylinder be insulated with fiberglass insulation. For the heat transfer rate with the insulation to be equal to the heat transfer rate found in part a, what should be the radial thickness of the insulation? Assume the inner and outer surface temperatures of the insulation are 80 °C and 20 °C, respectively.

7.56 Lacking an atmosphere to reject heat, rejection of heat from space ships is by radiation. A radiator in a space ship consists of tubes welded to a copper plate of thickness t. If the tubes are at a temperature T_f and the equivalent sky temperature is T_s, derive the differential equation for the temperature distribution in the copper plate. Write the boundary conditions.

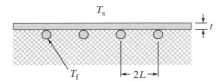

FIGURE P7.56

7.57 In some buildings space heating is accomplished by embedding heating elements in the floors or ceilings. Consider a room that is 4-m wide, 5-m long, and 2.5-m high. The ceiling is at 40 °C, and the floor and walls are at 18 °C. Determine the radiative heat transfer rate from the ceiling. The emissivity of all the surfaces is 0.9.

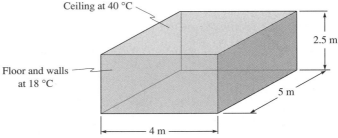

FIGURE P7.57

7.58 Reconsider Problem 7.57 with the floor at 30 °C and the side walls at 18 °C. If all surfaces are black, compute the heat transfer rate from the ceiling and the side walls.

7.59 Re-solve Problem 7.58 if the emissivity of each surface is 0.8.

7.60 Re-solve Problem 7.58 if the floor is black and the emissivity of the remaining surfaces is 0.8.

7.61 A furnace is in the form a very long, semicylindrical enclosure with a radius of 2 m. The base is divided into two parts by the axis of the cylinder. One half of the base 1 is maintained at 400 °C and the other half 2 at 100 °C. The emissivity of all the surfaces is 0.7. Find the radiative heat flux from surface 1. The cylindrical surface is perfectly insulated.

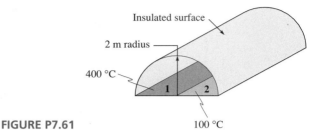

FIGURE P7.61

7.62 Consider two rectangular surfaces facing each other as shown. The surfaces are 30-cm wide and 20-cm high and are 10-cm apart. Surface 1, with an emissivity of 0.8, is maintained at 60 °C, and surface 2, with an emissivity of 0.6, is at 80 °C. If they are in an environment at 0 K, determine the radiant energy transfer rate from each surface.

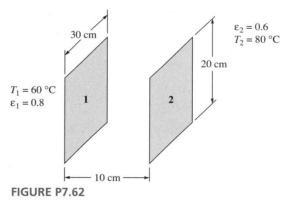

FIGURE P7.62

7.63 Re-solve Problem 7.62 if both surfaces are black.

7.64 Re-solve Problem 7.62 if the environment is at 20 °C and the surfaces are black.

7.65 Re-solve Problem 7.62 if the environment is at 20 °C.

7.66 To determine the emissivity of the surface of a material, a hollow cylinder of that material, 2-cm O.D. and 20-cm long, is heated to 130 °C by an electrical heater that fits snugly inside the cylinder. The power input to the heater is found to be 15.8 W. Another cylinder of identical dimensions with a known emissivity of 0.98 required 26.3 W to maintain its temperature at 130 °C. The axes of the cylinders are horizontal; the surrounding air and surfaces are at 25 °C. The heat transfer rate from the end surfaces may be neglected as steps to reduce such heat transfer have been taken.
(a) Determine the convective heat transfer coefficient using the data of the surface with the known emissivity.

(b) Having determined the convective heat transfer coefficient, compute the radiation heat transfer rate from the test specimen and the emissivity of the test surface.

(c) How does the convective heat transfer coefficient determined from the test data compare with the heat transfer coefficient obtained from the applicable correlation for natural convection heat transfer coefficient?

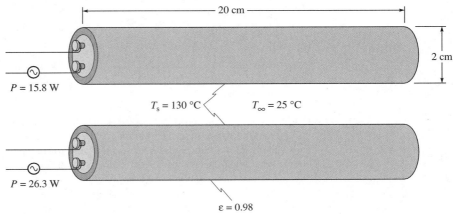

FIGURE P7.66

7.67 A 10-cm diameter sphere is mounted concentrically inside a 20-cm diameter spherical shell. The space between the two spheres is completely evacuated. The inside surface is heated to a temperature of 120 °C. Under steady conditions the temperature of the outer sphere is found to be 30 °C. The emissivity of both surfaces of the outer sphere is 0.95, and it is suspended in air at 20 °C. Determine the emissivity of the surface of the inner sphere if the convective heat transfer coefficient is 8 W/m^2 °C. Neglect conductive heat transfer through the supporting bracket.

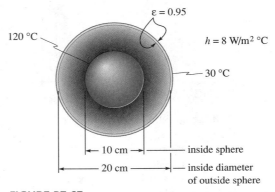

FIGURE P7.67

7.68 An ice rink is 35-m long and 20-m wide. The enclosure is 60-m long, 50-m wide, and 15-m high. The surface of the rink is at −8 °C and the enclosure is at 16 °C. Assuming the emissivity of the enclosure to be 0.8, estimate the heat transfer rate from the surroundings to the ice by radiation.

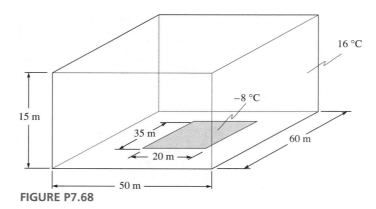

FIGURE P7.68

7.69 Liquid hydrogen is stored in a 30-cm diameter spherical container. The temperature of the outer surface of the container is 22 K and the surrounding air and surfaces are at 300 K. To reduce the heat transfer rate to the sphere, it is enclosed in two concentric spheres, one of 33-cm diameter and another of 36-cm diameter. The spaces between the spheres are completely evacuated. Determine the radiative heat transfer rate to the sphere, if the emissivity of all the surfaces of the spherical shells is 0.1 and the emissivity of the surface of the container is 0.3. Neglect convection.

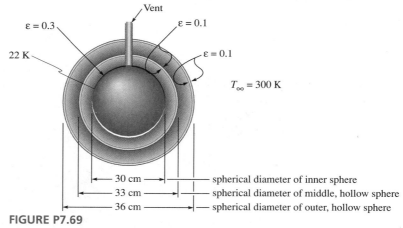

FIGURE P7.69

7.70 Liquid nitrogen is transferred in a long 2-cm diameter tube whose surface temperature is 80 K. It is enclosed by a 4-cm diameter coaxial tube that is at 260 K. The space between the two cylinders is completely evacuated. Determine the radiative heat transfer rate to the inner tube per meter length if the emissivity of the inner tube surface is 0.3 and the emissivity of both surfaces of the outer tube is 0.7.

7.71 In Problem 7.70 a thin 3-cm diameter coaxial tube with an emissivity of 0.1 of both surfaces is inserted. The outer tube surface is at 260 K. What is the radiative heat transfer rate to the inner tube?

7.72 The inner surface of the ceiling of a furnace ($\varepsilon = 0.85$) is maintained at 300 °C by an embedded heating element. The furnace is 1-m wide, 2-m long, and 50-cm high and is open to a large room at the ends. The room air and the walls are at 30 °C. The bottom surface ($\varepsilon = 0.7$) is at 100 °C and the side walls are insulated. Determine the radiative heat transfer rate from the top surface and the heat transfer rate to the bottom surface.

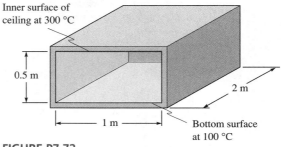

FIGURE P7.72

7.73 A 20-cm diameter heated circular disk, maintained at 400 °C, is placed in front of a perfectly insulated 40-cm diameter hemisphere. The disk is at the center of the base of the hemisphere. The disk and the hemisphere are kept in a large room with the surface of the disk facing the room perfectly insulated. If the emissivity of the disk surface is 0.7 and the room is at 20 °C, find the heat transfer rate from the disk and the temperature of the hemisphere.

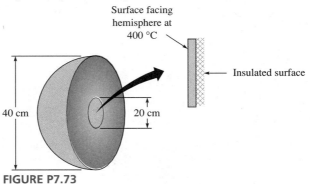

FIGURE P7.73

7.74 Two 20-cm diameter coaxial disks are 5 cm apart. The temperature of one surface is 200 °C and its emissivity is 0.8. The other surface, with an emissivity of 0.5, is at 400 °C. What are the radiative heat transfer rates from each surface if they are placed in a large room at 20 °C?

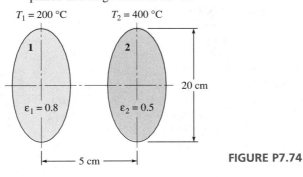

FIGURE P7.74

7.75 Re-solve Problem 7.74 if both the disks are black.

7.76 The temperature of the two end surfaces of an evacuated hollow cylinder are $T_1 = 300\ °C$ and $T_2 = 100\ °C$. The cylindrical surface is perfectly insulated. The diameter and the length of the cylinder are 10 cm and 8 cm, respectively. Considering the cylinder as a three-surface enclosure determine the heat transfer rate from surface 1 to surface 2 ($\varepsilon_1 = 0.5$ and $\varepsilon_2 = 0.7$).

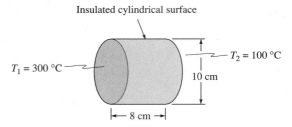

FIGURE P7.76

7.77 In Problem 7.76 a thin 10-cm diameter disk is placed at a distance of 4-cm from one of the end surfaces. The emissivities of the two surfaces of this disk are 0.6 and 0.8. Calculate the heat transfer rate from surface 1. What is the heat transfer rate from surface 1 if $T_1 = 400\ °C$ and $T_2 = 200\ °C$?

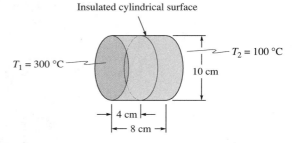

FIGURE P7.77

7.78 Combustion gases maintain the outer surface of 20-cm diameter duct at 400 °C. The duct is to be enclosed by a number of coaxial cylindrical ducts with a radial gap of 0.5 cm between successive ducts. The emissivity of every surface is 0.8. Determine the temperature of the outermost duct, and the heat transfer rate to the surroundings per meter length of the duct if the total number of ducts is
(a) 2
(b) 3
(c) 4
The temperature of the surroundings is 20 °C. Neglect convective heat transfer.

7.79 Re-solve Problem 7.78 if the emissivities of the outer surfaces of the innermost and outermost ducts are 0.8, and the emissivity of all the other surfaces is 0.2.

7.80 It is proposed to attach a rectangular plate, 6-m long, 1-m wide, and 12-mm thick, to a space ship to reject heat to the surroundings. The surface of the plate is maintained at 80 °C. The emissivity of the plate is 0.9. The sky temperature is 4 K. Determine the radiative heat heat transfer rate from the plate when
(a) The space ship is on the dark side of the earth.
(b) One surface of the plate receives solar radiation at 1360 W/m².
Assume that the plate "sees" very little of the space ship.

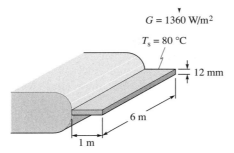

FIGURE P7.80

7.81 A thin rectangular plate is attached to a side of a space ship to reject heat to the surrounding. The edge of the plate attached to the space ship is at T_b and the effective sky temperature is T_∞. If the emissivity of the surface of the plate is ε, derive the differential equation for the steady-state temperature distribution $T(x)$ when
(a) The space ship is on the dark side of the earth.
(b) One surface receives solar radiation of G W/m².
Assume that the plate "sees" very little of the space ship.

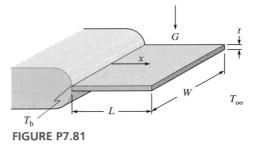

FIGURE P7.81

7.82 An array of rectangular plates is attached to a side of a space ship as shown. The spacing between the plates s is much less than the width W. As the length L is small, the plates and the base of the space ship to which the plates are attached are at a uniform temperature of 325 K. The effective sky temperature is 0 K. The emissivity of the surfaces is 0.8. Determine the heat transfer rate from a pair of adjacent plates if $s = 20$ mm, $W = 3$ m, $t = 3$ mm, and $L = 40$ mm.

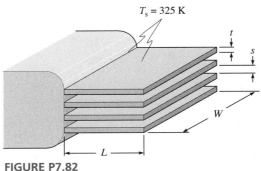

FIGURE P7.82

7.83 It is proposed to generate Refrigerant 113 vapor by radiant heating. Saturated R-113 flows through passages drilled inside a plate that is 60-cm long and 30-cm wide.

The plate, with an emissivity of 0.8, is at 50 °C. A rectangular heating element, also 60-cm long and 30-cm wide, is placed directly opposite to the plate at a distance of 20 cm. The heating element is at 300 °C and has an emissivity of 0.7. The temperature of the surroundings is 25 °C. Compute

(a) The radiative heat transfer rate to the plate.

(b) The radiative heat transfer rate from the heating element.

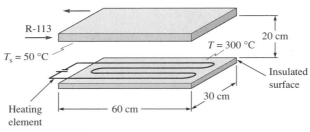

FIGURE P7.83

7.84 Re-solve Problem 7.83 if both the plate and the heating element are black.

7.85 In Problem 7.83 perfectly insulated, nonconducting side plates connecting the plate and the heating element are added. Determine the heat transfer rate from the heating element and the temperature of the side plates.

7.86 Re-solve Problem 7.85 if all surfaces are black.

7.87 The cylindrical surface of a hollow cylinder is insulated. One end surface ($\varepsilon_1 = 0.8$) is at a temperature of 500 °C. The net radiative heat transfer rate to the other end surface ($\varepsilon_3 = 0.4$) is 830 W. The cylinder has a diameter of 60 cm and a length of 20 cm. Assuming that the cylinder is perfectly evacuated and considering it as a three-surface enclosure, determine the temperatures of the cylindrical surface 2 and end surface 3.

7.88 A very long cylinder 1 (15-cm diameter, $\varepsilon = 0.9$) is maintained at 500 °C. It is completely enclosed by a thin, aluminum cylindrical sheet 2 (60-cm diameter). The emissivity of the aluminum surface facing the cylinder is 0.25 and the emissivity of the outer surface is 0.1. The assembly is placed in a large room maintained at 20 °C. Determine the radiative heat transfer rate from the cylinder per meter length and the temperature of the aluminum sheet. Neglect convective heat transfer. Compare the heat transfer rate with the heat transfer rate if the aluminum sheet is removed.

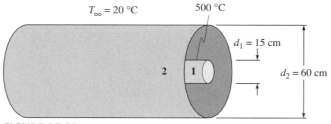

FIGURE P7.88

7.89 A 20-cm diameter sphere 1 is placed concentrically inside a 30-cm diameter sphere with the space between the two spheres completely evacuated. If $T_1 = 300\,°C$, $\varepsilon_1 = 0.8$, $T_2 = 20\,°C$, and $\varepsilon_2 = 0.4$, determine
(a) The heat transfer rate from sphere 1
(b) $F_{2\text{-}2}$

7.90 Re-solve Problem 7.89 if a spherical shell (25-cm diameter), made of a thin metal sheet with both surfaces having an emissivity of 0.05, is placed concentrically with the first sphere.

7.91 In Problem 7.36, the cavity at $120\,°C$ ($\varepsilon = 0.6$) is enclosed by a 3-cm diameter disk. The emissivity of both disk surfaces is 0.2; the surroundings are at $20\,°C$. Determine the heat transfer rate from the cavity and the temperature of the disk. Neglect convection.

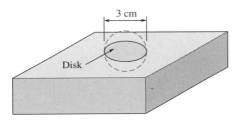

FIGURE P7.91, P7.92, P7.93

7.92 In Problem 7.36, the cavity (at $120\,°C$, $\varepsilon = 0.6$) is enclosed by a hemisphere (3-cm diameter emissivity of 0.2), and the surroundings are at $20\,°C$. Neglecting convection, determine the heat transfer rate from the cavity and the temperature of the enclosing hemisphere.

7.93 In Problem 7.92, in addition to the hemisphere, a disk is placed over the cavity ($T = 120\,°C$ and $\varepsilon = 0.6$). The emissivity of the disk and hemispherical surfaces is 0.2. Determine the heat transfer rate from the cavity and the temperatures of the hemisphere and the disk. The temperature of the surroundings is $20\,°C$. Neglect convection.

7.94 The skin temperature of the human body is approximately $33\,°C$. The convective heat transfer coefficient is estimated to be 8 $W/m^2\,°C$. If the room temperature varies in the range $20\,°C$ to $25\,°C$, determine a suitable total heat transfer coefficient (sum of heat transfer coefficients due to convection and radiation) to compute the heat transfer rate from the exposed surface of the human body.

7.95 A thermocouple is used to measure the temperature of air inside a large reservoir. The thermocouple indicates a temperature of $315\,°C$. The reservoir surface temperature is $290\,°C$. The convective heat transfer coefficient associated with the thermocouple junction is 20 $W/m^2\,°C$. What is the true temperature of the air? The emissivity of the thermocouple junction surface is 0.9.

  What steps would you suggest to reduce the difference between the temperature indicated by the thermocouple and the temperature of the air?

7.96 In Problem 7.95 if the temperature of the air is $400\,°C$, and the temperature of the reservoir surface is $380\,°C$, what is the temperature that the thermocouple indicates? The solution to the problem is facilitated by using the radiative heat transfer coefficient.

7.97 Re-solve Problem 7.41 employing the radiative heat transfer coefficient. The radiation heat transfer coefficient may be assumed to be constant at its average value over every 50 °C difference in the temperature of the block.

7.98 A thin horizontal, black plate is suspended outdoors. On a clear day the effective sky temperature is approximately 228 K. The convective heat transfer coefficient between the surface of the plate and the surrounding air is 20 W/m² °C. The underside of the plate is perfectly insulated. What is the maximum air temperature below which a drop of water placed on the plate will freeze?

7.99 Re-solve Problem 7.98 if the underside of the plate is not insulated but exposed to the surrounding air. The ground temperature is 10 °C.

CHAPTER

8

RADIATION II

In Chapter 7, consideration of radiation was limited to gray surfaces separated by a nonparticipating medium. A brief discussion of wavelength dependent properties was provided, and view factor and radiation intensity were introduced. In this chapter we build on these concepts and

- *Consider the wavelength dependent properties in greater detail*
- *Define total hemispherical emissivity and absorptivity for surfaces with wavelength dependent properties and illustrate their use*
- *Explain how view factors are computed and prove the view factor relations given in Equations 7.2.4a through d*
- *Show one method for determining the view factor between two infinite strips*
- *Elaborate on the electrical analog for solving radiation exchange among diffuse gray surfaces in an enclosure*
- *Present the radiosity method for the solution of radiation exchange among diffuse gray surfaces in an enclosure*
- *Show the equivalence of the electrical analog and the radiosity method*
- *Introduce radiation in gases*
- *Present a method for computing radiation heat transfer between gases and the bounding surface when they are at uniform but different temperatures*

8.1 RADIATIVE PROPERTIES OF SURFACES

In Chapter 7 the radiative properties—absorptivity, reflectivity, and emissivity—were assumed to be direction independent and, in most cases, wavelength independent. The results obtained with such simplifying assumptions are adequate to obtain an engineering estimate of radiative heat transfer rates in many cases. However, there are cases where the direction or wavelength dependency of the properties is significant. We now consider the direction and wavelength dependence of radiative properties of opaque surfaces.

8.1.1 Emissivity

Consider the radiant energy leaving an infinitesimal surface dA in a specified direction and contained in an infinitesimal solid angle $d\omega$ as shown in Figure 8.1.1 (for

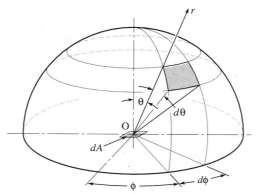

FIGURE 8.1.1 An Infinitesimal Surface Radiating Energy in a Specified Direction in an Infinitesimal Solid Angle

a brief introduction to solid angle see Section 7.5). The angles θ and ϕ in Figure 8.1.1 are known as the cone angle and azimuthal angle, respectively.

The emitted power d^4Q_e from dA in the solid angle $d\omega$ in the wave band λ to $\lambda + d\lambda$ is given by

$$d^4Q_e = I_{\lambda e}(\theta, \phi, T) \, dA \cos\theta \, d\omega \, d\lambda \tag{8.1.1}$$

where $I_{\lambda e}(\theta, \phi, T)$ is the emitted monochromatic radiation intensity from the surface in the direction (θ, ϕ) at wavelength λ and temperature T.

Spectral, directional emissivity

The radiation intensity, as indicated, is not only a function of the temperature of the surface but also of the wavelength and direction. The *spectral* (frequency or wavelength dependent), *directional emissivity* of the surface is defined as

$$\varepsilon_\lambda(\theta, \phi, T) = \frac{I_{\lambda e}(\theta, \phi, T)}{I_{b\lambda}(T)} \tag{8.1.2}$$

where $I_{b\lambda}$ is the radiation intensity of a perfect emitter (a blackbody, which, by definition, is a diffuse emitter) given by Planck's distribution:

$$I_{b\lambda}(T) = \frac{2hc_o^2}{\lambda^5 \left[\exp\left(\dfrac{hc_o}{k\lambda T}\right) - 1 \right]} \tag{8.1.3}$$

where

h = Planck's constant = 6.62377×10^{-34} J s
k = Boltzmann's constant = 1.38026×10^{-23} J/K
c_o = velocity of light in vacuum = 2.997925×10^8 m/s

Comparing Equation 8.1.3 with Equation 7.1.2 (with $E_{b\lambda} = \pi I_{b\lambda}$), we note that

$c_1 = 2\pi h c_o^2 = 3.74047 \times 10^{-16}$ W m^2 = 3.74047×10^8 W μm^4/m^2

$c_2 = \dfrac{hc_o}{k} = 0.0143868$ m K = 1.43868×10^4 μm K

The azimuthal dependence of ε_λ of a plane surface is generally weak and is neglected. We then have

$$\varepsilon_\lambda(\theta, T) = \frac{I_{\lambda e}(\theta, T)}{I_{b\lambda}(T)} \tag{8.1.4}$$

Total directional emissivity

The *total directional emissivity* is given by

$$\varepsilon(\theta, T) = \frac{\int_0^\infty I_{\lambda e}(\theta, T)\, d\lambda}{\int_0^\infty I_{b\lambda}(T)\, d\lambda} = \frac{\int_0^\infty \varepsilon_\lambda(\theta, T) I_{b\lambda}(T)\, d\lambda}{I_b(T)} \tag{8.1.5}$$

Figures 8.1.2a and b show the directional dependence of the emissivity in polar coordinates for a few metals. Emissivity increases with increasing θ but decreases as θ approaches 90°.

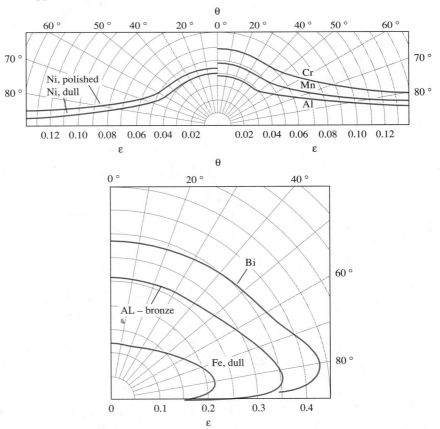

FIGURE 8.1.2 Directional Dependence of the Monochromatic Emissivity of a Few Metallic Surfaces.
The angle θ is from the normal to the surface. Reproduced by permission from Eckert, E.R.G.

Equation 8.1.5 defines the directional emissivity averaged over the wavelength. Similarly, spectral or monochromatic emissivity averaged over the direction is defined as the ratio of the total energy emitted in the wave band λ to $\lambda + d\lambda$ to a hemisphere surrounding the infinitesimal area to the energy emitted by a blackbody to a hemisphere over the same wave band. Employing spherical coordinates,

$$\varepsilon_\lambda(T) = \frac{\int_0^{2\pi} \int_0^{\pi/2} I_{\lambda e}(\theta, T) \cos\theta \sin\theta \, d\theta \, d\phi}{\int_0^{2\pi} \int_0^{\pi/2} I_{b\lambda}(T) \cos\theta \sin\theta \, d\theta \, d\phi}$$

$$= \frac{2\pi \int_0^{\pi/2} \varepsilon_\lambda(\theta, T) I_{b\lambda} \cos\theta \sin\theta \, d\theta}{\pi I_{b\lambda}(T)}$$

or

$$\varepsilon_\lambda(T) = 2 \int_0^{\pi/2} \varepsilon_\lambda(\theta, T) \cos\theta \sin\theta \, d\theta \tag{8.1.6}$$

Total hemispherical emissivity

The *total hemispherical emissivity* is defined as the ratio of the total energy emitted by the element in all directions over all wavelengths to the total energy emitted by a blackbody at the same temperature. Thus,

$$\varepsilon(T) = \frac{\int_0^\infty \int_0^{2\pi} \int_0^{\pi/2} I_{\lambda e}(\theta, \phi, T) \cos\theta \sin\theta \, d\theta \, d\phi \, d\lambda}{\int_0^\infty \int_0^{2\pi} \int_0^{\pi/2} I_{b\lambda}(T) \cos\theta \sin\theta \, d\theta \, d\phi \, d\lambda}$$

$$= \frac{\int_0^\infty \int_0^{2\pi} \int_0^{\pi/2} \varepsilon_\lambda(\theta, \phi, T) I_{b\lambda}(T) \cos\theta \sin\theta \, d\theta \, d\phi \, d\lambda}{\int_0^\infty \int_0^{2\pi} \int_0^{\pi/2} I_{b\lambda}(T) \cos\theta \sin\theta \, d\theta \, d\phi \, d\lambda}$$

With $I_{b\lambda}(T) = [E_{b\lambda}(T)]/\pi$, and the denominator $= E_b(T)$, we obtain

$$\varepsilon(T) = \frac{\int_0^\infty \int_0^{2\pi} \int_0^{\pi/2} \varepsilon_\lambda(\theta, \phi, T) \dfrac{E_{b\lambda}(T)}{\pi} \cos\theta \sin\theta \, d\theta \, d\phi \, d\lambda}{E_b(T)} \tag{8.1.7}$$

In the right-hand side of Equation 8.1.7, the numerator represents the emissive power of the surface. Assuming the emissivity to be independent of the azimuthal angle ϕ, Equation 8.1.7 becomes

$$\varepsilon(T) = \frac{2 \int_0^\infty \int_0^{\pi/2} \varepsilon_\lambda(\theta, T) E_{b\lambda}(T) \cos\theta \sin\theta \, d\theta \, d\lambda}{E_b(T)} \tag{8.1.8a}$$

The basic emission characteristic is $\varepsilon_\lambda(\theta, \phi, T)$. The other quantities, $\varepsilon(\theta, T)$, $\varepsilon_\lambda(T)$, and $\varepsilon(T)$ are derived quantities. If $\varepsilon_\lambda(\theta, \phi, T)$ is known, the emitted energy can be determined starting with Equation 8.1.1.

If a value of ε_λ averaged over all the directions is used, we obtain the total hemispherical emissivity given by

$$\varepsilon(T) = \frac{\int_0^\infty \varepsilon_\lambda(T)E_{b\lambda}(T)d\lambda}{E_b(T)} \qquad \text{(8.1.8b)}$$

Integrating Equation 8.1.8a over all wavelengths and denoting $\int_0^\infty \varepsilon_\lambda(\theta, T)E_{b\lambda}\, d\lambda$ by $\varepsilon(\theta, T)E_b$ we obtain

$$\varepsilon(T) = 2\int_0^{\pi/2} \varepsilon(\theta, T) \cos\theta \sin\theta\, d\theta \qquad \text{(8.1.8c)}$$

EXAMPLE 8.1.1 A surface has directional emissivity (averaged over all wavelengths) of 0.8 for $0 < \theta < 60°$ and 0.2 for $60° < \theta < 90°$ as shown in Figure 8.1.3. Determine

(a) The total hemispherical emissivity.
(b) The emitted flux from a surface at a temperature of 400 K in a cone with a half angle of 30° as shown in the figure.

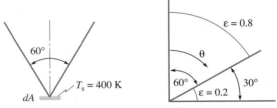

FIGURE 8.1.3 Sketch for Example 8.1.1

Given

$0 < \theta < 60°$ $\qquad \varepsilon = 0.8$ $\qquad T_s = 400\ \text{K}$
$60° < \theta < 90°$ $\qquad \varepsilon = 0.2$

Find

(a) The total hemispherical emissivity
(b) The emitted energy flux in a cone with half angle $= 30°$

SOLUTION

(a) The total hemispherical emissivity is determined from Equation 8.1.8c:

$$\varepsilon = 2\int_0^{\pi/2} \varepsilon(\theta, T) \cos\theta \sin\theta\, d\theta$$

$$= 2\left[\int_0^{\pi/3} 0.8 \cos\theta \sin\theta\, d\theta + \int_{\pi/3}^{\pi/2} 0.2 \cos\theta \sin\theta\, d\theta\right]$$

$$= 2\left[0.8\left(\frac{-\cos 2\theta}{4}\right)\Big|_0^{\pi/3} + 0.2\left(\frac{-\cos 2\theta}{4}\right)\Big|_{\pi/3}^{\pi/2}\right] = \underline{0.65}$$

(b) To find the emitted energy flux from the surface over all wavelengths in a cone with half angle $\pi/6$ radian, we use Equation 8.1.1:

$$d^4 Q_e = I_{\lambda e}(\theta, \phi, T) \, dA \cos\theta \sin\theta \, d\theta \, d\phi \, d\lambda$$

Dividing by dA, with $I_{\lambda e}(T) = \varepsilon_\lambda(\theta, \phi, T) I_{b\lambda}(T)$ and $dQ_e/dA = E$,

$$d^3 E = \varepsilon_\lambda(\theta, \phi, T) I_{b\lambda}(T) \cos\theta \sin\theta \, d\theta \, d\phi \, d\lambda$$

Integration with respect to ϕ and λ yields

$$dE = \int_0^\infty \int_0^{2\pi} \varepsilon_\lambda(\theta, \phi, T) I_{b\lambda}(T) \cos\theta \sin\theta \, d\theta \, d\phi \, d\lambda$$

$$= 2\pi\varepsilon(\theta, T) I_b(T) \cos\theta \sin\theta \, d\theta$$

The energy flux emitted into the cone can now be computed.

$$E_{0\text{-}\pi/6} = 2\pi I_b(T) \int_0^{\pi/6} \varepsilon(\theta, T) \cos\theta \sin\theta \, d\theta$$

$$= 2\pi I_b(T) 0.8 \left(\frac{-\cos 2\theta}{4} \right) \Bigg|_0^{\pi/6} = 0.2\pi I_b(T) = 0.2 E_b(T)$$

$$= 0.2\sigma T^4 = 0.2 \times 5.67 \times 10^{-8} \times 400^4 = \underline{290.3 \text{ W/m}^2}$$

In some cases wavelength dependency of the emission characteristics is of greater significance than directional dependency. For example, to increase the efficiency of solar collectors, absorber plates are specially coated so that the surfaces have a high absorptivity for short wavelengths (less than 2 µm) and low emissivity at long wavelengths (greater than 2 µm). As most of the solar energy is in wavelengths less than 2 µm, the plate absorbs a high fraction of the incident solar energy. The emitted energy from the absorber plate, which is at around 350 K, is mostly in the long wavelengths and with a low emissivity at long wavelengths the emitted energy is a small fraction of the blackbody emissive power. Thus, by absorbing most of the incident energy and emitting only a small fraction of the maximum possible energy, the efficiency of the collector is increased.

If directional dependency is averaged over all directions or if the radiation properties are direction independent (diffuse surface)

$$\varepsilon_\lambda(T) = \frac{I_\lambda(T)}{I_{b\lambda}(T)} = \frac{E_\lambda(T)}{E_{b\lambda}(T)} \tag{8.1.9}$$

The emissive power in this case is given by

$$E(T) = \int_0^\infty E_\lambda(T) \, d\lambda = \int_0^\infty \varepsilon_\lambda(T) E_{b\lambda}(T) \, d\lambda \tag{8.1.10}$$

Spectral, directional absorptivity

8.1.2 Absorptivity

In a manner similar to emissivity, the *spectral, directional absorptivity* is defined as

$$\alpha_\lambda(\theta, \phi, T) = \frac{I_\lambda(\theta, \phi, T_s)|_{\text{absorbed}}}{I_{\lambda,i}(\theta, \phi, T_s)} \tag{8.1.11}$$

where the denominator represents the incident monochromatic radiation intensity in the direction (θ, ϕ) from a source at T_s, and the numerator represents the absorbed intensity (in the same direction). Denoting the absorbed energy by Q_{abs}, the energy absorbed by an infinitesimal surface is given by

$$d^4Q_{abs} = \alpha_\lambda(\theta, \phi, T)I_{\lambda,i}(\theta, \phi, T_s) \, dA \cos \theta \, d\omega \, d\lambda$$

Spectral, hemispherical absorptivity

Incident intensity depends on the characteristics of the sources and their temperatures. As a surface may be surrounded by many surfaces with different monochromatic emissivities and at different temperatures, the complexity involved in including the directional dependency is obvious. Since including the directional dependence of α_λ leads to complex equations and the directional dependence is weak, in most engineering applications, the directional dependence of the absorptivity is accounted for by using a value averaged over all the directions.[1] Employing such an averaged value, the *spectral hemispherical absorptivity* is defined as

$$\alpha_\lambda(T) = \frac{dG_{abs}}{dG}$$

where dG_{abs} is the energy flux absorbed in the wave band λ to $\lambda + d\lambda$ and dG is the incident energy flux in the wave band λ to $\lambda + d\lambda$. If the source temperature is T_s,

$$G_\lambda(T_s) = \int_0^{2\pi} \int_0^{\pi/2} I_{\lambda,i}(\theta, \phi, T_s)) \cos \theta \sin \theta \, d\theta \, d\phi$$

and

$$dG = G_\lambda(T_s)d\lambda$$

$$dG_{abs} = \alpha_\lambda(T)G_\lambda(T_s)d\lambda$$

$$G_{abs} = \int_0^\infty \alpha_\lambda(T)G_\lambda(T_s)d\lambda$$

Total hemispherical absorptivity

Defining the *total hemispherical absorptivity* as the ratio of the total absorbed energy to the total incident energy, we obtain

$$\alpha(T, T_s) = \frac{G_{abs}(T_s)}{G(T_s)} = \frac{\displaystyle\int_0^\infty \alpha_\lambda(T)G_\lambda(T_s)d\lambda}{G(T_s)} \tag{8.1.12a}$$

Equation 8.1.12a is similar to Equation 8.1.8b but with one very important difference. In Equation 8.1.1b, $E_{b\lambda}$ and E_b depend on the temperature of the surface only and, hence, $\varepsilon(T)$ has a unique value for a given surface. In Equation 8.1.12a, G_λ and G depend not only on the source temperature but also on the source characteristics. Since α_λ is dependent on the surface temperature and G_λ is dependent on the source temperature and monochromatic emissivity, the averaged value of the absorptivity

---

[1]There are exceptions to this. For example, in determining the temperature of a surface by a radiometer, the directional dependence is significant. The sensor of the instrument is oriented normal to the incident beam. The normal absorptivity, which is equal to the normal emissivity, should be used in such a case.

α is a function of both the surface temperature and the source temperature and the source surface properties. Although $\alpha_\lambda = \epsilon_\lambda$, depending on the source, the value of α and ε can be different. There is no unique value for α. From Equations 8.1.8b and 8.1.12a it is easily seen that $\varepsilon(T)$ equals $\alpha(T, T_s)$ when either ε_λ is independent of λ, i.e., if the surface is gray, or the spectral distribution of the incident energy is the same as that of a blackbody at the temperature of the surface.

Of particular interest for solar energy utilization is the absorptivity of a surface for solar insolation (irradiation from the sun). *The total absorptivity for solar insolation is known as solar absorptivity.* The solar absorptivity of a surface is defined as

$$\alpha_s(T) = \frac{\int_0^\infty \alpha_\lambda G_{\lambda s} d\lambda}{G_s} \tag{8.1.12b}$$

where

$G_{\lambda s} d\lambda$ = incident solar energy flux in the wave band λ to $\lambda + d\lambda$
G_s = incident solar energy flux
$\alpha_s(T)$ = solar absorptivity at the temperature of the surface T

To evaluate the integral in Equation 8.1.12a or b, the spectral distribution of the incident radiation, G_λ, must be known.

EXAMPLE 8.1.2

A specially coated surface has an emissivity of 0.8 for $0 < \lambda < 2$ µm and 0.1 for $\lambda > 2$ µm as shown in Figure 8.1.4. Determine

(a) The solar absorptivity α_s (assume the sun is a blackbody at 5760 K).
(b) The absorptivity of the surface exposed to incident radiation from a blackbody at 1000 K.
(c) The total hemispherical emissivity of the surface at 300 K and 800 K.

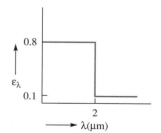

FIGURE 8.1.4 Wavelength Dependent Emissivity of the Special Surface

Given

$0 < \lambda < 2\ \mu$ $\varepsilon_\lambda = 0.8$
$2\ \mu m < \lambda < \infty$ $\varepsilon_\lambda = 0.1$

Find

(a) α_s (solar absorptivity)
(b) α for incident radiation from a blackbody at 1000 K
(c) ε at 300 K and 800 K

SOLUTION

(a) The solar absorptivity is given by Equation 8.1.12b.

$$\alpha_s = \frac{\int_0^\infty \alpha_\lambda G_{\lambda s} d\lambda}{G_s} = \int_0^\infty \alpha_\lambda \frac{G_{\lambda s}}{G_s} d\lambda$$

where the subscript s denotes the source, in this case, the sun. By Kirchoff's Law, the spectral absorptivity equals the spectral emissivity. It has two distinct values, 0.8 for $0 < \lambda < 2$ μm, and 0.1 for 2 μm $< \lambda < \infty$. Denoting the values in the range $0 < \lambda < 2$ μm by subscript 1 and those in the range 2 μm $< \lambda < \infty$ by subscript 2,

$$\alpha_s = \int_0^{\lambda_1} \alpha_1 \frac{G_{\lambda s}}{G_s} d\lambda + \int_{\lambda_1}^\infty \alpha_2 \frac{G_{\lambda s}}{G_s} d\lambda$$

$$= \frac{\alpha_1}{G_s} \int_0^{\lambda_1} G_{\lambda s} \, d\lambda + \frac{\alpha_2}{G_s} \int_{\lambda_i}^\infty G_{\lambda s} \, d\lambda$$

$$= \alpha_1 \frac{G_1}{G_s} + \alpha_2 \frac{G_2}{G_s}$$

To find G_1/G_s and G_2/G_s, construct a spherical shell around the sun, containing the surface (Figure 8.1.5). Denoting the surface area of the sun by A_s and the surface area of the spherical shell by A.

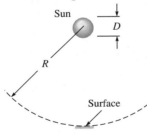

FIGURE 8.1.5 **A Sphere of Radius *R* (Distance Between the Sun and the Earth) Constructed Around the Sun**

The total energy leaving the sun $= E_{bs}A_s$

Energy received by the spherical surface $=$ energy leaving the sun

Therefore,

$$\text{Energy flux received by the spherical surface} = \frac{E_{bs}A_s}{A} = G_s$$

Similarly, the energy flux from the sun in the wave band 0 to λ_1 is given by

$$G_1 = \frac{E_{bs,0-\lambda_1}A_s}{A}$$

From the expressions for G_s and G_1 we get

$$\frac{G_1}{G_s} = \frac{E_{bs,0-\lambda_1}}{E_{bs}} \qquad \frac{G_2}{G_s} = \frac{G_s - G_1}{G_s} = 1 - \frac{G_1}{G_s}$$

Since the sun is considered as a blackbody at 5760 K, $\lambda_1 T = 11\,520$ μm K. From the software RF (or Table 7.1.2),

$$\left. \frac{E_{bs,0-\lambda_1}}{E_{bs}} \right|_{5760\ K} = 0.9392 = \frac{G_1}{G_s}$$

$$\frac{G_2}{G_s} = 1 - 0.9392 = 0.0608$$

Substituting the values for G_1/G_s and G_2/G_s,

$$\alpha_s = \alpha_1 \frac{G_1}{G_s} + \alpha_2 \frac{G_2}{G_s} = 0.8 \times 0.9392 + 0.1 \times 0.0608 = \underline{0.7574}$$

(b) In a manner similar to part a, the total hemispherical absorptivity for irradiation from a blackbody at 1000 K is given by

$$\alpha = \alpha_1 \frac{G_1}{G_s} + \alpha_2 \frac{G_2}{G_s}$$

where the subscript s denotes the source, a blackbody at 1000 K. From the software RF (or Table 7.1.2) for $\lambda_1 T = 2000$ μm

$$\frac{G_1}{G_s} = \left. \frac{E_{b,0-\lambda_1}}{E_b} \right|_{1000\ K} = 0.06675$$

$$\frac{G_2}{G_s} = 1 - \frac{G_1}{G_2} = 1 - 0.06675 = 0.9333$$

$$\alpha = 0.8 \times 0.06675 + 0.1 \times 0.9333 = \underline{0.1467}$$

(c) Total hemispherical emissivity of the surface at 300 K is found by employing Equation 8.1.8b,

$$\varepsilon = \frac{E}{E_b} = \int_0^\infty \frac{E_\lambda}{E_b}\, d\lambda = \int_0^\infty \varepsilon_\lambda \frac{E_{b\lambda}}{E_b}\, d\lambda = \int_0^{\lambda_1} \varepsilon_1 \frac{E_{b\lambda}}{E_b}\, d\lambda + \int_{\lambda_1}^\infty \varepsilon_2 \frac{E_{b\lambda}}{E_b}\, d\lambda$$

$$= \varepsilon_1 \frac{E_{b,0-\lambda_1}}{E_b} + \varepsilon_2 \frac{E_{b,\lambda 1-\infty}}{E_b}$$

At 300 K, $\lambda_1 T = 600$ μm K, and from the software (or Table 7.1.2),

$$\frac{E_{b,0-\lambda_1}}{E_b} \cong 0 \quad \text{and} \quad \frac{E_{b,\lambda_1-\infty}}{E_b} = 1$$

which gives

$$\underline{\varepsilon = 0.1}$$

At 800 K, $\lambda_1 T = 1600$ μm K,

$$\frac{E_{b,0-\lambda_1}}{E_b} = 0.01973 \quad \text{and} \quad \frac{E_{b,\lambda_1-\infty}}{E_b} = 1 - 0.01973 = 0.9803$$

The emissivity of the surface at 800 K is then given by

$$\varepsilon = 0.8 \times 0.01973 + 0.1 \times 0.9803 = 0.1138$$

COMMENT

The ratio of the solar absorptivity to emissivity at 300 K is 7.6. Also see comments on page 578.

EXAMPLE 8.1.3 The solar constant (the incident energy on a surface normal to the solar beam at the outer edge of the atmosphere) is 1353 W/m². Based on this value and assuming that the sun behaves as a blackbody, the temperature of the surface of the sun is 5760 K. Figure 8.1.6 shows a solar collector to be used on a space shuttle. Two surfaces

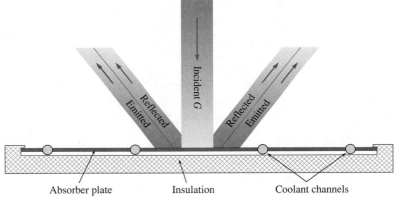

FIGURE 8.1.6 A solar collector receives solar energy, and the incident energy from the sky is negligible, i.e., the surroundings are assumed to be at 0 K.

are being considered for the absorber surface of the collector, a black surface and a specially coated surface (Figure 8.1.7) with an emissivity of 0.95 for 0 μm < λ < 2 μm and 0.1 for 2 μm < λ < ∞. A coolant cools the collector surface and maintains the temperature of the surface at 340 K. The collector surface is normal to the solar

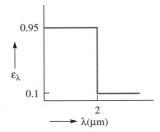

FIGURE 8.1.7 Monochromatic Emissivity of the Specially Coated Surface

beam. The thick absorber plate is made of a high thermal conductivity material and, with a small spacing between the coolant tubes, the temperature of the surface is uniform. Determine

(a) The heat transfer rate to the coolant for each surface.
(b) The temperature of each surface when there is no flow of the coolant.

Given

G_s (incident energy flux from the sun) $= 1353$ W/m^2
Temperature of the surface of the sun $= 5760$ K
Two surfaces: Black surface
                Specially coated surface with
                    $0 < \lambda < 2$ μm $\varepsilon_\lambda = 0.95$ 2 μm $< \lambda < \infty$ $\varepsilon_\lambda = 0.1$

Find

(a) q_{coolant} with T_s (temperature of the surface) $= 340$ K
(b) T_s (temperature of the surface) with $q_{\text{coolant}} = 0$

ASSUMPTIONS

1. Because the collector is mounted on a space shuttle and the surroundings are virtually a vacuum, there is no convective heat transfer from the surface to the surroundings. The heat transfer from the surface exposed to the sun is by radiation and from the surface adjacent to the coolant by convection.
2. The surface receives energy only from the sun, i.e., the surroundings are at 0 K. The view factor to the surrounding ≈ 1.
3. As the surface is normal to the solar beam, the incident solar energy flux is 1353 W/m^2.
4. Steady state

SOLUTION

(a) From an energy balance, *per unit surface area of the absorber plate* and per unit time, in steady state,

$$\text{Net energy transfer to the absorber plate} = 0$$

Therefore,

$$\begin{aligned} &\text{Radiant energy incident on the surface} \\ &- \text{Radiant energy leaving the surface} \\ &- \text{Heat transfer to the coolant} = 0 \end{aligned}$$

As the surroundings of the surface, except the sun, are assumed to be at 0 K, the radiant energy reaching the surface is from the sun only. Energy leaving the surface consists of the energy emitted by the surface, the reflected part of the incident energy, and the heat transfer to the coolant.

$$\begin{aligned} \text{Heat transfer to the coolant} \\ = \text{Incident energy} - \text{Reflected energy} - \text{Emitted energy} \\ = \text{Absorbed energy} - \text{Emitted energy} \end{aligned}$$

Black surface: $\varepsilon = \alpha = 1, \rho = 0$

Absorbed energy $= G = 1353$ W/m^2
At $T_s = 340$ K, emitted energy $= E_b = 5.67 \times 10^{-8} \times 340^4 = 757.7$ W
Heat transfer to the coolant $= 1353 - 757.7 = \underline{595.3 \text{ W/m}^2}$

Special surface:

Absorbed energy $= \alpha_1 G_{0\text{-}2\mu m} + \alpha_2 G_{2\mu m\text{-}\infty}$, where $\alpha_1 = 0.95$ and $\alpha_2 = 0.1$

From the results in Example 8.1.2,

$$\frac{G_{0\text{-}2\mu m}}{G} = \frac{E_{b,0\text{-}2\mu m}}{E_b}\bigg|_{sun}$$

With the temperature of the sun $= 5760$ K, from the software RF (or Table 7.1.2), for $\lambda T = 11\ 520\ \mu m\ K$,

$$\frac{E_{b,0\text{-}2\mu m}}{E_b}\bigg|_{sun} = 0.9392 = \frac{G_{0\text{-}2\mu m}}{G}$$

$$\frac{G_{2\mu m\text{-}\infty}}{G} = 1 - 0.9392 = 0.0608$$

Absorbed energy $= \left(\alpha_1 \dfrac{G_{0\text{-}2\mu m}}{G} + \alpha_2 \dfrac{G_{\mu m2\text{-}\infty}}{G}\right) G$

$$= (0.95 \times 0.9392 + 0.1 \times 0.0608) \times 1353 = 0.8983 \times 1353$$

$$= 1215.4 \text{ W/m}^2$$

(Here, by definition, solar absorptivity of the surface $= 0.8983$.)

Energy emitted by the surface at 340 K $= \varepsilon_1 E_{b,0\text{-}2\mu m} + \varepsilon_2 E_{b,2\mu m\text{-}\infty}$

For $T_s = 340$ K and for $\lambda = 2\ \mu m$, $\lambda T = 680\ \mu m\ K$. From Table 7.1.1,

$$\frac{E_{b,0\text{-}2\mu m}}{E_b} \cong 0 \quad \text{and} \quad \frac{E_{b,2\mu m\text{-}\infty}}{E_b} \cong 1$$

$$\text{Emitted energy} = (0.95 \times 0 + 0.1 \times 1) \times E_b$$
$$= 0.1 E_b = 0.1 \times 757.7 = 75.8 \text{ W/m}^2$$

(The emissivity of the surface for a surface temperature of 340 K is 0.1.)

Heat transfer rate to the coolant $=$ Absorbed energy $-$ Emitted energy
$$= 1215.4 - 75.8 = \underline{1139.6 \text{ W/m}^2}$$

By utilizing the specially coated surface, the heat transfer rate to the coolant is increased from 595.3 W/m^2 for a black surface to 1139.6 W/m^2 for the specially coated surface.

(b) If there is no heat transfer to the coolant, from an energy balance on the surface

Emitted energy $=$ Absorbed energy

For the black surface, $E = E_b = \sigma T_s^4$; therefore, $\sigma T_s^4 = 1353$.

$$T_s = \left(\frac{1353}{5.67 \times 10^{-8}}\right)^{1/4} = \underline{393 \text{ K}}$$

For the specially coated surface,

$$\varepsilon_1 E_{b,0\text{-}\lambda_1} + \varepsilon_2 E_{b,\lambda_1\text{-}\infty} = \alpha_1 G_{0\text{-}\lambda_1} + \alpha_2 G_{\lambda_1\text{-}\infty}$$

The right-hand side represents the absorbed energy, which has already been found to be 1215.4 W/m² in part a. The only unknown in the above equation is the surface temperature. For a given surface temperature both $E_{b,0-\lambda_1}$ and $E_{b,0-\lambda_2}$ are fixed and can be found from the f-values in Table 7.1.2 or the software (RF). But as there is no readily available, simple relation between $E_{b,0-\lambda_1}$ and $E_{b,0-\lambda_2}$, and the surface temperature T, the surface temperature is found by iteration. To find the equilibrium temperature of the surface, assume a temperature. From the software RF (or Table 7.1.2) determine the emitted energy. If the emitted energy is less than the absorbed energy, repeat the process with a higher temperature. The trial and error computations are carried out until a surface temperature for which the emitted energy equals the absorbed energy is found. In the solution to the problem, values of $f_{0-\lambda}$ are taken from the software RF and may be slightly different from those found by interpolating the values in Table 7.1.2.

To find the temperature of the surface, we have

$$\left(\varepsilon_1 \frac{E_{b,0-\lambda_1}}{E_b} + \varepsilon_2 \frac{E_{b,\lambda_1-\infty}}{E_b} \right) E_b = 1215.4$$

or

$$(\varepsilon_1 f_{0-\lambda_1} + \varepsilon_2 f_{\lambda_1-\infty})E_b = [\varepsilon_1 f_{0-\lambda_1} + \varepsilon_2(1 - f_{\lambda_1-\infty})] E_b = 1215.4$$

Assume a surface temperature of $T_s = 500$ K.

$$\lambda_1 T = 1000 \ \mu m \ K \qquad f_{0-\lambda_1} = 0.000321 \qquad E_b = 3544 \ W/m^2$$

Emitted energy = $[0.95 \times 0.000321 + (1 - 0.000321)0.1]3544 = 355.4 \ W/m^2$

The emitted energy for a surface temperature of 500 K is only 355.4 W/m², whereas the actual emitted energy is 1215.4 W/m². The assumed temperature of 500 K is too low. For the next trial, assume $T_s = 800$ K. Repeating the computations for the assumed temperature of 800 K, the emitted energy is

$$[0.95 \times 0.01973 + (1 - 0.01973)0.1]23\ 224 = 2712 \ W/m^2$$

The emitted energy in this case is higher than the absorbed energy and the assumed temperature of 800 K is too high. The surface temperature is, therefore, bracketed between 500 K and 800 K. With a few more trials, the correct temperature is found to be 673 K. At this temperature the emitted energy is 1220 W/m². The emitted energy is very close to the absorbed energy (within 0.37%) at this temperature and the equilibrium temperature is very close to 673 K. The surface temperature of 673 K is significantly greater than the temperature of a blackbody (393 K).

COMMENT

The solar absorptivity for the special surface is 0.8983 and the emissivity at 340 K is 0.1. The explanation for the increase in the heat transfer rate to the coolant with the special surface as compared with the black surface lies in this difference between the solar absorptivity and the emissivity. The black surface absorbs all the incident energy but also emits the maximum amount of energy at 340 K. The special surface absorbs 90% of the incident solar energy but emits only 10% of the energy from a black surface at the same temperature.

8.1.3 Reflectivity

Spectral, directional
reflectivity

The concepts developed for the spectral dependence of absorptivity (and emissivity) also apply to reflectivity. When radiant energy with an intensity $I_{\lambda,i}(\theta, \phi)$ reaches a surface, a part of it is reflected. The *spectral, directional reflectivity* at the surface temperature T, $\rho_\lambda(\theta, \phi, T)$, is defined as the ratio of the reflected intensity $I_{\lambda i}(\theta, \phi)|_{\text{reflected}}$ to the incident intensity $I_\lambda(\theta, \phi)$.

$$\rho_\lambda(\theta, \phi, T) = \frac{I_{\lambda,i}(\theta, \phi)\,|_{\text{reflected}}}{I_{\lambda,i}(\theta, \phi)} \tag{8.1.13}$$

In Equation 8.1.13 the term $I_{\lambda,i}(\theta, \phi)$ in the denominator represents the incident intensity and θ and ϕ are the incidence angles. A part of the incident intensity is reflected (Figure 8.1.8) and the numerator $I_{\lambda,i}(\theta, \phi, T|_{\text{reflected}}$ represents that part of $I_{\lambda,i}(\theta, \phi)$ that is reflected (and not the reflected intensity in the direction θ, ϕ).

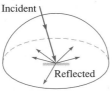

Incident

Reflected

FIGURE 8.1.8 The incident energy on a surface is reflected in many directions.

If the reflected intensity has no preferred direction regardless of the direction of the incident radiation, the reflection is termed *diffuse*. If the angle of incidence equals the angle of reflection, the reflection is *specular*. There are no perfectly diffuse or perfectly specular reflectors but reflection from highly polished, smooth surfaces tends to approach specular reflection; reflection from rough surfaces tend to be diffuse.

8.1.4 Transmissivity

The computation of the results of transmission of radiation through a semitransparent material is involved. The incident intensity is attenuated as it passes through the medium by absorption and scattering in the intervening medium. In addition there is emission at every point in the path of the radiation intensity. The intensity due to emission depends on the local temperature. The emitted intensity from a point in the medium is also attenuated as it traverses through the medium. Thus, the incident radiation intensity is attenuated due to absorption and scattering, but augmented due to emission. The combination of attenuation and augmentation leads to complex

Hemispherical,
spectral
transmissivity

equations. However, in many engineering applications involving incident radiation from a high temperature source or a medium whose temperature can be assumed to be uniform, acceptable results can be obtained by employing total hemispherical transmissivity. The *hemispherical, spectral transmissivity* is defined as

$$\tau_\lambda = \frac{dG_{\text{tr}}}{dG}$$

where

dG = incident radiant energy flux from all directions in the wave band λ to $\lambda + d\lambda$

dG_{tr} = energy transmitted in the same wave band

Total hemispherical transmissivity

The *total hemispherical transmissivity* is defined as $\tau = G_{tr}/G$ and is obtained from

$$\tau = \frac{\int_0^\infty \tau_\lambda G_\lambda d\lambda}{G} \tag{8.1.14}$$

From a radiant energy balance,

$$\alpha_\lambda + \rho_\lambda + \tau_\lambda = 1 \tag{8.1.15a}$$

If the properties are averaged over all wavelengths,

$$\alpha + \rho + \tau = 1 \tag{8.1.15b}$$

If the medium is opaque, $\tau_\lambda = \tau = 0$ and

$$\alpha_\lambda + \rho_\lambda = 1 \tag{8.1.16a}$$

$$\alpha + \rho = 1 \tag{8.1.16b}$$

EXAMPLE 8.1.4

An opaque surface, has an absorptivity of 0.8 in the wave band 0.4 μm < λ < 1 μm and receives solar energy in the outer reaches of the atmosphere. If the area of the surface is 0.01 m², determine

(a) The total rate of incident solar energy when the normal to the surface makes an angle of 30° with the solar beam (Figure 8.1.9a).
(b) The rate of energy absorbed in the wave band 0.4 μm < λ < 1 μm.

(Diameter of the sun = 1.39×10^9 m. Mean distance between the centers of the sun and the earth = 1.5×10^{11} m.)

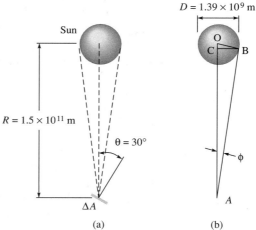

FIGURE 8.1.9 A surface, whose area is small enough to be considered as an infinitesimal area, receives solar insolation.

Given

$0 < \lambda < 1$ μm $\alpha_\lambda = 0.8$ $\Delta A = 0.01$ m²
Diameter of the sun = 1.39×10^9 m
Center-to-center distance between the sun and the earth = 1.5×10^{11} m

Find

(a) Incident solar energy for $\theta = 30°$

(b) Energy absorbed in the wave band $0.4\ \mu m < \lambda < 1\ \mu m$

ASSUMPTIONS

1. Sun behaves as a blackbody at 5760 K.
2. The radiant intensity from the sun is not attenuated.

SOLUTION

(a) From the definition of radiation intensity, the incident energy on the surface is given by

$$\Delta Q \approx I_i\ \Delta A\ \cos\theta\ \Delta\omega$$

where

ΔQ = the incident solar power on the surface

$\Delta\omega$ = solid angle subtended by the sun at dA

I_i = incident intensity = $I_{bi}(5760\ \text{K}) = [E_{bi}(5760\ \text{K})]/\pi = \sigma T_s^4/\pi$

T_s = temperature of the sun = 5760 K

$\theta = 30°$

ΔA = area of the surface = 0.01 m^2

To determine $\Delta\omega$, referring to Figure 8.1.9b,

$$\Delta\omega = \pi\ \frac{BC^2}{AC^2} = \pi\ \tan^2\phi$$

$$= \pi\ \frac{OB^2}{OA^2 - OB^2} = \frac{\pi \times (0.695 \times 10^9)^2}{(1.5 \times 10^{11})^2 - (0.695 \times 10^9)} = 6.744 \times 10^{-5}\ \text{sr}$$

$$\Delta Q = \frac{5.67 \times 10^{-8} \times 5760^4 \times 0.01 \times \cos 30}{\pi} \times 6.744 \times 10^{-5} = \underline{11.6\ \text{W}}$$

(b) The absorbed power in the wave band $0.4\ \mu m < \lambda < 1\ \mu m$ is given by

$$\Delta Q_{abs,\lambda_1-\lambda_2} = \alpha\ \Delta Q_{\lambda_1-\lambda_2} = \alpha\ \left(\frac{\Delta G_{0-\lambda_2}}{\Delta G} - \frac{\Delta G_{0-\lambda_1}}{\Delta G}\right) \Delta Q$$

$$= \alpha\ \left(\frac{I_{b,0-\lambda_2}}{I_b} - \frac{I_{b,0-\lambda_1}}{I_b}\right)_{sun} \Delta Q = \alpha\ \left(\frac{E_{b,0-\lambda_2}}{E_b} - \frac{E_{b,0-\lambda_1}}{E_b}\right)_{sun} \Delta Q$$

$\lambda_1 = 0.4\ \mu m$, $\lambda_1 T = 2304\ \mu m\ K$, and $\lambda_2 = 1\ \mu m$, $\lambda_2 T = 5760\ \mu m\ k$. From the software RF (or Table 7.1.2),

$$\left(\frac{E_{b,0-\lambda_1}}{E_b}\right)_{sun} = 0.1208 \qquad \left(\frac{E_{b,0-\lambda_2}}{E_b}\right)_{sun} = 0.7165$$

$$\Delta Q_{abs,\lambda_1-\lambda_2} = 0.8(0.7165 - 0.1208)11.6 = \underline{5.53\ \text{W}}$$

8.2 VIEW FACTORS

View factor was introduced in Section 7.2.2, and some of the relations involving view factors were presented in Equation 7.2.4. In this section we prove the relations given in Equation 7.2.4 and compute the view factor for one case.

Consider two gray, diffuse surfaces A_1 and A_2 separated by a nonparticipating medium. The radiosities of the surfaces are uniform at J_1 on A_1 and J_2 on A_2. To compute the view factor F_{1-2} consider two infinitesimal areas dA_1 and dA_2. The length of the line joining dA_1 and dA_2 is r. The angles between the normals to dA_1 and dA_2 and the line joining dA_1 and dA_2 are θ_1 and θ_2, respectively. Referring to Figure 8.2.1, if the intensity of radiation leaving surface A_1 is I_1, denoting the radiant power leaving A_1 by Q, the power leaving surface dA_1 and reaching dA_2 is

$$d^2Q = I_1 \, dA_1 \cos \theta_1 \, d\omega \qquad (8.2.1)$$

where $d\omega$ is the infinitesimal solid angle subtended by dA_2 at dA_1.

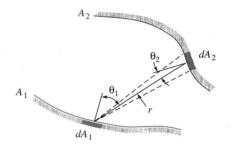

FIGURE 8.2.1 Radiative Energy Exchange Between Two Surfaces

From Equation 7.5.5, $I_1 = J_1/\pi$. The infinitesimal solid angle $d\omega = (dA_2 \cos \theta_2)/r^2$. Substituting these values, we obtain

$$d^2Q = \frac{J_1}{\pi} \, dA_1 \cos \theta_1 \, \frac{dA_2 \cos \theta_2}{r^2} \qquad (8.2.2)$$

Total energy leaving surface dA_1 is $J_1 \, dA_1$. Thus,

$$dF_{dA_1-dA_2} = \frac{\text{Power reaching } dA_2 \text{ from } dA_1}{\text{Total power leaving } dA_1} = \frac{(J_1 \, dA_1 \cos \theta_1 \, dA_2 \cos \theta_2)/\pi r^2}{J_1 \, dA_1}$$

or

$$\left\| \quad dF_{dA_1-dA_2} = \frac{\cos \theta_1 \cos \theta_2 \, dA_2}{\pi r^2} \quad \right\| \qquad (8.2.3)$$

In a similar manner, $dF_{dA_2-dA_1}$ is given by

$$dF_{dA_2-dA_1} = \frac{\cos \theta_1 \cos \theta_2 \, dA_1}{\pi r^2}$$

Reciprocity relation— infinitesimal areas

From the expressions for $dF_{dA_1-dA_2}$ and $dF_{dA_2-dA_1}$ one obtains the reciprocity relation (Equation 7.2.4a) for two infinitesimal areas.

$$\left\| \quad dA_1 \, dF_{dA_1-dA_2} = dA_2 \, dF_{dA_2-dA_1} \quad \right\| \qquad (8.2.4)$$

Only a part of the total power leaving A_1 reaches dA_2. From Equation 8.2.2, the part of the total power leaving A_1 and reaching dA_2 is

$$dQ = \int_{A_1} \frac{J_1 \cos \theta_1 \cos \theta_2 \, dA_1 \, dA_2}{\pi r^2}$$

From the definition of view factor, $dF_{A_1-dA_2}$ is

$$dF_{A_1-dA_2} = \frac{\text{Power received by } dA_2 \text{ from } A_1}{\text{Total power leaving } A_1}$$

The total power leaving surface A_1 is $J_1 A_1$ and $dF_{A_1-dA_2}$ is given by

$$\left\| \quad dF_{A_1-dA_2} = \frac{1}{A_1} \int_{A_1} \frac{\cos \theta_1 \cos \theta_2 \, dA_1 \, dA_2}{\pi r^2} \quad \right\| \qquad \textbf{(8.2.5a)}$$

Starting with infinitesimal area dA_2 the total power leaving dA_2 and reaching A_1 is $\int_{A_1} [(J_2 \cos \theta_1 \cos \theta_2)/(\pi r^2) \, dA_1 \, dA_2]$. The total power leaving surface dA_2 is $J_2 \, dA_2$. From the definition of the view factor

$$F_{dA_2-A_1} = \frac{1}{dA_2} \int_{A_1} \frac{\cos \theta_1 \cos \theta_2 \, dA_1 \, dA_2}{\pi r^2} \qquad \textbf{(8.2.5b)}$$

Reciprocity relation between a finite and an infinitesimal areas

Comparing Equations 8.2.5a and b, we obtain the reciprocity relation between a finite area and an infinitesimal area.

$$\left\| \quad A_1 \, dF_{A_1-dA_2} = dA_2 F_{dA_2-A_1} \quad \right\|$$

Finally, starting with Equation 8.2.2, the power reaching A_2 from A_1 is given by

$$Q = \int_{A_2} \int_{A_1} \frac{J_1 \cos \theta_1 \cos \theta_2 \, dA_1 \, dA_2}{\pi r^2}$$

The total power leaving surface $A_1 = J_1 A_1$. Hence, the view factor $F_{A_1-A_2}$ is obtained from

$$\left\| \quad F_{A_1-A_2} = \frac{1}{A_1} \int_{A_2} \int_{A_1} \frac{\cos \theta_1 \cos \theta_2 \, dA_1 \, dA_2}{\pi r^2} \quad \right\| \qquad \textbf{(8.2.6a)}$$

From the symmetry of the integrand in Equation 8.2.6a (or changing subscripts 1 and 2) we get

Reciprocity relation

$$F_{A_2-A_1} = \frac{1}{A_2} \int_{A_1} \int_{A_2} \frac{\cos \theta_2 \cos \theta_1 \, dA_2 \, dA_1}{\pi r^2} \qquad \textbf{(8.2.6b)}$$

Comparing Equations 8.2.6a and b, we have the reciprocity relation given in Equation 7.2.4a.

$$A_1 F_{A_1-A_2} = A_2 F_{A_2-A_1}$$

In the remaining part of the text, when denoting view factors between two surfaces, the subscript will be shortened to the numeral identifying the area without the subscript A. For example, F_{1-2} will denote the view factor of area A_1 with respect to area A_2.

Equations 7.2.4b and c can be derived from Equation 8.2.6a. To prove Equation 7.2.4b, consider a surface made up of two areas i and j and a second surface k (see Figure 7.2.1b). From Equation 8.2.6a

$$A_{ij}F_{ij\text{-}k} = \int_{A_k}\int_{A_{ij}} \frac{\cos\theta_{ij}\cos\theta_k \, dA_{ij} \, dA_k}{\pi r^2}$$

where

dA_{ij} = an infinitesimal area on ij

θ_{ij} = angle between the normal to the infinitesimal area dA_{ij} and the line joining dA_{ij} and dA_k

r = the distance between dA_{ij} and dA_k

Splitting the integral over A_i and A_j

$$\int_{A_k}\int_{A_{ij}} \frac{\cos\theta_{ij}\cos\theta_k \, dA_{ij} \, dA_k}{\pi r^2} = \int_{A_k}\int_{A_i} \frac{\cos\theta_i\cos\theta_k \, dA_i \, dA_k}{\pi r^2}$$
$$+ \int_{A_k}\int_{A_j} \frac{\cos\theta_j\cos\theta_k \, dA_j \, dA_k}{\pi r^2}$$

From the form of the integrals and the definition of view factors,

$$A_{ij}F_{ij\text{-}k} = A_iF_{i\text{-}k} + A_jF_{j\text{-}k}$$

Law of correspondence corners

Equation 7.2.4c can be derived in a similar manner.

The proof of the law of corresponding corners, Equation 7.2.4d is a little more complex in terms of the algebra involved. The derivation is taken from Love (1968). Consider two rectangular surfaces with a common edge shown in Figure 8.2.2. Referring to Figure 8.2.2, the law of corresponding corners states

$$A_1F_{1\text{-}2} = A_3F_{3\text{-}4}$$

To prove this from Equation 8.2.6, define x_A and y_A on plane A, and x_B and y_B on plane B with origin for both at P as shown in Figure 8.2.2. From Equation 8.2.6a

$$A_1F_{1\text{-}2} = \int_{A_2}\int_{A_1} \frac{\cos\theta_1\cos\theta_2 \, dA_1 \, dA_2}{\pi r^2}$$

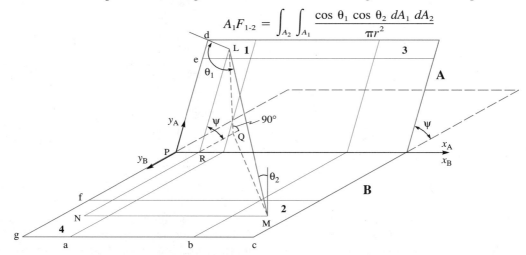

FIGURE 8.2.2 Two Rectangular Surfaces Sharing a Common Side

L and M are two arbitrary points on 1 and 2, respectively. Coordinates of L are $\{x_A, y_A\}$ and those of M are $\{x_B, y_B\}$. *LQ* is perpendicular to plane B extended. The angle $Q\hat{R}L$ is ψ. Thus,

$$r^2 = LM^2 = MQ^2 + LQ^2 = MN^2 + NQ^2 + LQ^2$$
$$= (x_B - x_A)^2 + (y_B + y_A \cos \psi)^2 + (y_A \sin \psi)^2$$

From triangle LMQ, $\cos \theta_2 = LQ/LM = y_A \sin \psi/r$. Similarly, $\cos \theta_1 = y_B \sin \psi/r$. The expression for $A_1 F_{1-2}$ can now be written as

$$A_1 F_{1-2} = \int_{y_B=f}^{g} \int_{x_B=b}^{c} \int_{y_A=e}^{d} \int_{x_A=0}^{a} \frac{y_A y_B \sin^2\psi \, dx_A \, dy_A \, dx_B \, dy_B}{\pi[(x_B - x_A)^2 + (y_B + y_A \cos \psi)^2 + (y_A \sin \psi)^2]}$$

The expression for $A_3 F_{3-4}$ has the same integral except that the limits on x_A and x_B are interchanged. As the integrand is unchanged, if x_A and x_B are interchanged, the integral is also unchanged. This proves the law of corresponding corners for two rectangular surfaces with a common side. In a similar manner it can be shown that the law of corresponding corners is valid for two rectangular surfaces facing each other (Figure 7.2.1e).

EXAMPLE 8.2.1

Determine the view factor between a very small disk of area dA_1 and a second coaxial disk of radius R located at a distance D as shown in Figure 8.2.3.

The view factor is given by Equation 8.2.5b, with subscripts 1 and 2 interchanged (denoting the distance between dA_1 and dA on the circular disk)

$$F_{1-2} \approx \frac{1}{dA_1} \int_{A_2} \frac{\cos \theta_1 \cos \theta_2 \, dA_1 \, dA_2}{\pi x^2} = \int_{A_2} \frac{\cos \theta_1 \cos \theta_2 \, dA_2}{\pi x^2}$$

From Figure 8.2.3, $\theta_1 = \theta_2 = \theta$ $\qquad dA_2 = r \, d\phi \, dr$

$$F_{1-2} = \int_0^{2\pi} \int_0^{R} \frac{\cos^2\theta \, r \, dr \, d\phi}{\pi x^2} = 2 \pi \int_0^{R} \frac{\cos^2\theta \, r \, dr}{\pi x^2}$$

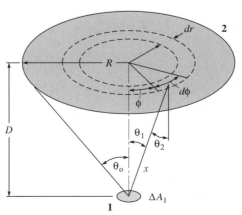

FIGURE 8.2.3 Elemental Disk ΔA_1 Coaxial with Disk A_2

From Figure 8.2.3, $r = D \tan \theta \qquad dr = D \sec^2\theta \; d\theta \qquad x = \dfrac{D}{\cos \theta}$

Substituting these values,

$$F_{1\text{-}2} = 2 \int_0^{\theta_0} \sin \theta \cos \theta \; d\theta = \sin^2\theta_0 = \underline{\dfrac{R^2}{(D^2 + R^2)}}$$

8.2.1 View Factors for Infinite Strips

The determination of view factors between two surfaces generally involves the evaluation of the integral given in Equation 8.2.6a. Some closed form solutions are available and a few of them are given in Table 7.2.1. For cases with no closed form solutions, the integrals are evaluated numerically. But, for two infinite surfaces generated by two lines (straight or curved) moving along two parallel lines, there is a simple way to compute the view factors. This method was developed by Hottel (1954) and is known as the method of *crossed and uncrossed strings*.

To understand how the method of crossed and uncrossed strings works, first consider an enclosure with three infinite surfaces, each one being plane or convex towards the enclosure. The three surfaces are generated by lines AB (curved), BC, and AC moving along parallel lines FG, HI, and DE, respectively (Figure 8.2.4).

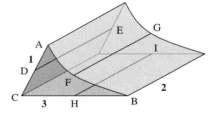

FIGURE 8.2.4 A Three-Surface Enclosure Consisting of Three Infinite, Parallel, Plane or Convex Surfaces

For each surface, $F_{i\text{-}i} = 0$ (i = 1, 2, 3). Applying the summation rule (Equation 7.2.4e) to each surface,

$$A_1 F_{1\text{-}2} + A_1 F_{1\text{-}3} = A_1$$
$$A_2 F_{2\text{-}1} + A_2 F_{2\text{-}3} = A_2 \qquad \text{(8.2.7)}$$
$$A_3 F_{3\text{-}1} + A_3 F_{3\text{-}2} = A_3$$

Employing the reciprocity relation for view factors, Equation 8.2.7 is recast as

$$A_1 F_{1\text{-}2} + A_1 F_{1\text{-}3} = A_1$$
$$A_1 F_{1\text{-}2} + A_2 F_{2\text{-}3} = A_2 \qquad \text{(8.2.8)}$$
$$A_1 F_{1\text{-}3} + A_2 F_{2\text{-}3} = A_3$$

We now have three equations in three unknowns, $F_{1\text{-}2}$, $F_{1\text{-}3}$, and $F_{2\text{-}3}$ (it is assumed that the areas of the three surfaces are known). Solving for $F_{1\text{-}2}$

$$F_{1\text{-}2} = \dfrac{A_1 + A_2 - A_3}{2A_1} \qquad \text{(8.2.9)}$$

Equation 8.2.9 is applied to find the view factor between two infinite strips, 1 and 2, as shown in Figure 8.2.5a. Dashed lines AED and BFC represent taut, uncrossed strings (they do not cross each other) joining the end points A and B on 1 and D and C on 2, respectively. AC and BD represent taut, crossed strings (they cross each other) joining the end points A to C and B to D. From repeated application of

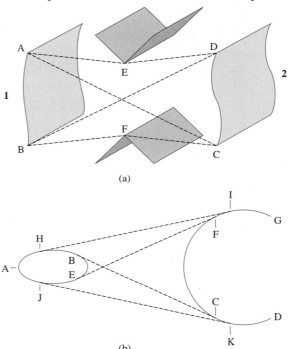

(a)

(b)

FIGURE 8.2.5 Illustration of Crossed and Uncrossed Strings
(a) Labels 1 and 2 represent two infinite, parallel strips with two other surfaces obstructing clear views between 1 and 2. (b) Crossed and uncrossed strings for an elliptical and partial cylindrical surface.

Equation 8.2.9 and the view factor relations given by Equation 7.2.4, it can be shown that

$$A_1 F_{1\text{-}2} = \frac{A_{\text{BD}} + A_{\text{AC}} - A_{\text{AED}} - A_{\text{BFC}}}{2} \qquad \textbf{(8.2.10)}$$

As the lengths of all the surfaces perpendicular to the plane of the paper are equal, and areas are proportional to the lengths of the lines generating the surfaces,

$$\overline{\text{AB}} F_{1\text{-}2} = \frac{\overline{\text{BD}} + \overline{\text{AC}} - \overline{\text{AED}} - \overline{\text{BFC}}}{2}$$

where $\overline{\text{AB}}$ represents the length along the arc AB on surface 1 and the terms on the right-hand side represent the lengths of the strings BD, AC, AED, and BFC. Thus,

$$\overline{AB}F_{1\text{-}2} = \frac{\begin{array}{c}\text{Sum the lengths of} \quad \text{Sum of the lengths of} \\ \text{the crossed strings} \; - \; \text{the uncrossed strings}\end{array}}{2} \qquad \textbf{(8.2.11)}$$

In the sketch each dashed line represents a string with one end of the string held at one of the end points of one surface and the other end of the string held against an end point on the second surface.

All the strings lie in the plane of a cross section and are held taut between the extremities of the surfaces. For example, to find the view factor between the elliptical surface and the partial cylindrical surface shown in Figure 8.2.5b, the crossed strings are ABCD and AEFG and the uncrossed strings are AHIG and AJKD.

EXAMPLE 8.2.2 The cross section of a rectangular duct is 16-cm wide and 12-cm high (Figure 8.2.6). Its length is much greater than the width or the height. Compute the view factors between a pair of opposite surfaces and a pair of adjacent surfaces.

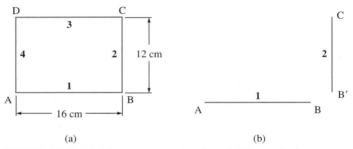

(a)   (b)

FIGURE 8.2.6 (a) A long, rectangular duct. (b) Two adjacent surfaces slightly displaced to clarify the application of the method.

ASSUMPTION

As the length is much greater than the width or the height, the duct is treated as infinitely long.

SOLUTION

$F_{1\text{-}3}$ (View Factor Between a Pair of Opposite Surfaces) From Equation 8.2.11,

$$F_{1\text{-}3} = \frac{\overline{AC} + \overline{BD} - \overline{AD} - \overline{BC}}{2\overline{AB}} = \frac{\overline{AC} - \overline{AD}}{\overline{AB}} = \frac{(12^2 + 16^2)^{1/2} - 12}{16} = \underline{0.5}$$

$F_{1\text{-}2}$ (View Factor Between a Pair of Adjacent Surfaces) To clarify the application of Equation 8.2.11, separate surfaces AB and BC as shown in Figure 8.2.6b. Apply Equation 8.2.11. The two surfaces are then brought together to form a pair of adjacent surfaces.

$$F_{1\text{-}2} = \frac{\overline{AB'} + \overline{BC} - \overline{AC} - \overline{BB'}}{2\,\overline{AB}}$$

As B′ is moved to B, with $\overline{BB'} = 0$, and $\overline{AB'} \rightarrow \overline{AB}$, we get

$$F_{1\text{-}2} = \frac{\overline{AB} + \overline{BC} - \overline{AC}}{2\,\overline{AB}} = \frac{16 + 12 - 20}{2 \times 16} = \underline{0.25}$$

Check:

$$F_{1\text{-}2} + F_{1\text{-}3} + F_{1\text{-}4} = 1 \qquad F_{1\text{-}2} = F_{1\text{-}4} \quad \text{and} \quad F_{1\text{-}2} = 0.5 \times (1 - F_{1\text{-}3}) = 0.25$$

EXAMPLE 8.2.3

A very long 4-cm diameter cylindrical surface, 1, is placed coaxially in front of a 10-cm diameter semicircular surface, 2 (Figure 8.2.7). Denoting the surroundings as 3, determine $F_{2\text{-}1}$, $F_{2\text{-}3}$, and $F_{2\text{-}2}$.

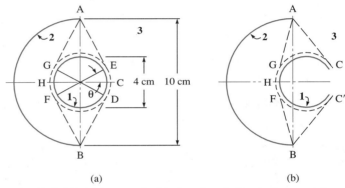

FIGURE 8.2.7 (a) Long, cylindrical surface in front of a coaxial, semicircular surface. (b) An incomplete circular surface whose axis is displaced from that of the semi-circular surface to clarify the method.

Given

$d_1 = 4$ cm $d_2 = 10$ cm

Find

$F_{2\text{-}1}$, $F_{2\text{-}3}$, and $F_{2\text{-}2}$

SOLUTION

Figure 8.2.7a depicts the configuration. To clarify the notion of crossed and uncrossed strings, the figure 8.2.7b shows an incomplete circular surface 1, whose axis is displaced from the axis of surface 2. As the cylindrical surface is completed and its axis is made to coincide with the axis of surface 2, we get the configuration shown at left. The uncrossed strings for the partial cylindrical surface (Figure 8.2.7b) are $\overline{AC}$ and $\overline{BC'}$. The crossed strings are $\overline{AGHC'}$ and $\overline{BFHC}$. As the cylindrical surface is completed and the two axes made to coincide, C and C′ collapse into a single point and the resulting configuration is a complete cylindrical surface coaxial with the semicylindrical surface as shown in Figure 8.2.7a. In Figure 8.2.7a, the uncrossed strings are $\overline{AEC}$ and $\overline{BDC}$. The crossed strings are $\overline{AGHFDC}$ and $\overline{BFHGEC}$.

Determine $F_{2\text{-}1}$ Representing the arc length AB by L_2, from Equation 8.2.11,

$$L_2 F_{2\text{-}1} = \frac{\overline{AGHFDC} + \overline{BFHGEC} - \overline{AEC} - \overline{BDC}}{2} \qquad (1)$$

The sum of the crossed and uncrossed strings in Equation 1 are given by

$$\overline{AGHFDC} + \overline{BFHGEC} = \overline{AG} + \overline{GH} + \overline{HFDC} + \overline{BF} + \overline{FH} + \overline{HGEC}$$

$$= 2\,\overline{AG} + 2\,\overline{GH} + 2\,\overline{HFDC}$$

$$\overline{AEC} + \overline{BDC} = \overline{AE} + \overline{EC} + \overline{BD} + \overline{DC} = 2\,\overline{AE} + 2\,\overline{EC}$$

$$= 2\,\overline{AG} + 2\,\overline{GH}$$

Hence, the right-hand side of Equation 1 is given by

$$\frac{\overline{AGHFDC} + \overline{BFHGEC} - \overline{AEC} - \overline{BDC}}{2} = \overline{HFDC} = \pi R_1$$

where R_1 is the radius of surface 1. $L_2 = \pi R_2$. Hence,

$$F_{2\text{-}1} = R_1/R_2 = 4/10 = \underline{0.4}$$

Determine $F_{2\text{-}3}$ To determine $F_{2\text{-}3}$, denote $\overline{AECDB}$ by 4. Then, $F_{2\text{-}3} = F_{2\text{-}4}$. By the summation rule

$$F_{4\text{-}1} + F_{4\text{-}2} = 1$$

By symmetry, $F_{1\text{-}4} = 0.5$. By the reciprocity rule

$$A_1 F_{1\text{-}4} = A_{4\text{-}1} F_{4\text{-}1} \qquad F_{4\text{-}1} = \frac{A_1}{A_4} F_{1\text{-}4} = \frac{L_1}{L_4} F_{1\text{-}4}$$

To evaluate L_4, we have

$$L_4 = 2(\overline{AE} + \overline{EC}) = 2(R_2 \cos\theta + R_1\theta)$$

$$F_{4\text{-}1} = \frac{2\pi R_1}{2(R_2 \cos\theta + R_1\theta)} F_{1\text{-}4}$$

$$\sin\theta = R_1/R_2 = 0.4 \qquad \theta = 0.4115 \text{ rad}$$

$$F_{4\text{-}1} = \frac{\pi \times 2}{5 \cos(0.4115) + 2 \times 0.4115} \times 0.5 = 0.5812$$

$$F_{4\text{-}2} = 1 - F_{4\text{-}1} = 1 - 0.5812 = 0.4188$$

$$F_{2\text{-}3} = F_{2\text{-}4} = \frac{A_4 F_{4\text{-}2}}{A_2} = \frac{L_4 F_{4\text{-}2}}{L_2} = \frac{2[5 \cos(0.4115) + 2 \times 0.4115]}{\pi \times 5} \times 0.4188$$

$$= \underline{0.2882}$$

$F_{2\text{-}2}$ Applying the summation rule $F_{2\text{-}1} + F_{2\text{-}2} + F_{2\text{-}3} = 1$. Therefore,

$$F_{2\text{-}2} = 1 - F_{2\text{-}1} - F_{2\text{-}3} = 1 - 0.4 - 0.2882 = \underline{0.3118}$$

View factors have been computed for a very large number of configurations. Algebraic expressions for view factors for many configurations and an extensive bibliography for view factors are available in Siegel and Howell (1992). A few of those expressions are given in Chapter 7.

Equation 8.2.6a can be applied to any two surfaces to find the view factor of one surface with respect to the other. The view factor has no parameters relating to radiation heat transfer and can be evaluated from purely mathematical principles. The concept of view factors is related to the radiant power reaching one surface from another. Equation 8.2.6a was arrived at as representing the ratio of the power reaching surface 2 from surface 1 to the power leaving surface 1 in all directions with the assumption that the radiosity of surface 1 is uniform. Thus, though the view factors can be evaluated without any reference to radiation, their use with a specific surface implies that the surface is diffuse and that the radiosity of the surface is uniform. We should, therefore, ensure that such an assumption of uniform radiosity of a surface is reasonably satisfied.

8.3 RADIATIVE HEAT TRANSFER AMONG GRAY, DIFFUSE, AND OPAQUE SURFACES IN AN ENCLOSURE

In Chapter 7 we studied radiative heat transfer among three surfaces forming an enclosure. Now, we consider radiative heat transfer among a number of surfaces forming an enclosure. Two methods are presented. The methods are restricted to enclosures satisfying four basic assumptions.

Radiative heat transfer among gray, diffuse, and opaque surfaces—

1. Each surface is at a uniform but not necessarily known temperature. If the temperature of a surface is not substantially uniform, it is divided into a number of smaller surfaces, each of which may be considered to be at a uniform temperature.
2. All surfaces are gray and opaque, i.e., the absorptivity and emissivity are independent of wavelength. Specifically $\alpha = \varepsilon$ and $\rho = 1 - \alpha$.
3. The surfaces are diffuse. The reflected energy has no preferred direction whatever be the direction of the incident energy. Similarly, the emitted energy has no preferred direction. With this assumption of diffuse surfaces, it is not possible to differentiate between the reflected and emitted energy. In most computations it is this sum of the reflected and emitted energy—radiosity—that is solved for.
4. Radiosity of each surface is uniform. In delineating the radiation surfaces it is important to closely approximate the condition of uniform radiosity. Recall that uniformity of temperature of a surface does not guarantee uniform radiosity. Uniformity of radiosity requires that the incident energy be uniform in addition to the surface temperature and emissivity being uniform.

In the solution of radiative heat transfer among gray surfaces forming a complete enclosure, either the temperature of or the heat transfer rate from each surface is known. We then wish to determine the heat transfer rate from those surfaces whose temperatures are known and the temperatures of those surfaces with known heat transfer rates. In addition to the surfaces forming the enclosure there may be surfaces within the enclosure. The two methods considered are the electrical network method, which has already been introduced in Chapter 7, and the radiosity method. The equivalence of the two methods is proved in Section 8.3.3.

8.3.1 Electrical Network Method

The electrical analogy [due to Oppenheim (1956)] has already been introduced in Section 7.3. The bases of this method are Equations 7.3.3 and 7.3.4. Consider an enclosure with N surfaces; the surfaces have uniform but different radiosities. The view factors for every pair of surfaces are known.

$$\left\| \quad q_i = \sum_{k=1}^{N} A_i F_{i\text{-}k}(J_i - J_k) \quad \right\| \tag{7.3.3}$$

$$\left\| \quad q_i = \frac{E_{bi} - J_i}{(1 - \varepsilon_i)/A_i \varepsilon_i} \quad \right\| \tag{7.3.4}$$

Electrical analogy and network

Combining Equations 7.3.3 and 7.3.4,

$$\left\| \quad \frac{E_{bi} - J_i}{(1 - \varepsilon_i)/A_i \varepsilon_i} + \sum_{k=1}^{N} \frac{J_k - J_i}{1/(A_i F_{i\text{-}k})} = 0 \quad \right\| \tag{8.3.1}$$

and

$$\left\| \quad q_i + \sum_{k=1}^{N} \frac{J_k - J_i}{1/(A_i F_{i\text{-}k})} = 0 \quad \right\| \tag{8.3.2a}$$

For a black surface

$$q_i - \sum_{k=1}^{N} \frac{E_{bi} - J_k}{1/(A_i F_{i\text{-}k})} = 0 \tag{8.3.2b}$$

Equations 8.3.1 and 8.3.2 are similar to those in an electrical resistance network. Equation 8.3.1 is equivalent to one of Kirchhoff's Laws: the algebraic sum of the currents to a node is zero. It also represents the energy balance on a surface: the algebraic sum of the heat transfer rates to a surface is zero. *Note that from Equation 8.3.2a, the radiosity of a surface with a specified heat transfer rate is independent of its emissivity.*

For surfaces with known temperatures, Equation 8.3.1 applies; for surfaces with known heat transfer rates, Equation 8.3.2a applies. Thus, by employing Equation 8.3.1 or Equation 8.3.2a as appropriate, we have N linear equations involving N unknown radiosities. The solution to the system of N simultaneous equations will yield the radiosities of each surface. From Equation 7.3.4, the unknown heat transfer rate from or the temperature of (emissive power) of each surface is found.

Equation 8.3.1 cannot be used for a black surface. If the temperature of a black surface is known, its radiosity, being equal to its emissive power, is also known and Equation 8.3.2b is applied. When applying Equation 8.3.1 or 8.3.2a for other surfaces, the emissive power of the black surface is substituted for its radiosity in the term $(J_i - J_k)/[1/(A_i F_{i\text{-}k})]$.

EXAMPLE 8.3.1

The long square duct in Figure 8.3.1 has three surfaces, 1, 2, and 3, maintained at uniform temperatures. Surface 4 is subjected to a uniform heat flux. Determine the net radiative heat fluxes from surfaces 1, 2, and 3 and temperature of surface 4. All surfaces are gray and diffuse.

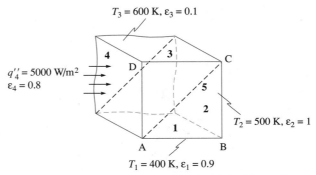

FIGURE 8.3.1 A Long Square Duct with Uniform (but Different) Temperatures of Three Surfaces and a Uniform Heat Flux on the Fourth Surface

Given

Long, square duct

$T_1 = 400$ K $\quad T_2 = 500$ K $\quad T_3 = 600$ K $\quad q_4'' = 5000$ W/m^2

$\varepsilon_1 = 0.9 \qquad\quad \varepsilon_2 = 1 \qquad\qquad \varepsilon_3 = 0.1 \qquad\quad \varepsilon_4 = 0.8$

Find

q_1'', q_2'', q_3'', and T_4

ASSUMPTIONS

1. The duct is a four-surface enclosure, i.e., the radiosity of each surface is uniform.
2. Medium in the duct does not participate in radiation
3. The four surfaces form a complete enclosure, i.e., the duct is very long and the radiant energy reaching the end surfaces is very small compared with the energy reaching each of the other surfaces forming the enclosure.
4. Negligible convection

SOLUTION

To solve for the radiosities, we need the view factors. With the assumption that the four surfaces form a complete enclosure, the various view factors are found either from basic principles relating to view factors, from Figure 7.2.9 for adjacent sides and from Figure 7.2.7 for opposite sides, by Hottel's method (Section 8.2.1), or by using the equations in Table 7.2.1 (or from the software). Here, the view factors are determined from the basic definition of view factor and the relations given in Equation 7.2.4.

Determine $F_{1\text{-}2}$ To find $F_{1\text{-}2}$, introduce the surface AC and consider the duct ABC.

Denoting the surface AC by 5, by symmetry

$$F_{5\text{-}1} = F_{5\text{-}2} = 1/2$$

From reciprocity relation, Equation 7.2.4a, $A_1 F_{1\text{-}5} = A_5 F_{5\text{-}1}$. Therefore,

$$F_{1\text{-}5} = \frac{A_5}{A_1} F_{5\text{-}1} = \frac{\sqrt{2}}{2} = \frac{1}{\sqrt{2}}$$

As surfaces 1, 2, and 5 form an enclosure, from Equation 7.2.4e (with $F_{1-1} = 0$)

$$F_{1-2} = 1 - F_{1-5} = 1 - \frac{1}{\sqrt{2}} = 0.2929$$

From symmetry, $F_{1-4} = F_{1-2} = 0.2929$

For the four-surface enclosure, from Equation 7.2.4e, with $F_{1-1} = 0$,

$$F_{1-3} = 1 - F_{1-2} - F_{1-4} = 1 - 2 \times 0.2929 = 0.4142$$

Since the duct is square, the view factor between any pair of adjacent surfaces is 0.2929 and between any pair of opposite surfaces it is 0.4142. Thus, all the view factors for the enclosure are known. We now proceed to apply Equation 8.3.1 or 8.3.2 to each surface.

Surface 1   Gray surface ($\varepsilon \neq 1$), known temperature; apply Equation 8.3.1.

$$\frac{E_{b1} - J_1}{(1 - \varepsilon_1)/A_1\varepsilon_1} + \frac{J_2 - J_1}{(1/A_1 F_{1-2})} + \frac{J_3 - J_1}{(1/A_1 F_{1-3})} + \frac{J_4 - J_1}{(1/A_1 F_{1-4})} = 0 \tag{1}$$

Surface 2   Black ($\varepsilon_2 = 1$), known temperature.

$$J_2 = E_{b2} \tag{2}$$

Surface 3   Gray surface ($\varepsilon_3 \neq 1$), known temperature; apply Equation 8.3.1.

$$\frac{E_{b3} - J_3}{(1 - \varepsilon_3)/A_3\varepsilon_3} + \frac{J_1 - J_3}{(1/A_3 F_{3-1})} + \frac{J_2 - J_3}{(1/A_3 F_{3-2})} + \frac{J_4 - J_3}{(1/A_3 F_{3-4})} = 0 \tag{3}$$

Surface 4   Known heat transfer rate; employ Equation 8.3.2a.

$$q_4 + \frac{J_1 - J_4}{(1/A_4 F_{4-1})} + \frac{J_2 - J_4}{(1/A_4 F_{4-2})} + \frac{J_3 - J_4}{(1/A_4 F_{4-3})} = 0 \tag{4}$$

As $A_1 = A_2 = A_3 = A_4 = A$, dividing each equation by A and substituting E_{b2} for J_2 for black surface 2, we obtain

$$\frac{E_{b1} - J_1}{(1 - \varepsilon_1)/\varepsilon_1} + \frac{E_{b2} - J_1}{1/F_{1-2}} + \frac{J_3 - J_1}{1/F_{1-3}} + \frac{J_4 - J_1}{1/F_{1-4}} = 0 \tag{5}$$

$$\frac{E_{b3} - J_3}{(1 - \varepsilon_3)/\varepsilon_3} + \frac{J_1 - J_3}{1/F_{3-1}} + \frac{E_{b2} - J_3}{1/F_{3-2}} + \frac{J_4 - J_3}{1/F_{3-4}} = 0 \tag{6}$$

$$q_4'' + \frac{J_1 - J_4}{1/F_{4-1}} + \frac{E_{b2} - J_4}{1/F_{4-2}} + \frac{J_3 - J_4}{1/F_{4-3}} = 0 \tag{7}$$

In Equations 5, 6, and 7,

$$E_{b1} = \sigma T_1^4 = 5.67 \times 10^{-8} \times 400^4 = 1451.5 \text{ W/m}^2$$

$$E_{b2} = \sigma T_2^4 = 5.67 \times 10^{-8} \times 500^4 = 3543.8 \text{ W/m}^2$$

$$E_{b3} = \sigma T_3^4 = 5.67 \times 10^{-8} \times 600^4 = 7348.3 \text{ W/m}^2$$

$$q_4'' = 5000 \text{ W/m}^2$$

Substituting the values and rearranging the equations,

$$10 J_1 - 0.4142 J_3 - 0.2929 J_4 = 14\,101 \qquad \textbf{(8)}$$

$$- 0.4142 J_1 + 1.111 J_3 - 0.2929 J_4 = 1854 \qquad \textbf{(9)}$$

$$- 0.2929 J_1 - 0.2929 J_3 + J_4 = 6468 \qquad \textbf{(10)}$$

Solving Equations 8, 9, and 10 for J_1, J_3, and J_4,

$$J_1 = 1843.1 \text{ W/m}^2 \qquad J_3 = 4555.2 \text{ W/m}^2 \qquad J_4 = 8342.1 \text{ W/m}^2$$

The heat transfer rates from surfaces whose temperatures are known are now determined through Equation 7.3.3 for any surface or Equation 7.3.4 for nonblack surfaces.

Surface 1 From Equation 7.3.4,

$$q_1'' = \frac{\varepsilon_1}{1 - \varepsilon_1} (E_{b1} - J_1) = \frac{0.9}{0.1} (1451.5 - 1843.1) = \underline{-3524 \text{ W/m}^2}$$

Surface 2 As surface 2 is black, apply Equation 7.3.3, to find q_2''.

$$q_2'' = F_{2\text{-}1}(E_{b2} - J_1) + F_{2\text{-}3}(E_{b2} - J_3) + F_{2\text{-}4}(E_{b2} - J_4)$$

$$= 0.2929(3543.8 - 1843.1) + 0.2929(3543.8 - 4555.2)$$

$$+ 0.4142(3543.8 - 8342.1)$$

$$= \underline{-1786 \text{ W/m}^2}$$

Surface 3 From Equation 7.3.4,

$$q_3'' = \frac{\varepsilon_3}{1 - \varepsilon_3} (E_{b3} - J_3) = \frac{0.1}{0.9} (7348.3 - 4555.2) = \underline{310.3 \text{ W/m}^2}$$

Surface 4 The heat transfer rate is already known and its temperature is to be found. From Equation 7.3.4,

$$E_{b4} - J_4 = \frac{1 - \varepsilon_4}{\varepsilon_4} q_4'' = \frac{0.2}{0.8} \times 5000 = 1250 \text{ W/m}^2$$

$$E_{b4} = 8432.1 + 1250 = 9592.1 \text{ W/m}^2 = \sigma T_4^4 \qquad T_4 = \underline{641.3 \text{ K}}$$

COMMENTS

1. A positive value of q_i indicates that the net radiative heat transfer is from the surface to the other surfaces of the enclosure. A negative value means that the net radiant heat transfer is to the surface.

2. If the emissivity of surface 4 is changed from 0.8 to a lower value of 0.4, keeping the net heat transfer rate per unit area at the same value (5000 W/m²), its radiosity is not changed. Its surface temperature is increased from 641.3 K to 727 K.

3. The algebraic sum of the net radiative heat transfer rates from all the surfaces must be zero. In this case, as the areas of all the surfaces are equal, the algebraic sum of the heat fluxes must be zero. Substituting the values of the heat flux,

$$\sum_{i=1}^{4} q_i'' = -3524 - 1786 + 310.3 + 5000 = -0.3 \text{ W/m}^2$$

The nonzero value is the result of round-off errors in the various stages of computations.

8.3.2 Radiosity Method

In the radiosity method, the method of setting up the equations for the unknown radiosities is different from the electrical network method. Once the radiosities are known, the heat transfer rates and temperatures of surfaces are found in the same manner as in the electrical network method. In both the electrical network method and the radiosity method, the radiosities are determined. The method of determining the radiosities in the two methods is different, however.

Consider an N-surface enclosure satisfying all the conditions enumerated in the introduction to Section 8.3. In addition to the surfaces forming the enclosure there are two surfaces, labelled 6 and 7, within the enclosure (Figure 8.3.2). As in the

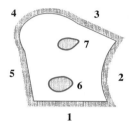

FIGURE 8.3.2 An Enclosure with N Gray, Diffuse, and Opaque Surfaces, Each with a Uniform Radiosity

electrical network method, it is assumed that the view factors between every pair of surfaces have been determined. Consider two surfaces represented by i and k. For surface i,

$$J_i = \varepsilon_i E_{bi} + \rho_i G_i$$

where G_i is the incident radiant energy flux on surface i from all the surfaces forming the enclosure. To find G_i, consider surface k.

Rate of energy leaving surface k and reaching i $= A_k F_{k-i} J_k$
From the reciprocity relation, Equation 7.2.4a, $A_k F_{k-i} = A_i F_{i-k}$
Therefore, the rate of energy leaving surface k and reaching i $= A_i F_{i-k} J_k$

Total rate of energy reaching surface i from all the surfaces of the enclosure is $\sum_{k=1}^{N} A_i F_{i-k} J_k$. In this expression, the summation includes the term $A_i F_{i-i} J_i$ (zero if surface i is plane or convex toward the enclosure).

Rate of radiant energy reaching surface i per unit area of the surface A_i is

$$G_i = \sum_{k=1}^{N} F_{i-k} J_k \tag{8.3.3}$$

Hence,

$$J_i = \varepsilon_i E_{bi} + \rho_i \sum_{k=1}^{N} F_{i-k} J_k \tag{8.3.4a}$$

or

$$\left\| \; (1 - \rho_i F_{i-i}) J_i - \rho_i \sum_{\substack{k=1 \\ k \neq i}}^{N} F_{i-k} J_k = \varepsilon_i E_{bi} \; \right\| \tag{8.3.4b}$$

For example, the application of Equation 8.3.4b for surface 4 (with known temperature) for the seven-surface enclosure is

$$(1 - \rho_4 F_{4-4})J_4 - \rho_4(F_{4-1}J_1 + F_{4-2}J_2 + F_{4-3}J_3 + F_{4-5}J_5 + F_{4-6}J_6 + F_{4-7}J_7) = \varepsilon_4 E_{b4}$$

From an energy balance on surface i,

$$q_i = A_i J_i - A_i G_i \quad \text{or} \quad \frac{q_i}{A_i} = J_i - G_i$$

Employing Equation 8.3.3 for G_i,

$$J_i - \sum_{k=1}^{N} F_{i-k}J_k = \frac{q_i}{A_i}$$

or

$$\left\| (1 - F_{i-i})J_i - \sum_{\substack{k=1 \\ k \neq i}}^{N} F_{i-k}J_k = \frac{q_i}{A_i} \right\| \tag{8.3.5}$$

Equation 8.3.4b for a surface with known temperature or Equation 8.3.5 for a surface with known heat transfer rate, is written for each surface. There are, then, N simultaneous, linear equations in the N unknown radiosities. Solution to these equations by any of the established methods yields the radiosity of each surface. As in the electrical network method, once the radiosities are determined, the heat transfer rate or the temperature of a surface is found from Equations 7.3.1 (or 7.3.4) or 7.3.3. If the temperature of a black surface is known, its radiosity, being equal to the black-body emissive power, is also known. In all equations the emissive power is substituted for the radiosity for a black surface. Also, for a black surface, Equation 8.3.4a (or b) is not required. If the temperature of a black surface is specified, Equation 8.3.5 for that surface is required only for computing the heat transfer rate from that surface; if the heat transfer rate from a black surface is specified, Equation 8.3.5 for that surface is required to compute the radiosities of other surfaces.

Both the electrical network method and the radiosity method are simple and require the solutions of N simultaneous, linear equations. The electrical network method may be simulated by an electrical circuit. The radiosities of the surfaces may be found by measuring the voltage at the appropriate points. By varying the internal resistances of the batteries, $(1 - \varepsilon_i)/A_i\varepsilon_i$, the effect of changing the emissivities of the different surfaces can be found. The radiosity method is a little easier to program on a digital computer. The HT software provided includes the program ''Simultaneous Equations,'' which can solve equations with up to 15 unknowns. With more than four surfaces, entering the values of the coefficients (more than 30) becomes tedious.

EXAMPLE 8.3.2

Rework Example 8.3.1 by the radiosity method.

ASSUMPTIONS

The same as for Example 8.3.1 are valid. The view factors determined in that example will be used in the solution.

SOLUTION

Surface 1   With the temperature known, employing Equation 8.3.4b, with $F_{1-1} = 0$,

$$J_1 - (1 - \varepsilon_1)(F_{1-2}J_2 + F_{1-3}J_3 + F_{1-4}J_4) = \varepsilon_1 E_{b1}$$

Substituting $\varepsilon_1 = 0.9$ and the view factors,

$$J_1 - 0.1(0.2929\, J_2 + 0.4141\, J_3 + 0.2929\, J_4) = 1306.4 \tag{1}$$

Surface 2   Surface 2 is black and its temperature = 500 K,

$$J_2 = E_{b2} = 3543.8 \text{ W/m}^2$$

Surface 3   The temperature is known; employ Equation 8.3.4b. With $F_{3-3} = 0$ and $\varepsilon_3 = 0.1$,

$$J_3 - 0.9(0.4142\, J_1 + 0.2929\, J_2 + 0.2929\, J_4) = 734.8 \tag{2}$$

Surface 4   Heat flux $q_4/A_4 = 5000$ W/m^2; employ Equation 8.3.5. With $F_{4-4} = 0$,

$$J_4 - (0.2929\, J_1 + 0.4142\, J_2 + 0.2929\, J_3) = 5000 \tag{3}$$

Substituting $J_2 = E_{b2} = 3543.8$ W/m^2 in Equations 1, 2, and 3, simplifying, and rearranging, we obtain

$$J_1 - 0.04142\, J_3 - 0.02929\, J_4 = 1410.2 \tag{4}$$

$$-0.37278\, J_1 + J_3 - 0.26361\, J_4 = 1669 \tag{5}$$

$$-0.2929\, J_1 - 0.2929\, J_3 + J_4 = 6467.8 \tag{6}$$

Solving for the radiosities,

$$J_1 = 1843.2 \text{ W/m}^2 \qquad J_3 = 4555.1 \text{ W/m}^2 \qquad J_4 = 8341.9 \text{ W/m}^2$$

These values are the same values (within round-off approximations) obtained by the electrical network method in Example 8.3.1. Employing Equation 7.3.4 for surfaces 1, 3, and 4, we get

$$q_1'' = \frac{0.9}{0.1}\,(1451.5 - 1843.2) = \underline{-3525 \text{ W/m}^2}$$

$$q_3'' = \frac{0.1}{0.9}\,(7348.3 - 4555.1) = \underline{310.3 \text{ W/m}^2}$$

From Equation 7.3.1, $\mathrm{T}_4 = \underline{641.3 \text{ K}}$

Employing Equation 8.3.5 for the black surface 2,

$$q_2'' = J_2 - (F_{2-1}J_1 + F_{2-3}J_3 + F_{2-4}J_4)$$

$$= 3543.8 - (0.2929 \times 1843.2 + 0.2929 \times 4555.1 + 0.4142 \times 8341.9)$$

$$= \underline{-1785 \text{ W/m}^2}$$

EXAMPLE 8.3.3 Figure 8.3.3 shows a hollow cylinder, 6-cm in diameter and 12-cm long. The cylindrical surface is maintained at 500 °C, one end surface at 300 °C, and the other end surface at 100 °C. Determine the radiative heat transfer rate from each surface if the emissivities of every surface is 0.8.

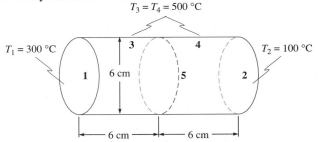

FIGURE 8.3.3 A Hollow Cylinder with Two End Surfaces and the Cylindrical Surface at Uniform Temperatures

Given

Diameter of cylinder $= 6$ cm $T_2 = 100\,°C$

Length of cylinder $= 12$ cm $T_3 = T_4 = 500\,°C$

$T_1 = 300\,°C$ $\varepsilon_1 = \varepsilon_2 = \varepsilon_3 = \varepsilon_4 = 0.8$

Find

$q_1, q_2, q_3,$ and q_4

ASSUMPTIONS

1. Heat transfer is from the inner surfaces only.
2. Medium inside the cylinder does not participate in radiation.
3. All surfaces are diffuse, gray.

SOLUTION

This problem has been solved as a three-surface enclosure in Example 7.3.5 (part a) with the end surfaces considered as two surfaces and the cylindrical surface as the third. To ascertain if the assumption of uniform radiosity of the cylindrical surface is appropriate, we re-solve the problem as a four-surface enclosure, considering the end surfaces as surfaces 1 and 2 and the cylindrical surface as two surfaces, 3 and 4, as shown in the figure. The effect of nonuniformity of the radiosity of the cylindrical surface is demonstrated.

To solve the problem, we first determine the view factors between the surfaces. From the results in Example 7.3.4,

$$F_{1\text{-}2} = 0.055728 = F_{2\text{-}1} \qquad F_{1\text{-}34} = 0.94427$$

To determine $F_{1\text{-}3}$, introduce an imaginary disk 5 as shown by the dotted lines in Figure 8.3.3. Determine $F_{1\text{-}5}$ and, considering the enclosure formed by the three surfaces 1, 3, and 5, determine $F_{1\text{-}3}$. For the two coaxial disks 1 and 5, the dimensionless radii (for finding view factor from equations in Table 7.2.1) are

$$R_1 = \frac{3}{6} = 0.5 = R_5$$

$$x = 1 + \frac{1 + R_5^2}{R_1^2} = 1 + \frac{1 + 0.5^2}{0.5^2} = 6$$

$$F_{1\text{-}5} = \frac{1}{2}\left\{x - \left[x^2 - 4\left(\frac{r_5^2}{r_1^2}\right)\right]^{1/2}\right\}$$

$$= \frac{1}{2}\left\{6 - \left[6^2 - 4\left(\frac{0.5}{0.5}\right)^2\right]^{1/2}\right\} = 0.1716$$

(The view factor can also be obtained from the software provided.)

For the enclosure formed by surfaces 1, 3, and 5, from Equation 7.2.4e,

$$F_{1\text{-}1} + F_{1\text{-}5} + F_{1\text{-}3} = 1$$

With $F_{1\text{-}1} = 0$, $\quad F_{1\text{-}3} = 1 - F_{1\text{-}5} = 1 - 0.1716 = 0.8284$
From Equation 7.2.4c, $\quad A_1 F_{1\text{-}34} = A_1 F_{1\text{-}3} + A_1 F_{1\text{-}4}$. Therefore,

$$F_{1\text{-}4} = F_{1\text{-}34} - F_{1\text{-}3} = 0.94427 - 0.8284 = 0.11587$$

By symmetry, $\quad F_{2\text{-}4} = F_{1\text{-}3} = 0.8284$ and $F_{2\text{-}3} = F_{1\text{-}4} = 0.11587$
From the reciprocity relation, $\quad A_3 F_{3\text{-}1} = A_1 F_{1\text{-}3}; F_{3\text{-}1} = (A_1/A_3)F_{1\text{-}3}$ or

$$F_{3\text{-}1} = \frac{\pi \times 0.03^2}{\pi \times 0.06 \times 0.06} \times 0.8284 = 0.2071 = F_{4\text{-}2}$$

By symmetry, $\quad F_{3\text{-}1} = F_{3\text{-}5} = 0.2071$
For the enclosure 135, $\quad F_{3\text{-}1} + F_{3\text{-}5} + F_{3\text{-}3} = 1$; hence,

$$F_{3\text{-}3} = 1 - 2 \times 0.2071 = 0.5858 = F_{4\text{-}4}$$

Employing reciprocity relation,

$$F_{4\text{-}1} = \frac{A_1}{A_4} F_{1\text{-}4} = \frac{\pi \times 3^2}{\pi \times 6 \times 6} \times 0.11587 = 0.028968 = F_{3\text{-}2}$$

For the four-surface enclosure, 1234, from Equation 7.2.4e,

$$F_{3\text{-}1} + F_{3\text{-}2} + F_{3\text{-}3} + F_{3\text{-}4} = 1$$

or

$$F_{3\text{-}4} = 1 - (F_{3\text{-}1} + F_{3\text{-}2} + F_{3\text{-}3}) = 0.17813 = F_{4\text{-}3}$$

Now that all the view factors have been determined, and, as the temperatures of all the surfaces are known, employing Equation 8.3.4b for each surface, we get the following.

Surface 1

$$(1 - \rho_1 F_{1\text{-}1})J_1 - \rho_1(F_{1\text{-}2}J_2 + F_{1\text{-}3}J_3 + F_{1\text{-}4}J_4) = \varepsilon_1 E_{b1}$$

$$\rho_1 = 1 - \varepsilon_1 = 1 - 0.8 = 0.2 = \rho_2 = \rho_3 = \rho_4$$

$$\varepsilon_1 E_{b1} = 0.8 \times 5.67 \times 10^{-8} \times 573.15^4 = 4894.9 \text{ W/m}^2$$

Substituting the various values,

$$J_1 - 0.011146 \, J_2 - 0.16568 \, J_3 - 0.023174 \, J_4 = 4894.9 \qquad \textbf{(1)}$$

Surface 2

$$(1 - \rho_2 F_{2\text{-}2})J_2 - \rho_2(F_{2\text{-}1}J_1 + F_{2\text{-}3}J_3 + F_{2\text{-}4}J_4) = \varepsilon_2 E_{b2}$$

Substituting the values of emissivity and view factors,

$$J_2 - 0.011146 \, J_1 - 0.023174 \, J_3 - 0.16568 \, J_4 = 879.4 \qquad \textbf{(2)}$$

Surface 3

$$(1 - \rho_3 F_{3\text{-}3})J_3 - \rho_3(F_{3\text{-}1}J_1 + F_{3\text{-}2}J_2 + F_{3\text{-}4}J_4) = \varepsilon_3 E_{b3}$$

In this case, note that $F_{3\text{-}3} \neq 0$. Substituting the known values,

$$0.88284 \, J_3 - 0.04142 \, J_1 - 0.0057936 \, J_2 - 0.035626 \, J_4 = 16\,208 \qquad \textbf{(3)}$$

Surface 4   In a manner similar to the computations for surface 3,

$$(1 - \rho_4 F_{4\text{-}4})J_4 - \rho_4(F_{4\text{-}1}J_1 + F_{4\text{-}2}J_2 + F_{4\text{-}3}J_3) = \varepsilon_4 E_{b4} \qquad \textbf{(4)}$$

$$0.88284 \, J_4 - 0.0057936 \, J_1 - 0.04142 \, J_2 - 0.035626 \, J_3 = 16\,208$$

Solving Equations 1, 2, 3, and 4 for the radiosities,

$$J_1 = 8641 \text{ W/m}^2 \qquad J_2 = 4648 \text{ W/m}^2$$
$$J_3 = 19\,579 \text{ W/m}^2 \qquad J_4 = 19\,424 \text{ W/m}^2$$

The heat transfer rates from the surfaces can now be determined by applying Equation 7.3.4 to each surface,

$$q_1 = \frac{\pi \times 0.06^2}{4} \times \frac{0.8}{0.2} \, (6119 - 8641) = \underline{-28.52 \text{ W}}$$

$$q_2 = \frac{\pi \times 0.06^2}{4} \times \frac{0.8}{0.2} \, (1099 - 4648) = \underline{-40.14 \text{ W}}$$

$$q_3 = \pi \times 0.06 \times 0.06 \times \frac{0.8}{0.2} \, (20\,260 - 19\,579) = \underline{30.81 \text{ W}}$$

$$q_4 = \pi \times 0.06 \times 0.06 \times \frac{0.8}{0.2} \, (20\,260 - 19\,424) = \underline{37.82 \text{ W}}$$

The heat transfer rate from the cylindrical surface is

$$q_3 + q_4 = 30.81 + 37.82 = \underline{68.63 \text{ W}}$$

(Note that the algebraic sum of the heat transfer rates from all the surfaces is -0.03 W, caused by round-off errors.)

COMMENTS

Now compare this solution with the solution obtained by considering it as a three-surface problem, Example 7.3.5 (note that the numbering of surfaces in the two problems is different),

Example 7.3.5 (Three-Surface)	This Solution (Four-Surface)
$q_1 = -28.39$ W	$q_1 = -28.53$ W
$q_2 = -40.24$ W	$q_2 = -40.14$ W
$q_3 = 68.66$ W	$q_{34} = 68.63$ W

The differences in the solution to the heat transfer rates in the two cases is quite small. We, therefore, conclude that the assumption of uniform radiosity for the cylindrical surface is justified. The same conclusion can be arrived at by comparing the radiosities of the two halves of the cylindrical surfaces in this solution to the radiosity of the cylindrical surface (considered as one surface) in Example 7.3.5. By considering the cylindrical surface as one surface, its radiosity was found to be 19 497.9 W/m². By treating it as two surfaces, the radiosities are found to be 19 424 W/m² and 19 579 W/m² with an average value of 19 503 W/m², which is only 0.026% higher than the value as a three-surface enclosure. The solution to this problem shows that to maintain the cylindrical surface at a uniform temperature, the heat transfer rate per unit area of the cylindrical surface should increase as we go from one end surface towards the other. Treating it as a four-surface enclosure, the heat flux on surface 3 [$= 30.81/(\pi \times 0.06 \times 0.06)$] is 2724.2 W/m² and the heat flux on surface 4 [$= 37.83/(\pi \times 0.06 \times 0.06)$] is 3345 W/m².

8.3.3 Equivalence of the Electrical Network and Radiosity Methods

Both the electrical network and the radiosity methods require that the radiosities of each surface forming the enclosure be determined. The only difference is the manner in which the equations are set up. We now demonstrate the equivalence of the two methods.

Figure 8.3.4 depicts an enclosure with N radiation surfaces. Rearranging Equation 8.3.4a for surface i,

$$J_i - (1 - \varepsilon_i) \sum_{k=1}^{N} F_{i\text{-}k} J_k = \varepsilon_i E_{bi} \qquad (8.3.6)$$

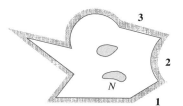

FIGURE 8.3.4 An *N*-surface Enclosure

Multiply Equation 8.3.6 by $A_i/(1 - \varepsilon_i)$ and rearrange,

$$\frac{A_i \varepsilon_i}{1 - \varepsilon_i} E_{bi} - \frac{A_i}{1 - \varepsilon_i} J_i + A_i F_{i\text{-}i} J_i + \sum_{\substack{k=1 \\ k \neq i}}^{N} A_i F_{i\text{-}k} J_k = 0$$

Add and subtract $\sum\limits_{\substack{k=1 \\ k\neq i}}^{N} A_i F_{i\text{-}k} J_i$. With $\sum\limits_{k=1}^{N} F_{i\text{-}k} = 1$, we get

$$\frac{A_i \varepsilon_i}{1 - \varepsilon_i} E_{bi} - \frac{A_i}{1 - \varepsilon_i} J_i + A_i J_i + \sum_{\substack{k=1 \\ k\neq i}}^{N} A_i F_{i\text{-}k}(J_k - J_i) = 0$$

or

$$\frac{E_{bi} - J_i}{(1 - \varepsilon_i)/(A_i \varepsilon_i)} + \sum_{\substack{k=1 \\ k\neq i}}^{N} \frac{J_k - J_i}{1/(A_i \varepsilon_i)} = 0$$

The last equation is the electrical analogy.

Two methods for the solution of heat exchange among gray, diffuse surfaces have been presented. These and other methods are described in Sparrow (1963) and Clark and Koyabashi (1974).

8.4 RADIATION IN AN ABSORBING, EMITTING MEDIUM

Radiation in gases and liquids

So far we have considered radiative heat transfer between surfaces separated by a medium that does not absorb or emit radiative energy. The assumption that a medium does not participate in radiation is valid if the medium consists of monatomic and/ or symmetric diatomic molecules. If air is a mixture of only N_2 and O_2, both of which are symmetric molecules, it may be considered to be a nonparticipating medium. If the air contains CO_2, H_2O vapor, or CO molecules, they may affect the radiative heat transfer by absorption at low temperatures and by absorption and emission at high temperatures. The effect of the presence of such gases depends on their concentrations. In combustion gases, the concentrations of CO_2 and H_2O vapor in the products of combustion are quite high and, hence, the effects of CO_2 and H_2O vapor are of engineering importance. In some cases even small concentrations of a participating gas may be quite significant. For example, the small amounts of CO_2, H_2O vapor, and O_3 (Ozone) that are present in the atmosphere have a significant effect on the solar radiation that reaches us.

Many liquids are opaque to infrared radiation

Many common liquids are opaque to radiation. The transmission of short wavelength radiation (visible spectrum) in some liquids is quite high but they are almost opaque to radiation in wavelengths greater than about 2 µm. Their temperatures usually do not exceed 500 °C in the liquid state. At a temperature of 500 °C the blackbody emissive power in the wave band 0–2 µm is less than 1% of the blackbody emissive power. Therefore, emission from the interior of such liquids is ignored. Unless the source temperature is quite high, the incident radiation is absorbed by a thin layer of the liquid at the surface and the liquids are essentially opaque to radiation. Recall the effect of radiation in film boiling in increasing the vapor generation due to radiative heat transfer from the heated surface to the liquid surface at the edge of the vapor layer. The thickness of the water layer S (mm) required to reduce the radiation intensity to 1% of the incident intensity for various wavelengths is given in the following table.

λ (μm)	S (mm)	λ (μm)	S (mm)
2.0	0.7	3.6	0.26
2.2	2.8	3.8	0.41
2.4	0.9	4.0	0.32
2.6	0.3	5.0	0.15
2.8	0.008	6.0	0.02
3.0	0.004	7.0	0.08
3.2	0.01	8.0	0.085
3.4	0.06	9.0	0.08
		10.0	0.07

The values in the table show that water in thicknesses exceeding 1 mm can be considered opaque for radiation in wavelengths greater than 2.4 μm. At source temperatures less than 400 °C most of the incident radiant energy is in wavelengths exceeding 3 μm; therefore, water can be considered opaque. Hence, in analyzing heat transfer to water in tubes, radiation is not considered separately as it contributes to what we have considered as conduction.

Gases do not absorb and emit at all wavelengths but in a few well-defined, discrete wavelengths, known as wave bands. Their emissions at normal temperatures are not high but become significant at elevated temperatures. Their absorption may be significant at all temperatures. For example, at the outer reaches of the atmosphere, solar radiation has approximately the same spectral distribution as a blackbody at 5760 K. The attenuation of the solar intensity in different wave bands by ozone (O_3), water vapor, and carbon dioxide is clearly seen in Figure 8.4.1. Water vapor has strong absorption in the wave bands centered around 1.4 μm, 1.9 μm, 2.4 μm, 2.7 μm, and 6.3 μm and carbon dioxide centered around 1.8 μm, 2.2 μm, 2.6 μm, 2.8 μm, 4.0 μm, 4.6 μm, 9 μm, and 19 μm. The effect of even small quantities of ozone

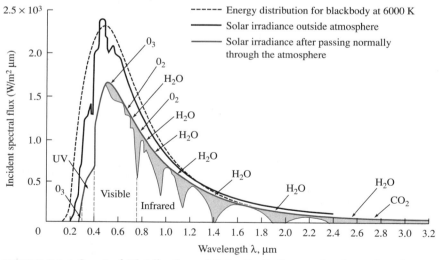

FIGURE 8.4.1 Spectral Distribution of Solar Radiation at the Outer Reaches of the Atmosphere and the Surface of the Earth
After Thekaekara, M.P. (1965), Survey of the Literature on the Solar Constant and the Spectral Distribution of Solar Radiant Flux, NASA SP-74.

Solar energy
attenuation in the
atmosphere

in absorbing ultraviolet radiation from the sun is also clearly seen.[2] Solar energy drops off rapidly for wavelengths greater than about 6 μm (more than 99.5% of the energy from a blackbody at 5760 K is in wavelengths less than 6 μm, see Table 7.1.2).

When a gas or mixture of gases participates in radiation, analysis of radiative heat transfer is complex. The complexity arises out of many factors. First, gases emit and absorb in a few well-defined wave bands. This means that although the surfaces surrounding a gas are gray, the heat transfer rate has to be determined considering the wavelength dependent properties of the gas. Second, the emission depends on the temperature of the gas. If the temperature of the gas varies spatially, the temperature variation has to be taken into account when determining the emitted energy from the gases. Third, in addition to emission and absorption scattering may be significant. Scattering is the change in the direction of propagation of the radiation intensity due to reflection, refraction, and diffraction. Scattering may be caused by the presence of foreign particles, fog, and fluidized beds. Scattering can also be caused by the gas molecules themselves (Rayleigh scattering). Fourth, gaseous radiation is a volumetric phenomenon. Every volumetric element of the gas participates in radiation. The effect of absorption, emission, and scattering in every volumetric element has to be considered.

To appreciate the complexity in determining the radiative heat transfer rates in a participating medium consider an enclosure containing a gas (Figure 8.4.2). The incident radiant energy flux at a point P on the bounding surface is given by

$$G = \int_\lambda \int_\omega I_\lambda \cos \theta \, d\omega \, d\lambda \qquad \textbf{(8.4.1)}$$

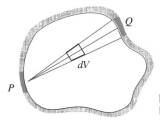

FIGURE 8.4.2 A Gas Participating in Radiation Inside an Enclosure

The radiation intensity I_λ arriving at P in the direction QP is the result of the intensity from the surface at Q in the direction QP (after attenuation by the gas in the path QP) and the intensities due to emission by the gas volume elements such as dV along the path QP (after attenuation in the gas between dV and P). The surface intensity is attenuated by the gas in the path QP. The attenuation depends on the wavelength, the temperature and concentration of the gas, and the path length QP. To determine the intensity due to the emission by the gas, the emitted intensity from every gaseous element in the path QP has to be considered. The intensities from different elements are different if the temperatures and/or the composition of the gas vary. The emitted intensity from an element such as dV along QP is also attenuated by the gas between

[2]Ozone depletion in the atmosphere leading to increased exposure to ultraviolet rays is being seriously discussed now. Higher dosages of UV rays to people lead to an increase in the incidence of skin cancer.

dV and P. To determine the irradiation G in Equation 8.4.1, the intensity along every path such as QP in the enclosure has to be considered.

The complexity of analyzing radiative heat transfer with a participating medium is clear. In many engineering applications, we seek a reasonable approximate solution; simplifying assumptions are employed to arrive at such approximate solutions. One such approximation to radiation heat transfer between a surface and a participating gaseous medium, the mean beam length approximation, is considered in this chapter. To understand the simplifying assumptions, we need an appreciation for the manner in which the absorption and emission in gases are evaluated and how the radiant energy is attenuated by absorption and augmented by emission. In the following section some of the radiative properties of gases are considered.

8.4.1 Radiation Properties of Gases

Absorptivity

Absorptivity of gases

We now consider the absorption of radiation emanating from a surface at T_s by a gas at T_g. A planar layer of the gas has a thickness ds^* and is located at a distance s^* from the surface as shown in Figure 8.4.3. The gas layer receives monochromatic

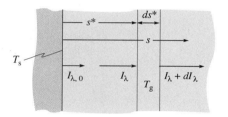

FIGURE 8.4.3 **An Opaque Surface Adjacent to a Gas with Temperature Varying with the Distance from the Surface**

radiation of intensity I_λ. As the intensity passes through the gas layer, it decreases due to absorption. The attenuation in the intensity is given by Beer's Law. The law states that the decrease in the intensity in a medium due to absorption, dI_λ, is proportional to the product of the intensity (I_λ) at that point and the thickness (ds^*) of the medium, i.e.,

$$dI_\lambda = -k_\lambda I_\lambda \, ds^* \qquad \textbf{(8.4.2)}$$

The coefficient of proportionality in Equation 8.4.2, k_λ, is the monochromatic absorption coefficient. It is a strong function of the wavelength and varies with temperature and pressure of the gas. Integrating Equation 8.4.2, the radiation intensity $I_\lambda(s)$ at s is given by

$$I_\lambda = c_1 \exp\left(-\int_0^s k_\lambda(s^*) \, ds^*\right) \qquad \textbf{(8.4.3)}$$

If $I_{\lambda,0}$ is the radiation intensity at the boundary surface located at $s = 0$, $c_1 = I_{\lambda,0}$ and it follows that

$$I_\lambda = I_{\lambda,0} \exp\left(-\int_0^s k_\lambda(s^*) \, ds^*\right) \qquad \textbf{(8.4.4)}$$

If the monochromatic absorptivity α_λ at the gas temperature T_g is defined as the fraction of the incident intensity that is absorbed in the path s to the incident intensity $I_{\lambda,0}$, we have

$$\alpha_\lambda(T_g) = \frac{I_{\lambda,0} - I_\lambda}{I_{\lambda,0}} = 1 - \exp\left(-\int_0^s k_\lambda(s^*)\, ds^*\right)$$

If the monochromatic absorption coefficient, k_λ, is constant,

$$\alpha_\lambda(T_g) = 1 - \exp(-k_\lambda s) \tag{8.4.5a}$$

The total absorptivity of the gas for radiation emanating from a surface at T_s is given by

$$\alpha(T_g, T_s) = \frac{\displaystyle\int_0^\infty \alpha_\lambda(T_g)I_{\lambda,0}\, d\lambda}{\displaystyle\int_0^\infty I_{\lambda,0}\, d\lambda} \tag{8.4.5b}$$

In Equation 8.4.5b α_λ depends on T_g and $I_{\lambda,0}$ depends on the source monochromatic emissivity and temperature T_s. Therefore, α defined by Equation 8.4.5b is a function of both the gas temperature T_g and the source temperature T_s.

Emissivity

Emissivity of gases

To determine the emissivity of a gas, consider a black-surfaced hemisphere of radius s, surrounding a black infinitesimal area, dA, at the base of the hemisphere at O (Figure 8.4.4). The hemisphere is filled with a gas whose emissivity is to be found.

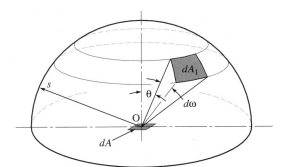

FIGURE 8.4.4 A Black Hemisphere Filled with a Gas and an Infinitesimal Surface at the Center of the Base of the Hemisphere
The gas, the hemisphere, and the infinitesimal surface are all at the same temperature.

The gas, the hemisphere, and the infinitesimal surface are in thermal equilibrium at T_g. Infinitesimal area, dA_1, is on the hemisphere subtending a solid angle $d\omega$ at O. The intensity reaching dA through the solid angle $d\omega$ from dA_1 is given by

$$I_\lambda = I_{\lambda 1} + I_{\lambda e}$$

where

$I_{\lambda 1}$ = intensity reaching dA from dA_1

$I_{\lambda e}$ = intensity reaching dA from the gas in the solid angle $d\omega$,
   due to emission from the gas and self-absorption

From Equation 8.4.4, $I_{\lambda 1}$ is given by

$$I_{\lambda 1} = I_{b\lambda} \exp \left(- \int_0^s k_\lambda \, ds \right)$$

where $I_{b\lambda}$ is the blackbody, monochromatic intensity for the surface dA_1.

As the temperature of the gas is uniform, its density is uniform and, therefore, k_λ is constant and

$$I_{\lambda 1} = I_{b\lambda} \exp(-k_\lambda s)$$

The energy reaching dA from the gas in the solid angle $d\omega$ and from the infinitesimal surface dA_1, in the wave band (λ to $\lambda + d\lambda$) is given by

$$d^3Q = I_\lambda \, dA \cos \theta \, d\omega \, d\lambda = I_\lambda \, dA \cos \theta \, \frac{dA_1}{r^2} \, d\lambda$$

where I_λ is the sum of the intensity due to emission from the gas and the intensity reaching dA from dA_1 and Q is the rate of total incident radiant energy reaching the base. Integration with respect to dA_1 (see Section 7.5 for integration) yields

$$d^2Q = I_\lambda \pi \, dA \, d\lambda = = (I_{\lambda 1} + I_{\lambda e})\pi \, dA \, d\lambda = [I_{b\lambda} \exp(-k_\lambda s) + I_{\lambda e}] \, \pi \, dA \, d\lambda$$

d^2Q represents the power reaching dA in the wave band λ to $\lambda + d\lambda$ and absorbed by dA. The emitted power from $dA = E_{b\lambda} \, dA \, d\lambda = \pi I_{b\lambda} \, dA \, d\lambda$.

In equilibrium conditions, the power absorbed by dA equals the power emitted by dA. Hence,

$$\pi I_{b\lambda} \, dA \, d\lambda = [I_{b\lambda} \exp(-k_\lambda s) + I_{\lambda e}]\pi \, dA \, d\lambda$$

Therefore,

$$I_{\lambda e} = I_{b\lambda} [1 - \exp(-k_\lambda s)]$$

We define the emissivity of the gas as

$$\varepsilon_\lambda(T_g) = \frac{I_{\lambda e}}{I_{b\lambda}} = 1 - \exp(-k_\lambda s)$$

Comparing the expression for $\varepsilon_\lambda(T_g)$ with Equation 8.4.5a, $\varepsilon_\lambda(T_g) = \alpha_\lambda(T_g)$.

The *total emissivity* of the gas is obtained from

$$\varepsilon(T_g) = \frac{\displaystyle\int_0^\infty I_{\lambda e} \, d\lambda}{I_b(T_g)} = \frac{\displaystyle\int_0^\infty \varepsilon_\lambda I_{b\lambda} \, d\lambda}{I_b(T_g)}$$

With $I_{b\lambda} = E_{b\lambda}/\pi$, and $I_b = E_b/\pi$,

$$\varepsilon(T_g) = \frac{\int_0^\infty \varepsilon_\lambda E_{b\lambda}(T_g)\, d\lambda}{E_b(T_g)}$$ (8.4.6a)

The absorptivity is obtained from Equation 8.4.5b

$$\alpha(T_g, T_s) = \frac{\int_0^\infty \alpha_\lambda(T_g) I_{\lambda,0}(T_s)\, d\lambda}{\int_0^\infty I_{\lambda,0}(T_s)\, d\lambda}$$ (8.4.5b)

In Equation 8.4.5b, $I_{\lambda,0}$ is the source intensity. If the source is a gray surface, the source emissivity is constant and

$$I_{\lambda,0} = \varepsilon I_{b\lambda}(T_s)$$

$$\int_0^\infty I_{\lambda,0}\, d\lambda = \varepsilon I_b(T_s)$$

For a gray surface source, Equation 8.4.5b simplifies to

$$\alpha(T_g, T_s) = \frac{\int_0^\infty \alpha_\lambda(T_g) I_{b\lambda}(T_s)\, d\lambda}{I_b(T_s)}$$ (8.4.6b)

EXAMPLE 8.4.1 The monochromatic absorptivity of carbon dioxide at 10 atm with a thickness of 38.8 cm) is shown in Figure 8.4.5. The spectral absorptivity is approximated as constant in defined wave bands. The approximated values are

Wave band (μm)	Absorptivity
1.8–2.2	0.32
2.6–3.0	0.92
4.0–5.0	0.96
9.0–20	0.86

Determine the total emissivity of carbon dioxide at temperatures of 1000 K and 1500 K. Assume that the values given for the absorptivity are valid at these temperatures.

SOLUTION

The total emissivity as defined by Equation 8.4.6a is

$$\varepsilon(T_g) = \frac{\int_0^\infty \varepsilon_\lambda E_{b\lambda}(T_g)\, d\lambda}{E_b(T_g)}$$ (1)

As the monochromatic emissivity is equal to the monochromatic absorptivity, employing the values of α_λ for ε_λ in the four wave bands, the emissive power of carbon dioxide, given by the numerator of Equation 1 is

$$E = \varepsilon_1 \int_{\lambda_1}^{\lambda_2} E_{b\lambda}(T_g)\, d\lambda + \varepsilon_2 \int_{\lambda_3}^{\lambda_4} E_{b\lambda}(T_g)\, d\lambda + \cdots$$ (2)

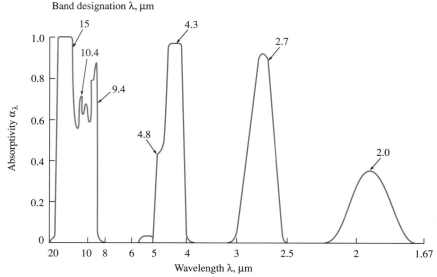

FIGURE 8.4.5 The Spectral Absorptivity of CO_2 at 830 K, 10 atm, and for a Path Length of 38.8 cm Through the Gas
Reproduced by permission from *Thermal Radiation Heat Transfer*, 3rd ed., by Siegel, R., and Howell, J.R. Washington, D.C.: Hemisphere, 1992.

where

$$\lambda_1 = 1.8 \ \mu m, \qquad \lambda_2 = 2.2 \ \mu m \qquad \varepsilon_1 = 0.32$$
$$\lambda_3 = 2.6 \ \mu m \qquad \lambda_4 = 3 \ \mu m \qquad \varepsilon_2 = 0.92 \cdots$$

$$\varepsilon(T_g) = \frac{\varepsilon_1 \int_{\lambda_1}^{\lambda_2} E_{b\lambda}(T_g) \, d\lambda}{E_b(T_g)} + \frac{\varepsilon_2 \int_{\lambda_3}^{\lambda_4} E_{b\lambda}(T_g) \, d\lambda}{E_b(T_g)} + \cdots$$

$$\varepsilon(T_g) = \varepsilon_1 f_{\lambda_1 - \lambda_2} + \varepsilon_2 f_{\lambda_3 - \lambda_4} + \cdots \tag{3}$$

where

$$f_{\lambda_1 - \lambda_2} = \int_{\lambda_1}^{\lambda_2} \frac{E_{b\lambda}}{E_b} \, d\lambda = \int_0^{\lambda_2} \frac{E_{b\lambda}}{E_b} \, d\lambda - \int_0^{\lambda_1} \frac{E_{b\lambda}}{E_b} \, d\lambda = f_{0 - \lambda_2} - f_{0 - \lambda_1}$$

Denoting the lower and upper values of the wavelengths by λ_L and λ_U, the values of $f_{0 - \lambda_U} - f_{0 - \lambda_L}$ are available in Table 7.1.2 and in the software (RF). The values in the following table are obtained from the software.

λ_L (μm)	λ_U (μm)	$f_{0 - \lambda_U} - f_{0 - \lambda_L}$ (1500 K)	$f_{0 - \lambda_U} - f_{0 - \lambda_L}$ (1500 K)
1.8	2.2	0.0616	0.1348
2.6	3.0	0.0901	0.1019
4.0	5.0	0.1529	0.0966
9.0	20.0	0.0956	0.0361

Inserting the values from the table in Equation 3

$$\varepsilon(1000\ \text{K}) = 0.32 \times 0.0616 + 0.92 \times 0.0901 + 0.96 \times 0.1529$$
$$+ 0.86 \times 0.0956 = \underline{0.3316}$$

$$\varepsilon(1500\ \text{K}) = 0.32 \times 0.1348 + 0.92 \times 0.1019 + 0.96 \times 0.0966$$
$$+ 0.86 \times 0.0361 = \underline{0.2607}$$

COMMENTS

1. The emissivity of CO_2 is mainly in the infrared region ($\lambda > 1.8\ \mu m$). As the temperature is increased, the fraction of the blackbody emissive power in the infrared region decreases and, hence, the total emissivity of CO_2 decreases as the temperature is increased.

2. By Kirchhoff's Law the emissivity equals the absorptivity for incident radiation from a blackbody at the same temperature. Hence, the absorptivity of carbon dioxide is 0.3316 for radiation from a blackbody at 1000 K and 0.2607 for radiation from a blackbody at 1500 K.

3. In arriving at the properties it was assumed that the emissivity of carbon dioxide at 830 K would also be valid at 1000 K and 1500 K. The absorptivity varies with temperature and the spectral values at 1000 K and 1500 K should be used for more accurate values of the properties.

Transmissivity

Transmissivity of gases

In most cases transmissivity in gases is finite (greater than zero) and it cannot be neglected. Scattering in gases is very small and is usually neglected in engineering applications. We then have

$$\alpha_\lambda + \tau_\lambda = 1$$

At a given temperature, if the total absorptivity is used,

$$\alpha + \tau = 1 \tag{8.4.7}$$

Consider a gas between two parallel, black plates as shown in Figure 8.4.6. The gas and the plates are at uniform temperatures of T_g and T_p, respectively. The emis-

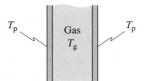

FIGURE 8.4.6 A Gas Between Two Parallel Plates at Different Temperatures

sivity of the gas is evaluated from Equation 8.4.6a and depends only on the gas temperature T_g. To find the energy emitted by one of the plates and absorbed (or transmitted) by the gas as a fraction of the total energy emitted by the plate, the absorptivity of the gas should be evaluated from Equation 8.4.6b. As the monochromatic absorptivity depends on the gas temperature T_g and the irradiation depends on the plate temperature T_p, the absorptivity (and the transmissivity) is a function of both the gas and the plate temperatures.

8.4.2 Energy Balance with Gaseous Radiation

Consider a gas at a uniform temperature, T_g, inside a black, spherical surface at temperature T_s, Figure 8.4.7. An infinitesimal black surface dA_1 is located at the

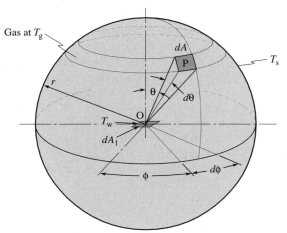

FIGURE 8.4.7 A gas at a uniform temperature T_g is inside a spherical surface at a uniform temperature T_s. An infinitesimal area dA_1 at T_w is located at the center of the sphere.

center of the sphere at a temperature T_w. The monochromatic absorptivity of the gas with a thickness of a planar layer equal to the radius of the sphere can be evaluated from Equation 8.4.5a. In gases the absorptivity and emissivity are finite only within narrow wave bands. As an approximation, they may be considered to be consatnt at a reasonable average value within each band. Such an approximation is known as band approximation. In Example 8.4.1 the emissivity of the gas was computed with a band approximation. With a band approximation, the total emissivity of the gas, ε, can be evaluated at T_g by the method illustrated in Example 8.4.1. If the irradiation to the gas is from a black surface at T_s, $\alpha(T_g, T_s)$ is evaluated from Equation 8.4.6b. If the emissivity and absorptivity of the gas are known, we can compute the energy reaching dA_1 due to emission from the gas and the bounding surface. Consider the gas in the conical element with a base area dA on the hemispherical surface subtending a solid angle $d\omega$ at dA_1. Consider only the upper surface of dA_1. The intensity of radiation reaching dA_1 along PO is expressed as

$$I = I_s + I_g$$

where

I_s = intensity due to emission from the surface at T_s
I_g = intensity due to emission from the gas at T_g

From Equation 8.4.5b,

$$\alpha(T_g, T_s) = \frac{\int_0^\infty \alpha_\lambda(T_g)I_{b\lambda}(T_s) \, d\lambda}{I_b(T_s)}$$

Therefore,

$$I_s = \int_0^\infty \tau_\lambda(T_g)I_{b\lambda}(T_s) \, d\lambda = \int_0^\infty [1 - \alpha_\lambda(T_g)]I_{b\lambda}(T_s) \, d\lambda$$

$$= [1 - \alpha(T_g, T_s)]I_b(T_s) = \tau(T_g, T_s)I_b(T_s)$$

Similarly, the intensity reaching dA_1 due to emission in the gas along PO is given by

$$I_g = \int_0^\infty \varepsilon_\lambda(T_g)I_{b\lambda}(T_g) \, d\lambda = \varepsilon(T_g)I_b(T_g)$$

From Equation 8.4.6a,

$$\varepsilon(T_g) = \frac{\int_0^\infty \varepsilon_\lambda(T_g)I_{b\lambda}(T_g) \, d\lambda}{I_b(T_g)}$$

Combining the expressions for I_s and I_g, we obtain

$$I = \tau_g(T_s, T_g)I_b(T_s) + \varepsilon_g(T_g)I_b(T_g)$$

The total energy reaching the upper surface of dA_1 is given by (G is the incident radiant energy flux)

$$G \, dA_1 = \int I \, dA_1 \cos \theta \, d\omega$$

The integral is to be evaluated over the entire upper hemisphere (see Section 7.5 for details). The final result is

$$G \, dA_1 = \pi I \, dA_1 = \pi[\tau_g(T_g, T_s)I_b(T_s) + \varepsilon_g(T_g)I_b(T_g)] \, dA_1$$

With $\pi I_b = E_b$, we get

$$G \, dA_1 = [\tau_g(T_g, T_s)E_b(T_s) + \varepsilon_g(T_g)E_b(T_g)] \, dA_1$$

As the surface dA_1 is black, all the incident energy is absorbed by it. The energy leaving $dA_1 = E_b(T_w) \, dA_1$.

The net radiant heat transfer from dA_1 = The energy leaving dA_1
                                         − The energy absorbed by dA_1

$$\| \quad dq = [E_b(T_w) - \tau_g(T_g, T_s)E_b(T_s) - \varepsilon_g(T_g)E_b(T_g)] \, dA_1 \quad \| \qquad \textbf{(8.4.8)}$$

EXAMPLE 8.4.2 A black surface 1 with an area of 1 cm^2 is at 500 K (Figure 8.4.8). It is at the center of a black, 38.8 cm diameter spherical surface that is at 1000 K. The sphere is filled with CO_2 at 1500 K, 10 atm pressure. The emissivity of CO_2 is shown in Figure 8.4.5. Determine the radiant heat transfer rate to one side of the surface 1.

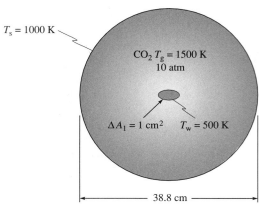

FIGURE 8.4.8 A Small Area 1 at the Center of Sphere Filled with CO_2

Given

Gas: CO_2 at 10 atm pressure, 1500 K $T_s = 1000$ K

All surfaces are black $T_w = 500$ K

Diameter of hemisphere = 38.8 cm

Find

q (to the surface at the center of the hemisphere)

SOLUTION

Surface 1 with an area of 1 cm² inside a 38.8-cm diameter sphere is approximated as an elemental area. As both the sphere and CO_2 are at uniform temperatures and the surfaces 1 and sphere are black, from Equation 8.4.8,

$$\Delta q_1 = - [E_b(T_w) - \varepsilon_g(T_g)E_b(T_g) - \tau_g(T_s, T_g)E_b(T_s)]\Delta A_1$$

where

$$\Delta q_1 = \text{radiant heat transfer rate to } \Delta A_1$$
$$T_w = \text{temperature of 1 (500 K)}$$
$$T_g = \text{temperature of } CO_2 \text{ (1500 K)}$$
$$T_s = \text{temperature of the spherical surface (1000 K)}$$
$$\varepsilon_g(T_g) = \text{emissivity of } CO_2 \text{ at 1500 K}$$
$$\tau_g(T_g, T_s) = \text{transmissivity of } CO_2 \text{ at } T_g \text{ for irradiation from a source at } T_s$$

From the results of Example 8.4.1,

$$\varepsilon_g(1500 \text{ K}) = 0.2607$$

$$\tau_g(T_g, T_s) = 1 - \alpha(T_g, T_s)$$

From Equation 8.4.6b

$$\alpha(T_g, T_s) = \frac{\displaystyle\int_0^\infty \alpha_\lambda(T_g)I_{b\lambda}(T_s) \, d\lambda}{I_b(T_s)}$$

Assuming $\alpha_\lambda(830\ \text{K}) \approx \alpha_\lambda(1000\ \text{K})$, with $\alpha_\lambda = \varepsilon_\lambda$ from the results of Example 8.4.1, $\alpha(T_g, T_s) = 0.3316$

$$\tau_g(T_g, T_s) = 1 - 0.3316 = 0.6684$$

$E_b(500\ \text{K}) = 3544\ \text{W/m}^2$
$E_b(1000\ \text{K}) = 5.67 \times 10^4\ \text{W/m}^2$
$E_b(1500\ \text{K}) = 2.87 \times 10^5\ \text{W/m}^2$

Substituting these values into the expression for Δq_1,

$$\Delta q_1 = -(3544 - 0.2607 \times 2.87 \times 10^5 - 0.6684$$
$$\times 5.67 \times 10^4)\ 1 \times 10^{-4} = \underline{10.92\ \text{W}}$$

If only the upper surface of 1 participates in radiant heat transfer, it should be cooled by removing 10.92 W. To maintain the temperature of surface 1 at 500 K the actual heat removal from 1 is the sum of the radiative and convective heat transfer rates to 1.

8.4.3 Energy Exchange Between an Enclosure and a Gas

Radiative heat transfer from gases to a bounding black surface

In the previous section, we determined the heat transfer rate from an infinitesimal surface. We were able to evaluate the exact heat transfer rate for the restricted case since the configuration was particularly simple. The same principles can be applied to a more general case but the solution to the resulting equations are quite involved. In this section we present a simple method given by Hottel (1954). The method can be applied to determine the heat transfer rate from a gas at a uniform temperature to the enclosing black surface, also at a uniform but different temperature from that of the gas.

Consider a gas at T_g enclosed by a black surface at T_s. As the surface is black, all the energy emitted by the gas and reaching the surface is absorbed by it. The surface also emits energy and a part of that energy is absorbed by the gas. The heat transfer rate from the gas to the surface is then given by the total energy emitted by the gas (and reaching the surface) less the total amount of energy emitted by the surface and absorbed by the gas. To determine these two quantities, consider two infinitesimal areas dA_1 on the enclosing surface and dA_2 on the gaseous surface adjacent to the solid surface, as shown in Figure 8.4.9.

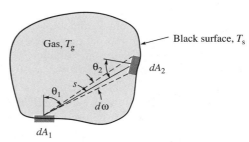

FIGURE 8.4.9 A Gas at a Uniform Temperature T_g Inside a Black Enclosure, Which Is at a Uniform Temperature T_s

Infinitesimal area dA_2 subtends a solid angle $d\omega$ at dA_1. Denoting the rate of incident energy on the solid surface from the gas by Q_{ge}, the incident rate of energy from the gas in the solid angle $d\omega$ is

$$d^2 Q_{ge} = \varepsilon_g(s, T_g) I_b(T_g) \, dA_1 \cos \theta_1 \, d\omega \qquad (8.4.9)$$

where

$\varepsilon_g(s, T_g) = \alpha_g(s, T_g); \qquad \alpha_\lambda(s) = 1 - \exp(-k_\lambda s)$

$\varepsilon_g(s, T_g)$ is evaluated at T_g by the method illustrated in Example 8.4.1.

$I_b(T_g) = $ blackbody radiation intensity at T_g

$s = $ distance between dA_1 and dA_2

$d\omega = $ solid angle subtended by dA_2 on $dA_1 = (dA_2 \cos \theta_2)/s^2$

Therefore, the total rate of energy incident on dA_1 due to gaseous emission is

$$dQ_{ge} = \int_{A_s} \frac{\varepsilon_g(s, T_g) I_b(T_g) \cos \theta_1 \cos \theta_2 \, dA_2 \, dA_1}{s^2} \qquad (8.4.10)$$

Equation 8.4.10 represents the radiant energy emitted by the gas and incident on dA_1, with the integration carried out with respect to dA_2. To get the total incident energy on the entire solid surface due to emission from the gas, we must integrate Equation 8.4.10 with respect to dA_1, over the surface area A_s. Denoting the total incident energy on the solid surface due to emission by the gas by Q_{ge},

$$Q_{ge} = \int_{A_s} \int_{A_s} \frac{\varepsilon_g(s, T_g) I_b(T_g) \cos \theta_1 \cos \theta_2 \, dA_2 \, dA_1}{s^2}$$

With $I_b = E_b/\pi$,

$$Q_{ge} = E_b(T_g) \int_{A_s} \int_{A_s} \frac{\varepsilon_g(s, T_g) \cos \theta_1 \cos \theta_2 \, dA_1 \, dA_2}{\pi s^2} \qquad (8.4.11)$$

Express the total rate of energy emitted by the gas to the surrounding surface by the relation

$$Q_{ge} = E_b(T_g) \varepsilon_g(L, T_g) A_s, \qquad (8.4.12)$$

Comparing Equation 8.4.11 and Equation 8.4.12,

$$\varepsilon_g(L, T_g) = \frac{1}{A_s} \int_{A_s} \int_{A_s} \frac{\varepsilon_g(s, T_g) \cos \theta_1 \cos \theta_2 \, dA_1 \, dA_2}{\pi s^2} \qquad (8.4.13)$$

Mean beam length　　The term $\varepsilon_g(L, T_g)$ in Equation 8.4.13 can be given a physical interpretation. The expression $\varepsilon_g(L, T_g) E_b(T_g)$ represents the average emissive power of the gas. The product of this average emissive power and the surface area gives the total energy emitted by the gas. Think of Equation 8.4.13 as the integrated mean emissivity of the gas, which can be equated to the emissivity of the gas of planar thickness L. Thickness L is the *mean beam length* of the gas. Employing Equation 8.4.13, the mean beam length of the gas can be determined for different enclosures at different temperatures of the gas. If the mean beam length is known, the emissivity can be computed. The average emissive power of the gas and, therefore, the total emitted energy from the gas can be computed.

The radiation properties of gases considered so far are on a volumetric basis. However, the absorptivity and emissivity depend on the mass of the gas in the volume. At a defined temperature of the gas, the mass is proportional to the density, which in turn is proportional to the pressure. Therefore, the emissivity is a function of the mean beam length and the pressure P or the product PL.

The mean beam length for a few geometries, as given by Siegel and Howell (1992), are given in Table 8.4.1. In the table, $L_{e,0}$ represents the mean beam length for $\alpha L_e \rightarrow 0$ and L_e for average conditions.

TABLE 8.4.1 Mean Beam Length for Radiation from a Gas at Uniform Temperature

	Characteristic Length	$L_{e,0}$	L_e
Hemisphere radiating to element at center	Radius R	R	R
Sphere radiating to its surface	Diameter D	$0.6667D$	$0.65D$
Circular cylinder of height equal to diameter radiating to element at center of base	Diameter D	$0.77D$	$0.71D$
Circular cylinder of infinite height radiating to convex bounding surface	Diameter D	D	$0.95D$
Circular cylinder of semi-infinite height radiating to element at center of base	Diameter D	D	$0.9D$
Circular cylinder of semi-infinite height radiating to entire base	Diameter D	$0.81D$	$0.65D$
Circular cylinder of height equal to diameter radiating to entire surface	Diameter D	$0.6667D$	$0.6D$
Cylinder of infinite height and semicircular cross section radiating to element at center of plane rectangular surface	Radius R	—	$1.26R$
Infinite slab of gas radiating to element on one face	Slab thickness D	$2D$	$1.8D$
Infinite slab of gas radiating to both bounding planes	Slab thickness D	$2D$	$1.8D$
Cube radiating to a face	Edge X	$0.6667X$	$0.6X$
Arbitrary enclosure, gas radiating to bounding surface	$4V/A_s$	—	$3.6\,V/A_s$

V = volume; A_s = surface area

Once we know the mean beam length, we can calculate the emissivity for different values of PL at different temperatures. Such curves of emissivity for CO_2 and H_2O vapor are given in Figures 8.4.10a and 8.4.11a. Values obtained from these figures can be used when either CO_2 or H_2O vapor is the only gas participating in radiation and when the total pressure is 1 atm. If the total pressure is different from 1 atm, the value of the emissivities from Figure 8.4.10a and 8.4.11a should be multiplied by the pressure correction factors C_{CO_2} and C_{H_2O} given in Figures 8.4.10b and 8.4.11b. In all cases, the partial pressure of the gas (which, in conjunction with the gas temperature determines the density of the gas) is used in computing PL. When both CO_2 and H_2O vapor are present, the emissivity of each gas corresponds to its

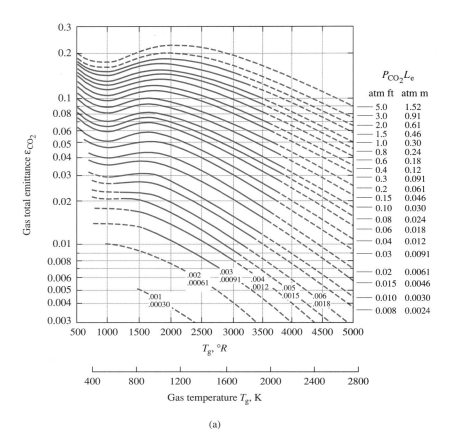

(a)

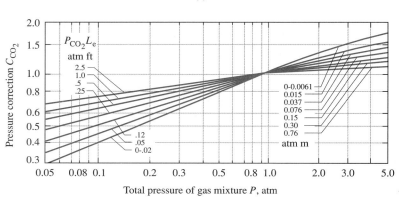

(b)

FIGURE 8.4.10 (a) Total emissivity of carbon dioxide in a mixture having a total pressure of 1 atm. (b) Pressure correction for CO_2 total emissivity for values of total pressure other than 1 atm. [Hottel and Egbert (1942). Used with permission of the American Institute of Chemical Engineers.]

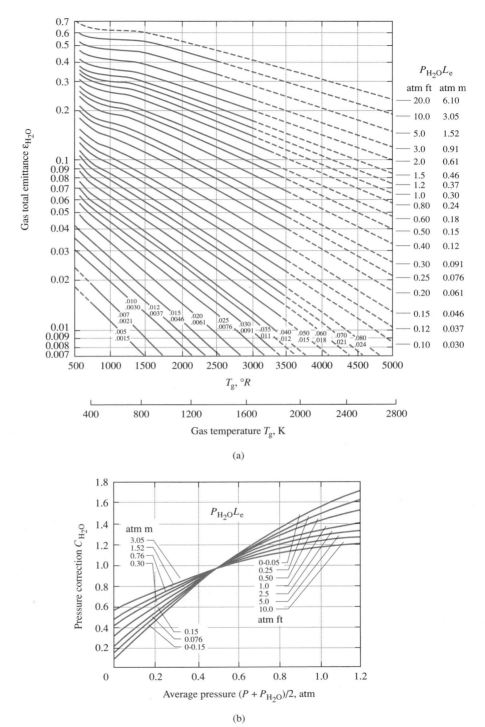

FIGURE 8.4.11 (a) Total emissivity of water vapor in a mixture at a pressure of 1 atm. (b) Pressure correction for the emissivity of water vapor when the total pressure is other than 1 atm. [Hottel and Egbert (1942). Used with permission of the American Institute of Chemical Engineers.]

partial pressure, but the effective emissivity of the mixture is not merely the sum of the emissivities of each component; the sum of the emissivities of the components must be corrected. With the correction $\Delta\varepsilon$ the total emissivity is obtained as

$$\varepsilon_g = C_{H_2O}\varepsilon_{H_2O} + C_{CO_2}\varepsilon_{CO_2} - \Delta\varepsilon \qquad (8.4.14)$$

where $\Delta\varepsilon$ is a correction factor resulting from the overlap of bands of emission for each component. The correction factor $\Delta\varepsilon$ is given in Figure 8.4.12.

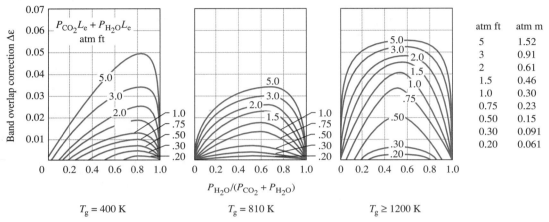

FIGURE 8.4.12 **Correction for Total Emissivity When Both CO_2 and Water Vapor Are Present in the Mixture**
[Hottel and Egbert (1942). Used with permission of the American Institute of Chemical Engineers.]

We may apply the same reasoning to compute the absorptivity. However there is one more factor to be considered—the temperature and wavelength dependence of α; these are taken into account by computing the emissivity at the temperature of the solid surface, and correcting it by Equations 8.4.15, 8.4.16 and 8.4.17.

$$\alpha_g = \alpha_{CO_2} + \alpha_{H_2O} - \Delta\alpha \qquad (8.4.15)$$

$$\alpha_{CO_2} = C_{CO_2}\left(\frac{T_g}{T_s}\right)^{0.65}\varepsilon_{CO_2}\left(T_s, P_{CO_2}L\frac{T_s}{T_g}\right) \qquad (8.4.16)$$

$$\alpha_{H_2O} = C_{H_2O}\left(\frac{T_g}{T_s}\right)^{0.45}\varepsilon_{H_2O}\left(T_s, P_{H_2O}L\frac{T_s}{T_g}\right) \qquad (8.4.17)$$

$$\Delta\alpha = \Delta\varepsilon$$

In computing the absorptivity of the gas, evaluate the emissivities at the solid surface temperature T_s, corresponding to $PL(T_s/T_g)$, where P is the partial pressure of the gas.

Having found the emissivity and absorptivity of the gas, the energy transfer from the solid surface to the gas is determined if the solid surface is black. The total rate of energy emitted by the gas is

$$Q_{ge} = \varepsilon_g(T_g)A_sE_b(T_g) \qquad (8.4.18)$$

The total amount of energy emitted by the surrounding black surface and absorbed by the gas is

$$Q_{abs} = \alpha_g(T_g, T_s)A_sE_b(T_s) \qquad (8.4.19)$$

The net radiative heat transfer rate from the gas to the solid surface is then given by

$$q = Q_{ge} - Q_{abs} = [\varepsilon_g(T_g)E_b(T_g) - \alpha_g(T_g, T_s)E_b(T_s)]A_s \qquad (8.4.20)$$

Equation 8.4.20 can be used for black surfaces only. If the enclosing surface is a gray surface (with emissivity less than 1), the computation of the heat transfer rate becomes more complex. However, if the emissivity of the surface is greater than 0.8, Hottel (1954) suggests that the value given in Equation 8.4.20 be modified as

$$q_{gray} = q_{black}\frac{\varepsilon + 1}{2} \qquad (8.4.21)$$

EXAMPLE 8.4.3

A cylindrical furnace, 2-m diameter and 2.3-m long (Figure 8.4.13), is fired by natural gas containing 70% methane and 30% ethane by volume. Fifty percent excess air is supplied for complete combustion and temperature control. If the combustion gases are at 1 atm pressure at a temperature of 1500 °C and the walls of the furnace are at 1200 °C, estimate the heat transfer rate from the gas to the furnace walls if

(a) The furnace walls are black.
(b) The emissivity of the furnace walls is 0.85.

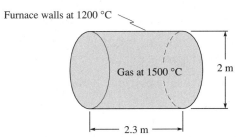

Furnace walls at 1200 °C

Gas at 1500 °C   2 m

2.3 m

FIGURE 8.4.13 A cylindrical furnace is 2-m diameter and 2.3-m long. The gases in the furnace are at 1500 °C and the walls at 1200 °C.

Given

Gas: Combustion products—70% methane, 30% ethane with 50% excess air
$T_g = 1500$ °C $T_s = 1200$ °C

Find

q if (a) $\varepsilon = 1$ (b) $\varepsilon = 0.85$

ASSUMPTIONS

1. Temperature of the combustion gases is uniform.
2. Complete combustion
3. Combustion products behave as ideal gas mixtures.

SOLUTION

Determine the partial pressure of CO_2 and H_2O. With 50% excess air the equation of reaction for 100 kmoles of natural gas is

$$70\ CH_4 + 30\ C_2H_6 + 1.5(245\ O_2 + 3.76 \times 245\ N_2) \rightarrow 130\ CO_2$$
$$+ 230\ H_2O + 122.5\ O_2 + 1381.8\ N_2$$

$$P_{CO_2} = \frac{130}{130 + 230 + 122.5 + 1381.8} = 0.06973\ \text{atm}$$

$$p_{H_2O} = \frac{230}{1864.3} = 0.1234\ \text{atm}$$

We will compute the mean beam length for the furnace by employing the last entry for an arbitrary enclosure in Table 8.4.1.

$$L_e = 3.6\ \frac{V}{A_s} = 3.6 \times \pi \times 1^2 \times \frac{2.3}{\pi \times 2 \times 2.3 + 2 \times \pi \times 1^2} = 1.25\ \text{m}$$

From Table 8.4.1, for a cylinder of infinite height radiating to the convex bounding surface, the mean beam length is $0.95\ D$. For a circular cylinder of height equal to the diameter, radiating to the entire surface, the mean beam length is $0.6\ D$. As the furnace being studied has a height of 1.15 times the diameter, the mean beam length should be between $0.6\ D$ and $0.95\ D$, i.e., between 1.2 m and 1.9 m but closer to 1.2 m. The computed value of 1.25 m is within these limits and close to 1.2 m as expected.

Determine the emissivity of the gas.

For H_2O, $p_{H_2O}L_e = 0.1234 \times 1.25 = 0.154$ atm m
From Figure 8.4.11a, for $T_g = 1773$ K, $\varepsilon_{H_2O} = 0.09$

As the total pressure is 1 atm, no pressure correction is needed.

For CO_2, $p_{CO_2}L_e = 0.06973 \times 1.25 = 0.0872$ atm m
From Figure 8.4.10a, for $T_g = 1773$ K, $\varepsilon_{CO_2} = 0.065$

There is no pressure correction for CO_2 also as the total pressure is 1 atm. To find the correction for total emissivity when both H_2O and CO_2 are present, use Figure 8.4.12.

$$p_{CO_2}L_e + p_{H_2O}L_e = 0.154 + 0.0872 = 0.241\ \text{atm m}$$

$$\frac{p_{H_2O}}{p_{CO_2} + p_{H_2O}} = \frac{0.154}{0.154 + 0.0872} = 0.638$$

From Figure 8.4.11, for $T_g > 1200$ K, $\Delta\varepsilon = 0.025$,

$$\varepsilon_g = \varepsilon_{H_2O} + \varepsilon_{CO_2} - \Delta\varepsilon = 0.09 + 0.065 - 0.025 = 0.13$$

Determine the absorptivity of the gas corresponding to $T_s = 1473$ K.

For H_2O, $p_{H_2O}L(T_s/T_g) = 0.154 \times 1473/1773 = 0.128$ atm m

$$\varepsilon_{H_2O}(1473\ \text{K},\ 0.128\ \text{atm m}) = 0.1$$

For $P = 1$ atm, $C_{H_2O} = 1$. From Equation 8.4.17,

$$\alpha_{H_2O} = \left(\frac{1773}{1473}\right)^{0.45} \times 0.1 = 0.11$$

For CO_2, $p_{CO_2} L(T_s/T_g) = 0.0872 \times 1473/1773 = 0.072$ atm m

$$\varepsilon_{CO_2} (1473 \text{ K}, 0.072 \text{ atm m}) = 0.09$$

With $C_{CO_2} = 1$, from Equation 8.4.16,

$$\alpha_{CO_2} = \left(\frac{1773}{1473}\right)^{0.65} \times 0.09 = 0.1$$

With $\Delta\alpha = \Delta\varepsilon = 0.025$,

$$\alpha_g = \alpha_{H_2O} + \alpha_{CO_2} - \Delta\alpha = 0.11 + 0.1 - 0.025 = 0.185$$

(a) The total heat transfer rate from the gas to the walls is obtained from Equation 8.4.20.

$$A_s = 2 \times \frac{\pi d^2}{4} + \pi dL = 2 \times \frac{\pi \times 2^2}{4} + \pi \times 2 \times 2.3 = 20.73 \text{ m}^2$$

$$q = [\varepsilon_g(T_g)E_b(T_g) - \alpha_g(T_g, T_s)E_b(T_s)]A_s$$
$$= (0.13 \times 5.67 \times 10^{-8} \times 1773^4 - 0.185 \times 5.67 \times 10^{-8} \times 1473^4)$$
$$\times 20.73 = \underline{4.86 \times 10^5 \text{ W}}$$

(b) If the emissivity of the walls is 0.85, from Equation 8.4.21,

$$q_{gray} = q_{black} \frac{\varepsilon + 1}{2} = 4.86 \times 10^5 \times \frac{0.85 + 1}{2} = \underline{4.5 \times 10^5 \text{ W}}$$

8.5 PROJECTS

PROJECT 8.5.1 Consider a fin of rectangular cross section for heat rejection in a space ship (Figure 8.5.1). The effective sky temperature is 0 K. The fin is made of plain carbon-steel and the base of the fin is at 400 K. Assuming that the fin sees very little of the space ship, determine the heat transfer rate from the fin to the surroundings.

ASSUMPTIONS

1. Steady state
2. Constant properties
3. One-dimensional temperature distribution, i.e., $T = T(x)$

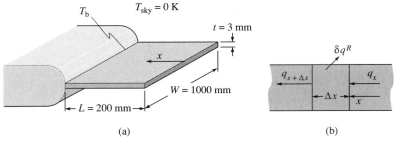

(a)   (b)

FIGURE 8.5.1 A Fin on a Space Ship
(a) Details of fin. (b) Element for analysis.

SOLUTION

In steady state the heat transfer rate from the fin to the surroundings by radiation must be equal to the heat transfer rate to the base of the fin by conduction. The expressions for the heat transfer rates are

$$q_f = kA_x \frac{dT}{dx}\bigg|_{x=L} \tag{1}$$

$$q_f = \int_0^L \sigma\varepsilon P T^4 \, dx + A_x \sigma\varepsilon T^4_{x=L} \tag{2}$$

where

A_x = area of cross section perpendicular to the x-direction
P = perimeter of the area of cross section
ε = surface emissivity

Equation 1 gives the conduction heat transfer to the fin at the base and Equation 2 is the radiative heat transfer from the fin surface to the surroundings. Both the expressions require that we find the temperature distribution $T(x)$. To find $T(x)$, consider an element of the fin shown in Figure 8.5.1b. For the element in steady state the net heat transfer rate is zero and

$$q_x - q_{x+\Delta x} - \delta q_R = 0 \tag{3}$$

where q_x and $q_{x+\Delta x}$ are the conductive heat transfer rates at x and $x + \Delta x$, respectively, and δq_R is the radiative heat transfer rate from the surface of the element exposed to the surroundings. As the fin sees very little of the space ship, it is assumed that the view factor of the surface of the element with respect to the sky is unity.

$$\delta q_R = P\,\Delta x\,\sigma\varepsilon T^4$$

Substituting the above expression for δq_R into Equation 3,

$$q_x - q_{x+\Delta x} - P\,\Delta x\,\sigma\varepsilon T^4 = 0$$

Dividing by Δx and taking the limit as $\Delta x \to 0$

$$-\frac{dq_x}{dx} - P\sigma\varepsilon T^4 = 0$$

Substituting $q_x = -kA(dT/dx)$, dividing by kA_x, we obtain

$$\frac{d^2T}{dx^2} - \frac{P\sigma\varepsilon}{kA_x} T^4 = 0 \qquad \textbf{(4a)}$$

Boundary conditions:

$$x = L \qquad T = T_b \qquad \textbf{(4b)}$$

$$x = 0 \qquad k\frac{dT}{dx} = \sigma\varepsilon T^4 \qquad \textbf{(4c)}$$

The solution to the nonlinear differential equation is given in Love (1968). Here we consider two approximate methods. In the first method Equations 4a and 4c are linearized employing the concept of radiative heat transfer coefficient. In the second method Equation 4a is numerically integrated.

Linearizing Equation 4a and Equation 4c The equations are linearized by using radiative heat transfer coefficient (Section 7.4). The radiative heat transfer coefficient (with $T_{sky} = 0$) is

$$h_R = \sigma\varepsilon T_{av}^3 \qquad \textbf{(5)}$$

A reasonable T_{av} is the arithmetic mean of T_b and $T(x = 0)$. But $T(x = 0)$ is yet to be determined. It is determined by iteration.

$$\frac{d^2T}{dx^2} - \frac{h_R P}{kA_x} T = 0 \qquad \textbf{(6a)}$$

Boundary conditions:

$$x = L \qquad T = T_b \qquad \textbf{(6b)}$$

$$x = 0 \qquad k\frac{dT}{dx} = h_R T \qquad \textbf{(6c)}$$

Denoting $(h_R P)/(kA_x)$ by m^2, the solution to Equation 6a is

$$T = c_1 \cosh mx + c_2 \sinh mx \qquad \textbf{(7a)}$$

Using boundary conditions 6b and 6c and solving for c_1 and c_2

$$c_1 = \frac{T_b}{\cosh mL + (h_R/km) \sinh mL} \qquad \textbf{(7b)}$$

$$c_2 = c_1 \frac{h_R}{km} \qquad \textbf{(7c)}$$

The iteration process consists of

1. Assuming a value of $T(x = 0)$ denoted by T_o
2. Determining h_R with $T_{av} = (T_b + T_o)/2$
3. Determining T_o ($= c_1$) from Equation 7a
4. Repeating the process until two successive values of T_o are sufficiently close to each other.

1. Assume $T_o = 300$ K.

$$T_{av} = \frac{400 + 300}{2} = 350 \text{ K}$$

From Table A1 the thermal conductivity of plain carbon-steel at 350 K is 58.6 W/m K. From Table A9 emissivity of plain carbon-steel is 0.81.

2. Determine h_R with T_{av}.

$$h_R = \sigma \varepsilon T_{av}^3 = 5.67 \times 10^{-8} \times 0.81 \times 350^3 = 1.97 \text{ W/m}^2 \text{ °C}$$

3. Determine T_o from Equation 7b.

$$P = 2.006 \text{ m} \qquad A_x = 0.003 \text{ m}^2$$

$$m = \left(\frac{h_R P}{k A_x}\right)^{1/2} = \left(\frac{1.97 \times 2.006}{58.6 \times 0.003}\right)^{1/2} = 4.74 \text{ m}^{-1}$$

$$mL = 4.74 \times 0.2 = 0.948 \qquad \frac{h_R}{km} = \frac{1.97}{58.6 \times 4.74} = 0.00709$$

$$T_o = c_1 = \frac{400}{\cosh 0.948 + 0.00709 \sinh 0.948} = 268 \text{ K}$$

4. As the computed value of T_o (= 268 K) is much lower than the assumed value of 300 K, repeat the computations with 268 K. The sequence of values are given in the following table.

T_o (K)	300	268	282	276	278.5	277.5
T_{av} (K)	350	334	341	338	339.3	338.8
h_R (W/m^2 °C)	1.97	1.71	1.82	1.77	1.79	1.785
k (W/m °C)	58.6	59.2	58.9	59.1	59.0	59.0
m (m^{-1})	4.74	4.396	4.55	4.48	4.508	4.497
mL	0.948	0.879	0.909	0.896	0.902	0.899
$h_R/km \times 10^3$	7.09	6.574	6.797	6.702	6.741	6.725
$T_o = c_1$ (K)	268	282	276	278.5	277.5	277.9

In the last column the computed value of T_o is only 0.4 K (in 278 K) higher than the value used for computing h_R. No further iteration is needed. The values in the last column are used to compute the heat transfer rate.

$$c_1 = 277.9 \text{ K} \qquad c_2 = 0.006725 \times 277.9 = 1.869 \text{ K}$$

$$q_f = kA_x \frac{dT}{dx}\bigg|_{x=L} = kA_x m (c_1 \sinh mL + c_2 \cosh mL)$$

$$= 59.0 \times 0.003 \times 4.497(277.9 \times \sinh 0.899 + 1.869 \times \cosh 0.899)$$

$$= \underline{229 \text{ W}}$$

COMMENTS

1. Fin efficiency: The maximum heat transfer rate is obtained by setting the fin temperature equal to the base temperature (= 400 K).

$$q_{\text{f,max}} = A_s \sigma \varepsilon T_b^4$$

$$= (2.006 \times 0.2 + 0.003 \times 1)5.67 \times 10^{-8} \times 0.81 \times 400^4$$

$$= 475 \text{ W}$$

$$\eta = \frac{229}{475} = 0.48$$

2. Note that the fin efficiency may also be defined by determining $q_{\text{f,max}}$ from a perfect conductor ($k \to \infty$) *and* a perfect emitter ($\varepsilon = 1$). On this basis $\eta = 0.48 \times \varepsilon = 0.48 \times 0.81 = 0.39$.

3. Heat transfer rate from the free end surface denoted by q_o is given by

$$q_o = A_x \sigma \varepsilon T_o^4 = 0.003 \times 5.67 \times 10^{-8} \times 0.81 \times 277.9^4 = 0.82 \text{ W}$$

Compared with the total heat transfer rate (= 229 W) the heat transfer rate from the end surface is negligibly small. The free end may, therefore, be considered as perfectly insulated for all practical purposes.

Numerical Solution   The second method is to integrate Equation 4a numerically. With reference to Figure 8.5.2 the nodal equation is

$$q_{i-1} + q_{i+1} - q_R = 0 \tag{8}$$

$$kwt \frac{T_{i-1} - T_i}{\Delta x} + kwt \frac{T_{i+1} - T_i}{\Delta x} - \sigma \varepsilon 2 \Delta x \, PT_i^4 = 0$$

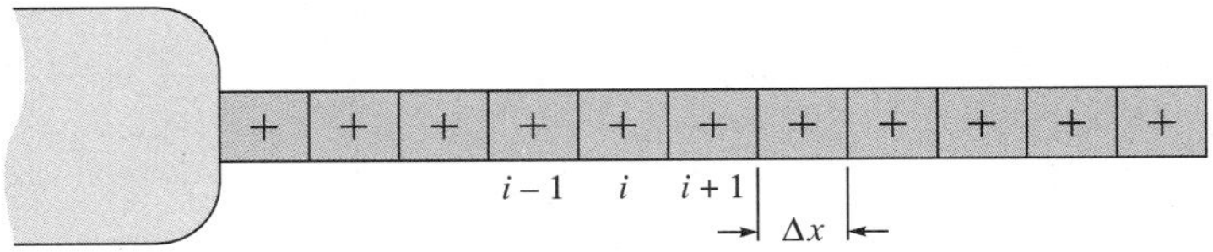

FIGURE 8.5.2 Nodes for Numerical Integration

The nodal equation is written for every node. Although the number of equations equals the number of the unknowns T_is, a direct solution for the unknown nodal temperatures is not easily available because of the nonlinear term containing T_i^4. The nonlinear term is linearized by using the radiative heat transfer coefficient to express the radiative heat transfer term.

$$\sigma \varepsilon 2 \, \Delta x \, PT_i^4 = h_{Ri} 2 \, \Delta x \, PT_i \qquad h_{Ri} = \sigma \varepsilon T_i^3$$

where h_{Ri} is the nodal radiative heat transfer coefficient. The nodal equation is then written as

$$T_{i-1} - \left(2 + \frac{h_{Ri} 2 \, \Delta x \, P}{kwt} \right) T_i + T_{i+1} = 0 \tag{9}$$

The solution proceeds in the following manner.

1. Assume reasonable values for the nodal temperatures.
2. Compute the nodal radiative heat transfer coefficient.
3. Write the nodal equation for each node, Equation 9.

4. Solve the resulting system of linear, simultaneous equations. A direct solution is obtained by employing the Tri-Diagonal Matrix Algorithm. See Section 3.3.3 for details of the algorithm and Appendix 7 for a typical FORTRAN program.
5. Update the values of h_{Ri} employing the computed values of the nodal T_i.
6. Repeat computations until two successive sets of the nodal temperatures are within a predefined convergence criterion.

The numerical computations are left as an exercise to the student.

An alternate method to numerically integrate Equation 4a is by the well-established Runge-Kutta procedure.

PROJECT 8.5.2 In Project 8.5.1 the heat transfer rate from a single fin was determined. In many applications an array of fins is used to increase the heat transfer rate from a surface. Consider an array of vertical, painted, plain carbon-steel fins of rectangular cross section as shown in Figure 8.5.3. The convective heat transfer coefficient is 5.5 W/m² °C.

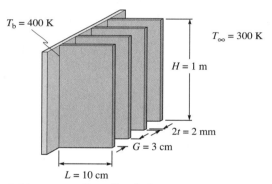

FIGURE 8.5.3 Details of Fins

The fins are 1-m high, 2-mm thick, and 10-cm long. They are spaced 3-cm apart. The emissivities of the painted surface and the base are very close to unity. The surface to which the fins are attached is at 400 K and the ambient air and surrounding surfaces are at 300 K. Determine the total heat transfer rate from one fin.

The inclusion of radiative heat transfer results in nonlinear equations, which are solved by numerical methods. Assuming the number of fins to be large, the temperature distribution in all the fins (except the outermost fins) is identical. We, therefore, determine the temperature distribution in and heat transfer rate from one interior fin.

ASSUMPTIONS

1. All fins and the base are black.
2. Steady state
3. Constant properties
4. The height is much greater than the gap G and, for determining the view factors, the fins are considered to be infinitely long in the vertical direction.

5. The boundary layer thickness is less than half the gap. Assuming laminar boundary layer, employ Equation 14 of Example 5.3.4. With $T_s = 400$ K and $T_\infty = 300$ K, $Gr_L = 7.69 \times 10^9$ and $\delta_{max} \approx 12$ mm. As the maximum boundary layer thickness is less than half the spacing between the fins, it is assumed that the convective heat transfer rate is given by

$$\delta q_c = h(T_s - T_\infty) \, dA$$

where T_∞ is 300 K.

Because of the symmetry of the fins, only one-half the thickness of each fin need be considered. An energy balance on a node i on fin A yields (Figure 8.5.4)

$$q_{i-1} + q_{i+1} - \delta q_{ic} - \delta q_{iR} = 0 \tag{1}$$

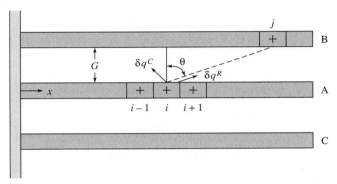

FIGURE 8.5.4 Energy Balance on a Node

where

$$q_{i-1}, q_{i+1} = \text{conductive heat transfer rate from nodes } i - 1 \text{ and } i + 1 \text{ to node } i$$
$$\delta q_{ic} = \text{convective heat transfer rate from node } i$$
$$\delta q_{iR} = \text{radiative heat transfer rate from node } i$$

The term δq_{iR} is the sum of the radiative heat transfer rates from node i to each node on fin B, the base (at T_b) and the surroundings. Each term is evaluated separately.

Denoting the heat transfer rate from node i to node j on B by $\delta^2 q_{i\text{-}j,R}$

$$\delta^2 q_{i\text{-}j,R} = \delta A_i \, \delta F_{i\text{-}j}(E_{bi} - E_{bj}) \tag{2}$$

where δA_i is the surface area of element $i = \Delta x \, H$; H is the height of the fin.

As the gap G between A and B is much smaller than the height H, fins A and B are assumed to be infinitely long in the vertical direction. The view factor $\delta F_{i\text{-}j}$ is evaluated by Hottel's crossed-uncrossed string method (Figure 8.5.5).

Uncrossed strings:

$$LM = NP = [(x_j - x_i)^2 + G^2]^{1/2}$$

Crossed strings:

$$MN = \left\{ \left[\left(x_j + \frac{\Delta x}{2} \right) - \left(x_i - \frac{\Delta x}{2} \right) \right]^2 + G^2 \right\}^{1/2} = [(x_j - x_i + \Delta x)^2 + G^2]^{1/2}$$

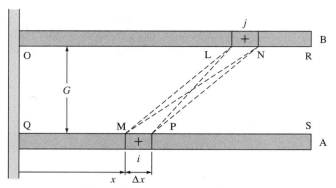

FIGURE 8.5.5 Two Elements on A and B for Evaluation of $\delta F_{i\text{-}j}$

Similarly,

$$\text{LP} = [(x_j - x_i - \Delta x)^2 + G^2]^{1/2}$$

$$\delta F_{i\text{-}j} = \frac{\overline{\text{LP}} + \overline{\text{MN}} - 2\,\overline{\text{LM}}}{2}$$

$$= \frac{[(x_j - x_i + \Delta x)^2 + G^2]^{1/2} + [(x_j - x_i - \Delta x)^2 + G^2]^{1/2} - 2[(x_j - x_i)^2 + G^2]^{1/2}}{2}$$

The total radiative heat transfer rate from node i to fin B is given by

$$\delta q_{i\text{-}B,R} = \delta A_i \sum_{j=1}^{N} (E_{bi} - E_{bj})\,\delta F_{i\text{-}j} \tag{3}$$

where N is the total number of nodes on the fin.

The view factor from i to the base b is similarly evaluated and is given by

$$\delta F_{i\text{-}b} = \frac{\{[x_1 - (\Delta x/2)]^2 + G^2\}^{1/2} + \Delta x - \{[x_i + (\Delta x/2)]^2 + G^2\}^{1/2}}{2} \tag{4}$$

$$\delta q_{i\text{-}b} = \delta A_i\,\delta F_{i\text{-}b}(E_{bi} - E_{bb}) \qquad E_{bb} = \sigma T_b^4$$

Finally, the radiative heat transfer rate from the node to the surroundings is

$$\delta q_{surr,R} = \delta A_i\,\delta F_{i\text{-}surr}(E_{bi} - E_{bsurr}) \qquad E_{bsurr} = \sigma T_\infty^4 \tag{5}$$

$$\delta F_{i\text{-}surr} = \frac{\{[L - x_i - (\Delta x/2)]^2 + G^2\}^{1/2} + \Delta x - \{[L - x_i + (\Delta x/2)]^2 + G^2\}^{1/2}}{2}$$

where L is the length of the fin.

The convective heat transfer rate from the node is

$$\delta q_{i,c} = \delta A_i h (T_i - T_\infty) \tag{6}$$

The conductive heat transfer rates are

$$q_{i-1} = kHt\,\frac{T_{i-1} - T_i}{\Delta x} \qquad q_{i+1} = kHt\,\frac{T_{i+1} - T_i}{\Delta x} \tag{7}$$

Substituting Equations 2, 3, 4, 5, 6, and 7 into Equation 1 and rearranging, we obtain

$$T_{i+1} + T_{i-1} - \left(2 + \frac{h\,\Delta x^2}{kt}\right) T_i - \frac{\Delta x^2 \sigma}{kt}$$

$$\left[\delta F_{i\text{-}b}(T_i^4 - T_b^4) + \sum_{j=1}^{N} (T_i^4 - T_j^4)\delta F_{i\text{-}j} + \delta F_{i\text{-}surr}(T_i^4 - T_\infty^4)\right] = 0 \quad \textbf{(8)}$$

Equation 8 is written for every node. Note that T_j on fin B equals T_j in fin A. There are N such equations in the N unknown T_i. To solve the equations, the terms containing T_i^4 are linearized by using the radiative heat transfer coefficient. For example, the term $\sigma(T_i^4 - T_j^4)$ is written as

$$\sigma(T_i^4 - T_j^4) = h_{R,ij}(T_i - T_j) \qquad h_{R,ij} = \sigma(T_i^2 + T_j^2)(T_i + T_j)$$

The procedure for solving for the unknown temperature is

1. Assume reasonable values for the unknown temperatures.
2. Compute the radiative heat transfer coefficients $h_{R,ij}$.
3. Linearize Equation 8 by substituting the computed radiative heat transfer coefficients.
4. Solve for the unknown temperatures.
5. If the computed T_i are different from those used for computing $h_{R,ij}$, recompute $h_{R,ij}$ with the latest available T_i and repeat the computations.
6. Compute the total heat transfer rate

$$q = \Sigma(\delta q_{ic} + \delta q_{iR})$$

PROJECT 8.5.3 A 2-cm diameter, horizontal steam pipe is used as a space heater (Figure 8.5.6). The temperature of the outer surface of the pipe is 400 K and the surrounding air and surfaces are at 293 K. To increase the heat transfer rate to the surroundings, it is proposed to wrap it with a material of thermal conductivity 1 W/m K. There is, of course, skepticism about increasing the heat transfer rate with a material that has a low thermal conductivity. Compute the sum of convective and radiative heat transfer rate for wrapping thickness in steps of 4 mm and plot the heat transfer rate per meter

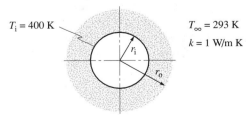

$T_i = 400$ K $T_\infty = 293$ K
                  $k = 1$ W/m K

FIGURE 8.5.6 Steam Pipe Wrapped with a Material of Low Thermal Conductivity

length of the pipe against the outer radius of the pipe. Find the thickness that yields the maximum heat transfer rate. Assume that the temperature of the outer surface of the pipe is 400 K for all thicknesses of the wrapping material. The emissivity of the surface of the wrapping material is 0.8. Assume natural convection from the outer surface.

The heat transfer rate is found by solving for the outer surface temperature. The outer surface temperature is found by equating the conductive heat transfer rate across the wrapping material to the sum of the convective and radiative heat transfer rate from the outer surface of the material. Thus,

$$\frac{2\pi k L (T_i - T_s)}{\ln(r_o/r_i)} = 2\pi r_o L [h(T_s - T_\infty) + \sigma\varepsilon(T_s^4 - T_\infty^4)] \tag{1}$$

The left-hand side of Equation 1 is the conductive heat transfer rate; the right-hand side is the sum of the convective and radiative heat transfer rates. The convective heat transfer coefficient h depends on the as yet unknown T_s. The outer surface temperature is determined by iteration. One method of iteration follows:

1. Assume a value for T_s.
2. Compute h from the software (CC).
3. Employing Equation 1, solve for T_s.
4. If the computed value of T_s is significantly different from the value used for computing h, repeat steps 2 and 3 using the latest available value of T_s for computing h.

After determining T_s for a given value of r_o determine the heat transfer rate from the left-hand side of Equation 1. Repeat computations for different values of r_o. Computations are left as an exercise.

SUMMARY

In Chapter 7 the discussion of radiation was essentially limited to gray, diffuse, and opaque surfaces with a nonparticipating medium separating the surfaces. In this chapter the *wavelength* and *direction dependence* of the radiative properties of surfaces with emphasis on *emissivity* and *absorptivity* are considered. View factors are presented in greater detail and an introduction to radiation in gases is presented.

In many cases the wavelength dependence of the properties is more significant than the direction dependence. The *total hemispherical emissivity* of a surface (with negligible direction dependence) is given by

$$\varepsilon(T) = \frac{\int_0^\infty \varepsilon_\lambda(T)E_{b\lambda}(T)\,d\lambda}{E_b(T)} \qquad T = \text{surface temperature}$$

$$\alpha(T, T_s) = \frac{\int_0^\infty \alpha_\lambda(T)G_\lambda(T_s)\,d\lambda}{G(T_s)} \qquad T_s = \text{source temperature}$$

At a specified temperature the total hemispherical emissivity has a unique value as it depends only on the surface temperature. The total hemispherical absorptivity depends not only on the surface temperature but also on the source radiation characteristics and, therefore, does not have a unique value.

The concept of view factors has been elaborated and the view factor relations stated in Chapter 7 (Equation 7.2.4) are derived from the mathematical expressions for the view factor between two surfaces. One method for computing the view factors between two infinitely long surfaces—the method of crossed and uncrossed strings—is presented. Although view factors can be computed from purely mathematical considerations, they are used with diffuse, opaque surfaces with uniform radiosities.

Two methods for computing the radiative heat transfer rates from gray, diffuse, and opaque surfaces in an N-surface enclosure are provided. In both methods the radiosities of surfaces are found from the solution to a set of linear, simultaneous equations.

In the *electrical analogy* an energy balance on the surfaces yields

Specified surface temperature
$$\frac{E_{bi} - J_i}{(1 - \varepsilon_i)/A_i\varepsilon_i} + \sum_{\substack{k=1 \\ k \neq 1}}^{N} \frac{J_k - J_i}{1/(A_iF_{i\text{-}k})} = 0$$

Specified heat flux
$$q_i + \sum_{\substack{k=1 \\ k \neq 1}}^{N} \frac{J_k - J_i}{1/(A_iF_{i\text{-}k})} = 0$$

Black surfaces
$$q_i + \sum_{\substack{k=1 \\ k \neq 1}}^{N} \frac{J_k - E_{bi}}{1/(A_iF_{i\text{-}k})} = 0$$

We, then, have N simultaneous, linear equations in N unknowns. The solution to the equations yield the radiosities of the surfaces. For nonblack surfaces, from the relation

$$q_i = \frac{A_i\varepsilon_i}{1 - \varepsilon_i}(E_{bi} - J_i)$$

the heat transfer rate (for specified surface temperature) or the surface temperature (for specified heat flux) is computed. The temperature of a black surface with a specified heat flux is found from the calculated surface emissive power. The heat transfer rate from a black surface with specified surface temperature is found from

$$q_i = \sum_{k=1}^{N} A_iF_{i\text{-}k}(E_{bi} - J_k)$$

In the *radiosity method* the procedure for setting up the equations for the solutions to the surface radiosities is different. The resulting equations are

Specified surface temperature
$$(1 - \rho_iF_{i\text{-}i})J_i - \rho_i \sum_{\substack{k=1 \\ k \neq i}}^{N} F_{i\text{-}k}J_k = \varepsilon_iE_{bi}$$

Specified heat flux
$$(1 - F_{i\text{-}i})J_i - \sum_{\substack{k=1 \\ k \neq i}}^{N} F_{i\text{-}k}J_k = \frac{q_i}{A_i}$$

For a black surface with specified temperature, its radiosity, being equal to its emissive power, is known and its emissive power is substituted for its radiosity in all the equations. The heat flux of surfaces with specified temperature or the temperature for surfaces with specified heat flux is found in the same way as in the electrical network method. The equivalence of the two methods is shown.

Solutions to radiative heat transfer rates with an *absorbing, emitting medium* are complex. Many common liquids such as water and oils are essentially opaque to radiation. Many gases, such as oxygen and nitrogen, are practically nonparticipating but some gases such as carbon dioxide, water vapor, and carbon monoxide, even in small concentrations, have a significant effect by absorption at room temperatures and by absorption and emission at elevated temperatures.

An introduction to the radiation properties of gases is provided. A method for computing the heat transfer rate from a gas at a uniform temperature completely enclosed by a black surface (or a gray surface with $\varepsilon > 0.8$) also at a uniform but different temperature is presented. The concept of *mean beam length* is discussed and its use in finding the heat transfer rate from a gas is illustrated. Mean beam lengths for a few configurations are given in Table 8.4.1. Charts for finding the emissivity and absorptivity of CO_2 and H_2O vapor in conjunction with the mean beam length are included. The effective emissivity and absorptivity of a mixture of gases containing CO_2 and H_2O vapor are given by

$$\varepsilon_g = C_{H_2O}\varepsilon_{H_2O} + C_{CO_2}\,\varepsilon_{CO_2} - \Delta\varepsilon$$
$$\alpha_g = \alpha_{CO_2} + \alpha_{H_2O} - \Delta\alpha$$
$$\alpha_{H_2O} = C_{H_2O}\left(\frac{T_g}{T_s}\right)^{0.45}\varepsilon_{H_2O}\left(T_s, P_{H_2O}L\frac{T_s}{T_g}\right)$$
$$\alpha_{CO_2} = C_{CO_2}\left(\frac{T_g}{T_s}\right)^{0.65}\varepsilon_{CO_2}\left(T_s, P_{CO_2}L\frac{T_s}{T_g}\right)$$
$$\Delta\alpha = \Delta\varepsilon$$

Black surface
$$q = [\varepsilon_g(T_g)E_b(T_g) - \alpha_g(T_g, T_s)E_b(T_s)]A_s$$

Gray surface ($\varepsilon > 0.8$)
$$q_{gray} = q_{black}\,\frac{\varepsilon + 1}{2}$$

REFERENCES

Brewster, M.Q. (1991). *Thermal Radiative Transfer and Properties*. New York: Wiley.

Clark, J.A., and Korybalski, M.E. (1974). Algebraic methods for the calculation of radiation heat exchange in an enclosure. *Warme- und Stoffugertragung,* 7:31.

Edwards, D.K. (1981). *Radiation Heat Transfer Notes*. Washington, D.C.: Hemisphere.

Howell, J.R. (1982). *A Catalog of Radiation Configuration Factors*. New York: McGraw-Hill.

Hottel, H. (1954). Radiant heat transmission. In *Heat Transmission,* ed. McAdams, W.C. New York: McGraw-Hill.

Hottel, H.C., and Egbert, R.B. (1942). Radiant Heat Transmission from Water Vapor, *AIChE* Transactions, 38:531.

Hottel, H.C., and Sarofim, A.F. (1967). *Radiative Heat Transfer*. New York: McGraw-Hill.

Love, T.J. (1968). *Radiative Heat Transfer*. Columbus, OH: Merril.

Planck, M. (1959). *The Theory of Heat Radiation.* New York: Dover.

Siegel, R., and Howell, J.R. (1992). *Thermal Radiation Heat Transfer,* 3rd ed. Washington, D.C.: Hemisphere.

Sparrow, E.M. (1963). On the calculation of radiant interchange between surfaces. In *Modern Developments in Heat Transfer,* ed. Ibele, W. New York: Academic Press.

Sparrow, E.M., and Cess, R.D., (1978). *Radiation Heat Transfer,* aug. ed. New York: McGraw-Hill.

Thekaekara, M.P. (1965). Survey of the Literature on the Solar Constant and the Spectral Distribution of Solar Radiant Flux. NASA SP-74.

REVIEW QUESTIONS

(a) Define monochromatic intensity of radiation.

(b) Define total hemispherical emissivity.

(c) Kirchhoff's Law states that under equilibrium conditions, $\varepsilon_\lambda = \alpha_\lambda$. For a surface with wavelength dependent properties, explain why the total hemispherical emissivity can be different from the total hemispherical absorptivity.

(d) For a nongray surface, explain why, at a fixed surface temperature, there is no unique value of the total absorptivity for the surface.

(e) A black surface has the highest absorptivity of all surfaces. Explain why a surface that has high emissivity for short wavelengths (< 2 μm) and low emissivity for long wavelengths (> 2 μm) is preferred to a black surface in solar collectors.

(f) A greenhouse has an enclosure that has a high transmissivity at short wavelengths and very low transmissivity (nearly opaque) for long wavelengths. Why does a greenhouse get warmer than the surrounding air during clear days? Will it have a similar effect during clear nights?

(g) View factors as given by Equation 8.2.6a can be evaluated without any knowledge of radiation. State the condition to be satisfied for the view factors to be used in radiation heat transfer.

(h) State the law of corresponding corners.

(i) What is the difference between the reciprocity relation and the law of corresponding corners for view factors?

(j) In gaseous radiation, why is its total emissivity evaluated at its temperature and why is its total absorptivity affected by the temperature of the bounding surfaces?

PROBLEMS

In all cases involving solar energy, assume the sun behaves as a black surface at 5760 K

8.1 A special surface has wavelength dependent properties, approximated as shown in the figure. Find the hemispherical emissivity of the surface at 400 K and its total absorptivity for incident radiation from a blackbody at 6000 K.

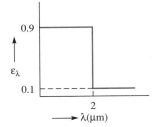

FIGURE P8.1

8.2 The emissivity of a surface is shown in Figure P8.2. Determine the emissive power of the surface if the surface is maintained at
   (a) 300 K
   (b) 1200 K

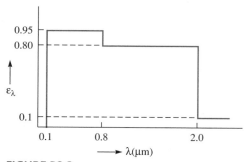

FIGURE P8.2

8.3 If the surface in Problem 8.2 receives solar radiation of 900 W/m^2, calculate the solar energy absorbed by the surface and the solar absorptivity for the surface.

8.4 For the surface in Problem 8.1, determine the solar absorptivity.

8.5 One possible method for heating a space shuttle is by solar energy. A panel with properties of the surface in Problem 8.1 is attached to a space shuttle. The solar radiation on a surface normal to the solar beam is 1350 W/m^2. Determine the equilibrium temperature of the panel if
   (a) The panel is normal to the solar beam.
   (b) The panel is inclined at 30° to the solar beam. Assume a sky temperature of 0 K.

8.6 The emissivity of a 30-cm diameter circular disk is approximated as $\varepsilon_\lambda = 0.9$ for $0 < \lambda < 6$ μm and $\varepsilon_\lambda = 0.1$ for 6 μm $< \lambda$. The disk is completely enclosed by a 30-cm diameter black hemisphere. If the disk is at 600 K and the hemisphere is at 400 K, determine the radiative heat transfer rate from the disk to the hemisphere.

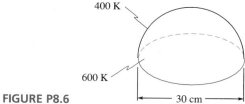

FIGURE P8.6

8.7 The monochromatic emissivity of the absorber plate of a solar collector is modeled as,

$$0 < \lambda < 2 \text{ μm} \qquad \varepsilon_\lambda = 0.95$$
$$2 \text{ μm} < \lambda < \infty \qquad \varepsilon_\lambda = 0.1$$

The absorber plate receives solar energy at 900 W/m^2. The collector is kept at 330 K by circulating water through tubes attached to the surface. The surrounding air is at 300 K. The convective heat transfer coefficient between the surface and the surrounding air is 4 W/m^2 °C. The effective surrounding temperature is 300 K. The collector is 1-m wide and 2-m long. Determine the heat transfer rate to the coolant. Assume uniform surface temperature.

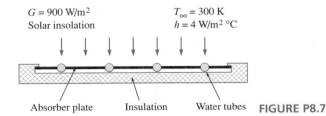

Absorber plate   Insulation   Water tubes   **FIGURE P8.7**

8.8 In Problem 8.7 determine the heat transfer rate to the cooling water if the absorber plate is black.

8.9 In Problem 8.7 determine the equilibrium temperature of the absorber plate if the flow of the cooling water is stopped and if
   (a) The properties of the plate are those given in Problem 8.7.
   (b) After prolonged use, the plate behaves as a black surface.

8.10 The monochromatic reflectivity of a black-chrome-on-nickel surface is modeled as,

$0 < \lambda < 1.5 \ \mu m$ $\rho_\lambda = 0.8$
$1.5 \ \mu m < \lambda < 3 \ \mu m$ $\rho_\lambda = 0.3$
$3 \ \mu m < \lambda < \infty$ $\rho_\lambda = 0.9$

Determine
   (a) The total absorptivity of the surface for
      (i) Solar radiation.
      (ii) Irradiation from a black surface at 1000 K.
   (b) The total emissivity of the surface at 400 K.

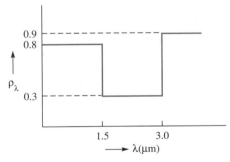

FIGURE P8.10

8.11 In Problem 8.10 determine the total absorptivity of the surface if the source for irradiation is a similar surface at 1000 K and has the same wavelength dependent properties.

8.12 The transmissivity of a nonreflecting glass is modeled as

$0 < \lambda < 0.2 \ \mu m$ $\tau_\lambda = 0$
$0.2 \ \mu m < \lambda < 3 \ \mu m$ $\tau_\lambda = 0.92$
$3 \ \mu m < \lambda < \infty$ $\tau_\lambda = 0$

The glass is used as a cover plate in a solar collector. The black absorber plate is heated by the solar insolation, and the plate is maintained at 330 K by a coolant as shown. For a solar insolation of 900 W/m², determine per unit area of the absorber plate,
   (a) The solar energy transmitted by the cover plate.
   (b) The energy transmitted by the cover plate due to emission from the absorber plate.

(c) Neglecting convection, the equilibrium temperature of the cover plate (assume that the surroundings are at 300 K).

(d) The heat flux to the cooling water.

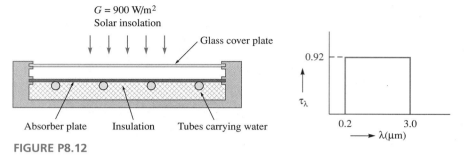

FIGURE P8.12

8.13 The wavelength integrated emissivity of a surface is direction dependent as shown in the figure. Determine the total emissivity of the surface.

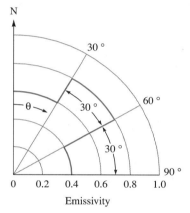

FIGURE P8.13

8.14 Show that the radiant energy emitted by a diffuse, gray surface of area dA in a cone with its apex at the element (and the axis along the normal to the surface) is given by, $dE = \varepsilon E_b \sin^2\theta \, dA$, where 2θ is the angle of the cone.

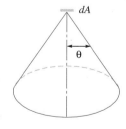

FIGURE P8.14

8.15 A radiation thermometer detects the incident radiant energy centered around a specified wavelength. The instrument is calibrated to indicate the temperature of a blackbody that has the same spectral intensity as the incident energy. If such a thermometer, constructed to detect the radiation intensity centered around 1 μm is pointed towards a source at 1000 K, what is the temperature indicated by the instrument if

(a) The surface is black and the medium separating the instrument and the surface does not participate in radiation?

(b) The surface is gray with an emissivity of 0.8 and the medium between the instrument and the surface does not participate in radiation?

(c) The surface is gray with an emissivity of 0.8 and absorptivity of the medium between the instrument and the surface at 1 μm is 0.05?

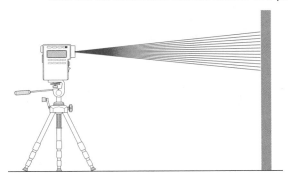

FIGURE P8.15

8.16 Consider a long extended surface of constant area of cross section and perimeter in outer space where the heat transfer is by radiation only. If the effective sky temperature is T_{sky} and the temperature at the base is T_B, derive the differential equation for the temperature distribution.

8.17 A circular surface, 2-cm diameter, has emissivity shown in Problem 8.1. It receives radiant energy from a black coaxial circular disk of radius 2 m at 2000 K placed at a distance of 3 m from the surface. Determine the energy absorbed by the surface.

8.18 The solar irradiation on a panel is 900 W/m². If the back of the panel (the surface not exposed to the sun) is perfectly insulated, find the equilibrium temperature of the panel if

(a) The panel surface is perfectly black.

(b) The surface has emissivity shown in Problem 8.2.

Assume negligible convection heat transfer and an effective sky temperature of 250 K.

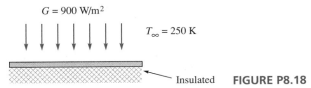

Insulated   **FIGURE P8.18**

8.19 Re-solve Problem 8.18 taking into consideration convection heat transfer to the surrounding air. The air is at 10 °C and the convective heat transfer coefficient is 8 W/m² K.

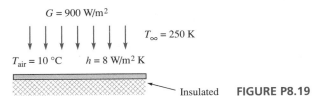

Insulated   **FIGURE P8.19**

8.20 Roof coverings are available with different radiation properties. Three such coverings have properties as shown. What is your choice for the covering
(a) In tropical climates?
(b) In cold climates?

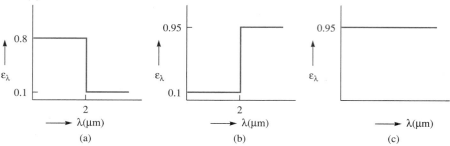

FIGURE P8.20

8.21 In space ships heat rejection to surroundings is by radiation. A radiator in a space ship has an absorptivity of 0.1 for $0 < \lambda < 2$ μm and 0.95 for 2 μm $< \lambda$. The surface is maintained at 600 K. What is the net heat flux from the surface when solar irradiation on the surface is 1200 W/m²? Assume a sky temperature of 0 K.

8.22 Re-solve Problem 8.21 if the special surface is replaced by a black surface.

8.23 It is proposed to design a radiation thermometer. The thermometer has a sensor whose output is dependent on the incident radiation in the wave band 3 μm to 10 μm. It is calibrated to detect the temperature of a black surface. It has a view of 1° and the collecting surface is a 3-mm diameter circular disk. Determine the energy incident on the instrument from a black wall at 500 K that is at a distance of 3 m. What are the factors that influence the temperature indicated by the instrument even if the wall is kept at a uniform temperature?

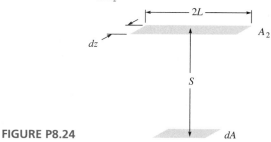

FIGURE P8.23

8.24 A small elemental area dA faces a narrow, rectangular strip A_2, of length $2L$ and width dz and separated by a distance S. Find $dF_{dA\text{-}A_2}$.

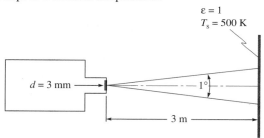

FIGURE P8.24

8.25 Determine the view factor of a cylinder of radius R, which is parallel to another cylinder of the same radius with the axes S apart if the axial length of the cylinders is very much greater than the distance between the two cylinders.

8.26 Determine the view factor, $F_{1\text{-}2}$, for a long cylinder 1, of diameter 20 cm, and a cylinder 2, of diameter 10 cm, if the distance between the two axes is 40 cm. The axes are parallel.

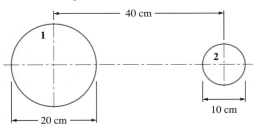

FIGURE P8.26

8.27 Determine $F_{1\text{-}2}$ for two long semicircular surfaces as shown.

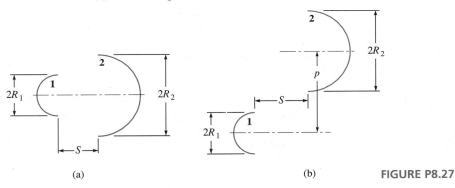

(a)   (b)   **FIGURE P8.27**

8.28 A furnace is constructed in the form of a tunnel. End surface 1 is electrically heated and maintained at 1200 K. Cylindrical surface 2, also electrically heated, is at 1100 K. Cylindrical surface 3 is perfectly insulated. End surface 4 is the load. Treating the furnace as a four-surface enclosure, find the radiant heat transfer rate to the load when its temperature is
(a) 300 K
(b) 600 K
(c) 900 K
All surfaces are black.

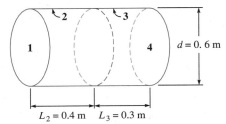

FIGURE P8.28

8.29 Employing the electrical analogy, find q_4 in Problem 8.28, if $\varepsilon_1 = 0.7$, $\varepsilon_2 = 1$, and $\varepsilon_4 = 0.8$.

8.30 Re-solve Problem 8.29 employing the radiosity method.

8.31 A room is to be heated by heaters embedded in the ceiling. The room is 4-m wide, 5-m long and 2.5-m high. The ceiling is maintained at 50 °C. Two adjacent walls are perfectly insulated and the other two adjacent walls are at 15 °C. The floor is at 30 °C. Employing the electrical analogy determine the radiant heat transfer rate from each surface treating the room as a four-surface enclosure. The emissivity of each surface is 0.8.

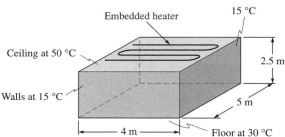

FIGURE P8.31

8.32 Re-solve Problem 8.31 employing the radiosity method.

8.33 Using the radiosity method, find the heat transfer rates in Problem 8.31 if the emissivity of each surface is increased from 0.8 to 0.9.

8.34 In Problem 8.31, instead of controlling the temperature of the ceiling, the heat transfer rate from the ceiling is maintained at 3500 W. Determine the temperature of the ceiling and the heat transfer rates to the floor and the side walls at 15 °C.

8.35 The radius of a long, semicylindrical furnace is 1 m. It is fabricated by welding four sheets (labeled 1, 2, 3, and 4) as shown. A heat flux of 20 kW/m² is imposed on the cylindrical surface. When the temperature of the diametral base is 400 K, determine the temperature of the cylindrical surface and the heat flux to the base surfaces 3 and 4. $\varepsilon_1 = \varepsilon_2 = 0.7$, $\varepsilon_3 = \varepsilon_4 = 0.9$.

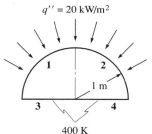

FIGURE P8.35

8.36 In the furnace in Problem 8.35, on a closer examination it is found that due to a mix-up in fabrication, $\varepsilon_1 = \varepsilon_4 = 0.7$ and $\varepsilon_2 = \varepsilon_3 = 0.9$. Considering the furnace as a four-surface enclosure and employing the electrical analogy, find T_1, T_2, q_3'', and q_4''.

8.37 Re-solve Problem 8.36 employing the radiosity method.

8.38 A long furnace is 2-m wide, and 1-m high. The top surface 1 ($\varepsilon_1 = 0.8$) is at 900 K, and the bottom surface 2 ($\varepsilon_2 = 0.7$) is at 400 K. The side walls are perfectly insulated. Determine the heat flux from the top surface, considering the furnace as

(a) A three-surface enclosure consisting of the top surface, the bottom surface, and the side walls.

(b) A four-surface enclosure consisting of the top surface, bottom surface, top half of the side walls, and bottom half of the side walls.

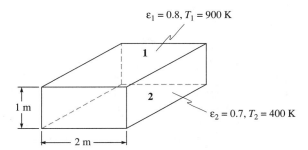

$\varepsilon_1 = 0.8, T_1 = 900$ K

1

2

$\varepsilon_2 = 0.7, T_2 = 400$ K

1 m

2 m

FIGURE P8.38

8.39 Re-solve Problem 8.38a for a furnace with a finite length of 6 m with the surroundings at 300 K.

8.40 As an approximate model, CO_2 may be considered to emit and absorb in four wave bands, 1.8 μm–2.2 μm, 2.6 μm–3.0 μm, 4.0 μm–5.0 μm, and 9 μm–20 μm. For a thick layer of CO_2, its monochromatic emissivity is unity in the four wave bands. Determine its total emissivity at 500 K, 1000 K, and 2000 K.

8.41 Re-solve Problem 8.40 if the monochromatic emissivity of CO_2 is as given in Example 8.4.1.

8.42 Determine the total absorptivity of CO_2 for solar radiation
   (a) If it is sufficiently thick that the monochromatic absorptivity is unity in the four wave bands in Example 8.4.1.
   (b) If the monochromatic absorptivities are as given in Example 8.4.1.

8.43 A thin, black 1-cm diameter disk is placed at the center of a black hemisphere filled with CO_2 at 1200 K. The surface of the hemisphere is at 1000 K. If the monochromatic emissivity of CO_2 is as given in Example 8.4.1, estimate the radiant energy absorbed by the disk.

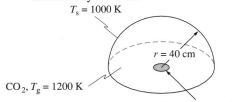

$T_s = 1000$ K

$r = 40$ cm

$CO_2, T_g = 1200$ K

1-cm diameter disk   **FIGURE P8.43**

8.44 A furnace is in the form of a cube, with each edge 1-m long. The furnace is filled with 30% CO_2, and 70% N_2 by volume. The gases are pressurized to 1.5 atm and maintained at 1000 K. The black wall surfaces are at 800 K. What is the radiant heat transfer rate to the walls?

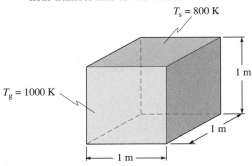

$T_s = 800$ K

1 m

1 m

$T_g = 1000$ K

1 m

FIGURE P8.44

8.45 The furnace of a coal-fired boiler is 8-m long, 6-m wide, and 6-m high. The products of combustion (at atmospheric pressure) contain 14% CO_2, 4.8% H_2O vapor, and the balance N_2 by volume. The temperature of the gases in the furnace is 1100 K. The walls of the combustion chamber are at 700 K and have an emissivity of 0.9. Determine the radiant heat transfer rate from the gases to the walls.

8.46 A plate heat exchanger consists of a series of black, parallel plates. Flue gases at atmospheric pressure flow between alternate pairs of plates, and water in between the remaining alternate passages. Flue gases, containing CO_2 at 0.08 atm and H_2O at 0.1 atm, are at 600 °C. The width and height of the plates are much greater than the spacing between them. The plates are at 250 °C. What is the net radiative heat flux to the plates?

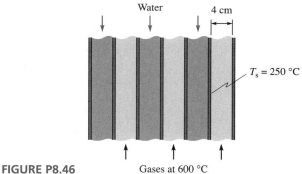

FIGURE P8.46

8.47 A domestic water heater is heated by combustion gases flowing in a 10-cm diameter tube. The tube is very long. The products of combustion contain 8.6% CO_2, 15.2% H_2O, and the balance N_2. The average temperature of the gases is 800 °C, and the surface of the tube is at 200 °C. Determine the radiant heat transfer rate per meter length of the tube, if the emissivity of the inner surface of the tube is 0.9.

8.48 Re-solve Problem 8.47 if the length of the tube is 1.5 m.

8.49 The products of combustion (per kilomole of fuel oil) in a boiler contain 7.36 kmol of carbon dioxide, 6.34 kmol of water vapor, and 46.22 kmol of nitrogen and oxygen. The products are at an average temperature of 1400 °C. The walls of the furnace are lined with water pipes. The temperature of the surface of the pipes is 300 °C. The furnace is 4-m wide, 10-m long, and 4-m high. Assuming the walls to be black, determine the radiant heat transfer rate to the water pipes.

9

HEAT EXCHANGERS

There are many engineering applications requiring the transfer of thermal energy to or from a fluid stream. In the majority of such applications, heat transfer is by convection between two fluids separated by a solid surface. In some special cases radiative heat transfer to the surroundings may be significant, however. Equipment to accomplish such heat transfer to or from a fluid stream are known as heat exchangers. In this chapter we consider only heat exchangers that transfer thermal energy from one fluid stream to another fluid stream. We now

- *Give an introduction to different types of heat exchangers*
- *Elaborate on the overall heat transfer coefficient*
- *Discuss the concept of logarithmic mean temperature difference and introduce heat exchanger effectiveness*
- *Show how the overall heat transfer coefficient is used in the design of heat exchangers*
- *Show how to estimate the performance of heat exchangers under conditions different from those for which they were designed, i.e., performance under off-design conditions.*

9.0 INTRODUCTION

In heat exchangers heat transfer occurs from one or more fluids. If there are two fluids, they may be separated by a solid surface. If they are not separated by a solid surface but come in contact with each other, the heat exchanger is known as a direct contact heat exchanger. Here, we deal with those heat exchangers in which heat is transferred from one fluid to another separated by a solid surface. Such heat exchangers find applications in many fields. They are used in power plants as boilers, condensers, feedwater heaters, economizers, superheaters, and air heaters; in refrigeration and air conditioning equipment as condensers and evaporators; in processing dairy and food products; in cooling engine cooling water in automobile radiators; in cooling computers, lasers; and in many other applications. Some of the applications are illustrated in Figure 9.0.1.

Problems related to heat exchangers fall into two categories:

1. Design of a heat exchanger to accomplish a defined heat transfer rate (sizing)
2. Estimation of the performance of a heat exchanger under conditions other than those for which it was designed (rating)

739

Sizing of a heat exchanger

Design of a Heat Exchanger To Accomplish a Defined Heat Transfer Rate   In this class of problems, the required heat transfer rate and the conditions in the heat exchanger are known. We then proceed to choose the type of heat exchanger and determine its size (surface area). For example, in the dairy industry the mass rate of milk to be heated for pasteurizing from an initial known temperature to the desired temperature and the fluid that will heat the milk, say, steam, may be specified. In the design of air conditioning equipment, the mass rate of the refrigerant to be condensed, the inlet and exit conditions of the refrigerant to the condenser, and the fluid to which heat is transferred (water or air) are known. In such cases, the heat transfer surface areas to accomplish the defined tasks have to be computed.

Rating of a heat exchanger

Performance of Heat Exchangers under Off-Design Conditions   When heat exchangers are designed for specific conditions, the actual conditions under which they operate may be different from those assumed in the design. For example, a refrigeration condenser may be designed to operate with air available at 20 °C, but the unit may operate in locations where the temperature may be higher or lower. If an air-cooled condenser is operated with multiple fans to force the air through the condensers, the heat transfer rate with one or more fans out of action may be required. Similarly, a heat exchanger in a computer or an automobile operate under different conditions depending on the locale where they are used. In all such cases, one should know how the heat exchanger will perform so that the conditions under which the equipment will operate satisfactorily may be specified. Another example is that of estimating the performance of a heat exchanger designed for one purpose and used for another.

9.1   TYPES OF HEAT EXCHANGERS

Classification of heat exchangers

Several types of heat exchangers are used in practice. Heat exchangers can be classified into five categories.

1. **Transfer Mechanism**
   (a) In direct contact type of heat exchangers, the two fluids are in contact with each other. An example of such a heat exchanger is the cooling tower, where water is cooled by air in direct contact with it. But in most cases, heat exchangers are of the indirect type where the two fluids are separated by a solid surface. Only indirect type heat exchangers are considered here.
   (b) Heat transfer may be by convection in single phase fluids, by phase change, or by radiation.
2. **Surface Compactness.** The classification of heat exchangers on the basis of surface compactness, i.e., the ratio of the heat transfer surface to the volume, is somewhat arbitrary. A heat exchanger with a surface area of more than about 700 m^2/m^3 is classified as a compact heat exchanger [Shah and Mueller (1985)].
3. **Construction.** Heat exchanger construction may be of the shell-and-tube type, the plate type, or with extended surfaces.
4. **Flow Arrangement.** The two fluids may flow in the same direction (parallel flow), in opposite directions (counterflow), in mutually perpendicular directions (cross flow), or in a combination of these types.

(a)

(b)

(c)

FIGURE 9.0.1 Different Types of Heat Exchangers
(a) Integral water tube boiler. Combustion gases transfer heat to water in the boiler by radiation and convection. Courtesy of Babcock & Wilcox, Akron, OH. (b) A heat exchanger to cool neon and highly corrosive fluorine gas in a laser, dissipates 60 kW, and measures 0.6 m × 1.2 m × 7.5 cm. Courtesy of Lytron Inc. Woburn, MA. (c) A heat exchanger on an Air Cushion Landing craft with a capacity of 100 kW to cool transmission oil. Courtesy of Lytron Inc. Woburn, MA.

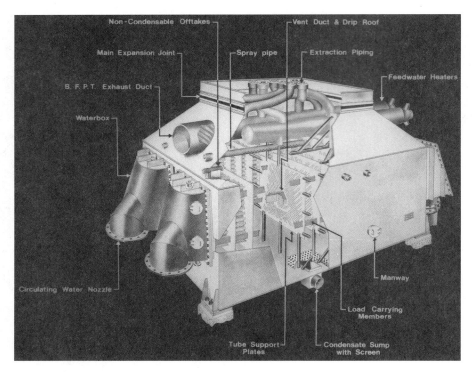

(d)

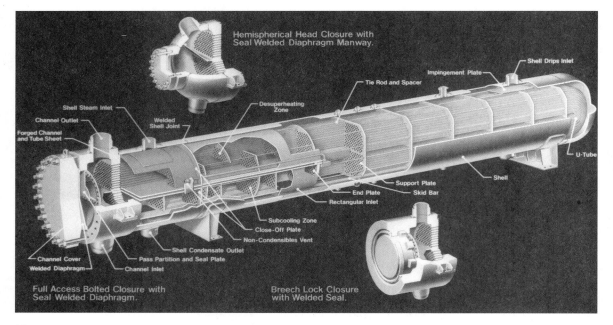

(e)

FIGURE 9.0.1 Different Types of Heat Exchangers (cont'd)

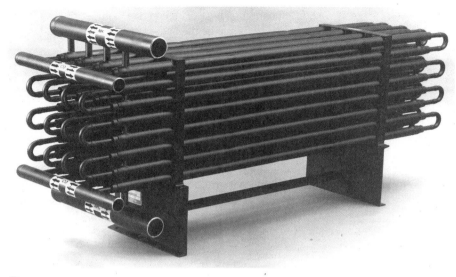

(f)

FIGURE 9.0.1 Different Types of Heat Exchangers (cont'd)
(d) A shell-and-tube steam condenser for a power plant. Condenser tubes
vary in length from 6 m–15 m (20 ft–50 ft) with surface areas from 4500
m^2–93 000 m^2 (50 000 ft^2–100 000 ft^2). The overall dimensions of a
condenser vary from 9 m–21 m long × 6 m–9 m wide × 9 m–12 m high
(30 ft–70 ft long × 20 ft–30 ft wide × 30 ft–40 ft high). Compare these
dimensions with those of a typical house which may be 40 ft–60 ft long ×
25 ft–30 ft wide × 10 ft–20 ft high. Courtesy of Senior Engineering Co.,
Los Angeles, CA. (e) A shell-and-tube feedwater heater for power plants.
The feedwater heater has three zones—one for desuperheating the
steam, a second one for condensation, and a third one for subcooling the
condensate. Courtesy of Senior Engineering Co., Los Angeles, CA.
(f) Manifolded tube-in-tube heat exchanger assembly for ammonia
refrigeration system heat recovery applications. Courtesy of Turbotec
Products, Windsor, CT. (g) A heat exchanger modular package containing
a radiator, air conditioning condenser, and oil cooler for automobile
applications. Courtesy of Modine Manufacturing Co., Racine, WI.

(g)

5. Number of Fluids. Two, three, or even four fluids may exchange thermal energy
with one another.

Yet another type of heat exchanger is the regenerator, which, usually, consists of an
energy storage material in the form of a solid matrix, or simply as a solid (such as
rocks or metal pieces), or a liquid. The material is alternately heated by the hot fluid
and cooled by the cool fluid. As the hot fluid flows over the storage material, the
fluid is cooled and the storage material is heated. Subsequently, when the cool fluid
flows over the storage material, the stored energy is transferred to the cool fluid. A
more detailed description of the regenerator is given in Section 9.5.

Classification of heat exchangers according to construction and flow type is the most common. Some of the heat exchangers classified according to this scheme are described in some detail here.

Double-pipe heat exchanger

A double-pipe heat exchanger, shown in Figure 9.1.1, is the simplest type. It consists of one tube inside another coaxial tube. If the fluids in the inner tube and the annulus flow in the same direction, the heat exchanger is termed parallel-flow type (Figure 9.1.1a). If the two fluids flow in opposite directions, it is of the counterflow type (Figure 9.1.1b). To increase the heat transfer rates, the inner pipe may be provided with extended surfaces, either on the inner surface or the outer surface or both surfaces. In some cases, three coaxial tubes are used either for two fluids or three fluids. To increase the heat transfer area without an excessive length, a number of double-pipe heat exchangers may be arranged in series—multiple tube pass as shown in Figure 9.1.1c. Double-pipe heat exchangers are widely used as they are easy to fabricate, but their capacities are limited because the heat transfer surface areas are not large. Figure 9.1.1d shows a double-pipe heat exchanger in the form of a coil.

Shell-and-tube heat exchanger

The shell-and-tube heat exchanger consists of a bundle of tubes enclosed in a cylindrical or rectangular shell (Figure 9.1.2). If both the fluids are single phase fluids, the inlet and exit are at the ends of the heat exchanger. If the fluid on the shell side changes phase, as in a steam condenser, the entry of the phase change fluid may be midway between the two ends (see the steam condenser in Figure 9.0.1d). The tube bundles are provided with baffles to give the desired flow pattern to the flow of the fluid on the shell side. If the fluid in the tubes flows from one end to the other end only once, it is known as a one-tube pass shell-and-tube heat exchanger. If the fluid in the tubes flows from one end to the other and back to the inlet end, it is known as a two-tube pass shell-and-tube heat exchanger. Four-tube pass and, in some cases, more than four-tube passes are also in use. Shell-and-tube heat exchangers are extensively used in power plants as feedwater heaters and condensers, in refrigeration applications as condensers and evaporators, and in chemical industrial applications. They have large ratios of heat transfer surface to volume and weight and are easy to fabricate in a wide range of capacities and sizes. They are also rugged, reasonably easy to maintain, and withstand high pressures and temperatures. A large bank of design data are available so that they can be easily designed. Fabrication facilities are widely available.

In both the double-pipe and the shell-and-tube heat exchangers, the two fluids flow along generally parallel lines either in the same or in opposite directions; baffles are introduced in shell-and-tube heat exchangers to induce cross flow in each section of the heat exchanger. In cross-flow heat exchangers, the two fluids flow in mutually perpendicular directions; cross-flow heat exchangers are extensively used if one of the fluids is a gas and the other is a liquid or a fluid undergoing phase change; for example, in refrigeration applications where the refrigerant is condensed in an air-cooled condenser. Other examples of cross-flow heat exchangers include superheaters in boilers, exhaust gas boilers in power plants with diesel engines, automobile radiators, and baseboard heaters employing hot water for space heating. They are also used as air-to-air heat exchangers for waste heat recovery in buildings.

In cross-flow heat exchangers, where the convective heat transfer coefficient on one or both sides is low, as in the case of a gas on one side and a liquid or a gas on

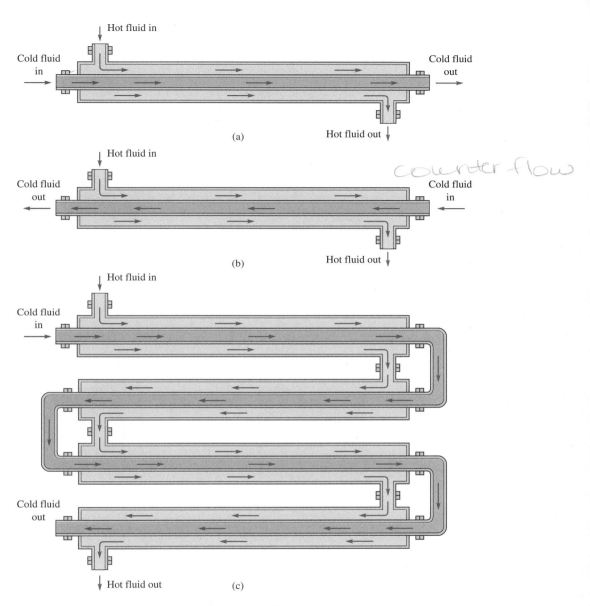

Hot fluid in

Cold fluid in →

Cold fluid out →

(a)

Hot fluid out ↓

Hot fluid in

Cold fluid out ←

← Cold fluid in

(b)

Hot fluid out ↓

counter flow

Hot fluid in

Cold fluid in →

Cold fluid out

↓ Hot fluid out

(c)

(d)

FIGURE 9.1.1 Different Arrangements for Double-Pipe Heat Exchangers

(a) Double-pipe parallel-flow heat exchanger. One fluid flows inside the inner tube and a second fluid flows in the annular region. Both fluids flow in the same direction. (b) Double-pipe counterflow heat exchanger. The direction of the flow of the fluid in the inner tube is opposite to that of the fluid in the annulus. (c) Four-tube pass double-pipe heat exchanger. (d) Double-tube condenser/evaporator for water source 4-ton heat pump application. Courtesy of Turbotec Products, Windsor, CT.

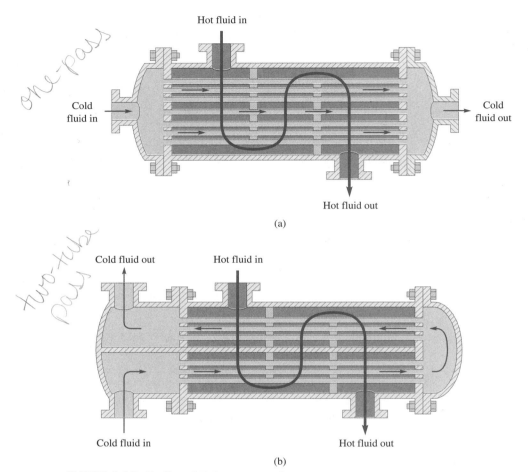

one-pass

two-tube pass

FIGURE 9.1.2 Shell-and-Tube Heat Exchangers
(a) One-tube pass shell-and-tube heat exchanger. The fluid in the tubes flows from one end to the other end only once. The effect of the baffles on the flow pattern is also shown. (b) Two-tube pass shell-and-tube heat exchanger. The fluid in the tubes flows from one end to the other and back to the inlet end.

Cross flow heat exchanger

the other, it is common to have plates attached to the tubes as shown in Figure 9.1.3a. These plates act as extended surfaces. Fins may also be provided on the inner surface of the tubes. The fluid between a pair of plates (outside of the tubes) cannot mix with the fluid between another pair of plates. If no plates are provided, the fluid in one region on the outside of the tubes has the possibility of mixing with the fluid in another region. In all cases the fluid in the pipes remains ''unmixed'' because each pipe is a separate channel. The arrangement with the plates is termed *both fluids unmixed* and the arrangement without the plates is termed *one fluid mixed and the other unmixed*. In cross-flow heat exchanges, the effective exterior heat transfer surface can be made quite large by using a large number of fins, as in an automobile radiator. However, with the use of such closely packed fins, the heat transfer surfaces are susceptible to fouling and obstruction of passages caused by the deposit of debris.

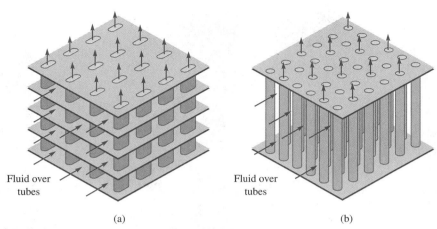

Fluid over tubes

Fluid over tubes

(a)

(b)

FIGURE 9.1.3 Cross-Flow Heat Exchangers
(a) Cross-flow heat exchanger, both fluids unmixed. The two streams of fluid flow in mutually perpendicular directions. As the fluids are constrained to flow in the tubes and between plates, the fluid in one region cannot mix with the fluid in another region and the arrangement is termed both fluids unmixed. (b) Cross-flow heat exchanger where the two fluids flow in mutually perpendicular directions. As the fluid in one region outside the tubes has the possibility of mixing with the fluid in another region, the arrangement is termed one fluid mixed and the other unmixed (fluid in the tubes).

Plate heat exchanger

A plate heat exchanger consists of a series of plates separated by gaskets between each pair of plates. One fluid flows between alternate pairs of plates, and a second fluid between the remaining passages between plates. A simple plate type heat exchanger is shown in Figure 9.1.4. The fluids between the plates may flow either in the same direction (parallel-flow) or in opposite directions (counterflow). By a suitable arrangement of gaskets, a plate heat exchanger can be used with more than two fluids. Plate heat exchangers have very high ratios of heat transfer surface to volume and the heat fluxes per unit temperature difference between the fluids on either side of the plates is high, making them particularly suitable for applications where the difference between the temperatures of the fluids is small. They are easily assembled, and cleaning and replacement of the plates are simple operations. Their capacity can be changed easily by adding or removing plates in the same frame. The disadvantages of plate heat exchangers are that they cannot withstand high pressures or temperatures. The operating pressure is limited to about 2.5 MPa and the temperature to about 250 °C, but usually to less than 150 °C, due to the limitations of the gasket materials. The pressure drops are higher in plate heat exchangers than in shell-and-tube heat exchangers. Plate heat exchangers are generally used as liquid-to-liquid heat exchangers with moderate to low viscosity liquids. They are not suitable for operation with gases or high viscosity liquids. They have to be specially designed if they are to be used as condensers. Recently, plate heat exchangers with welded or brazed joints in place of the gaskets have been introduced. With the elimination of the gaskets the possibility of the leakage of the fluids is reduced and the need to periodically replace the gaskets is eliminated. But such heat exchangers lack the flexibility of those with gaskets, and cleaning is more involved.

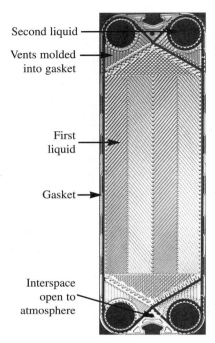

Second liquid

Vents molded into gasket

First liquid

Gasket

Interspace open to atmosphere

FIGURE 9.1.4 A Single-Pass Counterflow Plate Heat Exchanger.
Two fluids flow between alternate pairs of plates. Some of the larger units have heat transfer surface areas of up to 2500 m². The plates are generally of 0.5 mm thick stainless steel, with gaps of up to 5 mm between the plates. Size of the plates varies from 0.6 m × 0.1 m to 4 m × 1.25 m. Channel flow rates are from 0.05 m³/h to 12.5 m³/h. Courtesy of APV Crepaco Inc. Tonawanda, NY.

9.2 OVERALL HEAT TRANSFER COEFFICIENT

Overall heat transfer coefficient and its relation to convective and conductive resistances

To predict the heat transfer rate in a heat exchanger we need a coefficient relating the heat flux between the two fluid streams and the local differences in their temperatures similar to the convective heat transfer coefficient. The overall heat transfer coefficient, which was introduced in Section 2.4, provides such a coefficient. The overall heat transfer coefficient is defined by the relation

$$q'' = U(T_{f1} - T_{f2})$$

where T_{f1} and T_{f2} are the local reference temperatures of the fluids for defining the convective heat transfer coefficients on the two sides (Figure 9.2.1). In the case of

FIGURE 9.2.1 A Solid Surface of Thickness L Separates Two Fluids
Fluid 1 is at a temperature T_{f1} with an associated convective heat transfer coefficient of h_1. Fluid 2 is at T_{f2} with a convective heat transfer coefficient of h_2.

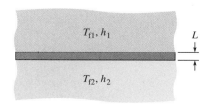

T_{f1}, h_1

L

T_{f2}, h_2

a single-phase fluid, the reference temperature is the bulk temperature. If the fluid is boiling or condensing, the reference temperature is usually the saturation temperature of the fluid.

If the two fluids are separated by a rectangular plate, from the thermal circuit we have

$$q'' = \frac{T_{f1} - T_{f2}}{1/h_1 + L/k + 1/h_2} = U(T_{f1} - T_{f2})$$

so that

$$U = \frac{1}{1/h_1 + L/k + 1/h_2} \tag{9.2.1}$$

where

T_{f1} and T_{f2} = the local reference temperatures for defining the convective heat transfer coefficients

h_1 and h_2 = the convective heat transfer coefficients

k = the thermal conductivity of the material of the plate

L = the thickness of the plate

The reciprocal of the convective heat transfer coefficient is the convective resistance, and L/k is the conductive resistance of the solid surface, both on a unit area basis. The application of the series thermal circuit implies that axial conduction in the solid (separating the fluids) is negligible. This is a reasonable approximation in most cases.

If the two fluids are separated by a cylindrical tube, the heat transfer rate per unit length of the pipe is given by (Figure 9.2.2)

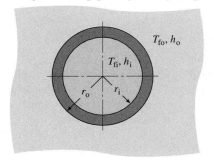

FIGURE 9.2.2 Two fluids at T_{fi} and T_{fo} are separated by a cylindrical tube with inner and outer radii of r_i and r_o, respectively. The heat transfer coefficients associated with the two fluids are h_i and h_o. The thermal conductivity of the material of the tube is k.

$$q' = \frac{T_{fi} - T_{fo}}{1/(2\pi h_i r_i) + \ln(r_o/r_i)/(2\pi k) + 1/(2\pi h_o r_o)} \tag{9.2.2}$$

The terms representing the convective and conductive resistances for a unit length of the pipe are easily recognized. The heat flux q'', can be defined on the basis of either the inner surface area or the outer surface area of the tube.

$$q_i'' = \frac{q'}{2\pi r_i} = \frac{T_{fi} - T_{fo}}{1/h_i + r_i\{[\ln(r_o/r_i)]/k\} + (1/h_o)(r_i/r_o)}$$

$$q_o'' = \frac{q''}{2\pi r_o} = \frac{T_{fi} - T_{fo}}{(1/h_i)\,r_o/r_i) + r_o\{[\ln(r_o/r_i)]/k\} + 1/h_o}$$

where q_i'' and q_o'' represent the heat fluxes based on the inner and outer surface areas of the tube, respectively.

The two expressions for the heat flux lead to the definition of the overall heat transfer coefficients based on the inner or outer surface area. With, $q_i'' = U_i(T_{fi} - T_{fo})$ and $q_o'' = U_o(T_{fi} - T_{fo})$ we obtain

$$U_i = \frac{1}{1/h_i + r_i\{[\ln(r_o/r_i)]/k\} + (1/h_o)(r_i/r_o)} \qquad (9.2.3)$$

$$U_o = \frac{1}{(1/h_i)(r_o/r_i) + r_o\{[\ln(r_o/r_i)]/k\} + 1/h_o} \qquad (9.2.4)$$

Although two overall heat transfer coefficients are defined, each is used in conjunction with the appropriate surface area, i.e., as the product of U_i *and the inside surface area, or* U_o *and the outside surface area. These products are equal to each other, and it does not matter whether the evaluation is made on the basis of one or the other surface area.* The heat transfer rate over an infinitesimal axial length dz, is given by

$$dq = U_i \, dA_i(T_{fi} - T_{fo}) = U_o \, dA_o(T_{fi} - T_{fo})$$

It is easily seen that, as $dA_i = 2\pi r_i \, dz$, and $dA_o = 2\pi r_o \, dz$, the values of $U_i \, dA_i$ and $U_o \, dA_o$ are equal to each other.

In many cases, the difference between the inner and outer surface areas is small and no distinction is made between the two. This is not true when heavy walled tubes of low thermal conductivity materials (for example, stainless steel in nuclear power plant applications) are used. In such cases axial conduction may also be significant.

A second observation is that, as the intent in these heat exchangers is to obtain the maximum heat transfer rate for a given temperature difference between the two fluids, the material separating the two fluids is chosen to reduce the conductive resistance. The conductive resistance is minimized by using a metal of high thermal conductivity and strength, permitting the use of very thin materials. In many applications the conductive resistance is negligibly small compared with the convective resistances. Usually, then, the conductive resistance is neglected.

Fouling resistance   After the equipment is in service for some time, the surfaces are no longer clean and additional resistances due to deposits on the surfaces are introduced. Such resistances are known as *fouling resistances*. They are influenced by the nature of the fluids and the flow conditions, the heat transfer surface, the length of service of the equipment, and so forth. Shah and Mueller (1985) have classified fouling according to its source.

- Scaling caused by deposits of salts
- Particulate fouling resulting from deposits of particles
- Chemical reaction within the fluids leading to surface deposits
- Corrosion fouling
- Biofouling from the accumulation of organisms on the surfaces
- Freezing fouling resulting from crystallization from a liquid phase on a subcooled surface

Many variables such as the temperature and velocity of the fluids, the properties of fluids, and the material of the surface of the heat exchanger affect the fouling resistance. The effect of each variable on the fouling resistance is not amenable to analytical computation. Hence, fouling resistances are estimated from experimental measurements. These additional resistances are added to the overall resistance ($= 1/U$) computed on the basis of clean surfaces. The overall heat transfer coefficient of heat exchangers that have been in service for some time is given by

$$\frac{1}{U} = \frac{1}{U_c} + R_f''$$

where

U = overall heat transfer coefficient in service
U_c = overall heat transfer coefficient with clean surfaces
R_f'' = fouling resistance for a unit area

Representative values of the fouling resistance, extracted from *Standards of Tubular Exchanger Manufacturers Association* (1970), are given in Table 9.2.1.

TABLE 9.2.1 Fouling Resistance

Fluid	$R_f''(\mathrm{m^2\ K/W})$
Sea water and treated boiler feedwater ($<50\,°C$)	0.0001
Sea water and treated boiler feedwater ($>50\,°C$)	0.0002
River water ($<50\,°C$)	0.0002–0.0001
Engine jacket water	0.0002
With the worst water in a heat exchanger	0.001
Fuel oil	0.0009
Transformer oil and engine lubricating oil	0.0002
Vegetable oils	0.0005
Refrigerating liquids and hydraulic fluids	0.0002
Steam	0.0001
Manufactured gas and engine exhaust gas	0.002
Compressed air	0.0004

EXAMPLE 9.2.1

A heat exchanger is made of 25-mm O.D., 3-mm thick wall, type 302 stainless steel tubes (Figure 9.2.3). Hot water flows in the tube with a velocity of 4 m/s, and air flows perpendicular to the axis of the tubes at 20 °C with a velocity of 10 m/s. At a section of the tube where the temperature of the water is 70 °C, determine the overall heat transfer coefficient based on

(a) The inner surface area.
(b) The outer surface area of the tubes.

Given

Fluid: Water	$T_b = 70\,°C$	$v_w = 4$ m/s
Fluid: Air	$T_a = 20\,°C$	$v_a = 10$ m/s
Tubes: 302 stainless steel	$d_i = 19$ mm	$d_o = 25$ mm

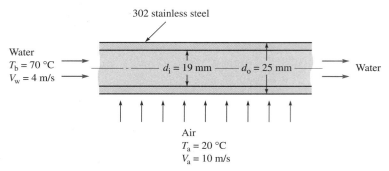

FIGURE 9.2.3 Water at 70 °C flows with a velocity of 4 m/s inside a 19-mm I.D. tube. Air at 20 °C flows with a velocity of 10 m/s perpendicular to the axis of the tube.

Find

(a) U_i
(b) U_o

ASSUMPTIONS

1. Steady state
2. Fully developed velocity and temperature profile for water
3. Air is at standard atmospheric pressure (101.325 kPa).

SOLUTION

To determine the overall heat transfer coefficient, Equations 9.2.3 and 9.2.4 are used. To use these equations, the values of h_i and h_o are needed. From the given information, h_i can be computed by employing one of the correlations for internal flows. To find the convective heat transfer coefficient associated with the outer surface, the outer surface temperature, which is not known, is needed. The surface temperature and the convective heat transfer coefficient should be found such that all appropriate relations are satisfied. An iterative procedure is adopted to find them. One possible iterative procedure is

1. Guess a reasonable outer surface temperature.
2. Compute h_o.
3. Compute the value of the outer surface temperature T_s, obtained from the energy balance on a unit length of the tube given by

$$2\pi r_o h_o (T_s - T_a) = \frac{T_b - T_a}{1/(2\pi r_i h_i) + [1/(2\pi k)][\ln(r_o/r_i] + 1/(2\pi r_o h_o)}$$

or

$$T_s = \frac{T_b - T_a}{(r_o/r_i)(h_o/h_i) + [(r_o h_o)/k][\ln(r_o/r_i)] + 1} + T_a \tag{1}$$

where

$$T_s = \text{outer surface temperature of the tube}$$
$$T_b = \text{bulk temperature of the water}$$
$$T_a = \text{temperature of the air}$$
$$r_i, r_o = \text{inner and outer radii of the tube}$$
$$k = \text{thermal conductivity of type 302 stainless steel} = 15.1 \text{ W/m K}$$

4. If the value of T_s computed from Equation 1 is sufficiently close to the value used in computing h_o, the correct value of h_o has been found. Otherwise, repeat steps 2–4 with the newly computed value of T_s.

Determine h_i

From the software CC (or Table A5) properties of water at the bulk temperature of 70 °C are

$\rho = 977.8 \text{ kg/m}^2$ $\qquad \mu = 3.933 \times 10^{-4} \text{ N s/m}^2$
$k = 0.662 \text{ W/m k}$ $\qquad \text{Pr} = 2.49$
$d_i = 25 - 6 = 19 \text{ mm}$

$$\text{Re}_d = \frac{\rho v_w d_i}{\mu} = \frac{977.8 \times 4 \times 0.019}{0.0003933} = 1.889 \times 10^5$$

Employing Equation 4.2.3b

$$\text{Nu}_d = 0.023(1.889 \times 10^5)^{0.8} \, 2.49^{0.3} = 502.8 = \frac{h_i d_i}{k}$$

$$h_i = \frac{502.8 \times 0.662}{0.019} = 17\,526 \text{ W/m}^2 \text{ K}$$

Determine h_o

To find the outside convective heat transfer coefficient, Equation 4.1.19, 4.1.20, or Equation 4.1.21 is used. To evaluate the properties at the film temperature, the surface temperature (which is not readily available) is required. The iterative procedure detailed in the beginning of the solution is used.

1. Guess a value for the outer surface temperature. As the forced convective heat transfer coefficient on the water side is much higher than the convective heat transfer coefficient on the air side, the outer surface temperature is expected to be closer to the water temperature than to the air temperature. A reasonable guess for the surface temperature is approximately 50 °C.

2. Compute h_o. For the assumed surface temperature, $T_f = 35$ °C. From the software CC (or Table A7), properties for air at 35 °C are

$\rho = 1.146 \text{ kg/m}^3;$ $\qquad \mu = 1.888 \times 10^{-5} \text{ N s/m}^2$
$k = 0.0267 \text{ W/m K}$ $\qquad \text{Pr} = 0.711$
$d = 0.025 \text{ m}$

$$\text{Re}_d = \frac{\rho v_a d_o}{\mu} = \frac{1.146 \times 10 \times 0.025}{1.888 \times 10^{-5}} = 15\,169$$

Employing Equation 4.1.21,

$$\mathrm{Nu}_d = 0.3 + \frac{0.62 \times 15\,169^{1/2} \times 0.711^{1/3}}{[1 + (0.4/0.711)^{2/3}]^{1/4}} = 60.16$$

$$h_\mathrm{o} = \mathrm{Nu}_d \frac{k}{d_\mathrm{o}} = 60.16 \times \frac{0.0267}{0.025} = 64.4 \text{ W/m}^2\,°\text{C}$$

3. Compute T_s. From Equation 1,

$$T_\mathrm{s} = \frac{T_\mathrm{b} - T_\mathrm{a}}{(r_\mathrm{o}/r_\mathrm{i})(h_\mathrm{o}/h_\mathrm{i}) + [(r_\mathrm{o}h_\mathrm{o})/k][\ln(r_\mathrm{o}/r_\mathrm{i})] + 1} + T_\mathrm{a} \qquad (1)$$

Substituting the values,

$$T_\mathrm{s} = \frac{70 - 20}{(0.0125/0.0095) \times (64.3/17526) + (0.0125 \times 64.3)/15.1[\ln(0.0125/0.0095)] + 1} + 20 = 69\,°\text{C}$$

4. The computed value of $T_\mathrm{s} = 69\,°\text{C}$ is significantly higher than the value used for computing h_o. Repeat steps 2–4.

With $T_\mathrm{s} = 69\,°\text{C}$, recomputing h_o, we obtain $h_\mathrm{o} = 64.2 \text{ W/m}^2\,°\text{C}$. With $h_\mathrm{o} = 64.2$ W/m^2 $°\text{C}$, we obtain $T_\mathrm{s} = 69\,°\text{C}$, which is equal to the value of $69\,°\text{C}$ used to compute h_o. Hence, the correct value of h_o is $64.2 \text{ W/m}^2\,°\text{C}$.

(a) To find U_i, compute each term in Equation 9.2.3.

$$\frac{1}{h_\mathrm{i}} = \frac{1}{17\,526} = 5.706 \times 10^{-5} \text{ m}^2\,°\text{C/W}$$

$$r_\mathrm{i}\frac{\ln(r_\mathrm{o}/r_\mathrm{i})}{k} = \frac{0.019}{2}\frac{\ln(25/19)}{15.1} = 1.727 \times 10^{-4} \text{ m}^2\,°\text{C/W}$$

$$\frac{1}{h_\mathrm{o}}\frac{r_\mathrm{i}}{r_\mathrm{o}} = \frac{1}{64.2}\frac{0.019}{0.025} = 1.184 \times 10^{-2} \text{ m}^2\,°\text{C/W}$$

$$U_\mathrm{i} = \frac{1}{5.706 \times 10^{-5} + 1.727 \times 10^{-4} + 1.184 \times 10^{-2}}$$
$$= \underline{82.9 \text{ W/m}^2\,°\text{C}}$$

(b) Employing Equation 9.2.4 to find U_o.

$$U_\mathrm{o} = \frac{1}{1/17\,526 \times 0.025/0.019 + 0.025/2 \times [\ln(25/19)]/15.1 + 1/64.2}$$
$$= \underline{63 \text{ W/m}^2\,°\text{C}}$$

COMMENTS

In the expression for U_o the conductive resistance (2.272×10^{-4} m^2 $°\text{C/W}$) is greater than the convective resistance on the inner surface (7.508×10^{-5} m^2 $°\text{C/W}$) as a heavy walled stainless steel tube with a low thermal conductivity is used, and the convective heat transfer coefficient associated with water is quite high. In this case the conductive resistance cannot be neglected in computing the value of the overall heat transfer coefficient. Notice how close U_o is to h_o. Why?

EXAMPLE 9.2.2

A long, horizontal, ⅜-in diameter, type K seamless copper tube is surrounded by saturated, dry steam at 100 °C (Figure 9.2.4). Water flows in the tube at a mass flow rate of 0.4 kg/s. Estimate the overall heat transfer coefficient based on the inner surface area at the section of the tube where the water temperature is 20 °C.

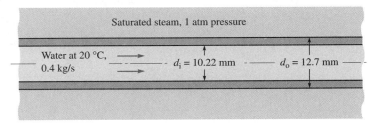

FIGURE 9.2.4 A Copper Tube, with Water Flowing Inside, Surrounded by Saturated Steam at 1 atm

Given

Fluid: Water $T_b = 20\ °C$ $\dot{m} = 0.4$ kg/s
Fluid: Saturated steam, 1 atm $T_{st} = 100\ °C$
Tube: Copper, ⅜-in diameter, type K

Find

U_i, the overall heat transfer coefficient based on the inner surface area

ASSUMPTIONS

1. Steady state
2. Fully developed velocity and temperature profile for water
3. Steam has no significant velocity over the tube.

SOLUTION

From the definition of the overall heat transfer coefficient, Equation 9.2.3,

$$\frac{1}{U_i} = \frac{1}{h_i} + \frac{r_i}{k} \ln \frac{r_o}{r_i} + \frac{1}{h_o} \frac{r_i}{r_o}$$

A ⅜-in diameter, type K steamless copper tubing has an outer diameter of 12.7 mm and an inside diameter of 10.22 mm with a wall thickness of 1.24 mm.

To find the overall heat transfer coefficient, the convective heat transfer coefficients on the inner and outer surfaces of the tube are needed. From the given information, h_i can be computed by employing one of the correlations for internal flows. To find the convective heat transfer coefficient associated with the condensation of steam on a horizontal tube, Equation 6.2.12 is used. But this equation requires the surface temperature, which is not known. The surface temperature should be found such that all appropriate relations are satisfied. For this purpose an iterative procedure similar to the one used in the solution of Example 9.2.1 is adopted. The suggested procedure for the solution is

1. Find h_i from Equation 4.2.3b or 4.2.1b, depending on the Reynolds number.
2. Assume a value of surface temperature T_s.

3. Find h_o from Equation 6.2.12.

4. On the basis of a unit length of the tube the surface temperature should satisfy the relation

Heat transfer rate from the steam to the tube
= Heat transfer rate from the steam to the water

$$2\pi r_o h_o (T_{st} - T_s) = \frac{T_{st} - T_b}{1/(2\pi r_i h_i) + 1(2\pi k)[\ln(r_o/r_i)] + 1/(2\pi r_o h_o)}$$

$$T_s = T_{st} - \frac{T_{st} - T_b}{(r_o/r_i)(h_o/h_i) + [(r_o h_o)/k][\ln(r_o/r_i)] + 1}$$

where

T_b = bulk temperature of the water
T_s = temperature of the outer surface of the tube
T_{st} = saturation temperature of the steam

5. If the computed value of T_s is close to the value employed in determining h_o, the correct value of T_s was used to find h_o. If the computed and the assumed values of T_s are far apart, steps 3 and 4 are repeated employing the newly computed value of T_s until the computed and assumed values are sufficiently close.

1. Compute h_i. At the bulk temperature of 20 °C, properties of water from the software **CC** (or Table A5) are
$\rho = 998.2$ kg/m^3 $\qquad \mu = 9.853 \times 10^{-4}$ N s/m^2
$k = 0.603$ W/m K $\qquad$ Pr = 6.839
$d = 0.01022$ m

$$\mathrm{Re}_d = \frac{4\dot{m}}{\pi d_i \mu} = \frac{4 \times 0.4}{\pi \times 0.01022 \times 9.853 \times 10^{-4}} = 50\,574$$

For $\mathrm{Re}_d = 50\,574$, employing Equation 4.2.3b with 0.4 as the exponent for Pr,

$$\mathrm{Nu}_d = 0.023\,\mathrm{Re}_d^{0.8}\,\mathrm{Pr}^{0.4} = 0.023 \times 50\,574^{0.8} \times 6.839^{0.4} = 287.7$$

$$h_i = \mathrm{Nu}_d \frac{k}{d} = 287.7 \times \frac{0.603}{0.01022} = 16\,974 \text{ W/m}^2 \text{ K}$$

2. Assume $T_s = 70$ °C.

3. Compute h_o. Properties to be used at $T_f = (70 + 100)/2 = 85$ °C, from software **CC** (or Table A5) are

$\rho = 968.6$ kg/m^3 $\qquad\qquad k = 0.673$ W/m K
$\mu = 3.254 \times 10^{-4}$ N s/m^2 $\qquad c_p = 4198$ J/kg K

From Table A6

$\rho_v(T_{st} = 100\text{ °C}) = 0.5977$ kg/m^3 $\qquad i_{fg} = 2257$ kJ/kg
d_o = outer diameter of the tube = 0.0127 m

All properties refer to those of liquid water, except ρ_v, which is the density of steam at 100 °C.

$$i'_{fg} = i_{fg} + 0.68\, c_p(T_{st} - T_s) = 2257 \times 10^3 + 0.68 \times 4198 \times (100 - 70)$$
$$= 2.343 \times 10^6 \text{ J/kg}$$

From Equation 6.2.12 (after increasing it by 20% to account for the effect of interfacial ripples),

$$h_o = 0.875 \left[\frac{g\,\rho(\rho - \rho_v)k^3 i'_{fg}}{\mu(T_{st} - T_s)d_o} \right]^{1/4}$$

$$= 0.875 \left[\frac{9.807 \times 968.6(968.6 - 0.5977)0.673^3 \times 2.343 \times 10^6}{3.254 \times 10^{-4}(100 - 70)0.0127} \right]^{1/4}$$

$$= 13\,274 \text{ W/m}^2 \text{ K}$$

4. Update the value of T_s. From Table A1, k (copper) = 401 W/m °C.

$$T_s = 100 - \frac{100 - 20}{[(0.0127 \times 13\,274)/(0.01022 \times 16\,974)] + (0.0127/2) \times (13\,274/401)[\ln(0.0127/0.01022)] + 1} = 60.3 \text{ °C}$$

5. Check if the computed value of T_s is close to the assumed value. The computed value of T_s, 60.3 °C, is a little less than the assumed value of 70 °C. Repeating steps 3 and 4, with an assumed value of 60.3 °C, we obtain $h_o = 12\,210$ W/m^2 °C and $T_s = 58.7$ °C (details of the computations are left as an exercise). As the computed and assumed values of T_s are now very close to each other, the value 12 210 W/m^2 °C is used for h_o.

Employing $h_i = 16\,974$ W/m^2 °C and $h_o = 12\,210$ W/m^2 °C

$$\frac{1}{U_i} = \frac{1}{h_i} + \frac{r_i}{k}\ln\frac{r_o}{r_i} + \frac{1}{h_o}\frac{r_i}{r_o} = \frac{1}{16\,974} + \frac{0.01022}{401 \times 2}\ln\frac{12.7}{10.22} + \frac{1}{12\,210} \times \frac{10.22}{12.7}$$
$$U_i = 7838 \text{ W/m}^2 \text{ °C}$$

COMMENTS

1. Based on the outer surface area, U_o is given by

$$\frac{1}{U_o} = \frac{1}{16\,974} \times \frac{12.7}{10.22} + \frac{0.0127}{2 \times 401}\ln\frac{12.7}{10.22} + \frac{1}{12\,210}$$

$$U_o = 6307 \text{ W/m}^2 \text{ °C}$$

(Note that to find the heat transfer rate, the overall heat transfer coefficient is used with the appropriate surface area, i.e., U_i with the inner surface area and U_o with the outer surface area, and that $U_i r_i = U_o r_o$. In fact, U_o can be computed from the relation $U_i r_i = U_o r_o$.)

2. If the conductive resistance is neglected, $U_i = 8011$ W/m^2 °C, which is only 2.2% higher than computed by including it. Similarly $U_o = 6447$ W/m^2 ° C, which is also 2.2% higher than that found by including the conduction resistance. If, in addition to this conduction resistance, the thickness of the tube is also neglected, the overall heat transfer coefficient is given by

$$\frac{1}{U} = \frac{1}{h_i} + \frac{1}{h_o} = \frac{1}{16\,974} + \frac{1}{12\,210} \quad \text{or} \quad U = 7102 \text{ W/m}^2 \text{ °C}$$

The value of the overall heat transfer coefficient is 9.4% lower than U_i and 12.6% higher than U_o.

From an examination of the approximations in computing the overall heat transfer coefficients, it is clear that neglecting the conduction resistance with a material of high thermal conductivity introduces less error than neglecting the thickness of the material of the tubes.

Examples 9.2.1 and 9.2.2 illustrate how the overall heat transfer coefficient is computed. However, there are many cases where the convective heat transfer coefficients are not easily computed. For instance, in a shell-and-tube heat exchanger with the introduction of baffles to create cross flow, the flow of the fluid on the shell side is neither totally parallel nor totally perpendicular to the axes of the tubes. There are no simple correlations for the computation of the convective heat transfer coef-

TABLE 9.2.2 Representative Values of the Overall Heat Transfer Coefficient

Fluids	U (W/m² °C)
Shell-and-Tube Heat Exchangers	
Water/Water	1220–2440
Ammonia/Water	1220–2440
Light organics ($\mu < 5 \times 10^{-4}$ N s/m²)/Water	370–730
Medium organics ($5 \times 10^{-4} < \mu < 10 \times 10^{-4}$ N s/m²)/Water	240–610
Heavy organics ($\mu > 10 \times 10^{-4}$ N s/m²)/Water	25–370
Gases/Water	10–240
Water/Brine	490–980
Steam/Water (feedwater heaters, condensers)	980–5700
Steam/Ammonia	980–3400
Steam/Light organics	490–980
Steam/Medium organics	240–490
Steam/Heavy organics	30–300
Steam/Gases	25–240
Light organics/Light organics	190–370
Medium organics/Medium organics	100–290
Heavy organics/Heavy organics	50–200
Air-Cooled Exchangers (with Large Finned Surfaces*	
Water	680–740
Lubricating oil	85–170
Flue gases	50–170
Air (values vary with air pressure)	114–511
Steam (condensing)	740–795
Refrigerant 12	340–455
Plate Heat Exchangers	
Water/Water	3000–7000

*with air on one side and the fluid from the table on the other

ficient in that case. Methods for computing the overall heat transfer coefficients in some of those cases are given in Rohsenow et al. (1985), Palen (1986), and Kakac et al. (1981). Table 9.2.2 gives representative values of the overall heat transfer coefficient in some common applications. The values are given only to indicate the range of the values of the overall heat transfer coefficient and are not to be taken as exact. The values for shell-and-tube heat exchangers and air-cooled exchangers are taken from Roshenow et al. (1985), and those for plate heat exchangers are from Palen (1986).

In cross-flow heat exchangers, if fins (plates) are used, the overall heat transfer coefficient should be based on the heat transfer rate to the fluid on the fin side as given by (see Section 2.6.2)

$$q = h_{\mathrm{o}}(A_{\mathrm{uf}} + \eta A_{\mathrm{f}})(T_{\mathrm{s}} - T_{\mathrm{f}})$$

where

h_{o} = convective heat transfer coefficient on the fin side, assumed to be uniform
A_{uf} = surface area of the tubes (not occupied by fins) exposed to the fluid
A_{f} = surface area of the fins
η = efficiency of the fins (see Section 2.6.2)
T_{s} = outer surface temperature of the tube, also the base temperature of the fins
T_{f} = temperature of the fluid on the fin side

In many cases the fins attached to the tube banks are not circular but rectangular. The efficiency of rectangular fins attached to circular tubes (Fig. 9.2.5) is available in Sparrow and Lin (1964) and Kuan et al. (1984).

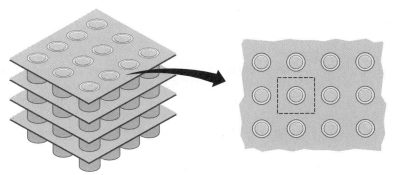

FIGURE 9.2.5 Plates attached to tube banks in cross-flow heat exchanger act as fins. The shape of the fin, shown in dotted lines, is that of a rectangular plate attached to a circular tube.

9.3 ANALYSIS OF HEAT EXCHANGERS

Recall that analysis of heat exchangers falls into two categories. One is the design of a heat exchanger when its duty, the heat transfer rate to be accomplished (i.e., the mass rates of flow of the fluids and their inlet and exit temperatures), is known and the heat transfer surface area of the heat exchanger is to be determined. In the second category, the inlet conditions of the two fluids (mass flow rates and inlet temperatures) and the details of the heat exchanger are specified and the performance of the heat exchanger (the heat transfer rate that can be obtained and, hence, the exit temperatures of the fluids) is to be evaluated. In both cases, it is assumed that the overall heat transfer coefficient is known. If the value of the overall heat transfer coefficient is available, the heat flux at any location can be determined if the bulk temperatures of the fluids on either side of the solid surface at that location are known. But, in general, the bulk temperatures of the fluids continuously vary in the direction of flow due to the heat transfer between them. Two methods to determine the heat transfer rate in a heat exchanger are presented. They are the *logarithmic mean temperature difference* (LMTD) method and the *number of transfer units* (NTU) method. In either method it is assumed that the overall heat transfer coefficient is available.

9.3.1 Logarithmic Mean Temperature Difference (LMTD) Method

Determining a mean temperature difference for computing the heat flux

It is convenient to define a suitable mean temperature difference such that the heat transfer rate in the heat exchanger can be computed by

$$q = UA_s \, \Delta T_m \qquad (9.3.1)$$

where

U = overall heat transfer coefficient
A_s = surface area associated with the overall heat transfer coefficient
ΔT_m = appropriate temperature difference that will yield the heat transfer rate from one fluid to the other

The intent of the LMTD method is to define ΔT_m in terms of the inlet and exit temperatures of the hot and cold fluids. The LMTD is derived for a counterflow heat exchanger. Its evaluation for other types of heat exchangers is discussed in the next section.

For purposes of analysis, consider a double-pipe, counterflow heat exchanger shown in Figure 9.3.1a. It is assumed that the cold fluid flows inside the tube and the hot fluid flows in the opposite direction in the annulus. Identical results are obtained if the hot fluid flows in the inner tube and the cold fluid in the annulus. The end where the cold fluid enters and the hot fluid exits the heat exchanger is arbitrarily designated as 1, and the other end as 2.

ASSUMPTIONS

1. Heat transfer is only between the two fluids and there is no heat transfer to the surroundings.
2. The overall heat transfer coefficient is uniform; i.e., it has the same value at every location. If the overall heat transfer coefficient varies, a suitable mean overall heat transfer coefficient is used.
3. Properties of the fluids are constant.
4. Steady state
5. No axial conduction in the pipes or in the fluids; i.e., in both the pipe and the fluid the heat transfer rates in the flow directions are negligible compared with the heat transfer in the transverse direction.

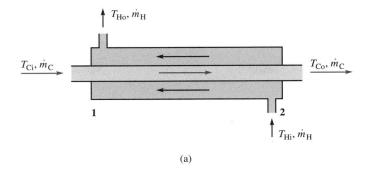

(a)

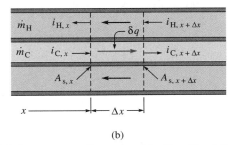

(b)

FIGURE 9.3.1 A Counterflow, Double-Pipe Heat Exchanger and Control Volume
(a) Double-pipe counterflow heat exchanger with the cold fluid flowing in the inner tube and the hot fluid in the annulus between the inner and outer tubes. The two fluids flow in opposite directions. (b) An elemental length of the double-pipe heat exchanger. The left end of the heat exchanger is taken to be the origin, and the x-coordinate is measured from the left end. The control volume is between the dotted lines.

Consider a small element of the heat exchanger, with x as the coordinate from end 1, Figure 9.3.1b. Applying the First Law of Thermodynamics to the cold and hot fluids in the control volume,

$$\delta q = \dot{m}_C(i_{C,x+\Delta x} - i_{C,x}) = \dot{m}_H(i_{H,x+\Delta x} - i_{H,x}) \qquad \textbf{(9.3.2a)}$$

where δq is the heat transfer rate from the hot fluid to the cold fluid. Here, i represents the specific enthalpy of the fluids. Subscript H represents the hot fluid and C the cold fluid. From the definition of the overall heat transfer coefficient δq is also given by

$$\delta q = U(A_{s,x+\Delta x} - A_{s,x})(T_H - T_C) \qquad \textbf{(9.3.2b)}$$

where

A_s = heat transfer surface area (of the inner pipe) measured from $x = 0$,
T_H, T_C = the local bulk temperatures of the hot and cold fluid in the control volume
U = the overall heat transfer coefficient associated with A_s.

Several cases arise.

Both T_H and T_C Are Uniform This is possible when both the hot and cold fluids undergo phase change, for example, when the hot fluid condenses and the cold fluid boils. Then $T_H - T_C$ is constant and we have

$$q'' = U(T_H - T_C) = \text{constant} \quad \text{and} \quad q = \int q'' \, dA_s = q''A_s = UA_s(T_H - T_C)$$

$$q = UA_s(T_H - T_C) \tag{9.3.3}$$

Thus, if the temperatures of the fluids entering the heat exchanger are known, the temperature difference between the two fluids is known everywhere and the heat transfer rate is also known. With both T_H and T_C spatially uniform, $\Delta T_m = T_H - T_C$.

Neglecting any subcooling of the condensing fluid or the superheating of the boiling liquid,

$$\text{Mass rate of boiling} = \frac{q}{i_{fg,b}} \quad \text{(boiling liquid)}$$

$$\text{Mass rate of condensation} = \frac{q}{i_{fg,c}} \quad \text{(condensing fluid)}$$

where

$i_{fg,b}$ = enthalpy of vaporization of the boiling fluid
$i_{fg,c}$ = enthalpy of vaporization of the condensing fluid

One of the Temperatures Is Uniform Let T_H be uniform (as when the hot fluid is condensing). The cold fluid is a single phase fluid and its temperature varies in the direction of its flow in the heat exchanger. Denoting the constant pressure specific heat of the cold fluid by c_C,

$$\delta q = \dot{m}_C(i_{C,x+\Delta x} - i_{C,x}) = \dot{m}_C c_C (T_{C,x+\Delta x} - T_{C,x})$$

$$= \dot{m}_C c_C \, \Delta T_C = U \, \Delta A_s (T_H - T_C) \quad (\Delta T_C = T_{C,x+\Delta x} - T_{C,x})$$

Separating the variables and taking the limit as $\Delta x \to 0$,

$$\frac{dT_C}{T_H - T_C} = \frac{U}{\dot{m}_C c_C} \, dA_s$$

Integrating this equation from inlet end 1 to exit end 2, (with T_H constant) we obtain

$$-\ln(T_H - T_C) \Big|_i^e = \frac{U}{\dot{m}_C c_C} A_s$$

or

$$\left\| \; \frac{T_H - T_C)_e}{(T_H - T_C)_i} = \exp\left(\frac{-UA_s}{\dot{m}_C c_C}\right) \; \right\| \tag{9.3.4}$$

If the inlet and exit temperatures of the cold fluid, and its mass flow rate are known, with $q = \dot{m}_C c_C (T_{Ce} - T_{Ci})$ where T_{Ci} and T_{Ce} indicate the inlet and exit temperature, of the cool fluid, we have

$$\dot{m}_C c_C = \frac{q}{T_{Ce} - T_{Ci}}$$

Substituting the above relation in Equation 9.3.4 and rearranging,

$$q = UA_s \frac{(T_H - T_C)_1 - (T_H - T_C)_2}{\ln[(T_H - T_C)_1/(T_H - T_C)_2]} = UA_s \frac{\Delta T_1 - \Delta T_2}{\ln(\Delta T_1/\Delta T_2)} \qquad (9.3.5)$$

where ΔT_1 and ΔT_2 represent the temperature differences between the hot and cold fluids at ends 1 and 2, respectively. In Figure 9.3.1 the end where the single-phase fluid enters is designated as end 1. Comparing Equation 9.3.5 with Equation 9.3.1, we obtain

$$\left\| \quad \Delta T_m = \frac{\Delta T_1 - \Delta T_2}{\ln(\Delta T_1/\Delta T_2)} \quad \right\| \qquad (9.3.6)$$

Note that in Equation 9.3.6 the same value of ΔT_m is obtained if the numeral designation of the ends is interchanged.

The same results are obtained when the cold fluid temperature T_C is constant (cold fluid boiling) and T_H varies from location to location. (*Note:* The temperature differences ΔT_1 and ΔT_2 are defined by the relation $\Delta T_1 = T_{H1} - T_{C1} = q_1''/U$ and $\Delta T_2 = T_{H2} - T_{C2} = q_2''/U$ where q_1'' and q_2'' are the heat fluxes at ends 1 and 2, respectively. These temperature differences are those that are used in conjunction with the overall heat transfer coefficient to obtain the heat flux. The actual temperatures of the fluids, particularly when phase change takes place, may be different. For example, when a vapor condenses, the overall heat transfer coefficient is based on the saturation temperature of the vapor. At the end where the condensate is drained, the actual temperature of the condensate is generally lower than the saturation temperature of the vapor, but it is the saturation temperature at both ends that should be used in computing the LMTD.)

The temperature profiles for the cold and hot fluids are sketched in Figure 9.3.2a.

Neither T_H nor T_C Is Uniform (No Phase Change Occurs).

$$\delta q = \dot{m}_C(i_{C,x+\Delta x} - i_{C,x}) = \dot{m}_C c_C \, \Delta T_C$$

$$\delta q = \dot{m}_H(i_{H,x+\Delta x} - i_{H,x}) = \dot{m}_H c_H \, \Delta T_H$$

$$\delta q = U(T_H - T_C) \, \Delta A_s$$

Hence, in the limit as $\Delta A_s \rightarrow 0$, with $dq = U(T_H - T_C) \, dA_s$

$$dT_H - dT_C = d(T_H - T_C) = dq \left(\frac{1}{\dot{m}_H c_H} - \frac{1}{\dot{m}_C c_C} \right)$$

$$= U(T_H - T_C) \left(\frac{1}{\dot{m}_H c_H} - \frac{1}{\dot{m}_C c_C} \right) dA_s \qquad (9.3.7)$$

The products $\dot{m}_C c_C$ and $\dot{m}_H c_H$ are known as capacity rates. Now two cases arise.

1. **Equal capacity rates, $\dot{m}_C c_C = \dot{m}_H c_H$.** In this case $d(T_H - T_C) = 0$ or $T_H - T_C$ = constant. Hence, the equation, $dq = U(T_H - T_C) \, dA_s$ can be integrated to give

$$q = U(T_H - T_C)A_s \qquad (9.3.8)$$

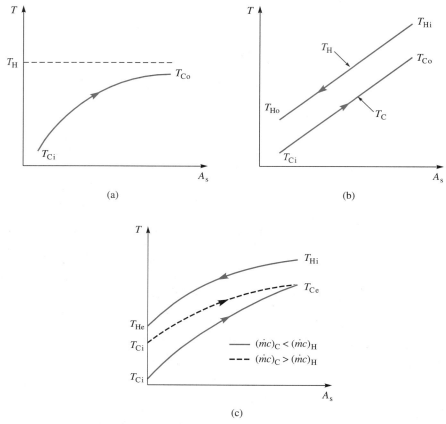

FIGURE 9.3.2 Temperature Profiles in a Counterflow Heat Exchanger
(a) Temperature profiles in a double-flow heat exchanger with T_H constant. (b) The temperature variation in the hot and cold fluids in a counterflow heat exchanger in which $\dot{m}_H c_H = \dot{m}_C c_C$. (c) Temperature profile in a double-pipe heat exchanger with two single-phase fluids ($m_C c_C \neq \dot{m}_H c_H$).

Both T_H and T_C vary, but $T_H - T_C$ = constant as shown in Figure 9.3.2b. In every element, the increase in the temperature of the cold fluid is exactly equal to the decrease in the temperature of the hot fluid. This condition can exist only in a counterflow heat exchanger. (In a parallel-flow heat exchanger the temperature of the hot fluid decreases and the temperature of the cold fluid increases in the direction of flow and the temperature difference between the hot and cold fluids continuously decreases.) With $\dot{m}_H c_H = \dot{m}_C c_C$, the temperature variation in the hot and cold fluids (Figure 9.3.2b) shows that we need to know the temperature difference between the hot and cold fluids at any one location, which is the temperature difference to be used in Equation 9.3.8. It is also the mean temperature difference. Generally, the inlet temperatures of the fluids and the exit temperature of one fluid are known. The heat transfer rate and the exit temperature of the other fluid can be determined from the energy balance.

$$q = \dot{m}_C c_C (T_{Co} - T_{Ci}) = \dot{m}_H c_H (T_{Hi} T_{Ho})$$

where the second subscript (i or o) represents inlet or outlet of the heat exchanger. The difference in temperatures of the two fluids at one of the ends is used in Equation 9.3.8 to find the surface area of the heat exchanger. The mean temperature difference to be used in Equation 9.3.1 is given by

$$\Delta T_m = T_H - T_C$$

The term $T_H - T_C$ can be evaluated at any location in the heat exchanger.

2. **Unequal Capacity Rates, $\dot{m}_C c_C \neq \dot{m}_H c_H$.** With $dq = U(T_H - T_C)\,dA_s$, Equation 9.3.7 can be rearranged as

$$\frac{d(T_H - T_C)}{T_H - T_C} = \left(\frac{1}{\dot{m}_H c_H} - \frac{1}{\dot{m}_C c_C}\right) U\,dA_s \tag{9.3.10}$$

Integrating Equation 9.3.10 from end 1 to end 2,

$$\ln\left[\frac{(T_H - T_C)_2}{(T_H - T_C)_1}\right] = \left(\frac{1}{\dot{m}_H c_H} - \frac{1}{\dot{m}_C c_C}\right) UA_s \tag{9.3.11}$$

With $q = \dot{m}_H c_H (T_{H,2} - T_{H,1}) = \dot{m}_C c_C (T_{C,2} - T_{C,1})$, eliminating $\dot{m}_H c_H$ and $\dot{m}_C c_C$, and rearranging Equation 9.3.11,

$$q = UA_s \frac{(T_H - T_C)_1 - (T_H - T_C)_2}{\ln[(T_H - T_C)_1/(T_H - T_C)_2]} = UA_s \frac{\Delta T_1 - \Delta T_2}{\ln(\Delta T_1/\Delta T_2)} \tag{9.3.12}$$

Comparing Equation 9.3.12 with Equation 9.3.1, we get

$$\left\|\ \Delta T_m = \frac{\Delta T_1 - \Delta T_2}{\ln(\Delta T_1/\Delta T_2)}\ \right\|$$

where

$\Delta T_1 = (T_H - T_C)_1 =$ the difference between the temperatures of the hot and cold fluid at end 1

$\Delta T_2 = (T_H - T_C)_2 =$ the temperature difference between the two fluids at end 2.

Either end may be designated as end 1. The temperature profile for this case is sketched in Figure 9.3.2c.

To summarize, when $T_H - T_C$ is uniform (either both T_H and T_C are uniform, as in the case with phase change on both sides of the heat exchanger, or when $\dot{m}_C c_C = \dot{m}_H c_H$ with no phase change on either side)

$$\|\ \ \Delta T_m = T_H - T_C\ \ \| \tag{9.3.13a}$$

In all other cases,

$$\left\|\ \Delta T_m = \frac{\Delta T_1 - \Delta T_2}{\ln(\Delta T_1/\Delta T_2)}\ \right\| \tag{9.3.13b}$$

The mean temperature difference defined by Equation 9.3.13b is known as the *logarithmic mean temperature difference* (LMTD). *Note:* Equations 9.3.3, 9.3.4, 9.3.6, and 9.3.13b are valid for a parallel-flow heat exchanger also for the same conditions. Their derivations are left as an exercise.

The heat transfer rate from one fluid to the other can also be determined by applying the law of energy balance for the two fluids.

$$q = \dot{m}(i_i - i_o)$$

where

q = heat transfer rate from the hot fluid to the cold fluid
$\dot{m}$ = mass rate of the hot fluid
i_i = specific enthalpy of the hot fluid at inlet to the heat exchanger
i_o = specific enthalpy of the hot fluid at outlet of the heat exchanger

For a constant density fluid (with a small pressure drop) or an ideal gas, $i_i - i_o = c_p(T_i - T_o)$. If the fluid changes phase, from saturated vapor to saturated liquid, i_i and i_o are the specific enthalpies of the fluid at the actual inlet and exit states of the fluid.

If the rate of heat transfer to be accomplished, the terminal temperatures of the fluids, and the overall heat transfer coefficient are known, the required surface area, A_s, can be determined directly from the relation $q = UA_s \Delta T_m$. However, if the inlet temperatures of the hot and cold fluids (without phase change in either fluid) and their mass rates of flow are known, for a given surface area and overall heat transfer coefficient, the heat transfer rate has to be determined by iteration in the LMTD method.

In determining the LMTD and the heat transfer rate, the assumption of uniform overall heat transfer coefficient should be satisfied. Such an assumption is a reasonable approximation in many situations but there is one case when it is not satisfied. In a flooded condenser, a vapor is totally condensed in one part of the condenser and the condensate is subcooled in another part of the same condenser where no condensation occurs. In such a condenser, the overall heat transfer coefficient in the condensing part is significantly different from the overall heat transfer coefficient in the subcooled part and the assumption of uniform overall heat transfer coefficient is not appropriate. The two parts should be analyzed separately, each with an overall heat transfer coefficient appropriate for the particular mode of heat transfer.

EXAMPLE 9.3.1 In an oil cooler for a diesel engine 0.1 kg/s of oil is to be cooled from 120 °C to 60 °C in a double-pipe heat exchanger with 0.1 kg/s of water available at 10 °C. The overall heat transfer coefficient is 400 W/m² °C. Determine the heat transfer surface area if the flow is

(a) Parallel.
(b) Counterflow.

(Specific heat of oil is 2131 J/kg °C.)

Given

Fluid: Oil $\dot{m}_o = 0.1$ kg/s $T_{oi} = 120$ °C $T_{oe} = 60$ °C
Fluid: Water $\dot{m}_w = 0.1$ kg/s $T_{wi} = 10$ °C
$U = 400$ W/m² °C

Find

(a) A_s for parallel flow arrangement
(b) A_s for counterflow arrangement

ASSUMPTIONS

1. Steady state
2. Constant properties for both fluids
3. No heat transfer other than between the two fluids
4. Uniform overall heat transfer coefficient
5. Incompressible fluids

SOLUTION

The heat transfer surface area is determined from (with subscripts o for oil and w for water)

$$q = UA_s \, \Delta T_m \tag{1}$$

$$\Delta T_m = \frac{\Delta T_1 - \Delta T_2}{\ln(\Delta T_1/\Delta T_2)} \tag{2}$$

From the First Law of Thermodynamics for incompressible fluids

$$q = [\dot{m}c_p(T_e - T_i)]_w = [\dot{m}c_p(T_i - T_e)]_o \tag{3}$$

Determine q and T_{we} from Equation 3, ΔT_m from Equation 2, and A_s from Equation 1.

(a) Parallel-flow arrangement. Determine T_{we} from Equation 3. Although the mean temperature of water is not known, from Table A5, the value of c_p for water ranges from 4195 J/kg °C at 10 °C to 4181 J/kg °C at 60 °C (the exit temperature of water cannot exceed the exit temperature of oil in a parallel flow arrangement) with a minimum value of 4175 J/kg °C at 40 °C. An initial estimate of the exit temperature can be obtained by using any value within this range. From such an estimate of the exit temperature a more appropriate value for the specific heat can be used to find the correct exit temperature of the water.

For the initial estimate, employing the value of 4176 J/kg °C for c_p at 30 °C,

$$T_{we} = T_{wi} + q/(\dot{m}c_p)_w$$

$$q = (\dot{m}c_p)_o(T_{oi} - T_{oe}) = 0.1 \times 2131(120 - 60) = 12\,786 \text{ W}$$

$$T_{we} = 10 + 12\,786/(0.1 \times 4176) = 40.6 \,°C$$

With an exit temperature of 40.6 °C the mean temperature is 25.3 °C. Employing the value of 4178 J/kg °C for c_p at 25.3 °C and recomputing the exit temperature of the water, we obtain

$$T_{we} = 10 + 12\,786/(0.1 \times 4178) = 40.6 \,°C$$

As the exit temperature equals the assumed value for evaluating the mean temperature for c_p, no further iteration is necessary. The parallel-flow arrangement with the exit temperatures for the fluids, is shown in Figure 9.3.3a. For the parallel-flow arrangement,

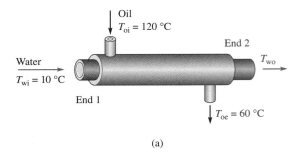

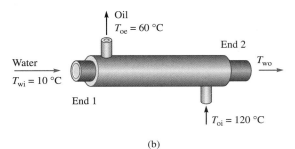

(a)

(b)

FIGURE 9.3.3 Double-Pipe Heat Exchanger
(a) Parallel-flow arrangement of the heat exchanger.
(b) Counterflow arrangement for the heat
exchanger.

$$\Delta T_m = \text{LMTD} = \frac{\Delta T_1 - \Delta T_2}{[\ln(\Delta T_1/\Delta T_2)]} = \frac{(120 - 10) - (60 - 40.6)}{\ln(120 - 10)/(60 - 40.6)]} = 52.2\ °C$$

$$A_s = \frac{q}{U\ \Delta T_m} = \frac{12\ 786}{400 \times 52.2} = \underline{0.612\ m^2}$$

(b) Counterflow arrangement. The exit temperature of the water determined from the energy balance is valid for the counterflow arrangement also but the temperature differences between the fluids at the two ends of the heat exchanger are different as shown in Figure 9.3.3b. As the terminal temperatures are different, the LMTD is also different. For the counterflow arrangement,

$$\Delta T_m = \text{LMTD} = \frac{\Delta T_1 - \Delta T_2}{\ln(\Delta T_1/\Delta T_2)} = \frac{(60 - 10) - (120 - 40.6)}{\ln[(60 - 10)/(120 - 40.6)]} = 63.6\ °C$$

$$A_s = \frac{q}{U\ \Delta T_m} = \frac{12\ 786}{400 \times 63.6} = \underline{0.503\ m^2}$$

COMMENT

The LMTD in the counterflow heat exchanger is higher than that in the parallel-flow heat exchanger. Consequently, the heat transfer surface area is 17.8% less than the area in the parallel-flow heat exchanger. A counterflow heat exchanger has the maximum value of ΔT_m for a given set of inlet and exit temperatures of the fluids. Hence, it requires the least heat transfer surface area. It is the most efficient type of heat exchanger, but operating conditions may require the use of other types of heat exchangers even though the surface area for such heat exchangers may be higher than the surface area with the counterflow arrangement.

9.3.2 Mean Temperature Difference for Other Types of Heat Exchangers

Logarithmic Mean
Temperature
Difference—
Correction factor

For a parallel-flow or a counterflow heat exchanger the mean temperature difference to be employed in determining the heat transfer rate is the LMTD given by Equation 9.3.13b. But many heat exchangers, such as shell-and-tube type and cross-flow type, are neither parallel-flow nor counterflow type. For example, in a two-tube pass, single-shell pass heat exchanger with baffles, shown schematically in Figure 9.3.4a, the overall direction of motion of the shell side fluid is from one end of the heat exchanger to the other. Because of the baffles, although the general direction of the flow of the shell side fluid is from one end to the other, the flow is actually transverse in sections between a pair of baffles. In cross-flow heat exchangers, the flows are

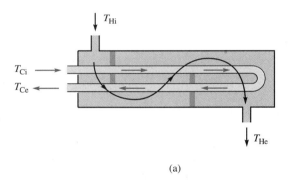

(a)

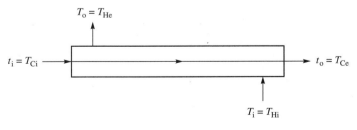

(b)

FIGURE 9.3.4 Shell-and-Tube Heat Exchanger and a Counterflow Heat Exchanger with Identical Terminal Temperatures
(a) A shell-and-tube heat exchanger with baffles. The flow of the fluid on the shell side is neither parallel-flow nor counterflow type.
(b) Counterflow heat exchanger with the same exit temperatures as in the shell-and-tube heat exchanger.

neither parallel nor counterflow but are mutually perpendicular. In such cases, where the heat exchanger is neither of the purely parallel-flow nor counterflow type, the determination of the mean temperature difference becomes complicated. The mean temperature difference is then obtained through a correction factor, F, defined by

$$F = \frac{\Delta T_m}{\text{LMTD}_{CF}}$$

In the expression for F, ΔT_m is the mean temperature difference for the actual heat exchanger, and the LMTD_{CF} in the denominator is the log mean temperature difference for a counterflow heat exchanger with the same terminal temperatures of the fluids as in the actual heat exchanger, shown in Figure 9.3.4b. For fixed terminal temperatures the LMTD is highest for counterflow heat exchanger and the maximum value of F is unity. The value of F has been determined for several types of heat exchangers and are presented in Figures 9.3.5 through 9.3.7. In the figures,

$$R = \frac{T_i - T_o}{t_o - t_i} \qquad P = \frac{t_o - t_i}{T_i - t_i}$$

$$\dot{m}_H c_H (T_i - T_o) = \dot{m}_C c_C (t_o - t_i)$$

Therefore,

$$R = \frac{\dot{m}_C c_C}{\dot{m}_H c_H}$$

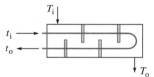

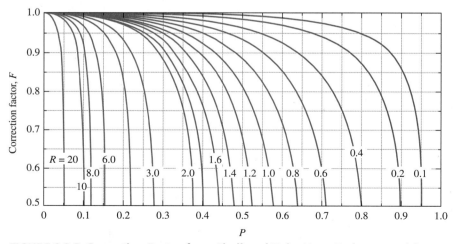

FIGURE 9.3.5 Correction Factor for a Shell-and-Tube Heat Exchanger with One Shell and Any Multiple of Two-Tube Passes. Reproduced from Bowman, et al. (1940), with the permission of the ASME.

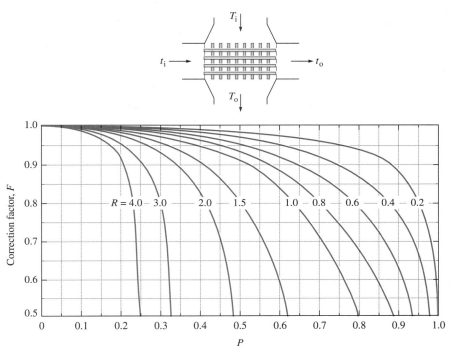

FIGURE 9.3.6 Correction Factor for a Single-Pass, Cross-Flow Heat Exchanger with Both Fluid Unmixed. Reproduced from Bowman, et al. (1940), with the permission of the ASME.

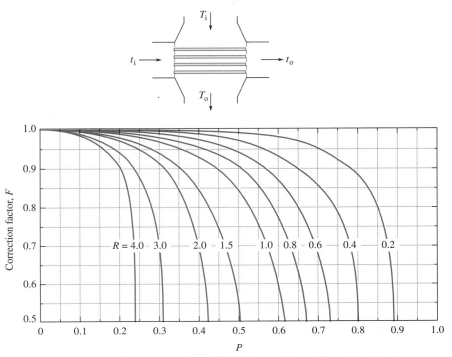

FIGURE 9.3.7 Correction Factor for a Single-Pass, Cross-Flow Heat Exchanger with One Fluid Mixed and the Other Unmixed. Reproduced from Bowman, et al. (1940), with the permission of the ASME.

The quantities $\dot{m}_H c_{CH}$ and $\dot{m}_C c_C$ are the capacity rates and their ratio R is the capacity ratio. $T_i - t_i$ is the maximum temperature difference that can be obtained in one of the fluids in a counterflow heat exchanger of infinite area and $t_o - t_i$ is actual temperature difference in one of the fluids. Multiplying the numerator and the denominator by the capacity rate $\dot{m}_C c_C$ or $\dot{m}_H c_H$, we see that P is the ratio of the actual heat transfer rate to the maximum possible heat transfer rate and represents the thermal effectiveness of the heat exchanger. The correction factor F for various types of heat exchangers is available in the HE module of the software.

For heat exchangers other than the parallel-flow or counterflow type, first determine the exit temperatures of the fluids [the temperatures that are appropriate in the equation $dq = U(T_H - T_C)\,dA$. The LMTD for a counterflow heat exchanger with these exit temperatures is then determined. The actual ΔT_m for the heat exchanger is calculated by finding the value of F from Figures 9.3.5–9.3.7.

EXAMPLE 9.3.2 Engine oil flowing at 0.5 kg/s is to be cooled in a two-tube pass shell-and-tube heat exchanger from 80 °C to 40 °C. Water at 20 °C enters the heat exchanger at a flow rate of 1 kg/s. If the overall heat transfer coefficient is 500 W/m² K, determine the required heat transfer area. A two-tube pass shell-and-tube heat exchanger is schematically shown in Figure 9.3.8.

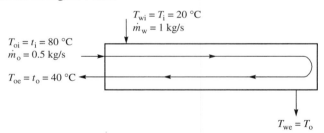

FIGURE 9.3.8 Shell-and-Tube Heat Exchanger in Example 9.3.1

Given

Fluid: Oil $\dot{m}_o = 0.5$ kg/s $T_{oi} = 80\,°C$ $T_{oe} = 40\,°C$
Fluid: Water $\dot{m}_w = 1$ kg/s $T_{wi} = 20\,°C$
$U = 500$ W/m² °C

Find

A_s (heat transfer surface area)

ASSUMPTIONS

1. Steady state
2. Constant properties
3. Oil has the properties of unused engine oil.
4. Uniform overall heat transfer coefficient
5. No heat transfer from the shell to the surroundings

SOLUTION

The surface area is found from the relation (with subscripts o for oil and w for water)

$$q = UA_s \, \Delta T_m \tag{1}$$

where

$$\Delta T_m = F \, \mathrm{LMTD_{CF}} \tag{2}$$

$$\mathrm{LMTD_{CF}} = \frac{\Delta T_1 - \Delta T_2}{\ln(\Delta T_1/\Delta T_2)} \tag{3}$$

$$\Delta T_1 = T_{oi} - T_{we} \qquad \Delta T_2 = T_{oe} - T_{wi}$$

To find T_{we}, from the First Law of Thermodynamics,

$$q = [\dot{m}c_p(T_i - T_e)]_o = [\dot{m}c_p(T_e - T_i)]_w$$

Find F from the terminal temperatures.

Determine the exit temperature of the water from Equation 4.

$$q = [\dot{m}c_p(T_i - T_e)]_o = [\dot{m}c_p(T_e - T_i)]_w$$

From Table A4, c_p for oil at the mean temperature of 60 °C = 2047 J/kg K. Thus,

$$q = (\dot{m}c_p)_o(T_i - T_e)_o = 0.5 \times 2047(80 - 40) = 40\,940 \text{ W}$$

Anticipating a mean temperature of 25 °C for the water, from Table A5, c_p for water = 4178 J/kg K. Then,

$$(T_e - T_i)_w = \frac{q}{(\dot{m}c_p)_w} = \frac{40\,940}{1 \times 4178} = 9.8 \text{ °C}$$

$$T_{we} = 29.8 \text{ °C}$$

The mean temperature of the water is close to the anticipated mean temperature and the appropriate value of c_{pw} was used in computing the exit temperature of water.

A counterflow heat exchanger with the same inlet and exit temperatures as the actual heat exchangers is shown schematically in Figure 9.3.9. Compute the LMTD for the counterflow heat exchanger from Equation 3.

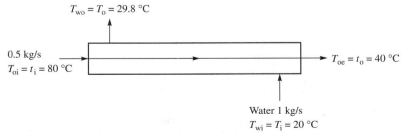

FIGURE 9.3.9 Counterflow Heat Exchanger with the Same Exit Temperatures of the Hot and Cold Fluid as in the Shell-and-Tube Heat Exchanger

$$\text{LMTD}_{\text{CF}} = \frac{\Delta T_1 - \Delta T_2}{\ln(\Delta T_1/\Delta T_2)} = \frac{(80 - 29.8) - (40 - 20)}{\ln[(80 - 29.8)/(40 - 20)]} = 32.8\,°\text{C}$$

Determine the value of F. Referring to Figure 9.3.5,

$$R = \frac{T_i - T_o}{t_o - t_i} = \frac{20 - 29.8}{40 - 80} = 0.245$$

$$P = \frac{t_o - t_i}{T_i - t_i} = \frac{40 - 80}{20 - 80} = 0.67$$

From Figure 9.3.5 corresponding to $R = 0.245$, $P = 0.67$, and $F \approx 0.93$, (From the HE software, $F = 0.933$)

$$\Delta T_m = F \times \text{LMTD}_{\text{CF}} = 0.93 \times 32.8 = 30.5\,°\text{C}$$

$$A = \frac{q}{U\,\Delta T_m} = \frac{40\,940}{500 \times 30.5} = \underline{2.68\ \text{m}^2}$$

LMTD method for rating

The LMTD method works well when, given the mass flow rates of the fluids, their inlet and exit temperatures, and the overall heat transfer coefficient, the surface area of a heat exchanger is to be determined. To determine the performance of a heat exchanger given $\dot{m}_C$, $\dot{m}_H$, T_{Ci}, T_{Hi}, U, and A, this method does not give a direct solution; the solution has to be obtained by a trial and error algorithm. However, in a heat exchanger where the temperature of one of the fluids is uniform, a direct solution is obtained by employing Equation 9.3.4 to determine the exit temperature of the fluid whose temperature changes. From the computed value of the exit temperature, the heat transfer rate is also obtained. If the temperatures of both the fluids are uniform, the mean temperature difference is readily available to compute the heat transfer rate.

Consider a heat exchanger whose surface area and overall heat transfer coefficients are known. For given mass flow rates and temperatures of the hot and cold fluids (with no phase change in either fluid and with unequal capacity rates) entering the heat exchanger, it is desired to determine the heat transfer rate when the mass flow rates and the inlet temperatures are not the same as those for which the heat exchanger was designed. The heat transfer rate is related to the difference in the inlet and exit temperatures of the fluids. But neither the exit temperatures nor the heat transfer rate is known. They are determined by iteration. One possible iteration algorithm follows:

1. Guess the exit temperature of one of the fluids.
2. From the guessed temperature determine the exit temperature of the other fluid from an energy balance.
3. Determine the LMTD for the corresponding counterflow heat exchanger.
4. Determine the correction factor, F.
5. Compute ΔT_m.
6. Check if the heat transfer rate obtained from $q = UA_s\,\Delta T_m$ is equal to that obtained from the energy balance. If the two heat transfer rates do not match, guess a new

temperature and repeat the process until the heat transfer rates calculated by both equations are close to each other.

This is a time consuming process, and it is desirable to have a method that would give a direct solution to this type of problem. The NTU (number of transfer units) method provides such a direct solution.

9.4 NUMBER OF TRANSFER UNITS (NTU)— EFFECTIVENESS METHOD

Effectiveness of a heat exchanger

The NTU method can be used both for the design of heat exchangers and to predict their performances under off-design conditions without any iteration. The *effectiveness of a heat exchanger, ε,* is defined as the ratio of the heat transfer rate in the actual heat exchanger to the maximum possible heat transfer rate between the two fluids. Such a maximum heat transfer occurs in a counterflow heat exchanger with an infinite area with the same mass flow rates and inlet temperatures as the actual heat exchanger.

$$\varepsilon = \frac{\text{Heat transfer rate in the actual heat exchanger}}{\text{Heat transfer rate with a counterflow heat exchanger of infinite area}}$$

$$\text{(9.4.1)}$$

In the NTU method, the maximum possible heat transfer rate is determined and related to the actual heat transfer rate through its effectiveness, ε. If the effectiveness of the heat exchanger is known, the actual heat transfer rate is readily determined.

The effectiveness of a heat exchanger is a function of two parameters, the number of transfer units $(UA)/(\dot{m}c_p)_{min}$, denoted by NTU, and the capacity ratio $(\dot{m}c_p)_{min}/(\dot{m}c_p)_{max}$, denoted by c_r.

$$NTU = \frac{UA}{(\dot{m}c_p)_{min}}$$

$$c_r = \frac{(\dot{m}c_p)_{min}}{(\dot{m}c_p)_{max}}$$

Maximum heat transfer rate— Counterflow heat exchanger with an infinite surface area

To compute the maximum heat transfer rate in a counterflow heat exchanger with an infinite surface area, consider two single-phase fluids exchanging heat in a counterflow heat exchanger. From an energy balance, with constant specific heats,

$$q = \dot{m}_C c_C (T_{Ce} - T_{Ci}) = \dot{m}_H c_H (T_{Hi} - T_{He})$$

where $\dot{m}$ is the mass flow rate and c is the specific heat at constant pressure. Subscript C and H refer to the cold and hot fluid, respectively. Denoting $\dot{m}_C c_C$ by C_C and $\dot{m}_H c_H$ by C_H, if $C_C > C_H$, then $T_{Hi} - T_{He} > T_{Ce} - T_{Ci}$. For this case, then, the maximum heat transfer rate is given by

$$q_{max} = C_H (T_{Hi} - T_{He})_{max}$$

With a given T_{Hi}, $(T_{Hi} - T_{He})_{max}$ is obtained when T_{He} reaches its minimum value. In a counterflow heat exchanger, the value of T_H continuously decreases and approaches the inlet temperature of the cold fluid. In the limiting case of a counterflow heat exchanger with an infinite surface area, T_{He} equals T_{Ci}. Thus, for this case,

$$q_{max} = C_H(T_{Hi} - T_{Ci})$$

For the case of $C_H > C_C$, by a similar logic as in the previous case, it is easily shown that

$$q_{max} = C_C(T_{Hi} - T_{Ci})$$

In either case, therefore,

$$q_{max} = C_{min}(T_{Hi} - T_{Ci})$$

The temperature profiles in a counterflow heat exchanger with infinite surface area are shown in Figure 9.4.1.

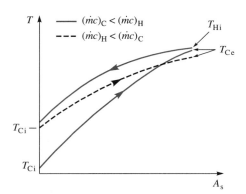

FIGURE 9.4.1 A Counterflow Heat Exchanger with a Large Surface Area If $(\dot{m}c)_C < (\dot{m}c)_H$, T_{Ce} approaches T_{Hi}. If $(\dot{m}c)_H < (\dot{m}c)_C$, T_{He} approaches T_{Ci}.

If the mass flow rates and the inlet temperatures of the two fluid streams are known, the maximum heat transfer rate can be determined. If the effectiveness of the heat exchanger is known, the actual heat transfer rate, and the exit temperatures of the fluids can be calculated. From the definition of the effectiveness we have

$$\varepsilon = \frac{q}{C_{min}(T_{Hi} - T_{Ci})}$$

or

$$q = \varepsilon C_{min}(T_{Hi} - T_{Ci})$$

Having determined the heat transfer rate, the exit temperatures are computed from the relation

$$q = C_C(T_{Ce} - T_{Ci}) = C_H(T_{Hi} - T_{He})$$

The effectiveness of various types of heat exchangers has been analyzed. Some of them are given in Table 9.4.1. Table 9.4.2 gives NTU in terms of the effectiveness. Figures 9.4.2 through 9.4.7 are graphical representations of ε for different types of heat exchangers. Equation 4 in Table 9.4.1 is from Eckert and Drake (1959). All other equations in the table are adapted from Shah and Mueller (1985). Table 9.4.2 is obtained from the expressions in Table 9.4.1.

The equations in Tables 9.4.1 and 9.4.2 have been programmed in the software HE.

TABLE 9.4.1 Formulae for Computing Heat Exchanger Effectiveness

$$c_r = \frac{(\dot{m}c_p)_{\min}}{(\dot{m}c_p)_{\max}} = \frac{C_{\min}}{C_{\max}} \qquad N = \text{NTU} = \frac{UA}{(\dot{m}c_p)_{\min}} \qquad \varepsilon = \frac{q}{q_{\max}}$$

Parallel-Flow

$$\varepsilon = \frac{1 - e^{-N(1 + c_r)}}{1 + c_r} \tag{1}$$

Counterflow

$$\varepsilon = \frac{1 - e^{-N(1 - c_r)}}{1 - c_r e^{-N(1 - c_r)}} \qquad (c_r \neq 1) \tag{2}$$

If $c_r = 1, \varepsilon = \dfrac{N}{1 + N}$

Shell-and-tube

One-shell pass, 2, 4, . . . tube passes

$$\varepsilon = \frac{2}{(1 + c_r) + a(1 + e^{-aN})/(1 - e^{-aN})} \tag{3a}$$

where

$$a = \sqrt{1 + c_r^2}$$

n-shell passes, $2n$, $4n$ . . . tube passes

$$\varepsilon = \left[\left(\frac{1 - \varepsilon_1 c_r}{1 - \varepsilon_1} \right)^n - 1 \right]^n \left[\left(\frac{1 - \varepsilon_1 c_r}{1 - \varepsilon_1} \right)^n - c_r \right]^{-1} \tag{3b}$$

ε_1 = effectiveness of a one-shell pass heat exchanger from Equation 3

Cross-Flow (single pass)

Both fluids unmixed

$$\varepsilon = 1 - e^{N^{0.22}\Gamma/c_r} \qquad \Gamma = e^{-c_r N^{0.78}} - 1 \tag{4}$$

$\dot{m}c_{p})_{\max}$(mixed), $\dot{m}c_{p})_{\min}$(unmixed)

$$\varepsilon = \frac{(1 - e^{-c_r \Gamma})}{c_r} \qquad \Gamma = 1 - e^{-N} \tag{5}$$

$\dot{m}c_{p})_{\min}$(mixed), $\dot{m}c_{p})_{\max}$(unmixed)

$$\varepsilon = 1 - e^{-\Gamma/c_r} \qquad \Gamma = 1 - e^{-c_r N} \tag{6}$$

Both fluids mixed

$$\varepsilon = \frac{N}{N/(1 - e^{-N}) + c_r N/(1 - e^{-Nc_r}) - 1} \tag{7}$$

All exchangers with $c_r = 0$

$$\varepsilon = 1 - e^{-N} \tag{8}$$

| TABLE 9.4.2 Formulae for Computing NTU

$$c_r = \frac{(\dot{m}c_p)_{min}}{(\dot{m}c_p)_{max}} = \frac{C_{min}}{C_{max}} \qquad N \text{ NTU} = \frac{UA}{(\dot{m}c_p)_{min}} \qquad \varepsilon = \frac{q}{q_{max}}$$

Parallel-flow

$$N = -\frac{\ln[1 - \varepsilon(1 + c_r)]}{1 + c_r} \qquad (1)$$

Counterflow

$$N = -\frac{1}{1 - c_r} \ln\left(\frac{1 - \varepsilon}{1 - \varepsilon c_r}\right) \qquad (2)$$

If $c_r = 1$,

$$N = \frac{\varepsilon}{1 - \varepsilon} \qquad (3)$$

Shell-and-Tube

One-shell pass, 2, 4, . . . tube passes

$$N = -\frac{1}{(1 + c_r^2)^{1/2}} \ln\left(\frac{b - 1}{b + 1}\right)$$

where

$$b = \frac{2/\varepsilon - (1 + c_r)}{(1 + c_r^2)^{1/2}} \qquad (4)$$

Cross-flow (single pass)

$\dot{m}c_p)_{max}$(mixed), $\dot{m}c_p)_{min}$(unmixed)

$$N = -\ln\left[1 + \frac{1}{c_r} \ln (1 - \varepsilon c_r)\right] \qquad (5)$$

$\dot{m}c_p)_{min}$(mixed), $\dot{m}c_p)_{max}$(unmixed)

$$N = -\frac{1}{c_r} \ln[1 + c_r \ln(1 - \varepsilon)] \qquad (6)$$

All exchangers with $c_r = 0$

$$N = -\ln(1 - \varepsilon) \qquad (7)$$

Both fluids unmixed
Both fluids mixed $\Big\}$ N determined by numerical methods

The methodologies for sizing and rating of heat exchangers by the NTU method are

1. T_{Hi}, T_{Ci}, $\dot{m}_H$, $\dot{m}_C$, U, and T_{He} (or T_{Ce}) known. Determine the surface areas A.
   (a) From $q = \dot{m}_C c_C(T_{Ce} - T_{Ci}) = \dot{m}_H c_H(T_{Hi} - T_{He})$, determine q and T_{Ce} (or T_{He}).
   (b) Compute $q_{max} = \dot{m}c)_{min}(T_{Hi} - T_{Ci})$.
   (c) Determine $c_r = (\dot{m}c)_{min}/(\dot{m}c)_{max}$ and $\varepsilon = q/q_{max}$.
   (d) From the values of c_r and ε, find NTU from Figures 9.4.2 through 9.4.7, or from the software, or from the equations in Table 9.4.2.
   (e) Determine $A_s = \text{NTU}(\dot{m}c)_{min}/U$
2. T_{Hi}, T_{Ci}, $\dot{m}_H$, $\dot{m}_C$, U, and A_s known. Determine q, and T_{He} and T_{Ce}.
   (a) Compute $(\dot{m}c)_H$, $(\dot{m}c)_C$, $\text{NTU} = UA_s/(\dot{m}c)_{min}$, and $c_r = (\dot{m}c)_{min}/(\dot{m}c)_{max}$.
   (b) From the values of NTU and c_r, find ε from Figures 9.4.2 through 9.4.6, or from the software, or from the equations in Table 9.4.1.
   (c) Compute $q_{max} = (\dot{m}c)_{min}(T_{Hi} - T_{Ci})$ and $q = \varepsilon q_{max}$.
   (d) Determine T_{He} and T_{Ce} from $q = \dot{m}_C c_C(T_{Ce} - T_{Ci}) = \dot{m}_H c_H(T_{Hi} - T_{He})$

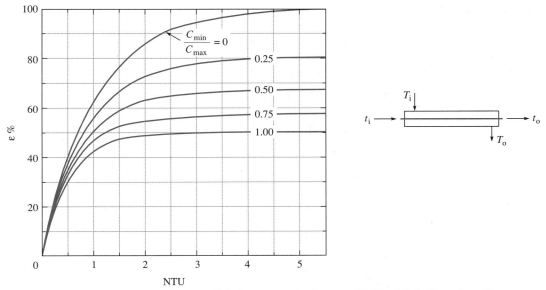

FIGURE 9.4.2 Effectiveness of a Parallel-Flow Heat Exchanger (Table 9.4.1, Equation 1)

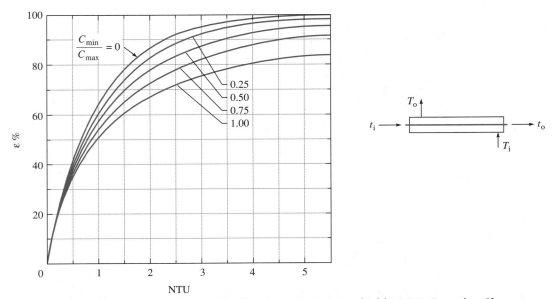

FIGURE 9.4.3 Effectiveness of a Counterflow Heat Exchanger (Table 9.4.1, Equation 2)

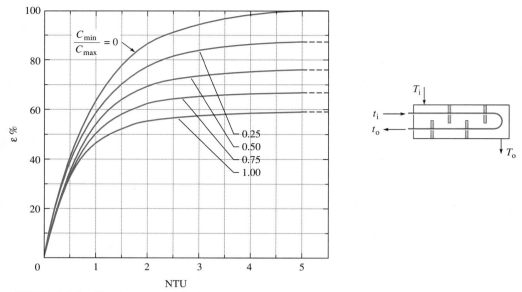

FIGURE 9.4.4 Effectiveness of a Shell-and-Tube Heat Exchanger with One-Shell Pass and Any Multiple Two-Tube Passes (two, four, etc.) (Table 9.4.1, Equation 3a)

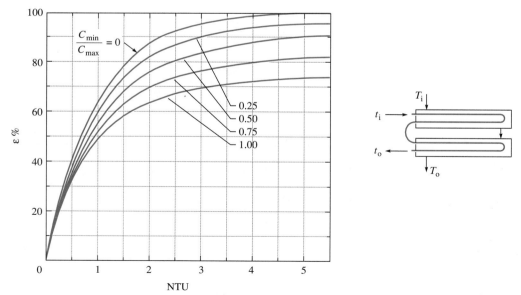

FIGURE 9.4.5 Effectiveness of a Shell-and-Tube Heat Exchanger with Two-Shell Passes and Any Multiple of Four-Tube Passes (Table 9.4.1, Equation 3b)

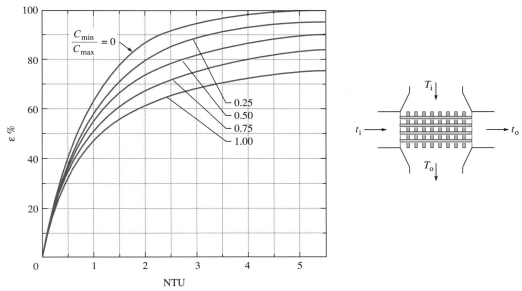

FIGURE 9.4.6 Effectiveness of Single-Pass, Cross-Flow Heat Exchanger with Both Fluids Unmixed (Table 9.4.1, Equation 4)

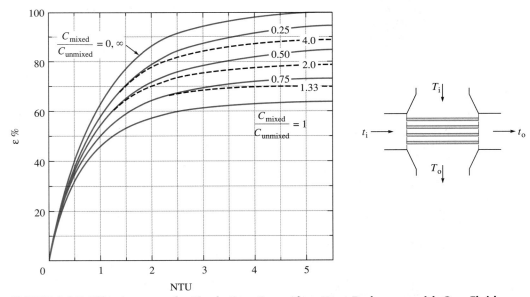

FIGURE 9.4.7 Effectiveness of a Single-Pass Cross-Flow Heat Exchanger with One Fluid Mixed and the Other Unmixed (Table 9.4.1, Equations 5 and 6)

EXAMPLE 9.4.1

Oil at 2 kg/s enters an oil cooler at 80 °C. The heat transfer surface area is 30 m², and it is estimated that the overall heat transfer coefficient for different flow rates of water are

Water flow rate (kg/s)	0.5	1.0	2.0	3.0
U(W/m² °C)	250	275	300	325

Estimate the water flow rate required to cool the oil to 35 °C if the inlet temperature of water is 15 °C. The oil cooler is of the shell-and-tube type with two-tube passes (Figure 9.4.8). The specific heat c_p(oil) = 2110 J/kg K.

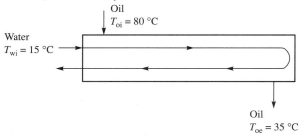

FIGURE 9.4.8 Shell-and-Tube Heat Exchanger to Cool Oil

Given

Fluid: Oil $\dot{m}_o$ = 2 kg/s T_{oi} = 80 °C
                c_{po} = 2110 J/kg °C T_{oe} = 35 °C

Fluid: Water U as a function of $\dot{m}_w$ T_{wi} = 15 °C

A_s = 30 m²

Find

$\dot{m}_w$ (the flow rate of water)

SOLUTION

From the given values of the mass flow rates of water and oil, the overall heat transfer coefficients, and surface area of the heat exchanger, compute c_r and NTU for each mass flow rate of water.

1. Find the effectiveness from the software (or Equation 3 in Table 9.4.1 or from Figure 9.4.4).
2. Compute the actual heat transfer rate $[q = (\dot{m}c_p)_{min}(T_{Hi} - T_{Ci})]$.
3. From the heat transfer rate, compute the exit temperature of the oil $[T_{oe} = T_{oi} - q/(\dot{m}c_p)_o]$.
4. Choose the value of the mass flow rate of water that yields an exit oil temperature closest to the given value of 35 °C.

For the computations, employ a reasonable value for c_{pw}. Anticipating water exit temperature of 30 °C, from Table A5, c_{pw}(22.5 °C) = 4180 kJ/kg °C. The computed values are given in the following table.

Water flow rate (kg/s)	0.5	1.0	2.0	3.0
$(\dot{m}c_p)_w$ (W/°C)	2090	4180	8360	12 540
$(\dot{m}c_p)_o$ (W/°C)	4220	4220	4220	4220
$\dfrac{(\dot{m}c_p)_{min}}{(\dot{m}c_p)_{max}}$	0.4953	0.9905	0.5048	0.3365
U (W/m² °C)	250	275	300	325
NTU $= UA/(\dot{m}c_p)_{min}$	3.589	1.974	2.133	2.31
ε (from software)	0.754	0.558	0.7015	0.7711
$q_{max} = (\dot{m}c_p)_{min}(T_{Hi} - T_{Ci})$ (W)	135 850	217 700	274 300	274 300
$q = \varepsilon \times q_{max}$ (W)	102 431	151 609	192 421	211 513
T_{oe} (°C)	55.7	44.1	34.4	29.9
T_{we} (°C)	64.0	51.3	38.0	31.9

From the tabulated values, the water flow rate should be 2 kg/s. The exit temperature of the oil can be increased to 35 °C by a slight reduction in the mass rate of flow of water. To appreciate the power of the NTU method, you may like to solve the problem by the LMTD method.

EXAMPLE 9.4.2 A plate heat exchanger is fabricated from 0.5-mm thick, type 304 stainless steel plates. The plates are 2-m high, 1-m wide, and the gap between the plates is 5 mm. Water enters the heat exchanger at 100 °C with a mean velocity of 1 m/s between the plates. Liquid Refrigerant 113 enters at 0 °C with a mean velocity of 0.8 m/s between the plates. The flow of the fluids is parallel to the 2-m dimension (Figure 9.4.9). Determine the overall heat transfer coefficient, the heat transfer rate across

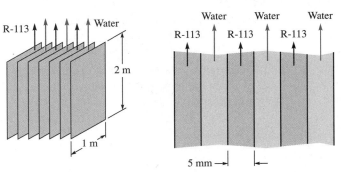

FIGURE 9.4.9 Plate Heat Exchanger with Parallel-Flow Arrangement.
Water and R-113 flow between alternate pairs of plates in the same direction.

each plate, and the exit temperatures of the water and refrigerant for parallel flow of the fluids.

Properties of R-113, extracted from the ASHRAE *Handbook of Fundamentals* (1989) are

Temperature (°C)	ρ (kg/m^3)	$\mu \times 10^6$ (N s/m^2)	c_p (J/kg K)	k (W/m K)
7	1603	885	933	0.0787
17	1580	763	946	0.0768
27	1557	664	958	0.0747
47	1511	516	984	0.0705

Given

Plate material: 304 stainless steel
Height = 2 m   Thickness = 0.5 mm   Width = 1 m
Fluid: Water $v_w = 1$ m/s $T_{wi} = 100\,°C$
Fluid: R-113 $v_R = 0.8$ m/s $T_{Ri} = 0\,°C$

Find

U, q_{plate}, and T_{we} and T_{Re}

ASSUMPTIONS

1. As the gap between the plates is very much less than the length, fully developed conditions prevail in the major part of the plates. Hence, correlations for fully developed internal flows are applicable.
2. Correlations for circular tubes can be used with the hydraulic mean diameter as the characteristic length. The temperature of the plate separating the fluids varies in the direction of flow, but the effect of such variation in computing the convective heat transfer coefficients is neglected and correlations that do not involve the plate temperature are employed.
3. Steady state
4. Uniform overall heat transfer coefficient

SOLUTION

The exit temperatures of the fluids are not known *a priori,* and the mean temperatures at which the fluid properties are to be evaluated are not known. The exit temperatures are found by iteration (with subscripts w for water and R for R-113).

1. Guess the exit temperatures of the fluids for purposes of evaluating the properties.
2. Evaluate the properties, the convective heat transfer coefficients, and the overall heat transfer coefficient.
3. Determine $\dot{m}_w c_w$, $\dot{m}_R c_R$, then compute $c_r = (\dot{m}c_p)_{\text{min}}/(\dot{m}c_p)_{\text{max}}$ and NTU $= UA/(\dot{m}c_p)_{\text{min}}$.
4. From the values of NTU and c_r, find ε.
5. Compute q from $q = \varepsilon q_{\text{max}} = \varepsilon(\dot{m}c_p)_{\text{min}}(T_{wi} - T_{Ri})$.
6. Find T_{we} and T_{re} from $q = \dot{m}_w c_{pw}(T_{wi} - T_{\text{we}}) = \dot{m}_R c_{PR}(T_{\text{Re}} - T_{Ri})$.
7. If the computed exit temperatures are different from those employed for determining the properties, repeat the computations using the computed exit temperatures for evaluating the properties. One or two such iterations will give the correct solution.

From Equation 9.2.1 the overall heat transfer coefficient for fluids separated by a rectangular plate is

$$\frac{1}{U} = \frac{1}{h_w} + \frac{L}{k} + \frac{1}{h_R}$$

where

h_w = convective heat transfer coefficient on the water side
h_R = convective heat transfer coefficient on the refrigerant side
k = thermal conductivity of the material of the plate
L = thickness of the plates

Evaluate properties at initially guested mean temperatures of 90 °C for water and 17 °C for R-113.

Determine h_w   The hydraulic mean diameter, d, for parallel plates is given by

$$d = 4A_c/P \approx 4wg/(2w) = 2g = 2 \times 0.005 = 0.01 \text{ m}$$

where

w = width of the plates
g = gap between the plates

(In computing the hydraulic mean diameter, the contribution of the side plates, with dimensions 2 m × 0.005 m, to the perimeter has been neglected as they do not participate in heat transfer. Also, the flow is essentially that between infinite parallel plates as the gap (5 mm) is much less than the width (1 m), and the end effects are negligible.)

Properties of water at 90 °C from the software CC (or Table A5) are

$\rho = 965.3 \text{ kg/m}^3$ $c_p = 4203 \text{ K/kg °C}$
$\mu = 3.074 \times 10^{-4} \text{ N s/m}^2$ $k = 0.676 \text{ W/m °C}$
$\text{Pr} = 1.91$

$$\text{Re}_d = \frac{\rho v_w d}{\mu} = \frac{965.3 \times 1.0 \times 0.01}{3.074 \times 10^{-4}} = 31\,406$$

Employing Equation 4.2.3b, with 0.3 as the exponent for the Prandtl number,

$$\text{Nu}_d = 0.023 \text{ Re}_d^{0.8} \text{ Pr}^{0.3} = 0.023 \times 31\,406^{0.8} \times 1.91^{0.3} = 110.6$$

$$h_w = \text{Nu}_d \frac{k}{d} = 110.6 \times \frac{0.676}{0.01} = 7477 \text{ W/m}^2 \text{ °C}$$

Determine h_R   Properties of R-113 at 17 °C are

$\rho = 1580 \text{ kg/m}^3$ $c_p = 946 \text{ J/kg °C}$
$\mu = 763 \times 10^{-6} \text{ N s/m}^2$ $k = 0.0768 \text{ W/m °C}$
$\text{Pr} = c_p\mu/k = 946 \times 763 \times 10^{-6}/0.0768 = 9.398$

$$\text{Re}_d = \frac{\rho v_r d}{\mu} = \frac{1580 \times 0.8 \times 0.01}{763 \times 10^{-6}} = 16\,566$$

Employing Equation 4.2.3b, with 0.4 for the exponent for the Prandtl number,

$$\text{Nu}_d = 0.023 \text{ Re}_d^{0.8} \text{ Pr}^{0.4} = 0.023 \times 16\,566^{0.8} \times 9.398^{0.4} = 133.8$$

$$h_R = \text{Nu}_d \frac{k}{d} = 133.8 \times \frac{0.0768}{0.01} = 1028 \text{ W/m}^2 \text{ °C}$$

Compute Overall Heat Transfer Coefficient From Table A1, k (304 Stainless steel) = 14.9 W/m °C. L is the thickness of the plates = 0.0005 m.

$$\frac{1}{U} = \frac{1}{h_w} + \frac{L}{k} + \frac{1}{h_R} = \frac{1}{7477} + \frac{0.0005}{14.9} + \frac{1}{1028}$$

$$U = \underline{877.1 \text{ w/m}^2 \text{ °C}}$$

As the exit temperatures of the fluids are not known, the NTU method is employed. From the symmetry of the arrangement of the heat exchanger, it can be considered as a number of independent heat exchangers in parallel, each one consisting of a pair of passages for the fluids separated by one heat transfer surface (i.e., one plate). The mass rate of flow of each fluid associated with one plate is half the mass flow rate in each passage as shown in Figure 9.4.10.

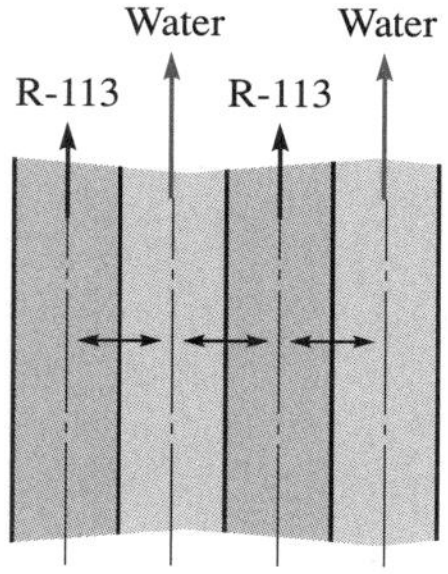

FIGURE 9.4.10 Sketch Showing the Fluids Exchanging Heat Across One Plate

Compute $(\dot{m}c_p)_w$, $(\dot{m}c_p)_r$, c_r and NTU

$$(\dot{m}c_p)_w = (\rho v A_c c_p)_w = 965.3 \times 1 \times 1 \times (0.005/2) \times 4203 = 10\,143 \text{ W/°C}$$

$$(\dot{m}c_p)_R = (\rho v A_c c_p)_R = 1580 \times 0.8 \times 1 \times (0.005/2) \times 946 = 2989 \text{ W/°C}$$

$$(\dot{m}c_p)_{min} = 2989 \text{ W/°C}$$

For each plate, the heat transfer surface area = A_s = 2 m². Therefore,

$$\text{NTU} = \frac{UA_s}{(\dot{m}c_p)_{min}} = \frac{877.1 \times 2}{2989} = 0.5869$$

$$c_r = (\dot{m}c_p)_{min}/(\dot{m}c_p)_{max} = 2989/10\,143 = 0.2947$$

Find ε From the software, for parallel flow heat exchangers, (or Equation 1, Table 9.4.1), for NTU = 0.5869 and c_r = 0.2947, ε = 0.4111.

Determine q

$$q = \varepsilon q_{max} = \varepsilon(\dot{m}c_p)_{min}(T_{Hi} - T_{Ci})$$
$$= 0.4111 \times 2989 \times (100 - 0) = 122\,880 \text{ W}$$

Find T_{Re} and T_{we}

$$T_{Re} = T_{Ri} + q/(\dot{m}c_p)_R = 0 + 122\,880/2989 = 41.1 \text{ °C}$$

$$T_{we} = T_{wi} - q/(\dot{m}c_p)_w = 100 - 122\,878/10\,143 = 87.9°C$$

Having determined the exit temperatures, compare the mean temperatures of the fluids with those assumed for computing the convective heat transfer coefficients.

For water, the mean temperature evaluated from the inlet and the computed exit temperature is 94 °C. For R-113, the mean temperature is 20.6 °C. These are slightly different from the assumed mean values of 90 °C and 17 °C at which the properties of the fluids were evaluated. A better estimate of the heat transfer coefficients may now be obtained by evaluating the properties at 94 °C for water, and 20.6 °C for R-113 and repeating the computations. The updated values are

$h_w = 7646$ W/m² °C $h_R = 1038$ W/m² °C $U = 886.7$ W/m² °C

$(\dot{m}c_p)_w = 10\ 124$ W/°C $(\dot{m}c_p)_R = 2986$ W/°C

$c_r = 0.2949$ NTU $= 0.5937$ $\varepsilon = 0.4143$

$q = 123\ 710$ W $T_{Re} = 41.4$ °C $T_{we} = 87.8$ °C

The mean values of the inlet and exit temperatures of each fluid are now very close to those employed in determining the overall heat transfer coefficient. Hence,

$$\text{Heat transfer rate} = \underline{123\ 710\ \text{W}}$$

$$\text{Exit temperature of R-113} = \underline{41.4\ ^\circ\text{C}}$$

$$\text{Exit temperature of water} = \underline{87.8\ ^\circ\text{C}}$$

COMMENT

If the flows were countercurrent,

$$\varepsilon = 0.4244, \quad q = 126\ 726\ \text{W}, \quad T_{re} = 42.4\ ^\circ\text{C}, \quad T_{we} = 87.5\ ^\circ\text{C}$$

EXAMPLE 9.4.3 In a gas-fired furnace it is proposed to recover waste heat from the flue gases to heat water in a double-pipe, counterflow heat exchanger. Water enters the inner, thin-walled 4-cm diameter pipe at 40 °C with a velocity of 0.5 m/s. As the gas side heat transfer coefficient is much less than the water side heat transfer coefficient, 16 axial copper fins of rectangular cross section are attached to the outer surface of the inner pipe. Each fin is 6-cm high (radial height) and 1-mm thick as shown in Figure 9.4.11. The combustion products enter the heat exchanger at 200 °C at a mass flow rate of 0.12 kg/s. The average heat transfer coefficient on the gas side is 115 W/m² °C. If

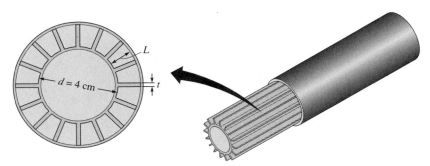

FIGURE 9.4.11 A Finned Double-Pipe Heat Exchanger

the length of the heat exchanger is 5 m, find the heat transfer rate and the exit temperatures of the gases and water. Assume that the properties of the gases are the same as those of air.

Given

Double-pipe heat exchanger with fins   Length = 5 m   d = 40 m
Fin material: copper $L = 6$ cm $t = 1$ mm
N (number of fins) = 16
Fluid: Water in inner pipe $T_{Ci} = 40\,°C$ $V_c = 0.5$ m/s
Fluid: Combustion gases $T_{Hi} = 200\,°C$ $\dot{m}_H = 0.12$ kg/s
$h_o = 115$ W/m² °C

Find

q, T_{He}, and T_{Ce}

ASSUMPTIONS

Fully developed velocity and temperature profiles for water, negligible pipe wall thickness

SOLUTION

Employing the NTU method,
1. Compute $c_r = (\dot{m}c)_{min}/(\dot{m}c)_{max}$.
2. Find h_i (convective heat transfer coefficient on the water side).
3. Compute U.
4. Determine NTU $= (UA)/\dot{m}c)_{min}$.
5. From the values of c_r and NTU find ε.
6. Find $q = \varepsilon q_{max} = \varepsilon(\dot{m}c)_{min}(T_{Hi} - T_{Ci})$.
7. Find the exit temperatures

$$T_{He} = T_{Hi} - \frac{q}{(\dot{m}_H c_H)} \qquad T_{Ce} = T_{Ci} + \frac{q}{(\dot{m}_c c_C)}.$$

Compute c_r. Employing properties of air at an estimated mean temperature of 450 K (177 °C) for gases, from Table A7,

$\rho = 0.785$ kg/m³ $c_H = 1021$ J/kg °C $\mu = 249.03 \times 10^{-7}$ N s/m²
$k = 0.0363$ W/m °C   Pr = 0.7

Thus,

$$\dot{m}_H c_H = 0.12 \times 1021 = 122.5 \text{ W/°C}$$

From Table A5, properties of water at an estimated mean temperature of 45 °C are

$\rho = 990.2$ kg/m³ $c_C = 4176$ J/kg °C $\mu = 577.4 \times 10^{-6}$ N s/m²
$k = 0.637$ W/m °C   Pr = 3.78

Thus,

$$\dot{m}_C = \rho V_c \pi d^2/4 = 990.2 \times 0.5 \times \pi \times 0.04^2/4 = 0.6222 \text{ kg/s}$$

$$\dot{m}_C c_C = 0.6222 \times 4176 = 2598 \text{ W/°C}$$

and

$$c_r = 122.5/2598 = 0.04715$$

Find h_i On the water side,

$$\text{Re}_d = \frac{990.2 \times 0.5 \times 0.04}{577.4 \times 10^{-6}} = 34\ 299$$

From Equation 4.2.3b ($n = 0.4$)

$$h_i = 0.023 \times 34\ 299^{0.8} \times 3.78^{0.4} \times \frac{0.637}{0.04} = 2649 \text{ W/m}^2 \text{ °C}$$

Compute U

For a definition of the overall heat transfer coefficient with fins, consider the heat transfer rate per unit length of the heat exchanger. (See Section 2.6 for a discussion of fin efficiency.)

$$q' = h_o \eta_f A_f'(T_H - T_s) + h_o A_{uf}'(T_H - T_s) \tag{1}$$

where

q' = heat transfer rate from the air to the inner tube per unit length
h_o = convective heat transfer coefficient on the air side
η_f = fin efficiency
A_f' = surface area of the fins per unit length = $2L_cN$
N = number of fins
L_c = corrected length of fins = $L + t/2$
A_{uf}' = area of the unfinned part of the tube per unit length = $\pi d - Nt$
T_H = local bulk temperature of the air (hot fluid)
T_s = temperature of the tube surface

The heat transfer rate from the air to the tube must be equal to the heat transfer rate from the tube to the water.

$$q' = h_i \pi d(T_s - T_C) \tag{2}$$

where T_C is the local temperature of the water (cold fluid).

Rearranging Equations 1 and 2,

$$T_H - T_s = \frac{q'}{h_o(\pi d - Nt + \eta_f 2L_c N)} \tag{3}$$

$$T_s - T_C = \frac{q'}{h_i \pi d} \tag{4}$$

Adding Equations 3 and 4 and rearranging, we obtain

$$q'' = \frac{q'}{\pi d} = \frac{T_H - T_C}{1/h_i + \{\pi d/[h_o(\pi d - Nt + \eta_f 2L_c N)]\}} = U(T_H - T_C)$$

From the above equation we get

$$\frac{1}{U} = \frac{1}{h_i} + \frac{\pi d}{h_o(\pi d - Nt + \eta_f 2L_c N)}$$

Find fin efficiency. For a rectangular fin

$$\eta_f = \frac{\tan h(mL_c)}{mL_c}$$

$$L_c = L + t/2 = 0.06 + 0.0005 = 0.0605 \text{ m}$$

$$m = \left(\frac{2h_o}{kt}\right)^{1/2}$$

From Table A1, k_{Cu} (copper) $= 401$ W/m °C

$$mL_c = \left(\frac{2 \times 115}{401 \times 0.001}\right)^{1/2} \times 0.0605 = 1.449$$

$$\eta_f = \tanh(1.449)/1.449 = 0.618$$

$$\frac{1}{U} = \frac{1}{2649} + \frac{\pi \times 0.04}{115(\pi \times 0.04 - 16 \times 0.001 + 0.618 \times 2 \times 0.0605 \times 16)}$$

$$U = 823.6 \text{ W/m}^2 \text{ °C}$$

Determine NTU

$$\text{NTU} = \frac{UA_s}{(\dot{m}c)_{min}} = \frac{823.6 \times \pi \times 0.04 \times 5}{122.5} = 4.224$$

(Note that as q'' is defined on the basis of the smooth tube, the area of the smooth tube is used in computing NTU.)

Find ε   From Equation 2, Table 9.4.1,

$$\varepsilon = \frac{1 - e^{-4.224(1-0.04715)}}{1 - 0.04175e^{-4.224(1-0.04715)}} = 0.9829$$

Determine q

$$q = \varepsilon q_{max} = 0.829 \times 122.5 \times (200 - 40) = \underline{19\ 265} \text{ W}$$

Find the exit temperatures

$$T_{He} = 200 - \frac{19\ 265}{122.5} = \underline{42.7\ °C} \qquad T_{Ce} = 40 + \frac{19\ 265}{2598} = \underline{47.4\ °C}$$

COMMENT

The temperature of the flue gases has been reduced to nearly the inlet temperature of the water. The heat exchanger behaves almost as an infinitely long heat exchanger. An exit temperature of 42.7 °C for the flue gases is acceptable for a gas furnace as natural gas does not contain sulphur. However, in oil furnaces, the sulphur in the oil leads to oxides of sulphur in the flue gases that combine with water vapor and condense when the temperature of the flue gases falls below the dew point temperature. The acids formed by the oxides of sulphur lead to rapid corrosion of the metals. When the fuel contains sulphur, the exit temperature of the flue gases should be higher than the dew point temperature to avoid such corrosion.

EXAMPLE 9.4.4 Derive the expression for the effectiveness of a counterflow heat exchanger (Figure 9.4.12) with both fluids being single-phase fluids.

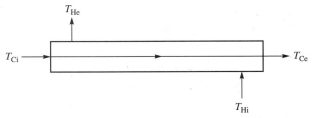

FIGURE 9.4.12 A Counterflow Heat Exchanger with Two Single-Phase Fluids

Consider the case for which $\dot{m}_H c_H = (\dot{m}c)_{min}$. Then,

$$\varepsilon = \frac{q}{q_{max}} = \frac{\dot{m}_H c_H (T_{Hi} - T_{He})}{\dot{m}_H c_H (T_{Hi} - T_{Ci})} = \frac{T_{Hi} - T_{He}}{T_{Hi} - T_{Ci}}$$

From energy conservation, with the assumption that heat transfer is only between the fluids (without any heat transfer to the surroundings),

$$C_C(T_{Ce} - T_{Ci}) = C_H(T_{Hi} - T_{He})$$

where C is $\dot{m}c$. For this case

$$\frac{C_{min}}{C_{max}} = \frac{C_H}{C_C} = c_r = \frac{T_{Ce} - T_{Ci}}{T_{Hi} - T_{He}}$$

From Equation 9.3.11, denoting the heat transfer surface area by A,

$$\ln\left[\frac{(T_{Hi} - T_{Ce})}{(T_{He} - T_{Ci})}\right] = \left(\frac{1}{C_H} - \frac{1}{C_C}\right) UA$$

or

$$\ln\frac{T_{He} - T_{Ci}}{T_{Hi} - T_{Ce}} = -\frac{UA}{C_H}\left(1 - \frac{C_H}{C_C}\right) = -\frac{UA}{C_{min}}\left(1 - \frac{C_{min}}{C_{max}}\right) \quad \textbf{(1)}$$
$$= -N(1 - c_r) = -x$$

where $x = (UA)/C_{min}[1 - (C_{min}/C_{max})] = -N(1 - c_r)$ and $N = $ NTU. Equation 9.3.11 can, therefore, be written as

$$\frac{T_{He} - T_{Ci}}{T_{Hi} - T_{Ce}} = e^{-x}$$

Now

$$\frac{T_{He} - T_{Ci}}{T_{Hi} - T_{Ce}} = \frac{(T_{Hi} - T_{Ci}) - (T_{Hi} - T_{He})}{T_{Hi} - T_{Ce}}$$
$$= \frac{T_{Hi} - T_{Ci}}{T_{Hi} - T_{Ce}}\left(1 - \frac{T_{Hi} - T_{He}}{T_{Hi} - T_{Ci}}\right)$$

or

$$\frac{T_{He} - T_{Ci}}{T_{Hi} - T_{Ce}} = \frac{T_{Hi} - T_{Ci}}{T_{Hi} - T_{Ce}}(1 - \varepsilon) \tag{2}$$

From $c_r = (T_{Ce} - T_{Ci})/(T_{Hi} - T_{He})$,

$$T_{Ce} = T_{Ci} + c_r(T_{Hi} - T_{He})$$

Substituting this expression for T_{Ce} in Equation 2 and dividing the numerator and the denominator by $(T_{Hi} - T_{Ci})$

$$\frac{T_{He} - T_{Ci}}{T_{Hi} - T_{Ce}} = \frac{(T_{Hi} - T_{Ci})(1 - \varepsilon)}{T_{Hi} - T_{Ci} - c_r(T_{Hi} - T_{He})} = \frac{1 - \varepsilon}{1 - c_r[(T_{Hi} - T_{He})/(T_{Hi} - T_{Ci})]} \tag{3}$$

$$= \frac{1 - \varepsilon}{1 - c_r\varepsilon}$$

Combining Equations 1 and 3,

$$\frac{T_{He} - T_{Ci}}{T_{Hi} - T_{Ce}} = e^{-x} = \frac{1 - \varepsilon}{1 - c_r\varepsilon}$$

Solving for ε we obtain

$$\varepsilon = \frac{1 - e^{-x}}{1 - c_r e^{-x}} = \frac{1 - e^{-N(1-c_r)}}{1 - c_r e^{-N(1-c_r)}}$$

If $c_r = 1$, the effectiveness from the above expression is undefined; ε is found as follows:

Let $c_r = 1 - y$. As $y \rightarrow 0$, $c_r \rightarrow 1$, and $y \ll 1$.

$$\varepsilon = \frac{1 - e^{-Ny}}{1 - (1 - y)e^{-Ny}}$$

As $Ny \ll 1$, $e^{-Ny} \approx 1 - Ny$.
Therefore,

$$\varepsilon = \frac{1 - (1 - Ny)}{1 - (1 - y)(1 - Ny)} = \frac{Ny}{y + Ny - Ny^2} = \frac{N}{1 + N - Ny}$$

As $y \rightarrow 0$, $\varepsilon = \frac{N}{1 + N}$

The application of the LMTD and NTU methods have been illustrated through several examples. In either case, the following salient points should be kept in mind.

1. If a fluid undergoes phase change, its $(\dot{m}c_p)$ value is infinite. Its temperature is uniform in the heat exchanger. The type of heat exchanger (whether it is parallel-flow, counterflow, or cross-flow type) makes no difference. The solution to the problem of design or estimating the off-design performance can be obtained by employing Equation 9.3.4.

2. If there is phase change in both fluids, the temperature difference between the fluids is constant everywhere. The solution is simple as the heat transfer rate is

directly related to the product of the surface area, the overall heat transfer coefficient, and the difference in the temperatures of the fluids. The problem reduces to one of determining the overall heat transfer coefficient.

3. In a counterflow heat exchanger with two single-phase fluids, the temperature difference between the fluids is uniform if the values of $(\dot{m}c_p)$ of the fluids are equal. If the heat transfer rate is known, the temperature difference between the fluids is determined by the application of the First Law of Thermodynamics. The problem reduces to one of finding the overall heat transfer coefficient.

4. In computing the overall heat transfer coefficient radiative heat transfer was neglected. Liquids are generally opaque to radiation and neglecting radiative heat transfer on the liquid side in computing the overall heat transfer coefficient is appropriate. With gases at high temperature radiative heat transfer may be significant, particularly if the gases contain carbon dioxide and water vapor.

5. In steam condensers operating at high vacuum, air leakage is inevitable. The presence of air leads to a reduction in the condensation heat transfer coefficient. To account for such a reduction in the heat transfer coefficient manufacturers of condensers have published graphs of the overall heat transfer coefficient as a function of the tube size, water velocity, and other appropriate variables.

9.5 REGENERATORS

In the heat exchangers studied so far, thermal energy was transferred directly from one fluid to another through a solid surface. In a regenerator heat transfer does not occur directly from one fluid to another but through a storage medium. The storage material may be a solid matrix with passages for fluid flow, solids (rocks or metal pieces), liquids (water or brine) or phase change materials (wax).

Stationary and rotary regenerators

With solids, liquids, or phase change materials, the operation is usually of the batch type. In such stationary regenerators all the storage material is heated for a certain time and the flow of the hot fluid ceases. The cold fluid then flows over the storage material and is heated, bringing the storage material to its original state. The flow of the cold fluid ceases and the flow of the hot fluid resumes. Such regenerators are common in solar energy applications (Figure 9.5.1a). During bright sunny days more solar energy than can be immediately used is available. The excess energy is stored in the storage material. During nights of cloudy days when the demand exceeds the available solar energy, the storage material supplies the required energy.

In rotary regenerators using a solid matrix (Figure 9.5.1b) the heat transfer from the hot fluid to the cool fluid is continuous but through the matrix. The drum containing the solid matrix rotates. As it rotates, the part of the matrix through which the hot gas flows is heated. When the heated section comes in contact with the cool fluid the cool fluid is heated.

Analysis of a Rotary Regenerator

Rotary regenerator analyses

Consider a rotary regenerator with axial plates as the energy storage medium as shown in Figure 9.5.2. A counterflow arrangement, usual with rotary regenerators, is assumed. Although the analysis is for a regenerator with axial plates, it is valid for other types of the matrix.

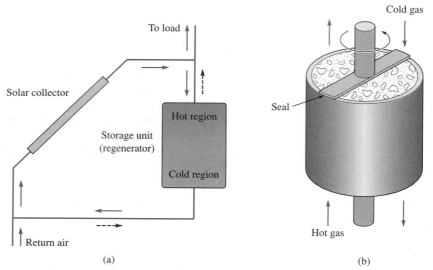

FIGURE 9.5.1 Stationary and Rotary Regenerators
(a) Stationary regenerator (schematic of air-heating solar-heating system). (b) Rotary regenerator

ASSUMPTIONS

1. Steady state
2. Uniform inlet and exit velocities of the hot and cold fluids
3. Constant properties of the fluids and the storage medium
4. Uniform convective heat transfer coefficients
5. Negligible axial heat transfer in the fluids and storage material
6. One-dimensional temperature distribution in the storage material with the temperature varying in the axial direction only. This assumption restricts the analysis to those cases where the Biot number ($=ht/2k$, t = thickness of the plate) is less than 0.1

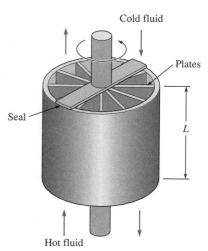

FIGURE 9.5.2 A Rotary Regenerator with Axial Plates

The nomenclature is as follows:

A = total surface area
A' = area of a plate exposed to the fluid
c = specific heat
M' = mass of one plate
M = total mass of the plates
$\dot{m}$ = mass rate of fluid flow between two plates
P = perimeter of the cross section of a plate perpendicular to the x-direction
t = thickness of the plate
ρ = density of the material of the plate

Subscripts:

C = cold fluid
H = hot fluid
m = material of the plate

Consider the hot and cold side of the regenerator at a distance x, measured from the inlet of the cold fluid as shown in Figure 9.5.3.

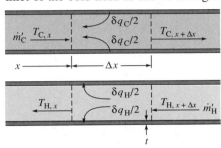

FIGURE 9.5.3 Elements in the Regenerator for Analysis

$$\dot{m}'_C c_C T_{C,x} + \delta q_C = \dot{m}'_C c_C T_{C,x+\Delta x} \tag{9.5.1}$$

$$\dot{m}'_H c_H T_{H,x+\Delta x} - \delta q_H = \dot{m}'_H c_H T_{H,x} \tag{9.5.2}$$

where

δq_C = convective heat transfer rate from the plate element to the cold fluid
δq_H = convective heat transfer rate from the hot fluid to the plate element

and

$$\delta q_C = P_C \, \Delta x h_C (T_m - T_C) \qquad \delta q_H = P_H \, \Delta x h_H (T_H - T_m)$$

where T_m is the mean temperature of the plate element.

Introducing the expressions for δq_C and δq_H into Equations 9.5.1 and 9.5.2, dividing by Δx, and taking the limit as $\delta x \to 0$,

$$\dot{m}'_C c_C \frac{\partial T_C}{\partial x} = h_C P_C (T_m - T_C) \tag{9.5.3}$$

$$\dot{m}'_H c_H \frac{\partial T_H}{\partial x} = h_H P_H (T_H - T_m) \tag{9.5.4}$$

As the hot and cold fluids flow over the plates, the plates are heated on the hot fluid side and cooled on the cold fluid side. From an energy balance on the plate elements, with $\Delta \tau$ representing the time for the fluid to flow from x to $x + \Delta x$,

$$-\rho c_{\mathrm{m}} P_{\mathrm{C}} t \; \Delta x \; \Delta T_{\mathrm{m}} = h_{\mathrm{C}} P_{\mathrm{C}} \; \Delta x (T_{\mathrm{m}} - T_{\mathrm{C}}) \; \Delta\tau \qquad (9.5.5)$$

$$\rho c_{\mathrm{m}} P_{\mathrm{H}} t \; \Delta x \; \Delta T_{\mathrm{m}} = h_{\mathrm{H}} P_{\mathrm{H}} \; \Delta x (T_{\mathrm{H}} - T_{\mathrm{m}}) \; \Delta\tau \qquad (9.5.6)$$

Dividing the above equations by $\Delta\tau$ and taking the limit as $\Delta\tau \to 0$,

$$-\rho c_{\mathrm{m}} P_{\mathrm{C}} t \; \frac{\partial T_{\mathrm{m}}}{\partial\tau} = h_{\mathrm{C}} P_{\mathrm{C}} (T_{\mathrm{m}} - T_{\mathrm{C}}) \qquad (9.5.7)$$

$$\rho c_{\mathrm{m}} P_{\mathrm{H}} t \; \frac{\partial T_{\mathrm{m}}}{\partial\tau} = h_{\mathrm{H}} P_{\mathrm{H}} (T_{\mathrm{H}} - T_{\mathrm{m}}) \qquad (9.5.8)$$

In time $\Delta\tau$, because of the rotation of the matrix, the fluid temperature changes with time but such changes in the temperature of the fluid with time is neglected; such an approximation restricts the development to gases.

Multiply Equations 9.5.3, 9.5.4, 9.5.7, and 9.5.8 by the length of the regenerator L.

$$L\dot{m}_{\mathrm{C}}' c_{\mathrm{C}} \; \frac{\partial T_{\mathrm{C}}}{\partial x} = h_{\mathrm{C}} A_{\mathrm{C}}' (T_{\mathrm{m}} - T_{\mathrm{C}}) \qquad (9.5.9)$$

$$L_{\mathrm{H}} \dot{m}_{\mathrm{H}}' c_{\mathrm{H}} \; \frac{\partial T_{\mathrm{H}}}{\partial x} = h_{\mathrm{H}} A_{\mathrm{H}}' (T_{\mathrm{H}} - T_{\mathrm{m}}) \qquad (9.5.10)$$

$$-c_{\mathrm{m}} M_{\mathrm{C}}' \; \frac{\partial T_{\mathrm{m}}}{\partial\tau} = h_{\mathrm{C}} A_{\mathrm{C}}' (T_{\mathrm{m}} - T_{\mathrm{C}}) \qquad (9.5.11)$$

$$c_{\mathrm{m}} M_{\mathrm{H}}' \; \frac{\partial T_{\mathrm{m}}}{\partial\tau} = h_{\mathrm{H}} A_{\mathrm{H}}' (T_{\mathrm{H}} - T_{\mathrm{m}}) \qquad (9.5.12)$$

There are no initial conditions as we are seeking the periodic steady-state solution. The conditions to be satisfied are

$$T_{\mathrm{C}}(x = 0) = T_{\mathrm{Ci}} \qquad T_{\mathrm{H}}(x = L) = T_{\mathrm{Hi}}$$

Starting the time at the instant the plate enters the hot fluid section,

$$T_{\mathrm{m,H}}(0) = T_{\mathrm{m,C}}(\tau') \qquad T_{\mathrm{m,H}}(\tau_{\mathrm{H}}) = T_{\mathrm{m,C}}(\tau_{\mathrm{H}})$$

where

$$\tau' = \text{period of rotation (time for one revolution of the matrix)}$$
$$\tau_{\mathrm{H}} = \text{plate residence time in the hot fluid region}$$
$$\tau' - \tau_{\mathrm{H}} = \tau_{\mathrm{C}} = \text{plate residence time in the cold fluid region}$$

The equations are numerically integrated and the results are given in dimensionless form. The dimensionless parameters are

	Hot Side	**Cold Side**
Dimensionless length	$\Lambda_{\mathrm{H}} = \dfrac{h_{\mathrm{H}} A_{\mathrm{H}}}{\dot{m}_{\mathrm{H}} c_{\mathrm{H}}}$	$\Lambda_{\mathrm{C}} = \dfrac{h_{\mathrm{C}} A_{\mathrm{C}}}{\dot{m}_{\mathrm{C}} c_{\mathrm{C}}}$
Dimensionless time	$\pi_{\mathrm{H}} = \dfrac{h_{\mathrm{H}} A_{\mathrm{H}} \tau_{\mathrm{H}}}{M_{\mathrm{H}} c_{\mathrm{m}}}$	$\pi_{\mathrm{C}} = \dfrac{h_{\mathrm{C}} A_{\mathrm{C}} \tau_{\mathrm{C}}}{M_{\mathrm{C}} c_{\mathrm{m}}}$
Dimensionless temperature	Fluid: $\theta_{\mathrm{f}} = \dfrac{T_{\mathrm{f}} - T_{\mathrm{Ci}}}{T_{\mathrm{Hi}} - T_{\mathrm{Ci}}}$	Storage material: $\theta_{\mathrm{m}} = \dfrac{T_{\mathrm{m}} - T_{\mathrm{Ci}}}{T_{\mathrm{Hi}} - T_{\mathrm{Ci}}}$

If a regenerator operates with equal capacity rates ($\dot{m}_C c_C = \dot{m}_H c_H$) it is termed a *balanced regenerator*. If, in addition, $A_H = A_C$ (i.e., $\tau_H = \tau_C$), the regenerator is said to be *symmetric*. For a balanced symmetric regenerator

$$\pi_C = \pi_H \qquad \Lambda_C = \Lambda_H$$

The effectiveness of a balanced regenerator is defined as

$$\eta_{reg} = \frac{T_{Hi} - T_{Ho}}{T_{Hi} - T_{Ci}} = \frac{T_{Co} - T_{Ci}}{T_{Hi} - T_{Ci}}$$

where T_{Ho} and T_{Co} are the average outlet temperatures of the hot and cold fluids, respectively. The relationship between η_{reg} and Λ with π as a parameter [from the numerical values in Schmidt and Willmott (1981)] is shown in Figure 9.5.4.

One of the difficult problems in the analysis and design of regenerators is determining the heat transfer coefficients. If the heat transfer coefficients, the inlet temperatures, and the geometry are known, the effectiveness of the regenerator can be determined from Figure 9.5.4. From the effectiveness the exit temperatures of the fluids and, therefore, the heat transfer rate are readily determined. The computations are illustrated in the following example.

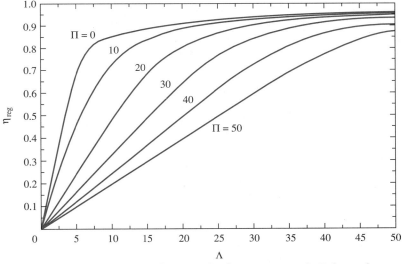

FIGURE 9.5.4 Relationship of η_{reg} and λ for a Symmetric Balanced Counterflow Regenerator

EXAMPLE 9.5.1 A 1-m diameter, 0.5-m high rotary regenerator consists of a total of 1600, 8-mm I.D., 9-mm O.D. plain carbon-steel tubes parallel to the axis of rotation as shown in Figure 9.5.5. Mass rates of flow of the hot gas is 0.4 kg/s and equals the mass rate of flow of the cold gas. The hot gas enters at 200 °C and the cold gas at 20 °C. The gases flow in opposite directions. The drum makes one revolution every two minutes. Determine the exit temperatures of the gases assuming the properties of the gases are those of air at atmospheric pressure.

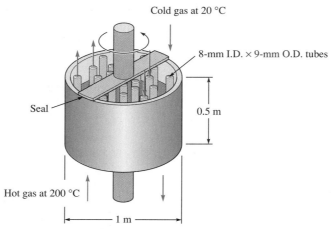

Cold gas at 20 °C

8-mm I.D. × 9-mm O.D. tubes

0.5 m

Seal

Hot gas at 200 °C

1 m

FIGURE 9.5.5 Rotary Regenerator with Axial Tubes as the Matrix

Given

Diameter = 1 m
Length = 0.5 m
Matrix: 1600 plain carbon-steel tubes.
        8 mm I.D., 9 mm O.D.
Hot fluid: $\dot{m}_H$ = 0.4 kg/s T_{Hi} = 200 °C
Cold fluid: $\dot{m}_C$ = 0.4 kg/s T_{Ci} = 20 °C

Find

T_{He} and T_{Ce}

ASSUMPTIONS

1. Steady state
2. Symmetric regenerator
3. Constant properties and, therefore, balanced regenerator
4. Fully developed flow of gases in the tubes

SOLUTION

The exit temperatures are determined by finding the value of η_{reg} from Figure 9.5.4. We need the values of Λ and π.

The residence time of the tubes in the hot and cold gases is 60 s each. The convective heat transfer coefficient is needed to compute Λ and π. We assume the convective heat transfer coefficients on both sides to be equal and based on the properties at 110 °C. Properties of atmospheric air at 110 °C from the software are

ρ = 0.9214 kg/m³ c_p = 1013 J/kg °C
μ = 2.22 × 10⁻⁵ N s/m² k = 0.0319 W/m °C
Pr = 0.704

There are 800 tubes on each side. Mass rate of flow per tube = 0.4/800 kg/s. Therefore,

$$\mathrm{Re}_d = \frac{4 \times 0.4/800}{\pi \times 0.008 \times 2.22 \times 10^{-5}} = 3585$$

From Equation 4.2.3d for fully developed flow, $\mathrm{Nu}_d = 11.1$ and

$$h = \frac{11.1 \times 0.0319}{0.008} = 44.3 \text{ W/m}^2 \text{ °C}$$

Surface area on the hot side is $A = \pi \times 0.008 \times 0.5 \times 800 = 10.05 \text{ m}^2$.

$$\Lambda = \frac{hA}{\dot{m}c} = \frac{44.3 \times 10.05}{0.4 \times 1013} = 1.099$$

From Table A1, properties of steel are

$$\rho_m = 7854 \text{ kg/m}^3 \qquad c_m(60 \text{ °C}) = 450 \text{ J/kg °C}$$

$$M = \text{mass of the tubes on the hot side}$$

$$= \frac{\pi(0.009^2 - 0.008^2)}{4} \times 0.5 \times 7854 \times 800 = 41.95 \text{ kg}$$

$$\pi = \frac{hA\tau}{M_m c_m} = \frac{44.3 \times 10.05 \times 60}{41.95 \times 450} = 1.42$$

From Figure 9.5.4, for $\Lambda = 1.1$ and $\pi = 1.42$, $\eta_{\mathrm{reg}} = 0.56$,

$$\eta_{\mathrm{reg}} = \frac{T_{\mathrm{Hi}} - T_{\mathrm{Ho}}}{T_{\mathrm{Hi}} - T_{\mathrm{Ci}}}$$

$$T_{\mathrm{Hi}} - T_{\mathrm{Ho}} = T_{\mathrm{Co}} - T_{\mathrm{Ci}} = 0.56 \times (200 - 20) = 100.8 \text{ °C}$$

$$T_{\mathrm{He}} = \underline{99.2 \text{ °C}} \qquad T_{\mathrm{Co}} = \underline{120.8 \text{ °C}}$$

$$q = \dot{m}c(T_{\mathrm{Hi}} - T_{\mathrm{Ho}}) = 0.4 \times 1013 \times (200 - 99.2) = \underline{4.08 \times 10^4 \text{ W}}$$

SUMMARY

One of the common applications of heat transfer is the design of or the determination of the off-design performance of devices in which heat transfer occurs from one fluid to another separated by a solid surface—heat exchangers. Different types of heat exchangers—double-pipe, shell-and-tube, cross-flow, plate, and regenerators—are described. A *double-pipe heat exchanger* consists of two coaxial pipes. A *shell-and-tube heat exchanger* has many tubes (in parallel) inside a shell. The number of tube passes in a shell-and-tube heat exchanger is the number of times the fluid in the tubes goes from one end of the shell to the other. In *cross-flow heat exchangers* the two fluids move in mutually perpendicular directions. *Plate heat exchangers* consist of a large number of plates separated by suitably shaped gaskets.

The *overall heat transfer coefficient* is the reciprocal of the thermal resistance for a unit area of the solid surface and is defined by

$$q'' = U(T_{\mathrm{f1}} - T_{\mathrm{f2}})$$

If the fluids are separated by rectangular plates,

$$\frac{1}{U} = \frac{1}{h_1} + \frac{L}{k} + \frac{1}{h_2}$$

If the fluids are separated by circular tubes

$$\frac{1}{U_i r_i} = \frac{1}{U_o r_o} = \frac{1}{h_i r_i} + \frac{1}{k} \ln\left(\frac{r_o}{r_i}\right) + \frac{1}{h_o r_o}$$

where U_i and U_o are the overall heat transfer coefficients based on the inner and outer surface areas, respectively. The overall heat transfer coefficient is used in conjunction with the surface area on which it is based and $U_i\, dA_i = U_o\, dA_o$. In most engineering applications the conductive resistance is negligible.

The heat transfer rate is given by

$$q = \dot{m}_H(i_{Hi} - i_{He}) = \dot{m}_C(i_{Ce} - i_{Ci}) = UA_s\, \Delta T_m$$

For constant density fluids and ideal gases,

$$i_2 - i_1 = c_p(T_2 - T_1)$$

For a *counterflow heat exchanger* with uniform T_H (condensing) and uniform T_C (boiling) or $\dot{m}_H c_H = \dot{m}_C c_C$,

$$\Delta T_m = T_H - T_C$$

In all other cases,
$$\Delta T_m = \frac{\Delta T_1 - \Delta T_2}{\ln(\Delta T_1/\Delta T_2)}$$

where ΔT_1 and ΔT_2 are the differences between the hot and cold fluid temperatures at ends 1 and 2, respectively.

For a *parallel-flow heat exchanger*

Uniform T_H and T_C, $\Delta T_m = T_H - T_C$

For all other cases, $\Delta T_m = \dfrac{\Delta T_1 - \Delta T_2}{\ln\,(\Delta T_1/\Delta T_2)}$

If only T_H or T_C is uniform, then for all types of heat exchangers with

Uniform T_H $\qquad \ln \dfrac{T_H - T_{Ce}}{T_H - T_{Ci}} = -\dfrac{UA_s}{\dot{m}_C c_C}$ $\qquad q = \dot{m}_C c_C(T_{Ce} - T_{Ci})$

Uniform T_C $\qquad \ln \dfrac{T_{He} - T_C}{T_{Hi} - T_C} = -\dfrac{UA_s}{\dot{m}_H c_H}$ $\qquad q = \dot{m}_H c_H(T_{Hi} - T_{He})$

The mean temperature differences for heat exchangers other than counterflow or parallel-flow heat exchangers is found from

$$\Delta T_m = \text{LMTD}_{CF} \times F$$

LMTD_{CF} is the LMTD for a counterflow heat exchanger with the same terminal temperatures as the actual heat exchanger and F is a correction factor found in Figures 9.3.5, through 9.3.7 or from the HE software.

The LMTD method yields the heat transfer surface area when the required heat transfer rate, the inlet conditions of the hot and cold fluids, and the overall heat transfer coefficient are known. But if the heat transfer surface area, inlet conditions of the hot and cold fluids, and the overall heat transfer coefficient are known, an iterative solution method is needed to determine the heat transfer rate and the exit conditions of the fluids. The *number of transfer units (NTU)* method can be used both to find the heat transfer surface area (with specified heat transfer rate, inlet conditions of the hot and cold fluids, and the overall heat transfer coefficient) and to determine the heat transfer rate and the exit conditions of the fluids (with known heat transfer surface area, inlet conditions of the hot and cold fluids, and the overall heat transfer coefficient).

The *effectiveness, ε,* of a heat exchanger is defined as the ratio of the heat transfer rate to the maximum attainable heat transfer rate in a counterflow heat exchanger with the same inlet conditions. That is,

$$\varepsilon = q/q_{max} \qquad q_{max} = (\dot{m}c)_{min}(T_{Hi} - T_{Ci})$$

The *effectiveness* is a function of the *NTU* and c_r.

$$\text{NTU} = \frac{UA_s}{(\dot{m}c)_{min}} \qquad c_r = \frac{(\dot{m}c)_{min}}{(\dot{m}c)_{max}}$$

The effectiveness of different heat exchangers is presented in Figures 9.4.4 through 9.4.7 and in Tables 9.4.1 and 9.4.2. To determine A_s for specified q, inlet conditions (c_r), and U, determine ε and find the required NTU that yields A_s. To determine q for specified inlet conditions (c_r), A_s, and U, compute NTU and find ε. From the value of ε, find q and the exit conditions of the fluids.

A brief introduction to regenerators is provided.

Software

Subprogram HE computes the LMTD correction factor, heat exchanger effectiveness with known values of NTU and other required values, and computes the NTU required for a given effectiveness.

REFERENCES

Bowman, R.A., Mueller, A.C., and Nagle, W.M. (1940) Mean temperature difference in design. *TRANS, ASME* 62:283.

Eckert, E.R.G., and Drake, R.M. (1959). *Heat and Mass Transfer.* New York: McGraw-Hill.

Hewitt, G.F., (1990). *Handbook of Heat Exchanger Design.* New York: Hemisphere Publishing.

Kakac, S., Shah, R.K., and Bergles, A.E. (1983). *Low Reynolds Number Flow Heat Exchangers.* New York: Hemisphere Publishing.

Kakac, S., Bergles, A.E., and Mayinger, F., eds. (1981). *Heat Exchangers—Thermal-Hydraulic Fundamentals.* New York: Hemisphere Publishing.

Kays, W.M., and London, A.L. (1984). *Compact Heat Exchangers,* 3d ed. New York: McGraw-Hill.

Kuan, D.Y., Aris, R., and Davis, H.T. (1984). Estimation of fin efficiencies of regular tubes arranged in circumferential fins. *Int. J. Heat Mass Transfer* 27:148.

Palen, J.W., ed. (1986). *Heat Exchanger Sourcebook.* New York: Hemisphere Publishing.

Rohsenow, W.M., Hartnett, J.P., and Ganic, E.N., eds. (1985). *Handbook of Heat Transfer Applications.* New York: McGraw-Hill.

Schmidt, F.W., and Willmott, A.J. (1981). *Thermal Energy Storage and Generation.* New York: McGraw-Hill.

Shah, R.K. (1981). Classification of heat exchangers. In *Heat Exchangers—Thermal-Hydraulic Fundamentals*. eds. Kakac, S., Bergles, A.E., and Mayinger, F. New York: Hemisphere Publishing.

Shah, R.K., and Mueller, A.C. (1985). Heat exchangers. In *Handbook of Heat Transfer Applications*. eds. Rohsenow, W.M., Hartnett, J.P., and Ganic, E.N., New York: McGraw-Hill.

Standards of Tubular Exchanger Manufacturers Association (1970), 5th ed. Tubular Exchanger Manufacturers Association, Inc., New York.

Sparrow, E.M., and Lin, S.H. (1964). Heat transfer characteristics of polygonal and plate fins. *Int. J. Heat Mass Transfer* 7:951.

REVIEW QUESTIONS

(a) Describe parallel-flow, counterflow, shell-and-tube, plate, and cross-flow heat exchangers.

(b) Define overall heat transfer coefficient.

(c) Define logarithmic mean temperature difference.

(d) State the conditions for which Equation 9.3.6 cannot be directly used to determine the mean temperature difference.

(e) How will you respond to the statement that it is better to arrange a steam condenser as a counterflow heat exchanger than as a parallel-flow heat exchanger?

(f) Your friend asserts that, in a heat exchanger, it is impossible for the exit temperature of the cold fluid to be greater than the exit temperature of the hot fluid when both fluids are single-phase fluids. What is your response?

(g) With at least one fluid being a single-phase fluid, express the maximum heat transfer rate in terms of the appropriate mass flow rates and inlet temperatures of the hot and cold fluids.

(h) Define the effectiveness of a heat exchanger.

(i) What are the advantages and disadvantages of shell-and-tube and plate heat exchangers.

(j) When fluids are separated by heavy walled tubes in a heat exchanger, the overall heat transfer coefficients based on the inside and outside surface areas of the tube have different values. Explain how the same solution is obtained for the heat transfer rate by using either of the two values.

(k) Give an example of a heat exchanger where the overall heat transfer coefficient can have significantly different values in different parts of the same heat exchanger.

PROBLEMS

9.1 Water flows inside a type 302 stainless steel tube (25-mm O.D., 3-mm wall thickness), and air flows over the tube, perpendicular to the axis of the tube. The inside and outside heat transfer coefficients are, respectively, 1500 W/m^2 K and 200 W/m^2 K. Determine the overall heat transfer coefficient based on the inside and outside surface area

(a) Neglecting the thickness of the tube and the conduction resistance.

(b) Neglecting the conduction resistance but including the wall thickness.

(c) Including the wall thickness and conduction resistance.

9.2 The heat transfer rate from a fin is equal to the heat transfer rate from a surface maintained at the base temperature with the surface area $= \eta A_f$, where η is the fin efficiency and A_f is the surface area of the fin (see Section 2.6.2). Consider a double-pipe heat exchanger with the inner pipe having an I.D. of r_1 and an O.D. of r_2. The outer surface is provided with longitudinal fins of rectangular cross section of thickness t and height H. The inside convective heat transfer coefficient is h_i and the outside heat transfer coefficient is h_o. Neglecting the conductive resistance, show that the overall heat transfer coefficient based on the inside surface area is given by

$$U_i = \frac{1}{A_i/[(A_o + \eta A_f)h_o] + 1/h_i}$$

where

A_i = inner surface area of the pipe

A_o = outer surface area of the pipe not occupied by the fins.

A_f = surface area of the fins

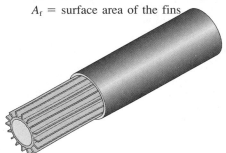

FIGURE P9.2

9.3 There are N longitudinal, rectangular fins in the double-pipe heat exchanger in Problem 9.2. The thermal conductivity of the material of the inner pipe is k. Including the conductive resistance of the wall of the inner pipe and assuming one-dimensional temperature distribution, show that

$$\frac{1}{U_i} = \frac{1}{h_i} + \frac{r_1}{k}\ln\left(\frac{r_2}{r_1}\right) + \frac{2\pi r_1}{h_o(2\pi r_2 - Nt + \eta 2NH)}$$

9.4 Water enters the 10-mm diameter inner tube of a double-pipe heat exchanger used to condense saturated steam at 80 °C. At a section of the tube where the water temperature is 20 °C, find the overall heat transfer coefficient for water velocities of 1 m/s and 5 m/s. Neglect wall thickness and conductive resistance of the tube.

9.5 A plate heat exchanger is used to heat cold water with hot water. Cold water enters with a velocity of 1 m/s between plates, and hot water enters with a velocity of 0.7 m/s between plates. The plates are made of 0.5-mm thick, type 304 stainless steel, and the gap between the plates is 5 mm. Estimate the overall heat transfer coefficient for a mean temperature of 20 °C for the cold water and 80 °C for the hot water.

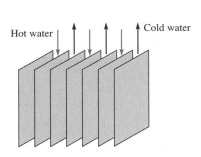

FIGURE P9.5

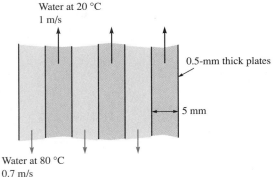

9.6 Water flows in a 12-mm I.D. and 15-mm O.D. copper tube with a velocity of 3 m/s. Air at 20 °C is in cross flow with a velocity of 15 m/s. Assume fully developed velocity and temperature profile in the tube. Based on the outer surface area of the tube, determine the overall heat transfer coefficient at the section of the tube where the water temperature is 60 °C.

9.7 In Problem 9.6, if the water is replaced by engine oil flowing at 2 m/s at 60 °C, determine the overall heat transfer coefficient.

9.8 In a test on a cross-flow heat exchanger with a heat transfer area of 1.7 m², the inlet and exit temperatures of the hot fluid are 90 °C and 60 °C, and those of the cold fluid are 10 °C and 50 °C, respectively. For the hot fluid, $\dot{m}c_p = 800$ W/K. Estimate the overall heat transfer coefficient
(a) If both fluids are unmixed.
(b) If the fluid in the tube is unmixed and the fluid on the outside of the tubes is mixed.

9.9 A Freon 12 condenser is constructed as a double-pipe heat exchanger with water flowing in the inner tube and saturated Freon at 50 °C condensing on the outer surface of the inner tube. The inner and outer diameters of the inner copper tube are 12 mm and 15 mm, respectively. Water flows inside the tube with a velocity of 2.5 m/s. Determine
(a) The overall heat transfer coefficient, U_i, at the section where the water temperature is 15 °C.
(b) The length of the tube required if the inlet and exit temperatures of the water are 10 °C and 20 °C, respectively.
(c) The mass rate of condensation of the Freon.
(Properties of Freon 12 at saturation temperature = 50 °C; density of vapor = 69.7 kg/m³; and $i_{fg} = 122.3$ kJ/kg.)

9.10 To recover waste heat from flue gases, a double-pipe heat exchanger is fabricated from a 4-cm diameter, thin-walled inner pipe. Twelve longitudinal fins of plain carbon-steel are welded to the outer surface of the pipe. The fins are of rectangular cross section with a radial height of 1.5 cm and a thickness of 0.5 mm. The inside convective heat transfer coefficient is 2800 W/m² °C and the average outside convective heat transfer coefficient is 110 W/m² °C. Determine the overall heat transfer coefficient based on the surface area of the smooth inner pipe.

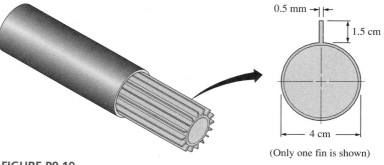

0.5 mm
1.5 cm
4 cm
(Only one fin is shown)

FIGURE P9.10

9.11 A heat exchanger is constructed with one bank of 10-mm I.D. and 13-mm O.D. commercial bronze tubes. Circular brass disks, 13-mm I.D. 40-mm O.D., and 0.5-

mm thick are soldered to the tube at a pitch of 5 mm. Water flows inside the tubes with a velocity of 2 m/s. Air at 20 °C is forced over the tubes and the convective heat transfer coefficient on the air side is 105 W/m² °C. Determine

(a) The overall heat transfer coefficient U_i based on the inner surface area of the tube, at a section where the temperature of the water is 80 °C.

(b) The temperature drop of the water per meter length of the tube at that section.

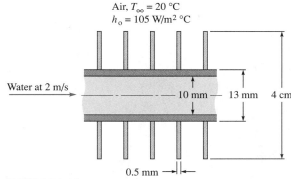

FIGURE P9.11

9.12 To heat air, a heat exchanger is made of 12-mm diameter thin-walled copper tube. The tube is made in sections of 1.5 m length of tubes as shown in Figure P9.12. Water enters the tube with a velocity of 2 m/s, at 80 °C. The tube is suspended in air with the plane of the assembly horizontal. The surrounding air is at 10 °C. Determine

(a) The overall heat transfer coefficient at a section of the tube where the temperature of the water is 75 °C.

(b) The total length of the tube if the exit temperature of the water is 70 °C.

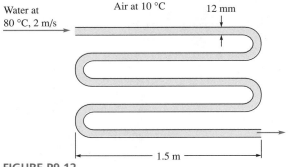

FIGURE P9.12

9.13 If the length of the tube in Problem 9.12 is 15 m, determine the exit temperature of the water.

9.14 In a counterflow heat exchanger 75 kg/s of oil is cooled from 80 °C to 40 °C with 100 kg/s of water entering at 15 °C. The heat exchanger has a heat transfer area of 275 m². Determine the overall heat transfer coefficient.

9.15 In Problem 9.14, if the heat exchanger is a two-tube pass shell-and-tube heat exchanger, what is the value of the overall heat transfer coefficient?

9.16 Show that the mean temperature difference for a parallel-flow heat exchanger is given by Equation 9.3.13b when there is no phase change in either fluid.

9.17 Water flows inside a thin, 25-mm diameter copper tube with a mass flow rate of 0.1 kg/s. Air at 20 °C, 200 kPa flows over the tube (cross-flow) with a velocity of 10 m/s.

(a) At a section of the tube where the water temperature is 55 °C, determine the overall heat transfer coefficient. Assume fully developed conditions.

(b) Assuming the overall heat transfer coefficient found in part a to be uniform, determine the length of the tube required to cool the water from 60 °C to 50 °C.

9.18 Engine oil flowing at 3 kg/s is to be cooled from 120 °C to 60 °C with 5 kg/s of water available at 20 °C. If the overall heat transfer coefficient is 300 W/m^2 °C, what should be the heat transfer area

(a) Of a counterflow heat exchanger?

(b) Of a two-tube pass shell-and-tube heat exchanger?

9.19 Steam is to be condensed in a double-pipe, horizontal heat exchanger with a 3-mm thick, 25-mm outside diameter stainless steel (type 316) tube mounted inside a coaxial pipe. Dry, saturated steam at 100 °C enters the annulus with a negligible velocity and is condensed on the outer surface of the inner tube. Water at 30 °C enters the tube at the rate of 0.6 kg/s.

(a) Estimate the overall heat transfer coefficient U_i at the section of the tube where the water temperature is 40 °C. Assume fully developed conditions for water.

(b) Determine the length of the tube required to condense 0.05 kg/s of steam assuming that the overall heat transfer coefficient found in part a is uniform.

9.20 A counterflow plate heat exchanger is used to heat cold water from 10 °C to 60 °C with hot water available at 100 °C. The velocities of the cold and hot water between the plates is 1 m/s and 0.7 m/s, respectively. The plates are made of 1-mm thick, type 304 stainless steel, and the gap between the plates is 5 mm. Determine the height of the plates. If the height of the plates is not to exceed 3 m, indicate how you would arrange the plates. Assume fully developed velocity and temperature profiles and that correlations for circular tubes are applicable with the diameter replaced by the hydraulic mean diameter.

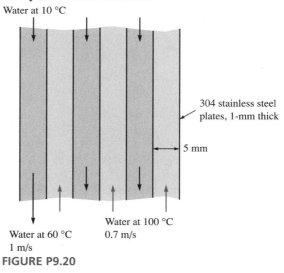

Water at 10 °C

304 stainless steel plates, 1-mm thick

5 mm

Water at 100 °C
0.7 m/s

Water at 60 °C
1 m/s

FIGURE P9.20

9.21 The exhaust gases from a diesel engine flow (cross-flow, unmixed) over a bank of tubes in the exhaust gas boiler. The mass flow rate of the exhaust gases is 21 000 kg/h. Saturated water (800 kPa) enters the boiler and leaves as saturated steam at 800 kPa. The gases enter at 400 °C and leave at 200 °C. Assume that the gases have the properties of air.

(a) What should be the surface area of the boiler if the overall heat transfer coefficient is 180 W/m² °C?

(b) What is the mass rate of steam produced in the boiler?

Exhaust gas boiler

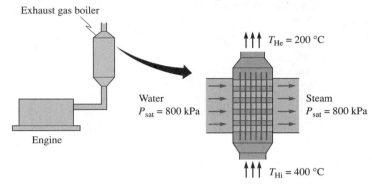

FIGURE P9.21

9.22 In a steam power plant 100 kg/s of feedwater is heated from 25 °C to 100 °C utilizing bleed steam available at a saturation temperature of 120 °C and a quality of 0.95. The overall heat transfer coefficient based on the mean temperature of water and the saturation temperature of the steam is 2500 W/m² K. The condensate leaves the heater at 100 °C. Determine the heat transfer surface area and the mass rate of condensation of steam.

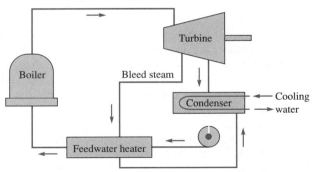

FIGURE P9.22

9.23 A double pipe heat exchanger is to be designed to cool 0.15 kg/s of oil from 150 °C to 30 °C with 0.2 kg/s of water available at 5 °C. Determine the heat transfer surface area if the overall heat transfer coefficient is 300 W/m² °C.

9.24 Saturated steam at 40 °C is to be condensed in a two-tube pass, one-shell pass condenser. Water enters the heat exchanger at 10 °C and exits at 20 °C. The mass rate of condensation of steam is 2.8 kg/s. Find the heat transfer area if the overall heat transfer coefficient is 3800 W/m² °C. Assume the condensate leaves the condenser at 30 °C.

9.25 The lubricating oil of a marine diesel engine is to be cooled from 120 °C to 80 °C with water. The heat transfer rate from the oil is 500 kW. The inlet and exit temperatures of the water are, respectively, 20 °C and 60 °C. The overall heat transfer coefficient is 300 W/m² K. Determine the surface area of the tubes of the heat exchanger

(a) For a parallel-flow heat exchanger.

(b) For a counterflow heat exchanger.

9.26 Re-solve Problem 9.25 for a two-tube pass shell-and-tube heat exchanger.

9.27 Re-solve Problem 9.25 for a cross-flow heat exchanger with one fluid mixed and the other unmixed.

9.28 A plate heat exchanger is to be designed for heating cold water with waste hot water for the conditions given below:

Mass flow rate of hot water	120 kg/s
Inlet temperature of hot water	50 °C
Mass flow rate of cold water	150 kg/s
Inlet temperature of cold water	22 °C
Exit temperature of cold water	36 °C
Gap between plates	5 mm
Thickness of plates	0.8 mm
Material of plates	Type 302 stainless steel

(a) Determine the overall heat transfer coefficient, limiting the water velocities between plates to 0.8 m/s.

(b) Determine the height and number of 1-m wide plates to be used for counterflow arrangement. Multiple pass arrangement may be used.

9.29 A supply of 0.1 kg/s of hot water at 85 °C is required. Water is available at 15 °C and is to be heated with dry saturated steam at 100 °C condensing on the outside of a 12-mm thin, horizontal tube. Determine the length of the tube required.

9.30 Re-solve Problem 9.29 if tubes of 10-mm, 12-mm, and 15-mm diameter are available. Support your choice of design with quantitative computations of length, pressure drop, and pumping power.

9.31 Ethylene glycol is to be cooled from 80 °C to 40 °C in a two-tube pass, one-shell pass heat exchanger with 2000 tubes. The heat exchanger is made of 1-cm diameter tubes of negligible wall thickness. The velocity of glycol is 0.1 m/s. The coolant is water entering at 15 °C and flowing on the shell side. The temperature rise of the water is limited to 10 °C and the heat transfer coefficient on the water side is 4800 W/m² °C.

(a) Assuming fully developed velocity profile, find the inside convective heat transfer coefficient at a location where the temperature of glycol is 60 °C and the overall heat transfer coefficient. Assume correlations for uniform surface temperature.

(b) If the overall heat transfer coefficient in the heat exchanger is uniform at the value found in part a, find the length of each tube.

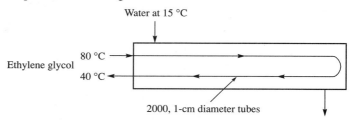

FIGURE P9.31

9.32 A one-tube pass steam condenser is made of type 1024 steel, 1-cm diameter and 4-m long horizontal tubes arranged in a square array of 20 × 20 tubes. The wall thickness of the tubes is negligible. Water enters the tubes at 12 °C with a velocity of 2 m/s. Dry, saturated steam enters at 50 °C.

(a) At a location where the water temperature is 18 °C, determine the inside convective heat transfer coefficient, the outside convective heat transfer coefficient, and the overall heat transfer coefficient.

(b) Assuming the overall heat transfer coefficient calculated in part a to be uniform, determine the heat transfer rate and the rate of condensation of steam.

9.33 Water flowing at 7 kg/s is to be cooled from 100 °C to 60 °C in a cross-flow heat exchanger with air entering the heat exchanger at a mass flow rate of 40 kg/s and an inlet temperature of 10 °C. If the overall heat transfer coefficient is 180 W/m² °C, estimate the heat transfer area. Assume one fluid mixed and the other unmixed.

9.34 Lubricating oil of an engine is to be cooled from 80 °C to 40 °C in a one-shell pass, two-tube pass heat exchanger. The mass flow rate of oil (c_p = 2100 J/kg K) is 1 kg/s. Water enters the heat exchanger at 10 °C with a mass flow rate of 2 kg/s. The overall heat transfer coefficient is 100 W/m² °C. Find the heat transfer surface area.

9.35 A one-shell pass, two-tube pass heat exchanger is used to cool oil (c_p = 2440 J/kg K) entering at 150 °C with water entering at 15 °C. The mass rates of flow of oil and water are, respectively, 28 000 kg/h and 15 000 kg/h. The overall heat transfer coefficient is 300 W/m² °C and the heat transfer surface area is 48 m². Determine the exit temperatures of the water and oil by

(a) The LMTD method.

(b) The NTU method.

9.36 Show that T_{He}, the exit temperature of the hot fluid in a heat exchanger is given by

Parallel-flow $\qquad T_{He} = T_{Ce} + (T_{Hi} - T_{Ci})e^{-UAb} \qquad b = \dfrac{1}{C_H} + \dfrac{1}{C_C}$

Counterflow $\qquad T_{He} = T_{Ci} + (T_{Hi} - T_{Ce})e^{-UAb} \qquad b = \dfrac{1}{C_H} - \dfrac{1}{C_C}$

$A = \begin{array}{l}\text{heat transfer}\\\text{surface area}\end{array} \qquad C_H = (\dot{m}c_p)_H \qquad C_C = (\dot{m}c_p)_C$

9.37 For a parallel-flow heat exchanger show that the correction factor, F, for determining the mean temperature difference is given by

$$F = \frac{R+1}{R-1}\frac{\ln[C1 - P)/(1 - PR)]}{\ln\{1/[1 - P(R+1)]\}}$$

where

$$R = \frac{\dot{m}_C c_C}{\dot{m}_H c_H} = \frac{T_{Hi} - T_{He}}{T_{Ce} - T_{Ci}} \qquad P = \frac{(T_{Ce} - T_{Ci})}{(T_{Hi} - T_{Ci})}$$

Hint: Express F given above in terms of the temperatures.

9.38 The feedwater heater in a steam power plant heats 100 kg/s of feedwater from 20 °C to 60 °C utilizing bleed steam available at a saturation temperature of 120 °C and a quality of 0.95. The condensate leaves the heater at 70 °C. The heater is made with 18-mm diameter tubes of negligible wall thickness. The velocity of the water is limited to 3 m/s. The tubes are arranged in a square array. Determine

(a) The overall heat transfer coefficient at a section where the water temperature is 40 °C.

(b) The length of the tubes if the heater is constructed as a single-pass, shell-and-tube heat exchanger.

9.39 Ethylene glycol flowing at 2 kg/s is to be cooled from 100 °C to 40 °C with water available at 20 °C in a coaxial, double-pipe heat exchanger. The mass rate of water is limited to 1 kg/s.
(a) Can this be accomplished in a parallel-flow arrangement? In a counter flow arrangement?
(b) What is the effectiveness of the heat exchanger?
(c) If the overall heat transfer coefficient is 500 W/m² K, determine the surface area required.

9.40 Water flowing at 2 kg/s is to be heated from 20 °C to 80 °C in a counterflow heat exchanger. Heating is accomplished with 4 kg/s of oil ($c_p = 2100$ J/kg °C) entering the heat exchanger at 100 °C. The overall heat transfer coefficient is 120 W/m² °C.
(a) Find the heat transfer surface area.
(b) During operation of the heat exchanger, the mass rate of flow of oil is reduced to 2.4 kg/s and the overall heat transfer coefficient to 105 W/m² °C. Determine the exit temperature of the water.

9.41 A steam condenser has a heat transfer area of 70 m². The overall heat transfer coefficient is 3000 W/m² °C. Dry saturated steam enters the condenser at a temperature of 100 °C and 375 kg/s of water flow through the condenser with an inlet temperature of 20 °C. Estimate
(a) The exit temperature of water.
(b) The mass of steam condensed per second. (Assume the condensate leaves as liquid at 80 °C.)
(c) The effectiveness of the heat exchanger.

9.42 Plain carbon-steel rods of 1-cm diameter are to be heated from 20 °C to 100 °C with air at 150 °C in cross flow as shown. The rods are being continuously fed to the heating chamber with a velocity of V m/s. If the length of the chamber is 10 m, determine the velocity V. The convective heat transfer coefficient is 50 W/m² °C. (Assume the temperature of the rods at any cross section is uniform.)

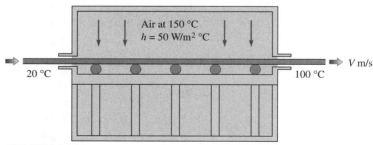

FIGURE P9.42

9.43 Plain carbon-steel rods of 1-cm diameter are to be heated from 20 °C to 80 °C with air at 150 °C flowing parallel to the axis of the rods as shown. The exit temperature of the air is 90 °C, and the steel rods are fed at a rate of 5 cm/s. The convective heat transfer coefficient is 50 W/m² °C. Determine the length L,

(a) If air motion is in the direction opposite to that of the rods.
(b) If the air motion is in the same direction as that of the rods.

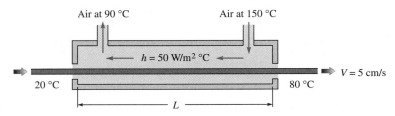

Counterflow arrangement shown

FIGURE P9.43

9.44 A condenser for an air-conditioning plant is designed to transfer 25 kW from the refrigerant. The condenser is of the cross-flow type with both fluids unmixed. The refrigerant is to be condensed at 70 °C. The ambient air temperature is 40 °C and the exit temperature of the air is limited to 45 °C. The overall heat transfer coefficient with a two-speed fan at the high speed is 140 W/m² °C. With the fan at the lower speed, the velocity of the air is reduced to 1/2 the velodity and the overall heat transfer coefficient is reduced to 110 W/m² °C. Determine the percentage decrease in the heat transfer rate.

9.45 Hot water flowing at 3 kg/s is to be cooled from 100 °C in a coaxial, double-pipe heat exchanger having a heat transfer surface area of 20 m², with 5 kg/s of cold water entering the heat exchanger at 20 °C. What is the minimum temperature of the hot water at exit if the overall heat transfer coefficient is 800 W/m² K?

9.46 A heat exchanger has been designed to condense 2.2 kg/s of saturated steam at 100 °C with water available at 15 °C. In the design, the temperature rise of the water is limited to 8 °C. During operation, the water flow rate is reduced to 50% of the designed mass flow rate. The reduction in the flow rate reduces the overall heat transfer coefficient to 70% of the value used in the design. What is the percentage reduction in the mass rate of condensation?

9.47 Air at 1.8 kg/s and water at 1.4 kg/s enter a heat exchanger with a surface area of 15 m². The inlet temperatures of air and water are, respectively, 20 °C and 80 °C. The overall heat transfer coefficient is 160 W/m² °C. Determine the exit temperatures of the air and water if the heat exchanger is of the
(a) Parallel-flow type.
(b) Counterflow type.
(c) Cross-flow type with air flowing over the tubes (mixed).

9.48 A heat exchanger to recover heat from the exhaust gases from a residential oil furnace, consists of a square array of twenty-five 25-mm diameter, 40-cm long in-line tubes of negligible wall thickness. The pitch (vertical and horizontal) is 40 mm. Combustion gases at 150 °C enter the heat exchanger with a velocity of 3 m/s. Air at 12 °C is forced through the tubes with a velocity of 5 m/s. Estimate
(a) The overall heat transfer coefficient.
(b) The heat transfer rate to the air, and the exit temperature of the air.
Assume the properties of the combustion products are those of air.

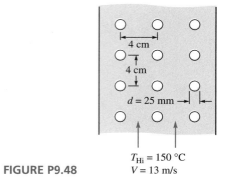

FIGURE P9.48

$T_{Hi} = 150\ °C$
$V = 13\ m/s$

9.49 In an industrial application, 2 kg/s of air are heated from 10 °C to 120 °C by passing it over a steady stream of carbon-steel pellets entering the heat exchanger at 800 °C at a mass flow rate of 2 kg/s. The specific heat of the material of the pellets is 2000 J/kg °C. An employee suggests that the exit temperature of the air can be increased by 50 °C by preheating the air from 10 °C to 60 °C before it enters the heat exchanger. As a consultant invited to evaluate the suggestion, write a brief report evaluating the employee's suggestion.

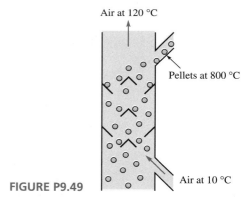

Air at 120 °C

Pellets at 800 °C

Air at 10 °C

FIGURE P9.49

9.50 A hot fluid [with $(\dot{m}c_p)_H = 800$ W/K] enters a heat exchanger at 80 °C. The inlet temperature of the cold fluid [with $(\dot{m}c_p)_C = 400$ W/K] is 10 °C. The surface area of the heat exchanger is 2 m². The overall heat transfer coefficient is 1200 W/m² °C. Determine the exit temperatures of the fluids and the heat transfer rates for
(a) Parallel-flow arrangement.
(b) Cross-flow arrangement with the cold fluid unmixed.

9.51 A room heater is made of one row of six 15-mm diameter, 2-m long tubes with fins on the outer surface. Hot water enters the tubes at 80 °C with a velocity of 1 m/s. Air is in cross flow over the heater. The inlet temperature of the air is 25 °C and its velocity in the duct is 3 m/s. The convective heat transfer coefficient on the air side based on the unfinned surface area of the tubes is [($q' =$ heat transfer rate per meter length of the tube $= h_o \pi d (T_s - T_a)$] is 200 W/m² °C. Find the heat transfer rate to the air and the exit temperature of the air.

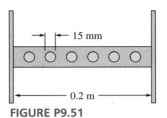 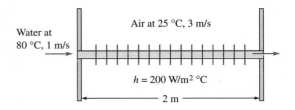

FIGURE P9.51

9.52 In super-computers, such as the Cray computers, the dissipation energy density is quite high. One method to cool the computers is described in Example 4.2.4. Another method is by forced convection with special dielectric liquids such as the Flourinert liquids developed by the 3M Company. The fluid is cooled in a heat exchanger and pumped back to the computer.

In one such computer the total energy dissipation is 50 kW. The coolant enters the computer at 20 °C and exits at 30 °C. The properties of the cooling fluid (FC-77, 3M Company) are

$$\rho = 1780 \text{ kg/m}^3 \qquad\qquad c_H = 1050 \text{ J/kg °C}$$
$$\nu = \mu/\rho = 0.8 \times 10^{-6} \text{ m}^2/\text{s} \qquad k = 0.063 \text{ W/m °C}$$

The fluid is to be cooled by water (on the shell side) in a two-tube pass shell-and-tube water exchanger. The diameter of the tubes is 12 mm and the wall thickness is negligible. The inlet and exit temperatures of the water are 12 °C and 18 °C, respectively. The convective heat transfer coefficient on the water side is 8000 W/m² °C. The velocity of the coolant in the tubes is 2 m/s.

(a) Determine the surface area and the length of the tubes.

(b) If the mass rate of flow of water is reduced by 30% and the corresponding heat transfer coefficient is 6500 W/m² °C, compute the inlet and exit temperatures of the coolant. (The heat dissipation in the computer and the water inlet temperature remain constant.)

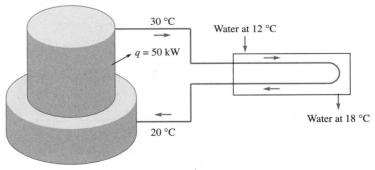

Computer

FIGURE P9.52

9.53 In a 10-m long double-pipe heat exchanger the inner, thin-walled tube has a diameter of 10 mm. Water at 120 °C, 0.2 kg/s, enters the inner tube and 4 kg/s of ethyl alcohol enters the annulus at 0 °C. The convective heat transfer coefficient associated with the ethyl alcohol is 800 W/m² °C. Estimate the maximum exit temperature of the alcohol.

9.54 A very long, double-pipe heat exchanger is used to heat water with oil. Details of the heat exchanger are given below:

Mass rate of flow of water	0.1 kg/s
Inlet temperature of water	10 °C
Mass rate of flow of oil	0.05 kg/s
Specific heat of oil	1800 J/kg °C
Inlet temperature of oil	120 °C
Overall heat transfer coefficient	300 W/m² °C

Determine the exit temperature of water and oil, the heat transfer rate, and the heat exchanger effectiveness for
(a) Parallel-flow arrangement.
(b) Counterflow arrangement.

9.55 The plates in a plate heat exchanger are 3-m high and 1.5-m wide. The heat exchanger is used to heat milk with hot water. The plates are made of 1-mm thick type 316 stainless steel. The gap between the plates is 5 mm. Water enters the heat exchanger at 90 °C with a velocity of 0.8 m/s between the plates. The inlet temperature of the milk is 5 °C, and the velocity of the milk between the plates is 0.4 m/s. Estimate the heat transfer rate and the exit temperatures of the milk and water if the heat exchanger is arranged as a
(a) Parallel flow heat exchanger
(b) Counterflow heat exchanger

Properties of milk: $\rho = 1030$ kg/m³ $c_p = 3850$ J/kg K
$k = 0.6$ W/m K $\mu = 2.12 \times 10^{-3}$ N s/m²

9.56 Re-solve Problem 9.55 if the fluids pass through the heat exchanger twice before exiting from it.

9.57 A one-tube pass shell-and-tube heat exchanger is designed to condense a fluid at its saturation temperature of 60 °C. The 12-mm diameter tubes (of negligible wall thickness) are 10-m long. Water enters the tubes at 10 °C with a velocity of 3 m/s. The exit temperature of the water is 30 °C. Estimate the convective heat transfer coefficient on the condensing side.

9.58 Starting with the definition of heat exchanger effectiveness, show that for a steam condenser

$$\varepsilon = 1 - e^{-\text{NTU}}$$

9.59 Starting with the definition of the effectiveness of a heat exchanger, show that for an infinitely long parallel-flow heat exchanger,

$$\varepsilon = \frac{1}{1 + c_r}$$

9.60 Derive the expression for the effectiveness of a parallel-flow heat exchanger for two single-phase fluids. See Table 9.4.1 for the expression.

9.61 A counterflow heat exchanger with a heat transfer area A_1 is utilized to cool a gas with a mass rate of flow of 10 000 kg/h from 150 °C to 100 °C with air entering the heat exchanger at 10 °C and leaving at 120 °C. It is desired to construct a heat exchanger with a heat transfer area A_2 so that the hot gas is cooled to 90 °C (instead of 100 °C). Assuming that the mass rates of flow of hot gas and air, the inlet temperatures, and the overall heat transfer coefficients are the same in both cases, determine the ratio, A_2/A_1.

9.62 The heat dissipation in a computer is 60 kW. It is cooled with a specially developed liquid (FC-77, 3M Company) with properties as given in Problem 9.52. The coolant is to be cooled with water available at 15 °C. The exit temperature of the water should not exceed 20 °C. The temperature of the coolant at exit from the computer is 35 °C. The temperature of the coolant at inlet to the computer should not be greater than 25 °C. Propose a suitable design of the heat exchanger. **(Design)**

9.63 In large commercial buildings, hospitals, and hotels energy resources can be conserved by recovering the waste heat from the exhaust air in cold weather. Design a plate heat exchanger to recover waste heat for the following conditions:

Mass rate of warm air	120 kg/s
Inlet temperature of warm air	22 °C
Mass rate of outside air	140 kg/s
Outside air temperature	−5 °C
Heat exchanger effectiveness	75% **(Design)**

9.64 Water flowing at 1.6 kg/s is to be heated from 15 °C to 85 °C by dry, saturated steam at atmospheric pressure. Thin-walled 12-mm diameter tubes are available. The tubes are to be arranged in a square array and the velocity of the water in the tubes is not to exceed 3 m/s. Making reasonable assumptions to compute the overall heat transfer coefficient design the condenser. **(Design)**

9.65 Design a steam condenser for the following conditions:

Mass of steam to be condensed	1000 kg/s of saturated steam at 40 °C
Condensate at exit	Saturated liquid
Water inlet temperature	20 °C
Water exit temperature	27 °C
Maximum velocity of water in tubes	2.5 m/s **(Design)**

9.66 A space heater is to be constructed in the form of a single tube provided with circular fins. Water at 80 °C enters the tube. Air at 10 °C is forced over the tube to give an average convective heat transfer coefficient of 40 W/m² °C. Propose a design to transfer 200 W to the air. Indicate
(a) The diameter of the tube.
(b) Flow velocity of the water.
(c) Dimensions of the fin to be made of aluminum.
(d) The length of the tube **(Design)**

9.67 Design a heat exchanger to condense 0.8 kg/s of saturated steam at 80 °C. Tubes of 10 mm, 12 mm, and 15 mm are available. Water, available at 10 °C, is the coolant. The temperature rise of the water is to be limited to 15 °C. The tubes may be assumed to be very thin. **(Design)**

9.68 Design a plate heat exchanger to cool 160 kg/s of water from 90 °C to 60 °C with cold water available at 20 °C. The temperature rise of the cold water is to be limited to 10 °C. The plates are to be made of type 304 stainless steel. The plates are available in heights of 2 m, 3 m, and 4 m with a width of 1.25 m, and heights of 1 m, 1.5 m, and 2 m with a width of 0.8 m. **(Design)**

9.69 Power Consultants Inc., in which you are an employee, has been awarded a contract for the design of a power plant. You are being assigned the preliminary design of the main condenser. The following information is available.

Mass rate of condensation of steam	40 kg/s
Pressure of steam at inlet	17.5 kPa
Velocity of cooling water	1–3 m/s
Type of condenser	Shell-and-tube
Maximum length of tubes	12 m
Temperature of cooling water at inlet	15 °C
Maximum temperature rise of cooling water	10 °C
Diameter of tubes	15 mm, 18 mm, 21 mm, and 25 mm

Design the condenser.   **(Design)**

9.70 Design an exhaust gas boiler to operate with a diesel engine with the following specifications.

Full load output of the engine	22 500 h.p.
Specific fuel consumption	170 g/bhp hr
Lower heating value of the fuel	42 000 kJ/kg
Air-fuel ratio	45
Temperature of exhaust gases at inlet to boiler	300 °C
Temperature of exhaust gases at exit from boiler	160 °C
Velocity of exhaust gases at inlet to boiler	3 m/s

Saturated liquid water enters the boiler at a pressure of 700 kPa. The mixture of water and steam is fed into an auxiliary boiler where the steam and water are separated. The exhaust gas boiler is a heat exchanger made of tubes with exhaust gases in cross flow. You may assume that the tube surface is at the saturation temperature of water.   **(Design)**

9.71 To improve the efficiency of gas turbines, a regenerator is used to recover the waste heat from the exhaust gases exiting the turbine. In the regenerator the compressed air is heated by the exhaust gases before it enters the combustion chamber. Raising the temperature of the air at inlet to the combustion chamber reduces the fuel requirements and increases the efficiency of the turbine. For a small gas turbine design a regenerator for the following conditions.   **(Design)**

Air flow	0.5 kg/s
State of air at compressor exit	230 °C, 540 kPa
State of gases at turbine exit	675 °C, 110 kPa
Regenerator efficiency	75%

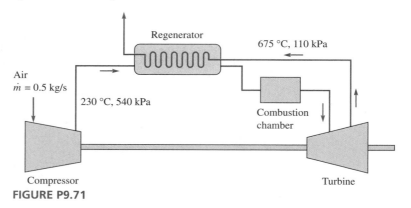

FIGURE P9.71

CHAPTER

10

MIXED-MODE HEAT TRANSFER

In the preceding chapters emphasis has been on a single mode of heat transfer. The isolation of each mode helps us to identify the mode of heat transfer and to develop the analytical and computational skills required to estimate the heat transfer by each mode. A few cases involving both conduction and convection were considered in the chapters on conduction, convection, and heat exchangers but the resulting equations were usually linear (except for the temperature dependence of the convective heat transfer coefficient, either because of temperature dependence of properties in forced convection or because of the temperature dependence of properties and the temperature difference in natural convection). When radiative heat transfer occurs in conjunction with conductive and (or) convective heat transfer, the resulting equations are nonlinear, usually requiring numerical techniques for their solution. Most practical problems involve multimode heat transfer. We now consider such cases with emphasis on radiative heat transfer. The algorithms for the numerical solutions depend on the nature of the equations. We illustrate some of the algorithms through examples. In this chapter we

- *Consider a few cases with heat transfer by more than one mode with emphasis on radiative heat transfer*
- *Illustrate a few algorithms to solve the resulting equations*
- *Give a brief introduction to solar collectors, which involve heat transfer by all three modes*

10.1 INTRODUCTION

In previous chapters emphasis was on analysis with only one mode of heat transfer, as an understanding of heat transfer by each mode is necessary before solving problems that involve heat transfer by more than one mode. However, most practical situations involve analysis of heat transfer by more than one mode. We have seen a few problems involving conductive and convective heat transfer in earlier chapters on conduction, convection, and heat exchangers. As both conductive and convective heat transfer rates are related to the difference in temperatures, the resulting equations are linear and, hence, the solution is straightforward unless the thermal conductivity or the convective heat transfer coefficient is strongly temperature dependent. If radiative heat transfer is also significant, the equations are nonlinear because radiative heat transfer is proportional to the difference in the fourth power of the temperatures. Radiative heat transfer usually occurs in combination with convective heat transfer when the space adjacent to the surfaces is filled with a gas. One exception is space applications where only conductive and radiative heat transfer occur simultaneously. If the convective heat transfer is by natural convection to gases or by forced convection to gases at high temperature levels, radiative heat transfer cannot be neglected. In this chapter we emphasize the solution to problems involving radiative heat transfer in addition to heat transfer by conduction and convection. Most such problems require some modeling.

In the majority of the cases involving multimode heat transfer, which require modeling, the resulting equations depend on the modeling. Therefore, no universal algorithm can be suggested for the solution to such problems. In many cases, particularly those involving radiative heat transfer in conjunction with conductive and convective heat transfer, numerical methods are needed for the solution to the resulting equations. The solutions may also need iterative methods. Each problem has to be considered separately. We will now present solutions to several such problems.

EXAMPLE 10.1.1 | ### Conduction, Convection, Radiation

A cylindrical, carbon resistance heating element, 20-cm long and 1-cm diameter, is placed inside a large duct (Figure 10.1.1). In the duct, air at 0 °C flows over the heating element, perpendicular to the axis of the heating element, with a velocity of 20 m/s. The resistance element is electrically heated with its outer surface maintained at 800 °C. The walls of the duct are at 20 °C. Determine

(a) The heat transfer rate from the element.
(b) The maximum temperature in the element.

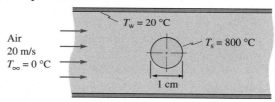

FIGURE 10.1.1 Air in Cross Flow over a Heated Carbon Element

Given

Material: Carbon	$L = 20$ cm	$d = 1$ cm	$T_s = 800\ °C$
Fluid: Air	$V = 20$ m/s	$T_\infty = 0\ °C$	$T_w = 20\ °C$

Find

(a) q
(b) T_{max}

ASSUMPTIONS

1. Steady state
2. Air does not participate in radiation.
3. Heat transfer from the end surfaces is negligible.

4. Air is at atmospheric pressure.

5. Constant thermal conductivity of the material of the heating element

6. Negligible blockage effect, i.e., the average air velocity is not significantly affected by the presence of the heating element.

SOLUTION

(a) The heat transfer rate from the element is by both convection to the surrounding air and radiation to the duct surface.

$$q = q_c + q_R$$

Heat transfer rate by each mode is determined separately.

To determine the convective heat transfer rate, we use an appropriate correlation for the convective heat transfer coefficient for flows over a cylinder. We use Equation 4.1.19, 4.1.20, or 4.1.21 depending on the Reynolds number. At a film temperature of 400 °C, properties of air from the software CC (or Table A7) are

$\rho = 0.524$ kg/m^3 $\qquad c_p = 1069.4$ J/kg K $\qquad k = 0.0497$ W/m K

$\mu = 3.268 \times 10^{-5}$ N s/m^2 $\qquad$ Pr $= 0.703$

$$Re_d = \frac{dV\rho}{\mu} = \frac{0.01 \times 20 \times 0.524}{3.268 \times 10^{-5}} = 3210$$

Employing Equation 4.1.21 for $Re_d < 20\,000$

$$Nu_d = 0.3 + \frac{0.62 \times 3210^{1/2} \times 0.703^{1/3}}{[1 + (0.4/0.703)^{2/3}]^{1/4}} = 27.71$$

$$h = Nu_d \frac{k}{d} = 27.71 \times \frac{0.0497}{0.01} = 137.8 \text{ W/m}^2 \text{ K}$$

$$q_c = hA_s(T_s - T_\infty) = 137.8 \times \pi \times 0.01 \times 0.2(800 - 0) = \underline{692.7 \text{ W}}$$

To compute the radiative heat transfer rate, we consider the heating element surface as one surface, completely enclosed in a much larger second surface, the duct surface. As the emissivity of the duct surfaces is not expected to be very small, from Equation 1.5.3,

$$q_R = \sigma A_s \varepsilon (T_s^4 - T_w^4)$$

From Table A10, the emissivity of carbon filament is 0.53. Therefore,

$$q_R = 5.67 \times 10^{-8} \times \pi \times 0.01 \times 0.2 \times 0.53(1073.2^4 - 293.2^4) = \underline{249.1 \text{ W}}$$

$$\text{Total heat transfer rate} = q_c + q_R = 692.7 + 249.1 = \underline{941.8 \text{ W}}$$

(b) To find the maximum temperature in the element, we consider the resistance heating as internal energy generation in the element.

ASSUMPTIONS

1. Uniform internal energy generation

2. Constant thermal conductivity

3. Negligible end effects

4. One-dimensional temperature distribution

SOLUTION

Total internal heat generation = 941.8 W; thus,

$$q''' = \text{internal energy generation per unit volume}$$

$$= \frac{941.8}{\pi(0.01^2/4)0.2} = 5.996 \times 10^7 \text{ W/m}^3$$

From Equation 2.7.1 (also Section 3.1.2) simplified to this steady, one-dimensional case,

$$\frac{d}{dr}\left(kr\frac{dT}{dr}\right) = -q'''r$$

Boundary conditions:

$$r = 0, \frac{dT}{dr} \text{ is finite or } T \text{ is finite}$$

$$r = R, T = T_s = 800 \,^\circ\text{C} \quad (R = \text{radius of the heating element})$$

Integrating once,

$$\frac{dT}{dr} = -\frac{q'''r}{2k} + \frac{c_1}{r}$$

At $r = 0$, dT/dr is finite and, hence, $c_1 = 0$.

Integrating again,

$$T = -\frac{q'''r^2}{4k} + c_2$$

Employing the boundary condition, $r = R$, $T = T_s = 800 \,^\circ\text{C}$

$$c_2 = T_s + \frac{q'''R^2}{4k}$$

The temperature distribution in the heating element is

$$T = -\frac{q'''r^2}{4k} + T_s + \frac{q'''R^2}{4k} = T_s + \frac{q'''R^2}{4k}\left(1 - \frac{r^2}{R^2}\right)$$

From the expression for the temperature distribution, it is clear that the maximum temperature is at $r = 0$. The maximum temperature is

$$T_{max} = T(r = 0) = T_s + \frac{q'''R^2}{4k}$$

From Table A3B, k (carbon filament) = 8.49 W/m K,

$$T_{max} = 800 + \frac{5.996 \times 10^7 \times 0.005^2}{4 \times 8.49} = \underline{844.1 \,^\circ\text{C}}$$

EXAMPLE 10.1.2 | **Conduction, Convection, Radiation**

Saturated steam at 200 °C flows in a thin-walled, 5-cm diameter horizontal tube. The tube is insulated with 5-cm thick rockwool and covered with a thin painted metal cladding ($\varepsilon = 0.8$). The surrounding air and surfaces are at 25 °C (Figure 10.1.2). Determine the heat loss per meter length of the pipe.

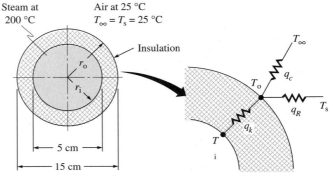

FIGURE 10.1.2 Conductive, Convective, and Radiative Heat Transfer from an Insulated Steam Pipe

Given

Fluid: Saturated steam	$T_{sat} = 200\ °C$
Fluid: Air	$T_\infty = 25\ °C$
Surroundings:	$T_s = 25\ °C$
Tube:	$2r_i = 5$ cm
Insulation: Rockwool	$2r_o = 15$ cm
Cladding: Painted steel	$\varepsilon = 0.8$

Find

q/L (heat loss per meter length)

ASSUMPTIONS

1. Condensation heat transfer coefficient with condensation of steam is large, and the convective resistance on the steam side is negligible compared with the conductive resistance of the insulation or the resistance associated with the convective and radiative heat transfer from the outer surface. Therefore, the tube surface is at 200 °C.
2. As the tube and the metal cladding are thin, their conductive resistances are negligible.
3. The room surface is much larger than the insulation surface.
4. Quiescent air

SOLUTION

Heat transfer is by conduction across the insulation and by convection and radiation from the cladding surface to the surrounding air and surface, respectively, as shown in Figure 10.1.2b. From an energy balance on the outer surface,

$$q_k = q_c + q_R \tag{1}$$

where

$$q_k = \frac{2\pi k L(T_i - T_o)}{\ln(r_o/r_i)}$$

$$q_c = h2\pi r_o L(T_o - T_\infty)$$

$$q_R = \sigma 2\pi r_o L\varepsilon(T_o^4 - T_s^4)$$

Substituting the expressions for the heat transfer rates into Equation 1, simplifying and rearranging, we obtain

$$\frac{k(T_i - T_o)}{r_o \ln(r_o/r_i)} = h(T_o - T_\infty) + \sigma\varepsilon(T_o^4 - T_s^4) \tag{2}$$

After solving Equation 2 for T_o, we obtain the heat transfer rate per meter length by conduction from $[2\pi k(T_i - T_o)]/[\ln(r_o/r_i)]$. Equation 2 is to be solved numerically. To solve for T_o, we need h. As the air is quiescent, convective heat transfer is by natural convection and in natural convection the heat transfer coefficient depends on T_o. Hence, T_o is determined by iteration. The following iteration scheme is adopted.

1. Assume a value for T_o.
2. Compute h.
3. Compute T_o from Equation 2.
4. If the computed value of T_o is different from the value used for computing h, go to step 2, recompute h employing the newly available value of T_o, and repeat the computations.

1. Assume $T_o = 60\,°C$.
2. Compute h with $T_o = 60\,°C$, $T_\infty = 25\,°C$, $d_o = 0.15$ m. From the software CC, properties of air at $T_f = (60 + 25)/2 = 42.5\,°C$, 101.325 kPa are

$$\rho = 1.119\ kg/m^3 \qquad \mu = 1.923 \times 10^{-5}\ N\ s/m^2 \qquad k = 0.0273\ W/m\,°C$$
$$\beta\ (at\ 298.15\ K) = 3.354 \times 10^{-3}\ K^{-1} \qquad Pr = 0.71$$

$$Ra_d = \frac{g\beta\rho^2(T_o - T_\infty)\,d^3\,Pr}{\mu^2}$$

$$= \frac{9.807 \times 3.354 \times 10^{-3} \times 1.119^2(60 - 25)0.15^3 \times 0.71}{(1.923 \times 10^{-5})^2} = 9.337 \times 10^6$$

From Equation 4.3.6c,

$$h = \left\{0.36 + \frac{0.518(9.337 \times 10^6)^{1/4}}{[1 + (0.559/0.71)^{9/16}]^{4/9}}\right\} \times \frac{0.0273}{0.15} = 4.0\ W/m^2\,°C$$

3. Compute T_o from Equation 2. From Table A3A, k (rockwool) $= 0.04$ W/m °C. From Equation 2,

$$\frac{0.04(473.2 - T_o)}{0.075 \times \ln(0.075/0.025)} = 4.0(T_o - 298.2)$$

$$+ 5.67 \times 10^{-8} \times 0.8(T_o^4 - 298.2^4)$$

or

$$0.48546(473.2 - T_o) = 4.0(T_o - 298.2) + 5.67 \times 10^{-8} \times 0.8(T_o^4 - 298.2^4)$$

Solving for T_o by a numerical technique, $T_\infty = 307.1$ K = 34 °C.

4. As the computed value of $T_o = 34$ °C is much less than 60 °C assumed for determining h, repeat steps 2 and 3 starting with $T_o = 34$ °C.

With $T_o = 34$ °C, the values are $T_f = 29.5$ °C, $Ra_d = 2.791 \times 10^6$, and $h = 2.88$ W/m² °C. From Equation 2, $T_o = 308.3$ K = 35.1 °C. The computed value of $T_o = 35.1$ °C is quite close to the value used for computing h (34 °C). Hence, $T_o = 35.1$ °C. The total heat transfer rate per meter length is given by

$$\frac{q_k}{L} = \frac{2\pi k(T_i - T_o)}{\ln(r_o/r_i)} = \frac{2 \times \pi \times 0.04(473.2 - 308.3)}{\ln(0.075/0.025)} = \underline{37.7 \text{ W/m}}$$

$$\frac{q_c}{L} = \pi \, d_o h(T_o - T_\infty) = \pi \times 0.15 \times 2.88(308.3 - 298.2) = 13.7 \text{ W/m}$$

$$\frac{q_R}{L} = \pi\sigma \, d_o\varepsilon(T_o^4 - T_s^4) = 5.67 \times 10^{-8} \times \pi \times 0.15 \times 0.8(308.3^4 - 298.2^4)$$
$$= 24.1 \text{ W/m}$$

COMMENTS

1. Note that the radiative heat transfer is greater than the convective heat transfer.
2. The solution to the problem can also be obtained by employing radiative heat transfer coefficient (Section 7.4)
3. The sum of the convective and radiative heat transfer rates/meter (37.8 W/m) is 0.1 W greater than the conductive heat transfer/meter length (37.7 W/m). The difference is due to round-off errors.

EXAMPLE 10.1.3 | **Conduction, Convection, Radiation**

Reconsider Example 7.3.7. The furnace arch is constructed with 40-cm thick firebricks, and there is no radiation shield (Figure 10.1.3). The inside surface temperature of the furnace is 1000 °C. Surrounding air and wall surface are at 35 °C. The

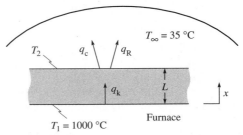

FIGURE 10.1.3 The inside surface of a furnace arch is at 1000 °C and the outside surface is exposed to surroundings at 35 °C.

arch may be considered as a rectangular top 6-m wide and 20-m long. The emissivity of the brick surfaces is 0.9. Determine the heat transfer rate from the arch to the surroundings,

(a) If convection is neglected.
(b) If convection is included.

Given

$T_1 = 1000\,°C = 1273.2\ K$
Material: Firebrick: $L = 0.4\ m$ Width $= 6\ m$ Length $= 20\ m$ $\varepsilon = 0.9$
Fluid: Air: $T_\infty = 35\,°C = 308.2\ K$
Surrounding surface: $T_s = T_\infty = 35\,°C = 308.2\ K$

Find

(a) q, if convection neglected
(b) q, if convection included

ASSUMPTIONS

1. Steady state
2. Surrounding air does not participate in radiation.
3. Constant properties
4. One-dimensional temperature distribution in the arch

SOLUTION

(a) To determine the heat transfer rate, from an energy balance on the outer surface, the conductive heat transfer rate to the outer surface equals the radiative heat transfer to the surroundings. The conductive heat transfer rate is given by

$$q_k = kA_x \frac{T_1 - T_2}{L} \tag{1}$$

where

$T_1 =$ temperature of the arch surface exposed to the hot gases
$T_2 =$ temperature of the outer surface of the arch
$L =$ thickness of the arch

Considering the surrounding surfaces to be much larger than the arch, the radiative heat transfer rate is given by

$$q_R = \sigma\varepsilon A_s(T_2^4 - T_\infty^4) \tag{2}$$

where T_∞ is the temperature of the surrounding surfaces.

Equating the two heat transfer rates and simplifying, we obtain (with $A_x = A_s$)

$$k\frac{T_1 - T_2}{L} = \sigma\varepsilon(T_2^4 - T_\infty^4) \tag{3}$$

The solution to Equation 3 yields T_2. The heat transfer rate is then calculated from either Equation 1 or Equation 2.

As the conductive heat transfer rate is proportional to the difference in the surface temperatures, $T_1 - T_2$, and the radiative heat transfer rate is proportional to the difference in the fourth powers of the surface temperatures, $T_2^4 - T_\infty^4$, the concept of thermal resistance and circuit cannot be applied. We then seek some other method of determining the heat transfer rate.

From Equation 3

$$T_2^4 + \frac{k}{L\sigma\varepsilon} T_2 - T_\infty^4 - \frac{k}{L\sigma\varepsilon} T_1 = 0 \tag{4}$$

The quartic Equation 4 can be solved for T_2 directly [see Spiegel (1968)] or by numerical methods. Here, we solve it by a numerical method—the bisection method. In this method any two values of T on either side of the root (or roots) are identified, say T_H and T_L, representing the higher and lower limits for T_2. Denoting the left-hand side of Equation 4 by $f(T)$, $f(T_H)$ and $f(T_L)$ will have opposite signs. The value of $f(T)$, for $T = (T_H + T_L)/2$ is then determined. If $f(T)$ and $f(T_L)$ have the same sign, i.e., $f(T) \times f(T_L) > 0$, the root should be between T and T_H. Then T_L is set equal to T. If $f(T)$ and $f(T_L)$ have opposite signs, i.e., $f(T) \times f(T_L) < 0$, then the root should be between T_L and T. T_H is then set equal to T. The process is repeated until $|f(T)|$ is less than an arbitrary, small value indicating that the solution has attained an acceptable degree of accuracy. To solve Equation 4, as T_2 cannot be higher than T_1 or less than T_∞, we set $T_H = T_1$, and $T_L = T_\infty$.

From Table A3B, k (firebricks) = 1 W/m K,

Substituting the values into Equation 4, and dividing it by 10^8,

$$\left(\frac{T_2}{100}\right)^4 + 48.99 \left(\frac{T_2}{100}\right) - 714.1 = 0 \tag{5}$$

Setting $T_H = 1273.2$ K, and $T_L = 308.2$ K, we get a sequence of values, some of which are given below (all temperatures have been divided by 100):

$T_L/100$ (K)	$T_H/100$ (K)	$T/100$ (K)	$f(T_L)$	$f(T)$	$f(T_L) \times f(T)$
3.082	12.732	7.907	-473	3582	< 0
3.082	7.907	5.494	-473	466	< 0
3.082	5.494	4.288	-473	-166	> 0
4.288	5.494	4.891	-166	97	< 0
$\vdots$	$\vdots$	$\vdots$	$\vdots$	$\vdots$	$\vdots$
4.691	4.693	4.692	-0.047	0.42	< 0
4.691	4.692	4.6915	-0.047	0.18	< 0

From the table it is clear that the value of T_2 is close to 469.1 K. From Equation 1 the heat transfer rate is given by

$$q = kA_x \frac{T_1 - T_2}{L} = 1 \times 6 \times 20 \times \frac{1273.2 - 469.1}{0.4} = \underline{\underline{241\ 200\ \text{W}}}$$

An alternate method of finding T_2 is to recast Equation 4 in the form,

$$\frac{T_2}{100} = \left[714.1 - 48.99 \left(\frac{T_2}{100}\right)\right]^{1/4} \tag{6}$$

Starting with any reasonable value of $T_2/100$ on the right-hand side, a new value of $T_2/100$ is computed. This is then used on the right-hand side again and the process repeated. When the absolute value of the difference between two successive values is within a pre-assigned convergence value, a good approximation to the correct value of T_2 is obtained.

Starting with $T_2/100 = 0$ on the right-hand side, the following sequence of values for $T_2/100$ are obtained:

5.169,   4.633,   4.698,   4.69,   4.691,   4.691

In this case a rapidly converging solution is obtained. There is no assurance that the convergence would be as rapid in all cases or even assurance of convergence.[1]

(b) Including the effects of convection, we have

$$kA_x \frac{T_1 - T_2}{L} = \sigma A_s \varepsilon (T_2^4 - T_\infty^4) + hA_s(T_2 - T_\infty) \tag{7}$$

With $A_x = A_s$, Equation 7 is simplified to

$$T_2^4 + \left(\frac{k}{\sigma \varepsilon L} + \frac{h}{\sigma \varepsilon} \right) T_2 - \left[\left(\frac{kT_1}{\sigma \varepsilon L} + \frac{hT_\infty}{\sigma \varepsilon} \right) + T_\infty^4 \right] = 0 \tag{8}$$

Equation 8 is solved for T_2 by iteration by recasting the equation in the form

$$T_2 = \left[\frac{k}{\sigma \varepsilon L} T_1 + \frac{h}{\sigma \varepsilon} T_\infty + T_\infty^4 - \left(\frac{k}{\sigma \varepsilon L} + \frac{h}{\sigma \varepsilon} \right) T_2 \right]^{1/4} \tag{9}$$

The value of the convective heat transfer coefficient has to be determined before Equation 9 can be solved for T_2. Assuming that the heat transfer from the outer surface of the arch to the surrounding air is by natural convection, the convective heat transfer coefficient is determined by one of the Equations 4.3.5a–4.3.5f.

For a horizontal rectangular surface, the characteristic length $L*$ is given by

$$L* = \frac{\text{surface area}}{\text{perimeter}} = \frac{6 \times 20}{2(6 + 20)} = 2.308 \text{ m}$$

For an initial estimate of the value of the convective heat transfer coefficient, assume $T_2 = 340$ K (66.8 °C). Then,

$$T_f = (340 + 308.2)/2 = 324.1 \text{ K (50.9 °C)}$$

From the software CC (for Table A7), for air at 50.9 °C,

$\rho = 1.0895$ kg/m^3 $c_p = 1008$ J/kg K
$\mu = 1.962 \times 10^{-5}$ N s/m^2 $\beta = 1/324.1$ K^{-1}
$k = 0.0279$ W/m K $Pr = 0.709$

---

[1]If, instead of recasting Equation 5 in the form of Equation 6, it is rearranged as

$$\frac{T_2}{100} = \frac{1}{48.99} \left[714.1 - \left(\frac{T_2}{100} \right)^4 \right]$$

and solved for T_2, we will find that a convergent solution is not obtained.

$$\text{Ra}_{L*} = \frac{9.807 \times 1.0895^2 (66.8 - 35)2.308^3}{324.1(1.962 \times 10^{-5})^2} \times 0.709 = 2.586 \times 10^{10}$$

From Equation 4.3.5b,

$$\text{Nu}_{L*} = 0.13 \, \text{Ra}_{L*}^{1/3} = 0.13(2.586 \times 10^{10})^{1/3} = 384.4$$

$$h = \frac{384.4}{2.308} \times 0.0279 = 4.65 \text{ W/m}^2 \text{ K}$$

Assuming the convective heat transfer coefficient does not vary very much with temperature, the value of T_2 is found by setting $h = 4.65$ W/m^2 K. After finding the value of T_2, the value of h is updated corresponding to the new value of the surface temperature and the process repeated, if necessary. Substituting all the known values, Equation 9 simplifies to

$$\frac{T_2}{100} = \left(994.8 - 140.1 \times \frac{T_2}{100} \right)^{1/4} \tag{10}$$

Solving Equation 10, $T_2 = 440.7$ K.

We now find a new value for the convective heat transfer coefficient corresponding to a surface temperature of 440.7 K. The new value, following the procedure already indicated for using Equation 4.3.5b, is 6.75 W/m^2 K. Employing $h = 6.75$ W/m^2 K in Equation 9, and solving for T_2, we get $T_2 = 430$ K. Repeating the procedure of finding h with updated values of T_2, and finding T_2, we get $h = 6.64$ W/m^2 K, $T_2 = 430.4$ K, which is very close to the previous value of T_2. The correct values of T_2 and h are, therefore, 430.4 K and 6.64 W/m^2 K, respectively. The heat transfer rate is then given by

$$\begin{aligned} q &= \sigma A_s \varepsilon (T_2^4 - T_\infty^4) + h A_s (T_2 - T_\infty) \\ &= 5.67 \times 10^{-8} \times 6 \times 20 \times 0.9(430.4^4 - 308.2^4) \\ &\quad + 6.64 \times 6 \times 20(430.4 - 308.2) = 154\,883 + 97\,369 = \underline{252\,252 \text{ W}} \end{aligned}$$

COMMENT

When convection is included the surface temperature goes down from 469.1 K to 430.4 K but the heat transfer rate increases slightly from 241 200 W to 252 250 W, an increase of 4.6%. By convection the heat transfer rate is 97 370 W, and by radiation it is 154 880 W. The radiative heat transfer rate is greater than by convection. It should also be pointed out that natural convection correlations obtained with still air far away from the heated surface were used for determining the convective heat transfer coefficient. In most cases, there is likely to be some air movement caused by external forces which will lead to a higher value of h.

| EXAMPLE 10.1.4 | **Conduction, Convection, Radiation** |

The temperature of air flowing in a duct is measured by a thermocouple. The thermocouple junction is attached to the bottom surface of the thermometer well as shown in Figure 10.1.4. The brass thermometer well is a hollow cylinder with an

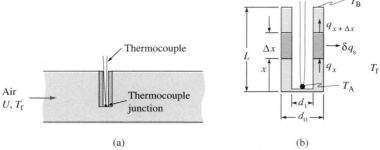

(a)   (b)

FIGURE 10.1.4 Thermocouple in a Thermocouple Well to Measure the Air Temperature
(a) Schematic of thermocopule and well. (b) Details of thermocouple well.

outer diameter d_o of 0.5 cm, an inner diameter d_i of 0.25 cm and a length L of 4 cm. The air flows with a velocity of 10 m/s at a pressure of 400 kPa. The thermocouple indicates an air temperature of 395 K. Due to heat losses, the temperature of the end of the well (T_B) is 375 K. Estimate the true temperature of the air.

Given

Fluid: Air	$U = 10$ m/s	$p = 400$ kPa	
Thermometer well:	$d_i = 0.25$ cm	$d_o = 0.5$ cm	$L = 4$ cm
Material: Brass			
$T_A = 395$ K	$T_B = 375$ K		

Find

T_f (air temperature)

ASSUMPTIONS

1. Negligible heat transfer from the inner surface of the well
2. Duct wall is at the same temperature as the air.
3. Duct surface is much larger than the well surface.
4. The thermocouple junction is at the same temperature as the end surface of the thermocouple well.
6. One-dimensional temperature distribution in the thermocouple well

SOLUTION

The temperature of the end surface T_A, is less than the air temperature due to conduction along the wall of the well. The variation in the temperature of the well is determined by an energy balance on an element shown shaded in Figure 10.1.4b.

$$q_x - q_{x+\Delta x} - \delta q_s = 0 \tag{1}$$

where δq_s is the sum of the convective and radiative heat transfer to the air and duct wall, respectively.

$$\text{Convective heat transfer rate to the air} = hP(T - T_f)\,\Delta x$$

where P = perimeter = πd_o

As the duct surface is much larger than the well, the radiative heat transfer (Equation 1.5.2) $= \sigma P \varepsilon(T^4 - T_s^4) \, \Delta x$. T_s is the duct surface temperature equal to the air temperature T_f. Substituting the expression for the convective and radiative heat transfer into Equation 1 we obtain

$$q_x - q_{x+\Delta x} - hP(T - T_f) \, \Delta x - \sigma P \varepsilon(T^4 - T_s^4) \, \Delta x = 0$$

Dividing by Δx and taking the limit as $\Delta x \to 0$, we obtain

$$-\frac{dq_x}{dx} - hP(T - T_f) - \sigma P \varepsilon(T^4 - T_s^4) = 0$$

Substituting $q_x = -kA_x \dfrac{dT}{dx}$, with $T_s = T_f$, we obtain the differential equation for the temperature distribution in the well.

$$\frac{d}{dx}\left(kA_x \frac{dT}{dx}\right) - hP(T - T_f) - \sigma P \varepsilon(T^4 - T_f^4) = 0 \tag{2}$$

We use the boundary conditions

$$x = 0, \, T = T_A \tag{3a}$$
$$x = L, \, T = T_B \tag{3b}$$

To solve nonlinear Equation 2, we linearize the radiative heat flux term by employing radiative heat transfer coefficient. The radiative heat transfer coefficient is defined by

$$\sigma \varepsilon(T^4 - T_f^4) = h_R(T - T_f)$$

where

$$h_R = \sigma \varepsilon(T^2 + T_f^2)(T + T_f)$$

Denoting $h + h_R = h_T$, $\theta = T - T_f$, and $m^2 = hP/kA_x$, Equation 2 and the associated boundary conditions are

$$\frac{d^2\theta}{dx^2} - m^2\theta = 0 \tag{4}$$

$$x = 0 \quad \theta = \theta_A = T_A - T_f \tag{4a}$$

$$x = L \quad \theta = \theta_B = T_B - T_f \tag{4b}$$

Solution to Equation 4 is

$$\theta = c_1 \cosh mx + c_2 \sinh mx$$

To satisfy the boundary conditions given by Equations 4a and 4b, we require that

$$c_1 = \theta_A \quad \text{and} \quad c_2 = \frac{\theta_A - \theta_B \cosh mL}{\sinh mL}$$

The temperature distribution is then given by

$$\theta = \theta_A \cosh mx + \frac{\theta_A - \theta_B \cosh mL}{\sinh mL} \sinh mx \tag{5}$$

The temperature distribution given by Equation 5 contains the unknown fluid temperature T_f. We need an additional condition to determine T_f. Such a condition is obtained from an energy balance on the surface at $x = 0$. Energy balance on the surface at $x = 0$ yields

$$k \frac{d\theta}{dx} = h_T \theta$$

or

$$km \frac{\theta_A - \theta_B \cosh mL}{\sinh mL} = h_T \theta_A \tag{6}$$

Substituting $T_A - T_f$ for θ_A and $T_B - T_f$ for θ_B in Equation 6 and solving for T_f, we obtain

$$T_f = \frac{[(h_T \sinh mL)/km]T_A - T_B + T_A \cosh mL}{\cosh mL - 1 + [(h_T \sinh mL)/km]} \tag{7}$$

The terms h_T and m (which depends on h_T) on the right-hand side of Equation 7 depend on the value of T_f, which is yet to be determined. It is determined by the following iteration scheme.

1. Assume a reasonable value of T_f.
2. Employing the value of T_f, determine h, h_R, and h_T.
3. Determine T_f from Equation 7.
4. Employing the value of T_f found in Equation 7 repeat steps 2 and 3 until the values in Steps 2 and 3 are within acceptable limits.

1. Assume $T_f = 400$ K $= 126.8$ °C
2. Both h and h_R depend on the well surface temperature, which varies from 395 K at $x = 0$ to 375 K at $x = L$. We will evaluate them at a mean surface temperature of 385 K (111.8 °C). From the software CC, properties of air at the film temperature of 392.5 K (119.3 °C), 400 kPa are

$\rho = 3.551$ kg/m^3 $\qquad \mu = 2.259 \times 10^{-5}$ ns/m^2
$k = 0.0326$ W/m °C $\qquad$ Pr $= 0.703$

$$\text{Re}_d = (\rho U\, d_o)/\mu = \frac{3.551 \times 10 \times 0.005}{2.259 \times 10^{-5}} = 7860$$

From Equation 4.1.21 for cross flow over a cylinder,

$$h = \left\{ 0.3 + \frac{0.62 \times 7860^{1/2} \times 0.703^{1/3}}{[1 + (0.4/0.703)^{2/3}]^{1/4}} \right\} \times \frac{0.0326}{0.005} = 281.4 \text{ W/m}^2 \text{ °C}$$

Evaluate h_R. From Table A9, the emissivity of oxidized brass is 0.66. For a well surface temperature of 385 K and a duct surface temperature of 400 K, h_R is given by

$$h_R = 5.67 \times 10^{-8} \times 0.66(385^2 + 400^2)(385 + 400) = 9.05 \text{ W/m}^2 \text{ °C}$$

$$h_T = 281.4 + 9.05 = 290.5 \text{ W/m}^2 \text{ °C}$$

From Table A1, k (brass) = 110 W/m °C

$$m = \left(\frac{h_T P}{k A_x}\right)^{1/2} = \left[\frac{290.5 \times \pi \times 0.005}{110 \times \pi(0.005^2 - 0.0025^2)/4}\right]^{1/2} = 53.07 \text{ per meter}$$

$$mL = 53.07 \times 0.04 \times 2.123$$

3. Substituting the values into Equation 7, we obtain $T_A = 400.8$ K.
4. As the computed value of T_A (= 400.8 K) and the value used to compute h_T in step 2 (= 400 K) are very close to each other, no further iteration is necessary.

$$T_f = \underline{400.8 \text{ K}}$$

COMMENTS

1. It just happened that we guessed a value very close to the correct value of T_f. If we had guessed $T_f = 420$ K in step 2, we would obtain $h = 280.9$ W/m² °C, $h_R = 9.8$ W/m² °C, $h_T = 290.7$ W/m² °C and $T_f = 400.8$ K. It is, therefore, clear that the iteration scheme will work for any reasonable starting value for T_f.
2. There is a 5.8 K difference between the indicated and the actual temperature of the air. Although many simplifying assumptions were made in arriving at the value of T_f, the computed temperature is much closer to the actual value than the indicated temperature.
3. With forced convection the radiative heat transfer coefficient is much less than the convective heat transfer coefficient and neglecting radiative heat transfer would yield almost the same value as the computed value.
4. Although the base temperature was assumed known, in practice, the base temperature has to be determined by finding the convective heat transfer coefficient associated with the base and performing an energy balance.

EXAMPLE 10.1.5 | **Conduction, Convection**

Cooling water enters a small steam condenser at 20 °C with a flow rate of 0.1 kg/s. Saturated steam at 100 °C condenses in the condenser (Figure 10.1.5). To measure the temperature of the cooling water at inlet, a thermocouple is to be attached to the outside surface of the 9-mm I.D., 12.5-mm O.D. cooling water copper tube. The outer surface of the thermocouple is insulated. If the thermocouple is placed too close to the inlet flange of the condenser, the indicated temperature will be different

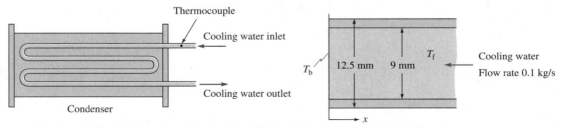

FIGURE 10.1.5 Sketch Showing the Cooling Water Tube Attached to the Condenser

from the actual temperature of the water in the tube due to conduction in the tube wall. How far away from the inlet flange should the thermocouple be placed if the indicated temperature is to be within 0.1 °C of the cooling water inlet temperature?

ASSUMPTIONS

To determine the location, where the temperature of the tube surface is 20.1 °C, the temperature distribution in the tube material is needed. Far away from the flange, the temperature of the tube is that of the cooling water, 20 °C. At the flange, because of the high convective heat transfer coefficient for condensing steam, the temperature of the tube surface just inside of the flange may be expected to be slightly less than 100 °C; we assume that it is 100 °C. This assumption leads to a slightly larger distance than when the true temperature is used, which in turn, leads to a more accurate reading by the thermocouple. At the point where the temperature of the tube is determined as 20.1 °C, the actual temperature will be close to 20.1 °C. The change in the temperature of the tube wall is due to conduction from the base and convection to the water and convection to the air outside. As the convective heat transfer on the air side is much smaller than on the water side, it will be neglected. It is assumed that the temperature distribution in the wall of the tube is one-dimensional, $T = T(x)$. It is also assumed that the water temperature is constant in this region at 20 °C. This latter assumption is equivalent to assuming that the heat transfer rate from the tube to the cooling water is negligible so far as the water temperature is concerned, but is significant in determining the temperature distribution in the tube. These assumptions should be checked later after the temperature distribution in the tube wall and the heat transfer rate to the water are determined. The tube is considered as an infinite fin with its base at 100 °C.

SOLUTION

For an infinite fin, the differential equation is (see Section 2.6.1)

$$\frac{d^2\theta}{dx^2} - m^2\theta = 0 \qquad m^2 = \frac{hp}{kA_x} \qquad \theta = T - T_f \qquad T_f = \text{water temperature}$$

with boundary conditions

$$\theta(x = 0) = \theta_b = T_b - T_f \qquad \theta(x \to \infty) = 0$$

The solution to the equation with the boundary conditions is $\theta = \theta_b e^{-mx}$.

To find the location where $\theta/\theta_b = 0.1/80$, we need the value of m. The value of the convective heat transfer coefficient is needed to determine m. From Table A5 properties of water at 20 °C are

$\mu = 9.853 \times 10^{-4}$ N s/m²   Pr = 6.83
$c_p = 4182$ J/kg K $k = 0.603$ W/m K

$$\text{Re}_d = \frac{\rho V d_i}{\mu} = \frac{4\dot{m}}{\pi d_i \mu} = \frac{4 \times 0.1}{\pi \times 0.009 \times 9.853 \times 10^{-4}} = 14\,358$$

Employing Equation 4.2.3b, $n = 0.4$ for heating,

$$\text{Nu}_d = 0.023 \times 14\,358^{0.8} \times 6.83^{0.4} = 105$$

Here the correlation suitable for uniform surface temperature has been used. But in the actual problem, neither the surface temperature nor the surface heat flux is uniform. As the intent is to find the location where the temperature has a defined value, the convective heat transfer coefficient is determined by using the correlation for uniform surface temperature. Recognizing that the value of the convective heat transfer coefficient is subject to an uncertainty of $\pm 10\%$, suitable allowance may be made in the final location of the thermocouple junction.

$$h = \frac{105 \times 0.603}{0.009} = 7035 \text{ W/m}^2 \text{ K}$$

From Table A1, k (copper) $= 401$ W/m K. Thus,

$$m^2 = \frac{hP}{kA_x} = \frac{h\pi d_i}{k\pi[(d_o^2 - d_i^2)/4]} = \frac{4h\, d_i}{k(d_o^2 - d_i^2)} = \frac{4 \times 7035 \times 0.009}{401(0.0125^2 - 0.009^2)}$$

$$= 8393/\text{meter}^2 \qquad m = 91.6/\text{meter}$$

The location where $\theta/\theta_b = 0.1/80$ is given by

$$x = -\frac{1}{m} \ln \frac{\theta}{\theta_b} = -\frac{1}{91.6} \ln \frac{0.1}{80} = 0.073 \text{ m}$$

The thermocouple must be located at a minimum distance of 7.3 cm from the flange. To take into account the uncertainty in the value of the convective heat transfer coefficient, it may be placed at a distance of 8 cm–10 cm from the flange.

Now, two assumptions that were originally made should be checked.

1. One-dimensional temperature distribution in the tube (see Section 2.6.1). Denoting the thickness of the tube by t,

$$\text{Bi} = ht/k = 7035 \times 0.00175/401 = 0.031$$

As the Biot number is less than 0.1, one-dimensional temperature distribution is acceptable.

2. Negligible change in the temperature of the water. Heat transfer rate from the tube to the water is

$$h = -kA_x \frac{d\theta}{dx}\bigg|_{x=0} = kA_x m \theta_b$$

$$= 401 \times \pi \frac{(0.0125^2 - 0.009^2)}{4} 91.6 \times 80 = 173.7 \text{ W}$$

$$(T_e - T_i)_{\text{water}} = \left(\frac{q}{\dot{m}c_p}\right)_{\text{water}} = \frac{173.7}{0.1 \times 4182} = 0.42 \,°\text{C}$$

The temperature of the water increases by 0.42 °C, from 20 °C to 20.42 °C. The difference between the maximum temperature of the tube surface and the water is reduced from 80 °C to 79.6 °C. This reduction in the base temperature has a negligible effect on the distance computed.

EXAMPLE 10.1.6 | Conduction, Convection, Radiation

The roof of a house is inclined at 45° to the vertical plane. It is modeled as a 10-mm thick plywood sheet with an emissivity of 0.9 (Figure 10.1.6). On a hot day the solar insolation is 800 W/m² and the ambient air temperature is 40 °C. The attic space is well ventilated and the roof is modeled as an inclined surface with natural convection on the inside surface. The outside convective heat transfer coefficient is 10 W/m² °C. The equivalent surrounding temperature is 20 °C. Determine

(a) The temperature of the surface of the roof exposed to the sun.
(b) The heat flux across the roof.

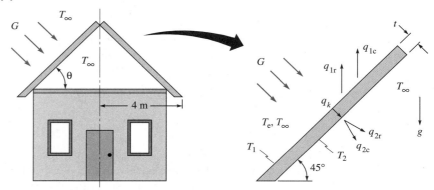

FIGURE 10.1.6 Heat Transfer by Different Modes on the Roof of a House

Given

Roof: 10-mm thick plywood, inclined at 45° to the horizontal, $\varepsilon = 0.9$
$G = 800 \ \text{W/m}^2$ $T_\infty = 40 \ ^\circ\text{C}$
$h_1 = 10 \ \text{W/m}^2 \ ^\circ\text{C}$ $T_e = 20 \ ^\circ\text{C}$

Find

(a) T_1
(b) q''

SOLUTION

Representing the conductive heat flux across the roof by q''_k from an energy balance on the upper surface of the roof, we have

$$\alpha G - q''_{1c} - q''_{1R} = q''_k$$

From an energy balance on the lower surface,

$$q''_k = q''_{2c} + q''_{2R}$$

The nomenclature employed is

$\quad G$ = solar insolation
$\quad \alpha$ = absorptivity = ε = 0.9
$\quad q''_{1R}$ = radiative heat flux from the upper surface = $\sigma\varepsilon(T_1^4 - T_e^4)$
$\quad T_1$ = temperature of upper surface
$\quad T_e$ = equivalent surrounding temperature

q_{1c}'' = convective heat flux from the upper surface = $h_1(T_1 - T_\infty)$
h_1 = outside convective heat transfer coefficient
q_k'' = conductive heat flux across the roof = $k(T_1 - T_2)/t$
k = thermal conductivity of plywood
T_2 = temperature of the lower surface of the roof
t = thickness of plywood
q_{2c}'' = convective heat flux from the lower surface = $h_2(T_2 - T_\infty)$
h_2 = inside convective heat transfer coefficient
q_{2R}'' = radiative heat flux from the lower surface = $\sigma\varepsilon(T_2^4 - T_{e2}^4)$
T_{e2} = inside surface temperature of the roof not exposed to the sun and the attic floor

Substituting the expressions for the heat fluxes, we have

$$\alpha G - h_1(T_1 - T_\infty) - \sigma\varepsilon(T_1^4 - T_e^4) = \frac{k(T_1 - T_2)}{t} \tag{1}$$

$$\frac{k(T_1 - T_2)}{t} = h_2(T_2 - T_\infty) + \sigma\varepsilon(T_2^4 - T_{e2}^4) \tag{2}$$

Solution to Equations 1 and 2 will give us the values of T_1 and T_2 and q_k''.

As the attic is well insulated, the temperatures of the inside air and the surfaces not exposed to the sun are assumed to be approximately equal to $T_\infty (T_{e2} \approx T_\infty)$. h_2 is found from Equation 4.3.1a or 4.3.1b, replacing g by $g \cos \theta$. Thus, the only parameter required for the solution is h_2. But the value of h_2 depends on T_2, which is yet to be determined. h_2 and T_2 are determined by iteration.

From Table A3A, k (plywood) = 0.115 W/m °C.
Determine h_2: Assume T_2 = 45 °C.
L = length of the roof = 4/cos 45° = 5.66 m.

From the software CC the properties of atmospheric air at T_f = (40 + 45)/2 = 42.5 °C are

ρ = 1.118 kg/m³ μ = 1.923 × 10⁻⁵ N s/m² k = 0.0273 W/m °C
β(313.2 K) = 3.193 × 10⁻³ K⁻¹   Pr = 0.71

$$\begin{aligned}
\mathrm{Ra}_L &= \frac{g(\cos\theta)\beta\rho^2\,\Delta T L^3}{\mu^2}\,\mathrm{Pr} \\
&= \frac{9.807 \times \cos 45° \times 3.193 \times 10^{-3} \times 1.118^2 \times 5 \times 5.66^3}{(1.923 \times 10^{-5})^2} \times 0.71 \\
&= 4.818 \times 10^{10}
\end{aligned}$$

From Equation 4.3.1b,

$$h_2 = \left\{0.825 + \frac{0.387 \times (4.818 \times 10^{10})^{1/6}}{[1 + (0.492/0.71)^{9/16}]^{8/27}}\right\}^2 \frac{0.0273}{5.66} = 2 \text{ W/m}^2 \text{ °C}$$

Substituting the known values into Equations 1 and 2, we obtain

$$0.9 \times 800 - 5.67 \times 10^{-8} \times 0.9(T_1^4 - 293.2^4)$$

$$- 10(T_1 - 313.2) = \frac{0.115(T_1 - T_2)}{0.1} \tag{3}$$

$$\frac{0.115(T_1 - T_2)}{0.1} = 2(T_2 - 313.2) + 5.67 \times 10^{-8} \times 0.9(T_2^4 - 313.2^4) \quad \textbf{(4)}$$

Solving Equations 3 and 4 numerically, we obtain

$$T_1 = 346.2 \qquad T_2 = 317.2 \text{ K}$$

The assumed value of 45 °C (318.2 K) to calculate h_2 is quite close to the computed value of T_2(317.2 K). No iteration is necessary.

$$q_k'' = \frac{0.115(346.2 - 317.2)}{0.1} = \underline{33.4 \text{ W/m}^2}$$

$$T_1 = \underline{346.2 \text{ K } (73 °C)}$$

10.2 SOLAR COLLECTORS

A solar collector is a good example of a component where heat transfer occurs by radiation, convection, and conduction. Some of the many applications of solar energy are for space heating, supplying hot water, heating swimming pools, crop drying, and power generation. Here, we consider only one component used extensively for the utilization of solar energy for residential use, the flat plate collector. Before going into the details of the collector, a brief introduction to solar radiation is given.

The sun is a star in our galaxy. It is a sphere of hot gases. Some of the data on the sun are given in Table 10.2.1.

TABLE 10.2.1 Some Data on the Sun

Diameter of sun	1.39×10^9 m
Mean distance between the sun and the earth	1.5×10^{11} m
Effective blackbody surface temperature of the sun	5760 K
Ratio of mass of sun to mass of earth	332 488
Total energy emitted by the sun ($E_b A_s$)	3.8×10^{26} W
Solar energy intercepted by the earth	1.7×10^{17} W
Solar constant	1353 W/m^2

The solar constant is the rate of incident solar energy per unit area of a surface perpendicular to the solar beam in the outer reaches of the atmosphere. As the solar energy passes through the atmosphere, its intensity is attenuated by scattering by the air molecules (Raleigh scattering); absorption by dust, water vapor, carbon dioxide, and ozone; and reflection and absorption in cloud layers. The extent of the attenuation depends on the height above sea level, the position of the sun (time of day and time of year), and dust and gas concentration in the atmosphere. On a clear day, approximately 70% of the solar constant (~ 1000 W/m^2) reaches the surface of the earth. About 99% of solar energy is in the wave band 0.2 μm–4 μm and 94% in the band 0.2 μm–2 μm.

Nomenclature In the discussion on solar collectors the following terms are used.

G_s = rate of solar insolation per unit area [rate of solar energy reaching a surface per unit area of the surface (W/m^2)]

$G_{\lambda s}$ = spectral solar insolation per unit area (W/m^2 μm)

$G_{\lambda s} \, d\lambda$ = solar insolation per unit area in the wave band λ to $\lambda + d\lambda$ (W/m^2)

E_{bs} = blackbody emissive power at the temperature of the sun, 5760 K (W/m^2)

$E_{b\lambda s}$ = blackbody monochromatic emissive power at the temperature of the sun, 5760 K (W/m^2 μm)

α_s = solar absorptivity (absorptivity for solar irradiation)

ε = infrared emissivity (emissivity at temperatures in the range 300 K–400 K)

τ_s = cover plate solar transmissivity (transmissivity for solar irradiation)

τ = cover plate infrared transmissivity (transmissivity for irradiation from a surface at temperatures in the range 300 K–400 K)

A solar collector is a heat exchanger in which a fluid is heated by a surface that absorbs the incident solar energy. A tube, exposed to the sun, in which a fluid is circulated, is a collector in its simplest form. In a flat plate collector the surface that absorbs the incident solar energy is a flat plate.

There are two types of solar collectors—the concentrating collector and the flat plate collector. The solar rays that reach us are essentially parallel as the distance between the sun and the earth is large in relation to the diameter of the sun. In a concentrating solar collector the solar beam is focused on a much smaller area by suitably shaped mirrors or lenses. There is no such concentration in a flat plate collector. We consider only flat plate collectors.

Two types of flat plate collectors for space heating, with air or water as the working fluid, are shown in Figures 10.2.1 and 10.2.2. A flat plate collector consists

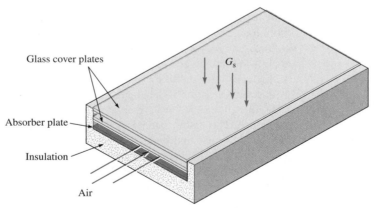

FIGURE 10.2.1 A Flat Plate Collector with Air as the Working Fluid
Two glass cover plates, essentially transparent to solar radiation, enclose an absorber plate. The incident solar energy heats the absorber plate. The absorber plate heats the air flowing between the absorber plate and the adjacent glass cover plate.

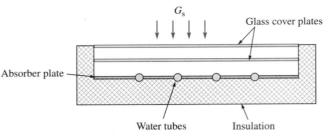

FIGURE 10.2.2 A Flat Plate Collector with Water as the Working Fluid
Two glass cover plates enclose the absorber plate. The absorber plate, heated by the incident solar energy, heats the water flowing in tubes attached to it.

of an absorber plate, which absorbs the incident solar energy. The absorber plate is covered by one or more glass cover plates. The glass cover plates are essentially transparent to the solar energy. Most of the solar energy is absorbed by the absorber plate, and its temperature increases. The glass cover plates are almost opaque to infrared radiation, and the emitted energy from the absorber plate (most of which is in the infrared region) does not pass through the glass cover plates. In dealing with solar collectors, terms such as infrared emissivity and transmissivity are used. Such terms usually refer to the infrared wave band, approximately 1 μm–100 μm. For a black surface at temperatures in the range 300 K–400 K, more than 99.8% of the total emitted energy is in the infrared wave band and a negligible amount in wavelengths less than 1 μm. In such cases, whether we define infrared emissivity as the emissivity in the infrared wave band or the emissivity of a surface at temperatures in the range 300 K–400 K makes very little difference.

In an air heating collector of the type shown in Figure 10.2.1, air is forced between the absorber plate and the glass cover plate next to it. The air is heated by the absorber plate. The air between the two topmost glass cover plates is stagnant (except for the motion caused by natural convection) and serves to reduce the heat losses from the collector to the surrounding air. In a water heating collector (Figure 10.2.2), tubes are attached to the absorber plate. The hot absorber plate heats the water circulating in the tubes attached to it.

A simple system for the supply of domestic hot water by solar collectors is shown in Figure 10.2.3. Water is circulated through the collectors mounted on the roof of the house. The water, heated in the collectors, passes through a coil in a holding tank in the house and heats the cooler water surrounding the coil in the tank. The cooler water flows back to the collector. The heated water from the holding tank flows through a water heater. If the temperature of the water at inlet to the heater is not high enough, it is further heated in the heater.

To analyze the operation of a collector, a knowledge of the radiation characteristics of the absorber plate and the glass cover plates is needed. A brief discussion of some of these characteristics follows.

Solar Absorptivity   The absorber plate of a collector receives solar energy from the sun radiating as a black body at 5760 K. The absorber plates generally operate at

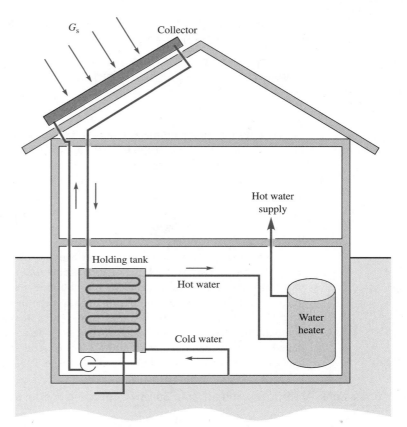

FIGURE 10.2.3 Schematic of Solar Collectors for Hot Water Supply
Hot water from the collector heats the cold water in holding tank. The
water from the holding tank goes through a water heater where the
water is further heated if necessary.

temperature less than 400 K (127 °C). A good absorber plate should absorb all the
incident solar energy but should emit very little energy. Most of the incident energy
from the sun (98%) is in the wave band 0.2 μm–3 μm. Most of the energy emitted
by the absorber plate is in the infrared region, 3 μm–100 μm (more than 99.8% for
operating temperatures around 400 K). Hence, the absorber plate should have a high
absorptivity at short wavelengths and low emissivity at intermediate and long wavelengths. For such special surfaces having wavelength dependent properties, the solar
absorptivity is defined by (see also Section 8.1.2)

$$\alpha_s = \frac{\int_0^\infty \alpha_\lambda G_{\lambda s}\, d\lambda}{G_s} \tag{10.2.1}$$

If the spectral absorptivity of a surface is known, its solar absorptivity is determined
from Equation 10.2.1. As the spectral distribution of the solar energy is fixed, the
solar absorptivity of a surface is also constant (except for minor variation due to the
temperature dependence of the spectral absorptivity of the surface).

Infrared Emissivity In a manner similar to the definition of the solar absorptivity, the infrared emissivity of a surface is defined by

$$\varepsilon = \frac{\int_0^\infty \varepsilon_\lambda E_{b\lambda} \, d\lambda}{E_b} \tag{10.2.2}$$

where both the monochromatic emissive power, $E_{b\lambda}$, and the blackbody emissive power, E_b, are evaluated at the temperature of the surface. The infrared emissivity of a surface is usually evaluated at surface temperatures in the range 300 K–400 K. At these temperatures, most of the emitted energy is in the wave band 3 μm to ∞ (99.8%), and only a negligible amount (0.2%) in the wave band 0 to 3 μm. At temperatures in the range 300 K–400 K, as only a negligible amount of emitted energy is in the wave band 0 to 3 μm, very little error is introduced whether the lower limit in Equation 10.2.2 is 0 or 3 μm.

Even if the spectral emissivity of a surface is fixed (temperature independent), its total infrared emissivity is a function of the temperature as the monochromatic blackbody emissive power of a surface is a function of its temperature. However, if the range of the temperature is not large, the variation in the total infrared emissivity is small. When values of the infrared emissivities are given without indicating the temperature at which it is evaluated, they may be considered as having been evaluated in the temperature range 300 K–400 K. The solar absorptivity and infrared emissivity of some special surfaces are given in Table 10.2.2.

TABLE 10.2.2 Solar Absorptivity and Infrared Emissivity of Some Special Surfaces

	α_s	ε	α_s/ε
Black nickel	0.87–0.96	0.07–0.12	7.3–13.7
Copper oxide	0.81–0.93	0.11–0.17	4.8–8.5
Black chrome	0.90–0.95	0.15–0.25	3.6–6.3
Lead sulfide (crystals)	0.89	0.20	4.45
Black paints	0.90–0.98	0.90–0.98	~1

Excerpted from Howell, J.R., Bannerot, R.B., and Vliet, G.C., *Solar-Thermal Systems* (New York: McGraw-Hill, 1982).

From the definition of the solar absorptivity and infrared emissivity, the solar energy flux absorbed and the radiant energy flux emitted by a special surface are given by

Absorbed solar energy flux = $\alpha_s G_s$ (W/m²)

Emitted energy flux = εE_b (W/m²)

Note that the solar absorptivity and infrared emissivity should be used only in conjunction with solar energy to evaluate absorption and to evaluate the emitted energy at temperatures in the range of 300 K–400 K.

EXAMPLE 10.2.1 A special opaque, diffuse surface has an emissivity of 0.9 for $0 < \lambda < 3$ μm and
0.1 for wavelengths > 3 μm (Figure 10.2.4). Calculate its
(a) Solar absorptivity.
(b) Infrared emissivity at 300 K, 400 K, and 500 K.

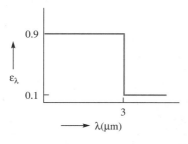

**FIGURE 10.2.4 Spectral Emissivity of a
Special Surface**

Given

$$0 < \lambda < 3 \text{ μm} \qquad \varepsilon_\lambda = 0.9$$
$$3 \text{ μm} < \lambda \qquad \varepsilon_\lambda = 0.1$$

Find

(a) α_s
(b) ε at $T = 300$ K, 400 K, and 500 K

SOLUTION

(a) From Equation 10.2.1, the solar absorptivity of the surface is given by

$$\alpha_s = \frac{\int_0^\infty \alpha_\lambda G_{\lambda s} \, d\lambda}{G_s}$$

From Kirchhoff's Law, $\alpha_\lambda = \varepsilon_\lambda$, and the spectral emissivity of the surface as shown
in Figure 10.2.4 can be used for the spectral absorptivity also. Denoting the quantities
in the wave band 0 to 3 μm by subscript 1 and those in the wave band 3 μm to ∞
by subscript 2,

$$\alpha_s = \alpha_1 \frac{\int_0^3 G_{\lambda s} \, d\lambda}{G_s} + \alpha_2 \frac{\int_3^\infty G_{\lambda s} \, d\lambda}{G_s}$$

where the limits for λ are in μm. From Example 8.1.2, $G_{\lambda s}/G_s = E_{b\lambda s}/E_{bs}$. Hence,

$$\alpha_s = \alpha_1 f_{0\text{-}\lambda_1} + \alpha_2 (1 - f_{0\text{-}\lambda_1})$$

where

$$f_{0\text{-}\lambda_1} = \frac{\int_0^3 E_{b\lambda s} \, d\lambda}{E_{bs}}$$

From the software RF (or Table 7.1.2), for $T = 5760$ K, $\lambda_1 = 3$ μm, $f_{0-\lambda_1} = 0.9786$

$$\alpha_s = 0.9 \times 0.9786 + 0.1(1 - 0.9786) = \underline{0.88}$$

(b) The infrared emissivity given by Equation 10.2.2 is

$$\varepsilon = \frac{\int_0^\infty \varepsilon_\lambda E_{b\lambda}\, d\lambda}{E_b} = \varepsilon_1 \frac{\int_0^3 E_{b\lambda}\, d\lambda}{E_b} + \varepsilon_2 \frac{\int_3^\infty E_{b\lambda}\, d\lambda}{E_b}$$

$$= \varepsilon_1 f_{0-\lambda_1} + \varepsilon_2 (1 - f_{0-\lambda_1})$$

1. $T = $ **300 K.** From the software RF (or Table 7.1.2), for $\lambda_1 = 3$ μm, $T = 300$ K, $f_{0-\lambda_1} = 0.00009$

$$\varepsilon(300 \text{ K}) = 0.9 \times 0.00009 + 0.1(1 - 0.00009) = \underline{0.1}$$

2. $T = $ **400 K.** For $\lambda_1 = 3$ μm, $T = 400$ K, $f_{0-\lambda_1} = 0.002135$

$$\varepsilon(400 \text{ K}) = 0.9 \times 0.002135 + 0.1(1 - 0.002135) = \underline{0.102}$$

3. $T = $ **500 K.** For $\lambda_1 = 3$ μm, $T = 500$ K, $f_{0-\lambda_1} = 0.01286$

$$\varepsilon(500 \text{ K}) = 0.9 \times 0.01286 + 0.1(1 - 0.01286) = \underline{0.11}$$

COMMENT

From this example, if the emissivity for wavelengths greater than 3 μm is fixed, the difference between the emissivities at 300 K and 400 K is very small, and quite small even in the range of 300 K to 500 K.

EXAMPLE 10.2.2 A plate attached to a space ship receives solar energy at 800 W/m² (although the solar constant is 1351 W/m², the actual insolation depends on the orientation of the surface to the solar beam). The underside of the plate is perfectly insulated (Figure 10.2.5). Assuming the sky temperature to be 0 K, determine

(a) The equilibrium temperature of the plate if the surface is (1) black and (2) black nickel.

(b) The heat flux to the cooling fluid flowing in the tubes attached to the plate if the surface is (i) black and (ii) black nickel. The cooling fluid maintains the plate at 320 K, in both cases.

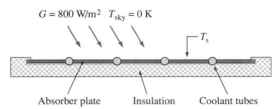

$G = 800$ W/m² $T_{sky} = 0$ K

T_s

Absorber plate   Insulation   Coolant tubes

FIGURE 10.2.5 Flat Plate Collector on a Space Ship

Given

$G_s = 800 \text{ W/m}^2$

Find

(a) T_s if $q = 0$ for
    (1) $\varepsilon_\lambda = \alpha_\lambda = 1$
    (2) Black nickel
(b) q'' if $T_s = 320 \text{ K}$ and if the surface is (1) black and (2) black nickel

ASSUMPTIONS

1. Steady state
2. For part a, as the plate is located in the outer reaches of the atmosphere, the heat transfer is by radiation only (no convection).
3. Negligible heat transfer across the insulation

SOLUTION

From an energy balance on the absorber plate,

Energy reaching the surface = Energy leaving the surface

or

Energy reaching the surface from the sun
= Reflected solar energy
+ Radiant energy transfer from the plate to the surroundings
+ Heat transfer to the coolant

(a) There is no heat transfer to the coolant and the energy balance simplifies to

Incident solar energy on the plate
− Reflected solar energy
= Radiant energy transfer from the plate to the surroundings

The left-hand side represents the absorbed solar energy. As the sky temperature is assumed to be 0 K, the radiant energy transfer from the surface to the surroundings per unit area of the surface is its emissive power. Hence, on a unit area basis,

Absorbed solar energy flux = Emissive power of the surface

1. For a black surface, $\alpha = \varepsilon = 1$, and the energy balance leads to

$$G_s = \sigma T_s^4$$

where T_s is the temperature of the plate surface. Then,

$$T_s = \left(\frac{G_s}{\sigma}\right)^{1/4} = \left(\frac{800}{5.67 \times 10^{-8}}\right)^{1/4} = \underline{344.6 \text{ K}}$$

2. For the black nickel surface,

Absorbed solar energy per unit area $= \alpha_s G_s$

Emissive power $= \varepsilon E_b(T_s) = \varepsilon \sigma T_s^4$

To find the absorbed solar energy, the solar absorptivity, α_s, is used. As it is anticipated that the temperature of the plate will be somewhat higher than 345 K, but not very much higher, the infrared emissivity is used to determine the emissive power of the surface. From Table 10.2.1, for a black nickel surface, $\alpha_s = 0.94$. As the temperature is expected to be somewhat higher than 300 K–400 K, the higher value of $\varepsilon = 0.12$ is used.

$$T_s = \left(\frac{\alpha_s G_s}{\varepsilon \sigma}\right)^{1/4} = \left(\frac{0.94 \times 800}{0.12 \times 5.67 \times 10^{-8}}\right)^{1/4} = \underline{576.6 \text{ K}}$$

(b) The energy balance on the plate should include the heat transfer to the coolant.

Absorbed solar energy flux
= Emissive power of the surface
+ Heat flux to the coolant

Therefore,

q'' = Heat flux to the coolant
 = Absorbed solar energy flux $-$ Emissive power of the surface
 = $\alpha_s G_s - \varepsilon \sigma T_s^4$

1. For the black surface, $\alpha_s = \varepsilon = 1$. With $T_s = 320$ K,

$$q'' = 800 - 5.67 \times 10^{-8} \times 320^4 = \underline{205 \text{ W/m}^2}$$

2. For the black nickel surface, $\alpha_s = 0.94$, $\varepsilon = 0.12$, $T_s = 320$ K, and

$$q'' = 0.94 \times 800 - 0.12 \times 5.67 \times 10^{-8} \times 320^4 = \underline{681 \text{ W/m}^2}$$

COMMENTS

1. The equilibrium temperature of the black nickel surface is 576.6 K and is higher than the 300 K–400 K range of temperature for which ε is given in Table 10.2.2. But, from the results of Example 10.2.1, we note that even at a temperature of 500 K the infrared emissivity of a surface does not appreciably differ from the value at 400 K if the spectral emissivity of the surface for $\lambda > 3$ μm is fixed. Therefore, the emissivity value used is acceptable.
2. Though the blackbody absorbs all the solar energy, it also emits the highest amount of energy at a given temperature. The special surface absorbs 94% of the incident solar energy, but at temperatures around 400 K, emits only 12% of the energy emitted by a black surface at the same temperature. Hence, to emit energy equal to the absorbed energy, the temperature of the special surface has to be significantly higher than that of the black surface.
3. If the plate is used as a solar collector operating at 320 K, the heat transfer to the coolant is 205 W/m^2 with a black surface and 681 W/m^2 with the special surface. If the efficiency of the solar collector is defined as the ratio of the heat transfer rate to the coolant to the rate of incident solar energy, the efficiency of the black surface is 25.7% and the efficiency of the special surface is as high as 85%. The advantages of using a specially coated surface for a solar collector is obvious. But such coatings are expensive, and, with time, many special coatings loose their selective properties.

Solar Transmissivity A flat plate collector is usually provided with one or more glass cover plates. The glass cover plates transmit most of the incident solar energy, while reducing both the transmission of the radiant energy emitted by the absorber plate, and the convective heat transfer from the collector to the surroundings.

The transmissivity of a glass plate depends not only on the material and the thickness of the plate but also on the orientation of the glass plate to the solar beam. The transmitted energy is the incident energy less the absorbed and reflected energy. When radiant energy strikes a glass plate, a part of it is reflected by the first surface and the rest reaches the second surface after being attenuated by absorption in the plate. A part of this energy reaching the second surface is reflected by it and this reflected energy reaches the first surface where, after some absorption, a part of it is again reflected. Thus, in computing the reflected energy, the multiple reflections and absorptions that occur within the glass plate should be considered. *In Figure 10.2.6 the angle of reflection is shown equal to the angle of incidence to illustrate multiple reflections and should not be construed as specular reflection.*

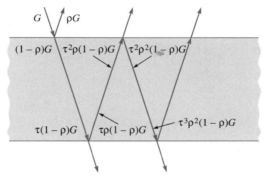

FIGURE 10.2.6 Sketch Showing Multiple Reflections and Absorptions from the Surfaces of a Glass Plate (Specular Reflection Is Not Implied)

The solar transmissivity of a cover plate is defined as

$$\tau_s = \frac{\int_0^\infty \tau_\lambda G_{\lambda s} \, d\lambda}{G_s} \tag{10.2.3a}$$

The infrared transmissivity of a cover plate is given by

$$\tau = \frac{\int_0^\infty \tau_\lambda E_{b\lambda} \, d\lambda}{E_b} \tag{10.2.3b}$$

where $E_{b\lambda}$ and E_b are evaluated at the temperature of the absorber plate, usually in the range 300 K–400 K.

Glass cover plates should have a very high solar transmissivity and very low infrared transmissivity. Most glass cover plates used in solar collectors are nearly opaque for radiation for wavelengths greater than 3 μm.

EXAMPLE 10.2.3 The spectral transmissivity of a 13-mm thick fused-silica glass is shown in Figure 10.2.7. Determine its

(a) Solar transmissivity.
(b) Infrared transmissivities at 300 K and 400 K.

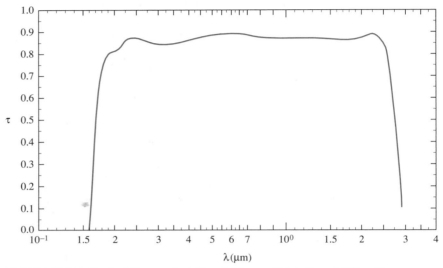

FIGURE 10.2.7 Typical Transmissivity of 13-mm Thick Fused-Silica Glass

SOLUTION

(a) The solar transmissivity is given by Equation 10.2.3a. Though the spectral transmissivity is a complicated function of the wavelength, from Figure 10.2.7, it is approximated as

$$\tau_\lambda = 0.88 \qquad 0.17 \ \mu m < \lambda < 2.5 \ \mu m$$
$$\tau_\lambda = 0 \qquad 0 < \lambda < 0.17 \ \mu m \text{ and } 2.5 \ \mu m < \lambda < \infty$$

With this approximation, Equation 10.2.3 reduces to

$$\tau_s = \frac{\int_{0.17}^{2.5} 0.88 G_{\lambda s} \, d\lambda}{G_s} = 0.88 \left(\int_0^{2.5} \frac{G_{\lambda s}}{G_s} \, d\lambda - \int_0^{0.17} \frac{G_{\lambda s}}{G_s} \, d\lambda \right)$$

But

$$\frac{G_{\lambda s}}{G_s} = \frac{E_{b\lambda}}{E_b} \bigg|_{5760 \ K} \qquad \text{(See Example 8.1.2)}$$

$$\tau_s = 0.88(f_{0-\lambda_2} - f_{0-\lambda_1}) \qquad \lambda_1 = 0.17 \ \mu m \qquad \lambda_2 = 2.5 \ \mu m \qquad T = 5760 \ K$$

From the software RF (or Table 7.1.2),

for $\lambda_2 T = 2.5 \times 5760 = 14\,400 \ \mu m \ K \qquad f_{0-\lambda_2} = 0.9655$
for $\lambda_1 T = 0.17 \times 5762 = 979.2 \ \mu m \ K \qquad f_{0-\lambda_1} = 0.0003$

and

$$\tau_s = 0.88(0.9655 - 0.0003) = \underline{0.85}$$

(b) From Equation 10.2.3b the infrared transmissivity is given by

$$\tau = \frac{\int_0^\infty \tau_\lambda E_{b\lambda} \, d\lambda}{E_b}$$

As $\tau_\lambda = 0$ for $\lambda < 0.17$ μm and $\lambda > 2.5$ μm,

$$\tau = 0.88 \int_{\lambda_1}^{\lambda_2} \frac{E_{b\lambda}}{E_b} \, d\lambda = 0.88(f_{0-\lambda_2} - f_{0-\lambda_1}) \qquad \lambda_1 = 0.17 \ \mu\text{m} \qquad \lambda_2 = 2.5 \ \mu\text{m}$$

1. $T = 300$ K.

$$\lambda_2 T = 2.5 \times 300 = 750 \ \mu\text{m K} \qquad f_{0-\lambda_2} \approx 0$$
$$\lambda_1 T = 0.17 \times 300 = 51 \ \mu\text{m K} \qquad f_{0-\lambda_1} \approx 0$$
$$\underline{\tau(300 \text{ K}) \approx 0}$$

2. $T = 400$ K.

$$\lambda_2 T = 2.5 \times 400 = 1000 \ \mu\text{m K} \qquad f_{0-\lambda_2} = 0.0003$$
$$\lambda_1 T = 0.17 \times 400 = 68 \ \mu\text{m K} \qquad f_{0-\lambda_1} \approx 0$$
$$\tau(400 \text{ K}) = \underline{0.88(0.0003 - 0) \approx 0}$$

COMMENT

From the values of solar and infrared transmissivity for this glass, it is clear that it has a high solar transmissivity and is opaque to infrared radiation.

Glass Cover Plates The solar transmissivity of fused-silica glass used in solar collectors is shown in Figure 10.2.7. From Figure 10.2.7, we observe that fused-silica glass has a high solar transmissivity in the wave band 0.17 μm–2.5 μm. More than 96% of the solar energy is in this narrow wave band. When more than one glass cover plate is used in a solar collector, the combined transmissivity is *approximately* given by the product of the transmissivities of the individual glass plates, i.e., $\tau = \tau_1 \times \tau_2 \times \tau_3 \times \cdots \times \tau_n$. The transmissivity of glass plates is a function of the angle of incidence of the radiant beam, but is approximately constant for angles up to 60° between the normal to the plate and the solar beam. Most collectors operate at angles less than 60°, and the effect of the angle of incidence of the solar beam may be neglected.

There is one further complication. In dealing with solar collectors, we are interested in estimating the energy absorbed by the plate. With one glass cover plate having a solar transmissivity of τ, the energy incident on the absorber plate is τG_s. The absorber plate absorbs $\tau \alpha G_s$, where α is the solar absorptivity of the plate, and reflects $\tau(1 - \alpha)G_s$, which reaches the glass plate (Figure 10.2.8). The glass plate reflects $\tau(1 - \alpha)\rho_d G_s$ back to the absorber plate (ρ_d is the diffuse reflectivity of the

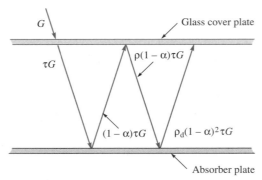

FIGURE 10.2.8 Solar Energy Absorption and Repeated Reflection and Absorption (Specular Reflection Not Implied)

glass plate). This process is repeated an infinite number of times, and the total absorbed energy is given by

$$\text{Absorbed energy} = \tau\alpha \sum_{n=0}^{\infty} [(1 - \alpha)\rho_d]^n G_s = \frac{\tau\alpha G_s}{1 - (1 - \alpha)\rho_d} = (\tau\alpha)_{\text{eff}} G_s$$

where

$$(\tau\alpha)_{\text{eff}} = \frac{\text{absorbed energy}}{\text{incident solar energy}} = \frac{\tau\alpha}{1 - (1 - \alpha)\rho_d}$$

Methods for computing ρ_d are available in Duffie and Beckman (1980). Duffie and Beckman also show that $1/[1 - (1 - \alpha)\rho_d] \approx 1.01$ so that $(\tau\alpha)_{\text{eff}}$ is only 1% greater than $\tau\alpha$, the product of τ and α. The value of τ is approximately 0.9 for a single cover plate, 0.8 for two cover plates, and 0.7 for three cover plates. The infrared emissivity of glass is approximately 0.88.

10.2.1 Performance of Flat Plate Collectors

The heat transfer rate to the working fluid in a solar collector is given by

$$q = \dot{m}c_p(T_e - T_i) \tag{10.2.4}$$

where

$\dot{m}$ = mass rate of flow of the working fluid
T_i, T_e = inlet and exit temperatures of the working fluid

The heat transfer rate and the fluid exit temperature depend on the absorber plate temperature and the convective heat transfer coefficient associated with the fluid in contact with the absorber plate. The absorber temperature, in turn, depends on the radiant energy absorbed by the plate and the heat transfer rate from the plate to the working fluid and to the surroundings.

Consider a flat plate collector with one glass cover plate. Water is the working fluid, which is heated by the collector. Water flows in tubes attached to the absorber plate. The surrounding temperature is T_a for the computation of both convective and radiative heat transfer from the cover plate to the surroundings. Given the solar irradiation, the heat transfer rate to the working fluid is to be determined.

In analyzing the performance of a collector, the following assumptions are made.

1. Steady state. Although solar insolation is time dependent, on clear days, the variation is slow enough that a steady-state assumption for short periods is a good approximation. However, if rapid changes occur due to changes in cloud cover, the steady-state assumption is not valid.
2. Uniform absorber plate temperature. Depending on the construction of the collector, there is some variation in the temperature of the plate. For the collector shown in Figure 10.2.9, temperature variations exist both in the direction of the flow of water and between the tubes. Such variations are neglected.

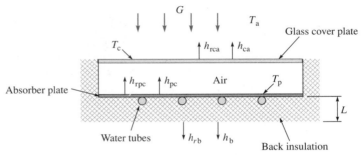

FIGURE 10.2.9 Sketch Showing a Water-Cooled Solar Collector with One Cover Plate
Water is circulated through tubes attached to the absorber plate.

3. Negligible temperature variation across the thickness of the cover plates
4. Glass cover plate is opaque to infrared radiation.
5. Solar absorptivity of the cover plate is negligible in so far as it affects the heat transfer to the surroundings. Solar absorptivity is significant in determining the effective transmissivity of the cover plate, but its effect on the temperature of the cover plate is negligible. The effect of infrared radiation from the adjacent absorber plate is more significant than the effect due to solar absorption.
6. The absorber plate is provided with back insulation, and the temperature distribution in the insulation is one-dimensional.
7. As the differences in temperatures between the cover plate, the absorber plate, and the surroundings are expected to be small, the radiative heat transfer can be computed by employing the radiative heat transfer coefficient (see Section 7.4).

From the above assumptions, the heat transfer in the collector can be analyzed by employing the thermal circuit. The thermal circuit is shown in Figure 10.2.10. The nomenclature is

h_{rb} = radiative heat transfer coefficient associated with the surface of the back insulation exposed to the surrounding surfaces at T_a.

h_b = convective heat transfer coefficient associated with the surface of the back insulation

h_{rpc} = radiative heat transfer coefficient between the cover plate and the absorber plate

h_{pc} = convective heat transfer coefficient associated with the air gap between the absorber plate and the cover plate

h_{rca} = radiative heat transfer coefficient between the cover plate and the surroundings at T_a

h_{ca} = convective heat transfer coefficient associated with the cover plate and the surrounding air

T_p = absorber plate temperature

T_c = cover plate temperature

T_s = temperature of the insulation surface exposed to the surrounding air

G_s = solar insolation on the cover plate

$(\tau\alpha)G_s$ = solar energy flux absorbed by the absorber plate

k = thermal conductivity of the insulating material

L = thickness of the insulation

q_u'' = useful heat flux to the working fluid

Figure 10.2.10a shows the individual resistances associated with the conductive, convective, and radiative heat transfers. Notice that as both convective and radiative heat transfers take place from the absorber plate to the cover plate, and from the cover plate to the surroundings, there are several resistances in parallel. The thermal circuit can be used with heat transfer by both convection and radiation, as the radiative heat transfer is computed by employing radiative heat transfer coefficient.

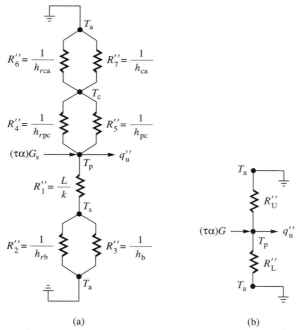

(a)   (b)

FIGURE 10.2.10 (a) The thermal circuit for the collector. (b) The equivalent circuit with effective resistances. The resistances are for a unit area of the absorber plate.

Using the radiative heat transfer coefficient permits us to relate both radiative and convective heat transfer rates to the difference in the appropriate temperatures. (If, instead of using the radiative heat transfer coefficient, the radiative heat transfer were computed exactly, it is related to the difference in the fourth powers of the temperatures of the surface and the surroundings, whereas the convective heat transfer is related to the difference in the temperatures. In such a case the thermal circuit cannot be used.) In Figure 10.2.10b, the various resistances have been replaced by the equivalent resistances R_L'' and R_U''.

$$R_L'' = \frac{R_2'' R_3''}{R_2'' + R_3''} + R_1''$$

$$R_U'' = \frac{R_4'' R_5''}{R_4'' + R_5''} + \frac{R_6'' R_7''}{R_6'' + R_7''}$$

From Figure 10.2.10b, we have

$$(\tau\alpha)G_s = \frac{T_p - T_a}{R_L''} + \frac{T_p - T_a}{R_U''} + q_u'' \qquad \textbf{(10.2.5a)}$$

In solar energy literature, the reciprocals of the resistances for a unit area (R_L'' and R_U'') are termed the bottom and top *heat loss coefficients,* and are denoted by U_L and U_T, respectively. The definition of the heat loss coefficient is similar to that of the overall heat transfer coefficient, except that the heat transfer is related to the difference in the temperatures of a surface and a fluid, instead of the difference in the temperatures of two fluids. Equation 10.2.5a is written as

$$(\tau\alpha)G_s = (U_L + U_T)(T_p - T_a) + q_u'' \qquad \textbf{(10.2.5b)}$$

If the temperature of the absorber plate, T_p, is known, q_u'', the heat flux to the working fluid, can be computed. In general, only the inlet temperature of the working fluid and the solar insolation, G_s, are known *a priori.* All other quantities are computed by iteration to satisfy the following relations:

$$q_u = \dot{m}c_p(T_e - T_i) \qquad \textbf{(10.2.6)}$$

where

q_u = heat transfer rate to the working fluid

$\dot{m}, T_i, T_e$ = mass rate of flow and inlet and exit temperatures of the working fluid, respectively

With the assumption of uniform temperature of the absorber plate, and assuming uniform convective heat transfer coefficient for the working fluid, from Equation 4.2.5b,

$$T_p - T_e = (T_p - T_i) \exp\left(-\frac{hA}{\dot{m}c_p}\right) \qquad \textbf{(10.2.7)}$$

where

h = convective heat transfer coefficient between the tube surface and the working fluid

A = tube surface area

From Equation 7.2.4b,

$$h_{rb} = \varepsilon_i \sigma (T_s^2 + T_a^2)(T_s + T_a)$$

$$h_{rca} = \varepsilon_g \sigma (T_c^2 + T_a^2)(T_c + T_a)$$

where

ε_i = emissivity of the insulation surface
ε_g = infrared emissivity of the glass cover plate

From Equation 2, Example 7.3.1 (with $A_1 = A_2$)

$$h_{rpc} = \frac{\sigma (T_p^2 + T_c^2)(T_p + T_c)}{(1/\varepsilon_p) + (1/\varepsilon_g) - 1}$$

where ε_p is the infrared emissivity of the absorber plate.

The various convective heat transfer coefficients should be evaluated from appropriate correlations for convective heat transfer coefficients. If convective heat transfers to the surrounding air from the insulation surface and the cover plate are by natural convection, the convective heat transfer coefficients are dependent on T_s, T_p, and T_c, which are not known *a priori*. An iterative scheme is adopted to find all the temperatures.

Both U_L and U_T are dependent on T_p only, although they require T_c and T_s, which are found by iteration. U_L and U_T can be computed for different values of T_p. With these values of U_L and U_T as functions of T_p, we proceed to find q_u for a given value of G_s. This also requires iteration. One possible iteration scheme is

1. Compute the convective heat transfer coefficient, h.
2. Compute U_L and U_T for different values of T_p.
3. Guess a value of T_p.
4. Find T_e from Equation 10.2.7.
5. Find q_u from Equation 10.2.6.
6. Compute q_u from Equation 10.2.5b

If the values of q_u from Equations 10.2.5b and 10.2.6 are close to each other, the value of T_p is correct. Otherwise, repeat steps 3 through 6. The iterative process for computing the top and bottom loss coefficients is illustrated in Example 10.2.4.

EXAMPLE 10.2.4

The specifications of a solar collector with one glass cover plate (Figure 10.2.11) are given in the following table.

Solar insolation	800 W/m²
Gap between absorber plate and glass cover plate (L)	30 min
Solar transmissivity of cover plate	0.88
Absorber plate emissivity (ε_p)	0.9
Infrared glass emissivity (ε_g)	0.9
Air and surrounding temperature (T_a)	0 °C
Absorber plate temperature (T_p)	90 °C
Angle of tilt of collector from horizontal (θ)	50°

Convective heat transfer coefficient associated with the surfaces exposed to surrounding air (h_{ca}, h_b) · · · · · · · · · · · · · · · 9 W/m² °C

Insulation thickness (t) · · · · · · · · · · · · · · · 75 mm

Insulation · · · · · · · · · · · · · · · Fiberglass

Collector length (H) · · · · · · · · · · · · · · · 1 m

Insulation is covered with a thin sheet of steel

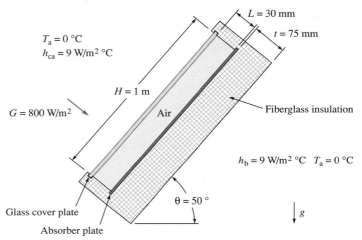

FIGURE 10.2.11 Flat Plate Collector with One Glass Cover Plate

Find

(a) The top heat loss coefficient.

(b) The bottom heat loss coefficient.

(c) The heat transfer rate to the working fluid per unit area of the absorber plate.

SOLUTION

(a) Top heat loss coefficient, U_T. For the collector with a single cover plate, the thermal circuit shown in Figures 10.2.10a and b are applicable. From Figure 10.2.10b,

$$\frac{1}{U_T} = \frac{1}{h_{rpc} + h_{pc}} + \frac{1}{h_{rca} + h_{ca}} \tag{1}$$

The heat transfer coefficients are dependent on the temperature of the glass cover plate, T_c, which is not known. T_c is determined by the following iteration scheme:

1. Guess a value of T_c

2. Determine the heat transfer coefficients and U_T

3. From the relation,

$$U_T(T_p - T_a) = (h_{rca} + h_{ca})(T_c - T_a) \tag{2}$$

compute T_c.

4. Using the newly computed value of T_c, repeat steps 2 and 3, until two consecutive values of T_c are sufficiently close to each other.

1. Assume $T_c = 40\ °C$.

2. Determine h_{rpc} and h_{rca}.

$$
\begin{aligned}
h_{rpc} &= \frac{\sigma(T_p^2 + T_c^2)(T_p + T_c)}{(1/\varepsilon_p) + (1/\varepsilon_g) - 1} \\
&= \frac{5.67 \times 10^{-8}(363.2^2 + 313.2^2)(363.2 + 313.2)}{(1/0.9) + (1/0.9) - 1} = 7.22\ \text{W/m}^2\ °C \\
h_{rca} &= \varepsilon_g\sigma(T_c^2 + T_a^2)(T_c + T_a) \\
&= 0.9 \times 5.67 \times 10^{-8}(313.2^2 + 273.2^2)(313.2 + 273.2) = 5.17\ \text{W/m}^2\ °C
\end{aligned}
$$

Determine h_{pc}, the convective heat transfer coefficient between the absorber plate and the glass cover plate across the 30-mm air gap. For the inclined cavity,

$$
\frac{H}{L} = \frac{1}{0.03} = 33.3
$$

As $H/L > 12$ and $\theta = 50°$, Equation 5.1.14 is used. Mean temperature of the two plates at $(90 + 40)/2 = 65\ °C$. Properties of air at $65\ °C$ from the software CC (or Table A7) are

$\rho = 1.044\ \text{kg/m}^3$ $\qquad c_p = 1009\ \text{J/kg K}$

$\mu = 2.025 \times 10^{-5}\ \text{N s/m}^2$ $\qquad k = 0.0289\ \text{W/m K}$

$\beta = 1/338.2\ \text{K}^{-1}$ $\qquad \text{Pr} = 0.708$

$$
\text{Ra}_L = \frac{9.807 \times 1.044^2(90 - 40)0.03^3}{338.2(2.025 \times 10^{-5})^2} \times 0.708 = 73\,670
$$

Employing Equation 5.1.14, with $\theta = 50°$,

$$
\begin{aligned}
\text{Nu}_L &= 1 + 1.44 \left[1 - \frac{1708}{73\,670 \times \cos(50)}\right]^* \left\{1 - \frac{1708 \times [\sin(1.8 \times 50)]^{1.6}}{73\,670 \times \cos(50)}\right\} \\
&\quad + \left\{\left[\frac{73\,670 \times \cos(50)}{5830}\right]^{1/3} - 1\right\}^* = 3.348
\end{aligned}
$$

$$
h_{pc} = 3.348 \times \frac{0.0289}{0.03} = 3.23\ \text{W/m}^2\ °C
$$

From Equation 1,

$$
\frac{1}{U_T} = \frac{1}{7.22 + 3.23} + \frac{1}{5.17 + 9}
$$

$$
U_T = 6.01\ \text{W/m}^2\ °C
$$

3. Compute T_c from Equation 2

$$
T_c = T_a + \frac{U_T(T_p - T_a)}{h_{rca} + h_{ca}} = 0 + \frac{6.01 \times 90}{5.17 + 9} = 38.2\ °C = 311.4\ \text{K}
$$

The newly computed value of T_c, $38.2\ °C$, is slightly lower than the assumed value of T_c, $40\ °C$. Recomputing the radiative and convective heat transfer coefficients with the updated value of $T_c = 38.2\ °C$,

$$h_{rpc} = 7.16 \text{ W/m}^2 \text{ °C} \qquad h_{rca} = 5.12 \text{ W/m}^2 \text{ °C}$$
$$h_{pc} = 3.25 \text{ W/m}^2 \text{ °C} \qquad U_T = 5.99 \text{ W/m}^2 \text{ °C}$$

With $U_T = 5.99$ W/m^2 °C, the value of T_c from Equation 2 is 38.2 °C. As the two successive values of T_c are equal to each other, the correct value of T_c for computing U_T is 38.2 °C, and

$$U_T = \underline{5.99 \text{ W/m}^2 \text{ °C}}$$

(b) Bottom heat loss coefficient, U_L. As the sheet steel covering the insulation is very thin, assume that its thermal resistance is negligibly small compared with that of the insulation. From Figure 10.2.10b,

$$\frac{1}{U_L} = \frac{t}{k} + \frac{1}{h_{rb} + h_b}$$

where t is 0.075 m.

From Table A3A, for fiberglass insulation, $k = 0.036$ W/m K, and $h_b = 9$ W/ m^2 °C.

To find h_{rb}, we need T_s, the temperature of the insulation surface that is exposed to the surrounding air. Its value is to be found by iteration (similar to finding T_c in determining U_T) such that it satisfies the relation,

$$U_L(T_p - T_a) = (h_{rb} + h_b)(T_s - T_a) \qquad \textbf{(3)}$$

For the insulation surface, because of the high thermal resistance of the insulation material, we anticipate that T_s is closer to T_a than to T_p. Assume $T_s = 10$ °C. Determine h_{rb} from the relation

$$h_{rb} = \varepsilon_i \sigma (T_s^2 + T_a^2)(T_s + T_a)$$

where ε_i is the emissivity of the steel.

We expect the steel cladding to be painted with $\varepsilon \approx 0.81$. Therefore,

$$h_{rb} = 0.81 \times 5.67 \times 10^{-8}(283.2^2 + 273.2^2)(283.2 + 273.2) = 3.96 \text{ W/m}^2 \text{ °C}$$

$$\frac{1}{U_L} = \frac{0.075}{0.036} + \frac{1}{3.96 + 9} \qquad U_L = 0.463 \text{ W/m}^2 \text{ °C}$$

With this initial estimate of U_L, from Equation 3,

$$T_s = T_a + \frac{U_L(T_p - T_a)}{h_{rb} + h_b} = 0 + \frac{0.463 \times 90}{3.96 + 9} = 3.2 \text{ °C}$$

Repeating the computations with $T_s = 3.2$ °C,

$$h_{rb} = 3.81 \text{ W/m}^2 \text{ °C} \qquad U_L = 0.463 \text{ W/m}^2 \text{ °C}$$

As the two consecutive values of U_L are equal, the correct value of T_s and U_L have been determined, and

$$U_L = \underline{0.463 \text{ W/m}^2 \text{ °C}}$$

(c) Heat transfer rate to the working fluid (water or air). The heat transfer rate from the absorber plate to the working fluid is given by Equation 10.2.5b.

$$q_u'' = (\tau\alpha)G_s - (U_L + U_T)(T_p - T_a)$$

For the collector, $(\tau\alpha)$ is approximated as the product of the transmissivity of the cover plate and the solar absorptivity of the absorber plate. From the specification of the absorber plate, it is assumed that the plate surface is gray with a solar absorptivity of 0.9. Hence,

$$(\tau\alpha) = 0.88 \times 0.9 = 0.792$$
$$q_u'' = 0.792 \times 800 - (0.463 + 5.99)(90 - 0) = \underline{52.8 \text{ W/m}^2}$$

COMMENT

The efficiency of this collector is: $\eta = 52.8/800 = 0.066$. The efficiency is low as the collector operating temperature of 90 °C is fairly high. Collectors with one glass cover plate would operate at lower temperatures and the efficiency of the collector would be higher. If the collector is to operate with an absorber plate temperature of 90 °C, two cover plates would be used. Although the use of two cover plates results in a lower value of the transmissivity of the cover plates, such a reduction in the transmissivity will be more than offset by a decrease in the top heat loss coefficient.

In the collector of Example 10.2.4, the bottom heat loss coefficient is less than 10% of the top heat loss coefficient. In determining it, we assumed that the heat transfer coefficient associated with the insulation surface exposed to air was equal to that for the glass cover plate. When solar collectors are mounted on the roofs of houses with little clearance between them and the roof, the heat transfer rate from the insulation surface is negligible as the air gap would be very small and the surface would be adjacent to the heated roof surface. In such cases, it is reasonable to neglect the heat transfer from the insulation surface.

An iterative method for determining the glass cover plate temperature of a solar collector with one cover plate is illustrated in Example 10.2.4. When more than one cover plate is used, the cover plate temperatures and loss coefficient are determined by an extension of such an iterative scheme. Such a scheme would involve guessing the temperature of each plate, and updating the temperatures after each iteration. The computational effort involved in such iterations justifies the use of a computer.

The absorber plate temperature was assumed to be uniform. There is variation in the plate temperature in an actual collector. The nonuniformity of the plate temperature depends on the flow pattern of the working fluid. If air is the working fluid, it is normally forced between the absorber plate and either the adjacent cover plate or a parallel plate on the insulation side. With such an arrangement, the absorber plate temperature varies in the direction of the flow of the air (why?). With water as the working fluid, the temperature of the absorber plate varies both in the direction of flow of the water and perpendicular to that direction between the tubes (being a minimum at the location of the tubes and a maximum midway between the tubes). A complete solution in such cases requires numerical procedures.

An estimate of the variation of the temperature in the direction perpendicular to the direction of the flow of water may be made by employing the principles presented in Section 2.6.

Example 10.2.5 The water tubes in the collector of Example 10.2.4 are 20 cm apart. The absorber plate is made of 1-mm thick copper sheet. The temperature of the plate adjacent to the water tube is 90 °C (Figure 10.2.12). Determine the maximum temperature of the plate.

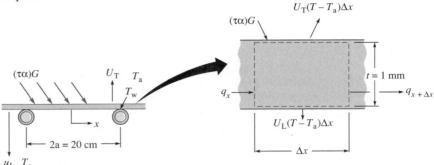

FIGURE 10.2.12 Collector with Water Tubes 2a (20 cm) Apart
The thickness of the plate is t (1 mm). The sketch on the right shows an element of the plate.

Given

$T(x = \pm a) = T_w = 90\,°C$ $2a = 20\ cm$
Material: Copper $t = 1\ mm$

Find

T_{max} (of the plate)

ASSUMPTIONS

1. Steady state
2. Variation of the temperature of the plate in the direction of the flow of water is negligible.
3. The top and bottom heat loss coefficient determined in Example 10.2.4 are applicable to the absorber plate. $U_T = 5.99\ W/m^2\,°C$ and $U_L = 0.463\ W/m^2\ C$.

SOLUTION

As the absorber plate is made of 1-mm thick copper plate, the temperature variation in the plate across the thickness of the plate is expected to be negligible. To verify that the temperature variation across the thickness of the plate can be neglected, we determine the Biot number with the heat transfer coefficient replaced by the heat loss coefficient. For copper, from Table A1, $k = 401\ W/m\ K$.

$$\mathrm{Bi} = \frac{(U_T + U_L)t}{k} = \frac{(5.99 + 0.463)0.001}{401} = 1.61 \times 10^{-5}$$

As $\mathrm{Bi} \ll 1$, the temperature variation across the thickness of the plate may be neglected.
    From an energy balance on the element shown on the right in Figure 10.2.12,

$$q_x - q_{x+\Delta x} - (U_T + U_L)W(T - T_a)\,\Delta x + (\tau\alpha)GW\,\Delta x = 0$$

where

W = length of the plate along the axis of the tubes
q = conductive heat transfer rate in the x-direction
$T = T(x)$ = temperature of the plate

Dividing by Δx and taking the limit as $\Delta x \to 0$,

$$-\frac{dq_x}{dx} - W(U_T + U_L)(T - T_a) + W(\tau\alpha)G_s = 0$$

Substituting $q_x = -kWt(dT/dx)$, dividing by (kWt), we obtain

$$\frac{d^2T}{dx^2} - \frac{(U_T + U_L)}{kt}(T - T_a) + \frac{(\tau\alpha)G_s}{kt} = 0$$

Denoting

$$(\tau\alpha)G_s = q'' \qquad \frac{(U_T + U_L)}{kt} = m^2 \qquad U_T + U_L = U$$

$$\frac{d^2T}{dx^2} - m^2\left(T - T_a - \frac{q''}{U}\right) = 0$$

With $T(x = \pm a) = T_w$ = temperature of the water tube, the solution to the differential equation is

$$T - T_a - \frac{q''}{U} = \frac{T_w - T_a - (q''/U)}{\cosh(ma)}\cosh(mx)$$

$$T_{max} = T(x = 0) = T_a + \frac{q''}{U} + \frac{T_w - T_a - (q''/U)}{\cosh(ma)}$$

The values are computed as follows

$q'' = (\tau\alpha)G_s = 0.88 \times 0.9 \times 800 = 633.6$ W/m^2
$U = U_T + U_L = 5.99 + 0.463 = 6.453$ W/m^2 °C
$T_w = 90$ °C $T_a = 0$ °C $a = 0.1$ m
$$m = \left(\frac{U}{kt}\right)^{1/2} = \left(\frac{6.453}{401 \times 0.001}\right)^{1/2} = 4.01 \text{ m}^{-1}$$

Then,

$$T_{max} = \frac{633.6}{6.453} + \frac{90 - (633.6/6.453)}{\cosh(4.01 \times 0.1)} = \underline{90.6 \text{ °C}}$$

COMMENT

The temperature of the plate varies from a minimum of 90 °C at the tube walls to a maximum of 90.6 °C. The heat transfer to the surroundings in Example 10.2.4 was computed with $T_p - T_a = 90$ °C. Considering the nonuniformity of the plate temperature, $(T_p - T_a)$ varies from 90 °C to 90.6 °C, a variation of less than 1%, which is acceptable in most cases involving convective heat transfer.

Generally, we are interested in the performance of a collector during a day when the solar insolation on the surface changes with time due to the changing angle between the solar beam and the collector surface. Although the solar insolation varies with time, the variation is slow and to determine the performance of a collector, the solar insolation may be assumed to be constant for short periods of time. The performance of the collector during such periods can be determined on the basis of the steady-state solution. Simulation programs that take into account not only the variation in the solar insolation during a day but also its variation during a year, the weather data relating the air temperature, and direction and velocity of wind for different geographical locations have been developed by a number of workers in the field. Some of the simpler algorithms to predict the performance of collectors are available in the references on solar energy given at the end of this chapter.

SUMMARY

Most heat transfer problems involve heat transfer by more than one mode. The methodologies to solve such problems depend on the nature of the problems, and each case has to be considered by itself. Many of them also involve iterative techniques and numerical solutions. Solutions to a few cases illustrating some of the iterative and numerical algorithms are obtained. Flat plate solar collectors are a good example where such multimode heat transfer takes place. A brief introduction to solar collectors is given. Some of the more significant radiation characteristics of the absorber plate and the glass cover plates used in flat plate collectors are considered in some detail. The use of special surfaces in collectors to increase the efficiency of collectors is demonstrated. The computations of the heat loss coefficients from a collector, and the heat transfer rate to the working fluid are illustrated.

Software

Multimode heat transfer problems involve a combination of conduction, convection, and radiation. In many cases iterative solutions requiring the evaluation of complex expressions, such as in convection correlations, are needed. Software programs TC, CC, RF, and HE assist in making rapid computations, eliminating errors, and reducing the tedium of repetitive calculations.

REFERENCES

Chapman, A.J. (1987). *Fundamentals of Heat Transfer*. New York: Macmillan.

Duffie, J.A., and Beckman, W.A. (1980). *Solar Engineering of Thermal Processes*. New York: Wiley.

Howell, J.R., Bannerot, R.B., and Vliet, G.C. (1982). *Solar-Thermal Systems*. New York: McGraw-Hill.

Incropera, F.P., and DeWitt, D.P. (1990). *Introduction to Heat Transfer*. New York: Wiley.

Lunde, P.J. (1980). *Solar Thermal Engineering*. New York: Wiley.

Rapp, D. (1981). *Solar Energy*. Englewood Cliffs, NJ: Prentice Hall.

Spiegel, M.P. (1968). *Mathematical Handbook for Formulas and Tables*. New York: McGraw-Hill.

PROBLEMS

10.1 The emissivity of a 20-cm diameter electric hot plate is 0.9. The surrounding air and surfaces are at 20 °C. If the power supply to the hot plate is 1000 W, determine its temperature. Assume that all heat transfer is from the upper surface only.

10.2 The temperature of the top surface of a reheat furnace (6-m wide and 15-m long) is 200 °C. The surrounding air and the walls are at 30 °C. The emissivity of the surface is 0.85. Compute the heat transfer rate from the surface.

10.3 In Problem 10.2, a radiation shield of a thin steel sheet is placed 20 cm above the furnace top surface. Compute the heat transfer rate from the furnace and the temperature of the radiation shield.

10.4 A solar collector is constructed by brazing tubes to an absorber plate as shown. Water flows in the tubes and maintains the temperature of the plate adjacent to the tubes at T_w. The incident solar energy is G. If the temperature of the surroundings is T_∞, derive the differential equation for the one-dimensional temperature distribution $T(x)$ in the plate. Both convective and radiative heat transfers are significant. The equivalent surrounding temperature for radiative heat transfer is T_e.

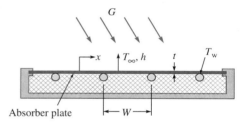

FIGURE P10.4   Absorber plate

10.5 A 1-cm thick insulation material ($k = 0.2$ W/m K, $\varepsilon = 0.85$) is attached to the bottom surface of a large, thin, horizontal steel plate ($\varepsilon = 0.8$). If the temperature of the steel plate is 600 °C, determine the ratio of the heat flux from the top surface to the surroundings to the heat flux across the insulation. The surrounding air and walls are at 30 °C and the convective heat transfer coefficient for both the top and bottom surfaces is 20 W/m² °C.

10.6 The roof of a reheat furnace, made of 40-cm thick firebricks ($k = 1$ W/m² °C), is 2-m wide and 4-m long. To reduce the heat transfer from the top surface, a radiation shield is placed 10 cm above the roof. The emissivity of both surfaces of the shield is 0.9. The inside surface of the firebricks is at 800 °C and the surroundings are at 40 °C. Calculate the heat transfer rate from the roof.

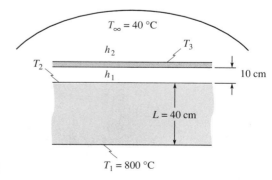

FIGURE P10.6

10.7 Re-solve Problem 10.6 if the firebrick is replaced by 15-cm thick zirconia.

10.8 The solar insolation on the flat, 20-cm thick concrete roof ($k = 0.663$ W/m °C) of a house is 800 W/m². The ambient surfaces and the inside air are at 35 °C. The equivalent sky temperature is 0 °C. The inside surfaces of the wall and floor are at 35 °C. The emissivity of the roof surface is 0.9. The outside and inside convective heat transfer coefficients are 15 W/m² °C and 10 W/m² °C, respectively. Assuming steady state, estimate the outside and inside surface temperatures of the roof and the heat flux across the roof.

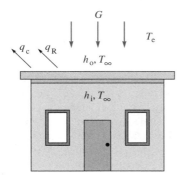

FIGURE P10.8

10.9 Re-solve Problem 10.8 if the roof is lined with a material having an emissivity of 0.1 in the wave band $0 < \lambda < 2$ μm, and 0.9 for wavelengths greater than 2 μm.

10.10 A horizontal, 20-cm diameter circular hot plate is rated at 1500 W. It consists of a 9-mm thick steel plate followed by a 6-mm thick insulating cement ($k = 1$ W/m K) and the electrical heating element. The underside of the heating element is well insulated. The temperature of the air and the walls of the room, in which the hot plate is located, is 30 °C. Determine the temperatures of the surface adjacent to the heating element and the outer steel surface if the heater is operated at
(a) Full-rated capacity.
(b) At half the rated capacity.

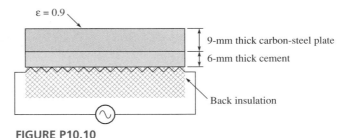

FIGURE P10.10

10.11 Steel parts are to be preheated to 930 °C in a furnace prior to carburizing. The interior of the furnace is 2-m wide, 3-m long, and 1.5-m high. The walls of the furnace are made of 20-cm thick firebricks ($k = 0.27$ W/m K) followed by 10-cm thick insulation ($k = 0.15$ W/m K). The outer wall is lined with a thin metal sheet with an emissivity of 0.3. The inner surface temperature of the furnace is 1000 °C and the surroundings are at 35 °C. Estimate the heat flux across the side walls.

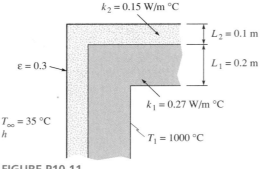

$k_2 = 0.15$ W/m °C

$L_2 = 0.1$ m

$L_1 = 0.2$ m

$\varepsilon = 0.3$

$k_1 = 0.27$ W/m °C

$T_\infty = 35$ °C
h

$T_1 = 1000$ °C

FIGURE P10.11

10.12 A space heater is made of 40-cm diameter stainless steel sphere (lightly oxidized), freely hanging from the ceiling (out of reach of people). The surface of the heater is maintained at 200 °C. The room temperature is 20 °C. Evaluate the heat transfer rate from the sphere and suggest a means of increasing the heat transfer rate without increasing the temperature of the surface.

10.13 The filament of a 75 W, evacuated, incandescent electric bulb is approximated as a gray body at 2800 K. The radiation properties of the glass bulb are

$0 < \lambda < 3$ μm $\qquad \tau_\lambda = 0.8, \rho_\lambda = 0$
3 μm $< \lambda < \infty \qquad \tau_\lambda = 0, \rho_\lambda = 0$

The bulb is freely suspended in a room which is at 30 °C. Determine the temperature of the glass bulb, approximating it as a 60-mm diameter sphere.

10.14 A spherical reservoir for storing liquid hydrogen is made of two concentric, spherical shells of thin stainless steel sheets. The diameters of the inner and outer sphere are 1 m and 1.04 m, respectively. The space between the two spheres is completely evacuated. Liquid hydrogen inside the container maintains the temperature of the inner sphere at 42 K. Assuming the container to be freely suspended in a large room maintained at 30 °C, determine the heat transfer rate to the liquid hydrogen.

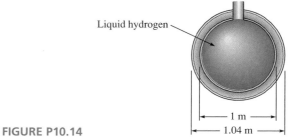

Liquid hydrogen

1 m

1.04 m

FIGURE P10.14

10.15 Re-solve Problem 10.14 if the outer sphere is covered with a 50-mm thick insulation material ($k = 0.06$ W/m K) enclosed inside a thin stainless steel cladding.

10.16 Due to a small leak between the two spheres in Problem 10.14, air enter the space between the two spheres. Determine the heat transfer rate to the liquid hydrogen.

10.17 On clear, cold nights, the effective sky temperature is considerably less than 0 °C.

As a result of radiative heat transfer, the temperature of some vegetation may fall below 0 °C even when the surrounding air temperature is greater than 0 °C. To estimate the surface temperature of vegetation under such conditions, consider a thin, freely suspended, black horizontal, plate (30 cm × 30 cm). The lower surface of the plate is assumed to be perfectly insulated. If the effective sky temperature is −40 °C, and the surrounding air temperature is 4 °C, determine the equilibrium temperature of the plate if

(a) The air is calm.

(b) The air has a velocity of 4 m/s parallel to one of the edges.

10.18 Re-solve part b of Problem 10.17 using a 8-cm diameter spherical shell (simulating an orange) instead of a horizontal surface.

10.19 A thermocouple junction is placed inside a duct to measure the temperature of the air flowing inside the duct. The junction is a 1-mm diameter sphere. The duct wall temperature is 380 °C and the air temperature as indicated by the thermocouple is 400 °C. The air velocity is 10 m/s. Assuming conduction in lead wires of the thermocouple to be negligible, find the corrected temperature of air if the emissivity of the surface of the thermocouple junction is

(a) 0.9

(b) 0.1

Suggest a method to reduce the error in measuring the air temperature without changing the emissivity of the surface.

10.20 The thermophysical properties of a thermocouple junction are

$$k = 20 \text{ W/m K} \qquad c_p = 400 \text{ J/kg K} \qquad \rho = 8500 \text{ kg/m}^3 \qquad \varepsilon = 0.9$$

The diameter of the spherical junction is 1 mm. It is placed in an air stream flowing at 10 m/s at 60 °C in a duct whose surface is also at 60 °C. If the initial temperature of the thermocouple is 20 °C, estimate the time taken for the junction to reach 59.5 °C

(a) Neglecting radiative heat transfer.

(b) Including radiative heat transfer. (Make a reasonable estimate of the radiative heat transfer.)

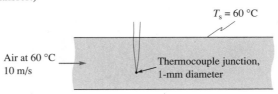

FIGURE P10.20

10.21 A 10-cm copper sphere, initially at a uniform temperature of 100 °C, is freely suspended in a large room. The temperature of the air is 20 °C. The emissivity of the surface of the sphere is 0.8. Determine the time for the temperature of the sphere to reach 25 °C. You may have to employ numerical methods, or radiative heat transfer coefficient for discrete changes in the temperature of the sphere.

10.22 The temperature of air flowing in a duct is to be measured by a thermocouple. The thermocouple is made by soldering two wires at the end of a 3-mm I.D., 1-mm wall plain carbon-steel tube. A 2-cm length of the tube is inserted into the duct. The temperature of the tube where it enters the duct is 25 °C. The air in the duct is at 70 °C, 400 kPa, and flows at 5 m/s. There is concern that due to conduction in the tube the temperature of the thermocouple junction at the end of the tube may be lower than the actual temperature of the air.

(a) Estimate the temperature that the thermocouple would indicate.

(b) If the indicated temperature is significantly different from the temperature of the air in the duct, propose methods to ensure that the indicated temperature is close to the actual temperature.

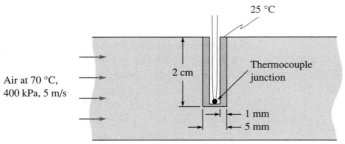

FIGURE P10.22

For Problems 10.23 to 10.30 on flat plate collectors with water as the working fluid, the specification of the collector are

Collector length	2.4 m
Collector width	1.2 m
Water tube inside diameter	1 cm
Pitch of water tubes	20 cm
Orientation of water tubes	Parallel to the 2.4-m dimension
Air space between absorber plate and cover plate	5 cm
Solar transmissivity of glass cover plate	0.85
Infrared emissivity of glass cover plate	0.88
Collector tilt angle with the horizontal	45°
Solar insolation	800 W/m^2

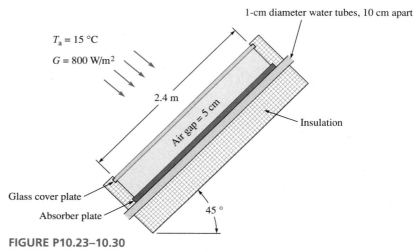

FIGURE P10.23–10.30

10.23 The flat plate collector is located in a place where the temperature of the surrounding air is 15 °C. The convective heat transfer coefficient associated with the surface of the cover plate exposed to the surrounding air is 6 W/m^2 °C. The absorber plate has an emissivity of 0.9 for all wavelengths. Determine the top heat loss coefficient for an absorber plate temperature of 60 °C.

10.24 Re-solve Problem 10.23 for an absorber plate temperature of 60 °C if the spacing between the absorber plate and the cover plate is changed to 3 cm.

10.25 The absorber plate of the collector is lined with a material whose emissivity is approximated as

$0 < \lambda < 2$ μm $\varepsilon_\lambda = 0.9$
2 μm $< \lambda < \infty$ $\varepsilon_\lambda = 0.15$

Compute the infrared emissivity of the surface at 350 K and its solar absorptivity.

10.26 By a special treatment of the surface of the absorber plate, the emissivity of the surface is modified to

$0 < \lambda < 2$ μm $\varepsilon_\lambda = 0.9$
2 μm $< \lambda < 5$ μm $\varepsilon_\lambda = 0.7$
5 μm $< \lambda < \infty$ $\varepsilon_\lambda = 0.15$

Determine the infrared emissivity of the surface at 350 K and its solar absorptivity.

10.27 The collector is operated in a location where the surrounding air temperature and the effective sky temperature are 15 °C. Due to a failure of the circulating water pump, the water in the collector tubes is stagnant. The pressure of the water in the tubes is 100 kPa and the emissivity of the absorber plate is 0.9. The convective heat transfer coefficient associated with the cover plate surface exposed to the atmospheric air is 6 W/m^2 °C. The heat loss through the back insulation is negligible. Determine if boiling of the water in the tubes occurs.

10.28 Re-solve Problem 10.27 if the absorber plate has a solar absorptivity of 0.85, infrared emissivity of 0.1, and the water is replaced by a mixture of water and ethylene glycol having a boiling point of 150 °C.

10.29 The collector with a gray absorber plate ($\varepsilon = 0.9$) is used to heat water. Determine the heat transfer rate to the circulating water if the temperature of the absorber plate is 60 °C and that of the surrounding air is 15 °C. There is a gentle breeze with a velocity of 3 m/s parallel to the 1.2-m dimension.

10.30 Re-solve Problem 10.29 if the absorber plate has a solar absorptivity of 0.9, and an infrared emissivity of 0.1.

10.31 The collector is modified to use air as the working fluid. Air flows with a velocity of 6 m/s parallel to the 2.4-m dimension, in a 2-cm deep channel on the underside of the absorber plate. The channel extends over the entire width of the collector. The inlet temperature of the air is 40 °C. The surroundings are at 15 °C. The convective heat transfer coefficient associated with the surface of the cover plate and the surrounding air is 6 W/m^2 °C. The solar absorptivity of the absorber plate is 0.9 and the infrared emissivity is 0.1. Assuming the absorber plate temperature to be uniform, compute
(a) The temperature of the absorber plate.
(b) The exit temperature of the air.

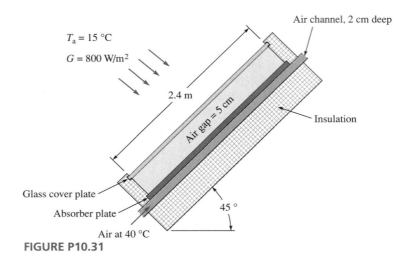

FIGURE P10.31

10.32 Design a handle for a wood stove door so that when the door reaches a temperature of 120 °C, one can comfortably use the handle to operate the door. **Design**

10.33 Design an experimental apparatus to measure the convective heat transfer coefficient in forced convection inside a tube under uniform heat flux conditions. The experimental setup should determine the heat transfer coefficient at several axial locations both for fully developed velocity and temperature profiles and fully developed velocity profile but developing temperature profile. The range of Reynolds numbers should be from 5000 to 100 000 using building air or water supply. The maximum flow rate of available air is 0.08 kg/s. **Design**

10.34 In an experiment setup it is required to condense 0.1 kg/s of trichloro-trifluoro ethane (R-113). The refrigerant enters the condenser as dry, saturated vapor at atmospheric pressure. Cooling water is available at 20 °C. The temperature rise of the cooling water is to be limited to 5 °C. Tubes of 10-, 12-, and 15-mm diameter are available. Total length of the condenser is to be limited to 2 m. Propose a suitable design of the condenser. **Design**

10.35 Containers with a volumetric capacity of 0.8 m³ are to be designed to transport liquid hydrogen. LH_2 is kept at a constant pressure (slightly above atmospheric pressure) by venting the container to release the vapor generated as a result of heat transfer. It is desired to limit the heat transfer rate such that no more than 5% of LH_2 is lost in a day when the container is full. Design a suitable container. **Design**

10.36 Re-solve Problem 10.35 for liquid nitrogen. **Design**

10.37 People like to have large windows so that they can enjoy a good view from inside buildings. It is also believed to provide better environment for work. However, the heat flux across a single-pane glass window is significantly greater than that across a well-insulated wall, leading to an increase in both cooling and heating loads. To reduce the heating and cooling load, it is proposed to use double-glazed windows. The glass panes are 6-mm thick. The glass used is opaque to infrared radiation. Propose a suitable air gap between the glass panes and determine the heat flux with and without the second pane.

To estimate convective heat transfer rates from the inside and outside surfaces, assume an inside air temperature of 20 °C and a convective heat transfer coefficient of 7 W/m² °C and an outside air temperature of -15 °C with a convective heat transfer coefficient of 12 W/m² °C. If the value of

$$\frac{g\beta\rho^2(T_1 - T_2)a^3}{\mu^2}\frac{c_p\mu}{k}\frac{a}{H} \ll 1$$

(T_1 and T_2 are the glass surface temperatures in the cavity; a = gap and H = height) heat transfer rate across the cavity will be essentially by conduction. Include the effect of radiative heat transfer.   **Design**

10.38 An electric heater is to be fabricated by sandwiching a heating element between two plates each 72-cm long and 36-cm wide. The heater is to transfer 4000 W when the plates are maintained at 40 °C above the ambient air temperature. The ambient air is drawn over the heater with a velocity not to exceed 5 m/s. Aluminum and copper rods of 3-mm diameter are available for use as extended surfaces. Rod spacing should not be less than $4d$ and weight and volume should be low. Propose a suitable design.   **Design**

10.39 A package has two sources of energy generation. Each source dissipates 200 W. The temperature of the sources may be assumed to be uniform and is to be limited to 60 °C. This is to be accomplished by installing suitable extended surfaces connecting the two sources and a fan to force the air over them at a free stream velocity of 5 m/s. Design the system of extended surfaces. Weight is an important consideration.   **Design**

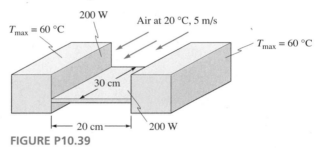

FIGURE P10.39

10.40 Design a space heater with a capacity of 2000 W. The heater is to use electric power available at 110 V. It is for residential use to supplement central heating. Identify the criteria for a successful design (greatest potential for selling) and propose a design.   **Design**

11

CONDUCTION—DIFFERENTIAL FORMULATION

In Chapters 2 and 3 on conduction, the application of differential equations was limited to one-dimensional temperature distribution. Approximate solutions to one-dimensional transient conduction problems by integral methods were presented in Chapter 3. In this chapter we

- *Derive the three-dimensional transient conduction equation in Cartesian coordinates*
- *Present solutions to two-dimensional steady state, and one-dimensional transient conduction problems by two techniques—the separation of variables and Laplace transforms and show how Figures 2.9.2 through 2.9.4 are obtained from one such solution*

The solutions to differential equations for one-dimensional, steady state conduction problems were obtained in Chapter 2. The extension of differential equations for the more general multidimensional, steady and transient problems and their solutions was delayed because the solution to the resulting partial differential equations require mathematical techniques beyond those used in the solution of ordinary differential equations. We begin with the derivation of the three-dimensional, transient conduction equation in Cartesian coordinates. With the assumption of constant properties, the resulting differential equation is linear. There are several methods for solving the linear equation. Two such methods are the separation of variables and Laplace transforms. The application of separation of variables for the solution of two-dimensional steady state, and one-dimensional, transient problems is illustrated through some examples. Laplace transforms are particularly suitable for the solution of transient problems; their use is also illustrated through some examples. However, the two methods are applicable only when the boundary and initial conditions satisfy certain conditions.

The solution to the differential equation is straightforward if the properties are constant and the solid is a rectangular slab, cylinder, or sphere. If the properties are temperature dependent, the differential equation is nonlinear; its solution is involved and, in some cases, there may be no closed-form solutions. In such cases, or if the geometry is irregular, solutions are obtained by numerical methods similar to the method discussed in Chapter 3.

A knowledge of orthogonal functions is necessary for an understanding of the technique of separation of variables. A brief introduction to the Sturm-Liouville equation, the solution of which leads to orthogonal functions, is given in Appendix 3. If you are not familiar with orthogonal functions, it is suggested that you go through Appendix 3.

11.1 DIFFERENTIAL EQUATION FOR THREE-DIMENSIONAL, TRANSIENT CONDUCTION WITH INTERNAL ENERGY GENERATION—CARTESIAN COORDINATES

Consider a rectangular solid (Figure 11.1.1) with different temperatures at the respective boundary surfaces with a volumetric internal energy generation. Because of the different temperatures on the boundaries and the internal energy generation, the temperature within the solid is three-dimensional. To find the differential equation to be satisfied by the temperature within the solid, $T(x, y, z, \tau)$, consider an element of dimensions Δx, Δy, and Δz located at (x, y, z) within the solid.

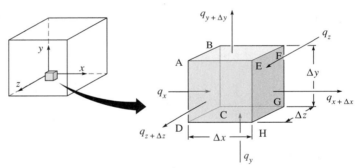

FIGURE 11.1.1 A Cubic Element Within the Solid with Dimensions Δx, Δy, and Δz Parallel to the Three Axes

Three-dimensional transient conduction equation with internal energy generation—Cartesian coordinates

The differential equation is obtained by applying the appropriate balance and particular laws to the element. We begin with the application of the First Law of Thermodynamics to the element whose mass is constant. Although the location of the element is fixed, the dimensions of the element may vary with time. Because of the unsteady temperature in the medium, the density of the medium changes. Although such changes in density for solids are quite small, they are considered in deriving the differential equation.

The First Law of Thermodynamics (balance of energy) applied to the element is

Rate of change of internal energy of the element
= Net rate of heat transfer to the element
− Net rate of work transfer from the element
+ Net rate of internal energy generation in the element

Each term in the energy balance is individually evaluated.

Internal energy of the element $= u \, dm$

where

$u =$ specific internal energy of the element
$dm =$ mass of the element $= \rho \, \Delta x \, \Delta y \, \Delta z$
$\rho =$ local density of the material of the element

Rate of change of internal energy of the element $= \dfrac{\partial}{\partial \tau} (\rho \, u \, \Delta x \, \Delta y \, \Delta z)$

Net rate of heat transfer to the element $= \delta q_x - \delta q_{x+\Delta x}$
$+ \, \delta q_y - \delta q_{y+\Delta y} + \delta q_z - \delta q_{z+\Delta z}$

Each term in the above expression is the heat transfer rate in the direction of a spatial coordinate across the appropriate elemental area. For example, referring to Figure 11.1.1, δq_x represents the heat transfer rate across ABCD and $\delta q_{x+\Delta x}$ the heat transfer rate across EFGH both in the x-direction. Surface EFGH is displaced a distance Δx from ABCD with constant y- and z-coordinates. Similarly, δq_y and $\delta q_{y+\Delta y}$ represent the heat transfer rates across DCGH and ABFE in the y-direction. Each term is now expressed in terms of the heat flux across the surface.[1]

$$\delta q_x = q''_x \, \Delta y \, \Delta z \qquad \delta q_{x+\Delta x} = q''_{x+\Delta x} \, \Delta y \, \Delta z$$

$$\delta q_y = q''_y \, \Delta x \, \Delta z \qquad \delta q_{y+\Delta y} = q''_{y+\Delta y} \, \Delta x \, \Delta z$$

$$\delta q_z = q''_z \, \Delta x \, \Delta y \qquad \delta q_{z+\Delta z} = q''_{z+\Delta z} \, \Delta x \, \Delta y$$

Combining the heat transfer terms across the six surfaces of the element,

Net heat transfer rate $= (q''_x - q''_{x+\Delta x}) \, \Delta y \, \Delta z$
$+ \, (q''_y - q''_{y+\Delta y}) \, \Delta x \, \Delta z + (q''_z - q''_{z+\Delta z}) \, \Delta x \, \Delta y$

Net rate of internal energy generation $= q''' \, \Delta x \, \Delta y \, \Delta z$

Rate of work transfer (boundary movement) from the element $= p \, \dfrac{\partial(\Delta x \, \Delta y \, \Delta z)}{\partial \tau}$

In most cases, the changes in pressure are negligible and, in such cases, with the assumption of constant pressure,

Rate of work transfer from the element $= \dfrac{\partial(p \, \Delta x \, \Delta y \, \Delta z)}{\partial \tau}$

Combining all the terms in the energy equation and transferring the work transfer term to the left side,

$$\frac{\partial}{\partial \tau} (\rho u \, \Delta x \, \Delta y \, \Delta z) + \frac{\partial}{\partial \tau} (p \, \Delta x \, \Delta y \, \Delta z) = (q''_x - q''_{x+\Delta x}) \, \Delta y \, \Delta z$$
$$+ (q''_y - q''_{y+\Delta y}) \, \Delta x \, \Delta z + (q''_z - q''_{z+\Delta z}) \, \Delta x \, \Delta y + q''' \, \Delta x \, \Delta y \, \Delta z$$

The two terms on the left side of the equation are rearranged as

---

[1]Strictly speaking, the heat flux is a vector, and its components should be specified by the direction and location. For example, $q''_{x+\Delta x}$ should be denoted by $q''_{x,x+\Delta x}$. This becomes cumbersome and, as there is no confusion in this case, it is denoted simply by $q''_{x+\Delta x}$. Note that at any given location there are three components of the heat flux vector: q''_x, q''_y, and q''_z.

$$\frac{\partial}{\partial \tau}\left[\rho\left(u + \frac{p}{\rho}\right)\Delta x\,\Delta y\,\Delta z\right] = \left(u + \frac{p}{\rho}\right)\frac{\partial}{\partial \tau}(\rho\,\Delta x\,\Delta y\,\Delta z)$$

$$+ \rho\,\Delta x\,\Delta y\,\Delta z\,\frac{\partial}{\partial \tau}\left(u + \frac{p}{\rho}\right)$$

But by mass conservation,

$$\frac{\partial}{\partial \tau}(\rho\,\Delta x\,\Delta y\,\Delta z) = 0$$

Thus, with $u + p/\rho = i$ (specific enthalpy), the energy equation is recast as

$$\rho\,\Delta x\,\Delta y\,\Delta z\,\frac{\partial i}{\partial \tau} = (q_x'' - q_{x+\Delta x}'')\,\Delta y\,\Delta z + (q_y'' - q_{y+\Delta y}'')\,\Delta x\,\Delta z$$

$$+ (q_z'' - q_{z+\Delta z}'')\,\Delta x\,\Delta y + q'''\,\Delta x\,\Delta y\,\Delta z$$

Dividing by $\Delta x\,\Delta y\,\Delta z$ and taking the limit as $\Delta x \to 0$, $\Delta y \to 0$ and $\Delta z \to 0$, and noting that

$$\operatorname*{Lim}_{\Delta x \to 0}\frac{q_x'' - q_{x+\Delta x}''}{\Delta x} = -\frac{\partial q_x''}{\partial x} = \frac{\partial}{\partial x}\left(k\frac{\partial T}{\partial x}\right)$$

we get

$$\rho\frac{\partial i}{\partial \tau} = \frac{\partial}{\partial x}\left(k\frac{\partial T}{\partial x}\right) + \frac{\partial}{\partial y}\left(k\frac{\partial T}{\partial y}\right) + \frac{\partial}{\partial z}\left(k\frac{\partial T}{\partial z}\right) + q'''$$

Now considering $i = i(p, T)$,

$$di = \left.\frac{\partial i}{\partial T}\right|_p dT + \left.\frac{\partial i}{\partial p}\right|_T dp$$

Therefore, at constant pressure, with $\partial i/\partial T|_p = c_p$, $di = c_p\,dT$, and $\partial i/\partial \tau = c_p(\partial T/\partial \tau)$, the final form of the energy equation is given by

$$\left\|\quad \rho\,c_p\frac{\partial T}{\partial \tau} = \frac{\partial}{\partial x}\left(k\frac{\partial T}{\partial x}\right) + \frac{\partial}{\partial y}\left(k\frac{\partial T}{\partial y}\right) + \frac{\partial}{\partial z}\left(k\frac{\partial T}{\partial z}\right) + q''' \quad\right\| \quad \textbf{(11.1.1)}$$

If the density of a solid is constant, $c_p = c_v$. For most solids the density is nearly constant and for such solids $c_p \approx c_v$. Because of the very small differences between c_p and c_v, only one value of the specific heat (c_p) of solids is given in the table of properties.

For a complete solution of the differential equation we need

1. One initial condition—the temperature distribution at some instant of time
2. Two boundary conditions in each of the three spatial directions. For a discussion of different boundary conditions, refer to Section 2.7.

If the thermal conductivity, specific heat, or internal energy generation is temperature dependent, the equation is nonlinear. If they are space-dependent only, the equation is linear with variable coefficients; if the properties are constant, the equation is linear with constant coefficients. We consider only cases with constant properties. With constant properties, dividing Equation 11.1.1 by k, and replacing $k/\rho c_p$ by α, the thermal diffusivity of the material, we obtain

$$\left\| \quad \frac{1}{\alpha} \frac{\partial T}{\partial t} = \frac{\partial^2 T}{\partial x^2} + \frac{\partial^2 T}{\partial y^2} + \frac{\partial^2 T}{\partial z^2} + \frac{q'''}{k} \quad \right\| \qquad \textbf{(11.1.2)}$$

Transient three-dimensional conduction equations in cylindrical and spherical coordinates are derived in a similar manner. The equation in cylindrical coordinates without internal energy generation can be obtained from Equation A5.10 and the equation in spherical coordinates from Equation A5.15 (in Appendix 5) by setting the velocities and the shear stresses equal to zero in each case.

11.2 STEADY, TWO-DIMENSIONAL TEMPERATURE DISTRIBUTION

We begin with exact solutions to steady, two-dimensional conduction problems by the technique of separation of variables. The technique of separation of variables can be applied if the following conditions are satisfied:

Solution by separation of variables, steady state—conditions to be satisfied

1. The differential equation and the boundary conditions are linear. To test for linearity, express each term of the equation as a product of the highest derivative in the term and the remaining parts of the term [$= c(\partial^n y/\partial x^n)$]. If any coefficient c is a function of the dependent variable (either explicitly or implicitly by containing a derivative of the variable), the equation is nonlinear. If one, or more, coefficient is a function of the independent variable only and none is a function of the dependent variable, the equation is linear with variable coefficients. If all the coefficients are constant the equation is linear with constant coefficients. For example, consider the terms $(\partial/\partial x)[k(\partial T/\partial x)]$:

$$\frac{\partial}{\partial x}\left(k \frac{\partial T}{\partial x} \right) = \frac{\partial k}{\partial x} \frac{\partial T}{\partial x} + k \frac{\partial^2 T}{\partial x^2}$$

If the material is homogeneous, $\partial k/\partial x = 0$. If the thermal conductivity is a function of temperature, the coefficient of $\partial^2 T/\partial x^2$ is a function of the dependent variable T and the resulting equation is nonlinear.

2. There is only one nonhomogeneity either in the differential equation or in the boundary conditions. If either the differential equation or a boundary condition contains a term that is either a function of the independent variable only or a constant, the equation is nonhomogeneous.

3. The solid is a rectangular slab, cylinder, or sphere.

(For a brief explanation of the limitations of separation of variables technique, see the additional note at the end of the solution of Example 11.3.2.)

If the system of equation and boundary conditions has more than one nonhomogeneity, the problem is split up into a number of simpler problems, each of which has only one nonhomogeneity. The solutions to these problems are then combined in a suitable manner by the principle of superposition. For a discussion of the method by the principle of superposition, consult one of the books given in the references at the end of the chapter. Here, we deal only with those problems that lead to a linear differential equation with constant coefficients satisfying the three conditions given above.

The solution is assumed to be a combination of several functions, each being a function of only one independent variable. The partial differential equation is then reduced to a system of ordinary differential equations, each containing only one of the functions. The boundary conditions for each differential equation are also functions of only one independent variable. Usually, in the case of two-dimensional, steady state problems, the dependent variable is expressed as the product of two functions, each being a function of only one independent variable. However, in some cases, it may be necessary to express the original function as the product of two functions and the sum of one or more functions; each is a function of only one independent variable.

Some Frequently Encountered Integrals—Separation of Variables

$$\int \sin^2 ax \; dx = \frac{x}{2} - \frac{\sin 2ax}{4a} = \frac{1}{2a} (xa - \sin ax \cos ax)$$

$$\int \cos^2 ax \; dx = \frac{x}{2} + \frac{\sin 2ax}{4a} = \frac{1}{2a} (xa + \sin ax \cos ax)$$

$$\int x \sin ax \; dx = \frac{\sin ax}{a^2} - \frac{x \cos ax}{a}$$

$$\int x \cos ax \; dx = \frac{\cos ax}{a^2} + \frac{x \sin ax}{a}$$

$$\int x^2 \sin ax \; dx = \frac{2x}{a^2} \sin ax + \left(\frac{2}{a^3} - \frac{x^2}{a} \right) \cos ax$$

$$\int x^2 \cos ax \; dx = \frac{2x}{a^2} \cos ax + \left(\frac{x^2}{a} - \frac{2}{a^3} \right) \sin ax$$

$$\int \sin ax \sin bx \; dx = \frac{\sin(a - b)x}{2(a - b)} - \frac{\sin(a + b)x}{2(a + b)} \qquad a \neq b$$

$$\int \sin ax \cos bx \; dx = - \frac{\cos(a - b)x}{2(a - b)} - \frac{\cos(a + b)x}{2(a + b)} \qquad a \neq b$$

$$\int \cos ax \cos bx \; dx = \frac{\sin(a - b)x}{2(a - b)} + \frac{\sin(a + b)x}{2(a + b)} \qquad a \neq b$$

EXAMPLE 11.2.1 Four surfaces of a rectangular slab are maintained at uniform but different temperatures as shown in Figure 11.2.1a. Three surfaces are maintained at T_o and the fourth surface is at a different temperature T_1. The surfaces perpendicular to the z-axis are perfectly insulated. Determine the temperature distribution in the slab.

ASSUMPTIONS

1. Steady state
2. Constant properties
3. No internal energy generation
4. As the temperatures of each surface are uniform and there is no temperature variation in the z-direction, the temperature distribution is two-dimensional, i.e., $T = T(x, y)$.

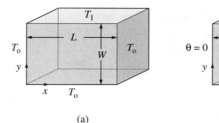

(a)   (b)

FIGURE 11.2.1 A Rectangular Solid with Two-Dimensional Temperature Distribution
(a) Boundary temperatures have finite, non-zero values.
(b) With a change of variable, only one boundary temperature has a non-zero value.

SOLUTION

For the steady, two-dimensional temperature distribution without internal energy generation, Equation 11.1.2 simplifies to

$$\frac{\partial^2 T}{\partial x^2} + \frac{\partial^2 T}{\partial y^2} = 0 \tag{1}$$

$$T(0, y) = T_o \tag{2a}$$

$$T(L, y) = T_o \tag{2b}$$

$$T(x, 0) = T_o \tag{2c}$$

$$T(x, W) = T_1 \tag{2d}$$

Equation 1 is a linear, homogeneous differential equation with constant coefficients, and the four boundary conditions represented by Equation 2 are all nonhomogeneous. To obtain a solution by the separation of variables technique, there should be only one nonhomogeneous boundary condition. In this problem, three of the nonhomogeneities in the boundary conditions can be removed by a change in the dependent variable. Define $\theta = T - T_o$. With this new variable, Equation 1 and the associated boundary conditions are

$$\frac{\partial^2 \theta}{\partial x^2} + \frac{\partial^2 \theta}{\partial y^2} = 0 \tag{3}$$

$$\theta(0, y) = 0 \tag{4a}$$

$$\theta(L, y) = 0 \tag{4b}$$

$$\theta(x, 0) = 0 \tag{4c}$$

$$\theta(x, W) = \theta_1 = T_1 - T_o \tag{4d}$$

Now there is only one nonhomogeneous boundary condition, Equation 4d. As the conditions for the application of the separation of variables technique have been satisfied, we proceed to find the solution by that method. The method consists of assuming that the dependent variable can be expressed as a product of two functions, each function being a function of only one independent variable. You may question the reason for such an assumption. One answer is that, if such a function satisfies the differential equation and the associated boundary conditions, we expect it to be the solution of the equation.

Now, assume $\theta(x, y) = f_1(x)f_2(y)$.

$$\frac{\partial\theta}{\partial x} = f_2 f_1' \qquad \frac{\partial^2\theta}{\partial x^2} = f_2 f_1''$$

$$\frac{\partial\theta}{\partial y} = f_1 f_2' \qquad \frac{\partial^2\theta}{\partial y^2} = f_1 f_2''$$

The primes denote differentiation with respect to the variable of the function, $f_1' = df_1/dx$, and so forth. Substituting the derivatives in terms of the functions into Equation 3, we get

$$f_2 f_1'' + f_1 f_2'' = 0$$

Dividing by $f_1 f_2$ and rearranging,

$$\frac{f_1''}{f_1} = -\frac{f_2''}{f_2} \tag{5}$$

In terms of the functions f_1 and f_2, the boundary condition Equation 4a becomes

$$f_1(0)f_2(y) = 0$$

If $f_2(y) = 0$, then $\theta = 0$ everywhere, and this is not an acceptable solution as boundary condition 4d cannot be satisfied. Thus, we require that $f_2(y) \neq 0$, except possibly, at some specified location. This means that the boundary condition can be satisfied only if $f_1(0) = 0$. By a similar logic we obtain

$$f_1(0) = 0 \tag{6a}$$

$$f_1(L) = 0 \tag{6b}$$

$$f_2(0) = 0 \tag{6c}$$

$$\theta(x, W) = \theta_1 \tag{6d}$$

Note that boundary condition 4d can be satisfied only by θ, i.e., by $f_1 f_2$.

In Equation 5, the left side of the equation is, at most, a function of x and the right side, at most, a function of y. Equation 5 should be satisfied for any value of $0 \le x \le L$ and $0 \le y \le W$. To satisfy the equality for any arbitrary values of x and y, we require the left side and the right side of Equation 5 to be equal to a constant. Thus, we require that

$$\frac{f_1''}{f_1} = -\frac{f_2''}{f_2} = -\lambda^2 \tag{7}$$

λ^2 is a real, positive constant yet to be determined. If, instead of taking the constant as $(-\lambda^2)$, we choose the constant to be λ^2, the boundary conditions cannot be satisfied (try it). The square of a real number is taken for convenience as will be evident in the solution. *Values of λ are known as the characteristic values or eigenvalues.* Equation 7 is written as

$$f_1'' + \lambda^2 f_1 = 0 \tag{8}$$

$$f_1(0) = 0 \tag{8a}$$

$$f_1(L) = 0 \tag{8b}$$

Solution to Equation 8 is

$$f_1 = a_1 \sin \lambda x + a_2 \cos \lambda x$$

From Equation 8a, $a_2 = 0$. From Equation 8b, $0 = a_1 \sin \lambda L$. If $a_1 = 0$, $\theta = 0$ everywhere, which is unacceptable. Hence, we require that $\sin \lambda L = 0$. This condition is satisfied if the constant λ is such that

$$\lambda L = \pi, \, 2\pi, \, 3\pi, \, \cdots, \, n\pi, \, \cdots$$

In Equation 7, if $\lambda = 0$, $f_1'' = 0$ and, therefore, $f_1 = a + bx$, where a and b are constants. Boundary condition 8a requires that $a = 0$. Boundary condition 8b requires that $bL = 0$ or $b = 0$. Hence, $\lambda = 0$ is not valid for this case.

As λ can take any value $n\pi/L$, $n = 1, 2, 3, \cdots, \infty$, we have

$$\lambda_n = \frac{n\pi}{L} \qquad n = 1, 2, 3, \cdots, \infty$$

(Why are negative values of n not included in the sequence?)

Thus, Equation 8 is satisfied for an infinite series of values of λ. For any one such value of $\lambda_m = m\pi/L$, where m is any positive integer, the function f_2 should satisfy

$$f_{2m}'' - \lambda_m^2 f_{2m} = 0 \tag{9}$$

Solution to Equation 9 is

$$f_{2m} = b_m \sinh \lambda_m y + c_m \cosh \lambda_m y$$

Boundary condition Equation 6c requires that $c_m = 0$. Thus, for any particular value of $\lambda_m (= m\pi/L)$ the corresponding solution is

$$f_{1m} f_{2m} = a_m \sin(\lambda_m x) b_m \sinh(\lambda_m y)$$

Boundary condition 6d is satisfied only by the general solution. The general solution is given by a linear combination of all solutions so that

$$\theta = \sum_{m=1}^{\infty} d_m \sin \lambda_m x \sinh \lambda_m y \qquad \lambda_m = \frac{m\pi}{L}$$

where $a_m b_m$ has been replaced by a single constant d_m. The constants d_m in the series are evaluated by using the last boundary condition, Equation 6d.

$$\theta(x, W) = \theta_1 = \sum_{m=1}^{\infty} d_m \sin \lambda_m x \sinh \lambda_m W \tag{10}$$

To evaluate the constant d_n, n representing one of the valid values for m, in Equation 10, multiply both sides of the equation by $\sin \lambda_n x$ and integrate between $x = 0$ and $x = L$.

$$\sum_{m=1}^{\infty} \int_0^L d_m \sin \lambda_n x \sin \lambda_m x \sinh \lambda_m W \, dx = \int_0^L \theta_1 \sin \lambda_n x \, dx \tag{11}$$

The left side of Equation 11 represents the right side of Equation 10 after every term of that equation has been multiplied by $\sin(\lambda_n x)$ and integrated between 0 and L. But

Equation 8 is a special case of the Sturm-Liouvelle equation (see Appendix 3) and, hence, every term on the left side of Equation 11 for which $m \neq n$ vanishes, and the only nonzero term is that for which $m = n$. Hence,

$$d_n = \frac{1}{\sinh \lambda_n W} \frac{\int_0^L \theta_1 \sin \lambda_n x \, dx}{\int_0^L \sin^2 \lambda_n x \, dx} \qquad \lambda_n = \frac{n\pi}{L}$$

$$\int_0^L \sin \lambda_n x \, dx = -\frac{1}{\lambda_n} \cos \lambda_n x \Big|_0^L = -\frac{L}{n\pi} (\cos n\pi - 1) = -\frac{L}{n\pi} [(-1)^n - 1]$$

$$\int_0^L \sin^2 \lambda_n x \, dx = \left[\frac{x}{2} - \frac{\sin(2\lambda_n x)}{4\lambda_n} \right] \Big|_0^L = \frac{L}{2}$$

In evaluating the integrals, $\lambda_n L$ has been set equal to $n\pi$. The constant d_n is given by

$$d_n = \frac{2}{n\pi} [1 - (-1)^n] \frac{\theta_1}{\sinh \lambda_n W}$$

where $d_n = 0$ for even values of n and $d_n = 4\theta_1/(n\pi \sinh \lambda_n W)$ for odd values of n. This can be expressed as

$$d_{2n+1} = \frac{4\theta_1}{(2n+1)\pi \sinh[(2n+1)\pi(W/L)]} \qquad n = 0, 1, 2, \cdots, \infty$$

The final solution is

$$\theta = T - T_o = \frac{4\theta_1}{\pi} \sum_{n=0}^{\infty} \frac{1}{2n+1} \frac{\sin[(2n+1)\pi(x/L)] \sinh[(2n+1)\pi(y/L)]}{\sinh[(2n+1)\pi(W/L)]} \tag{12}$$

Equation 12 is a convergent series and gives the temperature distribution at any location (x, y). A few isotherms—lines of constant temperature—are shown in Figure 11.2.2.

Having determined the temperature distribution, let us try to obtain the heat transfer rate from Fourier's Law. For example, the heat transfer rate across surface at $x = L$ in the x-direction, per unit length in the z-direction is given by

$$q_x(x = L)$$

$$= \int q_x'' \, dA_x = \int_0^W \left(-k \frac{\partial \theta}{\partial x} \Big|_{x=L} \right) dy$$

$$= -k \int_0^W \frac{\partial}{\partial x} \left[\frac{4\theta_1}{\pi} \sum_{n=0}^{\infty} \frac{1}{2n+1} \frac{\sin[(2n+1)\pi(x/L)] \sinh[(2n+1)\pi y/L]}{\sinh[(2n+1)\pi(W/L)]} \right]_{x=L} dy$$

$$= -\frac{4k\theta_1}{L} \sum_{n=0}^{\infty} \int_0^W \frac{\cos[(2n+1)\pi] \sinh[(2n+1)\pi(y/L)]}{\sinh[(2n+1)\pi(W/L)]} dy$$

$$= \frac{4k\theta_1}{\pi} \sum_{n=0}^{\infty} \frac{1}{(2n+1)} \frac{\cosh[(2n+1)\pi(W/L)] - 1}{\sinh[(2n+1)\pi(W/L)]} \tag{13}$$

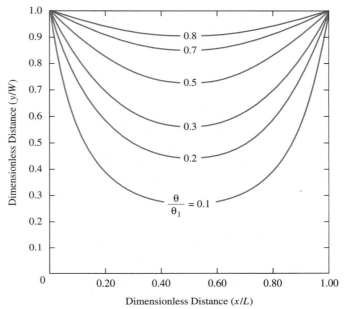

FIGURE 11.2.2 **Isotherms from Equation 12**

An examination of Equation 13 shows that as $n \to \infty$,

$$\frac{\cosh[(2n + 1)\pi(W/L)] - 1}{\sinh[(2n + 1)\pi(W/L)]} \to 1$$

and the series $\sum_{1}^{\infty} 1/(2n + 1)$ is not convergent. Hence, Equation 13 is not convergent and cannot be used to find the heat transfer rate. It may be pointed out that while the Fourier series solution may be convergent, term by term differentiation of the series may result in a divergent series. Integration of the series term by term results in a more rapidly converging series.

EXAMPLE 11.2.2 The heat flux to the base of a rectangular fin attached to a plate (Figure 11.2.3) is given by

$$q_b'' = q_o''(1 + a \cos by)$$

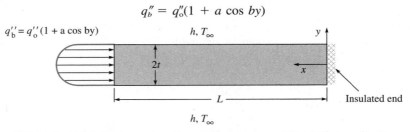

FIGURE 11.2.3 **A Rectangular Fin with an Imposed Heat Flux at its Base**

The temperature of the fluid surrounding the fin is T_∞. The convective heat transfer coefficient associated with the fin surface is h. The thickness of the fin is $2t$ and its length is L. The width of the fin is W, which is much greater than t. Determine the temperature distribution in the fin.

ASSUMPTIONS

1. Steady state
2. As $W \gg t$, end effects in the z-direction are negligible. The temperature distribution is two-dimensional, $T = T(x, y)$.
3. Constant properties
4. Uniform convective heat transfer coefficient
5. The heat transfer rate from the tip is negligible, i.e., it can be treated as insulated [the concept of corrected length (Section 2.6.4) has been used].

SOLUTION

The differential equation for the steady, two-dimensional temperature distribution in the fin and the associated boundary conditions are

$$\frac{\partial^2 T}{\partial x^2} + \frac{\partial^2 T}{\partial y^2} = 0 \tag{1}$$

$$x = 0, y, \qquad \frac{\partial T}{\partial x} = 0 \tag{2a}$$

$$x = L, y, \qquad k\frac{\partial T}{\partial x} = q_o''(1 + a \cos by) \tag{2b}$$

$$x, y = t, \qquad k\frac{\partial T}{\partial y} + h(T - T_\infty) = 0 \tag{2c}$$

$$x, y = -t, \qquad k\frac{\partial T}{\partial y} - h(T - T_\infty) = 0 \tag{2d}$$

By the symmetry of the differential equation and the boundary conditions in the y-direction, one of the boundary conditions in the y-direction can be replaced by $\partial T/\partial y(x, 0) = 0$. We replace Equation 2d by $\partial T/\partial y(x, 0) = 0$ in the following development.

Equation 1 is linear and homogeneous. There are now two linear, nonhomogeneous boundary conditions, Equations 2b and c [Equation 2d is replaced by $\partial T/\partial x(x, 0) = 0$]. Recall that solution by separation of variables cannot be obtained with more than one nonhomogeneity in the system of equations and boundary conditions. Equation 2b cannot be made nonhomogeneous. Equation 2c can be made homogeneous by a change in variable. Define $\theta = T - T_\infty$. The differential equation and the boundary conditions for θ are

$$\frac{\partial^2 \theta}{\partial x^2} + \frac{\partial^2 \theta}{\partial y^2} = 0 \tag{3}$$

$$x = 0, y, \qquad \frac{\partial \theta}{\partial x} = 0 \tag{4a}$$

$$x = L, y, \qquad k\frac{\partial\theta}{\partial x} = q_o''(1 + a \cos by) \qquad \text{(4b)}$$

$$x, y = t, \qquad k\frac{\partial\theta}{\partial y} + h\theta = 0 \qquad \text{(4c)}$$

$$x, y = 0, \qquad \frac{\partial\theta}{\partial y} = 0 \qquad \text{(4d)}$$

To solve for $\theta(x, y)$, assume

$$\theta(x, y) = f_1(x)f_2(y) \qquad \text{(5)}$$

Substituting Equation 5 into Equation 3, dividing the resulting equation by $f_1(x)f_2(y)$, and rearranging,

$$\frac{f_1''}{f_1} = -\frac{f_2''}{f_2} \qquad \text{(6)}$$

The boundary conditions are

$$x = 0, \qquad f_1' = 0 \qquad \text{(7a)}$$

$$x = L, y, \qquad k\frac{\partial\theta}{\partial x} = q_o''(1 + a \cos by) \qquad \text{(7b)}$$

$$y = t, \qquad kf_2' + hf_2 = 0 \qquad \text{(7c)}$$

$$y = 0, \qquad f_2' = 0 \qquad \text{(7d)}$$

In Equation 6, the left side is independent of y, and the right side is independent of x. As the two sides are equal to each other for all permissible values of x and y, we require that

$$\frac{f_1''}{f_1} = -\frac{f_2''}{f_2} = \lambda^2 \qquad \text{(8)}$$

Equation 8 is now written as two equations.

$$f_1'' - \lambda^2 f_1 = 0 \qquad \text{(9a)}$$

$$f_2'' + \lambda^2 f_2 = 0 \qquad \text{(9b)}$$

$$x = 0, \qquad f_1' = 0 \qquad \text{(10a)}$$

$$x = L, \qquad k\frac{\partial\theta}{\partial x} = q_o''(1 + a \cos by) \qquad \text{(10b)}$$

$$y = t, \qquad kf_2' + hf_2 = 0 \qquad \text{(10c)}$$

$$y = 0, \qquad f_2' = 0 \qquad \text{(10d)}$$

As the boundary conditions are homogeneous in the y-direction, the eigenvalues are obtained by solving the differential equation in that direction. Solution to Equation 9b is

$$f_2 = c_1 \cos \lambda y + c_2 \sin \lambda y$$

From boundary condition Equation 10d, $c_2 = 0$. From boundary condition, Equation 10c,

$$-k\lambda \sin \lambda t + h \cos \lambda t = 0$$

or

$$\lambda t \tan \lambda t = \frac{ht}{k} = \text{Bi} \tag{11}$$

There are an infinite number of roots for Equation 11. The first six roots for different values of the Biot number, Bi, are given in Appendix 2.

For any particular value of $\lambda = \lambda_m$, the solution to Equation 9a is

$$f_{1m} = d_{1m} \cosh \lambda_m x + d_{2m} \sinh \lambda_m x$$

From boundary condition Equation 10a, $d_{2m} = 0$. The last boundary condition, Equation 10b, can be satisfied only by the complete solution for θ. The complete solution is given by the linear combination of all individual solutions.

$$\theta = \sum_{m=1}^{\infty} e_m \cosh \lambda_m x \cos \lambda_m y \tag{12}$$

In Equation 12, the product, $c_{1m}d_{1m}$ has been replaced by e_m. To find e_m, we make use of the boundary condition Equation 10b.

$$\sum_{m=1}^{\infty} e_m \lambda_m \sinh \lambda_m L \cos \lambda_m y = \frac{q_o''}{k}(1 + a \cos by) \tag{13}$$

To determine e_n, (for a particular value of $m = n$), multiply both sides of Equation 13 by $\cos \lambda_n y$, and integrate with respect to y between the limits, $y = 0$, and $y = t$. Equation 9b is a special case of Equation A3.1 (Appendix 3). Hence, its solutions are orthogonal. All integrals on the left side for which $m \neq n$ vanish, and only the integral for which $m = n$ has a nonzero value. Thus,

$$e_n = \frac{\displaystyle\int_0^t (q_o''/k)(1 + a \cos by) \cos \lambda_n y \, dy}{\lambda_n \displaystyle\int_0^t \sinh \lambda_n L \cos^2 \lambda_n y \, dy} \tag{14}$$

$$\int_0^t \cos \lambda_n y \, dy = \frac{\sin \lambda_n t}{\lambda_n}$$

$$\int_0^t \cos by \cos \lambda_n y \, dy = \frac{\sin(b + \lambda_n)t}{2(b + \lambda_n)} + \frac{\sin(b - \lambda_n)t}{2(b - \lambda_n)}$$

$$\int_0^t \cos^2 \lambda_n \, y \, dy = \frac{1}{2\lambda_n}(\lambda_n t + \sin \lambda_n t \cos \lambda_n t)$$

$$e_n = \frac{q_o''}{k} \frac{[(2 \sin \lambda_n t)/\lambda_n] + \{[a \sin(b + \lambda_n)t]/(b + \lambda_n)\} + \{[a \sin(b - \lambda_n)t]/(b - \lambda_n)\}}{\sinh \lambda_n L(\lambda_n t + \sin \lambda_n t \cos \lambda_n t)}$$

$$\tag{15}$$

The temperature distribution is

$$\theta = \sum_{n=1}^{\infty} e_n \cosh \lambda_n x \cos \lambda_n y$$

with e_n given by Equation 15

COMMENT

The heat flux at the base is not likely to be a cosine function. Any profile of the heat flux can be expressed as a Fourier series consisting of an infinite number of terms of sines and cosines. In many cases a good approximation of the actual profile can be obtained from a limited number of the terms in the series. The solution to the heat flux being equal to any particular term can be found by the method indicated in the above problem. By suitably combining the solutions of the temperature profile with the heat flux being equal to each term in the series, the temperature profile satisfying the actual heat flux profile is obtained.

In Examples 11.2.1 and 11.2.2, the two-dimensional temperature distribution in a rectangular slab is determined. In a similar manner, the two-dimensional temperature distribution in a cylinder, where the temperature is a function of the radial and axial coordinates, may be found by applying the separation of variables technique. However, the solution in the r-direction involves Bessel functions instead of the trigonometric function in Cartesian geometry. Bessel functions and their use in series solutions are not considered in this book. The separation of variables technique can also be extended to the determination of three-dimensional temperature distribution.

11.3 TRANSIENT, ONE-DIMENSIONAL TEMPERATURE DISTRIBUTION

Separation of variables and Laplace transforms— transient temperature distribution

Separation of variables technique can also be employed to obtain solutions to transient conduction problems in finite geometry. An alternate method is to use Laplace transforms. Laplace transforms are particularly well suited for solving one-dimensional transient problems. Here we demonstrate the use of Laplace transforms to two cases where the inverse transform can be found from a table of transforms. Solutions to one-dimensional, transient conduction problems by the application of the two techniques—separation of variables and Laplace transforms—are illustrated through examples. Some of the properties of Laplace transforms and a table of transforms that are useful in the solution of conduction problems are given in Appendix 4.

EXAMPLE 11.3.1   A steel slab of thickness $2L$ is heated to a uniform temperature T_i in a furnace. It is then quenched in a stream of water at T_w (Figure 11.3.1). If the Biot number hL/k is large, it signifies that the convective resistance is much smaller than the conductive resistance and that the difference between the temperatures of the surface and the fluid is very small. Thus, with the convective heat transfer coefficient with a stream

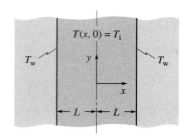

FIGURE 11.3.1 A slab initially at a uniform temperature is cooled by a stream of water, which maintains the exposed surfaces at a constant temperature.

of water being large, we expect that the Biot number is large and that the surface temperature is approximately equal to the water temperature. With the surface temperature at T_w for $\tau > 0$, find an expression for the one-dimensional, transient temperature distribution in the slab.

ASSUMPTIONS

1. The dimensions of the slab in the y- and z-directions are much larger than $2L$ and the temperature distribution can be approximated as one-dimensional in the x-direction, i.e., $T = T(x, \tau)$. End effects in the y- and z-directions are neglected.
2. Constant properties

SOLUTION

Because of the symmetry, the origin is taken at the midplane. With the assumption of constant properties, Equation 11.1.2, specialized to one-dimensional, transient temperature distribution without internal energy generation, simplifies to

$$\frac{1}{\alpha} \frac{\partial T}{\partial \tau} = \frac{\partial^2 T}{\partial x^2} \tag{1}$$

The initial and boundary conditions are

$$T(x, 0) = T_i \tag{2a}$$

$$T(L, \tau) = T_w \tag{2b}$$

$$T(-L, \tau) = T_w \tag{2c}$$

To solve Equation 1 by the method of separation of variables, the boundary conditions in Equation 2 are made homogeneous by a change of variable. Let $\theta = T - T_w$. Then,

$$\frac{1}{\alpha} \frac{\partial \theta}{\partial \tau} = \frac{\partial^2 \theta}{\partial x^2} \tag{3}$$

$$\theta(x, 0) = \theta_i = T_i - T_w \tag{4a}$$

$$\theta(L, \tau) = 0 \tag{4b}$$

$$\theta(-L, \tau) = 0 \tag{4c}$$

In the spirit of the separation of variables technique, it is assumed that the solution for θ is a product of one function of time and another function of space.

$$\theta = f_1(\tau) f_2(x)$$

Substituting the product solution in Equation 3, and dividing it by $f_1 f_2$, we obtain

$$\frac{1}{\alpha} \frac{f_1'}{f_1} = \frac{f_2''}{f_2} = -\lambda^2 \tag{5}$$

The primes indicate differentiation with respect to the appropriate variables. In Equation 5, $(1/\alpha)(f_1'/f_1)$ is a function of time only and f_2''/f_2 is a function of x only. They can be equal to each other for arbitrary values of x and τ only when each is equal to a constant. Each of them is set equal to a constant, $-\lambda^2$, the separation constant. The reason for making the constant negative is that, if it were positive, it would predict the temperature to increase indefinitely with time, which is contrary to the physics of the problem. Equation 5 can be split into two differential equations.

$$f_1' + \alpha\lambda^2 f_1 = 0 \tag{6a}$$

$$f_2'' + \lambda^2 f_2 = 0 \tag{6b}$$

From the initial and boundary conditions 4a, b, and c,

$$\theta(x, 0) = \theta_i \tag{7a}$$

$$f_2(L) = 0 \tag{7b}$$

$$f_2(-L) = 0 \tag{7c}$$

Solution to Equation 6b is

$$f_2 = a \cos \lambda x + b \sin \lambda x \tag{8}$$

From boundary conditions 7b and c,

$$0 = a \cos \lambda L + b \sin \lambda L \tag{9}$$

$$0 = a \cos \lambda L - b \sin \lambda L \tag{10}$$

From Equations 9 and 10, $b = 0$ and, as $a \neq 0$, for a meaningful solution[2]

$$\cos \lambda L = 0$$

or

$$\lambda L = \pi/2,\ 3\pi/2,\ \cdots$$

The eigenvalues are given by

$$\lambda_m = (2m + 1) \frac{\pi}{2L} \qquad m = 0, 1, 2, 3, \cdots, \infty$$

Corresponding to any specific value of m, from Equation 6a,

$$\frac{f_{1m}'}{f_{1m}} = -\alpha\lambda_m^2$$

whose solution is

$$f_{1m} = c_m \exp(-\alpha\lambda_m^2 \tau)$$

---

[2]To satisfy Equations 9 and 10, the constant a can be set equal to zero. In that case $\partial\theta/\partial x\ (x = 0) = b \cos(\lambda x)_{x=0}$, which is not equal to zero. But by symmetry of the equation and the boundary conditions, we require that $\partial\theta/\partial x(x = 0) = 0$. Hence, $b = 0$ and $a \neq 0$.

To satisfy initial condition Equation 7a, we take the most general solution, a linear combination of all possible solutions.

$$\theta(x, \tau) = \sum_{m=0}^{\infty} d_m \exp(-\alpha\lambda_m^2\tau) \cos \lambda_m x \tag{11}$$

where $d_m = a_m c_m$. From the initial condition, 7a, we have

$$\theta(x, 0) = \theta_i = \sum_{m=0}^{\infty} d_m \cos \lambda_m x \tag{12}$$

To evaluate the constant d_m for a specific value of $m = n$, we note that Equation 6b is a special case of the Sturm-Liouville equation and the functions satisfying the differential equation and the homogeneous boundary conditions are orthogonal. Multiply both sides of Equation 12 by $\cos \lambda_n x$ and integrate between $x = -L$ and $x = L$. From the properties of orthogonal functions, every integral

$$\int_{-L}^{+L} \cos \lambda_m x \cos \lambda_n x \, dx,$$

for which $m \neq n$, vanishes. Only the integral for which $m = n$ yields a nonzero value. Thus,

$$d_n = \frac{\theta_i \displaystyle\int_{-L}^{+L} \cos \lambda_n x \, dx}{\displaystyle\int_{-L}^{+L} \cos^2 \lambda_n x \, dx}$$

$$\int_{-L}^{+L} \cos \lambda_n x \, dx = \frac{2 \sin \lambda_n L}{\lambda_n} = \frac{2 \sin[(2n + 1)(\pi/2)]}{\lambda_n} = \frac{2(-1)^n}{\lambda_n}$$

$$\int_{-L}^{+L} \cos^2 \lambda_n x \, dx = \left(\frac{x}{2} + \frac{\sin 2\lambda_n x}{4\lambda_n}\right)\bigg|_{-L}^{+L} = L$$

[because $2\lambda_n L = (2n + 1)\pi$, sin of $2\lambda_n L = 0$.] Thus, $d_n = [2(-1)^n]/(\lambda_n L)$. The final solution is

$$\theta = \theta_i \sum_{n=0}^{\infty} \frac{2(-1)^n}{\lambda_n L} \exp(-\lambda_n^2\alpha\tau) \cos(\lambda_n x) \tag{13}$$

$$\lambda_n L = (2n + 1)\pi/2$$

As $\tau \to \infty$, $\theta \to 0$, i.e., the temperature of the slab approaches the temperature of the surroundings with increasing time.

The total heat transfer from the slab from $\tau = 0$ to any instant τ is calculated as follows. The heat flux at the surface $(x = L)$ is given by

$$q''_{x=L} = -k \frac{\partial\theta}{\partial x}\bigg|_{x=L} = k\theta_i \sum_{n=0}^{\infty} \frac{2(-1)^n}{L} \exp(-\lambda_n^2\alpha\tau) \sin \lambda_n L$$

With $\sin \lambda_n L = \sin(2n + 1)(\pi/2) = (-1)^n$,

$$q''_{x=L} = k\theta_i \sum_{n=0}^{\infty} \frac{2 \exp(-\lambda_n^2\alpha\tau)}{L}$$

The total heat transfer rate from both surfaces, q, is given by

$$q = 2A_x q''_{x=L} = 2A_x k\theta_i \sum_{n=0}^{\infty} \frac{2 \exp(-\lambda_n^2 \alpha \tau)}{L}$$

where A_x is the cross-sectional area perpendicular to the x-direction. Denoting the total heat transfer from the slab from $\tau = 0$ to τ by Q,

$$Q = \int_0^{\infty} q \, d\tau = 2A_x k\theta_i \sum_{n=0}^{\infty} \int_0^{\tau} \frac{2 \exp(-\lambda_n^2 \alpha \tau)}{L} \, d\tau$$

$$= 2A_x k\theta_i \sum_{n=0}^{\infty} \frac{2}{\lambda_n^2 \alpha L} [1 - \exp(-\lambda_n^2 \alpha \tau)]$$

Substituting $\lambda_n^2 = (2n + 1)(\pi/2L)$

$$Q = \frac{2A_x k\theta_i L}{\alpha} \sum_{n=0}^{\infty} \frac{8}{(2n + 1)^2 \pi^2} [1 - \exp(-\lambda_n^2 \alpha \tau)]$$

and

$$\frac{2A_x kL}{\alpha} = 2A_x L\rho c = \rho c V$$

where c is the specific heat of the solid and V is the total volume of the solid. Denoting $\rho c V \theta_i$ by Q_o

$$\frac{Q}{Q_o} = \sum_{n=0}^{\infty} \frac{8}{(2n + 1)^2 \pi^2} [1 - \exp(-\lambda_n^2 \alpha \tau)]$$

Also

$$\sum_{n=0}^{\infty} \frac{1}{(2n + 1)^2} = \frac{\pi^2}{8}$$

so that

$$\frac{Q}{Q_o} = 1 - \sum_{n=0}^{\infty} \frac{8}{(2n + 1)^2 \pi^2} \exp(-\lambda_n^2 \alpha \tau)$$

As $\tau \to \infty$, $Q \to Q_o$. Thus, Q_o, the difference in the enthalpy of the solid between the initial and final states, represents the total heat transfer when the solid goes from the initial to the final state.

In Example 11.3.1 two parallel surfaces were maintained at a uniform temperature of T_w. A more often encountered boundary condition is that the surfaces are exposed to a fluid at T_∞ with an associated convective heat transfer coefficient h. The fluid temperature T_∞ and the convective heat transfer coefficient h are both uniform and constant. It is required to find $T(x, \tau)$ and the heat flux at the surface. This is the problem considered in Section 2.9.1. We find the solution to the problem and demonstrate how the solution in dimensionless variables is used to produce Figure 2.9.2.

EXAMPLE 11.3.2 A steel slab of thickness $2L$ is at a uniform temperature T_i (Figure 11.3.2). It is then cooled by exposing two parallel surfaces to a fluid at T_∞ with an associated heat transfer coefficient h. Find the transient temperature distribution $T(x, \tau)$.

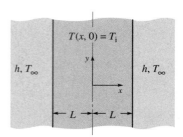

FIGURE 11.3.2 A large slab of thickness $2L$, initially at a uniform temperature, is exposed to a fluid adjacent to it.

ASSUMPTION

1. The dimensions of the slab in the y- and z-directions are much greater than its thickness $2L$, and the temperature distribution is one-dimensional in the x-direction.
2. Constant properties

SOLUTION

As we will be using dimensionless variables for the solution to the problem, we denote the dimensional coordinates by $\hat{x}$ and $\hat{y}$, and the dimensionless coordinates by x and y.

Taking the origin at the midplane, the differential equation and the initial and boundary conditions are

$$\frac{1}{\alpha} \frac{\partial T}{\partial \tau} = \frac{\partial^2 T}{\partial \hat{x}^2} \tag{1}$$

Initial condition:
$$T(\hat{x}, 0) = T_i \tag{2}$$

Boundary conditions. Surfaces at $\hat{x} = \pm L$ are exposed to convective heat transfer. Thus,

$$-k \frac{\partial T}{\partial \hat{x}} (L, \tau) = h[T(L, \tau) - T_\infty] \tag{3}$$

$$k \frac{\partial T}{\partial \hat{x}} (-L, \tau) = h[T(-L, \tau) - T_\infty] \tag{4}$$

From the symmetry of the problem (the differential equation and the initial and boundary conditions being symmetric about the midplane), we replace one of the boundary conditions, Equation 4, by $\partial T / \partial \hat{x}(0, \tau) = 0$. We now use appropriate dimensionless variables to solve this problem. Define

$$\theta = \frac{T - T_\infty}{T_i - T_\infty} \qquad x = \frac{\hat{x}}{L} \qquad \text{Fo} = \frac{\alpha \tau}{L^2}$$

In terms of these dimensionless variables, the differential equation and the associated initial and boundary conditions are

$$\frac{\partial \theta}{\partial Fo} = \frac{\partial^2 \theta}{\partial x^2} \tag{5}$$

$$\theta(x, 0) = 1 \tag{6}$$

$$\frac{\partial \theta}{\partial x}(0, Fo) = 0 \tag{7}$$

$$\frac{\partial \theta}{\partial x}(1, Fo) + Bi\ \theta(1, Fo) = 0 \qquad Bi = \frac{hL}{k} \tag{8}$$

Note that the solution for θ is identical for all cases that have the same Biot number, Bi. Hence, θ is dependent on only one parameter, Bi (in addition to the dimensionless distance x, and dimensionless time Fo). The advantage of employing dimensionless variables is that the solution to all problems with the same Bi is identical whatever may be the individual values of h, k, L, T_i, and T_∞.

To find the solution for θ by separation of variables, assume

$$\theta = f_1(x)f_2(Fo)$$

Substitution of $f_1(x)f_2(Fo)$ in Equation 5 and dividing it by $f_1(x)f_2(Fo)$, we obtain

$$\frac{f_1''}{f_1} = \frac{f_2'}{f_2} = -\mu^2 \tag{9}$$

As f_1''/f_1 cannot be a function of Fo and f_2'/f_2 cannot be a function of x, we set each of them equal to a constant. The separation constant is $-\mu^2$ (Why is the constant negative?). Substitution of $f_1(x)f_2(Fo)$ in Equations 6, 7, and 8 yields

$$f_1(x)f_2(0) = 1$$

$$f_1'(0)f_2(Fo) = 0$$

$$f_1'(1)f_2(Fo) + Bi\ f_1(1)f_2(Fo) = 0$$

In the first boundary condition, it is clear that $f_2(Fo)$ cannot be equal to zero because, if it were zero, then θ will be zero everywhere. Hence, we require that $f_1'(0) = 0$. A similar logic applied to the other boundary condition yields the following set of initial and boundary conditions,

$$f_1(x)f_2(0) = 1 \tag{10}$$

$$f_1'(0) = 0 \tag{11}$$

$$f_1'(1) + Bi\ f_1(1) = 0 \tag{12}$$

Equation 9 results in

$$\frac{f_1''}{f_1} = -\mu^2 \tag{13}$$

$$\frac{f_2'}{f_2} = -\mu^2 \tag{14}$$

Solution to Equation 13 is

$$f_1(x) = a \cos \mu x + b \sin \mu x \tag{15}$$

Applying boundary condition, Equation 11, $b = 0$. Using boundary condition, Equation 12,

$$-a \mu \sin \mu + \text{Bi } a \cos \mu = 0$$

As $a \neq 0$ (why?), we require

$$\mu \sin \mu - \text{Bi} \cos \mu = 0$$

or

$$\mu \tan \mu = \text{Bi} \tag{16}$$

Solution of Equation 16 gives a series of values for μ. The first six roots of the equation for different values of Bi are tabulated in Appendix 2. Corresponding to a given value of μ in the series, say μ_m, the solution to Equation 14 is

$$f_{2m} = d_m \exp(-\mu_m^2 \text{Fo})$$

so that with $f_{1m}(x) = a_m \cos \mu_m x$ and $f_{2m}(\tau) = d_m \exp(-\mu_m^2 \text{ Fo})$, one of the possible solutions to the differential equation satisfying the boundary conditions is $a_m \cos(\mu_m x) d_m \exp(-\mu_m^2 \text{ Fo})$. The general solution to the equation is then given by a linear combination of all such solutions.

$$\theta = \sum_{m=1}^{\infty} c_m \cos \mu_m x \exp(-\mu_m^2 \text{ Fo}) \tag{17}$$

where $a_m d_m$ has been replaced by c_m. To satisfy the initial condition we require

$$1 = \sum_{m=1}^{\infty} c_m \cos \mu_m x \tag{18}$$

Equation 13 is a special case of the Sturm-Liouville system (see Appendix 3), and the coefficient c_m can be evaluated by utilizing the orthogonality of the functions. To evaluate the coefficient c_n, (for $m = n$), multiply both sides of Equation 18 by $\cos \mu_n x$ and integrate the equation from $x = 0$ to $x = 1$. From the principle of orthogonality, we have

$$\int_0^1 \cos \mu_m x \cos \mu_n x \, dx \qquad = 0 \text{ for } m \neq n$$

$$= \text{constant for } m = n$$

Hence,

$$c_n = \frac{\displaystyle\int_0^1 \cos \mu_n x \, dx}{\displaystyle\int_0^1 \cos^2 \mu_m x \, dx}$$

$$\int_0^1 \cos \mu_n x \, dx = \frac{\sin \mu_n}{\mu_n}$$

$$\int_0^1 \cos^2 \mu_n x \, dx = \frac{1}{2} + \frac{\sin \mu_n \cos \mu_n}{2\mu_n}$$

so that

$$c_n = \frac{2 \sin \mu_n}{\mu_n + \sin \mu_n \cos \mu_n}$$

The solution for θ is, therefore, given by

$$\theta = 2 \sum_{n=1}^{\infty} \frac{\sin \mu_n}{\mu_n + \sin \mu_n \cos \mu_n} \exp(-\mu_n^2 \, Fo) \cos \mu_n x \qquad (19)$$

For small values of Fo the convergence of the series in Equation 19 is slow and a very large number of terms have to be evaluated to obtain acceptable accuracy. With Laplace transforms it is possible to obtain rapidly converging series for both small and large times.

We now have the solution to the transient temperature distribution in a semi-infinite rectangular solid, initially at a uniform temperature, and two parallel surfaces exposed to a fluid. Using Equation 19, the computed values of θ for three values of $x = 0$, 0.5, and 1.0 are shown in Figure 11.3.3a for Bi = 2. Figure 11.3.3b shows the temperature distribution for Fo = 0.5 for three different values of Bi. From Figure 11.3.3b it is clear that for Bi < 0.1 the assumption of spatially uniform temperature is appropriate. Thus, if Bi < 0.1, lumped analysis may be used. The temperature distribution for a slab with Bi = 1000 at Fo = 0.005 is given in Table 11.3.1.

From the values in Table 11.3.1 it is clear that the surface temperature approximately equals the fluid temperature at Fo = 0.005 but the temperature in a substantial part of the slab (up to 80% of the slab) has hardly changed. Representative values of dimensional time corresponding to Fo = 0.005 are given in Table 11.3.2.

Recall that the dimensionless temperature at the surface is the ratio of the difference in the surface temperature and the fluid temperature to the difference in the initial uniform temperature of the slab and the fluid temperature. Therefore, a dimensionless surface temperature of 0.008 indicates that the surface temperature is very nearly equal to the fluid temperature. From Table 11.3.2 the time required for the dimensionless surface temperature of 0.008 is less than 0.05 s for a slab thickness of 2 cm ($L = 1$ cm) and less than 1 s for aluminum and copper slabs of 20-cm thickness ($L = 10$ cm) but 3.6 s for bronze. The values indicate that the surface reaches the fluid temperature very quickly at large Biot numbers. Therefore, the solution can be used for a specified surface temperature by prescribing a sufficiently large Biot number for the fluid. From the table it is also clear that one method of achieving a prescribed surface temperature is to use a fluid that yields a large Biot number with a high convective heat transfer coefficient; high convective heat transfer coefficients can be obtained with boiling or condensation of fluids.

How are Heisler charts (Figures 2.9.2, 2.9.3, and 2.9.4) produced?

Figure 11.3.3b shows the effect of Biot number on the spatial variation of the temperature. For Bi = 5, there is a significant difference in the midplane temperature and the surface temperature. This temperature difference is slightly reduced if the Biot number is reduced to 2. If the Biot number is reduced to 0.1 the temperature difference is so small as to be negligible. Figure 11.3.3a is a part of Figure 2.9.2.

Solutions similar to Equation 1 for a cylinder and sphere have also been obtained. From the solutions the one-dimensional transient temperature distribution charts shown in Figures 2.9.2 through 2.9.4 are produced. The program TRANS on disk

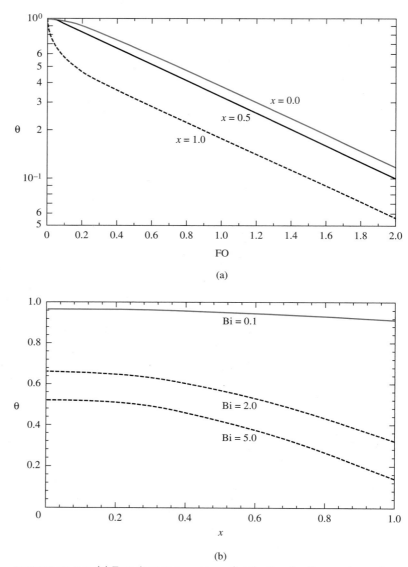

FIGURE 11.3.3 (a) Transient temperature distribution for three values of dimensionless distance for Bi = 2.0. (b) Temperature distribution for Fo = 0.5 for three values of the Biot number.

TABLE 11.3.1 Temperature Distribution in a Semi-Infinite Rectangular Slab: Bi = 1000, Fo = 0.005

x	0.0	0.1	0.5	0.7	0.8	0.9	1.0
θ	1.0	1.0	1.0	0.9974	0.9556	0.6875	0.008

TABLE 11.3.2 Representative Dimensional Times Corresponding to Fo = 0.005			
		τ(s)	
	$L = 0.01$m	$L = 0.1$ m	$L = 0.2$ m
Aluminum ($\alpha = 97.1 \times 10^{-6}$ m²/s	0.005	0.51	2.1
Copper ($\alpha = 117 \times 10^{-6}$ m²/s)	0.004	0.43	1.7
Bronze ($\alpha = 14 \times 10^{-6}$ m²/s)	0.036	3.6	14.3

yields the temperature distribution for a rectangular plate, cylinder, or sphere for different values of the dimensionless distance (from 0 to 1 in steps of 0.1) at a defined value of Fo or the temperature as a function of Fo (from 0 to 1.5 in steps of 0.1) at a defined value of x. Thus, from the software we can construct the transient temperature distribution charts. Also, HT.EXE subprogram Transient Conduction I can compute the temperature at defined values of the dimensionless distance and Fo. To compute the temperature, the eigenvalues are needed. The computations of the eigenvalues for rectangular and spherical geometry are fast, but the computations of the eigenvalues for cylindrical geometry (requiring the computations of Bessel functions) take a much longer time on a PC without a math coprocessor.

In the software the dimensionless temperature for Fo $\geq$ 0.05 is obtained. With the limited number of terms used in the software, convergence is not satisfactory for Fo < 0.05.

Equation 2.9.5a, which is a one-term approximation, is the first term in the series solution; for example, it is the first term in Equation 19 for rectangular geometry.

Note on Separation of Variables

Separation of variables— constraints

It should be noted that the technique of separation of variables works only under certain conditions. Briefly, some of them are

1. The technique will be successful only for rectangular slabs, cylinders, and spheres with finite dimensions. In Cartesian coordinates the solutions involve trigonometric functions. In cylindrical coordinates, the solutions contain Bessel functions, and in spherical coordinates the solutions contain Legendre polynomials.
2. The equations should be linear, second order, with, usually, constant coefficients.
3. In the directions in which we anticipate characteristic values (eigenvalues), the boundary conditions should be homogeneous, i.e., of the type,

$$ay' + by = 0$$

The boundary condition includes $y' = 0$ ($b = 0$) and $y = 0$ ($a = 0$).

4. There should be only one nonhomogeneity, either in the equation or in one of the boundary conditions. The nonhomogeneous boundary condition should be of the type

$$ay' + by = f(x)$$

5. If there is more than one nonhomogeneity, it is possible to apply the separation of variables technique by splitting the problem into an appropriate number of simpler problems, each of which has only one nonhomogeneity. The solutions to these simple problems may then be combined by the principle of superposition.

For a more detailed discussion of the technique, as applied to conduction heat transfer, the reader may refer to the books in the References at the end of this chapter.

As one of the limitations of the separation of variables technique is that the medium should be finite, it is not suitable when the medium is infinite. In many transient problems, where the medium is infinite, the Laplace transform technique is more suitable. The application of Laplace transforms is illustrated through two examples. It is assumed that the reader has a working knowledge of Laplace transforms. Only those cases for which solutions can be obtained from the table of inverse transforms (given in Appendix 4) are considered here. Inverse transforms given in Appendix 4 are useful in many conduction problems that can be solved by Laplace transforms.

Laplace transforms— transient temperature distribution

EXAMPLE 11.3.3

A very long bar of constant cross section is initially at a uniform temperature T_i. The peripheral surfaces are perfectly insulated. At time $\tau = 0$, the temperature of the surface at $x = 0$ (Figure 11.3.4) is changed to T_o and maintained at that value. Determine the transient temperature distribution.

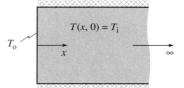

FIGURE 11.3.4 The temperature of one surface of a semi-infinite slab, initially at a uniform temperature, is suddenly changed and maintained at the new value.

ASSUMPTIONS

1. Constant properties
2. As the slab is initially at a uniform temperature, and the surface at $x = 0$ is at a uniform temperature; the temperature is one-dimensional, i.e., $T = T(x, \tau)$.

SOLUTION

The differential equation and the associated boundary conditions for this problem, with $\theta = T - T_i$ are

$$\frac{\partial^2 \theta}{\partial x^2} - \frac{1}{\alpha} \frac{\partial \theta}{\partial \tau} = 0 \tag{1}$$

The initial and boundary conditions are given by

$$\theta(x, 0) = 0 \tag{2a}$$

$$\theta(0, \tau) = \theta_o = T_o - T_i \tag{2b}$$

$$\theta(x \to \infty, \tau) = 0 \tag{2c}$$

Taking the Laplace transform of Equation 1 with respect to time and using the initial condition, Equation 2a (see Appendix 4),

$$\frac{d^2\overline{\theta}}{dx^2} - \frac{s}{\alpha}\,\overline{\theta} = 0 \tag{3}$$

The transforms of the boundary conditions, Equation 2b and c, are

$$\overline{\theta}\,(0,\,s) = \frac{\theta_o}{s} \tag{4}$$

$$\overline{\theta}\,(x \to \infty,\,s) = 0 \tag{5}$$

Note that there is no initial condition for the transformed equation as the initial condition, Equation 2a, is absorbed in the differential equation, Equation 3. The solution to Equation 3 is

$$\overline{\theta}(x,\,s) = c_1 \exp\left(-\sqrt{\frac{s}{\alpha}}\,x\right) + c_2 \exp\left(\sqrt{\frac{s}{\alpha}}\,x\right)$$

Boundary condition 5, yields $c_2 = 0$. Boundary condition 4 gives $c_1 = \theta_o/s$, and the complete solution in the transformed plane is

$$\overline{\theta}(x,\,s) = \theta_o\,\frac{\exp[-\sqrt{(s/\alpha)}\,x]}{s}$$

From entry 30 of the Table of Transforms in Appendix 4, the inverse transform is given by

$$\theta(x,\,t) = \theta_o\,\text{erfc}\left[\frac{x}{2(\alpha t)^{1/2}}\right]$$

where $\text{erfc}(z) = 1 - \text{erf}(z)$. The tabulated values of $\text{erf}(z)$ are available in Appendix 2 and in the software EF.

EXAMPLE 11.3.4 Re-solve Example 11.3.3 with a uniform heat flux imposed at the surface at $x = 0$ (Figure 11.3.5).

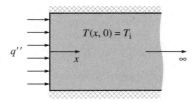

FIGURE 11.3.5 One surface of a semi-infinite slab, initially at a uniform temperature, is suddenly subjected to a uniform, constant heat flux.

ASSUMPTIONS

1. Constant properties
2. As the bar is initially at a uniform temperature, and the surface and the surface at $x = 0$ is subjected to a uniform heat flux for $\tau > 0$, the temperature is one-dimensional, i.e., $T = T(x, \tau)$.

SOLUTION

Define $\theta = T - T_i$. The differential equation and the initial and boundary conditions are

$$\frac{\partial^2 \theta}{\partial x^2} - \frac{1}{\alpha} \frac{\partial \theta}{\partial \tau} = 0 \tag{1}$$

$$\theta(x, 0) = 0 \tag{2a}$$

$$\frac{\partial \theta}{\partial x} (0, \tau) = -\frac{q''}{k} \tag{2b}$$

$$\theta(x \rightarrow \infty, \tau) = 0 \tag{2c}$$

Taking the transform of Equation 1 and using the initial condition, Equation 2a,

$$\frac{d^2 \bar{\theta}}{dx^2} - \frac{s}{\alpha} \bar{\theta} = 0 \tag{3}$$

The transformed boundary conditions are

$$\frac{d\bar{\theta}}{dx} (0, s) = -\frac{q''}{ks} \tag{4a}$$

$$\bar{\theta}(x \rightarrow \infty, s) = 0 \tag{4b}$$

Note that there is no initial condition in the transformed plane as the initial condition is already used in arriving at Equation 3.

The solution to Equation 3 is

$$\bar{\theta}(x, s) = c_1 \exp\left(-\sqrt{\frac{s}{\alpha}} x\right) + c_2 \exp\left(\sqrt{\frac{s}{\alpha}} x\right) \tag{5}$$

From boundary condition, Equation 4b, $c_2 = 0$. From boundary condition, Equation 4a,

$$c_1 = \frac{q''}{ks(s/\alpha)^{1/2}}$$

The solution for $\bar{\theta}$ is then given by

$$\theta = \frac{q''}{k} \frac{\exp[-\sqrt{(s/\alpha)} \, x]}{(s/\alpha)^{1/2} s} \tag{6}$$

From entry 31 in the table of transforms in Appendix 4, the inverse transform of Equation 6, is

$$\theta = 2 \frac{q''}{k} \left[\left(\frac{\alpha\tau}{\pi}\right)^{1/2} \exp\left(-\frac{x^2}{4\alpha\tau}\right) - \frac{x}{2} \, \text{erfc}\left(\frac{x}{2(\alpha\tau)^{1/2}}\right) \right] \tag{7}$$

This is the exact solution used in comparing the approximate solution in Example 3.5.2.

Comments on Examples 11.3.3 and 11.3.4   The solids in both the example problems are taken to be semi-infinite in the x-direction. Thus, we may wonder where these solutions are applicable. These solutions may be applied to slabs of finite thickness in a limited time domain if one of the surfaces is subjected to the boundary conditions employed in these examples. This will be illustrated through an example.

EXAMPLE 11.3.5

Consider the process of end-milling a plastic sheet. It is required to find the temperature at the end of the milling cutter. A thermocouple is placed at a distance x_o from the cutting end (Figure 11.3.6). We wish to determine the temperature at the cutting end from the temperature indicated by the thermocouple.

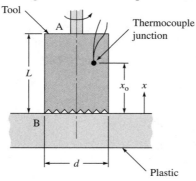

FIGURE 11.3.6 A thermocouple is installed to estimate the temperature of the cutting surface of an end mill used to machine a plastic sheet.

SOLUTION

To relate the temperature indicated by the thermocouple to the temperature at the cutting end of the cutter, we need to model the heat transfer process. In a milling machine, the pressure of the cutter may be assumed to be constant with respect to time. The process of milling is accompanied by energy transfer to the cutter at the end surface. The process is treated as one of uniform heat flux.[3] Further, to get an estimate of the temperature at the end of the cutter, the heat transfer from the cylindrical surface is neglected.

With these assumptions, the process is modeled as one of uniform heat flux at the end with the cylindrical surface insulated. Initially, the temperature of the cutter

---

[3] Consider the work done on an infinitesimal ring of radial length dr. Force normal to the ring = $p2\pi r\, dr$, where p = pressure. Frictional force is given by $\mu p2\pi r\, dr$, μ = coefficient of friction. If the angular velocity of the tool is ω, the rate of work done by the element is given by $\mu p2\pi\omega r^2\, dr$. Hence, the rate of work done per unit area of the mill is given by $\mu p\omega r$. For uniform wear we require that $\mu p\omega r$ = constant. Assuming the friction coefficient to be constant, with constant ω, pr = constant = c. If the total axial force is F, then $F = \int_0^R p2\pi r\, dr = 2\pi cR$ or $c = F/(2\pi R)$. The total work transfer is $\int_0^R \mu p2\pi\omega r^2 dr = \int_0^R \mu c2\pi r\omega\, dr = \mu c2\pi\omega R^2/2 = \mu F\omega R/2$. Assuming that all the work transfer manifests itself as heat transfer at the surface of the cutter, and, as the work transfer is uniform, the heat transfer at the surface is also uniform and is given by $q'' = \mu F\omega/(2\pi R)$. As the thermal conductivity of the plastic is considerably less than that of the tool steel, the plastic sheet is approximated as an insulating material.

is uniform at T_i. Though the cutter is finite in dimensions, it can be treated as an infinite cylinder until the temperature of surface A differs significantly from T_i. What constitutes significant change is somewhat subjective and is discussed later.

With the above assumptions, the solution given by Equation 7 in Example 11.3.4 can be used. Dividing Equation 7 by $q''L/k$, it is recast as

$$\frac{\theta}{(q''L/k)} = \frac{2}{L}\left(\frac{\alpha\tau}{\pi}\right)^{1/2} \exp\left(-\frac{x^2}{L^2}\frac{L^2}{4\alpha\tau}\right) - \frac{x}{L} \operatorname{erfc}\left[\frac{L}{2(\alpha\tau)^{1/2}}\frac{x}{L}\right] \tag{1}$$

In Equation 1 the denominator on the left side represents the temperature difference across a cylinder of length L, one end of which is subjected to a uniform heat flux q'', with the cylindrical surface perfectly insulated. Taking $q''L/k$ as the characteristic temperature difference, θ_R, we may treat the cutter (between surfaces A and B) as infinite if θ/θ_R is sufficiently small, say, 0.02. For tool steel, we use the properties of 0.2% C, 1.02% Cr, and 0.15% V. We expect the temperature to be high and we use the properties at 400 K. From Table A1, properties of the tool steel are

$$\rho = 7836 \text{ kg/m}^3 \qquad c_p = 492 \text{ J/kg K} \qquad k = 46.8 \text{ W/m K}$$

$$\alpha = \frac{k}{\rho c_p} = \frac{46.8}{7836 \times 492} = 12.14 \times 10^{-6} \text{ m}^2 \text{ s}$$

We seek a value of τ for which $\theta/\theta_R = 0.02$. Equation 1 is solved by iteration. With $x/L = 1$, for $(\alpha\tau)^{1/2}/L = 0.4$, $\theta/\theta_R = 0.0175$. If the axial length of the cutter, L, is 10 cm, for $\alpha = 12.14 \times 10^{-6}$ m²/s, $(\alpha\tau)^{1/2}/L = 0.4$ for $\tau = 132$ s, or a little more than 2 minutes. Thereafter, the assumption of an infinite slab leads to increasing error in the computed temperature distribution.

EXAMPLE 11.3.6   In Example 11.3.5, 90 s after starting the cutting, the thermocouple, located at a distance of 1 cm from the cutting surface, measures a temperature of 60 °C. What is the temperature at the cutting surface? The initial uniform temperature of the cutter is 20 °C.

SOLUTION

Denoting the temperature excess at x_o by θ_o (Figure 11.3.6) and that at the tip ($x = 0$) by θ_t, from Equation 7 of Example 11.3.4 we get

$$\frac{\theta_o}{\theta_t} = \frac{2(\alpha\tau/\pi)^{1/2} \exp[-x_o^2/(4\alpha\tau)] - x_o \operatorname{erfc}\{x_o/[2(\alpha\tau)^{1/2}]\}}{2(\alpha\tau/\pi)^{1/2}}$$

For $\tau = 90$ s and $x_o = 0.01$ m, we get, $\theta_o/\theta_t = 0.755$. Then, $\theta_o = 60 - 20 = 40$ °C or $\theta_t = 40/0.755 = 53$ °C.

Temperature at the tip = 53 + 20 = <u>73 °C</u>

SUMMARY

In this chapter, the three-dimensional, transient conduction equation is derived in Cartesian coordinates. The technique of solving two-dimensional, steady state and one-dimensional conduction equation by separation of variables was illustrated through examples. The limitations of the technique are indicated. Solutions to one-dimensional, transient conduction equation by Laplace transforms are presented for a few cases for which the inverse transforms are readily available in Appendix 4.

REFERENCES

Arpaci, V.S. (1966). *Conduction Heat Transfer.* Reading, MA: Addison Wesley.

Carslaw, H.S., and Jaeger, J.C. (1959). *Conduction of Heat in Solids,* 2d ed. Oxford: Oxford University Press.

Churchill, R.V. (1972). *Operational Mathematics,* 3d ed. New York: McGraw-Hill.

Kutataladze, S.S. (1963). *Fundamentals of Heat Transfer.* New York: Academic Press.

Myers, G.E. (1971). *Analytical Methods in Conduction Heat Transfer.* New York: McGraw-Hill.

Ozisik, M.N. (1968). *Boundary Value Problems of Heat Conduction.* Scranton, PA: International Textbook.

Ozisik, M.N. (1980). *Heat Conduction.* New York: Wiley.

Schneider, P.J. (1955). *Conduction Heat Transfer.* Reading, MA: Addison-Wesley.

Poulikakos, D. (1964). *Conduction Heat Transfer.* Englewood Cliffs, NJ: Prentice-Hall.

PROBLEMS

11.1 Show that the one-dimensional, transient conduction equation for the temperature distribution in the radial direction in cylindrical coordinates, with constant properties of the medium, is

$$\frac{\partial T}{\partial \tau} = \alpha \left[\frac{1}{r} \frac{\partial}{\partial r} \left(r \frac{\partial T}{\partial r} \right) \right]$$

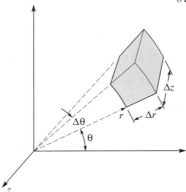

FIGURE P11.1

11.2 Show that the one-dimensional, transient conduction equation for the temperature distribution in the radial direction in spherical coordinates, with constant properties of the medium, is

$$\frac{\partial T}{\partial \tau} = \alpha \left[\frac{1}{r^2} \frac{\partial}{\partial r} \left(r^2 \frac{\partial T}{\partial r} \right) \right]$$

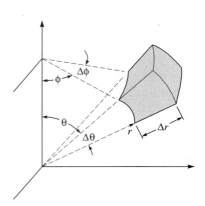

FIGURE P11.2

11.3 Theory of fire proof walls. Consider a wall, one side of which is lined with combustible material such as wood. It is required to find the time taken for the combustibles to attain a defined temperature. To obtain an estimate of the time, the wall is modeled as a large rectangular slab, of thickness L, initially at a uniform temperature T_i. The temperature of one side of the wall is suddenly raised to T_o. The heat transfer rate from the other side of the wall is assumed to be negligibly small. Find an expression for $T(x, t)$.

11.4 In Problem 11.3, find τ_o for the adiabatic surface to reach a temperature of 150 °C if the wall is made of concrete or stone. The walls are 30-cm thick and the initial uniform temperature of the walls is 20 °C. The temperature of the surface exposed to the fire reaches 800 °C within a very short time.

α (concrete) $= 6 \times 10^{-7}$ m²/s $\qquad$ α (stone) $= 1.3 \times 10^{-6}$ m²/s

11.5 Consider a rectangular slab of thickness L. The temperatures of the boundaries of the slab are

$$T(x, 0) = T_1 \qquad T(0, y) = T(L, y) = T(x, y \to \infty) = T_2$$

Show that the two-dimensional temperature distribution is given by

$$\theta = \frac{T - T_2}{T_1 - T_2}$$

$$= \frac{4}{\pi} \left(e^{-\pi y/L} \sin \pi \frac{x}{L} + \frac{1}{3} e^{-3\pi y/L} \sin 3\pi \frac{x}{L} + \frac{1}{5} e^{-5\pi y/L} \sin 5\pi \frac{x}{L} \cdots \right)$$

The series can be expressed as

$$\theta = \frac{2}{\pi} \tan^{-1} \left[\frac{\sin(x/L)}{\sin(y/L)} \right]$$

Employing the closed form solution for θ, plot isotherms corresponding to $\theta = 0.2$ and 0.8.

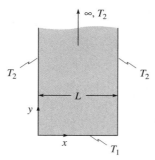

FIGURE P11.5

11.6 Show that the differential equation for the one-dimensional, transient temperature distribution in a rectangular fin with a uniform heat transfer coefficient is

$$\frac{1}{\alpha}\frac{\partial T}{\partial \tau} = \frac{\partial^2 T}{\partial x^2} - m^2(T - T_\infty) \qquad m^2 = hP/(kA_x)$$

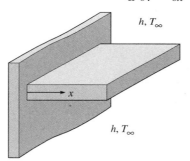

FIGURE P11.6

11.7 Find the two-dimensional, steady temperature distribution, $T(x, y)$, in a fin of rectangular cross section surrounded by a fluid. The temperature of the fluid and the convective heat transfer coefficient are uniform. Assume that the heat transfer rate from the free end is negligible and that the base is maintained at a uniform temperature T_b.

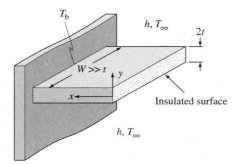

FIGURE P11.7

11.8 Re-solve Problem 11.7 if the fin is infinitely long.
11.9 Re-solve Problem 11.8 if, instead of the base being maintained at a uniform temperature, it is subjected to a uniform heat flux q_o''.

11.10 Reconsider Example 11.2.1. If, instead of the surface at $y = W$ being maintained at a uniform temperature, it is maintained at a temperature given by $T = T_o + a \sin bx$, where a and b are constants, determine $T(x, y)$. Comment on the solution for the special case of $b = \dfrac{p\pi}{L}$, where p is an integer.

11.11 Re-solve Problem 11.7 if, instead of the base being maintained at a uniform temperature, it is maintained at

$$T(L, y) = T_o + a \cos by$$

11.12 Consider a rectangular fin of thickness $2t$, and length L. The temperature of the base is approximated as

$$0 \le y \le t \qquad T = T_o - by$$
$$-t \le y < 0 \qquad T = T_o + by$$

Find an expression for the two-dimensional temperature distribution assuming the heat transfer from the free end to be negligible.

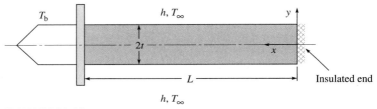

FIGURE P11.12

11.13 Re-solve Problem 11.7 if the heat transfer coefficients on the upper and lower surfaces are uniform but different.

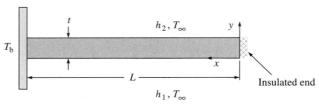

FIGURE P11.13

11.14 From the solution to Example 11.3.3 show that the heat flux to the surface at $x = 0$ is given by

$$q'' = \frac{k\theta_o}{(\pi\alpha\tau)^{1/2}} = \left(\frac{k\rho c_p}{\pi\tau}\right)^{1/2} \theta_o$$

11.15 Lord Kelvin made an estimate of the age of earth from the measured temperature gradient at the surface of the earth, which is approximated as 0.03602 °C/m. He modeled the earth as a semi-infinite rock (thermal diffusivity $= 1.178 \times 10^{-6}\,\text{m}^2/\text{s}$) at a uniform temperature of 3870 °C. At time, $\tau = 0$, the temperature of the surface of the earth was changed to 0 K and maintained at that value. What was his estimate of the age of the earth?

11.16 Heat pumps are used to heat houses with the ground as the source. To estimate the rate of energy available to such heat pumps, consider a heat exchanger consisting of a number of pipes parallel to the surface of the earth. The heat exchanger is modeled as a large plate buried deep in the ground. The initial uniform temperature of the soil is T_i and at time $\tau = 0$, the temperature of the plate is changed to T_o and maintained at that value. For a unit change in the temperature ($T_i - T_o = 1\ °C$) estimate the heat flux after 1 day, 10 days, and 60 days for
(a) Soil A: $k = 2.1$ W/m K $\rho = 1632$ kg/m^3 $c_p = 1880$ J/kg °C
(b) Soil B: $k = 0.69$ W/m K $\rho = 1600$ kg/m^3 $c_p = 1254$ J/kg °C

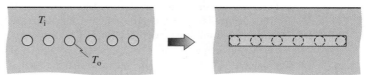

FIGURE P11.16

11.17 In winter the temperature of the soil decreases and there is the danger of freezing of water pipes that are buried in the soil. To prevent such freezing, the pipes are buried sufficiently deep so that the temperature of soil adjacent to the pipes does not fall below 0 °C. To estimate the safe depth for laying water pipes, assume that the soil temperature prior to the onset of winter is uniform at 10 °C. A cold spell in the winter, suddenly lowers the temperature of the surface of the soil to -10 °C, and maintains it at that temperature continuously for 60 days. If the thermal diffusivity of the soil is 0.15×10^{-6} m^2/s, determine the safe depth.

11.18 In Example 11.3.3, the transient, one-dimensional temperature distribution in a semi-infinite solid was determined. Instead of the temperature of the surface being suddenly changed to T_o, a more realistic condition is that the surface is suddenly exposed to a fluid at T_∞, with an associated heat transfer coefficient h. Show that the $T(x, \tau)$ is given by

$$\frac{\theta}{\theta_\infty} = \text{erfc } \eta - exp(\beta x + \alpha \beta^2 \tau)\ \text{erfc}[\eta + \beta(\alpha\tau)^{1/2}]$$

$$\theta = T - T_i \qquad \theta_\infty = T_\infty - T_i \qquad \eta = \frac{x}{2(\alpha\tau)^{1/2}} \qquad \beta = \frac{h}{k}$$

11.19 In Problem 11.17 we assumed that the temperature of the surface of the earth remained constant for a number of days. A more appropriate assumption is that the mean surface temperature oscillates with time. Assume that the surface temperature variation is given by

$$T_s - T_m = \Delta T \sin \omega\tau$$

where T_m is the mean surface temperature and ΔT is the amplitude of variation of the temperature. We anticipate that the interior temperature will also oscillate with a period ω but with a time lag and a reduced amplitude as we go deeper into the earth. Therefore, a reasonable assumption for the temperature variation in the interior of the earth is

$$T - T_m = ce^{-ax} \sin(\omega\tau + bx)$$

By substituting the temperature distribution into the differential equation for the one-dimensional, transient temperature show that

$$T - T_{\mathrm{m}} = \Delta T \exp\left[-\left(\frac{\omega}{2\alpha}\right)^{1/2} x\right] \sin\left[\omega\tau - \left(\frac{\omega}{2\alpha}\right)^{1/2} x\right]$$

11.20 In Problem 11.19, at a particular location the mean temperature of the surface of the earth is 7 °C and the amplitude of the variation of the annual temperature is 15 °C, i.e., the minimum and maximum surface temperatures are −8 °C and 22 °C, respectively. If the maximum temperature occurs on August 10, determine the minimum soil temperature at depths of 1 m and 2 m and the dates on which they occur. The thermal diffusivity of the soil is 4.9×10^{-7} m²/s. Assume sinusoidal variation of the surface temperature with a period of one year.

11.21 Consider a spherical ball of radius, R, initially at a uniform temperature T_{i}, suddenly exposed to condensing steam at T_∞ at $\tau = 0$. The convective heat transfer coefficient for condensing steam is very high and the surface temperature is very close to T_∞ for $\tau \geq 0$. The differential equation for $T(r, \tau)$ is (see Problem 11.2)

$$\frac{\partial T}{\partial \tau} = \alpha \left[\frac{1}{r^2}\frac{\partial}{\partial r}\left(r^2 \frac{\partial T}{\partial r}\right)\right]$$

Substituting $\theta = (T_\infty - T)r$, show that the differential equation transforms to

$$\frac{\partial \theta}{\partial \tau} = \alpha \frac{\partial^2 \theta}{\partial r^2}$$

Employing the technique of separation of variables, show that

$$\frac{T_\infty - T}{T_\infty - T_{\mathrm{i}}} = 2 \sum_{n=1}^{\infty} (-1)^{n+1} e^{-\lambda_n^2 \alpha\tau}\frac{\sin \lambda_n r}{\lambda_n r}$$

$$\lambda_n = \frac{n\pi}{R}, \, n = 1, 2, \cdots \qquad \theta_{\mathrm{i}} = T_\infty - T_{\mathrm{i}}$$

11.22 A 10-cm diameter stainless steel (type 316) ball, initially at a uniform temperature of 20 °C is suddenly exposed to steam. The condensing steam maintains the temperature of the surface of the sphere at 100 °C. Estimate the temperature of the center of the sphere 1 minute and 10 minutes after it is exposed to the steam.

11.23 In Problem 11.22, how long should the ball be exposed to the steam before the temperature of the center of the ball reaches 90 °C?

11.24 A semi-infinite solid is initially in equilibrium with the surroundings at T_∞. At $\tau = 0$, the surface of the solid is subjected to a net radiant heat flux of q''. The convective heat transfer coefficient is h. Find the transient, one-dimensional temperature distribution in the solid.

12

CONVECTION—DIFFERENTIAL FORMULATION

Approximate solutions by integral techniques to a few convection problems are presented in Chapter 5. We now consider the differential form of the energy equation in forced convection. The solution to a convective heat transfer problem requires the solution to the associated fluid flow problem. An introduction to the differential equations for the velocity and temperature distribution in a fluid and solutions for the equations for a few cases are given in this chapter. We now

■ *Derive the two-dimensional, steady-state differential equations for mass conservation, momentum balance, and energy balance for a fluid in motion*

■ *Present some exact solutions for incompressible fluids*

■ *Derive the approximate forms of the equations near a solid boundary—the boundary layer differential equations*

■ *Show how the momentum and energy boundary layer partial differential equations are transformed to ordinary differential equations and how the resulting ordinary differential equations are solved*

12.0 INTRODUCTION

A few approximate solutions to convection problems by integral methods are presented in Chapter 5. In this chapter, differential equations for convection problems are derived and their application illustrated through a few examples. The solutions to the differential equations for convective heat transfer problems require the associated fluid velocity field. The equations for the velocity and temperature field may be either uncoupled or coupled. In forced convection, if the fluid properties are assumed to be independent of temperature, the mass conservation and momentum balance equations, which yield the velocity field, are uncoupled from the energy equation. In natural convection, where the motion of the fluid is caused by temperature differences and temperature dependent densities, the equations of motion and energy are coupled and have to be solved simultaneously. In this chapter we limit our discussions to forced convection problems.

In general, the differential equations of fluid flow are nonlinear. Analytical solutions to the equations are available only for a limited number of cases. Because of the difficulty of solving the nonlinear system of equations, and the availability of high-speed digital computers, numerical solutions to a large number of fluid flow problems have been developed since the 1960s. The solutions to fluid flow equations are further complicated by the possible onset of turbulence. Differential equations for laminar flow have been well established, but equations for turbulent flows that are valid for all situations are yet to be defined. Many phenomenological and statistical theories of turbulence have been proposed, but no single theory has been satisfactory in all situations.

The intent of this chapter is to emphasize the conceptual understanding of the derivation of the differential equations for laminar flows and to solve them for a few simple cases. Hence, the discussion is limited to two-dimensional flows. To obtain the solutions to the velocity and temperature fields, the basic laws of mass conservation, linear momentum balance, and energy conservation have to be satisfied. The equations resulting from the application of these laws to a fluid are derived for a differential control volume in Cartesian coordinates and their solutions illustrated. The concept of boundary layers is then discussed; a solution to the boundary layer equations for convection on a flat plate is presented. The complete set of equations for unsteady, three-dimensional cases in Cartesian, cylindrical, and spherical coordinates are given in Appendix 5. The nomenclature adopted in the development of the equations is

$$e = \text{specific energy } [u + (v^2/2)]$$
$$g_x, g_y = \text{component of body forces, per unit mass in the } x\text{- and } y\text{-directions,}$$
$$\text{respectively}$$
$$t = \text{time}$$
$$p = \text{pressure}$$
$$q''_x, q''_y = \text{heat flux in the } x\text{- and } y\text{-direction, respectively}$$
$$u = \text{specific internal energy}$$
$$v_x = \text{velocity component in the } x\text{-direction}$$
$$v_y = \text{velocity component in the } y\text{-direction}$$
$$x, y = \text{Cartesian coordinates}$$
$$\mu = \text{dynamic viscosity}$$
$$\nu = \text{kinematic viscosity} = \mu/\rho$$
$$\rho = \text{density}$$
$$\tau = \text{stress}$$

In this chapter time is denoted by t and stresses by τ because the common symbol for stresses in fluids is τ.

The fluid is assumed to be homogeneous with temperature independent properties. Only two-dimensional flows are considered, i.e., the velocity has only two components in the x- *and* y-*directions,* v_x *and* v_y, *and that each is, at most, a function of* x, y, *and* t *only. As none of the variables is a function of the* z-*coordinate, it is sufficient to consider a unit depth of the fluid in the* z-*direction.*

12.1 MASS BALANCE (CONSERVATION OF MASS)

Consider a differential control volume (C.V.) of dimensions Δx, Δy, and unit depth in the z-direction as shown in Figure 12.1.1.

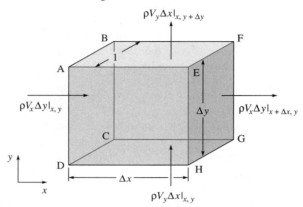

FIGURE 12.1.1 **Control Volume for Mass Conservation**

For the control volume, the law of mass balance is

Rate of change of mass in the C.V.
+ Net rate of mass flow of the fluid out of the C.V.
= 0

where

Mass in the C.V.	$\rho \, \Delta x \, \Delta y$
Rate of change of mass in the C.V.	$(\partial/\partial t)(\rho \, \Delta x \, \Delta y)$
Rate of mass entering the C.V. across surface ABCD	$\rho v_x \, \Delta y \vert_{x,y}$
Rate of mass leaving the C.V. across surface EFGH	$\rho v_x \, \Delta y \vert_{x+\Delta x, y}$
Rate of mass entering the C.V. across surface CDHG	$\rho v_y \, \Delta x \vert_{x,y}$
Rate of mass leaving the C.V. across surface BAEF	$\rho v_y \, \Delta x \vert_{x,y+\Delta y}$

Mass balance

Combining all the terms we get

$$\frac{\partial}{\partial t} (\rho \, \Delta x \, \Delta y) - \rho v_x \, \Delta y \vert_{x,y} + \rho v_x \, \Delta y \vert_{x+\Delta x, y} - \rho v_y \, \Delta x \vert_{x,y} + \rho v_y \, \Delta x \vert_{x,y+\Delta y} = 0$$

Dividing the equation by $\Delta x \, \Delta y$ and taking the limit as $\Delta x \to 0$ and $\Delta y \to 0$, we obtain

$$\left\Vert \; \frac{\partial \rho}{\partial t} + \frac{\partial \rho v_x}{\partial x} + \frac{\partial \rho v_\nu}{\partial y} = 0 \; \right\Vert \tag{12.1.1}$$

Equation 12.1.1 can be recast as

$$\frac{\partial \rho}{\partial t} + v_x \frac{\partial \rho}{\partial x} + v_y \frac{\partial \rho}{\partial y} + \rho \left(\frac{\partial v_x}{\partial x} + \frac{\partial v_y}{\partial y} \right) = 0$$

Now, ρ is a function of x, y, and t and a differential change in the density of the fluid is given by

$$d\rho = \frac{\partial \rho}{\partial t} \, dt + \frac{\partial \rho}{\partial x} \, dx + \frac{\partial \rho}{\partial y} \, dy$$

$d\rho$ represents the change in density that an observer would measure in time dt; during the time interval dt, the observer moves distances of dx and dy. The rate of increase of the density of the fluid that the observer would measure is obtained by dividing the above relation by dt,

$$\frac{d\rho}{dt} = \frac{\partial \rho}{\partial t} + \frac{\partial \rho}{\partial x}\frac{dx}{dt} + \frac{\partial \rho}{\partial y}\frac{dy}{dt}$$

The derivatives dx/dt and dy/dt represent the velocity components of the observer. For a stationary observer the velocities are zero and the rate of change of density of the fluid is given by the partial derivative of the density with respect to time, $\partial \rho / \partial t$. If the observer moves with the fluid packet, then dx/dt and dy/dt represent the velocity components of the fluid in the x- and y-directions. Then, $d\rho/dt$ is the time rate of change of density of an identifiable mass of fluid; i.e., the rate of change of density that an observer moving with the same velocity as the fluid realizes. Such a rate of change of density is known as the *substantial derivative* or the *material derivative of the density* and is denoted by $D\rho/Dt$.

Substantial (material) derivative

$$\frac{D\rho}{Dt} = \frac{\partial \rho}{\partial t} + v_x \frac{\partial \rho}{\partial x} + v_y \frac{\partial \rho}{\partial y}$$

The total derivative $d\rho/dt$ is replaced by $D\rho/Dt$ to indicate that the left side of the equation is the substantial derivative. Similarly, the substantial derivative of a fluid property is obtained from the relation

$$\frac{Db}{Dt} = \frac{\partial b}{\partial t} + v_x \frac{\partial b}{\partial x} + v_y \frac{\partial b}{\partial y}$$

where b represents any property of the fluid such as pressure, momentum, internal energy, enthalpy, entropy, and so on.

Equation 12.1.1 is the two-dimensional form of the differential equation of mass balance for any fluid. For an incompressible fluid, the density, ρ, is constant and Equation 12.1.1 simplifies to

Continuity equation

$$\left\| \; \frac{\partial v_x}{\partial x} + \frac{\partial v_y}{\partial y} = 0 \; \right\| \tag{12.1.2}$$

Equations 12.1.1 and 12.1.2 are also known as *continuity equations*.

12.2 MOMENTUM BALANCE

The law of linear momentum balance for a C.V. is

In any specified direction,

Momentum balance

        Rate of change of momentum of the fluid within the C.V.

    + Net rate of momentum flow out of the C.V.

    = Sum of all the forces on the fluid in the C.V.

Consider a C.V. (Figure 12.2.1) with dimensions Δx and Δy with a unit depth in the z-direction. We derive the momentum equation in the x-direction. *Whenever the term* momentum *is used in the development that follows, it refers to the momentum in the positive* x-*direction.*

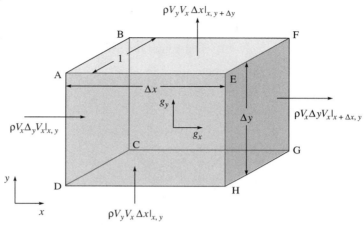

FIGURE 12.2.1 Control Volume Showing the x-Components of Momentum Flows at C.V. Surfaces

Mass of the fluid in the C.V. $\qquad\qquad\qquad\qquad\qquad$ $\rho\,\Delta x\,\Delta y$

Momentum of the fluid in the C.V. in the x-direction $\qquad$ $\rho v_x\,\Delta x\,\Delta y$
at any instant of time

Rate of change of x-momentum of the fluid in the C.V. $\qquad$ $\dfrac{\partial}{\partial t}\,(\rho v_x\,\Delta x\,\Delta y)$

To evaluate the rate of momentum flow out of the C.V. it should be recognized that mass is crossing each surface of the C.V.; the mass crossing each surface has momentum in the x-direction associated with it. To determine the rate of momentum flow out of the C.V., we find the mass rate of flow across each surface, and multiply it by the x-component of the velocity at that surface. Referring to Figure 12.2.1, consider surfaces ABCD and EFGH.

Rate of mass flow across surface ABCD $\qquad$ $\rho v_x\,\Delta y|_{x,y}$

The velocity component in the positive x-direction is v_x and as the mass is flowing into the C.V. across ABCD,

Rate of momentum flow into the C.V. $\qquad$ $\rho v_x^2\,\Delta y|_{x,y}$

Similarly,

Rate of mass flow out of the C.V. across EFGH $\qquad$ $\rho v_x\,\Delta y|_{x+\Delta x,y}$

Rate of momentum flow out across EFGH $\qquad$ $\rho v_x^2\,\Delta y|_{x+\Delta x,y}$

Net rate of momentum flow out of the C.V. $\qquad$ $\rho v_x^2\,\Delta y|_{x+\Delta x,y} - \rho v_x^2\,\Delta y|_{x,y}$
across ABCD and EFGH in the x-direction

Now consider the surfaces CDHG and BAEF:

Rate of mass flow into the C.V. across CDHG $\qquad$ $\rho v_y\,\Delta x|_{x,y}$

The fluid crossing CDHG has a velocity component v_x in the x-direction. The x-momentum flow per unit mass of the fluid crossing CDHG is v_x.

Rate of momentum flow into the C.V. across CDHG $\qquad \rho v_y v_x \, \Delta x|_{x,y}$

Similarly,

Rate of x-momentum flow out of the C.V. across BAEF $\qquad \rho v_y v_x \, \Delta x|_{x,y+\Delta y}$

Combining the momentum flows across BAEF and CDHG,

Rate of x-momentum flow out of the C.V. across BAEF and CDHG
$\rho v_y v_x|_{x,y+\Delta y} - \rho v_y v_x|_{x,y}$

Combining all the terms, the net rate of x-momentum flow out of the C.V. is

$$\rho v_x^2 \, \Delta y|_{x+\Delta x,y} - \rho v_x^2 \, \Delta y|_{x,y} + \rho v_y v_x|_{x,y+\Delta y} - \rho v_y v_x|_{x,y}$$

The left side of the x-momentum balance is given by

$$\frac{\partial}{\partial t} (\rho v_x \, \Delta x \, \Delta y) + \rho v_x^2 \, \Delta y|_{x+\Delta x,y} - \rho v_x^2 \, \Delta y|_{x,y} + \rho v_y v_x|_{x,y+\Delta y} - \rho v_y v_x|_{x,y}$$

Now consider the forces in the x-direction acting on the fluid in the C.V. The forces on the fluid in the C.V. are of two types—surface forces that act on the surfaces of the C.V., and body forces that act on every element of the fluid within the C.V. An example of surface forces is the shear stress on a surface. Gravitational and magnetic forces are body forces. The only type of body force that is considered here is that due to gravity. Figure 12.2.2 shows the C.V. with all the surface stresses and

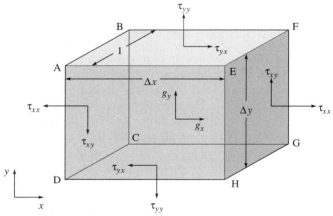

FIGURE 12.2.2 Sketch Showing the Sign Convections for the Stresses on an Element

body forces. Surface forces are subdivided into normal and tangential forces. Normal forces on a surface are due to the pressure of the fluid and viscous stresses normal to the surface. The tangential stresses (also termed shear stresses) are due to viscous effects only.

Sign Conventions for Surface Stresses Surface stresses are denoted by two subscripts. The first subscript indicates the direction of the surface, i.e., the axis parallel to the normal to the surface. The second subscript indicates the direction of the stress. If the outward normal to a surface is in the positive direction of a coordinate axis, the stresses on that surface are considered positive if the directions of the stresses are also in the positive directions of the coordinates. If the outward normal to a surface is in the negative direction of an axis, the stresses in the decreasing (negative) directions of the axes are considered to be positive. For example, the shear stress on the surface EFGH is positive in the y-increasing direction as the normal to the surface EFGH is in the x-increasing (or positive) direction. The shear stress on the surface ABCD is positive if it is in the negative y-direction as the outward normal to ABCD is in the negative x-direction. Stress τ_{xx} indicates that it is a stress on a surface that has an outward normal parallel to the x-axis and that the stress is also parallel to the x-axis; it, therefore, represents a normal stress. τ_{yx} indicates that it is a stress on a surface with an outward normal parallel to the y-direction and that the direction of the stress is parallel to the x-axis—a shear stress. The stresses indicated in Figure 12.2.2 follow this convention.

Surface stresses— sign convention

Now consider the forces, in the x-direction, on surfaces EFGH and ABCD. They are given by

$$\tau_{xx}\, \Delta y \big|_{x+\Delta x, y} - \tau_{xx}\, \Delta y \big|_{x,y}$$

The first term (the product of the stress and the area of the surface on which it is acting) is the force in the x-direction on surface EFGH. As the surface stress τ_{xx} is in the positive x-direction, the force is also positive. The second term represents the x-direction force on ABCD due to the stress τ_{xx} on that surface. As the outward normal to ABCD is along the negative x-direction, τ_{xx} is also along the negative x-direction; thus the force due to τ_{xx} on ABCD is in the negative x-direction. Similarly, the component of the forces on AEFB and DHGC in the x-direction are given by

$$\tau_{yx}\, \Delta x \big|_{x,y+\Delta y} - \tau_{yx}\, \Delta x \big|_{x,y}$$

Let g_x and g_y represent the components of the gravitational forces in the x- and y-directions per unit mass of the fluid.

x-component of the gravitational force on the fluid in the C.V. $\rho g_x\, \Delta x\, \Delta y$

Combining all the forces on the element in the x-direction, the momentum balance yields,

$$\frac{\partial}{\partial t}(\rho v_x\, \Delta x\, \Delta y) + \rho v_x^2\, \Delta y \big|_{x+\Delta x, y} - \rho v_x^2\, \Delta y \big|_{x,y} + \rho v_y v_x \big|_{x,y+\Delta y} - \rho v_y v_x \big|_{x,y}$$

$$= \tau_{xx}\, \Delta y \big|_{x+\Delta x, y} - \tau_{xx}\, \Delta y \big|_{x,y} + \tau_{yx}\, \Delta x \big|_{x,y+\Delta y} - \tau_{yx}\, \Delta x \big|_{x,y} + \rho g_x\, \Delta x\, \Delta y$$

Divide the equation by $\Delta x\, \Delta y$ and take the limit as the differential quantities approach zero. With

$$\lim_{\Delta x \to 0} \frac{\rho v_x^2 \big|_{x+\Delta x, y} - \rho v_x^2 \big|_{x,y}}{\Delta x} = \frac{\partial \rho v_x^2}{\partial x}$$

and similarly for other derivatives, we obtain

$$\frac{\partial \rho v_x}{\partial t} + \frac{\partial}{\partial x}(\rho v_x^2) + \frac{\partial}{\partial y}(\rho v_x v_y) = \frac{\partial \tau_{xx}}{\partial x} + \frac{\partial \tau_{yx}}{\partial y} + \rho g_x \qquad \textbf{(12.2.1)}$$

Equation 12.2.1 can be simplified by splitting each term on the left side into two terms, and utilizing Equation 12.1.1.

$$
\underbrace{\frac{\partial \rho v_x}{\partial t}}_{\longleftrightarrow} \qquad \underbrace{\frac{\partial}{\partial x}(\rho v_x^2)}_{\longleftrightarrow} \qquad \underbrace{\frac{\partial}{\partial y}(\rho v_x v_y)}_{\longleftrightarrow}
$$

$$
v_x \frac{\partial \rho}{\partial t} + \rho \frac{\partial v_x}{\partial t} + \rho v_x \frac{\partial v_x}{\partial x} + v_x \frac{\partial \rho v_x}{\partial x} + \rho v_y \frac{\partial v_x}{\partial y} + v_x \frac{\partial \rho v_y}{\partial y}
$$

$$
= \rho \frac{\partial v_x}{\partial t} + \rho v_x \frac{\partial v_x}{\partial x} + \rho v_y \frac{\partial v_x}{\partial y} + v_x \left(\frac{\partial \rho}{\partial t} + \frac{\partial \rho v_x}{\partial x} + \frac{\partial \rho v_y}{\partial y} \right)
$$

But from mass balance (Equation 12.1.1) the terms in parentheses vanish and the simplified momentum balance in the x-direction is

$$
\left\| \quad \rho \frac{\partial v_x}{\partial t} + \rho v_x \frac{\partial v_x}{\partial x} + \rho v_y \frac{\partial v_x}{\partial y} = \frac{\partial \tau_{xx}}{\partial x} + \frac{\partial \tau_{yx}}{\partial y} + \rho g_x \quad \right\| \qquad \textbf{(12.2.2a)}
$$

In a similar manner, the momentum balance in the y-direction is

$$
\left\| \quad \rho \frac{\partial v_y}{\partial t} + \rho v_x \frac{\partial v_y}{\partial x} + \rho v_y \frac{\partial v_y}{\partial y} = \frac{\partial \tau_{xy}}{\partial x} + \frac{\partial \tau_{yy}}{\partial y} + \rho g_y \quad \right\| \qquad \textbf{(12.2.2b)}
$$

Equations 12.2.a and b are applicable to all fluids in two-dimensional flow. For many fluids such as air and water, the stresses can be related to the velocity components as

Newtonian fluids

$$
\tau_{xx} = -p + 2\mu \frac{\partial v_x}{\partial x} + \lambda \left(\frac{\partial v_x}{\partial x} + \frac{\partial v_y}{\partial y} \right) \qquad \textbf{(12.2.3a)}
$$

$$
\tau_{yx} = \mu \left(\frac{\partial v_x}{\partial y} + \frac{\partial v_y}{\partial x} \right) = \tau_{xy} \qquad \textbf{(12.2.3b)}
$$

$$
\tau_{yy} = -p + 2\mu \frac{\partial v_y}{\partial y} + \lambda \left(\frac{\partial v_x}{\partial x} + \frac{\partial v_y}{\partial y} \right) \qquad \textbf{(12.2.3c)}
$$

where

μ = dynamic viscosity of the fluid
λ = second coefficient of viscosity
p = thermodynamic pressure

Stress relations given in Equation 12.2.3 are not applicable to all fluids but only to a class of fluids. Fluids for which stresses are given by Equations 12.2.3a, b, and c are known as *Newtonian fluids*. Further, for such fluids $\lambda = -2/3 \, \mu$. Substituting Equation 12.2.3 into Equation 12.2.2, we obtain

$$
\rho \frac{\partial v_x}{\partial t} + \rho v_x \frac{\partial v_x}{\partial x} + \rho v_y \frac{\partial v_x}{\partial y} = -\frac{\partial p}{\partial x} + 2 \frac{\partial}{\partial x} \left(\mu \frac{\partial v_x}{\partial x} \right)
$$

$$
- \frac{2}{3} \frac{\partial}{\partial x} \left[\mu \left(\frac{\partial v_x}{\partial x} + \frac{\partial v_y}{\partial y} \right) \right] + \frac{\partial}{\partial y} \left[\mu \left(\frac{\partial v_x}{\partial y} + \frac{\partial v_y}{\partial x} \right) \right] + \rho g_x \qquad \textbf{(12.2.4)}
$$

Equation 12.2.4 can be simplified for an incompressible fluid with constant viscosity. As ρ is constant, from Equation 12.1.2, $[(\partial v_x/\partial x) + (\partial v_y/\partial y)]$ is zero. Further, with constant viscosity,

$$2\frac{\partial}{\partial x}\left(\mu\frac{\partial v_x}{\partial x}\right) + \frac{\partial}{\partial y}\left[\mu\left(\frac{\partial v_x}{\partial y} + \frac{\partial v_y}{\partial x}\right)\right] = \mu\left(\frac{\partial^2 v_x}{\partial x^2} + \frac{\partial^2 v_x}{\partial y^2}\right) + \mu\frac{\partial}{\partial x}\left(\frac{\partial v_x}{\partial x} + \frac{\partial v_y}{\partial y}\right)$$
$$= \mu\left(\frac{\partial^2 v_x}{\partial x^2} + \frac{\partial^2 v_x}{\partial y^2}\right)$$

For an incompressible fluid with constant viscosity, the simplified form of Equation 12.2.4 is

Navier-stokes equations

$$\left\|\quad\frac{\partial v_x}{\partial t} + v_x\frac{\partial v_x}{\partial x} + v_y\frac{\partial v_x}{\partial y} = -\frac{1}{\rho}\frac{\partial p}{\partial x} + v\left(\frac{\partial^2 v_x}{\partial x^2} + \frac{\partial^2 v_x}{\partial y^2}\right) + g_x\quad\right\| \quad \textbf{(12.2.5)}$$

Equation 12.2.5 has been obtained by dividing Equation 12.2.4 by ρ and replacing μ/ρ by v, the kinematic viscosity of the fluid. From an analogy with Equation 12.2.5, the y-momentum equation for an incompressible fluid with constant properties is

$$\left\|\quad\frac{\partial v_y}{\partial t} + v_x\frac{\partial v_y}{\partial x} + v_y\frac{\partial v_y}{\partial y} = -\frac{1}{\rho}\frac{\partial p}{\partial y} + v\left(\frac{\partial^2 v_y}{\partial x^2} + \frac{\partial^2 v_y}{\partial y^2}\right) + g_y\quad\right\| \quad \textbf{(12.2.6)}$$

The derivation of Equation 12.2.6 is similar to that of Equation 12.2.5 and is left as an exercise. Equations 12.2.5 and 12.2.6 are known as *Navier-Stokes equations* for the two-dimensional flow of an incompressible, Newtonian fluid.

12.3 ENERGY BALANCE (THERMAL ENERGY + KINETIC ENERGY)— FIRST LAW OF THERMODYNAMICS

The derivation of the energy equation is, perhaps, more involved than any other derivation, requiring patience and perseverance. However, by the end of the derivation the meaning of the different terms and the restrictions that apply to the equation become clear.

For a C.V., (Figure 12.3.1) the law of balance of energy (First Law of Thermodynamics) is

Energy balance

Rate of change of energy of the fluid in the C.V.
+ Net rate of energy convected out of the C.V.
= Net rate of heat transfer to the fluid in the C.V.
+ Net rate of work transfer to the fluid in the C.V.

In stating the First Law of Thermodynamics, it has been assumed that the rate of internal energy generation is zero. The expression ''convected energy'' refers to the energy associated with the mass crossing the boundaries of the C.V. The specific energy of a substance is the sum of the internal and kinetic energies. We have the option of considering the work transfer due to gravitational force as potential energy and including it in the specific energy or considering such work transfer in the rate of work transfer term. Here we choose to consider the effect of gravity force in the rate of work transfer term. Each term in the statement of the law is now evaluated.

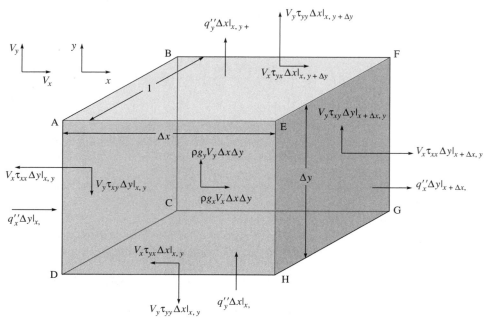

FIGURE 12.3.1 Sketch Showing the Work Transfer Due to Surface and Body Forces and Conductive Heat Transfer Terms

Rate of Change of Energy in the Control Volume

Mass of fluid in the C.V. at any instant $\qquad \rho\, \Delta x\, \Delta y$

Energy associated with the fluid in the C.V. $\qquad \rho e\, \Delta x\, \Delta y$

In the above equation, e is the energy per unit mass of the fluid and is the sum of the specific internal energy and kinetic energy.

Rate of change of energy in the C.V.

$$\frac{\partial}{\partial t}\,(\rho e\, \Delta x\, \Delta y) \tag{12.3.1}$$

Rate of Energy Convected Out of the Control Volume

Energy convection

Mass rate of flow of fluid into C.V. across ABCD $\qquad \rho v_x\, \Delta y|_{x,y}$

Rate of energy convected into the C.V. across ABCD $\qquad \rho v_x e\, \Delta y|_{x,y}$

Similarly,

Rate of energy convected out of the C.V. across EFGH $\qquad \rho v_x e\, \Delta y|_{x+\Delta x, y}$

Rate of energy convected into the C.V. across CDHG $\qquad \rho v_y e\, \Delta x|_{x,y}$

Rate of energy convected out of the C.V. across BAEF $\qquad \rho v_y e\, \Delta x|_{x,y+\Delta y}$

Combining all the terms we obtain

Rate of convected energy out of the C.V.

$$= \rho v_x e\, \Delta y|_{x+\Delta x, y} - \rho v_x e\, \Delta y|_{x,y} + \rho v_y e\, \Delta x|_{x,y+\Delta y} - \rho v_y e\, \Delta x|_{x,y} \tag{12.3.2}$$

Net Rate of Heat Transfer to the Fluid in the Control Volume Heat transfer occurs at the boundaries of the C.V. The heat transfer rate across each surface is the product of the heat flux (in the positive direction of the appropriate axis) and the area of the surface.

Heat conduction

Heat transfer rate across ABCD to the fluid $q_x'' \Delta y|_{x,y}$

Heat transfer rate across EFGH from the fluid $q_x'' \Delta y|_{x+\Delta x,y}$

Similarly, $q_y'' \Delta x|_{x,y}$ and $q_y'' \Delta x|_{x,y+\Delta y}$ are the heat transfer rates across surfaces CDHG and BAEF to and from the fluid in the C.V. Then, combining the terms,

Net rate of heat transfer conducted into the C.V.

$$= q_x'' \Delta y|_{x,y} - q_x'' \Delta y|_{x+\Delta x,y} + q_y'' \Delta x|_{x,y} - q_y'' \Delta x|_{x,y+\Delta y} \qquad \textbf{(12.3.3)}$$

Work transfer

Net Rate of Work Transfer to the Control Volume To evaluate the work transfer, *forces exerted on the fluid in the C.V. by the surroundings are multiplied by the components of velocities of the fluid in the direction of forces at the points of application of the forces.* If a force and the velocity of the fluid where the force is exerted are in the same direction, the work transfer to the fluid in the C.V. is positive; if they are in opposite directions, the work transfer is negative. Computations of work transfer by surface forces are illustrated for two surfaces, ABCD and EFGH, and the work transfer across other surfaces are determined by inspection of the terms so found.

First consider the surface EFGH. Normal force exerted by the surroundings on EFGH is $\tau_{xx}\Delta y$ and is in the positive x-direction. Velocity component in the x-direction is v_x. Hence, the work transfer to the C.V. by the action of the normal force on EFGH is $\tau_{xx}v_x \, \Delta y|_{x+\Delta x,y}$. Both τ_{xx} and v_x are in the same direction (positive x-direction) and the work transfer to the C.V. is positive. The tangential force is $\tau_{xy}\Delta y$ and is in the positive y-direction. The velocity component in the y-direction is v_y. Work transfer resulting from the action of the tangential force is $\tau_{xy}v_y \, \Delta y|_{x+\Delta x,y}$. As both τ_{xy} and v_y are in the same direction, the work transfer due to the tangential force on EFGH is positive.

Now consider the surface ABCD. The normal force is $\tau_{xx} \Delta y$ and is in the negative x-direction. The velocity component is v_x and is in the positive x-direction. As the force and the velocity component are in opposite directions, the work transfer is negative, i.e., work is done by the C.V. on the surroundings. The work transfer to the C.V. due to the normal force on ABCD is $-\tau_{xx}v_x\Delta y|_{x,y}$. The tangential force is $\tau_{xy} \Delta y$ and is in the negative y-direction. The velocity component in the y-direction, v_y, is in the positive y-direction. As the directions of the force and velocity are opposite to each other, the work transfer is negative. The work transfer to the C.V. due to the tangential force on ABCD is $-\tau_{xy}v_y\Delta y|_{x,y}$. Thus, the work transfer terms associated with the forces on surfaces EFGH and ABCD are

Work transfer due to the normal force on EFGH $\tau_{xx}v_x \, \Delta y|_{x+\Delta x,y}$

Work transfer due to the tangential force on EFGH $\tau_{xy}v_y \, \Delta y|_{x+\Delta x,y}$

Work transfer due to the normal force on ABCD $-\tau_{xx}v_x \, \Delta y|_{x,y}$

Work transfer due to the tangential force on ABCD $-\tau_{xy}v_y \, \Delta y|_{x,y}$

The work transfer to the C.V. due to the surface forces on BAEF and CDHG are evaluated in a similar manner and they are

Work transfer due to the normal force on BAEF $\qquad \tau_{yy}v_y \, \Delta x|_{x,y+\Delta y}$

Work transfer due to the tangential force on BAEF $\qquad \tau_{yx}v_x \, \Delta x|_{x,y+\Delta y}$

Work transfer due to the normal force on CDHG $\qquad -\tau_{yy}v_y \, \Delta x|_{x,y}$

Work transfer due to the tangential force on CDHG $\qquad -\tau_{yx}v_x \, \Delta x|_{x,y}$

Work Transfer Due to Gravitational Force

Gravitational force on the fluid (in the C.V.) in the x-direction is $\rho g_x \, \Delta x \, \Delta y$ and is in the positive x-direction. The work transfer due to the gravitational force in the x-direction is $\rho g_x v_x \, \Delta x \, \Delta y$. Similarly, the work transfer due to the gravitational force in the y-direction is $\rho g_y v_y \, \Delta x \, \Delta y$.

Summing up all the work transfer terms,

Net rate of work transfer to the C.V.

$$
\begin{aligned}
= \; & \tau_{xx}v_x \, \Delta y|_{x+\Delta x,y} - \tau_{xx}v_x \, \Delta y|_{x,y} + \tau_{xy}v_y \, \Delta y|_{x+\Delta x,y} - \tau_{xy}v_y \, \Delta y|_{x,y} \\
& + \tau_{yy}v_y \, \Delta x|_{x,y+\Delta y} - \tau_{yy}v_y \, \Delta x|_{x,y} + \tau_{yx}v_x \, \Delta x|_{x,y+\Delta y} - \tau_{yx}v_x \, \Delta x|_{x,y} \\
& + \rho g_x v_x \, \Delta x \, \Delta y + \rho g_y v_y \, \Delta x \, \Delta y
\end{aligned}
\tag{12.3.4}
$$

Combining Equations 12.3.1, 12.3.2, 12.3.3, and 12.3.4, dividing by $\Delta x \, \Delta y$ and taking the limit as Δx and Δy approach zero, we obtain

$$
\begin{aligned}
\frac{\partial}{\partial t}(\rho e) + \frac{\partial}{\partial x}(\rho v_x e) + \frac{\partial}{\partial y}(\rho v_y e) = \; & -\frac{\partial}{\partial x}(q_x'') - \frac{\partial}{\partial y}(q_y'') + \frac{\partial}{\partial x}(\tau_{xx}v_x) \\
& + \frac{\partial}{\partial x}(\tau_{xy}v_y) + \frac{\partial}{\partial y}(\tau_{yy}v_y) \\
& + \frac{\partial}{\partial y}(\tau_{yx}v_x) + \rho g_x v_x + \rho g_y v_y
\end{aligned}
\tag{12.3.5}
$$

Rearrange the left side of Equation 12.3.5 as

$$
e\frac{\partial \rho}{\partial t} + \rho\frac{\partial e}{\partial t} + e\left[\frac{\partial}{\partial x}(\rho v_x) + \frac{\partial}{\partial y}(\rho v_y)\right] + \rho v_x\frac{\partial e}{\partial x} + \rho v_y\frac{\partial e}{\partial y}
$$

From mass conservation, Equation 12.1.1

$$
e\frac{\partial \rho}{\partial t} + e\left[\frac{\partial}{\partial x}(\rho v_x) + \frac{\partial}{\partial y}(\rho v_y)\right] = 0
$$

With this simplification, after expanding the terms, such as $\tau_{xy}v_y$, on the right side, Equation 12.3.5 is written as

$$
\begin{aligned}
\rho\frac{\partial e}{\partial t} + \rho v_x\frac{\partial e}{\partial x} + \rho v_y\frac{\partial e}{\partial y} = \; & -\frac{\partial}{\partial x}(q_x'') - \frac{\partial}{\partial y}(q_y'') + v_x\frac{\partial \tau_{xx}}{\partial x} + \tau_{xx}\frac{\partial v_x}{\partial x} + v_y\frac{\partial \tau_{xy}}{\partial x} \\
& + \tau_{xy}\frac{\partial v_y}{\partial x} + v_y\frac{\partial \tau_{yy}}{\partial y} + \tau_{yy}\frac{\partial v_y}{\partial y} + v_x\frac{\partial \tau_{yx}}{\partial y} + \tau_{yx}\frac{\partial v_x}{\partial y} \\
& + \rho g_x v_x + \rho g_y v_y
\end{aligned}
\tag{12.3.6}
$$

In Equation 12.3.6 the terms

$$\tau_{xx} \frac{\partial v_x}{\partial x} + \tau_{xy} \frac{\partial v_y}{\partial x} + \tau_{yy} \frac{\partial v_y}{\partial y} + \tau_{yx} \frac{\partial v_x}{\partial y}$$

represent the work done by the surface forces in deforming the fluid. The deformation work transfer contributes to the increase in internal energy. The terms

$$v_x \frac{\partial \tau_{xx}}{\partial x} + v_y \frac{\partial \tau_{xy}}{\partial x} + v_y \frac{\partial \tau_{yy}}{\partial y} + v_x \frac{\partial \tau_{yx}}{\partial y}$$

represent the work done by the force imbalance, which accelerates the fluid, and thus, increases the kinetic energy of the fluid. Equation 12.3.6 is further simplified with the aid of Equation 12.2.2a and b. Multiply Equation 12.2.2a by v_x and Equation 12.2.2b by v_y, add the two equations, and obtain

$$\rho \frac{\partial}{\partial t} (w^2/2) + \rho v_x \frac{\partial}{\partial x} (w^2/2) + \rho v_y \frac{\partial}{\partial y} (w^2/2)$$

Transport of kinetic energy

$$= v_x \frac{\partial \tau_{xx}}{\partial x} + v_x \frac{\partial \tau_{yx}}{\partial y} + \rho v_x g_x$$

$$+ v_y \frac{\partial \tau_{xy}}{\partial x} + v_y \frac{\partial \tau_{yy}}{\partial y} + \rho v_y g_y \quad \textbf{(12.3.7)}$$

In Equation 12.3.7, $w^2 (= v_x^2 + v_y^2)$ is twice the kinetic energy of the fluid per unit mass. This equation can be viewed as the equation for the transport of kinetic or mechanical energy. The effect of the work transfer due to the force imbalance and gravitational force in increasing the kinetic energy of the fluid is clear from Equation 12.3.7. Subtracting Equation 12.3.7 from Equation 12.3.6, and noting that $e = u + w^2/2$, where $u =$ specific internal energy,

$$\rho \frac{\partial u}{\partial t} + \rho v_x \frac{\partial u}{\partial x} + \rho v_y \frac{\partial u}{\partial y} = - \frac{\partial}{\partial x} (q_x'') - \frac{\partial}{\partial y} (q_y'') + \tau_{xx} \frac{\partial v_x}{\partial x} + \tau_{xy} \frac{\partial v_y}{\partial x}$$

$$+ \tau_{yy} \frac{\partial v_y}{\partial y} + \tau_{yx} \frac{\partial v_x}{\partial y} \quad \textbf{(12.3.8)}$$

Equation 12.3.8 is applicable to both compressible and incompressible fluids. We now specialize it to incompressible Newtonian fluids.

From Fourier's law

$$q_x'' = -k \frac{\partial T}{\partial x} \qquad q_y'' = -k \frac{\partial T}{\partial y}$$

For incompressible fluids, from Equations 12.2.3a, b, and c,

$$\tau_{xx} = -p + 2\mu \frac{\partial v_x}{\partial x} \qquad \tau_{yy} = -p + 2\mu \frac{\partial v_y}{\partial y} \qquad \tau_{xy} = \tau_{yx} = \mu \left(\frac{\partial v_x}{\partial y} + \frac{\partial v_y}{\partial x} \right)$$

Now, $u = u(T, v)$ and

$$du = \left. \frac{\partial u}{\partial T} \right|_v dT + \left. \frac{\partial u}{\partial v} \right|_T dv$$

where v is the specific volume. For an incompressible fluid $v =$ constant and with $\partial u/\partial T|_v = c_v$, for a homogeneous fluid

$$\frac{\partial u}{\partial t} = c_v \frac{\partial T}{\partial t} \qquad \frac{\partial u}{\partial x} = c_v \frac{\partial T}{\partial x} \qquad \frac{\partial u}{\partial y} = c_v \frac{\partial T}{\partial y}$$

Substituting the above relations into Equation 12.3.8, we obtain

$$\rho c_v \frac{\partial T}{\partial t} + \rho v_x c_v \frac{\partial T}{\partial x} + \rho v_y c_v \frac{\partial T}{\partial y} = \frac{\partial}{\partial x}\left(k \frac{\partial T}{\partial x}\right) + \frac{\partial}{\partial y}\left(k \frac{\partial T}{\partial y}\right) - p\left(\frac{\partial v_x}{\partial x} + \frac{\partial v_y}{\partial y}\right)$$
$$+ 2\mu\left(\frac{\partial v_x}{\partial x}\right)^2 + 2\mu\left(\frac{\partial v_y}{\partial y}\right)^2 + \mu\left(\frac{\partial v_x}{\partial y} + \frac{\partial v_y}{\partial x}\right)^2$$

(12.3.9)

In Equation 12.3.9 the term $-p\left[(\partial v_x/\partial x) + (\partial v_y/\partial y)\right]$ represents the reversible compression work. Note that for an incompressible fluid the compression work is zero.

From Equation 12.1.2 for an incompressible fluid, $(\partial v_x/\partial x) + (\partial v_y/\partial y) = 0$. Assuming constant thermal conductivity, the energy equation is written as

$$\left\|\begin{array}{c} \dfrac{\partial T}{\partial t} + v_x \dfrac{\partial T}{\partial x} + v_y \dfrac{\partial T}{\partial y} = \dfrac{k}{\rho c_v}\left(\dfrac{\partial^2 T}{\partial x^2} + \dfrac{\partial^2 T}{\partial y^2}\right) \\[4mm] + \dfrac{1}{\rho c_v}\,\mu\left[2\left(\dfrac{\partial v_x}{\partial x}\right)^2 + 2\left(\dfrac{\partial v_y}{\partial y}\right)^2 + \left(\dfrac{\partial v_x}{\partial y} + \dfrac{\partial v_y}{\partial x}\right)^2\right] \end{array}\right\|$$

(12.3.10)

Viscous dissipation   The expression associated with the viscosity

$$\phi = 2\left(\frac{\partial v_x}{\partial x}\right)^2 + 2\left(\frac{\partial v_y}{\partial y}\right)^2 + \left(\frac{\partial v_x}{\partial y} + \frac{\partial v_y}{\partial x}\right)^2$$

is known as the dissipation function. The function, $\mu\phi$, represents the viscous dissipation or the work done by the viscous forces. As all the terms in ϕ are positive, the viscous dissipation is always positive. The viscous dissipation represents the irreversible work transfer resulting from the viscosity and leads to an increase in the entropy.

Equation 12.3.8 can be rearranged by eliminating the specific internal energy in favor of specific enthalpy. Recasting Equation 12.3.8 in terms of the internal energy

$$\rho \frac{\partial u}{\partial t} + \rho v_x \frac{\partial u}{\partial x} + \rho v_y \frac{\partial u}{\partial y} = \frac{\partial}{\partial x}\left(k \frac{\partial T}{\partial x}\right) + \frac{\partial}{\partial y}\left(k \frac{\partial T}{\partial y}\right) - p\left(\frac{\partial v_x}{\partial x} + \frac{\partial v_y}{\partial y}\right)$$
$$+ 2\mu\left[\left(\frac{\partial v_x}{\partial x}\right)^2 + 2\mu\left(\frac{\partial v_y}{\partial y}\right)^2 + \mu\left(\frac{\partial v_x}{\partial y} + \frac{\partial v_y}{\partial x}\right)^2\right]$$

(12.3.11)

From Equation 12.1.1 we have

$$-p\left(\frac{\partial v_x}{\partial x} + \frac{\partial v_y}{\partial y}\right) = \frac{p}{\rho}\left(\frac{\partial \rho}{\partial t} + v_x \frac{\partial \rho}{\partial x} + v_y \frac{\partial \rho}{\partial y}\right)$$

$$\frac{\partial \rho}{\partial t} = -\rho^2 \frac{\partial}{\partial t}\left(\frac{1}{\rho}\right) \Rightarrow \frac{p}{\rho}\frac{\partial \rho}{\partial t} = -p\rho \frac{\partial}{\partial t}\left(\frac{1}{\rho}\right)$$

$$\frac{\partial}{\partial t}\left(\frac{p}{\rho}\right) = p\frac{\partial}{\partial t}\left(\frac{1}{\rho}\right) + \frac{1}{\rho}\frac{\partial p}{\partial t} \Rightarrow \frac{p}{\rho}\frac{\partial \rho}{\partial t} = -p\rho \frac{\partial}{\partial t}\left(\frac{1}{\rho}\right) = -\rho \frac{\partial}{\partial t}\left(\frac{p}{\rho}\right) + \frac{\partial p}{\partial t}$$

Similarly,

$$\frac{p}{\rho} v_x \frac{\partial \rho}{\partial x} = -\rho v_x \frac{\partial}{\partial x}\left(\frac{p}{\rho}\right) + v_x \frac{\partial p}{\partial x} \qquad \frac{p}{\rho} v_y \frac{\partial \rho}{\partial y} = -\rho v_y \frac{\partial}{\partial y}\left(\frac{p}{\rho}\right) + v_y \frac{\partial p}{\partial y}$$

Thus,

$$-p\left(\frac{\partial v_x}{\partial x} + \frac{\partial v_y}{\partial y}\right) = -\rho\left[\frac{\partial}{\partial t}\left(\frac{p}{\rho}\right) + v_x \frac{\partial}{\partial x}\left(\frac{p}{\rho}\right) + v_y \frac{\partial}{\partial y}\left(\frac{p}{\rho}\right)\right]$$
$$+ \left(\frac{\partial p}{\partial t} + v_x \frac{\partial p}{\partial x} + v_y \frac{\partial p}{\partial y}\right)$$

Substituting the above relation into Equation 12.3.11, and transferring

$$-\rho\left[\frac{\partial}{\partial t}\left(\frac{p}{\rho}\right) + v_x \frac{\partial}{\partial x}\left(\frac{p}{\rho}\right) + v_y \frac{\partial}{\partial y}\left(\frac{p}{\rho}\right)\right]$$

to the left side, we obtain

$$\rho\left[\frac{\partial}{\partial t}\left(u + \frac{p}{\rho}\right) + v_x \frac{\partial}{\partial x}\left(u + \frac{p}{\rho}\right) + v_y \frac{\partial}{\partial y}\left(u + \frac{p}{\rho}\right)\right] = \frac{\partial}{\partial x}\left(k\frac{\partial T}{\partial x}\right)$$
$$+ \frac{\partial}{\partial y}\left(k\frac{\partial T}{\partial y}\right) + \left(\frac{\partial p}{\partial t} + v_x \frac{\partial p}{\partial x} + v_y \frac{\partial p}{\partial y}\right) + \mu\phi$$

The quantity $u + (p/\rho)$ is the specific enthalpy, i. Therefore, the above equation is rewritten as

$$\rho\left(\frac{\partial i}{\partial t} + v_x \frac{\partial i}{\partial x} + v_y \frac{\partial i}{\partial y}\right) = \frac{\partial}{\partial x}\left(k\frac{\partial T}{\partial x}\right) + \frac{\partial}{\partial y}\left(k\frac{\partial T}{\partial y}\right)$$
$$+ \left(\frac{\partial p}{\partial t} + v_x \frac{\partial p}{\partial x} + v_y \frac{\partial p}{\partial y}\right) + \mu\phi \qquad \textbf{(12.3.12)}$$

Recall that

$$\frac{\partial i}{\partial t} + v_x \frac{\partial i}{\partial x} + v_y \frac{\partial i}{\partial y} = \frac{Di}{Dt}$$

and

$$\frac{\partial p}{\partial t} + v_x \frac{\partial p}{\partial x} + v_y \frac{\partial p}{\partial y} = \frac{Dp}{Dt}$$

where Di/Dt and Dp/Dt are the time rate of change of the specific enthalpy and pressure of an identifiable mass of the fluid. Hence, denoting the change in the specific enthalpy of an infinitesimal mass of the fluid by Di, from thermodynamic relations we have [Van Wylen and Sonntag (1985)]

$$\rho\, Di = \rho c_p\, DT + (1 - \beta T)\, Dp$$

where β is the volumetric thermal expansion coefficient $[= -1/\rho(\partial\rho/\partial T)_p]$. Therefore,

$$\rho\frac{Di}{Dt} = \rho c_p\frac{DT}{Dt} + (1 - \beta T)\frac{Dp}{Dt}$$

Introducing the above relation into Equation 12.3.12, we obtain

$$\left\|\; \frac{\partial T}{\partial t} + v_x \frac{\partial T}{\partial x} + v_y \frac{\partial T}{\partial y} = \alpha \left(\frac{\partial^2 T}{\partial x^2} + \frac{\partial^2 T}{\partial y^2} \right) + \frac{\beta T}{\rho c_p} \frac{Dp}{Dt} + \frac{\nu}{c_p} \phi \;\right\|$$

(12.3.13)

For a constant density fluid, $\beta = 0$. If viscous dissipation is negligible, $\phi = 0$. Then, Equation 12.3.13 simplifies to

$$\left\|\; \frac{\partial T}{\partial t} + v_x \frac{\partial T}{\partial x} + v_y \frac{\partial T}{\partial y} = \alpha \left(\frac{\partial^2 T}{\partial x^2} + \frac{\partial^2 T}{\partial y^2} \right) \;\right\|$$ (12.3.13a)

The differential equations for mass, momentum, and energy balance for two-dimensional flows of incompressible fluids have been derived in Cartesian coordinates. Similar equations can be derived for three-dimensional flows. Three-dimensional, transient differential equations for the balance of mass, momentum, and energy in Cartesian, cylindrical, and spherical coordinates are given in Appendix 5. The momentum equations in Appendix 5 are known as Navier-Stokes equations.

Solutions to the nonlinear equations of motion are quite involved. There are no known general solutions to the system of equations. Only a few problems have closed-form solutions.

The viscous dissipation terms represent the increase in internal energy due to the work done on the fluid by viscous forces (conversion of mechanical energy to thermal energy). Such a process is analogous to frictional work. In many engineering applications, the dissipation effect is not particularly significant compared with the effect of heat transfer. An estimate of the effect of the dissipation function will now be obtained.

Consider the fully developed, steady, laminar flow of an incompressible fluid in a circular tube, with the z-axis coinciding with the axis of the tube. In such a flow, the axial component of the velocity of the fluid is a function of the radius only, $v_z = v_z(r)$ and $v_r = v_\theta = 0$. From the energy equation for an incompressible fluid in cylindrical coordinates (see Appendix 5), for steady state, we get

$$\frac{\partial T}{\partial z} = \frac{1}{v_z} \frac{k}{\rho c_p} \left[\frac{1}{r} \frac{\partial}{\partial r} \left(r \frac{\partial T}{\partial r} \right) + \frac{\partial^2 T}{\partial z^2} \right] + 2 \frac{1}{v_z} \frac{\mu}{\rho c_p} \left(\frac{\partial v_z}{\partial r} \right)^2$$

The term $k(\partial^2 T/\partial z^2)$ representing the axial conduction is negligible if Re Pr $\gg 1$, a condition that is satisfied in many cases. In such cases, the first group of terms on the right-hand side, $(1/v_z)(k/\rho c_p)\{(1/r)(\partial/\partial r)[r(\partial T/\partial r)]\}$, represents the rate of increase of the temperature in the z-direction due to r-direction conductive heat transfer to the fluid, and the last term represents the rate of increase of temperature in the z-direction due to viscous dissipation.

To compare the increase in the temperature of the fluid (in the direction of flow) due to heat conduction and viscous dissipation, we *estimate* the value of $\partial T/\partial z$ due to viscous dissipation by

$$\left. \frac{\partial T}{\partial z} \right|_\phi = 2 \frac{\mu}{\rho c_p v_z} \left(\frac{\partial v_z}{\partial r} \right)^2$$

(12.3.14)

For laminar flow of a fluid in a circular pipe, *for an estimate,* v_z in the denominator of Equation 12.3.14 is taken as the mean velocity (based on mass flow rate), which is $v_c/2$, where v_c is the axial velocity at the center line of the pipe. $\partial v_z/\partial r$ is approximated by v_c/R, R being the radius of the pipe. Substituting these values into Equation 12.3.14 and denoting the effect of viscous dissipation by $\partial T/\partial z|_\phi$, we estimate

$$\left.\frac{\partial T}{\partial z}\right|_\phi \approx \frac{4\mu}{\rho c_p v_c}\left(\frac{v_c}{R}\right)^2 = \frac{4\mu v_c}{\rho c_p R^2} = 4\,\frac{\rho v_c D}{\mu c_p}\left(\frac{\mu}{\rho}\right)^2\frac{4}{D^3} = 32\,\frac{\rho v_m D}{\mu c_p}\,\frac{\nu^2}{D^3}$$

$$= 32\,\text{Re}\,\frac{\nu^2}{c_p D^3}$$

The upper limit of Re for laminar flow is approximately 2000. For atmospheric air at 300 K ($c_p = 1007$ J/kg K, $\nu \approx 15.72 \times 10^{-6}$ m²/s) flowing in a 6-mm diameter tube,

$$\left.\frac{\partial T}{\partial z}\right|_\phi \approx = \frac{32 \times 2000(15.72 \times 10^{-6})^2}{1007 \times 0.006^3} = 0.07\ °\text{C/m}$$

In cases where conductive heat transfer is significant, the temperature change due to such conductive heat transfer is of the order of 2 °C/m–10 °C/m. Compared with the change in the temperature of the air due to heat transfer, the change in temperature due to viscous dissipation is quite small and is usually neglected. For water the value of $\partial T/\partial z|_\phi$ under similar conditions is much smaller, of the order of 5×10^{-5} °C/m. Thus, viscous dissipation is not significant in heat transfer applications in gases and many liquids but can become significant if the fluid is highly viscous or if the velocities are very high. For a discussion on viscous dissipation, see also Section 5.1.6.

12.3.1 Determination of the Convective Heat Transfer Coefficient

Convective heat transfer coefficient from analytical solutions

We have obtained the differential equations for the velocity and temperature distributions in a fluid in forced flow, with heat transfer. How is the solution to the equations related to the convective heat transfer coefficient? The solution to the energy equation gives the temperature distribution in the fluid, $T(x, y, z, t)$. From the no-slip boundary condition at the wall, the fluid adjacent to the solid surface is stationary. The heat flux from the wall to the stationary fluid layer is given by Fourier's law, $-k(\partial T/\partial n)|_{\text{wall}}$, where n is the outward normal to the wall. If T_s is the temperature of the wall, found by evaluating $T(x, y, z, t)$ at the wall, and T_{ref} is the reference temperature for defining the convective heat transfer coefficient, then,

$$q_w'' = -k\left.\frac{\partial T}{\partial n}\right|_{\text{wall}} = h(T_s - T_{\text{ref}})$$

or

$$h = \frac{-k(\partial T/\partial n)|_{\text{wall}}}{T_s - T_{\text{ref}}}$$

12.4 SOME EXACT SOLUTIONS

The application of the differential equations for mass, momentum, and energy balance will now be illustrated through some examples. First, we consider those cases where exact solutions can be found. Later, we examine how the momentum and energy equations are approximated to yield boundary layer equations, solutions to which are much simpler than solutions for the complete equations.

Heat Transfer in Circular Tubes

Fully developed velocity and temperature profiles—pipe flow

First, we consider the internal, laminar flow of an incompressible fluid in the fully developed velocity and temperature profile region with heat transfer from the tube wall to the fluid (Figure 12.4.1). The properties of the fluid are assumed to be constant. The concept of fully developed profiles (both velocity and temperature) was introduced in Chapter 4. By fully developed temperature profile we mean that an

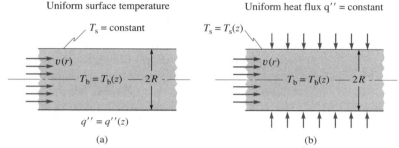

FIGURE 12.4.1 Fluid with Fully Developed Velocity Profile Flowing Inside a Heated Tube
(a) Uniform surface temperature. (b) Uniform surface heat flux.

appropriately defined dimensionless temperature of the fluid is a function of only the radius of the tube and is independent of the axial coordinate. For internal flows in tubes, define the dimensionless temperature, θ, as

$$\frac{T_s(z) - T(r, z)}{T_s(z) - T_c(z)} = \theta(r) \tag{12.4.1}$$

where

T_s = temperature of the wall surface
T = temperature of the fluid
T_c = temperature of the fluid at the axis of the tube
r = radial coordinate
z = axial coordinate

The condition given by Equation 12.4.1, where the dimensionless temperature is independent of the axial coordinate, is satisfied (both for uniform surface temperature and uniform heat flux) beyond some distance from the point where the heat transfer to the fluid begins. We now examine some of the consequences of θ being independent of z.

In both cases of uniform surface temperatures and uniform heat flux, the ratio of the difference between the surface temperature and the centerline temperature to the difference between the surface temperature and the bulk temperature, T_b is constant (the proof is left as an exercise, Problem 12.7).

$$\frac{T_s - T_c}{T_s - T_b} = \text{constant} \tag{12.4.2}$$

As θ is a function of the radius only (and is independent of the axial coordinate, z) it follows that its derivative with respect to the radius evaluated at the surface of the pipe (radius R) is independent of z.

$$\left.\frac{\partial\theta}{\partial r}\right|_{r=R} \neq f(z)$$

As neither T_s nor T_c is a function of the radius, from Equation 12.4.1,

$$\left.\frac{\partial\theta}{\partial r}\right|_{r=R} = -\frac{(\partial T/\partial r)_{r=R}}{T_s - T_c} \neq f(z)$$

But from Fourier's law, if q_w'' is the heat flux from the pipe to the fluid, then, $q_w'' = k(\partial T/\partial r)|_{r=R}$. Also, for fully developed velocity and temperature profiles, from Equation 12.4.2, $T_s - T_c = c(T_s - T_b)$, where c is a constant. Thus,

$$-\frac{q_w''}{k(T_s - T_b)} \neq f(z)$$

Defining the convective heat transfer coefficient for flow in tubes as

$$h = \frac{q_w''}{T_s - T_b}$$

we have

$$\frac{h}{k} \neq f(z) \tag{12.4.3}$$

Thus, the conclusion is that for a fluid with constant properties, in the fully developed temperature profile region, the convective heat transfer coefficient is independent of the axial coordinate z; i.e., it is constant. In arriving at Equation 12.4.3, no restriction as to whether the tube surface is at a uniform temperature or subjected to a uniform heat flux is made. Hence, it is valid for both cases. As Equation 12.4.1, the starting point for the inequality given by Equation 12.4.3, is not valid in the entrance region. Equation 12.4.3 is not applicable in the entrance region. In the entrance region, the convective heat transfer coefficient is a function of the axial coordinate.

Now consider the case of uniform wall heat flux. In this case both h and q_w'' are uniform. Hence, with $T_s - T_b = q_w''/h$,

$$T_s - T_b = \text{constant} \tag{12.4.4}$$

or

$$\frac{dT_s}{dz} = \frac{dT_b}{dz} \tag{12.4.5}$$

From the application of the First Law of Thermodynamics for an elemental length of the tube in the axial direction,

$$\dot{m}c_pT_b|_{z+\Delta z} - \dot{m}c_pT_b|_z = q_w''2\pi R\,\Delta z$$

$$\frac{dT_b}{dz} = \frac{q_w''2\pi R}{\dot{m}c_p} = \text{constant} \qquad (12.4.6a)$$

From Equations 12.4.1 and 12.4.4, with $T_s - T_c = \text{constant}$, $(d/dz)(T_s - T) = 0$ and $dT_s/dz = \partial T/\partial z$. Hence,

$$\frac{\partial T}{\partial z} = \frac{dT_s}{dz} = \frac{dT_b}{dz} \qquad (12.4.6b)$$

Now, for the case of uniform surface temperature, $T_s = \text{constant}$. Expanding Equation 12.4.1 [with $T_s - T_c = c(T_s - T_b)$],

$$-\frac{\partial T/\partial z}{T_s - T_b} - \frac{T_s - T}{(T_s - T_b)^2}\left(-\frac{dT_b}{dz}\right) = 0$$

or

$$\frac{\partial T}{\partial z} = \frac{T_s - T}{T_s - T_b}\left(\frac{dT_b}{dz}\right) \qquad (12.4.7)$$

EXAMPLE 12.4.1 An incompressible fluid is in steady laminar flow with fully developed velocity and temperature profiles in a horizontal, circular tube of radius R (Figure 12.4.2). A uniform heat flux, q_w'' is imposed on the surface of the tube. Determine (a) the velocity profile, (b) friction factor, and (c) the convective heat transfer coefficient (based on the bulk temperature). Assume constant properties.

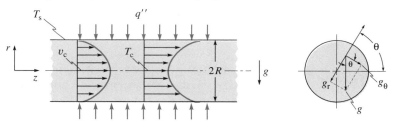

FIGURE 12.4.2 Fully Developed Velocity and Temperature Profiles in an Incompressible Fluid Flowing in a Circular Tube

SOLUTION

(a) With a fully developed velocity profile, the dimensionless axial velocity is independent of the axial coordinate and is a function of the radius only, i.e., $v_z/v_c = f(r)$. The axial velocity is then given by, $v_z = v_c f(r)$, where v_c is the velocity at the center line. Under steady-state conditions, the mass rate of flow of the fluid is constant. As a consequence,

$$\dot{m} = \int_0^R \rho v_z 2\pi r\,dr = \int_0^R \rho v_c f(r)2\pi r\,dr = \text{constant}$$

As the density is constant (incompressible fluid), it follows that v_c cannot be a function of z and is, therefore, constant. If v_c is constant, the axial velocity v_z is also a function of the radius only. We further assume that there is no swirl in the flow, i.e., the tangential velocity, v_θ, is zero everywhere. With $v_z = v_z(r)$, from mass balance, Equation A5.6 (Appendix 5), it follows that

$$\frac{\partial}{\partial r}(rv_r) = 0 \quad \text{or} \quad rv_r = f(z) \text{ at any cross section}$$

As $v_r = 0$ at the tube surface $(r = R)$, it follows that $f(z) = 0$ and, hence, v_r is zero every where. With v_r and v_θ being zero, and $v_z = v_z(r)$, from momentum balance, Equations A5.7 and A5.8, we have

r-component
$$\frac{\partial p}{\partial r} = -\rho g \sin \theta \tag{1}$$

θ-component
$$\frac{\partial p}{\partial \theta} = -\rho r g \cos \theta \tag{2}$$

At any given axial location, z, the change in pressure is given by

$$dp_z = \frac{\partial p}{\partial r} dr + \frac{\partial p}{\partial \theta} d\theta = -\rho g \sin \theta \, dr - \rho g r \cos \theta \, d\theta \tag{3}$$

Integrating Equation 3, at some value of z, from $(0, \theta_o, z)$ to (r, θ, z), and denoting the pressure at the center line, $p(r = 0, z) = p_o(z)$.

$$p(r, \theta, z) - p_o(z) = -\int_{0,\theta_o}^{r,\theta_o} \rho g \sin \theta_o \, dr - \int_{r,\theta_o}^{r,\theta} \rho r g \cos \theta \, d\theta$$
$$= -\rho r g \sin \theta$$

Here the integration is from the center line to r at constant θ $(= \theta_o)$ and along an arc with constant r, with θ changing from θ_o to θ.[1] Thus,

$$p(r, \theta, z) = p_o(z) - \rho r g \sin \theta \tag{4}$$

From the z-momentum balance, Equation A5.9,

$$-\frac{\partial p}{\partial z} + \mu \left[\frac{1}{r} \frac{d}{dr} \left(r \frac{dv_z}{dr} \right) \right] = 0 \tag{5}$$

Here, $g_z = 0$ for a horizontal pipe. From Equation 4, $\partial p/\partial z = dp_o/dz$. Thus, we get

$$\frac{1}{r} \frac{d}{dr} \left(r \frac{dv_z}{dr} \right) = \frac{1}{\mu} \frac{dp_o}{dz} \tag{6}$$

In Equation 6, the left side is at most a function of r and the right side a function of z. As Equation 6 is true for arbitrary values of r and z, the equality holds good only if each of these terms is a constant, i.e., dp_o/dz is constant.

---

[1]You may recognize that Equation 4 may be interpreted as the variation of pressure, at a given value of z, in a static fluid. $r \sin \theta$ represents the elevation of the point above the axis of the tube.

One of the boundary conditions for the axial velocity (the no-slip condition at the solid surface) is

$$v_z(r = R) = 0 \tag{7}$$

The second boundary condition is not immediately apparent. The second boundary condition is found from an examination of the solution and the physics of the problem. With dp_o/dz = constant, integrating Equation 6 twice,

$$v_z = \frac{1}{4\mu} \frac{dp_o}{dz} r^2 + c_1 \ln r + c_2 \tag{8}$$

Because $v_z(r = 0)$ is finite, c_1 is zero. Equation 7 gives

$$c_2 = -\frac{1}{4\mu} \frac{dp_o}{dz} R^2 \tag{9}$$

The velocity profile is then given by

$$v_z = -\frac{1}{4\mu} \frac{dp_o}{dz} R^2 \left(1 - \frac{r^2}{R^2}\right) \tag{10}$$

Note that for v_z to be positive dp_o/dz must be negative, i.e., the flow is in the direction of decreasing pressure. The term $-(1/4\mu)(dp/dz)R^2$, is the center line velocity of the fluid, v_c. Equation 10 can be recast as

$$v_z = v_c \left(1 - \frac{r^2}{R^2}\right) \tag{11}$$

which is the equation of a parabola. The parabolic velocity distribution in a pipe is applicable only for laminar flows. The velocity distribution in turbulent flows is quite different from the parabolic distribution.

We define an average velocity v_m based on the mass flow rate as that value of uniform velocity that gives the same mass flow rate as the actual mass flow rate. With the velocity profile given by Equation 11,

$$\rho\pi R^2 v_m = \int_0^R \rho v_c \left(1 - \frac{r^2}{R^2}\right) 2\pi r \, dr$$

which yields

$$v_m = \frac{v_c}{2} \tag{12}$$

The mean velocity based on the mass flow rate is one-half of the maximum velocity that occurs at the center line.

(b) The shear stress at the wall is given by

$$\tau_w = -\mu \left.\frac{dv_z}{dr}\right|_{r=R} = 2\mu \frac{v_c}{R}$$

Defining the Fanning friction factor, $C_F = \tau_w/(\rho v_m^2/2)$,

$$C_F = \frac{4\mu v_c}{\rho v_m^2 R} = \frac{16\mu}{\rho v_m d} = \frac{16}{Re_d} \qquad Re_d = \frac{\rho v_m d}{\mu}$$

Another common way of defining the friction factor is based on the pressure drop. The Darcy friction factor f is defined as

$$f = d \frac{-\partial p_o/\partial z}{\rho v_m^2/2} \qquad v_c = 2v_m = -\frac{1}{4\mu}\frac{dp_o}{dz}R^2$$

Substituting $dp_o/dz = (32\mu v_m)/d^2$, the friction factor is

$$f = \frac{64}{\mathrm{Re}_d} \tag{13}$$

(c) Having determined the velocity profile, we now determine the convective heat transfer coefficient in the fully developed temperature profile region. It has already been shown that, with constant properties, in the fully developed temperature profile region, the convective heat transfer coefficient is constant. With the velocity profile given by Equation 11, the energy balance in cylindrical coordinates from Appendix 5, [neglecting viscous dissipation ($\mu\phi$), with $v_r = v_\theta = 0$, $\partial T/\partial\theta = 0$, $\beta = 0$ for ρ = constant] is

$$v_z \frac{\partial T}{\partial z} = \alpha \left[\frac{1}{r}\frac{\partial}{\partial r}\left(r\frac{\partial T}{\partial r} \right) + \frac{\partial^2 T}{\partial z^2} \right] \tag{14}$$

From Equations 12.4.6a and b, $\partial T/\partial z = dT_b/dz$ = constant and, hence, $\partial^2 T/\partial z^2 = 0$. Substituting Equation 11 for v_z, Equation 14 simplifies to

$$\frac{\partial}{\partial r}\left(r\frac{\partial T}{\partial r} \right) = \frac{v_c}{\alpha}\left(1 - \frac{r^2}{R^2} \right)r\frac{dT_b}{dz} \tag{15}$$

Integrating Equation 15 once, in the r-direction, at a specified axial location

$$r\frac{\partial T}{\partial r} = \frac{v_c}{\alpha}\left(\frac{r^2}{2} - \frac{r^4}{4R^2} \right)\frac{dT_b}{dz} + c_1$$

As $\partial T/\partial r$ is finite at $r = 0$, $c_1 = 0$. Integrating once again,

$$T = \frac{v_c}{\alpha}\left(\frac{r^2}{4} - \frac{r^4}{16R^2} \right)\frac{dT_b}{dz} + c_2$$

Denoting the centerline temperature by T_c, $T(r = 0) = T_c = c_2$. (c_2 has been replaced by T_c for convenience and to give a physical interpretation for c_2. T_c is still not known.)

$$T - T_c = \frac{v_c}{\alpha}\left(\frac{r^2}{4} - \frac{r^4}{16R^2} \right)\frac{dT_b}{dz} \tag{16}$$

If the surface temperature at the particular axial location is $T(r = R) = T_s$,

$$T_s - T_c = \frac{v_c}{\alpha}\frac{3}{16}R^2\frac{dT_b}{dz} \tag{17}$$

Dividing Equation 16 by Equation 17, we obtain the dimensionless temperature profile,

$$\frac{T - T_c}{T_s - T_c} = \frac{4}{3}\frac{r^2}{R^2}\left(1 - \frac{r^2}{4R^2} \right) \tag{18}$$

Employing Equation 16, we determine the bulk temperature T_b from the relation (see Section 4.2)

$$T_b = \frac{\int_{A_c} \rho v_z c_p T \, dA_c}{\int_{A_c} \rho v_z c_p \, dA_c} = \frac{\int_0^R v_z T 2\pi r \, dr}{\int_0^R v_z 2\pi r \, dr}$$

where A_c is the cross-sectional area of the tube perpendicular to its axis.

$$T_b = \frac{\int_0^R v_c \left[1 - (r^2/R^2) \right] \left\{ \dfrac{v_c}{\alpha} [(r^2/4) - (r^4/16R^2)] dT_b/dz + T_c \right\} r \, dr}{\int_0^R v_c [1 - (r^2/R^2)] r \, dr}$$

which gives

$$T_b = \frac{v_c}{\alpha} \frac{7}{96} R^2 \frac{dT_b}{dz} + T_c \tag{19}$$

From Equation 16, the heat flux to the fluid from the surface is given by

$$k \left. \frac{\partial T}{\partial r} \right|_{r=R} = q_w'' = k \frac{v_c}{\alpha} \frac{R}{4} \frac{dT_b}{dz}$$

From the definition of the convective heat transfer coefficient (with the bulk temperature as the reference temperature),

$$h = \frac{q_w''}{T_s - T_b} = \frac{k(v_c/\alpha)(R/4)(dT_b/dz)}{T_s - T_b} \tag{20}$$

Subtracting Equation 19 from Equation 17, we obtain

$$T_s - T_b = \frac{v_c}{\alpha} \frac{3}{16} R^2 \frac{dT_b}{dz} - \frac{v_c}{\alpha} \frac{7}{96} R^2 \frac{dT}{dz}$$

or

$$T_s - T_b = \frac{11}{96} \frac{v_c}{\alpha} R^2 \frac{dT_b}{dz} \tag{21}$$

Substituting Equation 21 into Equation 20, $h = (24/11)(k/R)$ or

$$Nu_d = \frac{hd}{k} = \frac{48}{11} = 4.364 \tag{22}$$

This is the value obtained by Sellers et al. (1956) from the exact solution and given in Equation 4.2.2b.

We now consider a case where viscous dissipation is significant.

EXAMPLE 12.4.2

An incompressible fluid is in steady flow between two horizontal, semi-infinite, parallel plates at a distance $2a$ apart (Figure 12.4.3). The plates are maintained at uniform temperatures, the upper plate at T_1 and the lower plate at T_2. In the fully developed velocity and temperature profile region, determine the heat flux to each plate. Viscous dissipation is to be included.

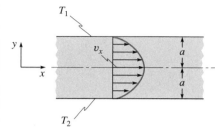

FIGURE 12.4.3 **An Incompressible Fluid in Steady Flow Between Two Parallel Plates, Each of Which Is Maintained at a Uniform Temperature**

ASSUMPTIONS

1. Steady state
2. Constant properties
3. Laminar flow

SOLUTION

In the fully developed region, $v_x = v_x(y)$, $v_y = v_z = 0$, $T = T(y)$. Following the development given in Example 12.4.1, it may be shown that

$$v_x = v_c \left(1 - \frac{y^2}{a^2} \right) \tag{1}$$

where

$v_c = (1/2\mu)(dp/dx)a^2$ = center line velocity, dp/dx = constant
v_m = mean velocity based on the mass flow rate = 2/3 v_c

In the fully developed temperature profile region, $T = T(y)$ and $\partial T/\partial x = \partial T/\partial z = 0$, and with $v_y = v_z = 0$, energy Equation A5.5 simplifies to

$$\alpha \frac{d^2T}{dy^2} + \frac{\mu}{\rho c_p} \left(\frac{dv_x}{dy} \right)^2 = 0 \tag{2}$$

$$T(y = a) = T_1; \quad T(y = -a) = T_2$$

From Equation 1, $dv_x/dy = -(2v_c y)/a^2$ and $\alpha = k/(\rho c_p)$. Equation 2 becomes

$$\frac{d^2T}{dy^2} = -\frac{\mu}{k}\frac{4v_c^2}{a^4} y^2$$

Integrating twice,

$$T = -\frac{1}{3}\frac{\mu}{k}\frac{v_c^2}{a^4} y^4 + c_1 y + c^2$$

Using the boundary conditions, $T(a) = T_1$ and $T(-a) = T_2$,

$$T_1 = -\frac{1}{3}\frac{\mu}{k}\frac{v_c^2}{a^4}a^4 + c_1 a + c_2 \qquad T_2 = -\frac{1}{3}\frac{\mu}{k}\frac{v_c^2}{a^4}a^4 - c_1 a + c_2$$

Solving for the constants of integration,

$$c_1 = \frac{T_1 - T_2}{2a} \qquad c_2 = \frac{T_1 + T_2}{2} + \frac{1}{3}\frac{\mu}{k}v_c^2$$

The temperature profile is now given by

$$T = -\frac{1}{3}\frac{\mu v_c^2}{k}\frac{y^4}{a^4} + \frac{T_1 - T_2}{2}\frac{y}{a} + \frac{T_1 + T_2}{2} + \frac{1}{3}\frac{\mu}{k}v_c^2 \tag{3}$$

Equation 3 can be recast as

$$T = \frac{T_1 + T_2}{2} + \frac{T_1 - T_2}{2}\frac{y}{a} + \frac{1}{3}\frac{\mu}{k}v_c^2\left(1 - \frac{y^4}{a^4}\right) \tag{3a}$$

Without viscous dissipation, $(T_1 + T_2)/2 + (T_1 - T_2)/2(y/a)$ is the temperature distribution with $(T_1 + T_2)/2$ representing the midplane temperature T_c. Equation 3a can, therefore, be interpreted as the sum of the temperature distribution without viscous dissipation and the effect of viscous dissipation $(\mu v_c^2/3k)(1 - y^4/a^4)$.

Equation 3 can be recast in dimensionless form in different ways depending on the characteristic temperature difference chosen. The characteristic temperature difference may be either $T_1 - T_2$, the difference in the temperature of the two plates, or $(\mu v_c^2/3k)$, which represents the maximum increase in the temperature caused by viscous dissipation. Choosing $T_1 - T_2$ as the reference temperature difference and a as the reference length, the dimensionless form of the temperature distribution is

$$\frac{T - [(T_1 + T_2)/2]}{T_1 - T_2} = \frac{1}{3}\frac{\mu v_c^2}{k(T_1 - T_2)}(1 - \eta^4) + \frac{1}{2}\eta \tag{4}$$

or

$$\theta = \frac{1}{3}\,\text{Ec Pr}\,(1 - \eta^4) + \frac{1}{2}\eta \tag{5}$$

where

$$\theta = \text{dimensionless temperature} = \frac{T - [(T_1 + T_2)/2}{T_1 - T_2}$$

$$\text{Ec Pr} = \frac{\mu v_c^2}{k(T_1 - T_2)} = \frac{\mu c_p}{k}\frac{v_c^2}{c_p(T_1 - T_2)}$$

$$\text{Pr} = \text{Prandtl number} = \frac{\mu c_p}{k}$$

$$\text{Ec} = \text{Eckert number} = \frac{v_c^2}{c_p(T_1 - T_2)}$$

$$\eta = \text{dimensionless coordinate} = \frac{y}{a}$$

The heat flux to the plates is given by

$$q_1'' = -k \left.\frac{dT}{dy}\right|_{y=a} = k \left(\frac{4}{3} \frac{\mu v_c^2}{ka} - \frac{T_1 - T_2}{2a} \right) \tag{6}$$

$$q_2'' = k \left.\frac{dT}{dy}\right|_{y=-a} = k \left(\frac{4}{3} \frac{\mu v_c^2}{ka} + \frac{T_1 - T_2}{2a} \right) \tag{7}$$

Assuming that $T_1 > T_2$, the heat flux to plate 2 is positive under all conditions. In Equation 7, the first term on the right side represents the heat flux to the plate due to viscous dissipation and the second term due to the difference in temperatures of the two plates. The heat flux to plate 1, given by Equation 6, can be negative, zero, or positive if

$$\frac{4}{3} \frac{\mu v_c^2}{ka} \underset{>}{\overset{<}{=}} \frac{T_1 - T_2}{2a} \tag{8}$$

Thus, even if $T_1 > T_2$, heat transfer can take place to both plates from the fluid if $(4/3)(\mu v_c^2/ka) > (T_1 - T_2)/2a$ or Ec Pr > 3/8.

Equation 5 is appropriate if $T_1 - T_2$ is not very small compared with $(\mu v_c^2/3k)$, which represents the temperature rise due to viscous dissipation, i.e., $(T_1 - T_2)/(\mu v_c^2/k)$ is not very small. The parameter $(T_1 - T_2)/(\mu v_c^2/k)$ can be recast as $[c_p(T_1 - T_2)/v_c^2](k/c_p\mu)$ or $(Ec\ Pr)^{-1}$. Hence, Equation 5 is suitable if Ec Pr is much less than 1 when viscous dissipation is not significant. However, if viscous dissi-

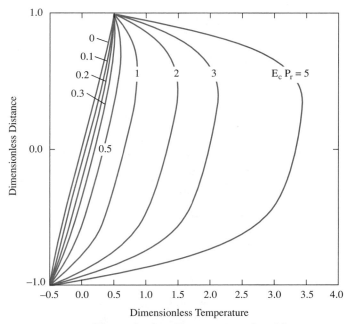

FIGURE 12.4.4 **Dimensionless Temperature in a Viscous Fluid in Steady Flow Between Two Parallel Plates Maintained at Different Uniform Temperatures**

pation is significant, i.e., Ec Pr is not much less than 1, and, in particular, if $T_1 = T_2$, the appropriate dimensionless temperature is obtained by taking $(\mu v_c^2/3k)$ as the characteristic temperature difference. For $T_1 = T_2$ the dimensionless temperature distribution is given by

$$\frac{T - T_1}{(1/3)(\mu v_c^2/k)} = 1 - \eta^4 \tag{9}$$

Figure 12.4.4 shows temperature profiles for different values of Pr Ec when $T_1 - T_2$ is not small. It is clear from the plots that for Ec Pr < 0.1, heat transfer is from plate 1 to the fluid (temperature decreasing from plate 1 towards plate 2). For Ec Pr $= 1$, 2, and 3 heat transfer takes place from the fluid to both plates. In all cases, heat transfer to plate 2 is positive.

12.5 BOUNDARY LAYERS

We have considered two examples of exact solutions to the Navier-Stokes equations and indicated the difficulties in solving the nonlinear equations. We now consider a simplification to the Navier-Stokes equations, applicable near solid boundaries, leading to the boundary layer equations. The concept of boundary layers was introduced in Chapter 4 when discussing convective heat transfer. In this section we study how the Navier-Stokes equations are simplified and solved near boundaries. The boundary layer equations are also nonlinear but they can be solved in a large number of cases.

Consider the flow of a fluid over an airfoil (Figure 12.5.1). No analytical solution to the Navier-Stokes equations for this case is available, but if we neglect viscous terms, the equations can be solved. From the solution to the inviscid flow (flows of fluids without viscosity—an idealization similar to the motion of one solid over another without friction) the pressure distribution around the airfoil can be deter-

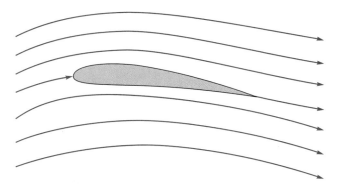

FIGURE 12.5.1 Flow over an Airfoil
Pressure around the airfoil and the lift force on it can be determined by neglecting viscous forces, but to compute the drag force the effect of viscosity has to be taken into account.

mined. The lift on the airfoil can then be determined from the known pressure distribution. The lift so computed will be in good agreement with the actual lift. But as viscous effects are ignored, the resulting solution predicts zero viscous drag, which is unrealistic.

In the beginning of this century Ludwig Prandtl simplified the Navier-Stokes equations near solid boundaries. He hypothesized that viscous effects are significant only in a thin layer of the fluid adjacent to the solid boundaries and that the momentum equations can be simplified in that region. The hypothesis led to the conclusion that some of the terms in the Navier-Stokes equations are much smaller than others and he discarded such terms and obtained the simplified form of the equations applicable near solid boundaries. The resulting approximate momentum equations for a flat plate were solved by one of his students, H. Blasius. It took some time for other people working on fluid mechanics problems to accept Prandtl's revolutionary idea. When significant experimental results were presented to validate the approximate solutions, the simplifications were accepted. The approximate equations led to great advances in analytical solutions to fluid mechanics problems, which were otherwise unsolvable.

In the following sections we derive the boundary layer equations for the flow of a fluid over a flat plate. Although the application of boundary layer equations is illustrated for one of the simplest cases, the flow of a fluid over a flat plate, the approximations have been used for the solutions of a large variety of problems.

12.5.1 Velocity Boundary Layer Equations

Consider the steady flow of an incompressible, Newtonian fluid flowing parallel to a thin, semi-infinite plate (a plate that is very large in the z-direction; if the plate is not thin, a separation region exists at the leading edge). The fluid approaches the plate with a uniform velocity (Figure 12.5.2). The free stream velocity, i.e., the

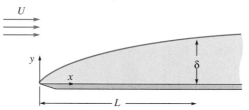

FIGURE 12.5.2 A fluid approaches a flat plate with a uniform velocity U. Viscous effects are confined to a thin region near the boundary, the boundary layer of thickness δ. Outside the boundary layer, viscous effects are negligible.

velocity far from the plate, is U. Viscous effects are significant in a thin layer adjacent to the plate, the boundary layer of thickness δ. The boundary layer is so thin that the flow outside it is not affected by its presence, i.e., the flow outside the boundary layer is the same as that of an inviscid fluid flowing over the plate. In this case the flow of a fluid outside the boundary layer has the solution

$$v_x = U = \text{constant} \qquad p = \text{constant}$$

The mass conservation and momentum balance equations within the boundary layer for the two-dimensional flow of an incompressible fluid are

$$\frac{\partial u}{\partial x} + \frac{\partial v}{\partial y} = 0 \tag{12.5.1}$$

$$u \frac{\partial u}{\partial x} + v \frac{\partial u}{\partial y} = -\frac{1}{\rho}\frac{\partial p}{\partial x} + \nu \left(\frac{\partial^2 u}{\partial x^2} + \frac{\partial^2 u}{\partial y^2}\right) \tag{12.5.2}$$

$$u \frac{\partial v}{\partial x} + v \frac{\partial v}{\partial y} = -\frac{1}{\rho}\frac{\partial p}{\partial y} + \nu \left(\frac{\partial^2 v}{\partial x^2} + \frac{\partial^2 y}{\partial y^2}\right) \tag{12.5.3}$$

where

u = velocity component along the x-direction
v = velocity component along the y-direction
p = pressure
ν = kinematic viscosity = μ/ρ

Prandtl's logic was that in Equations 12.5.2 and 12.5.3, if viscous effects are important only in a thin, boundary layer, some terms are not significant compared with others, and such terms can be neglected. The boundary layer thickness is assumed to be small. One may ask, How small is small, or more appropriately, small compared with what? If we consider the balance equations at a distance L from the leading edge, δ, the boundary layer thickness, may be assumed to be small if $\delta/L \ll 1$.

Velocity (momentum) boundary layer equations

To get an appreciation for the relative magnitudes of the various terms, we recast the equations in dimensionless form at $x = L$. As we are considering the equations at $x = L$, we take the reference length as L. Reference velocity is U. The dimensionless variables are then given by

$x^* = x/L$ $\qquad$ $y^* = y/L$
$u^* = u/U$ $\qquad$ $v^* = v/U$
$p^* = p/(\rho U^2)$ $\qquad$ $\text{Re} = UL/\nu$

Substituting these dimensionless variables into Equations 12.5.1 through 12.5.3 we obtain

$$\frac{\partial u^*}{\partial x^*} + \frac{\partial v^*}{\partial y^*} = 0 \tag{12.5.4}$$

$$u^* \frac{\partial u^*}{\partial x^*} + v^* \frac{\partial u^*}{\partial y^*} = -\frac{\partial p^*}{\partial x^*} + \frac{1}{\text{Re}} \left(\frac{\partial^2 u^*}{\partial x^{*2}} + \frac{\partial^2 u^*}{\partial y^{*2}}\right) \tag{12.5.5}$$

$$u^* \frac{\partial v^*}{\partial x^*} + v^* \frac{\partial v^*}{\partial y^*} = -\frac{\partial p^*}{\partial y^*} + \frac{1}{\text{Re}} \left(\frac{\partial^2 v^*}{\partial x^{*2}} + \frac{\partial^2 v^*}{\partial y^{*2}}\right) \tag{12.5.6}$$

Order of magnitude

In the above equations, $x^* \approx 1$ in the vicinity of $x = L$ and in the free stream $u^* = 1$. Consider these equations for such values of the variables for which $\text{Re} \gg 1$. In Equation 12.5.4, u^* changes from 1 to 0 near the surface and, hence, with $x^* \approx 1$, $\partial u^*/\partial x^* \sim O(1)$. The symbol $O(1)$ denotes "order of 1" (neither much greater nor much smaller than 1). The y-component of the velocity changes from 0 at $y = 0$ to v at $y = \delta$. Therefore, $\partial v/\partial y \sim v/\delta$ or $\partial v^*/\partial y^* \sim v^*/(\delta/L)$. To maintain mass balance, which we wish to satisfy under all conditions, in Equation 12.5.4 we require that both $\partial u^*/\partial x^*$ and $\partial v^*/\partial y^*$ be of the same order of magnitude. This leads us to the conclusion that $v^*/(\delta/L) \sim O(1)$ or $v^* \sim O(\delta/L)$. With a similar order of magnitude analysis, it can be shown that in Equation 12.5.5

$$u^* \frac{\partial u^*}{\partial x^*} \sim 1 \times 1 = O(1) \qquad v^* \frac{\partial u^*}{\partial y^*} \sim \frac{\delta}{L}\frac{1}{\delta/L} = O(1)$$

$\partial p^*/\partial x^*$ will be considered later.

$$\frac{\partial^2 u^*}{\partial x^{*2}} \sim O(1) \qquad \frac{\partial^2 u^*}{\partial y^{*2}} \sim O\left(\frac{1}{(\delta/L)^2}\right)$$

In Equation 12.5.6, the order of magnitudes for the terms are

$$u^* \frac{\partial v^*}{\partial x^*} \sim 1\frac{\delta/L}{1} = O(\delta/L) \qquad v^* \frac{\partial v^*}{\partial y^*} \sim \frac{\delta}{L}\frac{\delta/L}{\delta/L} = O(\delta/L)$$

$$\frac{\partial^2 v^*}{\partial x^{*2}} = \frac{\delta/L}{1} = O(\delta/L) \qquad \frac{\partial^2 v^*}{\partial y^{*2}} \sim \frac{\delta/L}{(\delta/L)^2} = O\left(\frac{1}{\delta/L}\right)$$

The order of magnitude of the various terms in the momentum balance equations are shown along with the equations.

$$u^* \frac{\partial u^*}{\partial x^*} + v^* \frac{\partial u^*}{\partial y^*} = -\frac{\partial p^*}{\partial x^*} + \frac{1}{Re}\left(\frac{\partial^2 u^*}{\partial x^{*2}} + \frac{\partial^2 u^*}{\partial y^{*2}}\right) \qquad \textbf{(12.5.5)}$$
$$1 \; \frac{1}{1} \qquad \frac{\delta}{L}\frac{1}{(\delta/L)} \qquad\qquad\qquad\qquad \frac{1}{1} \qquad \frac{1}{(\delta/L)^2}$$

$$\frac{\partial v^*}{\partial x^*} \quad v^* \frac{\partial v^*}{\partial y^*} \; \frac{\partial p^*}{\partial y^*} + \frac{1}{Re}\left(\frac{\partial^2 v^*}{\partial x^{*2}} \quad \frac{\partial^2 v^*}{\partial y^{*2}}\right) \qquad \textbf{(12.5.6)}$$
$$\frac{(\delta/L)}{1}\frac{1}{1} \quad \frac{\delta}{L}\frac{(\delta/L)}{(\delta/L)} \qquad\qquad \frac{(\delta/L)}{1} \quad \frac{(\delta/L)}{(\delta/L)^2}$$

In the x-momentum equation, Equation 12.5.5, as $(\delta/L)^2 \ll 1$, $\partial^2 u^*/\partial x^{*2} \ll \partial^2 u^*/\partial y^{*2}$ and, hence, $\partial^2 u^*/\partial x^{*2}$ can be neglected in the equation. Further, to retain the effect of viscous stresses, we require that $(1/Re)(\partial^2 u^*/\partial y^{*2})$ be of the same order of magnitude as the other terms in the equation, i.e., $(1/Re)(\partial^2 u^*/\partial y^{*2}) \sim O(1)$. This leads us to the conclusion that Re $(\delta/L)^2 \sim O(1)$ or that $\delta/L \sim 1/Re^{1/2}$. Is it not amazing that without solving the complicated equations we have come to the conclusion that the dimensionless thickness of the boundary layer, δ/x, varies inversely as the square root of Re_x? Note that δ/L is much less than 1 if $Re^{1/2}$ is much greater than 1; hence, this conclusion is valid only when the original assumption that $Re^{1/2} \gg 1$ is satisfied.

From the y-momentum equation, Equation 12.5.6 with Re $(\delta/L)^2 \sim O(1)$, Re $(\delta/L) \sim 1/(\delta/L)$, or Re $(\delta/L) \gg 1$. $u^*(\partial v^*/\partial x^*)$ and $v^*(\partial v^*/\partial y^*)$ are of $O(\delta/L)$ and the viscous terms are of $O(\delta/L)$ or less. Hence, all the terms in the equation are very small, which requires that $\partial p^*/\partial y^*$ also be small, i.e., it can be neglected. This implies that the pressure variation within the boundary layer in the y-direction is negligible or that the pressure in the boundary layer at any given value of x is equal to the pressure at the edge of the boundary layer at that location. In the free stream (outside the boundary layer) $u = U$ and $v = 0$ and, as viscous effects are negligible in the free stream, from Equation 12.5.4, the pressure at the edge of the boundary layer is constant and $\partial p/\partial x$ is zero. In the boundary layer as $\partial p/\partial y$ is zero, $\partial p/\partial x$ is also zero. Thus, we come to the conclusion that

$$\frac{\delta}{L} \sim \frac{1}{\mathrm{Re}^{1/2}} \qquad \frac{\partial p^*}{\partial x^*} = 0 \qquad \frac{\partial p^*}{\partial y^*} \sim 0$$

When these approximations are employed, the simplified, dimensional form of the equations is

$$\frac{\partial u}{\partial x} + \frac{\partial v}{\partial y} = 0 \tag{12.5.7}$$

$$u\frac{\partial u}{\partial x} + v\frac{\partial u}{\partial y} = \nu\frac{\partial^2 u}{\partial y^2} \tag{12.5.8}$$

$$0 = -\frac{1}{\rho}\frac{\partial p}{\partial y} \tag{12.5.9}$$

For a solution of the equations, we require

For u: one boundary condition in the x-direction and two in the y-direction
For v: one boundary condition in the y-direction

The boundary conditions are

$$u(0, y) = U \qquad u(x, 0) = 0 \qquad u(x, y \to \infty) = U \qquad v(x, 0) = 0$$

Although some simplification of the equations has been achieved by the boundary layer approximations, the solutions to the resulting equations are still quite involved. In 1908, by introducing a new variable, Blasius transformed the partial, nonlinear differential equations into a single ordinary differential equation. We now present the derivation of the ordinary differential equation.

In solving Equations 12.5.7, 12.5.8, and 12.5.9, the two coordinates, x and y, are combined into a single variable, η, in such a way that the velocities are functions of only this single variable. Both x and y may vary but they vary in such a way that for equal values of η, the velocities are also equal. Such a variable is known as a *similarity variable*. With some manipulation [White (1974)], η is found to be

Similarity variable

$$\eta = y\left(\frac{U}{\nu x}\right)^{1/2} \tag{12.5.10}$$

It is easily seen that η is dimensionless. From Equation 12.5.7, at fixed values of x, we have

$$\frac{\partial v}{\partial y} = -\frac{\partial u}{\partial x} \qquad \text{or} \qquad v = -\int_0^y \frac{\partial u}{\partial x}\, dy$$

Substituting the above expression for v in Equation 12.5.8,

$$u\frac{\partial u}{\partial x} - \left(\int_0^y \frac{\partial u}{\partial x}\, dy\right)\frac{\partial u}{\partial y} = \nu\frac{\partial^2 u}{\partial y^2} \tag{12.5.11}$$

For convenience we express the dimensionless velocity, u/U, as the derivative of a function of η. Note that as we have already utilized the mass conservation equation by expressing v in terms of u, we have only the momentum balance equation to solve. Let us try.

$$\frac{u}{U} = f'(\eta) \quad \text{or} \quad u = Uf'(\eta) \tag{12.5.12}$$

With u being given by Equation 12.5.12 and η by Equation 12.5.10, we have (by the chain rule for differentiation),

$$\frac{\partial u}{\partial x} = \frac{du}{d\eta}\frac{\partial \eta}{\partial x} = Uf''(\eta)\left[-\frac{y}{2}\left(\frac{U}{v}\right)^{1/2}\frac{1}{x^{3/2}}\right] = -\frac{U}{2x}\eta f''(\eta) \tag{12.5.13}$$

$$\frac{\partial u}{\partial y} = Uf''(\eta)\frac{\partial \eta}{\partial y} = Uf''(\eta)\left(\frac{U}{vx}\right)^{1/2} \tag{12.5.14a}$$

$$\frac{\partial^2 u}{\partial y^2} = Uf'''(\eta)\left(\frac{U}{vx}\right) \tag{12.5.14b}$$

$$v = -\int_0^y \frac{\partial u}{\partial x}\,dy = \frac{U}{2x}\int_0^y \eta f''(\eta)\,dy$$

The integration in the above equation is at fixed x. Hence, with $\eta = y(U/vx)^{1/2}$, *at fixed x*,

$$d\eta = \left(\frac{U}{vx}\right)^{1/2}dy \qquad dy = \left(\frac{vx}{U}\right)^{1/2}d\eta$$

$$v = \frac{1}{2}\frac{U}{x}\int_0^\eta \eta f''(\eta)\,dy = \frac{1}{2}\left(\frac{vU}{x}\right)^{1/2}\int_0^\eta \eta f''(\eta)\,d\eta$$

$$= \frac{1}{2}\left(\frac{vU}{x}\right)^{1/2}[\eta f'(\eta) - f(\eta) + f(0)]$$

In the expression for v, the constant $f(0)$ can be absorbed in $f(\eta)$ as we are dealing with only the derivatives of the function for u and, therefore, neither u nor v is affected by absorbing $f(0)$ in $f(\eta)$. To satisfy the conditions that $v(x, y = 0) = 0$ or $v(\eta = 0) = 0$, we require that $f(0) = 0$. Thus,

$$v = \frac{1}{2}\left(\frac{vU}{x}\right)^{1/2}[\eta f'(\eta) - f(\eta)] \tag{12.5.15}$$

Substituting Equations 12.5.13, 12.5.14a and b, and 12.5.15 into Equation 12.5.8,

$$Uf'\left[-\frac{U}{2x}\eta f''\right] + \frac{1}{2}\left(\frac{vU}{x}\right)^{1/2}\left(\eta f' - f\right)U\left(\frac{U}{vx}\right)^{1/2}f'' = v\frac{U^2}{vx}f'''$$

Simplifying and rearranging the above equation, we obtain

$$f''' + \frac{1}{2}ff'' = 0 \tag{12.5.16}$$

The boundary conditions are

$$u(x, y \to \infty) = u(\eta \to \infty) = U \quad \text{or} \quad f'(\eta \to \infty) = 1 \qquad \textbf{(12.5.17)}$$

$$u(x, y = 0) = u(\eta = 0) = 0 \qquad \text{or} \quad f'(0) = 0 \qquad \textbf{(12.5.18)}$$

$$u(x = 0, y) = u(\eta \to \infty) = U \quad \text{or} \quad f'(\eta \to \infty) = 1 \qquad \textbf{(12.5.19)}$$

$$v(x, y = 0) = v(\eta = 0) = 0 \qquad \text{or} \quad f(0) = 0 \qquad \textbf{(12.5.20)}$$

We also require that $v(x, y \to \infty) = v(\eta \to \infty) = 0$. From Equation 12.5.15 this requires that with $f'(\eta \to \infty) = 1, f(\eta \to \infty) \to \eta$. It should also be noted that the two boundary conditions for u at $y \to \infty$ and $x = 0$, given by Equations 12.5.17 and 12.5.19 merge into a single boundary condition in terms of η. This is a necessary condition as Equation 12.5.16 permits only three independent boundary conditions.

The numerical solution to Equation 12.5.16 with its associated boundary conditions was first given by Blasius, and is known as the *Balsius solution*. The equation constitutes a two-point boundary value problem in which the boundary conditions are specified at two different locations. As $f''(0)$ is not available, the marching process from $\eta = 0$ for the integration of the equation cannot be applied directly. One method to obtain the solution is to integrate Equation 12.5.16 with two values of $f''(0)$, c_1 and c_2. If the corresponding values of $f'(\infty)$ are b_1 and b_2, a new value of $f''(0)$ that is anticipated to yield the value of 1 for $f'(\infty)$ is determined from the relation

$$c_3 = c_2 + \frac{c_2 - c_1}{b_2 - b_1} (1 - b_2)$$

The computations are repeated until the value of $f''(0)$ for which $f'(\infty)$ approaches unity is found. Although we speak of very large values of η, we find that the functions reach their asymptotic values for values of η around 9 so that it is only necessary to carry out the integration up to a value of 10 for η. One of the programs on the software disk (BLVEL.EXE) numerically integrates Equation 12.5.16 by such an algorithm employing the 4th order Runge-Kutta method. It stores the values of η, f, f', and f'' in a file, VEL, on the disk for later use in the computation of the temperature profile. The results of the integration obtained from the program are given in Table 12.5.1.

The edge of the boundary layer is arbitrarily defined as that point at which the velocity is approximately 99% of the free stream velocity or that value of η that yields $f'(\eta) \cong 0.99$. From Table 12.5.1, it is seen that $u/U = f'(\eta) = 0.99$ for $\eta = 4.91$. However, η is usually taken as 5 at the edge of the boundary layer giving

$$\eta_{y=\delta} = 5 = \delta \left(\frac{U}{vx} \right)^{1/2}$$

or

$$\frac{\delta}{x} = \frac{5}{\text{Re}^{1/2}} \qquad \textbf{(12.5.21)}$$

where Re = Reynolds number = Ux/v.

The shear stress at the wall is given by $\mu \, \partial u/\partial y \mid_{y=0}$ or from Equation 12.5.14a

$$\tau_{\text{w}} = \mu U \left(\frac{U}{vx} \right)^{1/2} f''(0)$$

TABLE 12.5.1 Solution to Equation 12.5.16 (From Program BL.EXE in the Software)

$$\eta = y \left(\frac{U}{\nu x} \right)^{1/2} \qquad f' = \frac{u}{U}$$

η	$f(\eta)$	$f'(\eta)$	$f''(\eta)$
0.0	0.000000	0.000000	0.332057
0.2	0.006641	0.066408	0.331984
0.4	0.026560	0.132764	0.331470
0.6	0.059735	0.198937	0.330079
0.8	0.106108	0.264709	0.327389
1.0	0.165572	0.329780	0.323007
1.2	0.237949	0.393776	0.316589
1.4	0.322982	0.456262	0.307865
1.6	0.420321	0.516757	0.296663
1.8	0.529518	0.574758	0.282931
2.0	0.650024	0.629766	0.266752
2.2	0.781193	0.681310	0.248351
2.4	0.922290	0.728982	0.228092
2.6	1.072506	0.772455	0.206455
2.8	1.230977	0.811510	0.184007
3.0	1.396808	0.846044	0.161360
3.2	1.569095	0.876081	0.139128
3.4	1.746950	0.901761	0.117876
3.6	1.929525	0.923330	0.098086
3.8	2.116030	0.941118	0.080126
4.0	2.305746	0.955518	0.064234
4.2	2.498040	0.966957	0.050520
4.4	2.692361	0.975871	0.038973
4.6	2.888248	0.982683	0.029484
4.8	3.085321	0.987790	0.021871
5.0	3.283274	0.991542	0.015907
5.2	3.481868	0.994246	0.011342
5.4	3.680919	0.996155	0.007928
5.6	3.880291	0.997478	0.005432
5.8	4.079882	0.998375	0.003648
6.0	4.279621	0.998973	0.002402
6.2	4.479457	0.999363	0.001550
6.4	4.679357	0.999612	0.000981
6.6	4.879296	0.999768	0.000608
6.8	5.079260	0.999864	0.000370
7.0	5.279239	0.999922	0.000220
7.4	5.679220	0.999975	0.000074
7.8	6.079215	0.999993	0.000023
8.2	6.479213	0.999998	0.000006
8.6	6.879212	1.000000	0.000002
9.0	7.279212	1.000000	0.000000
9.4	7.679212	1.000000	0.000000

Defining the Fanning friction factor C_F by $C_F = \tau_w/(\rho U^2/2)$, with $f''(0) = 0.332$ from Table 12.5.1,

$$C_F = \frac{0.664}{\sqrt{\text{Re}}}$$

(12.5.22)

12.5.2 Temperature Boundary Layer

Consider an incompressible fluid flowing with a uniform free steam velocity U parallel to a flat plate (Figure 12.5.3). The free stream temperature is T_∞ and the plate is maintained at a uniform temperature of T_s. Viscous dissipation is negligible. We wish to find the temperature profile and the heat transfer coefficient.

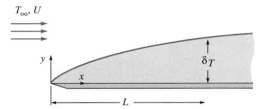

FIGURE 12.5.3 A fluid approaches with a uniform velocity U and uniform temperature T_∞. The thickness of the temperature boundary layer is δ_T.

Specializing the energy equation A5.5 (constant density and negligible viscous dissipation),

$$u\frac{\partial T}{\partial x} + v\frac{\partial T}{\partial y} = \alpha\left(\frac{\partial^2 T}{\partial x^2} + \frac{\partial^2 T}{\partial y^2}\right)$$

(12.5.23)

Employing the dimensionless variables (see also Section 12.5.1)

$$x^* = x/L \qquad\qquad y^* = y/L \qquad u^* = u/U \qquad v^* = v/U$$
$$\theta = (T_s - T)/(T_s - T_\infty) \qquad \text{Re} = UL/\nu \qquad \text{Pr} = \nu/\alpha$$

$$u^*\frac{\partial\theta}{\partial x^*} + v^*\frac{\partial\theta}{\partial y^*} = \frac{1}{\text{Re Pr}}\left(\frac{\partial^2\theta}{\partial x^{*2}} + \frac{\partial^2\theta}{\partial y^{*2}}\right)$$

$$\begin{array}{ccccc} 1 & 1 & \dfrac{\delta}{L}\dfrac{1}{(\delta_T/L)^2} & & 1 \quad \dfrac{1}{(\delta_T/L)^2} \end{array}$$

where

δ = velocity boundary layer thickness
δ_T = temperature boundary layer thickness

Note that the dimensionless temperature θ satisfies the boundary conditions $\theta(x, 0) = 0$, $\theta(x, y \to \infty) = 1$, and $\theta(0, y) = 1$

Temperature (energy) boundary layer equation

The order of magnitudes of the different terms are given under each term. These are obtained with a logic similar to the one used in estimating the order of magnitude of the terms in the momentum equation. For the temperature boundary layer we conclude that $(\delta_T/L)^2 = O(1/\text{Re Pr})$. Therefore, for $\text{Re Pr} \gg 1$, $\partial^2\theta/\partial x^2 \ll \partial^2\theta/\partial y^2$. With dimensional variables the energy equation for the temperature boundary layer is given by

$$u\frac{\partial\theta}{\partial x} + v\frac{\partial\theta}{\partial y} = \alpha\frac{\partial^2\theta}{\partial y^2} \qquad (12.5.24)$$

with

$$\theta(0, y) = 1 \qquad \theta(x, 0) = 0 \qquad \theta(x, \infty) = 1 \qquad (12.5.25)$$

Employing the same dimensionless variables used in developing the velocity boundary layer equation, boundary layer Equations 12.5.8 and 12.5.24 with $\partial p/\partial x = 0$, are

$$u^*\frac{\partial u^*}{\partial x^*} + v^*\frac{\partial u^*}{\partial y^*} = \frac{1}{\text{Re}}\frac{\partial^2 u^*}{\partial x^{*2}} \qquad (12.5.8a)$$

$$u^*(0, y^*) = 1 \qquad u^*(x^*, 0) = 0 \qquad u^*(x^*, \infty) = 1$$

$$u^*\frac{\partial\theta}{\partial x^*} + v^*\frac{\partial\theta}{\partial y^*} = \frac{1}{\text{Re Pr}}\frac{\partial^2\theta}{\partial x^{*2}} \qquad (12.5.24a)$$

$$\theta(0, y^*) = 1 \qquad \theta(x^*, 0) = 0 \qquad \theta(x^*, \infty) = 1$$

Note the similarity between Equation 12.5.24a and Equation 12.5.8a and the associated boundary conditions for the dimensionless velocity and temperature. If $\text{Pr} = 1$, the solution for θ is, $\theta = u^*$, as substitution of u^* for θ in Equation 12.5.24a satisfies the equation and the boundary conditions. This leads to the Reynolds analogy (see Section 4.2).

A transformation of Equation 12.5.24 in terms of the similarity variable, $\eta = y(U/vx)^{1/2}$ and $u = Uf'(\eta)$ yields

$$\theta'' + \frac{\text{Pr}}{2}f\theta' = 0 \qquad (12.5.26)$$

$$\theta(\eta = 0) = 0 \qquad \theta(\eta = \infty) = 1 \qquad (12.5.27)$$

Solution to Equation 12.5.26 has to be obtained by integrating it numerically. There are several methods for the solution of this two-point boundary value problem where the values of θ are known at $\eta = 0$ and $\eta = \infty$ but the value of $\theta'(0)$ is not known. The difficulty of $\theta'(0)$ not being known is overcome by making use of the linearity of Equation 12.5.26 (it is linear with a variable coefficient, f). In one method, two solutions θ_1 and θ_2 are found and linearly combined to satisfy the equation and the boundary conditions. Let

$$\theta = \theta_1 + c\theta_2$$

where c is a constant to be determined. The two functions θ_1 and θ_2 are defined such that

$$\theta_1'' + \frac{\text{Pr}}{2}f\theta_1' = 0 \qquad\qquad \theta_2'' + \frac{\text{Pr}}{2}f\theta_2' = 0$$
$$\theta_1(0) = 0 \qquad \theta_1'(0) = 1 \qquad\qquad \theta_2(0) = 0 \qquad \theta_2'(0) = -1$$

With the specified boundary conditions, both θ_1 and θ_2 are found from numerical integration. The constant c is now determined such that the boundary condition $\theta(\infty) = 1$ is satisfied.

$$\theta_1(\infty) + c\,\theta_2(\infty) = 1$$

or

$$c = \frac{1 - \theta_1(\infty)}{\theta_2(\infty)}$$

Having found the value of c, $\theta'(0)$ is found from

$$\theta'(0) = \theta_1'(0) + c\,\theta_2'(0)$$

It turns out that from the structure of the equations, $\theta_2 = -\theta_1$ and $\theta_2(\infty) = -\theta_1(\infty)$ so that we need to integrate Equation 12.5.26 only once with $\theta_1'(0) = 1$. The correct value of $\theta'(0)$ is then obtained from the relation

$$\theta'(0) = \frac{1}{\theta_1(\infty)}$$

Having found the correct value of $\theta'(0)$, a second integration of Equation 12.5.26 gives the temperature profile. The HT.EXE module ''Flat Plate Heat Transfer'' provides the value of $\theta'(0)$ for any value of Pr between 0.0001 and 3000 by the above method. Results of the integration of the energy equation from the program for Prandtl numbers of 0.7 and 10 are given in Tables 12.5.2 and 12.5.3 and shown in Figure 12.5.4. The figure also shows the velocity profile. Note how close the velocity profile is to the temperature profile for Pr $= 0.7$. The values of $\theta'(0)$ and δ_T for different values of Pr are given in Table 12.5.4. The value of η when θ is 0.99 is considered to be the temperature boundary layer thickness, δ_T. Note that the value of η for which $\theta = 0.99$ depends on Pr and except for Pr $= 1$, the value is different from the value of η (≈ 5) for which $u/U = 0.99$. The integration was stopped when $\theta'(\eta)$ was less than 10^{-6}.

Denoting the velocity boundary layer thickness by δ, from Table 12.5.1 for $u/U = 0.99$, $\delta/x^{1/2}(U/v)^{1/2} = 4.91$. From the values of $\delta_T/x^{1/2}(U/v)^{1/2}$ for $\theta = 0.99$ in Table 12.5.4, the values of δ_T/δ are given in Table 12.5.4. For Pr $= 1$ the temperature boundary layer thickness equals the velocity boundary layer thickness. The effect of increasing the thermal conductivity of the fluid (reducing the value of the Prandtl number) is to increase the temperature boundary layer thickness. Thus, for liquid metals (Pr $\ll 1$) the temperature boundary layer thickness is much greater than the velocity boundary layer thickness. On the other hand, for oils (Pr $\gg 1$) the temperature boundary layer thickness is much less than the velocity boundary layer thickness.

From the numerical solution, the temperature boundary layer thickness is approximated by

$$\frac{\delta_T}{\delta} = \mathrm{Pr}^{-n}$$

TABLE 12.5.2 Solution to Equation 12.5.26 for Pr = 0.7, η = 5.66 for θ = 0.99

η	θ(η)	θ′(η)
0.0	0.000000	0.292657
0.2	0.058529	0.292615
0.4	0.117028	0.292300
0.6	0.175413	0.291444
0.8	0.233551	0.289782
1.0	0.291256	0.287064
1.2	0.348291	0.283062
1.4	0.404382	0.277582
1.6	0.459216	0.270475
1.8	0.512457	0.261651
2.0	0.563760	0.251087
2.2	0.612779	0.238836
2.4	0.659190	0.225026
2.6	0.702699	0.209863
2.8	0.743063	0.193618
3.0	0.780096	0.176612
3.4	0.843775	0.141764
3.8	0.893680	0.108196
4.2	0.930831	0.078342
4.6	0.957058	0.053738
5.0	0.974593	0.034888
5.4	0.985690	0.021428
5.8	0.992333	0.012446
6.2	0.996096	0.006837
6.6	0.998111	0.003551
7.0	0.999133	0.001744
7.4	0.999622	0.000810
7.8	0.999844	0.000356
8.2	0.999939	0.000148
8.6	0.999977	0.000058
9.0	0.999992	0.000022
9.4	0.999997	0.000008
9.8	0.999999	0.000003
10.15	1.000000	0.000001

where $n = \frac{1}{2}$ for Pr < 0.1 and $n = \frac{1}{3}$ for Pr > 0.1.

The heat flux from the plate to the fluid is given by

$$q''_w = -k\left.\frac{\partial T}{\partial y}\right|_{y=0} = k\left.\frac{\partial \theta}{\partial y}\right|_{y=0}(T_s - T_\infty)$$

$$h = \frac{q''_w}{T_s - T_\infty} = \left.\frac{d\theta}{d\eta}\right|_{\eta=0}\left(\frac{U}{\nu x}\right)^{1/2}k$$

$$\mathrm{Nu}_x = \frac{hx}{k} = \left.\frac{d\theta}{d\eta}\right|_{\eta=0}\left(\frac{Ux}{\nu}\right)^{1/2} = \mathrm{Re}_x^{1/2}\left.\frac{d\theta}{d\eta}\right|_{\eta=0}$$

TABLE 12.5.3 Solution to Equation 12.5.26 for Pr = 10, η = 2.17 for θ = 0.99

η	$\theta(\eta)$	$\theta'(\eta)$
0.0	0.000000	0.727762
0.2	0.145481	0.726253
0.4	0.289860	0.715186
0.6	0.430328	0.685834
0.8	0.562551	0.632037
1.0	0.681405	0.552431
1.2	0.782121	0.452039
1.4	0.861561	0.341885
1.6	0.919164	0.236030
1.8	0.957104	0.146955
2.0	0.979541	0.081564
2.2	0.991324	0.039915
2.4	0.996762	0.017046
2.6	0.998947	0.006293
2.8	0.999704	0.001991
3.0	0.999929	0.000535
3.2	0.999985	0.000122
3.4	0.999998	0.000023
3.6	1.000000	0.000004
3.75	1.000000	0.000001

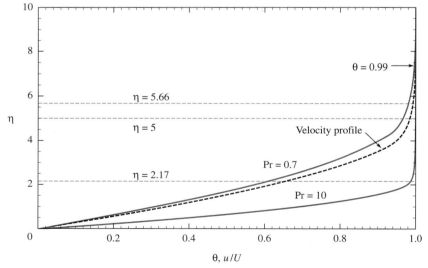

FIGURE 12.5.4 Dimensionless temperature profiles for Prandtl numbers of 0.7 and 10 and dimensionless velocity profile
The dimensionless velocity boundary layer thickness is 5 and the dimensionless temperature boundary layer thicknesses are 5.66 for Pr = 0.7 and 2.17 for Pr = 10.

TABLE 12.5.4 Values of $\theta'(0)$ for Different Values of the Prandtl Number from the Numerical Integration of Equation 12.5.26

$\delta_T = \eta(\theta = 0.99)$, $c = \theta'(0)/Pr^n$, $n = .5$ for $Pr < 0.1$ and $n = 1/3$ for $Pr \geq 0.1$

Pr	$\theta'(0)$	$\delta_T \left(\dfrac{U}{\nu x}\right)^{1/2}$	c		δ_T/δ
0.0001	5.588×10^{-3}	365.05	0.558	$n = \frac{1}{2}$	74.4
0.0005	0.012350	164.2	0.552		33.44
0.001	0.017316	116.45	0.548		23.72
0.005	0.037392	52.8	0.528		10.75
0.01	0.051588	37.7	0.516		7.68
0.05	0.10511	17.6	0.47		3.58
0.1	0.14003	12.85	0.301	$n = \frac{1}{3}$	2.62
0.5	0.25928	6.45	0.327		1.31
0.7	0.29266	5.66	0.33		1.15
1.0	0.33202	4.91	0.332		1.0
2.5	0.45582	3.51	0.336		0.715
5.0	0.57668	2.75	0.337		0.56
10.0	0.72813	2.17	0.338		0.442
20.0	0.91837	1.72	0.338		0.35
50.0	1.24721	1.26	0.339		0.257
100.0	1.57168	1.0	0.339		0.204
200.0	1.98030	0.79	0.339		0.161
400.0	2.49493	0.63	0.339		0.128
600.0	2.85582	0.55	0.339		0.112
800.0	3.14306	0.50	0.339		0.102
1000.0	3.38556	0.46	0.339		0.094

From the relations shown on page 943, as $x \to 0$, $h_x \to \infty$. The reason for such an unrealistic value of h_x is that the results are being used in a region where they are not applicable. The boundary layer simplifications are valid only for $Re \gg 1$ and $Re\ Pr \gg 1$. But as $x \to 0$ both Re and $Re\ Pr$ approach zero and the boundary layer simplifications are no longer valid.

From Table 12.5.4 the effect of Prandtl number on the temperature boundary layer thickness is clear. As the Prandtl number increases, the temperature boundary layer thickness decreases. Although Table 12.5.4 includes values of Prandtl numbers less than 0.001, the Prandtl numbers for liquid metals even at high temperatures of around 1000 °C (Prandtl numbers for liquid metals decrease with increasing temperature) are rarely less than 0.001.

In the relation $Nu = Re_x^{1/2}\ \theta'(0)$, $\theta'(0)$ depends on Pr. The relation can be recast as $Nu_x = cRe_x^{1/2}$, where $c = \theta'(0)/Pr^n$. Employing the values of $\theta'(0)$, the computed values of c are also given in Table 12.5.4. For $Pr > 5$, with $n = \frac{1}{3}$ the value of the constant is 0.339; for $Pr \approx 1$ its value is 0.332. For very low values of Pr with $n = \frac{1}{2}$, the value is around 0.5. The accepted value of c for very low values of Pr is 0.564, which is the limiting value of the constant when $Pr \to 0$. When the Prandtl number is very low (approaching zero) the velocity boundary layer thickness is much smaller than the temperature boundary layer thickness. The effect of the velocity boundary layer can be neglected and, in integrating the energy equation, the velocity may be assumed to be uniform (see Problem 12.14).

The suggested correlations for the Nusselt numbers for different ranges of the Prandtl numbers are

Nusselt number based on the local heat transfer coefficient

$$\text{Nu}_x = \frac{h_x x}{k} = \begin{cases} 0.564 \ \text{Re}_x^{1/2} \ \text{Pr}^{1/2} & \text{Pr} < 0.05 & \textbf{(12.5.28)} \\ 0.332 \ \text{Re}_x^{1/2} \ \text{Pr}^{1/3} & 0.6 < \text{Pr} < 10 & \textbf{(12.5.29)} \\ 0.339 \ \text{Re}_x^{1/2} \ \text{Pr}^{1/3} & \text{Pr} > 10 & \textbf{(12.5.30)} \end{cases}$$

Nusselt number based on the average heat transfer coefficient

$$\text{Nu}_L = \frac{h_L L}{k} = \begin{cases} 1.13 \ \text{Re}_L^{1/2} \ \text{Pr}^{1/2} & \textbf{(12.5.28a)} \\ 0.664 \ \text{Re}_L^{1/2} \ \text{Pr}^{1/3} & \textbf{(12.5.29a)} \\ 0.678 \ \text{Re}_L^{1/2} \ \text{Pr}^{1/3} & \textbf{(12.5.30a)} \end{cases}$$

Flat plate correlations from solution to boundary layer equations

---

EXAMPLE 12.5.1   Air at 20 °C flows parallel to a flat plate with a free stream velocity of 5 m/s (Figure 12.5.5). The plate is maintained at 80 °C. At a distance of 50 cm from the leading edge and 5 mm from the plate, find x- and y-components of the velocity and temperature of the air and the heat flux from the plate at $x = 0.5$ m.

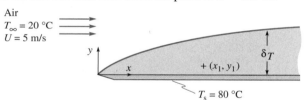

FIGURE 12.5.5 Air at 20 °C flows parallel to a plate maintained at 80 °C.

Given

$x_1 = 0.5$ m $T_\infty = 20 \ °C$

$y_1 = 0.005$ m $T_s = 80 \ °C$

$U = 5$ m/s

Find

u, v, T at $x = 0.5$ m, $y = 5$ mm

q_x'' at $x = 0.5$ m

SOLUTION

From software CC, property of air at 40 °C, 101.325 kPa are

$\rho = 1.093 \ \text{kg/m}^3$ $\mu = 1.957 \times 10^{-5} \ \text{N s/m}^2$

$k = 0.0278 \ \text{W/m °C}$ $\text{Pr} = 0.709$

$$\text{Re}_x = \rho U x / \mu = \frac{1.093 \times 5 \times 0.5}{1.957 \times 10^{-5}} = 1.396 \times 10^5$$

As $\text{Re}_x < 5 \times 10^5$ the boundary layer is laminar and the solutions given in Tables 12.5.1 and 12.5.4 are applicable.

$$u = Uf'(\eta) \qquad \eta = y_1 \left(\frac{U}{\nu x_1}\right)^{1/2}$$

$$\nu = \mu/\rho = \frac{1.9574 \times 10^{-5}}{1.093} = 1.791 \times 10^{-5} \text{ m}^2/\text{s}$$

$$\eta = 0.005 \left(\frac{5}{1.791 \times 10^{-5} \times 0.5}\right)^{0.5} = 3.736$$

From Table 12.5.1 at $\eta = 3.736, f'(\eta) = 0.9354$

$$u = 0.9354 \times 5 = \underline{4.68 \text{ m/s}}$$

From Equation 12.5.15

$$v = 0.5 \left(\frac{\nu U}{x}\right)^{1/2} \left[\eta f'(\eta) - f(\eta)\right]$$

At $\eta = 3.736, f'(\eta) = 0.9354$ and $f(\eta) = 2.056$ and

$$v = 0.5 \left(\frac{1.791 \times 10^{-5} \times 5}{0.5}\right)^{1/2} (3.736 \times 0.9354 - 2.056) = \underline{0.00963 \text{ m/s}}$$

As the Prandtl number is slightly greater than 0.7, we use the values of θ available for $\text{Pr} = 0.7$ to get a good estimate for the value of the temperature. From Table 12.5.2 at $\eta = 3.736, \theta = 0.8887$. Thus,

$$\frac{T_s - T}{T_s - T_\infty} = \theta \qquad \frac{80 - T}{80 - 20} = 0.8857 \qquad T = \underline{26.9 \,°\text{C}}$$

From Table 12.5.4, for $\text{Pr} = 0.7, \theta'(0) = 0.2927$

$$h_x = \theta'(0) \left(\frac{\rho U}{\mu x}\right)^{1/2} k = 0.2927 \left(\frac{5 \times 1.093}{1.957 \times 10^{-5} \times 0.5}\right)^{1/2}$$

$$\times 0.0278 = 6.08 \text{ W/m}^2 \,°\text{C}$$

$$q_x''(x = 0.5 \text{ m}) = 6.08(80 - 20) = \underline{364.8 \text{ W/m}^2}$$

COMMENT

Note that the value of v is 0.00963 m/s, whereas the value of u is 4.68 m/s. Throughout the boundary layer, the value of v is very much less than the free stream velocity of 5 m/s.

SUMMARY

In this chapter, two-dimensional differential equations for mass (Equation 12.1.2), momentum (Equations 12.2.5 and 12.2.6), and energy balance (Equation 12.3.10) for an incompressible fluid are derived. The equations are applicable to those fluids that satisfy the stress-rate-of-strain relations given by Equations 12.2.3a, b, and c; such fluids are known as Newtonian fluids. The momentum equations are known as Navier-Stokes equations. The three-dimensional form of the equations in Cartesian, cylindrical, and spherical coordinates are given in Appendix 5.

Exact solutions to the differential equations for the velocity and temperature distribution are obtained for a few cases. The effect of viscous dissipation is also illustrated.

Boundary layer equations (the simplified form of the Navier-Stokes equations), valid near solid boundaries for high Reynolds numbers, for mass balance (Equation 12.5.7), momentum balance (Equations 12.5.8 and 12.5.9), and thermal energy balance (Equation 12.5.24) are derived. The partial differential boundary layer equations are transformed to ordinary differential equations (Equations 12.5.16 and 12.5.26) through a similarity variable. The solutions to the ordinary differential equations are obtained by numerical methods. One method of integrating the boundary layer energy equation (Equation 12.5.26) is described and the numerical values found from that method. Exact solutions to the limiting cases with $\mathrm{Pr} \ll 1$ can be obtained through Laplace transforms.

Software

The disk contains two programs that give the velocities and the temperatures in the boundary layer when a fluid flows parallel to a flat plate. BLVEL.EXE numerically integrates Equation 12.5.16 employing the procedure described on page 938. The integration is carried out with steps of 0.01 for η. For η greater than 9.5 there is no change in the velocity and the integration stops at $\eta = 10$. Values of $f(\eta)$, $f'(\eta)$, and $f''(\eta)$ are printed on a printer at intervals specified by the user. VEL01 and VEL05 are values of $f(\eta)$ obtained from BL.EXE.

HT.EXE module "Flat Plate Heat Transfer" numerically integrates Equations 12.5.16 and 12.5.26. The program yields $\theta'(0)$ and values of η at which θ is around 0.99 so that the temperature boundary layer thickness can be approximated. It can be used to obtain the convective heat transfer coefficient for any fluid (Prandtl number between 0.0001 and 3000) flowing parallel to a flat plate.

REFERENCES

Arpaci, V.S. (1984). *Convection Heat Transfer.* Englewood Cliffs: Prentice Hall.

Batchelor, G.K. (1967). *Fluid Dynamics.* New York: Cambridge University Press.

Bejan, A. (1984). *Convection Heat Transfer.* New York: Wiley.

Bird, R.B., Stewart, W.E., and Lightfoot, E.N. (1960). *Transport Phenomena.* New York: Wiley.

Burmeister, L.C. (1993). *Convective Heat Transfer,* 2d ed. New York: Wiley.

Eckert, E.R.G., and Drake, R.M. (1972). *Analysis of Heat and Mass Transfer.* New York: McGraw-Hill.

Kakac, S., Shah, R.M., and Aung, W. (1987). *Handbook of Single-Phase Convective Heat Transfer.* New York: Wiley.

Kays, W.M., and Crawford, M.C. (1993). *Convective Heat and Mass Transfer,* 3d ed. New York: McGraw-Hill.

Schlichting, H. (1968). *Boundary Layer Theory,* 6th ed. New York: McGraw-Hill.

Sellars, J.R., Tribus, M., and Klein, J.S. (1956). Heat transfer to laminar flows in a round tube or flat conduit: The Graetz problem extended. *TRANS. ASME:* 78:441.

Whitaker, S. (1983). *Fundamentals Principles of Heat Transfer.* Malabar, FL: Robert E. Krieger.

White, F.M. (1974). *Viscous Fluid Flow.* New York: McGraw-Hill.

PROBLEMS

12.1 Starting from a differential control volume, with $v_z = 0$, show that the two-dimensional mass conservation equation in cylindrical coordinates (r, θ) is

$$\frac{\partial \rho}{\partial t} + \frac{1}{r} \frac{\partial}{\partial r} (\rho r v_r) + \frac{1}{r} \frac{\partial}{\partial \theta} (\rho v_\theta) = 0$$

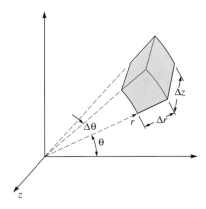

FIGURE P12.1

12.2 Starting from a differential control volume, neglecting viscous dissipation, show that the two-dimensional energy equation, in cylindrical coordination with $T = T(r, \theta)$, is

$$\rho c_v \left(\frac{\partial T}{\partial t} + v_r \frac{\partial T}{\partial r} + \frac{v_\theta}{r} \frac{\partial T}{\partial \theta} \right) = \frac{1}{r} \frac{\partial}{\partial r} \left(kr \frac{\partial T}{\partial r} \right)$$
$$+ \frac{1}{r} \frac{\partial}{\partial \theta} \left(\frac{k}{r} \frac{\partial T}{\partial \theta} \right) - \frac{p}{r} \left[\frac{\partial}{\partial r} (r v_r) + \frac{\partial v_\theta}{\partial \theta} \right]$$

12.3 A liquid film steadily drains down the outside surface of a vertical, cylindrical rod of radius r_1. After some distance, the outer radius of the film reaches a constant value of r_2. Assuming negligible shear stress at the interface of the film and the surrounding air, find an expression for the velocity distribution and the mass flow rate of the liquid. Assume $(r_2 - r_1) \ll r_1$

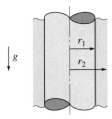

FIGURE P12.3

12.4 A plate (ρ = 2700 kg/m^3), 15-cm high and 3-mm thick, falls vertically in an oil (ρ = 860 kg/m^3, μ = 0.03 N s/m^2) and reaches a constant velocity after falling some distance. Employing the boundary layer relations for the shear stress, estimate the terminal, constant velocity.

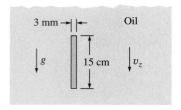

3 mm

Oil

g

15 cm

v_z

FIGURE P12.4

12.5 Consider the fully developed, laminar flow of an incompressible fluid in the annular space between two coaxial cylinders of radii R_1 and R_2. With $v_z = v_z(r)$, $v_\theta = v_r = 0$ and dp/dz = constant,
(a) Find an expression for $v_z(r)$.
(b) Determine the mass flow rate of the liquid.
(c) Find the shear stresses τ_{w1} and τ_{w2} on the inner and outer surfaces.
(d) Find a relation between dp/dz and the wall shear stresses.

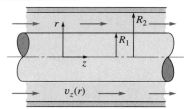

r

R_2

R_1

z

$v_z(r)$

FIGURE P12.5

12.6 Air (ρ = 1.1774 kg/m^3, μ = 1.846 × 10^{-5} N s/m^2) flows parallel to a flat plate with a free stream velocity of 2 m/s. At 1 m from the leading edge, employing values from Table 12.5.1, find v_x and v_y at a distance of $\delta/2$ from the plate (δ = boundary layer thickness).

12.7 Prove Equation 12.4.2

12.8 A fluid is in steady flow down an inclined plate that is maintained at a uniform temperature T_s. The thickness of the fluid film is δ. The free surface of the fluid is maintained at T_o. The shear stress at the free surface is negligible. Assuming fully developed velocity and temperature profiles (including viscous dissipation),
(a) Show that

$$v_x = \frac{\rho g \delta^2 \sin \theta}{2\mu} \left(1 - \frac{y^2}{\delta^2}\right)$$

(b) Show that

$$\frac{T - T_o}{T_s - T_o} = \frac{\rho^2 g^2 \delta^4 \sin^2 \theta}{12\mu k(T_s - T_o)} \left(\frac{y}{\delta} - \frac{y^4}{\delta^4}\right) + \frac{y}{\delta}$$

(c) Determine the location of the maximum temperature and show that the bulk temperature is

$$\frac{T_b - T_o}{T_s - T_o} = \frac{3}{8} + \frac{27}{1120} \frac{\rho^2 g^2 \delta^4 \sin^2 \theta}{\mu k(T_s - T_o)}$$

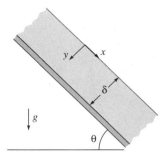

FIGURE P12.8

12.9 Consider the flow of a fluid between two horizontal, parallel plates at a distance of a apart. One plate is stationary and the other moves with a constant velocity, U.

(a) Show that, if $\partial p/\partial x = 0$, under steady conditions the velocity distribution in the fluid is $u = U(y/a)$.

(b) If both the plates are maintained at the same uniform temperature T_o, show that, with viscous dissipation, the maximum temperature is given by

$$T_{max} = T_o + \frac{\mu U^2}{8k}$$

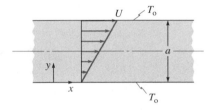

FIGURE P12.9

12.10 A 5-cm diameter shaft rotates at 3600 rpm inside a bearing. The bearing clearance is 0.1 mm and the length of the bearing is 6 cm. The lubricating oil has a viscosity of 0.0725 N s/m^2 and a thermal conductivity of 0.14 W/m K. Estimate

(a) The maximum temperature of the oil if (1) both the bearing and the shaft are at 60 °C, and (2) if the bearing is at 60 °C and the heat transfer to the shaft is negligible.

(b) The power required to rotate the shaft.

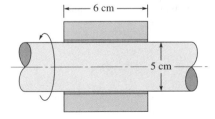

FIGURE P12.10

12.11 A liquid flows between two parallel plates, which are at a distance $2a$ apart, with a fully developed velocity profile. The velocity distribution is given by

$$v_x = v_c \left(1 - \frac{y^2}{a^2}\right)$$

The two plates are subjected to a uniform heat flux q_w''. Show that in the fully developed temperature profile region, with negligible viscous dissipation,

$$\frac{ha}{k} = \frac{35}{17} \qquad \left(h = \frac{q_w''}{T_s - T_b} \right)$$

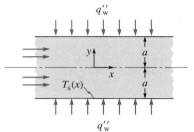

FIGURE P12.11

12.12 A liquid metal flows between two parallel plates. At $x = 0$, $T = T_i$. The upper plate is insulated and the lower plate is at T_s for $x > 0$. Assuming the velocity to be uniform, find an expression for $T(x, y)$. Assume negligible axial conduction and viscous dissipation.

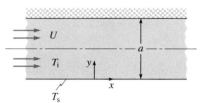

FIGURE P12.12

12.13 A highly viscous, incompressible fluid fills the annulus between two very long, vertical, coaxial cylinders. The inner cylinder rotates at a constant angular velocity, ω, and the outer cylinder is stationary. Under steady state conditions, neglecting end effects, and assuming $v_r = v_z = 0$ and $v_\theta = v_\theta(r)$ show that

$$\frac{v_\theta}{\omega r_i} = \frac{1}{(r_o/r_i)^2 - 1} \left(\frac{r_o^2}{r r_i} - \frac{r}{r_i} \right)$$

Derive the temperature distribution if the inner and outer cylinders are maintained at a uniform temperature of T_1.

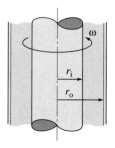

FIGURE P12.13

12.14 A liquid metal with a very low Prandtl number approaches a flat plate with a uniform velocity, U, and uniform temperature, T_∞. The plate is at a uniform temperature T_s. As Pr $\ll$ 1, the velocity boundary layer thickness is very much smaller than the temperature boundary layer thickness. Hence, the velocity of the fluid may be approximated as uniform everywhere. The boundary layer energy equation for such a flow is

$$U \frac{\partial T}{\partial x} = \alpha \frac{\partial^2 T}{\partial y^2}$$

Define $\theta = T_\infty - T$. Taking the Laplace transform of the equation and the boundary conditions in the x-direction, show that

$$\theta = \theta_s \; \text{erfc} \left[\frac{y}{2(\alpha x/U)^{1/2}} \right] \qquad \theta_s = T_\infty - T_s$$

Derive Equation 12.5.28.

12.15 In Problem 12.14, if, instead of the plate being maintained at a uniform temperature, it is subjected to a uniform heat flux q_o'' (following the same procedure as in Problem 12.14) show that

$$\theta = -\frac{q_o''}{k} \left[2(ax/\pi)^{1/2} \; exp[-y^2/(4ax)] - y \; \text{erfc} \left(\frac{y}{2(ax)^{1/2}} \right) \right]$$

$$\text{Nu}_x = 0.886(\text{Re}_x \; \text{Pr})^{1/2}$$

where $a = (\alpha/U)^{1/2}$.

TABLES OF PROPERTIES

Table A1 Thermophysical Properties of Metals
Table A2 Thermophysical Properties of Miscellaneous Solids
Table A3A Thermophysical Properties of Building Materials (300 K)
Table A3B Thermophysical Properties of Miscellaneous Materials
Table A4 Thermophysical Properties of Saturated Liquids
Table A5 Thermophysical Properties of Saturated Water
Table A6 Thermophysical Properties of Saturated Steam
Table A7 Thermophysical Properties of Gases at Atmospheric Pressure
Table A8 Thermophysical Properties of Selected Liquids at Atmospheric Pressure
Table A9 Normal, Total Emissivities of Surfaces, Metals
Table A10 Normal, Total Emissivities of Surfaces, Nonmetals

References for Tables of Properties

ASHRAE Handbook of Fundamentals (1985, 1989). American Society of Heating, Refrigerating and Air Conditioning Engineers, Atlanta, GA.

Chapman, A.J. (1987). *Fundamentals of Heat Transfer.* New York: Macmillan.

Eckert, E.R.G., and Drake, R.M. (1972). *Analysis of Heat and Mass Transfer.* New York:McGraw-Hill.

Incropera, F.P., and De Witt, D.P. (1990). *Fundamentals of Heat and Mass Transfer,* 3d ed. New York: Wiley.

Jasper, J.J. (1972). The surface tension of pure liquid compounds. *J. Physical and Chemical Reference Data,* (4).

Raznejevic, K. (1976). *Handbook of Thermodynamic Tables and Charts.* New York: McGraw-Hill.

Rosenhow, W.M., Hartnett, J.P., and Ganic, E.N. (1985). *Handbook of Heat Transfer Fundamentals.* New York: McGraw-Hill.

Sparrow, E.M., and Cess, R.D. (1978). *Radiation Heat Transfer* (Aug. Ed.). New York: McGraw-Hill.

Touloukian, Y.S., and Ho, C.Y., eds. (1972). *Thermophysical Properties of Matter.* Vol. 1, *Thermal Conductivity of Metallic Solids;* Vol. 2, *Thermal Conductivity of Nonmetallic Solids.* New York: Plenum.

Van Wylen, G., and Sonntag, R.E. (1985). *Fundamentals of Classical Thermodynamics,* 3d ed. New York: Wiley.

White, F.M. (1988). *Heat and Mass Transfer.* Reading, MA: Addison-Wesley.

TABLE A1 Thermophysical Properties of Metals

Material	Properties at 300 K					Properties at Various Temperatures (K), k/c_p[‡]					
	T^*_m[†] (K)	ρ (kg/m³)	c_p (J/kg K)	k (W/m K)	$\alpha \times 10^6$ (m²/s)	100	200	400	600	800	1000
Aluminum pure	993	2702	903	237	97.1	302	237	240	231	218	
						482	798	949	1033	1146	
Aluminum alloy 2024-T6	775	2770	875	177	73	65	163	186	186		
						473	787	925	1042		
Beryllium	1550	1850	1825	200	59.2	990	301	161	126	106	90.8
						203	1114	2191	2604	2823	3018
Cadmium	594	8650	231	96.8	48.4	203	99.3	94.7			
						198	222	242			
Chromium	2118	7160	449	93.7	29.1	159	111	90.9	80.7	71.3	65.4
						192	384	484	542	581	616
Cobalt	1769	8862	421	99.2	26.6	167	122	85.4	67.4	58.2	52.1
						236	379	450	503	550	628
Copper (pure)	1358	8933	385	401	117	482	413	393	379	366	352
						252	356	397	417	433	451
Bronze (commercial) (90% Cu, 10% Al)	1293	8800	420	52	14		42	52	59		
							785	460	545		
Bronze (phosphor) (89% Cu, 11% Sn)	1104	8780	355	54	17		41	65	74		
Brass (cartridge) (70% Cu, 30% Zn)	1188	8530	380	110	33.9	75	95	137	149		
							360	395	425		
Constantan (55% Cu, 45% Ni)	1493	8920	384	23	6.71	17	19				
						237	362				
Germanium	1211	5360	322	59.9	34.7	232	96.8	43.2	27.3	19.8	17.4
						190	290	337	348	357	375
Gold	1336	19 300	129	317	127	327	323	311	298	284	270
						109	124	131	135	140	145
Iridium	2720	22 500	130	147	50.3	172	153	144	138	132	126
						90	122	133	138	144	153
Iron (pure)	1810	7870	447	80.2	23.1	134	94	69.5	54.7	43.3	32.8
						216	384	490	574	680	975
Iron (99.75% Pure)		7870	447	72.7	20.7	95.6	80.6	65.7	53.1	42.2	32.3
						215	384	490	574	680	975

[†]T^*_m = melting point.
[‡]Upper row is thermal conductivity, k, and lower row is specific heat, c_p.

Composition	Melting Point (K)	ρ (kg/m³)	c_p (J/kg·K)	k (W/m·K)	α·10⁶ (m²/s)	100 K k	100 K c_p	200 K k	200 K c_p	400 K k	400 K c_p	600 K k	600 K c_p	800 K k	800 K c_p	1000 K k	1000 K c_p
Iron, cast (4% C)		7272	420	52	17												
Carbon Steels																	
Plain Carbon (AISI 1010) 0.08–0.13% C, 0.3–0.6% Mn		7854	434	60.5	17.7					56.7	487	48	559	39.2	685	30	1169
Carbon-Silicon (<1% Mn, 0.1% < Si < 0.6%)		7817	446	51.9	14.9					49.8	501	44	582	37.4	699	29.3	971
Carbon-Mn-Si 1–1.65% Mn, 0.1–6% Si		8131	434	41	11.6					42.2	487	39.7	559	35	685	27.6	1090
Chromium Steels																	
0.18% C, 0.65% Cr, 0.23% Mo, 0.6% Si		7822	444	37.7	10.9					38.2	492	36.7	575	33.3	688	26.9	969
0.16% C, 1% Cr, 0.54% Mo, 0.39% Cr		7858	442	42.3	12.2					42	492	39.1	575	34.5	688	27.4	969
0.2% C, 1.02% Cr, 0.15% V		7836	443	48.9	14.1					46.8	492	42.1	575	36.6	688	28.2	969
Stainless Steels																	
AISI 302		8055	480	15.1	3.91					17.3	512	20	559	22.8	585	25.4	606
AISI 304	1670	7900	477	14.9	3.95	9.2	272	12.6	402	16.6	515	19.8	557	22.6	582	25.4	611
AISI 316		8238	468	13.4	3.48					15.2	504	18.3	550	21.3	576	24.2	602
AISI 347		7978	480	14.2	3.71					15.8	513	18.9	559	21.9	585	24.7	606
Lead	601	11 340	129	35.3	24.1	39.7	118	36.7	125	34	132	31.4	142				
Magnesium	923	1740	1024	156	87.6	169	649	159	934	153	1074	149	1170	146	1267		
Molybdenum	2894	10 240	251	138	53.7	179	141	143	224	134	261	126	275	118	285	112	295
Nickel (pure)	1728	8900	444	90.7	23	164	232	107	383	80.2	485	65.6	592	67.6	530	71.8	562
Nichrome (80% Ni, 20% Cr)	1672	8400	420	12	3.4					14	480	16	525	21	545		
Inconel (73% Ni, 15% Cr, 6.7% Fe)	1665	8510	439	11.7	3.1	8.7		10.3	372	13.5	473	17	510	20.5	546	24	626

TABLE A1 (continued)

Material	T*_m† (K)	ρ (kg/m³)	c_p (J/kg K)	k (W/m K)	α × 10⁶ (m²/s)	100	200	400	600	800	1000
Palladium	1827	12 020	244	71.8	24.5	76.5	71.6	73.6	79.7	86.9	94.2
						168	227	251	261	271	281
Platinum (pure)	2045	21 450	133	71.6	25.1	77.5	72.6	71.8	73.2	75.6	78.7
						100	125	136	141	146	152
Platinum alloy 60% Pt, 40% Rh	1800	16 630	162	47	17.4			52	59	65	69
Rhodium	2236	12 450	243	150	49.6	186	154	146	136	127	121
						147	220	253	274	293	311
Silicon	1685	2330	712	148	89.2	884	264	98.9	61.9	42.2	31.2
						259	556	790	867	913	946
Silver	1235	10 500	235	429	174	444	430	425	412	396	379
						187	225	239	250	262	277
Tantalum	3269	16 600	140	57.5	24.7	59.2	57.5	57.8	58.6	59.4	60.2
						110	133	144	146	149	152
Thorium	2023	11 700	118	54	39.1	59.8	54.6	54.5	55.8	56.9	56.9
						99	112	124	134	145	156
Tin	505	7310	227	66.6	40.1	85.2	73.3	62.2			
						188	215	243			
Titanium	1953	4500	522	21.9	9.32	30.5	24.5	20.4	19.4	19.7	20.7
						300	465	551	591	633	675
Tungsten	3660	19 300	132	174	68.3	208	186	159	137	125	118
						87	122	137	142	145	148
Uranium	1460	19 070	116	27.6	12.5	21.5	25.1	29.6	34	38.8	43.9
						94	108	125	146	176	180
Vanadium	2192	6100	489	30.7	10.3	35.8	31.3	31.3	33.3	35.7	38.2
						258	430	515	540	563	597
Zinc	693	7140	389	116	41.8	117	118	111	103		
						297	367	402	436		
Zirconium	2125	6570	278	22.7	12.4	33.2	25.2	21.6	20.7	21.6	23.7
						205	264	300	322	342	362

†T*_m = melting point.
‡Upper row is thermal conductivity, k, and lower row is specific heat, c_p.

TABLE A2 Thermophysical Properties of Miscellaneous Solids

Material	Properties at 300 K					Properties at Various Temperatures (K), k/c_p‡					
	T_m^*† (K)	ρ (kg/m³)	c_p (J/kg K)	k (W/m K)	$\alpha \times 10^6$ (m²/s)	100	200	400	600	800	1000
Aluminum oxide (sapphire)	2323	3970	765	46	15.1	450	82	32.4 / 940	18.9 / 1110	13 / 1180	10.5 / 1225
Aluminum oxide (polycrystalline)	2323	3970	765	36	11.9	133	55	26.4 / 940	15.8 / 1110	10.4 / 1180	7.85 / 1225
Boron	2573	2500	1105	27.6	9.99	190	52.5	18.7 / 1490	11.3 / 1880	8.1 / 2135	6.3 / 2350
Boron fiber epoxy	590	2080									
k paral. to fiber				2.29		2.1	2.23	2.28			
k perp. to fiber				0.59		0.37	0.49	0.6			
c_p			1122			364	757	1431			
Carbon (amorphous)	1500	1950		1.6		0.67	1.18	1.89	2.19	2.37	2.53
Carbon (diamond)		3500	509	2300		10 000 / 21	4000 / 194	1540 / 853			
Graphite pyrolytic	2273	2210									
k, paral. to layer				1950		4970	3230	1390	892	667	534
k, perp. to layer				5.7		16.8	9.23	4.1	2.68	2.01	1.6
c_p			709			136	411	992	1406	1650	1793
Graphite fiber epoxy	450	1400									
k, paral. to fiber				11.1		5.7	8.7	13			
k, pepr. to fiber				0.87		0.46	0.68	1.1			
c_p			935			337	642	1216			
Pyroceram Corning 9606	1623	2600	808	3.98	1.89	5.25	4.78	3.64 / 908	3.28 / 1038	3.08 / 1122	2.96 / 1197
Silicon Carbide	3100	3160	675	490	230			/ 880	/ 1050	/ 1135	87 / 1195

†T_m^* = melting point
‡Upper row is thermal conductivity, k, and lower row is specific heat, c_p.

TABLE A2 (continued)

Material	T_m^*† (K)	Properties at 300 K				Properties at Various Temperatures (K), k/c_p‡					
		ρ (kg/m³)	c_p (J/kg K)	k (W/m K)	$\alpha \times 10^6$ (m²/s)	100	200	400	600	800	1000
Silicon dioxide quartz	1883	2650									
$\quad k$, paral. to c axis				10.4		39	16.4	7.6	5	4.2	
$\quad k$, perp. to c axis				6.21		20.8	9.5	4.7	3.4	3.1	
$\quad c_p$			745					885	1075		
Silicon dioxide (fused silica)	1883	2220	745	1.38	0.834	0.69	1.14	1.51	1.75	2.17	2.87
								905	1040	1105	1155
Silicon nitride	2173	2400	691	16	9.65			13.9	11.3	9.88	8.76
							578	778	937	1063	1155
Sulfur	392	2070	708	0.21	0.14	0.17	0.19				
						403	606				
Thorium dioxide	3573	9110	235	13	6.1			10.2	6.6	4.7	3.68
								255	274	285	295
Titanium dioxide polycrystalline	2133	4157	710	8.4	2.8			7.01	5.02	3.94	3.46
								805	880	910	930

†T_m^* = melting point
‡Upper row is thermal conductivity, k, and lower row is specific heat, c_p.

TABLE A3A Thermophysical Properties of Building Materials (300 K)

Material	ρ (kg/m³)	k (W/m K)	c_p (J/kg K)
Building Board			
Asbestos cement board	1920	0.576	1010
Gypsum or plasterboard	800	0.176	1090
Hardboard			
Medium density	800	0.105	1300
High density	880	0.118	1340
High density, standard tempered	1008	0.114	1340
Particle board			
Low density	592	0.078	1300
Medium density	800	0.135	1300
High density	1000	0.170	1300
Underlayment	640	0.110	1220
Plywood (Douglas fir)	544	0.115	1220
Vegetable fiberboard			
Sheathing, regular density (12.7 mm)	288	0.055	1300
Sheathing, intermediate density (19.05 mm)	352	0.059	1300
Sound-deadening board (7.94 mm)	240	0.053	1260
Tiles, acoustic	288	0.058	590
Laminated paperboard	480	0.072	1380
Wood subfloor (19 mm)		0.116	1380
Insulating Materials			
Blanket and batt (Values of k and ρ are typical, and depend on how they are compressed)			
Blanket (fiber, slag, glass)	12	0.046	
Boards and slabs			
Cellular glass	136	0.050	750
Cement fiber slabs	410	0.070	
Cork, impregnated boards	155	0.041	
Glass fiber, organic bonded	100	0.036	960
Expanded perlite, organic bonded	16	0.052	1260
Expanded polystyrene	28.8	0.036	1220
Expanded polystyrene, smooth skin	40	0.029	1220
Expanded polystyrene, molded beads	16	0.037	
Interior finish (plank, tile)	240	0.050	1340
Mineral fiber with resin binder	240	0.042	710
Mineral fiberboard, wet felted			
Core or roof insulation	265	0.049	
Acoustic tile	288	0.050	800
Acoustic tile	336	0.053	
Mineral fiberboard, wet molded			
Acoustic tile	368	0.060	590
Polyurethane, cellular	24	0.023	1590
Rockwool	224	0.04	
Wood or cane fiberboard			
Acoustic tile		0.058	

TABLE A3A **(continued)**

Material	ρ (kg/m^3)	k (W/m K)	c_p (J/kg K)
Loose fill			
Cellulose insulation	45	0.042	1380
Mineral fiber (rock, slag, glass)	30	0.056	710
Perlite, expanded	50	0.042	
Vermicullite	125	0.068	1340
Field applied			
Polyurethane foam	35	0.024	
Ureaformaldhyde foam	15	0.036	
Spray cellulosic fiber base	60	0.039	
Masonry Materials			
Cement mortar	1856	0.720	
Gypsum-fiber concrete	816	0.239	880
Lightweight aggregate	1280	0.360	
Perlite, expanded	480	0.102	1340
Sand and gravel aggregate (dried)	2240	1.296	920
Sand and gravel aggregate (not dried)	2240	1.728	
Stucco	1856	0.720	
Masonry Units			
Bricks, common	1920	0.720	
Bricks, face	2080	1.296	
Stone, lime or sand		1.800	800
Plastering Material			
Cement plaster			
Sand aggregate	1860	0.720	835
Gypsum plaster			
Lightweight aggregate	720	0.230	
Perlite aggregate	720	0.216	1340
Sand aggregate	1680	0.806	840
Vermicullite aggregate	720	0.245	
Hardboard siding	640	0.215	1170
Woods (12% Moisture Content)			
Hardwoods			
Birch	700	0.170	
Maple	670	0.162	
Ash	640	0.160	
Softwoods			1630
California redwood	425	0.115	
Douglas fir	550	0.152	
Hem-fir, spruce-pine-fir	450	0.120	
Southern cypress	508	0.131	
West Coast woods, cedars	400	0.110	

TABLE A3B Thermophysical Properties of Miscellaneous Materials

Material	ρ (kg/m^3)	k (W/m K)	c_p (J/kg K)
Asphalt	2110	0.740	920
Bakelite	1270	0.233	1590
Bark	400	0.055	
Bituminous coal	1100	0.174	
Carbon filamet		8.490	
Carborundum		0.212	
Chalk	2290	0.830	900
Charcoal (wood)	240	0.050	840
Chrome brick	3200	1.2	710
Clay, 47% moisture by vol.	1495	1.675	
Clay, dry	1550	0.930	
Coal (anthracite)	1370	0.238	1090
Coke (powdered)	990	0.95	1500
Concrete, air dried	500	0.186	
Concrete, air dried	2000	0.896	
Concrete, completely dried	500	0.128	
Concrete, completely dried	2000	0.663	
Cork	86	0.048	2030
Cotton (fiber)	1500	0.042	1340
Earth, dry		0.134	
Ebonite	1200	0.167	2010
Fat		0.174	
Felt	330	0.05	
Fireclay brick	1790	1	829
German Silver	8730	33.0	400
Glass			
Window	2800	0.700	
Crown	2500	1.047	
Lead	3400	0.847	
Mirror	2550	0.802	
Pyrex	2230	1.0	840
Granite	2750	3.5	
Gravel, as filling material	1750	0.86	
Graphite, powder	700	1.186	
Guttaperche		0.200	
Gypsum, plaster	800	0.395	
Gypsum, plaster	1200	0.663	
Ice, -100 °C.	928	3.489	1460
Ice, 0 °C.	917	2.21	2093
Leather, excised	0.209		
Lime		0.123	
Limestone, calcium carbonate	2650	2.21	
Linoleum	535	0.081	
Linoleum	1180	0.186	
Magnesia (powdered)	796	0.61	980
Marble	2750	2.7	
Mica	2900	0.523	

TABLE A3B (continued)

Material	ρ (kg/m³)	k (W/m K)	c_p (J/kg K)
Mortar, plaster	1600	0.663	
Mortar, plaster	2200	1.396	
Paper	930	0.13	1300
Paraffin	890	0.265	1300
Plastics			
Acetal (Delrin, Celcon)	1450	0.23	1465
Acrylics (Plexiglass)	1190	0.19	1465
Fluoroplastics (Teflon–TFE)	2200	0.25	1050
Fluoroplastics (15% glass fiber)	2220	0.45	
Fluoroplastics (15% graphite)	2190	0.46	
Fluoroplastics (60% bronze)	3970	0.46	
Polycarbonate (Lexan, Tuffak, Zelux)	1200	0.19	1260
Polymides (nylon)	1145	0.29	
Polypropylene	910	0.12	1925
Polyster (fiberglass)	1395	0.15	1170
Polyvinyl chloride (PVC, vinyl)	1470	0.1	840
Polystyrene	1050	0.157	
Polystyrene (expanded)	28.8	0.036	
Porcelain	2300	1.5	
Pumice stone	800	0.33	
Rubber, Buna	1200	0.39	
Rubber, hard normal	1200	0.15	
Rubber, natural	1150	0.279	
Sand, average value	1750	0.93	
Sandstone, moist	2250	1.291	
Sawdust	190	0.05	
Silica brick	2000	1.105	
Snow, freshly fallen	100	0.598	
Snow, (0 °C)	500	2.2	
Stone, natural dense		2.908	
Stone, natural porous		1.745	
Styrofoam		0.033	
Sugar	1600	0.582	
Tar (bituminous)	1200	0.71	
Tar (pitch)	1100	0.88	2500
Vaseline		0.174	
Wood, teak	720	0.14	
Wood felt	330	0.052	
Wool, fabric	120	0.06	
Wool, glass	52	0.038	657
Zirconia, cloth	900	0.15	

TABLE A4 Thermophysical Properties of Saturated Liquids

T (°C)	ρ (kg/m^3)	c_p (J/kg K)	k (W/m K)	$\alpha \times 10^8$ (m^2/s)	$\mu \times 10^3$ (N s/m^2)	$\nu \times 10^6$ (m^2/s)	Pr	$g\beta/\nu^2 \times 10^8$ (1/m^3 K)
Ammonia								
−40	692	4467	0.546	17.8	0.281	0.406	2.28	1050
−20	667	4509	0.546	18.2	0.254	0.381	2.09	1310
0	640	4635	0.540	18.2	0.239	0.373	2.05	1510
20	612	4798	0.521	17.8	0.220	0.359	2.02	1810
40	581	4999	0.493	17.0	0.198	0.340	2.00	2340
Ethyl alcohol								
−40	823	2037	0.186	11.10	4.810	5.840	52.7	2.9
−20	815	2124	0.179	10.30	2.830	3.470	33.6	8.4
0	806	2249	0.174	9.60	1.770	2.200	22.9	21.2
20	789	2395	0.168	8.89	1.200	1.520	17.0	45.4
40	772	2572	0.162	8.16	0.834	1.080	13.2	93.1
60	755	2781	0.156	7.43	0.592	0.784	10.6	181.0
80	738	3026	0.150	6.72	0.430	0.583	8.7	335.0
Ethelyne glycol								
0	1131	2295	0.254	9.79	65.10	57.50	588	0.0192
20	1117	2386	0.257	9.64	21.40	19.20	199	0.1730
40	1101	2476	0.259	9.50	9.57	8.69	91	0.8440
60	1088	2565	0.262	9.39	5.17	4.75	51	2.8200
80	1078	2656	0.265	9.26	3.21	2.98	32	7.1800
100	1059	2750	0.267	9.17	2.15	2.03	22	15.5000
Glycerin								
0	1276	2261	0.282	9.83	10 603.6	8310	84 700	7.103E-07
10	1270	2319	0.284	9.65	3810.0	3000	31 000	5.450E-06
20	1264	2386	0.286	9.45	1491.5	1180	12 500	3.523E-05
30	1258	2445	0.286	9.29	629.0	500	5380	1.962E-04
40	1252	2512	0.286	9.14	275.4	220	2450	1.013E-03
50	1245	2583	0.287	8.93	186.8	150	1630	2.180E-0e
Mercury								
0	13 628	140.3	8.20	429.9	1.690	0.124	0.0288	
20	13 579	139.4	8.69	460.6	1.548	0.114	0.0249	3774.2
50	13 506	138.6	9.40	502.2	1.405	0.104	0.0207	1650.7
100	13 385	137.3	10.51	571.6	1.242	0.0928	0.0162	2073.2
150	13 264	136.5	11.49	635.4	1.131	0.0853	0.0134	2453.8
200	13 145	157.0	12.34	690.8	1.054	0.0802	0.0116	2775.8
250	13 025	135.7	13.07	740.6	0.996	0.0765	0.0103	3050.8
316	12 847	134.0	14.02	815	0.865	0.0673	0.0083	3941.9
Methyl chloride								
−30	1016.5	1492	0.202	13.37	0.319	0.314	2.35	
−20	999.4	1504	0.196	13.01	0.309	0.309	2.38	
−10	981.5	1519	0.187	12.57	0.300	0.306	2.43	

TABLE A4 (continued)

T (°C)	ρ (kg/m³)	c_p (J/kg K)	k (W/m K)	$\alpha \times 10^8$ (m²/s)	$\mu \times 10^3$ (N s/m²)	$\nu \times 10^6$ (m²/s)	Pr	$g\beta/\nu^2 \times 10^8$ (1/m³ K)
0	962.4	1538	0.178	12.13	0.291	0.302	2.49	
10	942.4	1560	0.171	11.66	0.280	0.297	2.55	
20	923.3	1586	0.163	11.12	0.271	0.293	2.63	
30	903.1	1616	0.154	10.58	0.260	0.288	2.72	
40	883.1	1650	0.144	9.96	0.248	0.281	2.83	
50	861.2	1689	0.133	9.21	0.236	0.274	2.97	
Refrigerant 12								
−40	1516	885	0.069	5.14	0.424	28.0	5.4	2520
−20	1457	907	0.071	5.38	0.343	23.5	4.4	3730
0	1396	935	0.073	5.59	0.298	21.4	3.8	5040
20	1329	966	0.073	5.66	0.262	19.7	3.5	6540
40	1254	1002	0.069	5.46	0.240	19.1	3.5	8640
Unused Engine Oil								
0	899	1796	0.147	9.11	3850.00	4280.0	47 038.1	3.5E-06
20	888	1880	0.145	8.72	800.0	901.0	10 372.4	7.9E-05
40	876	1964	0.144	8.34	212.00	242.0	2891.4	0.00111
60	864	2047	0.140	8.00	72.50	83.9	1060.1	0.00939
80	852	2131	0.138	7.69	32.00	37.5	494.1	0.0477
100	840	2219	0.137	7.38	17.10	20.3	277.0	0.165
120	829	2307	0.135	7.10	10.20	12.4	174.3	0.448
140	817	2395	0.133	6.86	6.53	8.0	117.6	1.09
160	806	2483	0.132	6.63	4.49	5.6	84.5	2.26

TABLE A5 Thermophysical Properties of Saturated Water

T (°C)	ρ (kg/m³)	c_p (J/kg K)	k (W/m K)	$\mu \times 10^6$ (N s/m²)	$\nu \times 10^8$ (m²/s)	Pr	$\beta \times 10^6$ (1/K)
0.01	999.8	4226	0.569	1760.4	176.08	13.07	−44.82
5	999.9	4206	0.578	1499.6	149.97	10.91	24.39
10	999.6	4195	0.587	1290.9	129.14	9.23	89.39
15	999.0	4187	0.595	1122.6	112.37	7.90	150.04
20	998.2	4182	0.603	985.3	98.71	6.83	206.34
25	997.0	4178	0.611	872.4	87.50	5.97	258.30
30	995.6	4176	0.618	778.6	78.20	5.26	305.91
35	993.9	4175	0.625	700.0	70.43	4.68	349.17
40	992.2	4175	0.631	633.7	63.87	4.19	388.08
45	990.2	4176	0.637	577.4	58.31	3.78	422.65

TABLE A5 (continued)

T (°C)	ρ (kg/m³)	c_p (J/kg K)	k (W/m K)	$\mu \times 10^6$ (N s/m²)	$\nu \times 10^8$ (m²/s)	Pr	$\beta \times 10^6$ (1/K)
50	988.0	4178	0.643	529.2	53.56	3.44	465.72
55	985.7	4179	0.648	487.7	49.48	3.14	494.90
60	983.2	4181	0.653	451.8	45.95	2.89	523.95
65	980.5	4184	0.658	420.6	42.90	2.67	552.85
70	977.7	4187	0.662	393.3	40.23	2.49	581.61
75	974.9	4190	0.666	369.3	37.88	2.32	610.22
80	971.8	4194	0.670	348.1	35.82	2.18	638.69
85	968.6	4198	0.673	325.4	33.60	2.03	667.02
90	965.3	4202	0.676	307.4	31.84	1.91	695.20
95	961.9	4206	0.679	291.0	30.25	1.80	723.24
100	958.3	4211	0.681	276.0	28.81	1.71	751.14
105	954.8	4217	0.683	262.4	27.49	1.62	778.89
110	951.0	4224	0.685	250.0	26.29	1.54	806.50
115	947.2	4228	0.686	238.6	25.19	1.47	833.97
120	943.1	4232	0.687	228.1	24.18	1.40	861.30
125	939.1	4241	0.688	218.4	23.26	1.35	888.48
130	934.8	4250	0.688	209.5	22.41	1.29	915.51
135	930.5	4254	0.688	201.2	21.63	1.24	942.41
140	926.1	4257	0.688	193.6	20.90	1.20	969.16
145	921.6	4263	0.688	186.5	20.24	1.16	995.76
150	916.9	4270	0.687	179.9	19.62	1.12	1022.23
155	912.2	4277	0.686	173.7	19.05	1.08	1048.55
160	907.4	4285	0.685	168.0	18.52	1.05	1074.72
165	902.5	4313	0.683	162.6	18.02	1.03	1100.76
170	897.3	4340	0.681	157.6	17.56	1.00	1126.65
175	892.2	4368	0.679	152.9	17.14	0.98	1152.39
180	886.9	4396	0.677	148.5	16.74	0.96	1178.00
185	881.5	4438	0.674	144.3	16.37	0.95	1203.45
190	876.0	4480	0.671	140.4	16.03	0.94	1228.77
195	870.4	4491	0.668	136.7	15.71	0.92	1253.94
200	864.7	4501	0.665	133.2	15.41	0.90	1278.97
210	852.8	4560	0.657	126.9	14.87	0.88	1328.60
220	840.3	4605	0.649	121.1	14.41	0.86	1377.66
230	827.3	4690	0.639	116.0	14.02	0.85	1426.14
240	813.6	4731	0.628	111.4	13.70	0.84	1474.05
250	799.2	4857	0.616	107.7	13.47	0.85	1521.38
260	784.0	4982	0.603	103.8	13.24	0.86	1568.14
270	767.9	5030	0.589	99.9	13.01	0.85	1614.33
280	750.7	5234	0.574	96.0	12.79	0.88	1659.94
290	732.3	5445	0.558	92.1	12.58	0.90	1704.98
300	712.5	5694	0.540	88.3	12.40	0.93	1749.45

TABLE A6 Thermophysical Properties of Saturated Steam

T (°C)	Pressure (kPa)	v (m³/kg)	i_{fg} (kJ/kg)	c_p (J/kg K)	k (W/m K)	$\mu \times 10^6$ (N s/m²)	Pr
0.01	0.611	206.140	2501.3	1854	0.0182	8.02	0.817
5.00	0.872	147.120	2489.6	1857	0.0183	8.22	0.834
10.00	1.228	106.380	2477.7	1860	0.0186	8.42	0.842
15.00	1.705	77.930	2465.9	1863	0.0189	8.62	0.849
20.00	2.339	57.790	2454.1	1867	0.0193	8.82	0.853
25.00	3.169	43.360	2442.3	1871	0.0195	9.02	0.865
30.00	4.246	32.890	2430.5	1875	0.0196	9.22	0.882
35.00	5.628	25.220	2418.6	1880	0.0201	9.42	0.881
40.00	7.384	19.520	2406.7	1886	0.0204	9.62	0.889
45.00	9.593	15.260	2394.8	1892	0.0207	9.82	0.897
50.00	12.349	12.030	2382.7	1900	0.0210	10.02	0.906
55.00	15.758	9.568	2370.7	1908	0.0213	10.22	0.915
60.00	19.940	7.671	2358.5	1917	0.0217	10.42	0.920
65.00	25.03	6.197	2346.2	1926	0.0220	10.62	0.930
70.00	31.19	5.042	2333.8	1937	0.0223	10.82	0.939
75.00	38.58	4.131	2321.4	1949	0.0226	11.02	0.950
80.00	47.39	3.407	2308.8	1963	0.0230	11.22	0.957
85.00	57.83	2.828	2296.0	1977	0.0233	11.42	0.969
90.00	70.14	2.361	2283.2	1993	0.0237	11.62	0.977
95.00	84.55	1.982	2270.2	2010	0.0241	11.82	0.986
100.00	101.35	1.673	2257.0	2029	0.0245	12.02	0.995
110.00	143.27	1.210	2230.2	2072	0.0248	12.42	1.038
120.00	198.53	0.8919	2202.6	2121	0.0249	12.80	1.091
130.00	270.1	0.6685	2174.2	2178	0.0254	13.17	1.129
140.00	361.3	0.5089	2144.7	2243	0.0258	13.54	1.177
150.00	475.8	0.3928	2114.3	2316	0.0263	13.90	1.224
160.00	617.8	0.3071	2082.6	2398	0.0272	14.25	1.256
170.00	791.7	0.2428	2049.5	2492	0.0282	14.61	1.291
180.00	1002.1	0.1941	2015.0	2598	0.0298	14.96	1.304
190.00	1254.4	0.1565	1978.8	2715	0.0304	15.30	1.366
200.00	1553.8	0.1274	1940.7	2837	0.0317	15.65	1.400
210.00	1906.2	0.1044	1900.7	2990	0.0331	15.99	1.445
220.00	2318	8.619E-02	1858.5	3154	0.0346	16.34	1.490
230.00	2795	7.158E-02	1813.8	3333	0.0363	16.70	1.534
240.00	3344	5.976E-02	1766.5	3542	0.0381	17.07	1.587
250.00	3973	5.013E-02	1716.2	3782	0.0401	17.45	1.646
260.00	4688	4.221E-02	1662.5	4058	0.0423	17.84	1.711
270.00	5499	3.564E-02	1605.2	4387	0.0447	18.26	1.792
280.00	6412	3.017E-02	1543.6	4782	0.0475	18.76	1.888
290.00	7436	2.557E-02	1477.1	5273	0.0506	19.29	2.010
300.00	8581	2.167E-02	1404.9	5900	0.0540	19.92	2.176
310.00	9856	1.835E-02	1326.0	6699	0.0583	20.75	2.384
320.00	11 274	1.549E-02	1238.6	7791	0.0637	21.88	2.676

Ⓐ $\mu \times 10^7$ = value

μ = value $\times 10^{-7}$

TABLE A7 Thermophysical Properties of Gases at Atmospheric Pressure

T (K)	ρ (kg/m³)	c_p (J/kg K)	$\mu \times 10^7$ (N s/m²)	$\nu \times 10^6$ (m²/s)	k (W/m K)	$\alpha \times 10^6$ (m²/s)	Pr
Air							
100	3.530	1032	70.46	2.00	0.0092	2.53	0.788
150	2.354	1012	104.01	4.42	0.0138	5.77	0.765
200	1.765	1007	133.62	7.57	0.0181	10.18	0.743
250	1.412	1006	160.42	11.36	0.0222	15.65	0.726
300	1.177	1007	184.99	15.72	0.0262	22.08	0.712
350	1.009	1009	207.77	20.60	0.0297	29.17	0.706
400	0.883	1014	229.03	25.95	0.0331	36.95	0.702
450	0.785	1021	249.03	31.74	0.0363	45.36	0.700
500	0.706	1030	267.93	37.95	0.0395	54.32	0.699
550	0.642	1040	285.89	44.54	0.0426	63.81	0.698
600	0.588	1051	303.02	51.50	0.0456	73.77	0.698
650	0.543	1063	319.41	58.81	0.0484	83.84	0.701
700	0.504	1075	335.14	66.45	0.0513	94.54	0.703
750	0.471	1087	350.28	74.41	0.0541	105.78	0.703
800	0.441	1099	364.89	82.68	0.0570	117.43	0.704
850	0.415	1110	379.02	91.25	0.0597	129.50	0.705
900	0.392	1121	392.71	100.11	0.0624	141.80	0.706
950	0.372	1131	405.99	109.24	0.0649	154.40	0.708
1000	0.353	1141	418.89	118.65	0.0673	167.14	0.710
1000	0.353	1141	418.89	118.65	0.0673	167.14	0.710
1100	0.321	1159	442.00	137.71	0.0719	193.17	0.713
1200	0.294	1175	465.00	158.05	0.0760	219.76	0.720
1300	0.272	1189	488.00	179.69	0.0797	246.86	0.729
1400	0.252	1199	509.00	201.84	0.0831	274.99	0.734
1500	0.235	1205	530.00	225.18	0.0863	304.30	0.740
Ammonia							
300	0.689	2158	101.5	14.7	0.0247	16.6	0.887
320	0.645	2170	109.0	16.9	0.0272	19.4	0.870
340	0.606	2192	116.5	19.2	0.0293	22.1	0.872
360	0.572	2221	124.0	21.7	0.0316	24.9	0.872
380	0.541	2254	131.0	24.2	0.0340	27.9	0.869
400	0.514	2287	138.0	26.9	0.0370	31.5	0.853
420	0.489	2322	145.0	29.7	0.0404	35.6	0.833
440	0.466	2357	152.5	32.7	0.0435	39.6	0.826
460	0.446	2393	159.0	35.7	0.0463	43.4	0.822
480	0.427	2430	166.5	39.0	0.0492	47.4	0.822
500	0.410	2467	173.0	42.2	0.0525	51.9	0.813
520	0.394	2504	180.0	45.7	0.0545	55.2	0.827
540	0.380	2540	186.5	49.1	0.0575	59.7	0.824
560	0.371	2577	193.0	52.0	0.0606	63.4	0.827
580	0.353	2613	199.5	56.5	0.0638	69.1	0.817

TABLE A7 (continued)

T (K)	ρ (kg/m^3)	c_p (J/kg K)	$\mu \times 10^7$ (N s/m^2)	$\nu \times 10^6$ (m^2/s)	k (W/m K)	$\alpha \times 10^6$ (m^2/s)	Pr
Argon							
200	2.434	523.6	160.0	6.6	0.0124	9.8	0.674
250	1.947	522.2	195.0	10.0	0.0152	14.9	0.672
300	1.623	521.6	227.0	14.0	0.0177	20.9	0.669
350	1.391	521.2	257.0	18.5	0.0201	27.8	0.666
400	1.217	521.0	285.0	23.4	0.0223	35.2	0.665
450	1.082	520.9	312.0	28.8	0.0244	43.3	0.665
500	0.974	520.8	337.0	34.6	0.0264	52.0	0.664
550	0.885	520.7	360.0	40.7	0.0283	61.4	0.662
600	0.811	520.6	383.0	47.2	0.0301	71.2	0.662
700	0.695	520.6	425.0	61.1	0.0336	92.8	0.658
800	0.609	520.5	464.0	76.2	0.0369	116.0	0.655
900	0.541	520.5	501.0	92.6	0.0398	141.0	0.654
1000	0.487	520.5	535.0	109.9	0.0427	168.0	0.652
Carbon Dioxide							
280	1.9022	830	140	7.4	0.0152	9.63	0.765
300	1.7730	851	149	8.4	0.0166	11.0	0.766
320	1.6609	872	156	9.4	0.0181	12.5	0.754
340	1.5618	891	165	10.6	0.0197	14.2	0.746
360	1.4743	908	173	11.7	0.0212	15.8	0.741
380	1.3961	926	181	13.0	0.0228	17.6	0.737
400	1.3257	942	190	14.3	0.0243	19.5	0.737
450	1.1782	981	210	17.8	0.0283	24.5	0.728
500	1.0594	1020	231	21.8	0.0325	30.1	0.725
550	0.9625	1050	251	26.1	0.0366	36.2	0.721
600	0.8826	1080	270	30.6	0.0407	42.7	0.717
650	0.8143	1100	288	35.4	0.0445	49.7	0.712
700	0.7564	1130	305	40.3	0.0481	56.3	0.717
750	0.7057	1150	321	45.5	0.0517	63.7	0.714
800	0.6614	1170	337	51.0	0.0551	71.2	0.716
Carbon Monoxide							
200	1.708	1045	127	7.5	0.0175	9.8	0.763
250	1.366	1048	154	11.3	0.0214	15.0	0.753
300	1.138	1051	178	15.6	0.0252	21.1	0.743
350	0.976	1056	201	20.5	0.0288	28.0	0.735
400	0.854	1060	221	25.9	0.0323	35.7	0.727
450	0.759	1065	241	31.8	0.0355	43.9	0.723
500	0.683	1071	260	38.0	0.0386	52.8	0.720
550	0.621	1077	277	44.6	0.0416	62.2	0.717
600	0.569	1084	294	51.7	0.0444	72.0	0.718
700	0.488	1099	325	66.6	0.0497	92.7	0.718
800	0.427	1114	354	82.9	0.0549	115.0	0.718

TABLE A7 (continued)

T (K)	ρ (kg/m^3)	c_p (J/kg K)	$\mu \times 10^7$ (N s/m^2)	$\nu \times 10^6$ (m^2/s)	k (W/m K)	$\alpha \times 10^6$ (m^2/s)	Pr
900	0.379	1128	381	100.4	0.0596	139.0	0.721
1000	0.342	1142	406	119.0	0.0644	165.0	0.720
Helium							
200	0.2440	5197	15.0	61	0.115	91	0.676
250	0.1952	5197	17.5	90	0.134	154	0.680
300	0.1627	5197	19.9	122	0.150	177	0.690
350	0.1394	5197	22.1	159	0.165	228	0.698
400	0.1120	5197	24.3	199	0.180	283	0.703
450	0.1085	5197	26.3	243	0.195	345	0.702
500	0.0976	5197	28.3	290	0.211	417	0.695
550	0.0887	5197	30.2	340	0.229	497	0.684
600	0.0813	5197	32.0	393	0.247	584	0.673
700	0.0697	5197	35.5	509	0.278	767	0.663
800	0.0610	5197	38.8	637	0.307	968	0.657
900	0.0542	5197	42.0	775	0.335	1190	0.652
1000	0.0488	5197	45.0	923	0.363	1430	0.645
Hydrogen							
200	0.1229	13 540	6.8	55	0.128	77	0.717
250	0.0983	14 070	7.9	80	0.156	113	0.713
300	0.0819	14 320	8.9	109	0.182	155	0.705
350	0.0702	14 420	9.9	142	0.203	201	0.705
400	0.0614	14 480	10.9	178	0.221	249	0.714
450	0.0546	14 500	11.8	217	0.239	302	0.719
500	0.0492	14 510	12.7	259	0.256	359	0.721
550	0.0447	14 520	13.6	304	0.274	422	0.722
600	0.0410	14 540	14.5	354	0.291	489	0.724
700	0.0351	14 610	16.1	459	0.325	634	0.724
800	0.0307	14 710	17.7	576	0.360	797	0.723
900	0.0273	14 840	19.2	703	0.394	1080	0.723
1000	0.0246	14 900	20.7	842	0.428	1160	0.724
Nitrogen							
100	3.4388	1070	68.8	2.00	0.0096	2.6	0.768
150	2.2594	1050	100.6	4.45	0.0139	5.9	0.759
200	1.6883	1043	129.2	7.65	0.0183	10.4	0.736
250	1.3488	1042	154.9	11.48	0.0222	15.8	0.727
300	1.1233	1041	178.2	15.86	0.0259	22.1	0.716
350	0.9625	1042	200.0	20.78	0.0293	29.2	0.711
400	0.8425	1045	220.4	26.16	0.0327	37.1	0.704
450	0.7485	1050	239.6	32.01	0.0358	45.6	0.703
500	0.6739	1056	257.7	38.24	0.0389	54.7	0.700
550	0.6124	1065	274.7	44.86	0.0417	63.9	0.702

TABLE A7 (continued)

T (K)	ρ (kg/m³)	c_p (J/kg K)	$\mu \times 10^7$ (N s/m²)	$\nu \times 10^6$ (m²/s)	k (W/m K)	$\alpha \times 10^6$ (m²/s)	Pr
600	0.5615	1075	290.8	51.79	0.0446	73.9	0.701
700	0.4812	1098	321.0	66.71	0.0499	94.4	0.706
800	0.4211	1220	349.1	82.90	0.0548	116	0.715
900	0.3743	1146	375.3	100.30	0.0597	139	0.721
1000	0.3368	1167	399.9	118.70	0.0647	165	0.721
Oxygen							
100	3.9450	962	76.4	1.94	0.0092	2.4	0.796
150	2.5850	921	114.8	4.44	0.0138	5.8	0.766
200	1.9300	915	147.5	7.64	0.0183	10.4	0.737
250	1.5420	915	178.6	11.58	0.0226	16.0	0.723
300	1.2840	920	207.2	16.14	0.0268	22.7	0.711
350	1.1000	929	233.5	21.23	0.0296	29.0	0.733
400	0.9620	942	258.2	26.84	0.0330	36.4	0.737
450	0.8554	956	281.4	32.90	0.0363	44.4	0.741
500	0.7698	972	303.3	39.40	0.0412	55.1	0.716
550	0.6998	988	324.0	46.30	0.0441	63.8	0.726
600	0.6414	1003	343.7	53.59	0.0473	73.5	0.729
700	0.5498	1031	380.8	69.26	0.0528	93.1	0.744
800	0.4810	1054	415.2	86.32	0.0589	116.0	0.743
900	0.4275	1074	447.2	104.60	0.0649	141.0	0.740
1000	0.3848	1090	477.0	124.00	0.0710	169.0	0.733
Water Vapor (Steam)							
380	0.5863	2060	127.1	21.68	0.0246	20.4	1.060
400	0.5542	2014	134.4	24.25	0.0261	23.4	1.040
450	0.4902	1980	152.5	31.11	0.0299	30.8	1.010
500	0.4405	1985	170.4	38.68	0.0339	38.8	0.998
550	0.4005	1997	188.4	47.04	0.0379	47.4	0.993
600	0.3652	2026	206.7	56.60	0.0422	57.0	0.993
650	0.3380	2056	224.7	66.48	0.0464	66.8	0.996
700	0.3140	2085	242.6	77.26	0.0505	77.1	1.000
750	0.2931	2119	260.4	88.84	0.0549	88.4	1.000
800	0.2739	2152	278.6	101.7	0.0592	100.0	1.010
850	0.2579	2186	296.9	115.1	0.0637	113.0	1.020

TABLE A8 Thermophysical Properties of Selected Liquids at Atmospheric Pressure

Liquid	T_b† (K)	ρ (kg/m³)	i_{fg} (kJ/kg)	c_p (J/kg K)	μ‡ $\times 10^6$ (N s/m²)²	k (W/m K)
Acetic acid	391.7	1049	405.0	2180	1222	0.17
Acetone	329.4	791	532.4	2150	331	0.176
Alcohol-ethyl	351.7	789.2	854.8	2840	1194	0.182
Alcohol-methyl	338.1	791.3	1100.0	2510	592.8	0.215
Aniline	457.5	1021	434.0	2140	4467	0.173
Benzene	353.3	879	394.0	1720	653	0.147
n-Butyl alcohol	390.7	811	591.5	2350	2950	0.15
Ca-chloride brine						
20% by weight		1180		3110	2000	0.574
Carbon tetra-chloride	349.9	1590	195.0	842	967	0.11
Ethyl chloride	285.5	897.8	385.9	1540		0.31
Formic acid	373.9	1219	502.0	2200	29.7	0.18
Glycerine	454	1261			17800	0.195
Heptane	371.6	684	321.0	2220	409	0.128
Hydrogen (normal)	20.4	70.7	445.7	9410		
Isobutyl alcohol	381.2	801	579.0	486	3910	0.14
Kerosene	477	820		2000	2480	0.15
Linseed oil		920			42 900	
Nitric acid	359.2	1512	628.0	1700	910	0.28
Nitorgen	77.4	808.6	198.6	2060		
Octane	398.9	703	306.3	2100	562	0.15
Sodium chloride brine						
20% by weight	378	1150		3110	1570	0.583
10% by weight	375.1	1070		3620	1180	0.593
Sulfuric acid and water						
100% by weight	560.9	1833		1400	22 000	
90% by weight	533.2	1816		1600	25 000	0.38
Toulene	383	867	363.0	1690	587	0.16
Turpentine	423	863	286.0	1700	546	0.13
Water	373	998.2	2257.0	4180	988	0.602

†Boiling point at atmospheric pressure.
‡μ at 20 °C.

TABLE A9 Normal, Total Emissivity of Surfaces, Metals

Material	Temperature (K)	Emissivity (ϵ)
Aluminum		
Polished	300–900	0.04–0.06
Rough plate	310	0.07
Commercial sheet	400	0.09
Heavily oxidized	400–800	0.20–0.33
Anodized	300–400	0.82–0.76
Antimony, polished	310–530	0.28–0.31

TABLE A9 (continued)

Material	Temperature (K)	Emissivity (ϵ)
Brass		
Highly polished	500–650	0.03–0.04
Polished	350	0.09
Dull plate	300–600	0.22
Oxidized	450–800	0.66
Chromium, polished	300–1400	0.08–0.27
Cobalt, unoxidized	530–810	0.13–0.23
Copper		
Highly polished	300	0.02
Polished	300–500	0.04–0.05
Commercial sheet	300	0.15
Oxidized	600–1000	0.5–0.8
Gold, highly polished	300–1000	0.03–0.06
Inconel		
X, stably oxidized	500–1145	0.55–0.78
B, stably oxidized	500–1225	0.32–0.55
X and B, polished	420–590	0.20
Iron		
Pure polished	366	0.06
Bright etched	423	0.13
Bright abrased	293	0.24
Red rusted	293	0.61
Hot rolled	293	0.77
Hot rolled	400	0.60
Heavily crusted	293	0.85
Heat-resistant oxidized	353	0.61
Heat-resistant oxidized	473	0.64
Cast iron, bright	366	0.21
Cast iron, oxidized	366	0.61
Black iron oxide	366	0.56
Wrought iron (polished)	400	0.28
Lead		
Polished	310–530	0.05–0.08
Gray, oxidized	310	0.28
Oxidized at 865 K	310	0.63
Magnesium (polished)	310–530	0.07–0.13
Mercury, pure and clean	310–365	0.10–0.12
Molybdenum		
Polished	310–530	0.06–0.08
Polished	810–1365	0.11–0.18
Filament	810–3030	0.08–0.29
Monel		
Repeated heating & cooling	505–1170	0.45–0.70
Oxidized at 866 K	475–865	0.41–0.46
Polished	310	0.17
Nickel		
Polished	500–1200	0.07–0.17

TABLE A9 (continued)

Material	Temperature (K)	Emissivity (ϵ)
Nickel		
Oxidized	450–1000	0.37–0.57
Wire	530–1365	0.10–0.19
Platinum		
Electrolytic	530–810	0.06–0.10
Filament	310–1365	0.04–0.19
Oxidized at 865 K	530–810	0.07–0.11
Polished	500–1500	0.06–0.18
Strip	810–1360	0.12–0.14
Wire	475–1645	0.07–0.18
Silver		
Polished or deposited	310–810	0.01–0.03
Oxidized	310–810	0.02–0.04
Stainless Steel		
Polished	300–1000	0.17–0.30
Lightly oxidized	600–1000	0.30–0.40
Highly oxidized	600–1000	0.70–0.80
Steel		
Commercial sheet	500–1200	0.20–0.32
Heavily oxidized	300	0.81
Mild, polished	420–755	0.14–0.32
Polished	310–530	0.07–0.10
Sheet, ground	1200	0.55
Sheet, rolled	310	0.66
Tin, bright	310–360	0.04–0.06
Tungsten		
Polished	300–2500	0.03–0.29
Filament	810–1365	0.11–0.16
Filament	3500	0.39
Zinc		
Pure, polished	310–530	0.02–0.03
Oxidized at 1200 K	1200	0.11
Galvanized, gray	310	0.28
Galvanized, fairly bright	310	0.23
Dull	310–530	0.21

TABLE A10 Normal, Total Emissivity of Surfaces, Nonmetals

Material	Temperature (K)	Emissivity (ϵ)
Aluminum oxide	600–1500	0.69–0.41
Asbestos	300	0.96
Brick		
Common	300	0.95
Fireclay	1200	0.75
Ordinary, refractory	1350	0.59
White, refractory	1350	0.29

TABLE A10 (continued)

Material	Temperature (K)	Emissivity (ϵ)
Carbon		
Filament	2000	0.53
Lampsoot	310	0.95
Clay, fired	365	0.91
Cloth	300	0.8
Concrete, rough	300	0.94
Glass, window	300	0.93
Glass, pyrex	300–1200	0.82–0.62
Glass, pyroceram	300–1500	0.85–0.57
Gypsum	310	0.80–0.90
Ice	273	0.97
Limestone	500–700	0.95–0.83
Magnesium oxide	400–800	0.69–0.55
Marble	300	0.94
Masonry	300	0.8
Mica	310	0.75
Mortar, lime		0.92
Paints		
Aluminum	300	0.45
Black, lacquer, shiny	300	0.88
Enamel	293	0.85–0.95
Lacquer white	273	0.925
Lampblack	325	0.94
Lacquer black matte	353	0.97
Oils, all colors	300	0.94
White, acrylic	300	0.90
Red primer	300	0.93
Varnish, dark glossy	311	0.89
Paper, white	300	0.95–0.98
Paper, other colors	300	0.92–0.94
Paper, roofing	300	0.91
Plaster, white	300	0.93
Porcelain, glazed	300	0.93
Quartz, rough, fused	300	0.93
Rubber, soft	300	0.86
Rubber, hard	300	0.93
Sand	300	0.90
Silicon carbide	600–1500	0.86
Skin, human	300	0.95
Snow	273	0.85
Soil, earth	300	0.95
Soot	300–500	0.95
Teflon	300–500	0.85–0.92
Water, 0.1 mm or more thick	273–373	0.95
Wood, beech	300	0.94
Wood, oak	300	0.90

MATHEMATICAL FUNCTIONS AND TABLES

HYPERBOLIC FUNCTIONS

$$\sinh x = \frac{e^x - e^{-x}}{2} \qquad \cosh x = \frac{e^x + e^{-x}}{2} \qquad \tanh x = \frac{\sinh x}{\cosh x} = \frac{e^x - e^{-x}}{e^x + e^{-x}}$$

$$\cosh^2 x - \sinh^2 x = 1 \qquad \sinh(x + y) = \sinh x \cosh y + \cosh x \sinh x$$
$$\cosh(x + y) = \cosh x \cosh y + \sinh x \sinh y$$

$$\sinh(-x) = -\sinh x \qquad \cosh(-x) = \cosh x \qquad \tanh(-x) = -\tanh x$$

$$\sinh(0) = 0 \qquad \cosh(0) = 1 \qquad \tanh(0) = 0$$

$$\sinh(\infty) = \infty \qquad \cosh(\infty) = \infty \qquad \tanh(\infty) = 1$$

$$\frac{d}{dx}(\sinh ax) = a \cosh ax \qquad \int \sinh ax \, dx = \frac{1}{a} \cosh ax$$

$$\frac{d}{dx}(\cosh ax) = a \sinh ax \qquad \int \cosh ax \, dx = \frac{1}{a} \sinh ax$$

no sign change {

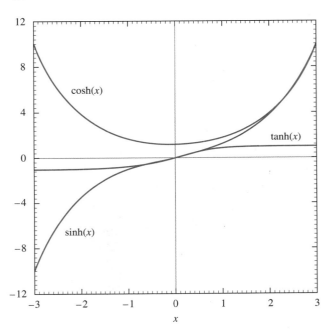

TABLE A11 Roots of $\mu \tan \mu = Bi$

Bi	μ_1	μ_2	μ_3	μ_4	μ_5	μ_6
0.001	0.031618	3.141911	6.283344	9.424884	12.566450	15.708027
0.002	0.044706	3.142229	6.283504	9.424990	12.566530	15.708091
0.006	0.077382	3.143501	6.284140	9.425415	12.566848	15.708345
0.010	0.099834	3.144773	6.284776	9.425839	12.567166	15.708600
0.020	0.140952	3.147946	6.286367	9.426900	12.567962	15.709236
0.040	0.198676	3.154273	6.289545	9.429020	12.569553	15.710509
0.060	0.242526	3.160574	6.292720	9.431140	12.571143	15.711782
0.080	0.279126	3.166849	6.295891	9.433258	12.572734	15.713055
0.100	0.311053	3.173097	6.299059	9.435376	12.574323	15.714327
0.200	0.432841	3.203935	6.314846	9.445948	12.582265	15.720685
0.300	0.521791	3.234090	6.330539	9.456492	12.590194	15.727036
0.400	0.593242	3.263550	6.346133	9.467005	12.598111	15.733381
0.500	0.653271	3.292310	6.361620	9.477486	12.606013	15.739719
0.600	0.705066	3.320366	6.376997	9.487932	12.613901	15.746050
0.700	0.750558	3.347720	6.392258	9.498342	12.621774	15.752372
0.800	0.791034	3.374376	6.407398	9.508714	12.629629	15.758685
0.900	0.827401	3.400339	6.422413	9.519045	12.637467	15.764990
1.000	0.860334	3.425618	6.437298	9.529334	12.645287	15.771285
1.200	0.917846	3.474169	6.466665	9.549780	12.660868	15.783844
1.400	0.966548	3.520126	6.495472	9.570038	12.676367	15.796360
1.600	1.008421	3.563601	6.523697	9.590094	12.691775	15.808829
1.800	1.044857	3.604717	6.551323	9.609939	12.707088	15.821247
2.000	1.076874	3.643597	6.578334	9.629560	12.722299	15.833611
2.500	1.142227	3.731838	6.643121	9.677580	12.759847	15.864265
3.000	1.192459	3.808762	6.703956	9.724027	12.796648	15.894513
3.500	1.232272	3.876052	6.760880	9.768813	12.832636	15.924313
4.000	1.264592	3.935162	6.814010	9.811878	12.867756	15.953626
4.500	1.291341	3.987322	6.863515	9.853194	12.901962	15.982417
5.000	1.313838	4.033568	6.909595	9.892752	12.935221	16.010659
6.000	1.349553	4.111618	6.992352	9.966672	12.998813	16.065395
7.000	1.376615	4.174638	7.064031	10.033914	13.058442	16.117689
8.000	1.397816	4.226362	7.126281	10.094916	13.114132	16.167459
9.000	1.414865	4.269445	7.180560	10.150187	13.165991	16.214677
10.000	1.428870	4.305802	7.228110	10.200262	13.214185	16.259361
15.000	1.472916	4.425495	7.395901	10.389796	13.407760	16.447369
20.000	1.496129	4.491480	7.495412	10.511670	13.541976	16.586395
25.000	1.510452	4.533017	7.560313	10.594730	13.637777	16.690108
30.000	1.520167	4.561495	7.605690	10.654324	13.708547	16.769057
35.000	1.527190	4.582209	7.639092	10.698912	13.762524	16.830531
40.000	1.532503	4.597943	7.664660	10.733415	13.804845	16.879444
45.000	1.536662	4.610294	7.684839	10.760852	13.838817	16.919129
45.000	1.536662	4.610294	7.684839	10.760852	13.838817	16.919129
50.000	1.540006	4.620246	7.701159	10.783164	13.866634	16.951884
60.000	1.545051	4.635287	7.725921	10.817204	13.909368	17.002622
70.000	1.548676	4.646113	7.743804	10.841911	13.940587	17.039975
80.000	1.551406	4.654276	7.757317	10.860641	13.964354	17.068554
90.000	1.553537	4.660650	7.767885	10.875320	13.983032	17.091093

TABLE A11 (continued)

Bi	μ_1	μ_2	μ_3	μ_4	μ_5	μ_6
100.000	1.555245	4.665765	7.776374	10.887130	13.998090	17.109307
200.000	1.562982	4.688949	7.814927	10.940924	14.066948	17.193005
500.000	1.567661	4.702983	7.838306	10.973631	14.108956	17.244285
1000.000	1.569227	4.707681	7,846136	10.984590	14.123045	17.261500
10 000.000	1.570639	4.711918	7.853196	10.994475	14.135753	17.277032
100 000.000	1.570781	4.712342	7.853903	10.995464	14.137026	17.278587
500 000.000	1.570793	4.712380	7.853966	10.995552	14.137139	17.278725

TABLE A12 Bessel Function of the First Kind $J_o(x)$

x	0.00	0.01	0.02	0.03	0.04	0.05	0.06	0.07	0.08	0.09
0.00	1.0000	1.0000	0.9999	0.9998	0.9996	0.9994	0.9991	0.9988	0.9984	0.9980
0.10	0.9975	0.9970	0.9964	0.9958	0.9951	0.9944	0.9936	0.9928	0.9919	0.9910
0.20	0.9900	0.9890	0.9879	0.9868	0.9857	0.9844	0.9832	0.9819	0.9805	0.9791
0.30	0.9776	0.9761	0.9746	0.9730	0.9713	0.9696	0.9679	0.9661	0.9642	0.9623
0.40	0.9604	0.9584	0.9564	0.9543	0.9522	0.9500	0.9478	0.9455	0.9432	0.9409
0.50	0.9385	0.9360	0.9335	0.9310	0.9284	0.9258	0.9231	0.9204	0.9177	0.9149
0.60	0.9120	0.9091	0.9062	0.9032	0.9002	0.8971	0.8940	0.8909	0.8877	0.8845
0.70	0.8812	0.8779	0.8745	0.8711	0.8677	0.8642	0.8607	0.8572	0.8536	0.8500
0.80	0.8463	0.8426	0.8388	0.8350	0.8312	0.8274	0.8235	0.8195	0.8156	0.8116
0.90	0.8075	0.8034	0.7993	0.7952	0.7910	0.7868	0.7825	0.7783	0.7739	0.7696
1.00	0.7652	0.7608	0.7563	0.7519	0.7473	0.7428	0.7382	0.7336	0.7290	0.7243
1.10	0.7196	0.7149	0.7101	0.7054	0.7006	0.6957	0.6909	0.6860	0.6810	0.6761
1.20	0.6711	0.6661	0.6611	0.6561	0.6510	0.6459	0.6408	0.6356	0.6305	0.6253
1.30	0.6201	0.6149	0.6096	0.6043	0.5990	0.5937	0.5884	0.5830	0.5777	0.5723
1.40	0.5669	0.5614	0.5560	0.5505	0.5450	0.5395	0.5340	0.5285	0.5230	0.5174
1.50	0.5118	0.5062	0.5006	0.4950	0.4894	0.4838	0.4781	0.4725	0.4668	0.4611
1.60	0.4554	0.4497	0.4440	0.4383	0.4325	0.4268	0.4210	0.4153	0.4095	0.4038
1.70	0.3980	0.3922	0.3864	0.3806	0.3748	0.3690	0.3632	0.3574	0.3516	0.3458
1.80	0.3400	0.3342	0.3284	0.3225	0.3167	0.3109	0.3051	0.2993	0.2934	0.2876
1.90	0.2818	0.2760	0.2702	0.2644	0.2586	0.2528	0.2470	0.2412	0.2354	0.2297
2.00	0.2239	0.2181	0.2124	0.2066	0.2009	0.1951	0.1894	0.1837	0.1780	0.1723
2.10	0.1666	0.1609	0.1553	0.1496	0.1440	0.1383	0.1327	0.1271	0.1215	0.1159
2.20	0.1104	0.1048	0.0993	0.0937	0.0882	0.0827	0.0773	0.0718	0.0664	0.0609
2.30	0.0555	0.0502	0.0448	0.0394	0.0341	0.0288	0.0235	0.0182	0.0130	0.0077
2.40	0.0025	−0.0027	−0.0079	−0.0130	−0.0181	−0.0232	−0.0283	−0.0334	−0.0384	−0.0434
2.50	−0.0484	−0.0533	−0.0583	−0.0632	−0.0681	−0.0729	−0.0778	−0.0826	−0.0873	−0.0921
2.60	−0.0968	−0.1015	−0.1062	−0.1108	−0.1154	−0.1200	−0.1245	−0.1291	−0.1336	−0.1380
2.70	−0.1424	−0.1469	−0.1512	−0.1556	−0.1599	−0.1641	−0.1684	−0.1726	−0.1768	−0.1809
2.80	−0.1850	−0.1891	−0.1932	−0.1972	−0.2012	−0.2051	−0.2090	−0.2129	−0.2167	−0.2205
2.90	−0.2243	−0.2280	−0.2317	−0.2354	−0.2390	−0.2426	−0.2462	−0.2497	−0.2532	−0.2566
3.00	−0.2601	−0.2634	−0.2668	−0.2701	−0.2733	−0.2765	−0.2797	−0.2829	−0.2860	−0.2890
3.10	−0.2921	−0.2951	−0.2980	−0.3009	−0.3038	−0.3066	−0.3094	−0.3122	−0.3149	−0.3176
3.20	−0.3202	−0.3228	−0.3253	−0.3278	−0.3303	−0.3328	−0.3351	−0.3375	−0.3398	−0.3421

TABLE A12 (continued)

x	0.00	0.01	0.02	0.03	0.04	0.05	0.06	0.07	0.08	0.09
3.30	−0.3443	−0.3465	−0.3486	−0.3507	−0.3528	−0.3548	−0.3568	−0.3587	−0.3606	−0.3625
3.40	−0.3643	−0.3661	−0.3678	−0.3695	−0.3711	−0.3727	−0.3743	−0.3758	−0.3773	−0.3787
3.50	−0.3801	−0.3815	−0.3828	−0.3841	−0.3853	−0.3865	−0.3876	−0.3887	−0.3898	−0.3908
3.60	−0.3918	−0.3927	−0.3936	−0.3944	−0.3953	−0.3960	−0.3967	−0.3974	−0.3981	−0.3987
3.70	−0.3992	−0.3997	−0.4002	−0.4007	−0.4011	−0.4014	−0.4017	−0.4020	−0.4022	−0.4024
3.80	−0.4026	−0.4027	−0.4027	−0.4028	−0.4027	−0.4027	−0.4026	−0.4025	−0.4023	−0.4021
3.90	−0.4018	−0.4015	−0.4012	−0.4008	−0.4004	−0.4000	−0.3995	−0.3990	−0.3984	−0.3978
4.00	−0.3971	−0.3965	−0.3958	−0.3950	−0.3942	−0.3934	−0.3925	−0.3916	−0.3907	−0.3897
4.10	−0.3887	−0.3876	−0.3865	−0.3854	−0.3842	−0.3831	−0.3818	−0.3806	−0.3793	−0.3779
4.20	−0.3766	−0.3752	−0.3737	−0.3722	−0.3707	−0.3692	−0.3676	−0.3660	−0.3644	−0.3627
4.30	−0.3610	−0.3593	−0.3575	−0.3557	−0.3539	−0.3520	−0.3501	−0.3482	−0.3463	−0.3443
4.40	−0.3423	−0.3402	−0.3381	−0.3360	−0.3339	−0.3318	−0.3296	−0.3274	−0.3251	−0.3228
4.50	−0.3205	−0.3182	−0.3159	−0.3135	−0.3111	−0.3087	−0.3062	−0.3037	−0.3012	−0.2987
4.60	−0.2961	−0.2936	−0.2910	−0.2883	−0.2857	−0.2830	−0.2803	−0.2776	−0.2749	−0.2721
4.70	−0.2693	−0.2665	−0.2637	−0.2609	−0.2580	−0.2551	−0.2522	−0.2493	−0.2464	−0.2434
4.80	−0.2404	−0.2374	−0.2344	−0.2314	−0.2283	−0.2253	−0.2222	−0.2191	−0.2160	−0.2129
4.90	−0.2097	−0.2066	−0.2034	−0.2002	−0.1970	−0.1938	−0.1906	−0.1874	−0.1841	−0.1809

TABLE A13 Bessel Function of the First Kind $J_1(x)$

x	0.00	0.01	0.02	0.03	0.04	0.05	0.06	0.07	0.08	0.09
0.00	0.0000	0.0050	0.0100	0.0150	0.0200	0.0250	0.0300	0.0350	0.0400	0.0450
0.10	0.0499	0.0549	0.0599	0.0649	0.0698	0.0748	0.0797	0.0847	0.0896	0.0946
0.20	0.0995	0.1044	0.1093	0.1142	0.1191	0.1240	0.1289	0.1338	0.1386	0.1435
0.30	0.1483	0.1531	0.1580	0.1628	0.1676	0.1723	0.1771	0.1819	0.1866	0.1913
0.40	0.1960	0.2007	0.2054	0.2101	0.2147	0.2194	0.2240	0.2286	0.2332	0.2377
0.50	0.2423	0.2468	0.2513	0.2558	0.2603	0.2647	0.2692	0.2736	0.2780	0.2823
0.60	0.2867	0.2910	0.2953	0.2996	0.3039	0.3081	0.3124	0.3166	0.3207	0.3249
0.70	0.3290	0.3331	0.3372	0.3412	0.3452	0.3492	0.3532	0.3572	0.3611	0.3650
0.80	0.3688	0.3727	0.3765	0.3803	0.3840	0.3878	0.3915	0.3951	0.3988	0.4024
0.90	0.4059	0.4095	0.4130	0.4165	0.4200	0.4234	0.4268	0.4302	0.4335	0.4368
1.00	0.4401	0.4433	0.4465	0.4497	0.4528	0.4559	0.4590	0.4620	0.4650	0.4680
1.10	0.4709	0.4738	0.4767	0.4795	0.4823	0.4850	0.4878	0.4904	0.4931	0.4957
1.20	0.4983	0.5008	0.5033	0.5058	0.5082	0.5106	0.5130	0.5153	0.5176	0.5198
1.30	0.5220	0.5242	0.5263	0.5284	0.5305	0.5325	0.5344	0.5364	0.5383	0.5401
1.40	0.5419	0.5437	0.5455	0.5472	0.5488	0.5504	0.5520	0.5536	0.5551	0.5565
1.50	0.5579	0.5593	0.5607	0.5620	0.5632	0.5644	0.5656	0.5667	0.5678	0.5689
1.60	0.5699	0.5709	0.5718	0.5727	0.5735	0.5743	0.5751	0.5758	0.5765	0.5772
1.70	0.5778	0.5783	0.5788	0.5793	0.5798	0.5802	0.5805	0.5808	0.5811	0.5813
1.80	0.5815	0.5817	0.5818	0.5818	0.5819	0.5818	0.5818	0.5817	0.5816	0.5814
1.90	0.5812	0.5809	0.5806	0.5803	0.5799	0.5794	0.5790	0.5785	0.5779	0.5773
2.00	0.5767	0.5761	0.5754	0.5746	0.5738	0.5730	0.5721	0.5712	0.5703	0.5693
2.10	0.5683	0.5672	0.5661	0.5650	0.5638	0.5626	0.5614	0.5601	0.5587	0.5574
2.20	0.5560	0.5545	0.5530	0.5515	0.5500	0.5484	0.5468	0.5451	0.5434	0.5416

TABLE A13 (continued)

x	0.00	0.01	0.02	0.03	0.04	0.05	0.06	0.07	0.08	0.09
2.30	0.5399	0.5381	0.5362	0.5343	0.5324	0.5305	0.5285	0.5265	0.5244	0.5223
2.40	0.5202	0.5180	0.5158	0.5136	0.5113	0.5091	0.5067	0.5044	0.5020	0.4996
2.50	0.4971	0.4946	0.4921	0.4895	0.4870	0.4843	0.4817	0.4790	0.4763	0.4736
2.60	0.4708	0.4680	0.4652	0.4624	0.4595	0.4566	0.4536	0.4507	0.4477	0.4446
2.70	0.4416	0.4385	0.4354	0.4323	0.4291	0.4260	0.4228	0.4195	0.4163	0.4130
2.80	0.4097	0.4064	0.4030	0.3997	0.3963	0.3928	0.3894	0.3859	0.3825	0.3790
2.90	0.3754	0.3719	0.3683	0.3647	0.3611	0.3575	0.3538	0.3502	0.3465	0.3428
3.00	0.3391	0.3353	0.3316	0.3278	0.3240	0.3202	0.3164	0.3125	0.3087	0.3048
3.10	0.3009	0.2970	0.2931	0.2892	0.2852	0.2813	0.2773	0.2733	0.2694	0.2654
3.20	0.2613	0.2573	0.2533	0.2492	0.2452	0.2411	0.2370	0.2330	0.2289	0.2248
3.30	0.2207	0.2165	0.2124	0.2083	0.2042	0.2000	0.1959	0.1917	0.1876	0.1834
3.40	0.1792	0.1751	0.1709	0.1667	0.1625	0.1583	0.1541	0.1500	0.1458	0.1416
3.50	0.1374	0.1332	0.1290	0.1248	0.1206	0.1164	0.1122	0.1080	0.1038	0.0996
3.60	0.0955	0.0913	0.0871	0.0829	0.0788	0.0746	0.0704	0.0663	0.0621	0.0580
3.70	0.0538	0.0497	0.0456	0.0414	0.0373	0.0332	0.0291	0.0250	0.0210	0.0169
3.80	0.0128	0.0088	0.0047	0.0007	−0.0033	−0.0074	−0.0114	−0.0153	−0.0193	−0.0233
3.90	−0.0272	−0.0312	−0.0351	−0.0390	−0.0429	−0.0468	−0.0507	−0.0546	−0.0584	−0.0622
4.00	−0.0660	−0.0698	−0.0736	−0.0774	−0.0811	−0.0849	−0.0886	−0.0923	−0.0960	−0.0996
4.10	−0.1033	−0.1069	−0.1105	−0.1141	−0.1177	−0.1212	−0.1247	−0.1282	−0.1317	−0.1352
4.20	−0.1386	−0.1421	−0.1455	−0.1489	−0.1522	−0.1556	−0.1589	−0.1622	−0.1654	−0.1687
4.30	−0.1719	−0.1751	−0.1783	−0.1814	−0.1845	−0.1876	−0.1907	−0.1938	−0.1968	−0.1998
4.40	−0.2028	−0.2057	−0.2086	−0.2115	−0.2144	−0.2173	−0.2201	−0.2229	−0.2256	−0.2284
4.50	−0.2311	−0.2337	−0.2364	−0.2390	−0.2416	−0.2442	−0.2467	−0.2492	−0.2517	−0.2541
4.60	−0.2566	−0.2589	−0.2613	−0.2636	−0.2659	−0.2682	−0.2704	−0.2726	−0.2748	−0.2770
4.70	−0.2791	−0.2812	−0.2832	−0.2852	−0.2872	−0.2892	−0.2911	−0.2930	−0.2949	−0.2967
4.80	−0.2985	−0.3003	−0.3020	−0.3037	−0.3054	−0.3070	−0.3086	−0.3102	−0.3117	−0.3132
4.90	−0.3147	−0.3161	−0.3175	−0.3189	−0.3202	−0.3216	−0.3228	−0.3241	−0.3253	−0.3264

TABLE A14 Error Function

$$\operatorname{erf}(x) = \frac{2}{\pi^{1/2}} \int_0^x e^{-t^2}\, dt = \frac{2}{\pi^{1/2}}\left(x - \frac{x^3}{3 \cdot 1!} + \frac{x^5}{5 \cdot 2!} - \frac{x^7}{7 \cdot 3!} + \cdots \right)$$

$$\operatorname{erfc}(x) = 1 - \operatorname{erf}(x)$$

x	0.00	0.01	0.02	0.03	0.04	0.05	0.06	0.07	0.08	0.09
0.00	0.000000	0.011283	0.022565	0.033841	0.045111	0.056372	0.067622	0.078858	0.090078	0.101281
0.10	0.112463	0.123623	0.134758	0.145867	0.156947	0.167996	0.179012	0.189992	0.200936	0.211840
0.20	0.222703	0.233522	0.244296	0.255023	0.265700	0.276326	0.286900	0.297418	0.307880	0.318283
0.30	0.328627	0.338908	0.349126	0.359279	0.369365	0.379382	0.389330	0.399206	0.409009	0.418739
0.40	0.428392	0.437969	0.447468	0.456887	0.466225	0.475482	0.484655	0.493745	0.502750	0.511668
0.50	0.520500	0.529244	0.537899	0.546464	0.554939	0.563323	0.571616	0.579816	0.587923	0.595936
0.60	0.603856	0.611681	0.619411	0.627046	0.634586	0.642029	0.649377	0.656628	0.663782	0.670840
0.70	0.677801	0.684666	0.691433	0.698104	0.704678	0.711156	0.717537	0.723822	0.730010	0.736103

| TABLE A14 (continued)

$$\text{erf}(x) = \frac{2}{\pi^{1/2}} \int_0^x e^{-t^2}\, dt = \frac{2}{\pi^{1/2}} \left(x - \frac{x^3}{3 \cdot 1!} + \frac{x^5}{5 \cdot 2!} - \frac{x^7}{7 \cdot 3!} + \cdots \right)$$

$$\text{erfc}(x) = 1 - \text{erf}(x)$$

x	0.00	0.01	0.02	0.03	0.04	0.05	0.06	0.07	0.08	0.09
0.80	0.742101	0.748003	0.753811	0.759524	0.765143	0.770668	0.776100	0.781440	0.786687	0.791843
0.90	0.796908	0.801883	0.806768	0.811564	0.816271	0.820891	0.825424	0.829870	0.834232	0.838508
1.00	0.842701	0.846810	0.850838	0.854784	0.858650	0.862436	0.866144	0.869773	0.873326	0.876803
1.10	0.880205	0.883533	0.886788	0.889971	0.893082	0.896124	0.899096	0.902000	0.904837	0.907608
1.20	0.910314	0.912956	0.915534	0.918050	0.920505	0.922900	0.925236	0.927514	0.929734	0.931899
1.30	0.934008	0.936063	0.938065	0.940015	0.941914	0.943762	0.945561	0.947312	0.949016	0.950673
1.40	0.952285	0.953852	0.955376	0.956857	0.958297	0.959695	0.961054	0.962373	0.963654	0.964898
1.50	0.966105	0.967277	0.968413	0.969516	0.970586	0.971623	0.972628	0.973603	0.974547	0.975462
1.60	0.976348	0.977207	0.978038	0.978843	0.979622	0.980376	0.981105	0.981810	0.982493	0.983153
1.70	0.983790	0.984407	0.985003	0.985578	0.986135	0.986672	0.987190	0.987691	0.988174	0.988641
1.80	0.989091	0.989525	0.989943	0.990347	0.990736	0.991111	0.991472	0.991821	0.992156	0.992479
1.90	0.992790	0.993090	0.993378	0.993656	0.993923	0.994179	0.994426	0.994664	0.994892	0.995111
2.00	0.995322	0.995525	0.995719	0.995906	0.996086	0.996258	0.996423	0.996582	0.996734	0.996880
2.10	0.997021	0.997155	0.997284	0.997407	0.997525	0.997639	0.997747	0.997851	0.997951	0.998046
2.20	0.998137	0.998224	0.998308	0.998388	0.998464	0.998537	0.998607	0.998674	0.998738	0.998799
2.30	0.998857	0.998912	0.998966	0.999016	0.999065	0.999111	0.999155	0.999197	0.999237	0.999275
2.40	0.999311	0.999346	0.999379	0.999411	0.999441	0.999469	0.999497	0.999523	0.999547	0.999571
2.50	0.999593	0.999614	0.999635	0.999654	0.999672	0.999689	0.999706	0.999722	0.999736	0.999751
2.60	0.999764	0.999777	0.999789	0.999800	0.999811	0.999822	0.999831	0.999841	0.999849	0.999858
2.70	0.999866	0.999873	0.999880	0.999887	0.999893	0.999899	0.999905	0.999910	0.999916	0.999920
2.80	0.999925	0.999929	0.999933	0.999937	0.999941	0.999944	0.999948	0.999951	0.999954	0.999956
2.90	0.999959	0.999961	0.999964	0.999966	0.999968	0.999970	0.999972	0.999973	0.999975	0.999976
3.00	0.999978	0.999979	0.999981	0.999982	0.999983	0.999984	0.999985	0.999986	0.999987	0.999988
3.10	0.999988	0.999989	0.999990	0.999990	0.999991	0.999992	0.999992	0.999993	0.999993	0.999994
3.20	0.999994	0.999994	0.999995	0.999995	0.999995	0.999996	0.999996	0.999996	0.999996	0.999997
3.30	0.999997	0.999997	0.999997	0.999998	0.999998	0.999998	0.999998	0.999998	0.999998	0.999998
3.40	0.999998	0.999999	0.999999	0.999999	0.999999	0.999999	0.999999	0.999999	0.999999	0.999999
3.50	0.999999	0.999999	0.999999	0.999999	0.999999	0.999999	1.000000	1.000000	1.000000	1.000000
3.60	1.000000	1.000000	1.000000	1.000000	1.000000	1.000000	1.000000	1.000000	1.000000	1.000000

APPENDIX 3

ORTHOGONAL FUNCTIONS, STURM-LIOUVILLE EQUATION

A sequence of functions $\phi_i(x)$, $i = 1, 2, 3, \ldots$ is said to be orthogonal in the interval (a, b) with respect to the weight function $w(x)$, if

$$\int_a^b (w(x)\phi_n(x)\phi_m(x) \, dx = \begin{cases} 0 \text{ if } n \neq m \\ \text{constant if } n = m \end{cases}$$

provided $w(x)$ does not change sign in the interval $a < x < b$.

Consider the differential equation

$$[r(x)\phi']' + [q(x) + \lambda w(x)]\phi = 0 \qquad \textbf{(A3.1)}$$

with the general homogeneous, linear boundary conditions,

$$a_1\phi(a) + a_2\phi'(a) = 0 \qquad \textbf{(A3.2)}$$
$$b_1\phi(b) + b_2\phi'(b) = 0$$

where a_1, a_2, b_1, and b_2 are constants.

The differential equation (Equation A3.1) with the boundary conditions A3.2 is known as the Sturm-Liouville system. Note that the boundary condition $\phi(a) = 0$ and $\phi'(a) = 0$ are obtained as special cases of A3.2 by setting $a_2 = 0$ or $a_1 = 0$.

In Equation A3.1 primes denote differentiation with respect to x. λ is a constant; it may take any value in a sequence $\lambda_1, \lambda_2, \ldots \lambda_n$, $n = 1, 2, \ldots$ and, corresponding to a given value of λ_n, we have a solution $\phi_n(x)$ satisfying the boundary conditions given by Equation A3.2. It will now be shown that the functions ϕ_n's are orthogonal in the interval (a, b) with weight function $w(x)$.

Let ϕ_m and ϕ_n be two such solutions corresponding to λ_m and λ_n. As these functions satisfy Equation A3.1, we have

$$[r(x)\phi_m']' + [q(x) + \lambda_m w(x)]\phi_m = 0 \qquad \textbf{(A3.3)}$$
$$[r(x)\phi_n']' + [q(x) + \lambda_n w(x)]\phi_n = 0 \qquad \textbf{(A3.4)}$$

Multiply Equation A3.3 by ϕ_n and Equation A3.4 by ϕ_m and subtract the former from the latter,

$$(\lambda_n - \lambda_m)w\,\phi_n\phi_m = [r\phi_m']'\phi_n - [r\phi_n']'\phi_m$$

Add and subtract $r\phi_m'\phi_n'$ to the right-hand side and rearrange,

$$(\lambda_n - \lambda_m)w\,\phi_n\phi_m = \frac{d}{dx}[r(\phi_m'\phi_n - \phi_m\phi_n')]$$

Integrate the above equation from a to b,

$$(\lambda_n - \lambda_m) \int_a^b w\phi_n\phi_m \, dx = \int_a^b \frac{d}{dx} [r\phi_m'\phi_n - \phi_m\phi_n')] \, dx$$
$$= [r(\phi_m'\phi_n - \phi_m\phi_n')]|_a^b$$

(A3.5)

As the functions ϕ_m and ϕ_n satisfy the boundary conditions, Equation A3.2, we have

$$a_1\phi_g(a) + a_2\phi_g'(a) = 0 \qquad b_1\phi_g(b) + b_2\phi_g'(b) = 0$$

where $g = m$ or n.

Consider the right side of Equation A3.5, denoted by RS,

$$RS = r \{[\phi_m'(b)\phi_n(b) - \phi_m(b)\phi_n'(b)] - [\phi_m'(a)\phi_n(a) - \phi_m(a)\phi_n'(a)]$$

If $a_1 = b_1 = 0$, $\phi_m'(a) = \phi_m'(b) = \phi_n'(a) = \phi_n'(b) = 0$ and, hence, RS = 0.
If $a_2 = b_2 = 0$, $\phi_m(a) = \phi_m(b) = \phi_n(a) = \phi_n(b) = 0$ and again RS = 0.

In the general case of each of a_1, a_2, b_1, and b_2 being nonzero,

$$\phi_m'(b) = -\frac{b_1}{b_2} \phi_m(b) \quad \text{and} \quad \phi_n'(b) = -\frac{b_1}{b_2} \phi_n(b)$$

Substituting these we have

$$\phi_m'(b)\phi_n(b) - \phi_m(b)\phi_n'(b) = -\frac{b_1}{b_2} \phi_m(b)\phi_n(b) - \phi_m(b)\left(-\frac{b_1}{b_2} \phi_n(b)\right) = 0$$

Similarly, $\phi_m'(a)\phi_n(a) - \phi_m(a)\phi_n'(a) = 0$. Thus, in every case the right side of Equation A3.5 is zero. Hence,

$$(\lambda_n - \lambda_m) \int_a^b w\phi_n\phi_m \, dx = 0$$

But as $\lambda_n \neq \lambda_m$,

$$\int_a^b w\phi_n\phi_m \, dx = 0$$

If $\lambda_n = \lambda_m$ ($n = m$), the equality $(\lambda_n - \lambda_m) \int_a^b w\phi_n\phi_m \, dx = 0$ is satisfied but the integral $\int_a^b w\phi_n\phi_m \, dx$ is now $\int_a^b w\phi_n^2 \, dx$. ϕ_n^2 is always positive. From the original constraint, $w(x)$ does not change sign in the interval (a, b). The integrand $w\phi_n^2$ has the same sign throughout the interval and, hence, the integral in the same interval cannot be zero. Thus, for $n \neq m$, $\int_a^b w\phi_n^2 \, dx \neq 0$ but is a constant.

LAPLACE TRANSFORMS

Definition The Laplace transform of $f(t)$, $f(t)$ being piecewise continuous in any interval $t_1 \le t \le t_2$, $t_1 > 0$, is defined by

$$\mathcal{L}\{f(t)\} = \bar{f}(s) = \int_0^\infty e^{-st}f(t)\ dt \tag{A4.1}$$

Some frequently needed properties of Laplace Transforms, which can be derived from the definition of the transform, Equation A4.1, are given below.

1. $\mathcal{L}[af(t) + bg(t)] = a\bar{f}(s) + b\bar{g}(s)$ $\hspace{2cm}$ (A4.2)

2. $\mathcal{L}\left[\dfrac{df(t)}{dt}\right] = s\bar{f}(s) - f(0+)$ $\hspace{2cm}$ (A4.3)

    $\mathcal{L}\left[\dfrac{d^2f(t)}{dt^2}\right] = s^2\bar{f}(s) - sf(0+) - f'(0+)$ $\hspace{1cm}$ (A4.4)

3. $\mathcal{L}\left[\displaystyle\int_0^t f(\tau)\ d\tau\right] = \dfrac{1}{s}\bar{f}(s)$ $\hspace{2cm}$ (A4.5)

4. $\mathcal{L}[f(at)] = \dfrac{1}{a}\bar{f}\left(\dfrac{s}{a}\right)$ $\hspace{2cm}$ (A4.6)

5. $\mathcal{L}[e^{-at}f(t)] = \bar{f}(s + a)$ $\hspace{2cm}$ (A4.7)

6. $\mathcal{L}\left[\displaystyle\int_0^t f(t - \tau)g(\tau)\ d\tau\right] = \bar{f}(s)\bar{g}(s)$ $\hspace{1.5cm}$ (A4.8)

The integral in Equation A4.8 is known as the convolution integral of $f(t)$ and $g(t)$. Some of the transforms required in solving heat transfer problems are given in Table A15.

TABLE A15 Table of Laplace Transforms	
$f(s)$	$f(t)$
1. $\dfrac{1}{s}$	1
2. $\dfrac{1}{s^2}$	t
3. $\dfrac{1}{s^k}$	$\dfrac{t^{k-1}}{(k-1)!}$
4. $\dfrac{1}{s^{1/2}}$	$\dfrac{1}{(\pi t)^{1/2}}$
5. $\dfrac{1}{s^{3/2}}$	$2\left(\dfrac{t}{\pi}\right)^{1/2}$
6. $\dfrac{1}{s^{k+1/2}}$	$\dfrac{2^k}{\pi^{1/2}(2k-1)!!}\,t^{k-1/2}$
7. $\dfrac{1}{s^\nu}$	$\dfrac{t^{\nu-1}}{\Gamma(\nu)}$
8. $s^{1/2}$	$-\dfrac{1}{2\pi^{1/2}t^{3/2}}$
9. $s^{3/2}$	$\dfrac{3}{4\pi^{1/2}t^{5/2}}$
10. $s^{k-1/2}$	$\dfrac{(-1)^k(2k-1)!!}{2^k\pi^{1/2}t^{k+1/2}}$
11. $s^{n-\nu}$	$\dfrac{t^{\nu-n-1}}{\Gamma(\nu-n)}$
12. $\dfrac{1}{s+\alpha}$	$e^{-\alpha t}$
13. $\dfrac{1}{(s+\alpha)(s+\beta)}$	$\dfrac{e^{-\beta t}-e^{-\alpha t}}{\alpha-\beta}$

14. $\dfrac{1}{(s+\alpha)^2}$ $te^{-\alpha t}$

15. $\dfrac{1}{(s+\alpha)(s+\beta)(s+\gamma)}$ $\dfrac{(\gamma-\beta)e^{-\alpha t}+(\alpha-\gamma)e^{-\beta t}+(\beta-\alpha)e^{-\gamma t}}{(\alpha-\beta)(\beta-\gamma)(\gamma-\alpha)}$

16. $\dfrac{1}{(s+\alpha)^2(s+\beta)}$ $\dfrac{e^{-\beta t}-e^{-\alpha t}[1-(\beta-\alpha)t]}{(\beta-\alpha)^2}$

17. $\dfrac{1}{(s+\alpha)^3}$ $\dfrac{1}{2}t^2 e^{-\alpha t}$

18. $\dfrac{1}{(s+\alpha)^k}$ $\dfrac{t^{k-1}e^{-\alpha t}}{(k-1)!}$

19. $\dfrac{s}{(s+\alpha)(s+\beta)}$ $\dfrac{\alpha e^{-\alpha t}-\beta e^{-\beta t}}{\alpha-\beta}$

20. $\dfrac{s}{(s+\alpha)^2}$ $(1-\alpha t)e^{-\alpha t}$

21. $\dfrac{s}{(s+\alpha)(s+\beta)(s+\gamma)}$ $\dfrac{\alpha(\beta-\gamma)e^{-\alpha t}+\beta(\gamma-\alpha)e^{-\beta t}+\gamma(\alpha-\beta)e^{-\gamma t}}{(\alpha-\beta)(\beta-\gamma)(\gamma-\alpha)}$

22. $\dfrac{s}{(s+\alpha)^2(s+\beta)}$ $\dfrac{[\beta-\alpha(\beta-\alpha)t]e^{-\alpha t}-\beta e^{-\beta t}}{(\beta-\alpha)^2}$

23. $\dfrac{s}{(s+\alpha)^3}$ $t\left(1-\dfrac{1}{2}\alpha t\right)e^{-\alpha t}$

24. $\dfrac{\alpha}{s^2+\alpha^2}$ $\sin \alpha t$

25. $\dfrac{s}{s^2+\alpha^2}$ $\cos \alpha t$

26. $\dfrac{\alpha}{s^2-\alpha^2}$ $\sinh \alpha t$

27. $\dfrac{s}{s^2-\alpha^2}$ $\cosh \alpha t$

TABLE A15 (continued)

$f(s)$	$f(t)$
28. e^{-qx}	$\dfrac{x}{2(\pi a t^3)^{1/2}}\, e^{-x^2/4at}$
29. $\dfrac{e^{-qx}}{q}$	$\left(\dfrac{a}{\pi t}\right)^{1/2} e^{-x^2/4at}$
30. $\dfrac{e^{-qx}}{s}$	$\mathrm{erfc}\left[\dfrac{x}{2(at)^{1/2}}\right]$
31. $\dfrac{e^{-qx}}{qs}$	$2\left(\dfrac{at}{\pi}\right)^{1/2} e^{-x^2/4at} - x\,\mathrm{erfc}\left[\dfrac{x}{2(at)^{1/2}}\right]$
32. $\dfrac{e^{-qx}}{s^2}$	$\left(t + \dfrac{x^2}{2a}\right)\mathrm{erfc}\left[\dfrac{x}{2(at)^{1/2}}\right] - x\left(\dfrac{t}{\pi a}\right)^{1/2} e^{-x^2/4at}$
33. $\dfrac{e^{-qx}}{s^{1+n/2}}$	$(4t)^{n/2}\, i^n\,\mathrm{erfc}\left[\dfrac{x}{2(at)^{1/2}}\right]$
34. $\dfrac{e^{-qx}}{s^{3/4}}$	$\dfrac{1}{\pi}\left(\dfrac{x}{2ta^{1/2}}\right)^{1/2} e^{-x^2/8at} K_{1/4}\left(\dfrac{x^2}{8at}\right)$
35. $\dfrac{e^{-qx}}{q+\beta}$	$\left(\dfrac{a}{\pi t}\right)^{1/2} e^{-x^2/4at} - \alpha\beta e^{\beta x + a\beta^2 t}\,\mathrm{erfc}\left(\dfrac{x}{2(at)^{1/2}} + \beta(at)^{1/2}\right)$
36. $\dfrac{e^{-qx}}{q(q+\beta)}$	$a e^{\beta x + a\beta^2 t}\,\mathrm{erfc}\left(\dfrac{x}{2(at)^{1/2}} + \beta(at)^{1/2}\right)$
37. $\dfrac{e^{-qx}}{s(q+\beta)}$	$\dfrac{1}{\beta}\,\mathrm{erfc}\left[\dfrac{x}{2(at)^{1/2}}\right] - \dfrac{1}{\beta}e^{\beta x + a\beta^2 t}\,\mathrm{erfc}\left[\dfrac{x}{2(at)^{1/2}} + \beta(at)^{1/2}\right]$
38. $\dfrac{e^{-qx}}{qs(q+\beta)}$	$2\left(\dfrac{at}{\pi}\right)^{1/2} e^{-x^2/4at} - \dfrac{(1+\beta x)}{\beta^2}\,\mathrm{erfc}\left[\dfrac{x}{2(at)^{1/2}}\right] + \dfrac{1}{\beta^2}e^{\beta x + a\beta^2 t}\,\mathrm{erfc}\left[\dfrac{x}{2(at)^{1/2}} + \beta(at)^{1/2}\right]$
39. $\dfrac{e^{-qx}}{q^{n+1}(q+\beta)}$	$\dfrac{a}{(-\beta)^n}e^{\beta x + a\beta^2 t}\,\mathrm{erfc}\left[\dfrac{x}{2(at)^{1/2}} + \beta(at)^{1/2}\right] - \dfrac{a}{(-\beta)^n}\sum_{r=0}^{n-1}[-2\beta(at)^{1/2}]^r\, i^r\,\mathrm{erfc}\left[\dfrac{x}{2(at)^{1/2}}\right]$

40. $\dfrac{e^{-qx}}{(q+\beta)^2}$

$-2\beta\left(\dfrac{a^3t}{\pi}\right)^{1/2}e^{-x^2/4at} + a(1 + \beta x + 2a\beta^2 t)e^{\beta x + a\beta^2 t}\,\text{erfc}\left[\dfrac{x}{2(at)^{1/2}} + \beta(at)^{1/2}\right]$

41. $\dfrac{e^{-qx}}{s(q+\beta)^2}$

$\dfrac{1}{\beta^2}\text{erfc}\left[\dfrac{x}{2(at)^{1/2}}\right] - \dfrac{2}{\beta}\left(\dfrac{at}{\pi}\right)^{1/2}e^{-x^2/4at} - \dfrac{1}{\beta^2}(1 - \beta x - 2a\beta^2 t)e^{\beta x + a\beta^2 t}\,\text{erfc}\left[\dfrac{x}{2(at)^{1/2}} + \beta(at)^{1/2}\right]$

42. $\dfrac{e^{-qx}}{s-\gamma}$

$\dfrac{1}{2}e^{\gamma t}\left\{e^{-x(\gamma/a)^{1/2}}\text{erfc}\left[\dfrac{x}{2(at)^{1/2}} - (\gamma t)^{1/2}\right] + e^{x(\gamma/a)^{1/2}}\text{erfc}\left[\dfrac{x}{2(at)^{1/2}} + (\gamma t)^{1/2}\right]\right\}$

43. $\dfrac{e^{-qx}}{q(s-\gamma)}$

$\dfrac{1}{2}e^{\gamma t}\left(\dfrac{a}{\gamma}\right)^{1/2}\left\{e^{-x(\gamma/a)^{1/2}}\text{erfc}\left[\dfrac{x}{2(at)^{1/2}} - (\gamma t)^{1/2}\right] - e^{x(\gamma/a)^{1/2}}\text{erfc}\left[\dfrac{x}{2(at)^{1/2}} + (\gamma t)^{1/2}\right]\right\}$

44. $\dfrac{e^{-qx}}{(s-\gamma)^2}$

$\dfrac{1}{2}e^{\gamma t}\left\{\left[t - \dfrac{x}{2(at)^{1/2}}\right]e^{-x(\gamma/a)^{1/2}}\text{erfc}\left[\dfrac{x}{2(at)^{1/2}} - (\gamma t)^{1/2}\right] + \left[t + \dfrac{x}{2(at)^{1/2}}\right]e^{x(\gamma/a)^{1/2}}\text{erfc}\left[\dfrac{x}{2(at)^{1/2}} + (\gamma t)^{1/2}\right]\right\}$

45. $\dfrac{e^{-qx}}{(s-\gamma)(q+\beta)}$
$\gamma \neq a\beta^2$

$\dfrac{1}{2}e^{\gamma t}\left\{\dfrac{a^{1/2}}{a^{1/2}\beta + \gamma^{1/2}}e^{-x(\gamma/a)^{1/2}}\text{erfc}\left[\dfrac{x}{2(at)^{1/2}} - (\gamma t)^{1/2}\right] + \dfrac{a^{1/2}}{a^{1/2}\beta - \gamma^{1/2}}e^{x(\gamma/a)^{1/2}}\text{erfc}\left[\dfrac{x}{2(at)^{1/2}} + (\gamma t)^{1/2}\right]\right\}$
$- \dfrac{a\beta}{a\beta^2 - \gamma}e^{\beta x + a\beta^2 t}\text{erfc}\left[\dfrac{x}{2(at)^{1/2}} + \beta(at)^{1/2}\right]$

46. $e^{x/s} - 1$

$\left(\dfrac{x}{t}\right)^{1/2}I_1[2(xt)^{1/2}]$

47. $\dfrac{1}{s}e^{x/s}$

$I_0[2(xt)^{1/2}]$

48. $\dfrac{1}{s^\nu}e^{x/s}$

$\left(\dfrac{t}{x}\right)^{(\nu-1)/2}I_{\nu-1}[2(xt)^{1/2}]$

49. $K_0(qx)$

$\dfrac{1}{2t}e^{-x^2/4at}$

50. $\dfrac{1}{s^{1/2}}K_{2\nu}(qx)$

$\dfrac{1}{2(\pi t)^{1/2}}e^{-x^2/8at}K_\nu\left(\dfrac{x^2}{8at}\right)$

TABLE A15 (continued)

$f(s)$	$f(t)$
51. $s^{\nu/2-1}K_\nu(qx)$	$x^{-\nu}a^{\nu/2}2^{\nu-1}\displaystyle\int_{x^2/4at}^\infty e^{-u}u^{\nu-1}\,du$
52. $s^{\nu/2}K_\nu(qx)$	$\dfrac{x^\nu}{a^{\nu/2}(2t)^{\nu+1}}e^{-x^2/4at}$
53. $[s-(s^2-x^2)^{1/2}]^\nu$	$\nu\dfrac{x^\nu}{t}I_\nu(xt)$
54. $e^{x[(s+\alpha)^{1/2}-(s+\beta)^{1/2}]}-1$	$\dfrac{x(\alpha-\beta)e^{-(\alpha+\beta)t/2}I_1\left[\frac{1}{2}(\alpha-\beta)t^{1/2}(t+4x)^{1/2}\right]}{t^{1/2}(t+4x)^{1/2}}$
55. $\dfrac{e^{x[s-(s+\alpha)^{1/2}(s+\beta)^{1/2}]}}{(s+\alpha)^{1/2}(s+\beta)^{1/2}}$	$e^{-(\alpha+\beta)(t+x)/2}I_0\left[\frac{1}{2}(\alpha-\beta)t^{1/2}(t+2x)^{1/2}\right]$
56. $\dfrac{e^{x[(s+\alpha)^{1/2}-(s+\beta)^{1/2}]}}{(s+\alpha)^{1/2}(s+\beta)^{1/2}[(s+\alpha)^{1/2}+(s+\beta)^{1/2}]^{2\nu}}$	$\dfrac{t^{\nu/2}e^{-(\alpha+\beta)t/2}I_\nu\left[\frac{1}{2}(\alpha-\beta)t^{1/2}(t+4x)^{1/2}\right]}{(\alpha-\beta)^\nu(t+4x)^{\nu/2}}$
57. $\dfrac{\sinh sx}{s\sinh sa}$	$\dfrac{x}{a}+\dfrac{2}{\pi}\displaystyle\sum_{n=1}^\infty\dfrac{(-1)^n}{n}\sin\dfrac{n\pi x}{a}\cos\dfrac{n\pi t}{a}$
58. $\dfrac{\sinh sx}{s\cosh sa}$	$\dfrac{4}{\pi}\displaystyle\sum_{n=1}^\infty\dfrac{(-1)^n}{2n-1}\sin\dfrac{(2n-1)\pi x}{2a}\sin\dfrac{(2n-1)\pi t}{2a}$
59. $\dfrac{\cosh sx}{s\sinh as}$	$\dfrac{t}{a}+\dfrac{2}{\pi}\displaystyle\sum_{n=1}^\infty\dfrac{(-1)^n}{n}\cos\dfrac{n\pi x}{a}\sin\dfrac{n\pi t}{a}$
60. $\dfrac{\cosh sx}{s\cosh sa}$	$1+\dfrac{4}{\pi}\displaystyle\sum_{n=1}^\infty\dfrac{(-1)^n}{2n-1}\cos\dfrac{(2n-1)\pi x}{2a}\cos\dfrac{(2n-1)\pi t}{2a}$
61. $\dfrac{\sinh sx}{s^2\sinh sa}$	$\dfrac{xt}{a}+\dfrac{2a}{\pi^2}\displaystyle\sum_{n=1}^\infty\dfrac{(-1)^n}{n^2}\sin\dfrac{n\pi x}{a}\sin\dfrac{n\pi t}{a}$
62. $\dfrac{\sinh sx}{s^2\cosh sa}$	$x+\dfrac{8a}{\pi^2}\displaystyle\sum_{n=1}^\infty\dfrac{(-1)^n}{(2n-1)^2}\sin\dfrac{(2n-1)\pi x}{2a}\cos\dfrac{(2n-1)\pi t}{2a}$

63. $\dfrac{\cosh sx}{s^2 \sinh sa}$ → $\dfrac{t^2}{2a} + \dfrac{2a}{\pi^2}\sum_{n=1}^{\infty} \dfrac{(-1)^n}{n^2}\cos\dfrac{n\pi x}{a}\left(1 - \cos\dfrac{n\pi t}{a}\right)$

64. $\dfrac{\cosh sx}{s^2 \cosh sa}$ → $t + \dfrac{8a}{\pi^2}\sum_{n=1}^{\infty} \dfrac{(-1)^n}{(2n-1)^2}\cos\dfrac{(2n-1)\pi x}{2a}\sin\dfrac{(2n-1)\pi t}{2a}$

65. $\dfrac{\cosh sx}{s^3 \cosh sa}$ → $\dfrac{1}{2}(t^2 + x^2 - a^2) - \dfrac{16a^2}{\pi^3}\sum_{n=1}^{\infty} \dfrac{(-1)^n}{(2n-1)^3}\cos\dfrac{(2n-1)\pi x}{2a}\cos\dfrac{(2n-1)\pi t}{2a}$

66. $\dfrac{\sinh x\sqrt{s}}{\sinh a\sqrt{s}}$ → $\dfrac{2\pi}{a^2}\sum_{n=1}^{\infty} (-1)^n n\, e^{-n^2\pi^2 t/a^2}\sin\dfrac{n\pi x}{a}$

67. $\dfrac{\cosh x\sqrt{s}}{\cosh a\sqrt{s}}$ → $\dfrac{\pi}{a^2}\sum_{n=1}^{\infty} (-1)^{n-1}(2n-1)e^{-(2n-1)^2\pi^2 t/4a^2}\cos\dfrac{(2n-1)\pi x}{2a}$

68. $\dfrac{\sinh x\sqrt{s}}{\sqrt{s}\cosh a\sqrt{s}}$ → $\dfrac{2}{a}\sum_{n=1}^{\infty} (-1)^{n-1}e^{-(2n-1)^2\pi^2 t/4a^2}\sin\dfrac{(2n-1)\pi x}{2a}$

69. $\dfrac{\cosh x\sqrt{s}}{\sqrt{s}\sinh a\sqrt{s}}$ → $\dfrac{1}{a} + \dfrac{2}{a}\sum_{n=1}^{\infty} (-1)^n e^{-n^2\pi^2 t/a^2}\cos\dfrac{n\pi x}{a}$

70. $\dfrac{\sinh x\sqrt{s}}{s \sinh a\sqrt{s}}$ → $\dfrac{x}{a} + \dfrac{2}{\pi}\sum_{n=1}^{\infty} \dfrac{(-1)^n}{n} e^{-n^2\pi^2 t/a^2}\sin\dfrac{n\pi x}{a}$

71. $\dfrac{\cosh x\sqrt{s}}{s \cosh a\sqrt{s}}$ → $1 + \dfrac{4}{\pi}\sum_{n=1}^{\infty} \dfrac{(-1)^n}{2n-1} e^{-(2n-1)^2\pi^2 t/4a^2}\cos\dfrac{(2n-1)\pi x}{2a}$

72. $\dfrac{\sinh x\sqrt{s}}{s^2 \sinh a\sqrt{s}}$ → $\dfrac{xt}{a} + \dfrac{2a^2}{\pi^3}\sum_{n=1}^{\infty} \dfrac{(-1)^n}{n^3}(1 - e^{-n^2\pi^2 t/a^2})\sin\dfrac{n\pi x}{a}$

73. $\dfrac{\cosh x\sqrt{s}}{s^2 \cosh a\sqrt{s}}$ → $\dfrac{1}{2}(x^2 - a^2) + t - \dfrac{16a^2}{\pi^3}\sum_{n=1}^{\infty} \dfrac{(-1)^n}{(2n-1)^3} e^{-(2n-1)^2\pi^2 t/4a^2}\cos\dfrac{(2n-1)\pi x}{2a}$

Sources: Arpaci, V.S., *Conduction Heat Transfer* (Reading, MA: Addison-Wesly, 1966). Spiegel, Murray R., *Mathematical Handbook of Formulas and Tables* (New York: McGraw-Hill, 1968).

Notes:

I and K are the modified Bessel functions of the second kind.

$q = (s/a)^{1/2}$; a and x are positive and real. k and n are finite integers; v is a fractional number; $1 \cdot 2 \cdot 3 \cdots n = n!$; $1 \cdot 3 \cdot 5 \cdots (2n-1) = (2n-1)!!$

$n\Gamma(n) = \Gamma(n+1)$ $\Gamma(1) = 1$ $\Gamma(v)\Gamma(v-1) = \dfrac{\pi}{\sin v\pi}$ $\Gamma(\tfrac{1}{2})\,\pi^{1/2}$

APPENDIX 5

EQUATIONS OF MOTION AND ENERGY EQUATION FOR A NEWTONIAN FLUID WITH CONSTANT ρ AND μ (NATURAL CONVECTION NEGLECTED)

CARTESIAN COORDINATES

Mass Conservation

$$\frac{\partial v_x}{\partial x} + \frac{\partial v_y}{\partial y} + \frac{\partial v_z}{\partial z} = 0 \tag{A5.1}$$

x-Momentum

$$\rho \left(\frac{\partial v_x}{\partial t} + v_x \frac{\partial v_x}{\partial x} + v_y \frac{\partial v_x}{\partial y} + v_z \frac{\partial v_x}{\partial z} \right) = -\frac{\partial p}{\partial x} + \rho g_x + \mu \left(\frac{\partial^2 v_x}{\partial x^2} + \frac{\partial^2 v_x}{\partial y^2} + \frac{\partial^2 v_x}{\partial z^2} \right) \tag{A5.2}$$

y-Momentum

$$\rho \left(\frac{\partial v_y}{\partial t} + v_x \frac{\partial v_y}{\partial x} + v_y \frac{\partial v_y}{\partial y} + v_z \frac{\partial v_y}{\partial z} \right) = -\frac{\partial p}{\partial y} + \rho g_y + \mu \left(\frac{\partial^2 v_y}{\partial x^2} + \frac{\partial^2 v_y}{\partial y^2} + \frac{\partial^2 v_y}{\partial z^2} \right) \tag{A5.3}$$

z-Momentum

$$\rho \left(\frac{\partial v_z}{\partial t} + v_x \frac{\partial v_z}{\partial x} + v_y \frac{\partial v_z}{\partial y} + v_z \frac{\partial v_z}{\partial z} \right) = -\frac{\partial p}{\partial z} + \rho g_z + \mu \left(\frac{\partial^2 v_z}{\partial x^2} + \frac{\partial^2 v_z}{\partial y^2} + \frac{\partial^2 v_z}{\partial z^2} \right) \tag{A5.4}$$

Energy

$$\rho c_p \left(\frac{\partial T}{\partial t} + v_x \frac{\partial T}{\partial x} + v_y \frac{\partial T}{\partial y} + v_z \frac{\partial T}{\partial z} \right) = \frac{\partial}{\partial x} \left(k \frac{\partial T}{\partial x} \right) + \frac{\partial}{\partial y} \left(k \frac{\partial T}{\partial y} \right) + \frac{\partial}{\partial z} \left(k \frac{\partial T}{\partial z} \right) + \mu \phi$$

$$+ T\beta \left(\frac{\partial p}{\partial t} + v_x \frac{\partial p}{\partial x} + v_y \frac{\partial p}{\partial y} + v_z \frac{\partial p}{\partial z} \right)$$

$$\phi = 2 \left[\left(\frac{\partial v_x}{\partial x} \right)^2 + \left(\frac{\partial v_y}{\partial y} \right)^2 + \left(\frac{\partial v_z}{\partial z} \right)^2 \right]$$

$$+ \left[\left(\frac{\partial v_y}{\partial x} + \frac{\partial v_x}{\partial y} \right)^2 + \left(\frac{\partial v_z}{\partial y} + \frac{\partial v_y}{\partial z} \right)^2 \right.$$

$$\left. + \left(\frac{\partial v_z}{\partial x} + \frac{\partial v_x}{\partial z} \right)^2 \right] \tag{A5.5}$$

CYLINDRICAL COORDINATES

Mass Conservation

$$\frac{1}{r}\frac{\partial}{\partial r}(rv_r) + \frac{1}{r}\frac{\partial v_\theta}{\partial \theta} + \frac{\partial v_z}{\partial z} = 0 \tag{A5.6}$$

r-Momentum

$$\rho\left(\frac{\partial v_r}{\partial t} + v_r\frac{\partial v_r}{\partial r} + \frac{v_\theta}{r}\frac{\partial v_r}{\partial \theta} - \frac{v_\theta^2}{r} + v_z\frac{\partial v_r}{\partial z}\right) = -\frac{\partial p}{\partial r} + \rho g_r$$
$$+ \mu\left[\frac{\partial}{\partial r}\left(\frac{1}{r}\frac{\partial}{\partial r}(rv_r)\right) + \frac{1}{r^2}\frac{\partial^2 v_r}{\partial \theta^2} - \frac{2}{r^2}\frac{\partial v_\theta}{\partial \theta} + \frac{\partial^2 v_r}{\partial z^2}\right] \tag{A5.7}$$

θ-Momentum

$$\rho\left(\frac{\partial v_\theta}{\partial t} + v_r\frac{\partial v_\theta}{\partial r} + \frac{v_\theta}{r}\frac{\partial v_\theta}{\partial \theta} + \frac{v_r v_\theta}{r} + v_z\frac{\partial v_\theta}{\partial z}\right) = -\frac{1}{r}\frac{\partial p}{\partial \theta} + \rho g_\theta$$
$$+ \mu\left[\frac{\partial}{\partial r}\left(\frac{1}{r}\frac{\partial}{\partial r}(rv_\theta)\right) + \frac{1}{r^2}\frac{\partial^2 v_\theta}{\partial \theta^2} + \frac{2}{r^2}\frac{\partial v_r}{\partial \theta} + \frac{\partial^2 v_\theta}{\partial z^2}\right] \tag{A5.8}$$

z-Momentum

$$\rho\left(\frac{\partial v_z}{\partial t} + v_r\frac{\partial v_z}{\partial r} + \frac{v_\theta}{r}\frac{\partial v_z}{\partial \theta} + v_z\frac{\partial v_z}{\partial z}\right) = -\frac{\partial p}{\partial z} + \rho g_z$$
$$+ \mu\left[\frac{1}{r}\frac{\partial}{\partial r}\left(r\frac{\partial v_z}{\partial r}\right) + \frac{1}{r^2}\frac{\partial^2 v_z}{\partial \theta^2} + \frac{\partial^2 v_z}{\partial z^2}\right] \tag{A5.9}$$

Energy

$$\rho c_p\left(\frac{\partial T}{\partial t} + v_r\frac{\partial T}{\partial r} + \frac{v_\theta}{r}\frac{\partial T}{\partial \theta} + v_z\frac{\partial T}{\partial z}\right) = \frac{1}{r}\frac{\partial}{\partial r}\left(rk\frac{\partial T}{\partial r}\right) + \frac{1}{r}\frac{\partial}{\partial \theta}\left(\frac{k}{r}\frac{\partial T}{\partial \theta}\right)$$
$$+ \frac{\partial}{\partial z}\left(k\frac{\partial T}{\partial z}\right) + \mu\phi + T\beta\left(\frac{\partial p}{\partial t} + v_r\frac{\partial p}{\partial r} + \frac{v_\theta}{r}\frac{\partial p}{\partial \theta} + v_z\frac{\partial p}{\partial z}\right) \tag{A5.10}$$

$$\phi = 2\left[\left(\frac{\partial v_r}{\partial r}\right)^2 + \left(\frac{1}{r}\frac{\partial v_\theta}{\partial \theta} + \frac{v_r}{r}\right)^2 + \left(\frac{\partial v_z}{\partial z}\right)^2\right] + \left[r\frac{\partial}{\partial r}\left(\frac{v_\theta}{r}\right) + \frac{1}{r}\frac{\partial v_r}{\partial \theta}\right]^2$$
$$+ \left[\frac{1}{r}\frac{\partial v_z}{\partial \theta} + \frac{\partial v_\theta}{\partial z}\right]^2 + \left[\frac{\partial v_r}{\partial z} + \frac{\partial v_z}{\partial r}\right]^2$$

SPHERICAL COORDINATES

Mass Conservation

$$\frac{1}{r^2}\frac{\partial}{\partial r}(r^2 v_r) + \frac{1}{r\sin\theta}\frac{\partial}{\partial \theta}(v_\theta\sin\theta) + \frac{1}{r\sin\theta}\frac{\partial v_\phi}{\partial \phi} = 0 \tag{A5.11}$$

r-Momentum

$$\rho\left(\frac{\partial v_r}{\partial t} + v_r\frac{\partial v_r}{\partial r} + \frac{v_\theta}{r}\frac{\partial v_r}{\partial\theta} + \frac{v_\phi}{r\sin\theta}\frac{\partial v_r}{\partial\phi} - \frac{v_\theta^2 + v_\phi^2}{r}\right) = -\frac{\partial p}{\partial r} + \rho g_r$$

$$+ \mu\left(\nabla^2 v_r - \frac{2v_r}{r^2} - \frac{2}{r^2}\frac{\partial v_\theta}{\partial\theta} - \frac{2v_\theta\cot\theta}{r^2} - \frac{2}{r^2\sin\theta}\frac{\partial v_\phi}{\partial\phi}\right)$$

(A5.12)

θ-Momentum

$$\rho\left(\frac{\partial v_\theta}{\partial t} + v_r\frac{\partial v_\theta}{\partial r} + \frac{v_\theta}{r}\frac{\partial v_\theta}{\partial\theta} + \frac{v_\phi}{r\sin\theta}\frac{\partial v_\theta}{\partial\phi} + \frac{v_r v_\theta}{r} - \frac{v_\phi^2\cot\theta}{r}\right) = -\frac{1}{r}\frac{\partial p}{\partial\theta} + \rho g_\theta$$

$$+ \mu\left(\nabla^2 v_\theta + \frac{2}{r^2}\frac{\partial v_r}{\partial\theta} - \frac{v_\theta}{r^2\sin^2\theta} - \frac{2\cos\theta}{r^2\sin^2\theta}\frac{\partial v_\phi}{\partial\phi}\right)$$

(A5.13)

φ-Momentum

$$\rho\left(\frac{\partial v_\phi}{\partial t} + v_r\frac{\partial v_\phi}{\partial r} + \frac{v_\theta}{r}\frac{\partial v_\phi}{\partial\theta} + \frac{v_\phi}{r\sin\theta}\frac{\partial v_\phi}{\partial\phi} + \frac{v_\phi v_r}{r} + \frac{v_\theta v_\phi}{r}\cot\theta\right) = -\frac{1}{r\sin\theta}\frac{\partial p}{\partial\phi}$$

$$+ \rho g_\phi + \mu\left(\nabla^2 v_\phi - \frac{v_\phi}{r^2\sin^2\theta} + \frac{2}{r^2\sin\theta}\frac{\partial v_r}{\partial\phi} + \frac{2\cos\theta}{r^2\sin^2\theta}\frac{\partial v_\theta}{\partial\phi}\right)$$

(A5.14)

Energy

$$\rho c_p\left(\frac{\partial T}{\partial t} + v_r\frac{\partial T}{\partial r} + \frac{v_\theta}{r}\frac{\partial T}{\partial\theta} + \frac{v_\phi}{r\sin\theta}\frac{\partial T}{\partial\phi}\right) = \frac{1}{r^2}\frac{\partial}{\partial r}\left(r^2 k\frac{\partial T}{\partial r}\right) + \frac{1}{r\sin\theta}\frac{\partial}{\partial\theta}\left(\frac{k\sin\theta}{r}\frac{\partial T}{\partial\theta}\right)$$

$$+ \frac{1}{r\sin\theta}\frac{\partial}{\partial\phi}\left(\frac{k}{r\sin\theta}\frac{\partial T}{\partial\phi}\right) + \mu\phi + T\beta\left(\frac{\partial p}{\partial t} + v_r\frac{\partial p}{\partial r} + \frac{v_\theta}{r}\frac{\partial p}{\partial\theta} + \frac{v_\phi}{r\sin\theta}\frac{\partial p}{\partial\phi}\right)$$

(A5.15)

$$\phi = 2\left[\left(\frac{\partial v_r}{\partial r}\right)^2 + \left(\frac{1}{r}\frac{\partial v_\theta}{\partial\theta} + \frac{v_r}{r}\right)^2 + \left(\frac{1}{r\sin\theta}\frac{\partial v_\phi}{\partial\phi} + \frac{v_r}{r} + \frac{v_\theta\cot\theta}{r}\right)^2\right]$$

$$+ \left[r\frac{\partial}{\partial r}\left(\frac{v_\theta}{r}\right) + \frac{1}{r}\frac{\partial v_r}{\partial\theta}\right]^2 + \left[\frac{1}{r\sin\theta}\frac{\partial v_r}{\partial\phi} + r\frac{\partial}{\partial r}\left(\frac{v_\phi}{r}\right)\right]^2$$

$$+ \left[\frac{\sin\theta}{r}\frac{\partial}{\partial\theta}\left(\frac{v_\phi}{\sin\theta}\right) + \frac{1}{r\sin\theta}\frac{\partial v_\theta}{\partial\phi}\right]^2$$

$$\nabla^2 = \frac{1}{r^2}\frac{\partial}{\partial r}\left(r^2\frac{\partial}{\partial r}\right) + \frac{1}{r^2\sin\theta}\frac{\partial}{\partial\theta}\left(\sin\theta\frac{\partial}{\partial\theta}\right) + \frac{1}{r^2\sin^2\theta}\left(\frac{\partial^2}{\partial\phi^2}\right)$$

APPENDIX 6

DIMENSIONAL ANALYSIS

To get an appreciation of dimensionless variables, consider the one-dimensional, steady-state temperature distribution in a flat plate (Figure A6.1).

$$T = T_1 - (T_1 - T_2)\frac{x}{L} \tag{A6.1}$$

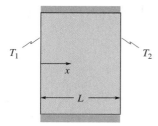

FIGURE A6.1

If we want to represent $T(x)$ in a graphical manner, we have to consider the vast number of cases arising from different values of T_1, T_2, and L. If Equation A6.1 is to be experimentally validated for different values of the variables in the equation, we have to show the data for each set of variables on a different graph.

Now consider an alternate form of Equation A6.1.

$$\frac{T_1 - T}{T_1 - T_2} = \frac{x}{L} \tag{A6.2}$$

Equation A6.2 represents the temperature distribution given by Equation A6.1 in a slightly different manner. However, all the different sets of values for the variables can be represented by a single graph by Equation A6.2. The experimental values for all slabs can now be shown on a single graph and compared with the values predicted by Equation A6.2 (Figure A6.2).

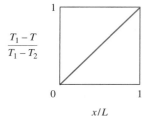

FIGURE A6.2

In recasting Equation A6.1 into Equation A6.2, we have employed a dimensionless temperature $(T_1 - T)/(T_1 - T_2)$ and a dimensionless distance x/L. Instead of using the man-made unit length (foot, meter, etc.), we seek a length that is characteristic to the problem. In this case the thickness, L, is such a length and all other length dimensions are measured in terms of this length L. Similarly, the unit of temperature is the difference in the temperature of the two surfaces ($= T_1 - T_2$) and all other temperatures are measured relative to T_1 with $T_1 - T_2$ as the characteristic temperature difference.

Compared with the dimensional form Equation A6.1, the dimensionless form Equation A6.2 has two main advantages. The number of parameters is reduced to two $[(T_1 - T)/T_1 - T_2)$ and x/L] in Equation A6.2 compared with five (T_1, T_2, T, x, L) in Equation A6.1. The second advantage is that, irrespective of the system of units used, the dimensionless parameters have the same value.

Dimensionless parameters can be found by different methods. Two methods, the power-product method and the Π Theorem are briefly illustrated. Both methods require that the independent variables affecting the dependant variable be known.

POWER-PRODUCT METHOD

Consider the dependent variable y, which is a function of the independent variables, $x_1, x_2, \ldots, x_n$. In a narrow range of the variables, it is generally possible to express the functional relationship between the variables in the form

$$y = px_1^{a_1}x_2^{a_2} \cdots x_n^{a_n}$$

where p is a dimensionless constant. By the principle of dimensional homogeneity, we require that the two sides of the equation have the same dimensions. This requirement yields as many equations as the number of dimensions involved in the equation. Equating the exponents of each dimension on both sides of the equation, we get m equations in n exponents, where m is the number of primary dimensions. We may then solve for any chosen m exponents in terms of the remaining $n - m$ exponents. This will then yield $n - m$ dimensionless variables.

As an example consider the flow of a fluid over a flat plate. With heat transfer to the fluid, it is known that the value of the local convective heat transfer coefficient (h) is a function of the velocity (U), properties of the fluid—density (ρ), thermal conductivity (k), constant pressure specific heat (c_p), viscosity (μ)—and the distance from the leading edge (x). We represent the relationship between h and the remaining variables in the form

$$h = h(U, \rho, k, c_p, \mu, x)$$

In a narrow range of the variables the functional form the relationship can be represented as

$$h = pU^a\rho^bk^cc_p^d\mu^ex^f$$

where p is a dimensionless constant. Here we employ dimensions of length, L; mass, M; time, T; and temperature, θ as the primary dimensions. Substituting the dimensions of the variables,

$$MT^{-3}\theta^{-1} = p\,(LT^{-1})^a (ML^{-3})^b (MLT^{-3}\theta^{-1})^c (L^2T^{-2}\theta^{-1})^d (ML^{-1}T^{-1})^e L^f$$

From the principle of dimensional homogeneity, we require that the exponents of each of the dimensions on either side of the equation should be equal.

L: $\quad 0 = a - 3b + c + 2d - e + f$ \hfill (1)

M: $\quad 1 = b + c + e$ \hfill (2)

T: $\quad -3 = -a - 3c - 2d - e$ \hfill (3)

θ: $\quad -1 = -c - d$ \hfill (4)

Out of the six unknown exponents, with four equations we can solve for any four of them in terms of the remaining two. The choice of exponents to be solved for is arbitrary. Let us solve for a, b, d, and f in terms of c and e.

From Equation 4, $\quad d = 1 - c$

From Equation 2, $\quad b = 1 - c - e$

From Equation 3, $\quad a = 3 - 3c - 2 + 2c - e = 1 - c - e$

From Equation 1, $\quad f = -a + 3b - c - 2d + e = -c - e$

Substituting these,

$$h = pU^{1-c-e}\rho^{1-c-e}k^c c_p^{1-c}\mu^e x^{-c-e}$$

Rearrangement gives

$$\frac{h}{U\rho c_p} = p\left(\frac{k}{U\rho c_p x}\right)^c \left(\frac{\mu}{U\rho x}\right)^e \tag{5}$$

Here the exponents, c and e, must be permitted to take suitable values depending on the functional form of the relationship. Their values may change in different ranges of the variables. To satisfy dimensional homogeniety, we require that each group be dimensionless (verify that each group is dimensionless by substituting the dimensions of each of the variables in the group). We have, thus, identified three dimensionless groups:

$$\frac{h}{U\rho c_p}, \quad \frac{k}{U\rho c_p x} \quad \text{and} \quad \frac{\mu}{U\rho x}$$

The dimensionless parameters that have been identified are different from those used in Equation 4.1.1 for the same situation. The dimensionless groups used in Equation 4.1.1 can be obtained from those that have been obtained above. As multiplication or division of one dimensionless variable by another yields a third dimensionless parameter, we can obtain those used in Equation 3.1.1 as,

$$\frac{h}{U\rho c_p} \times \frac{U\rho c_p x}{k} = \frac{hx}{k}$$

$$\frac{U\rho c_p x}{k} \times \frac{\mu}{U\rho x} = \frac{c_p \mu}{k}$$

Thus, we obtain the three dimensionless groups, hx/k, $U\rho x/\mu$, and $c_p\mu/k$ employed in Equation 4.1.1.

Instead of solving for a, b, d, and f in terms of c and e, we may solve for b, c, e, and f in terms of a and d. We then obtain $b = a$, $c = 1 - d$, $e = d - a$, $f = a - 1$. With this combination of exponents, we get

$$\frac{hx}{k} = p \left(\frac{U\rho x}{\mu} \right)^a \left(\frac{c_p \mu}{k} \right)^d$$

In this case we obtain directly the dimensionless variables used in Equation 4.1.1.

Π THEOREM

If, in a functional relationship of the different variables, there are n variables and m primary dimensions, there will be $(n - m)$ Π terms, each of the Π terms being dimensionless (here Π is used to designate a dimensionless group and has no connection with the ratio of the circumference of a circle to its radius). To form the dimensionless variables, choose m dimensional variables that do not form a dimensionless group among themselves. With these as the primary variables, form dimensionless groups by combining one of the remaining dimensional variable (to them) at a time.

To illustrate the methodology, consider the case of the convective heat transfer coefficient just examined. The dimensional variables and their dimensions are listed below:

h $MT^{-3}\theta^{-1}$ U LT^{-1} ρ ML^{-3}
k $MLT^{-3}\theta^{-1}$ c_p $L^2 T^{-2}\theta^{-1}$ μ $ML^{-1}T^{-1}$
x L

Here, there are seven (n) variables and four (m) dimensions. Hence, we expect three $(n - m)$, dimensionless variables. As there are four dimensions, we choose four variables that do not form a dimensionless variable among themselves. From the list of variables and their dimensions, we may choose U, k, μ, and x as the primary variables that do not form a dimensionless group among themselves. Note that, because the temperature dimension appears in only one variable, it is impossible to eliminate the temperature dimension however the four variables are combined. Thus, it is guaranteed that these four cannot form a dimensionless group. We add to these four the remaining variables, h, ρ, and c_p, one at a time and form a dimensionless group. The groups will be designated as Π_1, Π_2 and Π_3.

$$\Pi_1: U, k, \mu, x, h \quad \text{or} \quad \Pi_1 = U^a k^b \mu^c x^d h$$

Substituting the dimensions of each of the variables for the group,

$$(LT^{-1})^a (MLT^{-3}\theta^{-1})^b (ML^{-1}T^{-1})^c L^d (MT^{-3}\theta^{-1})$$

As this group is to be dimensionless, the exponent of each dimension should be zero.

L: $a + b - c + d = 0$
M: $b + c + 1 = 0$
T: $-a - 3b - c - 3 = 0$
θ: $-b - 1 = 0$

Solving for a, b, c, and d, we get $a = 0$, $b = -1$, $c = 0$, $d = 1$. Hence,

$$\Pi_1 = \frac{hx}{k}$$

The second dimensionless group, Π_2, is formed from u, k, μ, x, and ρ.

$$\Pi_2 = U^a k^b \mu^c x^d \rho$$

Substituting the dimensions for each variable and equating the exponent of each dimension to zero, we get

L: $a + b - c + d - 3 = 0$
M: $b + c + 1 = 0$
T: $-a - 3b - c = 0$
θ: $-b = 0$

TABLE A16 Dimensionless Parameters

Parameter	Definition	Physical Interpretation
Bi (Biot)	$\dfrac{hL}{k}$	Ratio of internal conduction resistance to external convective resistance
Bo (Bond)	$\dfrac{g(\rho_L - \rho_v)L^2}{\sigma}$	Ratio of gravitational force to surface tension force
Ec (Eckert)	$\dfrac{v^2}{c_p(T_s - T_\infty)}$	Ratio of kinetic energy to the enthalpy difference in the boundary layer
Fo (Fourier)	$\dfrac{\alpha t}{L^2}$	Dimensionless time
Gr (Grashof)	$\dfrac{g\beta(T_s - T_\infty)L^3}{v^2}$	Ratio of buoyancy to viscous forces
Ja (Jakob)	$\dfrac{c_p(T_s - T_{sat})}{i_{fg}}$	Ratio of sensible energy change to latent heat
Ma (Mach)	v/c	Ratio of velocity to local sonic velocity
Nu (Nusselt)	$\dfrac{hL}{k_f}$	Dimensionless temperature gradient
Pe (Peclet)	Re Pr	
Pr (Prandtl)	$\dfrac{v}{\alpha}$	Ratio of momentum and thermal diffusivities
Ra (Raleigh)	Gr Pr	
Re (Reynolds)	$\dfrac{\rho VL}{\mu}$	Ratio of inertia force to viscous force
St (Stanton)	$\dfrac{Nu}{Re\ Pr}$	
We (Weber)	$\dfrac{\rho V^2 L}{\sigma}$	Ratio of inertia force to surface tension force

Solving, we get $a = 1$, $b = 0$, $c = -1$, $d = 1$, so that

$$\Pi_2 = \frac{U\rho x}{\mu}$$

The last group, Π_3 is formed from U, k, μ, x, and c_p.

$$\Pi_3 = U^a k^b \mu^c x^d c_p$$

Again, substituting the dimensions and equating the exponents of each dimension to zero,

L: $a + b - c + d + 2 = 0$
M: $b + c = 0$
T: $-a - 3b - c - 2 = 0$
θ: $-b - 1 = 0$

Solving, $a = 0$, $b = -1$, $c = 1$, and $d = 0$,

$$\Pi_3 = \frac{c_p \mu}{k}$$

Some of the more common dimensionless parameters used in heat transfer are given in Table A16 (see previous page).

APPENDIX 7

FORTRAN PROGRAMS

PROGRAM FOR EXAMPLE 3.3.2

```
      REAL T(50,50)
C     ΔX AND ΔY NEED NOT BE EQUAL
C     TH = THICKNESS OF SLAB
C     W = WIDTH OF SLAB
C     IEND AND JEND ARE THE LAST NODES IN THE X- AND Y-DIRECTIONS
C     SUPPLY VALUES OF W, TH, IEND AND JEND
      W = 8
      TH = 6
      IEND = 20
      JEND = 15
      EPS = 1E-6
C       DEFINE BOUNDARY TEMPERATURES
      DO 10 I = 2, IEND - 1
      T(I,1) = 100
10    T(I,JEND) = 60
      DO 20 J = 2 TO JEND - 1
      T(1,J) = 150
20    T(IEND,J) = 0
C       SUPPLY INITIAL GUESSED TEMPERATURES
      DO 30 J = 2, JEND - 1
      DO 30 I = 2, IEND - 1
30    T(I,J) = T(1,J) - (T(1,J) - T(IEND,J))/(IEND-1)
      DX = W/(IEND-1)
      DY = TH/(JEND-1)
      DYX = DY/DX
      DXY = DX/DY
C     UPDATE TEMPERATURES AT INTERIOR NODES AND TEST FOR CONVERGENCE AT
C     EACH NODE. AVOID ENDLESS LOOP BY SPECIFYING THE MAXIMUM NUMBER OF
C     ITERATIONS, JX. IF IND = 0 EVERYWHERE THE SOLUTION HAS CONVERGED.
      IEND = 0
      TMAX = T(1,2) - T(IEND,2)
      DO 200 JX = 1, 10000
      DO 100 J = 2, JEND-1
```

```
      DO 100 I = 2, IEND-1
      TEMP = T(I,J)
      T(I,J) = DYX * (T(I-1,J) + T(I+1,J)) + DXY * (T(I,J-1) + T(I,J+1))
      T(I,J) = 0.5 * T(I,J)/(DYX + DXY)
C     CHECK FOR CONVERGENCE
      IF (ABS((T(I,J) - TEMP)/TMAX).GT.EPS) THEN IND = IND + 1
100   CONTINUE
C      IF IND = 0 THE SOLUTION HAS CONVERGED. GO TO OUTPUT
C      IF IND > 0 CONTINUE ITERATION
      IF (IND.EQ.0) THEN GO TO 300
      IND = 0
200   CONTINUE
      C STARTING FROM STATEMENT 300 WRITE A PROGRAM FOR OUTPUT OF T(I,J)
300   C STATEMENTS FOR OUTPUT OF THE TEMPERATURES AT VARIOUS NODES
      END
```

TRI-DIAGONAL MATRIX ALGORITHM (TDMA)—PROGRAM TO SOLVE EXAMPLE 3.3.3

```
C     PROGRAM TO SOLVE EXAMPLE 3.3.2 BY TDMA
C     A, B, C, AND D ARE THE COEFFICIENTS IN THE MATRIX
C     P AND Q ARE DETERMINED BY ITERATION
C     IEND AND JEND ARE THE LAST NODES IN THE X- AND Y-DIRECTIONS
C     TO = DIFFERENCE BETWEEN BASE AND SURROUNDING TEMPERATURES
C     K = THERMAL CONDUCTIVITY
C     H AND HE ARE THE CONVECTIVE HEAT TRANSFER COEFFICIENTS
C     N = NUMBER OF ITERATIONS
C     XL = LENGTH OF FIN
C     TH = HALF THE THICKNESS OF THE FIN
      REAL A(20,20), B(20,20), C(20,20), D(20,20), T(20,20), P(20), Q(20)
      L = 0.2
      TH = 0.005
      TO = 80
      H = 6
      HE = 6
      K = 16
      EPS = 1E-6
      INT = 10
      IEND = 11
      JEND = 5
      DX = L/(IEND-1)
      DY = TH/(JEND-1)
      DXDY = DX/DY
      DYDX = DY/DX
```

```
C        INITIALIZE TEMPERATURE DISTRIBUTION
         DO 10 I = 1, IEND
         DO 10 J = 1, JEND
         T (I,J) = TO - TO*(I-1)/(IEND-1)
10       CONTINUE
C        INCLUDE STATEMENTS FOR THE OUTPUT OF THE INITIAL TEMPERATURE GUESS
C        COMPUTE COEFFICIENTS. ALL D(I,J) ARE DETERMINED WHEN EXECUTING TDMA AS
C        THEY DEPEND ON THE TEMPERATURE OF THE NEIGHBORING ROWS
         J = 1
         DO 30 I = 2, IEND - 1
         A(I,J) = 0.5*DYDX
         B(I,J) = -DYDX - DXDY
         C(I,J) = 0.5*DYDX
30       CONTINUE
         A(2,1) = 0
         DO 40 J = 2, JEND - 1
         DO 40 I = 2, IEND - 1
         A(I,J) = DYDX
         IF (I.EQ.2) THEN A(I,J) = 0
         B(I,J) = -2*(DYDX+DXDY)
         C(I,J) = DYDX
40       CONTINUE
         J = JEND
         DO 50 I = 2, IEND - 1
         A(I,J) = 0.5*DYDX
         IF (I.EQ.2) THEN A(I,J) = 0
         B(I,J) = -(DYDX + DXDY + H*DX/K)
         C(I,J) = 0.5*DYDX
50       CONTINUE
         I = IEND
         J = 1
         A(I,J) = DYDX
         B(I,J) = -(DYDX + HE*DY/K + DXDY)
         C(I,J) = 0
         I = IEND
         DO 60 J = 2, JEND-1
         A(I,J) = DYDX
         B(I,J) = -(DYDX + DXDY + HE*DY/K)
         C(I,J) = 0
60       CONTINUE
         I = IEND
         J = JEND
         A(I,J) = DYDX
         B(I,J) = -(DYDX + HE*DY/K + DXDY + H*DX/K)
         C(I,J) = 0
C        ALL A(I,J), B(I,J) AND C(I,J) HAVE BEEN COMPUTED
         INT = 1000
```

```
        IW = INT
        DO 70 N = 1,30000
C       START TDMA ROW BY ROW
        DO 80 J = 1 TO JEND
C       COMPUTE D(I,J)
        IF (J.EQ.1) THEN
        I = 2
        D(I,J) = - DXDY * T(I,J+1) - 0.5 * DYDX * TO
        DO 315 I = 3, IEND
315     D(I,J) = - DXDY * T(I,J+1)
        ELSE IF (J.GT.1 .AND. J.LT.JEND) THEN
        I = 2
        D(I,J) = -DXDY * (T(I,J-1) + T(I,J+1)) - DXDY*TO
        DO 325 I = 3, IEND-1
325     D(I,J) = -DXDY * (T(I,J-1) + T(I,J+1))
        I = IEND
        D(I,J) = - 0.5 * DXDY * (T(I,J-1) + T(I,J+1))
        ELSE IF (J.EQ.JEND) THEN
        I = 2
        D(I,J) = - DXDY * T(I,J-1) - 0.5 * DYDX * TO
        DO 335 I = 3, IEND
335     D(I,J) = - DXDY * T(I,J-1)
        END IF
C       ALL D(I,J) HAVE BEEN COMPUTED FOR A PARTICULAR VALUE OF J. COMPUTE
C       T(I,J) FOR THE PARTICULAR ROW DENOTED BY THE VALUE OF J
        DO 210 I = 2, IEND
        P(I) = - C(I,J)/(A(I,J)*P(I-1) + B(I,J))
        Q(I) = (D(I,J) - A(I,J)*Q(I-1))/(A(I,J)*P(I-1) + B(I,J))
        IF (I.EQ.2) THEN
        P(I) = - C(I,J)/B(I,J)
        Q(I) = D(I,J)/B(I,J)
        END IF
210     CONTINUE
C       DETERMINE T(I,J) FOR THE PARTICULAR ROW BY BACK SUBSTITUTION
        TEMP = T(IEND,J)
        T(IEND,J) = Q(IEND)
        IF (ABS((TEMP-T(IEND,J))/TO).GT.EPS THEN IND = IND + 1
        DO 220 I = IEND-1, 2, -1
        TEMP = T(I,J)
        T(I,J) = P(I) * T(I+1,J) + Q(I)
        IF (ABS((TEMP - T(I,J))/TO).GT.EPS) THEN IND = IND + 1
220     CONTINUE
80      CONTINUE
C       ALL T(I,J) HAVE BEEN COMPUTED. TEST FOR CONVERGENCE
        IF (IND.EQ.0) GO TO 100
C       WRITE INTERMEDIATE RESULTS IF DESIRED
70      CONTINUE
```

```
C      IF THE NUMBER OF ITERATIONS FOR CONVERGENT SOLUTION IS NOT ADEQUATE,
C      INCLUDE A COMMENT TO INDICATE THAT CONVERGENT SOLUTION WAS NOT OBTAINED
C      STATEMENT TO SHOW THAT CONVERGENT SOLUTION IS NOT OBTAINED
       GO TO 1000
100
C      WRITE RESULTS FOR CONVERGENT SOLUTION
C      INCLUDE STATEMENT FOR THE OUTPUT OF CONVERGED SOLUTION
1000  END
```

SUBROUTINE FOR TRIDIAGONAL MATRIX ALGORITHM

```
C      A,B,C ARE THE ELEMENTSINEACH ROW ANDD IS THECONSTANT ON THE RIGHT
C        SIDE OF THE EQUATION
C      N = NUMBER OF EQUATIONS,NOTTOEXCEED500. IF N EXCEEDS 500 MODIFY
C        THE DIMENSION STATEMENT FOR A, B, C, D, AND T
C        T IS THE UNKNOWN VARIABLE
C      A, B, C, AND D ARE COMPUTEDINTHECALLINGPROGRAM. THE SUBROUTINE
C        SOLVES THE EQUATIONS AND PASSES THE VALUES OF T TO THE CALLING PROGRAM
C        CALLING STATEMENT: CALL TDMA(A,B,C,D,N,T)
       SUBROUTINE TDMA (A, B, C, D, N, T)
       DIMENSION A(500), C(500), D(500), T(500), P(500), Q(500)
C        COMPUTE P(I) AND Q(I)
       P(1) = - C(1)/B(1)
       Q(1) = D(1)/B(1)
       DO 10 I = 2, N
       P(I) = - C(I)/(A(I)*P(I-1) + B(I))
       Q(I) = (D(I) - A(I)*Q(I-1)/(A(I)*P(I-1) + B(I))
10     CONTINUE
C        COMPUTE T(I) BY BACK SUBSTITUTION
       T(N) = Q(N)
       DO 20 I = N-1, 1, -1
       T(I) = P(I)*T(I+1) + Q(I)
20     CONTINUE
       END
```

PROGRAM FOR THE INTEGRATION OF EQUATION 12.5.16

```
C      INTEGRATION OF EQ.12.5.16 BY 4TH ORDER RUNGE-KUTTA
C      YA = F YB = FPRIME YC = FDOUBLE-PRIME
C       Z =
C      REAL YA(1001)
C      OPEN FILE ''VEL'' FOR WRITING THE OUTPUT REQUIRED FOR INTEGRATING
C        EQ.12.5.26
       DIMENSION YA(1001)
```

```
      DZ = 0.01
C     INTEGRATION IS STOPPED AT η = 10 i.e. FOR IMAX = 1001 (Z = 10)
C     ZGR3 IS THE INITIAL GUESS OFF''(0)
      ZGR3 = 0.4
C     DZGR3 IS THE INCREMENT OF ZGR3 AFTER THE FIRST SET OF COMPUTATIONS
      DZGR3 = 0.2
      EPS = 1E−6
C     N1T GIVES THE NUMBER OF ITERATIONS
C     1P IS THE INTERVAL AT WHICH VALUES OF Z, YA, YB, AND YC ARE PRINTED
      PRINT*',    Z           F        FP           FPP'
      IP = 20
      IPRINT = 21
      NIT = 1
      IMAX = 1001
      IND = 0
C     WHEN CONVERGED SOLUTION IS OBTAINED IND IS SET TO 1
      DO 10 NIT = 1, 11
      Z = 0
      YA(1) = 0
      YB = 0
      YC = ZGR3
      IF (IND.EQ.1) THEN
C     INCLUDE STATEMENT FOR THE OUTPUT OF Z, YA(1), YB AND YC AT Z = 0. THE
C     VALUES OF Z YA AND YB SHOULD ALSO BE ON FILE ''VEL'' FOR INTEGRATION OF
C     THE ENERGY EQUATION
      PRINT100,Z,YA(1),YB,YC
      END IF
      DO 500 I = 2, IMAX
C     COMMENCE INTEGRATION
      P1 = YB
      Q1 = YC
      R1 = −0.5*YA(I−1)*YC
      P2 = YB + Q1*DZ*0.5
      Q2 = YC + R1*DZ*0.5
      R2 = −0.5*(YA(I−1) + P1*DZ*0.5)*(YC + R1*DZ*0.5)
      P3 = YB + Q2*DZ*0.5
      Q3 = YC + R2*DZ*0.5
      R3 = −0.5*(YA(I-1) + P2*DZ*0.5)*(YC + R2*DZ*0.5)
      P4 = YB + Q3*DZ
      Q4 = YC + R3*DZ
      R4 = −0.5*(YA(I−1) + PE*DZ)*(YC + R3*DZ)
      YA(I) = YA(I−1) + (P1 + 2*P2 + 2*P3 + P4)*DZ/6
      YB = YB + (Q1 + 2*Q2 + 2*Q3 + Q4)*DZ/6
      YC = YC + (R1 + 2*R2 + 2*R3 + R4)*DZ/6
      Z = Z + DZ
      IF (IND.EQ.1.AND.I.EQ.IPRINT) THEN
      PRINT 100,Z,YA(I),YB,YC
```

```
100     FORMAT (4(F10.6', '))
        IPRINT = IPRINT + IP
        END IF
C       INCLUDE STATEMENT FOR PRINTING RESULTS. THE FUNCTIONS ARE REQUIRED FOR
C       THE INTEGRATION OF THE ENERGY EQUATION AND THE VALUES OF Z, YA AND YB
C       SHOULD BE STORED IN THE FILE ''VEL''.
500     CONTINUE
C       IF CONVERGENT SOLUTION IS OBTAINED QUIT
        IF (IND.EQ.1) GO TO 1000
        DIFF = 1 - YB
        IF (ABS(DIFF).LT.EPS) THEN
        IND = 1
        GO TO 10
        END IF
        IF (NIT.EQ.1) THEN
        DIFFT = DIFF
        ZGRT = ZGR3
        ZGR3 = ZGR3 + DZGR3
        GO TO 10
        END IF
C       COMPUTE NEW VALUE ZGR3
        DZGR3 = (ZGR3 - ZGRT)*DIFF/(DIFFT-DIFF)
        ZGRT = ZGR3
        DIFFT = DIFF
        ZGR3 = ZGR3 + DZGR3
10      CONTINUE
C       INDICATE THAT CONVERGENT SOLUTION IS NOT OBTAINED
1000    END
```

PROGRAM FOR THE INTEGRATION OF EQUATION 12.5.26

The program first computes the functions f, f', and f'' starting with a value of 0.3320578 for $f''(0)$. The integration starts with an assumed value of 1 for $\theta'(0)$. The output is $\theta'(0) = 1/\theta(\infty)$, where $\theta(\infty)$ is obtained when θ' is less than 10^{-6}. If a temperature profile is desired, the integration should be performed a second time with the correct value of $\theta'(0)$.

For values of Prandtl number less than 0.01 the program takes some time to run, particularly without a math coprocessor.

```
C       FIRST INTEGRATE EQ.12.5.16 BY 4TH ORDER RUNGE-KUTTA
C       YA = F YB = FPRIME YC = FDOUBLE-PRIME
C       Z = η
C       REAL YA(1001)
        DIMENSION YA(1001), YB(1001)
        DZ = 0.01
C       INTEGRATION IS STOPPED AT Z = 10 i.e. for IMAX = 1001 (Z = 10)
C       ZGR3 IS THE INITIAL GUESS OF F''(0)
```

```
      ZGR3 = 0.3320578
C     DZGR3 IS THE INCREMENT OF ZGR3 AFTER THE FIRST SET OF COMPUTATIONS
      DZGR3 = 0.00001
      EPS = 1E-6
C     NIT GIVES THE NUMBER OF ITERATIONS
C     IP IS THE INTERVAL AT WHICH VALUES OF Z, YA, YB, AND YC ARE PRINTED
      NIT = 1
      IMAX = 1001
      IND = 0
C     WHEN CONVERGED SOLUTION IS OBTAINED IND IS SET TO 1
      DO 10 NIT = 1, 11
      Z = 0
      YA(1) = 0
      YB(1) = 0
      YC = ZGR3
      DO 500 I = 2, IMAX
C     COMMENCE INTEGRATION
      P1 = YB(I-1)
      Q1 = YC
      R1 = -0.5*YA(I-1)*YC
      P2 = YB(I-1) + Q1*DZ*0.5
      Q2 = YC + R1*DZ*0.5
      R2 = -0.5*(YA(I-1) + P1*DZ*0.5)*(YC + R1*DZ*0.5)
      P3 = YB(I-1) + Q2*DZ*0.5
      Q3 = YC + R2*DZ*0.5
      R3 = -0.5*(YA(I-1) + P2*DZ*0.5)*(YC + R2*DZ*0.5)
      P4 = YB(I-1) + Q3*DZ
      Q4 = YC + R3*DZ
      R4 = -0.5*(YA(I-1) + P3*DZ)*(YC + R3*DZ)
      YA(I) = YA(I-1) + (P1 + 2*P2 + 2*P3 + P4)*DZ/6
      YB(I) = YB(I-1) + (Q1 + 2*Q2 + 2*Q3 + Q4)*DZ/6
      YC = YC + (R1 + 2*R2 + 2*R3 + R4)*DZ/6
      Z = Z + DZ
500   CONTINUE
C     IF CONVERGENT SOLUTION IS OBTAINED QUIT
      IF (IND.EQ.1) GO TO 1000
      DIFF = 1 - YB(1001)
      IF (ABS(DIFF).LT.EPS) THEN
      IND = 1
      GO TO 10
      END IF
      IF (NIT.EQ.1) THEN
      DIFFT = DIFF
      ZGRT = ZGR3
      ZGR3 = ZGR3 + DZGR3
      GO TO 10
      END IF
```

```
C      COMPUTE NEW VALUE ZGR3
        DZGR3 = (ZGR3 — ZGRT)*DIFF/(DIFFT—DIFF)
        ZGRT = ZGR3
        DIFFT = DIFF
        ZGR3 = ZGR3 + DZGR3
10      CONTINUE
C      INDICATE THAT CONVERGENT SOLUTION IS NOT OBTAINED
1000  PRINT,NIT,YA(1), YB(1001), ZGR3
        CONTINUE
C       VALUES OF F AND FPRIME HAVE BEEN COMPUTED AND STORED AS YA(I) AND YB(I)

C      INTEGRATION OF EQ.12.2.26 — 4TH ORDER RUNGE-KUTTA
C      ENTER PRANDTL NUMBER
        PRINT*, 'ENTER PRANDTL NUMBER'
        READ*, PR
        DZ = 0.01
40      EPS = 1E-6
C      TGR = TEMPERATURE GRADIENT AT THE PLATE
        TGR = 1.0
        IND = 0
        IP = 0
C      IND = 1 INDICATES THAT THE BOUNDARY VALUES ARE SATISFIED
        IMAX = 25000
15      T = 0.0
        Z = 0
C      BEGIN INTEGRATION
        DO 30 I = 2, IMAX
        IF (I.GT.1001) THEN
        FT = FT + DZ
        FP = 1
        ELSE
        FT = YA(I-1)
        FP = YB (I-1)
        END IF
        GR = TGR
        P1 = GR
        Q1 = —0.5 * PR * FT * GR
        P2 = GR + Q1 * 0.5 * DZ
        Q2 = — 0.5 * PR * (FT + FP * 0.5 * DZ) * (GR + Q1 * 0.5 * DZ)
        P3 = GR + Q2 * 0.5 * DZ
        Q3 = — 0.5 * PR * (FT + FP * 0.5 * DZ) * (GR + Q2 * 0.5 * DZ)
        P4 = GR + Q3 * DZ
        Q4 = — 0.5 * PR * (FT + FP * DZ) * (GR + Q3 * DZ)
        T = T + (P1 + 2 * P2 + 2 * P3 + P4) * DZ/6
        TGR = GR + (Q1 + 2 * Q2 + 2 * Q3 + Q4) * DZ/6
        Z = Z + DZ
        IF (TGR.LT.EPS) GO TO 60
```

```
30      CONTINUE
60       TGR = 1/T
         PRINT 200,PR
200     FORMAT(' PRANDTL NUMBER = ',F8.4)
         PRINT 210, TGR
210     FORMAT (' TEMPERATURE GRADIENT AT PLATE = ',F7.5)
         PRINT *,''
         PRINT *,''
         PRINT *, 'ENTER PRANDTL NUMBER. TO QUIT ENTER 0'
         READ *, PR
         IF (PR.GT.0) GO TO 40
         END
```

INDEX

HT: Heat Transfer Computations
to Accompany Engineering Heat Transfer by N. V. Suryanarayana
ISBN: 0–314–01093–9

0-314-01093-9

90000

9 780314 010933